HANDBOOK OF BIOLOGICAL EFFECTS OF ELECTROMAGNETIC FIELDS

THIRD EDITION

Bioengineering and Biophysical Aspects of Electromagnetic Fields

HANDBOOK OF BIOLOGICAL EFFECTS OF ELECTROMAGNETIC FIELDS

THIRD EDITION

Bioengineering and Biophysical Aspects of Electromagnetic Fields

EDITED BY

Frank S. Barnes
University of Colorado-Boulder
Boulder, CO, U.S.A.

Ben Greenebaum
University of Wisconsin-Parkside
Kenosha, WI, U.S.A.

CRC Taylor & Francis
Taylor & Francis Group
Boca Raton London New York

CRC is an imprint of the Taylor & Francis Group, an informa business

CRC Press
Taylor & Francis Group
6000 Broken Sound Parkway NW, Suite 300
Boca Raton, FL 33487-2742

Printed in the United States of America on acid-free paper
10 9 8 7 6 5 4 3 2 1

International Standard Book Number-10: 0-8493-9539-9 (Hardcover)
International Standard Book Number-13: 978-0-8493-9539-0 (Hardcover)

Visit the Taylor & Francis Web site at
http://www.taylorandfrancis.com

and the CRC Press Web site at
http://www.crcpress.com

Preface

We are honored to have been asked to carry on the tradition established by Dr. Postow and the late Dr. Polk in the first two editions of the *Handbook of Biological Effects of Electromagnetic Fields*. Their editions of this handbook were each recognized as the authoritative standards of their time for scientists working in bioelectromagnetics, the science of electromagnetic field effects on biological systems, and for others seeking information about this field of research.

In revising and updating this edition of the *Handbook of Biological Effects of Electromagnetic Fields*, we have expanded the coverage to include more material on diagnostic and therapeutic applications. At the same time, in updating and expanding the previous editions' coverage of the basic science and studies related to the possible biological effects of the electromagnetic fields, we have added new material on the related physics and chemistry as well as reviews of the recent developments in the setting standards for exposure limits. Following the previous edition's lead, we have charged the authors of the individual chapters with providing the reader, whom we imagine is fairly well founded in one or more of the sciences underlying bioelectromagnetics but perhaps not in the others or in the interdisciplinary subject of bioelectromagnetics itself, with both an introduction to their topic and a basis for further reading. We asked the chapter authors to write what they would like to be the first thing they would ask a new graduate student in their laboratory to read. We hope that this edition, like its two predecessors, will be useful to many as a reference book and to others as a text for a graduate course that introduces bioelectromagnetics or some of its aspects.

As a "handbook" and not an encyclopedia, this work does not intend to cover all aspects of bioelectromagnetics. Nevertheless, taking into account the breadth of topics and growth of research in this field since the last edition, we have expanded the number of topics and the number of chapters. Unavoidably, some ideas are duplicated in chapters, sometimes from different viewpoints that could be instructive to the reader; and different aspects of others are presented in different chapters. The increased amount of material has led to the publication of the handbook as two separate, but inter-related volumes: *Biological and Medical Aspects of Electromagnetic Fields (BMA)* and *Bioengineering and Biophysical Aspects of Electromagnetic Fields (BBA)*. Because there is no sharp dividing line, some topics are dealt with in parts of both volumes. The reader should be particularly aware that various theoretical models, which are proposed for explaining how fields interact with biological systems at a biophysical level, are distributed among a number of chapters. No one model has become widely accepted, and it is quite possible that more than one will in fact be needed to explain all observed phenomena. Most of these discussions are in the *Biological and Medical* volume, but the *Bioengineering and Biophysics* volume's chapters on electroporation and on mechanisms and therapeutic applications, for example, also have relevant material. Similarly, the chapters on biological effects of static magnetic fields and on endogenous electric fields in animals could equally well have been in the *Biological and Medical* volume. We have tried to use the index and cross-references in the chapters to direct the reader to the most relevant linkages, and we apologize for those we have missed.

Research in bioelectromagnetics stems from three sources, all of which are important; and various chapters treat both basic physical science and engineering aspects and the biological and medical aspects of these three. Bioelectromagnetics first emerged as a separate scientific subject because of interest in studying possible hazards from exposure to electromagnetic fields and setting exposure limits. A second interest is in the beneficial use of fields to advance health, both in diagnostics and in treatment, an interest that is as old as the discovery of electricity itself. Finally, the interactions between electromagnetic fields and biological systems raise some fundamental, unanswered scientific questions and may also lead to fields being used as tools to probe basic biology and biophysics. Answering basic bioelectromagnetic questions will not only lead to answers about potential electromagnetic hazards and to better beneficial applications, but they should also contribute significantly to our basic understanding of biological processes. Both strong fields and those on the order of the fields generated within biological systems may become tools to perturb the systems, either for experiments seeking to understand how the systems operate or simply to change the systems, such as by injecting a plasmid containing genes whose effects are to be investigated. These three threads are intertwined throughout bioelectromagnetics. Although any specific chapter in this work will emphasize one or another of these threads, the reader should be aware that each aspect of the research is relevant to a greater or lesser extent to all three.

The reader should note that the chapter authors have a wide variety of interests and backgrounds and have concentrated their work in areas ranging from safety standards and possible health effects of low-level fields to therapy through biology and medicine to the fundamental physics and chemistry underlying the biology. It is therefore not surprising that they have different and sometimes conflicting points of view on the significance of various results and their potential applications. Thus authors should only be held responsible for the viewpoints expressed in their chapters and not in others. We have tried to select the authors and topics so as to cover the scientific results to date that are likely to serve as a starting point for future work that will lead to the further development of the field. Each chapter's extensive reference section should be helpful for those needing to obtain a more extensive background than is possible from a book of this type.

Some of the material, as well as various authors' viewpoints, are controversial, and their importance is likely to change as the field develops and our understanding of the underlying science improves. We hope that this volume will serve as a starting point for both students and practitioners to come up-to-date with the state of understanding of the various parts of the field as of late 2004 or mid-2005, when authors contributing to this volume finished their literature reviews.

The editors would like to express their appreciation to all the authors for the extensive time and effort they have put into preparing this edition, and it is our wish that it will prove to be of value to the readers and lead to advancing our understanding of this challenging field.

Frank S. Barnes
Ben Greenebaum

Editors

Frank Barnes received his B.S. in electrical engineering in 1954 from Princeton University and his M.S., engineering, and Ph.D. degrees from Stanford University in 1955, 1956, and 1958, respectively. He was a Fulbright scholar in Baghdad, Iraq, in 1958 and joined the University of Colorado in 1959, where he is currently a distinguished professor. He has served as chairman of the Department of Electrical Engineering, acting dean of the College of Engineering, and in 1971 as cofounder/director with Professor George Codding of the Political Science Department of the Interdisciplinary Telecommunications Program (ITP).

He has served as chair of the IEEE Electron Device Society, president of the Electrical Engineering Department Heads Association, vice president of IEEE for Publications, editor of the *IEEE Student Journal* and the *IEEE Transactions on Education*, as well as president of the Bioelectromagnetics Society and U.S. Chair of Commission K—International Union of Radio Science (URSI). He is a fellow of the AAAS, IEEE, International Engineering Consortium, and a member of the National Academy of Engineering.

Dr. Barnes has been awarded the Curtis McGraw Research Award from ASEE, the Leon Montgomery Award from the International Communications Association, the 2003 IEEE Education Society Achievement Award, Distinguished Lecturer for IEEE Electron Device Society, the 2002 ECE Distinguished Educator Award from ASEE, The Colorado Institute of Technology Catalyst Award 2004, and the Bernard M. Gordon Prize from National Academy of Engineering for Innovations in Engineering Education 2004. He was born in Pasadena, CA, in 1932 and attended numerous elementary schools throughout the country. He and his wife, Gay, have two children and two grandchildren.

Ben Greenebaum retired as professor of physics at the University of Wisconsin–Parkside, Kenosha, WI, in May 2001, but was appointed as emeritus professor and adjunct professor to continue research, journal editing, and university outreach projects. He received his Ph.D. in physics from Harvard University in 1965. He joined the faculty of UW–Parkside as assistant professor in 1970 following postdoctoral positions at Harvard and Princeton Universities. He was promoted to associate professor in 1972 and to professor in 1980. Greenebaum is author or coauthor of more than 50 scientific papers. Since 1992, he has been editor in chief of *Bioelectromagnetics*, an international peer-reviewed scientific journal and the most cited specialized journal in this field. He spent 1997–1998 as consultant in the World Health Organization's International EMF Project in Geneva, Switzerland. Between 1971 and 2000, he was part of an interdisciplinary research team investigating the biological effects of electromagnetic fields on biological cell cultures. From his graduate student days through 1975, his research studied the spins and moments of radioactive nuclei. In 1977 he became a special assistant to the chancellor and in 1978, associate dean of faculty (equivalent to the present associate vice chancellor position). He served 2 years as acting vice chancellor (1984–1985 and 1986–1987). In 1989, he was appointed as dean of the School of Science and Technology, serving until the school was abolished in 1996.

On the personal side, he was born in Chicago and has lived in Racine, WI, since 1970. Married since 1965, he and his wife have three adult sons.

Contributors

Frank S. Barnes Department of Electrical and Computer Engineering, University of Colorado, Boulder, Colorado

Paolo Bernardi Department of Electronic Engineering, University of Rome, Rome, Italy

Martin Bier Department of Physics, East Carolina University, Greenville, North Carolina

Jon Dobson Institute for Science and Technology, Keele University, Stoke-on-Trent, U.K. and Department of Materials Science and Engineering, University of Florida, Gainesville, Florida

Stefan Engström Department of Neurology, Vanderbilt University, Nashville, Tennessee

Camelia Gabriel Microwave Consultants Ltd, London, U.K.

Ben Greenebaum University of Wisconsin–Parkside, Kenosha, Wisconsin

Kjell Hansson Mild National Institute for Working Life, Örebro University, Örebro, Sweden

William T. Joines Department of Electrical and Computer Engineering, Duke University, Durham, North Carolina

Sven Kühn Foundation for Research on Information Technologies in Society (IT'IS Foundation), Swiss Federal Institute of Technology (ETH), Zurich, Switzerland

Niels Kuster Foundation for Research on Information Technologies in Society (IT'IS Foundation), Swiss Federal Institute of Technology (ETH), Zurich, Switzerland

A.R. Liboff Center for Molecular Biology and Biotechnology, Florida Atlantic University, Boca Raton, Florida

James C. Lin Department of Electrical and Computer Engineering and Department of Bioengineering, University of Illinois, Chicago, Illinois

Qing H. Liu Department of Electrical and Computer Engineering, Duke University, Durham, North Carolina

Richard Nuccitelli Department of Electrical and Computer Engineering, Old Dominion University, Norfolk, Virginia

Tsukasa Shigemitsu Department of Biomedical Engineering, University of Tokyo, Tokyo, Japan

Shoogo Ueno Department of Biomedical Engineering, University of Tokyo, Tokyo, Japan

James C. Weaver Massachusetts Institute of Technology, Cambridge, Massachusetts

Gary Ybarra Department of Electrical and Computer Engineering, Duke University, Durham, North Carolina

Table of Contents

Introduction

Charles Polk*

Revised for the 3rd Edition by Ben Greenebaum

Much has been learned since this handbook's first edition, but a full understanding of biological effects of electromagnetic fields has is to be achieved. The broad range of what must be studied has to be a factor in the apparent slow progress toward this ultimate end. The broad range of disciplines involved includes basic biology, medical science and clinical practice, biological and electrical engineering, basic chemistry and biochemistry, and fundamental physics and biophysics. The subject matter ranges over characteristic lengths and timescales from, at one extreme, direct current (dc) or $\sim 10^4$ km-wavelengths, multimillisecond ac fields and large, long-lived organisms to, at the other extreme, submillimeter wavelength fields with periods below 10^{-12} s and subcellular structures and molecules with subnanometer dimensions and characteristic times as short as the 10^{-15} s or less of biochemical reactions.

This chapter provides an introduction and overview of the research and the contents of this handbook.

0.1 Near Fields and Radiation Fields

In recent years it has become, unfortunately, a fairly common practice—particularly in nontechnical literature—to refer to the entire subject of interaction of electric (E) and magnetic (H) fields with organic matter as biological effects of nonionizing radiation, although fields that do not vary with time and, for most practical purposes, slowly time-varying fields do not involve radiation at all. The terminology had its origin in an effort to differentiate between relatively low-energy microwave radiation and high-energy radiation, such as UV and x-rays, capable of imparting enough energy to a molecule or an atom to disrupt its structure by removing one or more electron\s with a single photon. However, when applied to dc or extremely low-frequency (ELF), the term "nonionizing radiation" is inappropriate and misleading.

A structure is capable of efficiently radiating electromagnetic waves only when its dimensions are significant in comparison with the wavelength λ. But in free space $\lambda = c/f$, where c is the velocity of light in vacuum (3×10^8 m/s) and f is the frequency in hertz (cycles/s); therefore the wavelength at the power distribution frequency of 60 Hz, e.g., is 5000 km, guaranteeing that most available human-made structures are much smaller than one wavelength.

The poor radiation efficiency of electrically small structures (i.e., structures whose largest linear dimension $L \ll \lambda$ can be illustrated easily for linear antennas. In free space the radiation resistance, R_r of a current element, i.e., an electrically short wire of length ℓ carrying uniform current along its length [1], is

$$R_r = 80\pi^2 \left(\frac{\ell}{\lambda}\right)^2 \tag{0.1}$$

*Deceased.

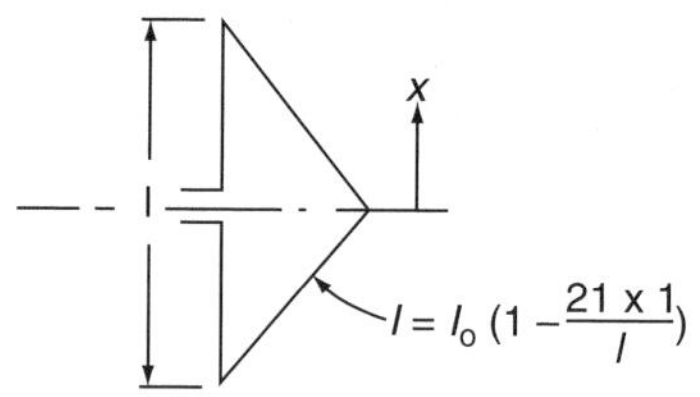

FIGURE 0.1
Current distribution on short, thin, center-fed antenna.

whereas the R_r of an actual center-fed radiator of total length ℓ with current going to zero at its ends, as illustrated in Figure 0.1, is

$$R_r = 20\pi^2 \left(\frac{\ell}{\lambda}\right)^2 \tag{0.2}$$

Thus, the R_r of a 0.01 λ antenna, 50 km long at 60 Hz, would be 0.0197 Ω. As the radiated power $P_r = I^2 R_r$ where I is the antenna terminal current, whereas the power dissipated as heat in the antenna wire is $I^2 R_d$; when I is uniform, the P_r will be very much less than the power used to heat the antenna, given that the ohmic resistance R_d of any practical wire at room temperature will be very much larger and R_r. For example, the resistance of a 50-km long, 1/2-in. diameter solid copper wire could be 6.65 Ω. At dc, of course, no radiation of any sort takes place, as acceleration of charges is a condition for radiation of electromagnetic waves.

The second set of circumstances, which guarantees that any object subjected to low-frequency E and H fields usually does not experience effects of radiation, is that any configuration that carries electric currents sets up E and H field components which store energy without contributing to radiation. A short, linear antenna in free space (short electric dipole) generates, in addition to the radiation field E_r, an electrostatic field E_s and an induction field E_i. Neither E_s nor E_i contribute to the P_r [2,3]. Whereas E_r varies as $1/r$, where r is the distance from the antenna, E_i varies as $1/r^2$, and E_s as $1/r^3$. At a distance from the antenna of approximately one sixth of the wavelength ($r = \lambda/2\pi$), the E_i equals the E_r, and when $r \ll \lambda/6$ the E_r quickly becomes negligible in comparison with E_i and E_s. Similar results are obtained for other antenna configurations [4]. At 60 Hz the distance $\lambda/2\pi$ corresponds to about 800 km and objects at distances of a few kilometers or less from a 60-Hz system are exposed to nonradiating field components, which are orders of magnitude larger than the part of the field that contributes to radiation.

A living organism exposed to a static (dc) field or to a nonradiating near field may extract energy from it, but the quantitative description of the mechanism by which this extraction takes place is very different than at higher frequencies, where energy is transferred by radiation:

1. In the near field the relative magnitudes of E and H are a function of the current or charge configuration and the distance from the electric system. The E field may be much larger than the H field or vice versa (see Figure 0.2).
2. In the radiation field the ratio the E to H is fixed and equal to 377 in free space, if E is given in volt per meter and H in ampere per meter.
3. In the vicinity of most presently available human-made devices or systems carrying static electric charges, dc, or low-frequency (<1000 Hz) currents, the E and H fields will only under very exceptional circumstances be large enough to produce heating effects inside a living object, as illustrated by Figure 0.3. (This statement assumes that the living object does not form part of a conducting path

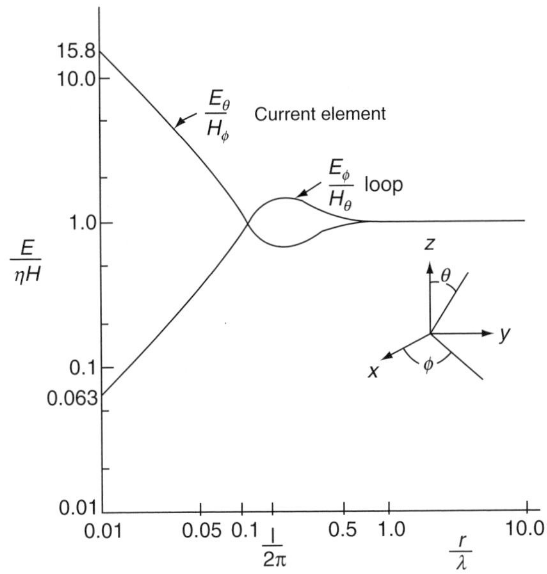

FIGURE 0.2
Ratio of E to H field (divided by wave impedance of free space $\eta = 377\ \Omega$) at $\theta = 90°$; for electric current element at origin along z-axis and for electrically small loop centered at the origin in x–y plane.

that permits direct entrance of current from a wire or conducting ground.) However, nonthermal effects are possible; thus an E field of sufficient magnitude may orient dipoles, or translate ions or polarizable neutral particles (see Chapter 3 and Chapter 4 in *BBA**).

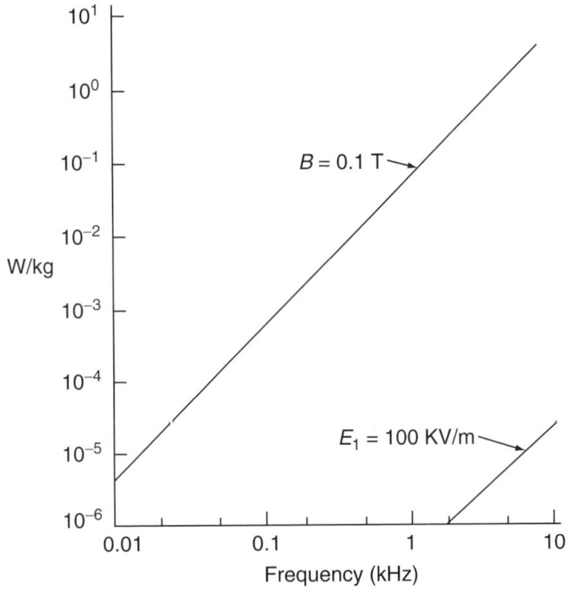

FIGURE 0.3
Top line: Eddy current loss produced in cylinder by sinusoidally time-varying axial H field. Cylinder parameters are conductivity $\sigma = 0.1\,\mathrm{S/m}$, radius 0.1 m, density $D = 1100\,\mathrm{kg/m^3}$, RMS magnetic flux density $0.1\,\mathrm{T} = 1000\,\mathrm{G}$. Watt per kilogram $= \sigma B^2 r^2 w^2/8D$; see Equation 0.15 and use power per volume $= J^2/\sigma$, *Lower line:* Loss produced by 60-Hz E field in Watt per kilogram $= \sigma\, {E_{int}}^2/D$, where external field E_1 is related to E_{int} by Equation 0.9 with $\varepsilon_2 = \varepsilon_0 \times 10^5$ at 1 kHz and $\varepsilon_0 = 8 \times 10^4$ at 10 kHz.

**BBA*: *Bioengineering and Biophysical Aspects of Electromagnetic Fields* (ISBN 0-8493-9539-9); *BMA*: *Biological and Medical Aspects of Electromagnetic Fields* (ISBN 0-8493-9538-0).

4. With radiated power it is relatively easy to produce heating effects in living objects with presently available human-made devices (see Chapter 10 in *BBA* and Chapter 5 in *BMA*). This does not imply, of course, that all biological effects of radiated radio frequency (RF) power necessarily arise from temperature changes.

The results of experiments involving exposure of organic materials and entire living organisms to static *E* and ELF *E* fields are described in *BBA*, Chapter 3. Various mechanisms for the interaction of such fields with living tissue are also discussed there and in *BBA*, Chapter 5. In the present introduction, we shall only point out that one salient feature of static (dc) and ELF *E* field interaction with living organisms is that the external or applied *E* field is always larger by several orders of magnitude than the resultant average internal *E* field [5,6]. This is a direct consequence of boundary conditions derived from Maxwell's equations [1–3].

0.2 Penetration of Direct Current and Low-Frequency Electric Fields into Tissue

Assuming that the two materials illustrated schematically in Figure 0.4 are characterized, respectively, by conductivities σ_1 and σ_2 and dielectric permittivities ε_1 and ε_2, we write *E*-field components parallel to the boundary as E_P and components perpendicular to the boundary as $E_\perp$. For both static and time-varying fields

$$E_{P1} = E_{P2} \tag{0.3}$$

and for static (dc) fields

$$\sigma_1 E_{\perp 1} = \sigma_2 E_{\perp 2} \tag{0.4}$$

as a consequence of the continuity of current (or conservation of charge). The orientations of the total *E* fields in media 1 and 2 can be represented by the tangents of the angles between the total fields and the boundary line

$$\tan\theta_1 = \frac{E_{\perp 1}}{E_{P1}}, \quad \tan\theta_2 = \frac{E_{\perp 2}}{E_{P2}} \tag{0.5}$$

From these equations it follows that

$$\tan\theta_1 = \frac{\sigma_2}{\sigma_1}\frac{E_{\perp 1}}{E_{P1}} = \frac{\sigma_2}{\sigma_1}\frac{E_{\perp 2}}{E_{P2}} = \frac{\sigma_2}{\sigma_1}\tan\theta_2 \tag{0.6}$$

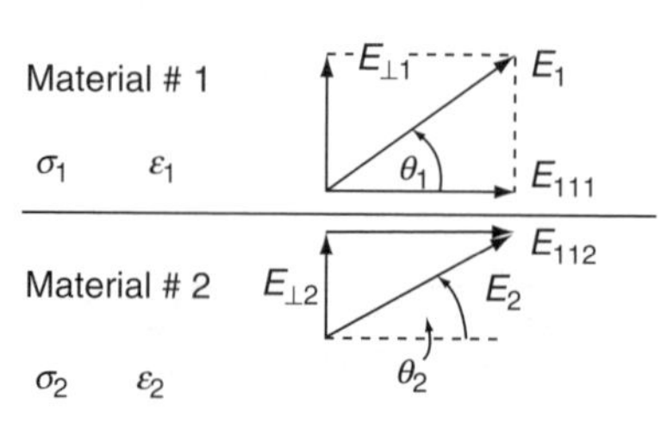

FIGURE 0.4
Symbols used in description of boundary conditions for *E*-field components.

If material 1 is air with conductivity [7] $\sigma_1 = 10^{-13}$ S/m and material 2 a typical living tissue with $\sigma_2 \approx 10^{-1}$ S/m (compare Chapter 3 in *BBA*), $\tan\theta_1 = 10^{12}\tan\theta_2$, and therefore even if the field in material 2 (the inside field) is almost parallel to the boundary so that $\theta_2 \cong 0.5^\circ$ or $\tan\theta_2 \approx (1/100)$, $\tan\theta_1 = 10^{10}$ or $\theta_1 = (\pi/2 - 10)^{-10}$ radians. Thus an electrostatic field in air, at the boundary between air and living tissue, must be practically perpendicular to the boundary. The situation is virtually the same at ELF although Equation 0.4 must be replaced by

$$\sigma_1 E_{\perp 1} - \sigma_2 E_{\perp 2} = -\mathrm{j}\omega\rho_\mathrm{s} \tag{0.7}$$

and

$$\varepsilon_1 E_{\perp 1} - \varepsilon_2 E_{\perp 2} = \rho_\mathrm{s} \tag{0.8}$$

where $\mathrm{j} = \sqrt{-1}$, ω is the radian frequency ($= 2\pi \times$ frequency), and ρ_s is the surface charge density. In Chapter 3 in *BBA* it is shown that at ELF the relative dielectric permittivity of living tissue may be as high as 10^6 so that $\varepsilon_2 = 10^6\ \varepsilon_0$, where ε_0 is the dielectric permittivity of free space $(1/36\ \pi)\ 10^{-9}$ F/m; however, it is still valid to assume that $\varepsilon_2 \leq 0^{-5}$. Then from Equation 0.7 and Equation 0.8

$$E_{\perp 1} = \frac{\sigma_2 + \mathrm{j}\omega\varepsilon_2}{\sigma_1 + \mathrm{j}\omega\varepsilon_1} E_{\perp 2} \tag{0.9}$$

which gives at 60 Hz with $\sigma_2 = 10^1$ S/m, $\sigma_1 = 10^{-13}$ S/m, $\varepsilon_2 \approx 10^{-5}$ F/m, and $\varepsilon_1 \approx 10^{-11}$ F/m

$$E_{\perp 1} = \frac{10^{-1} + \mathrm{j}_4 10^{-3}}{10^{-13} + \mathrm{j}_4 10^{-9}} E_{\perp 2} \approx \frac{\sigma_2}{\mathrm{j}\omega\varepsilon_1} = -\mathrm{j}\,(2.5 \times 10^7) E_{\perp 2} \tag{0.10}$$

This result, together with Equation 0.3 and Equation 0.5, shows that for the given material properties, the field in air must still be practically perpendicular to the boundary of a living organism: $\tan\theta_1$: $2.5(10^7)\tan\theta_2$.

Knowing now that the living organism will distort the E field in its vicinity in such a way that the external field will be nearly perpendicular to the boundary surface, we can calculate the internal field by substituting the total field for the perpendicular field in Equation 0.4 (dc) and Equation 0.9 (ELF). For the assumed typical material parameters we find that in the static (dc) case

$$\frac{E_\text{internal}}{E_\text{external}} \approx 10^{-12} \tag{0.11}$$

$$\rho_\mathrm{f} = \frac{3(\sigma_2\varepsilon_1 - \sigma_1\varepsilon_2)E_0}{2\sigma_1 + \sigma_2} \cos\vartheta\ \mathrm{C/m^2}$$

and for 60 Hz

$$\frac{E_\text{internal}}{E_\text{external}} \approx 4(10^{-8}) \tag{0.12}$$

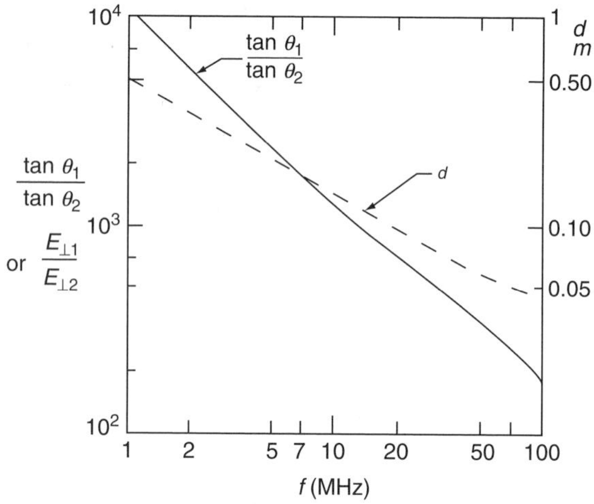

FIGURE 0.5
Orientation of E-field components at air–muscle boundary (or ratio of fields perpendicular to boundary); depth (d) at which field component parallel to boundary surface decreases by approximately 50% ($d = 0.6938$).

Thus, a 60-Hz external field of 100 kV/m will produce an average E_{internal} field of the order of 4 mV/m.

If the boundary between air and the organic material consists of curved surfaces instead of infinite planes, the results will be modified only slightly. Thus, for a finite sphere (with ε and σ as assumed here) embedded in air, the ratios of the internal field to the undisturbed external field will vary with the angle θ and distance r as indicated in Figure 0.5, but will not deviate from the results indicated by Equation 0.7 and Equation 0.8 by more than a factor of 3 [3,8]. Long cylinders ($L \ll r$) aligned parallel to the external field will have interior fields essentially equal to the unperturbed external field, except near the ends where the field component perpendicular to the membrane surface will be intensified approximately as above (see Chapter 9 and Chapter 10 in this volume).

0.3 Direct Current and Low-Frequency Magnetic Fields

Direct current H fields are considered in more detail in the Chapter 3, Chapter 5, and Chapter 8 in *BBA*. ELF H fields are considered in various places, including Chapter 5 and Chapter 7 in *BBA* and Chapter 2 and Chapter 11 in *BMA*. As the magnetic permeability μ of most biological materials is practically equal to the magnetic permeability μ_0 of free space, $4\pi(10^{-7})$ H/m, the dc or ELF H field "inside" will be practically equal to the H field "outside." The only exceptions are organisms such as the magnetotactic bacteria, which synthesize ferromagnetic material, discussed in Chapter 8 of *BBA*. The known and suggested mechanisms of interaction of dc H fields with living matter are:

1. Orientation of ferromagnetic particles, including biologically synthesized particles of magnetite.
2. Orientation of diamagnetically or paramagnetically anisotropic molecules and cellular elements [9].
3. Generation of potential differences at right angles to a stream of moving ions (Hall effect, also sometimes called a magnetohydrodynamic effect) as a result of the magnetic force $F_m = qvB \sin \theta$, where q is the the electric charge, v is the

velocity of the charge, B is the magnetic flux density, and $\sin\theta$ is the sine of the angle θ between the directions v and B. One well-documented result of this mechanism is a "spike" in the electrocardiograms of vertebrates subjected to large dc H fields.

4. Changes in intermediate products or structural arrangements in the course of light-induced chemical (electron transfer) reactions, brought about by Zeeman splitting of molecular energy levels or effects upon hyperfine structure. (The Zeeman effect is the splitting of spectral lines, characteristic of electronic transitions, under the influence of an external H field; hyperfine splitting of electronic transition lines in the absence of an external H field is due to the magnetic moment of the nucleus; such hyperfine splitting can be modified by an externally applied H field.) The magnetic flux densities involved not only depend upon the particular system and can be as high as 0.2 T (2000 G) but also <0.01 mT (100 G). Bacterial photosynthesis and effects upon the visual system are prime candidates for this mechanism [10,11].
5. Induction of E fields with resulting electrical potential differences and currents within an organism by rapid motion through a large static H field. Some magnetic phosphenes are due to such motion [12].

Relatively slow time-varying H fields, which are discussed in the basic mechanisms and therapeutic uses chapters (Chapter 5 of *BBA* and Chapter 11 in *BMA*), among others, may interact with living organisms through the same mechanisms that can be triggered by static H fields, provided the variation with time is slow enough to allow particles of finite size and mass, located in a viscous medium, to change orientation or position where required (mechanism 1 and 2) and provided the field intensity is sufficient to produce the particular effect. However, time-varying H fields, including ELF H fields, can also induce electric currents into stationary conducting objects. Thus, all modes of interaction of time-varying E fields with living matter may be triggered by time-varying, but not by static, H fields.

In view of Faraday's law, a time-varying magnetic flux will induce E fields with resulting electrical potential differences and "eddy" currents through available conducting paths. As very large external ELF E fields are required (as indicated by Equation 0.9 through Equation 0.12) to generate even small internal E fields, many human-made devices and systems generating both ELF E and H fields are more likely to produce physiologically significant internal E fields through the mechanism of *magnetic* induction.

The induced voltage V around some closed path is given by

$$V = \oint E \cdot \mathrm{d}\ell = -\iint \frac{\partial B}{\partial t}\,\mathrm{d}s \tag{0.13}$$

where E is the induced E field. The integration $\oint E\,\mathrm{d}\ell$ is over the appropriate conducting path, $\partial B/\partial t$ is the time derivative of the magnetic flux density, and the "dot" product with the surface element, ds, indicates that only the component of $\partial B/\partial t$ perpendicular to the surface, i.e., parallel to the direction of the vector ds, enclosed by the conducting path, induces an E field. To obtain an order-of-magnitude indication of the induced current that can be expected as a result of an ELF H field, we consider the circular path of radius r, illustrated by Figure 0.6. Equation 0.13 then gives the magnitude of the E field as

$$E = \frac{\omega B r}{2} \tag{0.14}$$

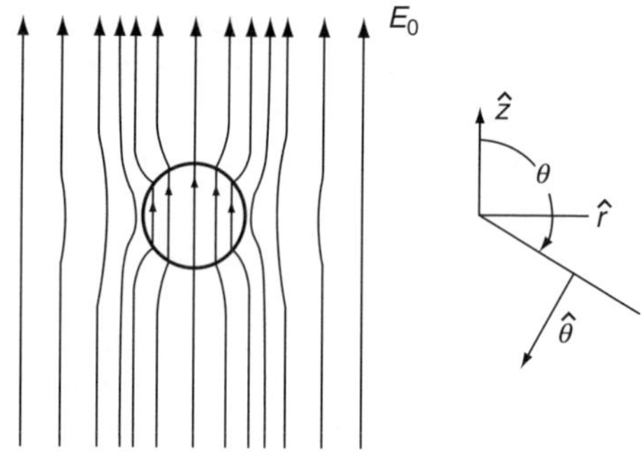

FIGURE 0.6
E field when sphere of radius R, conductivity σ_2, and dielectric permittivity ε_2 is placed into an initially uniform static field ($E = 2E_0$) within a medium with conductivity σ_1 and permittivity ε_1. The surface charge density is $\rho_r = \frac{3(\sigma_2\varepsilon_1 - \sigma_1\varepsilon_2)E_0}{2\sigma_1 + \sigma_2}\cos\theta\ \mathrm{C/m^2}$.

$$r < R \quad \bar{E} = \frac{3\sigma_1 E_0}{2\sigma_1 + \sigma_2}\hat{z} \qquad \varepsilon_2, \sigma_2$$

$$r < R \quad \bar{E} = E_0 \cos\theta\left[1 + \frac{2R^3(\sigma_2 \propto \sigma_1)}{r^3(2\sigma_1 + \sigma_2)}\right]\hat{r} \qquad \varepsilon_1, \sigma_1$$

$$- E \sin\theta\left[1 - \frac{R^3(\sigma_2 \propto \sigma_1)}{r^3(2\sigma_1 + \sigma_2)}\right]\hat{\theta}$$

where ω is the $2\pi f$ and f is the frequency. The magnitude of the resulting electric current density J in ampere per square meter is*

$$J = \sigma E = \frac{\sigma\omega B r}{2} \tag{0.15}$$

where σ is the conductivity along the path in Siemens per meter. In the SI (Systeme Internationale) units used throughout this book, B is measured in tesla ($T = 10^4$ G) and r in meters. Choosing for illustration a circular path of 0.1 m radius, a frequency of 60 Hz, and a conductivity of 0.1 S/m, Equation 0.14 and Equation 0.15 give $E = 18.85$ B and $J = 1.885$ B. The magnetic flux density required to obtain a current density of 1 mA/m^2 is 0.53 mT or about 5 G. The E field induced by that flux density along the circular path is 10 mV/m. To produce this same 10 mV/m E_{internal} field by an external 60 Hz E_{external} field would require, by Equation 0.12, a field intensity of 250 kV/m.

As the induced voltage is proportional to the time rate of change of the H field (Equation 0.13), implying a linear increase with frequency (Equation 0.14), one would expect that the ability of a time-varying H field to induce currents deep inside a conductive object would increase indefinitely as the frequency increases; or conversely, that the magnetic flux density required to induce a specified E field would decrease linearly with frequency, as indicated in Figure 0.7. This is not true however, because the displacement current density $\partial D/\partial t$, where $D = \varepsilon E$, must also be considered as the frequency increases. This leads to the wave behavior discussed in Part III, implying that at sufficiently high frequencies the effects of both external E and H fields are limited

*Equation 0.15 neglects the H field generated by the induced eddy currents. If this field is taken into account, it can be shown that the induced current density in a cylindrical shell of radius r and thickness Δ is given by $\Delta r < 0.01$ m^2/[$1 + j\Delta r/\delta^2$], where $H_0 = B_0/\mu_0$ and δ is the skin depth defined by Equation 0.17 below. However, for conductivities of biological materials ($\sigma < 5$ s/m) one obtains at audio frequencies $\delta > 1$ m and as for most dimensions of interest $\Delta r < 0.01$ m^2 the term $j\Delta r/\delta^2$ becomes negligible. The result $-jrH_0/\delta^2$ is then identical with Equation 0.15.

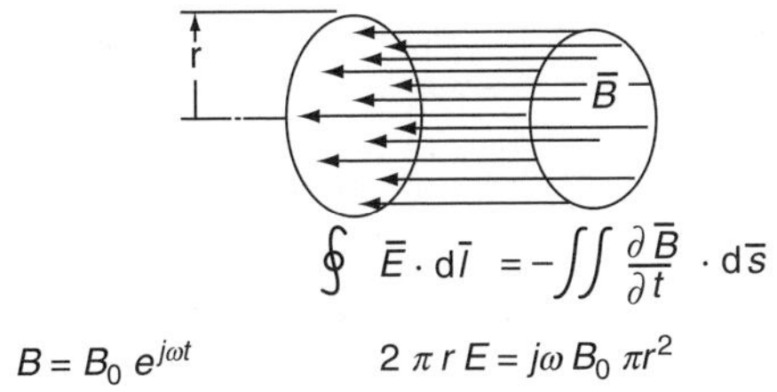

FIGURE 0.7
Circular path (loop) of radius r enclosing uniform magnetic flux density perpendicular to the plane of the loop. For sinusoidal time variation $B = B_0 e^{j\omega t}$.

by reflection losses (Figure 0.8 through Figure 0.10) as well as by skin effect [13], i.e., limited depth of penetration d in Figure 0.5.

0.4 RF Fields

At frequencies well below those where most animals and many field-generating systems have dimensions of the order of one free space wavelength, e.g., at 10 MHz where $\lambda = 30$ m, the skin effect limits penetration of the external field. This phenomenon is fundamentally different from the small ratio of internal to external E fields described in Equation 0.4 (applicable to dc) and Equation 0.9.

Equation 0.9 expresses a "boundary condition" applicable at all frequencies, but as the angular frequency ω increases (and in view of the rapid decrease with frequency of the dielectric permittivity ε_2 in biological materials—see Chapter 3 of *BBA*, the ratio of the normal component of the external to the internal E field at the boundary decreases

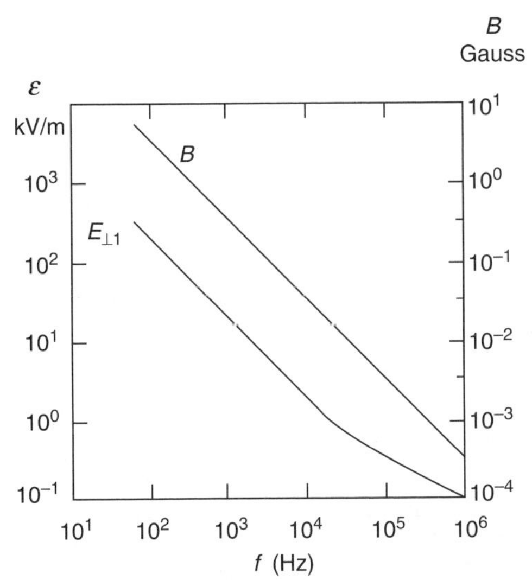

FIGURE 0.8
External E and H field required to obtain an internal E field of 10 mV/m (conductivity and dielectric permittivity for skeletal muscle from Foster, K.R., Schepps, J.L., and Schwan, H.P. 1980. *Biophys. J.*, 29:271–281. H-field calculation assumes a circular path of 0.1-m radius perpendicular to magnetic flux).

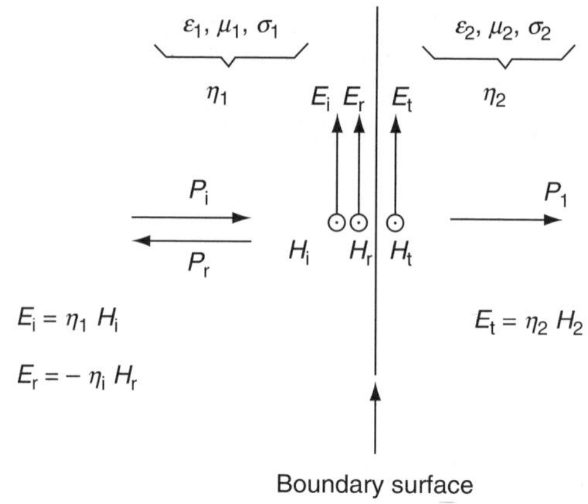

FIGURE 0.9
Reflection and transmission of an electromagnetic wave at the boundary between two different media, perpendicular incidence; P_i = incident power, P_r = reflected power, P_t = transmitted power.

with increasing frequency. This is illustrated by Figure 0.10 where $\tan\theta_1/\tan\theta_2$ is also equal to $E_{\perp 1}/E_{\perp 2}$ in view of Equation 0.3, Equation 0.5, and Equation 0.9. However, at low frequencies the total field inside the boundary can be somewhat larger than the perpendicular field at the boundary; and any field variation with distance from the boundary is not primarily due to energy dissipation, but in a homogeneous body is a consequence of shape. At RF, on the other hand, the *E* and *H* fields of the incoming

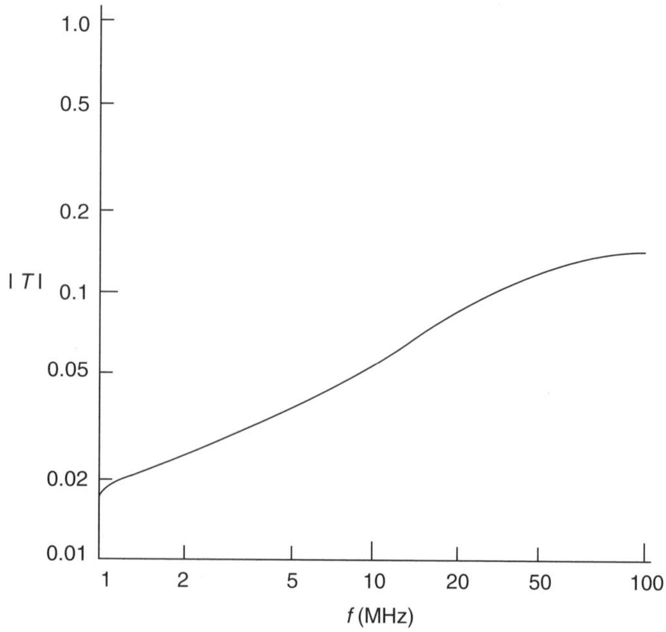

FIGURE 0.10
Magnitude of transmission coefficient T for incident E field parallel to boundary surface. $T = E_t/E_i$: reflection coefficient $r = E_r/E_i = T\text{-}1$. Γ and T are complex numbers; ε_r and σ for skeletal muscle from Chapter 3 in *BBA*.

electromagnetic wave, after reflection at the boundary, are further decreased due to energy dissipation. Both E and H fields decrease exponentially with distance from the boundary

$$g(z) = Ae^{-\frac{z}{\delta}} \tag{0.16}$$

where $g(z)$ is the field at the distance z and A is the magnitude of the field just *inside* the boundary.

As defined by Equation 0.16 the skin depth δ is the distance over which the field decreases to $1/e$ (= 0.368) of its value just *inside* the boundary. (Due to reflection, the field A just inside the boundary can already be very much smaller than the incident external field; see Figure 0.8 and Figure 0.9.)

Expressions for δ given below were derived [2,3,13,14] for plane boundaries between infinite media. They are reasonable accurate for cylindrical structures if the ratio of radius of curvature to skin depth (r_0/δ) is larger than about five [13]. For a good conductor

$$\delta = \frac{1}{\sqrt{\pi f \mu \sigma}} \tag{0.17}$$

where a good conductor is one for which the ratio p of conduction current, $J = \sigma E$, to displacement current, $\partial D/\partial t = \varepsilon\,(\partial E/\partial t) = j\omega\varepsilon E$ is large:

$$p = \frac{\sigma}{\omega\varepsilon} \gg 1 \tag{0.18}$$

Since for most biological materials p is of the order of one ($0.1 < p < 10$) over a very wide frequency range (see Chapter 3 of *BBA*), it is frequently necessary to use the more general expression [13]

$$\delta = \frac{1}{\omega\left[\frac{\mu\varepsilon}{2}(\sqrt{1+p^2} - 1\right]^{1/2}} \tag{0.19}$$

The decrease of field intensity with distance from the boundary surface indicated by Equation 0.16 becomes significant for many biological objects at frequencies where $r_0/\delta \geq 5$ is not satisfied. However, the error resulting from the use of Equation 0.16 and Equation 0.17 or Equation 0.19 with curved objects is less when $z < \delta$. Thus at $z = 0.693\,\delta$, where $g(z) = 0.5$ A from Equation 0.16 and Equation 0.17, the correct values of $g(z)$, obtained by solving the wave equation in cylindrical coordinates, differs only by 20% (it is 0.6 A) even when r_0/δ is as small as 2.39 [14]. Therefore, Figure 0.10 shows the distance $d = 0.693\,\delta$, at which the field decreases to half of its value just inside the boundary surface, using Equation 0.19 with typical values for σ and ε for muscle from Figure 0.11. It is apparent that the skin effect becomes significant for humans and larger vertebrates at frequencies >10 MHz.

Directly related to skin depth, which is defined for fields varying sinusoidally with time, is the fact that a rapid transient variation of an applied magnetic flux density constitutes an exception to the statement that the dc H field inside the boundary is equal to the H field outside. Thus, from one viewpoint one may consider the rapid application or removal of a dc H field as equivalent to applying a high-frequency field during the switching period, with the highest frequencies present of the order of $1/\tau$, where τ is the rise time of the applied step function. Thus, if $\tau < 10^{-8}$ s, the skin effect will be important during the transient period, as d in Figure 0.5 is <5 cm above 100 MHz. It is also possible to calculate directly the magnetic flux density inside a conducting cylinder as a function of radial position r and time t when a magnetic pulse is applied in the axial

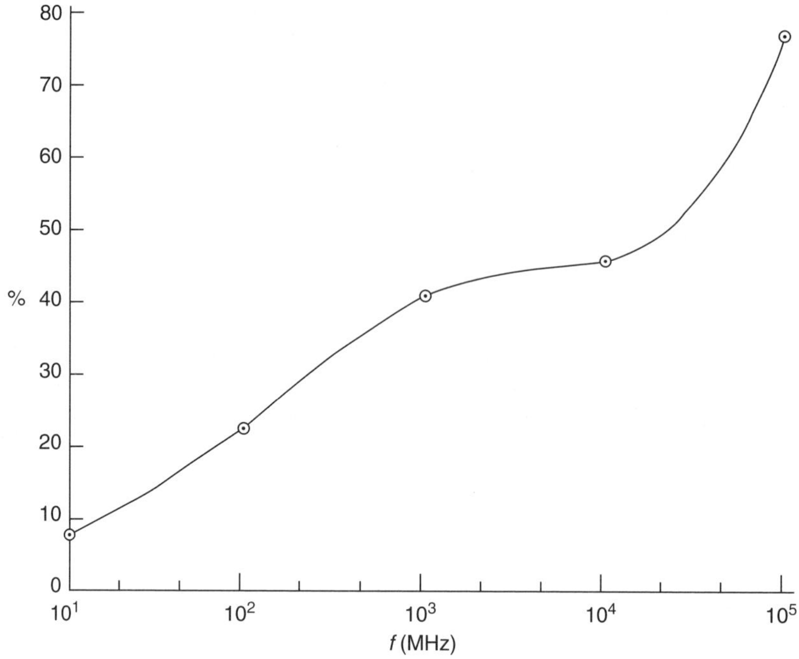

FIGURE 0.11
Ratio of transmitted to incident power expressed as percent of incident power. Air–muscle interface, perpendicular incidence (Equation 0.31, Table 0.1).

direction [15,16]. Assuming zero rise time of the applied field B_0, i.e., a true step function, one finds that the field inside a cylinder of radius a is

$$B = B_0\left[1 - \sum_{k=1}^{\infty} J_0\left(r\frac{v_k}{a}\right)e^{-t/T_k}\right] \tag{0.20}$$

where $J_0\ (r\ v_k/a)$ is the zero-order Bessel function of argument $r\ v_k/a$ and the summation is over the nulls of J_0 designated v_k (the first four values of v_k are 2.405, 5.520, 8.654, and 11.792).* T_k is the rise time of the kth term in the series and is given by

$$T_k = \frac{\mu_0 \sigma a^2}{v_k} \tag{0.21}$$

As v_k increases, the rise time decreases and therefore the longest delay is due to the first term in the summation with $k = 1$

$$T_1 = \frac{\mu_0 \sigma a^2}{2.405} \tag{0.22}$$

For a cylinder with 0.1 m radius and a conductivity $\sigma \approx 1$ S/m, which is a typical value for muscle between 100 and 1000 MHz, Equation 0.22 gives $T_1 = 2.6 \times 10^{-8}$ s. This finite rise time (or decay time in case of field removal) of the internal H field may be of some importance when pulsed H fields are used therapeutically [17]. It might also be used

*This result is based on solution of $\partial B/\partial t = (1/\mu_0)\nabla^2 B$, which is a consequence of Ampere's and Farraday's laws when displacement is disregarded. Equations 0.20 to 0.22 are therefore only correct when $p \gg 1$.

to measure noninvasively the conductivity of biological substances *in vivo* through determination of the final decay rate of the voltage induced into a probe coil by the slowly decaying internal field after the applied field is removed [16].

The properties of biological substances in the intermediate frequency range, above ELF (>300 Hz), and below the higher RFs, where wave behavior and skin effect begin to be important (~20 MHz), are discussed in Chapter 3 of *BBA*. However, many subsequent chapters are concerned with biological effects at dc and ELF frequencies below a few kilohertz, while others deal primarily with the higher RFs, >50 MHz. One reason for this limited treatment of the intermediate frequency range is that very little animal data are available for this spectral region in comparison with the large number of experiments performed at ELF and microwave frequencies in recent years.* Another reason is that most electrical processes known to occur naturally in biological systems—action potentials, EKG, EEG, ERG, etc.—occur at dc and ELF frequencies. Therefore, one might expect some physiological effects from external fields of appropriate intensity in the same frequency range, even if the magnitude of such fields is not large enough to produce thermal effects. As illustrated by Figure 0.3 and Figure 0.7, most E fields below 100 kHz set up by currently used human-made devices, and most H fields below 10 kHz except the very strongest, are incapable of producing thermal effects in living organisms, excluding, of course, fields accompanying currents directly introduced into the organism via electrodes. Thus, the frequencies between about 10 and 100 kHz have been of relatively little interest because they are not very likely to produce thermal or other biological effects. On the other hand, the higher RFs are frequently generated at power levels where enough energy may be introduced into living organisms to produce local or general heating. In addition, despite skin effect and the reflection loss to be discussed in more detail below, microwaves modulated at an ELF rate may serve as a vehicle for introducing ELF fields into a living organism of at least the same order of magnitude as would be introduced by direct exposure to ELF. Any effect of such ELF-modulated microwaves would, of course, require the existence of some amplitude-dependent demodulation mechanism to extract the ELF from the microwave carrier.

Among the chapters dealing with RF, Chapter 10 and Chapter 11 of *BBA* give the necessary information for establishing the magnitude of the fields present in biological objects: (1) experimental techniques and (2) analytical methods for predicting field intensities without construction of physical models made with "phantom" materials, i.e., dielectric materials with properties similar to those of living objects which are to be exposed. As thermal effects at microwave frequencies are certainly important, although one cannot assume *a priori* that they are the only biological effects of this part of the spectrum, and as some (but not all) thermal effects occur at levels where the thermoregulatory system of animals is activated, thermoregulation in the presence of microwave fields is discussed in Chapter 5 of *BMA*, as well as in Chapter 10 of *BBA*. Not only are the therapeutic applications of microwaves based upon their thermal effects, but also the experimental establishment of possible nonthermal effects at the threshold of large scale tissue heating in particular living systems and also requires thorough understanding of thermoregulatory mechanisms. The vast amount of experimental data obtained on animal systems exposed to microwave is discussed in Chapter 3 and Chapter 4 in *BMA*. Both nonmodulated fields and modulated fields, where the type of modulation had no apparent effect other than modification of the average power level, are considered. These chapters and the Chapter 9 in *BMA* are considered to be very new extension of experiments into exposures to ultra-short and to ultra-high power pulses.

*Though this statement was written in for the second edition in 1995, it continues to be true in 2005—Ben Greenebaum.

At the higher RFs, the external E field is not necessarily perpendicular to the boundary of biological materials (see Figure 0.4 and Figure 0.10), and the ratio of the total external E field to the total internal field is not given by Equation 0.9. However, the skin effect (Equation 0.16 through Equation 0.19) and reflection losses still reduce the E field within any biological object below the value of the external field. As pointed out in Chapter 3, dielectric permittivity and electrical conductivity of organic substances both vary with frequency. At RF, most biological substances are neither very good electrical conductors nor very good insulators, with the exception of cell membranes, which are good dielectrics at RF but at ELF can act as intermittent conductors or as dielectrics and are ion-selective [18–20]). The ratio p (Equation 0.18) is neither much smaller nor very much larger than values shown for typical muscle tissue [21,22] in Table 0.1.

Reflection loss at the surface of an organism is a consequence of the difference between its electrical properties and those of air. Whenever an electromagnetic wave travels, from one material to another with different electrical properties, the boundary conditions (Equation 0.3 and Equation 0.8) and similar relations for the H field require the existence of a reflected wave. The expressions for the reflection coefficient

$$\Gamma = \frac{E_r}{E_i} \tag{0.23}$$

and the transmission coefficient

$$T = \frac{E_t}{E_i} \tag{0.24}$$

become rather simple for loss-free dielectrics ($p \ll 1$) and for good conductors ($p \gg 1$). As biological substances are neither the most general expressions for Γ and T, applicable at plane boundaries, are needed [3,13]. For perpendicular incidence, illustrated by Figure 0.8,

$$\Gamma = \frac{\eta_2 - \eta_1}{\eta_2 + \eta_1} \tag{0.25}$$

$$T = \frac{2\eta_2}{\eta_2 + \eta_1} = 1 + \Gamma \tag{0.26}$$

TABLE 0.1

Ratio p of Conduction Current to Displacement as a Function of Frequency

f (MHz)	σ	ε_r	$p = \frac{\sigma}{\omega \varepsilon_0 \varepsilon_r}$
1	0.40	2000	3.6
10	0.63	160	7.1
100	0.89	72	2.2
10^3	1.65	50	0.59
10^4	10.3	40	0.46
10^5	80	6	2.4

where η_1 and η_2 are the wave impedances, respectively, of mediums 1 and 2. The wave impedance of a medium is the ration of the E to the H field in a plane wave traveling through that medium; it is given by [13]

$$\eta = \left(\frac{j\omega\mu}{\sigma + j\omega\varepsilon} \right)^{1/2} \tag{0.27}$$

Clearly Γ and T are in general complex numbers, even when medium 1 is air for which Equation 0.27 reduces to the real quantity $\eta_0 = \sqrt{\mu_0/\varepsilon_0}$, because medium 2, which here is living matter, usually has a complex wave impedance at RFs.

The incident, reflected, and transmitted powers are given by [13]

$$P_i = R_1 |E_i|^2 \frac{1}{\eta_1^*} = \frac{|E_i|^2}{|\eta_1|^2} R_1 \tag{0.28}$$

$$P_r = R_1 |E_r|^2 \frac{1}{\eta_1^*} = \frac{|E_r|^2}{|\eta_1|^2} R_1 \tag{0.29}$$

$$P_t = R_1 |E_t|^2 \frac{1}{\eta_2^*} = \frac{|E_t|^2}{|\eta_2|^2} R_2 \tag{0.30}$$

where the E fields are effective values ($E_{\text{eff}} = E_{\text{peak}}/\sqrt{2}$) of sinusoidal quantities, R_1 signifies ''real part of,'' η^* is the complex conjugate of η, and R_1 and R_2 are the real parts of η_1 and η_2. If medium 1 is air, $\eta_1 = R_1 = 377\ \Omega$, it follows from Equation 0.23, Equation 0.24, and Equation 0.28 through Equation 0.30 and conservation of energy that the ratio of the transmitted to the incident real power is given by

$$\frac{P}{P_1} = |T|^2 \frac{\eta_1 \eta_2^* + \eta_1^* \eta_2}{2|\eta_2|^2} = 1 - \frac{P_r}{P_i} = 1 - |\Gamma|^2 \tag{0.31}$$

The magnitude of the transmission coefficient T for the air–muscle interface over the 1- to 100-MHz frequency range is plotted in Figure 0.9, which shows that the magnitude of the transmitted E field in muscle tissue is considerably smaller than the E field in air. The fraction of the total incident power that is transmitted (Equation 0.31) is shown in Figure 0.11, indicating clearly that reflection loss at the interface decreases with frequency. However, for deeper lying tissue this effect is offset by the fact that the skin depth δ (Equation 0.19) also decreases with frequency (Figure 0.12) so that the total power penetrating beyond the surface decreases rapidly.

In addition to reflection at the air–tissue boundary, further reflections take place at each boundary between dissimilar materials. For example, the magnitude of the reflection coefficient at the boundary surface between muscle and organic materials with low-water content, such as fat or bone, is shown in Table 0.2.

The situation is actually more complicated than indicated by Figure 0.9 and Figure 0.11, because the wave front of the incident electromagnetic wave may not be parallel to the air–tissue boundary. Two situations are possible: the incident E field may be polarized perpendicular to the plane of incidence defined in Figure 0.13 (perpendicular polarization, Figure 0.13a) or parallel to the plane of incidence (parallel polarization, Figure 0.13b). The transmission and reflection coefficients [8] are different for the two types of polarization and also become functions of the angle of incidence α_1:

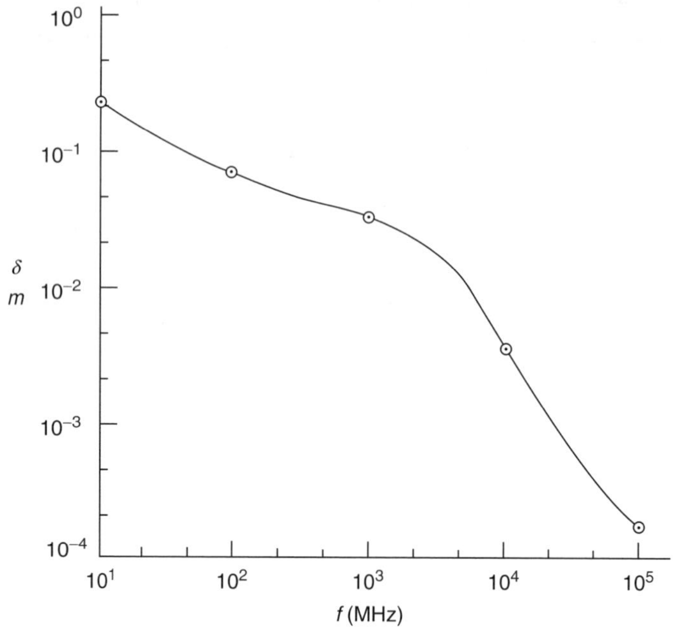

FIGURE 0.12
Electromagnetic skin depth in muscle tissue from plane wave expression (Equation 0.19, Table 0.1).

$$\text{Perpendicular polarization}\begin{cases} T_\perp = \dfrac{2\eta_2\cos\alpha_1}{\eta_2\cos\alpha_1 + \eta_1\cos\alpha_2} \\ \Gamma_\perp = \dfrac{\eta_2\cos\alpha_1 - \eta_1\cos\alpha_2}{\eta_2\cos\alpha_1 + \eta_1\cos\alpha_2} \end{cases} \qquad (0.32),(0.33)$$

$$\text{Parallel polarization}\begin{cases} T_\text{p} = \dfrac{2\eta_2\cos\alpha_1}{\eta_2\cos\alpha_2 + \eta_1\cos\alpha_1} \\ \Gamma_\text{p} = \dfrac{\eta_1\cos\alpha_1 - \eta_2\cos\alpha_2}{\eta_2\cos\alpha_2 + \eta_1\cos\alpha_1} \end{cases} \qquad (0.34),(0.35)$$

where α_2 is given by the generalized Snell's law (when both the media have the magnetic permeability of free space) by

TABLE 0.2

Reflection Coefficient "Capital Gamma" for Low–Water-Content Materials

	Fat or Bone		
***f* (MHz)**	**σ (S/m)**	**ε_r**	**Muscle[a]–Fat (Γ)**
10^2	0.048	7.5	0.65
10^3	0.101	5.6	0.52
10^4	0.437	4.5	0.52

[a] σ and ε_r for muscle from Table 0.1.

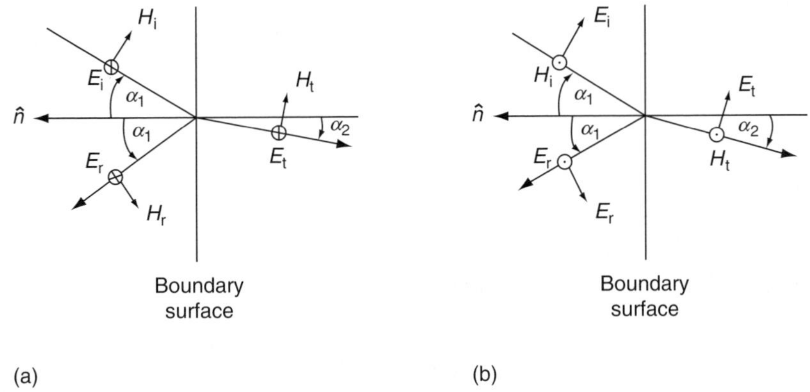

FIGURE 0.13
Oblique incidence of an electromagnetic wave at the boundary between two different media. (a) Perpendicular polarization (*E* vector perpendicular to plane of incidence); (b) parallel polarization (*E* vector parallel to plane of incidence). The plane of incidence is the plane formed by the surface normal (unit vector **n** and the direction of the incident wave); ⊗ indicates a vector into the plane of the paper; ⊙ indicates a vector out of the plane of the paper. The orientation of the field vectors in the transmitted field is shown for loss-free dielectrics. For illustration of the transmitted wave into a medium with finite conductivity, where the wave impedance η_2 becomes a complex number, see Stratton, J.A., *Electromagnetic Theory*, McGraw-Hill, New York, 1941, p. 435.

$$\sin\alpha_2 = \frac{\sqrt{\varepsilon_1}}{\sqrt{\varepsilon_2 - \mathrm{j}\frac{\sigma_2}{\omega}}} \tag{0.36}$$

so that $\cos\alpha_2 = \sqrt{1-\sin^2\alpha_2}$ is a complex number unless $\rho_2 = (\sigma_2/\omega\varepsilon_2) = 1$.

As illustration, the variation with angle of incidence of the transmission coefficient for parallel polarization at the air–muscle interface at 10 MHz, is shown in Figure 0.14. It is apparent that the transmitted field is not necessarily maximized by perpendicular incidence in the case of parallel polarization. Furthermore, whenever $p \approx 1$ or $p > 1$ (see Table 0.1, above), α_2 is complex, which causes the waves entering the tissue to be inhomogeneous—they are not simple plane waves, but waves where surfaces of constant phase and constant amplitude do not coincide [3,23]; only the planes of constant amplitude are parallel to the boundary surface.

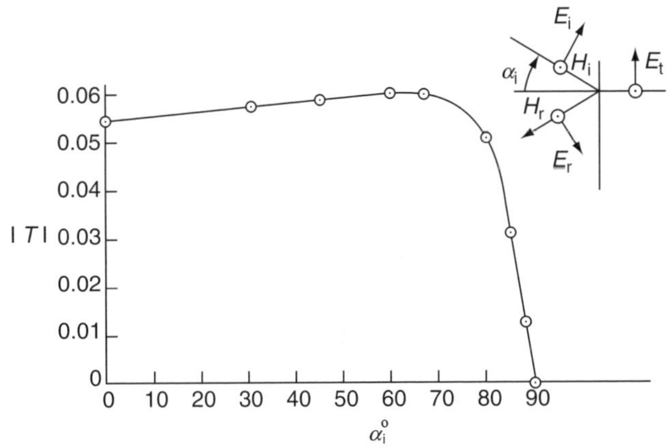

FIGURE 0.14
Magnitude of complex transmission coefficient for parallel polarization versus angle of incidence α_1 at 10 MHz (*E* field in plane of incidence, *H* field parallel to boundary plane; $\sigma_2 = 0.7\,\mathrm{S/m}$, $\varepsilon_{r2} = 150$, $T = E_t/E_r$).

Analytical solutions for nonplanar structures taking into account size and shape of entire animals have been given [24] and are also described in the RF modeling Chapter 10 of *BBA*.

0.5 Biophysical Interactions of Fields: Ionization, Ionizing Radiation, Chemical Bonds, and Excitation

RF fields can be characterized as nonionizing radiation. By this we mean that there is not enough energy in a single quantum of energy, hf, to ionize an atom or a molecule at RFs, where h is Planck's constant and f is the frequency. By comparison radiation in the UV or x-rays often lead to ionization. It is desirable to begin by reviewing the differences between ionizing and nonionizing radiations, to explain ionization phenomena and also to discuss related excitation phenomena, which require less energy than ionization. Then a number of the proposed models concerning atomic or molecular-level interactions of fields will be introduced. A number of these theories will be discussed and their predictions compared with experimental results in later chapters, including Chapter 5 through Chapter 7 and Chapter 9 in *BBA*; Chapter 9 and Chapter 11 in *BMA*. Heating, cell excitation, electroporation, and other results of high-intensity fields have been accepted as explanations for many bioelectromagnetic phenomena. For low-intensity exposure, however, no theory is widely accepted as a general explanation for bioelectromagnetic phenomena, and few specific phenomena have accepted explanations. It is quite possible that no general explanation exists and that more than one mechanism of interaction between fields will be found to be operating, depending on the situation. Binhi's book [25] contains a good summary of most recent theoretical proposals, including comparisons with data and critiques of their strong and weak points, as well as his own theory.

We note first that the energy of electromagnetic waves is quantized with the quantum of energy (in joules) being equal to Planck's constant ($h = 6.63 \times 10^{-34}$ J s) times the frequency. This energy can also be expressed in electron volts, i.e., in multiples of the kinetic energy acquired by an electron accelerated through a potential difference of 1 eV (1 eV $\approx 1.6 \times 10^{-19}$ J). Energy quanta for a few frequencies are listed in Table 0.3.

Quantized energy can "excite" molecules; appropriate frequencies can couple to vibrational and rotational oscillation; and if the incident energy quantum has sufficient magnitude it can excite other changes in the electron configuration, such as changing an electron to another (unoccupied) energy level or tearing an electron away from one of the constituent atoms, the latter process called as ionization. The energy required to remove one electron from the highest energy orbit of a particular chemical element is called its "ionization potential." Typical ionization potentials are of the order 10 eV; for example, for the hydrogen atom it is 13.6 eV and for gaseous sodium it is 5.1 eV. As chemical binding forces are essentially electrostatic, ionization implies profound chemical changes. Therefore ionization by any outside agent of the complex compounds that make up a living system leads to profound and often irreversible changes in the operation of that system.

Table 0.3 shows that even the highest RF (millimeter waves) has quantum energies well below the ionization potential of any known substance; thus one speaks of nonionizing radiation when referring to electromagnetic waves below UV light frequencies. Ionizing radiation includes UV and higher frequency electromagnetic waves (x-rays, γ-rays).

TABLE 0.3

Wave and Quantum Characteristics of Various Types of Radiation

Name of Radiation or Application	Frequency (Hz)	Wavelength (m)	Energy of 1 Quantum of Radiation (eV)
UHF TV	7×10^8	0.43	2.88×10^{-6}
Microwave radar	10^{10}	3×10^{-2}	4.12×10^{-5}
Millimeter wave	3×10^{11}	1×10^{-3}	1.24×10^{-3}
Visible light	6×10^{14}	5×10^{-7}	2.47
Ionizing UV	10^{16}	3×10^{-4}	41.2
Soft x-ray	10^{18}	3×10^{-10}	4120
Penetrating x-ray	10^{20}	3×10^{-12}	4.12×10^5

This explanation of the difference between ionizing and nonionizing radiation should not imply that nonionizing electromagnetic radiation cannot have profound effects upon inorganic and organic substances. As excitation of coherent vibrational and rotational modes requires considerably less energy than ionization, it could occur at RF; this will be discussed in later chapters. In addition, many other possible biological effects require energies well below the level of ionizing potentials. Examples are tissue heating, dielectrophoresis, depolarization of cell membranes, mechanical stress due to piezoelectric transduction, or dielectric saturation, resulting in the orientation of the polar side chains of macromolecules and leading to the breaking of hydrogen bonds. These and other mechanisms will be discussed by the authors of several chapters (see especially Chapter 5 through Chapter 7 of *BBA* and Chapter 9 of *BMA*), who will also give estimates of rates at which energy must be delivered to produce particular effects.

Returning to the discussion of ionization, it is important to note that ionization of a chemical element can be brought about not only by absorption of electromagnetic energy, but also by collision either with foreign (injected) atoms, molecules, or subatomic particles of the requisite energy, or by sufficiently violent collision among its own atoms. The latter process constitutes ionization by heating, or thermal breakdown of a substance, which will occur when the kinetic energy of the colliding particles exceeds the ionization potential. As the average thermal kinetic energy of particles is related to temperature [26] by $W = kT$ where k is Boltzmann's constant ($= 1.38 \times 10^{-23}$ J/K), we find that the required temperature is

$$1.38(10^{-23})T \approx 5\ \text{eV} \approx (5)1.6(10^{-19})\,\text{J}$$

$$T \approx 5(10^4)\,\text{K}$$

which is about twice the temperature inside a lightning stroke [27] and orders of magnitude higher than any temperature obtainable from electromagnetic waves traveling through air.

Actually, initiation of lightning strokes is an example of ionization by collision with injected energetic particles. The few free electrons and ions always present in the air due to ionization by cosmic rays are accelerated by the E fields generated within clouds to velocities corresponding to the required ionization energy. Only when the field is large enough to impart this energy over distances shorter than the mean free path of the free electrons or ions at atmospheric pressure can an avalanche process take place: an accelerated electron separates a low-energy electron from the molecule with which it collides and in the process loses most of its own energy; thus, one high-energy free electron is exchanged for two free low-energy electrons and one positive ion. Both the electrons are in turn accelerated again by the field, giving them high kinetic energy before they collide with neutral molecules; their collision produces four free electrons and the multiplication process continues. The breakdown field strength for air at atmospheric pressure is approximately 3×10^6 V/m, implying a mean free path of electrons

$$\Delta\ell \approx [5\,\text{eV}/3 \times 10^6\,\text{V/m}] \approx 10^{-6}\,\text{m}$$

However, this model is not entirely accurate because the actual mean free path corresponds to energies of the order of 0.1 eV, which is only sufficient to excite vibrational modes in the target molecule. Apparently such excitation is sufficient to cause ionization if the collision process lasts long enough [28].

Except for some laboratory conditions where a sufficiently high potential difference can be applied directly across a biological membrane to bring about its destruction, collisional ionization is generally not a factor in the interaction of electromagnetic waves with tissue: The potential difference required for membrane destruction [29] is between 100 nV and 300 mV, corresponding to a field strength of the order of 2×10^7 V/m, assuming a membrane thickness ($d = 100$ Å; $E = V/d$). However, there is a third mechanism of ionization that is particularly important in biological systems. When a chemical compound of the type wherein positive and negative ions are held together by their electrostatic attraction, such as the ionic crystal NaCl, is placed in a suitable solvent, such as H_2O, it is separated into its ionic components. The resulting solution becomes an electrolyte, i.e., an electrically conducting medium in which the only charge carriers are ions.

In this process of chemical ionization, the Na^+ cations and Cl^- anions are separated from the original NaCl crystal lattice and individually surrounded by a sheet of solvent molecules, the "hydration sheath." If the solvent is H_2O, this process is called "hydration," or more generally, for any solvent, "solvation."

A dilute solution of NaCl crystals in H_2O is slightly cooler than the original constituents before the solvation process, indicating that some internal energy of the system was consumed. Actually energy is consumed in breaking up the original NaCl bonds and some, but less, is liberated in the interaction between the dipole moment of the solvent molecule (H_2O in our example) and the electric charges on the ions. Thus, solvents with higher relative dielectric constant ε_r, indicating higher inherent electric dipole moment per unit volume (P), solvate ions more strongly ($\varepsilon_r = 1 + P/[\varepsilon_o E]$, where E is the electric field applied during the measurement of ε_r). For example, H_2O with $\varepsilon_r \approx 80$ solvates more strongly than methanol with $\varepsilon_r \approx 33$. For biological applications it is worth noting that solvation may affect not only ionic substances, but also polar groups, i.e., molecular components which have an inherent dipole moment, such as—C=O, —NH, or $—NO_2$. Details of the process are discussed in texts on electrochemistry [30,31].

In biological processes not only chemical ionization and solvation of ionic compounds, but also all kinds of chemical reaction take place. One of the central questions in the study of biological effects of E and H fields is therefore not only whether they can cause or influence ionization, but also whether they can affect—speed up, slow down, or modify—any naturally occurring biologically important chemical reaction.

In Table 0.4 typical energies for various types of chemical bonds are listed. For comparison the thermal energy per elementary particle at 310 K is also shown. Complementing the numbers in Table 0.4 one should also point out that:

1. The large spread in the statistical distribution of energies of thermal motion guarantees that at physiological temperatures some molecules always have sufficient energy to break the strongest weak bonds [32].

TABLE 0.4

Bond and Thermal Energies

Type of Bond	Change in Free Energy (Binding Energy) kcal/mol	eV/Molecule
Covalent	50–100	2.2–4.8
Van der Waals	1–2	0.04–0.08
Hydrogen	3–7	0.13–0.30
Ionic[a]	5	0.2
Avg. thermal energy at 310 K	0.62	0.027

[a]For ionic groups of organic molecules such as COO^-, $NH3_3^-$ in aqueous solution.

2. The average lifetime of a weak bond is only a fraction of a second.
3. The weak binding forces are effective only between the surfaces in close proximity and usually require complementary structures such as a (microscopic) plug and hole, such as are thought to exist, for instance, between antigen and antibody [33].
4. Most molecules in aqueous solution form secondary bonds.
5. The metabolism of biological systems continuously transforms molecules and therefore also changes the secondary bonds that are formed.

Comparison of the last columns in Table 0.3 and Table 0.4 shows that millimeter waves have quantum energies, which are only about one order of magnitude below typical Van der Waals energies (waves at a frequency of 10^{12} Hz with a quantum energy of 0.004 eV have a wavelength of 0.3 mm and can still be classified as millimeter waves). One might expect therefore that such waves could initiate chemically important events, such as configurational changes, by e.g., multiple transitions between closely spaced vibrational states at successively high-energy levels [46].

Energies associated with transition from one to another mode of rotation of a diatomic molecule are given by $W = \ell(\ell+1)A$ [26,33], where $\ell = 0, 1, 2, 3 \ldots$ and $A = 6 \times 10^{-5}$ eV; thus an electromagnetic wave with a frequency as low as 29 GHz—still in the microwave region—can excite a rotational mode. Vibrational modes of diatomic molecules [26,33] correspond to energies of the order of 0.04 eV, requiring excitation in the IR region. Vibrational frequencies in a typical H-bonded system [34] are of the order of 3000 GHz; however, attenuation at this frequency by omnipresent free H_2O may prevent any substantial effect [34].

Kohli et al. [34] predict that longitudinal and torsional modes of double helical DNA should not be critically damped at frequencies >1 GHz, although relaxation times are of the order of picoseconds, and Kondepudi [36] suggests the possibility of an influence of millimeter waves at approximately 5×10^{11} Hz upon oxygen affinity of hemoglobin due to resonant excitation of heme plane oscillations. Although Furia et al. [37] did not find resonance absorption at millimeter waves in yeast, such was reported by Grundler et al. [38,47]. The latter experiment has been interpreted [39,40] as supporting Fröhlich's theory of cooperative phenomena in biological systems. That theory postulates "electric polarization waves" in biological membranes which are polarized by strong biologically generated [18] fields (10^7 V/m). Fröhlich [41,42] suggests that metabolically supplied energy initiates mechanical vibrations of cell membranes. The frequency of such vibrations is determined by the dimensions and the elastic constants of the membranes;

based on an estimate of the sound velocity in the membrane of 10^3 m/s and a membrane thickness of 100 Å (equal to one half wavelength) one obtains a frequency of $5(10^{10})$ Hz. Individual molecules within and outside the membrane may also oscillate, and frequency estimates vary between 10^9 Hz for helical RNA [43] and 5×10^{13} Hz for hydrogen-bonded amide structures [44]. As the membranes and molecules involved are strongly polarized, the mechanically oscillating dipole electromagnetic fields that are able to transmit energy, at least in some situations, over distances much larger than the distance to the next adjacent molecule.

Electromagnetic coupling of this type may produce long-range cooperative phenomena. In particular, Fröhlich [45] has shown that two molecular systems may exert strong forces upon each other when their respective oscillation frequencies are nearly equal, provided the dielectric permittivity of the medium between them is strongly dispersive or excitation is supplied by pumping, i.e., by excitation at the correct frequency from an external source. The mechanism is nonlinear in the sense that it displays a steplike dependence on excitation intensity. Possible long-range effects may be, for example, attraction between enzyme and substrate [42]. These and related topics have been discussed in detail by Illinger [34] and are reviewed in the present volume in Chapter 11 and Chapter 5 of *BBA*.

References

1. Jordan, E.C., *Electromagnetic Waves and Radiating Systems*. Prentice-Hall, Englewood Cliffs, NJ, 1950.
2. Schelkunoff, S.A., *Electromagnetic Waves*, D Van Nostrand, New York, 1943, p. 133.
3. Stratton, J.A., *Electromagnetic Theory*, McGraw-Hill, New York, 1941, p. 435.
4. Van Bladel, J., *Electromagnetic Fields*, McGraw-Hill, New York, 1964, p. 274.
5. Kaune, W.T. and Gillis, M.F., General properties of the interaction between animals and ELF electric fields, *Bioelectromagnetics*, 2, 1, 1981.
6. Bridges, J.E. and Preache, M., Biological influences of power frequency electric fields—a tutorial review from a physical and experimental viewpoint, *Proc. IEEE*, 69, 1092, 1981.
7. Iribarne, J.V. and Cho, H.R., *Atmospheric Physics*, D. Reidel, Boston, 1980, p. 134.
8. Zahn, M., *Electromagnetic Field Theory, A Problem Solving Approach*, John Wiley & Sons, New York, 1979.
9. Raybourn, M.S., The effects of direct-current magnetic fields on turtle retina in vitro, *Science*, 220, 715, 1983.
10. Schulten, K., Magnetic field effects in chemistry and biology, *Festkörperprobleme/Advances in Solid State Physics*, Vol. 22, Heyden, Philadelphia, 1982, p. 61.
11. Blankenship, R.E., Schaafsma, T.J., and Parson, W.W., Magnetic field effects on radical pair intermediates in bacterial photosynthesis, *Biochim. Biophys. Acta*, 461, 297, 1977.
12. Sheppard, A.R., Magnetic field interactions in man and other mammals: an overview, in *Magnetic Field Effect on Biological Systems*, Tenforde, T.S., Ed., Plenum Press, New York, 1979, p. 33.
13. Jordan, E.C., *Electromagnetic Waves and Radiating Systems*, Prentice-Hall, Englewood Cliffs, NJ, 1950, p. 132.
14. Ramo, S., Whinnery, J.R., and Van Duzer, T., *Fields and Waves in Communication Electronics*, John Wiley & Sons, New York, 1965, p. 293.
15. Smyth, C.P., *Static and Dynamic Electricity*, McGraw-Hill, New York, 1939.
16. Bean, C.P., DeBlois, R.W., and Nesbitt, L.B., Eddy-current method for measuring the resistivity of metals, *J. Appl. Phys.*, 30(12), 1959, 1976.
17. Bassett, C.A.L., Pawluk, R.J., and Pilla, A.A., Augmentation of bone repair by inductively coupled electromagnetic fields, *Science*, 184, 575, 1974.

18. Plonsey, R. and Fleming, D., *Bioelectric Phenomena*, McGraw-Hill, New York, 1969, p. 115.
19. Houslay, M.D. and Stanley, K.K., *Dynamics of Biological Membranes*, John Wiley & Sons, New York, 1982, p. 296.
20. Wilson, D.F., Energy transduction in biological membranes, in *Membrane Structure and Function*, Bittar, E.D., Ed., John Wiley & Sons, New York, 1980, p. 182.
21. Johnson, C.C. and Guy, A.W., Nonionizing electromagnetic wave effects in biological materials and systems, *Proc. IEEE*, 60, 692, 1972.
22. Schwan, H.P., Field interaction with biological matter, *Ann. NY Acad. Sci.*, 303, 198, 1977.
23. Kraichman, M.B., *Handbook of Electromagnetic Propagation in Conducting Media*, NAVMAT P-2302, U.S. Superintendent of Documents, U.S. Government Printing Office, Washington, D.C., 1970.
24. Massoudi, H., Durney, C.H., Barber, P.W., and Iskander, M.F., Postresonance electromagnetic absorption by man and animals, *Bioelectromagnetics*, 3, 333, 1982.
25. Binhi, V.N., *Magnetobiology: Understanding Physical Problems*, Academic Press, London, 473 pp.
26. Sears, F.W., Zemansky, M.W., and Young, H.D., *University Physics*, 5th ed., Addison-Wesley, Reading, MA, 1976, p. 360.
27. Uman, M.A., *Lightning*, McGraw-Hill, New York, 1969, p. 162.
28. Coelho, R., *Physics of Dielectrics for the Engineer*, Elsevier, Amsterdam, 1979, p. 155.
29. Schwan, H.P., Dielectric properties of biological tissue and biophysical mechanisms of electromagnetic field interaction, in *Biological Effects of Nonionizing Radiation*, Illinger, K.H., Ed., ACS Symposium Series 157, American Chemical Society, Washington, D.C., 1981, p. 121.
30. Koryta, J., *Ions, Electrodes and Membranes*, John Wiley & Sons, New York, 1982.
31. Rosenbaum, E.J., *Physical Chemistry*, Appleton-Century-Crofts, Education Division, Meredith Corporation, New York, 1970, p. 595.
32. Watson, J.D., *Molecular Biology of the Gene*, W.A. Benjamin, Menlo Park, CA, 1976, p. 91.
33. Rosenbaum, E.J., *Physical Chemistry*, Appleton-Century-Crofts, Education Division, Meredith Corporation, New York, 1970, p. 595.
34. Illinger, K.H., Electromagnetic-field interaction with biological systems in the microwave and far-infrared region, in *Biological Effects of Nonionizing Radiation*, Illinger, K.H., Ed., ACS Symposium Series 157, American Chemical Society, Washington, D.C., 1981, p. 1.
35. Kohli, M., Mei, W.N., Van Zandt, L.L., and Prohofsky, E.W., Calculated microwave absorption by double-helical DNA, in *Biological Effects of Nonionizing Radiation*, Illinger, K.H., Ed., ACS Symposium Series 157, American Chemical Society, Washington, D.C., 1981, p. 101.
36. Kondepudi, D.K., Possible effects of 10^{11} Hz radiation on the oxygen affinity of hemoglobin, *Bioelectromagnetics*, 3, 349, 1982.
37. Furia, L., Gandhi, O.P., and Hill, D.W., Further investigations on resonant effects of mm-waves on yeast, Abstr. 5th Annu. Sci. Session, Bioelectromagnetics Society, University of Colorado, Boulder, June 12 to 17, 1983, 13.
38. Grundler, W., Keilman, F., and Fröhlich, H., Resonant growth rate response of yeast cells irradiated by weak microwaves, *Phys. Lett.*, 62A, 463, 1977.
39. Fröhlich, H., Coherent processes in biological systems, in *Biological Effects of Nonionizing Radiation*, Illinger, K.H., Ed., ACS Symposium Series 157, American Chemical Society, Washington, D.C., 1981, p. 213.
40. Fröhlich, H., What are non-thermal electric biological effects? *Bioelectromagnetics*, 3, 45, 1982.
41. Fröhlich, H., Coherent electric vibrations in biological systems and the cancer problem, *IEEE Trans. Microwave Theory Tech.*, 26, 613, 1978.
42. Fröhlich, H., The biological effects of microwaves and related questions, in *Advances in Electronics and Electron Physics*, Marton, L. and Marton, C., Eds., Academic Press, New York, 1980, p. 85.
43. Prohofsky, E.W. and Eyster, J.M., Prediction of giant breathing and rocking modes in double helical RNA, *Phys. Lett.*, 50A, 329, 1974.
44. Careri, J., Search for cooperative phenomena in hydrogen-bonded amide structures, in *Cooperative Phenomena*, Haken, H. and Wagner, W., Eds., Springer-Verlag, Basel, 1973, p. 391.
45. Fröhlich, H., Selective long range dispersion forces between large systems, *Phys. Lett.*, 39A, 153, 1972.

46. Barnes, F.S. and Hu, C.-L.J., Nonlinear Interactions of electromagnetic waves with biological materials, in *Nonlinear Electromagnetics*, Uslenghi, P.L.E., Ed., Academic Press, New York, 1980, p. 391.
47. Grundler, W., Keilmann, F., Putterlik, V., Santo, L., Strube, D., and Zimmermann, I., Nonthermal resonant effects of 42 GHz microwaves on the growth of yeast cultures, in *Coherent Excitations of Biological Systems*, Frölich, H. and Kremer, F., Eds., Springer-Verlag, Basel, 1983, p. 21.

1

Environmental and Occupationally Encountered Electromagnetic Fields

Kjell Hansson Mild and Ben Greenebaum

CONTENTS

1.1 Introduction

We encounter electromagnetic (EM) fields every day, both naturally occurring and man-made fields. This leads to exposure both in our homes as well as in our various workplaces, and the intensity of the fields varies substantially with the situation. Quite high exposure can occur in some of our occupations as well as our personal activities, for instance, in trains, where the extremely low-frequency (ELF) magnetic field can reach rather high levels. The frequency of the fields we are exposed to covers a wide range, from slowly changing static fields to the gigahertz range.

In this chapter, we give an overview of the fields we encounter in various situations.

1.2 Direct Current and ELF (0–3000 Hz) EM Fields

1.2.1 Naturally Occurring Fields

The most obvious naturally occurring field is the Earth's magnetic field, known since ancient times. The total field intensity diminishes from the poles, with a high of 67 μT at the south magnetic pole and a low of about 30 μT near the equator. In South Brazil, an area with flux densities as low as about 24 μT can be found. Indeed, the angle of the Earth's field to the horizontal (inclination) varies, primarily with latitude, ranging from very small near the equator to almost vertical at high latitudes. More information is available in textbooks (see, e.g., Dubrov [1]) and in databases available on the Web (see, e.g., the U.S. National Geophysical Data Center [2]).

However, the geomagnetic field is not constant, but is continuously subject to more or less strong fluctuations. There are diurnal variations, which may be more pronounced during the day and in summer than at night and in winter (see, e.g., König et al. [3]). There are also short-term variations associated with ionospheric processes. When the solar wind brings protons and electrons toward the Earth, phenomena like the Northern Lights and rapid fluctuations in the geomagnetic field intensity occur. The variation can be rather large; the magnitude of the changes can sometimes be up to 1 μT on a timescale of several minutes. The variation can also be very different in two fairly widely separated places because of the atmospheric conditions. There is also a naturally occurring direct current (DC) electric field at the surface of the Earth in the order of 100–300 V/m (Earth's surface negative) in calm weather and can be 100 kV/m in thunderstorms, caused by atmospheric ions [4].

EM processes associated with lightning discharges are termed as atmospherics or "sferics" in short. They consist mostly of waves in the ELF (strictly speaking 30–300 Hz but usually taken in the bioelectromagnetics literature to extend from 0 to 3000 Hz) and very low-frequency (VLF) ranges (3–30 kHz) (see König et al. [3]). Each second about 100 lightning discharges occur globally, and in the United States one cloud-to-ground flash occurs about every second, averaged over the year [3]. The ELF and VLF signals travel efficiently in the waveguide formed by the Earth and the ionosphere and can be detected many thousands of kilometers from the initiating stroke. Since 1994, several experiments studying the effects of short-term exposure to simulated 10-kHz sferics have been performed at the Department of Clinical and Physiological Psychology at the University of Giessen, Germany [5,6]. In the ELF range, very low-intensity signals, called Schumann resonances, also occur. These are caused by the ionosphere and the Earth's surface acting as a resonant cavity, excited by lightning [3,7] (see also http://www.oulu.fi/~spaceweb/textbook/schumann.html). These cover the low-frequency spectrum, with broad peaks of diminishing amplitude at 7.8, 14, 20, and 26 Hz and higher frequencies. Higher-frequency fields, extending into the microwave region, are also present in atmospheric or intergalactic sources. These fields are much weaker, usually by many

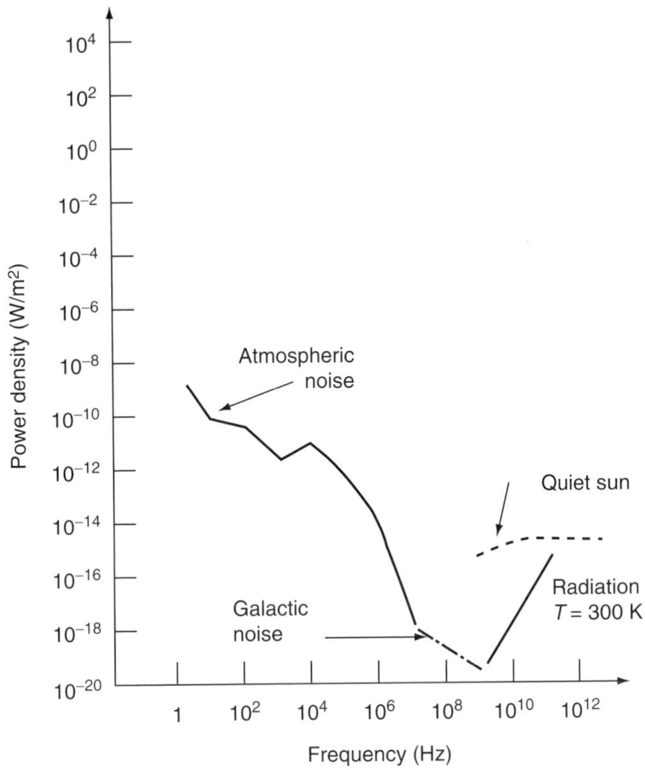

FIGURE 1.1
Power density from natural sources as a function of frequency. (Data from Smith, E. *Proceedings of the IEEE Symposium on Electromagnetic Compatibility*. Institute of Electrical and Electronic Engineering, Piscataway, NJ, 1982. Graph adapted from Barnes, F.S. *Health Phys*. 56, 759–766, 1989. With permission.)

orders of magnitude, than those caused by human activity (compare Figure 1.1 and subsequent tables and figures in this chapter).

1.2.2 Artificial DC and Power Frequency EM Fields in the Environment

1.2.2.1 DC Fields

Although alternate current (AC) power transmission is facilitated by the availability of transformers to change voltages, DC is also useful, especially since high-power, high-efficiency solid-state electronic devices have become available. Overland high-voltage DC lines running at up to ±1100 kV are found in Europe, North America, and Asia [8] (see also, e.g., http://www.answers.com/topic/high-voltage-direct-current, accessed on August 17, 2005). Electric and magnetic fields near these lines are essentially the same as those for AC lines running at the same voltages and currents, which are discussed below. Because potentials on the cables do not vary in time and there are only two DC conductors (+ and −) instead of the three AC phases, the DC electric fields and space charge clouds of air ions that partially screen them are somewhat different from those near AC transmission lines, though the general features are the same, especially for positions away from the lines. Electric fields, corona, and air ions are discussed further in the AC transmission line section below; see also Refs. [9,10].

For transfer of electric power between countries separated by sea, undersea power cables are especially useful, since their higher capacity causes increased losses in AC.

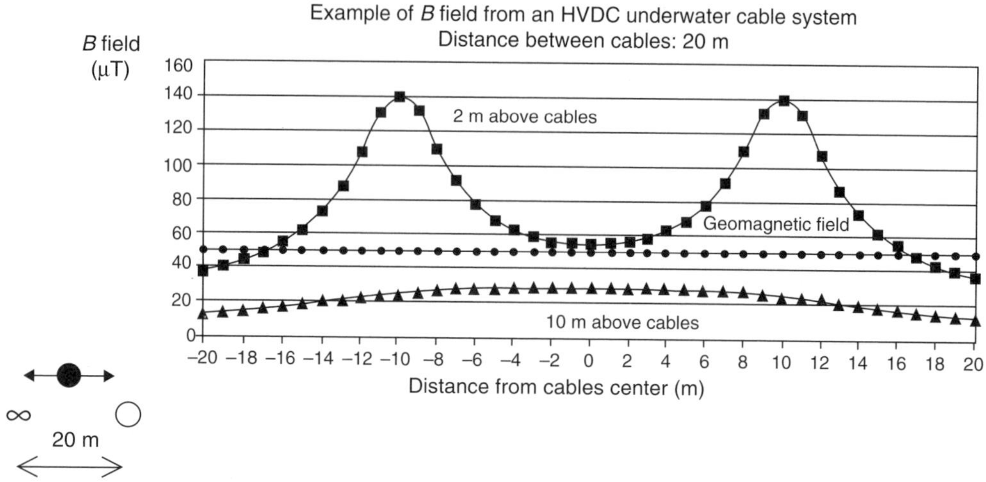

FIGURE 1.2
Predicted DC magnetic field from a high-voltage DC cable with the return cable placed at a distance of 20 m. The current in the cable was assumed to be 1333 A, which is the maximum design current. (From Hansson Mild, K. In Matthes, R., Bernhardt, J.H., and Repacholi, M.H., Eds. Proceedings from a joint seminar, *International Seminar on Effects of Electromagnetic Fields on the Living Environment*, of ICNIRP, WHO, and BfS, Ismaning, Germany, October 4–5, 1999, pp. 21–37. With permission.)

Examples are cables between Sweden and Finland, Denmark, Germany, and Gotland, a Swedish island in the Baltic Sea. Under construction at present is a cable from Sweden to Poland (SwePol). In these cables DC is used, and the ELF component of the current is less than a few tenths of a percent. The maximum current in these cables is slightly above 1000 A, and the estimated normal load is about 30% or 400 A. Depending on the location of the return path, the DC magnetic field will range from a maximum disturbance of the geomagnetic field (with a return through water) to a minimal disturbance (with a return through a second cable as close as possible to the feed cable). With a closest distance of 20 m between the cables, the predicted field distribution can be seen in Figure 1.2, immediately above the cables (2 m), practically the same value as that obtained for a single wire. When the distance between cables is increased beyond 20 m, the distortion at a given distance rises above that of Figure 1.2. Since the cables are shielded, no electric field will be generated outside the cable. For a more detailed discussion of the fields associated with this technique, the reader is referred to the paper by Koops [11].

Few other DC fields from human activity are broadly present in the environment, though very short-range DC fields are found near permanent magnets, usually ranging from a few tenths of a millitesla to a few millitesla at the surface of the magnet and decreasing very rapidly as one moves away. Occupationally encountered DC fields are discussed below.

1.2.2.2 High-Voltage AC Power Lines

The electric and magnetic fields from high-voltage power lines have been figuring for a long time in the debate on the biological effects of EM fields. Although the AC power systems in the Americas, Japan, the island of Taiwan, Korea, and a few other places are 60 Hz, while most of the rest of the world is 50 Hz, the frequency difference has no effect on high-voltage transmission line fields. In the early days of bioelectromagnetics research, the electric field was considered the most important part, and measurements of field strengths were performed in many places. Figure 1.3 shows an example of such

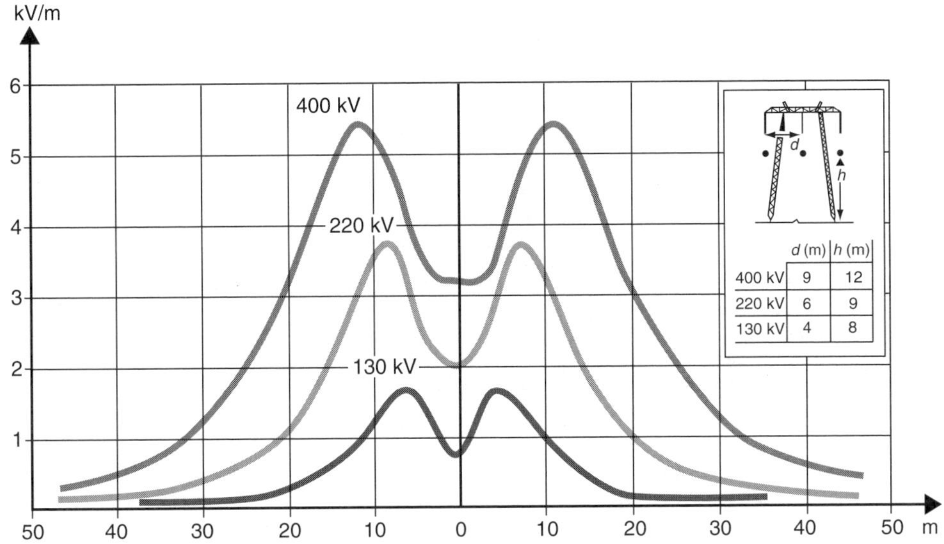

FIGURE 1.3
Electric field from three different high-voltage power lines as a function of the distance from the center of the line. In the inset the distance between the phases as well as the height above ground of the lines are given. (From Hansson Mild, K. In Matthes, R., Bernhardt, J.H., and Repacholi, M.H., Eds. Proceedings from a joint seminar, *International Seminar on Effects of Electromagnetic Fields on the Living Environment*, of ICNIRP, WHO, and BfS, Ismaning, Germany, October 4–5, 1999, pp. 21–37. With permission.)

measurements from three different types of lines: 400, 220, and 130 kV lines, respectively. The field strength depends not only on the voltage of the line but also on the distance between the phases and the height of the tower. The strongest field can be found where the lines are closest to the ground, and this usually occurs midway between two towers. Here, field strengths up to a few kilovolts per meter can be found. Since the guidelines of the International Commission on Non-Ionizing Radiation Protection (ICNIRP) [12] limit public exposure to 5 kV/m and there is no time averaging for low-frequency fields, people walking under high-voltage power lines may on some occasions be exposed in excess of existing international guidelines.

Because electric fields are well shielded by trees, buildings, or other objects, research in the 1970s and 1980s did not turn up any major health effects (see, e.g., Portier and Wolf [13]), and because of the epidemiological study by Wertheimer and Leeper [14] (see also Chapter 6 on ELF epidemiology in this volume), attention turned from electric to magnetic fields in the environment. The magnetic field from a transmission line or any other wire depends on the current load carried by the line, as well as the distance from the conductors; in Figure 1.4 calculations of the magnetic flux density from several different types of transmission lines are shown. There is a very good agreement between the theoretical calculation and the measured flux density in most situations. The flux density from two-wire power lines is directly proportional to the electric current, generally inversely proportional to the square of the distance to the power line for distances greater than several times the distance between the phase lines, and directly proportional to the distance between the phase wires. For three and six-wire systems the fields decrease more rapidly with distance at a rate that is dependent on the phase sequences and the spacing between the wires. For most lower-voltage lines, around 10–20 kv, the distance at which the B field falls below 0.2 μT is generally less than 10 m; this distance still depends on current and the spacing of the wires.

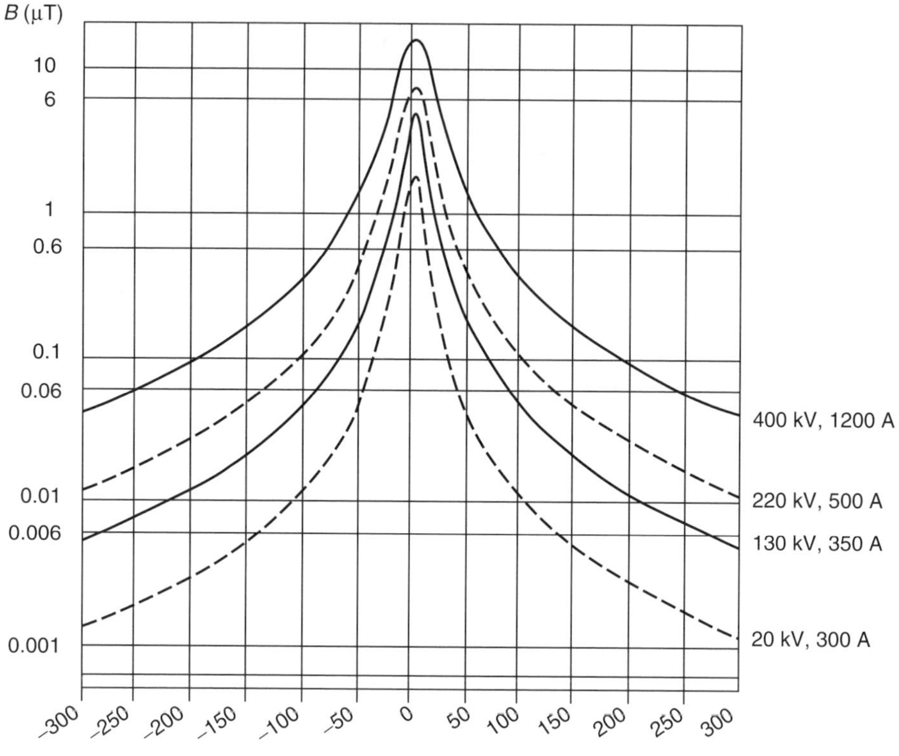

FIGURE 1.4
Magnetic flux density from different high-voltage power lines at a distance (in meters) from the center of the line. The currents in the lines are the maximum values allowed and are given to the right in the figure. (Figure courtesy of Swedish National Institute for Working Life.)

The electric or magnetic field vector from a single AC conductor displays a sinusoidal waveform, oscillating back and forth through zero intensity in a single direction determined by the observation position with respect to the wire, ignoring any small distortions due to harmonics, etc. However, near a three-phase high-voltage transmission line, the electric and magnetic field vectors from the group of conductors, which are at some distance from each other and whose individual sinusoidal variations are out of phase, rotate in space as well as change in magnitude, but their magnitude never decreases exactly to zero [15]. This so-called elliptical polarization may or may not have a different biological significance than the single conductor's ''plane polarization.''

Several approaches have been used for reducing the magnetic field from a line, and in Figure 1.5 some examples are given. Instead of hanging the three phases at the same height and in parallel, the lines can be arranged in a triangular form, thereby reducing the distance between the phases and thus also the flux density. The reduction is of the order of about 1.6. An even greater reduction is obtained if the so-called split phase arrangement is used. Here, five lines are used. One phase is placed in the center, and the other two phases are split into two lines each, which are placed diagonally (see Figure 1.4). The reduction is almost tenfold.

When high voltage is present, there is a possibility of the insulation breaking down, causing a catastrophic discharge—a spark; lightning is an obvious example. There is also the more common possibility of very minor discharges occurring, in which one or a relatively small number of molecules near the high-voltage element become

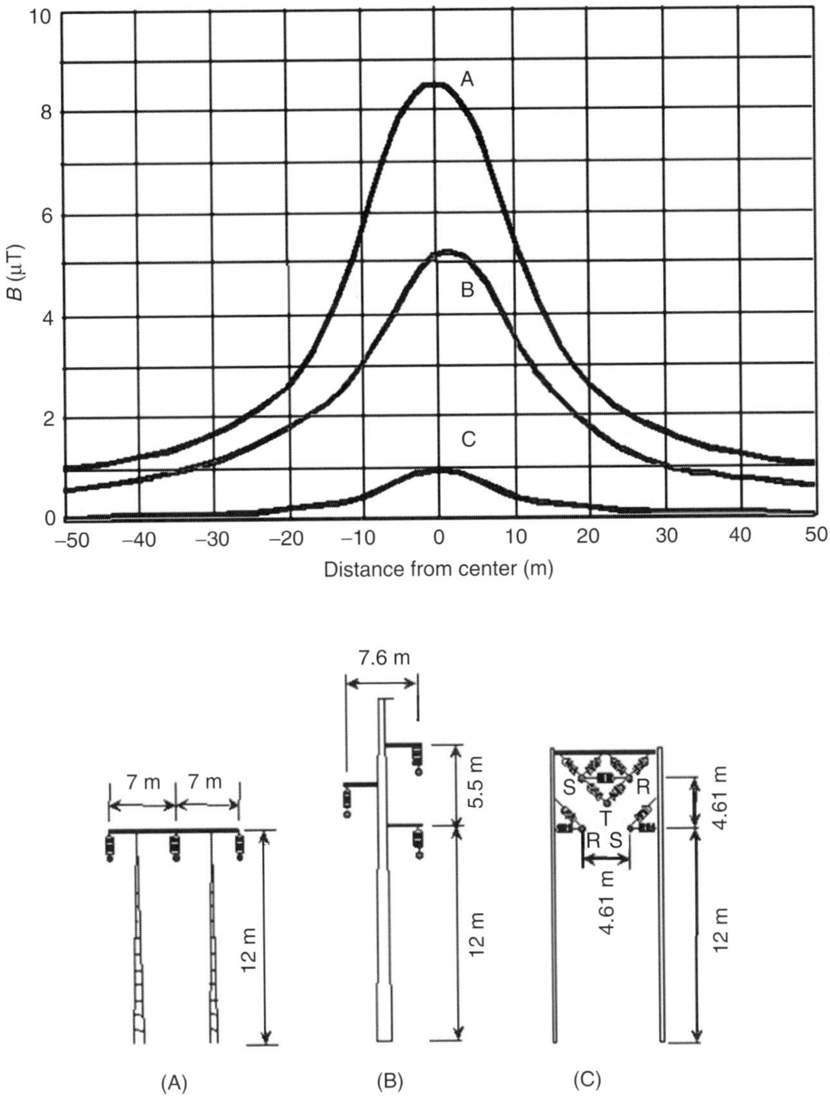

FIGURE 1.5
Examples of reduction of the magnetic flux density from a 220-kV line with a maximum phase current of 500 A. In (A) the normal configuration is used and the maximum flux density is about 8 μT, and in (B) a delta arrangement is used which gives a reduction to about 5 μT maximum under the line. In (C) the split-phase arrangement is used leading to a maximum value of only 1 μT. (Figure courtesy of Swedish National Institute for Working Life.)

ionized; this is often called a *corona*, since in extreme cases a small glow can be seen near parts of the high-voltage system. Corona discharge can also occur at grounded objects near a high voltage and is more likely to occur at more pointed objects; this is the principle of the lightning rod. Minor corona damage has been observed on pine tree needles very close to a 1200 V transmission line [16]. (No other environmental damage to plants or animals from either fields or corona has been found [17].) The resulting ions screen the electric field of the transmission line cables to varying extents, because their number depends on a variety of factors, including humidity, dust, rain, and wind [9,10]. While a hypothesis has been put forward that ions from power lines make small airborne particles, particularly those carrying naturally occurring radioactive atoms, more likely to enter and remain in the lungs and cause cancer or various other diseases [18], it has not found much acceptance.

1.2.2.3 Exposure in Homes

Although Wertheimer and Leeper [14] initially used transmission and distribution line sizes and configurations as surrogates for estimating magnetic field exposure from transmission lines, it quickly became apparent that the correlation was not very good and that sources of exposure inside the home were at least as important, unless the home was very close to a transmission line [13]. Several studies have explored the exposure to ELF electric and magnetic fields in homes in different countries. Deadman et al. [19] investigated the exposure of children in Canada. A logging device was used, which recorded the fields during two consecutive 24-h periods. For 382 children up to the age of 15 they found an arithmetic mean (AM) of the magnetic field of 0.121 μT with a range of 0.01–0.8 μT. The corresponding values for the electric field were AM 14.4 V/m, range 0.82–64.7 V/m. Hansson Mild et al. [20] compared the ELF fields in Swedish and Norwegian residential buildings. The overall mean values were as follows: E fields 54 V/m (SD = 37) and 77 V/m (SD = 58) in Sweden and Norway, respectively; the corresponding values for B fields were 40 nT (SD = 37) and 15 nT (SD = 17). Table 1.1 shows additional comparisons.

Mccurdy et al. [21] measured women's exposure in the United States by using personal magnetic field exposure meters that were worn during a working day or a day at home. The geometric mean of the time-weighted average for the working day was 0.138 μT with a range of 0.022–3.6 μT, and for the homemakers the corresponding values were 0.113 μT, range 0.022–0.403 μT.

In the meta-analysis by Ahlbom et al. [22] on childhood cancer and residential magnetic fields, it was stated that 99.2% of the population resided in homes with $B \leq 0.4$ μT.

Exposure varies widely in time, according to the time of day and the season. One may be outdoors, far from any field sources at one time, indoors near an operating appliance at another, riding in an electric transit vehicle at some other time, and so forth. Sample exposure values for an individual, recorded as a function of time over a 24-h period in spring and summer, are shown in Figure 1.6.

TABLE 1.1

Comparison of Personal Exposure and Background Fields

Country		Geometric Mean of Personal Exposure (nT)	Geometric Mean of Long-Term Background Field (nT)	Ratio, Personal Mean to Background Mean
United States	Adults at home	134	58	2:3
	Adults at home, not in bed	111	74	1:5
	Children, residential	96	99	1:0
	Children, at home	96	67	1:4
Canada	Children at home	117	107	1:1
	Adults at home	133		1:2
U.K.	Adults at home	54	37	1:5
	Adults	42	29	1:5

Source: From Swanson, J. and Kaune, W.T. *Bioelectromagnetics* 20, 244–254, 1999.

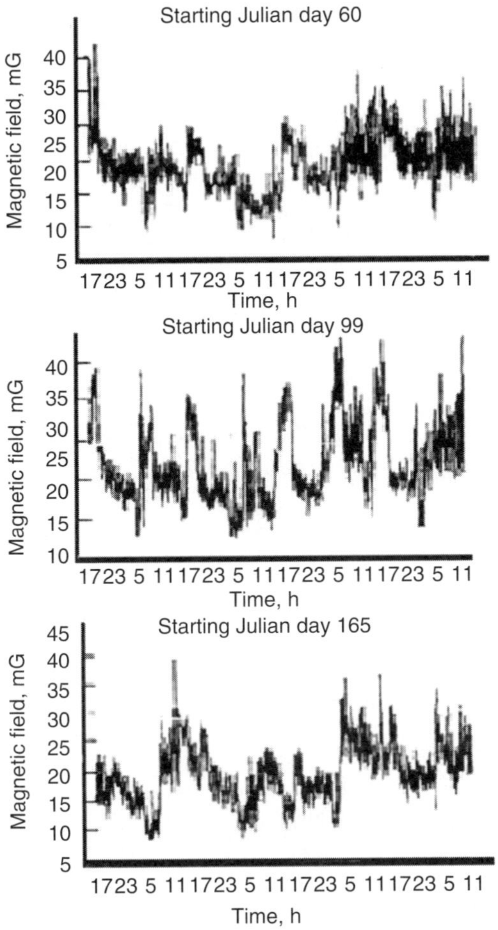

FIGURE 1.6
An individual's measured magnetic field exposure over the course of a day. Note that 1 mG = 0.1 μT. (From Koontz, M.D., Mehegan, L.L., Dietrich, F.M., and Nagda, N.L. Assessment of Children's Long Term Exposure to Magnetic Fields [The Geomet Study]. Final Report TR-101406, Research Project 2966-04, Electric Power Research Institute, Palo Alto, CA, 1992. With permission.)

Since the three-phase systems used for electrical distribution are dimensioned for sinusoidal fields, the harmonic content can create problems. Today we may find large stray currents, usually resulting from unbalanced currents between phases, in water pipes, ventilation systems, concrete reinforcement mesh, etc., and the current flowing also contains these harmonics. Figure 1.7 gives an example of a measurement of a current flowing in a cable in a large apartment building, and Figure 1.8 shows the corresponding Fourier frequency analysis. The magnetic field in the building thus also has these harmonic components. Often, the largest stray currents, which generate large domestic fields, are due to errors in wiring that violate the building code [23] or to a poorly planned wiring layout that has currents flowing in open loops instead of both wires of a circuit being laid next to each other in the same conduit [24].

From Figure 1.6 through Figure 1.8, as well as the data in the rest of this chapter, it is easy to see that average field strength is far from being the only parameter that is needed to characterize electric or magnetic field exposure. Other parameters include frequency or frequencies present (or the related parameters, the rise and fall times of up-and-down excursions or "transients"), numbers and height of transients, number of

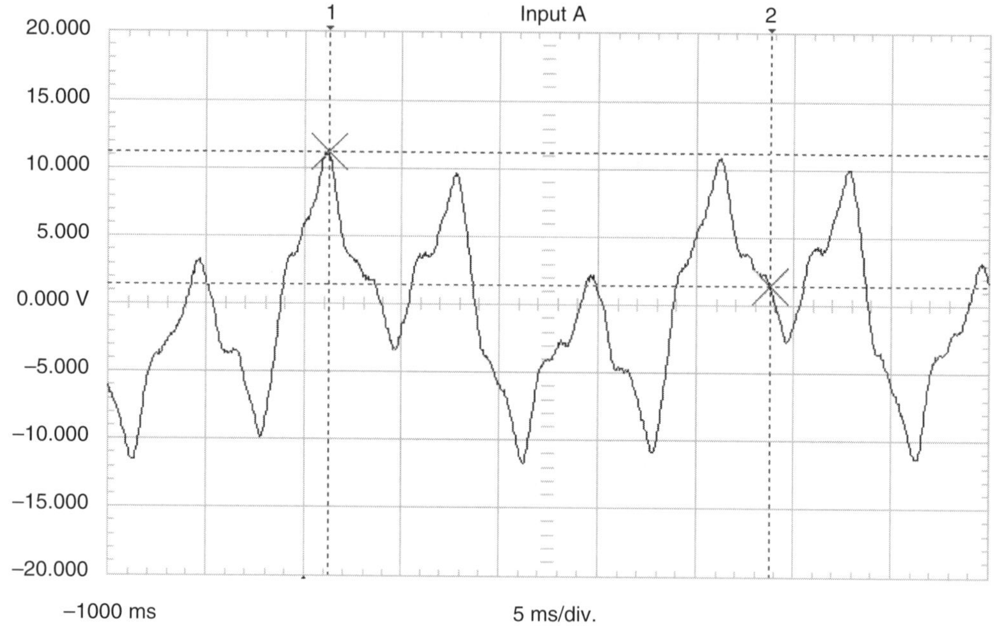

FIGURE 1.7
Stray current wave shape in the 50 Hz power delivery cable in an office building. The peak to peak current is of the order 20 A. (Figure courtesy of Swedish National Institute for Working Life.)

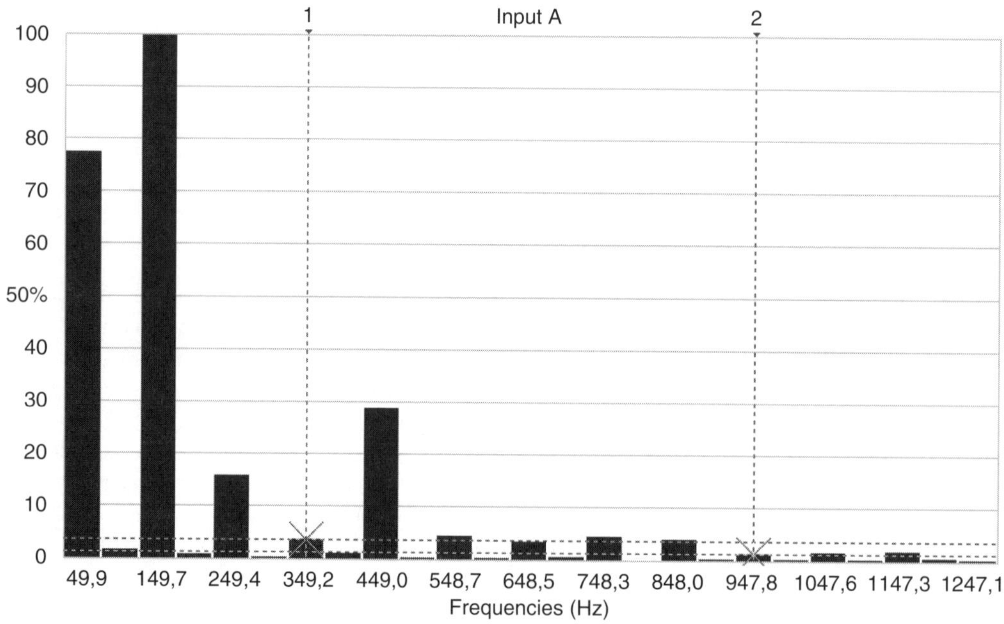

FIGURE 1.8
The Fourier spectrum of the wave shape in Figure 1.5. Note the high 150 Hz (third harmonic) component. (Figure courtesy of Swedish National Institute for Working Life.)

times the field exceeds or falls below a certain fraction of its average value, whether both DC and time-varying fields are present, relative direction of multiple fields, etc. As discussed elsewhere (e.g., the Introduction and chapters such as Chapter 5, Chapter 6, and Chapter 9 in this volume and Chapter 5 through Chapter 8 and Chapter 11 in *BMA*), it is not clear, in most cases, which one or group of these parameters is related to a particular biological effect. To date, average field strength is the most commonly used parameter, partly because it is the most easily obtainable summary of exposure over an extended period. For a given frequency range, average field strength is related to some other parameters, such as fraction of time over a certain threshold, but not to others, such as number of transients per hour. For further discussion of various parameters and their interrelationships, see, for example, Refs. [25,26].

Most measurements have been done in detached houses, even though many city dwellers live in apartment buildings. In apartment buildings, the current in the wiring in the ceiling of one unit, for instance, for ceiling lamps, may most strongly affect the magnetic field level of the unit above. Also, some apartment buildings have an electric substation in the basement, where a transformer reduces the medium-voltage distribution line power to 110 or 220 V for domestic use. The low-voltage conductors of the substation may carry substantial currents and create magnetic fields up to several tens of microtesla directly above the substation; reduction through placing conductors away from the substation ceiling and shielding with aluminum plates is possible [27].

In the United States and Canada, though not in other countries, the neutral wire of the AC power distribution system is required to be physically connected to the earth (grounded) at regular intervals to avoid injury from electric shocks; building wiring systems' neutral wires must also be grounded, often by connection to the buried water pipe as it enters the building. Unbalanced loading of the system can produce currents in the ground system, sometimes including currents that leave one residence through the grounding system and return to the power grid through another, which further contributes to the residential magnetic fields [28,29].

Kavet and colleagues [30–32] have proposed that effects observed in children, which epidemiology has associated with domestic magnetic fields, are in fact due to small shocks that arise due to potential differences that build up between the water tap and the grounded drain of a tub. Shocks received in the bath can still induce in a small child's body current densities of a magnitude known to induce a biological effect. This alternative hypothesis is still under investigation.

1.2.2.4 Electrical Appliances

The United States, Japan, Canada, and some other countries use 110 V_{rms} AC for basic electrical power, while most of the rest of the world uses 230 V. Since transmission and distribution voltages in the two types of system are about the same, only differences due to appliances or building wiring would be expected. For a given power consumption and similar design, 110 V appliances draw twice as much current and create twice as strong a local magnetic field, although their local electric fields are half as strong. However, because both types of field fall off rapidly with increasing distance from the appliance and metal appliance cabinets shield electric fields, measurements of exposure to magnetic fields have not yielded great differences between the two systems (see Table 1.1). Measurements of magnetic fields from a sample of various appliances show that the fields have a rapid falloff with distance from the device [33]. Very close, the values may exceed international guidelines, but at a distance of 0.5–1 m the fields are seldom higher than few tenths of a microtesla. In general, it can be said that the more power the equipment uses, the higher the magnetic field. Table 1.2 presents some representative values from 110 V appliances.

TABLE 1.2

Ratios (B_{on}/B_{off}) of Magnetic Fields Measured with Appliances Turned On and Off

Appliance	Measurement Location	6–54 Hz	B_{on}/B_{off} 54–606	606–3066 Hz	8–200 kHz
Hair dryer	5 cm from nozzle	8.3	57	76	11
	10 cm from nozzle	3.2	17	31	—
	15 cm from nozzle	2.1	7.9	16	—
	25 cm from nozzle	1.4	3.1	6.1	—
Headset playing music	Forehead	1.5	1.0	1.2	—
	Center of head	<1.0	1.0	1.2	1.1
	Above ear	2.0	1.0	2.5	—
	Sternum	1.4	1.0	1.1	—
	Hip	2.7	1.0	2.6	—
Home sewing machines	Front of abdomen	2.7	2.8	2.4	1.5
	Left side of abdomen	1.7	2.0	1.8	—
	Right side of abdomen	1.7	1.6	1.5	—
Motorized clock	10 cm from clock	<1.0	13	4.3	1.1
	25 cm from clock	<1.0	4.0	<1.0	—
	50 cm from clock	<1.0	1.7	<1.0	—
	100 cm from clock	<1.0	1.0	1.0	—
	At subject's head	<1.0	1.4	1.0	—
Electronic clock	10 cm from clock	<1.0	4.7	1.5	1.5
	25 cm from clock	<1.0	1.8	1.0	—
	50 cm from clock	<1.0	1.1	<1.0	—
	100 cm from clock	1.1	1.0	1.0	—
	At subject's head	1.1	1.1	1.0	—

Note: B_{off} was estimated using linear interpolation at those measurement locations where it was not directly measured.

Source: From Kaune, W.T., Miller, M.C., Linet, M.S., Hatch, E.E., Kleinerman, R.A., Wacholder, S., Mohr, A.H., Tarone, R.E., and Haines, C. *Bioelectromagnetics* 23, 14–25, 2002.

Vistnes [34] recently gave some examples of flux densities near 220 V appliances. Of special interest may be a clock radio, which because of bad electrical design may give rise to exposure of the order of 100 μT close to the equipment. Since people are likely to place a clock radio very close to the pillow, the head may be exposed to quite a large magnetic field, exceeding the normal levels in the house.

The general range of magnetic and electric field magnitudes at various distances from transmission lines, local distribution lines, and appliances is shown in Figure 1.9.

Most modern electrical appliances are equipped with an electronically switched power supply in which an electronic circuit replaces the old-style transformer. This means that the current is no longer a pure sinusoidal 50- or 60-Hz signal but contains harmonics. The current used by a low-energy 50-Hz fluorescent lamp is illustrated in Figure 1.10, and the Fourier analysis is shown in Figure 1.11 indicating all the harmonics. Higher harmonics and transients (fast spikelike excursions) are also generated by motor-driven appliances and those run by vibrating mechanisms using make-and-break switching contacts, such as older electric shavers or doorbells (Table 1.2).

The magnetic field in different infant incubators used in hospital nurseries varied between 0.23 and 4.4 μT, with an arithmetic average of 1.0 μT [35]. Most of these values are considerably higher than the exposure that can be measured in residential areas close to transmission lines. The technology to reduce the exposure is at hand and can be easily applied.

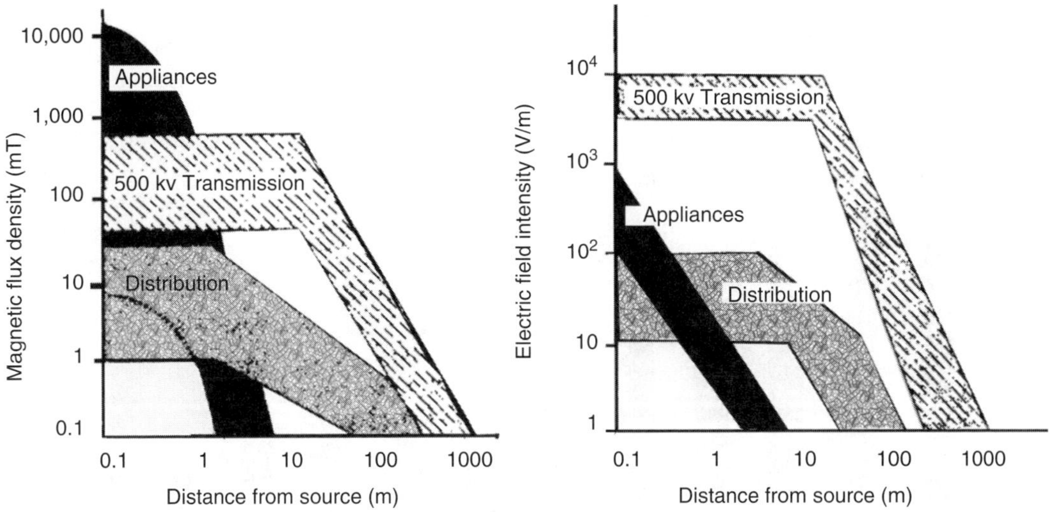

FIGURE 1.9
Magnetic flux density (left) and electric field strength (right) as a function of distance from transmission lines, local distribution lines, and appliances. (From U.S. Office of Technology Assessment. Biological Effects of Power Frequency Electric and Magnetic Fields. U.S. Government Printing Office, Washington, DC, Background Paper OTA-BP-E-53, 1989.)

Occupational exposure from handheld electrical appliances can be quite high. This is mainly equipment that is held close to the body and that uses high power, such as drills and circle saws. These devices usually have adjustable speed, which is done through the switched power supply. Values for the magnetic field of the order 100–200 μT are not uncommon, and in order to show compliance with standards the measurements have to take into account the harmonic contents of the waveform.

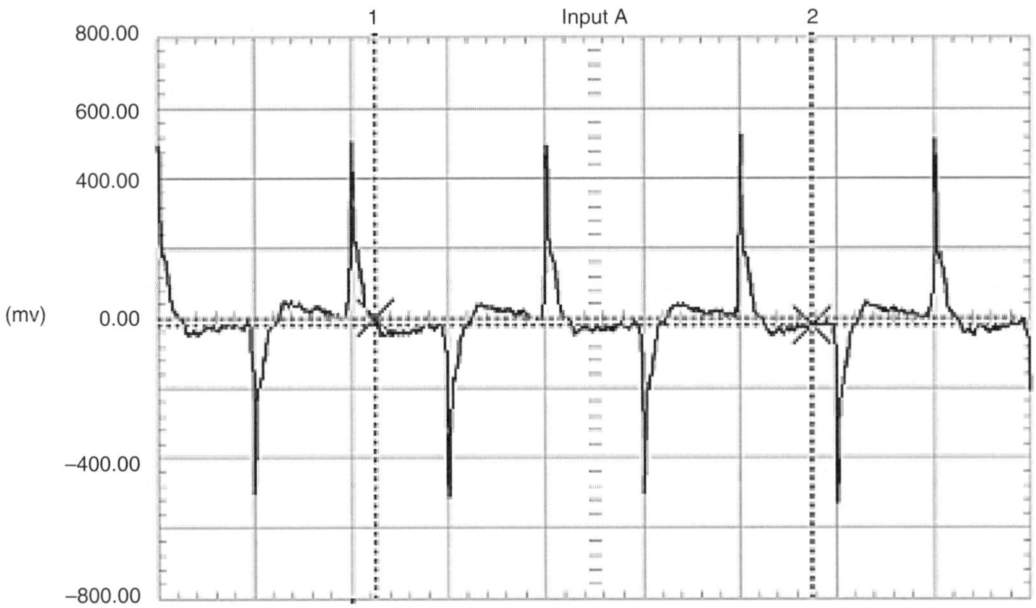

FIGURE 1.10
Wave shape of the current to a low-energy fluorescent lamp. The timescale is 10 ms/div. (Figure courtesy of Swedish National Institute for Working Life.)

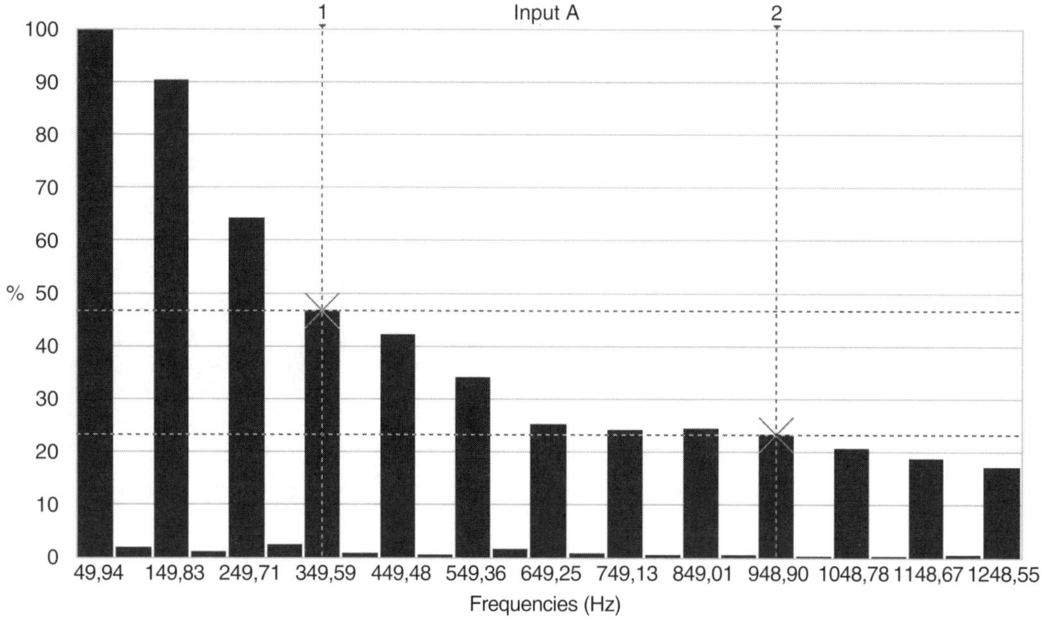

FIGURE 1.11
The Fourier spectrum of the wave shape in Figure 1.7. Note the high 150-Hz component. (Figure courtesy of Swedish National Institute for Working Life.)

1.2.2.5 ELF Fields in Occupational Settings

Wertheimer and Leeper [14] were not only the first to publish evidence in support of increased childhood cancer risk with magnetic field exposure, but they also pointed to increased cancer risk in occupations with high magnetic field exposure. Since then, hundreds of studies have looked into this problem, and the assessment of workers' exposure has been debated. There are studies where individual estimates of the exposure have been made for male [36] and females [37]. For workday means, the 25th, 50th, and 75th percentiles were 0.13, 0.17, and 0.27 μT, respectively, for males, and the corresponding values for females were almost similar: 0.14, 0.17, and 0.23 μT. The study on exposure of males investigated the 1000 most common occupations in Sweden, and the study on female exposure included 61 job categories. Table 1.3 shows additional estimates for various professions.

Sewing machines—Near sewing machines increased magnetic fields can be found, and depending on the type of machines used the values differ. The mean average value logged during some working hours is of the order of several tenths of a microtesla [38].

Welders—Among the occupations where quite high exposure exists, electric arc welders are a prominent example. They handle cables carrying hundreds of amperes very close to their bodies. The welder normally grasps the cable, and it sometimes also is in contact with other parts of the body, for instance, it might be draped over the shoulder. Depending of the technique used—DC or AC, type of rectification, etc.—the ELF magnetic field varies, but several studies report values in the range of tens to hundreds of microtesla [39]. Skotte and Hjøllund [40] found a mean of 21 μT for a full-shift average workday of manual metal arc welders. During the actual welding, the *B* field can be up to several millitesla.

The frequency content of the signal can be rather complex. In one of the most common situations the welding equipment is connected to a three-phase outlet, and the current for the weld is thus three-phase full-wave rectified. This means that we have first a DC

TABLE 1.3

EMF Exposures in Common Environments

Environment	Median[a] Exposure	Top 5th Percentile	Environment	Median[a] Exposure	Top 5th Percentile
Office Building			*Machine Shop*		
Support staff	0.6	3.7	Machinist	0.4	6.0
Professional	0.5	2.6	Welder	1.1	24.6
Maintenance	0.6	3.8	Engineer	1.0	5.1
Visitor	0.6	2.1	Assembler	0.5	6.4000
School			Office staff	0.7	4.7
Teacher	0.6	3.3	*Grocery Store*		
Student	0.5	2.9	Cashier	2.7	11.9
Custodian	1.0	4.9	Butcher	2.4	12.8
Administrative staff	1.3	6.9	Office staff	2.1	7.1
Hospital			Customer	1.1	7.7
Patient	0.6	3.6	—	—	—
Medical staff	0.8	5.6			
Visitor	0.6	2.4			
Maintenance	0.6	5.9			

Note: Magnetic fields are measured in milligauss (mG); 1 mG = 0.1 μT.
[a]The median of four measurements. For this table, the median is the average of the two middle measurements.
Source: National Institute for Occupational Safety and Health. From Portier, C.J. and Wolfe, M.S., Eds. Assessment of Health Effects from Exposure to Power-Line Frequency Electric and Magnetic Fields. NIH Publication 98-3981, National Institute of Health Sciences, Research Triangle Park, NC, 1998 (accessed April. 7, 2005, at http://www.niehs.nih.gov/emfrapid/html/WGReport/WorkingGroup.html).

component and on that a large AC ripple with main frequency 300 Hz (50 Hz power system), but it also has harmonics at 600, 900, 1200 Hz, etc. A newer type of equipment has a pulsed DC (50–200 Hz pulse frequency) as a base with a 53 kHz current applied between the pulses. This leads to frequencies in the current equal to the pulse frequency and its harmonics and also 53 kHz and harmonics. It is a very complex situation to evaluate with respect to compliance with guidelines, because of the complexity of the signal.

Since in many cases, high exposure results from the cables being very close to the body, much can be done to reduce the exposure of the welder by carefully arranging the workstation to keep the cables away from the body. By placing the welding machine on the right-hand side of the worker (if right-handed) and seeing that the return cable is as close as possible to the current cable, the exposure can be reduced by one order of magnitude.

Induction heaters Induction heating is used for heating metals for purposes that include surface or deep hardening, welding, melting, soft soldering, brazing, annealing, tempering, and relieving stress. The frequency can be from 50 Hz to the low megahertz range, depending on the desired skin depth and purpose. Since high currents are used, the leakage magnetic field can be substantial. At the operator's position, values of the order of 0.5–8 μT are common, and the maximum field near the coil, where, for instance, the hands can be exposed, can reach several hundreds of microtesla. The field strength is in many cases high compared with recommended limits [12].

Railway workers—Engine drivers of AC electric engines experience rather high magnetic field exposure. The intensity depends of several factors, one of them being the age of the engine. Nordensson et al. [41] (see also Refs. [42,43]) found that drivers of Swedish model RC engines were exposed to flux densities of the order of 10–100 μT. The older

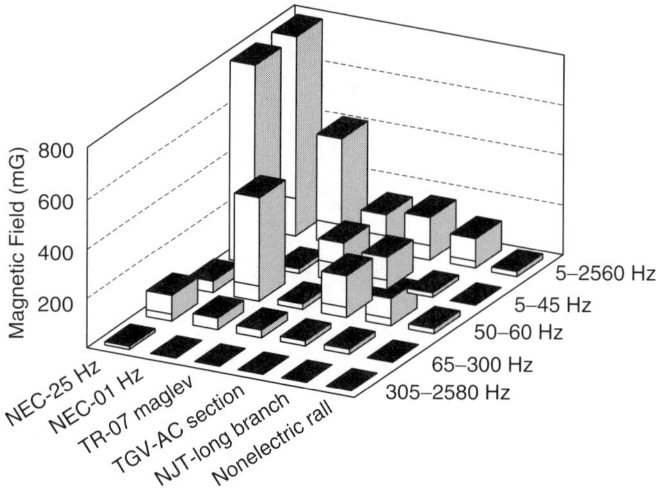

FIGURE 1.12
Maximum (top of bar) and average (horizontal bar) magnetic fields in various frequency bands in the passenger compartment of several intercity rail systems. NEC = U.S. Amtrak Northeast Corridor (Washington, DC, to Boston, MA), which has both 25- and 60-Hz segments; TR-07 = German Transrapid maglev system; TGV = French "Train a Grande Vitesse," AC-powered segment of Paris-Tours line; NJT = New Jersey Transit, NJ Coast Line Long Branch section. (From Bernardi, A., Fraser-Smith, A.C., and Villard, O.G., Jr. *IEEE Trans. Electromagn. Compat.* 31, 413–417, 1989.)

models of engines had the higher values. The mean average values for a full workday ranged from 2 to 15 μT. The main input power frequency is $16\frac{2}{3}$ Hz, and this frequency was dominant at idle, but at full power, harmonics up to 150 Hz existed. Wenzl [44] measured the exposure of rail maintenance workers in the United States and found peak values ranging from 3.4 to 19 μT, and the time-weighted average was in the range 0.3–1.8 μT. Chadwick and Lowes [45] have examined the exposure of passengers on trains in the U.K., and they found static magnetic flux densities up to several microtesla. The alternating field was also substantial in some locations and reached up to 15 mT at floor level. However, none of the whole-body alternating magnetic flux densities approached the National Radiological Protection Board (NRPB) investigation levels.

Trains operating on DC, such as in the Washington, DC, and San Francisco, CA, transit systems, also produce time-varying fields in the passenger compartments, particularly below 5 Hz [46,47]. Figure 1.12 shows field intensity in various frequency bands in the passenger compartment of several representative electric rail systems and a nonelectric one. Interestingly, the figure shows that an experimental magnetic levitation (maglev) system does not exhibit substantially different field levels [48].

Electrochemical plants—In factories producing, for instance, aluminum, copper, or chloride through electrochemical processes, very high DC currents are used, often of the order of tens of kiloamperes. The DC current is obtained through rectification of the incoming three-phase AC power. Often there is still a substantial AC component of the current and hence an AC magnetic field. Measurements have shown broadband ELF measurements of the order of 10–50 μT, with many different frequencies present that need to be taken into account in the evaluation of the exposure situation. Typically, a 50 Hz component can be present, because of unbalance between the three phases, and the full-wave rectification gives 300, 600, and 900 Hz components. The exposure guidelines can often be exceeded in some locations in the plants, and special requirements may be needed to reduce the exposure. DC fields in these smelters are often on the order of several millitesla, with peaks of at least 20–30 mT; up to 70 mT has been reported [49,50].

1.2.2.6 Internal ELF Fields Induced by External and Endogenous Fields

Because the bodies of humans, other animals, and even plants contain ionic solutions and because cell cultures, as well as many one-celled and other organisms such as fish or the roots of plants, live in conductive media, external exposure to electric or time-varying magnetic fields can produce internal fields, which can be quite different than the unperturbed external fields.

In an electric field, as discussed in Chapter 3 and Chapter 4 in this volume on properties of materials, the conductivity and dielectric constants of tissue are quite different from those of air or vacuum, creating a layer of charge due to polarization at the surface of the body, which decreases the internal field, often by many orders of magnitude. For a human standing in the ELF electric field below a high-voltage transmission line, the field inside the body may be only 10^{-6} of the external field. The shape of the body also affects the amount of polarization. Since a standing humans body has more of a "lightning rod" shape than a crouching rat, a rat must be exposed to a much lower external field to achieve an equivalent internal electric field. A squatting human will experience lower and the rearing rat, higher fields. The body shape and foot area also affect the average current densities in various body locations because of the external electric field. Figure 1.13 illustrates these differences [51]. As shown in the figure, current densities increase in areas of smaller cross section, for example, the human neck or leg, and closer to the ground, for example, the upper and lower human torso. When calculated without averaging across a cross section, current densities are higher near a junction point; for instance, they are higher and more horizontal at the armpit than in the middle of the chest area [52].

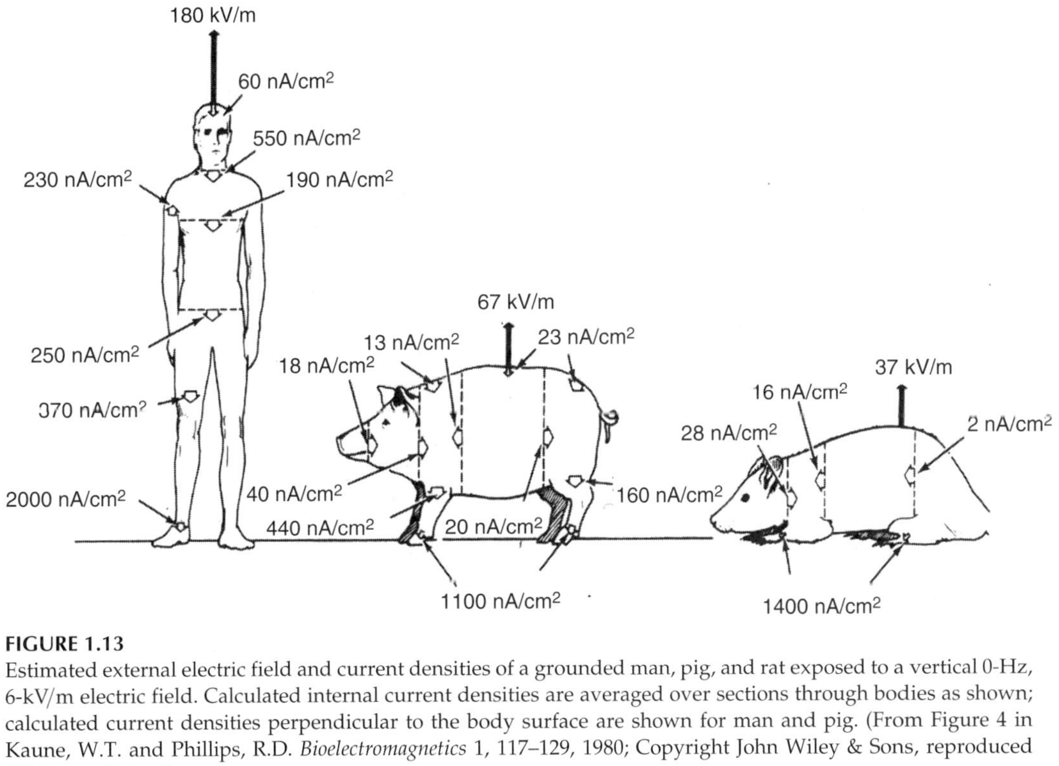

FIGURE 1.13
Estimated external electric field and current densities of a grounded man, pig, and rat exposed to a vertical 0-Hz, 6-kV/m electric field. Calculated internal current densities are averaged over sections through bodies as shown; calculated current densities perpendicular to the body surface are shown for man and pig. (From Figure 4 in Kaune, W.T. and Phillips, R.D. *Bioelectromagnetics* 1, 117–129, 1980; Copyright John Wiley & Sons, reproduced With permission.)

It is important to recognize that electric fields and current densities such as those in Figure 1.13 are averages, whether across the whole cross section of the body or a limb or across a localized region. Fields vary greatly across very small distances when one examines them at dimensions on the order of a cell or a molecule; this is called *microdosimetry*. Forming a good picture at this level of fields from either endogenous or external sources is an unsolved but very important problem. Chapter 5 in this volume on basic mechanisms discusses this issue further.

An external magnetic field's value is little changed as it enters a biological system, whether the human body or cells in culture, since the average biological magnetic susceptibilities are very close to those of air or vacuum (see Chapter 3 and Chapter 4 in this volume on magnetic properties of materials). However, the internal electric fields and currents induced in the body according to Faraday's Law are strongly determined by bodys (or specimen) shape, electric conductivity, and orientation with respect to the field. Table 1.4 gives some comparisons between the current induced in a human by the ELF magnetic fields generated in various situations and the external vertical 60 Hz electric field needed to produce the same current densities.

As discussed further in several chapters in this volume, especially Chapter 2 on endogenous fields, Chapter 5 on the basic interactions of fields and biological systems, and Chapter 7 on noise, as well as in the various discussions of models of field–biological system interaction, an externally applied field is unlikely to cause a biological effect unless the part of the biological system with which the field interacts is able to distinguish the external field from the internal electric fields and currents that are an integral part of the system. Exactly how to formulate the aspects of the endogenous field or current

TABLE 1.4

Magnetically Induced Total Body Current and Current Densities and Vertical 60-Hz Electric Field Inducing Equivalent Currents

Source	Current (μA)	Current Density (A/m²)	Electric Field (kV/m)
Sinusoidal waveforms			
Cord-connected household appliance	20–500	0.5–12[a]	1.5–38
Man in 8-kV/m electric field	120	3[a]	8
Electric blanket (not low field)	7–25	2–40[b]	0.5–1.7
Man in 0.16-kV/m electric field	2.2	0.05[a]	0.16
Nonsinusoidal waveforms—medical devices			
Electric anesthesia device (100-Hz square wave)	10,000	71,000[c]	670
Pacemaker electrode in myocardium[d,e]	6,000	20,000	400
Pacemaker electrode implanted in abdomen[d,f]	6,000	300	400

[a]Through 40-cm² ankle.
[b]0.63 cm from electric wire in blanket.
[c]Next to electrode.
[d]Peak pulse current $\sim 10^{-3}$-s duration, repeated every 0.8 s.
[e]Electrode area, 0.3 cm².
[f]Electrode area, 20 cm².
Source: After Bridges, J.E. and Preache, M. *Proc. IEEE* 69, 1092–1120, 1981. (Slightly modified from Table 1.3, Chapter 2 of second edition.)

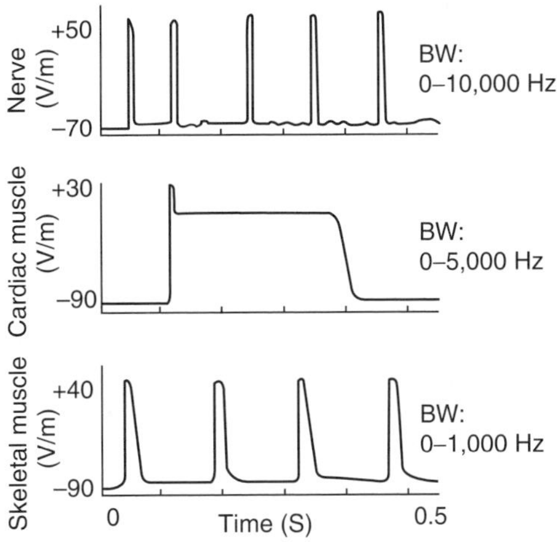

FIGURE 1.14
Typical time course and amplitudes of time-varying membrane potentials (V_m) of various cells. BW is the equivalent frequency bandwidth containing the main Fourier components of each voltage excursion. (From H. Wachtel, University of Colorado, private communication, copyright 1992; reprinted with permission.)

density that should be compared with the local field or currents in a particular situation is still an open research question; for example, over what region (how many molecules or cells) and over what range of frequencies (very narrow or broad) does the biological system average?

These endogenous fields range from the normal ~50–100 mV DC transmembrane potentials of most cells (negative in animals, sometimes positive in plants) to the relatively rapid pulses of nerve cell depolarization or repolarization spikes and the less rapid pulses of, for instance, muscle cells (see Figure 1.14 for examples). They also include the very large and often highly local and hence very nonuniform fields because of local charge densities on some macromolecules or changes in the double ion layer next to a membrane because of the inclusion of a protruding structure, such as a channel, at a particular location (see, e.g., diagrams in Chapter 5 in this volume on basic interactions of fields and biological systems).

1.3 EM Fields at Intermediate and Radio Frequencies (3 kHz to 300 MHz)

1.3.1 Electronic Article Surveillance

Many libraries and stores are currently equipped with electronic article surveillance systems, which generate EM fields ranging from ELF to radio frequencies (RFs). Sometimes, employees spend long periods of time close to parts of these systems, and they might therefore be exposed to strong EM fields, well in excess of the reference levels (RLs). RLs are the levels of easily measured fields in air that exposure-limiting regulations or guidelines state may, but do not necessarily, create health questions in those exposed to them. Above RLs, more precise measurements or calculations are needed for each specific situation to determine whether limits for the actual fields inside the body, the basic restrictions (BRs), are exceeded (for a discussion of RLs and BRs, see Chapter 8 in *BMA*) on standards in [82]).

Kjellsson et al. [53] made measurements on two different systems, one used in a shop and one in a library. The coils in the library at the exit were working at 920 Hz, and the flux density at the center of the coils was of the order of 10 μT, compared with the RL for the general public of 6.25 μT. When the books are returned to the library, the electronic tag in the book has to be activated before the book is put back into the library. The system used for deactivating operates at 50 Hz, and at close range the flux densities were of the order of some millitesla, thus again above the RL for occupational exposure at 500 μT. In the store the signal was a mixture of 17 Hz and 6.25 kHz. The flux densities at the lower frequency were of the order of 200–300 μT, and at the higher frequency the values were well in excess of 100 μT. The RLs at 6.25 kHz for the general public is 6.25 μT, so the RLs are clearly exceeded, and more in-depth analysis is needed to see if the BRs are still not violated. See also Harris et al. [54] for a technical description of these devices.

Both at the shop and in the library there are situations and places where the general public as well as the employees are exposed to magnetic fields that exceed the RL. Logging the exposure during a work shift at the shop also showed that the cashiers are exposed to magnetic fields generated by the transmitter signal of the detection gates, which were often activated by customers during the work shift studied. The cashiers are therefore a highly exposed category of personnel and hence are of interest to include in epidemiological studies. Eskelinen et al. [55] have published similar results and conclusions in a previous study.

In the library not only the detection gates but also the activators and deactivators contribute to the high magnetic fields. One problem is the handheld activator used by the employees. Because of its small size, it is possible to hold the activator against almost every part of the body. If for example, a user is wearing a pacemaker and is not aware of the magnetic fields, the worst case could lead to disturbances in the pacemaker.

1.3.2 EM Fields from Video Display Terminals

A very commonly used appliance emitting EM fields is the computer video display terminal (VDT). Flat panel displays are increasingly replacing the cathode ray tube (CRT)-based VDTs at present, but a great many CRTs are still in use. There are five different types of fields present in the vicinity of the CRT: an electrostatic field, VLF electric and magnetic fields at the horizontal sweep frequencies, and various ELF electric and magnetic fields at the screen refreshing rate, related to the power frequency. These are considered separately in the next few sections. Measurements show that the equivalent electrostatic surface potential on the screen can reach up to 20 kV for some VDTs, and the ELF electric fields in front of the VDT at a distance of 0.5 m range from a few to tens of volts per meter, although most of the time it is not distinguishable from the office background 50 or 60 Hz electric field. The ELF magnetic field can reach a few tenths of a microtesla, and close to the tube the values are up to a few microtesla. The VLF electric fields range from a few to tens of volts per meter, and the corresponding magnetic field is of the order of a few tenths of a microtesla at 0.5 m in front of the VDT. The VLF B field time derivative ranges from a few to a few hundreds of millitesla per second (see Ref. [56]).

In Figure 1.15 a schematic drawing of a CRT-based VDT is shown with the deflection coils. Since these coils are to move the electron beam horizontally, from left to right, the magnetic field affecting the electrons has to be vertically directed; therefore, the coils are in horizontal planes above and below the neck of the tube as shown in Figure 1.15. The stray field from these coils at the operator's position is mainly vertical. The sawtooth wave shape of the magnetic field is also schematically shown in Figure 1.10. The times, 3 and 30 μs, given in the figure are typical examples of values found on some VDTs. Also shown in the figure is the time derivative of the signal. Since the vertical deflection requires a horizontal magnetic field, the coils for the vertical deflection (omitted in the drawing in

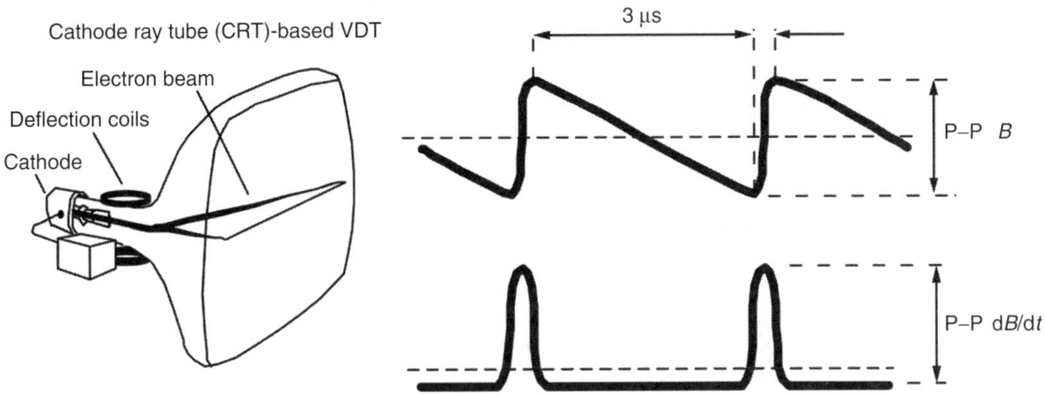

FIGURE 1.15
Schematic drawing of CRT-based VDT. To the right, examples of an idealized waveform of the VLF magnetic field and its time derivative are given. (Figure courtesy of Swedish National Institute for Working Life.)

Figure 1.15) lie in vertical planes on each side of the neck of the tube, giving a horizontal field and, thus, a corresponding stray field. The field from these coils has a frequency equal to the refresh rate, usually equal to or a small harmonic of the line frequency, and has a sawtooth wave shape. The electric field experienced by the operator is usually due to charge accumulating on the screen and is directed between the screen and the operator's body.

The time derivative, dB/dt, is usually measured only for the maximum field component in front of the VDTs. The median values found are 0.63 and 15 mT/s for the ELF and VLF frequency ranges, respectively. The corresponding maximum values are 1.9 and 101 mT/s for the ELF and VLF fields. For the ELF magnetic field in front of the VDT the median value was 0.21 μT. This is then a combination of the general magnetic field level in the office (median value 0.07 μT) and the emission from the VDT [56].

Flat panel displays for computers are presently based primarily on liquid crystal technology, particularly in portable computers. Compared with VDTs they use much lower internal voltages and currents and do not use magnetic field deflection, leading to the presumption that users' exposures are quite low. Even with a flat panel display, it or various sources inside the computer may expose portions of an operator's body to some fields that can approach few tenths of a microtesla in the ELF and VLF regions, especially for a portable computer that is held on the knees. At the time of writing, any values of the fields from laptop computers or other flat panel displays do not seem to have been published in the peer-reviewed literature. Swedish and Australian trade union tests on several laptop computers have shown magnetic field emissions below their internal certification limits for exposure (see Swedish levels in Table 1.5), initially established for VDTs. The Australians found two exceptions: a peak of up to 10 mG for up to 5 s during the program loading and a 0.01–0.16-μT fluctuating field at 30 cm, emanating from the computer transformer (frequencies not given) [57]. Computers with flat panel displays that have ungrounded power supplies, including laptops running either on batteries or on an ungrounded charger, can generate considerable static charges on their screens, producing DC potentials that exceed the Swedish unions' specifications. Older displays using cold-cathode backlighting sometimes also exceed the specifications for ELF or intermediate RF electric fields (Y. Hamnerius, Chalmers University, private communication, 2005).

Computers equipped for local area RF networking, such as "Wi-Fi" systems, and hubs for such networks emit and receive fields in the same general ranges of frequency and

TABLE 1.5

Swedish Tjänstemännens Central Organisation (TCO) Certification Limits for Computer Display Terminal Fields

Frequency Band	Quantity	Field Strength	Measurement Distance From Screen
DC	Electric potential	±0.5 kV	0
Band I	Electric field	≤10 V/m	30 and 50 cm
5 Hz–2 kHz	Magnetic field	≤200 nT	30 and 50 cm
Band II	Electric field	≤1 V/m	30 and 50 cm
20–400 kHz	Magnetic field	≤25 nT	30 and 50 cm

Source: TCO'03 Displays Flat Panel Displays, version 2.01 (translated by Y. Hamnerius), January. 2, 2004 (http://www.tcodevelopment.com).

intensity and are subject to the same sorts of limits as other portable RF devices, such as portable or cell phones. These are discussed later.

CRT-based VDTs use magnetic deflection of the electron beam, and therefore an external magnetic field can cause jitter and flicker on the screen. Sandström et al. [58] applied both 50 and 60 Hz magnetic fields to monitors with different refresh rates. The distortion was detectable from 0.6 up to 1.1 μT for seven monitors investigated. Background magnetic fields higher than 0.5 μT are not uncommon in offices; Sandström et al. [56] found that 5% of the workplaces measured exceeded this value. The ability to detect jitter depends, among other things, on the frequency of the disturbance. Since the frequency of the jitter oscillation is equal to the difference between the frequency of applied magnetic field and the refresh rate, the sensitivity to the external magnetic field will be different with different frequencies of the applied magnetic field.

1.3.3 RF Transmissions

1.3.3.1 Shortwave Transmission

High-power shortwave transmitters (approximately 2–25 MHz) are used for international broadcasts. Often the power supplied to the antenna system can be several hundreds of kilowatts. The antenna systems used are most often movable log-periodic or steerable curtain-type antennas.

Measurements of field strengths from such transmitters have been recently presented by Altpeter et al. [59] in connection with their study on health effects on people living near a station. The magnetic field values ranged from tens of milliamperes per meter for those at a distance of 500 m from the antenna to some tenths of a milliampere per meter for those at a distance of a few kilometers. In a study of leukemia and residence near a high-power shortwave transmitter in Italy, Michelozzi et al. [60] reported spot measurements of the electric field in some of the closest houses to be between 3 and 20 V/m. (Under far-field conditions 3 V/m corresponds to a magnetic field of 8 mA/m and 20 V/m to 53 mA/m.) For a review of measured field strengths see also Mantiply et al. [61].

1.3.3.2 FM Radio and TV Transmission

Exposures to RF fields have been occurring for as long as we have had radio broadcasting. Since the antenna towers are usually quite high and the emissions are directed for reaching a long distance, exposure levels near the towers are minimal. The Environmental Protection Agency has done field strength measurements in the United States, and an

TABLE 1.6

Electric Field Strengths

Source	Distance 1 km	10 km	50 km
(a) Combined electric field strengths from large radio and TV transmitters and one type of base station for mobile phones[a]			
Radio	1	0.1	0.02
UHF TV	5	0.5	0.1
(b) Electric field strengths from radio base stations for Mobile Communication			
	10 m	**100 m**	**1000 m**
Base station	7	0.7	0.07

Note: E field in V/m.

[a]Radio: 60 kW ERP; TV1: 60 kW ERP; TV2: 1000 kW ERP. $S = PG/(4\pi R^2) = E^2/120\pi$. (ERP is the effective radiated power, also expressed as PG, where P is the total power to the antenna and G is the gain; R is the distance to the antenna.)

overview is given by Mantiply et al. [61]. At a distance, the fields can be estimated from calculations using the far-field formula, and some examples are given in Table 1.6. The input power is often highest for UHF TV broadcasting. The TV signal consists of an amplitude-modulated video signal and a frequency-modulated audio signal. In Sweden, for instance, 30 kW of power is used for the video signal and about 5 kW for the audio. With the antenna gain this gives an effective radiated power (ERP) of 1000 kW.

There have not been many radio and TV towers built during the last 10 y or so, although a number have been moved or upgraded, e.g., moved to taller buildings for greater range, but we have seen an increase in the number of terrestrial channels available. New ways of transmitting information have also come into play with digital radio and TV. However, with these the output RF power may be lower; instead, this technique uses a much larger frequency bandwidth.

The total power density of RF exposure of the public has undeniably increased during the last 10 y. Tell and Mantiply [62] reported in 1980 that the average level in the United States from FM radio and TV transmitters was about 50 μW/m^2. Today, exposure due to such sources is about the same [50], but with mobile phone base stations added, the total power density is now often found to be on the order of 100 μW/m^2 or so (see Figure 1.16). Despite the proliferation of mobile telephone systems, most of the power is still due to broadcasting (Figure 1.11). Although the total level of RF exposure of the general public has increased during the last few years, it should be clearly stated that at all distances the RF field levels on the ground from base stations are well within the international guidelines for RF exposure of the general public.

1.3.3.3 Wireless Communication Systems (Base Stations, Personal Wireless Devices Such as Cellular Telephones and Pagers)

1.3.3.3.1 *Mobile Phone Base Stations*

Current mobile telephone systems operate at frequencies between 800 and 2100 MHz, but some older systems operating near 450 MHz are still in use (see Table 1.7). With the rapid increase in the use of mobile phones, the number of base stations has also increased. The phones operate by communicating with a nearby base station, which is a low-powered

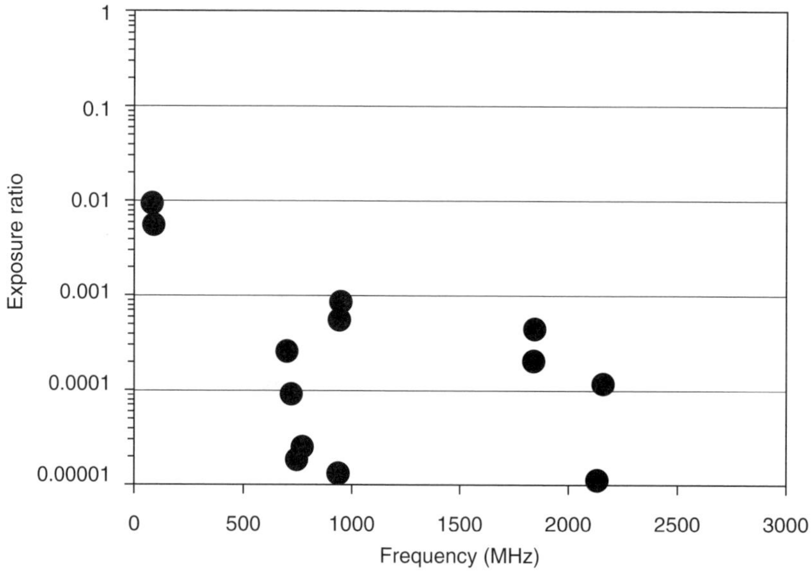

FIGURE 1.16
Power and frequency of environmental RF fields at a sample urban location in Sweden, plotted at each frequency as a comparison to the ICNIRP guideline power limit for personal exposure at the same frequencies, which is set equal to 1 (for plot of limits as a function of frequency, see Chapter 8 in *BMA* on standards in Ref. [82]). (From Hamnerius, Y. COST 281 Workshop 2004/09, Workshop on RF Exposure Assessment, Paris, September 20–21, 2004. With permission.)

radio transmitter, typically mounted on a tower or the roof of a building, that relays calls between the user and the telephone system. In Nordic countries the market penetration is over 100%; that is, there are more mobile phone subscribtions than people; many have more than one phone, and practically everone in the working population has access to a mobile phone. The number of base stations in Sweden is estimated at over 20,000. In 2000,

TABLE 1.7
Frequencies and Modulation Characteristics of Various Mobile Telephone Systems

System	Carrier Frequency	Modulation	Multiplexing Type
TETRA	380–470 MHz	17 Hz	TDMA
MIRS	806–821 MHz	11.1 Hz	TDMA
NADC	824–849 MHz	50 Hz	TDMA
CDMA	824–849 MHz	800 Hz	Multichannel
Analog	824–849 MHz	FM	None
GSM	890–915 MHz	217 Hz	TDMA
PDC	929.2 MHz/1.5 GHz	50 Hz	TDMA
Iridium	1616–1626 MHz	11 Hz	TDMA
CDMA	1765 MHz	800 Hz	Multichannel
PCS	1805–1880 MHz	217 Hz	TDMA
GSM	1800 MHz	217 Hz	TDMA
GSM	1900 MHz	217 Hz	TDMA
NMTS	1920–2170 MHz	100 Hz–1.5 kHz FDD	

Notes: TDMA = Time division multiple access; conversations use different time slices of shared channels. CDMA = code division multiple access; digital identifiers signal phone when the conversation is using one of many closely spaced channels.
Source: Data courtesy of M. Swicord, private communication, 2005.

the United States had more than 82,000 cellular base station sites, some with more than one base station in operation, while there were some 20,000 base stations sites in the U.K. [63].

The antenna system used for base stations comprises often either omnidirectional (whip) antennas, which radiate in all directions in the horizontal plane, or directional (panel) antennas, which radiate energy primarily from their front surfaces. Most commonly employed is a sectorized panel arrangement with three sets of directional transmitting and receiving antennas, oriented 120° apart.

The antennas have a high gain, giving a narrow beam in the vertical direction but a quite wide one in the horizontal direction. In the main beam of the antenna, that is, several meters or more directly in front of and at the same height as the antenna, the intensity of the beam (i.e., the power density) decreases as the inverse square of the distance from it. The RF exposure a person receives from a base station thus depends both on the distance from the antenna and on the angle with respect to the direction of the main beam. At ground level, the signal is relatively weak near the base of the antenna tower since the main beam is passing directly overhead. The ground level signal characteristically increases with distance from the tower to a maximum at between 10 and 100 m from the base of the tower and then decreases at still greater distances. Panel antennas only radiate significant amounts of energy in the forward direction. Thus, for panel antennas mounted facing outward on building parapets, the exposure is low for people on the rooftop or in rooms below the antenna. (See further the 1999 report on cellular telephones and health by the U.K. House of Commons Committee on Science and Technology [64].)

In rural areas, where there is a lower user density, base stations are typically a couple of kilometers apart. In towns and cities where there are more users the cells are smaller and transmitter base stations can be as little as a few hundred meters apart. The actual levels of transmitted power vary widely between urban and rural areas. Since a longer reaching distance is wanted in rural areas, the output power tends to be greater, whereas the large increase in the number and density of base stations within the cities will make the cells smaller and thereby also permit a reduction of output power from the surrounding base stations. Therefore, the levels of public exposure to RF energy from any system will not increase in proportion to the number of its base stations in an area.

In Table 1.6b an example is given of the calculation of electric field strength under far-field conditions. The values given are for the main beam, and since the beam would not be aimed at the ground in the near zone, the theoretical values are presumably overestimating the field strength for close distances. Measurements of field strength from base stations have recently been reported by Thuroczy et al. [65] and Hamnerius and Uddmar [66]. They found levels encountered within about 20 m from the antenna in the range of a few to some tens of volts per meter. However, at distances where most people would be exposed, the field strength is down to tens of millivolts per meter within cities, and in rural area values as low as a few millivolts per meter can be measured. For an estimation of maximum encountered levels, the formula used in Table 1.6b above may be used, and as can be seen from this table values in the range of tens of millivolts per meter are estimated for a distance of 1000 m from the station.

1.3.3.3.2 *Handheld Mobile Phones*

The technical details of analog and digital systems used for mobile phones have been described in detail [67,68], and only a short background is given here. The Nordic Mobile Telephone (NMT) system operates at 900 MHz with a continuous carrier wave. The maximum output from the handheld NMT900 phones is 1 W. The NMT phones have their output power regulated through the base station in two levels, 0.1 or 1 W; the closer to the station, the lower is the output power likely to be.

To make more efficient use of the available frequencies, various schemes that take advantage of the relatively slow way speech transmits information have allowed more than one conversation to occur in the same frequency channel. Some of these systems include pulsing the RF signal at various frequencies, often in the ELF frequency region, or otherwise modulating it. The result is a departure from a pure, continuous sinusoidal field. In the digital Global System for Mobile Communications (GSM), which is in very common use worldwide, the information is sent in pulses with a repetition rate of 217 Hz. The pulse length and repetition frequency give a duty cycle of 1/8. The maximum output power is 2 W, which gives a time-averaged value of 0.25 W maximum. The GSM system also provides a battery-saving function, which in practice reduces the output power to about half of the maximum. The output power is also regulated from the base station, from a maximum of 2 W down to a minimum of 20 mW, depending on the strength of the signal received at the phone; with the newest phones sold today an even lower value of 5 mW is used. The mean output power is thus normally well below 0.1 W. Frequencies and coding patterns of some other current or proposed mobile telephone signal systems are indicated in Table 1.7.

Different models of phones have different specifications for the antenna design position and physical dimensions, for instance, a dipole antenna or a helical antenna. Kuster [69] measured 16 different European digital phones and found a very wide variation in the SAR values. The phone giving the lowest value, when averaged over time and 10 g of tissue, had a specific absorption rate (SAR) of 0.28 W/kg in the user's head, and the one with the highest value had an SAR of 1.33 W/kg, all normalized to an antenna input power of 0.25 W, which is the maximal value for a GSM phone. If the averaging was done over 1 g of tissue, the span was from a low of 0.42 W/kg to a high of 2.0 W/kg. Anger [70] reported the measurement of SAR on 21 different phones, and his result is similar. The SAR ranged from 0.3 to 1.7 W/kg over 10-g tissues. He also reported the telephone communication power value, that is, how much of the output power can be used for communication, and he found values ranging from only about 5% in the lowest phone to just less than 50% in the "best" phone. Thus, more than half of the output power from the phone is lost because of mismatch between the phone and the antenna, and some is deposited as SAR in the user.

These SAR measurements were done under normal user conditions. However, when the phone is slightly tilted toward the head, Kuster [69] shows that the value can go up from 0.2 to 3.5 W/kg. Thus, for different phones under maximal output, we have a factor of about 5 between the extremes, and to this the personal handling of the phone gives a factor of tenfold or more. It should be noted that all given values are the maximum SAR values found, regardless of the anatomical location. An equal weight is given to the values, independent of whether they are obtained on the external ear, in the middle or inner ear, or behind the ear. In the future, it will be necessary to make the comparison at the same anatomical location. Presumably then, the values as given by Kuster [69] might differ even more.

Taken together there is a rather large uncertainty in estimating the actual SAR that depends on the specific situation, including a factor of 100 from the distance to the base station and at least a factor 10–50 depending on the make and model of the telephone and the personal style of use.

The currents from the battery also give rise to magnetic fields near the phone. For GSM phones magnetic flux densities of a few microtesla near the phone have been measured [66,71]. The fields are pulsed DC fields with a frequency of 217 Hz. For the NMT phones the magnetic field from the battery current can be regarded as pure DC. Jokela et al. [72] recently measured seven different GSM phones and examined the frequency content of the magnetic pulse, and he found that a considerable amount is found in the low kilohertz range. It was even found that some phones exceeded the ICNIRP guideline reference values when the multiple frequency formula was applied, but calculations show that the BRs were not exceeded.

1.3.4 RF EM Fields in Industrial Settings (RF Dielectric Heaters, Worker Exposure to Broadcast Systems)

1.3.4.1 RF Sealers

Operators of RF plastic sealers (RF operators) are an occupational category highly exposed to RF with frequencies around 27 MHz. RFs are used to produce heat to seal, for instance, plastics for tarpaulins, tents, rain clothes, and covers. Around an RF plastic sealer (RF sealer), both a magnetic and an electric field will be present, and close to the machine the coupling between the two is complex since far-field conditions are not at hand.

During RF exposure, a current is induced in the body, the magnitude of which is dependent on many factors such as the electric and magnetic field strength, the polarization of the field, and the grounding conditions. See further Wilén et al. [73], who tested different techniques to measure the induced current in grounded as well as ungrounded conditions.

Within the category of RF sealer operators, many different exposure situations exist, which can roughly be divided into two main types: tarpaulin workers and readymade clothing workers. Tarpaulin workers often stand in front of the RF sealer and will experience whole-body exposure, which often causes a high current to pass through the ankles. Readymade clothing workers often sit down in front of the RF sealer, often without perfect contact with the ground, but with their hands close to the electrode.

To get a good weld, both the welding time and the total power can be adjusted. The combination of the two will produce enough energy to seal the plastic. Typical welding times range from 1 to 10 s, depending on the material being sealed. The total exposure time will also vary between different RF-sealing processes; in the same amount of time, more welds are made in the readymade clothing industry than in the tarpaulin industry. For more information about the welding process see Refs. [74–76].

In a recent study, Wilén et al. [77] studied the exposure among RF sealer operators. The mean values of the calculated 6-min spatially averaged field strengths, in line with ICNIRP recommendations, are 107 V/m and 0.24 A/m, respectively. The maximum measured field strengths were 2 kV/m and 1.5 A/m, respectively. The induced current in ankles varied depending on the work situation, with a mean value of 101 mA and a maximum measured value of 1 A. In total, 16 out of the 46 RF plastic sealers measured exceeded the ICNIRP guidelines.

1.3.4.2 Occupational Exposure from Broadcasting and Radars

Workers in the fields of communication and radar are only exposed to low-level field strength in most situations. However, when climbing FM or TV towers, for instance, accidental exposure can be intense if proper safety precautions are not taken to keep exposure below the regulated limits. The same is true when someone is working near transmitter cabinets with the interlocks defeated and doors open. In these instances, the exposure, usually inadvertent, can be substantial.

Radar is used to detect the presence and directions of aircraft, ships, or other usually movable objects. The systems usually operate in the frequency range from 300 MHz to 15 GHz depending on the purpose of their use. The output power can range from a few milliwatts from police radar to several kilowatts for large air surveillance systems. A short description of various systems can be found, for example, on the WHO Fact Sheet No. 226 [78]. Here, as an example, a short description is given of one usual system where environmental exposure may occur. For civilian air traffic control, radars operating at 1305 MHz are used. The systems use a pulsed field with pulse duration of 2 μs and repetition rate of 625 pps. The peak power is 1.8 MW, and with the antenna gain this gives 3590 MW peak ERP in the maximum direction. Calculations using the far-field formula then give a peak electric field at 5 km of 65 V/m (equivalent power density of 11 W/m^2). However, with the pulsing this comes down to a time-averaged power density of 1.4 $\mu W/cm^2$, and since the antenna is rotating at 9 rpm and the lobe width is 1.2° this is then further reduced to 3×10^{-5} $\mu W/cm^2$. For a review of measurements on different systems see Mantiply et al. [61]

In all known exposure situations of exposure of the general public to EMF from radar systems, the level is below the limiting values given in international guidelines. However, occupational exposure in the near field can occur, and strict adherence to safety instructions is needed, in form of both engineering controls, for example, interlocks, and administrative controls, for instance, strong training and enforcement of safety policies.

1.3.4.3 Exposure in Medical Applications

When medical uses of EM devices are considered, a distinction is made between exposure of patients and medical workers operating the equipment. The balance between any possible risks and the anticipated benefit of exposure of the patient has to be considered by consultation between the patient and the prescribing physician; as a result, exposures are often in excess of limits for workers or the general public. However, operator exposure is governed by the usual rules for occupational exposures; hence, only operator exposure is considered here.

One of the earliest applications of RF energy was shortwave diathermy, which usually operates at 27 MHz. Usually, unshielded electrodes are used, and this may lead to high stray fields. The person operating the equipment, therefore, may be exposed to high field strengths, and the time-averaged guideline for exposure may be exceeded for both RF electric (E) and magnetic (H) fields, depending on the working conditions at hand. Microwave diathermy at 2.45 GHz is also being used in therapy. In this case, the antenna is such that the beam can be directed toward the area to be treated, but the microwave beam may easily be directed toward the patients. Although the beam may not be exposed to the therapist, it may be exposed to the people in the immediate vicinity, including people waiting in the next room. It is therefore necessary to take this into account when planning the space where the microwave diathermy treatment is to be performed.

Electrosurgical units are commonly used in operating suites. They employ RF energy for cutting and coagulation and typically operate at frequencies from about 0.5 MHz to a few megahertz. The energy is supplied to the cutting tool via unshielded cable, which may pass close to the arm and hand of the surgeon, resulting in exposure of operating room personnel to RF energy. The electric and magnetic field near the active lead may be quite high, from a few hundred volts per meter to several kilovolts per meter [79,80]. The induced current in the tissue is of the same order of magnitude as the basic guideline limit for these frequencies, about 5 A/m^2. Since the exposure is rather high, precautionary action would be wise, and shielded wires should be used for the active lead.

Operator and general public exposure to static and RF fields may occur in connection with the use of magnetic resonance imaging (MRI) and nuclear magnetic resonance spectrometers. MRI systems use strong DC magnets, typically from 0.05 to about 3 T, as well as rapidly changing gradient magnetic fields with time derivatives (dB/dt) typically from 1 to 3 mT/ms (1–3 T/s). The RF fields (10–100 MHz) are low and almost fully contained within the patient enclosure, and the RF exposure of the operators is negligible. Inside treatment rooms, near MRI equipment, the maximum exposure level is about 1 T in front of the magnet, and nurses and technicians staying with patients can be exposed to magnetic flux densities up to 200 mT, approaching the protection guideline. When professional activities take place very close to or inside the magnet's tube, workers can be exposed to higher fields (up to 1–2 T), for example, when assisting the patient, plugging in RF cables of treatment coils, or device cleaning. Reduction of the workers' exposure level is possible if MRI device design is required to include operation with workers no less than 0.5 m from the magnet. Workers' training should also include methods for exposure reduction.

Transcranial magnetic stimulation (TMS) has been introduced recently as a noninvasive and focal stimulation tool for the study of connectivity of brain regions, localization of functions, and pathophysiology of neuropsychiatric disorders and a therapeutic intervention method in the treatment of chronic depression. A high-intensity, fast magnetic field pulse produces a cortical stimulus through the induction of locally confined eddy currents; therefore, tailored coil arrangements can be used to achieve controlled local levels of stimulus. The TMS makes use of magnetic fields that can have intensities of up to 1 T with pulse durations in the range of ~0.05 to 0.2 ms. The resulting time derivative of the field can be several tens of kilotesla per second. This impulsive field transient is able to induce a rapid depolarization of the nerve cells within a volume of about 5 mm^3 at the cortical level.

In some instances, the operator holds the TMS transducer coil in place during exposure. Measurements of the leakage magnetic field at different distances from the coils of the transducer show that the intensity of the field decays proportionally to $1/r^3$ (r = distance). The regulations set limits aimed at avoiding excitation of the central nervous system, while TMS or repetitive TMS aims at just reaching the level of local exposure high enough to produce cortical excitations in patients. For the pulse trains in use, one group found a pulse spacing of ~0.3 ms and about 72-μs active pulse width, which gives an equivalent frequency of about 3.5 kHz. For this the limit value is about 1 T/s, and this is transgressed at distances of about 0.7 m from the surface of the transducer's coils under normal treatment conditions [81].

1.4 Conclusion

EM fields, both natural and of human origin, are ubiquitous. Fields of human origin are primarily a result of technological developments that did not begin until late in the 19th century. In general, the natural fields in the environment are much smaller than those inside organisms; natural environmental fields are also usually smaller than fields of human origin at the same frequency. Inside an organism, naturally occurring charges, currents, and fields in cells, tissues, and organs are very important physiologically, and electric charges and magnetic moments are crucial factors in determining molecular structure and chemical reaction rates. Since organisms, including humans, evolved in the natural fields alone, it is not clear how their adaptation to artificial ones might affect them. The other chapters of this handbook explore this question.

References

1. Dubrov, A.P. *The Geomagnetic Field and Life. Geomagnetobiology*. Plenum Press, New York, 1978.
2. U.S. National Geophysical Data Center. http:// www.ngdc.noaa.gov

3. König, H.L., Krueger, A.P., Lang, S., and Sönning, W. *Biologic Effects of Environmental Electromagnetism*. Springer-Verlag, New York, 1981.
4. National Academy Press. *Studies in Geophysics: The Earth's Electrical Environment*. National Academy Press, Washington, DC, 1986.
5. Schienle, A., Stark, R., Kulzer, R., Klöpper, R., and Vaitl, D. Atmospheric electromagnetism: individual differences in brain electrical response to simulated sferics. *Int. J. Pschyophysiol.* 21, 177–188, 1996.
6. Schienle, A., Stark, R., and Vaitl, D. Electrocortical responses of headache patients to the simulation of 10 kHz sferics. *Int. J. Neurosci.* 97, 211–224, 1999.
7. Campbell, W.C. Geomagnetic pulsations. In Matsushita, S. and Campbell, W.H., Eds. *Physics of Geomagnetic Phenomena*. Academic Press, New York, 1967, pp. 822–909.
8. Hingorani, N.G. High-voltage DC transmission: a power electronics workhorse. *IEEE Spectr.* 33(4): 63–72, 1996.
9. Kaune, W.T., Gilis, M.F., and Weigel, R.J. Analysis of air ions in biological exposure systems, including HVDC electric power transmission lines, in rooms containing generators, and near exposed humans and animals. *J. Appl. Phys.* 54, 6274–6283, 1983.
10. Fews, A.P., Wilding, R.J., Keitch, P.A., Holden, N.K., and Henshaw, D.L. Modification of atmospheric DC fields by space charge from high-voltage power lines. *Atmos. Res.* 63, 271–289, 2002.
11. Koops, F.B.J. Electric and magnetic fields in consequence of undersea power cables. In Matthes, R., Bernhardt, J., and Repacholi, M., Eds. *Effects of Electromagnetic Fields on the Living Environment*. Proceedings from International Seminar, Ismaning, Germany, October. 4–5, 1999, ICNIRP 10/2000, pp. 189–210.
12. ICNIRP. Guidelines for limiting exposure to time-varying electric, magnetic, and electromagnetic fields (up to 300 GHz). *Health Phys.* 74(4), 494–522, 1998.
13. Portier, C.J. and Wolfe, M.S., Eds. Assessment of Health Effects from Exposure to Power-Line Frequency Electric and Magnetic Fields. NIH Publication 98-3981, National Institute of Health Sciences, Research Triangle Park, NC, 1998 (accessed April 7, 2005, at http://www.niehs.nih.gov/emfrapid/html/WGReport/WorkingGroup.html).
14. Wertheimer, N. and Leeper, E. Electrical wiring configurations and childhood cancer. *Am. J. Epidemiol.* 109, 273–284, 1979.
15. Deno, D.W. Transmission line fields. *IEEE Trans. Power Appl. Syst.* PAS-95, 1600–1611, 1976.
16. Rogers, L.E., Beedlow, P.A., Carlile, D.W., Ganok, K.A., and Lee, J.M. Environmental Studies of a 1100-kV Prototype Transmission Line: An Annual Report for the 1984 Study Period. Prepared by Battelle Pacific Northwest Laboratories for Bonneville Power Administration, Portland, OR, 1984.
17. Lee, J.M., Pierce, K.S., Spiering, C.A., Stearns, R.D., and VanGinhoven, G. *Electrical and Biological Effects of Transmission Lines: A Review*. Bonneville Power Administration, Portland, OR, 1996.
18. Fews, A.P., Henshaw, D.L., Keitch, P.A., Close, J.J., and Wilding, R.J. Increased exposure to pollutant aerosols under high voltage powerlines. *Int. J. Radiat. Biol.* 75, 1505–1521, 1999.
19. Deadman, J.E., Armstrong, B.G., McBride, M.L., Gallagher, R., and Thériault, G. Exposures of children in Canada to 60 Hz magnetic and electric fields. *Scand. J. Work Environ. Health* 1999, 25(4), 368–375.
20. Hansson Mild, K., Sandström, M., and Johnsson, A. Measured 50 Hz electric and magnetic fields in Swedish and Norwegian residential buildings. *IEEE Trans. Instrum. Meas.* 45(3), 710–714, 1996.
21. Mccurdy, A.L., Wijnberg, L., Loomis, D., Savitz, D., and Nylander-French, L. Exposure to extremely low frequency magnetic fields among working women and homemakers. *Ann. Occup. Hyg.* 45, 643–650, 2001.
22. Ahlbom, A., Day, N., Feychting, M., Roman, E., Skinner, J., Dockerty, J., Linet, M., McBride, M., Michaelis, J., Olsen, J.H., Tynes, T., and Verkasalo, P.K. A pooled analysis of magnetic fields and childhood leukemia. *Br. J. Cancer* 83, 692–698, 2000.
23. Adams, J., Bitler, S., and Riley, K. Importance of addressing National Electrical Code® violations that result in unusual exposure to 60 Hz magnetic fields. *Bioelectromagnetics* 25, 102–106, 2004.

24. Moriyama, K. and Yoshitomi, K. Apartment electrical wiring: a cause of extremely low frequency magnetic field exposure in residential areas. *Bioelectromagnetics* 26, 238–241, 2005.
25. Zhang, J., Nair, I., and Morgan, M.G. Effects function simulation of residential appliance field exposures. *Bioelectromagnetics* 18, 116–124, 1997.
26. Verrier, A., Souques, M., and Wallet, F. Characterization of exposure to extremely low frequency magnetic fields using multidimensional analysis techniques. *Bioelectromagnetics* 26, 266–274, 2005.
27. Forsgren, P.G., Berglund, A., and Hansson Mild, K. Reduktion av lågfrekventa magnetiska fält i nya och befintliga anläggningar för eldistribution. Arbetsmiljöinstitutets undersökningsrapport 35, 1994 (in Swedish), 20 pp.
28. von Winterfeldt, D. and Trauger, T. Managing electromagnetic fields from residential electrode grounding systems: a predecision analysis. *Bioelectromagnetics* 17, 71–84, 1996.
29. Kaune, W.T., Dovan, T., Kavet, R.I., Savitz, D.A., and Neutra, R.R. Study of high- and low-current-configuration homes from the 1988 Denver childhood cancer study. *Bioelectromagnetics* 23, 177–188, 2002.
30. Kavet, R. and Zaffanella, L.E. Contact voltage measured in residences: implications to the association between magnetic fields and childhood leukemia. *Bioelectromagnetics* 23, 464–474, 2002.
31. Bridges, J.E. Non-perceptible body current ELF effects as defined by electric shock safety data. *Bioelectromagnetics* 23, 542–544, 2002.
32. Kavet, R., Zaffanella, L.E., Pearson, R.L., and Dallapiazza, J. Association of residential magnetic fields with contact voltage. *Bioelectromagnetics* 25, 530–536, 2004.
33. Kaune, W.T., Miller, M.C., Linet, M.S., Hatch, E.E., Kleinerman, R.A., Wacholder, S., Mohr, A.H., Tarone, R.E., and Haines, C. Magnetic fields produced by hand held hair dryers, stereo headsets, home sewing machines, and electric clocks. *Bioelectromagnetics* 23, 14–25, 2002.
34. Vistnes, A.I. Electromagnetic fields at home. In Brune, D., Hellborg, R., Persson, B.R.R., and Pääkkönen, R., Eds. *Radiation at Home, Outdoors and in Workplace*. Scandinavian Science Publisher, Oslo, Norway, ISBN 82-91833-02-8, 2001, pp. 286–305.
35. Söderberg, K.C., Naumburg, E., Anger, G., Cnattingius, S., Ekbom, A., and Feychting, M. Childhood leukemia and magnetic fields in infant incubators. *Epidemiology* 13(1), 45–49, 2002.
36. Floderus, B., Persson, T., and Stenlund, C. Magnetic-field exposures in the workplace: reference distribution and exposures in occupational groups. *Int. J. Occup. Environ. Health* 2(3), 226–238, 1996.
37. Deadman, J.E. and Infante-Rivard, C. Individual estimation of exposures to extremely low frequency magnetic fields in jobs commonly held by women. *Am. J. Epidemiol.* 155(4), 368–378, 2002.
38. Kelsh, M.A., Bracken, T.D., Sahl, J.D., Shum, M., and Ebi, K.L. Occupational magnetic field exposures of garment workers: results of personal and survey measurements. *Bioelectromagnetics* 24, 316–326, 2003.
39. Stuchly, M.A. and Lecuyer, D.W. Exposure to electromagnetic fields in arc welding. *Health Phys.* 56, 297–302, 1989.
40. Skotte, J.H. and Hjøllund, H.I. Exposure to welders and other metal workers to ELF magnetic fields. *Bioelectromagnetics* 18, 470–477, 1997.
41. Nordensson, I., Hansson Mild, K., Järventaus, H., Hirvonen, A., Sandström, M., Wilén, J., Blix, N., and Norppa, H. Chromosomal aberrations in peripheral lymphocytes of train engine drivers. *Bioelectromagnetics* 22, 306–315, 2001.
42. Ptitsyna, N.G., Villoresi, G., Kopytenko, Y.A., Tyasto, M.I., Kopytenko, E.A., Iucci, N., Voronov, P.M., and Zaitsev, D.B. Magnetic field environment in ULF range (0–5 Hz) in urban areas: man-made and natural fields in St Petersburg (Russia). In Bersani, F., Ed. *Electricity and Magnetism in Biology and Medicine*. Kluwer Academic/Plenum Publishers, New York, 1999, pp. 279–282.
43. Ptitsyna, N.G., Kopytenko, Y.A., Villoresi, G., Pfluger, D.H., Ismaguilov, V., Iucci, N., Kopytenko, E.A., Zaitzev, D.B., Voronov, P.M., and Tyasto, M.I. Waveform magnetic field survey in Russian DC and Swiss AC powered trains: a basis for biologically relevant exposure assessment. *Bioelectromagnetics* 24, 546–556, 2003.
44. Wenzl, T.B. Estimating magnetic field exposures of rail maintenance workers. *Am. Ind. Hyg. Assoc. J.* 58, 667–671, 1997.

45. Chadwick, P. and Lowes, F. Magnetic fields on British trains. *Ann. Occup. Hyg.* 42, 331–335, 1998.
46. Fraser-Smith, A.C. and Coates, D.B. Large amplitude ULF electromagnetic fields from BART. *Radio Sci.* 13, 661–668, 1978.
47. Bernardi, A., Fraser-Smith, A.C., and Villard, O.G., Jr. Measurement of BART magnetic fields with an automatic geomagnetic pulsation index generator. *IEEE Trans. Electromagn. Compat.* 31, 413–417, 1989.
48. Dietrich, F.M., Feero, W.E., and Jacobs, W.L. Safety of High Speed Guided Ground Transportation Systems: Final Report to the U.S. Federal Railroad Administration. U.S. Government Printing Office, Washington, DC, DOT/FRA/ORDL-93-07, DOT-VNTSC-TRA-93-13, 1993.
49. NIOSH. NIOSH Health Hazard Evaluation Report: Alumax of South Carolina, Centers for Disease Control and Prevention. National Institute of Occupational Safety and Health, Cincinnati, OH, 1994.
50. Von Kaenel, R. et al. The determination of the exposure to electromagnetic fields in aluminum electrolysis. In Mannweiler, U., Ed. *Light Metals 1994*. The Minerals, Metals & Materials Society, Warrendale, PA, 1994, pp. 253–260.
51. Kaune, W.T. and Phillips, R.D. Comparison of the coupling of grounded humans, swine and rats to vertical, 60-Hz electric fields. *Bioelectromagnetics* 1, 117–129, 1980.
52. Tenforde, T.S. and Kaune, W.T. Interaction of extremely low frequency electric and magnetic fields with humans. *Health Phys.* 53, 585–606, 1987.
53. Kjellsson, N., Sandström, M., Stensson, O., and Hansson Mild, K. Measurements of magnetic fields generated by electronic article surveillance systems. In Kostarakis, P., Ed. *Proceedings* of the 2nd International Workshop on Biological Effects of EMFs, Rhodes, Greece, October 7–11, 2002, ISBN 960-86733-3-X, pp. 937–941.
54. Harris, C., Boivin, W., Boyd, S., Coletta, J., Kerr, L., Kempa, K., and Aronow, S. Electromagnetic field strength levels surrounding electronic article surveillance (EAS) systems. *Health Phys.* 78(1), 21–27, 2000.
55. Eskelinen, T., Toivonen, T., Jokela, K., Keskinen, K., and Juutilainen, J. Occupational exposure to magnetic fields from EAS devices. In Proceedings of the EBEA 2001 5th International Congress of the European BioElectromagnetics Association, Helsinki, Finland, September 6–8, 2002, pp. 147–148.
56. Sandström, M., Hansson Mild, K., Stenberg, B., and Wall, S. A survey of electric and magnetic fields among VDT operators in offices. *IEEE Trans. EMC* 35(8), 394–397, 1993.
57. ACTU. *ACTU Guidelines for Screen Based Work*. Publ. D. No. 40/1998ACTU OHS UNIT, Australian Council of Trade Unions, Melbourne, 1998 (accessed April 2005 at http://www.ohsrep.org.au/data/files/general/SBW98.rtf).
58. Sandström, M., Hansson Mild, K., Sandström, M., and Berglund, A. External power frequency magnetic field induced jitter on computer monitors. *Behav. Inf. Technol.* 12(6), 359–363, 1993.
59. Altpeter, E.S., Krebs, Th., Pfluger, D.H., von Känel, J., Blattmann, R., Emmenegger, D., Cloetta, B., Rogger, U., Gerber, H., Manz, B., Coray, R., Baumann, R., Staerk, K., Griot, Ch., and Abelin, Th. Study on Health Effects of the Shortwave Transmitter Station of Schwarzenburg, Berne, Switzerland. Major report on behalf of the Federal Office of Energy, BEW Publication Series Study No. 55, August. 1995.
60. Michelozzi, P., Capon, A., Kirchmayer, U., Fusco, D., and Ancona, C. Risk of leukemia and residence near high-power radio-transmitters in Italy. *Epidemiology* 10, S116, 1999.
61. Mantiply, E.D., Pohl, K.R., Poppell, S.W., and Murphy, J.A. Summary of measured radiofrequency electric and magnetic fields (10 kHz to 30 GHz) in general and work environment. *Bioelectromagnetics* 18, 563–577, 1997.
62. Tell, R.A. and Mantiply, E.D. Population exposure to VHF and UHF broadcast radiation in the United States. *Proc. IEEE* 80, 6–12, 1980.
63. World Health Organization. Electromagnetic Fields and Public Health: Mobile Telephones and Their Base Stations. WHO International EMF Project Fact Sheet 193, Geneva, June. 2000 (accessed September 30, 2005, at http://www.who.int/mediacentre/factsheets/fs193/en).
64. Clark, R., et al. Cellular Phones and Health. Report House of Commons, No. 489, U.K. House of Commons Committee on Science and Technology, London, September 22, 1999 (see also http://www.parliament.uk/commons/selcom89/s&thome.htm).

65. Thuroczy, G., Janossy, J., and Nagy, N. Radiofrequency (RF) exposure of mobile communications in Hungary and evaluation relevant to EU and national standard: base stations and handy devices. Manuscript for NATO Workshop, Slovenia, October. 1998.
66. Hamnerius, Y. and Uddmar, T. Microwave exposure from mobile phone and base stations. Presented at Mobile Telephones and Health, Mobitel conference, Gothenburg, Sweden, September 16–17, 1999.
67. Bach Andersen, J., Johansen, C., Frölund Pedersen, G., and Raskmark, P. On the Possible Health Effects Related to GSM and DECT Transmissions. Report to the European Commission, Aalborg University, Denmark, 1995.
68. McKinlay, A.F., Andersen, J.B., Bernhardt, J.H., Grandolfo, M., Hossman, K.-A., van Leeuwen, F.E., Hansson Mild, K., Swerdlow, A.J., Verschaeve, L., and Veyret, B. Possible health effects related to the use of radiotelephones. Proposal for a research programme by a European Commission Expert Group, 1996.
69. Kuster, N. Review of dosimetry and near-field measurement techniques for human exposure evaluations and bioexperiments. In Legris, C., Ed. *Communication Mobile-Effects*, Biologique. CADAS, Acadèmie des Sciences, Paris, 2001, pp. 45–69.
70. Anger, G. SAR och utstrålad effekt för 21 mobiltelefoner. Statens Strålskyddsinstitut, SSI Rapport, 2002.
71. Linde, T. and Hansson Mild, K. Measurement of low frequency magnetic fields from digital cellular telephones. *Bioelectromagnetics* 18, 184–186, 1997.
72. Jokela, K, Purananen, L., and Sihvonen, A.-P. Assessment of the magnetic field exposure due to the battery current of digital mobile phones. *Health Phys.* 86, 56–66, 2004.
73. Wilén, J., Hansson Mild, K., Paulsson, L.E., and Anger, G. Induced current measurements in whole body exposure condition to radio frequency electric fields. *Bioelectromagnetics* 22(8), 560–567, 2001.
74. Eriksson, A. and Hansson Mild, K. Radiofrequency electromagnetic leakage fields from plastic welding machines. Measurements and reducing measures. *J. Microwave Power* 20, 95–107, 1985.
75. Williams, P. and Hansson Mild, K. *Guidelines for Measurement of RF Welders*. Arbetsmiljöinstitutet, Solna, Sweden, 1991.
76. International Labour Office. *Safety in the Use of Radiofrequency Dielectric Heaters and Sealers: A Practical Guide*. Occupational Safety and Health Series 71. International Labour Office Bureau of Publications, Geneva, 1998.
77. Wilén, J., Hörnsten, R., Sandström, M., Bjerle, P., Wiklund, U., Stensson, O., Lyskov, E., and Hansson Mild, K. Electromagnetic field exposure and health among RF plastic sealer operators. *Bioelectromagnetics* 25, 5–15, 2004.
78. World Health Organization. Electromagnetic Fields and Public Health: Radars and Human Health. Fact Sheet No. 226, World Health Organization, Geneva, 1999 (accessed May 2005 at http://www.who.int/mediacentre/factsheets/fs226/en/).
79. Paz, J.D., Milliken, R., Ingram, W.T., Frank, A., and Atkin, A. Potential ocular damage from microwave exposure during electrosurgery: dosimetric survey. *J. Occup. Med.* 29, 580–583, 1987.
80. Hansson Mild, K., Ly, C., Liljestrand, B., Sandström, M., and Foster, K.R. Dosimetric studies on electrosurgical units. Poster presentation at BEMS 23rd Annual Meeting, St. Paul, MN, 2001.
81. Figueroa Karlström, E., Stensson, O., Lundström, R., and Hansson Mild, K. Therapeutic staff exposure to magnetic field pulses during TMS/rTMS treatments. *Bioelectromagnetics* 27, 156–158, 2006.
82. van Deventer, E., Simunic, D., Repacholi, M., EMF Standards for Human Health, In Barnes, F.S., and Greenebaum, B., Eds., *Biological and Medical Aspects*, Taylor & Francis, Boca Raton, FL, 2006, Chapter 8.

2

Endogenous Electric Fields in Animals

Richard Nuccitelli

CONTENTS

2.1 Introduction

In a volume presenting the biological effects of electromagnetic fields it is appropriate to review the information we have regarding endogenous electric fields in the body. After all, imposed electromagnetic fields may augment the naturally occurring ones, so a complete understanding of the possible effects of imposed fields requires consideration of those electric fields already present. Here, I will provide a brief overview of the direct current (DC) endogenous fields that have been best characterized in animals and will touch on the evidence that these electric fields are required for the function of various cellular and organ systems. Other well-known variable fields that are generated by various electrically excitable organs such as the heart (electrocardiogram), brain (electroencephalogram), and eye (electrooculogram) will not be covered here. Another very comprehensive review of endogenous fields that will be of interest has appeared quite recently in *Physiological Reviews* by McCaig et al. (2005), and another review of the roles of such fields in development was presented by Levin (2003).

In order to put these endogenous fields in perspective for the reader, I would like to summarize here their main characteristics. Unlike much of the material considered in this volume, these endogenous fields are small and very slowly changing. Endogenous fields typically fall into the 10- to 100-V/m range and are generally very steady, DC fields generated by the flow of ionic currents through cells and embryos. This can be compared with the much higher fields required to electroporate cells (3 V/cell diameter or 3×10^5 V/m for a 10-μm-diameter cell), which are usually only applied for a short time on the order of a millisecond.

2.1.1 Sources of Endogenous Electric Fields

In order to generate an electric field in the body, a voltage generator or power source is required. There are two main sources of such a power source in living systems: (1) the plasma membrane surrounding every cell in the body and (2) the epithelium that surrounds every organ in the body as well as the entire body itself in the form of the skin. The plasma membrane forms the defining boundary for every cell and is a lipid bilayer with many embedded transporter proteins whose main function is to control the movement of molecules inside or outside of the cell. One of these transporter proteins is the Na^+/K^+-ATPase, which is responsible for maintaining two ion concentration gradients across the plasma membrane (high internal $[K^+]$ and low internal $[Na^+]$). The K^+ concentration gradient, in combination with a large number of K^+ channels in the plasma membrane, results in the outward diffusion of K^+. This outward movement of K^+ leaves behind the anion that was associated with maintaining electroneutrality and thereby separates charge across the membrane. This separation of charge generates a voltage difference or membrane potential (inside negative). This membrane potential is used for a wide variety of cellular functions, from capturing nutrients to signaling the occurrence of important events, such as sperm–egg fusion in many egg types or light absorption in retinal rods. Indeed, we expend more than half of our energy in maintaining this voltage across all of the cells in our brain and kidney, where excitatory events and Na^+-dependent transport constitute major functions (Clausen et al., 1991).

Voltage differences are also found across all epithelial layers, and this is called the transepithelial potential (TEP). Both organs and embryos are surrounded by one or more monolayers of cells called an epithelium. The outer epithelium belongs to the organ while the plasma membrane to the cell, and epithelia pump ions across themselves to generate the TEP. One clear difference between these two voltage sources (plasma membrane and epithelial layer) is their opposite polarity. Whereas the plasma membrane potential is usually negative on the inside with respect to the outside, the TEP is usually positive on the inside or basal side of the epithelial monolayer with respect to the outside or apical side of the monolayer.

This polarity difference results from the opposite flow of charge across these layers. The plasma membrane of most animal cells generates a potential difference across itself that is negative on the interior. This results from the outward diffusion of K^+ that separates charge across the membrane. In contrast, most animal epithelia are composed of highly polarized cells in which Na^+ channels are localized to the apical end of the cell and both the K^+ channels and the Na^+/K^+ ATPases are localized at the basal end. This polarized distribution of these transport proteins leads to Na^+ influx at the apical end and both Na^+ and K^+ efflux at the basal end to drive positive charge to the interior of the epithelial

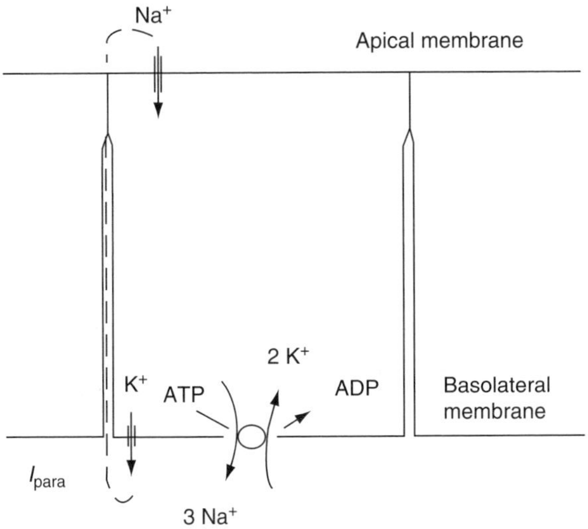

FIGURE 2.1
Diagram of a typical epithelial cell in a monolayer with Na^+ channels localized on the apical plasma membrane and K^+ channels localized on the basolateral membranes along with the Na^+/K^+-ATPase. This asymmetric distribution of ion channels generates a transcellular flow of positive current that must flow back between the cells through the paracellular pathway (I_{para}). This current flow generates a TEP that is positive on the basolateral side of the monolayer.

layer. However, this current cannot flow freely in the extracellular medium because these polarized epithelial cells are attached to each other with both tight and adherent junctions. As the apical–basal transcellular current flows back extracellularly, it must follow a pathway between the cells called the paracellular pathway (Figure 2.1). Current flow through this pathway encounters a high-resistance region at the tight junctions near the apical end, and the current flow through this resistance leads to a TEP that is positive on the basal side of the monolayer with respect to the apical side. The TEP will be proportional to the resistance of this paracellular pathway, and typical values for this TEP range from 15 to 60 mV, basal side positive.

It is this TEP that is the driving force for most endogenous ionic currents in embryos and adults. This voltage across epithelia will drive current out of regions of low resistance where there has been a break in the epithelium (wounds) or where tight junction resistance is low, such as along the primitive streak (Jaffe and Stern, 1979; Winkel and Nuccitelli, 1989), the posterior intestinal portal (Hotary and Robinson, 1990) in chick or mouse embryos, or at the forming limb bud in amphibian, chick, and mouse embryos (Borgens et al., 1983; Robinson, 1983; Altizer et al., 2001). This "leakage current" will in-turn generate a lateral electric field along its path that will be proportional to the resistivity in that region. This electric field results from Ohm's law in a conductive medium,

$$E = \rho J$$

where J is the current density and ρ is the local resistivity.

The earliest measurements of the leakage current associated with wounds were made more than a century ago. DuBois-Reymond (1843) used a unique galvanometer that he built himself with more than two miles of wire and measured about 1 μA flowing out of a cut in one of his fingers. This was confirmed in 1849 and 1910 by other investigators, and the history of these measurements is presented in a scholarly review by Vanable (1991). More modern techniques have also been used to study this wound current as discussed. Direct measurements of electric fields *in situ* have been made, and I will discuss them next.

2.1.2 Methods for Measuring Endogenous Electric Fields

2.1.2.1 Self-Referencing Probe

Lionel Jaffe and I developed a technique for exploring transcellular ionic currents, called the vibrating or self-referencing probe (Jaffe and Nuccitelli, 1974). This instrument vibrates a small platinum sphere between two points about 10 μm apart at about 300 Hz and measures the voltage between those points using signal averaging to improve the signal-to-noise ratio. In a conducting medium where most cells find themselves, a voltage difference can exist only where there is a current flowing through the medium. Ionic currents entering or leaving cells can be readily detected by measuring the voltage they generate as they flow through the extracellular medium, and the past 30 years of research on more than 30 cell types has revealed that most cells have an asymmetrical distribution of ion channels that naturally leads to a transcellular current density on the order of 1–10 $\mu A/cm^2$ (Nuccitelli, 1988, 1990). Most epithelia that have been studied exhibit extracellular current densities on the order of 10–100 $\mu A/cm^2$ flowing through the organ or embryo with which they are associated. This technique detects the current that is flowing outside the cell or tissue, and the exact electric field that is generated by this current when it flows inside the cell or tissue can only be estimated based on tissue resistivity. It is more accurate to directly measure the electric fields in the tissue or cells as described here.

2.1.2.2 Microelectrode Techniques for Measuring Endogenous Electric Fields

The classic approach to these measurements is to use KCl-filled glass microelectrodes to penetrate the outer epithelium and measure the voltage just beneath it in several positions. Such electrodes are typically connected via a Ag–AgCl junction to a very high-input impedance preamplifier, so that they do not drain current from the system under study, and their tips are small (on the order of 0.1–1 μm) to minimize tissue damage (Wallis, 1993). If ionic currents are flowing within a tissue, these currents will generate an electric field that can be detected as the difference in potential at various sites along the current path. We will see below that this is the most commonly used approach to measure intraembryonic electric fields.

2.1.2.3 Voltage-Sensitive Fluorescent Dyes for Measuring Endogenous Electric Fields

Another popular technique for measuring the voltage across lipid membranes uses lipophilic fluorescent dyes whose fluorescence is voltage sensitive (Loew, 1992; Loew et al., 2002). For some of these dyes, their fluorescence intensity is dependent on their position within the lipid bilayer, which is in turn influenced by the membrane potential drop across this region. Membrane potential changes can be monitored by measuring the fluorescence intensity of these dyes, and differences in the membrane potential of cells making up an organ or embryo can also be detected. One example of the use of this approach, which will be discussed below, is to provide information about electric fields within sheets of cells that are electrically coupled via gap junctions in the chick embryo.

2.2 Measurements of Endogenous Extracellular Electric Fields

2.2.1 Amputated Limbs

Among the earliest direct measurements of endogenous extracellular electric fields were those made in the regenerating amphibian limb (McGinnis and Vanable, 1986). Upon amputation, the skin battery of the amphibian limb drives 10–100 $\mu A/cm^2$ out of the cut end of the limb stump. This current flow generates an electric field within the limb tissue that is 60 mV/mm near the lesion during the first hours after amputation, and this field drops to about 25 mV/mm within 6 h as the healing process leads to an increase in the resistance of the wound.

Electric fields of this magnitude have been found to stimulate the growth of neurons into the limb via galvanotropism, and the presence of enhanced nerve in limbs has been correlated with enhanced regeneration (Borgens et al., 1979). The study of nerve galvanotropism has a long and fascinating history that is thoroughly reviewed by McCaig et al. (2005). Briefly, most neurons exhibit sensitivity to imposed electric fields by either bending their outgrowth direction toward the negative pole of the field or in the case of a neural ganglion, exhibiting a higher density of outgrowths on the side of the ganglion facing the negative pole. This galvanotropism may play a role in the guidance of neurons to their targets during development and has also been found to play an important role in regeneration.

2.2.2 Embryonic Electric Fields Beneath the Skin

Direct measurements of electric fields have been made in both avian and amphibian embryos during normal development. Hotary and Robinson (1990) used both the self-referencing probe and microelectrodes to first detect the transembryonic current in the 2- to 4-day-old chick embryo and then measure the electric field that the transembryonic current generates beneath the epidermis. They measured current entering much of the epidermis during stage 14 of development with the outward current focused mainly at the posterior intestinal portal, where up to 105 $\mu A/cm^2$ was measured (Figure 2.2). One would expect that this large anterior–posterior current would generate an internal electric field, so they then used microelectrodes to measure the TEP along this axis. Here, they measured electric fields of 5–20 mV/mm.

They then proceeded to test the hypothesis that these fields are important for normal development by perturbing them (Figure 2.3). They implanted a glass capillary that was filled with either conductive saline agar or nonconducting glass, used as a control, through the ectoderm at the dorsal trunk of the embryo (Hotary and Robinson, 1992). The conductive capillary allowed large currents of about 5 $\mu A/cm^2$ to leak out of the embryo. While the control embryos developed quite normally, most embryos with the implanted conducting capillary exhibited abnormalities in posterior structures where the endogenous electric field is normally the largest but was reduced by the capillary shunt. Perturbing the normal voltage pattern within the embryo resulted in striking tail abnormalities; and an investigation of a genetic mutant, rumpless, that exhibits similar tail abnormalities led to a very interesting correlation. They found that most rumpless mutants exhibited a much lower transembryonic current density and lower electric field within the embryo and those mutants that exhibited a normal electric field pattern also exhibited normal development. Therefore, they found a good correlation between the internal electric field and the normal posterior development. These observations certainly support the hypothesis that the endogenous field is important for the development of posterior structures.

Studies of endogenous fields have also been carried out on the stage 14–21 developing axolotl embryo (Metcalf et al., 1994; Shi and Borgens, 1995). Current is driven out of the lateral walls of the neural folds and the blastopore and enters most of the rest of the embryo's body surface (Figure 2.4). Measurements of the TEP indicate an internal, caudally negative electrical field beneath the neural plate ectoderm. The magnitude of the endogenous field is on the order of 10–20 mV/mm (Figure 2.5). When these embryos

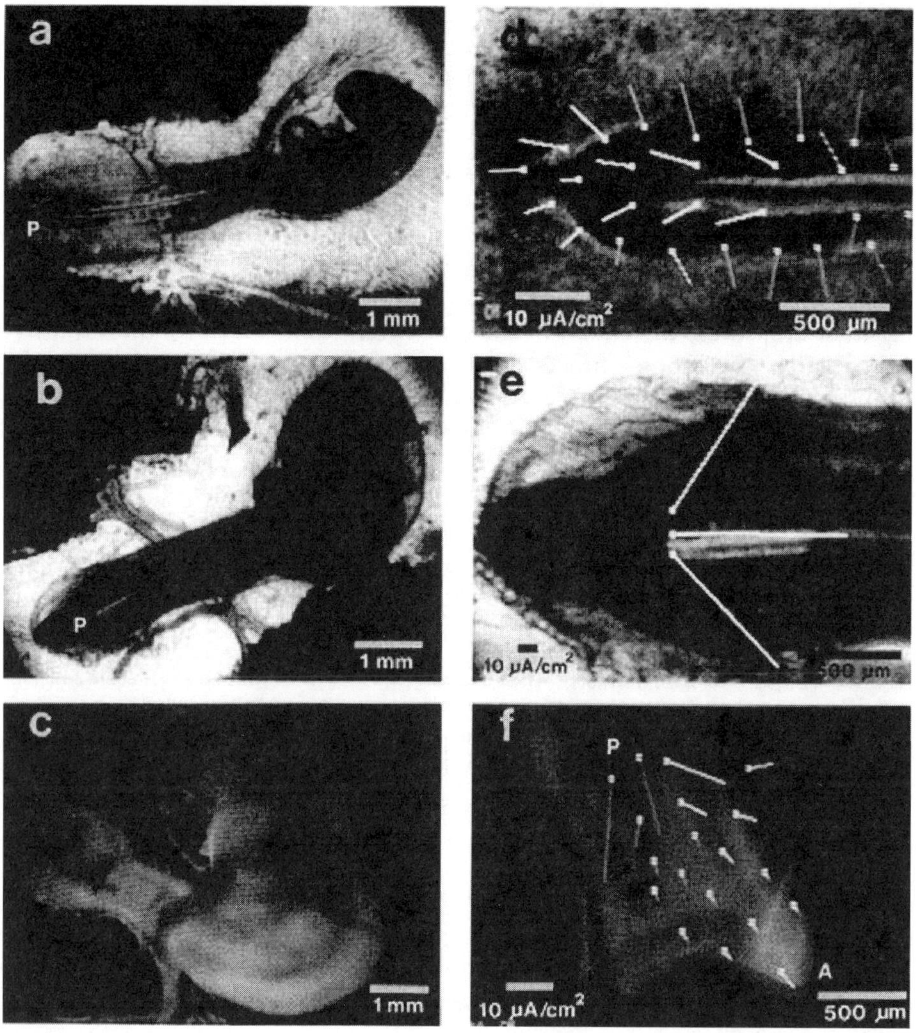

FIGURE 2.2
Ventral surface of three chick embryos at stages 14 (a and d), 17 (b and e), and 20 (c and f). Low-magnification views of the whole embryo are shown in (a)–(c), while (d)–(f) show the current pattern around the posterior intestinal portals of the embryos. Current vectors are represented by lines originating at a dot that indicates the position of the self-referencing probe when the measurement was made. The direction of the vector line away from the dot indicates the direction of current flow at that point, and the length of the line is proportional to the current density. At stage 14 (d), all vectors point toward the posterior intestinal portal or the lateral walls of the midgut. The three vectors shown at stage 17 (e) indicate large currents of about 100 $\mu A/cm^2$ leaving the posterior intestinal portal. At stage 20 (f), outward currents were also found at the posterior intestinal portal. Current densities were much lower by this stage. Note the inward current at the anterior intestinal portal (A). (From Hotary, K.B. and Robinson, K.R. (1990). *Dev. Biol. 140*, 149–160. With permission.)

were placed into an external electric field designed to modify the internal field, abnormalities were observed that depended on the developmental stage (Metcalf and Borgens, 1994). Gastrula-stage embryos exhibited normal development after exogenous field exposure, indicating that the imposed field does not harm the embryo in some nonspecific way. In contrast, neurula-stage embryos exhibited developmental abnormalities when exposed to similar electric fields of 25–75 mV/mm. These data support the hypothesis that the natural electric field within the embryo influences normal morphogenesis.

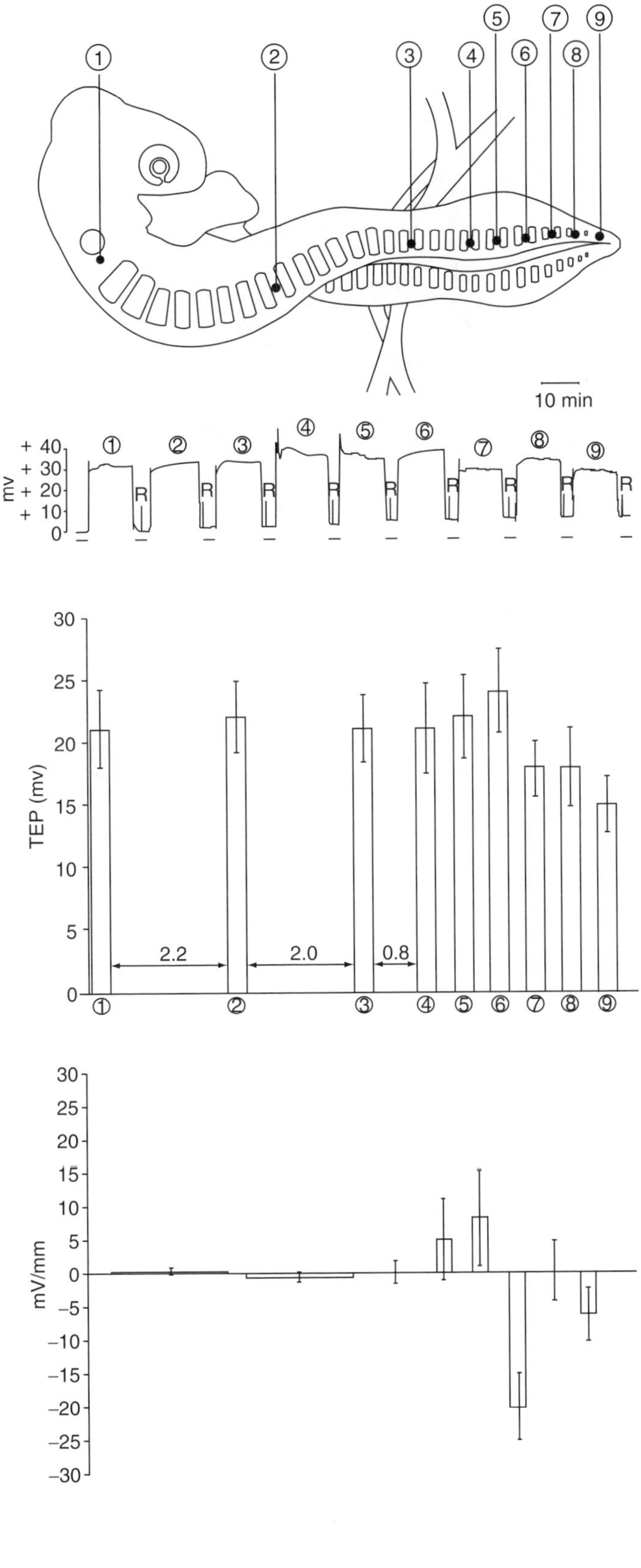

FIGURE 2.3
Stage 17 chick embryo TEP measured in nine rostral–caudal positions. Numbered measurement positions are shown in the drawing in the upper part of the figure. Below the drawing is a chart recording, tracing, and showing the TEP at the different measurement positions. At each numbered peak (corresponding to the positions shown in the drawing) the integument of the embryo was impaled and a stable positive potential was measured. Times at which the embryo was not impaled are indicated by a solid line below the recording. The upper bar chart shows the TEP at each position. The numbers below each bar correspond to the measurement positions indicated. The numbers between bars indicate the average distance (in mm) between each position. Where this is not indicated, the average distance is 0.3 mm. The lower chart shows the average voltage gradient between each consecutive position. Error bars indicate the standard error of the mean, $N = 6$. A steep voltage gradient was found between positions 6 and 7. (From Hotary, K.B. and Robinson, K.R. (1991). *Dev. Biol.* 140, 149. With permission.)

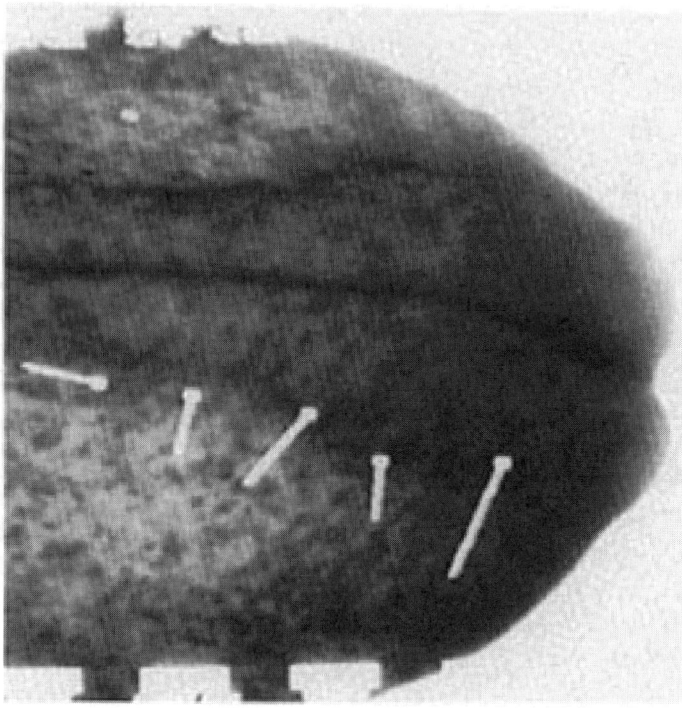

FIGURE 2.4
Neural fold currents in a stage 18 axolotl embryo measured with a two-dimensional self-referencing probe. Current vectors are displayed as a line originating at a dot that marks the measurement position. The direction of current flow from the dot is denoted by the line direction, and its length is proportional to the current density. Note the outwardly directed currents at the edge of the cranial neural folds. (From Metcalf, M.E.M., Shi, R.Y., and Borgens, R.B. (1994). *J. Exp. Zool. 268,* 307–322. With permission.)

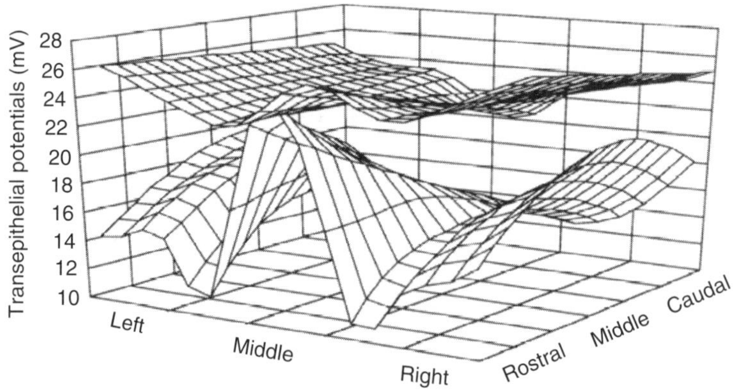

FIGURE 2.5
Summary of three-dimensional plot of TEPs at stage 15/16 (bottom) and stage 18/19 (top) axolotl embryos. The overall increase in the magnitude of TEPs at stage 18/19 is real, permitting these two views to be presented adjacent to each other. Note that only a hint of the characteristic voltage gradients beneath the ectoderm is evident at stage 18/19, and the potentials are not statistically different from one another. The embryo is essentially isopotential within the extracellular domain of the neural plate near the climax of neurulation. (Reproduced from Shi R.L and Borgens R.B. (1995) Dev. Dyn. *202,* 101–114, 1995. With permission.)

2.2.3 Fields Associated with the Neural Tube

The neural tube forms during development as a folding over of neural plate epithelium, inverting the normal polarity of this layer. The apical end of the neural plate becomes the inside of the tube so that transport of Na^+ occurs from the inside to the outside of the tube and the trans-tube potential is inside negative. This potential has been measured in both frog and axolotl to be as large as −90 mV (Hotary and Robinson, 1991; Shi and Borgens, 1994). Neuroblasts within the wall of the tube are therefore exposed to very large electric fields, since the wall is only about 50 μm thick and the 90 mV across this distance generates a field of roughly 1800 mV/mm. There is little doubt that such a large field will influence the migration and sprouting of these neuroblasts.

2.2.4 Fields Associated with Epithelial Wounds

As noted above, the earliest measurements of the electrical phenomena associated with wounds did not measure the electric field itself but rather the current flowing out of the wound. More modern techniques have also been used to study this wound current. The leakage current that is driven out of epithelia in low-resistance regions has been measured using the vibrating probe technique (Jaffe and Nuccitelli, 1974) in several systems. One of the earliest such measurements was a current as large as 100 $\mu A/cm^2$ leaving the stumps of regenerating newt limbs (Borgens et al., 1977). Similar measurements have also been made on fingertip amputation currents in humans (Illingworth and Barker, 1980), where up to 30 $\mu A/cm^2$ was detected leaving the accidentally amputated stump for about 3 wk. These currents will certainly generate electric fields just beneath the epidermis that will be proportional to the resistivity encountered in the tissue. The range of human tissue resistivity spans 200–1000 Ω cm (Faes et al., 1999), so these currents would be expected to generate an electric field within the tissue of about 10–100 mV/mm.

However, since this tissue resistivity can vary substantially as a function of cell density and tissue anatomy, it is always more reliable to measure these fields directly in the tissue rather than estimating them based on the transembryonic current density. This has been accomplished in four different wound types in skin and cornea. The classic approach to these measurements is to use KCl-filled glass microelectrodes to penetrate the outer epithelium and measure the voltage just beneath it in several positions along a line leading away from the wound. However, for skin measurements, another approach is more common. This method is to measure the potential gradient just beneath the stratum corneum on the surface of the epidermis, either with surface electrodes or by other means. The field generated by the current flowing between the upper surface of the epidermis and the stratum corneum is often larger than that generated below the epidermis because of the higher resistivity of that upper region. The range of field strengths measured in the four cases in the literature is surprisingly small, between 40 and 200 mV/mm (Table 2.1). The field direction is a function of position. Beneath and within the epidermis the field polarity has the negative pole at the wound center, and above the epidermis the wound current is flowing in the opposite direction so that the positive pole is at the wound (Figure 2.6).

These wound fields have some useful properties for signaling. First, they appear immediately upon wounding since the TEP is continuously present to drive current out of any low-resistance region as soon as it is formed. Second, the lateral electric field illustrated in Figure 2.6, that is generated by the wound current, will persist until the resistance increases as the wound heals. Thus, we have a signal that is immediate and persistent. These are ideal properties for a physiological signal to stimulate wound healing. If the epithelial cells forming the epidermis were able to detect such electric

TABLE 2.1

Endogenous Electric Fields Measured near Wounds

Species	Tissue	Wound Type	*E* Field (mV/mm)	Reference
Bovine	Cornea	Cut	42	Chiang et al., 1992; Sta Iglesia and Vanable, 1998
Notophthalmus viridescens	Digit	Digit tip amputation	40	McGinnis and Vanable, 1986; Chiang et al., 1989; Iglesia et al., 1996
N. viridescens	Limb stump	Amputation	7–50	McGinnis and Vanable, 1986
Guinea pig	Skin	Small cut	100–200	Barker et al., 1982

fields, they would be able to initiate wound healing immediately upon wounding. This is in fact the case, as discussed in the next section.

Measurements of endogenous electric fields near epithelial wounds have been made in three different systems. The first was a skin wound in the guinea pig (Barker et al., 1982). The transepidermal potential was measured in several locations lateral to a skin wound (Figure 2.7). At the wound itself, there is no epidermis, so the transcutaneous potential is zero, whereas about 1 mm away, the transcutaneous potential exhibited the normal value of 50–70 mV. The steepest voltage gradient was found immediately adjacent to the wound edge where values as high as 150 mV/mm were measured. The second direct measurement of the electric field near a wound was made in two regions of the newt limb. The electric field adjacent to an amputated digit was measured with microelectrodes and found to be about 40 mV/mm (McGinnis and Vanable, 1986; Chiang et al., 1989; Iglesia et al., 1996). This is very similar to the field near an amputated limb of the newt of 7–50 mV/mm (McGinnis and Vanable, 1986). The third direct measurement was made on the bovine cornea, and the magnitude of the electric field was 42 mV/mm (Chiang et al., 1992; Sta Iglesia and Vanable, 1998).

The fields in the cornea and the newt digit have been found to play a role in wound healing. When the field strength associated with the wound is modified, the rate of wound healing changes. The newt's wound healing rate can be optimized under normal

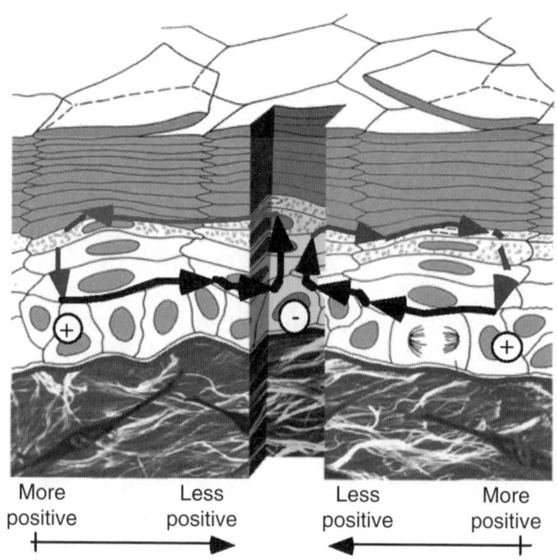

FIGURE 2.6
Generation of skin wound electric fields. Unbroken skin maintains a "skin battery" or TEP, generated by the apical influx of Na^+ and basolateral efflux of K^+. When there is a wound, the potential drives current flow through the newly formed low-resistance pathway, generating an electric field whose negative vector points toward the wound center at the lower portion of the epidermis and away from the wound on the upper portion below the stratum corneum.

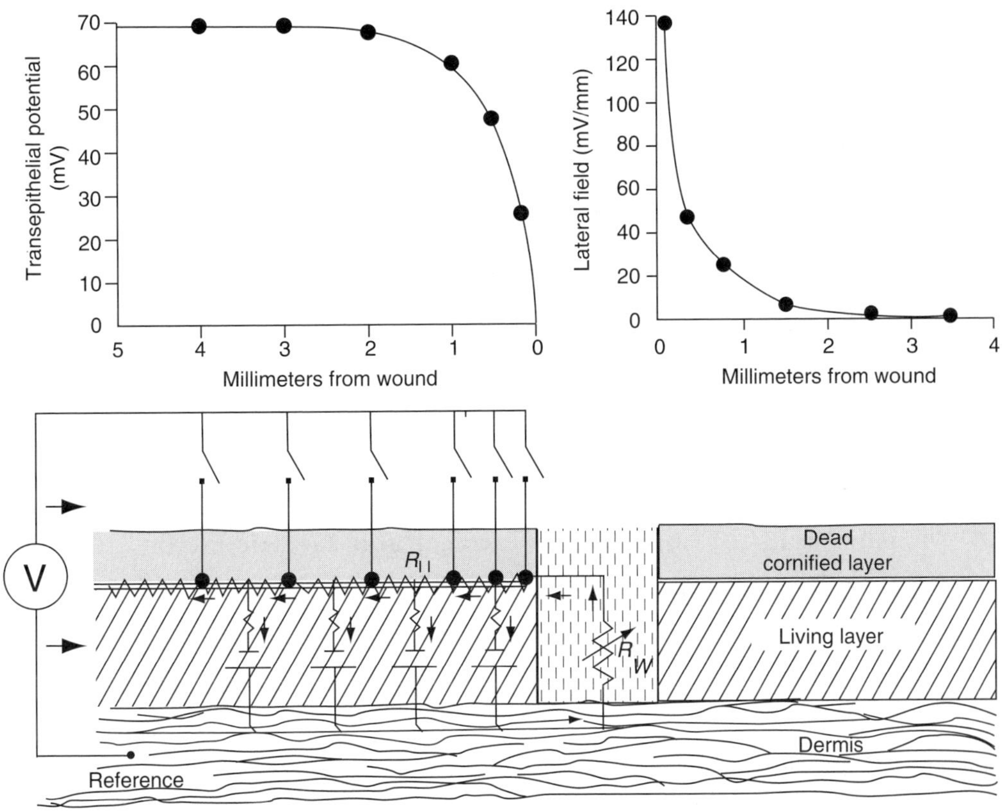

FIGURE 2.7
The electric field near a skin wound in a guinea pig. Bottom: diagram of the method used. A metal probe penetrated the stratum corneum to contact the epidermis at various distances from the wound site, and the surface voltage at each location was recorded with respect to the potential within the dermis. This voltage is the TEP and is plotted in the upper left of the figure. The voltage at the wound must be zero because the epidermal layer has been broken there, and far from that region, the TEP is 70 mV. The electric field at a given location is given by the voltage difference between sampling sites divided by the distance between those sites and is plotted on the upper right of the figure.

conditions but cannot be speeded up, removing the endogenous field does slow down the rate of wound healing by about 25%. However, the rate of bovine corneal wound healing can be enhanced by increasing the electric field strength. Epithelization was fastest in wounds with the field strength raised to 80 mV/mm, more than twice the normal field strength present in wounds maintained in Hanks' solution alone. Epithelization decreased, however, when the field strength was increased to 120 mV/mm. A similar pattern was also observed when the field's polarity was reversed. Decreasing the field strength by submersion of the lesions or by treating the lesions with the Na^+ channel blocker, benzamil, significantly retarded healing. In addition, an increase in the field strength of lesions treated with Na^+ depleted Hanks' solution, by the addition of DC, increased the rate of epithelization. These observations suggest that the endogenous electric field plays a role in the normal wound healing process.

More recent studies of rat cornea have reported a similar correlation between the wound healing rate and the electric field strength. The mammalian cornea exhibits an internally positive TEP of 30–40 mV. Na^+ and K^+ are transported out of the tear fluid and into the epithelium while Cl^- is transported in the opposite direction. A wound in this epithelium exhibits the same current pattern as skin (Figure 2.6), and a laterally oriented electric field is generated by this wound current with the cathode at the wound site. Recent work from McCaig, Zhao, and colleagues has indicated a strong correlation between the rate of wound healing in the rat cornea and the electric field strength lateral to the wound (Zhao et al., 1996; Song et al., 2004). They have used drugs to either increase the trans-corneal potential or decrease it. Drugs that increased this voltage difference also increased the rate of wound healing, while drugs that reduced this voltage difference also slowed down the rate of wound healing. In addition, when current was injected to restore and amplify the endogenous electric field in bovine cornea, the wound healing rate increased (Sta Iglesia and Vanable, 1998). These data provide very strong support for the hypothesis that these electric fields near corneal wounds play an important role in influencing the healing of these wounds.

2.3 Measurements of Endogenous Intracellular Electric Fields

2.3.1 Nurse Cell Complex in Insects

In the silk moth oocyte-nurse cell complex or the ovarian follicle, the oocyte cytoplasm is about 10 mV more positive than that of the nurse cell cytoplasm despite their connection by a broad cytoplasmic bridge. Woodruff and Telfer have published several studies of this system in which they show that the polarized transport of fluorescently labeled proteins between nurse cells and oocyte depends on the charge of these proteins in a manner consistent with intercellular electrophoresis driven by the voltage gradient across the cytoplasmic bridge of the silk moth and other insects (Jaffe and Woodruff, 1979; Woodruff and Telfer, 1980; Woodruff and Cole, 1997; Cole and Woodruff, 2000). Since these nurse cell complexes arise from incomplete cell division they must be considered electrically as a single cell that is generating a significant voltage gradient across a small region.

2.3.2 Development of Left–Right Polarity in Chick and Frog

The most recent study of the role of electric fields in development utilized both frog and chick embryos. The generation of left–right asymmetry in these systems was found to depend on both functioning gap junctions and voltage difference between blastomeres in very early stages of development (Levin and Mercola, 1998, 1999; Levin et al., 2002). This work started with a pharmacological screen to identify drugs that interfered with the development of left–right patterning. The most effective drugs were the ones that interfered with K^+ and H^+ ion fluxes. Lansoprazole and chromanol 293B were the most effective and these block the H^+ pump and a specific K^+ channel (KvLQT-1), respectively. Next most effective was SCH28080 and omeprazole, which block the H^+/K^+-ATPase, followed by $BaCl_2$, which blocks all K^+ channels.

They further tested the hypothesis that these ion fluxes were important for left–right patterning by injecting mRNA for either the alpha or beta subunits of the H^+/K^+-ATPase or the K^+ channel into the frog egg. They found that heterotaxia was induced when either was injected but by far the largest increase, 37%, occurred when all three mRNAs were injected. This suggests that overexpression of these molecules, which can influence or perturb ion concentration gradients of K^+ or H^+, disrupted the normal left–right patterning of the embryo.

In order to learn more about the mechanism through which H^+/K^+-ATPase influenced patterning, they used *in situ* hybridization to show an asymmetrical distribution of the K^+/H^+-ATPase. It was concentrated in the right ventral blastomere of the 4-cell stage frog embryo. the addition of Both K^+ channels and K^+/H^+-ATPase should hyperpolarize cells making them more negative than their neighbors. This voltage difference between the blastomeres of the 4-cell stage embryo could be used to segregate low molecular weight determinants through gap junctions to achieve asymmetric gene expression. They further tested this hypothesis by measuring the membrane potential of cells near the primitive streak during early chick development. They used an anionic fluorescent dye, $DiBAC_4$, whose distribution depended on membrane potential to show that the early chick embryo exhibited a voltage gradient across the primitive streak. The left side of the primitive streak was 10–20 mV more positive than right side and this difference was inhibited by the K^+ channel blocker, $BaCl_2$, and the H^+/K^+-ATPase inhibitor, omeprazole. Since these drugs also partially inhibited the development of left–right patterning, the membrane potential difference appears to play a role in this aspect of development. This paper is the first to demonstrate a role for membrane potential differences between blastomeres in early vertebrate pattern formation.

2.4 Methods for Modifying Endogenous Electric Fields

Modifying endogenous fields within embryos or cells is really quite challenging. The only way to generate a relatively steady field in a conducting medium is to pass current through it. However, the outer membranes or epithelia surrounding the targets here make this very hard to accomplish. These structures have high resistance to current flow so that slowly varying fields placed outside the cell or embryo will not effectively penetrate to the inside but will mainly generate current flow around the outside of the cells or embryos. Because of this there have been precious few attempts to modify endogenous fields, but I will discuss some of them.

The only exception to this field penetration problem is the application of ultrashort pulses in the nanosecond domain that have risen times faster than the charging time of biological membranes (Schoenbach et al., 2004). These pulses penetrate beyond the plasma membrane into the interior of cells and tissues but are present only for very short times. It is conceivable that by combining millions of these short pulses per second, one could significantly modify the DC electric field within cells and tissues. This remains to be explored.

2.4.1 Passing Current between Electrodes

If two electrodes can be placed within a single cell or in different regions of an embryo, current can be passed between them and a well-defined electric field can be generated. However, this requires penetrating the outer membrane or epithelium and may damage this outer layer. The longer the field is applied, the more likely that some damage will occur due to vibrations or embryo movement. This approach has been used with some success in the regenerating tip of newt limbs (Chiang et al., 1991; Iglesia et al., 1996), where current was passed along the limb by inserting one electrode into a slit made at the knee and the current delivery electrode was placed distal to the wounded digit tip. The electric field at the newt limb could be manipulated in this way, and elimination of this field significantly slowed down the rate of wound healing.

2.4.2 Low-Resistance Shunts

One way to perturb the natural transembryonic current pattern is to place a low-resistance pathway or shunt into a resistive membrane or epithelium. The endogenous fields will drive current out through this shunt, and this new current flow will definitely perturb the endogenous field within the tissue. This technique was used with striking results by Robinson's group in the chick embryo as described above (Hotary and Robinson, 1994).

2.4.3 Placing Tissues in an External Electric Field

While DC electric fields cannot penetrate into the conducting cytoplasm of a cell or tissue, they have been found to perturb normal development at certain stages in amphibians (Metcalf and Borgens, 1994). In addition, there is a very extensive old literature in which the polarity of both development and regeneration could be strongly influenced by imposed DC electric fields (Jaffe and Nuccitelli, 1977; Levin, 2003). One interpretation of these observations is that the imposed field has its effective target along the exterior of the embryo via lateral electrophoresis of membrane glycoproteins.

The only external electric field that can penetrate into the interior of cells is the one that rises faster than the charges within the cell can redistribute (Schoenbach et al., 2004). Such a pulsed field could certainly modify internal electric fields for the extremely short duration of the pulse. Perhaps by using either a large field strength or multiple pulses, such brief pulsed fields might have a significant effect on endogenous fields. This relatively new area deserves future investigation.

2.5 Summary

All plasma membranes and epithelia generate voltage differences across themselves. These batteries are the power sources that drive ionic currents through cells, tissues, and organisms. These currents will generate internal electric fields as they traverse tissue, and such fields can do work through electrophoresis of charged molecules within and between cells as well as in the plane of the plasma membrane. Cases in which such fields have been measured include the following:

1. Two- to 4-day-old chick embryos generate a 20-mV/mm field near the posterior intestinal portal that is important for normal development of posterior structures.
2. Stage 14–21 amphibian embryos generate similar internal electric fields, and modifying these fields during neurulation but not gastrulation results in developmental abnormalities.
3. Neural tubes in amphibians generate an internally negative voltage difference of as much as 90 mV across the wall of the tube, and cells in this region are exposed to fields as large as 1800 mV/mm.
4. Mammalian skin wounds generate 150-mV/mm fields just below the stratum corneum, and corneal epidermal wounds exhibit fields of 40 mV/mm lateral to wounds.
5. The development of left–right asymmetry in frog and chick embryos utilizes an electric field between blastomeres that is generated by an asymmetrical distribution of the K^+/H^+-ATPase among the blastomeres.

In nearly all of these well-documented examples, the electric field plays a critical role in either the development of the organism or the wound healing and regeneration of adult structures. Because of this, imposed electric fields may be utilized to perturb or influence normal development. One intriguing possible perturbation is the use of imposed electric fields to enhance the rate of wound healing, particularly in cases where the normal healing process is slower than normal, as observed in chronic wounds.

References

Altizer, A.M., Moriarty, L.J., Bell, S.M., Schreiner, C.M., Scott, W.J., and Borgens, R.B. (2001). Endogenous electric current is associated with normal development of the vertebrate limb. *Dev. Dyn. 221*, 391–401.

Barker, A.T., Jaffe, L.F., and Vanable, J.W., Jr. (1982). The glabrous epidermis of cavies contains a powerful battery. *Am. J. Physiol. 242*, R358–R366.

Borgens, R.B., Vanable, J.W., Jr., and Jaffe, L.F. (1977). Bioelectricity and regeneration: large currents leave the stumps of regenerating newt limbs. *Proc. Natl. Acad. Sci. USA 74*, 4528–4532.

Borgens, R.B., Vanable, J.W., Jr., and Jaffe, L.F. (1979). Small artificial currents enhance *Xenopus* limb regeneration. *J. Exp. Zool. 207*, 217–226.

Borgens, R.B., Rouleau, M.F., and DeLanney, L.E. (1983). A steady efflux of ionic current predicts hind limb development. *J. Exp. Zool. 228*, 491–503.

Chiang, M., Cragoe, E.J., Jr., and Vanable, J.W., Jr. (1989). Electrical fields in the vicinity of small wounds in *Notophthalmus viridescens* skin. *Biol. Bull. 176*(S), 179–183.

Chiang, M., Cragoe, E.J., Jr., and Vanable, J.W., Jr. (1991). Intrinsic electric fields promote epithelialization of wounds in the newt, *Notophthalmus viridescens. Dev. Biol. 146*, 377–385.

Chiang, M., Robinson, K.R., and Vanable, J.W., Jr. (1992). Electrical fields in the vicinity of epithelial wounds in the isolated bovine eye. *Exp. Eye Res. 54*, 999–1003.

Clausen, T., Van, H.C., and Everts, M.E. (1991). Significance of cation transport in control of energy metabolism and thermogenesis. *Physiol Rev. 71*, 733–774.

Cole, R.W. and Woodruff, R.I. (2000). Vitellogenic ovarian follicles of Drosophila exhibit a charge-dependent distribution of endogenous soluble proteins. *J. Insect Physiol. 46*, 1239–1248.

DuBois-Reymond, E. (1843). Vorlaufiger abrifs einer untersuchung uber den sogenannten frosch-strom und die electomotorischen fische. *Ann. Phys. U. Chem. 58*, 1.

Faes, T.J., van der Meij, H.A., de Munck, J.C., and Heethaar, R.M. (1999). The electric resistivity of human tissues (100 Hz–10 MHz): a meta-analysis of review studies. *Physiol. Meas. 20*, R1–R10.

Hotary, K.B. and Robinson, K.R. (1990). Endogenous electrical currents and the resultant voltage gradients in the chick embryo. *Dev. Biol. 140*, 149–160.

Hotary, K.B. and Robinson, K.R. (1991). The neural tube of the *Xenopus* embryo maintains a potential difference across itself. *Dev. Brain Res. 59*, 65–73.

Hotary, K.B. and Robinson, K.R. (1992). Evidence of a role for endogenous electrical fields in chick embryo development. *Development 114*, 985–996.

Hotary, K.B. and Robinson, K.R. (1994). Endogenous electrical currents and voltage gradients in *Xenopus* embryos and the consequences of their disruption. *Dev. Biol. 166*, 789–800.

Iglesia, D.D.S., Cragoe, E.J., Jr., and Vanable, J.W., Jr. (1996). Electric field strength and epithelization in the newt (*Notophthalmus viridescens*). *J. Exp. Zool. 274*, 56–62.

Illingworth, C.M. and Barker, A.T. (1980). Measurement of electrical currents emerging during the regeneration of amputated fingertips in children. *Clin. Phys. Physiol. Meas. 1*, 87–89.

Jaffe, L.F. and Nuccitelli, R. (1974). An ultrasensitive vibrating probe for measuring extracellular currents. *J. Cell Biol. 63*, 614–628.

Jaffe, L.F. and Nuccitelli, R. (1977). Electrical controls of development. *Annu. Rev. Biophys. Bioeng. 6*, 445–475.

Jaffe, L.F. and Stern, C.D. (1979). Strong electrical currents leave the primitive streak of chick embryos. *Science 206*, 569–571.

Jaffe, L.F. and Woodruff, R.I. (1979). Large electrical currents traverse developing Cecropia follicles. *Proc. Natl. Acad. Sci. USA 76*, 1328–1332.

Levin, M. (2003). Bioelectromagnetics in morphogenesis. *Bioelectromagnetics 24*, 295–315.

Levin, M. and Mercola, M. (1998). Gap junctions are involved in the early generation of left–right asymmetry. *Dev. Biol. 203*, 90–105.

Levin, M. and Mercola, M. (1999). Gap junction-mediated transfer of left–right patterning signals in the early chick blastoderm is upstream of Shh asymmetry in the node. *Development 126*, 4703–4714.

Levin, M., Thorlin, T., Robinson, K.R., Nogi, T., and Mercola, M. (2002). Asymmetries in H^+/K^+-ATPase and cell membrane potentials comprise a very early step in left–right patterning. *Cell 111*, 77–89.

Loew, L.M. (1992). Voltage-sensitive dyes: measurement of membrane potentials induced by DC and AC electric fields. *Bioelectromagnetics 13*(Suppl. 1), 179–189.

Loew, L.M., Campagnola, P., Lewis, A., and Wuskell, J.P. (2002). Confocal and nonlinear optical imaging of potentiometric dyes. *Methods Cell Biol. 70*, 429–452.

McCaig, C.D., Rajnicek, A.M., Song, B., and Zhao, M. (2005). Controlling cell behavior electrically: current views and future potential. *Physiol. Rev. 85*, 943–978.

McGinnis, M.E. and Vanable, J.W., Jr. (1986). Electrical fields in *Notophthalmus viridescens* limb stumps. *Dev. Biol. 116*, 184–193.

Metcalf, M.E.M. and Borgens, R.B. (1994). Weak applied voltages interfere with amphibian morphogenesis and pattern. *J. Exp. Zool. 268*, 323–338.

Metcalf, M.E.M., Shi, R.Y., and Borgens, R.B. (1994). Endogenous ionic currents and voltages in amphibian embryos. *J. Exp. Zool. 268*, 307–322.

Nuccitelli, R. (1988). Ionic currents in morphogenesis. *Experientia 44*, 657–666.

Nuccitelli, R. (1990). The vibrating probe technique for studies of ion transport. In *Noninvasive Techniques in Cell Biology*, J.K. Foskett and S. Grinstein, eds. New York: Wiley-Liss, pp. 273–310.

Robinson, K.R. (1983). Endogenous electrical current leaves the limb and prelimb region of the Xenopus embryo. *Dev. Biol. 97*, 203–211.

Schoenbach, K.H., Joshi, R.P., Kolb, J.F., Chen, N., Stacey, M., Buescher, E.S., Beebe, S.J., and Blackmon, P. (2004). Ultrashort electrical pulses open a new gateway into biological cells. *Proc. IEEE 92*, 1122–1137.

Shi, R. and Borgens, R.B. (1994). Embryonic neuroepithelial sodium transport, the resulting physiological potential, and cranial development. *Dev. Biol. 165*, 105–116.

Shi, R. and Borgens, R.B. (1995). Three-dimensional gradients of voltage during development of the nervous system as invisible coordinates for the establishment of embryonic pattern. *Dev. Dyn. 202*, 101–114.

Song, B., Zhao, M., Forrester, J., and McCaig, C. (2004). Nerve regeneration and wound healing are stimulated and directed by an endogenous electrical field *in vivo. J. Cell Sci. 117*, 4681–4690.

Sta Iglesia, D.D. and Vanable, J.W., Jr. (1998). Endogenous lateral electric fields around bovine corneal lesions are necessary for and can enhance normal rates of wound healing. *Wound Repair Regen. 6*, 531–542.

Vanable, J.W., Jr. (1991). A history of bioelectricity in development and regeneration. In *A History of Regeneration Research*, C.E. Dinsmore, ed. Cambridge: Cambridge University Press, pp. 151–177.

Wallis, D.I. (1993). *Electrophysiology: A Practical Approach.* Dublin: IRL Press.

Winkel, G.K. and Nuccitelli, R. (1989). Large ionic currents leave the primitive streak of the 7.5-day mouse embryo. *Biol. Bull. 176*, 110–117.

Woodruff, R.I. and Cole, R.W. (1997). Charge dependent distribution of endogenous proteins within vitellogenic ovarian follicles of *Actias luna. J. Insect Physiol. 43*, 275–287.

Woodruff, R.I. and Telfer, W.H. (1980). Electrophoresis of proteins in intercellular bridges. *Nature 286*, 84–86.

Zhao, M., Agius-Fernandez, A., Forrester, J.V., and McCaig, C.D. (1996). Directed migration of corneal epithelial sheets in physiological electric fields. *Invest. Ophthalmol. Vis. Sci. 37*, 2548–2558.

3

Dielectric Properties of Biological Materials

Camelia Gabriel

CONTENTS

3.1 Introduction

At some level of organization, all matter consists of charged entities held together by various atomic, molecular, and intermolecular forces. The effect of an externally applied electric field on the charge distribution is specific to the material; the dielectric properties are a measure of that effect; they are intrinsic properties of matter used to characterize nonmetallic materials. Biological matter has free and bound charges; an applied electric field will cause them to drift and displace, thus inducing conduction and polarization currents. Dielectric spectroscopy is the science that relates the dielectric properties to the underlying microscopic mechanisms of polarization and conduction. These dielectric phenomena are determined by and are informative about the structure and composition of the material. Consequently, knowledge of the dielectric properties is of practical importance in all fields of science where electromagnetic fields impinge or are used to probe or process matter. It is equally important in biomedical fields such as electrophysiology, where endogenous bioelectric sources provide signals that are sensed through various body tissues and are affected by their dielectric properties.

The past decade has seen a dramatic increase in the exposure of people to electromagnetic fields from wireless telecommunication devices and infrastructure. This situation sparked large research programs on the assessment and quantification of exposure of people and on the biological effects resulting from the exposure. Information on the dielectric properties of tissues is vital to these studies, for the computation of exposure metrics and the provision of a mechanistic explanation for biological effects. To satisfy the need of current research activity, this chapter will review the dielectric data for body tissue and the underlying mechanisms of interaction at the cellular, subcellular, and molecular levels. In doing so, we will draw on the authoritative article by Foster and Schwan (1996), published in the second edition of this book, which goes a long way toward establishing dielectric spectroscopy as a powerful tool for mechanistic studies.

Another area of scientific activity in the last decade revolved around the formulation of standard procedures for the experimental assessment of human exposure from electromagnetic sources, mostly telecommunication radio transceivers and their accessories. This created the need to formulate and measure the dielectric properties of tissue equivalent material and made dielectric measurement and the assessment of the associated uncertainty part of the compliance testing procedure. This chapter will deal with the fundamental issues that need to be established if dielectric measurement is to become a routine but accurate laboratory procedure.

It is not possible to review the dielectric properties and the polarization mechanisms in biological material without singling out the contribution to this field of Herman P. Schwan, whose name is associated with all the major findings over the past 50 years (Foster, 2002). Indeed, his 1957 review of the bulk electrical properties of cells and cell suspensions is one of the earliest and most studied texts on the subject. There are other reviews by Pethig (1979), Stuchly (1979), Schwan and Foster (1980), Pethig and Kell (1987), and Foster and Schwan (1989). The fundamental aspects and a more extensive treatment of the theoretical aspects of this subject can be found in books by Cole (1972), Grant et al. (1978), Schanne and Ceretti (1978), Pethig (1979), and more recently, Craig (1995), Roussy and Pearce (1995), and Grimnes and Martinsen (2000).

For most biological materials, the magnetic permeability is close to that of free space (i.e., diamagnetic), which implies that there is no direct interaction with the magnetic component of electromagnetic fields at low field strengths. However, this position is now changing following relatively recent reports of the presence of magnetite in human nervous tissue (see, e.g., Dobson and Grassi, 1996), which suggest that magnetite may provide a mechanism for direct interaction of external magnetic fields with the human central nervous system. The role of these strongly magnetic materials in organisms is only just beginning to be unraveled. This subject is elaborated in the second part of next chapter.

3.2 Dielectric Properties—Molecular Origin

From a historical perspective, the dielectric properties of materials were first observed experimentally by Faraday in the 1830s as a change in the capacity of an empty capacitor when a material is introduced inside it. Faraday introduced the term *specific inductive capacity* to describe the ratio of the capacities of the filled and empty capacitor. This quantity is now known as the permittivity and is denoted by ε. It is a fundamental property of nonmetallic or dielectric materials. Under quasistatic conditions, the capacitance C_0 of a perfect capacitor of area A and plate separation d changes to $C > C_0$:

$$C = \varepsilon\varepsilon_0 = \varepsilon \frac{\varepsilon_0 A}{d} \tag{3.1}$$

where ε is the relative permittivity of the material (dimensionless), ε_0 is the permittivity of free space (8.8542×10^{-12} F/m), and the product $\varepsilon\varepsilon_0$ is the absolute permittivity. The increase in capacity is due to the additional charge density induced by the field in the material. The field is said to have polarized the medium; polarizability or the ability of the material to polarize is the main determinant of its dielectric properties.

This section will start with the fundamental concepts of the interaction of homogenous matter with static fields and proceed, in steps, to heterogenous mixtures and their dynamic response to time-varying fields leading to the dielectric properties of tissues.

3.2.1 Quasi-Static Response

Considering the simple case of a monomolecular material, three main interaction mechanisms are possible: electronic, atomic, and molecular polarization. Electronic polarization is the shift of electrons, in the direction of the field, from their equilibrium position with respect to the positive nuclei. Atomic polarization is the relative displacement of atoms or atom groups relative to each other. The orientation of permanent or induced molecular dipoles, when present, is known as molecular polarization. The total polarizability α_T is the sum of the contribution of, in this case, all three processes, termed α_e, α_a, and α_d.

In view of the relative importance of molecular orientation processes in defining the total polarization ($\alpha_e + \alpha_a < \alpha_d$) and hence the dielectric properties, it is usual to differentiate between the polar and nonpolar materials when these properties are considered.

For ideal nonpolar materials, the relationship $\varepsilon = n^2$, where n is the optical refractive index, holds true. When a dielectric material becomes polarized by the application of an external electric field E, the dipole moment of the constituent molecules is given by

$$\overline{m} = \alpha_T \overline{E}_l \tag{3.2}$$

where E_l is the local field acting on the molecules. The dipole moment per unit volume of the material P increases the total displacement flux density D, defined from the relationship $D = \varepsilon_0 E$ in vacuum and $D = \varepsilon_0 \varepsilon E$ in a medium of relative permittivity ε. The latter expression may also be written as

$$D = \varepsilon_0 E + P \tag{3.3}$$

The dependence of P on E can take several forms, the simplest and most common being a scalar proportionality:

$$P = \varepsilon_0 \chi E \tag{3.4}$$

where $\chi = \varepsilon - 1$ is the relative dielectric susceptibility. This simple relationship is valid for a perfect isotropic dielectric, at low or moderate field intensities and at static or quasistatic field frequencies.

If the material contains N dipoles per unit volume, then

$$P = N\alpha_T E_l \tag{3.5}$$

and

$$\varepsilon - 1 = \frac{N\alpha_T}{\varepsilon_0} \frac{E_l}{E} \tag{3.6}$$

The molecular description of the permittivity requires that the relationship between the microscopic and the macroscopic field intensities be known. In most cases, there is no exact solution to this problem, only more or less good approximations that hold within the confines of the assumptions and simplifications made, as will be briefly illustrated for typical classes of materials.

3.2.2 Permittivity of Low-Pressure Gases

At low pressures the molecules are far apart from each other, and their interaction with each other may be assumed to be negligible in comparison with the macroscopic field intensity E. Under these conditions $E_l \approx E$ and

$$\varepsilon - 1 = \frac{N\alpha_T}{\varepsilon_0} \tag{3.7}$$

The relative permittivity of a nonpolar gas is very close to 1, typically of the order of 1.0001 at atmospheric pressure.

3.2.3 Permittivity of Liquids and Dense Gases

When the intermolecular interactions are such that $E_l \neq E$, the local field must be estimated. One approach is to consider a spherical region inside the dielectric that is large compared to the size of a molecule; the field inside it is estimated for nonpolar materials to be

$$E_l = \left(\frac{\varepsilon + 2}{3}\right) E \tag{3.8}$$

which yields

$$\frac{3(\varepsilon - 1)}{\varepsilon + 2} = \frac{N}{\varepsilon_0} \alpha_T \tag{3.9}$$

The above expression is known as the Claussius–Mossoti–Lorentz formulation. It is not always valid, such as when the density of the material corresponds to $N = 3\,\varepsilon_0/\alpha_T$. An alternative formulation, valid when the molecules are polarizable point dipoles of permanent moment μ, was provided by Onsager:

$$\frac{(\varepsilon - n^2)(2\varepsilon + n^2)}{\varepsilon(n^2 + 2)^2} = \frac{N\mu^2}{9kT\varepsilon_0} \tag{3.10}$$

where k is the Boltzman constant and T is the absolute temperature.

Debye separated out the contribution to the total polarization of the permanent dipole from those associated with electronic and atomic displacements and arrived at the following relationship

$$\frac{\varepsilon - 1}{\varepsilon + 2} = \frac{N}{3\varepsilon_0}(\alpha + \mu^2/3kT) \tag{3.11}$$

While Onsager and Debye used semistatistical techniques to estimate the local field, others like Kirkwood, and later Fröhlich, used statistical methods to obtain a rigorous expression of the permittivity after taking local interactions into consideration and obtained

$$\frac{(\varepsilon - 1)(2\varepsilon + 1)}{3\varepsilon} = \frac{N}{\varepsilon_0}(\alpha + g\mu^2/3kT) \tag{3.12}$$

This is Kirkwood's equation for permittivity in which g is known as the Kirkwood correlation parameter, introduced to account for the effect of local ordering in the material. Fröhlich's theory gives

$$\frac{(\varepsilon - n^2)(2\varepsilon + n^2)}{\varepsilon(n^2 + 2)^2} = \frac{Ng\mu^2}{9kT\varepsilon_0} \tag{3.13}$$

which, except for the correlation parameter g, is identical to Onsager's equation.

This brief outline of the dielectric theory gives an idea of the nature of the electric field interaction problems and of the various techniques used to partially solve them under static field conditions. The solutions hold for slow time-varying fields as long as there is a quasistatic state. References to the original work by Debye (1929), Kirkwood (1936), Onsager (1936), and Fröhlich (1955) are given in Böttcher and Bordewijk (1978) and other well-known texts (Hill et al., 1969; Jonscher, 1983).

3.3 Time and Frequency Dependence of the Dielectric Response

Much of the interest in the dielectric properties of biological materials is concerned with their response to time-varying electric fields. This can be explained by the same macroscopic variables used for the quasi-static state except for the introduction of a time dependence for the excitation and response. The general discussion will assume sinusoidal fields and linear and isotropic responses, nonsinusoidal fields and material anisotropy, and nonlinearity being special cases.

3.3.1 Time-Dependent Polarization—Impulse Response—Kramers–Krönig Relations

The following relationship holds irrespective of the polarization mechanism:

$$P(t) = D(t) - \varepsilon_0 E(t) \tag{3.14}$$

For an ideal dielectric material with no free charge, the polarization follows the pulse with a delay determined by the time constant of the polarization mechanism. Assuming a rate process, which is that the rate of polarization is proportional to the constantly decreasing number of unpolarized units, the simplest expression for the polarization is obtained from the solution of the first-order differential rate equation with constant coefficients and time constant τ, giving

$$P(t) = P(1 - \mathrm{e}^{-t/\tau}) \tag{3.15}$$

The decay of polarization is also an exponential function

$$P(t) = \mathrm{e}^{-t/\tau} \tag{3.16}$$

For a linear system, the response to a unit-step electric field is the impulse response $f(t)$ of the system. The response of the system to a time-dependent field can be obtained from summation in a convolution integral of the impulses corresponding to a sequence of elements making up the electric field. For a harmonic field and a causal, time-independent system, the Fourier transform exists and yields

$$P(\omega) = \varepsilon_0 \chi(\omega) E(\omega) \tag{3.17}$$

indicating that the dielectric susceptibility $\chi(\omega)$ is the Fourier transform of $f(t)$. In general, the susceptibility is a complex function reflecting the fact that it informs on the magnitude and phase of the polarization with respect to the polarizing field

$$\chi(\omega) = \chi' - \mathrm{j}\chi'' \tag{3.18}$$

The real and imaginary parts of $\chi(\omega)$ can be obtained from the separate parts of the Fourier transform:

$$\chi'(\omega) = \int_{-\infty}^{+\infty} f(t)\cos(\omega t)\,dt = \int_{0}^{+\infty} f(t)\cos(\omega t)\,dt$$
$$\chi''(\omega) = \int_{-\infty}^{+\infty} f(t)\sin(\omega t)\,dt = \int_{0}^{+\infty} f(t)\sin(\omega t)\,dt \tag{3.19}$$

The limit of integration can be changed from $-\infty$ to 0 since $f(t)$ is causal.

The impulse response $f(t)$ defines the dielectric response, and conversely, knowledge of the complex susceptibility allows the determination of the impulse response by carrying out the reverse transformation, which gives $f(t)$ in terms of either $\chi'(\omega)$ or $\chi''(\omega)$:

$$f(t) = (2/\pi)\int_{0}^{+\infty} \chi'(\omega)\cos(\omega t)\,d\omega$$
$$f(t) = (2/\pi)\int_{0}^{+\infty} \chi''(\omega)\sin(\omega t)\,d\omega \tag{3.20}$$

Eliminating $f(t)$ from the above equations gives an expression of $\chi'(\omega)$ in terms of $\chi''(\omega)$ and vice versa. Thus, there is a relationship between the real and imaginary parts of the complex susceptibility of any material, such that knowledge of either enables the other to be calculated. The expressions of real and imaginary parts of the susceptibility or permittivity in terms of each other are known as the Kramers–Krönig relations and have been derived as

$$\chi'(\omega) = \chi'(\infty) + \frac{2}{\pi}\int_{0}^{\infty} \frac{u\chi''(u) - \omega\chi''(\omega)}{u^2 - \omega^2}\,du$$
$$\chi''(\omega) = \frac{2}{\pi}\int_{0}^{\infty} \frac{\chi'(u) - \chi'(\omega)}{u^2 - \omega^2}\,du \tag{3.21}$$

where u is a variable of integration. Recalling the relationship between relative dielectric susceptibility and relative permittivity $\chi = \varepsilon - 1$, the permittivity is a complex function given by

$$\hat{\varepsilon}(\omega) = \varepsilon'(\omega) - j\varepsilon''(\omega) = (1 + \chi'(\omega) - j\chi''(\omega)) \tag{3.22}$$

Thus, the Kramers–Krönig relations relate ε' to the complete spectrum of ε'' and vice versa. A clear account of their derivation can be found in Jonscher (1983).

3.3.2 Permittivity of a Polar Substance—The Debye Equation

When a step field E is applied to a polar dielectric material, the electronic and atomic polarizations are established almost instantaneously compared to the time scale of the molecular orientation; the total polarization reaches a steady state as a first-order process characterized by the time constant of the dipolar rotation. When the field is removed, the process is reversed; electronic and atomic polarizations subside first, followed by a relatively slow decay in dipolar polarization (Figure 3.1). The time constant τ depends on the physical process, in this case the rotational dynamics of the dipole determined by the size, shape, and intermolecular relations of the molecules. If P_∞ and P_0 are the instantaneous and steady-state polarization, respectively, then the total polarization for a first-order process characterized by a time constant τ is

$$P = P_\infty + (P_0 - P_\infty)(1 - \mathrm{e}^{-t/\tau}) \tag{3.23}$$

In time-varying fields the permittivity is a complex function originating from the magnitude and phase shift of the polarization with respect to the polarizing field:

$$\hat{\varepsilon} = \varepsilon' - \mathrm{j}\varepsilon'' = \varepsilon' - \mathrm{j}\sigma/\omega\varepsilon_0 \tag{3.24}$$

The real part ε' is a measure of the induced polarization per unit field and the imaginary part ε'' is the out-of-phase loss factor associated with it. The loss factor can also be represented by a conductivity term $\sigma = \omega\varepsilon_0\varepsilon''$ where ω is the angular frequency. The SI unit of conductivity is siemens per meter (S/m).

The frequency response of the first-order system is obtained from the Laplace transformation, which provides the relationship known as the Debye equation:

$$\hat{\varepsilon} = \varepsilon_\infty + \frac{(\varepsilon_0 - \varepsilon_\infty)}{1 + \mathrm{j}\omega\tau} = \varepsilon' - \mathrm{j}\varepsilon'' \tag{3.25}$$

The limiting values of the permittivity, ε_s and ε_∞, are known as static and infinite permittivity, respectively. The relaxation time τ corresponds to a relaxation frequency $f_r = 1/2\pi\tau$. For a highly associated liquid such as water, the static permittivity can be expressed in terms of molecular parameters in accordance with the discussions in the previous section as

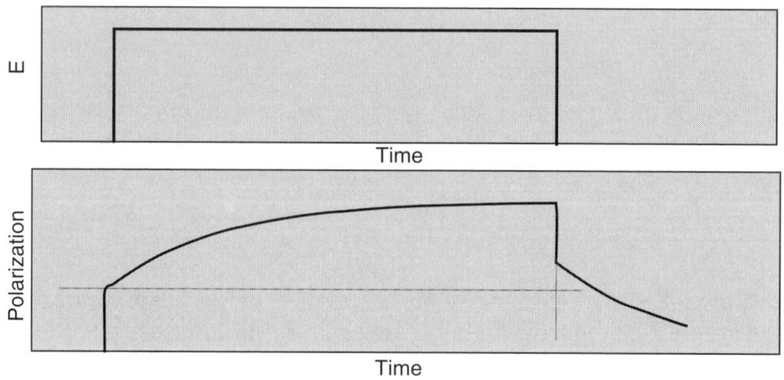

FIGURE 3.1
Schematic graph of dielectric response of a polar material expressed as polarization in the time domain in response to the step onset and removal of a polarizing electric field E. The nearly instantaneous rise in polarization, indicated by the horizontal dotted line, is P_∞; the vertical dotted line marks the instant and equivalent drop in polarization occurs when the field is removed.

$$\varepsilon_s = \varepsilon_\infty + \frac{Ng\mu^2}{2kT\varepsilon_0} \tag{3.26}$$

The relaxation time may be identified with the time constant of the molecular polarization and expressed in terms of molecular parameters. If η is the viscosity, then for a spherical molecule of radius a

$$\tau = 4\pi a^3\eta/kT \tag{3.27}$$

For most polar materials, though not for water, ε_∞ corresponds to the optical permittivity and is equal to the square of optical refractive index n of the medium:

$$\varepsilon_\infty = n^2 \tag{3.28}$$

The dielectric properties of polar molecules vary with temperature; in general, both ε_s and τ decrease with increasing temperature.

As with the charge density and polarization, the time dependence of the current density J and σ, the current density per unit field, also follows a first-order law such that

$$J/E = \sigma_\infty + (\sigma_s - \sigma_\infty)(1 - e^{-t/\tau}) \tag{3.29}$$

This transforms into the conductivity equivalent of the Debye equation:

$$\hat{\sigma} = \sigma_\infty + \frac{(\sigma_s - \sigma_\infty)}{1 + j\omega\tau} \tag{3.30}$$

Figure 3.2 shows the variation in the permittivity, loss factor, and conductivity with frequency for a single time constant relaxation; such behavior pertains to an idealized monomolecular polar substance with no residual frequency-independent conductivity, that is $\sigma_s = 0$. The best, if not the only, example of such material is pure water as will be discussed later.

At the relaxation frequency, the permittivity is halfway between its limiting values and the loss factor at its highest. In the case of a single time constant as described in Figure 3.2, the conductivity is halfway between its limiting values at the relaxation frequency.

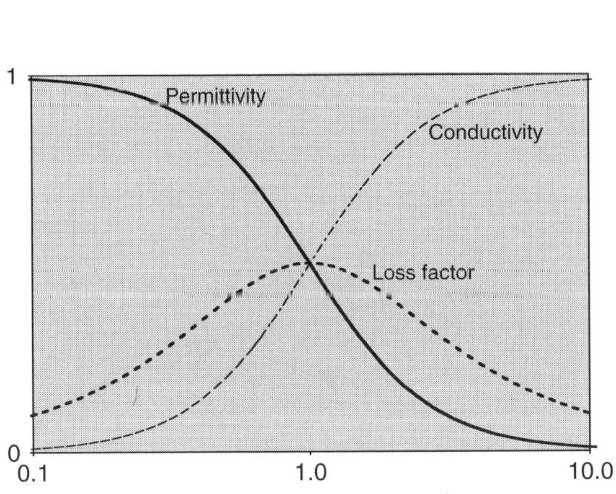

FIGURE 3.2
Normalized permittivity $(\varepsilon' - \varepsilon_\infty)/(\varepsilon_s - \varepsilon_\infty)$, loss factor $\varepsilon''/(\varepsilon_s - \varepsilon_\infty)$, and conductivity $\omega\varepsilon_0\,\varepsilon''/(\varepsilon_s - \varepsilon_\infty)$ for a single time constant relaxation plotted against f/f_r.

3.3.3 Nonpolar Molecules

The permittivity of nonpolar materials is virtually constant throughout the frequency range. In general, the temperature dependence is not significant. The static and optical values of the permittivity are almost identical, hence the Maxwell relation $\varepsilon = n^2$ holds true throughout the frequency and temperature range.

3.4 Observed Responses of Real Systems—Conduction—Multiple Relaxations—The Universal Law

In the previous section we described the expected behavior of idealized materials; we now need to deal with the observed responses of real systems. Few materials exhibit single relaxation time dispersions as in the Debye model; real materials depart from this ideal behavior to a greater or lesser extent depending on the complexity of the underlying mechanisms. To describe these responses we need to introduce the concepts of multiple dispersions and distribution of relaxation time. Moreover, biological materials exhibit conduction as well as polarization mechanisms; this needs to be taken into consideration in describing their dielectric response.

3.4.1 Conduction

The Debye expression does not include the effect of conduction currents as would arise from, for example, the drift of free ions in static fields. If σ_s is the static conductivity, the Debye expression becomes

$$\hat{\varepsilon} = \varepsilon_\infty + \frac{(\varepsilon_s - \varepsilon_\infty)}{1 - j\omega\tau} - \frac{j\sigma_s}{\omega\varepsilon_0} \tag{3.31}$$

In terms of real and imaginary parts we have

$$\begin{aligned} \varepsilon' &= \varepsilon_\infty + \frac{(\varepsilon_s - \varepsilon_\infty)}{1 + (\omega\tau)^2} \\ \varepsilon'' &= \frac{\sigma_s}{\omega\varepsilon_0} + \frac{(\varepsilon_s - \varepsilon_\infty)\omega\tau}{1 + (\omega\tau)^2} \end{aligned} \tag{3.32}$$

The total conductivity σ is given by

$$\sigma = \omega\varepsilon_0\varepsilon'' = \sigma_s + \frac{(\varepsilon_s - \varepsilon_\infty)\varepsilon_0\omega^2\tau}{1 + (\omega\tau)^2} \tag{3.33}$$

The total conductivity is thus made of two terms corresponding to the residual static conductivity and polarization losses. In practice, it is only possible to measure the total conductivity of a material; σ_s is obtained from data analysis or by measurement at frequencies corresponding to $\omega\tau \ll 1$ where the dipolar contribution to the total conductivity is negligible.

3.4.2 Multiple Relaxation Models—Distribution of Relaxation Times—Fractional Power Law Behavior

The occurrence of multiple interaction processes or the presence of more than one molecular conformational state or type of polar molecule may cause the dielectric behavior of a substance to exhibit multiple relaxation time dispersions. Deviation from Debye behavior may also indicate a polarization process whose kinetics are not first order or the presence of a complex intermolecular interaction. Models are needed to analyze the dielectric spectra of complex systems to unravel the underlying interaction mechanisms.

The simplest case is that of a dielectric response arising from multiple first-order processes; in this case the dielectric response will consist of multiple Debye terms to correspond to the polarization processes such that

$$\hat{\varepsilon} = \varepsilon_\infty + \frac{\Delta\varepsilon_1}{1 - j\omega\tau_1} + \frac{\Delta\varepsilon_2}{1 - j\omega\tau_2} + \cdots \tag{3.34}$$

where $\Delta\varepsilon_n$ corresponds to the limits of the dispersion characterized by time constant τ_n. If the relaxation times are well separated such that $\tau_1 \ll \tau_2 \ll \tau_3 \ll \cdots$, a plot of the dielectric properties as a function of frequency will exhibit clearly resolved dispersion regions.

If, as is quite often the case, the relaxation times are not well separated, the material will exhibit a broad dispersion encompassing all the relaxation times. In the limit of a continuous distribution of relaxation times, the multiple Debye expression would be

$$\hat{\varepsilon} = \varepsilon_\infty + (\varepsilon_s - \varepsilon_\infty) \int_0^\infty \frac{\rho(\tau)\,d\tau}{1 - j\omega\tau} \tag{3.35}$$

where

$$\int_0^\infty \rho(\tau)\,d\tau = 1 \tag{3.36}$$

The above equations can be used to represent all dielectric dispersion data, provided an appropriate distribution function $\rho(\tau)$ is available. Conversely, it should also be possible, at least in principle, to invert dielectric relaxation spectra to determine $\rho(\tau)$ directly; however, this is not easily achievable in practice. More commonly, one has to assume a distribution to describe the frequency dependence of the dielectric properties observed experimentally. The choice of distribution function should depend on the cause of the multiple dispersions in the material. For example, one can assume a Gaussian distribution as is known to occur for other physical characteristics (Figure 3.3) would be

$$\rho(t/\tau) = \frac{b}{\sqrt{\pi}} e^{-b^2[\ln(t/\tau)]^2} \tag{3.37}$$

where τ is the mean relaxation time. The shape of the Gaussian function depends on the parameter b; it reduces to the delta function when b tends to infinity and becomes very broad when b decreases; the area under the curve remains the same as required by the normalization condition. Incorporated into the expression for complex permittivity, it produces an expression that cannot be solved analytically, which makes it impractical for experimental data analysis.

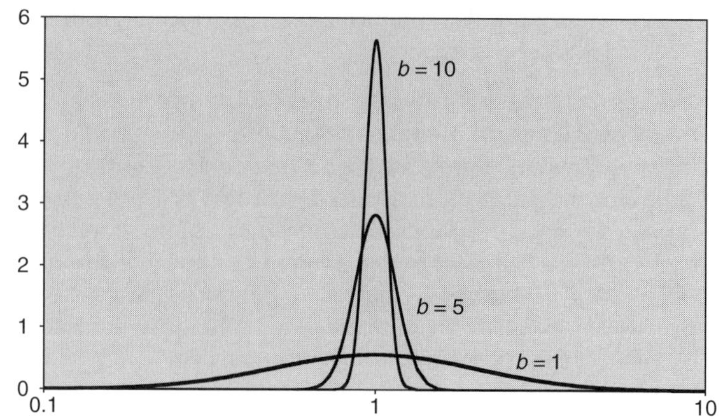

FIGURE 3.3
Gaussian distribution function as a function of t/τ.

Numerous empirical distribution functions or models have been proposed to model the experimental data without elaboration of the underlying mechanisms. One of the most commonly used models, a modified version of the Debye expression, was proposed in 1941 by Cole and Cole and is widely known as the Cole–Cole model:

$$\hat{\varepsilon} = \varepsilon_\infty + \frac{(\varepsilon_s - \varepsilon_\infty)}{1 - (j\omega\tau)^{1-\alpha}} = \varepsilon' - j\varepsilon'' \tag{3.38}$$

In it, α is a distribution parameter in the range $1 > \alpha \geq 0$; for $\alpha = 0$, the model reverts to the Debye equation. The real and imaginary parts are

$$\begin{aligned} \varepsilon' &= \varepsilon_\infty + \frac{(\varepsilon_s - \varepsilon_\infty)[1 - (\omega\tau)^{1-\alpha}\sin(\alpha\pi/2)]}{1 + (\omega\tau)^{2(1-\alpha)} + 2(\omega\tau)^{1-\alpha}\sin(\alpha\pi/2)} \\ \varepsilon'' &= \frac{(\varepsilon_s - \varepsilon_\infty)(\omega\tau)^{1-\alpha}\cos(\alpha\pi/2)}{1 + (\omega\tau)^{2(1-\alpha)} + 2(\omega\tau)^{1-\alpha}\sin(\alpha\pi/2)} \end{aligned} \tag{3.39}$$

Eliminating $\omega\,\tau$ from the above equations gives

$$\left(\varepsilon' - \frac{(\varepsilon_s + \varepsilon_\infty)}{2}\right)^2 + \left(\varepsilon'' + \frac{(\varepsilon_s + \varepsilon_\infty)}{2}\cot\frac{(1-\alpha)\pi}{2}\right)^2 = \left(\frac{\varepsilon_s - \varepsilon_\infty}{2}\operatorname{cosec}\frac{(1-\alpha)\pi}{2}\right)^2 \tag{3.40}$$

indicating that a plot of ε' against ε'' is a semicircle with its center below the real axis. For $\alpha = 0$, the Debye equivalent of the above equation is

$$\left(\varepsilon' - \frac{(\varepsilon_s + \varepsilon_\infty)}{2}\right)^2 + \varepsilon''^2 = \left(\frac{\varepsilon_s - \varepsilon_\infty}{2}\right)^2 \tag{3.41}$$

which indicates that ε' against ε'' is a semicircle with its center on the real axis (Figure 3.4); these semicircle plots are known as Cole–Coles.

The distribution function that corresponds to the Cole–Cole model is

$$\rho(t/\tau) = \frac{1}{2\pi}\,\frac{\sin(\alpha\pi)}{\cos h[(1-\alpha)\ln(t/\tau)] - \cos(\alpha\pi)} \tag{3.42}$$

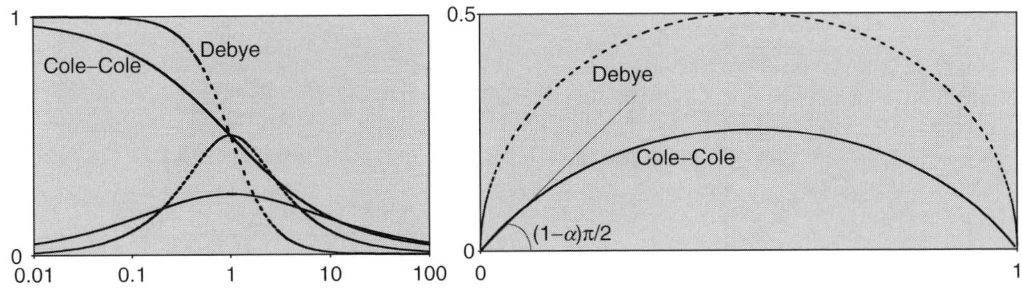

FIGURE 3.4
Left: Frequency dependence of normalized permittivity $(\varepsilon' - \varepsilon_\infty)/(\varepsilon_s - \varepsilon_\infty)$ and loss factor $\varepsilon''/(\varepsilon_s - \varepsilon_\infty)$ against frequency, normalized to the relaxation frequency, for a Debye and Cole–Cole with $\alpha = 0.4$. Right: Plot of normalized permittivity against loss factor showing a semicircle with its center on the real axis in the case of the Debye and an arc of a semicircle with its center below the real axis in the case of the Cole–Cole; the apex of the arc corresponds to the mean relaxation frequency.

Here again, τ is the mean relaxation time. As with the Gaussian, this distribution is logarithmically symmetrical about t/τ (Figure 3.5).

In 1951, Davidson and Cole proposed another variant of the Debye equation in which an exponent β is applied to the whole denominator:

$$\hat{\varepsilon} = \varepsilon_\infty + \frac{(\varepsilon_s - \varepsilon_\infty)}{(1 - j\omega\tau)^\beta} \tag{3.43}$$

which gives

$$\begin{aligned} \varepsilon' &= \varepsilon_\infty + (\varepsilon_s - \varepsilon_\infty)\cos(\beta\phi)(\cos\phi)^\beta \\ \varepsilon'' &= (\varepsilon_s - \varepsilon_\infty)\sin(\beta\phi)(\cos\phi)^\beta \end{aligned} \tag{3.44}$$

where $\phi = \arctan(\omega\,\tau)$. The corresponding distribution of relaxation times is

$$\rho(t/\tau) = \frac{1}{\pi}\left(\frac{t}{\tau - t}\right)^\beta \sin(\pi\beta) \tag{3.45}$$

When $\beta = 1$, the model reverts to the Debye equation. A plot of the real and imaginary parts of the model presents a skewed arc, similar to the Debye plot at low-frequencies, where it intercepts the abscissa at $\pi/2$, but at high frequencies the tangent to the arc is $\beta\ \pi/2$ (Figure 3.6).

The distribution function is shown graphically in Figure 3.7. It has a singularity at $t/\tau = 1$ and returns zero at $t > \tau$.

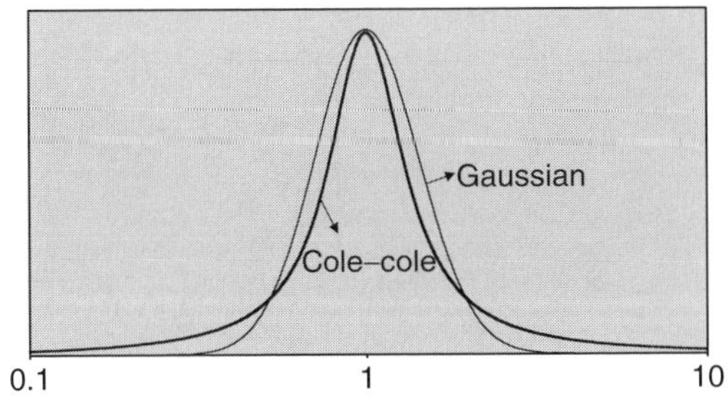

FIGURE 3.5
Gaussian distribution with $b = 2$ and Cole–Cole distribution with $\alpha = 0.09$ as a function of t/τ.

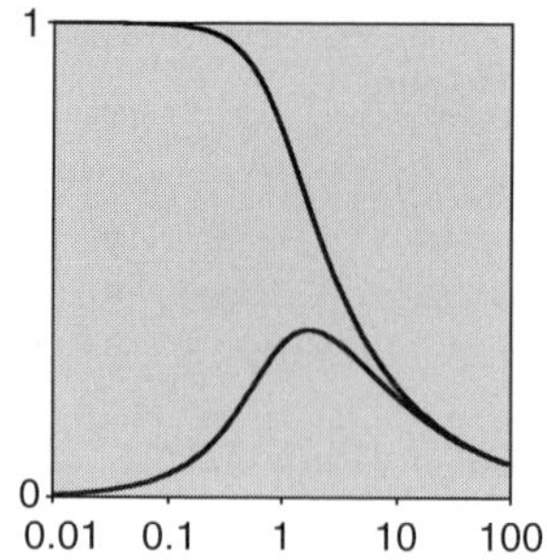

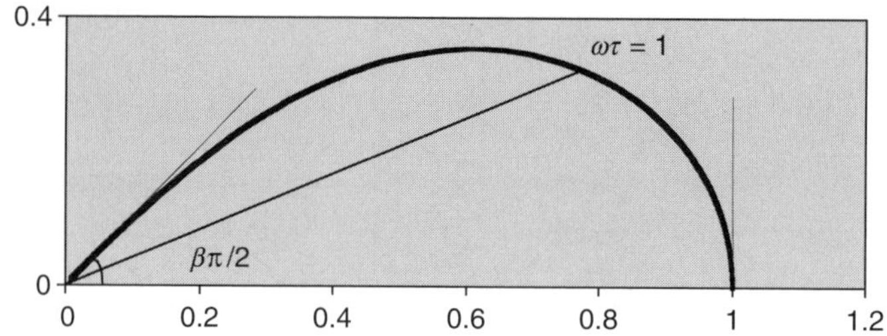

FIGURE 3.6
Left: Frequency dependence of normalized permittivity $(\varepsilon' - \varepsilon_\infty)/(\varepsilon_s - \varepsilon_\infty)$ and loss factor $\varepsilon''/(\varepsilon_s - \varepsilon_\infty)$ for a Cole–Davidson model with $\beta = 0.5$. Right: Plot of normalized loss factor against permittivity showing the characteristic Cole–Davidson skewed arc where the maximum in ε'' does not correspond with $\omega\tau = 1$; this point is found at the interception of the bisector of the high-frequency limiting angle with the data plot.

Another expression, sometimes used to model dielectric data, is the Havriliak–Negami relation (Havriliak and Negami, 1966). It combines the variations introduced in both the Cole–Cole and the Cole–Davidson models, giving

$$\hat{\varepsilon} = \varepsilon_\infty + \frac{(\varepsilon_s - \varepsilon_\infty)}{(1 - (j\omega\tau)^{(1-\alpha)})^\beta} \tag{3.46}$$

with real and imaginary parts:

$$\begin{aligned} \varepsilon' &= \varepsilon_\infty + \frac{(\varepsilon_s - \varepsilon_\infty)\cos(\beta\phi)}{1 + 2(\omega\tau)^{(1-\alpha)}\sin(\alpha\pi/2) + (\omega\tau)^{2(1-\alpha))^{\beta/2}}} \\ \varepsilon'' &= \frac{(\varepsilon_s - \varepsilon_\infty)\sin(\beta\phi)}{1 + 2(\omega\tau)^{(1-\alpha)}\sin(\alpha\pi/2) + (\omega\tau)^{2(1-\alpha))^{\beta/2}}} \end{aligned} \tag{3.47}$$

in which $\phi = \arctan\{[(\omega\tau)^{(1-\alpha)}\cos(\alpha\pi/2)]/[1 + (\omega\tau)^{(1-\alpha)}\sin(\alpha\pi/2)]\}$ and the corresponding distribution of relaxation times is

$$\rho(t/\tau) = \frac{1}{\pi}\frac{(t/\tau)^{\beta(1-\alpha)}\sin(\beta\theta)}{(t/\tau)^{2(1-\alpha)} + 2(t/\tau)^{(1-\alpha)}\cos(\pi(1-\alpha)) + 1)^{(\beta/2)}} \tag{3.48}$$

where $\theta = \arctan\{[(\sin(1-\alpha)\pi)/((t/\tau) + \cos(1-\alpha)\pi)]\}$.

The Cole–Cole plot of the Havriliak–Negami model is an asymmetric curve intercepting the real axis at different angles at high and low-frequencies (Figure 3.8). The distribution of relaxation times is also asymmetric (Figure 3.9).

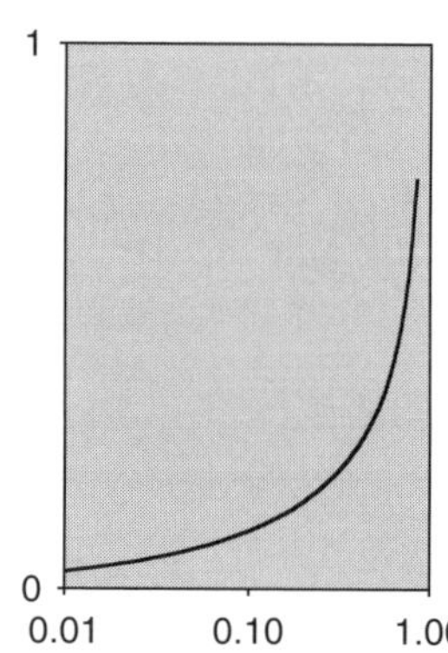

FIGURE 3.7
Cole–Davidson distribution with $\beta = 0.5$ as a function of t/τ.

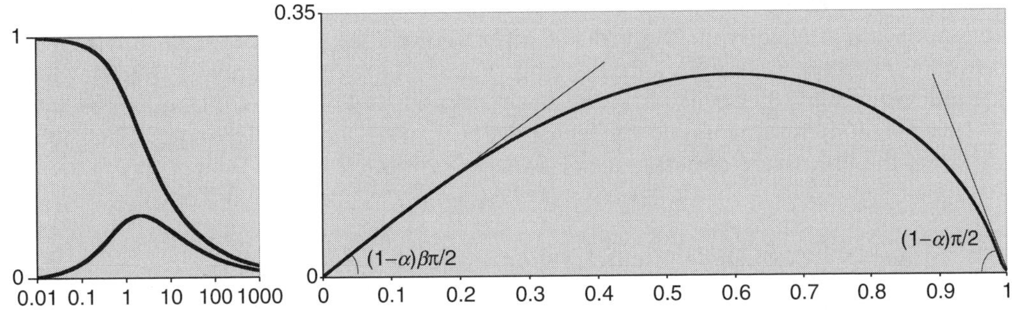

FIGURE 3.8
Left: frequency dependence of normalized permittivity $(\varepsilon' - \varepsilon_\infty)/(\varepsilon_s - \varepsilon_\infty)$ and loss factor $\varepsilon''/(\varepsilon_s - \varepsilon_\infty)$ for a Havriliak–Negami model with $\alpha = 0.2$ and $\beta = 0.5$. Right: Cole–Cole plot of the same data. As with the Cole-Davidson plot, the $\omega\tau = 1$ point is found at the interception of the bisector of the high-frequency limiting angle with the data plot.

Havriliak–Negami expressions revert to their Cole–Cole, Cole–Davidson, and Debye equivalents at the limiting values of β, α, and α and β, respectively. In principle, this should be the model of choice for dielectric data analysis. In practice, it is not widely used to describe the dielectric properties of biological material, as will be discussed later. It is important to recall that these empirical distribution functions lack mechanistic justification; however, they do serve a useful purpose in enabling the parametrization of the experimental data, albeit with very limited clarification of the underlying mechanisms.

Another limitation of this type of analysis is the possibility of obscuring multirelaxation processes, particularly the presence of a small amplitude dispersion following in the high-frequency tail end of a much larger principal one. This point is well illustrated by Wei and Sridhar (1993); they point out that a graphical representation of the parameter $\sigma'' = \omega \varepsilon_0(\varepsilon' - \varepsilon_\infty)$ versus $\sigma' = \omega\varepsilon_0\varepsilon''$ provides a more sensitive visualization of multirelaxation processes.

The Debye model and its many variations, including those described in this section, have been widely used over more than half a century primarily because they lend themselves to simple curve-fitting procedures. In particular, the Cole–Cole model is

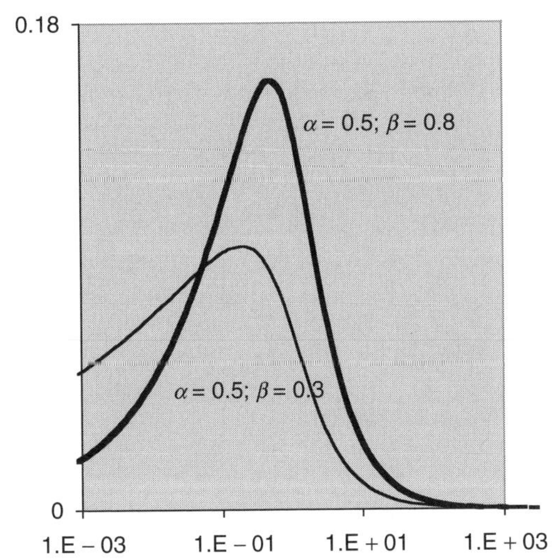

FIGURE 3.9
Havriliak–Negami distribution as a function of t/τ showing the effect of β for a given value of α.

used almost as a matter of course in the analysis of the dielectric properties of biological materials. Mathematically, at the limit of high frequencies, the Cole–Cole function simplifies to a fractional power law, that is, both ε' and ε'' are proportional to $(\omega\tau)^{(\alpha\,-1)}$. This fractional power law behavior is at the basis of what is known as the universal law of dielectric phenomena developed by Jonscher, Hill, and Dissado (Jonscher, 1983) for the analysis of the frequency dependence of dielectric data.

3.4.3 Universal Law of Dielectric Relaxation

Jonscher and his collaborators (Hill and Jonscher, 1983; Dissado and Hill, 1989) collated and analyzed extensive dielectric data obtained from numerous sources, pertaining to a wide range of materials, measured over a broad range of temperatures and frequencies. Their aim was to observe how dielectrics behave rather than presume a model for their frequency dependence; they studied the data on a log–log scale to better recognize the presence of a power law dependence, if present. Figure 3.10 shows plots for Debye and non-Debye responses where ω_p is the loss peak radial frequency.

Very few materials exhibit a pure Debye behavior where, at frequencies in excess of ω_p, the logarithmic slopes for $\varepsilon'(\omega)$ and $\varepsilon''(\omega)$ are -2 and -1, respectively, which is a Kramers–Krönig compatible result. However, for most materials a power law dependence of the type ω^{n-1}, with $n \neq 0$, applies for both $\varepsilon'(\omega)$ and $\varepsilon''(\omega)$. This is in compliance with the Kramers–Krönig relations, which require that at frequencies exceeding ω_p, both parameters follow the same frequency dependence, making the ratio $\varepsilon''(\omega)/\varepsilon'(\omega)$ frequency independent. Under such conditions, the ratio of energy dissipated to energy stored per radian of sinusoidal excitation is constant. The universal law can be summarized by the following frequency dependencies for the normalized complex permittivity:

$$\text{for } \omega < \omega_p, \quad \varepsilon''(\omega) \approx \omega^m \quad \text{and} \quad \varepsilon'(\omega) \approx 1 - \varepsilon''(\omega) \tag{3.49}$$

$$\text{for } \omega > \omega_p, \quad \varepsilon''(\omega) \approx \omega^{n-1} \quad \text{and} \quad \varepsilon'(\omega) \approx \varepsilon''(\omega) \approx \omega^{n-1} \tag{3.50}$$

Observation of the experimental data showed that ω_p is temperature dependent and follows an Arrhenius function:

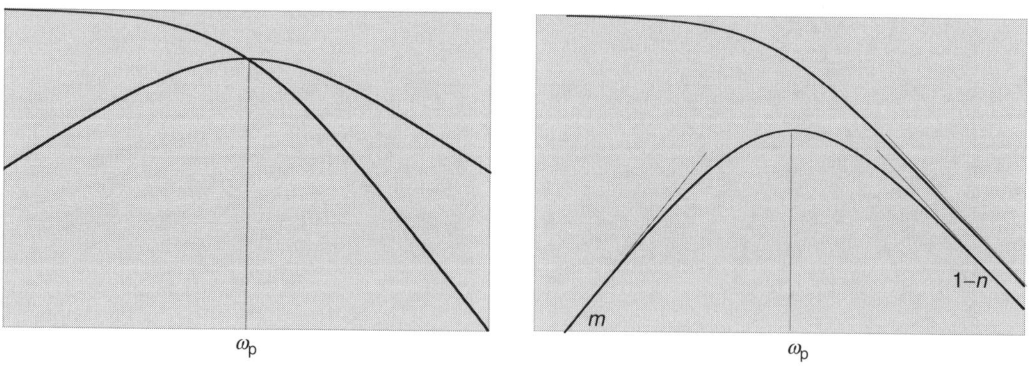

FIGURE 3.10
Log–log plot of the normalized permittivity and loss factor against frequency for a Debye-type behavior (left) and a non-Debye response (right).

$$\omega_p = Ae^{-W/kT} \tag{3.51}$$

and the functional form for $\varepsilon''(\omega)$ is

$$\varepsilon''(\omega) = \frac{A}{(\omega/\omega_p)^{1-n} + (\omega_p/\omega)^m} \tag{3.52}$$

The values for $\varepsilon'(\omega)$ can then be determined numerically from the Kramer–Krönig relations.

Although the features of the dielectric spectra of most materials can be described using this approach, there is no theoretical justification for it. This makes it yet another empirical model, albeit a very general and mathematically elegant one.

3.4.4 Combined Response Model

A model that combines features from Debye-type and universal dielectric response behavior was proposed by Raicu (1999). In the course of modeling broad dielectric dispersions, as is often observed in the dielectric spectrum of biological materials, Raicu found that neither approach was good enough over a wide frequency range. He proposed the following very general function

$$\hat{\varepsilon} = \varepsilon_\infty + \frac{\Delta}{[(j\omega\tau)^\alpha + (j\omega\tau)^{1-\beta}]^\gamma} \tag{3.53}$$

where α, β, and γ are real constants in the range [0,1], τ is the characteristic relaxation time, and Δ is a dimensional constant, which becomes the dielectric increment $(\varepsilon_s - \varepsilon_\infty)$ when $\alpha = 0$, and the above expression reverts to the Havriliak–Negami model, which further reduces to the Debye, Cole–Cole, or Cole–Davison models with an appropriate choice of the α, β, and γ parameters. For $\gamma = 1$, it reverts to Jonsher's universal response model; in the special case where $\gamma = 1$ and $\alpha = 1 - \beta$, it becomes

$$\hat{\varepsilon} = \varepsilon_\infty + \left(j\frac{\omega}{S}\right)^{\beta-1} \tag{3.54}$$

which is known as the constant phase angle model (Dissado, 1990). In this expression S is a scaling factor given by $S = (\Delta/2)^{1/(1-\beta)}\tau^{-1}$. The above expression was successfully used to model the dielectric spectrum of a biological material over five frequency decades from 10^3 Hz to 10^8 Hz.

3.5 Dielectric Properties of Biological Materials—Main Components

Tissue is a heterogenous material containing water, dissolved organic molecules, macromolecules, ions, and insoluble matter. The constituents are highly organized in cellular and subcellular structures forming macroscopic elements and soft and hard tissues. The presence of ions plays an important role in the interaction with an electric field, providing means for ionic conduction and polarization effects. Ionic charge drift creates conduction currents and also initiates polarization mechanisms through charge accumulation at structural interfaces, which occur at various organizational levels. Their dielectric properties will thus reflect contributions to the polarization from both structure and composition. In this section, the contribution of each of the components will be determined individually and then collectively, leading to the formulation of models for the dielectric response of biological tissue.

3.5.1 Water

Water is a constituent of all living things; it is the environment in which body electrolytes and biomolecules reside and interact. Knowledge of its properties must precede the study of the more complex system. Many of the physical properties peculiar to water are due to its molecular asymmetry, polar nature, and ability to hydrogen bond, which are all interrelated. Water is described as an associated liquid because of its intermolecular hydrogen bonding. One practical reason for emphasizing the study of water in this chapter is its increasing use as a reference liquid, that is, a material of well-known dielectric properties. Consequently, it is often used as a standard for the calibration and testing of dielectric measuring procedures.

The dielectric properties of water are among the most studied and reported in the literature. Over the past decades, many experimental studies have been carried out to determine the dielectric properties of water over wide frequency and temperature ranges. These include Haggis et al. (1952), Lane and Saxton (1952), Hastead and El Sabeh (1953), Grant et al. (1957), Grant and Shack (1967), Grant and Sheppard (1974), Schwan et al. (1976), Grant et al. (1981), Hastead et al. (1985), Kaatze (1986, 1988), Buckmaster (1990), and Buchner et al. (1998). A comprehensive list of references and a historical overview of the subject can be found in Ellison et al. (1996). Other notable reviews were carried out by Kaatze (1989) and Liebe et al. (1991).

Data up to 100 GHz exhibit a near-perfect Debye dispersion with fairly well-defined parameters. Table 3.1 gives the Debye parameters for water at 20°C from three relatively recent reviews. Kaatze (1989) used a Debye expression to model his own extensive experimental data covering −4°C to 60°C and 1 to 57 GHz in addition to what he considered to be credible data from other sources.

Liebe et al. (1991) gathered extensive static and high-frequency data. For frequencies up to 100 GHz and temperatures from 0°C to 30°C, the data were a very good fit to the Debye function. However, including data at higher frequency somewhat reduced the goodness of the fit, suggesting the possible presence of a much smaller secondary dispersion in the hundreds of gigahertz range. The next logical step was then to use a two-Debye model. This proved a good fit to all experimental data up to 1 THz, thus confirming the presence of a small, high-frequency dispersion, probably due to some subtle molecular mechanism.

TABLE 3.1

Debye Parameters for Pure Water at 20°C

Review	ε_s	τ (ps)	ε_∞
Kaatze (1989)	80.2	9.47	5.2
Liebe et al. (1991)	80.1	9.35–9.39	5.3–5.4
Buchner et al. (1998)	80.2	9.32–9.52	5.9–6.0

Notes: For Liebe et al. (1991), τ and ε_∞ values are those of the single-Debye model (<100 GHz) and of the principal dispersion in the two-Debye model (up to 1 THz). Buchner et al. (1998) provide upper and lower bounds for τ and ε_∞ of the principal relaxation of a two-Debye model.

This secondary dispersion, centered around 670 GHz, brought down the high-frequency permittivity from 5.4 to 3.3 and made practically no impact on the characteristics of the principal dispersion, which were almost unchanged (Table 3.1). Liebe et al. (1991) extended the model to the far infrared (30 THz) by accounting for two near-infrared resonance absorption terms.

The most recent and comprehensive analysis of the dielectric properties of water is provided by Ellison et al. (1996), who critically reviewed the literature spanning the late 19th and most of the 20th centuries. With respect to the static permittivity, they obtained a function $\varepsilon_s = a\,e^{-b}$ with $a = 87.85306$ and $b = 0.00456992$, which predicts the value of ε_s at a given temperature to well within the limits of experimental accuracy for the range $-35°C < t < 100°C$. All high-frequency data (up to 1 THz) that met their selection criteria are tabulated. They stopped short of formulating models for the frequency dependence of the data; instead, they invited comments from the scientific community prior to the determination of what would probably be the ultimate model and spectral parameters for the dielectric properties of pure water, a finding that will greatly benefit this field of study. Already, other researchers have used this extensive survey. For example, Buchner et al. (1998) reported values for τ and ε_∞ of the principal water dispersion (Table 3.1) by fitting a two-Debye model to combined experimental data from Ellison et al. (1996) and other, more recent, studies (Barthel et al., 1995).

It is evident from Table 3.1 that the static permittivity and the relaxation time are fairly well-defined, less so the infinite permittivity. Fortunately, this parameter has little impact on the dielectric data in the gigahertz range because its value is only a small percentage of the permittivity in that frequency range. This relatively large uncertainty highlights the fact that even this most studied, pure substance is not a perfect reference liquid and that much remains to be done in the characterization of the dielectric properties of water at terahertz frequencies. Figure 3.11 is a plot of the dielectric properties of water at 20° tabulated by Ellison et al. (1996).

In biological materials, water is a solvent for salts, protein, nucleic acids, and smaller molecules. It is therefore important to study the effect of solutes on its dielectric response.

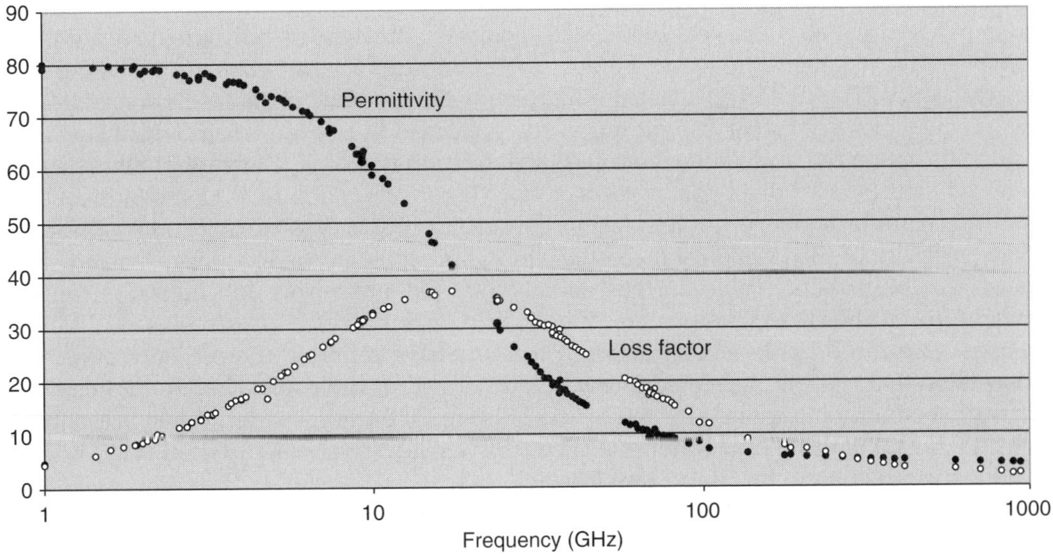

FIGURE 3.11
Experimental data from numerous sources reviewed and tabulated by Ellison et al. (1996).

3.5.2 Carbohydrates

In quantitative terms carbohydrates are not major constituents of animal cells; they are present at the surface of the cell membrane and are known to play a role in cellular communications. They are responsible for the gel consistency that gives certain body fluids such as vitreous humor and synovial fluid cushioning or lubricating properties. They are important constituents of certain tissues such as cartilage, tendon, and ligament. In terms of molecular structure, they vary in complexity and molecular size; they have in common the presence of one or more hydroxyl groups and the ability to hydrogen bond with each other or with water molecules. In aqueous solution they modify the principal dispersion of water to an extent that depends on the nature and concentration of the organic radical. In general, the dispersion is likely to be broader than a Debye, the static permittivity lower, and the relaxation time longer than for pure water, as observed and reported by Bateman and Gabriel (1987).

3.5.3 Proteins and Other Macromolecules

Protein constitutes the bulk of the organic matter in the body. Proteins are described as biopolymers, each molecule being a sequence of amino acids folded into a specific three-dimensional structure enclosing its hydrophobic sites within it. The surface has polar, hydrophilic groups with an affinity to bind water molecules from its surrounding aqueous environment. Part of the function of a protein resides in its structure; if the structure unfolds the protein is said to be denatured and is no longer functional. A good model for a globular protein in solution is that of a cluster of organic matter surrounded by a layer of strongly bound water; the solvent is referred to as free water to differentiate it from bound water. The size of the cluster depends on the molecular weight of the protein, which is typically of the order of tens or hundreds of thousands, that is, significantly larger than a water molecule. In an aqueous environment, most biological macromolecules including proteins act like polar molecules with permanent or induced dipole moment the magnitude of which depends on the molecular structure, configuration, and size.

Dielectric spectroscopy is therefore an important tool in the study of these molecular properties (Bateman et al., 1990, 1992). Typically, the dielectric dispersion of a protein will be in the megahertz frequency range, corresponding to a time constant of the order of microseconds. The dielectric spectrum of an aqueous globular protein solution will have two dispersion regions corresponding to the polarization of the protein and water molecules; the larger the protein the more clearly defined they will be. Conventionally, they are referred to as β and γ dispersions, respectively (Figure 3.12). Figure 3.12 shows a conceptual spectrum of a binary, protein–water system. In practice, to maintain the conformational stability of the biological molecules, inorganic ions, in the form of dissolved salts, must also be present. Table 3.2 has actual data, gathered from the literature, on the magnitude of the dielectric increment and the relaxation time for proteins of different shapes, sizes, and dipole moments. Many authors have reported the presence of a small dispersion that is attributed to bound water, described as molecules that are more or less strongly bound or otherwise affected by the presence of organic matter. When present, the spectral region of the bound water is termed δ dispersion. The book by Grant et al. (1978) is a good introduction to this important topic.

Larger biopolymers such as DNA, whose molecular weight may be of the order of several million, have more complex dielectric spectra with dispersions extending from kilohertz to megahertz. The elucidation of the polarization mechanisms responsible for

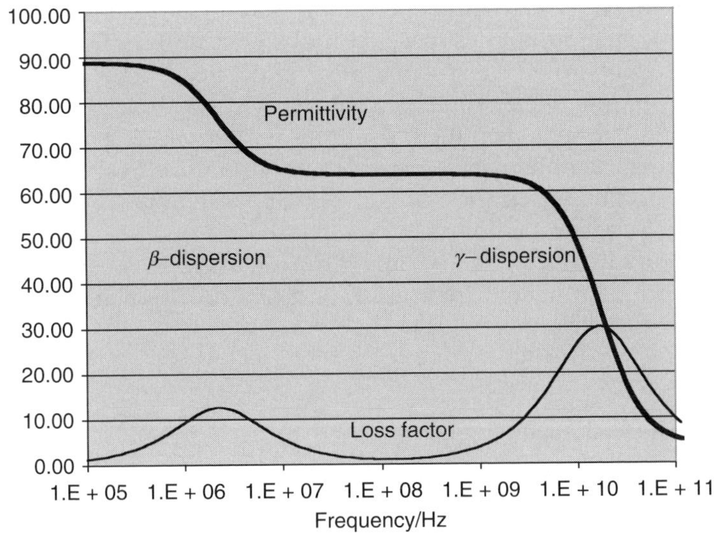

FIGURE 3.12
Conceptual representation of the dielectric spectrum of an aqueous protein solution. In practice, the two dispersions may overlap. The nearest to this picture is the complex permittivity spectrum of an aqueous solution of 1,2-dimyristoyl-L-3-phosphatidycholine reported by Kaatze and Giese (1980).

the dielectric spectrum of aqueous DNA solution is an area of active research. A good introduction to the subject is the book on biopolymers by Takashima (1989).

There is more than just academic interest in the study of biopolymers. Advances in nanotechnology are such that biological macromolecules are being considered as possible nanoscale electronic devices for fast information processing and transfer, a quest that will keep theoreticians and experimentalists busy for a long time.

3.5.4 Electrolytes

Electrolytes in the form of sodium, potassium, calcium, magnesium, chloride, and other ions play an important role in the function of biological systems. Many vital processes depend on a subtle balance being established between the concentration of electrolytes

TABLE 3.2
Dielectric Parameters of Various Proteins at 25°C

Protein	Mol wt. ($\times 10^3$)	$\Delta\varepsilon$	μ (D)	$\tau \times 10^8$ (s)	a/b
Myoglobin	17	0.15	170	2.9	—
β-Lactoglobulin (in 0.25 M glycine)	40	1.51	730	15, 5.1	4
Ovalbumin	44	0.10	250	18, 4.7	5
Horse carboxyhemoglobin	67	0.33	480	8.4	1.6
Horse serum albumin	70	0.17	380	36, 7.5	6
Horse serum pseudoglobulin	142	1.08	1100	250, 28	9

Notes: a/b is the axial ratio that determines the shape of the molecule. Where the shape deviates significantly from the spherical, two relaxation times are observed. The dielectric increment is $\Delta\varepsilon$, the dipole moment μ is given in Debye unit (1 D = 3.33×10^{-30} cm)
Source: From Foster KR, Schwan HP. 1989. *Crit Rev Biomed Eng* 17(1): 25–104. With permission.

inside and outside the cell. The cell membrane is, to a great extent, impermeable to the passive exchange of ions but allows directed movement under physiological control. In terms of dielectric properties, electrolytes have two effects. The direct effect, already mentioned, is the production of ohmic currents and energy loss in the system. This has the effect of making the static conductivity finite with a value commensurate with the ionic concentration and mobility. There are also important indirect effects whereby ionic charges contribute to the polarization of a biological system. One is interfacial polarization, whereby charge accumulation occurs at interfaces that are impermeable to ions. Another polarization mechanism is the ionic diffusion in electrical double layers adjacent to charged surfaces. Conduction, interfacial, and ion diffusion phenomena contribute significantly to the dielectric spectra of tissue.

3.5.5 Dielectric Dispersions in Tissue

The dielectric spectrum of a biological tissue (spleen at 37°C) is given in Figure 3.13 as an example of the response of a high water content tissue. Three main dispersion regions are immediately obvious and are referred to as α, β, and γ dispersions. The dispersions are rather broad, indicating the possible overlap of discrete relaxations arising from the polarization mechanisms encountered in the complex biological environment. Ionic conductivity contributes significantly to the loss factor, obliterating its features, and it is more

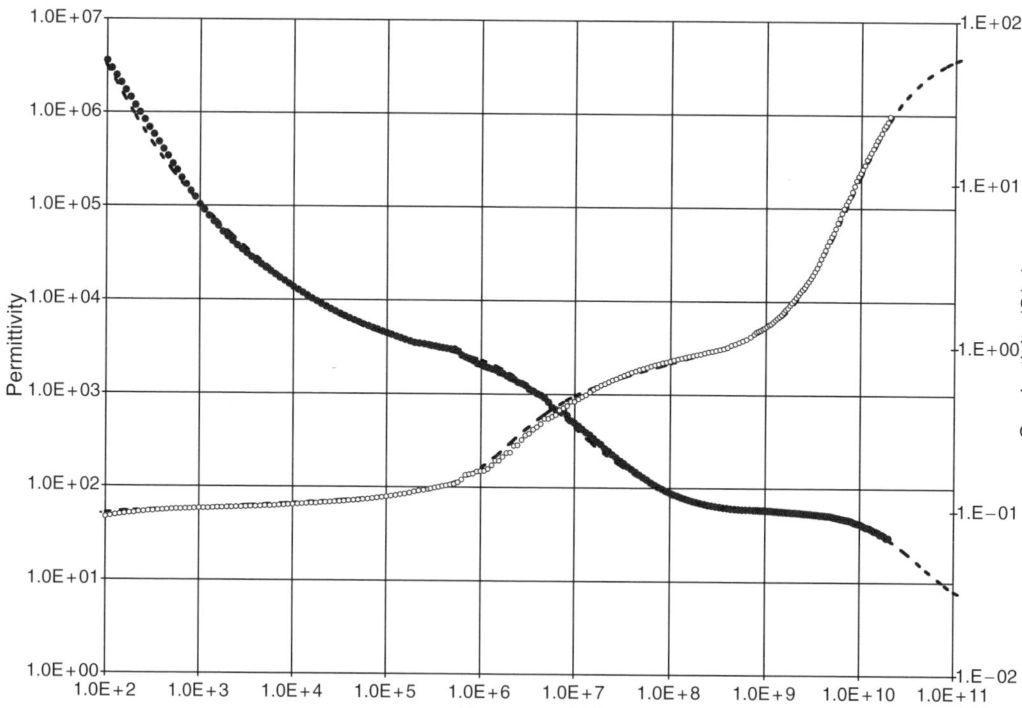

FIGURE 3.13
Dielectric spectrum of a high water content tissue (spleen at 37°C); experimental data are from Gabriel et al. (1996); dotted line is a best fit to a model of four Cole–Coles and a conductivity term.

informative to express the dielectric properties of tissues as permittivity and conductivity as in Figure 3.13.

3.5.5.1 α Dispersion

The α or low-frequency dispersion is characterized by very high permittivity values and can be ascribed, at least partially, to counterion diffusion effects. Such large dispersions are predicted by theories of ionic diffusion in heterogenous media. While it is not possible to model the complexity of a tissue, simpler mixture models predict dielectric increments of the right order of magnitude. Other mechanisms were postulated to contribute to the α dispersion; many relate to interactions in the vicinity of the cell membrane. The cell membrane is a complex, dynamic structure comprising a phospholipid bilayer. The lipid, hydrophobic ends of the phospholipids form a middle layer; the hydrophilic groups cover the inner and outer surfaces. Embedded in the bilayer are proteins, transport organelles, and ionic channels that operate under physiological control. The ionic balance between the intra- and extracellular media maintains a 60- to 70-mV potential difference between them of about 10 kV/mm across the membrane, assuming a 6- to 7-nm thickness. Membrane-related mechanisms that are thought to contribute to the α dispersion include the charging of intracellular membrane-bound organelles and a frequency dependence in the impedance of the cell membrane itself. An important reason for the uncertainty in the understanding of this dispersion is the paucity of error-free dielectric data in its frequency range. The α dispersion has a very large permittivity increment. The corresponding decrement in conductivity is small; this, however, does not contravene the principle of causality and the Kramers–Kronig relations, which predict a change in conductivity of about 0.005 S/m for a 10^6 increment in permittivity and relaxation frequency of 100 Hz.

3.5.5.2 β Dispersion

The β dispersion occurs at intermediate frequencies and originates mostly from the capacitive charging of the cellular membranes and those of membrane-bound intracellular bodies. This phenomenon, also known as interfacial polarization, has been studied theoretically and experimentally. It was established experimentally that damage to the cell membrane changes the features of the β dispersion. Numerous biomedical applications are based on the variation of the parameters of the β dispersion with pathological conditions involving changes in cell physiology and morphology. Tissue with directed, anisotropic cellular structure would exhibit an anisotropic dielectric response in the frequency range of the β dispersion.

Modeling the electrodynamics of a simplified tissue-like system, for example, suspensions of spherical inclusions in conductive media, has established theoretical grounds for the presence of the β dispersions. It enables the computation of an effective permittivity of similar order of magnitude to the β dispersion. For a system of concentric shells in conductive media (Hanai et al., 1988; Irimajiri et al., 1991), it predicts the presence of dielectric dispersions equal in number to the number of interfaces. These interfacial polarizations are boundary effects that occur in addition to other polarizations that may occur in the components of the system. The multiplicity of mehanisms goes a long way toward explaining why the β dispersions in tissue are rather broad.

3.5.5.3 γ Dispersion

The γ dispersion is due to the dipolar polarization of tissue water. At frequencies in excess of a few hundred megahertz, where the response of tissue water is the dominant

TABLE 3.3

Dielectric Parameters of Water Dispersion in Tissues Obtained by Analysis of Experimental Results at 37°C

Tissue	ε_s	τ (ps)	α	σ (S/m)
Bone (cortex)	14.9	13.8	0.26	0.092
Bone (section)	22.1	14.4	0.22	0.208
Cartilage	43.6	12.8	0.27	0.58
Cornea	53.0	8.72	0.13	1.05
Lens (cortex)	52.1	9.18	0.11	0.72
Lens (nucleus)	38.1	11.3	0.20	0.33
Retina	67.3	7.25	0.05	1.42
Brain (gray)	55.5	7.76	0.12	1.03
Brain (white)	37.0	8.04	0.24	0.47
Cerebellum	50.2	8.52	0.09	0.89
Dura	49.2	9.63	0.14	0.77
Brain stem	34.6	8.45	0.20	0.47
Tongue (*in vivo*)	57.7	9.12	0.08	0.63
Aqueous humor	74.2	6.81	0.01	1.83
Water	74.1	6.2	0.0	>0.0001

Source: From Gabriel et al. (1996c).

mechanism, the complex permittivity may be expressed as Cole–Cole plus a conductivity term to simulate the dipolar dispersion of water and the contribution of the electrolytes; thus,

$$\hat{\varepsilon}(\omega) = \varepsilon_\infty + \frac{\varepsilon_s - \varepsilon_\infty}{1 + (j\omega\tau)^{1-\alpha}} + \frac{\sigma}{j\omega\varepsilon_0}$$

where σ is the conductivity due to ionic currents and to the lower-frequency polarization mechanisms. Table 3.3 gives the parameters of the γ dispersion of tissues modeled to the above expression. The water content of the tissues considered ranges from >95% for vitreous humor and >85% for retina to <20% for cortical bone. The correlation between ε_s and tissue water content is an obvious and expected result. The value of the distribution parameter α is significant for most tissues and negligible for body fluids (as for aqueous humor, for example). The mean relaxation time τ is generally longer than the value for water, indicating a restriction in the rotational ability of at least some of the tissue water molecules. The lengthening of the relaxation time of water in biological material is a well-studied hypothesis; the effect is common to most organic solutes, is known to increase with solute concentration (Grant et al., 1981; Bateman et al., 1990), and has previously been observed in tissues (Gabriel et al., 1983).

3.5.5.4 δ Dispersion

Tissues and other biological materials may exhibit dispersions other than the three main ones. The δ dispersion, identified in some protein solutions between the β and γ, dispersions may also occur in tissue in the hundreds of megahertz range; when present, its magnitude is small compared to the adjacent ones. Possible mechanisms include the dipolar relaxation of bound water, relaxation of small dipolar segments or side chains of biological molecules, and counterion diffusion along small regions of the charged surface. Under these conditions it is difficult to isolate and, in view of the multiplicity of possible mechanisms, difficult to interpret. It is often treated as the tail end of the β dispersion or a broadening of the γ dispersion.

3.5.6 Effective Complex Permittivity of a Heterogenous System

Where adequate experimental data are available, the complex permittivity of a tissue can be quite adequately modeled with four Cole–Coles and a static conductivity term:

$$\hat{\varepsilon}(\omega) = \varepsilon_{\infty} + \sum_{n=1}^{4} \frac{\Delta\varepsilon_n}{1 + (j\omega\tau_n)^{(1-\alpha_n)}} + \sigma_s/j\omega\varepsilon_0 \tag{3.55}$$

This is a descriptive model, imparting no definite information on the polarization mechanisms; the measured dielectric properties represent the bulk response of the tissue. Assigning effective parameters to a heterogenous medium is equivalent to treating it as homogenous where these parameters are concerned; in this case, its structural components are much finer than the wavelength of the field probing it.

The derivation of a general formula for the effective permittivity of a system in terms of those for its constituents is based on the theory of the transport properties of mixtures (Reynolds and Hough, 1957). For example, in the case of a binary mixture where a medium with permittivity ε_1 has inclusions of permittivity ε_2 and assuming that v_1 andv_2 are their respective volume fractions such that $v_1 + v_2 = 1$, then the average electric displacement D is given by

$$D = v_1 D_1 + v_2 D \tag{3.56}$$

and the average electric field is given by

$$E = v_1 E_1 + v_2 E_2 \tag{3.57}$$

The effective permittivity ε of the mixture is given by

$$D = \varepsilon\varepsilon_0 E \tag{3.58}$$

and for each component, $D_1 = \varepsilon_1\varepsilon_0\ E_1$ and $D_2 = \varepsilon_2\varepsilon_0 E_2$. Equation 3.56 and Equation 3.57 give

$$\varepsilon = \varepsilon_1 v_1 f_1 + \varepsilon_2 v_2 f_2 \tag{3.59}$$

where $v_1 f_1 + v_2 f_2 = 1$, $f_1 = E_1/E$, and $f_2 = E_2/E$, which gives two general formulations for the effective permittivity of the mixture:

$$\varepsilon = \varepsilon_1 + (\varepsilon_2 - \varepsilon_1) v_2 f_2 \tag{3.60}$$

or

$$(\varepsilon - \varepsilon_1) v_1 f_1 + (\varepsilon - \varepsilon_2) v_2 f_2 = 0 \tag{3.61}$$

The theoretical problem that needs to be resolved for specific mixtures boils down to the determination of the field ratios f_2 or f_2 and f_1. Theoretically, the two formulations are equivalent but, when approximations have to be made for the values of the field ratios, this is no longer true. Most of the published mixture equations differ from one another in the approximations considered appropriate. As for which of the two formulations to use as a starting point, it would seem reasonable to use Equation 3.60 for the case of sparse inclusions in a continuous medium and Equation 3.61 when the volume fractions of the two components are comparable. In the general case, the dielectric properties (in Equation 3.60 and Equation 3.61) are complex. However, under static or quasistatic conditions, the mixture equations hold for either permittivity or conductivity.

Maxwell (1891) was the first to characterize the field ratios for a system of sparse spherical inclusions in a homogenous medium under static field conditions and obtained:

$$\frac{\sigma - \sigma_1}{\sigma + 2\sigma_1} = \nu_2 \frac{\sigma_2 - \sigma_1}{\sigma_2 + 2\sigma_1} \tag{3.62}$$

where the subscripts 1 and 2 refer to the suspending medium and inclusions, respectively. In terms of permittivity, Equation 3.62 is known as the Rayleigh formula; in its complex form it is attributed to Wagner and commonly known as the Maxwell–Wagner equation. Other well-known mixture equations for spherical inclusions include:

Böttcher equation:

$$\frac{\varepsilon - \varepsilon_1}{3\varepsilon} = \nu_2 \frac{\varepsilon_2 - \varepsilon_1}{\varepsilon_2 + 2\varepsilon_1} \tag{3.63}$$

Bruggeman equation:

$$\left(\frac{\varepsilon - \varepsilon_1}{\varepsilon_2 - \varepsilon_1}\right)^3 + (1 - \nu_2)\frac{\varepsilon}{\varepsilon_1} = 1 \tag{3.64}$$

Looyenga equation:

$$\varepsilon^{1/3} = \nu_1 \varepsilon_1^{1/3} + \nu_2 \varepsilon_2^{1/3} \tag{3.65}$$

To first-order approximation, the above mixture equations revert to the same expression irrespective of the formulation of the problem and of the technique used to solve it. This is because the conditions of infinite dilution and the spherical shape enable an almost exact solution to the field ratio to be obtained.

Other formulations exist for different shape inclusions such as oblate and prolate spheroids. The subject has been reviewed by, among others, Van Beek (1967), Hanai (1968), Dukhin (1971), and more recently, Greffe and Grosse (1992), Sihvola and Lindell (1992), and Tinga (1992).

It is possible to extend mixture equations to multiple inclusions by using an iterative procedure (Tamasanis, 1992). For example, if $\varepsilon(\varepsilon_1, \varepsilon_2, \nu_2)$ is the effective permittivity of a binary mixture of background of permittivity ε_1, inclusion of permittivity ε_2, and volume fraction v_2, then a mixture with two types of inclusions identified with subscripts 2 and 3, respectively, can be described as a binary mixture of background permittivity $\varepsilon(\varepsilon_1, \varepsilon_2, v_2/1 - v_3)$ and inclusion of permittivity ε_3 and volume fraction v_3. The effective permittivity of such a mixture will be

$$\varepsilon(\varepsilon_1; \varepsilon_2, v_2; \varepsilon_3, v_3) \approx \varepsilon\left[\varepsilon\left(\varepsilon_1; \varepsilon_2, \frac{v_2}{1 - v_3}\right); \varepsilon_3, v_3\right] \tag{3.66}$$

The contributions of the different types of inclusion are added recursively in order of increasing density.

Where the assumptions used in their derivation can be approximated in the tissue model, mixture equations can be used to analyze the dielectric properties of biological materials in terms of their constituents. A few examples are given here to illustrate their application.

If the ionic conductivities of the suspending phase and that of a protein solution are known, say, by measurement at a frequency below the protein β dispersion, and if an assumption is made about the conductivity of the protein molecules, an appropriate mixture equation can then be used to determine the volume fraction of the inclusions, which, in this case, is the hydrated protein. In turn, this enables the amount of bound water to be calculated given that the fraction of anhydrous protein is known. Bull and Breese (1969) followed this approach and calculated the bound water for a variety of proteins. They evaluated the water fraction to be 0.6 g/g of protein. Pauly and Schwan (1966), following a similar procedure, estimated the conductivity of the human erythrocyte to be 0.518 S/m at 25°C, compared to the value of 1.45 S/m calculated from the known ionic composition of the cell. They attributed the difference partly to excluded volume by the protein-bound water and partly to decreased ionic mobility due to hydrodynamic effects.

Bound water in biological systems including tissue was estimated by applying mixture equations in the near-plateau region between the β and γ dispersions of the permittivity spectrum. Assuming the tissue to be a suspension of hydrated organic matter in an electrolyte solution that is little affected by the presence of the organic matter, the measured permittivity at 1 GHz is a reasonable estimate of the effective static permittivity. Knowledge of the permittivity of the suspending medium and an estimate of the permittivity of the organic matter enables the volume fraction of the inclusion to be calculated. Comparison with the known organic content provides an estimate of bound water (e.g., Grant et al., 1984; Kaatze, 1990; Schaefer et al., 2003).

3.6 Dielectric Relaxation Mechanisms in Heterogenous Media

The description of a material as heterogenous is a matter of scale; in the context of dielectric relaxation, it refers to electrical heterogeneity or the presence of electrical boundaries or interfaces. Boundary conditions at and around the interfaces gives rise to dielectric dispersions quite apart from dipolar-type dispersions that occur in the surrounding media. In biological materials, cellular membranes provide such interfaces; their presence is associated with two major dispersion regions in the dielectric spectra of tissues, namely, α and β dispersions originating mainly from interfacial polarization and ionic diffusion effects. The main mechanisms giving rise to these phenomena will be briefly discussed in this section.

3.6.1 Interfacial Polarization

Interfacial polarization is due to the charging of interfaces between conducting media and is an important mechanism of interaction in biological material. The basic principles of this phenomenon are best illustrated in simple models first before discussing their occurrence in biological materials.

3.6.1.1 Interface between Two Media

The simplest model is that of an interface between two media, for example, two slabs of thickness d_1 and d_2 in contact with each other with their interface perpendicular to an external electric field (Figure 3.14a). If the static permittivity and conductivity of the two materials are ε_1, σ_1 and ε_2, σ_2, respectively, the boundary condition on the electric field component normal to the interface gives

$$E_1\varepsilon_1 = E_2\varepsilon_2 \tag{3.67}$$

If the current densities j_1 and j_2 are equal, there will be no charge accumulation at the interface; this, however, is hardly ever the case. The ratio of current densities at the interface is

$$j_1/j_2 = \sigma_1 E_1/\sigma_2 E_2 = \sigma_1\varepsilon_2/\sigma_2\varepsilon_1 \tag{3.68}$$

Therefore, if $\sigma_1\,\varepsilon_2 \neq \sigma_1\,\varepsilon_2$, the interface will be charged at a rate that is proportional to the difference between j_1 and j_2.

The effective permittivity ε and conductivity σ of the system are calculated from its effective capacitance. With the field across the interface, this is equivalent to capacitances in series combination; thus,

$$\frac{d_1 + d_2}{\varepsilon - j\sigma/\omega\varepsilon_0} = \frac{d_1}{\varepsilon_1 - j\sigma_1/\omega\varepsilon_0} + \frac{d_2}{\varepsilon_2 - j\sigma_2/\omega\varepsilon_0} \tag{3.69}$$

This can be rearranged into a Debye type expression with a relaxation time of

$$\tau = \varepsilon_0\frac{\varepsilon_1 d_2 + \varepsilon_2 d_1}{\sigma_1 d_2 + \sigma_2 d_1} \tag{3.70}$$

and limiting values for low and high frequencies

$$\varepsilon_s = \frac{(\varepsilon_2\sigma_1 - \varepsilon_1\sigma_2)^2(d_1 + d_2)d_1 d_2}{(\varepsilon_1 d_2 + \varepsilon_2 d_1)(\sigma_1 d_2 + \sigma_2 d_1)^2} + \varepsilon_\infty \tag{3.71}$$

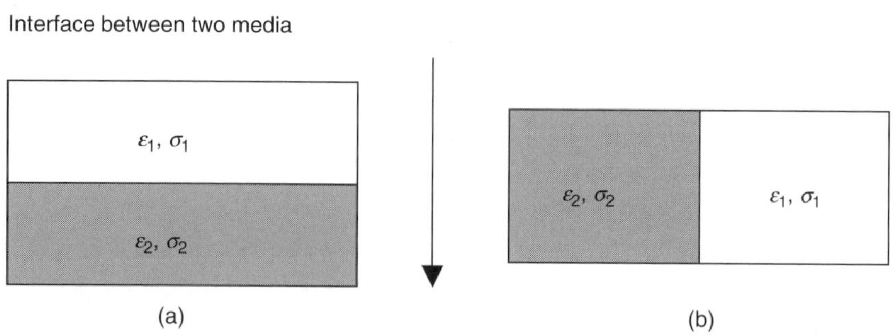

FIGURE 3.14
Two-media system; the arrow gives the direction of the electric field: (a) interface at right angle to the field and (b) interface along the field.

$$\sigma_s = (d_1 + d_2)\frac{\sigma_1\sigma_2}{\sigma_1 d_2 + \sigma_2 d_1} \tag{3.72}$$

$$\varepsilon_\infty = \frac{(d_1 + d_2)\varepsilon_1\varepsilon_1}{(\varepsilon_1 d_2 + \varepsilon_2 d_1)} \tag{3.73}$$

The polarization of the effective capacitance occurs in addition to any polarization process within the constituent phase, in which case the dielectric spectrum of the composite system will reflect the multiple dispersions.

If the field is along the interface (Figure 3.14b), no interfacial dispersion is observed, the effective permittivity is $\hat{\varepsilon} = (\hat{\varepsilon}_1 + \hat{\varepsilon}_2)/2$.

3.6.1.2 Suspension of Spheroids

The simplest system in this category is a dilute suspension of spherical inclusions in a continuum. Its effective complex permittivity is given by the formulation known as the Maxwell–Wagner equation, where the subscripts 1 and 2 refer to the suspending medium and inclusions, respectively:

$$\frac{\hat{\varepsilon} - \hat{\varepsilon}_1}{\hat{\varepsilon} + 2\hat{\varepsilon}_1} = v_2\frac{\hat{\varepsilon}_2 - \hat{\varepsilon}_1}{\hat{\varepsilon}_2 + 2\hat{\varepsilon}_1} \tag{3.74}$$

This equation can be rearranged in the form of a dispersion equation with the following parameters:

$$\begin{aligned}
\varepsilon_\infty &= \varepsilon_2\frac{2\varepsilon_2 + \varepsilon_1 - 2v_2(\varepsilon_2 - \varepsilon_1)}{2\varepsilon_2 + \varepsilon_1 + v_2(\varepsilon_2 - \varepsilon_1)} \\
\varepsilon_s - \varepsilon_\infty &= \frac{9(\varepsilon_2\sigma_1 - \varepsilon_1\sigma_2)^2 v_2(1 - v_2)}{[2\varepsilon_2 + \varepsilon_1 + v_2(\varepsilon_2 - \varepsilon_1)][2\sigma_2 + \sigma_2 + v_2(\sigma_2 - \sigma_1)]^2} \\
\sigma_s &= \sigma_2\frac{2\sigma_s + \sigma_2 - 2v_2(\sigma_2 - \sigma_1)}{2\sigma_2 + \sigma_2 + v_2(\sigma_2 - \sigma_1)} \\
\sigma_s - \sigma_\infty &= \frac{9(\sigma_2\varepsilon_1 - \sigma_1\varepsilon_2)^2 v_2(1 - v_2)}{[2\sigma_2 + \sigma_1 + v_2(\sigma_2 - \sigma_1)][2\varepsilon_2 + \varepsilon_2 + v_2(\varepsilon_2 - \varepsilon_1)]^2} \\
\tau &= \varepsilon_0\frac{2\varepsilon_2 + \varepsilon_1 - v_2(\varepsilon_2 - \varepsilon_1)}{2\sigma_2 + \sigma_2 + v_2(\sigma_2 - \sigma_1)}
\end{aligned} \tag{3.75}$$

The dispersion will occur when $\sigma_1\varepsilon_2 \neq \sigma_1\varepsilon_2$, which is practically always the case. The magnitude of the dispersion depends on the differences in dielectric parameters between the two phases.

In the context of mixture theory, sparse means $v_2 \leq 0.2$; to model the effective permittivity of more concentrated suspensions, it is necessary to take into consideration interparticle interactions. This becomes a very complex model; there are no rigorous solutions even for the relatively simple case of identical spherical inclusions. The Bruggeman–Hanai Equation 3.64 was formulated taking account of some interaction between particles and is therefore better suited than the Maxell–Wagner equation to model more concentrated suspension. In its complex form, it can be shown to predict the occurrence of a dispersion with the following limiting parameters:

$$\left(\frac{\varepsilon_\infty - \varepsilon_2}{\varepsilon_1 - \varepsilon_2}\right)\left(\frac{\varepsilon_1}{\varepsilon_s}\right)^{1/3} = 1 - v_2$$
$$\varepsilon_s\left(\frac{3}{\varepsilon_s - \sigma_2} - \frac{1}{\sigma_s}\right) = 3\left(\frac{\varepsilon_1 - \varepsilon_2}{\sigma_1 - \sigma_2} + \frac{\varepsilon_2}{\sigma_s - \sigma_2}\right) - \frac{\varepsilon_2}{\sigma_2}$$
$$\left(\frac{\varepsilon_s - \sigma_2}{\sigma_1 - \sigma_2}\right)\left(\frac{\sigma_1}{\sigma_s}\right)^{1/3} = 1 - v_2$$
$$\sigma_\infty\left(\frac{3}{\varepsilon_\infty - \varepsilon_2} - \frac{1}{\varepsilon_\infty}\right) = 3\left(\frac{\sigma_1 - \sigma_2}{\varepsilon_1 - \varepsilon_2} + \frac{\sigma_2}{\varepsilon_\infty - \varepsilon_2}\right) - \frac{\sigma_1}{\varepsilon_1} \tag{3.76}$$

In these expressions ε_s, ε_∞, σ_∞ are the limiting values of the corresponding parameters. The dispersion is characterized by a distribution of relaxation times.

The validity of the model has been verified experimentally for mixtures of known composition and geometry (Hanai et al., 1982; Ishikawa et al., 1982). The dispersion is broader than a single time constant because of the interactions between particles. Moreover, if the components of the heterogenous system exhibit molecular dielectric dispersion of their own, then these intrinsic dispersions will also appear, together with the interfacial dispersion, in the complete frequency spectrum of the system.

Fricke (1955), Sihvola and Kong (1988), Sihvola and Lindell (1992), and many others have extended the model to the more general case of a suspension of spheroids with any combination of axial ratios. Ultimately, the outcome is equivalent to introducing a parameter to account for the shape. The Maxwell–Wagner equation becomes

$$\frac{\hat{\varepsilon} - \hat{\varepsilon}_1}{\hat{\varepsilon} + F\hat{\varepsilon}_1} = v_2\frac{\hat{\varepsilon}_2 - \hat{\varepsilon}_1}{\hat{\varepsilon}_2 + F\hat{\varepsilon}_1} \tag{3.77}$$

where F is the shape factor, equal to 2 for spheres, which reverts to Equation 3.74. In cases where the shape of the inclusion is not known, limiting values for the effective permittivity can be obtained using the shape factors in the extreme cases of infinitely long thin rods and infinitely thin circular disks.

An interesting case is that of ellipsoids with their axes aligned in the same direction. The permittivity would be different depending on the direction of the field; the mixture would be electrically anisotropic and the effective permittivity is represented by a tensor.

Another case of practical interest is that of layered spherical inclusions. This situation is required when modeling cellular structures surrounded by a membrane of finite thickness. Solutions for the effective permittivity of this model were provided by many researchers in this field (e.g., Schwan, 1957; Zhang et al., 1983; Grosse, 1988). These authors applied two mixture models, once to the concentric bodies, thus obtaining an effective permittivity for the inclusions, and then treating the mixture as a suspension of homogenous spheres. The parameters of the dispersion could be expressed in terms of the physical dimensions and electrical characteristics of the cell and cell membrane; simplified versions of these expressions are reported by Foster and Schwan (1989). Sihvola (1989) and Irimajiri et al. (1991), among others, extended the treatment to several concentric shells by using a recursive technique. These complex models are more relevant to the study of biological systems and to the understanding of the interactions at the cellular level. They are not sufficiently developed for the quantitative characterization of the interfacial polarization in biological systems, but do provide an insight into the factors that determine its characteristics. An example from the recent literature is the modeling of the dielectric response of heart tissue by Schaefer et al. (2002). The model is a function of the cell shape, electrical cell coupling and polarization of cell membranes, and intracellular structure. It describes heart cells and subcellular organelles as rotational ellipsoids filled with electrolyte enclosed by an isolating membrane and is capable of reproducing the main features of the dielectric spectrum of heart tissue.

In recent years, statistical methods using probabilistic descriptions of the physical mixture in terms of a spatial density function have been developed to provide realistic bounds for the effective permittivity of mixtures. This approach, developed by, among others, Bergman (1978) and Milton (2002), provides an analytic integral representation of the effective permittivity $\hat{\varepsilon}$ of an arbitrary binary mixture in terms of a spatial density function $g(x)$:

$$\frac{\hat{\varepsilon} - \hat{\varepsilon}_1}{\hat{\varepsilon}_1} = \int_0^1 \frac{g(x)\,\mathrm{d}x}{x + \hat{\varepsilon}_1/\hat{\varepsilon}_2 - \hat{\varepsilon}_1}$$

where, as before, the subscripts 1 and 2 refer to continuum and dispersed phases, respectively, and the integration is over all possible positions. Depending on the choice of distribution function $g(x)$, it is possible for the above equation to revert to some of the well-known binary mixture equations. Recursive application is possible; modeling biological systems remains challenging.

A new tool for the study of mixtures, including biological materials, has evolved with the development of increasingly powerful numerical modeling packages for the propagation of electromagnetic fields in complex structures from full solutions of Maxwell's equations. With structures being defined at the nanoscale, the characterization of fields within cells and subcellular structures appears to be within reach (Gimsa and Wachner, 1998, 1999, 2001a,b; Bianco et al., 2000; Sebastian et al., 2001; Munoz et al., 2003).

3.6.2 Counterion Polarization Effects

Another important polarization phenomenon in electrically biological materials originates from ionic diffusion effects near charged surfaces and the formation of counterion or electric double layers. The distribution of ions in the vicinity of charged interfaces is subject to concentration and electric field gradients; an equilibrium is reached with the ions continuously distributed over the volume of the electrolyte solution. The time constant associated with this mechanism is longer than that of the Maxwell–Wagner effect, it is of the form L^2/D, where L is the length over which diffusion occurs and D is a diffusion coefficient (Schwarz, 1962).

Counterion phenomena are difficult to analyze rigorously; they involve coupled electrodynamic and hydrodynamic mechanisms. The theories are complex, but good reviews are available as an introduction to the subject (Dukhin, 1971; Dukhin and Shilov, 1974; Fixman, 1980, 1983; Mandel and Odijk, 1984). Relatively simple models that provide exact solutions have been proposed (Grosse and Foster, 1987; Grosse, 1988), whereby coupled differential equations for the ion concentrations and current densities are obtained for a macroscopic sphere of radius a in an ionic medium. Their solution yields a broad, asymmetrical, low-frequency dispersion. The time constant of this dispersion is a^2/D, where D is the diffusion coefficient of ions in the bulk electrolyte.

To visualize the effect, consider the motion of an ion in the bulk electrolyte near the particle, it will be conducted away or excluded depending on whether its sign is the same or opposite that of the ions in the counterion layer. Thus, for an ion in the electrolyte, the particle acts either as a good conductor or as an insulator depending on its charge compared to that of the counterion. A cloud of charge accumulates within a Debye length of the charged surface; for physiological saline (0.15 N NaCl) the Debye length is very small, <1 nm, resulting in a very large induced capacitance and hence a large permittivity dispersion.

3.7 Dielectric Properties of Tissue—State-of-Knowledge

Research into the dielectric properties of biological materials and their variation with frequency has been ongoing for most of the past century. Early studies went a long way toward understanding and establishing the principles of interaction and the corresponding features of the highly frequency-dependent spectrum of a tissue. In the last few decades, the research was driven, above all, by the need to establish a credible database of dielectric properties of all body tissues for use in electromagnetic dosimetry studies, where the object is to quantify the exposure of people to external electromagnetic fields from knowledge of the effective internal fields and currents induced in them. In these studies, tissues are characterized by their measured dielectric properties. In the last decade, most dosimetric studies drew on data published in the scientific literature in 1996 and made widely available on the Internet thereafter (Gabriel et al., 1996; Gabriel and Gabriel, 1997).

3.7.1 1996 Database

The backbone of the 1996 database is a large experimental study providing data pertaining, almost exclusively, to excised animal tissue at 37°C. For most tissues, the characterization was over a wide frequency range, 10 Hz to 20 GHz, using three previously established experimental setups with overlapping frequency ranges. The following are some of its characteristics:

- The data are presented in the context of a review covering all relevant publications in the preceding half-century. By and large, the experimental data were well within the confines of corresponding values from the literature.
- The data showed good internal consistency, that is, good agreement between data obtained with different experimental setups in a common frequency range.
- Finally, an element of great practical importance, the dielectric spectra were parametrized using a multidispersion model consisting of four Cole–Cole terms and one ionic conductivity term. For each tissue, the parameters of the model enable the reconstruction of its spectrum, a procedure that could easily be incorporated in numerical studies to provide dielectric data that are broadly in line with the vast body of literature on the subject.
- The fact that the complex permittivity data could be fitted to Cole–Cole dispersions implies that they also agree with the Kramers–Kronig relation in accordance with the principle of causality for a linear system. This imparts another level of consistency to the data.

Examples are given here to illustrate the extent of the available data in the literature for certain tissues at certain frequencies in contrast to the scarcity of data elsewhere

(Figure 3.15 through Figure 3.17). No attempt is made at a quantitative or mechanistic analysis.

While useful, the 1996 database has several limitations, as pointed out by its authors:

- Most measurements were carried out on excised tissue, while data pertaining to live tissue would have been more relevant in bioelectromagnetics studies.
- For most tissues, the predictions of the model can be used with confidence for frequencies above 1 MHz because of the availability of supporting data in the literature.
- At lower frequencies, where the literature values are scarce and have larger than average uncertainties, the model should be used with caution in the knowledge that it provides a "best estimate" based on the then available knowledge. This is particularly important for tissues where there are no data to support its predictions.
- Electrode polarization, an inevitable source of error at low-frequencies, was not totally accounted for. It affects the data at frequencies below 100 Hz.
- Because of the geometry of the sampling probe, it was not possible to orient the field along and across directed structure to demonstrate the anisotropy of the dielectric properties.

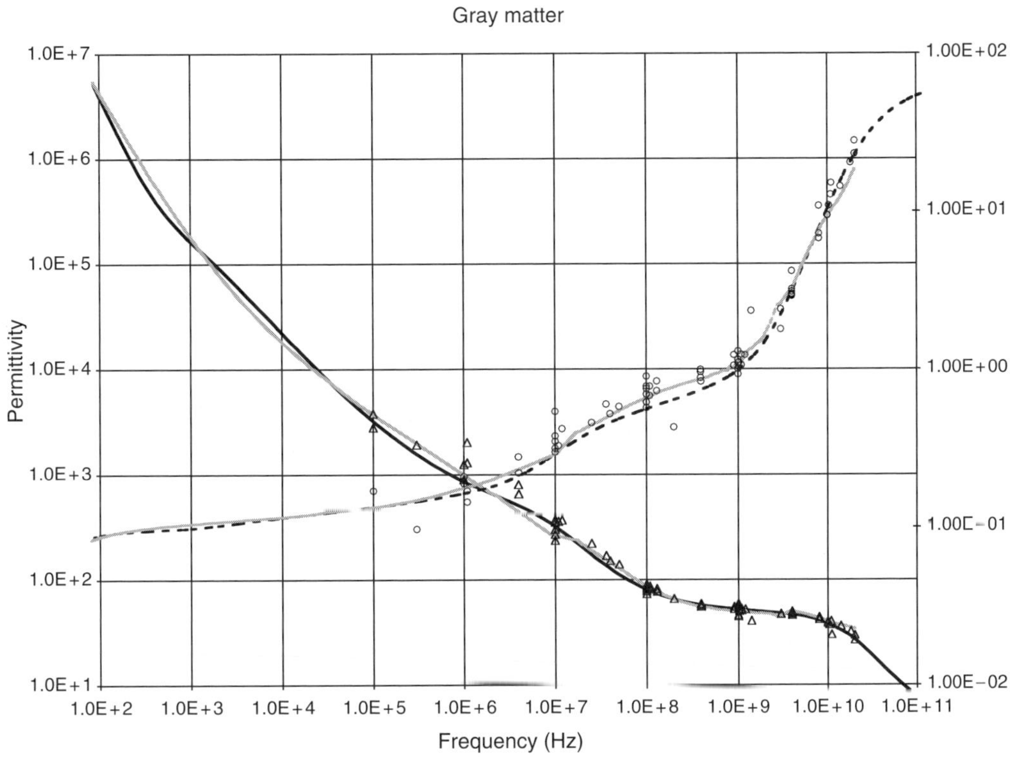

FIGURE 3.15
Permittivity and conductivity of gray matter at 37°C; gray lines are experimental data from Gabriel et al. (1996), triangles and circles are permittivity and conductivity values from the pre-1996 literature, black solid and dashed lines are the predictions of the model.

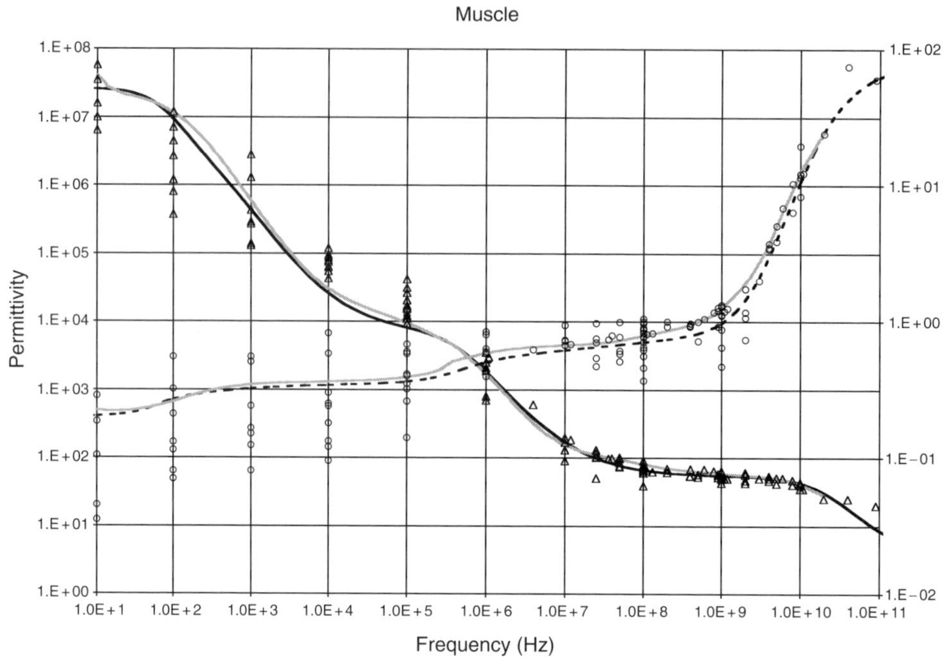

FIGURE 3.16
Permittivity and conductivity of skeletal muscle at 37°C; legend as in Figure 3.15. The very wide spectrum of data below 1 MHz is, at least partially, due to the anisotropy in the dielectric properties of muscle tissue. The literature data pertain to measurement along and across the muscle fibers and to measurements where the direction was not specified.

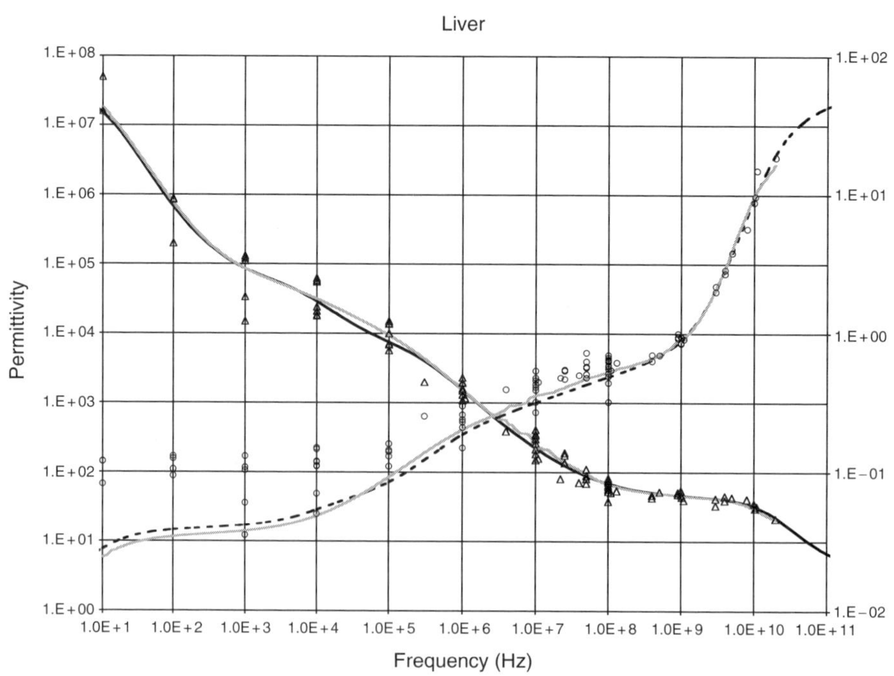

FIGURE 3.17
Permittivity and conductivity of liver tissue at 37°C; legend as in Figure 3.15. Liver tissue exhibits no significant anisotropy in its dielectric properties, but as with all tissue, the characteristics of the β dispersion and the static conductivity are sensitive to the viability and time after death when the measurements were made.

Evidently, much remains to be done, in particular with respect to reducing the uncertainty in the data and filling in the gaps identified. Ten years on, the recent literature is reviewed to update the state-of-knowledge on the subject.

3.7.2 Literature After 1996—A Brief Review

The review is carried out per tissue type or thematic underline. In some cases, data from recent studies are compared with the model in the 1996 database and with data from a recent study, where the dielectric properties of over 40 tissues were characterized *in vivo* and *in vitro* in the frequency range 10^2 to 10^4 MHz (Peyman et al., 2005).

3.7.2.1 Brain Tissue: Gray and White Matter

At microwave frequencies, three studies reported new data for brain tissue (Bao et al., 1997; Schmid et al., 2003a,b). Data tabulated by the authors are given in Figure 3.18. Data by Peyman et al. (2005) are in reasonable agreement with the database, while data by Bao and coworkers and Schmid and coworkers are higher for both permittivity and conductivity.

It is important to find a reason as to why carefully conducted studies, using adaptations of a conceptually similar experimental procedure, are still coming up with different results. In terms of explanation we note the handling of the sample by Bao et al., in which the whole brain is excised, immersed in saline, temperature regulated, and measured while immersed. The authors give good reasons for following this procedure.

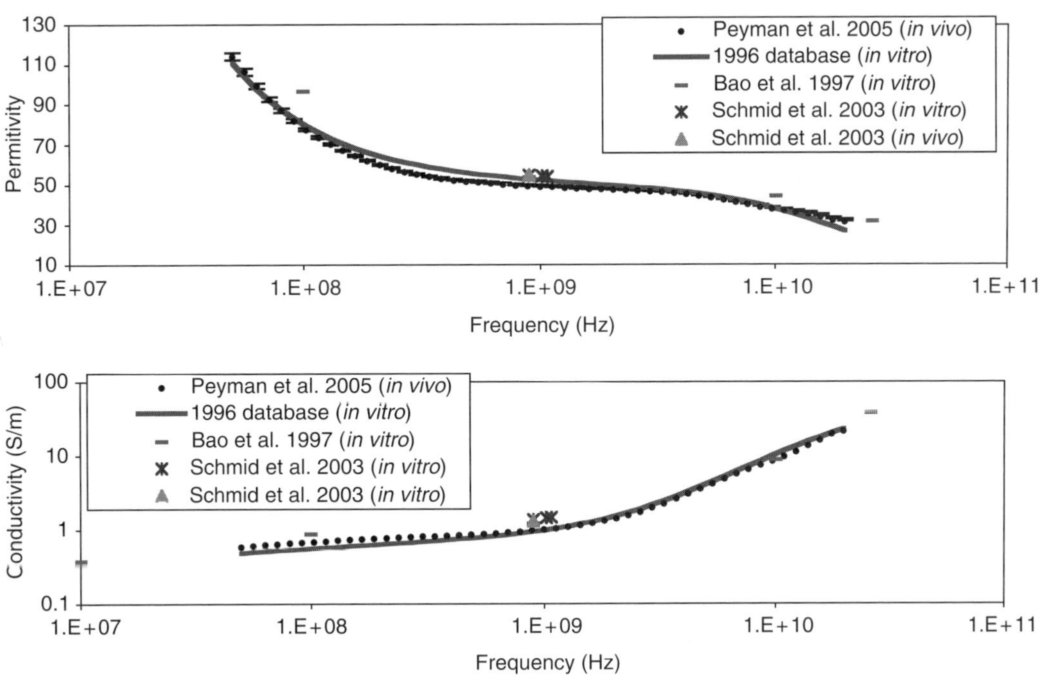

FIGURE 3.18
Dielectric properties of gray matter at 37°C. Data from recent studies compared to the prediction of the 1996 database.

It is inevitable, however, that one should expect their data to fall on the high side of average because of the presence of saline. Schmid et al. published data on porcine (*in vivo*) and human gray matter (*in vitro*). Their porcine gray matter data were obtained under conditions designed for the study of variation with time over a period spanning the time of death and beyond. Presumably, this is why their measurements were carried out over a narrow frequency range with the measurement probe held in position for the duration of the experiment (150 min). One might speculate that the amount of pressure used to maintain constant contact between the probe and the live brain could cause local oozing of fluid and higher conductivity values.

Conjecture apart, these data are valuable additions to the literature, but one must be cautious not to generalize on the basis of such limited data that measurement *in vitro* underestimates the dielectric properties of living tissues at microwave frequencies. Differences between tissue properties obtained *in vivo* and *in vitro* are to be expected at lower frequencies, in the range of α and β dispersions, in view of the sensitivity of their causal mechanism to the physiological state of the tissue. Differences between *in vivo* and *in vitro* data are much less likely in the frequency range of the γ dispersion, where water content is the most important determinant factor, and as recently reported by Stauffer et al. (2003), for liver tissue, and by Peyman et al. (2005), for many tissues including gray matter. Measurements *in vivo* are fraught with difficulties. For example, Burdette et al. (1986) measured the gray matter, *in vivo*, through the pia matter and directly beneath it. Of the two sets of data obtained, one is similar to that of Schmid et al., and the other is significantly lower.

In their human study, Schmid et al. measured the dielectric properties of gray matter in the frequency range of 800 MHz to 2450 MHz on 20 human brains immediately after excision. The measurements were carried out at room temperature in the range 18°C to 25°C and extrapolated to 37°C using experimentally determined thermal coefficients. Nevertheless, the dielectric properties at 900 MHz were in very good agreement with their data for porcine gray matter *in vivo*. It is understandably frustrating to realize that the bounds of uncertainty remain high when comparing data from different laboratories, even for those tissues that have been widely measured and reported.

Latikka et al. (2001) reported conductivity values at 50 kHz for gray matter (0.28 S/m), white matter (0.25 S/m), cerebrospinal fluid (1.25 S/m), and tumors (0.1 S/m to 0.43 S/m). They used a monopolar needle electrode during brain surgery on nine patients who had deep brain tumors. The technique is not geared toward making directed measurement and detecting anisotropy in the electrical properties, although those are anticipated on theoretical grounds and are known to be present in the hertz to kilohertz frequency range (Nicholson, 1965; Ranck et al., 1965; Yeldin et al., 1974; Nicholson and Freeman, 1975). Observed differences in the conductivity along and across the cellular structure were factors between 2 and 10 depending on the tissue. Clearly, this is an area of importance to electrophysiology, among other applications; it is also an area where data are scarce.

Peyman et al. (2001) reported variation in the dielectric properties of rat brain tissue as a function of age, at microwave frequencies. Their data pertained to the whole brain. The observed variation was ascribed, at least partially, to the change in the ratio of gray to white matter, which is known to occur throughout the developmental stage. In a recent study on porcine tissues (Peyman et al., 2005), they were able investigate gray and white matter separately. In this case, no variations were observed in the dielectric spectrum of gray matter, while statistically significant variations were observed in the dielectric spectrum of white matter (Figure 3.19). The observed variations are probably related to the process of myelination, which begins at birth and lasts to maturation. Similar variations were observed in the dielectric properties of the spinal cord.

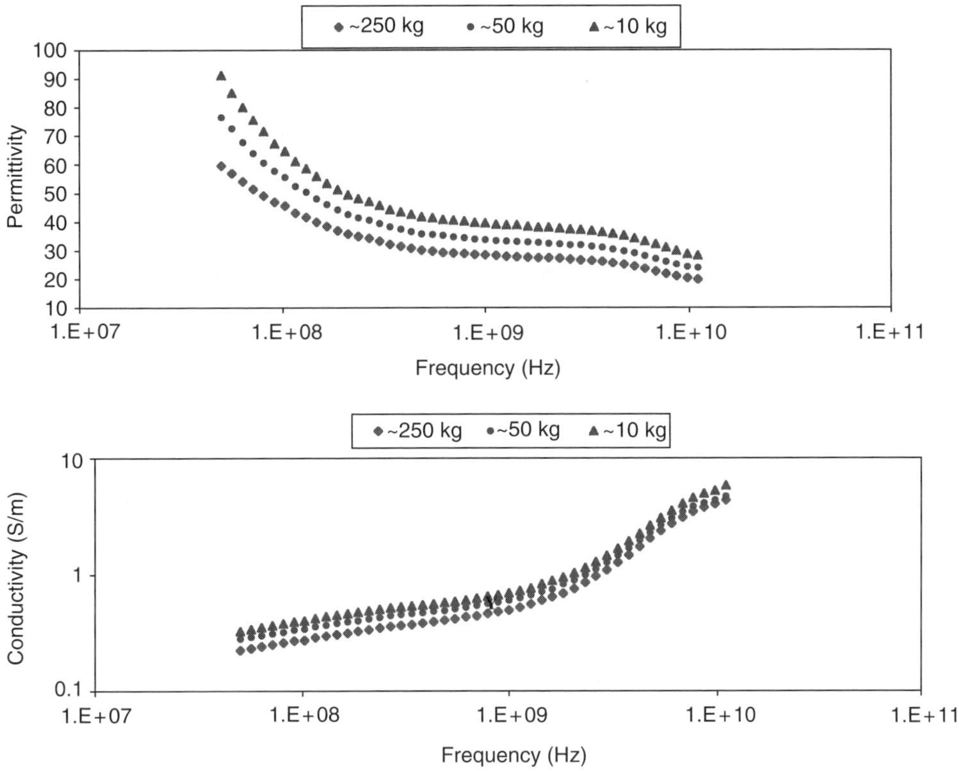

FIGURE 3.19
Permittivity and conductivity of white matter as a function of animal growth. The measurements were made *in vitro* at 37°C. The lowest permittivity and conductivity spectra pertain to a fully grown, 250-kg sow, the highest to a 10-kg piglet. (Data are from Peyman A, Holden S, Gabriel C. 2005. Dielectric Properties of Biological Tissues at Microwave Frequencies. Final technical report, MTHR Department of Health, U.K.)

3.7.2.2 Liver

Dielectric data for liver tissue were reported in several studies carried out under different conditions for a variety of reasons. For example, Riedel et al. (2003) developed a contact-free inductive measurement procedure and demonstrated the system by carrying out conductivity measurements on liver tissue between 50 and 400 kHz as a function of time after death. Stauffer et al. (2003) characterized the dielectric properties of normal and cancerous liver tissue in the frequency range of 0.3 to 3 GHz and reported higher permittivity and conductivity for tumor tissue. Chin and Sherar (2001) observed irreversible changes in the dielectric properties of liver tissue at 915 MHz because of excessive heating causing protein denaturation. Haemmerich et al. (2002) reported changes in the electrical resistivity of liver tissue during induced ischemia and postmortem. They observed increases in resistivity *in vivo* during occlusion. They analyzed the data in terms of intra- and extracellular resistance and cell membrane capacitance.

Valuable contributions by Raicu et al. (1998a,b) have provided data for rat liver tissue, measured *in vivo*, in the frequency range of 10^2 to 10^8 Hz. This is an eventful part of the dielectric spectrum of a tissue where contributions from interfacial and counterion interaction mechanisms occur. The measured data were corrected for electrode polarization and found to be in reasonable agreement with some previous studies (Surowiec et al., 1986; Foster and Schwan, 1996; Gabriel et al., 1996c). As expected, the data obtained traced a broad dielectric dispersion curve over the range of 10^3 to 10^8 Hz, suggestive of the involvement of widely distributed relaxation times. In the second of their 1998 papers, they provide a mechanistic analysis. A simple application of the Maxwell–Wagner interfacial polarization theory could not fully explain the observed dielectric behavior, especially at frequencies below 1 MHz. The Bruggeman–Hanai-type effective medium theory (EMT) was better, but not perfect, at simulating the observation at low-frequencies. A better simulation of the effective permittivity was obtained when second-order corrections, for possible dipole–dipole interaction (DDI) effects, were introduced to the classical EMT for a concentrated suspension of particles. Application of the new EMT-DDI model enabled reasonable estimates to be made of the following: effective size and shape of hepatic cell; specific capacitance for the plasma, nuclear, and mitochondrial membranes associated with the hepatocyte; and cytosolic as well as nucleoplasmic conductivities of physiological interest.

3.7.2.3 Muscle

Muscle tissues, be it skeletal, myocardial, lingual, or other, exhibit large anisotropy in their electrical properties. This is to be expected from the tissue structure and was observed at frequencies below 1 MHz (Epstein and Foster, 1983; Fallert et al., 1993). The static conductivity value measured along the muscle fiber may be up to an order of magnitude higher than when measured across. The α dispersion is more prominent and the β dispersion less defined in the longitudinal direction in accordance with the predictions of effective permittivity modeling of elongated structures. For example, Semenov et al. (2002) evaluated anisotropy of the myocardium using a cellular model of the myocardial tissue and concluded that at frequencies lower than 10 MHz, myocardial dielectric properties are highly anisotropic (up to a factor of 10). Reliable, low-frequency data are very scarce, and there is a wide range in what is available in the literature, partly due to the fact that many authors do not specify the measurement orientation. The situation is not helped by the fact that the apparent anisotropy depends on the measurement procedure, in particular the interelectrode distance in relation to the size of the muscle fiber. Some of the problems associated with obtaining good data at frequencies below 1 MHz have been described by Tsai et al. (2000, 2002) in the context of their *in vivo* measurement of swine myocardial resistivity. In their study, they report changes in the myocardial resistivity as a function of time after death. The postmortem resistivity at 1, 10, and 100 Hz increased to about three times their original *in vivo* value and at 500 kHz and 1 MHz increased less than 15%, 6 h after death.

Most recent studies on the electrical properties of muscle tissue focused on the differences between normal and ischemic or hypoxic tissue. One of the drivers is to investigate the possibility of using *in situ* impedance measurements to map the histological changes in tissue *in vivo*. There is also potential for noninvasive imaging provided that the electrical characteristics of both normal and scar tissue are well-defined. Miyauchi et al. (1999) and Schaefer et al. (1999) observed changes in the α and β dispersions of normal and ischemic skeletal muscle. Ischemia in myocardial muscle is a matter of clinical importance in the assessment of myocardial infarction and has been the subject of many dielectric investigations. Schwartzman et al. (1999) investigated the properties of the border zone, which were found to be intermediate between healthy and infarcted tissue in the case of chronically infarcted ventricular myocardium. Semenov et al. (2002) observed the dielectric properties of canine myocardium during acute ischemia and hypoxia to explore the potential of these observations for the clinical assessment of myocardial tissue using electrical impedance and microwave tomography. One of the problems identified is the need to know and take into consideration the tissue electrical anisotropy.

3.7.2.4 *Skin*

Skin is the interface of the body with environmental agents including electromagnetic fields; knowledge of its dielectric properties is of importance in the assessment of human exposure and in numerous biomedical applications. Data from *in vivo* measurements are now available, obtained using noninvasive, open-ended coaxial probes (Gabriel et al., 1996c; Gabriel, 1997; Raicu et al., 2000). However, the interpretation of such topical measurements as effective permittivity of the skin is far from straightforward. Lahtinen et al. (1997) and Alanen et al. (1998) advocate an analysis based on a quasistatic approximation of the fringing field of the probe penetrating a layered structure. Gabriel (1997) drew attention to the effect on the dielectric spectrum of the degree of hydration of the stratum corneum (Figure 3.20), which also affects penetration of the field into the tissue. Joining the discussion, Raicu et al. (2000) carried out *in vivo* measurements on dry skin and on skin moistened with physiological saline, in the frequency range of 100 Hz to 100 MHz. They analyzed the data using the dispersion model comprising a Debye-type and "universal" responses (Equation 3.53) and a conductivity term. Comparing the parameters of the model for dry and saline-moistened skin, they noted a fivefold increase in the dispersion magnitude. One possible explanation they provide is that the "effective penetration depth" increases and contributions from the innermost skin layers become evident. This statement appears paradoxical; in fact, it is due to the reduction in the layering effect; when moistened, the skin appears more homogenous and behaves like a high water content tissue.

Raicu et al. (2000) further speculate that an interfacial polarization originating from the stratum corneum–epidermis interface occurs in the case of dry skin, as suggested by Alanen et al. (1999), but not in that of the saline-moistened skin. In support of this argument, they refer to the change in one of the distribution parameters, β, which decreases from 0.152 to 0.076 after the skin is hydrated with aqueous NaCl solution. It therefore appears that topical measurement on dry (normal) skin *in vivo* may not be proportionately representative of the inner layers. On the other hand, the use of an aqueous coupling agent that is likely to hydrate the stratum corneum affects the results of the measurements. In practice, the use of a coupling agent gives more reproducible results and leads to better agreement between data from the recent literature, as is evident in Figure 3.21 through Figure 3.23, which contain such data (Gabriel, 1997; Ghodgaonkar et al., 2000; Raicu et al., 2000; Sunaga et al., 2002; Hwang et al., 2003; Petaja et al., 2003) and collectively cover the frequency range of 100 Hz to 100 GHz.

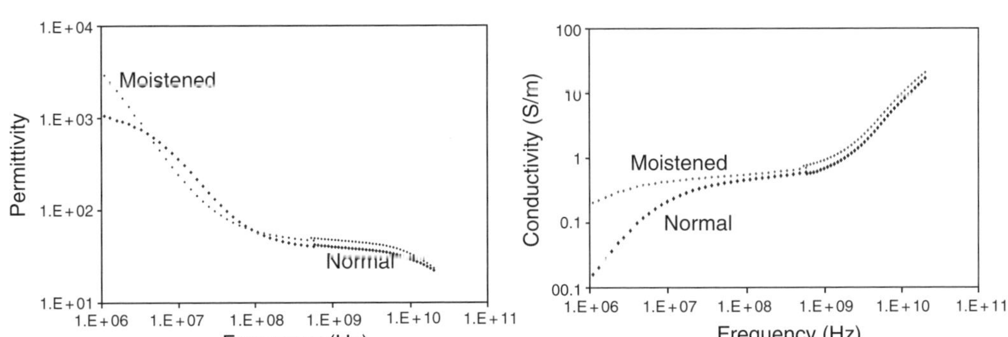

FIGURE 3.20
Permittivity and conductivity of skin (ventral forearm) illustrating the effect of moistening the skin on the dielectric spectra (Gabriel, 1996).

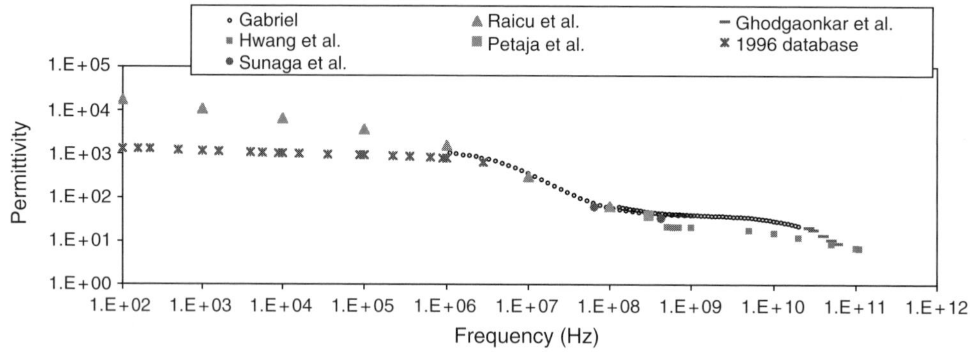

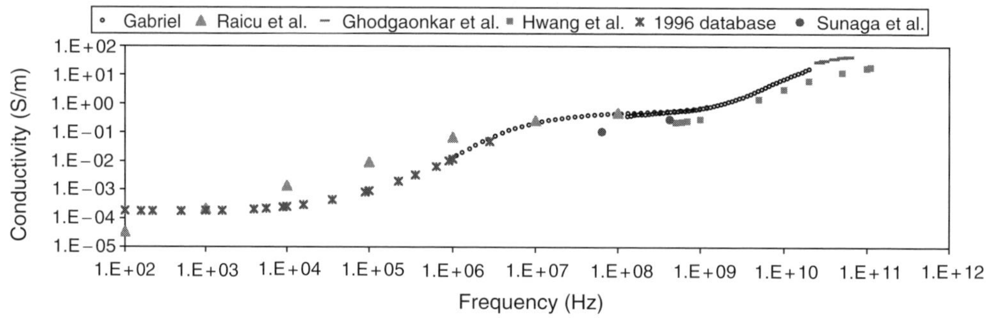

FIGURE 3.21
Permittivity and conductivity of skin (different parts of the body, excluding palms and soles). No moistening or contact gel was used. Different measurement techniques were used including open-ended coaxial probes of vastly different sizes.

The dielectric properties of skin have been widely investigated as monitors of various pathological conditions. Hayashi et al. (2005) investigated the dielectric properties of human skin *in vivo* at frequencies up to 10 GHz to monitor the progress of the healing process of skin burns using water content as the determinant factor. Their measurement technique, time domain spectroscopy and open-ended probe, is similar to that used by

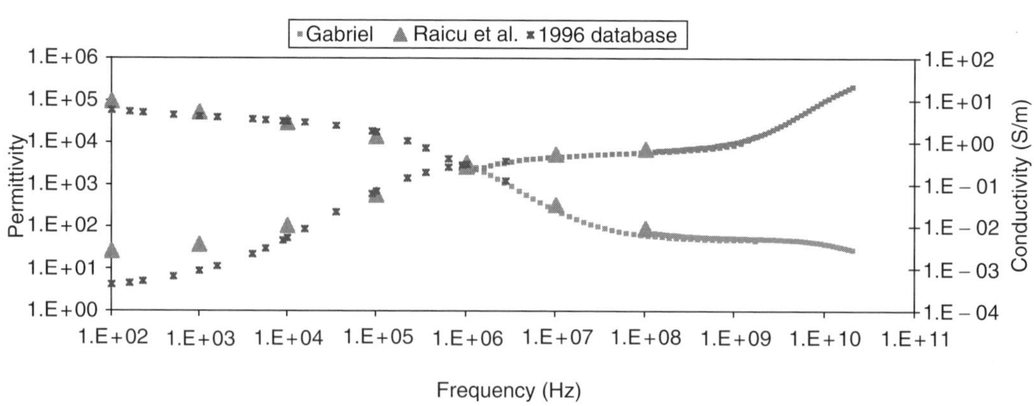

FIGURE 3.22
Permittivity and conductivity of skin (Raicu et al.: back of neck, moistened with physiological saline; Gabriel and database: ventral forearm, moistened with water). Open-ended coaxial probes of vastly different sizes were used.

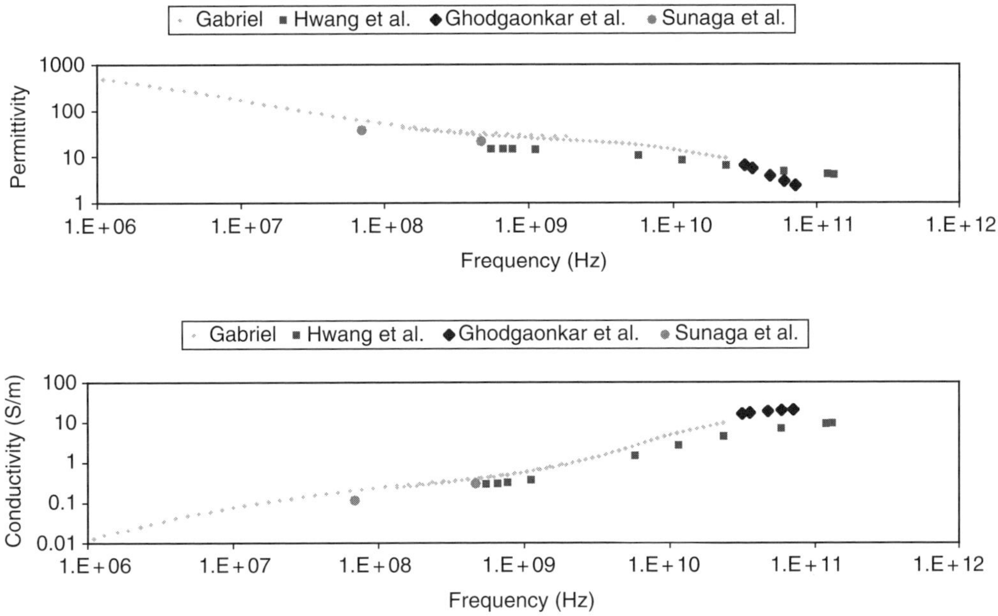

FIGURE 3.23
Permittivity and conductivity of palm, from recent studies.

Gabriel et al. (1997), who reported the dielectric spectra of normal and wounded tissue and ascribed the differences to water content.

Petaja et al. (2003) attempted to correlate the dielectric properties of skin at 300 MHz to body fluid changes after cardiac surgery and report limited success. Sunaga et al. (2002) investigated the variability in the dielectric properties of human skin of healthy volunteers, collagen disease patients, and dialysis patients over the frequency range of 1 to 450 MHz. No significant difference was detected in the dielectric properties among the three groups; some regional (abdomen, thigh, and forearm) dependence was observed. Marzec et al. (1999) measured the conductance and susceptance of soles and calves in leg skin from healthy controls and patients with ischemia in the frequency range of 100 Hz to 100 kHz. Ischemia was found to have no effect on the admittance at frequencies lower than 10 kHz, where the effect of the stratum corneum is dominant. Observed differences at frequencies in excess of 10 kHz are ascribed to ischemia in the underlying skin tissue.

Lindholm-Sethson et al. (1998) investigated the potential of using noninvasive skin impedance spectroscopy for the early detection of diabetic changes. They implemented a multivariate data analysis procedure to demonstrate how a regression model between the skin impedance and other diagnostic data for diabetic and control groups can be developed into a novel diagnostic tool for the early discovery of possible complications in diabetic patients. Statistical procedures are increasingly being applied to correlate dielectric parameters to structural or compositional elements of biological material, particularly in cases where there is a physical mechanism underpinning the effect that is obscured by noisy data (Kent et al., 2002).

The use of dielectric spectroscopy to monitor damage to the skin caused by ionizing radiation is an active field of research aiming at monitoring the side effects of clinical radiotherapy objectively and quantitatively. It appears that the changes to the skin during the acute stage cause both permittivity and conductivity to decrease (Tamura et al., 1994; Nuutinen et al., 1998), while the reverse happens when radiation-induced fibrosis finally sets in (Lahtinen et al., 1999). The initial decrease in permittivity, which also means a decrease in skin water, may be due to damage to skin capillaries resulting in swelling of the cytoplasm, with narrowing or occlusion of capillaries and a reduction in the effective microcirculation of the skin. In the long term, an increase in collagen and collagen-bound water is a likely explanation for the observed increase in the permittivity in line with a clinical indicator of subcutaneous fibrosis (Nuntineu, 1998).

3.7.2.5 Bone

Peyman et al. (2001) observed variation in the dielectric properties of rat skull bone as a function of developmental stage from neonate to 70 days old. They reported lower permittivity and conductivity values across the spectrum (100 MHz to 20 GHz). Similar results were recently reported for porcine skull, cortical bone, and bone marrow (Peyman et al., 2005).

3.7.2.6 Dielectric Properties of Cancerous Tissue

The pathological differences between normal and cancerous cells affect their composition and morphology and shape their dielectric spectrum. Interest in this field of study is driven by biomedical applications where such differences can be exploited for the treatment or diagnosis of cancer. Most of the clinical applications are based on hyperthermia, whereby electromagnetic energy is preferentially applied to the cancerous tissue, usually as an adjunct to radiotherapy. Electromagnetic hyperthermia was an active field of study in the 1980s, and it remains the domain of specialist medical centers. In contrast, there is a growing interest in applications geared toward the detection of cancerous regions using three-dimensional microwave tomography procedures and signal analysis (Hagness et al., 1998, 1999; Bulyshev et al., 2001; Wersebe et al., 2002). In cases where the suspect region is accessible for dielectric measurements, the procedure relies on the characterization and comparative analysis of the dielectric spectrum (Walker et al., 2000). The ultimate goal is to detect changes at the precancerous stage prior to their visibility by x-rays and to the emergence of serious clinical symptoms.

There is evidence that tumors have higher water content than the corresponding normal tissue; for certain types of tumor, such as breast carcinoma surrounded by fatty tissue, the difference could be considerable. In terms of dielectric properties, one would expect cancerous tissue to have higher permittivity and conductivity at microwave frequencies compared to normal tissue, as observed and reported by Schepps and Foster (1980), Foster and Schepps (1981), Rogers et al. (1983), and more recently, Stauffer et al. (2003).

Morphological changes affect the dielectric properties in the frequency range of the β dispersion and can be quite significant (Smith et al., 1986). Walker et al. (2000) used a finite element analysis to model the differences in impedivity between normal and precancerous cervical cells in the frequency range of 100 Hz to 10 MHz. Their results showed significant differences at frequencies lower than 10 kHz, basically in line with measurements carried out *in situ* with a four-electrode pencil probe. Polevaya et al. (1999) used time domain dielectric spectroscopy to study the differences between normal and malignant white blood cells. They used a Maxwell–Wagner mixture formulation and a double-shell cell model to determine differences in cellular and nuclear membrane characteristics between normal and malignant cells. Their detailed analysis reflects on some functional differences between membranes and provides some insight into the etiology of cancer.

The use of the electrical characteristics of tissue to understand, image, or treat cancerous tissue relies on the availability of good representative data across the spectrum but particularly in the frequency range of the β dispersion, where the changes are specific to the cellular transformation as well as the water content.

3.7.2.7 Conductivity of Tissue at Low-Frequency

There are limited, reliable dielectric data for body tissue at frequencies below 100 kHz. Some of the reasons relate to the dependence of the dielectric properties on the physiological state, degree of perfusion, time after death, and other biological parameters. There are also experimental difficulties, in particular electrode polarization, which is a major source of systematic error at frequencies below 100 Hz, even when precautions are taken to minimize its effects. Based on typical tissue dielectric data and a simple model for the electrode polarization, it is possible to estimate that it affects the permittivity more than the conductivity and that for body tissue, the conductive rather than the capacitive component dominates its electrical admittance (Schwan, 1992). For this reason, at extremely low-frequencies, the conductivity of tissue is considered real rather than complex, and the body is modeled as a resistive network the parameters of which are determined by the conductivity of the various tissues.

The conductivity of body tissue can be estimated by modeling on a cellular scale and applying appropriate mixture equations. Using this approach, Peters et al. (2001) evaluated the effective conductivity of several tissues such as cerebral cortex, liver, and blood. Such studies help to place upper and lower bounds on the conductivity values based on cellular parameters and knowledge of the conductivity of the phases.

Faes et al. (1999) carried out a meta-analysis of review studies (Geddes and Baker, 1967; Stuchly and Stuchly, 1980; Duck, 1990; Gabriel et al., 1996a,b) of tissue conductivity in the frequency range 100 Hz to 10 MHz. To make relative comparisons between different tissues, they calculated the mean and 95% confidence interval. They found large confidence intervals such that the conductivities of most high water content tissues (skeletal and cardiac muscle, kidney, liver, lung, and spleen) were not statistically different from one another at that level of significance. In contrast, blood has higher conductivity, while bone and fat have demonstrably lower conductivities. The insignificance of differences in high water content tissues could, of course, imply an equality of their conductivities, but it could also point at a large source of experimental variation that obscures real differences.

The conductivity of the body, or body parts, can be obtained by volume averaging using anatomical models and individual tissue conductivity as in Table 3.4, where the conductivity of the whole and the various parts of the body is given based on tissue conductivity in Gabriel (2000) and a voxel anatomical human model (Dimbylow, 1996).

At 50 Hz, the calculated conductivity values are comparable to the commonly used estimate of 0.2 S/m for the effective permittivity of a homogenous body model. When dielectric data become available, it would become imperative to reexamine the whole

TABLE 3.4

Conductivity (in S/m) of the Whole and Parts of the Body Obtained by Volume Averaging

Frequency	Whole Body	Head	Torso	Arm	Leg	Neck
50 Hz	0.22	0.25	0.22	0.19	0.20	—
10 kHz	0.28	0.28	0.26	—	0.24	0.22
100 kHz	0.29	0.30	0.33	—	0.24	0.24

Source: From Gabriel (2000).

question of whether or not there is sufficient justification for neglecting the contribution of the capacitive element of the body's electrical properties.

3.7.2.8 Nonlinear Dielectric Properties

The polarization mechanisms discussed so far occur from interactions with weak fields eliciting linear responses. At high field strength, nonlinear molecular and cellular polarization phenomena are predicted on theoretical grounds, on the basis of induced dipolar properties and classical electrodynamics. The threshold for initiating such effects is system and frequency dependent. As a general rule, as the frequency increases, so does the field level required to cause an effect. In general, at frequencies below the manifestation of the β dispersion, field strengths of the order of 10^6 V/m may be capable of initiating polarization mechanisms that affect the cellular function; higher fields may cause dielectric breakdown within the membrane, ultimately leading to cell destruction.

Under controlled conditions, high field strength, nonlinear effects are the focus of numerous applications in biotechnology. For example, dielectrophoresis, or the motion of particles caused by electrical polarization effects in nonuniform fields, is used to separate and manipulate cells. For a review of this subject, see Pethig (1996). Another nonlinear phenomenon, electroporation, is a consequence of the electrical breakdown of biological membranes, resulting in the formation of pores and a significant increase in the membrane permeability to external ions and molecules. Under controlled conditions, electroporation could be reversible and used to advantage in therapeutic, drug delivery applications. The basic principles can be found in mechanistic studies by Plickett and Weaver (1996), Prausnitz et al. (1999), and many others.

The hypothesis that weak fields may trigger nonlinear effects in cells has been investigated in theory and practice (Weaver and Astumian, 1990; Woodward and Kell, 1990). The generation of harmonics is one aspect that can be used to monitor their occurrence (Woodward and Kell, 1990); it also means that the dielectric properties are nonlinear, responding to harmonics as well as to the fundamental frequency. The study of harmonics is a subtle and clever tool; it has yet to prove its effectiveness in monitoring physiological responses to weak fields.

More recently, Balzano (2002) proposed to use a similar approach to test whether biological cells exhibit nonlinear responses to weak fields, at microwave frequencies. His idea hinges on the possibility of detecting a microwave signal at 1.8 GHz (second harmonic of 900 MHz) as weak as one microwave photon per cell per second. An experimental project based on this hypothesis is currently under way (see MTHR Web site: http://www.mthr.org.uk/research_projects/HO_funded_projects_excell.htm, accessed August 5, 2005). It will help to clarify the issue of nonlinear dielectric properties.

References

Alanen E, Lahtinen T, Nuutinen J. 1998. Variational formulation of open-ended coaxial line in contact with layered biological medium. *IEEE Trans Biomed Eng* 45(10): 1241–1248.

Alanen E, Lahtinen T, Nuutinen J. 1999. Penetration of electromagnetic fields of an open-ended coaxial probe between 1 MHz and 1 GHz in dielectric skin measurements. *Phys Med Biol* 44: N169–N176.

Balzano Q. 2002. Proposed test for detecting nonlinear responses in biological preparations exposed to RF energy. *Bioelectromagnetics* 23(4): 278–287.

Bao JZ, Lu ST, Hurt WD. 1997. Complex dielectric measurements and analysis of brain tissues in the radio and microwave frequencies. *IEEE Trans Microwave Theory Tech MTT* 45(10): 1730–1741.

Barthel J, Buchner R, Munsterer M. 1995. The dielectric properties of water in aqueous electrolyte solutions. In: *Chemistry Data Series 12*, Kreysa G, Ed., DECHEMA, Frankfurt.

Bateman JB, Gabriel C. 1987. Dielectric properties of aqueous glycerol and a model relating these to the properties of water. *J Chem Soc Faraday Trans* 2(83): 355–369.

Bateman JB, Gabriel C, Grant EH. 1990. Permittivity at 70 GHz of water in aqueous solutions of some amino acids and related compounds. *J Chem Soc Faraday Trans* 2(86): 3577–3583.

Bateman JB, Evans GF, Brown PR, Gabriel C, Grant EH. 1992. Dielectric properties of the system bovine albumin: urea: betaine in aqueous solutions. *Phys Med Biol* 37: 175.

Bergman DJ. 1978. The dielectric constant of a composite material. *Phys Rep* C43: 377.

Bianco B, Giordano S. 2003. Electrical characterization of linear and nonlinear random networks and mixtures. *Int J Circ Theor Appl* 31: 199–218.

Bianco B, Chiabrera A, Giordano S. 2000. DC-ELF characterization of random mixtures of piecewise nonlinear media. *Bioelectromagnetics* 21: 145–149.

Böttcher CJF, Bordewijk P. 1978. *Theory of Electric Polarization*. Elsevier, Amsterdam.

Buchner R, Hefter GT, May PM. 1998. Dielectric relaxation of aqeous NaCl solutions. *J Phys Chem A* 103: 1–9.

Buckmaster HA. 1990. Precision microwave complex permittivity measurements of high loss liquids. *J Electromagn Waves Appl* 4: 645–656.

Bull HB, Breese K. 1969. Electrical conductance of protein solutions. *J Colloid Interface Sci* 29: 492.

Bulyshev AE, Semenov SY, Souvorov AE, Svenson RH, Nazarov AG, Sizov YE, Tatsis GP. 2001. Computational modeling of three dimensional microwave tomography of breast cancer. *IEEE Trans Biomed Eng* 48(9): 1053–1056.

Burdette EC, Friederich PG, Seaman RL, Larsen LE. 1986. *In situ* permittivity of canine brain: regional variations and postmortem changes. *IEEE Trans Microwave Theory Tech MTT* 34(1): 38–49.

Chin L, Sherar M. 2001. Changes in dielectric properties of *ex vivo* bovine liver at 915 MHz during heating. *Phys Med Biol* 46: 197–211.

Cole KS. 1972. *Membrane, Ions and Impulses*. University of California Press, Berkeley.

Cole KS, Cole RH. 1941. Dispersion and absorption in dielectrics. I. Alternating current characteristics. *J Chem Phys* 9: 341–351.

Craig DQM. 1995. *Dielectric Analysis of Pharmaceutical Systems*. Taylor & Francis, London.

Davidson DW, Cole RH. 1951. Dielectric relaxation in glycerol, propylene glycol and *n*-propanol. *J Chem Phys* 19: 1484–1490.

Debye, P., Polar Molecules, Dover, Mincola, N.Y., 1929.

Dimbylow PJ. 1996. The development of realistic voxel phantoms for electromagnetic field dosimetry. In: *Proceedings of an International Workshop on Voxel Phantom Development*, National Radiological Protection Board, Chilton, U.K., 1–7.

Dissado LA. 1990. A fractal interpretation of the dielectric response of animal tissues. *Phys Med Biol* 35(11): 1487–1503.

Dissado LA, Hill RM. 1989. The fractal nature of the cluster model dielectric response functions. *J Appl Phys* 66(6): 2511–2524.

Dobson J, Grassi P. 1996. Magnetic properties of human hippocampl tissue: evaluation of artefact and contamination sources. *Brain Res Bull* 39: 255–259.

Duck FA. 1990. *Physical Properties of Tissue: A Comprehensive Reference Book*. Academic Press, London.

Dukhin SS. 1971. Dielectric properties of disperse systems. *Surface Colloid Sci* 3: 83.

Dukhin SS, Shilov VN. 1974. *Dielectric Phenomena and the Double Layer in Disperse Systems*. John Wiley & Sons, New York.

Ellison WJ, Lamkaouchi K, Moreau J. 1996. Water: a dielectric reference. *J Mol Liq* 68: 171–279.

Epstein BR, Foster KR. 1983. Anisotropy in the dielectric properties of skeletal muscle. *Med Biol Eng Comput* 21: 51–55.

Faes TJC, Meij HA, Munk JC, Heethaar RM. 1999. The electric resistivity of human tissues (100 Hz–10 MHz) a meta-analysis of review studies. *Physiol Meas* 20: R1–R10.

Fallert MA, Mirotznik MS, Bogen DK, Savage EB, Foster KR, Josephson ME. 1993. Myocardial electrical impedance mapping of ischemic sheep hearts and healing aneurisms. *Circulation* 87: 188.

Fixman M. 1980. Charged macromolecules in external fields. 1. The sphere. *J Chem Phys* 72: 5177.

Fixman M. 1983. Thin double layer approximation for electrophoresis and dielectric response. *J Chem Phys* 78: 1483.

Foster KR. 2002. Herman P. Schwan: a scientist and pioneer in biomedical engineering. *Annu Rev Biomed Eng* 4: 1–27.

Foster KR, Schepps JL. 1981. Dielectric properties of tumor and normal tissues at RF through microwave frequencies. *J Microwave Power* 16(2): 107–119.

Foster KR, Schwan HP. 1989. Dielectric properties of tissues and biological materials: a critical review. *Crit Rev Biomed Eng* 17(1): 25–104.

Foster KR, Schwan HP. 1996. Dielectric properties of tissues. In: *CRC Handbook of Biological Effects of Electromagnetic Fields*, Polk C, Postow E, Ed., CRC Press, Boca Raton, FL pp. 25–101.

Fricke HA. 1955. A mathematical treatement of the electric conductivity and capacity of disperse systems. II. The capacity of a suspension of conducting spheroids surrounded by a nonconducting membrane for a current of low frequency. *Phys Rev* 26: 678.

Fröhlich H. 1955. *Theory of Dielectrics*. Oxford University Press, Oxford.

Gabriel C. 1997. Comments on "Dielectric properties of the skin." *Phys Med Biol* 42: 1671–1674.

Gabriel C. and Gabriel S. 1997. http://www.brooks.af.mil/AFRL/HED/hedr/reports/dielectric/Report/Report.html

Gabriel C, Bentall RH, Grant EH. 1987. Comparison of the dielectric properties of normal and wounded human skin material. *Bioelectromagnetics* 8: 23–28.

Gabriel C, Gabriel S, Corthout E. 1996a. The dielectric properties of biological tissues: I. Literature survey. *Phys Med Biol* 41: 2231–2249.

Gabriel S, Lau RW, Gabriel C. 1996b. The dielectric properties of biological tissues: II. Measurements in the frequency range of 10 Hz to 20 GHz. *Phys Med Biol* 41: 2251–2269.

Gabriel S, Lau RW, Gabriel C. 1996c. The dielectric properties of biological tissues: III. Parametric models for the dielectric spectrum of tissues. *Phys Med Biol* 41: 2271–2293.

Gabriel C, Sheppard RJ, Grant EH. 1983. Dielectric properties of ocular tissues at 37°C. *Phys Med Biol* 28: 43–49.

Gabriel C., 2000. The dielectric properties of tissues. In: Radiofrequency radiation dosimetry and its relationship to the biological effects of electromagnetic fields. Editors: Klauengerg BJ and Miklavic D. Nato science series. *High Technology* (82): 75–84.

Geddes LA, Barker LE. 1967. The specific resistance of biological material—a compendium of data for the biomedical engineer and physiologist. *Med Biol Eng* 5: 271–293.

Ghodgaonkar DK, Gandhi Om P, Iskander MF. 2000. Complex permittivity of human skin *in-vivo* in the frequency band 26.5–60 GHz. *IEEE* Antennas and Propagation Society Symposium Proceedings, Salt Lake City, UT, USA, July 16–20, 2000, Vol. 2, pp. 1100–1103.

Gimsa J, Wachner D. 1998. A unified resistor–capacitor model for impedance, dielectrophoresis, electrorotation, and induced transmembrane potential. *Biophys J* 75: 1107–1116.

Gimsa J, Wachner D. 1999. A polarization model overcoming the geometric restrictions of the Laplace solution for spheroidal cells: obtaining new equations for field-induced forces and transmembrane potential. *Biophys J* 77: 1316–1326.

Gimsa J, Wachner D. 2001a. Analytical description of the transmembrane voltage induced on arbitrarily oriented ellipsoidal and cylindrical cells. *Biophys J* 81: 1888–1896.

Gimsa J, Wachner D. 2001b. On the analytical description of transmembrane voltage induced on spheroidal cells with zero membrane conductance. *Eur Biophys J* 30: 463–466.

Grant E, Shack R. 1967. Complex permittivity measurements at 8.6 mm wavelength over the temperature range 1–60 degrees centigrade. *Br J Appl Phys* 18: 1807–1814.

Grant EH, Sheppard RJ. 1974a. Dielectric relaxation in water in the neighbourhood of 4°C. *J Chem Phys* 60: 1792–1796.

Grant E, Buchanan T, Cook H. 1957. Dielectric behaviour of water at microwave frequencies. *J Chem Phys* 26: 156–161.
Grant EH, Sheppard RJ, South GP. 1978. *Dielectric Behaviour of Bilogical Molecules in Solution*. Clarendon Press, Oxford.
Grant EH, Szwarnowski S, Sheppard RJ. 1981. Dielectric properties of water in the microwave and infrared regions. In: *Biological Effects of Nonionising Radiation*, Illinger KH, Ed., American Chemical Society Symposium Series, Washington, DC, 157: 47–56.
Grant EH, McLean VER, Nightingale NRV and Gabriel C. 1984. Dielectric properties of water in biological solutions. In: *Interaction between Electromagnetic Fields and Cells*, Plessum Press, Newyork and London. Chiabrera A, Nicolini C, Schwan HP, Eds., NATO ASI Series, 97: 65–74.
Greffe JL, Grosse C. 1992. Static permittivity of emulsions. *Prog Electromagn Res* 41.
Grimnes S, Martinsen ØG, Eds. 2000. *Bioimpedance and Bioelectricity*. Academic Press, New York.
Grosse C. 1988. Permittivity of a suspension of charged spherical particles in electrolyte solution. II. Influence of the surface conductivity and asymmetry of the electrolyte on the low and high frequency relaxations. *J Phys Chem* 92: 3905–3910.
Grosse C, Foster KR. 1987. Permittivity of a suspension of charged spherical particle in electrolyte solution. *J Phys Chem* 91: 3073.
Haggis GH, Hasted JB, Buchanan TJ. 1952. The dielectric properties of water in solutions. *J Chem Phys* 20: 1452–1465.
Hagness SC, Taflove A, Bridges JE. 1998. Two-dimensional FDTD analysis of a pulsed microwave confocal system for breast cancer detection: fixed-focus and antenna-array sensors. *IEEE Trans Biomed Eng* 45(12): 1470–1479.
Hagness SC, Taflove A, Bridges JE. 1999. Three-dimensional FDTD analysis of a pulsed microwave confocal system for breast cancer detection: design of an antenna-array element. *IEEE Trans Antennas Propagat* 47(5): 783–791.
Hammerich D, Ozkan OR, Tsai JZ, Staelin ST, Tungjitkusolmun S, Mahvi DM, Webster JG. 2002. Changes in electrical resistivity of swine liver after occlusion and postmortem. *Med Biol Eng Comput* 40: 29–33.
Hanai T, Imakita T, Koizumi N. 1982. Analysis of dielectric relaxations of w/o emulsions in the light of theories of interfacial polarization. *Colloid Polym Sci* 260(11): 1029.
Hasted JB, El Sabeh. 1953. The dielectric properties of water in solutions. *Trans Faraday Soc* 49: 1003–1011.
Hasted JB, Husain SK, Frescura FAM, Birch JR. 1985. Far-infrared absorption in liquid water. *Chem Phys Lett* 118(6): 622–625.
Havriliak SJ, Negami S. 1966. A complete plane analysis of dispersion in some polymer systems. *J Polym Sci Polym Symp* 6: 99–117.
Hayashi Y, Miura N, Shinyashiki N, Yagihara S. 2005. Free water content and monitoring of healing processes of skin burns studied by microwave dielectric spectroscopy *in vivo*. *Phys Med Biol* 50: 599–612.
Hill NE, Vaughan WE, Price AH, Davies M. 1969. *Dielectric Properties and Molecular Behaviour*. Van Nostrand, London.
Hill RM, Jonscher AK. 1983. The dielectric behaviour of condensed matter and its many-body interpretation. *Contemp Phys* 24(1): 75–110.
Hwang H, Yim J, Cho J, Cheon C, Kwon Y. 2003. 110 GHz broadband measurement of permittivity on human epidermis using 1 mm coaxial probe. *IEEE MTT S Digest* 399–402.
Irimajiri A, Suzaki T, Asami K, Hanai T. 1991. Dielectric modeling of biological cells. Models and algorithm. *Bull Inst Chem Res Kyoto Univ* 69(4): 421–438.
Ishikawa A, Hanai T, Koizumi N. 1982. Dielectric properties of dextran gel Sephadex G-25 dispersed in aqueous phases. *Jpn J Appl Phys* 21(12): 1762.
Jonscher AK. 1983. *Dielectric Relaxation in Solids*. Chelsea Dielectrics Press, London.
Kaatze U. 1986. The dielectric spectrum of water in the microwave and near-millimetre wavelength region. *Chem Phys Lett* 132(3): 291–293.
Kaatze U. 1988. Complex permittivity of water as a function of frequency and temperature. *J Chem Eng* 34: 371–374.
Kaatze U. 1989. Complex permittivity of water as function of frequency and temperature. *J Chem Eng Data* 34: 371–374.

Kaatze U. 1990. On the existence of bound water in biological systems as probed by dielectric spectroscopy. *Phys Med Biol* 35(12): 1663–1681.

Kaatze U, Giese K. 1980. Dielectric relaxation spectroscopy of liquids: frequency domain and time domain experimental methods. *J Phys E: Sci Instrum* 13: 133–141.

Kent M, Peyman A, Gabriel C, Knight A. 2002. Determination of added water in pork products using microwave dielectric spectroscopy. *Food Control* 13: 43–149.

Kirkwood JG, 1936. On the Theory of Dielectric Polarization, *J Chem Phys*-Vol 4(9), 592–601.

Lahtinen T, Nuutinen J, Alanen E. 1997. Dielectric properties of the skin. *Phys Med Biol* 42: 1471–1472.

Lahtineu T, Nuntineu J, Alanen E, Turunen M, Nuortio L, Usenius T, Hopewell JW: Quantitative assessment of protein content in irradiated human skin. Int J Radiat Oncol Biol Phys 43: 635–638, 1999.

Lane JA, Saxton, JA. 1952. Dielectric dispersion in pure polar liquids at very high radio frequencies. *Proc R Soc Lond* 213: 400–408.

Latikka J, Kuurne T, Skola H. 2001. Conductivity of living intracranial tissues. *Phys Med Biol* 46: 1611–1616.

Liebe HJ, Hufford GA, Manabe T. 1991. A model for the complex permittivity of water at frequencies below 1 THz. *Int J Infrared Millimeter Waves* 12: 659–660.

Lindholm-Setson B, Han S, Ollmar S, Nicander I, Jonsson G, Lithner F, Bertheim U, Geladi P. 1998. Multivariate analysis of skin impedance data in long term type 1 diabetic patients. *Chemom Intell Lab Syst* 44: 381–394.

Mandel M, Odijik T. 1984. Dielectric properties of polyelectrolyte solutions. *Annu Rev Phys Chem* 35: 75.

Marzec E, Wachal K. 1999. The electrical properties of leg skin in normal individuals and in-patients with ischemia. *Bioelectrochem Bioenerg* 49: 73–75.

Maxwell JC. 1891. *A Treatise on Electricity and Magnetism*, 2nd ed., Vol. 1, chap. IX. Clarendon Press, Oxford.

Milton GW. 2002. *Theory of Composites*, Cambridge University Press, Cambridge.

Miyauchi T, Hirose H, Sasaki E, Hayashi M, Mori Y, Murakawa S, Takagi H, Yasuda H, Kumada Y, Iwata H. 1999. Predictability of dielectric properties for ischemic injury of the skeletal muscle before reperfusion. *J Surg Res* 86: 79–88.

Munoz San Martýn S, Sebastian JL, Sancho M, Miranda JM. 2003. A study of the electric field distribution in erythrocyte and rod shape cells from direct RF exposure. *Phys Med Biol* 48: 1649–1659.

Nicholson PW. 1965. Specific impedance of cerebral white matter. *Exp Neurol* 13: 386–401.

Nicholson C, Freeman JA. 1975. Theory of current source-density: analysis and determination of conductivity tensor for anuran cerebellum. *Neurophysiol* 38: 356–368.

Nuutinen J, Lahtinen T, Turunen M, Alanen E, Tenhunen M, Usenius T, Kolle R. 1998. A dielectric method for measuring early and late reactions in irradiated human skin. *Radiother Oncol* 47: 249–254.

Onsager L, 1936. Electric Moments of Molecules in Liquids *J. Am. Chem. Soc.* 58, 1486–1493.

Pauly H, Schwan HP. 1966. Dielectric properties and ion mobility in electrocytes. *Biophys J* 6: 621.

Petaja L, Nuutinen J, Uusaro A, Lahtinen T, Ruokonen E. 2003. Dielectric constant of skin and subcutaneous fat to assess fluid changes after cardiac surgery. *Physiol Meas* 24: 383–390.

Peters MJ, Stinstra JG, Hendriks M. 2001. Estimation of the electrical conductivity of human tissue. *Electromagnetics* 21: 545–557.

Pethig R. 1979. *Dielectric and Electronic Properties of Biological Materials*. John Wiley & Sons, New York.

Pethig R, Kell DB. 1987. The passive electrical properties of biological systems: their significance in physiology, biophysics, and biotechnology. *Phys Med Biol* 32(8): 933–970.

Pethig R. 1996. Dielectrophoresis: using inhomogeneous AC electrical fields to separate and manipulate cells. *Critical Reviews in Biotechnology* 16(4): 331–348.

Peyman A, Rezazadeh AA, Gabriel C. 2001. Changes in the dielectric properties of rat tissue as a function of age at microwave frequencies. *Phys Med Biol* 46: 1617–1629.

Peyman A, Holden S, Gabriel C. 2005. Dielectric Properties of Biological Tissues at Microwave Frequencies. Final technical report, MTHR Department of Health, U.K.

Polevaya Y, Erunolina I, Schlesinger M, Ginzburg BZ, Feldman Y. 1999. Time domain dielectric spectroscopy study of human cells II. Normal and maligant white blood cells. *Biochemic Biophysica Acta* 1419: 257–271.

Prausnitz MR, Pliquett U and Vanbever R (1999) Mechanistic studies of skin electroporation using biophysical methods in Electrically Mediated Delivery of Molecules to cells – Electrochemotherapy, Electrogenetherapy, and Transdermal Delivery by Electroporation, MJ Jaroszeski, R Gilbert, and R Heller, eds., Humana Press, Totowa, NJ, pp. 214–235.

Raicu 1999. Dielectric dispersion of biological matter: model combining Debye-type and "universal" response. *Phys Rev E* 60(4): 4677–4680.

Raicu V, Saibara T, Enzan H, Irimajiri A. 1998a. Dielectric properties of rat liver *in vivo*: analysis by modeling hepatocytes in the tissue architecture. *Phys Med Biol* 47: 333–342.

Raicu V, Saibara T, Irimajiri A. 1998b. Dielectric properties of rat liver *in vivo*: a non-invasive approach using an open-ended coaxial probe at audio/radio frequencies. *Phys Med Biol* 47: 325–332.

Raicu V, Kitagawa N, Irimajiri A. 2000. A quantitative approach to the dielectric properties of the skin. *Phys Med Biol* 45: L1–L4.

Ranck JB, BeMent SL. 1965. The specific impedance of the dorsal columns of cat: an anisotropic medium. *Exp Neurol* 11: 451–463.

Reynolds JA, Hough JM. 1957. Formulae for dielectric constant of mixtures. *Proc Phys Soc* 70: 769–775.

Riedel CH, Keppelen M, Nani S, Dössel O. 2003. Postmortem conductivity measurement of liver tissue using a contact free magnetic induction sensor. *EMBC IEEE* 3126–3129.

Rogers JA, Sheppard RJ, Grant EH. 1983. The dielectric properties of normal and tumour mouse tissue between 50 MHz and 10 GHz. *Br J Radiol* 56: 335–338.

Roussy G, Pearce JA. 1995. *Foundations and Industrial Applications of Microwaves and Radio Frequency Fields: Physical and Chemical Processes*. Wiley Publishers, New York.

Schaefer M, Kirlum HJ, Schlegel C, Gebhard MM. 1999. Dielectric properties of skeletal muscle during ischemia in the frequency range from 50 Hz to 200 MHHz. *Ann NY Acad Sci* 873: 59–64.

Schaefer M, Gross W, Ackemann J, Gebhard MM. 2002. The complex dielectric spectrum of heart tissue during ischemia. *Bioelectrochemistry* 58: 171–180.

Schaefer M, Gross W, Preuss M, Ackemann J, Gebhard MM. 2003. Monitoring of water content and water distribution in ischemic hearts. *Bioelectrochemistry* 61: 85–92.

Schanne OF, Ceretti ERP. 1978. *Impedance Measurements in Biological Cells*. Wiley, New York.

Schepps JL, Foster KR. 1980. UHF and microwave dielectric properties of normal and tumor tissues: variation in dielectric properties with tissue water content. *Phys Med Biol* 25: 1149–1159.

Schmid G, Neubauer G, Illievich UM, Alesch F. 2003a. Dielectric properties of porcine brain tissue in the transition from life to death at frequencies from 800 to 1900 MHz. *Bioelectromagnetics* 24: 413–421.

Schmid G, Neubauer G, Mazal PR. 2003b. Dielectric properties of human brain tissue measured less than 10 h post-mortem at frequencies from 800 to 2450 MHz. *Bioelectromagnetics* 24: 423–430.

Schwan HP. 1957. Electrical properties of tissue and cell suspensions. In: *Advances in Biological and Medical Physics*, Lawrence JH, Tobias CA, Eds., Academic Press, New York, 5: 147–209.

Schwan HP. 1992. Linear and nonlinear electrode polarization and biological materials. *Ann Biomed Eng* 20: 269–288.

Schwan HP, Foster KR. 1980. RF-field interactions with biological systems: electrical properties and biophysical mechanisms. *Proc IEEE* 68: 104–13.

Schwan HP, Sheppard RJ, Grant EH. 1976. Complex permittivity of water at 25°C. *J Chem Phys* 64: 2257–2258.

Schwartzman D, Chang I, Michele JJ, Mirotznik MS, Foster KR. 1999. Electrical impedance properties of normal and chronically infarcted left ventricular myocardium. *J Interv Card Electrophysiol* 3: 213–224.

Schwarz G. 1962. A theory of the low frequency dielectric dispersion of colloidal particles in eletrolyte solution. *J Phys Chem* 66: 2636.

Sebastian JL, Munoz S, Sancho M, Miranda JM. 2001. Analysis of the influence of the cell geometry, orientation and cell proximity effects on the electric field distribution from direct RF exposure. *Phys Med Biol* 46: 213–225.

Semenov SY, Svenson RH, Posukh VG, Nazarov AG, Sizov YE, Bulyshev AE, Souvorov AE, Chen W, Kasell J, Tatsis GP. 2002. Dielectrical spectroscopy of canine myocardium during acute ischemia and hypoxia at frequency spectrum from 100 kHz to 6 GHz. *IEEE Trans Med Imaging* 21(6): 703–707.

Sihvola AH. 1989. Self-consistency aspects of dielectric mixing theories. *IEEE Trans Geosci Remote Sensing* 27(4): 403.

Sihvola AH, Kong JA. 1988. Effective permittivity of dielectric mixtures. *IEEE Trans Geosci Remote Sensing* 26(4): 420.

Sihvola AH, Lindell IV. 1992. Polarizability modeling of heterogeneous media. *Prog Electromagn Res* PIER 06: 101–151.

Smith SR, Foster R, Wolf GL. 1986. Dielectric properties of VX-2 carcinoma versus normal liver tissue. *IEEE Trans Biomed Eng* 33(5): 522–524.

Stauffer PR, Rossetto F, Prakash M, Neuman DG, Lee T. 2003. Phantom and animal tissues for modelling the electrical properties of human liver. *Int J Hyperthermia* 19(1): 89–101.

Stuchly M. 1979. Interaction of radiofrequency and microwave radiation with living systems. *Radiat Environ Biophys* 16: 1.

Stuchly MA, Stuchly SS. 1980. Dielectric properties of biological substances—tabulated. *J Microwave Power* 15(1): 19–26.

Sunaga T, Ikehira H, Furukawa S, Shinkai H, Kobayashi H, Matsumoto Y, Yoshitome E, Obata T, Tanada S, Murata H, Sasaki Y. 2002. Measurement of the electrical properties of human skin and the variation among subjects with certain skin conditions. *Phys Med Biol* 47: N11–N15.

Surowiec AJ, Stuchly SS, Keaney M, Swarup A. 1986. *In vivo* and *in vitro* dielectric properties of feline tissues at low radio-frequencies. *Phys Med Biol* 31(8): 901–909.

Takashima S. 1989. Electrical properties of proteins. I. Dielectric relaxation. In: *Physical Principles and Techniques of Protein Chemistry*, Leach JS, Ed., Academic Press, New York.

Tamasanis D. 1992. Application of volumetric multiple scattering approximation to foliage media. *Radio Science (USA)* 27(6): 797–812.

Tamura T, Tenhunen M, Lahtinen T, Repo T, Schwan HP. 1994. Modelling of the dielectric properties of normal and irradiated skin. *Phys Med Biol* 39: 927–936.

Tinga WR. 1992. Mixture laws and microwave material interactions. *Prog Electromagn Res* 6: 1–40.

Tsai JZ, Cao H, Tungjitkusolmun S, Woo EJ, Vorperian VR, Webster JG. 2000. Dependance of apparent resistance of four-electrode probes on insertion depth. *IEEE Trans Biomed Eng* 47(1): 41–48.

Tsai JZ, Will JA, Stelle SHV, Cao H, Tungjitkusolmun S, Choy YB, Haemmerich D, Vorperian VR, Webster JG. 2002. *In-vivo* measurements of swine myocardial resistivity. *IEEE Trans Biomed Eng* 49(5): 472–483.

Van Beek LKH. 1967. Dielectric behaviour of heterogeneous systems, *Prog Dielectr* 7: 69–114.

Walker DC, Brown BH, Hose DR, Smallwood RH. 2000. Modelling the electrical impedivity of normal and prealignant cervical tissue. *Electron Lett*. 36(19): 1603–1604.

Weaver JC, Astumian RD. 1990. The response of living cells to very weak electric fields: the thermal noise limit. *Science* 247(4941): 459–462.

Wei YZ, Sridhar S. 1993. A new graphical representation for dielectric data. *J Chem Phys* 99(4): 3119–3124.

Wersebe A, Siegmann K, Krainick U, Fersis N, Vogel U, Claussen CD, Muller-Schimpfle M. 2002. Diagnostic potential of targeted electrical impedance scanning in classifying suspicious breast lesions. *Invest Radiol* 37(2): 65–72.

Woodward AM, Kell DB. 1990. On the nonlinear dielectric properties of biological systems *Saccharomyces cerevisiae. Bioelectrochem Bioenerg* 24: 83–100.

Yeldin M, Kwan H, Murphy JT, Nguyen-Hou H, Wong YC. 1974. Electrical conductivity in cat cerebellar cortex. *Exp Neurol* 43: 533–569.

Zaghloul H, Buckmaster HA. 1985. The complex permittivity of water at 9.356 GHz from 10 to 40 degrees C. *J Phys D: Appl Phys* 18: 2109–2118.

Zhang ZH, Sekine K, Hanai T, Koizuni N. 1983. Dielectric observations on polystrene microcapsules and the theoretical analysis with reference to interfractal polarization. *Colloid Polym Sci* 261: 381.

4

Magnetic Properties of Biological Material

Jon Dobson

CONTENTS

4.1 Introduction

Magnetism arises from the movement of electrical charges, such as the oscillation of electrons in a conducting wire, the flow of ions in an organism or the orbital, and spin motions of electrons in an atom. Since all materials contain moving subatomic particles (electrons and protons), all materials are, in some sense, "magnetic." In order to understand more about the role of magnetic materials in organisms, it is first necessary to examine what is meant by the various types of magnetic behavior of materials.

The magnetic properties of materials are dominated by the electron spin motion. In quantum mechanical terms, electrons may assume two possible spin states: spin $+1/2$ or spin $-1/2$. These may also be referred to as "spin up" and "spin down." The Pauli exclusion principle states that no two electrons may occupy the same energy state in an atom. This means that no two electrons may have the same set of values for the quantum numbers as they would then be indistinguishable. As electrons are added, they fill up each possible state in a given shell before filling the shell associated with the next higher energy state. The filling of the shells is governed by Schrödinger's wave equation and the quantum numbers.

Electrons are added to subshells in parallel spin configurations first according to Hund's rule. If all electrons are paired, there is no "spin" magnetic moment. These materials are still magnetic though, because of the electron's orbital motion. In most materials of biological origin, however, electron spin motion is cancelled, allowing electron orbital motion to dominate.

The spin structure of the transition series elements (iron in particular) is most important for the magnetic properties of biological materials. This is due to the presence of uncompensated spins in the 3d orbital, which gives rise to a spin magnetic moment. The spin moment is much stronger than the orbital moment and is aligned parallel to an applied field.

All materials, including biological materials, fall into one of the three categories of magnetic materials based on the spin and orbital motion of electrons: (1) diamagnetic, (2) paramagnetic, and (3) ferromagnetic.

4.2 Diamagnetic Materials

Diamagnets are materials in which all electron spins are paired (i.e., there are no uncompensated spins). Therefore, the magnetic properties of diamagnets are determined by the electron orbital motion.

According to Faraday's law, in the presence of an applied magnetic field there is an electric field induced that is acting on the orbiting electron. The induced electric field produces a torque on the electron, which gives the electron extra angular momentum. This extra angular momentum produces a magnetic moment, the sign of which is negative (i.e., antiparallel to the induced or applied magnetic field). Therefore, diamagnetic materials are repelled in magnetic fields (they have weak, negative magnetic susceptibility).

4.3 Paramagnetic Materials

Paramagnetic materials are those in which individual atoms, ions, or molecules have some number of uncompensated spins and thus a permanent net spin magnetic moment. As we have stated before, the spin moment is much larger than the orbital moment, so we would therefore expect that the behavior of paramagnetic materials when placed in a magnetic field will be governed by the behavior of the spin magnetic moments. This is indeed the case.

When paramagnetic substances are placed in an external magnetic field, the uncompensated spin moments tend to align, to some degree, parallel to the applied field direction (Figure 4.1). The magnetic energies involved in this alignment are relatively

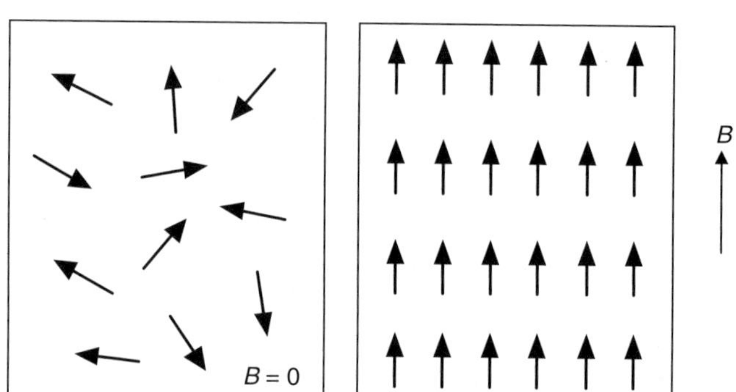

FIGURE 4.1
Orientation of spins in a paramagnet in the absence of a magnetic field ($B = 0$; left) and in the presence of a strong magnetic field (right)

small, and the energy associated with thermal agitation tends to work against the alignment, having a randomizing effect. The degree of alignment of the uncompensated spins with the applied magnetic field depends, therefore, on the strength of the field (the stronger the field, the greater the degree of alignment up to very high fields) and the temperature (the hotter the material, the lower the degree of alignment in the same applied field).

Since the spin moments in paramagnetic materials align with the applied field in this classical model, they add to it, so that the net effect is that these materials are attracted to a magnetic field (and they have a positive magnetic susceptibility). The linear temperature dependence of the magnetic susceptibility in paramagnetic materials was worked out by Pierre Curie and is known as Curie's law:

$$M/H = \chi = C/T$$

where M is the magnetization, H is the applied magnetic field, χ is the magnetic susceptibility, T is the temperature, and C is the Curie constant and is related to the magnetic properties of the material.

In paramagnetic materials, the individual dipole spin moments of the ions may be thought of as noninteracting (in other words, the magnetic moment of one atom has no effect on its neighboring atoms in the material). Because of the noninteraction of the magnetic moments, this fairly weak effect (paramagnetism) is lost upon removal of the external field. Therefore, when a paramagnetic material is not in an external magnetic field, the net magnetic moment in the material is zero because of the randomizing effects of thermal agitation.

4.4 Ferromagnetic Materials

As in paramagnets, in ferromagnetic materials, there are also uncompensated spins; however, these spins are coupled, giving rise to strong magnetic effects. Ferromagnetism may be thought of as a "group phenomenon," where groups of spin moments act in concert, whereas paramagnetism may be thought of as an "individual phenomenon," where the moment of one atom has little or no effect on the moment of neighboring atoms.

In ferromagnetic materials, as in paramagnets, the magnetic susceptibility is positive, and these materials acquire a positive magnetization when placed in an applied field because of alignment of the spin moments in the material with the field. Unlike paramagnets, however, the net magnetization is not lost upon removal of the field (as long as the material is below a certain temperature, which will be discussed in a moment), and the induced moment in the material may be very strong (Figure 4.2). This is to say that ferromagnetic materials exhibit "hysteresis."

The mechanism responsible for coupling of the spin magnetic moments in neighboring atoms in a material is due to quantum mechanical phenomena and is governed by the Pauli exclusion principle. The uncompensated spins in individual atoms of a ferromagnetic material may couple either directly (direct exchange) or through an intermediate anion—usually oxygen (superexchange). In ferromagnetic materials, this gives rise to a net magnetic moment because of the coupling of spins in a preferred orientation. Keep in mind that this coupling is quantum mechanical in nature and not purely due to magnetic forces acting between uncompensated spins in neighboring atoms.

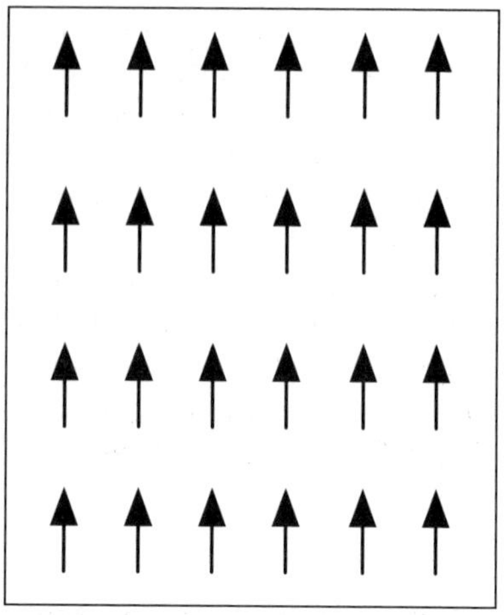

FIGURE 4.2
In ferromagnets, all spins are coupled and aligned, even in the absence of an applied field. This is the origin of remanent magnetization (i.e., hysteresis) in materials.

There are also special cases of ferromagnetism in which neighboring spins are coupled, but not necessarily in the same direction. We will examine two of these cases, antiferromagnetism and ferrimagnetism, as they are important to biological materials, though there are others.

4.4.1 Antiferromagnetism

For some ferromagnetic materials, the exchange coupling between neighboring lattice elements is such that the spins are aligned opposite to each other. This is called antiferromagnetism, and the exchange coupling arises from super exchange according to the Pauli exclusion principle and Hund's rule (Figure 4.3).

In this case the spin moments will still align themselves to an external applied field, only some will be parallel to the applied field and those exchange coupled to them will be antiparallel. This would normally give rise to a material with no net magnetic moment if for every spin up, it was coupled to a spin down. This is not, however, always the case. In some materials, there is a canted antiferromagnetic spin structure (Figure 4.3) or lattice defects and frustrated surface spins (in very fine particles), which can give rise to a net moment in the absence of an applied field.

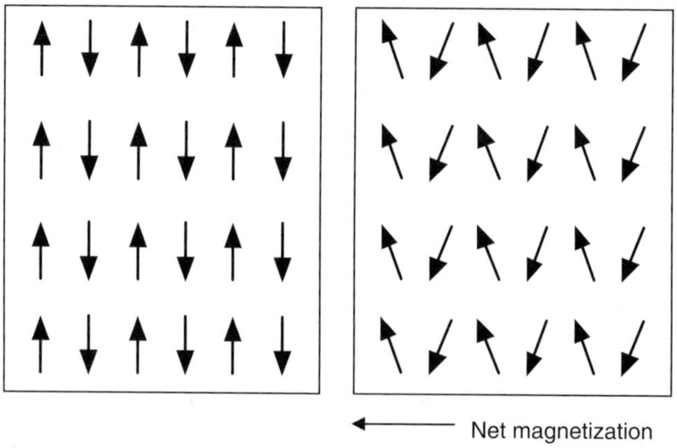

FIGURE 4.3
Pure antiferromagnetic behavior in which spins are coupled antiparallel to each other (left) and an antiferromagnet with a canted spin structure that gives rise to a net magnetization even in the absence of an applied field (right).

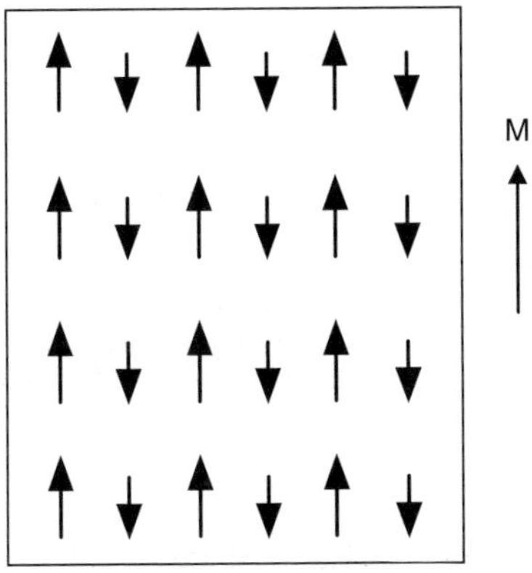

FIGURE 4.4
Electron spin configuration for a ferrimagnet. M = magnetization.

4.4.2 Ferrimagnetism

In addition to antiferromagnetic materials, it is also possible for the neighboring lattice subunits to have unequal numbers of uncompensated electrons coupled antiparallel to each other. This is the case for magnetite (Fe_3O_4), which contains both Fe^{2+} and Fe^{3+} in its lattice structure. The unequal distribution of the two neighboring iron ions gives rise to a net moment (again, even in the absence of an applied field) since one sublattice will have a magnetic moment of greater magnitude than the other, as shown in Figure 4.4. This type of material is called ferrimagnetic.

4.4.3 Temperature Dependence of Ferromagnetism

Since ferromagnetism results from the interaction of atomic moments in materials, there is an exchange energy associated with coupling of the spin moments. At room temperature, this exchange energy is much greater than the energy due to randomizing thermal effects (kT). If thermal energy exceeds the spin coupling (exchange) energy, the coupling breaks down, and the material behaves as a paramagnet. This temperature is dependent on the material and is called the Curie temperature (or, in the case of antiferromagnetic materials, the Néel temperature) (Figure 4.5).

Finally, materials that are superparamagnetic are generally very small (on the order of nanometers), and the electron spins may be coupled either parallel (ferromagnet) or

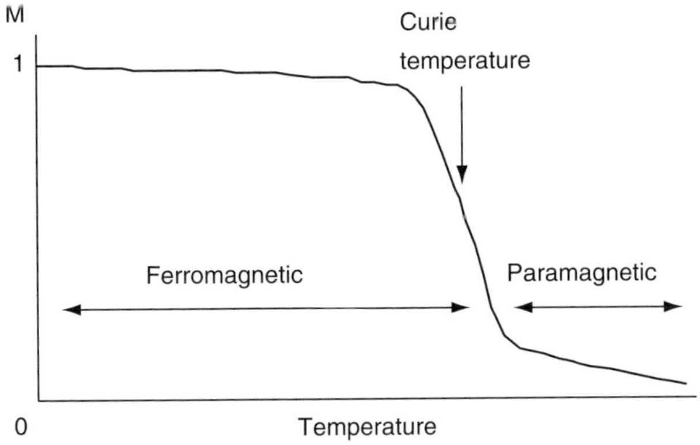

FIGURE 4.5
Magnetization as a function of temperature for a ferromagnet. Above the Curie temperature, spin coupling breaks down, and the material behaves as a paramagnet.

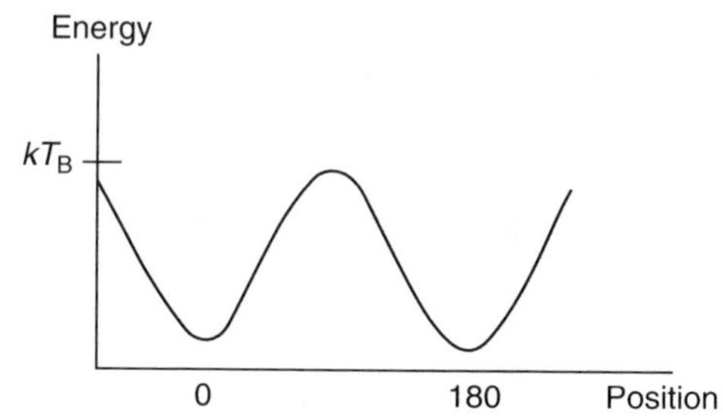

FIGURE 4.6
Schematic of the energy barriers in a superparamagnet. If thermal energy is above a given temperature—the blocking temperature (T_B)—the spins will be able to very rapidly flip between 0° and 180° along the easy axis of magnetization. If the particle becomes larger or the temperature is lowered, the height of the energy barrier constrains the orientation of the electron spins to one or the other of the easy axis energy wells.

antiparallel (ferri- or antiferromagnet). In the case of these materials, however, thermal considerations are dominant. Superparamagnetic materials are named as such because thermal energy causes them to behave—even though this is a special class of ferromagnetism—like a paramagnet. The difference is that, because of coupling of the spin moments, thermal energy causes the spins to flip rapidly as a group rather than individually, as is the case for paramagnetism. On a macroscopic level, the behavior of the two is very similar, except at low temperatures when the thermal energy is sufficiently reduced.

If the superparamagnetic particles are placed in an applied field, the energy of the field will cause the spins to align themselves parallel to it, just as with a normal ferromagnet or paramagnet (the degree of alignment depending on the strength of the field and the temperature—as with paramagnets). If the field is taken away, thermal fluctuation causes a "relaxation" of the spins, and they begin to flip rapidly between parallel and antiparallel orientations along the easy axis of magnetization (determined by magnetocrystalline or shape anisotropy). This has the effect of causing the material to appear paramagnetic when the magnetization is examined on timescales longer than the flipping frequency (generally $\sim 10^{-9}$ s). The remanent magnetization (which is the magnetization left in the material after removal of the field) decays with time according to the equation:

$$\tau = f_0 \exp\left(\frac{\nu M_s H_c}{2kT}\right)$$

where τ = relaxation time, f_0 = lattice vibration frequency, ν = grain volume, M_s = spontaneous magnetization, H_c = coercivity, k = Boltzmann's constant, and T = temperature.

The relaxation time depends on the height of the energy barrier—which is a function of grain volume and the magnetic properties of the material—and the amount of thermal energy (kT) required to overcome it (Figure 4.6). Using this equation, it is possible to calculate grain volumes above which the magnetization becomes stable or "blocked" (i.e., decays over very long periods of time at a given temperature).

4.5 Biological Magnets

For most biological materials, the magnetic permeability is close to that of free space (i.e., diamagnetic), which implies that there is no direct interaction with the magnetic component of electromagnetic fields at low field strengths. However, this is not the case for all biological materials.

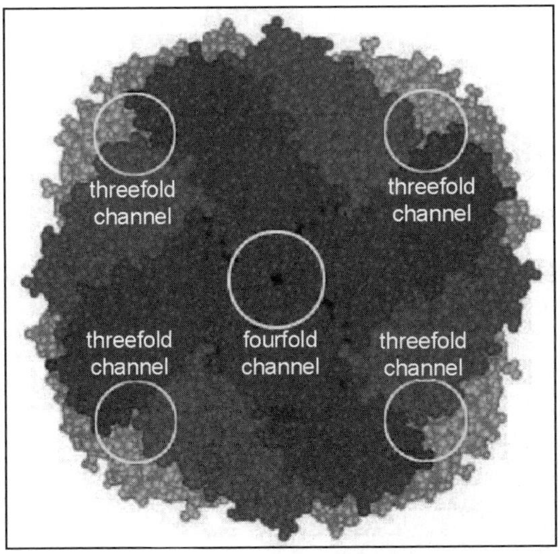

FIGURE 4.7 (See color insert following page 380.) Model of the ferritin protein showing the peptide subunits and iron transport channels. (From www.chemistry.wustl.edu/~edudev/LabTutorials/Ferritin/FerritinTutorial.html. With permission.)

Most magnetic (paramagnetic and ferromagnetic) materials in organisms are compounds of iron—and in particular, iron oxides. Virtually all organisms require iron to function normally. This is mainly due to its redox activity, which allows it to play an important role in energetic biochemical reactions. In organisms, iron is stored as the mineral ferrihydrite ($5Fe_2O_3 \cdot 9H_2O$) within the iron storage protein ferritin. It consists of a 12-nm hollow spherical protein shell made up of 24 subunits (Figure 4.7). The core of ferritin protein is 8 nm in diameter, and it can hold up to 4500 iron atoms in the form of ferrihydrite. Iron is transported into and out of the core through three- and fourfold channels in the shell. During transport, highly toxic Fe(II) is oxidized to Fe(III) for storage as ferrihydrite (Harrison and Arosio, 1996). The specific iron biochemistry of ferritin is complex and is not completely understood (e.g., Yang et al., 1998; Zhao et al., 2001).

Ferrihydrite is a superparamagnetic antiferromagnet at body temperature, and as such, its magnetic properties are potentially important for understanding the environmental consequences of electromagnetic field exposure, including exposure to strong fields within magnetic resonance imaging (MRI) scanners. In fact, the development of pulse sequences for MRI of ferritin is leading to novel ways of assessing iron concentrations in the liver and examining iron associated with neurodegenerative disorders such as Alzheimer's and Parkinson's diseases (Clarke and St. Pierre, 2000; Bartzokis et al., 2004; St. Pierre et al., 2004, 2005). These techniques rely on the magnetic fields generated by ferritin to produce contrast in ferrihydrite-rich tissue in much the same way as synthetic superparamagnetic iron oxide contrast agents do (for a review, see Pankhurst et al., 2003).

In addition to ferrihydrite in ferritin, in 1992 Joseph Kirschvink's group at the California Institute of Technology discovered that biogenic magnetite is produced in the human brain (Kirschvink et al., 1992). Later work demonstrated that this magnetic iron biomineral is present in several organs in the human body, including the heart, liver, and spleen (Schultheiss-Grassi et al., 1997).

Magnetite (Fe_3O_4)—a ferromagnetic iron oxide—is also known as lodestone and is a mineral more commonly associated with sedimentary rocks and volcanics. However, it is also found in many organisms. Probably, the most well-known example is the magnetotactic bacterium. These bacteria use chains of single-domain, biogenic magnetite arranged in chains in order to sense the geomagnetic field and use it for navigation—much like a

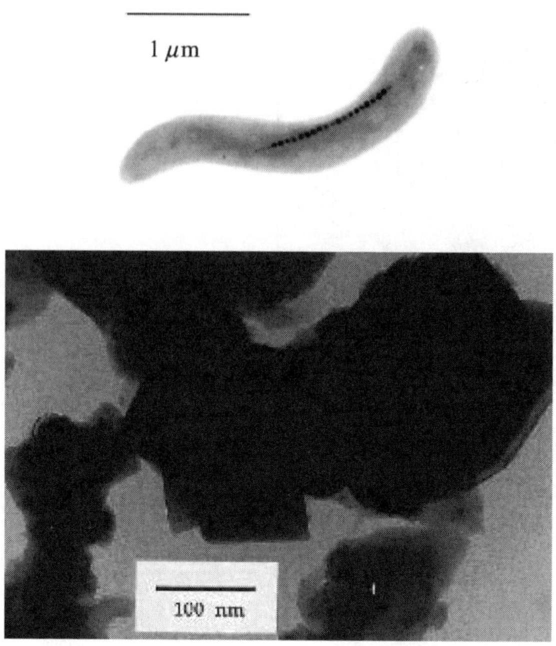

FIGURE 4.8
Transmission electron micrograph of the magnetotactic bacterium MS-1 (top). (From www.calpoly.edu/~rfrankel/mtbphoto.html) Biogenic magnetite extracted from the human hippocampus (bottom). (From Schultheiss-Grassi, PP, R Wessiken, and J Dobson (1999) *Biochim. Biophys. Acta* 1426: 212–216. With permission.)

compass needle (Blakemore, 1975) (Figure 4.8). Although the mechanism by which these organisms produce perfect magnetite crystals is not well understood, it has recently been shown that the process appears to be mediated by specific proteins (Arakaki et al., 2003). Since Blakemore's early discovery, magnetite has been found in a wide variety of animals from bacteria to humans, and in some cases, as with magnetotactic bacteria, it appears to be used for navigation (e.g., Walker and Bitterman, 1989; Walker et al., 1997; Wiltschko and Wiltschko, 2002; for a review, see Kirschvink and Hagadorn, 2000).

Although by 1992, magnetite was known to occur in many organisms, the discovery of biogenic magnetite in the human brain proved controversial. Further work in this area confirmed and extended Krischvink's results, examining not only the occurrence of magnetite in the brain but also its potential role in neurophysiological processes (Dunn et al., 1995; Dobson and Grassi, 1996; Schultheiss-Grassi and Dobson, 1999; Dobson 2001, 2002, 2004; Hautot et al., 2003; Collingwood et al., 2005) (Figure 4.8).

One area of research where the presence of biogenic magnetite in the brain may provide answers to some controversial questions is in the examination of potential mechanisms for the interaction of environmental electromagnetic fields with humans. Two theoretical models have been proposed to demonstrate how biogenic magnetite could act as a transducer of both low-frequency magnetic fields and radio frequency (RF) fields emitted from mobile phones and base stations (Kirschvink, 1992, 1996; Dobson and St. Pierre, 1996). These models rely on the fact that magnetite will couple strongly to the magnetic fields from electrical devices and, either through ferromagnetic resonance effects or through mechanical effects on membrane ion channels, can disrupt the normal functioning of cells in the brain.

Early tests of these models indicate that low-frequency, pulsed magnetic fields from mobile phones may have an influence on cellular activity (including cell death), whereas effects due to ferromagnetic resonance in RF fields are less clear (Cranfield et al., 2003a,b). In addition, magnetite contamination and the presence of biogenic magnetite in model organisms have also been highlighted as potential confounders that could influence the results of studies on the effects of environmental magnetic fields on organisms (Kobayashi and Kirschvink, 1995; Cranfield et al., 2004).

Although magnetite and ferrihydrite are two of the most ubiquitous magnetic materials in organisms, they are not the only ones. Greigite (Fe_7S_8) is a ferrimagnetic iron sulfide found in some iron-reducing bacteria (Posfai et al., 1998). It has a strong magnetic moment similar to magnetite and is thought to be produced as a by-product of iron reduction. Hematite (Fe_2O_3) and wüstite-like (FeO) iron phases also have recently been found within human ferritin (Quintana et al., 2004). And hemosiderin (FeOOH) is a goethite-like iron oxyhydroxide that is antiferromagnetic and is found primarily in pathogenic liver tissue (St. Pierre et al., 1998). In addition to these ferromagnetic materials, other ions and iron compounds, such as hemoglobin, are paramagnetic.

4.6 Magnetic Iron Compounds Related to Pathogenesis

Though iron is an important component of normally functioning organisms, in some cases, it can also be toxic. Because of its redox potential, Fe(II) generally has the potential to do more damage than oxidized Fe(III). For this reason, iron is primarily stored as Fe(III) within ferritin in organisms.

Disruption of the body's normal mechanisms for iron storage can lead to iron overload diseases such as hemochromatosis and β-thalassemia and the deposition of significant amounts of magnetic iron compounds (St. Pierre et al., 1998; Chua-anusorn et al., 2000). These diseases result in the formation of significant iron deposits, predominantly in the liver, which consist mainly of the iron biomineral hemosiderin (FeOOH). Hemosiderin is antiferromagnetic, but can occur as large particles in the body. As with ferrihydrite, both these antiferromagnets contain only Fe(III) and as such have a relatively weak magnetic moment primarily due to lattice defects and, for very fine particles, to frustrated spins on the particle's surface. As mentioned earlier, the presence of these magnetic iron compounds is being exploited for the development of noninvasive MRI monitoring of liver iron in order to track the effectiveness of chelating compounds that are used to treat these diseases (Clarke and St. Pierre, 2000; St. Pierre et al., 2004, 2005). In the brain, disruption of normal iron metabolism results from either disease or trauma and can often lead to pathogenesis (Beard et al., 1993). The accumulation of excess iron is known to induce epileptic activity, primarily as a result of intracranial bleeding due to head trauma. The model of iron-induced epilepsy was first introduced by Willmore et al. (1978), and the electrophysiological responses have been characterized in many studies since then (e.g., Ueda and Willmore, 2000; Engstrom et al., 2001). More recently, evidence shows that iron overload diseases may even predispose a person to epilepsy (Ikeda, 2001).

Although the initial seizure response to trauma and excess iron is relatively swift (Ueda et al., 1998), the long-term consequences of intracranial bleeding on the formation of various magnetic iron compounds in the brain are unclear. Little is known about the formation of iron compounds in the brain of epileptic patients, other than that there appears to be a relationship with an increase in iron (e.g., Willmore et al., 1978; Ueda and

Willmore, 2000; Engstrom et al., 2001). This increase may lead to increased iron loading in the ferritin core or the sequestration of free iron in another form, which could have significant consequences for disease progression. A preliminary correlation between magnetite particle packing geometry and epileptogenic tissue has been demonstrated; however, there is as yet no indication of an increase in magnetite biomineralization associated with temporal lobe epilepsy (Schultheiss-Grassi and Dobson, 1999).

The association of abnormal accumulations of iron with neurodegenerative disorders such as Alzheimer's, Parkinson's, and Huntington's diseases has been known for over 50 y (Goodman, 1953). Although this relationship has been studied extensively, it is not yet clear which iron compounds are present and what their role in these diseases is; however, evidence is mounting that there are likely synergistic mechanisms related to iron and amyloid-β (Aβ, the principal component of Alzheimer's plaques) and other disease-related proteins (e.g., Sayre et al., 2000; Rottkamp et al., 2001; Atamna and Frey, 2004]. Recent work seems to confirm that iron plays a role in disease progression as iron chelators appear to have neuroprotective effects in rat models of Parkinson's disease (Ben Shachar et al., 2004).

Various forms of iron may play a significant role in the biochemical processes that lead to the progression of neurodegenerative diseases. Although there is much speculation on that role, the primary mechanism is thought to be the result of oxidative stress—the generation of free radicals via the Fenton reaction (Connor and Menzies, 1995; Markesbery, 1997; Koppenol, 2001). Recently, Floor (2000) demonstrated the connection between high levels of iron in the basal ganglia and oxidative stress in Parkinson's patients. A disruption of iron metabolism and increased iron in the same region of the brain also has been implicated in Alzheimer's and Huntington's diseases (Bartzokis and Tishler, 2000). Furthermore, iron accumulation has been associated with microgliosis and correlated with increased damage to the CA1 region of the hippocampus via iron–zinc interactions in models of neurodegenerative diseases (Shoham and Youdim, 2000).

It should be noted, however, that some results have shown that the presence of oxidized nucleosides in neurons does not appear to be related to senile plaque material or neurofibrillary tangles in Alzheimer's (Nunomura et al., 1999). In some cases, free radical damage has even been reported to be reduced by Aβ deposition because of the inhibitory role of Aβ-related Zn^{2+} in H_2O_2-mediated toxicity (Cuajungco et al., 2000; Nunomura et al., 2000). More recently, it has been noted that in transgenic mouse models of Alzheimer's, neurogenesis was related to increased deposition of Aβ (Jankowsky et al., 2003). It has been postulated that this may be due to a possible neuroprotective role of Aβ in binding excess redox active iron, which is deposited because of the increased need for iron during neurogenesis (Dobson and Batich, 2004).

Recently, state-of-the-art superconducting quantum interference device (SQUID) magnetometry, nuclear forward scattering, and synchrotron x-ray fluorescence imaging have been employed to evaluate which specific iron compounds are present in these diseases and to map them to structures in the tissue (Hautot et al., 2003; Collingwood et al., 2005; Mikhailova et al., 2005). Early results of these studies indicate that several magnetic iron compounds are associated with these diseases—ferrihydrite (from ferritin), biogenic magnetite, α-iron, and hemosiderin. These results represent a major step toward our understanding of the origin and role of magnetic iron compounds in these diseases.

The role of the various magnetic materials in many organisms, particularly in humans, is only just beginning to be unraveled. After more than two decades of research on magnetic biominerals, it is clear that in many cases they play an important role in biological processes and may provide us with insights into mechanisms of interaction of environmental electromagnetic fields and organisms.

References

Arakaki, A, J Webb, and T Matsunaga (2003) A novel protein tightly bound to bacterial magnetite particles in *Magnetospirillum magneticum* strain AMB-1. *J. Biol. Chem.* 278: 8745–8750.

Atamna, H and WH Frey (2004) A role for heme in Alzheimer's disease: heme binds amyloid beta and has altered metabolism. *Proc. Natl. Acad. Sci. USA* 101: 11153–11158.

Bartzokis, G and TA Tishler (2000) MRI evaluation of basal ganglia ferritin iron and neurotoxicity in Alzheimer's and Huntington's disease. *Cell. Mol. Biol.* 46: 821–834.

Bartzokis, G, TA Tishler, IS Shin, PH Lu, and JL Cummings (2004) Brain ferritin iron as a risk factor for age at onset in neurodegenerative diseases. *Ann. NY Acad. Sci.* 1012: 224–236.

Beard, JL, JR Connor, and BC Jones (1993) Iron in the brain. *Nutr. Rev.* 51: 157–170.

Ben Shachar, D, N Kahana, V Kampel, A Warshawsky, and MBH Youdim (2004) Neuroprotection by a novel brain permeable iron chelator, VK-28, against 6-hydroxydopamine lession in rats. *Neuropharmacology* 46: 254–263.

Blakemore, RP (1975) Magnetotactic bacteria. *Science* 190: 377–379.

Chua-anusorn, W, KC Tran, J Webb, DJ Macey, and TG St. Pierre (2000) Chemical speciation of iron deposits in thalassemic heart tissue. *Inorg. Chim. Acta* 300–302: 932–936.

Clarke, P and TG St. Pierre (2000) Quantitative 1/T2 mapping of hepatic iron overload: a single spin echo imaging methodology. *Magn. Reson. Imaging* 18: 431–438.

Collingwood, JF, A Mikhailova, M Davidson, C Batich, WJ Streit, J Terry, and J Dobson (2005) *In-situ* characterization and mapping of iron compounds in Alzheimer's tissue. *J. Alzheimers Dis.* In press.

Connor, RJ and SL Menzies (1995) Cellular management of iron in the brain. *J. Neurol. Sci.* 134: 33–44.

Cranfield, CG, HG Wieser, JM Al Madan, and J Dobson (2003a) Evaluation of ferromagnetic transduction mechanisms for mobile phone bioeffects. *IEEE Trans. NanoBiosci.* 2: 40–43.

Cranfield, CG, HG Wieser, and J Dobson (2003b) Exposure of magnetic bacteria to simulated mobile phone type RF radiation has no impact on mortality. *IEEE Trans. NanoBiosci.* 2: 146–149.

Cranfield, CG, A Dawe, V Karloukovski, RE Dunin-Borkowski, D de Pomerai, and J Dobson (2004) Biogenic magnetite in the nematode *Caenorhabditis elegans*. *Proc. R. Soc. B (Suppl.) Biol. Lett.* 271: S436–S439.

Cuajungco, MP, LE Goldstein, A Nunomura, MA Smith, JT Lim, CS Atwood, X Huang, YW Farrag, G Perry, and AI Bush (2000) Evidence that the β-amyloid plaques of Alzheimer's disease represent the redox-silencing and entombment of Aβ by zinc. *J. Biol. Chem.* 275: 19439–19442.

Dobson, J (2001) Nanoscale biogenic iron oxides and neurodegenerative disease. *FEBS Lett.* 496: 1–5.

Dobson, J (2002) Investigation of age-related variations in biogenic magnetite levels in the human hippocampus. *Exp. Brain Res.* 144: 122–126.

Dobson, J (2004) Magnetic iron compounds in neurological disorders. *Ann. NY Acad. Sci.* 1012: 183–192.

Dobson, J and C Batich (2004) A potential iron-based mechanism for enhanced deposition of amyoid plaques due to cognitive stimulation in Alzheimer's disease. *J. Neuropathol. Exp. Neurol.* 63: 674–675.

Dobson, J and P Grassi (1996) Magnetic properties of human hippocampal tissue: evaluation of artefact and contamination sources. *Brain Res. Bull.* 39: 255–259.

Dobson, J and TG St. Pierre (1996) Application of the ferromagnetic transduction model to D.C. and pulsed magnetic fields: effects on epileptogenic tissue and implications for cellular phone safety. *Biochem. Biophys. Res. Commun.* 227: 718–723.

Dunn, JR, M Fuller, J Zoeger, J Dobson, F Heller, E Caine, and BM Moskowitz (1995) Magnetic material in the human hippocampus. *Brain Res. Bull.* 36: 149–153.

Engstrom, ER, L Hillered, R Flink, L Kihlstrom, C Lindquist, JX Nie, Y Olsson, and HC Silander (2001) Extracellular amino acid levels measured with intracerebral microdialysis in the model of posttraumatic epilepsy induced by intracortical iron injection. *Epilepsy Res.* 43: 135–144.

Floor, E (2000) Iron as a vulnerability factor in nigrostriatal degeneration in aging and Parkinson's disease. *Cell. Mol. Biol.* 46: 709–720.

Goodman, L (1953) Alzheimer's disease—a clinicopathologic analysis of 23 cases with a theory on pathogenesis. *J. Nerv. Ment. Dis.* 118: 97–130.

Harrison, PM and P Arosio (1996) The ferritins: molecular properties, iron storage function and cellular regulation. *Biochim. Biophys. Acta* 1275: 161–203.

Hautot, D, QA Pankhurst, N Khan, and J Dobson (2003) Preliminary evaluation of nanoscale biogenic magnetite and Alzheimer's disease. *Proc. R. Soc. Lond. B: Biol. Lett.* 270: S62–S64.

Ikeda, M (2001) Iron overload without the C282Y mutation in patients with epilepsy. *J. Neurol. Neurosurg. Psychiatry* 70: 551–553.

Jankowsky, JL, G Xu, D Fromholt, V Gonzales, and DR Borchelt (2003) Environmental enrichment exacerbates amyloid plaque formation in a transgenic mouse model of Alzheimer disease. *J. Neuropathol. Exp. Neurol.* 16: 1220–1227.

Kirschvink, JL (1992) Comments on "Constraints on biological effects of weak extremely-low-frequency electromagnetic fields." *Phys. Rev. A* 46: 2178–2184.

Kirschvink, JL (1996) Microwave absorption by magnetite: a possible mechanism for coupling non-thermal levels of radiation to biological systems. *Bioelectromagnetics* 17: 187–194.

Kirschvink JL and JW Hagadorn (2000) A grand unified theory of biomineralization. In Bäuerlein, E (ed.) *The Biomineralisation of Nano- and Micro-Structures.* Wiley-VCH Verlag GmbH, Weinheim, Germany, pp. 139–149.

Kirschvink, JL, A Kobayashi-Kirschvink, and BJ Woodford (1992) Magnetite biomineralization in the human brain. *Proc. Natl. Acad. Sci., USA* 89: 7683–7687.

Kobayashi, AK, JL Kirschvink, and MH Nesson (1995) Ferromagnetism and EMFs. *Nature* 374: 123.

Koppenol, WH (2001) The Haber–Weiss cycle 70 years later. *Redox Rep.* 6: 229–234.

Markesbery, WR (1997) Oxidative stress hypothesis in Alzheimer's disease. *Free Radic. Biol. Med.* 23: 134–147.

Mikhailova, A, M Davidson, JET Channel, Y Guyodo, C Batich, and J Dobson (2005) Detection, identification and mapping of iron anomalies in brain tissue using x-ray absorption spectroscopy. *J. R. Soc. Interface* 2: 33–37.

Nunomura, A, G Perry, MA Pappolla, R Wade, K Hirai, S Chiba, and MA Smith (1999) RNA oxidation is a prominent feature of vulnerable neurons in Alzheimer's disease. *J. Neurosci.* 19: 1959–1964.

Nunomura, A, G Perry, MA Pappolla, RP Friedland, K Hirai, S Chiba, and MA Smith (2000) Neuronal oxidative stress precedes amyloid-beta deposition in Down's syndrome. *J Neuropathol. Exp. Neurol.* 59: 1011–1017.

Pankhurst, QA, J Connoly, SK Jones, and J Dobson (2003) Applications of magnetic nanoparticles in biomedicine. *J. Phys. D* 36: R167–R181.

Posfai, M, PR Buseck, DA Bazylinski, and RB Frankel (1998) Reaction sequence of iron sulfide minerals in bacteria and their use as biomarkers. *Science* 280: 880–883.

Quintana, C, JM Cowley, and C Marhic (2004) Electron diffraction and high-resolution electron microscopy studies of the structure and composition of physiological and pathological ferritin. *J Struct. Biol.* 147: 166–78.

Rottkamp, CA, AK Raina, X Zhu, E Gaier, AI Bush, CS Atwood, M Chevion, G Perry, and MA Smith (2001) Redox-active iron mediates amyloid-β toxicity. *Free Radic. Biol. Med.* 30: 447–450.

Sayre, LM, G Perry, PLR Harris, YH Liu, KH Schubert, and MA Smith (2000) *In situ* oxidative catalysis by neurofibrillary tangles and senile plaques in Alzheimer's disease: a central role for bound transition metals. *J. Neurochem.* 74: 270.

Schultheiss-Grassi, PP and J Dobson (1999) Magnetic analysis of human brain tissue. *BioMetals* 12: 67–72.

Schultheiss-Grassi, PP, F Heller, and J Dobson (1997) Analysis of magnetic material in the human heart, spleen and liver. *BioMetals* 10: 351–355.

Schultheiss-Grassi, PP, R Wessiken, and J Dobson (1999) TEM observation of biogenic magnetite extracted from the human hippocampus. *Biochim. Biophys. Acta* 1426: 212–216.

Shoham, S and MBH Youdim (2000) Iron involvement in neuronal damage and microgliosis in models of neurodegenerative disease. *Cell. Mol. Biol.* 46: 743–750.

St. Pierre, TG, W Chua-anusorn, J Webb, DJ Macey, and P Pootrakul (1998) The form of iron oxide deposits in thalassemic tissues varies between different groups of patients: a comparison between Thai β-thalassemia/hemoglobin E patients and Australian β-thalassemia patients. *Biochim. Biophys. Acta* 1407: 51–60.

St. Pierre, TG, PR Clark, and W Chua-anusorn (2004) Single spin-echo proton transverse relaxometry of iron loaded liver. *NMR Biomed.* 17: 446–458.

St. Pierre, TG, PR Clark, W Chua-anusorn, AJ Fleming, GP Jeffrey, JK Olynyk, P Pootrakul, E Robins, and R Lindeman (2005) Non-invasive measurement and imaging of liver iron concentrations using proton magnetic resonance. *Blood* 105: 855–861.

Ueda, Y and LJ Willmore (2000) Sequential changes in glutamate transporter protein levels during Fe^{3+}-induced epileptogenesis. *Epilepsy Res.* 39: 201–209.

Ueda, Y, LJ Willmore, and WJ Triggs (1998) Amygdalar injection of $FeCl_3$ causes spontaneous recurrent seizures. *Exp. Neurol.* 153: 123–137.

Walker, MM and ME Bitterman (1989) Honeybees can be trained to respond to very small changes in geomagnetic field intensity. *J. Exp. Biol.* 145: 489–494.

Walker, MM, CE Diebel, CV Haugh, PM Pankhurst, JC Montgomery, and CR Green (1997) Structure and function of the vertebrate magnetic sense. *Nature* 390: 371–376.

Willmore, LJ, GW Sypert, and JB Munson (1978) Recurrent seizures induced by cortical iron injection—model of post-traumatic epilepsy. *Ann. Neurol.* 4: 329–336.

Wiltschko W and R Wiltschko (2002) Magnetic compass orientation in birds and its physiological basis. *Naturwissenschaften* 89: 445–452.

Yang, X, Y Chen-Barrett, P Arosio, and ND Chasteen (1998) Reaction paths of iron oxidation and hydrolysis in horse spleen and recombinant human ferritins. *Biochemistry* 37: 9743–9750.

Zhao, G, F Bou-Abdallah, X Yang, P Arosio, and ND Chasteen (2001) Is hydrogen peroxide produced during iron(II) oxidation in mammalian apoferritins? *Biochemistry* 40: 10832–10838.

5

Interaction of Direct Current and Extremely Low-Frequency Electric Fields with Biological Materials and Systems

Frank S. Barnes

CONTENTS

5.1 Introduction

The fact that electrical currents can affect the behavior of biological systems has been known for more than 2000 years. Electric shocks have been used to treat a wide variety of ailments since the eighteenth century. However, our knowledge of how these fields and the resulting currents influence biological systems is surprisingly incomplete. Electrical signals are clearly important in the control of biological processes and in carrying information from one part of the body to another. Nerve cells propagate electrical signals from sensors of pressure, temperature, light, sound, etc., to the brain and return control signals to muscles and other tissue. Yet, if we choose to stimulate these processes with external electrical inputs, we have a relatively limited understanding of how a given electrical signal will affect various biological organs; what the safe limits of exposure are (particularly overextended periods of time); and how electrical signals are carried across cell membranes, are propagated along nerves, or affect growth processes and cell division.

The purpose of this review is to bring together some of the physical concepts that underlie the interaction between electric fields and biological materials with the objective of providing background for determining safe levels of exposure and new applications for the use of electricity in therapy. This is the first step in a long chain of events that lead from externally applied electric field forces to significant biological changes. An objective of this chapter is to provide a background for some of the other chapters that cover both possible health effects and some therapeutic applications of electric and magnetic fields.

The approach that will be taken is to start with Maxwell's equations and couple them to the bulk electric and magnetic properties of the materials. These equations allow us to calculate the values of the electric and magnetic fields as a function of time and space given the values for the dielectric constant and the conductivity. The techniques for the solution of these equations for a wide variety of biological systems are covered in Chapter 10 in this volume by J. Lin and P. Bemardi and for imaging see Chapter 12 in this volume by W.T. Joines, O.H. Liu and G. Ybarra. From the solutions of these equations for the electric and magnetic fields, the conservation of energy, and charge, expressions can be obtained for the current density and other parameters of interest.

Next, the force equations are used to develop equations for the conductivity and the dielectric constant. This section is organized to begin at the lowest level of complexity by examining some of the forces that are exerted on charged particles in fluids, and it then proceeds to some of the effects of electric fields on chemical reaction rates and membranes. At the next level of complexity, some effects of externally applied electric fields on currents through membranes and membrane nonlinearities are described. Some effects of high-level fields and electroporation are described in Chapter 9 in *BMA* by Weaver and Chizmadzhev [112]. This is followed by a discussion of some long-term adaptive processes and secondary effects of current flow due to heating and by a description of a few effects on whole animals. This information is presented with the objective of specifying the general level or intensity of fields, currents, and temperatures where one can expect to observe a given class of biological responses. Much more data are presented in Chapter 3 in *BMA* by Anderson et al. [113]. The next section contains data on the levels of typical naturally occurring and man-made fields. This section also includes a comparison of some externally applied fields, fundamental noise levels, and signals generated in the body. J. Weaver and M. Bier present a more complete treatment of some of the noise sources in biological cells in Chapter 7 in this volume.

5.2 Maxwell's Equations and the Properties of Materials

Maxwell's equations can be written in both differential and integral form, and solutions to them under a variety of boundary conditions are extensively covered in Chapter 10 in this volume by James C. Lin and Paolo Bernardi. In differential form they are given by [1]

$$\vec{\nabla} \times \vec{H} = \vec{J} + \frac{\partial \vec{d}}{\partial t} \tag{5.1}$$

$$\vec{\nabla} \times \vec{E} = -\frac{\partial \vec{B}}{\partial t} \tag{5.2}$$

$$\vec{d} = \varepsilon_0 \vec{E} + \vec{p} \tag{5.3}$$

$$\vec{B} = \mu_0 \vec{H} + \vec{M}_B \tag{5.4}$$

where $\vec{H}$ is the magnetic field, $\vec{J}$ is the current density, $\vec{d}$ is the displacement vector, t is time, $\vec{E}$ is the electric field, $\vec{\nabla}$ is the partial differential operator, del, $\vec{B}$ is the magnetic flux density, ε_0 is the dielectric constant for free space, $\vec{p}$ is the electrical polarization per unit volume, μ_0 is the magnetic permeability in vacuum, and $\vec{M}_B$ is the magnetic polarization per unit volume. It is often convenient to expand the electrical polarization in a power series:

$$\vec{p} = \vec{p}_0 + \varepsilon_0 X_e \vec{E} + \varepsilon_0 X_2 \vec{E}^2 + \cdots \tag{5.5}$$

where $\vec{p}_0$ is the permanent polarization, X_e is the linear electric susceptibility, and X_2 is the quadratic coefficient. X_2 is significant for nonlinear optics and very large values of the electric field. For most of the exposure conditions described in this volume it is convenient to express the induced polarization in terms of the dielectric constant ε

$$\varepsilon = \varepsilon_0(1 + X_e) \tag{5.6}$$

Note that in anisotropic materials, both ε and X_e may be tensors and complex numbers. The imaginary part of ε is associated with the loss or dissipation of energy by the bound charges. The solutions to Maxwell's equations for $\vec{E}$ and $\vec{B}$ are given for a variety of boundary conditions and under the assumption that values for $\vec{J}$, $\vec{p}$, and $\vec{M}_B$ are known (given in Chapter 10 in this volume by James C. Lin and Paolo Bernardi). These fields, in turn, determine the forces on the components of the biological material.

5.3 Physics of the Interactions of Electric Fields with Biological Materials

Biological systems consist of complex physical subsystems. In an attempt to understand them, we will start at the most elementary level. Perhaps the simplest level—which is already surprisingly complicated—is the effect of electric fields on biological fluids. These fluids contain a large number of components, including ions, polar molecules such as water, proteins, lipids, hormones, and colloidal particles. Current flow in these fluids is given by the sum of the drift and diffusion currents for each component. At low current densities the system is linear; however, at moderate to high current densities nonlinearities are observed [2]. In addition, the fields can change the orientation of molecules with dipole moments, induce dipoles by distorting electron orbits, and change the relative positions of some of the atoms within the molecule. This, in turn, leads to changes in the dielectric constant. The next level of complexity involves the interaction of the fields with membranes that behave like porous solids for fields applied perpendicularly to their surface and like viscous liquids for fields in the plane of the membrane [3,4]. Membranes are inhomogenous so that different portions of them may be affected differently by the perturbing fields. Additionally, membranes are involved in active chemical reactions that change their porosity to various ions selectively. Both electrical potentials and chemical signals may change the membranes' conductivity by orders of magnitude and transmit signals across membranes. The next level of complexity occurs in the interactions between the biological fluids and the membranes in the presence of electric fields. Electric fields affect the selective transport of ions or molecules through the membrane. They change the buildup of charged ion layers at the surface and change the way new molecules are incorporated into the membrane or are bound to its surface. The result of changes in the transport of molecules or ions across cell membranes is changes in the performance of the cells and, in turn, of the organs of which they are a part. They can also lead to changes in the rate of exchange of electrons between molecules in the membrane and ions or molecules in the fluid [5]. For example, a biasing electric voltage across a pacemaker cell in the heart will change its firing rate and thus the pumping rate of the heart.

The fundamental law describing the forces on charged particles is given by

$$\vec{F} = q(\vec{E} + \vec{v} \times \vec{B}) \tag{5.7}$$

where $\vec{F}$ is the force, q is the charge on the particle, and $\vec{v}$ is the velocity of the particle. $\vec{E}$ and $\vec{B}$ are coupled by Maxwell's equations so that a time-varying magnetic field generates an $\vec{E}$ field and vice versa. This force may lead to ion currents and changes in the orientation of dipoles in molecules, and it may also lead to transitions between energy levels, to shifts in their spacing and induced dipole moments, $\vec{P}$. Additionally, if non-linearities or time-varying impedances are present, alternating current (AC) fields can be rectified to produce direct current (DC), frequencies at the second and higher harmonics, and sum and difference frequencies with biological or molecular oscillations. Direct magnetic field effects may occur through the term $\vec{v} \times \vec{B}$. For example, for a Na^+ moving at a thermal velocity of 4×10^2 m/sec in the earth's magnetic field of about 5×10^{-5} T, this term has a magnitude of 2×10^{-2} V/m, and the force is at right angles to the field and the velocity.

In addition to the forces on charged particles, electric fields can induce forces on polarizable atoms, molecules, ions, and molecules with dipole moments. To first order these forces are described by

$$\vec{F}_{\mathrm{d}} = (\vec{P}_0 \cdot \vec{\nabla})\vec{E} \tag{5.8}$$

$$\vec{F}_{\mathrm{L}} = \boldsymbol{\alpha} V(\vec{E} \cdot \vec{\nabla})\vec{E} \tag{5.9}$$

where $\vec{F}_{\mathrm{d}}$ is the force on a molecule with a permanent dipole moment $\vec{P}_0$, and $\vec{\nabla}\vec{E}$ is the gradient of the electric field. $\vec{F}_{\mathrm{L}}$ is the force on a molecule with an induced dipole moment $\vec{P}_{\mathrm{i}} = \boldsymbol{\alpha} V\vec{E}$, where α is the tensor polarizability and V is the volume [6]. Note that biological materials are highly inhomogenous and that there are large electric field gradients at the boundaries between the fluids and membranes. Additionally, for induced dipole moments, when the sign of the dipole reverses with an alternating field the force along the gradient of the field can be in a constant direction.

The current flow $\vec{J}_{\mathrm{i}}$ of a given molecule or ion, in molecules or ions per second per meter square, has both drift and diffusion components that may be given by

$$\vec{J}_{\mathrm{i}} = N_{\mathrm{i}}\mu\vec{F} + qD\vec{\nabla}N_{\mathrm{i}} \tag{5.10}$$

where N_{i} is the ion concentration, μ is the mobility in seconds per kilogram, D is the diffusion constant in meter square per second, and $\vec{\nabla}N_{\mathrm{i}}$ is the gradient of the concentration. For charged particles, the force has two components [7]:

$$\vec{F} = q\vec{E} + (\vec{M} \cdot \vec{\nabla}\vec{E}) \tag{5.11}$$

where $\vec{M}$ is the sum of the permanent and induced dipole moments.

The drift portion of the ion currents takes the form

$$\vec{J} = \sum q_i N_i \mu_i \vec{E} + \sum N_i \mu_i (\vec{M} \cdot \vec{\nabla}) \vec{E} \tag{5.12}$$

The conductivity $\sigma = \sum q_i N_i \mu_i$, and N_i is the concentration of each ion, μ_i is the mobility, and $\vec{M}_i$ is the dipole moment. Table 5.1 shows some typical values of mobility.

The conductivity of biological fluids such as blood, which contains cells, is in the vicinity of $\sigma = 0.6\,\mathrm{S/m}$, while for physiological saline it is approximately $1.4\,\mathrm{S/m}$. For more detailed material on the conductivities and dielectric constants of biological materials, see Chapter 3 in this volume by C. Gabriel. If the fluid channels between cells are relatively thick and the fluids are relatively good conductors, the channels tend to short circuit the voltages that might otherwise appear across membranes that typically have conductivities at least a thousand times smaller.

The total dielectric constant for a material includes the sum of the induced polarizabilities of the components and interaction terms between them. It is sometimes useful to think of the dielectric constant as a way of describing the fraction of the electric field that is shorted out by the bound charges. This can be shown by considering an ideal parallel plate capacitor where the outside plates are separated by a distance l. The capacity for these plates is given by

$$C_0 = \frac{\varepsilon_0 A}{l} \tag{5.13}$$

If an ideal thin metal plate of thickness w is inserted halfway between these plates, then the resulting capacity is given by

$$C = \frac{C_0}{(1 - w/l)} = \frac{\varepsilon A}{l} \tag{5.14}$$

The corresponding dielectric constant is given by

$$\varepsilon = \frac{\varepsilon_0}{(1 - w/l)} \tag{5.15}$$

In this example it is apparent that as the fraction of the field that is shorted out by the metal plate of thickness w increases, so does the effective dielectric constant. The electrons or ions forming an induced dipole moment can be thought of as doing the same thing. If there is a significant time lag for the movement of the charge, the effective value of w is reduced. The very large values of the dielectric constants of some tissues at low-frequencies can be thought of as the resulting motion of ions that are trapped inside highly

TABLE 5.1

Typical Values of Biological Ionic Mobilities

Particles	Mobilities	Ions	Mobilities
Proteins	$\mu = 10^{-10}$ to $10^{-8}\ \mathrm{m^2/V\,sec}$	Ca^{2+}	$\mu = 6.2 \times 10^{-8}\ \mathrm{m^2/V\,sec}$
Na^+	$\mu = 5.2 \times 10^{-8}\ \mathrm{m^2/V\,sec}$	Mg^+	$\mu = 5.4 \times 10^{-8}\ \mathrm{m^2/V\,sec}$
K^+	$\mu = 7.6 \times 10^{-8}\ \mathrm{m^2/V\,sec}$	Cr^+	$\mu = 1.9 \times 10^{-8}\ \mathrm{m^2/V\,sec}$

resistive membranes. At higher frequencies, the ions can no longer move fast enough to fully charge the surfaces of the membranes, and the effective dielectric constant for the tissue decreases. If energy is absorbed in inducing the dipole moments, then the dielectric constant becomes a complex number.

The forces applied by an electric field superimpose a drift velocity on the much larger random thermal velocity in the opposite directions for positively and negatively charged particles. These forces can lead to a redistribution of ions or molecules as a result of the differential mobilities and to an increase in the concentration of ions at interfaces. The average drift velocity $\vec{v}$ for a charged particle is given by

$$\vec{v} = \mu_{\mathrm{i}}\vec{E} \tag{5.16}$$

The separation of molecules as a result of the different velocities in a DC electric field is known as electrophoresis and is frequently used to identify large molecules or charged colloidal particles [8]. The separation of particles in an AC field gradient is known as dielectrophoresis [5].

For a spherical particle in a homogenous insulating fluid the mobility μ_{i} is given by

$$\mu_{\mathrm{i}} = q/6\pi\eta a \tag{5.17}$$

provided that the particle is significantly larger than the background particles of the fluid, where η is the viscosity of the fluid and a is the radius of the particle. In a conducting medium, counterions, or ions with a charge opposite to that of the particle, and molecules with dipole moments are attracted to it. They change the effective radius of the particle and then partially shield its charge. Additionally, small counterions may flow in the direction opposite to the particle motion, exerting a viscous drag. The theory for motion of a rigid sphere through a conducting liquid is complicated if all these effects are taken into account. Often some of the parameters, including the charge on the sphere, are not measurable. However, a relatively simple expression for the electrophoretic mobility is often used:

$$\mu_{\mathrm{i}} = \frac{\varepsilon_{\mathrm{i}}\zeta}{4\pi\eta} \tag{5.18}$$

where ε_{i} and η are the dielectric permittivity and the viscosity of the fluid (in kg/m sec), respectively, and ζ is the electrical potential drop from the particle surface across the bound fluid to the interface where the liquid begins to flow under the shear stress. Stated another way the ''zeta potential,'' ζ, is the potential at the surface boundary between the stationary fluid and the liquid that is moving with the particle. It should be noted that ζ is less than the total potential ψ across the charge double layer surrounding the charged particle. Also, note that water molecules bind to the ions, increase the effective diameter, and reduce the effective charge. This, in turn, makes the mobility less than that which might be expected at first from the atomic size and Stokes' law.

In a uniform AC field a charged particle oscillates about its mean position, and the electrical energy added to the solution is largely converted to heat. If there is a gradient in the field, as is to be expected in biological materials that are highly inhomogenous, then the gradient of the field can lead to a net charge displacement if the fields are large enough to lead to nonlinearities in the mobility or induced dipole moments. For large $\vec{E}$ the velocity saturates and mobility varies as

$$\mu_{\mathrm{i}} = \frac{\mu_0'}{\left|\vec{E}\right|} \tag{5.19}$$

and Equation 5.16 yields

$$|\vec{v}| = \mu_0'$$

Most biological systems are highly inhomogenous, and the induced currents will vary rapidly in space. For the case of an induced dipole moment and an ideal dielectric sphere with a permittivity ε_2 and a conductivity $\sigma_2 = 0$ in an ideal dielectric fluid with a dielectric permittivity ε_1 and a conductivity $\sigma_1 = 0$ and a nonuniform electric field prior to inserting the sphere, the force

$$\vec{F}_{\mathrm{L}} = \boldsymbol{\alpha} V(\vec{E}\cdot\vec{\nabla})\vec{E} = 4\pi a^2 \varepsilon_1 \left\{\frac{\varepsilon_2 - \varepsilon_1}{\varepsilon_2 + 2\varepsilon_1}\right\}\{(\vec{E}_1\cdot\vec{\nabla})\vec{E}_1\} \tag{5.20}$$

where $\vec{E}_1$ is the field in the fluid prior to insertion of the sphere. Written another way

$$\vec{F}_{\mathrm{L}} = (3/2)V\varepsilon_1 \left\{\frac{\varepsilon_2 - \varepsilon_1}{\varepsilon_2 + 2\varepsilon_1}\right\}\vec{\nabla}\left|\vec{E}_1\right|^2 \tag{5.21}$$

If we assume that the viscous drag on a spherical particle is given by Stokes' law, then

$$\vec{F}_{\mathrm{d}} = 6\pi a \eta \vec{v} \tag{5.22}$$

and the mobility μ_{i} is given by

$$\mu_{\mathrm{i}} = (2a^2/3\eta)\varepsilon_1 \left\{\frac{\varepsilon_2 - \varepsilon_1}{\varepsilon_2 + 2\varepsilon_1}\right\}\vec{\nabla}\vec{E}_1 \tag{5.23}$$

Dielectrophoresis may also be used for identifying molecules, and a more general treatment of the forces needs to take into account the conductivity or a complex dielectric constant for both the fluid and the particle [9].

For particles with dipole moments to change their distribution under the influence of an electric field gradient the force $\vec{F}$ must be large enough to overcome other forces. One of these forces that frequently must be overcome is due to osmotic pressure or diffusion. The osmotic pressure can be thought of as the force per unit area arising from diffusion or the random motion of the particles and is given by

$$\Pi = N_{\mathrm{i}} kT \tag{5.24}$$

where k is Boltzmann's constant and T is the absolute temperature [10]. The average differential force on a particle is proportional to the gradient of the osmotic pressure and is given by

$$\vec{F}_{\mathrm{os}} = -\frac{1}{N_{\mathrm{i}}}\vec{\nabla}\Pi = \frac{kT\vec{\nabla}N_{\mathrm{i}}}{N_{\mathrm{i}} - k\vec{\nabla}T} \tag{5.25}$$

If we consider the case of a spherical volume with a radial concentration gradient at constant temperature, the force is given by

$$\vec{F}_{\mathrm{os}} = -kT\frac{\Delta N_{\mathrm{i}}}{N_{\mathrm{i}}}\cdot\frac{r_0}{\Delta r} \tag{5.26}$$

where r_0 is the unit vector, ΔN_i is the incremental change in concentration, and Δr is the incremental change in distance. The maximum change is given by

$$\frac{\Delta N_i}{N_i} = 1 \tag{5.27}$$

when the presence or absence of a particle occurs at a distance $\Delta r = 2a$, where a is the particle radius. In this case, we get the maximum force

$$\left|\vec{F}_{os(max)}\right| = -\frac{kT}{2a} \tag{5.28}$$

To get an idea of the size of these forces, consider a particle of fat with $a = 1\,\mu m$ in water. The maximum osmotic pressure at $T = 300\,K$ is $|\vec{F}_{os(max)}| = 2 \times 10^{-13}\,N$. The dielectric constant for water is approximately $\varepsilon_1 = 80\varepsilon_0$ and for a fat particle, $\varepsilon_2 = 2\varepsilon_0$, where $\varepsilon_0 = 8.854 \times 10^{-12}\,F/m$. To get a dielectric force greater than the maximum osmotic force, we need a value of $\vec{\nabla} \mid \vec{E}_1 \mid^2 > 10^{12}\,V^2/m^3$. This is given approximately by a voltage of 100 V across a 5 mm gap when the $\vec{E}$ field goes from zero to a peak value of $5 \times 10^4\,V/m$ over the same gap. For a particle with a single charge in a uniform field, we would need a field of $E = 1.3 \times 10^4\,V/m$ to get an equal force.

The electric current densities generated by a concentration gradient are given by

$$\vec{J}_d = -qD\vec{\nabla}N_i \tag{5.29}$$

where D is the diffusion constant and is given by

$$D = \upsilon kT \tag{5.30}$$

where $\boldsymbol{\upsilon}$ is the hydrodynamic mobility with the dimensions of velocity/force and D has the dimensions of meter square per second. For rigid spherical particles of radius a, where $a \gg a_{H_2O}$, the Einstein–Stokes equation gives

$$D = \frac{kT}{6\pi\eta a} \tag{5.31}$$

This is only a first-order approximation because D varies slightly with concentration, departure of the molecule from a spherical shape, and other factors. η is the viscosity (in kg/m sec).

It is sometimes of interest to estimate the ratio of the drift to the diffusion current in order to estimate the level of the applied fields or the applied field gradients that lead to biological changes. This ratio is approximately given by [6]

$$\frac{\vec{J}_{i,diffusion}}{\vec{J}_{i,drift}} = \left(\frac{kT}{F_i}\right) \cdot \left(\frac{\Delta N_i}{N_i}\right) \tag{5.32}$$

where $\vec{F}_i$ is the force on the particle due to both the charge on the particle and the gradient of the field on the dipole moment. If we now assume that the maximum change in N_i goes from N_i in the solution to 0 at the membrane surface over a distance of the diameter of the molecule of N_i and that this is the same distance over which the field goes from the field in the fluid to the field in the membrane, then

$$\frac{\vec{J}_{i,\text{diffusion}}}{\vec{J}_{i,\text{drift}}} = \frac{kT}{W_i} \tag{5.33}$$

where W_i is the energy acquired by the particle moving through the field and its gradient. At room temperature the thermal energy $kT \cong 0.026\,\text{eV}$. A voltage drop from an externally applied source across the membrane liquid boundary on the order of 2×10^{-3} V would be required in order to make the drift current significant with respect to the total diffusion current under these assumptions. One way in which smaller drift currents and smaller voltages could be significant is if the ions with low velocities perpendicular to the membrane are the most important in binding to the membrane. Slow molecules stay close to the membrane for longer times, and these molecules are most affected by the applied forces [6].

For different boundary conditions the results will be quite different. If, for example, the boundary was nearly perfectly reflective, then the concentration gradient would be nearly 0, and so would the net diffusion current. Additionally, the gradient in the concentration may occur over a larger distance than the gradient of the electric field. At steady state the concentration at the membrane can be expected to increase until the diffusion current and the drift current balance each other so that the net current is equal to the rate at which the molecules are bound to the membrane or pass through it.

There are four forces that may become important when considering the interaction between two particles in a fluid. These are the osmotic diffusion force, the electrostatic force, the van der Waals force, and the hydration force [9]. These forces may all become important in considering the interaction between particles or bilipid membranes in an aqueous fluid. Electrostatic or coulomb forces between particles of like charge are repulsive. Because the charged particles attract free ions of the opposite sign—which produces a double layer—they are effectively shielded or are screened by the charged ions of the opposite sign when immersed in a conducting fluid. This force decays exponentially or

$$\vec{F}_c = \vec{F}_0 \exp -\frac{r}{\lambda_d} \tag{5.34}$$

where λ_d is known as the Debye screening length [11]

$$\lambda_d = \left[\frac{2q^2 n}{\varepsilon kT}\right]^{1/2} \tag{5.35}$$

where n is the density of the ion species doing the shielding, q is the charge and ε is the dielectric constant of the solution, k is Boltzmann's constant, and T is the absolute temperature. For physiological saline solution of approximately 0.14 M, the Debye length is approximately 0.83 nm [12]. Thus, the electrostatic forces are important only at very short ranges.

For like particles, the forces are repulsive at short distances (0.1 to 0.2 nm) and attractive at longer ranges. These forces may be thought of as being generated by transient electromagnetic fields because of fluctuations that occur as a result of thermal agitation or natural uncertainties in the position and momentum of the electrons and atomic nuclei.

If one thinks of the local transient fluctuations in terms of the underlying contributions from oscillations at all possible frequencies, it can be shown that the strength of the contributions due to the local fluctuations at a given frequency is proportional to the absorption of light at that frequency by the material. For an individual atom these forces fall off very rapidly as $1/r^7$ [7]. However, when they are integrated over the surface of a membrane, which is thick compared to an atomic layer, they are correlated over many atoms, as the wavelengths are large compared to an atomic diameter.

A calculation of these fields has been performed starting with quantum field theory [13]. The size of the forces and the rate at which they decay depend on the distance between the membranes and the difference of the bulk polarizability of the membrane and the aqueous gap in a complex way. All frequencies of the charge fluctuations contribute to the attraction, and each gives rise to a different relationship between energy and the distance of separation. For many simple cases and for a classical explanation of these forces and potential distributions, see Ref. [14].

One case of interest is for two membranes with a distance d_{w} across the aqueous gap between them. The thickness of the membranes is assumed to be large compared to the spacing. The force between these two membranes is approximately given by

$$\vec{F}_{\mathrm{w}} = \frac{H}{6d_{\mathrm{w}}^3} \tag{5.36}$$

where H is the Hamaker coefficient [9]. In a typical situation, the distances over which the van der Waals force is estimated to be important extend out to separations of 10 to 20 nm, which is substantially longer than a Debye length or the rate of falloff for the electrostatic or coulomb forces. The hydration forces are repulsive forces that rise extremely rapidly as the membrane bilayers approach a separation distance of approximately an atomic spacing. Experimentally, these forces can be expressed in the form [9]

$$\vec{F}_{\mathrm{H}} = \vec{F}_{\mathrm{H}_0} \exp -d_{\mathrm{w}}/\Lambda \tag{5.37}$$

where Λ is a scaling constant. In the case of egg phosphomonoesterase bilayers, $\vec{F}_{\mathrm{H}_0} \approx 7 \times 10^{-13}\ \mathrm{N/m^2}$ and $\Lambda = 0.256\,\mathrm{nm}$. This force may be important up to about 2 nm and is assumed to come about as a consequence of the work required to remove water from the hydrophilic surface of the membrane.

Long-range attractive forces have been observed between hydrophobic surfaces [15]. These forces are proportional to the contact area and may be the orders of magnitude larger than the van der Waals forces. They may also have decay lengths up to 25 nm. The magnitude and range of these forces depends on the temperature and the length of the surfactant's chain, and they appear only when the chains are in a fixed ordered state. These forces appear to be generated when the fields emanating from one surface induce a larger polarization in the other surface layer than in the intervening medium. Some relatively complex expressions for these attractive forces have been worked out [14].

These forces are important in self-organizing processes such as protein folding, ligand binding to hydrophobic receptor sites, and transformation of membrane structures. All these forces act over relatively short ranges in typical biological fluids. The ion densities are so large that charge neutrality is maintained everywhere except very close to charged surfaces.

The effect of an electric field, or an electric field gradient in a fluid, is to superimpose a small drift velocity on a relatively large random thermal velocity. For example, if we apply an electric field of 10^3 V/m to a Na^+ ion, we would expect a drift velocity of about 5×10^{-5} m/sec as compared to a thermal velocity of about 4×10^2 m/sec. For a protein, we would expect a drift velocity approximately one tenth of the speed of the Na^+ ion, although at higher fields. This means that if we are to transport proteins or other small charged particles over appreciable distances of a few millimeters, we can expect it to take minutes or longer. In the case of bacteria, we have measured drift velocities of 10^{-6} m/sec at 100 Hz in fields of about 10^4 V/m and gradients of 5×10^6 V/m^2 or about 0.2% of the velocity of the Na ions in the same field [5,16].

This drift velocity may also change chemical reaction rates if the rate is limited by the availability of one of the charged components. Consider the case of a chemical reaction that takes place in a homogenous fluid if the chemical reaction has the form

$$(\mathrm{A}) + (\mathrm{B}) \underset{k_2}{\overset{k_1}{\rightleftarrows}} (\mathrm{C}) \tag{5.38}$$

where (A) and (B) are the concentrations of the two input chemical reactants and k_1 is the reaction rate for A + B to C and k_2 is the rate for the back reaction of C to A and B. If we simplify the system and let k_2 be small, then the initial reaction rate may take the form [17,18]

$$k_1 = k(\mathrm{A})^n(\mathrm{B})^m \tag{5.39}$$

where n and m refer to the order of the reaction. In order to find the values of n and m, one can make the concentration of one of the reactants small so that it takes the form

$$k_1 = k(\mathrm{A}_0)^n(\mathrm{B})^m \tag{5.40}$$

where we have made the concentration (A_0) large enough so that it is approximately constant, and the changes in the reaction rate can be measured by varying (B). The value of k is given by

$$k = zpe^{-\Psi/RT} \tag{5.41}$$

where z is the collision frequency and p is the steric factor, which is <1 and reflects the fact that not all collisions occur with the right orientation of the molecules to react. ψ is the activation energy, R is the gas constant, and T is the absolute temperature. For many cases the collision frequency is proportional to the current density, and thus there are terms that are proportional to both the drift and the diffusion currents. For example, consider the case of an enzyme reaction on a charged substrate such as a biological membrane. The total current density for a given ion in the fluid incident on the membrane is given by Equation 5.10.

If the chemical reaction rate is limited by the number of ions arriving at the membrane surface with enough energy to overcome the barrier required to initiate the reaction, then a DC drift current may either add to or subtract from the diffusion current. If the field direction is such that it prevents the ion from reaching the surface, then the chemical reaction is blocked. If the direction is reversed, the rate can grow exponentially. These changes in chemical reaction rates with the direction of the electric field are likely to be responsible for changes in the growth and reabsorption of neuritis [19]. They are also likely to be involved in the mobility of cells such as leukocytes and fibroblasts [20,21].

An AC drift current will add to the diffusion current. For the AC fields, the drift current can be thought of as increasing the volume covered by a particle executing a random walk as a result of Brownian motion. Thus, in an asymmetrical environment an electric field oscillating in the x direction may increase the number of particles that will strike the y–z plane in a fixed period of time and can increase the chemical reaction rate for a catalytic reaction at the y–z plane. Seto and Hsieh show that AC fields as low as 5 V/m can increase enzyme reaction rates by a factor of 5 [22]. For a 60 Hz field, the peak-to-peak displacement for Ca^{2+} resulting from this field is estimated to be about 1.6 nm or about twice the thickness of the Debye layer. AC magnetic fields have also been shown to change the growth rate of corn roots at fields levels of 5×10^{-3} T. It is likely that these fields are inducing significant currents [23].

An additional mechanism by which AC or DC electric or magnetic fields can effect chemical reactions is by shifting the energy level and the distribution of particles in them. DC electric fields can shift the energy level by an amount that is given by the change in the dipole moment, $\Delta\overrightarrow{M}$, and in the polarizability, $\Delta\alpha$, associated with the transition. A DC field can either stretch or compress a dipole depending on its orientation with respect to the field, thus increasing or decreasing the energy levels for atoms or molecules with different orientations. It also can modify its rate of rotation, speeding it up when the dipole is pointed in the direction of the field and slowing it down when it is pointed away from the field [24]. In a vacuum the modifications of the rotational states have been worked out from the quantum mechanics for relatively simple molecules and can lead to a relatively complex set of allowed energy states that are a function of the applied field [23]. In a solid or a membrane the energy levels between the states corresponding to the different orientations shift, by different amounts, for each state as a function of the applied field. This is known as the Stark effect [25]. The frequency corresponding to this energy shift with a fixed orientation is given by

$$h\Delta f = -\Delta\overrightarrow{M} \cdot \vec{E} - \frac{1}{2}\vec{E} \cdot \Delta\alpha \cdot \vec{E} \tag{5.42}$$

where Δf is the frequency splitting between levels, $\Delta\vec{M}$ is the change in the dipole moment, $\Delta\alpha$ is the change in polarizability, and h is Plank's constant. These terms give the linear and quadratic Stark effects for a transition in a uniaxially oriented system. These energy levels will be inhomogenously broadened by the random orientation of the dipoles with respect to the applied field and by the thermal energy. Since the quantum of energy, hf, at microwave and lower frequencies is very much smaller than a quantum of thermal energy, low-lying energy levels are approximately equally populated. However, higher-energy states may be preferentially excited by chemical reactions or optical photons so that different excited states may contain different populations. These states may be further defined by the magnetic field and separated by a Zeeman splitting. Thus, radio- and low-frequency fields corresponding to energy separation between these states may excite transitions between levels that are separated by the Stark splitting and change the population distribution in these excited states. The transition rate between an excited molecule and its final product depends on the overlap between the energy levels of the two states. Thus, the application of an electric field can shift the energy levels so as to either increase or decrease this overlap and the transition rate. This has been discussed at length in Ref. [23] for the case of Zeeman splitting of the energy levels by a DC magnetic field and free radicals [26]. See also Chapter 6 in this volume by S. Engstrom.

5.4 Biological Amplification

Biological systems are not in a state of thermal equilibrium. In a typical cell, energy is supplied by hydrolysis of an ATP molecule, which leads to the pumping of three sodium ions out of a cell and two potassium ions into the cell. The net result is creation of potential difference between the inside and outside of a cell in the range of 50 to 100 mV so that the interior of the cell is at a negative potential with respect to the external environment [27]. This potential difference can be used to amplify a variety of external signals just as a typical electronic amplifier can use a small AC signal to convert DC energy into a larger AC signal. For example, the input from many dentric junctions can be summed in a pyramidal cell to trigger an action potential that is larger than any of the input signals [28]. The input from a single synaptic junction might change the cell resting potential by 0.5 to 1 mV, and 10 to 20 inputs might be required to fire an action potential of 50 to 100 mV [27]. Additionally, subthreshold inputs can lead to the release of neural transmitters that, in turn, can release from 2 to 10,000 Ca^{2+} ions from internal stores. These neural transmitters may remain bound to the postsynaptic membrane for up to 4 sec and reduce the threshold for the firing of successive pulses [29]. Feedback from the postsynaptic membrane to the presynaptic membrane can further reduce the firing potential for the synaptic junction.

Another mechanism of amplification that may be of interest is the extraction of energy from a high-frequency signal. If a low-frequency electric field is added to a higher-frequency field and is incident on a nonlinear reactance, then the low-frequency signal may be amplified parametrically. This mechanism for amplification is valuable in the optical region for the generation of tunable signal sources and for low-noise microwave amplifiers [1,30]. In a biological system the low-frequency signal might be generated by an ongoing process such as the heart and amplified by mixing with an external signal from a power line field.

An additional mechanism for amplifications is stochastic resonance. Stochastic resonance differs from the foregoing mechanisms of amplification in that the energy is extracted from the noise. Consider, for example, a small, externally applied sinusoidal electric field incident on an ion in a potential well. If the energy acquired from the external signal is not large enough to exceed the potential barrier, the ion stays trapped in the potential well. However, if noise is added to the system, then when the sum of the applied electric field and the noise are large enough to provide enough energy to exceed the height of the potential barrier, the ion may escape the potential well. This happens most frequently at the peaks of the applied electric field so that the signal is amplified at the applied frequency. Gains on the order of 20 to 30 dB and increases in the signal-to-noise ratio of 18 dB have been observed for stochastic resonance amplifiers [31]. For an extensive review of this subject and some application neuronal systems, see Gammaitoni [32]. For a bistable system, such as a pacemaker cell that is driven by both noise and a periodic signal, it has been shown that the signal-to-noise ratio can be enhanced by the addition of noise to a weak periodic signal and that power can be extracted from the noise. A strong periodic signal can be generated at signal-to-noise ratios <1 [31,33]. This phenomenon occurs when two energy states are separated by a barrier, and the probability of a transition increases exponentially with increasing noise power. For periodic signals that are insufficient to cause a transition over the barrier but periodically increase the energy of the particle, the transition rate at the signal frequency first increases with increased noise power up to some maximum. When the noise power is increased above this level, the output signal becomes more random (see also Chapter 9 in this volume by Weaver and Bier).

5.5 Effects of Electric Fields on Cell Membranes

Electric fields play a very important role in the normal biological functioning of membranes. Membranes are complex structures containing lipids, voltage-activated ion channels, and proteins. It would be surprising if externally applied fields did not affect the membrane behavior. First, an electric field exerts a mechanical force on a membrane by means of the force exerted on charges in the Debye layer on either side of it and on charged proteins that may protrude from the lipid bilayer. Note that although as a first approximation the membrane is often modeled as a smooth planar or spherical surface, it is highly inhomogenous, and the charges are sparsely disturbed. Thus, the field on a protein may be widely different from the average field. See Figure 5.1a. and b for a partial indication of a membrane and cell complexity. The effects from fields in the plane of the membrane, where large molecules such as proteins are free to move as in a viscous fluid, are significantly different from the effects of fields in the transverse direction, where the membrane components are bound in a layer typically 5 to 15 nm thick.

The field distribution incident on a particular part of a cell membrane is a function of its geometry, frequency, and the cells around it. As can be seen from the models in Figure 5.1d, a wide variety of environments may exist. The currents that flow through and along the membranes are dependent on the geometry and frequency. A variety of equivalent circuits have been used to model both the impedance of the membrane and the extracellular fluids. The simplest of these are a resistor and capacitor in parallel. At very low-frequencies a collection of cells can be modeled with resistors as indicated in Figure 5.1c.

The interiors of cells are normally negatively biased in relation to the surrounding fluid by 50 to 150 mV, which leads to average transverse electric fields up to tens of millions of volts per meter [34]. The effective membrane resistance (R_m) per unit area takes on values of 0.14 to 15 Ω/m^2 in the transverse direction. This corresponds to resistivities in the range of $\rho_m = 10^7$ to 10^9 Ω m. The relative dielectric constant for the membrane is typically in the range of 2 to 4. Both the surrounding fluid and the interior of a cell have resistivities ρ_f of about 2 Ω m and a relative dielectric constant of 50 to 80. This means that the cell membrane tends to shield the interior of a cell very effectively from externally applied fields at frequencies below a few kilohertz and becomes almost a short circuit in the multimegahertz region of the spectrum. In most cells the interior of the cell contains complex structures that are functions of time as the cell grows and divides. See Figure 5.1b.

Consider the case of an oversimplified hypothetical rectangular cell as shown in Figure 5.1c. At low-frequencies, an external field $\vec{E}$ causes a current density $\vec{J}_f = \vec{E}/\rho_f$ to flow in the external medium, where ρ_f is the resistivity of the fluid. The corresponding voltage drop is $V = \vec{E}L = \vec{J}_f\rho_f L$, which we can consider to be applied to the cell. This voltage is distributed across the cell length as

$$V = [\rho_m 2t + \rho_f(L - 2t)]|J_m| \tag{5.43}$$

where $\vec{J}_m$ is the current density through the cell and ρ_m is the resistivity of the membranes. Typical cell membrane thicknesses are 6 to 10 nm, and typical dimensions are 10 to 150 μm. Setting $L = 100$ μm and

$$\rho_m 2t = 10\,\Omega m^2 \tag{5.44}$$

$$V = (10\,\Omega m^2 + 2 \times 10^{-6}\,\Omega m^2)|J_m| \tag{5.45}$$

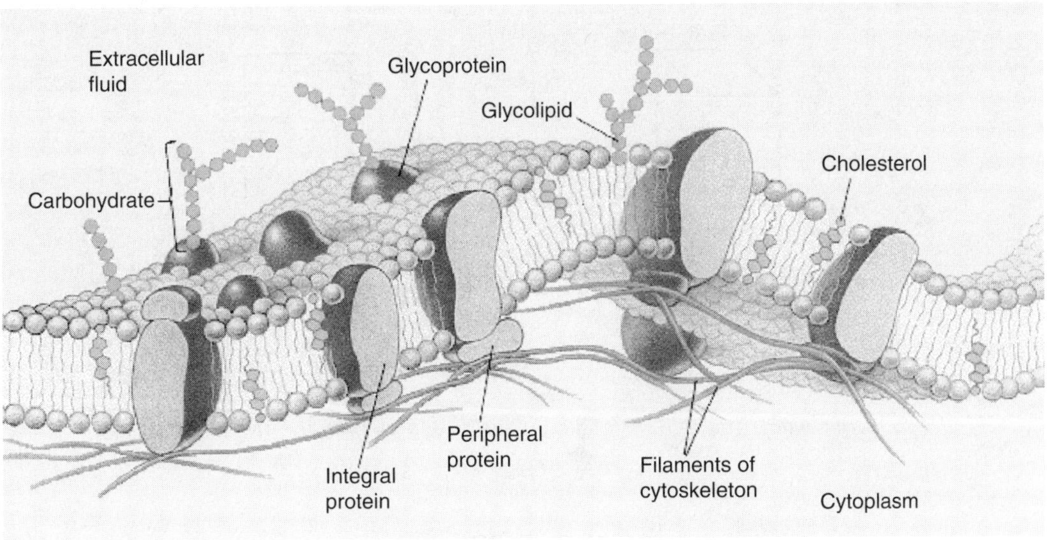

(a)

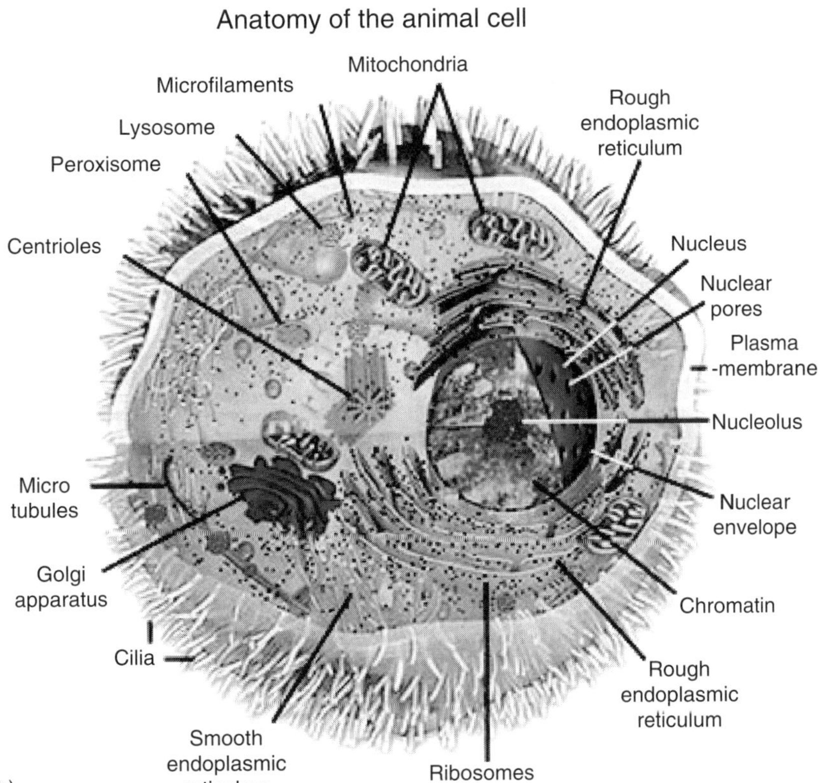

(b)

FIGURE 5.1 (See color insert following page 380.)
(a) Model of a cell membrane. (From Chiras, D., *Human Biology*, © 5th edition, 2006, Jones and Baretlett Publishers, Boston. With permission.) (b) Anatomy of the animal cell. (From *Molecular Expression*, http://microscopy.fsu.edu, accessed Sep. 30, 2005; drawing © M.W. Davidson and Florida State University. With permission.)
(continued)

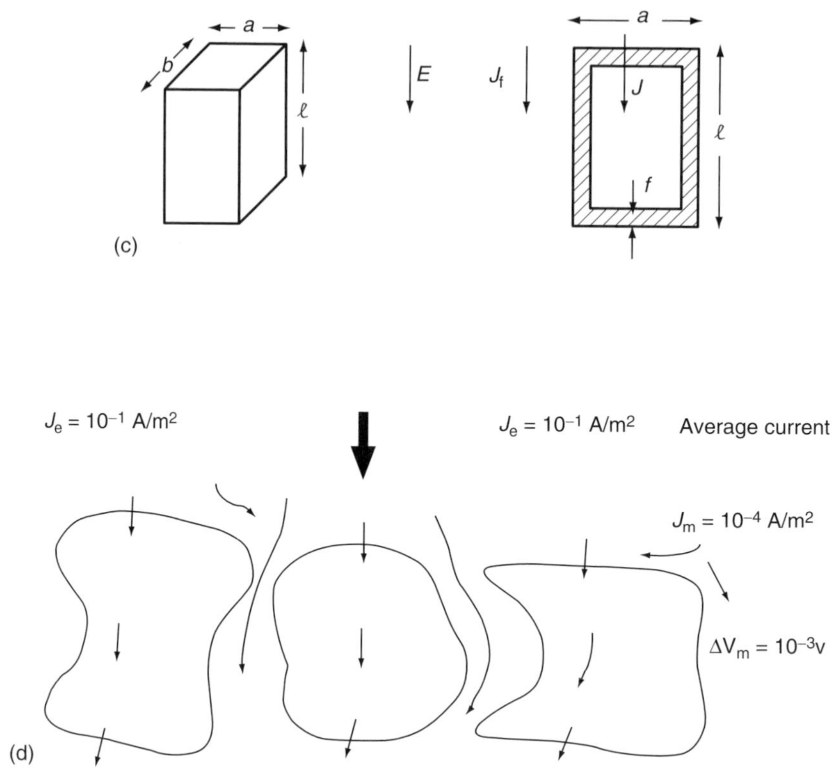

FIGURE 5.1 (continued) (See color insert following page 380.)
(c) Current distribution in a hypothetical rectangular cell. (d) Partition of 60-Hz currents through and around cells for an average current density $J = 10^2\ \mathrm{A/m^2}$. J_c is the current density between cells, and J_m is the current density through the membrane. (From Wachtel, H., private communication. With permission.)

This shows that essentially all of the transverse voltage drop occurs across the membrane at low-frequencies and the interior of the cell is almost completely shielded from external fields. A more complete theory for long cells that accounts for the internal resistance of the cell is given by Cooper et al. [35]. See also Chapter 11 by Pila [114]. Note that muscles, nerves, and a number of other cells may be much longer than 100 μm and may have dimensions in centimeters. Additionally, blood vessels form long, low-resistance paths that may concentrate currents due to externally applied fields. The anisotropic characteristics of cells are reflected in an anisotropy of the dielectric and conductive properties of tissue, which may give variations of as much as 10 to 1 in conductivity, depending on the direction of measurement relative to cell orientation [36]. A number of fish, including sharks, have been shown to use very long cells and to sum signals in both series and parallel to increase the voltage drop across a sensitive membrane in order to sense fields as low as 10^{-6} V/m [37–39]. The long cell may be thought of as an antenna that concentrates the field across a very thin, voltage-sensitive detector membrane. This membrane appears to have a built-in amplifier that allows detection of signals that are only a little above the natural electrical noise.

Membranes are not just simple linear resistors, but they usually are nonlinear and, in the case of nerve cells, are time varying as well. For a passive membrane in which the membrane potential is primarily determined by the concentration gradient of a single ion such as K^+, the Nernst equation predicts a diode-rectifying characteristic of the form [31]

$$I = I_0\left[\exp\left(\frac{V_m}{\eta V_T}\right) - 1\right] \tag{5.46}$$

where V_T is given by

$$V_T = \frac{kT}{q} = 0.026\,\text{V} \tag{5.47}$$

at $T = 300\,\text{K}$. q is the charge on the electron, V_m is the voltage across the membrane, I_o is the current for ideal back-biased current in amperes, and η is a dimensionless constant. Thus, for currents flowing through a membrane in one direction, the current is nearly constant, whereas for flow in the other direction the currect increases exponentially with voltage. In addition to passive currents, cells also use the energy from metabolic processes for the active transport of ions against the fields established by the concentration gradients. These processes are usually modeled as current sources and described as pumps [31]. A thermodynamic approach to pumping shows that ions can be pumped if they form a compound with a material that can flow through the membrane and that is created on one side of the membrane and destroyed on the other [40]. A large variety of models have been generated to characterize the effects of externally applied fields on the transport of ions through membranes [41–43]. However, the details of the pumping process are not well understood. In addition to ion transport, electrical fields can change the binding of ions or molecules to the membrane surface.

In the case of pacemaker cells, there are also feedback processes that lead to an oscillating membrane potential and a membrane resistance that is a function of time. The current flow for these cells is described empirically by the Hodgkin–Huxley equation [31]. An alternate approach that treats the nerve pulse like a plasma instability has been proposed by Triffet and Green and Vaccaro and Green [44,45]. Na^+ and K^+ currents are the dominant carriers for the propagation of nerve impulses along a cell. It is generally believed that the Na^+ and K^+ currents that flow through the membrane in opposite directions are carried through separate channels. Ca^{2+} ion currents are involved in the activation of at least a portion of the K^+ currents and are voltage gated. By activating the K^+ currents, the Ca^{2+} ions shorten the length of time the cell is depolarized and thus speed up the firing cycle [46]. A statistical approach to the formation of protein channels in the membrane by Baumann and Easton predicts many of the observed characteristics [47–49].

During the firing of a nerve cell, the Na^+ current pulse precedes the K^+ current pulse, which returns the cell to its resting potential [31]. The overall concentration balances are maintained by active ion pumps. Cl^- Mg^{2+}, and possibly OH^- and H^+ ions may also be involved in the current flow across a cell membrane.

The firing of a nerve cell typically involves voltage spikes of 10^{-1} V and peak current densities of $1.5\,\text{A/m}^2$. Changes in the firing rate can be induced by the injection of charge through a microelectrode of $<10^{-9}$ A for a few milliseconds. However, in cases where electrodes are used to stimulate muscles or to control epilepsy, the current is injected through a series of cell membrane fluid boundaries at a distance from the controlling nerve fiber. Thus, typical injected currents to produce behavioral changes in cells are in milliamperes, and current densities are $10\,\text{A/m}^2$ or higher.

For fields parallel to the plane of the membrane, it is possible to obtain electrophoresis or a rearrangement of charged particles. This has been shown by Poo in a striking fashion in cultured embryonic *Xenopus* myotomal muscle cells [50,51]. Receptors on the surface of the cell were labeled with a fluorescent dye and allowed to uniformly distribute themselves. Exposures to electric fields of 10^2 to $10^3\,\text{V/m}$ were sufficient to concentrate the fluorescent-labeled receptors on the side of the anode in about 10 min. After shutting off the field, diffusion returned the dye to its uniform distribution in about 2 h. This corresponds to an in-plane diffusion constant of about $3 \times 10^{-12}\,\text{m}^2/\text{sec}$. The force on the receptor molecules or particles in the membrane includes not only $q\vec{E}$ but also any viscous drag that may be generated by the flow of ions of the opposite sign moving along the surface in the opposite direction. The direction of motion for a given charged particle seems to depend on whether it has a larger or smaller zeta (ζ) potential than the potential across the charged double layer at the interface between the cell surface and the fluid (see Equation 5.16).

Additional work has shown that the distribution of acetylcholine (ACh) receptors is changed by external fields [48,49]. These receptors are concentrated on the cathode-facing surface of the cell in fields of $10^3\,\text{V/m}$ over a period of 30 min by literally rearranging channels already existing in the cell membrane. The concentration or clustering persists for at least 5 h after the field has been turned off, indicating that the clustering is relatively stable. Single-channel patch measurements show both a higher density of ACh channels in the clusters near the cathode and a longer mean duration of the pulses through the transmembrane channels. The length of the current pulse near the anode does not differ from the controls, indicating that the field itself does not have a direct effect on the channel kinetics. The lateral diffusion coefficient, D, of ACh receptors in the plasma membrane of cultured *Xenopus* embryonic muscle cells is estimated to be $2.6 \times 10^{-6}\,\text{m}^2/\text{sec}$ at 22°C [48,49]. Lateral concentration gradients in lipid monolayers have been shown to be induced by externally applied electric field gradients. For binary mixtures of dihydrocholesterol and dimyristoylphosphatidylcholine, the application of an electric field gradient at pressures below the critical pressure produces a liquid–liquid phase separation in a monolayer that is otherwise homogenous [52]. This separation occurs at field levels on the order of $10^7\,\text{V/m}$ and gradients of 10^9 to $10^{11}\,\text{V/m}^2$.

5.6 Nonlinear Effects of AC Fields on Cells

5.6.1 Introduction

The application of an AC electric field to nonlinear systems, which can be described by either a nonlinear resistance or capacitance, leads to at least partial rectification of the input signal and the generation of harmonics. If two or more signal frequencies are applied, it also leads to frequency mixing of the form

$$f_o = \pm m f_1 \pm n f_2 \tag{5.48}$$

where f_o is the output frequency, f_1 and f_2 are input frequencies, and m and n are integers. The rectified component of the AC current can, in turn, lead to ion accumulation at interfaces, which results in changes in ion concentration [53,54]. These changes in ion concentration, in turn, can affect biological function. Another important additional effect is the dependence of the dielectric constant on frequency. This leads to changes in the electric field distributions in tissue with frequency. Thus, both the electrophoretic and the dielectrophoretic forces become both size and frequency dependent. A third—possibly important—additional effect is the excitation of frequency-sensitive biological systems in a resonant manner. By driving systems near their resonant frequency, we may change the effective amplitude of the stimulating signal and change the frequency of the nerve cells firing.

In this section we will review the rectification process at the cell membrane in some detail. Additionally, we will show that cell nonlinearities lead to frequency-dependent effects such as injection phase locking of pacemaker cells. We will also briefly examine some problems associated with the exposure of cells to very low, extremely low-frequency (ELF) fields and the application of large ELF fields to biological systems.

5.6.2 Rectification by Cell Membranes

The rectification of currents flowing across membranes has been studied by many authors, beginning with Katz in 1949. Much of this work is referenced by Hayashi and Fishman in their paper on the inward rectifier K^+ channel kinetics [53].

For many passive cell membranes, an approximate relation for the transmembrane current can be derived from the Nernst equation as given in Equation 5.41 [31]. If we apply an AC signal across the membrane of the form $V_M = V_0 + V_1 \cos \omega t$, the resulting current can be approximated for small values of V_M (i.e., $qV_M < \eta kT$) by a Taylor series yielding

$$I = \frac{I_0}{\eta V_T}\left(V_0 + \frac{V_1^2}{4\eta V_T} + V_1 \cos \omega t + \frac{1}{4\eta V_T} V_0 V_1 \cos \omega t + \frac{V_1^2}{4\eta V_T} \cos 2\omega t + \cdots\right) \tag{5.49}$$

It is to be noted that the second term in the expression is the first approximation to the fraction of the applied AC voltage V_1 that yields a DC current component ΔI,

$$\Delta I = \frac{I_0}{4}\left(\frac{V_1}{\eta V_T}\right)^2 \tag{5.50}$$

or an offset voltage V_{DC} given by

$$V_{DC} \approx \frac{I_0}{4}\left(\frac{V_1}{\eta V_T}\right)^2 R_m \tag{5.51}$$

where R_m is the membrane impedance. This predicted voltage offset for an applied AC current has been measured by Montaigne and Pickard [55]. In their experiments, an AC signal was applied to a large plant cell by a strip line, and the measured voltage shift was obtained through microelectrodes located outside the applied AC fields. For an applied AC field of about 0.2 V, they measured a DC offset of 1 to 2 $\times$ 10^{-4} V. For frequencies above 2.5 kHz, the effects of the membrane capacitance must be taken into account, and the effective driving voltage is reduced to

$$(V_1)_{\text{eff}} = [\sqrt{2}\, a\sigma_e\, \vec{E}_{1\text{rms}}][(\sigma_e + aG)^2 + (a\omega C)^2]^{-1/2} \tag{5.52}$$

where a is the cell radius, σ_e, is the conductivity of the medium $\vec{E}_{1\text{rms}}$, is the electric field strength in the medium surrounding the cell, G is the membrane conductance per unit area, ω is the frequency, and C is the membrane capacitance per unit area [56]. This leads to the usual roll-off in the measured DC offset with increasing frequency. Note that the DC effect gets still smaller at higher frequencies (above 1 MHz) because of transit time limitations for ion flow across the membrane [57].

The relaxation times for a typical K^+ channel in an *Aplysia* membrane has been measured to be from 2 to 8 ms [51]. Rectification has also been demonstrated in thin lipid membranes [58]. In these systems, both the conductivity of the membrane and the ion concentration differences across it can be controlled. The Nernst equation was shown

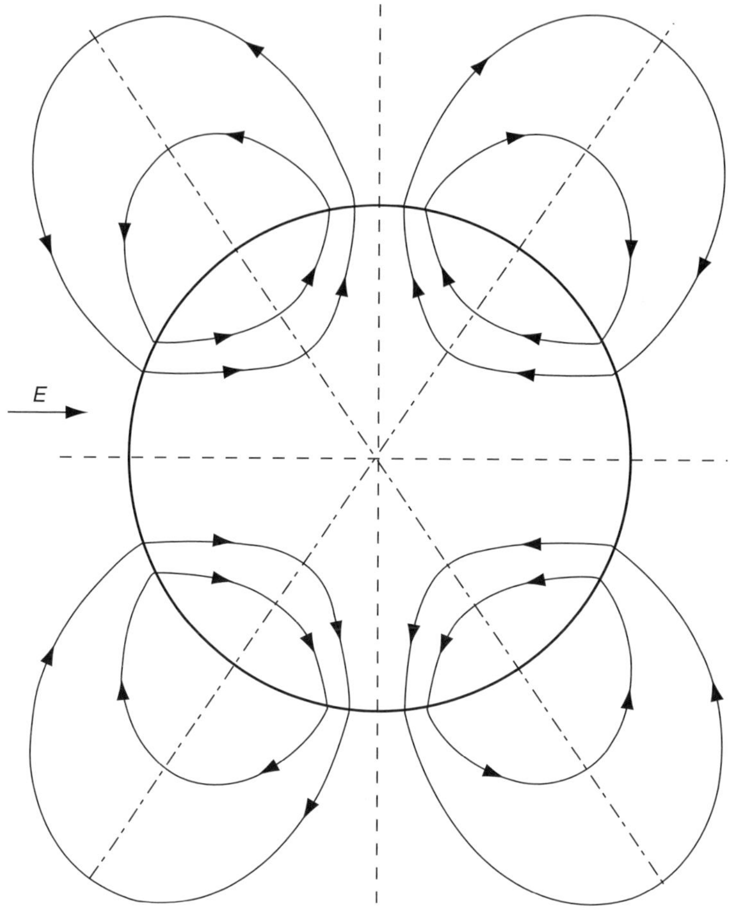

FIGURE 5.2
Induced DC current distribution in a spherical cell. (From Bisceglia and Pinto, I., personal communication, 1984. With permission.)

to apply to the I vs. V curve over a range of voltages from -60 to $+40$ mV. Depending on the ion concentration and membrane doping, the values of η ranged from 1 to 0.25.

A different treatment of the nonlinear response of passive cell membranes to an applied AC field has been carried out by Franceschetti and Pinto and by Casaleggio et al. [59,60]. Both these groups have expanded the Nernst equation in a Volterra series that takes into account memory of the preceding state of the cell. They have also treated the cell in spherical rather than planar geometry. The inclusion of a spherical cell requires that the total current into and out of the cell be equal to zero, and thus loops are formed circulating through the cell membrane (see Figure 5.2). All the theoretical treatments predict a DC component that varies as the square of the input signal V and tends to hyperpolarize the cell or make the interior of the cell more negative.

Cain has considered the effects of an AC field on nonlinearities of the nerve cell by numerical analysis of the Hodgkin–Huxley equation [61].* He applied a voltage

*Bisceglia and Pinto have applied a Volterra series expansion to the Hodgkin–Huxley equations. This approach gives an alternate method to Cain's of computing the current shifts resulting from applied AC signals [62].

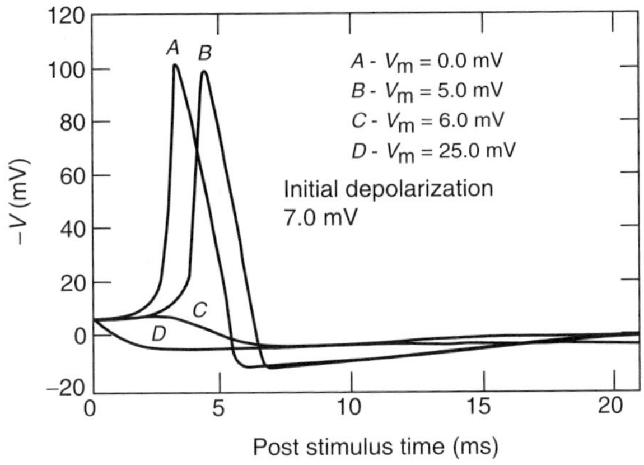

FIGURE 5.3
Computed membrane action potentials in response to an initial membrane depolarization of 7 mV for different values of V_m. Curves are solutions to the Hodgkin–Huxley equations. (From Cain, C.A., *Bioelectromagnetics*, 2, 23, 1981. With permission.)

$$V_m = V_0 + V_1 \cos \omega t[u(t) - u(t - \tau)] \tag{5.53}$$

across the membrane, where $u(t)$ and $u(t - \tau)$ are unit step functions that define an AC pulse of length τ. For the case where the AC frequency is large compared to the reciprocal of the pulse length, if a 7 mV depolarizing pulse is also applied to the membrane, the action potential is obtained as shown in Figure 5.3. Cain has assumed coefficients appropriate to the giant squid axon. Increasing V_1 first delays, and then suppresses, the action potential. If no depolarizing pulse is applied, the predicted changes in g_{Na} and g_K and the deviation V from the resting potential are as shown in Figure 5.4 for a 10 msec AC pulse with $V_1 = 25\,\mathrm{mV}$. Note that the applied AC frequency is assumed high enough not to be resolved in these figures. From these results, it is clear that AC signals can induce substantial changes in the operating characteristics of nerve cells at moderate to high levels of applied voltage. Although the appropriate coefficients were not measured in order to make a direct comparison between theory and experiments, Wachtel's results on *Aplysia* at frequencies above the lock-in range would appear to support Cain's theoretical predictions [63].

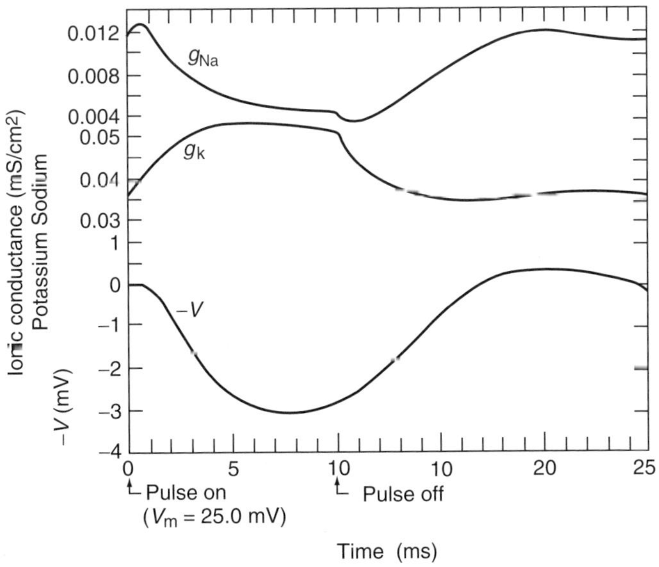

FIGURE 5.4
Response of model axon to a pulsed oscillating component of membrane electric field (10-ms pulse, $V_m = 25\,\mathrm{mV}$). The membrane potential and the sodium and potassium conductances are shown. These curves are solutions of the Hodgkin–Huxley equations. (Note: $1\,\mathrm{mS/cm^2} = 10\,\mathrm{S/m^2}$.) (From Cain, C.A., *Bioelectromagnetics*, 2, 23, 1981. With permission.)

Wachtel has made a series of measurements that demonstrate the nonlinear characteristics of pacemaker cells from *Aplysia* [61]. First, he measured the current input through a microelectrode that changed the firing rate of the cell. The current threshold for a minimum detectable change was approximately 6×10^{-10} A at frequencies between 0.8 and 1 Hz (see Figure 5.5). The natural firing rate for this cell is about 0.8 Hz, and an increasing current is required to synchronize the cell to the injected signal as the frequency deviates from the natural firing rate. A theory for injection locking of electronic oscillators predicts that the signal required for locking an oscillator to an external signal increases linearly as the difference between the two frequencies $\Delta\omega$ increases [64]. The signal required for lock-in according to this theory is given by

$$I_t \approx |A\Delta\omega|\ I \tag{5.54}$$

where I_t is the injected signal current and I is the peak unperturbed oscillator current. $A = \partial\phi/\partial\omega$ is the rate of change of phase with respect to frequency in the unperturbed oscillator. $\Delta\omega_0$ is equal to the difference between the frequency of the free-running oscillator and the injected signal. This expression is applicable as long as

$$\Delta\omega_0 \leq \frac{2\pi}{\tau} \tag{5.55}$$

where τ is the time constant for adjusting the gain of the circuit. The time constant τ for the *Aplysia* cells varied between 0.1 and 0.5 sec, and this corresponds to a maximum measured lock-in frequency of about 10 Hz. The results in Figure 5.5 show the threshold for one-to-one locking up to about 2 Hz. In the range from 2 to 10 Hz, Wachtel observed a lower threshold for subharmonic locking than one-to-one locking. At frequencies above 80 Hz, he observed a constant shift in the firing rate of the neuron in response to the injected transmembrane AC signal. The natural firing rate would be restored by also injecting a transmembrane DC signal equal in amplitude to about 1% of the peak-to-peak value of the AC current. This DC current was in the depolarizing direction, making the exterior of the cell more negative with respect to the cell cytoplasm to increase the firing rate (i.e., to restore it to its natural value). Apparently, the applied transmembrane AC current was partially rectified so as to hyperpolarize the membrane (making the interior of the cell more negative with respect to the external fluid). The details of how the applied field modifies the ion flow are only partially understood, but one characteristic is an increase in the conductivity for K^+, which increases its flow out of the cell. Wachtel also

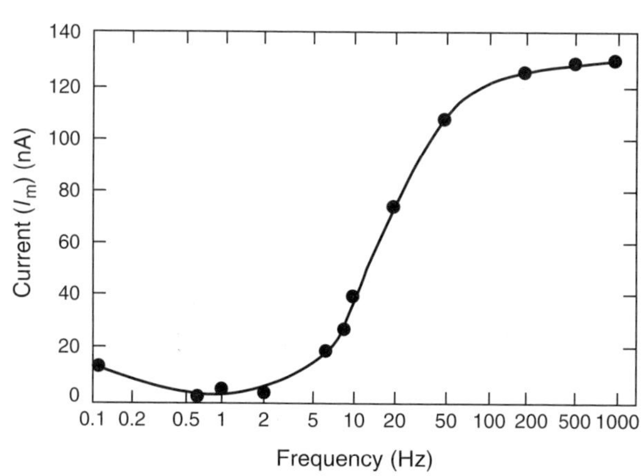

FIGURE 5.5
Intracellular (transmembrane) currents I_m (in nA) needed at different frequencies to produce firing-pattern changes (in a pacemaker neuron). Note that the detectable changes take on different forms at different frequencies. (From Wachtel, H., *Proceedings of the 18th Annual Hansford Life Science Symposium*, Technical Information Center, U.S. Department of Energy, Richland, WA, 132, 1978. With permission.)

injected low-frequency currents into the seawater surrounding the cell preparation through external electrodes [61]. In this case, the minimum current densities flowing in the vicinity of the cell preparation for injection locking were estimated to be about 10^{-2} A/m^2, and there was about a 30 to 1 variation between the maximum and minimum sensitivities for changes in angle between the applied field and the cells. At frequencies above 100 Hz, a minimum of about 0.35 A/m^2 was necessary to obtain a detectable change in firing rate.

These studies have been extended by Barnes et al., and injection locking at harmonic and subharmonics has been shown to occur. It is suggested that phase locking may provide a mechanism for narrow banding or time averaging so that a weak coherent signal may be distinguished from noise by a cell. In an electronic circuit model we showed we could phase lock an oscillator at signal-to-noise ratios <1 [65]. Extensive modeling of phase locking for a squid axon using two versions of the Hodgkin–Huxley equations has been carried out by Fohlmeister et al. [66]. They show that phase locking can occur for a wide variety of frequencies with AM-modulated signals at injected current densities greater than 0.1 A/m^2 [67]. For natural oscillation frequencies less than the externally applied signal, the system may be treated as a parametric process. For parametric amplification, a phase stability such that

$$\frac{d\Phi}{dt} < \Delta\omega - KV_s \tag{5.56}$$

is required for injection locking of the frequency of oscillation to an external signal, where $d\Phi/dt$ is the rate of change of the phase, $\Delta\omega$ is the frequency offset, K is the linear control characteristic in units of (2π Hz/V) and is closely related to the loop gain, and V_s is the injected signal [68]. Stated in words, this equation requires that the amplified signal, KV_s be large enough to correct for the random frequency fluctuations $d\Phi/dt$ generated by the noise for the system to become phase locked to a signal that is displaced by $\Delta\omega$ (see Figure 5.6 for some examples of injection locking of pacemaker cells to an external signal) [65]. An increase in the sensitivity to electromagnetic fields has also been shown in isolated frog hearts for signals that approach the natural resonant frequency or firing rate [69]. In these experiments, the firing rate of the heart was shown to increase as much as 30% when a signal in the vicinity of 10 to 20 V/m was applied through Ringer's solution to the isolated frog hearts at a frequency between 0.5 and 1 Hz. The natural firing rate of these excised hearts started out at approximately 1 Hz and dropped to about 0.5 Hz

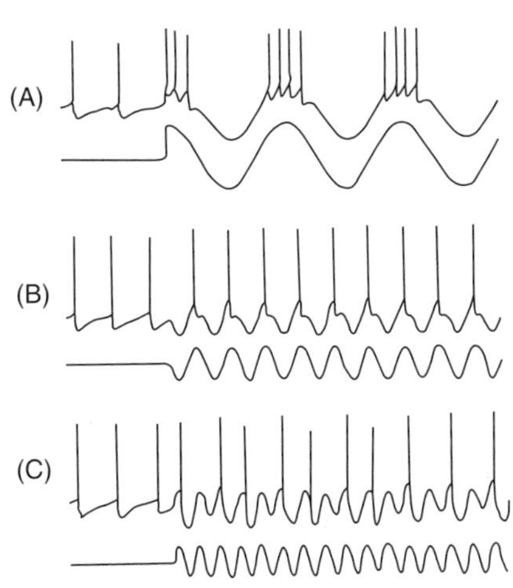

FIGURE 5.6
Examples of several modes of synchrony between an imposed ELF field and neuronal patterns. In each case the ELF current is shown below the transmembrane potential recording. (A) For ELF frequencies well below FR_0, several nerve impulses (spikes) are locked to each ELF half cycle. (B) ELF frequencies slightly above FR_0 are effective in phase locking the rise of neuronal spikes on a one-to-one basis. (C) For ELF frequencies several times greater than FR_0, phase locking can take the form of spikes occurring on alternate cycles (two-for-one synchrony). (From Barnes, F.S., *Bioelectromagn. Suppl.*, 1, 67–85, 1992. With permission.)

over a period of 2 h, where they remained stable for at least 5 h. To get a 30% increase in firing rate at 60 Hz, it was necessary to apply field strengths of 60 to 80 V/m. Thus, we have additional evidence that electric fields with repetition rates near the natural biological signaling frequencies are more likely to induce changes than those of higher frequencies and that signal strengths required for a given shift increase approximately linearly up to some cutoff, as shown in Figure 5.5.

For weak fields, it has been shown that cells can respond differently to signals that are both space and time coherent than they do for signals that look like the background noise. Litovitz et al. have shown that the application of 10 μT magnetic fields at either 55 or 65 Hz doubles the specific activity of ornithine decarboxylase (ODC) in L929 cells if the signals are coherent for periods of 10 sec or longer during the course of a 4 h exposure [70]. The applied signal and the corresponding ODC response as a function of the coherence time are shown in Figure 5.7. The ODC response of the cell can be fitted to an exponential curve of the form

$$(\mathrm{ODC}) = 1 + 1.26\left[1 - \exp\left(\frac{\tau_{\mathrm{coh}}}{\tau_{\mathrm{cell}}}\right)\right] \tag{5.57}$$

where τ_{coh} is the length of the time between shifts in frequency and the introduction of a random phase shift and τ_{cell} is the effective time constant of the cell [70]. τ_{cell} has a value of about 8 sec for these cells. If a spatially coherent noise signal with a power spectral density ranging from 30 to 90 Hz is superimposed on the coherent signal, the increased ODC response decreases with a decreasing signal-to-noise ratio and is less than 10% at a signal-to-noise ratio of 1 [71]. This work has been extended to show that temporally incoherent magnetic fields inhibit 60 Hz-induced changes in the ODC activity of developing chick embryos [72].

For the exposure geometry used in these experiments, the magnetic field induced a corresponding electric field of 4 μV/m. This signal is well below the calculated thermal noise field of 0.02 V/m for a 20 μm cell diameter. The combined results of the experiments cited above indicate that both space and time coherence may be used by cells to separate useful signals from larger natural background noise signals. For example, to get a significant biological response, some threshold number of channels or receptor molecules may need to be activated within a given period of time; this, in turn, requires nearly

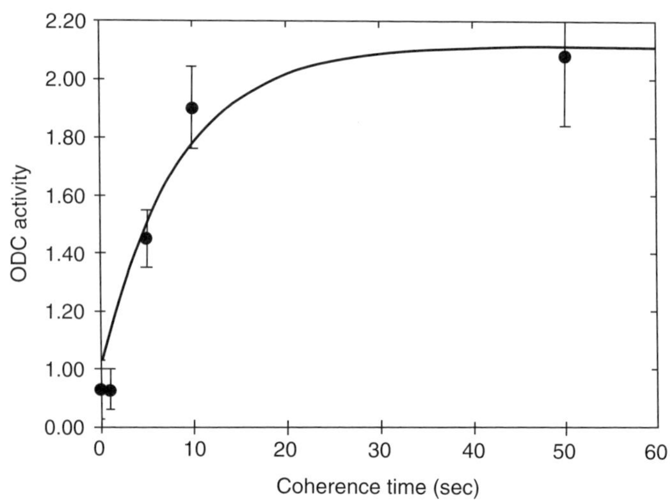

FIGURE 5.7
Plot of the enhancement of ODC activity (exposed/control) as a function of the coherence time, τ_{coh}, of the applied field. The solid line is the best fit to the mathematical function given by Equation 5.44, where τ_{cell} is found to be 8.2 sec. The experimental points shown represent a minimum of six different exposures. (© Academic Press; From Litovitz, T.A., Krause, D., and Mullins, J.M., *Biochem. Biophys. Res. Commun.*, 178, 3, 862, 1991. With permission.)

simultaneous activation over a significant fraction of the cell surface. Similar results have been obtained for developing chick embryos, where weak coherent signals lead to an increased incidence of abnormalities [73]. In this work, Litovitz and his colleagues show an increase in the incidence of abnormalities of approximately a factor of 3 for White Leghorn chicken embryos incubated in periodic magnetic fields with peak field strength of 1 μT (100 Hz repetition rate, 500 μs pulse duration, 2 μs rise, and decay times) when compared with the controls. This increased rate of the incidence of abnormalities was nearly eliminated with the addition of band-filtered noise with a spectrum running from 30 to 100 Hz and a root mean square value of 1 μT. Thus, Litovitz makes a strong case for a requirement of both space and time coherence for biological systems to detect signals below the natural noise environment.

A number of experiments indicate that at least two mechanisms are involved in the effects of low-level time-varying magnetic fields on membrane transport. The first of these is through Faraday's law or the induced electric field, which, in turn, induces electric currents. In these experiments, one would expect to get the same effects by introducing electric fields with electrodes at levels that induce the same current densities. The second group of experiments indicates that the background DC magnetic field is also important and that the combined effects of AC and DC magnetic fields are observed.

The initial experiments by Walleczek and Liburdy showed an enhanced uptake of Ca^{2+} in Con. A-activated rat thymocytes with exposures of 1 h to 60 Hz magnetic fields of 22 mT and induced current densities of 0.16 A/m^2 [74]. In this paper the exposure system consisted of concentric rings on cell culture plates, which, in turn, were placed in a water-cooled solenoid that produced a uniform magnetic field. This was followed by a group of experiments by Liburdy on Ca^{2+} transport across mitogen-activated lymphocyte membranes [75]. In these experiments, both the DC and the AC magnetic fields were controlled so that the DC geomagnetic field and the ambient 60 Hz fields were perpendicular to the exposed and control plates. The results show an increase in the Ca^{2+} influx during the plateau phase of the calcium signaling for Con. A-activated lymphocytes, which was a function of the induced electric field and which could be reproduced by applying the electric fields across the cells with a salt bridge at levels between 0.1 and 0.17 V/m. This corresponds to induced current levels of 0.168 to 0.28 A/m^2 in the fluid surrounding the cells, which had a conductivity, $\sigma = 1.68$ S/m, that is approximately a hundred times larger than the current densities observed around growing cells. Thus, the approximately 20% to 25% increase in the initial Ca^{2+} uptake is the result of a relatively large external current. In other experiments it was also shown that the response is dependent on the age of the animals from which cells are taken [76].

Most other reported experiments have not been done in a way to sort out the differences between possible direct effects of the magnetic fields and the induced electric fields. Yost and Liburdy [77] have also conducted experiments in the same system that show a direct dependence of the calcium uptake on the DC magnetic field.

The experiments by McLeod et al. [78] show both a frequency dependence and a dependence on the electric field strength across the cell membrane. They exposed neonatal bovine fibroblast cells to electric fields in culture through a media bridge. The fibroblasts populated a collagen matrix that enabled the cells to be grown with a dominant orientation and exposed to a well-defined current. An estimate of newly synthesized protein was made by measuring the incorporation of (^{3}H) proline into macromolecules after a 12 h exposure to current densities ranging from 10^{-3} to 10 A/m^2 and frequencies from 0.1 Hz to 1 kHz. The results in Figure 5.8 show an approximately 30% reduction in the ^{3}H counts with current densities as low as 10^{-2} A/m^2. This reduction is interpreted as a reduction in the incorporation of newly synthesized protein into the extracellular matrix rather than as a change in the cell number. The frequency specificity for this threshold is

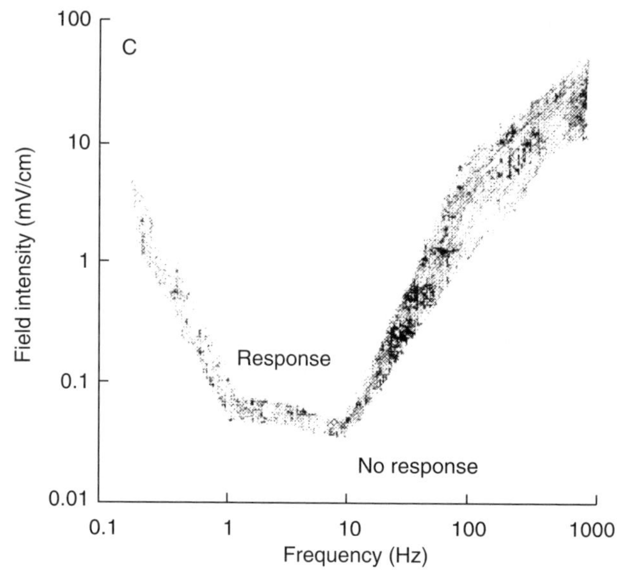

FIGURE 5.8
Minimum field intensity for a detectable response. Summary of results for all tested frequencies and current densities. Current densities were converted to peak field intensities by using the measured media resistivity of 65 Ω cm. The lower boundary of the gray region represents the highest field intensity at which no significant change in extracellular protein accumulation was detected; the upper boundary represents the lowest intensity evoking a statistically significant change ($n = 6$). (From McLeod, K.J., Lee, R.C., and Ehrlich, H.P., *Science*, 136, 1465–1469, 1987.)

shown in Figure 5.8; the peak sensitivity was recorded at 5×10^{-3} A/m^2 and 10 Hz. The corresponding peak electric field intensity was 4.5 mV/m. The fractional change in the (^{3}H) proline was nearly independent of the current density for increases in current density up to two orders of magnitude above 10^{-2} A/m^2. The cell membranes have a resistance many times higher than the resistance of the matrix as a whole. The cells are also asymmetric, with a ratio of major to minor axes of about 7 to 10. Thus, the current through the cell membranes would be expected to be at a maximum when the long axes of the cells are parallel to the applied field. For randomly oriented cells, current densities of 3 mA/m^2 produced no significant effect on the rate of proline incorporation. However, when the cells were oriented parallel to the electric field that was estimated at 2 mV/m and 10 Hz, a little more than a 30% reduction was observed. The estimated transmembrane potential was 0.5 μV. With the cell oriented perpendicular to the field, no significant change in proline incorporation was measured at 5 mA/m^2.

In addition to the nonlinear conductances associated with Na^+ and K^+ currents, membranes also exhibit nonlinear (i.e., potential dependent) and frequency-dependent capacitances and inductances. It is sometimes useful to think of these effects in terms of a phasor diagram as shown in Figure 5.9, where the electric field vector $\vec{E}$ is rotating at a velocity ω, and ϕ is the phase angle between $\vec{E}$ and the current density $\vec{J}$. If there is, for example, a fixed time delay between the field activation of a current gate and the current flow, then, depending on the frequency, $\vec{J}$ may be in any of the four quadrants and appear capacitive or inductive or even present a negative resistance to an external driving source.

Nonlinear inductive effects seem to be associated with the time delay for the onset of the K^+ currents under excitation in a typical excitable membrane, and they have been studied in the giant squid axon [79]. The nonlinear capacitive effects are difficult to measure at frequencies below a few kilohertz. Extra care needs to be exercised to minimize the series resistance and the end effects of the wire being used to measure the capacitance or inductance. Additionally, corrections must be made in the calculations of the membrane capacitance to take into account the appropriate variations in the frequency response that these terms introduce. However, when this is done, it can be shown that the membrane capacitance has both frequency- and voltage-dependent terms. The capacitance of giant squid axons is shown in Figure 5.10 and Figure 5.11 as a function of frequency and membrane voltage.

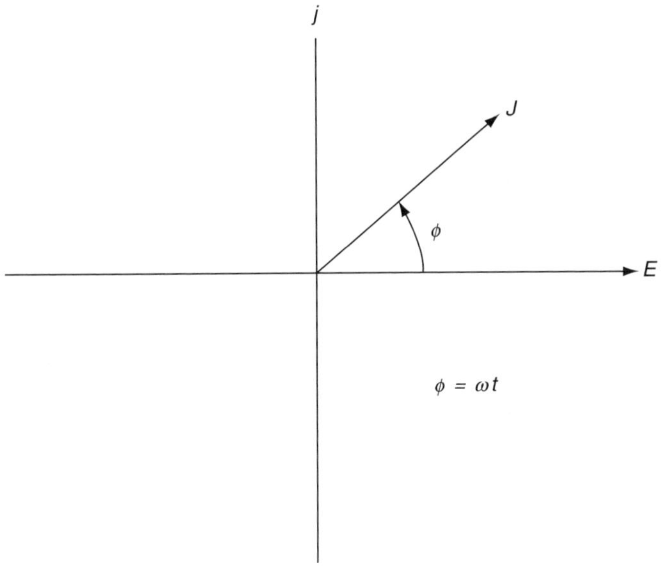

FIGURE 5.9
Steady-state vector characterization of electric fields $\vec{E}$ vs. current density $\vec{J}$ in polar form. ϕ is the phase angle between the sinusoidal electric field $\vec{E}$ and the resulting current density $\vec{J}$.

Variation of the capacitance of these membranes with frequency and amplitude differs from that of a simple bilipid membrane that has nearly constant capacitance. The variation appears to be associated with changes in the conformation of the proteins associated with the Na^+ conductance channels. Nonlinearity in conductance and capacitance can be induced into a bilipid membrane by the addition of Alamethicin. The nonlinear inductance or capacitance may also generate both sum and difference frequencies if two signals are applied. For the case of the single signal, a DC term is added to the current density that is proportional to membrane potential and the square of the applied AC signal [80]. The effects due to nonlinear membrane capacitance thus far observed are small. They

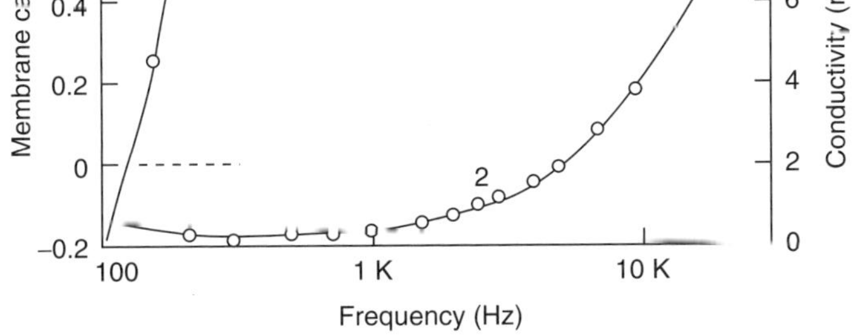

FIGURE 5.10
Membrane capacitance (Curve 1) and conductivity (Curve 2) of squid giant axon at various frequencies. Note the anomalous behavior at low-frequencies. (Note: 1 μF/cm^2.) (From Takashima, S., in *Biological Effects of Nonionizing Radiation*, ACS Symposium Series, No. 157, Illinger, K.H., Ed., 133–145, 1981.)

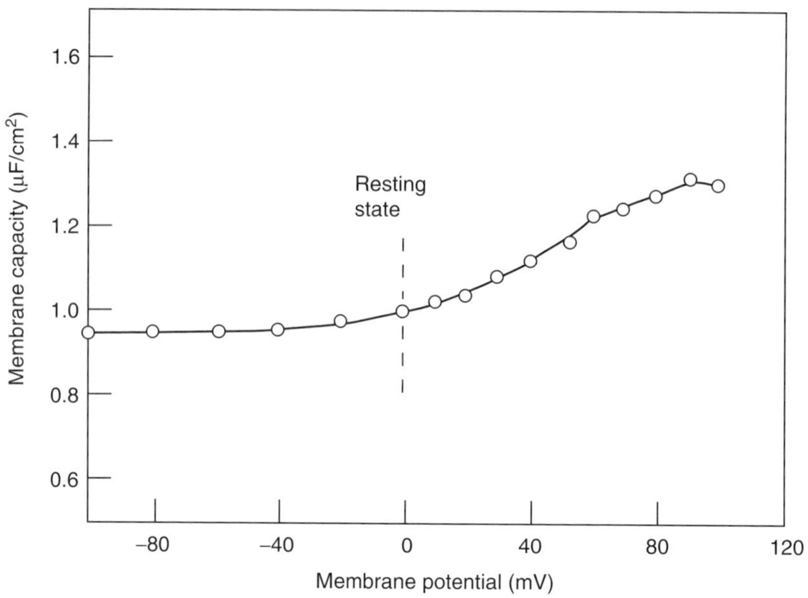

FIGURE 5.11
Membrane capacitance of squid giant axon at various membrane potentials. Membrane potential was shifted by injecting currents. The abscissa shows the actual potential across the membrane in millivolts. (From Takashima, S., in *Biological Effects of Nonionizing Radiation*, ACS Symposium Series, No. 157, Illinger, K.H., Ed., 133–145, 1981.)

appear likely to be more important in providing an understanding of the possible gating mechanism in membranes than as a mechanism for introducing rectification.

Another form of nonlinearity in the electrical response of cells comes about in what is often described as adaptive processes. For example, we found that repetitive exposures of pacemaker cells (taken from the ganglion of an *Aplysia)* to microwave pulses resulted in a decreasing reduction in the firing rate by successive pulses. This kind of change has also been shown to occur in neurons that have been conditioned with repetitive stimulation. Studies of conditioning have shown decreases in potassium ion conductance through membranes, thus raising the internal potential and enhancing the excitability [81,82]. The decrease in resistance between adjacent cells can occur in two ways. First, the resistance of gap junctions may be reduced by repetitive electrical stimulation, which increases the electrical coupling between the cells by up to 62%. Second, repetitive electrical stimulation can modify the chemical excitatory postsynaptic potential by amounts ranging from 31% to 140% [83]. This change is associated with the movement of protein kinase C from the interior of the cell into the membrane. An accompanying change in Ca^{2+} concentrations and the movement of a second messenger, diacylglycerol, into the membrane reduce the potassium ion flow. This enhanced excitability reduces the voltage or the charge required to initiate an action potential. If charge is transferred efficiently between cells, either actively or passively, cell length is effectively multiplied in the linear model by the number of cells in the chain; this, in turn, reduces the external electric field required to generate a given voltage across a terminating membrane (see Chapter 11 in *BMA* by A. Pilla).

An interesting speculation that is raised by these adaptive processes is whether or not a neural network can be trained to identify a repetitive signal such as 60 Hz in the presence of larger electric fields generated by the surrounding biological material. To test this hypothesis, we programmed a computer to simulate a neural network as shown in Figure 5.12 [65]. Using a backpropagation algorithm to adjust the connecting weights between neurons, a sigmoidal summing junction to model the neurons, and a pseudorandom noise generator,

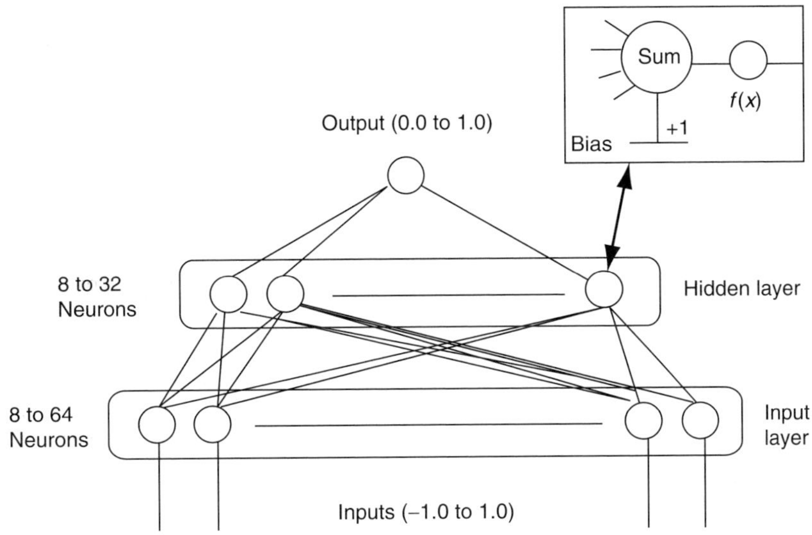

FIGURE 5.12
Backpropagation neural network. (From Barnes, F.S., *Bioelectromagn. Suppl.*, 1, 67–85, 1992. With permission.)

we measured the number of runs required to train the network to recognize a 60 Hz signal with 97% accuracy as a function of the input signal-to-noise ratio. The results in Figure 5.13 show that the training time increased from about 200 runs to about 1400 runs as the signal-to-noise ratio decreased from 1 to 0.001. The way the noise is presented to this network during the training makes a difference. For example, if you want the network to separate 59 Hz from 60 Hz, it helps to tell the network that 59 Hz is noise. This computer network model is clearly too simple to describe a biological nervous

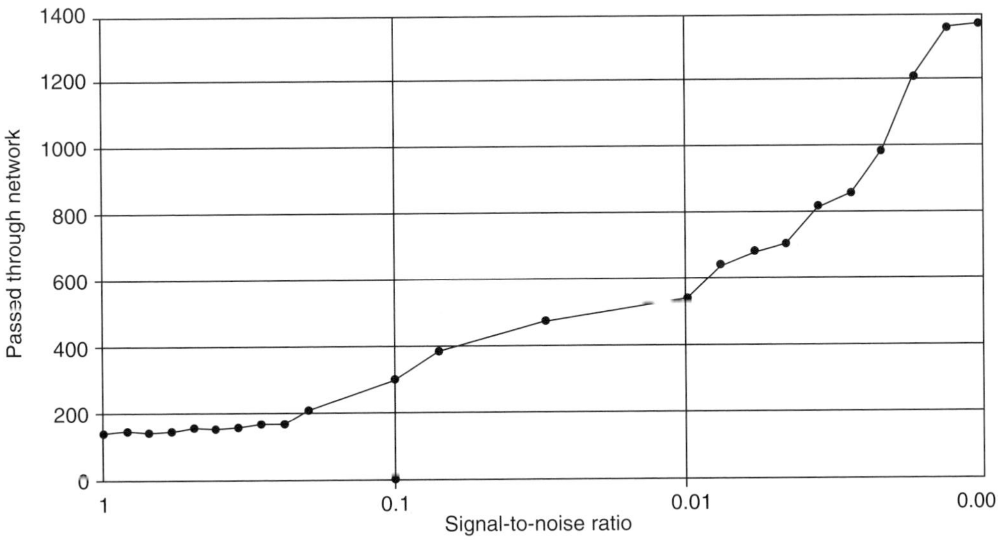

FIGURE 5.13
The learning response of a neural network with 64 input neurons, 8 neurons in the hidden layer, and 1 output neuron to a 60 Hz input signal and a pseudorandom noise signal with a decreasing signal-to-noise ratio. (From Barnes, F.S., *Bioelectromagn. Suppl.*, 1, 67–85, 1992. With permission.)

system, but it may provide a clue to one way in which a collection of cells may be able to respond to weak, externally applied electric fields, but a single cell would not.

5.7 Thermal Effects

One important effect of current flow due to electric fields is heating. The power input to a given volume of material can be expressed by $P' = I^2R$, where I is the total current and R is the resistance of the sample. For many calculations, a more useful expression is given by the power per unit volume or $P = \sigma E^2$, where σ is the conductivity, E is the electric field intensity. (For a more complete treatment of heating, see Chapter 10 and Chapter 12 of *BMA*.) The temperature rise resulting from this heat input is determined by the thermal capacity of the volume and the mechanisms for carrying the heat energy away. Typically, these thermal loss mechanisms include a combination of conduction and convection processes. For short current pulses, the heat dissipation is usually dominated by thermal conduction, and the basic equation for the rate of change of temperature is given by

$$\frac{\partial T}{\partial t} = \frac{P}{\rho' C_p} - \frac{T - T_0}{\tau_c} \tag{5.58}$$

where T is the temperature, T_0 is the initial temperature, t is time, and P is the power supplied per unit volume. ρ' is the density of the material (in kg/m^3), C_p is the specific heat under constant pressure, and τ_c is the thermal relaxation time.

If we consider a homogenous sphere of radius a immersed in an infinite fluid, the thermal conductive relaxation time is approximately given by

$$\tau_c = \frac{a^2}{4\overline{K}} \tag{5.59}$$

where $\overline{K}$ is the thermal diffusivity and is measured in meter square per second [84]. The thermal diffusivity is given by

$$\overline{K} = \frac{K'}{\rho' C_p} \tag{5.60}$$

where K' is the thermal conductivity (in cal/m sec °C), ρ' is the material density (in kg/m^3), and C_p is the thermal capacity (in cal/°C kg). If an applied current pulse is short compared to τ_c, the maximum temperature change is given by

$$\Delta T_{\max} = \left(\frac{3}{2\pi e}\right)^{3/2} \frac{\overline{H}}{\rho' C_p a^3} \tag{5.61}$$

where $\overline{H}$ is the total input energy in calories and e is the base of natural logarithms [84]. For current inputs that are long compared to the thermal relaxation time τ_c, the peak temperature is determined by a balance between the input power and the dissipation process controlled by conduction and convection. It is interesting to note that if we assume the thermal properties of water as a first approximation to various kinds of tissue, then τ_c for a sphere with a equal to 1 μm is a little less than 2 μs. Since a sphere has the smallest surface to volume ratio, Equation 5.60 gives an upper bound on τ_c, and Equation 5.61 gives an upper bound on the peak temperature excursion for small structures and pulses that are short compared to τ_c. Simply stated, it takes high power densities and large differential absorption coefficients to get significant differential temperature rises in small biological structures.

For situations where the volume involved is a cubic millimeter or larger, the thermal time constant is controlled by the amount of blood flowing through the volume. In these cases, temperatures may be more easily measured than calculated since a complicated thermal and electrical boundary value problem would have to be solved to calculate the temperature rise. This is particularly true since the viscosity η and other thermal and electrical parameters such as ρ, C_p, $\overline{K}$, etc. are functions of temperature. For example, C_p for an artificial bilipid membrane is shown in Figure 5.14 [85]. Another example of the importance of change in temperature is the conductivity of saline,

$$\sigma \approx C_1[10^{[(1/T)+\alpha](1/b)}] \times 10^{-4}\,\mathrm{S/m} \tag{5.62}$$

where C_1 is the concentration of NaCl in milligram equivalents per liter, T is the absolute temperature, $\alpha \simeq 6.23 \times 10^{-3}$ degrees^{-1}, and $b \simeq 1.4 \times 10^{-3}$ degrees^{-1} [86]. In the range around 37.5°C, this means that a 5°C change in temperature corresponds to a little less than 9% change in conductivity [86].

Changes in temperature are important, not only because they change transport properties such as viscosity, mobility, and the diffusion coefficient D, but also because they change chemical reaction rates. Typical biochemical reactions can be described by an equation of the form

$$\frac{dS}{dt} = -K'S \tag{5.63}$$

where S is the fraction of the material that has undergone the chemical reaction, t is the time, and K' is the reaction rate [87]. K' is often given by

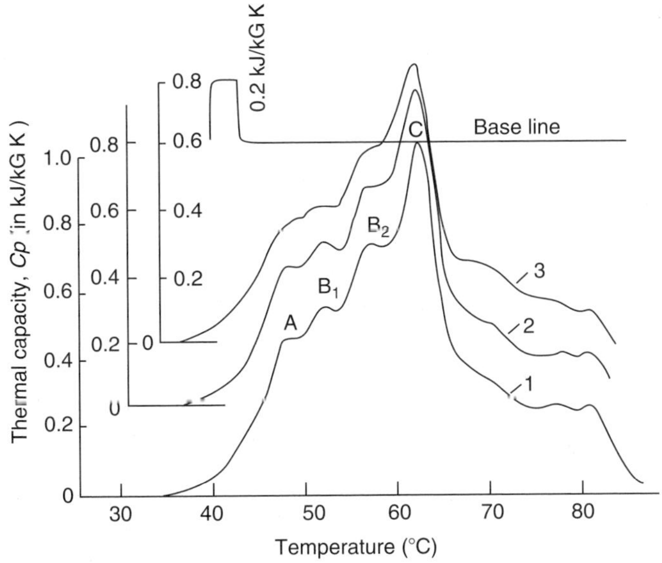

FIGURE 5.14
Differential changes in the heat capacity, C_p, of erythrocyte membranes as a function of temperature in 5 mmol/L sodium phosphate with pH 7.4 and a concentration of 5 mg protein per milliliter. The changes at A, B, B_2, and C correspond to changes in the structure of the membrane with temperature and are irreversible. Curve 1: intact membranes. Curve 2: irradiated at 330 mHz for 5 min (SAR 9 W/kg). Curve 3: irradiated at 300 mHz for 30 min (SAR W/kg). (From Shuyrou, V.L., Zhodan, G.G., and Akorv, I.G., Academy of Science, Institute of Biophysics, Pushchino, Moscow Region, Russia, personal communication, 1984. With permission.)

$$K' = \frac{kT}{h} \exp\left(\frac{+\Delta H' - T\Delta S'}{R'T}\right) \tag{5.64}$$

where k is the Boltzmann constant, T is the absolute temperature, H' is the free energy, S' is the entropy, h is Plank's constant, and R' is the gas constant. The significant feature is that the reaction rate K' varies exponentially with temperature, and $\Delta H'$ and $T\Delta S'$ are large numbers. Thus, very small changes in temperature can lead to big changes in chemical reaction rates.

In addition to chemical reaction rate changes, there may be changes in the binding of the proteins to cell membranes that lead to a shedding of proteins with a small increase in temperature. An exponential temperature dependence of the binding to membrane receptors is to be expected just as it is for chemical reactions [88].

A rule of thumb that the author uses to estimate whether or not significant biological changes are likely is to see if ΔT is $>10°C$ for 10^{-6} sec, 5°C for 1 sec, or 2°C for hours. If the ΔTs are larger, then they can be expected to lead to important changes in the biological system. Typical mammalian temperature regulatory systems will hold the internal body temperature constant to within +/−0.5°C.

In addition to the magnitude of the temperature change, it can be shown that the rate of temperature rise, dT/dt, is important and can induce current to flow across membranes. Changes in the firing rate of pacemaker cells from the ganglion of *Aplysia* have been induced by total temperature changes of as little as 1/10°C when the rates of change are about 1°C/sec [89]. This change of the firing rate corresponds to the injection of approximately 1 nA into the cell. By taking the time derivative of the Nernst equation, which describes the passive equilibrium potential across a membrane for a single ion, it can be shown that a current proportional to the temperature derivative is to be expected, or

$$I = -qV_1'C_1\left(\frac{\phi}{\phi_T}\right)\left(\frac{\dot{\phi}}{\phi} - \frac{\dot{T}}{T}\right) \tag{5.65}$$

where q is the charge of the ion, V_1' is the volume of the cell, C_1 is the concentration of ions inside the cell, ϕ is the resting potential, ϕ is given by $\phi_T = \frac{kT}{q}$, $\dot{\phi}$ is the derivative of the membrane potential with respect to time, T is the temperature, and $\dot{T}$ is the temperature derivative with respect to time [90].

Bol'shakov and Alekseyev [91] have observed similar changes in the firing rate of pacemaker cells taken from the large parietal ganglion of the central nervous system of *Limnea stagnalis*. In their experiments they observed a slow increase in temperature (1°C/min or slower) to increase the firing rate of the pacemaker cell and a rapid increase in temperature (0.1°C/sec or faster) to decrease or stop the firing. They ascribe these changes to changes in the Na^+ pump as the rapid temperature effect was completely blocked by adding ouabain to the solution. In addition to the changes in the Na^+ currents, Ca^+ currents have been shown to be sensitive to rapid changes in temperature [92]. The rate of rise has also been shown to be significant in exciting a brain slice from a mouse with pulses of 10^{-3} sec and peak temperature rises of less than 0.5°C [93].

Temperature rises also lead to thermal expansion, and rapid temperature rises lead to the generation of acoustic waves [94]. These acoustic waves, in turn, can affect stretch receptors in nerve cells and other tissue and thus generate a biological response that may be at a considerable distance from the electrical heating [95].

To get an idea of the magnitudes of both heating (as described in Equation 5.58) and the effect of the rate of rise (as given by Equation 5.65), consider the case of liver tissue with $\sigma = 0.14$ S/m and a field strength in the tissue of 2×10^3 V/m. The rate of temperature rise is approximately 13°C/sec assuming no conduction or convection heat losses and the thermal capacity of water. For this high field, a significant temperature rise occurs in about 1/2 sec. However, the rate of rise has been shown to be significant in exciting a brain slice from a mouse with pulses of 10^{-3} sec [96].

5.8 Natural Fields and Man-Made Fields

It is of interest to compare man-made with naturally occurring fields. First, we would like to know the approximate magnitudes of the fields that occur in nature outside man or the biological system of interest. Second, we would like to have values for the internal or physiological fields.

The natural electric fields at the surface of the earth have both DC and AC components [97]. One may think of the earth as a spherical capacitor where the surface is negatively charged with respect to an electrical conducting ionosphere that is about 50 km above the surface. This capacitor is being continuously charged by about 100 lightning strokes per second from thunderstorms worldwide. Since the atmosphere is a finite conductor, it also discharges with an *RC* time constant of about 18 sec. The result is an average electric field of about 130 V/m. This field is not uniform with height and typically falls off to 30 V/m at 1 km above the surface. The local values vary widely with temperature and humidity. In the Sahara during dust storms caused by winds in the dry season, a field of 1500 V/m has been measured with the polarity reversed from the normal. In thunderstorms, fields of up to 3000 V/m have been measured without lightning, and the polarity has been known to reverse in minutes. Storms as far as 50 km away have been shown to affect local fields. See Chapter 1 in this volume by Mild and Greenebaum for more details.

The atmosphere is a relatively poor conductor and as such will suspend a significant number of charged ions, dust particles, etc. This helps to contribute to local field variations of 20% to 50% over the course of the day and is a normal characteristic of our environment. The level of natural AC fields in the atmosphere falls very rapidly from a DC value of about 130 V/m [97]. The average value of the vertical component of the electric field above 1 Hz has a typical value of 10^{-4} V/m $\text{Hz}^{1/2}$. However, this value fluctuates widely with the time of day, the season of the year, and location. Additionally, the Schumann resonances impose multiple-cavity resonances on this spectrum with a periodicity of about 10 Hz. These resonances may be explained in terms of standing waves in a cavity formed by the earth and the atmosphere. These very low levels of the natural fields are one of the reasons why electronic communications in the ELF band are useful for ships at sea and submarines. However, because of the very low level of the natural atmospheric fields at frequencies above a few hertz, there is very little reason for biological organisms to develop natural protection against perturbations at these frequencies. It also means that biological systems could communicate internally at these frequencies using very low signal power levels and still maintain a good signal-to-noise ratio.

The signals generated within the body are the result of nerve firing and other cell activity. A typical nerve cell fires with an action potential of 50 to 100 mV and transmits a current pulse about 0.4 ms long [98]. The rise time for this current spike is approximately 0.1 ms, and the fall time is about 0.5 ms. Each pulse is followed by a refractory period that is typically on the order of 1 to 3 ms. The longitudinal fields along the exterior of a nerve cell membrane are estimated to have a maximum value of about 5×10^{-2} V/m during an action potential when the cell is surrounded by a relatively high conductivity fluid of 5 S/

m [98]. If we look at these signals closely, it will be noted that the interspike interval along any given nerve cell fluctuates in time. Additionally, variations in the beat-to-beat intervals for the ECG are random or chaotic, and the period can vary up to 30%. This is frequently seen, particularly at slow heart rates.

In looking at the natural fields in the body, we have two concerns. The first is how large an external signal takes to perturb the ongoing natural signal that is being used to communicate or control some biological process [99]. The second is how much of the signal field typically leaks away from active nerve fibers or bundles to form a background noise environment for surrounding tissue and processes. Regarding the first of these questions, it is interesting to look on the microscopic level at the electrical noise, i.e., the fluctuations that occur fundamentally as a result of the electrical process itself.

The first of several sources of noise that are always present is blackbody radiation, or Johnson noise, which is given by

$$P_{\mathrm{n}} = kTB \tag{5.66}$$

where P_{n} is the noise power, k is the Boltzmann constant, T is the absolute temperature, and B is the bandwidth [100–102]. The voltage equivalent of this noise power, which can be delivered to a matched load, or the mean squared voltage fluctuation $\overline{V}_{\mathrm{n}}^{2}$ across a resistance R, is given by [101,102]

$$\overline{V}_{\mathrm{n}}^{2} = 4kTBR \tag{5.67}$$

or by the mean squared current fluctuations

$$\overline{i}_{\mathrm{n}}^{2} = \frac{4kTB}{R} \tag{5.68}$$

Johnson noise applies to systems at thermodynamic equilibrium. Living systems are not at thermodynamic equilibrium. Thus, the foregoing expressions must be applied with caution to only those portions of biological systems where thermodynamic equilibrium is a good approximation. In the case of lasers, the spontaneous emission noise associated with the nonequilibrium population inversion of the energy levels can be obtained from Planck's radiation law by defining a negative temperature that assumes a Boltzmann distribution of atoms with N_2 atoms in the excited energy level E_2, which is greater than the N_1 atoms in the energy level E_1, such that

$$\frac{N_1}{N_2} = \exp\left[\frac{E_2 - E_1}{kT}\right] \tag{5.69}$$

In this case, the spontaneous emission noise $P_{\mathrm{n}} = h\upsilon B$, where h is Planck's constant and υ is the frequency of the radiation corresponding to a transition from E_2 to E_1 [100]. In those situations where the nonequilibrium characteristic may be described by an amplifier that can be modeled by a negative resistor or by energy storage in an inverted population distribution, the concept of a negative temperature may be a useful approach. Note that an equivalent temperature, T, is a convenient way to describe the energy distribution of a large number of particles. A much more complete description of nonequilibrium noise is given in Chapter 7 in this volume.

The second source of noise that is also present is the shot noise, which is given by

$$\overline{i}_{\mathrm{n}}^{2} = 2q\overline{I}_{\mathrm{DC}}B \tag{5.70}$$

where $\overline{i_n^2}$ is the mean-squared current fluctuation. This noise comes about because of the discreteness of the electronic charge q and the assumption that the motion of each charge is independent. With negative feedback, this noise may be reduced, as has been shown for space charge-limited diodes. Shot noise results in an AC fluctuation, $\overline{i_n^2}$, which is proportional to the average value of the current, $\bar{I}_{DC}$. A third source of noise is $(1/f)$ noise. This noise may be generated by many processes, some of which are described in Chapter 7 in this volume. $1/f$ noise can be synthesized from Gaussian noise by filtering it with a circuit that requires about one low pass state variable per decade for the period of time over which the model is used to generate noise with a power density spectrum $S(f) = (C/f^{\alpha})$, where C is a constant and α is a constant between 1 and 2 [103]. We can expect to find this kind of noise for processes that evolve with time and, or have memory. $1/f$ noise describes the power spectral density of the fluctuations at low-frequencies in such diverse phenomena as transistors, quartz crystal oscillators, the closing Dow Jones Averages for the stock market, and the weather. It is also generated by the flow of ion currents through an orifice and thus is a fundamental part of the transport of current through channels in membranes [102]. Measurements of the noise voltage across a 10 μm hole in a 6 μm Mylar film showed that for a wide range of ionic concentrations the voltage noise spectral density $S(f)$ is given by

$$\frac{S_{\phi}(f)}{\phi^2} = \frac{a}{bnr^3 f} \tag{5.71}$$

where b is a numerical geometric factor, n is the density of ions in the solution, r is the radius of the hole, a is a constant, and ϕ is the applied voltage. The data showed that $2.5 < a < 40$ with a mean value of 10 for a wide range of solutions including HCl, KCl, and AgNO, with concentrations from 0.05 to 5 mol.

For natural membranes, this noise has been shown to take the form of

$$S_E(f) = \frac{C_E}{f^{\alpha}} \tag{5.72}$$

where $0.7 < \alpha < 1.2$ with a mean close to $\alpha = 1$. For the frog node of Ranvier, the noise is a function of the membrane voltage as shown in Figure 5.15 [104]. The dominant source of this noise appears to be the K^+ current, and it has a minimum when the membrane is biased, so that this K^+ current is biased to zero.

To get an estimate of the size of these noise sources, let us consider a pacemaker cell from the abdominal ganglion of *Aplysia*. This cell fires 20-ms pulses at about 1 Hz/sec. It has a resting voltage of about 50 mV and a resistance R measured with a microelectrode between the inside of the cell and the surrounding solution of approximately $10^6\ \Omega$. If we assume a system bandwidth of 100 Hz and $T = 300$ K, the Johnson noise voltage would be $\bar{V}_n \approx 3 \times 10^{-6}$ V. This gives a resting potential-to-noise ($\bar{V}_n$) ratio of about 4×10^4. The peak current flow in these cells is estimated to be about 10^{-7} A, and thus the estimated shot noise current is $\bar{i}_n \approx 2 \times 10^{-12}$ A, and the ratio of the peak current to the noise current is about 2×10^4. We do not have the available value $S(f)$ for the *Aplysia*, $\bar{v}_{\phi} = \sqrt{S(f)B}$, where B is the bandwidth. If it is assumed that the maximum value of the noise is the same as that of the frog node of Ranvier, then for a bandwidth of 1 Hz we get $\bar{v}_{\phi} = 1.4 \times 10^{-5}$ V at a center frequency of 1 Hz from the curve for −50 mV in Figure 5.15. This is about a factor of 10^3 greater than the Johnson noise. It is likely that $(1/f)$ noise is the largest source of noise at the cell membranes for frequencies below 160 Hz [105,106].

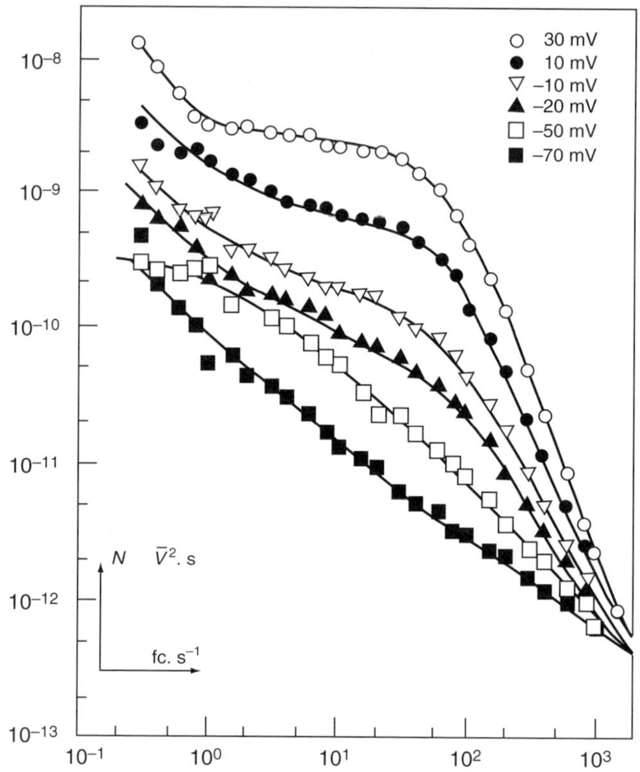

FIGURE 5.15
Voltage noise spectra of a frog node of Ranvier at different levels of membrane potential. (From Sichenga, E. and Verveen, A.A., in *Proceedings of the 1st European Biophysics Congress*, Vol. 5, Verlag Wiener Medizinischen Akademic, Vienna, Austria, 219, 1971. With permission.)

These fundamental sources of noise, which are generated by random fluctuations in the position of ions and their transport through channels, are spatially incoherent [73]. For many processes the important quantity in deciding whether or not an electrical signal is biologically important is the signal-to-noise ratio S/N where S is the power in the signal and N is the noise power. Typically, it is assumed that a signal-to-noise ratio of 1 is required for an externally applied signal to be detectable. In the foregoing discussion for both thermal and shot noise the movement of each charge was assumed to be statistically independent of each other. If for an externally applied signal the openings of the channels in a cell are excited in parallel and coherently and the noise is generated by incoherent random firing, then the signal-to-noise ratio increases with the square root of the number of channels. Similarly, the signal-to-noise ratio for a bundle of nerves would be expected to increase with the square root of the number of nerves for a signal applied externally to the whole bundle. Thus, a collection of cells can be expected to detect smaller signals than a single cell.

In addition to the electrical noise generated by currents and voltages that are part of the single-cell operation, electrical signals propagate through the body as a result of the incomplete confinement of electrical signals propagating in nerve cells. In a sense, these signals may be thought of as noise if they are not pertinent to the activity in that portion of the body through which they are propagating. If, on the other hand, they are used by tissue within the organism at some distance from the source, they must be thought of as signals. In the brain, the fraction of these signals that reach the scalp is called the electroencephalogram (EEG). The EEG is obtained by placing two or more conducting electrodes on the scalp and measuring the voltage between them. For electrodes placed 5 cm apart, the peak-to-peak voltages range up to 30 μV [107]. The author views this voltage as the integral of the vector sum of the leakage fields from the firing of the nerve cells in the brain between the two electrodes. Since there are a very large number of cells firing, most of the 50 mV signals from an isolated nerve are canceled by summing over many like cells firing at different times and by the attenuation caused by propagation through the tissue. Estimates of surface potential gradients along a nerve fiber range from 3×10^{-4} to 5×10^{-2} V/m, and the corresponding current densities external to the nerve cells range from 5×10^{-2} to 4 A/m^2 [107,108]. The EEG voltage has a strong periodic component (particularly during sleep) near 10 Hz, which is known as the alpha (α) wave. A peak amplitude of this component may be as large as 50 μV when measured at the surface of the scalp. It is interesting to note that the EEG signal contains significant information on the brain's activity, and a few individuals have been trained to control these signals so as to control a computer in way that allows them write messages.

At the surface of the chest, a signal may be recorded between two electrodes known as the ECG or EKG (electrocardiogram). This signal results from the highly coordinated firing of the cells in the heart and has a definite wave shape that is closely related to the operation of the heart. The peaks of the so-called R wave in this signal may range up to 2.5 mV and are typically 0.5 to 1.5 mV, depending on the placement of the electrodes, the amount of body fat, etc. The pulse repetition rate is usually in the range of 1 to 2 Hz, and the ''QRS spike'' of the typical cardiogram is 40 ms long. Again, the signal measured at the surface of the skin is the result of leakage from electrically active cells located at a distance. The estimated current density near the firing heart cell ranges up to 1 A/m^2 [109]. In this case the shape of the signal reaching the skin is so closely related to the activity of the heart that it provides detailed information on heart function.

One result of electric discharges in the atmosphere, as well as natural ionizing radiation, is the creation of small positive and negative ions in the atmosphere. In clean country or mountain air, the typical ion density is about 10^{10}/m^3 with an average ion lifetime of a few minutes [110]. When a hot dry wind is blowing, positive ions created by the shearing forces can increase in concentration significantly. It has been shown that increases in the negative ion concentration reduce the amount of serotonin (5-HT) in mice and rabbits, possibly by accelerating the enzymatic oxidation process [111]. A similar result has been demonstrated in the oxidation of cytochrome *c*. Positive ions appear to block monoamine oxidase action, thus raising the concentration of free 5-HT [111]. Changes in 5-HT levels produce significant changes in the central nervous system, with high levels of positive ions raising the anxiety levels under stress. Other effects of increased positive ion concentration include a decrease in the survival rate of mice exposed to a measured dose of influenza virus, while an increase of negative ions reduced the mortality rate [111].

The significance of these results is that it is relatively easy to change the ion concentration in air using high-voltage DC systems where a leakage current of 1 μA from a burr or other sharp point would correspond to the generation of about 10^{12} ions per second.

Relatively few high-voltage DC transmission lines are in use today for distribution of power. Because the shocks resulting from a short contact across a high DC voltage are so painful and obviously dangerous, these systems are nearly always shielded. Thus, one is rarely exposed to DC electric fields $>10^3$ V/m. An additional feature of this exposure is that air is such a good insulator that the DC currents flowing through the body in a noncontacting situation are very small, as explained in the Introduction. For example, 1000 V across a 1-cm gap would yield a current density of approximately 10^{-7} A/m^2 flowing across the air gap. Thus, the principal hazards from DC fields occur when parts of the body make contact with a conductor.

5.9 Discussion and Summary

In this review some of the physical mechanisms by which DC and time-varying electric fields affect biological systems are presented. A few typical values of electric field strengths and current densities that are known to affect the biological system are compared with those of natural fields and other forces. Some values of electric fields and their gradients that are shown to modify the currents and shift energy levels are given. These in turn are shown to modify chemical reaction rates, which can lead to changes in the growth cells and other characteristics of biological systems. It is hoped that this information will help the readers to make their own estimates of when a given exposure to electric fields will be significant in modifying biological systems and provide a basis for understanding some of the biological results presented in other chapters of this handbook.

Acknowledgments

The author wishes to express appreciation to Mikhail Zhadin, Howard Wachtel, Maria Stuchly, Ross Adey, Mike Marron, Elliot Postow, Charles Polk, and many students for their many helpful comments and suggestions and to Adam Sadoff for help in the preparation of the text. He also wishes to express his appreciation to the Office of Naval Research under Contract N00014-81-K-0387, the Mobile Manufactures Form, and the University of Colorado for financial support of his work in this area.

References

1. Yariv, A., *Optical Electronics in Modern Communications*, 5th edition, Oxford University Press, Oxford, UK, 1997.
2. Barnes, F.S. and Hu, C.L., Nonlinear interacations of electromagnetic waves with the biological materials, in *Nonlinear Electromagnetics*, Uslenghi, P.L.L., Ed., Academic Press, New York, 391–426, 1980.
3. Singer, S.J. and Nicolson, G.L, The fluid mosaic model of the structure of cell membranes, *Science*, 175, 720–731, 1972.
4. Poo, M., *In situ* electrophoresis or membrane components, in *Annual Review of Biophysics and Bioengineering*, Vol. 10, Mullins, L.J., Ed., Academic Press, New York, 245–276, 1981.
5. Kyriacous, D., *Modern Electroorganic Chemistry*, Springer-Verlag, Berlin, 1994.
6. Pohl, H.A., Dielectrophoresis, the behavior of neutral matter, in *Nonuniform Electric Fields*, Cambridge University Press, Cambridge, MA, 1978.
7. Kwon, Y. and Barnes, F.S., A theoretical study of the effects of RF fields in the vicinity of membranes, *Bioelectromagnetics*, 26, 2, 118–124, 2005.
8. Bier, M., Ed., *Electrophoresis: Theory, Methods, and Applications*, Vol. 2, Academic Press, New York, 1967.
9. Gascoyne, P.R. and Vykoukal, J.V., Dielectrophoresis-based sample handling in general-purpose programmable diagnostic instruments, *Proc. IEEE*, 92, 1, 22–41, 2004.
10. Rand, E.P., Interacting phospholipids bilayers: measured forces and induced structural changes, in *Annual Review of Biophysics and Bioengineering*, Vol. 10, Mullins, L.J., Ed., Academic Press, New York, 277–314, 1981.

11. Pethig, R., *Dielectric and Electronic Properties of Biological Materials*, John Wiley & Sons, New York, 184, 1979.
12. Parsegian, J.L., Long range physical forces in the biological milieu, in *Annual Review of Biophysics and Bioengineering*, Mullins, L.J., Ed., Academic Press, New York, 1973, 221.
13. Lifshitz, F.M., *Zh. Eksp. Teor. Fiz.*, 29, 95, 1955; *Sov. Phys. JETP* 2, 73, 1956.
14. Iralelachvili, J., *Intermolecular and Surface Forces*, 2nd edition, Academic Press, San Diego, CA, 1992.
15. Tasao, Y.T., Evan, D.F., and Wennerstrom, H., Long-range attractive force between hydrophobic surfaces observed by atomic force microscopy, *Science*, 262, 1993.
16. Barnes, F.S., Ginley, H., and Shulls, W., AC electric field effects on bacteria, presented at the Third Annual Conference, Bioelectromagnetics Society (BEMS), Washington, DC, 1981.
17. Zumdahl, S.S., *Chemistry*, 2nd edition, D.C. Health Co., Lexington, MA, 546, 1989.
18. Atkins, P.W., *Physical Chemistry*, 4th edition, W.H. Freeman Co., New York, Chapter 26, 1990.
19. McCaig, C.D., Spinal neurite reabsorption and regrowth *in vitro* depend on the polarity of an applied electric field, *Development*, 100, 31–41, 1987.
20. Nuccitelli, R., Physiological electric fields can influence cell motility, growth, and polarity, *Adv. Cell Biol.*, 2, 213–233, 1988.
21. Erickson, C.A. and Nuccitelli, R., Embryonic fibroblast motility and orientation can be influenced by physiological electric fields, *J. Cell Biol.*, 98, 296–307, 1984.
22. Seto, Y.J. and Hsieh, S.T., Electromagnetic induced kinetic effects on charged substrates in localized enzyme systems, *Biotechnology and Bioengineering*, XVIII, 813–837, 1976.
23. Muraji, M., Nishimura, M., Tatebe, W., and Fujii, T., Effect of alternating magnetic field on the growth of the primary root of corn, *IEEE Trans. Magn.*, 28, 4, 1996–2000, 1992.
24. Towns, C. and Schawlow, A., *Microwave Spectroscopy*, McGraw-Hill, New York, 255–283, 1955.
25. Bublitz, G.U. and Boxer, S.G., Stark spectroscopy: applications in chemistry, biology, and materials science, *Ann. Rev. Phys. Chem.*, 48, 213–242, 1997.
26. Steiner, U. and Ulrich, T., Magnetic field effects in chemical kinetics and related phenomena, *Chem. Rev.*, 89, 47–151, 1989.
27. Guyton, A.C., *Textbook of Medical Physiology*, 8th edition, W.B. Saunders, Philadelphia, PA, 47, 1991.
28. Barnes, F., A model for the detection of weak ELF electric and magnetic fields, *Bioelectrochem. Bioenerg.*, 47, 207–212, 1999.
29. Alkon, D. and Rasmussen, H., A spatial-temporal model of cell activation, *Science*, 26, 998–1004, 1989.
30. Louisell, W., *Coupled Mode and Parametric Electronics*, John Wiley & Sons, New York, 1960.
31. Moss, F., Stochastic resonance, *Ber. Bunsenges. Phys. Chem.*, 95, 3, 302–311, 1991.
32. Gammaitoni, L., Stochastic resonance, *Rev. Mod. Phys.*, 70, 1, 223–287, 1998.
33. McNamara, B. and Wiesenfeld, K., Theory of stochastic resonance, *Phys. Rev. A*, 39, 9, 4854–4869, 1989.
34. MacGregor, R.J. and Lewis, E.R., *Neural Modelling*, Plenum Press, New York, 1977.
35. Cooper, M.S., Miller, J.P., and Fraser, S.E., Electropheretic repatterning of charged cytoplasmic molecules within tissues coupled by gap junctions by externally applied electric fields, *Dev. Biol.*, 132, 170–188, 1989.
36. Epstein, B.R. and Foster, K.R., Anisotropy in the dielectric properties of skeletal muscle, *Med. Biol. Eng. Compnt.*, 21, 51, 1983.
37. Kalmijn, A.J., Electric and magnetic field detection in elamobranch fishes, *Science*, 281, 916–918, 1982.
38. Kalmijn, A.J., Detection of weak electric fields, in *Sensory Biology of Aquatic Animals*, Atema, E.J. et al., Eds., Springer-Verlag, Berlin, Chapter 6, 1987.
39. McCleave, J.D, Rommel, S., and Cathcart, A.S., Weak electric and magnetic fields in fish orientation, in *Orientation: Sensory Basis*, Vol. 188, Adler, H.E., Ed., New York Academy of Sciences, New York, 270, 1971.
40. Ransom, B.P. and Eyring, H., Membrane permeability and electrical potential, in *Ion Transport across Membranes*, Clarke, H.J. and Nachmansohn, D., Eds., Academic Press, New York, 103, 1954.

41. Tsong, T.Y., Liu, D.S., and Chauvin, R., Electroconformational coupling (ECC): an electric field induced enzyme oscillation for cellular energy and signal transductions, *Bioelectrochem. Bioenerg.*, 21, 319–331, 1989 (a section of *J. Electroanal. Chem.* and constituting Vol. 275, Elsevier Sequoia, S.A., Lausanne, Switzerland, 1985).
42. Liu, D.S., Astumian, R.D., and Tsong, T.Y., Activation of Na^+ and K^+ pumping modes of (Na, K)-ATPase by an oscillating electric field, *J. Biol. Chem.*, 265, 13, 7260–7267, 1990.
43. Horn, L.W., A novel method of the observation of membrane transporter dynamics, *Biophys. J.*, 64, 281–289, 1993.
44. Triffet, T. and Green, H.S., Information and energy flow in a simple nervous system, *J. Theor. Biol.*, 86, 3, 1980.
45. Vaccaro, S.R. and Green, H.S., Ionic processes in excitable membranes, *J. Theor. Biol.*, 81, 771, 1979.
46. Eckert, R. and Ewald, D., Residual calcium ions depress activation of calcium-dependent current, *Science*, 216, 730, 1982.
47. Baumann, G. and Easton, G., Modeling state-dependent sodium conductance data by memoryless random process, *Math. Biosci.*, 60, 265, 1982.
48. Schauf, C.L. and Baumann, G., Experimental evidence consistent with aggregation kinetics in sodium current of myxicola giant axons, *Biophys. J.*, 35, 707, 1981.
49. Baumann, G. and Eastman. G.S., Charge immobilization linked to inactivation in the aggregation model of channel gating, *J. Theor. Biol.*, 99, 249, 1982.
50. Poo, M., Rapid lateral diffusion of functional ACh receptors in embryonic muscle cell membrane, *Nature*, 295, 332, 1982.
51. Poo, M. and Lam, J.W., Lateral electrophoresis and diffusion of concanavalin A receptors in the membrane of embryonic muscle cell, *J. Cell Biol.*, 76, 483, 1978.
52. Lee, K.Y.C., Lingler, J.B., and McConnnell, H.M., Electric field-induced concentration gradients in lipid monolayers, *Science*, 263, 655, 1994.
53. Hayashi, H. and Fishman, H., Inward rectifier K^+ channel from analysis of the complex conductance of aplysia neuronal membrane, *Biophys. J.*, 53, 747–757, 1988.
54. Barnes, F.S. and Hu, C.L., Model for some non thermal effects of radio and microwave fields on biological membranes, *IEEE Trans. Microwave Theory Technol.*, 25, 742, 1977.
55. Montaigne, K. and Pickard, W.F., Offset of the vascular potential of characean cells in response to electromagnetic radiation over the range of 250 Hz to 250 kHz, *Bioelectromagnetics*, 5, 31, 1984.
56. Pickard, W.F. and Rosenbaum, F.J., Biological effects of microwaves at membrane level: two possible athermal electrophysiological mechanisms and a proposed experimental test, *Math. Biosci.*, 39, 235, 1978.
57. Pickard, W.F. and Barsoum, Y.H., Radio-frequency bioeffects at the membrane level: separation of thermal and athermal contributions in the Characeae, *J. Membrane Biol.*, 61, 39, 1981.
58. Kalkwarf, D.R., Frasco, D.L., and Brattain, W.H., Current rectification and action potentials across thin lipid membranes, in *Physical Principles of Biological Membranes*, Snell, F., Wolken, J., Iverson, G., and Lam, J., Eds., Gordon & Breach, New York, 1970.
59. Franceschetti, G. and Pinto, I., Cell membrane nonlinear response to applied electromagnetic field, *IEEE Trans. Microwave Theory Technol.*, 32, 7, 1984.
60. Casaleggio, A., Marconi, L., Morgavi, G., Ridella, S., and Rolando, C., Current flow in a cell, with a nonlinear membrane, stimulated by an electric field, personal communication, 1985.
61. Cain, C.A., Biological effects of oscillating electric fields: role of voltage sensitive ion channels, *Bioelectromagnetics*, 2, 23, 1981.
62. Bisceglia and Pinto, I., Volterra series solution of Hodgkin–Huxley equation, personal communication, 1984.
63. Wachtel, H., Firing-pattern changes and transmembrane currents produced by low frequency fields in pacemaker neurons, in *Proceedings of the 18th Annual Hansford Life Science Symposium*, Technical Information Center, U.S. Department of Energy, Richland, WA, 132–147, 1978.
64. Adler, R., A study of locking phenomena in oscillators, *Proc., IEEE* 34, 351, 1946.
65. Barnes, F.S., Smoller, A., and Sheppard, A., Injection locking of pacemaker cells, in *Proceedings of the 9th Annual Conference*, IEEE Engineering in Medicine and Biology Society, 1986.

66. Fohlmeister, J.F., Adelman, W.J., Jr., and Poppele, R.E., Excitation properties of the squid axon membrane and model systems with current stimulation, *Biophys. J.*, 30, 79–98.
67. Barnes, F.S., Some engineering models for interactions of electric and magnetic fields with biological systems, *Bioelectromagn. Suppl.*, 1, 67–85, 1992.
68. Kroupa., V.F., *Frequency Synthesis*, John Wiley & Sons, New York, 1973.
69. Koss, D.A. and Carstensen, E. L, Effects of ELF electric fields on the isolated frog heart, *IEEE Trans. Biomed. Eng.*, 30, 347, 1983.
70. Litovitz, T.A., Krause, D., and Mullins, J.M., Effect of coherence time of the applied magnetic field on ornithine decarboxylase activity, *Biochem. Biophys. Res. Commun*, 178, 3, 862–865, 1991.
71. Mullins, J.M., Krause, D., and Litovitz, T.A., Simultaneous application of a spatially coherent noise field blocks use response of cell cultures to a 60 Hz electromagnetic field, in *Proceedings of the 1992 World Congress on Bioelectromagnetism*, Bioelectroemagnetics Society, Orlando, FL, 1992.
72. Farrell, J.M., Barber, M., Krause, D., and Litovitz, T.A., The superposition of a temporally incoherent magnetic field inhibits 60 Hz-induced changes in the ODC activity of developing chick embryos, *Bioelectromagnetics*, 19, 53–56, 1998.
73. Litovitz, T.A., Montrose, C.J., Doinov, P., Brown, K.M., and Barber, M., Superimposing spatially coherent electromagnetic noise inhibits field induced abnormalities in developing chick embryos, *Bioelectromagnetics*, 15, 2, 105–113, 1994.
74. Walleczek, J. and Liburdy, R.P., Nonthermal 60 Hz sinusoidal magnetic-field exposure enhances $^{45}Ca^{2+}$ uptake in rat hymocytes: dependence on mitogen activation, *Fed. Eur. Biochem. Soc. Lett.*, 271, 1/2, 157–160, 1990.
75. Liburdy, R.P., Calcium signaling in lymphocytes and ELF fields, *Fed. Eur. Biochem. Soc.*, 301, 1, 53–59, 1992.
76. Liburdy, R.P., Biological interactions of cellular systems with time-varying magnetic fields, *Ann. N.Y. Acad. Sci.*, 649, 74–94, 1992.
77. Yost, M.G. and Liburdy, R.P., Time-varying and static magnetic fields act in combination to alter calcium signal transduction in the lymphocyte. *Fed. Eur. Biochem. Soc.*, 296, 2, 17–122, 1992.
78. McLeod, K, J., Lee, R.C., and Ehrlich, H.P., Frequency dependence of electric field modulation of fibroblast protein synthesis, *Science*, 136, 1465–1469, 1987.
79. Takashima, S., Non-linear properties of nerve membranes, *Biophys. Chem.*, 11, 447, 1980.
80. Berkowitz, G.C. and Barnes, F.S., The effects of nonlinear membrane capacity on the interaction of microwave and radio frequencies with biological materials, *IEEE Trans. Microwave Theory Technol.*, 27, 204, 1979.
81. Alkon, D.L., Memory storage and neural systems, *Sci. Am.*, 261, 42–50, 1989.
82. Alkon, D.L. and Rasmussen, H., A spatial temporal model of cell activation, *Science*, 26, 998–1004, 1988.
83. Yang, X., Korn, H., and Faver, A.S., Long-term potentiation of electronic coupling at mixed synapses, *Nature*, 348, 542–545, 1990.
84. Hu, C.L. and Barnes, F.S., The thermal chemical damage in biological materials under laser irradiation, *IEEE Trans. Biomed. Eng.*, 17, 220, 1970.
85. Shuyrou, V.L., Zhodan, G.G., and Akorv, I.G., Effects of 330 MHz radio frequency, Academy of Science, Institute of Biophysics, Pushchino, Moscow Region, Russia, personal communication, 1984.
86. Trautman, E.D. and Newbower, R.S., A practical analysis of the electrical conductivity of blood, *IEEE Trans. Biomed. Eng.*, 30, 141, 1983.
87. Johnson, F. H, Eyring, H., and Stover, B.J., *The Theory of Rate Processes in Biology and Medicine*, John Wiley & Sons, New York, 1974.
88. Liburdy, R.P. and Penn, A., Microwave bioeffect in erythrocyte are temperature and pO dependent action permeability and protein shedding occur at the membrane phase transition, *Bioelectromagnetics*, 5, 283–291, 1984.
89. Chalker, R., The effect of microwave absorption and associated temperature dynamics on nerve cell activity in Aplysia, M.S. thesis, University of Colorado, Boulder, 1982.
90. Barnes, F.S., Cell membrane temperature rate sensitivity predicted from the Nernst equation, *J. Bioelectromagn.*, 5, 113, 1964.

91. Bol'shakov, M.A. and Alekseyev, S.I., Change in the electrical activity of the pacemaker neurons of *L. stagnalis* with the rate of their heating, *Biophysics*, 31, 569–571, 1986.
92. Alekseyev, S.I., IL'in, V.I., and Tyazhelov, V.V., Effect of electromagnetic radiation in decimeter wavelength range on calcium current of mollusk neurons, *Biophysics*, 31, 2, 290–295, 1986.
93. Philippova, T.M., Novoselov, V.I., Bystrova., M.F., and Alekseev, S.I., Microwave effect on camphor binding to rat olfactory epithelium, *Bioelectromagnetics*, 9, 347–354, 1988.
94. Bushanam, S. and Barnes, F.S., Laser generated thermoelastic shock waves in liquids, *J. Appl. Phys.*, 46, 5, 2074–2082, 1975.
95. Mihran, R.T., Barnes, F.S., and Wachtel, H., Temporally-specific modification of myelinated axon excitability *in vitro* following a single ultrasound pulse, *Ultrasound Med. Biol.*, 16, 3, 297–309, 1990.
96. Ady, G., McNaughton, B.L., and Wachtel, H., A system for recording microwave effects on isolated mammalian brain slices, presented at the 5th Annual Conference, Bioelectromagnetics Society, Boulder, CO, 1983.
97. Polk, C., Sources, propagation amplitude and temporal variations of extremely low frequency (0–100 Hz) electromagnetic fields, in *Biological and Clinical Effects of Low-Frequency Magnetic and Electric Fields*, Llaurado, J.G., Sances, A., Jr., and Battocletti, J.H., Eds., Charles C Thomas, Springfield, IL, 21, 1974.
98. Plonsey, R., *Bioelectric Phenomena*, McGraw-Hill, New York, 1969.
99. Barnes, F.S. and Seyed-Madani, M., Some possible limits on the minimum electrical signals of biological significance, in *Mechanistic Approaches to Interactions of Electric and Electromagnetic Fields with Living Systems*, Blank, M. and Findl, E., Eds., Plenum Press, New York, 339–349, 1987.
100. Yariv, A, *Introduction to Optical Electronics*, Holt, Reinhart & Winston, New York, 1976.
101. MacDonald, D.K.C., *Noise and Fluctuations: An Introduction*, John Wiley & Sons, New York, 1962.
102. Beck, A.H.W., *Statistical Mechanics, Fluctuations, and Noise*, Edward Arnold, London, 1967.
103. Keshner, M.F., 1/f noise, *Proc. IEEE*, 70, 3, 212–218, 1982.
104. Sichenga, E. and Verveen, A.A., The dependence of the 1/f noise intensity of the diode of Ranvier on membrane potential, in *Proceedings of the 1st European Biophysics Congress*, Vol. 5, Verlag Wiener Medizinischen Akademic, Vienna, Austria, 219, 1971.
105. Dorset, D.L. and Fishman, H.M., Excess electrical noise during current flow through porous membranes separating ionic solutions, *J. Membrane Biol.*, 21, 291–301, 1975.
106. Verveen, A.A. and DeFelice, L.J., Membrane noise, *Prog. Biophys. Mol. Biol.*, 28, 189, 1974.
107. Nuney, P.L. and Katznelson, R. D., *Electric Fields of the Brain—The Neurosphysics of EEG*, Oxford University Press, New York, 1981.
108. Bernhardt, J., The direct influence of electromagnetic fields on nerve and muscle cells of man within the frequency range of 1 Hz to 30 MHz, *Radiat. Envriron. Biophys.*, 16, 309, 1979.
109. Verveen, A.A. and Derksen, H.E., Fluctuation phenomena in nerve membrane, *Proc. IEEE*, 56, 906, 1968.
110. Beckwith, J.R. and McGuire, L.B., *Basic Electrocardiography and Vector Cardiography*, Raven Press, New York, 1982.
111. Krueger, A. and Reed, E., Biological impact of small air ions, *Science*, 193, 1209, 1976.
112. Weaver, J. C. and Chizmadzhev, Y., Electroporation, *Biological and Medical Aspects*, Barnes, F.S. and Greenebaum, B., Eds., Taylor & Francis, Boca Raton, FL, Chapter 9, 2006.
113. Michaelson, S.M., Elson, E.C., Anderson, L.E., Interaction of non-modulated and pulse modulated radio frequency fields with living fields with living matter: experimental results, *Biological and Medical Aspects*, Barnes, F.S., and Greenebaum, B., Eds., Taylor & Francis, Boca Raton, FL, Chapter 3, 2006.
114. Pilla, A.A., Mechanisms and therapuetic applications of time-varying and static magnetic fields, *Biological and Medical Aspects*, Barnes, F.S., and Greenebaum, B., Eds., Taylor & Francis, Boca Raton, FL, Chapter 11, 2006.

6

Magnetic Field Effects on Free Radical Reactions in Biology

Stefan Engström

CONTENTS

6.1 Introduction

The physical chemistry of spin-correlated free radical pairs offers several mechanisms explaining how magnetic fields may influence biochemical processes. The mechanisms are classified on the basis of the dominating contribution to spin interconversion, and they cover a wide range of field strengths. Of particular interest is what is called the low-field mechanism, which has been extensively developed over the last decade and is now capable of explaining biological effects induced by magnetic fields well below 1 mT.

The principal mechanism behind the free radical mechanism was discovered in the physical problem of magnetic field dependence on positronium decay (Deutsch and Brown, 1952; Halpern, 1954). However, the development of the radical pair mechanism in chemistry has its roots in the work of Kaptein and Oosterhoff (1969), Closs (1969), and Brocklehurst (1969).

An ambitious survey of the literature up to its date of publication is the review of Steiner and Ulrich (1989), which lists some 775 references, 58 of which are themselves reviews on magnetokinetic phenomena. Another, now classic, reference on the subject is the book by Salikhov et al. (1984). McLauchlan and Steiner (1991) published a review including the possible mechanisms at lower fields. The review by Grissom (1995) explored the higher-field mechanisms with particular attention to the context of biological systems. A didactic paper geared toward the issues in biological systems is that of Brocklehurst and McLauchlan (1996). There are also some recent reviews on the free radical mechanism in general (Woodward, 2002) and with particular attention to biological systems (Brocklehurst, 2002).

6.2 Theoretical Background

A radical is an atom or a molecule with an unpaired electron. It tends to be highly reactive, a property that defines their best known roles in biology. A spin-correlated (or geminate) radical pair is typically created by hemolytic cleavage of a covalent bond, that is, each molecule retains one of the electrons that formed the chemical bond that was broken. The electron spins may remain correlated for a significant time (microseconds) after the pair's creation. As the radicals separate, the electron interaction term becomes small, and the electron states of the pair will fluctuate between antiparallel (singlet, or "S") and parallel (triplet, or "T") because of coherent spin evolution by hyperfine interactions between the electron spin and the nuclei. There is a chance of reencounter between the spin-correlated radicals. Reforming the bond is only permitted by quantum spin selection rules if the electron spins are oriented in the singlet state.

Magnetic fields can affect the electron state in the time before the reencounter; and if singlet and triplet pairs have different chemical fates, we have a basis for magnetic field effects on free radical chemistry (Figure 6.1). In exemplifying the process, we will assume that the radical pair is formed in the singlet state, the normal case for biologically relevant reactions (Eveson et al., 2000).

Many physical transduction mechanisms proposed to explain magnetic field effects in biology are very vulnerable to the randomizing effects of thermal noise at normal temperatures of living systems (Binhi and Savin, 2003). The free radical mechanism is uniquely resistant to this obstacle since it is a nuclear effect and is not strongly coupled to the thermal bath (Adair, 1999).

A quantum mechanical formulation of the free radical mechanism contains many possible contributions to the hamiltonian; Steiner and Ulrich (1989) provide a good categorization of the main components. The stochastic Liouville equation (SLE) is a tool

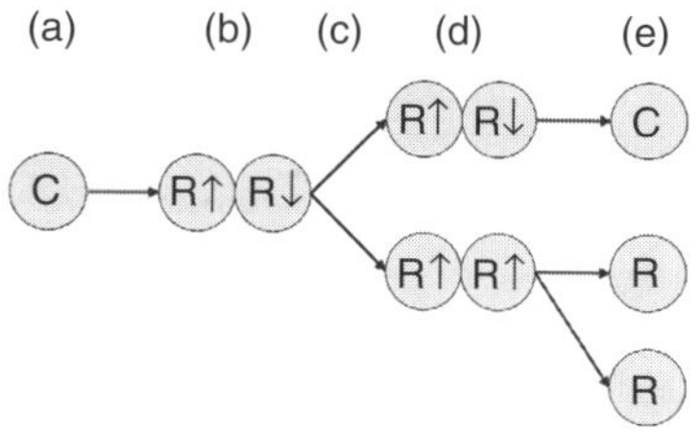

FIGURE 6.1
A molecule "C" (a) is split into two radicals (b). After diffusion and spin interconversion (c), the radicals may reencounter while still spin correlated (d). If the encounter occurs in the singlet state (e, top), the radicals may recombine. If the encounter occurs in the triplet state (e, bottom), recombination cannot occur, and the radicals will diffuse apart again and eventually lose their spin correlation. A magnetic field can influence this reaction by changing the rate of spin interconversion as long as the singlet and triplet products have different chemical fates.

for addressing the problem of simultaneous spin mixing and diffusion, but simplified models in which these two components are treated separately are often useful for the great reduction in problem complexity (Brocklehurst and McLauchlan, 1996). Recently, analytical results using a backward SLE have been presented (Pedersen and Christensen, 2004).

6.2.1 Hyperfine Interaction-Induced Singlet-to-Triplet Conversion

Hyperfine interactions between the spins of the electron and the nucleus cause the electron spins of the radicals to process and induce singlet-to-triplet conversion. The triplet state has a net magnetic moment, and in the presence of an external magnetic field the energy levels of the triplet states that have a moment aligned with the magnetic field (T_+ and T_-) will be separated by Zeeman splitting. As the applied field strength is increased, the T_+ and T_- energy levels will be shifted away from the singlet state so much that they are decoupled from the spin interconversion process between singlet and triplet states, and only the remaining triplet state (T_0), which has a magnetic moment that is oriented perpendicular to the field, is capable of participating in the spin conversion process. In this way the magnetic field can reduce the number of triplet states that can be converted into singlet states and subsequently reform the original chemical bond. This is the "normal" magnetic field effect on free radical chemistry. It becomes relevant for external fields larger than the effective field driving the hyperfine interaction mixing, typically 1–10 mT.

A way in which singlet-to-triplet conversion can be facilitated is when an applied field causes the T_- energy level to cross the nonmagnetic singlet level, which occurs when the Zeeman differential matches the electron exchange interaction energy (cf. Figure 6.2). This effect has been observed for fields as low as 6.6 mT (Werner et al., 1993), but it can in principle be observed for much lower fields if the radical pairs are fixed with an appropriate separation.

6.2.2 High-Field Regime: Spin Rephrasing through the Δg Mechanism

The product of the magnetic field and the Landé g-factor determines the precession of the unpaired electron spin, independent of the hyperfine contribution. If the two radicals have slightly different g-factors, this provides an additional source of spin conversion. Differences are usually quite small, so this mechanism typically becomes significant only for quite large fields, $B > 0.1$–1 T.

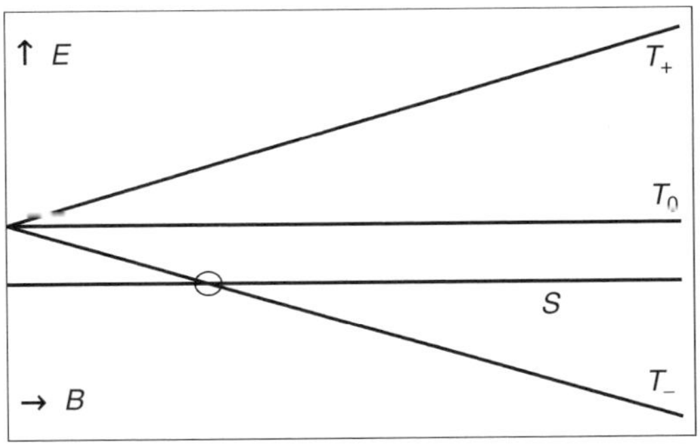

FIGURE 6.2
Energy level diagram for singlet and triplet states of a radical pair. At high fields, T_+ and T_- become completely disconnected from the singlet state, and only the T_0 state is left for singlet–triplet interconversion. At a specific B-field, the Zeeman split causes the T_- state to closely match the singlet state (circled), increasing the likelihood of transitions between the two states.

6.2.3 Low-Field Effect: $B < 1$ mT

At fields below the hyperfine interaction energy, it is still possible to see effects of external fields under certain circumstances. It was found by Brocklehurst (1976) that the selection rules of the hyperfine-induced spin mixing are more restrictive in zero field than when a field is applied (McLauchlan and Steiner, 1991). This becomes relevant, even for a very small field, as long as the coherence of the pair's state is maintained for long periods (100 ns to 1 μs). A helpful vector model to visualize this effect, along with some illustrative numerical examples, is given by Till et al. (1998).

The low-field effect (LFE) can theoretically produce a large (40%) drop in the singlet yield if the conditions are optimal (Timmel et al., 1998), but in practice only smaller effects attributed to this mechanism have been reported in the experimental literature.

Since long coherence times are required, it is necessary to understand spin relaxation effects and under what conditions they may be sufficiently long. Anisotropic hyperfine interactions provide noncoherent spin relaxation in solution, and it appears that the relaxation is slower in the low-field situation than has been generally thought (Fedin et al., 2001, 2003). It is becoming clear that understanding the local environment is crucial for evaluating LFEs (Eveson et al., 2000).

An important way to extend free radical magnetic field effects into the low-field region is to extend the lifetime of the spin-correlated pair. Integrating the magnetic field's influence over a longer time increases its ability to influence the spin evolution. Spin relaxation is not the primary problem here; rather, it is the reencounter probability that needs to be enhanced. There are several ways to achieve this:

1. Increased viscosity to restrict diffusion (Krissinel et al., 1999; Christensen and Pedersen, 2003; Kitahama et al., 2004)
2. Oppositely charged radicals that will oppose a tendency to diffuse apart (Adair, 1999)
3. Physically restricted mobility of the radicals through confining them to a surface or having the reaction taking place inside micelles, which are nanometer-scale compartments that form and reform on microsecond timescales and are able to confine the radical pairs and increase the reencounter probability (Eveson et al., 2000).

6.2.4 Free Radicals in Radio Frequency Fields

For higher-frequency fields ($f > 1$ MHz) the time modulation of the field starts to correspond to timescales present in the reaction dynamics, and a range of resonant interactions become available (Timmel and Hore, 1996). A detailed treatment of the coenzyme B_{12} system demonstrated resonant phenomena for relatively low-level radio frequency (RF) fields using a variety of mathematical techniques (Canfield et al., 1994, 1995).

6.3 General Characteristics of the Free Radical Mechanism

Free radical reactions are generally quite fast. There is evidence of picosecond reactions (Gilch et al., 1998; Musewald et al., 1999), but many known reactions occur over nanosecond timescales. If the free radical diffusion is constrained by micelles (Eveson et al.,

2000; Christensen and Pedersen, 2003) or by Coulomb attraction (Horiuchi et al., 2003), it may be possible to extend the radical pair lifetimes to hundreds of nanoseconds or even microseconds.

Since the free radical mechanism is practically instantaneous when compared to the timescale of time-varying magnetic fields in the extremely low-frequency regime (<300 Hz), one would expect that the observable output from a biological detection of the field should depend only on the time-averaged absolute field amplitude (Scaiano et al., 1995). However, if there is a downstream system that is able to decode the low-frequency signal, this statement is not necessarily true (Engstrom, 1997; Engstrom and Fitzsimmons, 1999).

6.3.1 Experimental Discrimination of Free Radical Models

The relative orientation of a static and a much smaller oscillating field will provide a discriminating test for a free radical-based model as long as the timescale of the applied field is long compared to the lifetime of the radical pair. In an isotropic system, such as a liquid suspension, described by this field situation, only the amplitude of the magnetic field is relevant. While the smaller parallel field adds linearly to the larger static field, the perpendicular component is effectively reduced by the calculation of vector length and will cause a much smaller variation in field amplitude (Engstrom, 1997).

For oriented systems in which the free radical chemistry steps have spatial preferences, it is possible to have angular dependence with respect to the angle of the applied field, $f(\theta)$. One can argue that some symmetry properties are very likely to be present in that kind of situation (Ritz et al., 2000). Polarity changes are not expected to be relevant, so $f(\theta + 180^\circ) = f(\theta)$. Furthermore, due to the isotropic distribution of nuclear spins in the initial state of the radical pair, one also expects that $f(180^\circ - \theta) = f(\theta)$.

If the free radical mechanism is active for low-frequency fields, one can also expect a response to RF fields in the same system, a property not expected by any other suggested physical transduction mechanisms for magnetic fields (Henbest et al., 2004). In a comparison between static fields in the range 0–2 mT, it was shown that a 300-μT, 5-MHz RF field applied perpendicularly or in parallel with the static field induced a response that was dependent on the magnitude of the static field (Henbest et al., 2004). This is qualitatively, if not quantitatively, similar to the observation that the magnetic sense of migratory birds can be disrupted by a 470-nT, 7-MHz field when the field is applied at an angle with the geomagnetic field (Ritz et al., 2004).

Magnetic isotope effects are another possible discriminating character of free radical effects, and it is possible to differentiate this signature from pure mass effects (Brocklehurst, 1997).

6.4 Free Radicals in Biology

6.4.1 Biological Transduction Mechanisms

Direct detection methods for free radicals have been developed in chemistry only relatively recently (Woodward, 2002). In biological systems, almost all experimental evidence for free radical involvement remains indirect. Generating hypotheses based on the signatures of free radical systems as outlined above are necessary to link free radical chemistry to magnetic field effects, but there is circumstantial evidence that free radical chemistry underlies some effects reported in the bioelectromagnetics literature.

It has been suggested that complex dynamical systems may have special sensitivity to magnetic field influences (Grundler et al., 1992; Walleczek, 1995). This idea has been elaborated in a series of theoretical models of oscillatory systems (Eichwald and Walleczek, 1996a,b; Kaiser, 1996). These models show that enzyme dynamics involving free radical chemistry may be frequency specific, although only for timescales comparable to or slower than the chemical kinetics of the system (Eichwald and Walleczek, 1997). Field amplitude can influence enzyme dynamics in some instances and this can be used to exert some control over enzyme systems (Eichwald and Walleczek, 1998).

The peroxidase–oxidase system has interesting, well-documented dynamical properties (Scheeline et al., 1997). Detailed modeling of this system has shown how a magnetic field-induced perturbation can affect its dynamical behavior (Eichwald and Walleczek, 1998; Moller and Olsen, 1999). The point of interaction in this system is suspected to be electron-transfer enzyme intermediates (Moller and Olsen, 2000; Moller et al., 2000).

Downstream effects from changes in chemistry must be taken into account to evaluate biological significance (Brocklehurst and McLauchlan, 1996). This can both facilitate detection (Walleczek, 1995) as well as introduce interventions that could block the biological significance of a physical detection event.

It may be relevant that other enzyme systems have been studied in detail without the specific intention of addressing free radicals as a possible mechanism. Examples include electric and magnetic field effects in ATPase (Blank and Soo, 2001b). Direct interactions with DNA have been suggested; and electron transfer reactions are proposed interaction targets (Blank and Goodman, 2000; Blank and Soo, 2001a), although not by mechanisms addressed here. Myosin phosphorylation is another enzyme system that has shown sensitivity to time-varying (Markov et al., 1993) and static magnetic fields (Markov and Pilla, 1994), as well as gradient-specific effects (Engstrom et al., 2002).

6.4.2 Role of Freely Diffusing Radicals

Free radicals observed in biology are most commonly oxygen or nitrogen based with an unpaired electron, leading to the terms reactive oxygen species (ROS) and reactive nitrogen species (RNS). A dominant role for these radicals is to act in immunological defense. They are secreted by macrophages and neutrophils and during attempts to kill bacteria, viruses, and tumors (Nathan, 1992). The highly reactive nature of the radicals also means that damage to normal cells is possible, and various defense mechanisms against this have evolved as well (Yu, 1994). This immunological weapon with checks and balances already suggests that there is a signaling system built around free radicals, but it seems that the ROS and RNS also have roles in intracellular cell signaling (Lander, 1997) as well as intercellular communication (Thannickal and Fanburg, 2000).

Consider a biochemical reaction producing a spin-correlated free radical pair in the singlet state. Depending on the specific mechanism at work, the ratio of singlet-to-triplet product at reencounter will be modified. This will increase or decrease the fraction of pairs that tend to recombine because of a reencounter finding the spins in singlet versus the triplet states. For the LFE, the triplet state is favored, and we would see an excess of escape product. The situation is the opposite for the ''normal'' field effect, in which the T_- and T_+ states are decoupled from the interconversion process, increasing the proportion of singlet reencounters leading to a larger amount of cage product and leaving fewer freely diffusing radicals. The Δg mechanism brings the T_- and T_+ states back into play and therefore again boosts the triplet reencounter escape products.

Given the wide involvement of free radicals in signaling and biological function, it is clear that there is the potential of both subtle and not-so-subtle effects on biological systems if we are able to alter the production of free radicals and thereby change the dynamics of already ongoing processes. The conventional wisdom regarding the deleterious effects of magnetic field effects on free radical recombination has been that more escape product means more radical-induced damage. This may be an oversimplification since the direct effects on cage or escape products are typically fairly small, certainly not larger than tens of percent, implying that drastic biological effects must involve downstream responses that amplify this relatively slight modulation. The answer may lie in the signaling properties of free radicals.

6.4.3 Animal Navigation Models Based on Free Radicals

The free radical mechanism was the first mechanism suggested as an explanation of avian use of magnetic fields for navigation (Schulten et al., 1978; Schulten, 1982). This model has since undergone several iterations of refinement (Ritz et al., 2000; Cintolesi et al., 2003). One interesting aspect of this work is a connection between photosensitivity and magnetoreception (Ritz et al., 2002). Dependence on light is a well-known feature of the avian magnetoreceptor (Deutschlander et al., 1999; Wiltschko et al., 2004a,b), but it has also appeared in other behavioral studies of animal magnetic field sensitivity (Prato et al., 1997, 1998).

The Ritz-Schulten model (Ritz et al., 2000) has an appealing geometrical application. Being integrated into the bird's retina, the suggested compass would appear as a modulation overlay on the bird's field of view. The mechanism operates through the so-called low field effect (LFE), based on a single nuclear spin, and operates near the limit of the theoretical sensitivity, despite omitting degrading effects such as the presence of multiple nuclear spins, dipolar effects, and various spin relaxation process that will start to become relevant for the long radical pair lifetimes (>100 ns) considered in the model.

Cryptochromes provide one possible source of free radicals in a spatially ordered system (Ritz et al., 2002). A recent theoretical model for avian magnetoreception develops that idea by investigating a flavin–tryptophan radical pair with a high degree of homology to the cryptochromes (Cintolesi et al., 2003). This multinucleus model is realistic in that it still manages to provide sensitivity to fields in the geomagnetic field range. Interestingly, it does not operate through the LFE described above (the multinucleus approach appears to remove most signs of that mechanism), but rather it depends on immobilized radicals and assumes that the free radical pair may have a lifetime up to 5 μs.

6.4.4 Coenzyme B_{12}-Dependent Reactions

Magnetic field effects in the coenzyme B_{12} are well explored experimentally with matching theoretical predictions (Harkins and Grissom, 1994; Grissom and Natarajan, 1997; Taoka et al., 1997). While most work on this model system has been concerned with intermediate and higher-field mechanisms, there are also detailed theoretical investigations suggesting that weak ($<100\ \mu$T), relatively low-frequency (<100 kHz), might be able to affect this system (Canfield et al., 1994, 1995, 1996).

6.4.5 Other Experimental Observations

The addition of iron ions or exposure to a 7-mT static magnetic field (SMF) did not affect the survival of rat lymphocytes *in vitro* when performed in isolation, but combined exposure led to a significant increase in cell death (Jajte et al., 2002). One possible explanation of this behavior is that the addition of iron ions enhanced levels of ROS and that the field exposure further promoted the creation of free radicals, leading to cell death by both apoptosis and necrosis. An experiment with a similar rationale used added $FeCl_2$ to stimulate ROS production, and a 930-MHz, 5-W/m^2 cell phone-generated field affected a biological marker for ROS production. It should be noted that the vacuum magnetic field associated with this exposure is quite low (approximately 0.14 μT) and the frequency is a relatively unexplored region for this mechanism.

Proliferation of chick fibroblasts was observed to be enhanced by a 100-Hz, 0.7-mT sinusoidal magnetic field (Katsir et al., 1998). In a follow-up study it was found that free radical scavengers (Katsir and Parola, 1998) suppressed this effect, suggesting that that free radicals may have a role in mediating the magnetic field effect on proliferation.

Genotoxic effects from intermediate static magnetic fields (250 mT) have been studied in *Escherichia coli* DNA, both *in vivo* and *in vitro* (Potenza et al., 2004). Free radical formation was stimulated, and the genetic damage was mapped as a function of exposure duration. *In vitro* experiments showed detectable genotoxic effects, but the *in vivo* assays did not, indicating that cellular protective responses may prevent damage in the intact system.

A reported effect on the oxidative burst in neutrophils by a 0.1-mT field was attributed to free radicals (Roy et al., 1995). In that study, the connection to free radicals lies in that the fluorescent probe used to study the neutrophil activity reacts specifically with free radical-derived oxidants that create the fluorescing compounds. Work in neutrophils in humans (Heine et al., 1999) using a much larger field (1.5 T) did not find any effects of magnetic fields on the respiratory burst of human neutrophils or on the production of radical species.

Phagocytosis was observed to be affected by 0.5–1.5-mT, 50-Hz sinusoidal magnetic fields (Simko et al., 2001). An attendant increase in superoxide production may be an indication that the field stimulated the system through a free radical process.

6.5 Conclusion

Free radical reactions are ubiquitous in biology, and recent developments of the low-field mechanisms (Timmel et al., 2001) and the consideration of detailed biochemical systems (Cintolesi et al., 2003) make this mechanism a contender for field effects down to geomagnetic field strengths. The physical transduction step is not vulnerable to thermal perturbations, a significant advantage over competing models.

This model does not produce large (factors >2) changes at the initial field detection step. Theoretical models and direct experimental observations in the low-field region typically operate around or below the 10% level, so we should expect the physical detection mechanism to need cooperation from downstream processes for biologically relevant detection of magnetic fields with free radicals as the starting point.

References

Adair RK (1999): Effects of very weak magnetic fields on radical pair reformation. *Bioelectromagnetics* 20:255–263.

Binhi VN, Savin AV (2003): Effects of weak magnetic fields on biological systems: physical aspects. *Physics Uspekhi* 46:259–291.

Blank M, Goodman R (2000): Stimulation of the stress response by low-frequency electromagnetic fields: possibility of direct interaction with DNA. *IEEE Transactions on Plasma Science* 28:168–172.

Blank M, Soo L (2001a): Electromagnetic acceleration of electron transfer reactions. *Journal of Cellular Biochemistry* 81:278–283.

Blank M, Soo L (2001b): Optimal frequencies for magnetic acceleration of cytochrome oxidase and Na,K-ATPase reactions. *Bioelectrochemistry* 53:171–174.

Brocklehurst B (1969): Formation of excited states by recombining organic ions. *Nature* 221:921.

Brocklehurst B (1976): Spin correlation in geminate recombination of radical ions in hydrocarbons. 1. Theory of magnetic-field effect. *Journal of the Chemical Society—Faraday Transactions II* 72:1869–1884.

Brocklehurst B (1997): Magnetic isotope effects in biology: a marker for radical pair reactions and electromagnetic field effects? *International Journal of Radiation Biology* 72:587–596.

Brocklehurst B (2002): Magnetic fields and radical reactions: recent developments and their role in nature. *Chemical Society Reviews* 31:301–311.

Brocklehurst B, McLauchlan KA (1996): Free radical mechanism for the effects of environmental electromagnetic fields on biological systems. *International Journal of Radiation Biology* 69:3–24.

Canfield JM, Belford RL, Debrunner PG, Schulten KJ (1994): A perturbation-theory treatment of oscillating magnetic-fields in the radical pair mechanism. *Chemical Physics* 182:1–18 (see also *Journal of Chemical Physics* 191:347, 1995).

Canfield JM, Belford RL, Debrunner PG, Schulten KJ (1995): A perturbation treatment of oscillating magnetic-fields in the radical pair mechanism using the Liouville equation. *Chemical Physics* 195:59–69.

Canfield JM, Belford RL, Debrunner PG (1996): Calculations of Earth-strength steady and oscillating magnetic field effects in coenzyme B-12 radical pair systems. *Molecular Physics* 89:889–930.

Christensen M, Pedersen JB (2003): On the validity of the one-particle diffusion model of geminate recombination in micelles. *Chemical Physics* 295:235–241.

Cintolesi F, Ritz T, Kay CWM, Timmel CR, Hore PJ (2003): Anisotropic recombination of an immobilized photoinduced radical pair in a 50-mu T magnetic field: a model avian photomagnetoreceptor. *Chemical Physics* 294:385–399.

Closs GL (1969): A mechanism explaining nuclear spin polarizations in radical combination reactions. *Journal of the American Chemical Society* 91:4552.

Deutsch M, Brown SC (1952): Zeeman effect and hyperfine splitting of positronium. *Physical Review* 85:1047–1048.

Deutschlander ME, Phillips JB, Borland SC (1999): The case for light-dependent magnetic orientation in animals. *Journal of Experimental Biology* 202:891–908.

Eichwald C, Walleczek J (1996a): Activation-dependent and biphasic electromagnetic field effects: model based on cooperative enzyme kinetics in cellular signaling. *Bioelectromagnetics* 17:427–435.

Eichwald C, Walleczek J (1996b): Model for magnetic field effects on radical pair recombination in enzyme kinetics. *Biophysical Journal* 71:623–631.

Eichwald C, Walleczek J (1997): Low-frequency-dependent effects of oscillating magnetic fields on radical pair recombination in enzyme kinetics. *Journal of Chemical Physics* 107:4943–4950.

Eichwald C, Walleczek J (1998): Magnetic field perturbations as a tool for controlling enzyme-regulated and oscillatory biochemical reactions. *Biophysical Chemistry* 74:209–224.

Engstrom S (1997): What is the time scale of magnetic field interaction in biological systems? *Bioelectromagnetics* 18:244–249.

Engstrom S, Fitzsimmons R (1999): Five hypotheses to examine the nature of magnetic field transduction in biological systems. *Bioelectromagnetics* 20:423–430.

Engstrom S, Markov MS, McLean MJ, Holcomb RR, Markov JM (2002): Effects of non-uniform static magnetic fields on the rate of myosin phosphorylation. *Bioelectromagnetics* 23:475–479.

Eveson RW, Timmel CR, Brocklehurst B, Hore PJ, McLauchlan KA (2000): The effects of weak magnetic fields on radical recombination reactions in micelles. *International Journal of Radiation Biology* 76:1509–1522.

Fedin MV, Purtov PA, Bagryanskaya EG (2001): Anisotropic hyperfine interaction-induced spin relaxation in a low magnetic field. *Chemical Physics Letters* 339:395–404.

Fedin MV, Purtov PA, Bagryanskaya EG (2003): Spin relaxation of radicals in low and zero magnetic field. *Journal of Chemical Physics* 118:192–201.

Gilch P, Pollinger-Dammer F, Musewald C, Michel-Beyerle ME, Steiner UE (1998): Magnetic field effect on picosecond electron transfer. *Science* 281:982–984.

Grissom CB (1995): Magnetic-field effects in biology—a survey of possible mechanisms with emphasis on radical-pair recombination. *Chemical Reviews* 95:3–24.

Grissom CB, Natarajan E (1997): Use of magnetic field effects to study coenzyme B-12-dependent reactions. *Vitamins and Coenzymes*, Pt K 281:235–247.

Grundler W, Kaiser F, Keilmann F, Walleczek J (1992): Mechanisms of electromagnetic-interaction with cellular-systems. *Naturwissenschaften* 79:551–559.

Halpern O (1954): Mangetic quenching of the positronium decay. *Physical Review* 94:904–907.

Harkins TT, Grissom CB (1994): Magnetic-field effects on B-12 ethanolamine ammonia-lyase—evidence for a radical mechanism. *Science* 263:958–960.

Heine J, Scheinichen D, Jaeger K, Herzog T, Sumpelmann R, Leuwer M (1999): Effect of magnetic resonance imaging on human respiratory burst of neutrophils. *FEBS Letters* 446:15–17.

Henbest KB, Kukura P, Rodgers CT, Hore PJ, Timmel CR (2004): Radio frequency magnetic field effects on a radical recombination reaction: a diagnostic test for the radical pair mechanism. *Journal of the American Chemical Society* 126:8102–8103.

Horiuchi M, Maeda K, Arai T (2003): Magnetic field effect on electron transfer reactions of flavin derivatives associated with micelles. *Applied Magnetic Resonance* 23:309–318.

Jajte J, Grzegorczyk J, Zmyslony M, Rajkowska E (2002): Effect of 7 mT static magnetic field and iron ions on rat lymphocytes: apoptosis, necrosis and free radical processes. *Bioelectrochemistry* 57:107–111.

Kaiser F (1996): External signals and internal oscillation dynamics: biophysical aspects and modelling approaches for interactions of weak electromagnetic fields at the cellular level. *Bioelectrochemistry and Bioenergetics* 41:3–18.

Kaptein R, Oosterhoff LJ (1969): Chemically induced dynamic nuclear polarization III (anomalous multiplets of radical coupling and disproportionation products). *Chemical Physics Letters* 4:214–216.

Katsir G, Parola AH (1998): Enhanced proliferation caused by a low frequency weak magnetic field in chick embryo fibroblasts is suppressed by radical scavengers. *Biochemical and Biophysical Research Communications* 252:753–756.

Katsir G, Baram SC, Parola AH (1998): Effect of sinusoidally varying magnetic fields on cell proliferation and adenosine deaminase specific activity. *Bioelectromagnetics* 19:46–52.

Kitahama Y, Wakasa M, Sakaguchi Y (2004): Viscosity dependence of the magnetic field effect due to the delta g mechanism. *Journal of Physical Chemistry A* 108:754–757.

Krissinel EB, Burshtein AI, Lukzen NN, Steiner UE (1999): Magnetic field effect as a probe of distance-dependent electron transfer in systems undergoing free diffusion. *Molecular Physics* 96:1083–1097.

Lander HM (1997): An essential role for free radicals and derived species in signal transduction. *FASEB Journal* 11:118–124.

Markov MS, Pilla AA (1994): Static magnetic-field modulation of myosix phosphorylation—calcium-dependence in 2 enzyme preparations. *Bioelectrochemistry and Biochergetics* 35(1–2):57–61.

Markov MS, Wang S, Pilla AA (1993): Effects of weak low frequency sinusoidal and DC magnetic fields on myosin phosphorylation in a cell-free preparation. *Bioelectrochemistry and Bioenergetics* 30:119–125.

McLauchlan KA, Steiner UE (1991): The spin-correlated radical pair as a reaction intermediate. *Molecular Physics* 73:241–263.

Moller AC, Olsen LF (1999): Effect of magnetic fields on an oscillating enzyme reaction. *Journal of the American Chemical Society* 121:6351–6354.

Moller AC, Olsen LF (2000): Perturbations of simple oscillations and complex dynamics in the peroxidase–oxidase reaction using magnetic fields. *Journal of Physical Chemistry B* 104:140–146.

Moller AC, Lunding A, Olsen LF (2000): Further studies of the effect of magnetic fields on the oscillating peroxidase–oxidase reaction. *Physical Chemistry Chemical Physics* 2:3443–3446.

Musewald C, Gilch P, Hartwich G, Pollinger-Dammer F, Scheer H, Michel-Beyerle ME (1999): Magnetic field dependence of ultrafast intersystem-crossing: a triplet mechanism on the picosecond time scale? *Journal of the American Chemical Society* 121:8876–8881.

Nathan C (1992): Nitric-oxide as a secretory product of mammalian-cells. *FASEB Journal* 6:3051–3064.

Pedersen JB, Christensen M (2004): The backward stochastic Liouville equation. *Journal of Physical Chemistry B* 108:9516–9523.

Potenza L, Cucchiarini L, Piatti E, Angelini U, Dacha M (2004): Effects of high static magnetic field exposure on different DNAs. *Bioelectromagnetics* 25:352–355.

Prato FS, Kavaliers M, Cullen AP, Thomas AW (1997): Light-dependent and -independent behavioral effects of extremely low frequency magnetic fields in a land snail are consistent with a parametric resonance mechanism. *Bioelectromagnetics* 18:284–291.

Prato FS, Kavaliers M, Thomas AW, Ossenkopp KP (1998): Modulatory actions of light on the behavioural responses to magnetic fields by land snails probably occur at the magnetic field detection stage. *Proceedings of the Royal Society of London Series B: Biological Sciences* 265:367–373.

Ritz T, Adem S, Schulten K (2000): A model for photoreceptor-based magnetoreception in birds. *Biophysical Journal* 78:707–718.

Ritz T, Dommer DH, Phillips JB (2002): Shedding light on vertebrate magnetoreception. *Neuron* 34:503–506.

Ritz T, Thalau P, Phillips JB, Wiltschko R, Wiltschko W (2004): Resonance effects indicate a radical-pair mechanism for avian magnetic compass. *Nature* 429:177–180.

Roy S, Noda Y, Eckert V, Traber MG, Mori A, Liburdy R, Packer L (1995): The phorbol 12-myristate 13-acetate (PMA)-induced oxidative burst in rat peritoneal neutrophils is increased by a 0.1 mT (60 Hz) magnetic-field. *FEBS Letters* 376:164–166.

Salikhov KM, Molin YuN, Sagdeev RZ, Buchachenko AL (1984): *Spin Polarization and Magnetic Effects in Radical Reactions*. Elsevier, Amsterdam.

Scaiano JC, Cozens FL, Mohtat N (1995): Influence of combined AC-DC magnetic-fields on free-radicals in organized and biological-systems—development of a model and application of the radical pair mechanism to radicals in micelles. *Photochemistry and Photobiology* 62:818–829.

Scheeline A, Olson DL, Williksen EP, Horras GA, Klein ML, Larter R (1997): The peroxidase–oxidase oscillator and its constituent chemistries. *Chemical Reviews* 97:739–756.

Schulten K (1982): Magnetic field effects in chemistry and biology. In Treusch J, ed. *Festkörperprobleme*, Vol. 22. Vieweg, Braunschweig, pp. 61–83.

Schulten K, Swenberg CE, Weller A (1978): Biomagnetic sensory mechanism based on magnetic-field modulated coherent electron-spin motion. *Zeitschrift fur Physikalische Chemie-Frankfurt* 111:1–5.

Simko M, Droste S, Kriehuber R, Weiss DG (2001): Stimulation of phagocytosis and free radical production in murine macrophages by 50 Hz electromagnetic fields. *European Journal of Cell Biology* 80:562–566.

Steiner UE, Ulrich T (1989): Magnetic-field effects in chemical-kinetics and related phenomena. *Chemical Reviews* 89:51–147.

Taoka S, Padmakumar R, Grissom CB, Banerjee R (1997): Magnetic field effects on coenzyme B-12-dependent enzymes: validation of ethanolamine ammonia lyase results and extension to human methylmalonyl CoA mutase. *Bioelectromagnetics* 18:506–513.

Thannickal VJ, Fanburg BL (2000): Reactive oxygen species in cell signaling. *American Journal of Physiology—Lung Cellular and Molecular Physiology* 279:L1005–L1028.

Till U, Timmel CR, Brocklehurst B, Hore PJ (1998): The influence of very small magnetic fields on radical recombination reactions in the limit of slow recombination. *Chemical Physics Letters* 298:7–14.

Timmel CR, Hore PJ (1996): Oscillating magnetic field effects on the yields of radical pair reactions. *Chemical Physics Letters* 257:401–408.

Timmel CR, Till U, Brocklehurst B, McLauchlan KA, Hore PJ (1998): Effects of weak magnetic fields on free radical recombination reactions. *Molecular Physics* 95:71–89.

Timmel CR, Cintolesi F, Brocklehurst B, Hore PJ (2001): Model calculations of magnetic field effects on the recombination reactions of radicals with anisotropic hyperfine interactions. *Chemical Physics Letters* 334:387–395.

Walleczek J (1995): Magnetokinetic effects on radical pairs: a paradigm for magnetic field interactions with biological systems at lower than thermal energy. *Electromagnetic Fields* 250:395–420.

Werner U, Kuhnle W, Staerk H (1993): Magnetic-field dependent reaction yields from radical-ion pairs linked by a partially rigid aliphatic chain. *Journal of Physical Chemistry* 97:9280–9287.

Wiltschko W, Gesson M, Stapput K, Wiltschko R (2004a): Light-dependent magnetoreception in birds: interaction of at least two different receptors. *Naturwissenschaften* 91:130–134.

Wiltschko W, Moller A, Gesson M, Noll C, Wiltschko R (2004b): Light-dependent magnetoreception in birds: analysis of the behaviour under red light after pre-exposure to red light. *Journal of Experimental Biology* 207:1193–1202.

Woodward JR (2002): Radical pairs in solution. *Progress in Reaction Kinetics and Mechanism* 27:165–207.

Yu BP (1994): Cellular defenses against damage from reactive oxygen species. *Physiological Reviews* 74:139–162.

7

Signals, Noise, and Thresholds

James C. Weaver and Martin Bier

CONTENTS

7.1 Signals, Detection, and Measurement

Measurement is a quantitative observation and well known to be of great importance to science. However, measurements involving biological systems are complicated by the complexity of cells and tissues, particularly if fields are expected to interact weakly and field-induced changes are found to be small. Some key parameters, for example, temperature coefficient of a measured quantity, may be inadequately characterized, and related quantities may be determined incompletely (e.g., measurement or modeling of the time-dependent temperature throughout the volume of the biological system being studied). Detection is a special case of measurement, that is, the measurement is so coarse that an observer can only distinguish between "signal" and "no signal."

Generally speaking, the smaller the change in an observed quantity (e.g., cell biomass) due to a stimulus (e.g., an applied electromagnetic field), the more difficult the experimental interpretation. There may be multiple candidate causes if small changes in biomass are found, for example, any of many growth-altering biochemical changes, unnoticed and uncharacterized temperature variations, or even changes in ambient light or mechanical vibration. In physical science a model can often be made of the experiment. This allows estimates of the influence of various quantities and parameters on the expected experimental outcome (change in observed quantity in response to a stimulus) and is valuable in the interpretation of experiments. Similar approaches to bioelectromagnetics should also be valuable.

Consider an illustrative measurement on a biological system: a population of microorganisms contained within a glass toroid. Application of an alternating magnetic field to a primary coil wrapped around one part of the toroid induces an alternating current. The induced current can be measured with a coil wound around part of the toroid. The induced current is related to the electrical conductivity of the aqueous electrolyte. The electrical conductivity of the extracellular medium changes when small, charged metabolites are excreted, and measurement of microbial metabolic activity can thus be accomplished electrically. First observed in 1899 by a nulling technique [1], electrical impedance detection of microorganisms has received significant attention as a measurement method [2,3]. A toroidal device has actually been explored as the basis for determining microbial activity [4], with metabolic acid production causing a change in extracellular ion concentration (activity) and therefore creating a change in the electrical conductivity of the extracellular medium. But complications may arise. If cytotoxic chemicals leach from the glass, there can be a time-dependent poisoning of microbial activity. Ambient temperature changes couple through the glass to create internal temperature variations that alter the conductivity. In short, because electrical conductivity change has more than one candidate cause, this measurement system lacks specificity. This also illustrates a basic challenge to measurement of effects of electromagnetic fields on biological systems, namely, demonstrating both a statistically significant change and convincing evidence that it is the field interaction with the biological system, not an associated competing influence, that is responsible for the observation.

7.2 Specificity

Specificity is a hallmark of biological interactions involving biochemicals. A cell contains a large number of coexisting molecules whose interactions are not spontaneous but are instead highly regulated. Enzymes can be highly specific in the reactions they catalyze. Antibodies and receptor–ligand binding are also often specific. However, interactions of electromagnetic fields with a biological system are rather general. Magnetic fields interact indirectly by inducing electric fields and directly through magnetically sensitive reactions [5] and through interactions with magnetic material. Such magnetic materials may be contaminant ferromagnetic particles in the human body [6] or they may be biologically synthesized magnetite granules [7,8]. Electric fields interact nonspecifically with charge and polarizable material. Thus, unlike ligand–receptor biochemical interactions, there are no molecular receptors that are highly specific for electromagnetic fields. Instead, magnetic and electric sensory systems interact broadly and can be regarded as nonspecific. Evolved sensory systems are rather special. To date, it appears that biological, electric and magnetic field reception is indeed accomplished by organized systems.

Lack of electromagnetic field specificity has important implications for interpreting experiments. If an experiment quantifies a change in an observed parameter, the cause of the change is not automatically known. Continuing the example of cell growth determination based on biomass measurement, if an increase in biomass (or cell number) is associated with a field exposure, then additional analysis is needed to determine whether this change is due directly to the field or is instead due to interfering influences such as temperature change or biochemical concentration changes.

The challenge of specificity is not limited to weakly interacting fields. Consider the case of strong, electroporating fields *in vivo*, for which the motivation is local tumor treatment or gene therapy. Strong fields can generate tissue movement by stimulating muscles and possibly also by bulk tissue polarization forces. Tissue motion can itself create membrane openings, and these can lead to biochemical transport [9–11]. Thus, observation of molecular uptake associated with electrical pulsing does not by itself show that electroporation is responsible. Specificity is an issue.

7.3 Signal-to-Noise Ratio

We adopt a recent discussion of the signal-to-noise ratio (S/N) for experiments with biological systems exposed to weakly interacting electromagnetic fields [12]. The observed quantity is x. For bioelectromagnetics experiments, examples of x include a local or spatially averaged transmembrane voltage change, temperature rise at a particular site, radioactivity of an incorporated unstable isotope, specific enzyme activity, intracellular calcium ion concentration, cell biomass, etc. Typically, experiments obtain data that can be characterized by their means and standard deviations, often presented as a bar chart. One bar of each bar pair represents the control result, and the other bar represents the exposed result. Each bar height represents the mean value, and the error bar is usually the standard deviation. (In some cases the error bars instead represent the standard error, that is, the uncertainty in the mean, rather than the standard deviation, but generally a report states which is being used.) Bar charts present a concise summary of an investigator's knowledge of the underlying natural distributions. The measured mean and standard deviation of the control distribution can be defined to be $\bar{x}_{\mathrm{con}}$ and σ_{con}, respectively. Similarly, $\bar{x}_{\mathrm{exp}}$ and σ_{exp} are the observed mean and standard deviation of the exposed distribution.

When repeating the same experiment and doing the same measurement many times over, one generally finds a Gaussian distribution of outcomes. This is because in a complex system there are many variables and sources of inaccuracy that are not under the control of the experimentalist. For the cumulative effect of all these imprecisions the central limit theorem becomes applicable. This theorem says that with many independent stochasticities involved, the outcome will be a Gaussian distribution [13]. As an example of this theorem in practice, do 100 coin tosses and record the number N of "heads." Repeat this experiment many times. The result will converge to a Gaussian distribution of N that is centered around 50.

The threshold for a field exposure effect occurs under conditions of detection, that is, the minimum change of x that is discernable using generally accepted statistical criteria. This is equivalent to determining whether or not the control statistical distribution and the exposed distribution are distinguishable (significantly different by accepted criteria). This requires sufficiently precise knowledge of the statistical distribution parameters. Increasing the number of determinations of the natural distribution generates more precise knowledge of its parameters. For example, if an investigator carries out a number, m_{con}, of determinations of x_{con} and another number, m_{exp}, of determinations of x_{exp}, then the empirically determined values can be reported as

$$x_{\mathrm{con}} = \bar{x}_{\mathrm{con}} \pm \frac{\sigma_{\mathrm{con}}}{\sqrt{m_{\mathrm{con}}}} \quad \text{and} \quad x_{\mathrm{exp}} = \bar{x}_{\mathrm{exp}} \pm \frac{\sigma_{\mathrm{exp}}}{\sqrt{m_{\mathrm{exp}}}} \tag{7.1}$$

The ratio $\sigma/\sqrt{m}$ actually represents the aforementioned standard error. Increasing the number of determinations reduces the standard error and the ensuing uncertainty in the mean. However, it does not decrease the standard deviation, σ_{exp}, of the underlying distribution, which is assumed unperturbed by the measurement process.

As the means $\bar{x}_{con}$ and $\bar{x}_{exp}$ become better known through more determinations, the potential distinguishability of the two distributions increases. The p-value of the experiment is often reported as a measure of this distinguishability. The p-value is the probability that the two means would be found to be as different as observed (or even more different) purely because of random variability. For example, $p = 0.01$ indicates that there is only a 1% chance that the difference between the control mean and the exposed mean would be due to the (assumed) random variability of the measured quantity, namely, the standard deviation [14]. After an investigator completes an experiment and finds a reasonably small p-value (.01 and .05 are widely used values), it is a common practice for the investigator to report that an effect due to the field exposure has occurred. However this assumes specificity, namely, that the field exposure rather than an associated competing influence is responsible. Indeed, a small p-value supports an effect of some sort but not necessarily one due to the field during the exposure. Additional analysis that considers other competing influences such as temperature variations, vibrations, and chemical concentration variations [15,16] is required for that conclusion.

Bioelectromagnetics experiments with weakly interacting fields typically involve determination of changes with respect to background values of, for instance, transmembrane voltage, fluorescence intensity, enzyme activity, or cell number. Observed changes in "exposed" relative to "control" are generally small. At the other extreme, strongly interacting fields create large changes with respect to background, for example, molecular uptake by electroporation (see Chapter 9 on electroporation in Ref. [127]). For the "weakly interacting" situation the uncertainties (error bars) are about the same for exposed and control. However, there is another figure of merit, distinct from the p-value, namely, an empirically determined signal-to-noise ratio $(S/N)_{obs}$, which is associated with the observation and which is presumed due to the underlying statistical distributions for the control and exposed cases. Classical detection theory shows that the associated distributions are expected to be Gaussians [17].

In continuation of a recent discussion [12], we consider the observed signal (S_{obs}) to be the difference between the control and the exposed means, and the observed noise (N_{obs}) as the standard deviation of the control distribution [17]. This yields

$$S_{obs} = \bar{x}_{exp} - \bar{x}_{con} \quad \text{and} \quad N_{obs} = \sigma_{con} \tag{7.2}$$

so that the empirically determined signal-to-noise ratio is the magnitude of

$$(S/N)_{obs} = \frac{\bar{x}_{exp} - \bar{x}_{con}}{\sigma_{con}} \tag{7.3}$$

Like the p-value $(S/N)_{obs}$ is a measure of the distinguishability of the two distributions. However, unlike the p-value, the signal-to-noise ratio is an inherent characteristic of the biological system, its environment, and a particular field exposure and does not depend on the number of determinations. In this view, S_{obs} is the observed change and is assumed to be a measure of the strength of the perturbation to the biological system by the field exposure. N_{obs} is a measure of the natural variability in the system for the conditions of the experiment. In the absence of an exposure, N_{obs} provides the appropriate scale to gauge the strength of S_{obs}.

$(S/N)_{obs}$ is based only on experimental determinations of x. However, in many cases the field exposure is believed to *indirectly* alter x. According to this general hypothesis, the field exposure affects one or more molecular-level biochemical processes through physical interactions. In this sense, the exposure is creating a "primary" molecular change, which is then amplified through a biochemical cascade that creates a downstream change. It is this downstream change that is eventually measured. The signal-to-noise ratio cannot be increased by the amplification process. Later in this chapter, we will describe how amplification generally adds noise to a signal.

7.4 Detection Criteria

The criterion $(S/N) \leq 0.1$ is a very conservative basis for ruling out a particular class of biophysical mechanism for a given field exposure. Similarly, the criterion $(S/N) \geq 10$ is a conservative basis for ruling in a candidate biophysical mechanism for a given exposure, retaining that biophysical mechanism hypothesis for further evaluation. This approach provides a quantitative basis for rejecting or accepting hypothetical biophysical mechanisms as candidate explanations for an experimental measurement. The traditional choice $(S/N) \approx 1$ is a useful but somewhat arbitrary dividing line, which indicates conditions for which an effect might appear. $(S/N) \leq 0.1$ and $(S/N) \geq 10$ provide criteria for stronger conclusions, allowing rejection or provisional retention of a biophysical mechanism hypothesis.

We should recognize that thresholds are defined by generally accepted statistical criteria. The widely used p-values of .01 and .05 are examples of such generally accepted statistical values. In the case of signal-to-noise ratios, a commonly accepted value is $(S/N) \approx 1$, where the approximately equal symbol denotes the imprecision. Specifically, if (S/N) (empirical or theoretical) exceeds 1, then the threshold is viewed as being exceeded. Similary, if (S/N) is less than 1, the response is interpreted as subthreshold. Clearly, it makes little sense to take a strong position if (S/N) is close to 1. But, as noted above, if the signal-to-noise ratio is significantly greater or less than 1, then some confidence can be attached to the result. In short, a threshold is imprecise but nevertheless a useful guide.

7.5 Equilibrium Noise

In this section we will examine how Brownian noise, the simple random motion of molecules due to thermal agitation, interferes with the coupling of an electromagnetic field to a biochemical system. Some organisms have evolved an ability to sense and effectively "measure" electric and magnetic fields. We will see that the thermal noise that a signal has to compete against sets fundamental limits on detectability. We will also see how evolution has come up with structures to optimize the signal-to-noise ratio in sensory perception.

Fish generally carry a small dipolar field relative to the water that they swim in. Sharks, skates, and rays have developed special organs to detect such fields [18,19] and they use this ability to pinpoint the position of their prey when they get close and the water is too turbulent to rely on smell. To be effective, the shark should be able to sense its prey instantaneously. So, for the signal not to be mistaken for Brownian noise and for Brownian noise not to be mistaken as a signal, a signal should carry an energy that is significantly larger than kT. kT constitutes the average energy in the thermal noise band [20]. This baseline criterion already works to explain some of the physiology of the electric sensing organs. The electric fields are picked up by the ampullae of Lorenzini. These ampullae terminate at pores in the skin around the fish's head. They are enclosed in a highly resistive material and are filled with a very conductive gel. The eventual setup is equivalent to an electrical wire with no voltage drop inside. These ampullae are, furthermore, well insulated against electrical noise that originates from the fish's own physiology. Two pores that are about 10 cm apart on the surface of the fish's head can, on the inside ends, be separated only by a few nanometers. A field of 500 nV/m can be detected. Two pores that are 10 cm apart on the surface could thus transfer 50 nV into a transmembrane potential.

By having a lot of ion channels that are sensitive to such small voltage variations, the thermal noise can be effectively averaged out. With N ion channels instead of just one, N times as much signal strength is picked up. The thermal noise at each channel is independent of that at any other channel. The noise is zero-average and the noise variances are added up for N channels. So the average noise amplitude will be only $\sqrt{N}$ times as large if N channels are involved instead of one. After detection, the fish has to amplify this signal to the millivolt range that the nervous system operates with. Amplifier noise constitutes a problem that builders of electric circuits have dealt with for decades. Amplifier noise is nonequilibrium noise, and we will discuss it in the next section. Over the past decade, researchers have built up a good and detailed understanding of the physiology [21] and physics [22,23] of the fishes' amplification system.

Many animal species have the ability to detect the geomagnetic field. Two mechanisms have been proposed for magnetosensitivity. The first mechanism involves chemical transitions that are sensitive to external magnetic fields. Upon excitation by light, many polyatomic molecules will start transiting between the singlet ground state, the singlet excited states, and the excited triplet state. The energy difference between a singlet ($\uparrow\downarrow$) state and a triplet ($\uparrow\uparrow$) state is affected by an external magnetic field. This energy difference is generally small for fields of the magnitude of the Earth's magnetic field. But the magnetism that living cells generate is even smaller. A magnetically sensitive reaction of this type is therefore not subject to significant thermal noise. However, a detection limit can be established by considering a model in which reacting product molecules can bind to receptors. There is an innate stochasticity in chemical reactions; rates represent an average behavior, and there is a Gaussian distribution around this average. This is called fundamental chemical noise, and we will come back to it later in this chapter. In this model, the average number of occupied receptors varies with the magnetic field, and the detection limits are set by this fundamental chemical noise [24]. The fact that many bird species actually need light for their magnetic compass to work is a strong indication that singlet–triplet transitions are involved in the navigation. Recently, additional evidence was found when it turned out that robins get disoriented when they are subjected to an RF magnetic field that oscillates at the singlet–triplet resonance frequency [5] (for more details, see Chapter 6 on free radical models).

The second mechanism that has been proposed to explain magnetosensitivity involves the small (<100 nm in diameter) granules of magnetite (Fe_3O_4). This material, also known as lodestone, is biochemically formed and has about 30% of the magnetic strength of pure Fe. In the 1970 s, it was discovered that certain microbes use single-domain magnetite granules, also called magnetosomes, as a kind of rudder to help them stay under water right at the interface between the water and the mud at the bottom. There is a force trying to align the magnetic granule(s) with the Earth's magnetic field, and the microbe thus "finds out" what its own orientation is relative to the inclination of the Earth's magnetic field [25,26]. For a single-domain magnetite granule of about 100 nm in diameter, the product μB of the magnetic moment μ and the Earth's magnetic field B amounts to about $5kT$. This $5kT$ alignment is sufficient to exceed the kT thermal agitation in the granule's rotation. In higher animals it appears that the granules are commonly embedded in biopolymers and lined up to form a rigid linear rod. Such an alignment effectively increases the magnetic moment and thereby the sensitivity to small variations in the magnetic field [27]. Indications are that there can be up to a million magnetite-containing cells in the brain of almost any animal. Even humans, who exhibit no apparent magnetosensitivity, have magnetite in their brain tissue [7,8].

The intensity of the Earth's magnetic field varies from 25 to 65 μT, and the direction varies from parallel to perpendicular to the Earth's surface. The magnetic sensitivity of, for instance, homing pigeons has been shown to be such that field variations smaller than 10 nT can be detected. With such a sensitivity the pigeon can use the change of the magnetic field vector to furnish itself a kind of global positioning system (GPS) [29]. Recent data indicate that some birds incorporate both magnetite and singlet–triplet chemistry in their magnetosense [5].

It is tempting to hypothesize that extremely low-frequency (ELF) radiation or microwave radiation could have a physiological effect through the interactions with magnetosomes. Cells produce their own electricity and concurrent electric noise. But there is no significant endogenous magnetic field noise. So the magnetic part of ELF radiation or microwave radiation would not have to compete against such endogenous biological noise. The average 24-h personal 60-Hz magnetic field due to house wiring, distribution lines, electric motors, etc., for individuals in the U.S. population is about 10^{-7} T [30], that is, orders of magnitude smaller than the earth's stationary magnetic field. Starting from this premise, the magnetosome in the cytoplasm was modeled as a damped harmonic oscillator with an external 60-Hz modulation [31]. The restoring force is the force pushing to align the magnetosome's moment with the earth's magnetic field, and the damping is due to the viscosity of the cytoplasm. The associated equation is easily solved. Using reasonable values for the involved parameters, it is was found that even with exposure to a 60-Hz field with an amplitude of 5 μT, the alternating field transfers an amount of energy to the magnetosome that is orders of magnitude smaller than kT. In other words, the thermal agitations in the rotation far overwhelm any "signal" from an ambient 60-Hz field. But subsequently, the legitimacy of a simple linear approximation was questioned [32]. It was pointed out that there are intricacies that make the viscosity of the cytoplasm, which determines the damping coefficient in the model, hard to specify. Most importantly, the possibility of many individual magnetosomes in a cell acting in concert should be considered. With N magnetosomes in a cell instead of just one, the signal-to-noise ratio is $\sqrt{N}$ times larger. The explanation for this apparent amplification is the same as with the aforementioned N ion channels in the shark's electroreception. An alternative model that includes such cooperativity leads to a signal-to-noise ratio that is well over unity with a 2-μT amplitude 60-Hz magnetic field [32]. However, almost nothing is currently known about how forces on magnetosomes are transduced into physiological signals. More solid estimates of detection thresholds can probably be derived only after such biophysical mechanisms are revealed.

Electric fields are also of interest. Close to a power line, a human can be exposed to an electric field of about 10 kV/m. Two steps have to be taken to get to an assessment of the transmembrane voltage that such an exposure leads to. First of all, living tissue is much more conducting than air. So, charge in the tissue will move and follow the external field until it is compensated. Depending on the amount of movable dipoles, different materials have different dielectric permittivities. The ratio between the internal field and the field in the air is [33,34]:

$$\frac{E_i}{E_0} \approx \varepsilon_0 \omega \rho_t \tag{7.4}$$

Here $\varepsilon_0 = 8.8 \times 10^{-12}\,\mathrm{C^2/(N\,m^2)}$ represents the dielectric permittivity of a vacuum, ω is the angular frequency ($2\pi f$), and ρ_t is the resistivity of the tissue. So for a frequency of about 100 Hz and with a typical tissue resistivity of about 1–2 Ω m, the attenuation factor for the field entering the body is found to be in a range of 10^{-8}–10^{-7}. Hence, most of the external field goes around the person in the way water in a river flows around a big rock. Once inside the tissue, an amplification at the cell membranes occurs again through the mechanism explained in the previous paragraph. For a spherical cell with a diameter of about $d = 10\,\mu\mathrm{m}$ in a field E, the voltage across the diameter will be $\Delta V = Ed$, and the eventual field in the membrane will be of the order of $E_{\mathrm{mem}} \approx E(d/h)$, where h is the thickness of the membrane. With $h \approx 5\,\mathrm{nm}$ we find an amplification factor of about a 1000. We thus find a net conversion factor of 10^{-5}–10^{-4} and an electric field of about 0.1–1.0 V/m across a membrane as a result of the 10-kV/m power line exposure. This leads to an ELF-induced potential difference of at most 10^{-8} V across the membrane. It should, however, be noted that muscle cells or nerve cells are cylindrically shaped and may have lengths in the millimeter or even centimeter range. When the imposed field is along the axis of the cylinder, there may be a conversion factor at the caps of the cylinder that is two to three orders of magnitude higher.

When a living cell is suddenly exposed to an external electric field, ions will start flowing in the conducting interior to compensate for this field. In a typical mammalian cell, it is generally within microseconds that ions have accumulated near the membrane to achieve a zero intracellular electric field. This means that stationary electric fields and ELF (<300 Hz) AC fields distribute over cell membranes. Power lines and high-voltage distribution stations have been the subject of a lot of public anxiety. The power grid operates at 60 Hz in the United States and at 50 Hz in most other countries, that is, well within the ELF regime.

The 10^{-8} V that we derived may appear small relative to, for instance, the transmembrane potential of about 0.1 V that is present in about every living cell. However, when we talk about detectability, this 10^{-8} V should first be compared to the transmembrane voltages due to Brownian motion. The thermal noise voltage across standard resistors was already detected in the 1920s [35]. A formula was subsequently derived by Nyquist [36]:

$$\langle dV^2 \rangle = 4kTRdf \tag{7.5}$$

This equation gives the average square voltage in a frequency window of width df. The noise is white, that is, it has the same intensity at all frequencies. Technically, this would lead to an absurdity. It would imply that the noise carries an infinite amount of energy. However, as Nyquist already pointed out, $\langle dV^2 \rangle$ starts vanishing when we get to high frequencies f where $hf \approx kT$. Here, h represents Planck's constant, $h = 6.6 \times 10^{-34}\,\mathrm{J\,sec}$. At these high frequencies, quantum physics takes over and makes $\langle dV^2 \rangle$ go to zero. Such high frequencies are not in our realm of interest.

What Nyquist had in mind for a resistor in his derivation was a Brownian gas of frequently colliding charge carriers. With a 5-nm cell membrane that consists of a lipid bilayer with embedded proteins, the charge carriers are small ions (Na^+, K^+, Cl^-, etc.).

The ions do not form a "gas" inside the membrane, and it is not *a priori* obvious that Nyquist's formalism would apply. The equilibrium noise current through a membrane that separates two ionic solutions is due to two-sided shot noise. Shot noise was first described by Schottky [37] in the context of vacuum amplifier tubes. It is due to the elementary charge being finite and the charge carriers making random "jumps." It can be shown that two-sided shot noise ultimately leads back again to Nyquist's Equation 7.5 [38,39]. Ultimately, Equation 7.5 is a manifestation of something much more general than Nyquist may have had in mind. What underlies Equation 7.5 is Einstein's fluctuation–dissipation theorem. This theorem says that the same random collisions that cause diffusion, thermal noise, or shot noise also cause dissipation, friction, or resistance. The theorem, moreover, makes this connection quantitative:

$$\beta = \frac{kT}{D} \tag{7.6}$$

For the motion of a macromolecule in a liquid, D is the diffusion coefficient and β is the coefficient of friction, that is, the ratio $\beta = F/v$, where F represents the pulling force and v represents the resulting average speed. But in the context of the current through a membrane, β represents the electrical resistance ($R = V/I$). For D we find $D = e^2 P_S c$ in the membrane electrical case. Here, P_S is the membrane permeability to the monovalent ion S that is responsible for the current, c represents the concentration of this ion on both sides of the membrane, and e is the elementary charge.

Electrically, a cell membrane can be modeled as in Figure 7.1a. A lipid bilayer membrane has a capacitance of about 1 $\mu F/cm^2$. The capacitance of an actual cell membrane is generally not much different. The resistance of a pure lipid bilayer depends on the ionic concentrations of the solutions on either side of the membrane. With these concentrations at biological levels the resistance of a lipid bilayer membrane can be as high as $10^9\ \Omega\,cm^2$. Because of the presence of ion channels, ion transporters, and ion pumps [40,41], an actual cell membrane has a resistance that is orders of magnitude smaller (typically about $10^3\ \Omega\ cm^2$). The resistance of a patch of membrane is inversely proportional to the area of that patch. So, in order to characterize a membrane, the approach is to measure the resistance through an actual patch and then multiply it with the surface area of that patch. That is why we give the resistance of a membrane in terms of $\Omega\ cm^2$.

The setup in Figure 7.1a is equivalent to the one in Figure 7.1b, that is, an ordinary RC circuit. When calculating the characteristic time, RC, of the circuit, the surface area cancels

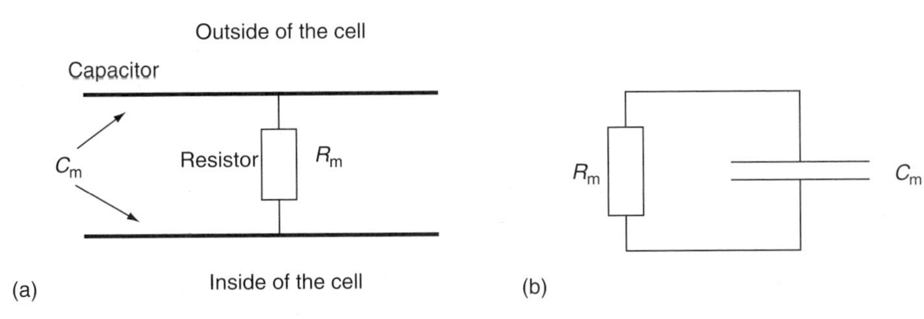

FIGURE 7.1
The electrical structure of the membrane is shown on the left. R_m and C_m are the resistance and capacitance between the inside and outside of the cell, respectively. The resistor also provides a thermal electromotive force. The equivalent circuit is shown on the right.

out. For a pure lipid bilayer the *RC* time constant can be of the order of minutes. But for a cell membrane it is of the order of milliseconds.

In our context, the resistor in Figure 7.1 is not just a resistor, but, following Nyquist (cf. Equation 7.5), also a white noise generator. At each frequency the resistor generates a harmonic oscillation. All these harmonic oscillations have the same amplitude. To evaluate the voltage across the capacitor, we thus have to analyze a simple *RC* circuit with an AC source. It has been argued that the high frequencies, that is, $f > (RC)^{-1}$, that are generated in the resistor do not have enough time to build up across the capacitor [42]. However, for low-frequencies, that is, $f < (RC)^{-1}$, changes are sufficiently slow for the capacitor to keep up and follow the voltage in the resistor. In this view the transmembrane voltage is the voltage across the capacitor, and the equilibrium noise is thus expected to occur mostly at low-frequencies. As mentioned before, the *RC* time of a cell membrane is of the order of milliseconds, and ELF fields thus operate in the $f < (RC)^{-1}$ regime where the noise is largest. A straightforward quantitative analysis shows that the low-frequency equilibrium noise far overwhelms any reasonable ambient power frequency field [42]. There would be no way to ever instantaneously detect such a field.

It was later put forward that everything that is happening in the cell membrane should, in the model of Figure 7.1, be imagined to happen inside the resistor [43]. Membrane proteins go through their catalytic cycle against a background of intramembrane noise. Inside the membrane means, in the context of Figure 7.1, inside the resistor. In this picture, the thermal noise voltage (cf. Equation 7.5) derives from a net electric field that results from inhomogeneities in the distribution of the charge carriers. Now at low-frequencies, the capacitor will be able to follow the imposed oscillation and effectively produce a field to counter the field generated inside the resistor (Figure 7.1a). This model thus leads to a vanishing net potential inside the membrane at low-frequency. At high frequency, the voltage changes in the resistor are too fast for the capacitor to keep up with. The capacitor will remain uncharged, and the thermal AC voltage will not be compensated for.

However, Figure 7.1 is no longer the appropriate model when we try to derive the intramembrane electric fields. For a cell of about 20 μm in diameter, the surface area amounts to about a billion square nanometers. The membrane is only about 5 nm thick, so the resistor resembles a very thin sheet. The lateral conductivity, that is, the conductivity from one place on the sheet to another, is very low. So at different spots on the sheet, different unrelated noise fields are generated. The more sensible model would therefore be one where the resistor in Figure 7.1 is cut up into millions of independent parallel resistors. Each of these resistors creates its own field. The capacitor plate corresponds to the conducting liquid on either side of the membrane, and it can be conceived of as having perfect lateral conductivity. So each resistor generates its own particular field, but they all experience the same field from the capacitor. With this model the noise gets very large. Not only there are more, say N, resistors producing noise. Each of these resistors has a resistance NR (N parallel resistors of resistance NR lead to a net resistance of R) and, according to Equation 7.5, thus produces more noise. Because the N parallel resistors that make up the resistance R are independent, they oscillate out of phase at each frequency f. As a result the parallel resistors end up pushing a lot of current in and out of each other. Most of the generated noise current thus remains intramembrane and never reaches the capacitor. The mathematics associated with this parallel setup is challenging, but an exact solution can be derived [39,44]. The capacitor, and therefore the *RC* time, plays no role in the intramembrane noise. The intramembrane noise is white and has an intensity that is many orders of magnitude larger than the noise that reaches the capacitor. What matters for biological function is actually the intramembrane noise. This, after all, is the noise that a membrane-embedded protein would "feel." The protein's catalytic cycle takes place against the background of such noise. The parallel setup model leads to a noise intensity that is much larger than that of the earlier models.

At first sight, all this extensive treatment of intramembrane noise may seem to have little to do with the two-sided shot noise that a membrane is subject to. However, when rigorously modeling the membrane as a thin sheet in an ionic solution, something similar to the overwhelming intramembrane noise is found. The ions that constitute the net charge on the membrane in Figure 7.1a move across the membrane–solution interface with an average speed of about 100 m/sec. This is just their thermal motion, and it is easily derived from $(1/2)mv^2 \approx kT$. This effectively causes laterally traveling electric pulses in the membrane. The noise intensity of these traveling electric pulses appears to be many orders of magnitude higher than the noise that is due to the shot noise-like membrane passages by the ions [39].

Current models of membrane noise thus lead to transmembrane voltage noise estimates that far exceed the strength of any reasonable magnitude ELF field-induced "signal." What the previous paragraphs lead up to is the conclusion that an ELF signal cannot be detected instantaneously.

However, under certain conditions and given enough time, even the smallest signal can get out of the noise band. The following example is meant to illustrate this. Consider the system depicted in Figure 7.2. Let the the resistance R represent a membrane patch. For simplicity, imagine that on either side of the resistor there is an infinite reservoir (i.e., a capacitor with infinite capacitance), so no net voltage can develop across the resistor. The average square charge $\langle q^2(t)\rangle$ that accumulates on either side of the membrane can be easily derived from Equation 7.5 and amounts to

$$\langle q^2(t)\rangle = \frac{2kT}{R}t \tag{7.7}$$

Again, there is an obvious analogy between Equation 7.7 and the well-known diffusion formula $\langle x^2(t)\rangle = 2Dt$, which describes the average square displacement of a particle with a diffusion coefficient D during a time interval of length t. The above formula clearly shows how, in an electrical context, kT/R plays the role of the diffusion coefficient D.

From Equation 7.7 we infer that for the accumulated charge as a function of time we have $|q_{\mathrm{Br}}(t)| \approx \sqrt{\langle q^2(t)\rangle} \propto \sqrt{t}$. The thermal noise-driven accumulation of any charged or uncharged molecule on either side of the membrane carries this $\sqrt{t}$ proportionality. The coupling of ELF electromagnetic fields to biochemical activity occurs mostly through

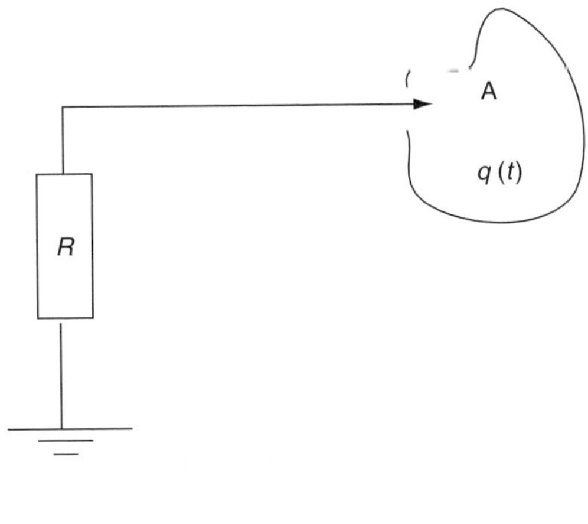

FIGURE 7.2
A resistor is connected to the ground and to an infinite reservoir A. The net voltage between the reservoirs remains zero. The situation is like the one in Figure 7.1 with the capacitor having infinite capacitance. Because of Brownian motion of electrons in the conduction band, there is a zero-average fluctuating current through the resistor. The net charge accumulating in the reservoir A is the result of these fluctuations in the same way that diffusive displacement is the result of random Brownian kicks. We have $\langle q^2(t)\rangle = 2(kT/R)t$ for the average square charge accumulation in time t.

membrane proteins. Membrane proteins whose conformational changes involve significant changes of the dipole moment are particularly sensitive. ELF fields can affect the catalytic rates of such proteins. So, for instance, electrogenic ion pumps [41], and also transporters or pumps that just carry a dipole, may have a slightly altered throughput in the presence of an ELF field. If there is no restoring force for a transported or pumped molecule, the accumulation will continue. The cumulative effect of the altered throughput will be a linear function of time. The excess charge that accumulates because of an ELF field thus follows $q_{\mathrm{ELF}} \propto t$.

Consequently, we see that on a small timescale the Brownian noise ($\propto \sqrt{t}$) will be stronger than the signal ($\propto t$). But there will always come a time $t = t_*$ when $|q_{\mathrm{B}r}(t_*)| = |q_{\mathrm{ELF}}(t_*)|$, and we then achieve S/N $= 1$. It depends on the values of the proportionality constants when t_* occurs. If molecular change is the measurement criterion, then it is only on timescales of the order of t_* that the effect becomes measurable. Estimates for t_* with realistic ELF exposure have been made [45] and have led to a timescale larger than the age of the universe.

7.6 Nonequilibrium Noise

In the previous section, we considered equilibrium noise. A living cell, however, constitutes a system that is far from equilibrium. Between the intracellular and extracellular solutions there is an electric potential difference of about 100 mV. For ions like Na^+, K^+, Cl^-, and Ca^{2+} there is a more than tenfold difference between intra- and extracellular concentration. The 100-mV transmembrane voltage over a width of about 5 nm implies a very strong field of tens of megavolts per meter.

The electrochemical potential across the cell membrane is an energy source for many processes [41]. The Na, Ca exchanger, for instance, is a membrane protein that picks up a sodium ion on the outside and then goes through a cycle in the course of which it drops the sodium ion off on the inside. The protein couples the energetically downhill movement of sodium to the uphill transport of calcium. In the course of the cycle a calcium ion is picked up on the inside and pumped, against the electrochemical potential, to the outside. The membrane potential is maintained by ATP-driven ion pumps. The most common of these is Na, K-ATPase. This is a membrane protein that, in the course of its catalytic cycle, hydrolyzes one ATP and uses the released energy to transport three sodium ions out of the cell and bring two potassium ions in.

Each working protein is like a small engine. A living cell contains millions of these engines: they are continuously converting energy from one form to another, and in the process, they are also generating heat, that is, dissipating energy. A living cell constitutes a far from equilibrium system, and the continuous transduction and dissipation of energy generates noise, which adds to the thermal, Brownian noise that was discussed in the previous section.

It would not be against the first law of thermodynamics (i.e., conservation of energy) if ion pumps were to extract heat from the environment and use it to power the maintenance of the transmembrane potential. This would, however, be in gross violation of the second law of thermodynamics. There are many equivalent formulations of the second law. The most common formulation is the proposition that every isolated system strives to increase and maximize its entropy. The teleological form of this formulation is somewhat bewildering. After all, most laws in science are formulated as conservation laws, for example, conservation of energy, or as causal laws, for instance, Newton's $F = ma$. However, after properly defining entropy, entropy maximization is often the easiest form of the second law to work with when dealing with macroscopic systems.

When going to the molecular realm, the second law can pose some challenging paradoxes. Consider, for instance, an ion channel in a cell membrane. Many ion channels rectify, that is, they pass current more easily in one direction than in the other. So the *I*–*V* characteristic is not a straight line through the origin, but it also has a curvature. Any frequency from the white spectrum of equilibrium noise should, in principle, be rectified. It thus might look like a rectifying ion channel could use zero-average equilibrium Nyquist noise to charge a battery. It would not work, of course. As pointed out above, it would be in violation of the second law. Thinking in the context of rectifying p–n junctions, solid-state physicists ran into this paradox long before ion channels were discovered. In 1950, L. Brillouin wrote a paper "Can the Rectifier Become a Thermodynamic Demon?" [46]. In this paper, he presents a short derivation to show that in a circuit with all components at the same temperature, no diode can rectify. He is aware that his case represents a special case of the so-called principle of detailed balance: "No system in thermal equilibrium in an environment at constant temperature spontaneously and of itself arrives in such a condition that any of the processes taking place in the system by which energy may be extracted, run in a preferred direction, without a compensating reverse process." The principle is a consequence of the second law [47,48] and, for our rectifier, basically states that there must, on average, be as much current in one direction as there is in the opposite direction.

In the *Feynman Lectures on Physics* [49] a ratchet and pawl system, originally thought up by Smoluchowski [50], is considered and eloquently discussed. The device operates as a mechanical rectifier (Figure 7.3) and essentially establishes the mechanical equivalent of Brillouin's paradox. The paradox is solved with the realization that the pawl must also be

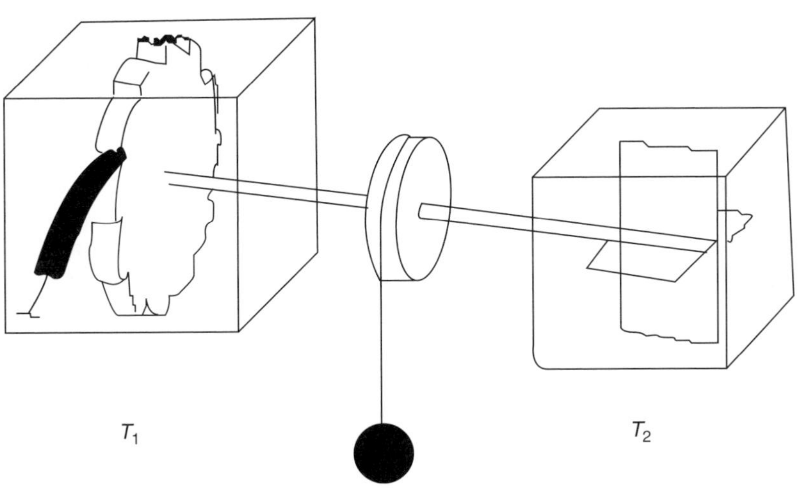

FIGURE 7.3
The mechanical thermal ratchet as it was originally conceived by Smoluchowski [50] and later discussed by Feynman et al. [49]. The device is small, and the paddle wheel in the right reservoir is moved by collisions of the molecules from the surrounding medium against the paddles. Because of the asymmetry of the teeth, the ratchet and pawl in the left reservoir allow motion in one direction and block it in the opposite direction. With the resulting net rotation it should be possible, in principle, to lift a weight. However, it would be in violation of the second law of thermodynamics to extract work from thermal fluctuations in the equilibrium situation, that is, $T_1 = T_2$. The solution of the paradox lies in the realization that the ratchet and pawl are also subject to thermal fluctuations if the system is small.

subject to thermal noise. The pawl involves a spring, and the spring will, at thermal equilibrium, exhibit a Boltzmann distribution over the accessible energy range. Even here the second law is involved, though on a deeper level. Given the macroscopic variables (e.g., temperature, concentration, pressure, etc.) there are still many possible molecular arrangements, that is, microstates, that correspond to that macrostate. For a fixed amount of energy the Boltzmann distribution is the energy distribution that has the most permutations [20]. It is therefore the most likely distribution. On the level of statistical mechanics, the second law can be formulated as the rule that given a macrostate, every microstate that corresponds to that macrostate has equal probability.

Second law issues can be subtle. The connection between statistics, entropy, information, and physical work still poses paradoxes that are hard to fathom. Books and articles still appear in which researchers are attempting to come to a fuller understanding and a better intuition [51,52]. At the scale of ion channels the simple invocation of detailed balance reveals little. An appropriate description is like the one Feynman gave for his mechanical ratchet and pawl: it involves Boltzmann distributions and Brownian motion. So it would simply be wrong to take any frequency from the white spectrum of equilibrium noise and model a rectifying ion channel as subject to this oscillation. The ion channel itself and its Brownian fluctuations have to be included in the description. At equilibrium, no part of a system can be "subject" to any other part. This is what detailed balance can be interpreted to mean.

However, when energy is dissipated, it is possible for one part of the system to impose its fluctuations on another part. When a rectifying ion channel is subject to nonequilibrium fluctuations, it will actually rectify the fluctuations and drive a net current. Consider, for instance, an electrogenic ion pump like Na,K-ATPase. As was mentioned before, this pump utilizes the energy of ATP hydrolysis to pump three sodium ions out and pump two potassium ions in. All this transport is against the electrochemical potential and requires about $15kT$ units of energy per stroke under physiological conditions. The power source is the hydrolysis of ATP, which under physiological conditions, releases about $20kT$ units of energy per cycle. It is the remaining $5kT$ that drives the process forward and that is ultimately released as heat. Na,K-ATPase is binding and releasing ions and thus generates fluctuating electric fields in its direct vicinity. For a nearby ion channel these fields can be conceived of as imposed because the $5kT$ that drives the Na,K-ATPase cycle is enough to overwhelm the small amount of energy ($<1kT$ [53]) necessary for the opening or closing of a channel. There is no feedback from the channel to the pump. The channel will rectify the fluctuations as a result, and a zero-average field can thus lead to net charge transport. In essence, the nonequilibrium fluctuations generated by the pump and imposed on the channel are part of the conversion of chemical energy, that is, the energy in ATP, to an electrochemical potential across the membrane.

So energy-dissipating, nonequilibrium oscillations and fluctuations are able to do work. ELF radiation from outside the organism can impose a varying field on an ion channel in much the same way that the nearby ion pump from the previous paragraph can impose a field on an ion channel. ELF radiation brings energy into the organism. Part of this energy will be dissipated to become heat, and part of it may be converted into chemical or electrical work. There is obviously no feedback from an ion channel back to the ELF source.

The selectivity of ion channels for the different kinds of ions is still hard to understand and model. But the rectification property is much easier to intuit (Figure 7.4). The channel is shaped like an asymmetric double cone, and charges in the lining of the channel are

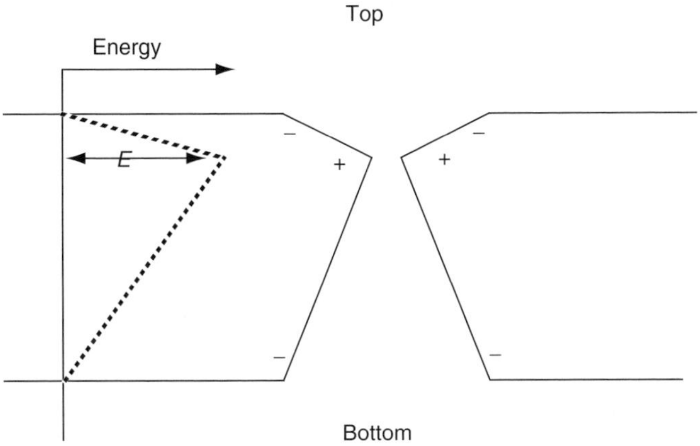

FIGURE 7.4
A simple continuum model of an ion channel imagined to be shaped like an asymmetric double cone. The energy profile on the left depicts the activation barrier that a positive ion going through the channel has to pass. The barrier has an obvious anisotropy.

indicated in the figure. A sodium or potassium ion that is going from the top to the bottom of the channel faces a rapid increase of the potential and then a slow decrease. A sodium or potassium ion that goes through in the opposite direction faces a slow increase and a fast subsequent decrease. A positive ion thus has a larger force to overcome when going from top to bottom than when going from bottom to top. Because of this, an imposed zero-average oscillation will lead to a net current [54]. As a matter of fact, any anisotropic potential shape along the length of the channel will rectify a zero-average harmonic field to lead to a net current [55]. Ion channels are proteins consisting of many amino acids, and anisotropy along the inside lining will be the rule rather than the exception.

The plethora of ratchet research in the late 1990s has made it clear that almost any zero-average oscillation or fluctuation imposed on a ratchet-like structure as in Figure 7.4 leads to a net current. Imagine, for instance, a temperature oscillation. With energy expressed in units of kT, a variation in temperature implies an oscillation of the barrier height E. Because of the difference in relaxation times on the slopes on either side of the barrier, a net current will result [55–58]. Recently, ever more examples have been found of nature exploiting ratchet effects for the purpose of regulation [59].

Researchers have meanwhile also succeeded in making artificial channels. Cone-shaped (and therefore anisotropic) channels form when a heavy ion is shot through an artificial membrane [60,61]. The I–V characteristic for the current of different types of ions has subsequently been recorded. It has even been experimentally shown that net charge transfer results when a zero-average field is imposed on such an artificial channel. The channel is thus made to behave like a kind of pump that converts an AC input into a DC output [62].

Imagine a number of identical anisotropic channels in a vesicle with an otherwise impermeable membrane. Next, put a large number of such vesicles in a beaker with an ionic solution. Any nonequilibrium fluctuation from the environment, or any "signal" for that matter, will now be picked up and converted into an electrochemical potential. The convection caused by a temperature gradient will heat up and cool down the vesicles and lead to their electrically charging up. The electric component of an ELF electromagnetic field will do the same thing. The beaker could thus be a battery that recharges by harnessing any incoming nonequilibrium fluctuation. This mechanism might, moreover, have played a role in the emergence of early prokaryotic life.

The Fourier spectrum of the noise that is associated with processes that dissipate energy is not white. Nonequilibrium noise appears to have higher amplitudes at lower frequencies; in other words, it exhibits an intensity that decreases with frequency. The so-called $1/f$ noise was first studied in the 1920s in the very nonequilibrium context of thermionic vacuum tube amplifiers [37]. In current scientific discourse the term "$1/f$ noise" actually applies to all noises that have spectral densities behaving like $1/f^{\alpha}$, where α ranges from about 0.5 to about 1.5. Especially in electrical devices, such noise is very commonly and easily observed. It is also known as "excess noise" or "flicker noise." In a log–log plot the $1/f^{\alpha}$ behavior usually extends over several frequency decades.

In the 1930s, it was proposed that the flicker noise originated from a variable number of electrons present in the conduction band. Electrons would shuttle between a free state and a bound state as in a chemical reaction. Let the relaxation time of that reaction be $1/\lambda$. This leads to a simple exponential relaxation $N(t) = N_0 \exp[-\lambda t]$ after any kind of fluctuation that has a magnitude N_0. The Fourier transform of the exponential decay is easily found:

$$F(\omega) = N_0 \int_{t=0}^{\infty} \exp\left[-(\lambda + \mathrm{i}\omega)t\right] \mathrm{d}t = \frac{N_0}{\lambda + \mathrm{i}\omega} \tag{7.8}$$

For the power spectral density, $S(\omega) = \|F(\omega)\|^2$, we find:

$$S(\omega) \propto \frac{1}{\lambda^2 + \omega^2} \tag{7.9}$$

where the proportionality constant involves the magnitudes of the fluctuations as well as the rates at which fluctuations occur. The power spectral density is a useful quantity as it describes how the energy in the noise is distributed over the different frequencies. $S(\omega)\mathrm{d}\omega$ is proportional to the amount of power that the noise carries between the frequencies ω and $\omega + \mathrm{d}\omega$. Equation 7.8 describes a so-called Lorentzian power spectrum. With a log scale for the frequency, the resulting curve is a sigmoid. At high ω, $S(\omega)$ behaves like $1/\omega^2$. As better data became available, it was found that a better fit was obtained when a distribution of infinitely many relaxation times was assumed [63]. Take, for instance, a uniform distribution of relaxation times between λ_1 and λ_2. With Equation 7.9 this leads to:

$$S(\omega) \propto \frac{1}{\lambda_2 - \lambda_1} \int_{\lambda_1}^{\lambda_2} \frac{1}{\lambda^2 + \omega^2} \mathrm{d}\lambda = \frac{1}{\omega(\lambda_2 - \lambda_1)} \left\{ \arctan\frac{\lambda_2}{\omega} - \arctan\frac{\lambda_1}{\omega} \right\} \tag{7.10}$$

It is easy to check that on $\lambda_1 < \omega < \lambda_2$ this $S(\omega)$ is approximately proportional to $1/(\omega(\lambda_2 - \lambda_1))$. This $S(\omega)$ is, moreover, roughly constant for $\omega < \lambda_1$ and drops off like $1/\omega^2$ when $\omega > \lambda_2$.

If we let, between λ_1 and λ_2, the relaxation rates contribute proportional to $\lambda^{-\beta}$, we can actually get any $1/f^{\alpha}$ dependence that we want, since

$$S(\omega) \propto \int_{\lambda_1}^{\lambda_2} \frac{1}{\lambda^{\beta}(\lambda^2 + \omega^2)} \mathrm{d}\lambda \propto \frac{1}{\omega^{1+\beta}} \quad \text{for } \lambda_1 < \omega < \lambda_2 \tag{7.11}$$

At $\omega < \lambda_1$, this spectrum would again flatten out.

In experimental practice with electrical resistors and amplifiers the $1/f^{\alpha}$ behavior has been observed to extend over more than six frequency decades with no noticeable flattening at low-frequency [64].

$1/f^{\alpha}$ spectra have been observed in nature in a wide variety of systems: electrocardiac waves [65], the variation of sea levels [66], tardiness at work [67], etc. An essential feature of $1/f^{\alpha}$ noise is that it exhibits self similarity, that is, if one magnifies both time and space with the appropriate factor, the noise pattern is indistinguishable from the original one. So the noise does not have a characteristic timescale or length scale. $1/f$ noise is often seen as a signature of the fractal character of nature.

In 1987, Bak et al. [68] proposed a model for a universal mechanism behind $1/f$ noise. In their landmark paper they illustrated the concept of "self-organized criticality" with a sandpile model. When a sandpile has an inclination steeper than a critical angle θ, avalanches will occur that bring the pile back to the critical angle. When sand is added to the pile in a random fashion, these avalanches do not exhibit a characteristic size, nor do they appear after regular time intervals. Instead, there are bigger avalanches that are relatively rare and smaller avalanches that occur more frequently. The size distribution follows a power law in the frequency f. For instance, in one day there can be one avalanche involving more than 1000 grains, 10 involving more than 100 grains, 100 avalanches involving more than 10 grains, and so on. The picture that emerges is one of a system that is sitting on the critical edge between two phases and is "organizing" avalanches to stay there [69]. The most commonly cited real-life example of self-organized criticality is the Gutenberg–Richter power law for earthquakes. It appears that every year, on average, there is one earthquake larger than magnitude 8, 10 earthquakes larger than magnitude 7, and 100 earthquakes larger than magnitude 6. Self-organized criticality is an attractive theory. It proposes a simple mechanism and predicts power laws that can be easily verified or falsified. It has been utilized in a wide variety of contexts [69]. It has, for instance, been applied to evolutionary theory [70] and has been used to explain frequency-size distributions of forest fires [71].

How truly universally applicable self-organized criticality is and to what extent its claims may be unwarranted are matters that are still very much under debate. The $1/f$ proportionality for earthquakes applies only between magnitudes 5 and 8. Even for the archetypal sandpile, things turn out to be more involved upon close inspection than self-organized criticality suggests. Accurate measurements [72–74] on real sandpiles showed that in many cases there is no $1/f$ pattern in the avalanches. It turns out that system parameters, like the grain size and the rate of sand addition, determine to a large extent what kind of spectrum eventually emerges. The entire concept of self-organized criticality collapses, of course, if fine tuning by the experimentalist is crucial for the $1/f$ spectrum to materialize. All in all, $1/f$ noise is not as universal as first thought, and the dynamics behind nonequilibrium noise are usually best unraveled with *ad hoc* models.

The node of Ranvier is where the action potential for myelinated nerve cells is generated [75]. There is a high concentration of ion channels in the node of Ranvier, and in the days before patch clamp, it was a good place to record membrane electrical activity. In the mid-1960s, Verveen and Derksen measured 5–10 min of cell membrane voltage noise at a Ranvier node of an unstimulated nerve cell [76,77]. The resulting power spectrum showed two decades, between 10 Hz and 1000 Hz, of $1/f$ noise (see Figure 7.5). This $1/f$ noise, they found, was much larger in magnitude than what Nyquist's $4kTR$ formula (cf. Equation 7.5) would predict. Following the explanation for $1/f$ noise in ordinary resistors (cf. Equation 7.10 and Equation 7.11), Verveen and Derksen suggested that an ion channel could, from time to time, get "clogged up." The wide distribution of waiting times (i.e., the λs in Equation 7.8 through Equation 7.11) before getting unclogged would then give rise to the $1/f$ spectrum [78].

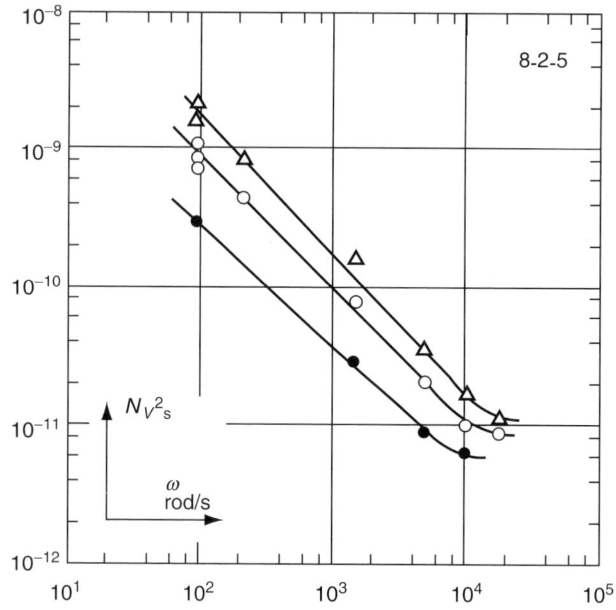

FIGURE 7.5
The voltage noise spectral densities from the frog node of Ranvier at room temperature at rest (open circles), at 10-mV depolarization (open triangles), and at 10-mV hyperpolarization (filled circles) [76,77]. (© Dutch Physiological Society. With permission.)

In the last two decades, single-channel recordings have shown how, even without stimulus, ion channels open and close repeatedly [40,75]. The kinetics behind the openings and closings is still very much a matter of debate. Modeling an ion channel as a two-state molecule with an open and a closed state and chemical steps with constant rates connecting these states appears not to account for the data in many cases. Electrophysiologists have commonly resorted to explaining the nonexponential distributions of open and closed times with a kinetic scheme that contains more than two states. With such an approach any distribution of open and closed times can always be fitted with a kinetic scheme [79–81]. It is just a matter of coming up with sufficiently many parameters (i.e., states and rates) to fit the data. In chemical kinetics the transitions are always assumed to be Markov transitions, that is, the probability of moving from a state 1 to a state 2 is constant and does not depend on the time that the molecule has been in state 1. A channel that is making such Markov transitions between a finite number of states always exhibits a power density spectrum that is a sum of Lorentzians. The number of characteristic times in the spectrum will always be one less than the number of states. If the characteristic times are sufficiently far apart, the power density spectrum will exhibit a number of identifiable plateaus when plotted on the customary logarithmic scale. The inflection points between the plateaus occur at the inverses of the characteristic times.

An alternative approach, foreshadowed by the aforementioned suggestion of Verveen and Derksen, has been to model the open to closed transition rates of an ion channel as time dependent, for instance $k(t) \propto t^{-\mu}$, where $0 < \mu < 1$ [82–85]. The exponent μ is taken to be smaller than unity to make $\int_{\tau}^{\infty} k(t)\, dt$ diverge for all $\tau > 0$ and thus guarantee the inevitability of an eventual transition. The proportionality $k(t) \propto t^{-\mu}$ leads to a decreasing transition probability, that is, the channel is "stabilizing in its openess," as more time is spent in the open state. There is ample justification for the use of open–closed transition rates that vary in time. A protein has many degrees of freedom and is subject to many equilibrium and nonequilibrium fluctuations. If an intramolecular rearrangement, like a transition between an open and a closed state, can be modeled as the crossing of an activation barrier, then that barrier will most likely not be fixed and stationary. A fluctuating barrier implies fluctuating open–closed transition rates. We could thus get the infinitely many relaxation rates that give rise to the $1/f$ power density spectra of Equation 7.10 and Equation 7.11. Under physiological conditions an ion channel is trafficking ions in an electric field of tens of megavolts per meter and comparable chemical gradients. This is a very nonequilibrium setup, and it has been conjectured that the channel operates as a self-organized critical structure. One authoritative textbook [65] states it as follows:

> A channel protein may be a self-organizing critical system. The channel protein consists of many pieces that interact with their neighbors. The energy added to the protein from the environment causes local strains that are spread throughout the structure. If these distortions spread faster than the time it takes for the structure to thermally relax, then the channel protein may be a self organizing critical system. If that is the case, then the fluctuations in the channel structure will be due to a global organization of the local interactions between many small interacting pieces of the channel protein. The fractal scaling would then be due to the fact the channel structure is poised at a phase transition between its open and closed conformational shapes.

Over the past few years increasing amounts of data have been gathered with ever more accurate technology. Recently, the $1/f$ power spectral density of a nerve cell that Verveen and Derksen discovered was more accurately rerecorded [86]. But through careful subsequent experimentation and computer simulations, these researchers were also able to show how the apparent $1/f$ result comes about as the sum of a number of Lorentzian contributions. Each type of channel has its own Lorentzian, and because of close characteristic times the sum of the individual sigmoids appears like a smoothly decreasing $1/f$ curve.

For a single channel, things often turn out to be much more intricate than simple $1/f$ versus Markov kinetics. In single-channel recordings of a bacterial ion channel it was found that actual channel openings and closings follow Markov kinetics and lead to Lorentzian contributions to the ultimate net power spectrum [87]. The $1/f$ noise that is present in the power spectrum originates from transitions between open states of a slightly different (about 1–5%) conductance. The rates of these miniconductance transitions appeared to be independent of the transmembrane voltage. The small transitions in conductance have been conjectured to be due to small clusters within the channel's structure moving in and out of the lining of the pore [87]. A cluster can cause a partial flow constriction when it sticks out into the pore. Following this idea, the apparent $1/f$ behavior can be attributed to many different clusters moving in and out with equally many different relaxation times. The voltage independence comes about because these clusters are either uncharged or the external electric field is somehow screened. Noise in synthetic channels has also been studied [88]. There it was found that potassium currents through a one-state, permanently open channel exhibit $1/f^2$ noise. An artificial channel that can open and close, on the other hand, was found to exhibit $1/f$ noise when the externally applied voltage is in the right regime. With this latter artificial channel there is good ground to attribute the open–closed transitions to the movement of "dangling ends" of polymers in the pore's lining. So the result supports the "moving cluster" for the mechanism behind $1/f$ noise in channels.

There appears to be no simple theory that can convincingly bring all manifestations of $1/f$ noise under one common denominator. All the indications are that an *ad hoc* approach to nonequilibrium noise phenomena is still the most fruitful one.

In ordinary resistors the amount of $1/f$ noise grows linearly with the dissipated power W. If we take $S(\omega)\,d\omega$ to denote the power in energy per unit of time (watts) in an interval $d\omega$, then we have for the power spectral density:

$$S(\omega) = \frac{gW}{f} \tag{7.12}$$

Here, g is a dimensionless constant the value of which depends on the type of resistor.

Nyquist noise is simple in that the net value of the resistance R fully determines the noise amplitude. With $1/f$ noise a more complex situation arises. Experimentally, the constant g (cf. Equation 7.12) turns out to be proportional to the volume-to-power ratio [89]. In Figure 7.6 the four resistors in design (b) are identical to the one resistor in design (a). It is obvious that (a) and (b) will have the same net resistance and therefore the same amount of Nyquist noise. Design (a), however, will exhibit four times as much $1/f$ noise as design (b). Design (b) is quieter because the energy dissipation is distributed over a larger volume. Generally, we have $g \propto 1/V$, where V denotes the resistor's volume. The gW in Equation 7.12 can be expressed as $g_e w$, where g_e is the g-value for a single elementary charge carrier in the resistor and w is the energy dissipated in the volume of such a single, independent charge carrier.

Pumps and carriers move ions one by one. Imagine a single pump or carrier that moves ions across the membrane at a rate ν. During a small time interval dt there is a probability $p = \nu\,dt$ that an ion is transported. We take dt to be sufficiently small so that the probability of more than one ion being transported during dt is negligible. We also take the duration of the catalytic cycle, that is, the "processing" time for an ion going through the membrane, to be negligible in comparison to the time between catalytic cycles. If we were not to make the latter assumption, we would simply have to multiply by the probability that an average channel is available for transport when we want to express the transport rate. For the average number of ions $\langle n\rangle_{dt}$ transported by the channel in time dt, we now have $\langle n\rangle_{dt} = 1\cdot p + 0\cdot(1-p) = p$. For the variance we have $\sigma_{dt} = \langle n^2\rangle - \langle n\rangle^2 = 1^2\cdot p - (1\cdot p)^2 = p(1-p)$. For M subsequent timesteps and $M\,dt = T$, the variances add up, and we have $\langle n\rangle_T = Mp$ and $\sigma_T = Mp(1-p)$. So the standard deviation, $\sqrt{\sigma_T}$, works out to be proportional to $\sqrt{M}$. Over time the standard deviation becomes more and more negligible compared to the average. For sufficiently small dt we can take $1-p$ to be equal to 1, and we then have a variance that equals the average.

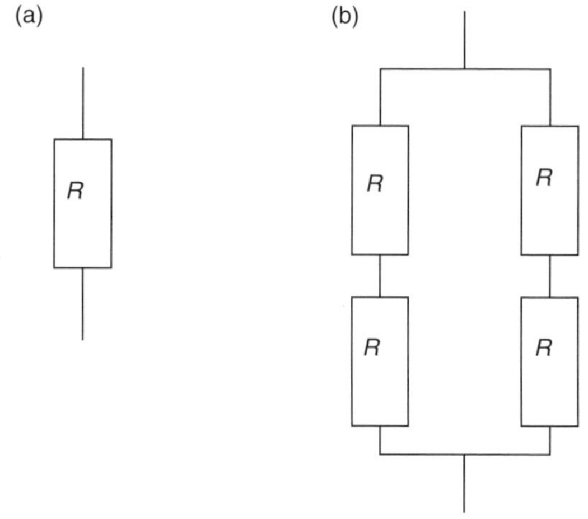

FIGURE 7.6
Design (a) and design (b) both have a net resistance R. They are different in that design (b) actually consists of two parallel resistors of $2R$. Designs (a) and (b) will exhibit the same amount of equilibrium noise, as it is only the net resistance (cf. Equation 7.5) that determines the equilibrium noise amplitude. It appears, however, that design (a) has four times as much power spectral density as design (b). The amount of nonequilibrium noise is proportional to the current density.

In the textbook by DeFelice [89] it is shown how the power spectral density of the process above amounts to $S(f) = 2\sigma_T/T = 2\nu$. The frequency-independent power spectrum can be intuited as follows. If the actual transport time through the membrane is small, then we can conceive of the transmembrane current as delta function-like pulses occurring at a rate ν. The Fourier transform of a Dirac delta function is a flat spectrum. In order to go from particle current to electrical current we must, to obtain the current power spectral density, multiply with the square of the charge of the involved ion. Taking this to be the elementary charge e, we find

$$S_i(f) = 2\nu e^2 = 2e\langle i\rangle \tag{7.13}$$

Here, $\langle i\rangle = e\nu$ denotes the average current through one pump.

Next, we let $\langle I\rangle$ denote the total transmembrane current due to pumps of a particular ion through the entire cell surface. We then have for the total current power spectral density:

$$S_I^{\mathrm{pu}}(f) = 2e\langle I\rangle \tag{7.14}$$

For a living cell in a steady state, for each kind of ion, there are just as many ions going in as there are going out, that is, there is just as much uphill transport through pumps as there is downhill flow through ion channels. So we have the same current $\langle I\rangle$ uphill as well as downhill. Below, we will first show that the downhill flow through the channels generates much more nonequilibrium noise than the uphill flow through the pumps. We will then show how the channel noise far exceeds the Johnson–Nyquist equilibrium noise.

Ion channels stay open for an average time of about $\tau_{\mathrm{op}} = 10^{-3}$ sec, and during that time there is a current of about 10^7 ions per sec (i.e., about 1 pA). So the equivalent of the elementary charge mentioned in the previous paragraph is now $N = 10^4$ ions.

However, before we blindly substitute Ne for e in Equation 7.14 to obtain the current power spectral density generated by the channel population of a cell, we have to take another source of variance into account. The channel-open time of about a millisecond is an average. If we view opening and closing of a channel as simple chemical steps between an open and a closed state, then the millisecond is the average of an exponential distribution of open times. For an exponential distribution the average open time equals the standard deviation in the open time. So the standard deviation ΔN equals N itself. The associated variance has to be added in. So we have $S_I^{\mathrm{ch}}(f) \approx 2\,\nu e^2\,(N^2 + (\Delta N)^2)$. The rate ν now represents the number of channel openings per unit of time. So we get for the current power spectral density produced by the channels:

$$S_I^{\mathrm{ch}} \approx 4Ne\langle I\rangle \tag{7.15}$$

So the channel contribution to the total nonequilibrium current noise is about 10^4 times as large as the contribution of the pumps. For the total current power spectral density, S_I^{noneq}, due to nonequilibrium currents, we thus neglect the pump contribution.

The above Equation 7.15 constitutes $S_I^{\mathrm{ch}}(0)$. At higher frequencies, when f approaches the average open time τ_{op} of the channel, $S_I^{\mathrm{ch}}(f)$ will decrease. The characteristic inverse time for a channel is $f_* = \tau_{\mathrm{op}}^{-1} + \tau_{\mathrm{cl}}^{-1}$, where τ_{cl} is the average closed time of the channel. With only one type of channel present in a cell, there will be a sigmoidally shaped, Lorentzian noise spectrum with an inflection point at $f = f_*$. With many types of channels for different kinds of ions present it is indeed possible to obtain a $1/f$ spectrum over several decades as the sum of Lorentzians [86].

To obtain the voltage power spectral density, we have to multiply $S_I^{noneq}(f)$ by R^2, where R represents the electrical resistance of the entire cell membrane. We let the cell surface area measure A. So we have $S_V^{noneq}(0) \approx 4\,Ne\,\langle I\rangle\,R^2$. The equilibrium Nyquist voltage noise across the same cell membrane is $S_V^{eq}(f) \approx 4kTR$. This is white noise and has the same strength at all frequencies. We thus obtain the following formula at $f = 0$ for the ratio $\theta(0)$ of the nonequilibrium noise and the equilibrium noise:

$$\theta(0) \approx \frac{Ne\langle I\rangle R}{kT} \tag{7.16}$$

The transmembrane current $\langle I\rangle$ is proportional to the cell surface A. The resistance R is inversely proportional to A. So eventually, the cell surface area and the cell geometry in general cancel out of the equation.

Data for the steady-state Na^+ flux through several types of cell membranes (e.g., rat soleus, sheep purkinje, squid axon, guinea pig auricles, and frog sartorius) are available [90]. That flux is about 50 pmol/(cm^2 sec). The vast majority of transmembrane ion transport is carried out by Na,K-ATPase, which transports two K^+ ions for every three Na^+ ions. We assume Na^+ and K^+ transport to therefore be about equal. After multiplying by Faraday's constant (the number of coulombs in a mole, i.e., about 10^5), we get a total current of about $\langle I\rangle \approx 10\ \mu A/cm^2$. The resistance of a cell membrane varies from 10^3 Ω cm^2 (squid axon) to 7×10^3 Ω cm^2 (mammalian cardiac cell). We thus find for $\theta(0)$ a value of about 5000.

Based on experimental data, it has been estimated that at 1 Hz the $1/f$ noise in the frog node of Ranvier is about a thousand times larger than thermal noise [91]. This is consistent with our estimate. Experimentally, it turns out that the power spectral density is constant from $f = 0$ up to about somewhere between 1 and 10 Hz [86,89]. At that point, the power spectral density starts to fall off as $1/f$. This means that we reach $\theta(f) = 1$, that is, the equilibrium and nonequilibrium noise being equal, somewhere near 10^4 Hz. Figure 7.5, in which the horizontal axis is in radians per second, indeed shows flattening between 10^3 and 10^4 Hz. At the 50- and 60-Hz power line frequencies, the nonequilibrium voltage noise is expected to exceed the equilibrium voltage noise by a factor of at least 10^2.

However, we should pause before taking the value of θ and employ it to incorporate nonequilibrium noise in the evaluation of a signal-to-noise ratio. As we saw earlier in this section, nonequilibrium noise may be a way to transduce energy from one stored form to another. So a signal can come in the form of a piece of nonequilibrium noise. With this gray area between signal and noise, it may no longer be straightforward to calculate a signal-to-noise ratio. There has been a natural selection toward high signal-to-noise ratios for signals whose detection has been important for the survival of the organism for many millennia. But what the signal-to-noise ratios and detection thresholds are for ELF and microwave radiation, which are relatively new phenomena in the environment, and how nonequilibrium noise figures in all of this is still open to conjecture and debate.

7.7 Chemical Noise

Chemical noise consists of both fundamental chemical noise (stochastic variations in net or accumulated amounts of a particular ion or molecule) and nonfundamental changes in chemical amount (molecular number) due to influences other than the applied field. We again adopt a recent discussion [12] in which an arbitrary biological system is considered.

In the discussion, attention focuses on weakly interacting fields that can create small chemical changes, but much of the approach is also relevant to strongly interacting fields, for example, those causing cell membrane electroporation (see Chapter 9 on electroporation in Ref. [127]). To begin, consider a small physical perturbation of the biological system due to the interaction of a local electromagnetic field, $\vec{F}_{\text{local}}(\vec{r}, t)$, that may vary from site to site within the volume of the biological system. In general, $\vec{F}_{\text{local}}$ may have a complicated dependence of its magnitude and direction on time and position, such that a formal prescription for calculating the field-induced molecular change due to an exposure is

$$\bar{n}_S = \int_{t=0}^{t=t_{\text{exp}}} \int_{\text{system volume}} J_0(t') f_{\text{bpm}}(\vec{F}_{\text{local}}(\vec{r},t'))\, dV\, dt' \tag{7.17}$$

We regard $\bar{n}_S$ as a molecular change signal. It is the primary consequence of the field exposure for the case where only one process, or one step in a cascade, is altered. Integration is carried out over the entire biological system volume and over the time comprising the exposure, t_{exp} (or control). This yields the accumulated, total chemical (molecular) change due to the applied field during the exposure. Other changes in the same ionic or molecular species may result from competing influences, for example, temperature variations, during t_{exp}.

The applied field interacts through one or more of a limited class of biophysical mechanisms. Here, *biophysical mechanism* means a class of interactions by which the field alters an ongoing biochemical rate (transport or reaction), with the rate arising from nonequilibrium processes dependent on metabolism. Examples of known biophysical mechanisms involving electric fields are heating (most biochemical processes have a nonzero temperature dependence), voltage-gated channels, electroconformational coupling of membrane enzymes, electroporation, and iontophoresis (mainly electrophoresis, but in some cases also electroosmosis). Examples involving magnetic fields are radical-pair reactions and twisting of magnetic material (magnetite or contaminant magnetic particles). As used here, a biophysical mechanism modulates an ongoing biochemical process, and both the coupling strength and the magnitude of the basal rate are important.

For a particular type of biophysical mechanism (bpm), the function $f_{\text{bpm}}\, (\vec{F}_{\text{local}}\, (t))$ describes the instantaneous alteration of the basal rate, J_0, which itself can vary in time [92]. The local field can be computed numerically at the tissue level (millimeter scale; see Chapter 11 on dosimetry, this volume) [93–99] and at the cellular level [100,101]. The time and position dependence of $\vec{F}_{\text{local}}\, (\vec{r},t)$ is often simple, namely, a constant magnitude (steady or DC) field, a constant amplitude periodic (AC) field, or at high frequencies a spatially decaying amplitude field due to power absorption. Environmental and occupational fields can be much more complicated, such that piecewise continuous representations may be needed.

If a weakly coupled physical perturbation alters the basal rate of a biochemical process (transport or reaction), the total chemical (molecular) change, expressed as the number of molecules, is

$$\bar{n} = \bar{n}_0 + \bar{n}_S \tag{7.18}$$

where $\bar{n}_0$ is the basal change during an exposure (sensing) time t_{exp}, and $\bar{n}_S$ is the (much smaller) molecular change due to the field exposure [22,45,92,102]. As noted above, $\bar{n}_S$ can be regarded as a molecular change signal. The basal process is far from equilibrium, driven by free-energy differences associated with metabolism. The largest field-induced molecular change occurs for a steady (DC) field exposure [22,102]:

$$\bar{n}_S = K_{\text{bpm,dc}} F_0 \, J_0 \, t_{\text{exp}} \tag{7.19}$$

where $K_{\text{bpm,dc}}$ describes the alteration of the basal rate by the steady field, here of magnitude F_0. Equation 7.19 is the DC version of the case of a weakly coupled periodic perturbation, previously described for the case of an extracellular electric field [45,92], namely,

$$\bar{n}_S = K_{\text{bpm,ac}} F_0^2 \, J_0 \, t_{\text{exp}} \tag{7.20}$$

where $K_{\text{bpm,ac}}$ describes the coupling that leads to rectification of the ongoing rate [45]. For basal rates with more complicated time dependence, Equation 7.17 may need to be evaluated numerically, but the same basic ideas apply. Equation 7.20 is valid for long exposures, involving a large number of cycles of the periodic field. Basal rates that can be altered by weakly interacting electromagnetic fields by definition involve small interaction energies, so that thermal fluctuations and chemical free-energy differences result in nonzero basal rates. A zero basal rate with an extremely large activation or interaction energy cannot, therefore, be expected to be changed to a measurable nonzero rate by a weakly interacting field.

A generalized, molecular change-based signal-to-noise ratio can be constructed by estimating the ratio of primary molecular change to the combined competing changes for the same molecular (ionic) species. We consider the simplest case of the field altering the rate at one step in a single pathway but note that in principle the present analysis can be extended to include multiple steps involving more than one biochemical pathway. We further assume that this biochemical has its rate through the pathway altered slightly by a physical perturbation, here an electromagnetic field. But competing influences can also alter the rate. Such influences include temperature variations, normal physiological concentration variations, changes in hormones and other regulating biochemicals, and mechanical perturbations of cells and tissues. Competing molecular changes can also be created by a background electromagnetic field, for example, normal electrical activity within the human body or by movement in the earth's magnetic field, interacting through the same biophysical mechanism. Such competition goes beyond fundamental chemical noise (molecular shot noise). Nonionizing influences can only modulate ongoing processes, and therefore such influences cannot (essentially by definition) introduce foreign molecules. This has the important consequence that competing molecular changes may arise from several sources (Table 7.1).

TABLE 7.1

Quantities Employed in Generalized Chemical Noise

Symbol	Molecular Change	Source
S	$\bar{n}_S$	Field-induced molecular change signal
N	$\sqrt{\bar{n}} \approx \sqrt{\bar{n}_0}$	Molecular shot noise (fundamental)
V	$\bar{n}_V$	Molecular change due to temperature variations
C	$\bar{n}_C$	Molecular change due to concentration variations
I	$\bar{n}_M$	Molecular change due to mechanical interference
B	$\bar{n}_B$	Molecular change due to background fields

A generalized signal-to-noise ratio $(S/N)_{\text{gen}}$, can thus be considered. The field-induced molecular change signal, S, is thereby quantitatively compared to the several sources of competing molecular changes for the same biochemical (molecule or ion), yielding

$$(S/N)_{\text{gen}} = \frac{S}{f_{\text{com}}(N,V,C,I,B)} \tag{7.21}$$

The various competing molecular changes, which may or may not be independent, are combined to give the total competing molecular change, f_{com}. Important simplifications can be made if the various competing molecular changes can be approximated as independent and random around their mean values. In this case, f_{com} can be approximated as

$$f_{\text{com}}(N,V,C,I,B) \approx [N^2 + V^2 + C^2 + I^2 + B^2]^{1/2} \tag{7.22}$$

Alternatively, emphasizing the changes in terms of numbers of molecules,

$$f_{\text{com}}(N,V,C,I,B) \approx [\bar{n}_0 + (\Delta n_V)^2 + (\Delta n_C)^2 + (\Delta n_M)^2 + (\Delta n_B)^2]^{1/2} \tag{7.23}$$

All significant sources of competing molecular change are directly relevant.

Consistent with experimental treatment of errors as random, here we consider the special case that all the important competing molecular changes can be approximated as independent and random variations around their mean value, as this allows the competing changes to be added in quadrature (Equation 7.22). This leads to a general molecular change-based signal-to-noise ratio that involves Gaussian distributions, namely,

$$(S/N)_{\text{gen}} \approx \frac{S}{[N^2 + V^2 + C^2 + I^2 + B^2]^{1/2}} \tag{7.24}$$

Each of these competing molecular changes is discussed briefly below, with reference to Table 7.1.

As indicated in Table 7.1, $N = \sqrt{\bar{n}} \approx \sqrt{\bar{n}_0}$ is the competing molecular change due to fundamental stochastic variations in biochemical reaction and transport processes [15,45,92], which provides a fundamental, minimum molecular change noise. Fundamental chemical noise is increasingly recognized as important to understanding other aspects of biological systems, such as the circadian clock [103,104], control of genetic circuits [105,106], and bacterial chemotaxis [107].

Temperature variations within the volume of the biological system are generally expected to result in altered rates. When integrated over the system volume and over the exposure time, a contribution to the end-point molecular change is expected, because most biochemical processes have nonzero temperature dependence. Thus, $V = \bar{n}_v$ is the resulting, competing molecular change due to temperature variations [15]. Human core body temperature has daily variations of more than 1°C [108–113], and there are even larger variations in the extremities. Often *in vitro* electric and magnetic field experiments use feedback control, for instance, in temperature-regulated exposure chambers, but these typically have variations greater than about 0.01°C at one or a few temperature measurement sites. Temperature variations within the biological system itself are often inferred, preferably by numerical models that can reasonably predict the temperature through the biological system by first predicting the specific absorption rate. During the exposure time, interfering temperature variation can be significant. To allow correction for temperature variations, the biological system should be characterized for its temperature sensitivity, and each particular apparatus should be characterized for its temperature variations for control and exposed conditions. As an example, an investigation first reporting athermal effects [114] was subsequently found to have temperature variation $\sim 0.1^{\circ}C$ at temperature measurement sites [115]. Without thermal modeling of the exposure systems and the biological systems, however, larger temperature changes away from the measurement site cannot be ruled out. It is the temperature change and variation over the entire volume containing cells (or other specimens) that need to be quantitatively understood. Temperature measurement at one site, typically somewhere along the perimeter or boundary of a temperature-regulated apparatus, is generally insufficient. The measured biochemical quantity should also be characterized for its temperature sensitivity for the biological system studied, so that the expected V can be determined, to address the basic specificity question of whether an observed change is due to the field or to temperature changes [116].

Changes in concentration of biochemicals involved in a process are well known to alter the rate of a process. Relevant chemical species include substrates, products, inhibitors, etc. The competing molecular change due to one or more interfering concentration changes is $C = \bar{n}_C$. In this case, a significant difference may exist for *in vitro* and *in vivo* experiments. Usually, only small, slow changes of chemical concentrations are expected *in vitro*, occurring, for example, through absorption or release of molecules (ions) from glass- and plasticware, spontaneous chemical decomposition, binding to cellular constituents, or evaporation. Uptake or release of interfering biochemicals from a biological preparation could be the predominant source, particularly if cells grow (taking up molecules) or die (releasing molecules). *In vivo* concentration variations are relatively large, because of normal physiologic variations. For example, Ca^{2+} concentration varies in humans by more than 1% over a day [117,118]. Unless buffered, these normal biochemical variations also compete with the field-induced molecular change.

Movement of tissue *in vivo* and vibration of an experimental apparatus containing a biological system can also create a mechanically induced molecular change, $I = \bar{n}_M$, that competes with a molecular change signal. In this case, the competing molecular change is due to interference of mechanical stress and strain [119,120], often present at high levels in living humans [11,121,122] but at low levels for *in vitro* experiments. *In vitro* apparatus can have quite different mechanical properties and isolation from ambient vibrations. Indeed, it has been found in some experiments that mechanical vibrations create effects larger than the field exposure [123]. Tissues *in vivo* experience significant mechanical deformation, but there is the least strain expected within the bone marrow and the brain [120]. This may be relevant to the "contact current hypothesis," which suggests that currents in the bone marrow may be important in exposures of children [124–126].

Background fields can, of course, also couple to biochemical processes through the same biophysical mechanisms as the applied field. Background field-induced molecular change competes, and is denoted by $B = \bar{n}_B$. Examples of background electromagnetic fields include the endogenous electrical fields generated within the body by cardiac, muscular, and neural activity and the sampling of different field values (local anomalies) in the ambient magnetic field as mobile humans move about in their environment. *In vitro* background fields will depend on the particular experimental environment, are usually small and constant, and are often measured.

7.8 Interpretation of Experiments

Specificity is fundamentally important to interpret experiments, as one wants to know what agent is responsible for the observed change(s). This is particularly relevant to experiments that find small changes in biological systems that are exposed to small electromagnetic fields. A basic challenge is to show that other influences are not responsible. Because it is well known that most biochemical processes have a significant temperature dependence, the approximate temperature sensitivity of the observed quantity should be determined or known, and some bound should be established for temperature drift or variations in the experiment. However, as already noted, even big changes can have more than one candidate cause. Both tissue electroporation and tissue movement can, for instance, underlie the changes in molecular uptake associated with large field pulses. For this reason, signal-to-noise ratio considerations should be preceeded by establishing field specificity, which can be much more difficult than observing a change associated with a field exposure. To establish specificity, a number of issues must be considered, many of which will be discussed next.

Many experiments determine quantities related to biochemical change. Exceptions are experiments that determine physical quantities such as voltages and currents, temperature changes, and magnetic particle rotation. However, electrical measurements are the most frequent physical measurement. These are incredibly important to systems of excitable cells, with experimental preparations ranging from isolated cells to electrophysiologic measurements on humans. Accumulation of charge might be measured, but probably as a voltage on a capacitance. Signal-to-noise ratio issues are still important, of course, but usually there is an important distinction: voltages and currents (rarely charge) are readily measured continuously.

A further distinction is that most experiments involving exposures to small fields use long exposure times (many seconds to hours or even days). Such experiments commonly determine biochemical quantities directly, for example, enzyme activity, or indirectly, for example, fluorescence emission from fluorescent indicators of intracellular calcium concentration. In this broad case, consideration of generalized chemical noise is relevant. Following the discussion in a recent paper [12], both the magnitude of the field perturbation and the nature and magnitude of chemical competition need to be understood. Such analysis should explicitly estimate the coupling to ongoing, far from equilibrium, metabolically driven biochemical processes and should quantitatively determine molecular changes due to competing influences. Only then can the analysis distinguish idealized conditions from *in vitro* conditions and *in vivo* conditions and then determine whether reported effects can be explained by known biophysical mechanisms.

In vivo there are several kinds of noise, and it is important to distinguish between them. Equilibrium noise comes about as a consequence of Brownian motion, that is, the random movement of molecules at finite temperature. At equilibrium, every degree of freedom takes on the same amount of energy, and equilibrium noise is therefore easy to evaluate. For a signal to exceed the equilibrium noise band, the quantitative criteria are often readily derived. Such baseline criteria can be useful when assessing the electroreception and magnetoreception that many organisms exhibit. But the nonequilibrium nature of life brings in nonequilibrium noise. When a primary molecular change is amplified through a biochemical cascade, "amplifier noise" is inevitable. Nonequilibrium noise appears whenever energy is dissipated, that is, when work is done. In many biological contexts the nonequilibrium noise is much more intense than the equilibrium noise. A serious complication is constituted by the fact that nonequilibrium noise, unlike equilibrium noise, is also able to perform work, that is, be a power source for an energetically uphill process. Many biological processes may rely on the energy transduction that can be accomplished through nonequilibrium noise. The analysis of such situations poses challenges as the noise may be a signal and the signal may be noise. There is no easy general "common denominator" theory for nonequilibrium noise like there is for equilibrium kT-noise.

In a recent discussion [12], it is argued that experimental measurements can be plausibly related quantitatively to an underlying primary molecular change because of a field exposure operating through a biophysical mechanism. It is further argued that only the uncertainty in this change propagates through biochemical amplification and therefore dominates the measurement uncertainty. A more complete approach would involve traditional, independent determination of the instrumental or assay error (quantitative characterization of the experimental measurement system). After removal of the "instrumental noise," the generalized signal-to-noise ratio $(S/N)_{gen}$, could be revised upward. This would allow interpretation (correction) of experimental error to estimate the uncertainty in the measured quantity itself. Assessment of combinations of biophysical mechanism models and particular exposure can then be carried out, using the most field-sensitive versions of theoretical models for the candidate biophysical mechanisms. The criterion $(S/N)_{gen} \leq 0.1$ is a very conservative basis for ruling out a particular class of biophysical mechanism for a given field exposure. Similarly, the criterion $(S/N)_{gen} \geq 10$ is a conservative basis for ruling in a candidate biophysical mechanism for a given exposure, retaining that biophysical mechanism hypothesis for further evaluation. This approach provides a quantitative basis for rejecting or accepting hypothetical biophysical mechanisms as candidate explanations for an experimental measurement.

The traditional choice $(S/N)_{gen} \approx 1$ is a useful but somewhat arbitrary dividing line, which indicates conditions for which an effect might appear. $(S/N)_{gen} \leq 0.1$ and $(S/N)_{gen} \geq 10$ provide criteria for stronger conclusions, allowing rejection or provisional retention of a biophysical mechanism hypothesis. This approach to interpreting experiments thus provides a general method for carrying out theoretical assessment of reported weak field exposure effects. This approach can distinguish relatively quiet *in vitro* conditions from *in vivo* conditions containing more and larger influences of competing molecular change. This in turn allows quantitative estimates of whether an *in vitro* result is relevant to *in vivo* conditions.

Acknowledgment

This work was supported partially by NIH grant RO1-GM63857.

References

1. G.N. Stewart. The changes produced by the growth of bacteria in the molecular concentration and electrical conductivity of culture media. *J. Exp. Med.*, 4:235–243, 1899.
2. R. Eden and G. Eden. *Impedance Microbiology*. Research Studies Press, Letchworth, Herts, U.K., 1984.
3. M. Wawerla, A. Stolle, B. Schalch, and H. Eisgruber. Impedance microbiology: applications in food hygiene. *J. Food Prot.*, 62:1488–1496, 1999.

4. J. McPhillips and N. Snow. Studies on milk with a new type of conductivity cell. *Aust. J. Dairy Technol.*, 3:192–196, 1958.
5. T. Ritz, P. Thalau, J.B. Phillips, R. Wiltschko, and W. Wiltschko. Resonance effects indicate a radical-pair mechanism for avian magnetic compass. *Nature*, 429:177–180, 2004.
6. D. Cohen. Ferromagnetic contaminants in the lungs and other organs of the body. *Science*, 180:745–748, 1973.
7. J.L. Kirschvink, A.K. Kirschvink, and B.J. Woodford. Magnetite biomineralization in the human brain. *Proc. Natl. Acad. Sci.*, 89:7683–7687, 1992.
8. J.R. Dunn, M. Fuller, J. Zoeger, J. Dobson, F. Heller, J. Ilammann, E. Caine, and B.M. Moskowitz. Magnetic material in the human hippocampus. *Brain Res. Bull.*, 36:149–153, 1994.
9. P.L. McNeil and S. Ito. Molecular traffic through plasma membrane disruptions of cells. *In Vivo J. Cell Sci.*, 96:549–556, 1990.
10. P.L. McNeil and R.A. Steinhardt. Loss, restoration, and maintanence of plasma membrane integrity. *J. Cell Biol.*, 137:1–4, 1997.
11. P.L. McNeil and M. Terasaki. Coping with the inevitable: how cells repair a torn surface membrane. *Nat. Cell Biol.*, 3:124–129, 2001.
12. T.E. Vaughan and J.C. Weaver. Molecular change signal-to-noise criteria for interpreting experiments involving exposure of biological systems to weakly interacting electromagnetic fields. *Bioelectromagnetics*, 26:305–322, 2005.
13. N.G. Van Kampen. *Stochastic Processes in Physics and Chemistry*. Elsevier, Amsterdam, 1992.
14. R.C. Duncan, R.G. Knapp, and M.C. Miller. *Introductory Biostatistics for the Health Sciences*. John Wiley & Sons, New York, 1983.
15. J.C. Weaver, T.E. Vaughan, and G.T. Martin. Biological effects due to weak electric and magnetic fields: the temperature variation threshold. *Biophys. J.*, 76:3026–3030, 1999.
16. J.C. Weaver. Understanding conditions for which biological effects of nonionizing electromagnetic fields can be expected. *Bioelectrochemistry*, 56:207–209, 2002.
17. C.W. Helstrom. *Statistical Theory of Signal Detection*, Second Edition. Pergamon Press, New York, 1968.
18. A.J. Kalmijn. Electro-perception in sharks and rays. *Nature*, 212:1232–1233, 1966.
19. A.J. Kalmijn. Electric and magnetic field detection in elasmobranch fish. *Science*, 218:916–918, 1982.
20. W.J. Moore. *Physical Chemistry*. Longman, London, 1972.
21. A.J. Kalmijn. Graded positive feedback in elasmobranch ampullae of lorenzini. In S.M. Bezrukov, Ed. *Unsolved Problems of Noise and Fluctuations: UPoN 2002: Third International Conference*, volume 665 of *Conference Proceedings*, pp. 133–141, American Institute of Physics, Melville, NY, 2003.
22. R.K. Adair, R.D. Astumian, and J.C. Weaver. Detection of weak electric fields by sharks, rays, and skates. *Chaos*, 8:576–587, 1998.
23. S.M. Bezrukov. Sensing nature's electric fields: ion channels as active elements of linear amplification. In S.M. Bezrukov, Ed. *Unsolved Problems of Noise and Fluctuations: UPoN 2002: Third International Conference*, volume 665 of *Conference Proceedings*, pp. 142–149, American Institute of Physics, Melville, NY, 2003.
24. J.C. Weaver, T.E. Vaughan, and R.D. Astumian. Biological sensing of small field differences by magnetically sensitive chemical reactions. *Nature*, 405:707–709, 2000.
25. R.P. Blakemore. Magnetotactic bacteria. *Science*, 190:377–379, 1975.
26. R. Blakemore. Magnetotactic bacteria. *Annu. Rev. Microbiol.*, 36:217–238, 1982.
27. J.L. Kirschvink, M.M. Walker, and C.E. Diebel. Magnetite-based magnetoreception. *Curr. Opin. Neurobiol.*, 11:462–467, 2001.
28. J.L. Kirschvink. Comment on "Constraints on biological effects of weak exteremely-low-frequency electromagnetic fields." *Phys. Rev. A*, 46:2178–2184, 1992.
29. M.M. Walker, T.E. Dennis, and J.L. Kirschvink. The magnetic sense and its use in long-distance navigation by animals. *Curr. Opin. Neurobiol.*, 12:735–744, 2002.
30. National Institute of Environmental Health Sciences (NIEHS). NIEHS Report on Health Effects from Exposure Power-Line Frequency Electric and Magnetic Fields. NIH Publication No. 99–4493, 1999.

31. R.K. Adair. Effects of ELF magnetic fields on biological magnetite. *Bioelectromagnetics*, 14:1–4, 1993.
32. C. Polk. Effects of extremely-low-frequency magnetic fields on biological magnetite. *Bioelectromagnetics*, 15:261–270, 1994.
33. R.K. Adair. Constraints on biological effects of weak extremely-low-frequency electromagnetic fields. *Phys. Rev. A*, 43:1039–1048, 1991.
34. K.R. Foster and H.P. Schwan. Dielectric properties of tissues and biological materials: a critical review. *CRC Crit. Rev. Bioeng.*, 17:25–104, 1989.
35. J.B. Johnson. Thermal agitation of electricity in conductors. *Phys. Rev.*, 32:97–109, 1928.
36. H. Nyquist. Thermal agitation of electric charge in conductors. *Phys. Rev.*, 32:110–113, 1928.
37. W. Schottky. Über spontane stromschwankungen in verschiedenen elektrizitätsleitern. *Ann. Phys. (Leipzig)*, 57:541–568, 1918.
38. R. Sarpeshkar, T. Delbrück, and C.A. Mead. White noise in MOS transistors and resistors. *IEEE Circuits Devices*, Nov.:23–29, 1993.
39. M. Bier. Gauging the strength of power frequency fields against membrane electrical noise. *Bioelectromagnetics*, 26:595–609, 2005.
40. B. Hille. *Ion Channels of Excitable Membranes*. Sinauer, Sunderland, MA, 1992.
41. P. Läuger. *Electrogenic Ion Pumps*. Sinauer, Sunderland, MA, 1991.
42. J.C. Weaver and R.D. Astumian. The response of living cells to very weak electric fields, the thermal noise limit. *Science*, 247:459–462, 1990.
43. W.T. Kaune. Thermal noises limit on the sensitivity of cellular membranes to power frequency electric and magnetic fields. *Bioelectromagnetics*, 23:622–928, 2002.
44. G. Vincze, N. Szasz, and A. Szasz. On the thermal noise limit of cellular membranes. *Bioelectromagnetics*, 26:28–35, 2005.
45. R.D. Astumian, J.C. Weaver, and R.K. Adair. Rectification and signal averaging of weak electric fields by biological cells. *Proc. Natl. Acad. Sci.*, 92:3740–3743, 1995.
46. L. Brillouin. Can the rectifier become a thermodynamical demon? *Phys. Rev.*, 78:627–628, 1950.
47. P.W. Bridgman. Note on the principle of detailed balancing. *Phys. Rev.*, 31:101–102, 1928.
48. R.C. Tolman. The principle of microscopic reversibility. *Proc. Natl. Acad. Sci.*, 11:436–439, 1925.
49. R.P. Feynman, R.B. Leighton, and M. Sands. *The Feynman Lectures on Physics*. Addison-Wesley, Reading, MA, 1966.
50. M. von Smoluchowski. Experimentell nachweisbare, der üblichen thermodynamik widersprechende molekularphänomene. *Phys. Zeitschr.*, 13:1069–1080, 1912.
51. P.W. Atkins. *The Second Law*. Scientific American Books, New York, 1984.
52. H.S. Leff and A.F. Rex, Eds. *Maxwell's Demon: Entropy, Information, Computing*. Princeton University Press, Princeton, NJ, 1990.
53. J. Howard and A.J. Hudspeth. Compliance of the hair bundle associated with gating of mechanoelectrical transduction channels in the bullfrog's saccular hair cell. *Neuron*, 1:189–199, 1988.
54. M. Magnasco. Forced thermal ratchets. *Phys. Rev. Lett.*, 71:1477–1480, 1993.
55. M. Bier. Brownian ratchets in physics and biology. *Contemp. Phys.*, 38:371–379, 1997.
56. R.D. Astumian and M. Bier. Fluctuation driven ratchets: molecular motors. *Phys. Rev. Lett.*, 72:1766–1769, 1994.
57. R.D. Astumian. Thermodynamics and kinetics of a brownian motor. *Science*, 276:917–922, 1997.
58. M. Bier and R.D. Astumian. Biased Brownian motion as the operating principle for microscopic engines. *Bioelectrochem. Bioenerg.*, 39:67–75, 1996.
59. I. Kosztin and K. Schulten. Fluctuation-driven molecular transport through an asymmetric membrane channel. *Phys. Rev. Lett.*, 93:238102-1–238102-4, 2004.
60. Z. Siwy, I.D. Kosińska, and A. Fuliński. On the validity of continuous modelling of ion transport through nanochannels. *Europhys. Lett.*, 67:683–689, 2004.
61. Z. Siwy, I.D. Kosińska, A. Fuliński, and C.R. Martin. Asymmetric diffusion through synthetic nanopores. *Phys. Rev. Lett.*, 94:048102-1–048102-4, 2005.
62. Z. Siwy and A. Fuliński. Fabrication of a synthetic nanopore ion pump. *Phys. Rev. Lett.*, 89:198103-1–198103-4, 2002.

63. J. Bernamont. Fluctuations in the resistance of thin films. *Proc. Phys. Soc.*, 49:138–139, 1937.
64. B. Pellegrini, R. Saletti, P. Terreni, and M. Prudenziati. $1/f^{\gamma}$ noise in thick-film resistors as an effect of tunnel and thermally activated emissions, from measures versus frequency and temperature. *Phys. Rev. B*, 27:1233–1243, 1983.
65. J.B. Bassingthwaighte, L.S. Liebovitch, and B.J. West. *Fractal Physiology*. Oxford University Press, New York, 1994.
66. C. Wunsch. Bermuda sea level in relation to tides, weather, and baroclinic fluctuations. *Rev. Geophys.*, 10:1–49, 1972.
67. M. Dishon-Berkovits and R. Berkovits. Work-related tardiness: lateness incident distribution and long-range correlations. *Fractals*, 5:321–324, 1997.
68. P. Bak, C. Tang, and K. Wiesenfeld. Self-organized criticality: an explanation of $1/f$ noise. *Phys. Rev. Lett.*, 59:381–384, 1987.
69. P. Bak. *How Nature Works*. Springer-Verlag, New York, 1996.
70. S. Kauffman. *Investigations*. Oxford University Press, New York, 2000.
71. B.D. Malamud, G. Morein, and D.L. Turcotte. Forest fires: an example of self-organized critical behavior. *Science*, 281:1840–1842, 1998.
72. H.M. Jaeger, C. Liu, and S.R. Nagel. Relaxation at the angle of repose. *Phys. Rev. Lett.*, 62:40–43, 1989.
73. S. Nagel. Instabilities in a sandpile. *Rev. Mod. Phys.*, 64:321–325, 1992.
74. M. Bretz, J.B. Cunningham, P.L. Kurczynski, and F. Nori. Imaging of avalanches in granular materials. *Phys. Rev. Lett.*, 69:2431–2434, 1992.
75. J.G. Nicholis, A.R. Martin, and B.G. Wallace, Eds. *From Neuron to Brain*. Sinauer, Sunderland, MA, 1992.
76. H.E. Derksen and A.A. Verveen. Fluctuations of resting neural membrane potential. *Science*, 151:1388–1389, 1966.
77. H.E. Derksen. Axon membrane voltage fluctuations. *Acta Physiol. Pharmacol. Neerl.*, 13:373–466, 1965.
78. H.E. Derksen and A.A. Verveen. Fluctuations in membrane potential of axons and the problem of coding. *Kybernetik* 2:152–160, 1965.
79. D. Colquhoun and A.G. Hawkes. On the stochastic properties of bursts of single ion channel openings and of clusters of bursts. *Philos. Trans. R. Soc. Lond.*, 300:1–59, 1982.
80. S.D. Silberberg and K.L. Magleby. Preventing errors when estimating single channel properties from the analysis of current fluctuations. *Biophys. J.*, 65:1570–1584, 1993.
81. D. Colquhoun, C. Hatton, and A.G. Hawkes. The quality of maximum likelihood estimation of ion channel rate constants. *J. Physiol. (Lond.)*, 547:699–728, 2003.
82. L.S. Liebovitch, J. Fishbarg, and J.P. Koniarek. Ion channel kinetics: a model based on fractal scaling rather than multistate Markov processes. *Math. Biosci.*, 84:37–68, 1987.
83. L.S. Liebovitch. Analysis of fractal ion channel gating kinetics: kinetic rates, energy levels, and activation energies. *Math. Biosci.*, 93:97–115, 1989.
84. L.S. Liebovitch. Testing fractal and Markov models of ion channel kinetics. *Biophys. J.*, 55:373–385, 1989.
85. I. Goychuk and P. Hänggi. Fractional diffusion modeling of ion channel gating. *Phys. Rev. E*, 70:051915-1–051915-9, 2004.
86. K. Diba, H.A. Lester, and C. Koch. Intrinsic noise in cultured hippocampal neurons: experiment and modeling. *J. Neurosci.*, 24:9723–9733, 2004.
87. S.M. Bezrukov and M. Winterhalter. Examining noise sources at the single molecule level: $1/f$ noise of an open maltoporin channel. *Phys. Rev. Lett.*, 85:202–205, 2002.
88. Z. Siwy and A. Fuliński. Origin of $1/f^{\alpha}$ noise in membrane channel currents. *Phys. Rev. Lett.*, 89:158101-1–158101-4, 2002.
89. L.J. DeFelice. *Introduction to Membrane Noise*. Plenum Press, New York, 1981.
90. O.M. Sejersted. Maintenance of Na,K-homeostasis by Na,K-pumps in striated muscle. *Prog. Clin. Biol. Res.*, 268B:195–206, 1988.
91. F.S. Barnes. Interaction of DC and ELF electric fields with biological materials and systems. In C. Polk and E. Postow, Eds. *Handbook of Biological Effects of Electromagnetic Fields*, pp. 103–147, CRC Press, Boca Raton, FL, 1996.

92. J.C. Weaver, T.E. Vaughan, R.K. Adair, and R.D. Astumian. Theoretical limits on the threshold for the response of long cells to weak ELF electric fields due to ionic and molecular flux rectification. *Biophys. J.*, 75:2251–2254, 1998.
93. M. Stuchly and T. Dawson. Interaction of low-frequency electric and magnetic fields with the human body. *Proc. IEEE*, 88:643–662, 2000.
94. M.A. Stuchly and O.P. Gandhi. Inter-laboratory comparison of numerical dosimetry for human exposure to 60 Hz electric and magnetic fields. *Bioelectromagnetics*, 21:167–174, 2000.
95. P.A. Mason, W.D. Hurt, T.J. Walters, J.A. D'Andrea, P. Gajšek, K.I. Ryan, D.A. Nelson, K.I. Smith, and J.M. Ziriax. Effects of frequency, permittivity and voxel size on predicted specific absorption rate values in biological tissue during electromagnetic-field exposure. *IEEE Trans. Microwave Theory Technol.*, 48:2050–2058, 2000.
96. P. Gajsek, T.J. Walters, W.D. Hurt, D.A. Nelson, and P.A. Mason. Empirical validation of SAR values predicted by FDTD modeling. *Bioelectromagnetics*, 23:37–48, 2002.
97. M. Nadeem, T. Thorlin, O.P. Gandhi, and M. Persson. Computation of electric and magnetic stimulation in human head using the 3-D impedance method. *IEEE Trans. Biomed. Eng.*, 50:900–907, 2003.
98. T.W. Dawson, K. Caputa, M.A. Stuchly, and R. Kavet. Comparison of electric fields induced in humans and rodents by 60-Hz contact currents. *IEEE Trans. Biomed. Eng.*, 50:744–753, 2003.
99. S.J. Allen, E.R. Adair, K.S. Mylacraine, W. Hurt, and J. Ziriax. Empirical and theoretical dosimetry in support of whole body radio frequency (RF) exposure in seated human volunteers at 220 MHz. *Bioelectromagnetics*, 26:440–447, 2005.
100. T.R. Gowrishankar and J.C. Weaver. An approach to electrical modeling of single and multiple cells. *Proc. Natl. Acad. Sci.*, 100:3203–3208, 2003.
101. D.A. Stewart, T.R. Gowrishankar, and J.C. Weaver. Transport lattice approach to describing cell electroporation: use of a local asymptotic model. *IEEE Trans. Plasma Sci*, 32:1696–1708, 2004.
102. P.C. Gailey. Membrane potential and time required for detection of weak signals by voltage-gated ion channels. *Bioelectromagnetics*, 20:102–109, 1999.
103. N. Barkai and S. Liebler. Circadian clocks limited by noise. *Nature*, 403:267–268, 2000.
104. M.B. Elowitz and S. Leibler. A synthetic oscillatory network. *Nature*, 403:335–338, 2000.
105. T.S. Gardner, C.R. Cantor, and J.J. Collins. Construction of a genetic toggle switch in *Escherichia coli*. *Nature*, 403:339–342, 2000.
106. A. Becskei and L. Serrano. Engineering stability in gene networks by autoregulation. *Nature*, 405:590–593, 2000.
107. T.S. Shimizu, S.V. Aksenov, and D. Bray. A spatially extended stochastic model of the bacterial chemotaxis signalling pathway. *J. Mol. Biol.*, 329:291–309, 2003.
108. H.T. Hammel. Regulation of internal body temperature. *Annu. Rev. Physiol.*, 30:641–710, 1968.
109. S.A. Rubin. Core temperature regulation of heart rate during exercise in humans. *J. Appl. Physiol.*, 62:1997–2002, 1987.
110. W.R. Keatinge, A.C. Mason, C.E. Millard, and C.G. Newstead. Effects of fluctuating skin temperature on thermoregulatory responses in man. *J. Physiol.*, 378:241–252, 1986.
111. K. Shiraki, S. Sagawa, F. Tajima, A. Yokota, M. Hashimoto, and G.L. Brengelmann. Independence of brain and tympanic temperatures in an unanesthetized human. *J. Appl. Physiol.*, 65:482–486, 1988.
112. P. Webb. Temperatures of skin, subcutaneous tissue, muscle and core in resting men in cold, comfortable and hot conditions. *Eur. J. Appl. Physiol.*, 64:471–476, 1992.
113. R.K. Adair. Biophysical limits on athermal effects of RF and microwave radiation. *Bioelectromagnetics*, 24:39–48, 2003.
114. D. de Pomerai, C. Daniells, H. David, J. Allan, I. Duce, M. Mutwakil, D. Thomas, P. Sewell, J. Tattersall, D. Jones, and P. Candido. Non–thermal heat-shock response to microwaves. *Nature*, 405:417–418, 2000.
115. D.I. de Pomerai, B. Smith, A. Dawe, K. North, T. Smith, D.B. Archer, I.R. Duce, D. Jones, and E.P.M. Candido. Microwave radiation can alter protein conformation without bulk heating. *FEBS Lett.*, 543:93–97, 2003.

116. Y.L. Zhao, P.G. Johnson, G.P. Jahreis, and S.W. Hui. Increased DNA synthesis in INIT/$10T_{1/2}$ cells after exposure to a 60 Hz magnetic field: a magnetic-field or a thermal effect? *Radiat. Res.*, 151:201–208, 1999.
117. B. Morrison, A. Shenkin, A. McLelland, D.A. Robertson, M. Barrowman, S. Graham, G. Wuga, and K.J. Cunningham. Intra-individual variation in commonly analyzed scrum constituents. *Clin. Chem.*, 25:1799–1805, 1979.
118. M.G. Weyer and H. Lommel. *LONG I: Eine Longitudinal-Studie über individuelle Normbereiche, individuelle Standardbereiche, statislische Normbereiche für Prävention und Früherkennung*. Verlag Kirchheim, Mainz, 1981.
119. T.E. Vaughan and J.C. Weaver. Energetic constraints on the creation of cell membrane pores by magnetic particles. *Biophys. J.*, 71:616–622, 1996.
120. T.E. Vaughan and J.C. Weaver. Molecular change due to biomagnetic stimulation and transient magnetic fields: mechanical interference constraints on possible effects by cell membrane pore creation via magnetic particles. *Bioelectrochem. Bioenerg.*, 46:121–128, 1998.
121. P.L. McNeil. Cell wounding and healing. *Am. Sci.*, 79:222–235, 1991.
122. D. Bansal, K. Miyake, S.S. Vogel, S. Groh, C.C. Chen, R. Williamson, P.L. McNeil, and K.P. Cambell. Defective membrane repair in dysferlin-deficient muscular dystrophy. *Nature*, 423:129–131, 2003.
123. U. Valtersson, K. Hanssson Mild, and M.-O. Mattsson. Uncharacterized physical parameters can contribute more than magnetic field exposure to ODC activity *in vitro*. In F. Bersani, Ed. *Electricity and Magnetism in Biology and Medicine*, pp. 449–452, Plenum Press, New York, 1999.
124. R. Kavet, L. Zaffanella, J. Daigle, and K. Ebi. The possible role of contact current in cancer risk associated with residential magnetic fields. *Bioelectromagnetics*, 21:538–553, 2000.
125. R. Kavet, L.E. Zaffenella, R.L. Pearson, and J. Dallaplazza. Association of residential magnetic fields with contact voltages. *Bioelectromagnetics*, 25:530–536, 2004.
126. R. Kavet. Contact current hypothesis: summary of results to date. *Bioelectromagnetics*, suppl. 7:S75–S85, in press.
127. J.C. Weaver, Y. Chizmadzhev, Electroporation. In F.S. Barnes and B. Greenebaum, *Biological and Medical Aspects*, Eds. Taylor & Francis, Boca Raton, FL, Chapter 9, 2006.

8

Biological Effects of Static Magnetic Fields

Shoogo Ueno and Tsukasa Shigemitsu

CONTENTS

8.1 Introduction

In ancient times, magnetism, especially permanent magnets, were a symbol of mystique because of the magnetic force that remains even after numerous attractions of iron. Although William Gilbert was called the pioneer of modern magnetics, he was surprised by the magnetism associated with living organisms. Magnetism, in contrast to transient static electricity, has been used as an explanation for various invisible effects and is expected to possess miraculous healing powers. Although people have been using magnetism for healing purposes without any scientific evidence, there have not been significant problems concerning the side effects of magnetism. As permanent magnets have improved in quality and achieved higher magnetic strength, magnetism has become more commonly used in modern medicine. For example, magnetism is commonly used in the correction of dentures.

In the society today there are many sources of electromagnetic fields. Humans are exposed daily to man-made and naturally originated fields. During the past decade, questions about whether the exposure to electromagnetic fields may be linked to adverse health effects have been raised. Although the interaction of electromagnetic fields with biological systems has been investigated, there are no biophysical mechanisms that can explain many of the observed biological effects of low-level of magnetic fields. Proposed mechanisms include effects on currents, direct forces on biomagnetic materials, effects on free radicals, ion cyclotron resonance, charge transfer processes, stochastic resonance, etc.

Recent developments in medical instrumentation such as magnetic resonance imaging (MRI) and transcranial magnetic stimulation (TMS) have raised questions as to whether or not strong (in the tesla range) magnetic fields influence human health. Medical applications of weak (1–200 mT) magnetic fields for the purpose of pain reduction and tissue healing have also been studied for many years. Therapeutic applications of permanent magnets and other magnetic devices have recently been expanded to various areas such as treatment of pain and diseases like rheumatoid arthritis and cancer.

The objective of this chapter is to describe some of the more recent information on biological effects and medical applications of static magnetic fields. This chapter consists of four sections. Each section has a comprehensive review of a recent topic of interest. A short summary of the mechanism of static magnetic field action on biological systems is described in the second section, which includes a brief review of well-known mechanisms that are discussed before mentioning the interactions of weak static magnetic fields with biological systems. The third section reviews and summarizes more recent *in vivo* and *in vitro* experimental results of the effects of static magnetic fields, including near-zero magnetic fields, geomagnetic fields, and MRI fields. The fourth section covers special topics including magnetic sensing, magnetite, and plant response to magnetic fields. In parallel with the comprehensive review of biological effects, the development and medical application of the static magnetic field phenomena are introduced in the fifth section and reviewed with emphasis on the applications that are currently under investigation.

8.2 Mechanisms of Biological Effects of Static Magnetic Fields

The biological effects of static magnetic fields are not well understood. Magnetic fields exert a force on moving charged particles at right angles to both the field and the velocity, $\vec{F} = q\vec{v} \times \vec{B}$, where $\vec{F}$ is the force, q is the charge, $\vec{v}$ is the velocity, and $\vec{B}$ is the magnetic flux density. In vacuum, the magnetic flux density is given by $\vec{B} = \mu_0 \vec{H}$, where $\vec{H}$ is the magnetic field strength and μ_0 is the magnetic permeability of a vacuum. (In this section we will be careful to distinguish between $\vec{B}$ and $\vec{H}$, although in subsequent sections we will be less careful, using "magnetic field" as a general term that often is used to describe flux density.) Additionally, these fields exert a torque on particles with both fixed and induced magnetic dipole moments. This torque tends to align the dipoles along the magnetic flux density and is given by $\vec{T} = \vec{M} \times \vec{H}$ where $\vec{M}$ is the magnetic dipole moment. For paramagnetic and diamagnetic materials, $\vec{M}$ is proportional to the magnetic flux density. The magnetic susceptibility, χ, is given by the equation, $\vec{M} = \chi \vec{H}$, where $\vec{H}$ is the local value of the magnetic field strength and χ is dimensionless. All materials may be mainly divided into three categories based on their susceptibility values. Materials with negative susceptibility are called diamagnetic. Materials with positive values are referred to as paramagnetic. Materials with large susceptibility include both ferro- and ferrimagnetic materials. Recognition of the role of diamagnetic, paramagnetic, and ferro- or ferrimagnetic materials in the body helps in the understanding the underlying mechanisms of biomagnetic effects.

Table 8.1 shows three types of well-known mechanisms of the biological effects of magnetic fields including time-varying magnetic fields. As shown in Table 8.1, there are two basic mechanisms of static magnetic fields: first, the magnetic torques on objects and second, the mechanical force effects. A radical pair mechanism is also proposed.

When biological materials or systems are exposed to a spatially homogenous magnetic field, they tend to rotate to a stable direction, which is determined by the anisotropy of magnetic susceptibility of the materials and magnetic torque acting on the materials, as described by the following equation:

$$T = -\frac{1}{2\mu_0} B^2 \Delta\chi \sin 2\theta$$

where B is the magnetic flux density, $\Delta\chi$ is the anisotropy of magnetic susceptability of the materials, θ is the angle between the direction of the magnetic field and the long axis of the materials, and μ_0 is the magnetic permeability of a vacuum. The magnetic orientation of diamagnetic materials such as fibrin and collagen can be observed and explained by this principle (Torbet et al., 1981).

Next, when biological materials or systems are exposed to a spatially inhomogenous magnetic field, the materials or systems tend to move along the direction of the steepest

TABLE 8.1

Well-Known Mechanisms of the Biological Effects of Magnetic Fields

1. Time-varying magnetic field	
Eddy currents $J = -\sigma \frac{B}{t}$	Nerve stimulation
Heat SAR $= \sigma \frac{E^2}{\rho}$	Thermal effects
2. Static magnetic fields	
a. Homogenous magnetic field	Magnetic orientation of biological cells
Magnetic torque	
$T = -\frac{1}{2\mu_0} B^2 \Delta \chi \sin 2\theta$	
b. Inhomogenous magnetic field	Parting of water by magnetic fields (Moses effect)
Magnetic force	
$F = \frac{\chi}{\mu_0} (\text{grad } B) B$	
3. Multiplication of magnetic fields and other energy	Yield effect of cage product and escape product
Photochemical reactions with radical pairs	
Singlet–triplet intersystem crossing	

gradient of magnetic force. The magnetic force acting on the materials is proportional to the multiplication of the magnetic flux density B, the gradient of the magnetic flux density B (grad B), and the magnetic susceptibility χ of the materials, as described by the following equation:

$$F = \frac{\chi}{\mu_0}(\text{grad}B)B$$

where μ_0 is the magnetic permeability in a vacuum.

Dramatic demonstrations of these forces can be seen when water is parted by magnetic fields using magnetic field exposures of 4–8 T with a gradient of 50 T/m (Ueno and Iwasaka, 1994a,b). Magnetic levitation of diamagnetic materials such as wood and other organic materials, is realized in magnetic fields of more than 20 T (Beaugnon and Tournier, 1991). This phenomenon can also be explained by the principles of magnetic force.

Oxygen is a paramagnetic molecule, and its behavior under magnetic field gradients has an important role both as a gas and when dissolved in solution. The blockage and disturbance of gas flow by magnetic fields have been observed (Ueno and Harada, 1987). This phenomenon, called a magnetic curtain, can be explained by the action of magnetic force on a paramagnetic molecule. The magnetic curtain in this example is a wall of oxygen or air. Experimentation with this principle has demonstrated the quenching of burning candle flames (Ueno, 1989), where the interception of oxygen by the magnetic curtain has a quenching effect on the flames. As a second example, when a human subject is positioned inside a space shielded by the magnetic curtain, one may expect the respiratory function, evaporation of water molecules, body temperature, blood circulation, and other physiological functions to be disturbed or modulated.

The effect of static magnetic fields on water vaporization rate was investigated (Nakagawa et al., 1999). This study demonstrated that the water vaporization rate was found to be significantly influenced when both air and oxygen were in the presence of magnetic fields, and the observed effect was dependent on the field–field gradient product rather than on the B field itself. It should be noted that this magneto-enhancement of vaporization might be the indirect cause of certain physiologic effects on living organisms.

For example, skin temperature decreases in rats exposed to magnetic fields have been observed (Ichioka et al., 2003). This observation can be explained in terms of the effects of the magnetic field on air convection. That is, high magnetic field gradients (135–140 T^2/m) push the diamagnetic water molecules toward the magnet bore. This increases the movement of water molecules in the air around the animal body and the vaporization rate. The heat of vaporization leads to a decrease in skin temperature. The decrease in the blood flow of the skin microcirculation is assumed to be a secondary change to the decrease in skin temperature.

Oxygen dissolved in water is also affected by magnetic fields. Changes in dissolved oxygen concentration were observed for magnetic field exposures on the order of 1 T (Ueno and Harada, 1982; Hirota et al., 2000; Kishioka et al., 2000). The changes in oxygen concentration are accelerated or regulated by gas transport of oxygen and water molecules inside and outside the water surface.

The magnetic field effects for large fields can be explained by these well-known mechanisms. With ever-increasing evidence indicating that weak static magnetic fields have profound effects on biological systems, a number of mechanisms for the action of these static magnetic fields, including geomagnetic fields, on biological systems have been proposed.

Possible biomagnetic and chemical effects can be expected when biological systems are exposed to both static magnetic fields and other forms of energy such as light and radiation (Ueno and Harada, 1986). Photochemical reactions produced by a radical pair intermediate can be expected to show magnetic field effects that arise from an electron Zeeman interaction, electron-nuclear hyperfine interaction, or a hyperfine interaction mechanism including an electron-exchange interaction in a radical pair intermediate (Hata, 1976; Schulten et al., 1976; Tanimoto et al., 1976; Nagakura and Molin, 1992; Natarajan and Grissom, 1996; Hayashi, 2004).

The magnetic field effect observed with radical pair recombination is one of the well-known mechanisms by which magnetic fields interact with biological systems. Throughout the past decades there have been several experimental results describing the effects of magnetic fields on radical pair recombination. Mohtat et al. (1998) examined the behavior of radical pairs derived by hydrogen abstraction of triplet benzophenone and some of its derivatives from bovine serum albumin, human serum albumin, and calf thymus DNA. The magnetic field strength was as high as 150 mT with durations as long as 10 μs. This result indicated that radical pair behavior is sensitive to magnetic fields, and this effect can be interpreted by using the theory of free radical recombination. Using the triplet state of benzophenone as a convenient source of pairs, Eveson et al. (2000) examined the effects of weak (>1 mT) magnetic fields on radical recombination reactions in micells. They found that the concentration of free radicals escaping from the micelle was both affected and depended on the conditions surrounding the radical pair.

Timmel and Till discussed the weak magnetic field effects on free radical recombination reactions (Till et al., 1998; Timmel et al., 1998). Vink and Woodward (2004) described the effects of a weak magnetic field, 21 mT, on the recombination reaction of neutral free radicals in isotropic solution.

Ritz et al. (2002) reviewed the physiological basis of animal magnetoreception. They suggested that there was a link between photoreception and magnetoreception, from their findings in behavioral and theoretical studies. Migratory birds have the ability to sense the geomagnetic field and use it as a source of compass information. The candidates for a biophysical mechanism of this magnetoreception are magnetite and magnetically sensitive chemical reactions in animals. Ritz et al. (2000) postulated the possibility that magnetoreception involves radical pair processes as a biophysical mechanism.

8.3 Experimental Studies on Static Magnetic Field Effects

This section focuses on *in vivo* and *in vitro* studies of the effects of static magnetic fields. It covers the field effects observed on behavior, the cardiovascular system, reproductive system, cellular and tissue development, the neuroendocrine system, and the magneto-mechanical system, utilizing molecular, cellular, tissue, and cell-free systems.

8.3.1 *In Vivo* Studies

8.3.1.1 Animal Behavior: Recognition and Analgesia

Scientific interest in behavioral changes has led to the development of a psychology of learning that studies the effects of various external stimuli, and this research has further led to the development of behavioral pharmacology to observe the effects of drugs on the central nervous system. There have been several investigations for studying the effects of magnetic fields on behavior and the central nervous system using techniques developed specifically in these research fields. Behavioral research directed toward the effects of magnetic exposure of living organisms mainly addresses two questions: whether magnetic fields are sensed and avoided and whether magnetic fields have any influence on the functions of leaning and memory. Magnetic field experiments designed to address both questions have been conducted using several indicators, such as open-field behavior, operant behavior, and spontaneous motor activity, as tools of observation and measurement. (See also Chapter 4 on behavioral effects by Johnston and D'Andrea.)

8.3.1.1.1 Animal Behavior and Recognition

Nikolskaya et al. (1996) investigated the influence of inhomogeneity of natural magnetic fields on rat cognition with regard to whether or not magnetic fields could serve as an informational factor for cognition. Under three natural magnetic field conditions, $37 \pm 2\,\mu T$ (condition 1; horizontal component (N–S) is 14 μT, and vertical 34 μT), 16–118 μT (condition 2), and 55–240 μT (condition 3), rats were subjected to a food-operant behavior study. All rats in conditions 2 and 3 were unable to form operant behavior, while rats in condition 1 demonstrated the behavior. Using the combination of an original behavioral model and a multiple alternative maze, an impact of Opilong, which is an analog of dermorphine and μ-receptor agonist, on rat sensitivity to $38 \pm 2\,\mu T$, in the static magnetic field has been investigated by the same authors (Nikolskaya et al., 1999). They concluded that chemical modulation of the opioid system in rats induced both an increased magnetic field sensitivity and an allowed perception of magnetic field parameters. In the following study (Nikolskaya and Echenko, 2002), it was reported that cognitive activity in the natural magnetic field of 38 μT caused an increase of ethanol intake in 34.8% of rats.

During the 1980s, Liboff (1985) argued favorably for the combined effects of static (DC) with extremely low-frequency (ELF) magnetic fields (see also Chapter 9 resonance phenomena by Liboff on). A surprising effect was observed at the ELF magnetic field frequency close to the cyclotron frequency of a calcium ion. Thomas et al. (1986) reported the disruption of operant behavior in rats after exposure to low-intensity magnetic fields. The protocol developed by them in the report has been reexamined by Stern et al. (1996) in a two-part experiment. In the first part, the vertical component of the static field was reduced to 0.0261 μT. In the second part, both the horizontal and the vertical components were matched to those used by Thomas et al. The results obtained in these experiments were found to be inconsistent with the results reported by Thomas et al. Effects of the combination of static (DC) and AC magnetic fields at the cyclotron frequency on rat open-field behavior have been investigated (Zhadin et al., 1999). Levels of locomotor and exploratory activities were decreased after exposure to DC and AC magnetic fields at the calcium cyclotron frequency, whereas field exposure at the magnesium cyclotron frequency increased levels of these activities.

Studies have been conducted to determine whether rats could acquire a two-choice discrimination based on a specified discrimination stimulus (Creim et al., 2002). The specified discriminative stimulus used in this study was tested both in ambient illumination as well as in a combination of an oscillatory field of 50 μT at 60 Hz and a static field of 26 μT. The results demonstrated that rats were able to discriminate between two-choice tasks easily during the period of changing illumination and that the presence or absence of the static and oscillatory fields had no observed effect on these findings.

Tsuji et al. (1996) evaluated physiological consummatory behavior by observing intakes of food and water and changes in the body weight using BALB/c mice exposed to magnetic field levels of 5 T for a period of 24 and 48 h. Exposure to a 5-T magnetic field for 48 h suppressed eating and drinking behavior. The decreased body weight, the increased blood urea nitrogen (BUN) level, and the slightly increased BUN–Cr ratio observed in this experiment might be attributed to the loss of body fluid secondary to decreased food and water intake.

For the purpose of observing changes caused by magnetic field exposure, MRI systems and application of their high-strength static magnetic fields have been useful and widely used to obtain intact images. There have been three reports describing the behavioral effects observed as a result of high-strength static magnetic field exposure of rats and mice. Using a conditioned taste aversion technique, it was shown that rats developed a conditioned taste aversion after exposure to a high magnetic field of 9.4 T for 30 min (Nolte et al., 1998). Following this report, similar experiments were carried out. Restrained rats and both unrestrained and restrained mice were exposed to magnetic fields of 7 and 14 T generated by superconducting magnets (Houpt et al., 2003). In the report, it was found that exposure of rats to high magnetic fields suppressed rearing and locomotor circling and induced conditioned taste aversion and expression of c-Fos in vestibular neclei. The rat's orientation in magnetic fields is a key factor for the direction of circling. Similar results were obtained with mice (Lockwood et al., 2003). All tested mice showed development of conditioned taste aversion, and a significant number showed tight circling and rearing suppression. Effects were observed more significantly in unrestrained mice than in restrained mice. Snyder et al. (2000) have identified brain stem regions that were activated by exposure to static magnetic field levels of 9.4 T for 30 min, of restrained rats, by using a c-Fos immunohistochemisty detection assay. Increased expression of c-Fos and neural activation in visceral and vestibular nuclei by magnetic field exposure have been reported. It has been suggested that the neural activation response might be a factor in promoting conditioned taste aversion learning.

Superconducting high magnetic field exposures of 7 T have been reported to have reduced the trehalase enzyme activity in honey bees (Kefuss et al., 1999). There were no changes found in the level of fatty acids, triacyglycerols, and steroids in this study.

8.3.1.1.2 Analgesia

Effects of a hypogeomagnetic environment with a flux density of 4 μT inside a Mu-metal box on stress-induced analgesia in C57 male mice have been investigated (Del Seppia et al., 2000). This study consisted of three consecutive parts: (1) maintaining the mice under various magnetic exposure conditions: hypogeomagnetic, altered magnetic field, and Earth's geomagnetic field of 46 μT for 90 min; (2) immobilizing the animals in a tube for 30 min under each exposure condition; and (3) recording nociceptive responses of the restraint-stressed mice as the latency of front-paw lifting to hot-plate stimulus. Stress-induced analgesia was significantly reduced in the animal group exposed to the hypogeomagnetic field, and this result was comparable with that in the mice exposed to altered magnetic fields or treated with prototypic opiate antagonist naloxone. It has been suggested that the exposure period in a hypogeomagnetic environment might be responsible for the inhibition of stress-induced analgesia. They also demonstrated that exposure to altered magnetic fields induce more rapid habituation to a novel environment (open field) (Del Seppia et al., 2003). The experiment was carried out to investigate effects of irregularly varying (<1 Hz), 20–70-μT or regular 37-Hz, 80-$\mu T_{p\text{-}p}$ magnetic fields with a 2-h exposure of mice. The nociceptive response was measured by a hot-plate test and showed that the nociceptive sensitivity was significantly greater in magnetically treated mice than in controls.

Various sensing-transduction mechanisms have been proposed to explain the biological effects of magnetic fields. Prato et al. (1996a). showed inhibitory effects of 60-Hz magnetic fields at levels of $299 \pm 1\ \mu T_{p\text{-}p}$ and static magnetic fields at levels of $78 \pm 1\ \mu T$ on opioid-mediated analgesia in the land snail, *Cepaea nemoralis*. It has been reported that the effects were dependent on the relative direction of both the weak static and the 60-Hz magnetic field, as well as on the presence of light. In a following study, indirect and direct mechanisms of ELF magnetic fields on an endogenous opioid peptide-mediated analgesic response have been proposed. It was shown that the energy transduction mechanism did not involve induced electric currents or magnetite (Prato et al., 1996b), and the results indicated that a direct magnetic field detection mechanism was consistent with the parametric resonance model. In a later study to reexamine the results obtained and further clarify the role of light, it was found that reduction of the opioid-induced analgesia by magnetic field exposure was enhanced by the presence of light. It was reported that the reduction rate of analgesia was not dependent on ELF frequency; however, the effect of the ELF magnetic field was in fact mediated by direct magnetic field detection.

In order to detect static magnetic field-induced functional changes in brain tissue, Veliks et al. (2004) carried out an investigation to identify the effects of static magnetic fields on rat brain structures using heart rate and heart rhythm as physiologic indicators. Rats put under ketamine–xylazine anesthesia were exposed to magnetic field levels of 100 mT for 15 min. Before and after the exposure, an electrocardiogram was recorded for analysis of heart rate and heart rhythm. Static magnetic fields were found to evoke changes in both heart rate and heart rhythm in 80% of the subject animals.

McLean et al. (2003b) examined the effects of inhomogenous static magnetic fields, alone or in combination with the chemical agent phenytoin (PHT), on audiogenic seizures (AGS) in DBA/2 mice. In experimental studies where the static magnetic fields ranged from 0.26 to 10.5 mT, with a field gradient ranging between 0.012 and 0.48 T/m, seizure severity decreased as the magnetic flux density and exposure duration period increased. It was found that the magnetic field pretreatment enhanced the effect of PHT, and it was also found that a static magnetic field alone had some anticonvulsant effects as well. Further investigation is required to clarify the anticonvulsant effects of magnetic fields in AGS.

8.3.1.2 Reproduction and Development

The effects of static magnetic fields, including MRI fields (static, gradient, and RF), on fertility, the developing embryo, and the fetus have been investigated. There have been a series of reports describing the effects of 10- and 35-day exposures to a magnetic field level of 0.7 T on mice, in terms of sperm motility, maturation, and production and morphological and developmental changes (Tablado et al., 1996, 1998, 2000). In the first report, mice were exposed to a magnetic field level of 0.7 T for time periods between 1 and 24 h/d, over 35 d. It was found that sperm motility, maturation, and production were not affected. Two years later, it was reported that the size of sperm heads was still intact; however, the animals that had undergone continuous exposure demonstrated increased sperm head abnormality. In a study by Tablado et al. of magnetic effects during developmental changes, experiments were carried out to investigate the *in utero* exposure effects from magnetic field levels of 0.5 to 0.7 T on testis and epididymis development in mice. After mating, female mice were exposed from day 7 of gestation until the day of birth. Results showed that there were no significant differences between exposed and sham-exposed animals in terms of body weight gain of dam, litter size, body weight of male pups, and testis–epididymus weight gain of pups up to 35 d of age. In addition, there were no detectable changes found during a histopathological evaluation of the testis and epididymis of pups.

Narra et al. (1996) reported the biological effects of a static magnetic field on spermatogenesis and embryogenesis in Swiss Webster mice. Male and pregnant female mice were exposed to a magnetic field level of 1.5 T for 30 min. There was no increase in sperm head shape abnormality and no reduction in testicular sperm numbers; however, a decreased survival rate of preimplantation embryos was indicated.

It has been reported that there were no harmful effects detected in a static magnetic field of 4.7 T on ICR mice (Okazaki et al., 2001). Pregnant ICR mice were exposed to the field from days 7.5–9.5 of gestation and sacrificed on day 18.5 of gestation. There were no significant differences in the incidence of prenatal death or malformations between the exposed and control groups. The investigation has been conducted for potential adverse effects of a 10-week exposure to a 9.4-T magnetic field on the development of male and female rats as well as their offspring (High et al., 2000). All rats were exposed to 9.4 T for 3 h twice a week with an intermittent duration of 5 week on, 2 week off, and 5 week on. The results demonstrate hematological, biochemical, pathological, and behavioral changes in both adults and their offspring; however, there were no adverse effects observed in either male or female adult rats, as well as in their offspring.

Effects of long duration and high magnetic field exposure on fetal growth and postnatal development in mice have been investigated (Magin et al., 2000). One group of mice was exposed for 9 h on day 9 or on day 12 postcoitus (or both) to MRI conditions (static magnetic field of 4 T with 5-T/s gradient and whole-body SAR of 0.2 W/kg at 170 MHz). A second group was exposed to a combination of ultrasound and MRI fields. There were no significant changes in fetal growth in animals exposed to MRI or ultrasound fields individually. The average fetal weight of animals exposed to combined fields was lower than that of the other group. Carnes et al. (1996) investigated the effects of static magnetic field levels of 4.7 T, which is equivalent to the level used in MRI, on fetus, adult growth, and testicular development in mice.The 8-h exposure was carried out on day 9 or day 12 of gestation (or both). Effects of the combined application of 1 MHz ultrasound on day 9 and MRI on day 12 was also investigated. The average fetal weight of the exposed groups was found to be less than that of the control groups. It was found that the postpartum death rate was higher after the MRI exposure, and sperm production was reduced; however, no changes were detected in embryonic death rate, sex ratio, body weight at day 50, spleen weight, and seminal vesicle weight.

Jove et al. (1999) studied the effects of static magnetic field levels of 18 and 36 mT on the development of chick embryo, including the pineal gland. They found that static magnetic fields affected the development and growth of embryos, and this effect was dependent not only on the intensity but also on the length of exposure time. Ruggiero et al. (2004) reported the effects of a 3-h exposure to a magnetic field level of 0.2 T generated by a clinically used MRI system on the angiogenesis of chick embryo. Angiogenesis was evaluated using a chick embryo chorioallantoic membrane assay. Results indicated that static magnetic field exposure inhibited angiogenesis in chick embryo. Effects of a uniform static magnetic field of 29 mT on cell migration and differentiation in the cerebellum of chick embryo have been studied (Espinar et al., 1997). The cerebella of chick embryos was exposed to the magnetic field on 6 d of incubation and sacrificed at day 13 of incubation. Results showed that static magnetic field exposure could induce irreversible developmental effects on cell migration and differentiation. Clear signs of cell degenerations and delay in the process of neuronal differentiation were indicated.

There have been a series of reports describing the effects of static magnetic field on fish embryos (Formicki and Winnicki, 1996, 1998; Winnicki et al., 1996; Formicki et al. 1997; Formicki and Perkowski, 1998). First, it was reported that during a prolonged transportation of fertilized salmonid eggs and trout (*Salmo trutta* L. and *Oncorhynchus mykiss*), a magnetic field of 4 mT lowered the mortality rate and enhanced the conservation condition (Winnicki et al., 1996). Effects of static magnetic fields ranging from 50 to 70 mT on the cardiac muscle activity of carp (*Cyprinus carpio* L.) embryos and larvae have been investigated (Formicki and Winnicki, 1996). An increase in heart rate was observed after exposure to these fields. Effects from both the strength and the direction of the geomagnetic field and static magnetic fields of 0.5 and 1 mT on the orientation of trout embryo

(*S. trutta*) and rainbow trout (*O. mykiss*) have also been investigated. Results have shown that the embryos were sensitive to the static magnetic fields, with a preference for a certain field direction. Further investigation has been conducted to clarify the response of fish embryos and larvae from the same type of trout to a static magnetic field (Formicki and Winnicki, 1998). Eggs were placed in magnetic fields during the period from fertilization to hatching. The exposure of eggs resulted in slower embryonic development; thus, eggs were incubated for a prolonged period of time. Embryos of rainbow trout (*O. mykiss*) and trout (*S. trutta*) exhibited orientation both in the natural magnetic field and in artificial magnetic fields of 0.5 and 1 mT. In a supplementary paper, Formicki and Perkowski (1998) showed the effects of static magnetic fields of 5 and 10 mT on the gas exchange in rainbow trout (*O. mykiss*) embryos. An increase of oxygen uptake was observed. It was determined that the impact of magnetic field exposure on the respiratory system was significant in the periods of advanced morphogenesis.

It has been reported that the exposure to static magnetic fields of 10–100 mT could alter the early embryonic development in two species of sea urchin embryo, *Lytechinus pictus* and *Strongylocebtrotus purpuratus* (Levin and Ernst, 1997). Results suggested that static magnetic fields delayed the onset of mitosis in both species. A static magnetic field of 30 mT caused an eightfold increase in the incidence of exogastrulation in one species. Static magnetic field effects of 4 and 8 mT and effects of exposure to 60-Hz AC magnetic fields of 6 and 8 mT on the first division of two sea urchin embryos, *Sphaerechinus granularis* and *Paracentrotus lividus*, have been studied (Pagnac et al., 1998). No differences were found in the time of the first cleavage in both exposures.

In a static magnetic field of 35 mT, the fruit fly (*Drosophila melanogaster*) was reared through several generations to investigate the width and length variability of both wings (Stamenkovic-Radak et al., 2001). In the second-generation exposure, there was a significant difference between sexes in the context of directional change of wing size variability. In the sixth generation, the differences in wing size between the magnetic field exposure group and control groups showed the same directional pattern in both sexes. Pan and Liu (2004) reported the effects of high static magnetic fields of 9.4 and 14.1 T on the hatching behavior of fresh mosquito eggs. Hatching was delayed, and the delay time depended on the intensity of the magnetic field.

Effects of exposure to high static magnetic field with levels up to 16.7 T on the first three cleavages of *Xenopus laevis* embryos have been investigated (Denegre et al., 1998). Results have indicated that cleavage furrows aligned parallel to the magnetic field. The most significant effects on the second and third cleavages have been observed in homogenous fields, not in the gradient fields. It has been suggested that this phenomenon was caused by interactions between the magnetic fields and the diamagnetic materials in the *Xenopus* embryo. Further studies have shown that after exposure to static magnetic fields of 17–22 T during either or both of the first two cell cycles, the third cell cycle mitotic apparatus at metaphase could be induced, and the third cleavage furrows aligned perpendicular to their nominal orientations (Valles, 2002; Valles et al., 2002). Valles et al. (1997) investigated the magnetic field gradient levitation (MFGL) as a technique for simulating low gravity for biological systems.They levitated living biological specimens, embryos of the frog *X. laevis*, using a large inhomogenous magnetic field. MFGL of embryos reduced the body forces and gravity-induced stress on them.

8.3.1.3 Circulatory System Effects

Recently, effects of static magnetic field applications on the circulatory system have been reviewed in experimental animals (Tenforde, 2005) and in humans (Chakeres and de Vocht, 2005; Crozier and Liu, 2005; van Rongen, 2005). Most of the studies have been related to MRI systems, and it has been reported that in MRI-related studies using strengths of up to 8 T, there were little or no significant changes in cardiovascular and circulatory parameters. In contrast, it was found that moderate-intensity static magnetic fields ranging from 1 to 350 mT can have significant circulatory system effects, most notably on cutaneous microcirculation and arterial blood pressure.

8.3.1.3.1 Microcirculation

Ohkubo and Xu (1997) studied the acute effects of 1–10-mT static magnetic fields applied for 10 min on the microcirculatory changes in conscious rabbits using a rabbit ear chamber (REC) and microphotoelectric plethysmography (MPPG). The static magnetic fields induced biphasic changes in vasomotion in a non-dose-dependent manner. Static magnetic fields suppressed vasomotion when the vascular tone was high, while the same level fields enhanced vasomotion when the tone was low. To clarify these effects, the same group, Okano et al. (1999), investigated the effect of a 1-mT static magnetic field for 10 min on microcirculation in conscious rabbits using pharmacological manipulation together with REC and MPPG. The results showed that the static magnetic field enhanced vasodilatation and increased vasomotion under norepinephrine, which induced high vascular tone and, in contrast, induced vasoconstriction and decreased vasomotion under acetylcholine, which induced low vascular tone.

These studies are strongly supported by another independent study that examined the effect of a moderate-intensity static magnetic field on microcirculation, suggesting that the static magnetic field could modulate the biphasic responses of relatively smaller-diameter blood vessels (arterioles). Morris and Skalak (2005) studied the acute effect of a 70-mT static magnetic field for 15 min on the diameter of microvessels in rat skeletal muscle placed under pentobarbital anesthesia, using intravital microscopy. This study suggested that the static magnetic field could modulate microvascular tone in a restorative fashion, thereby acting to normalize the tone.

Gmitrov and colleagues reported the acute effects of 250–350 mT static magnetic field exposures for a time period of up to 80 min on microcirculation within cutaneous tissue of the rabbit ear lobe placed under pentobarbital anesthesia using REC and MPPG (Gmitrov and Ohkubo, 1999a,b; Gmitrov et al., 2002). They demonstrated that application of static magnetic fields to the carotid sinus baroreceptor region for a period of 65–80 min increased microcirculation. The static magnetic fields suppressed or recovered a Ca^{2+} channel blocker, verapamil, which reduced microcirculation.

Xu et al. (1998) observed the subchronic effects of a 180-mT static magnetic field, with length of exposure times for up to 4 weeks, on cutaneous microcirculation in conscious rabbits, using REC and MPPG. The static magnetic field significantly increased the long-lasting vasodilatation and enhanced the vasomotion. The same authors found that exposure to static magnetic fields at 1 mT or higher for 10 min enhanced microcirculation and increased peak blood velocity (Xu et al., 2000).

Mayrovitz et al. (2001, 2005) investigated the effects of static magnetic fields with strengths as high as 100 mT and exposure periods of up to 36 min on skin blood flow in humans, using a laser-Doppler flowmeter. They showed that the static magnetic fields have no significant effect on the normal, unstressed circulation or vasoconstrictive response in conjunction with skin temperature. Steyn et al. (2000) also indicated that there are no effects observed on blood flow in horses after exposure to static magnetic fields.

In a series of studies that observed the effects of ultrastrong (>5 T) static magnetic field exposure, Ichioka et al. (1998, 2000) investigated the acute effect of an 8-T static magnetic field exposure for a period of 5 min on blood flow in rat by using a laser-Doppler flowmeter and thermistor-derived measurements.They demonstrated that blood flow and skin temperature decreased during the field exposure, through the movement of water vapor over the animal and the decreased humidity in the air.

8.3.1.3.2 Blood Pressure

Gmitrov and colleagues studied the influence of both a 350-mT static magnetic field and geomagnetic field activity on mean arterial blood pressure (MAP) in pentobarbital-anesthetized rabbits (Gmitrov and Ohkubo, 2002; Gmitrov et al., 2002). Application of the static magnetic field to the baroreceptor region for 65–80 min decreased MAP. In testing geomagnetic field applications, they found that there was a positive correlation of this field's activity with MAP, and this result implied that magnetic storms could increase the incidence of severe cardiovascular events.

Okano and Ohkubo (2001) examined the acute effect of a 1-mT static magnetic field applied for 30 min on pharmacologically altered blood pressure in conscious rabbits. It was found that (1) the static magnetic field reduced the vasodilatation effect from enhanced vasomotion and antagonized the reduction of blood pressure under a Ca^{2+} channel blocker, nicardipine, which induced low vascular tone, and (2) the static magnetic field attenuated vasoconstriction and suppressed the elevation of blood pressure while under the influence of a nitric oxide (NO) synthase inhibitor, L-NAME, which induced high vascular tone. However, two of their experiments, which were carried out under normal conditions without pharmacological drugs, showed that static magnetic fields did not induce any significant effects on hemodynamics and blood pressure (Okano and Ohkubo, 2001, 2003a). With regard to these undetectable effects, Muehsam and Pilla (1996) speculated that physiologically significant bioresponses to therapeutic signals appear to occur only when the physiologic state of the target system is far from homeostasis.

In contrast to the experiments done without pharmacological manipulation, Okano and Ohkubo (2003a) found that exposure to a 5.5-mT static magnetic field for 30 min caused the suppression of norepinephrine- or L-NAME-induced vasoconstriction and hypertension in rabbits. Furthermore, they tested exposures of 5–10-mT static magnetic fields for a period of several weeks on the development of hypertension in spontaneously hypertensive rats (Okano and Ohkubo, 2003b; Okano et al., 2005a). Experimental results indicated that the static magnetic fields suppressed and retarded the development of hypertension because of the reduction in plasma levels of both angiotesin II and aldosterone together with lower levels of NO metabolites (NO_x).

In addition, the antihypotensive effects of static magnetic fields on reserpine-induced hypotensive rats were investigated (Okano et al., 2005b). The result suggested that exposure to a 25-mT static magnetic field for several weeks suppressed the reserpine-induced hypotension and bradykinesia through the inhibition of norepinephrine depletion.

Saunders (2005) commented that most of these studies were undertaken in the context of the potential therapeutic effects of static magnetic field on various disorders. Further studies with some independent replications are required even if the effects of static magnetic field on both blood flow and blood pressure indicate possible medical applications.

8.3.1.4 Neuroendocrine, Visual, and Neurophysiological Systems

Effects of static magnetic field exposures of 0.05 μT to 80 mT, and 7 T on the level of melatonin in rat have been examined (Kroeker et al., 1996). The first experimental exposure using field strengths of up to 80 mT for 12 h/d and 8 d showed no significant changes in night-time pineal and serum melatonin levels, as did the second experimental exposure using 7 T for 45 min. The visual system of the fruit fly (*D. melanogaster*) was investigated after an exposure to a zero magnetic field (Creanga et al., 2002). Adults from pupae maintained in a zero magnetic field for 20 h were used for the electroretinogram. A significant increase in sensitivity of neural cells from the first optic ganglion was indicated.

Osuga and Tatsuoka (1999) tested the effects of a 1.5-T static magnetic field by using an MRI system application on neuroconduction in a partially active nerve in the bullfrog (*Rana catesbeiana*). The action potential and nerve impedance measurements indicated that a field strength of 1.5 T had no effect on neuroconduction; therefore, it was determined that neuroconduction in damaged nerves was not affected by the exposure.

8.3.1.5 Magneto-Mechanical Systems

Testorf et al. (2002) studied the influence of homogenous static magnetic fields of 8 and 14 T on melanophore aggregation in black tetra (*Gymnocorymbus ternetzi*). The result showed no significant field effects on the aggregation after exposure to magnetic fields.

Effects of a 0.2-T static magnetic field on a normal human neuronal cell culture, FNC-B4, has been investigated with MCF-7 and WEHI-3 cells as controls (Pacini et al., 1999b). FNC-B4 cells changed their morphology after the exposure. Cells became elongated and formed vortexes, while controls did not show any alteration. The morphological changes in MRC-5 fibroblasts were evaluated as well (Pate et al., 2003). The cells were screened for cell mobility, cell distribution, and cellular morphology (size, shape, lysis, and background). These cells were exposed to both a static magnetic field and a pulsating magnetic field for a period of 0, 24, 48, and 72 h. Although the static magnetic field-exposed cells showed cell membrane damage and morphological change, as well as other interesting findings that were included in the report, this report may not be useful because it did not provide essential dosimetric data, such as strength of the field.

Danielyan et al. (1999) examined the effects of a 0.2-T static magnetic field on binding of ouabain-H^3, which is a specific inhibitor of Na^+-K^+-ATP-ase, in normal glandular breast tissue and in cancerous breast tissue. The static magnetic field-induced decrease of binding was considered as evidence for the dehydration effect of the field. This study has indicated that the static magnetic field tested could influence the cancer cell's metabolism through cell hydration changes. They investigated the effects of a 0.2-T static magnetic field on the hydration of rat tissues (Danielyan and Ayrapetyan, 1999). They assumed that the target for magnetic field action was the structured water of the cell. Decreases in hydration and adaptation of brain, liver, and spleen and an increase in the case of kidney were observed.

8.3.1.6 Musculoskeletal System

Yan et al. (1998) investigated the effects of static magnetic fields on bone formation of rat femurs. They implanted magnetized samarium cobalt rods with a field strength of 180 mT into rat femurs. The bone mineral density (BMD) and bone calcium content were measured 12 weeks after implantation. Results indicated that the femurs adjacent to the magnetized specimens had significantly higher BMD and calcium content. However, BMD and calcium content levels were found to be normal in both magnetized and unmagnetized specimen groups. The same research group further studied the effects of a 180-mT static magnetic field on bone formation, using an ischemic rat femur model (Xu et al., 2001). It was reported that the enhancement of the femoral bone formation was due to the improved blood circulation in the femur.

Satow et al. (2001) observed the effect of a 0.65-T static magnetic field on muscle tension in the neuromuscular preparation of the sartorius muscle of bullfrog (*R. catesbeiana*). The tension development was obtained by stimulation of the sciatic nerve or of the sartorius muscle itself for a duration of 30 min. A decrease in muscle tension was observed. The results indicated that application of the static magnetic field was responsible for tension development.

8.3.2 Tissue, Molecular, and Cellular Studies

8.3.2.1 DNA and Chromatin

The exposure of isolated rat lymphocytes to a static magnetic field of 7 mT for 3 h did not increase the number of damaged cells (Zmyslony et al, 2000). Although incubation with 10 μg/ml $FeCl_2$ did not cause DNA damage, the number of damaged cells increased when the $FeCl_2$-incubated lymphocytes were simultaneously exposed to the field. A hypothesis for these observations was that the number of reactive oxygen species generated by iron ions in cells might increase after the exposure to the magnetic field (Jajte et al., 2002).

Binhi et al. (2001) have reported the effect of a weak static magnetic field on *Escherichia coli* K12 AB1157 cells, by using anomalous viscosity time dependence (AVTD) assay methods. The AVTD changes were found when the cells were exposed to static magnetic field levels up to 110 μT. These results were consistent with the calculations of individual rotations of the ion–protein complexes Ca^{2+}, Mg^{2+}, and Zn^{2+}, provided that all complexes rotated at the same speed. They suggested that the rotation for all ion–protein complexes is on the same carrier, such as DNA.

The effect of a zero magnetic field on the conformation of chromatin in human VH-10 fibroblasts and lymphocytes was investigated by the AVTD method (Belyaev et al., 1997). A decrease in the AVTD peaks was observed within 40–80 min of exposure to fibroblasts, and this decrease was transient, disappearing 120 min after the beginning of exposure. A similar effect of zero field was observed when cells were exposed for 20 min and kept at an ambient field. They concluded that both zero field and γ-rays caused hypercondensation and decondensation of chromatin. Zero field effects were more significant in the beginning of the G_1-phase than in the G_0-phase in human lymphocytes.

Okuda et al. (1998) evaluated the effects of a 6.34-T static magnetic field on the instability of microsatellite repetitive sequences in DNA mismatch repair (MMR)-proficient and MMR-deficient cell lines, HeLa S3, and HCT116, respectively. After exposure to the field, both cell lines exhibited no significant microsatellite sequence changes. This result suggested that the static magnetic field might not induce the genetic changes in microsatellite sequences.

8.3.2.2 Cell Growth, Cell Proliferation, and Cell Cycle

Potenza et al. (2004b) showed that *E. coli* cell growth and gene expression were affected by a static magnetic field exposure level of 300 mT. Cell proliferation at the stationary phase was increased by exposure to those cells growing in a modified medium culture containing glutamic acid; however, cell proliferation was not affected in those cells growing in traditional Luria–Bertani (LB) medium. Gene expression differences were estimated by differential display assays using arbitrary primers, and four genes were found to be responsive to the static magnetic field. One clone, expressed only in the exposed cells, corresponded to a putative transposase. Potenza et al. suggested that the static magnetic field exposure might stimulate transposition activity.

Stansell et al. (2001) reported that antibiotic (piperacillin) resistance of the clinically isolated *E. coli* was increased by the heterogenous static magnetic field exposure level of

8–60 mT for 45 min. They suggested the observation may be unique to the particular strain of *E. coli* or the specific antibiotic used. They did not suggest any mechanistic implication for this observation.

Poiata et al. (2003) reported zero magnetic field effect on the antibiotic resistance of *E. coli* strains isolated from human subjects. They used 26 *E. coli* strains and 5 different antibiotics, ampicillin, ceftazidine, tetracycline, ofloxacin, and kanamycin. Approximately one third of the tested strains was sensitive to the zero field treatment. Minimum inhibitory concentrations (MICs) of each antibiotic for some strains were decreased by zero magnetic field exposure, while the MICs for other strains were increased by the exposure. Their mechanistic hypothesis, based on magnetic particles, did not support these observations.

Ruiz-Gomez et al. (2004) showed that growth of the haploid yeast strain *Saccharomyces cerevisiae*, a eukaryotic cell, was not affected by exposure to static magnetic field levels of 0.35 and 2.45 mT.

Effects of a 0.2-T static magnetic field, generated by MRI alone or in combination with vitamin D treatment, on cell damage and proliferation in the human breast cancer cell MCF-7, human neuronal cell FNC-B4, and murine leukemia cell WEHI-3 have been investigated (Pacini et al., 1999a). Three-hour exposures to the 0.2 T field had no effect on the cell colony formation number in all three cell lines. Results demonstrated that [^{3}H] thymidine incorporation level decreased in MCF-7 and FNC-B4 cells, while no changes were observed in WEHI-3. It was also demonstrated that the treatment of cells using vitamin D had a permanent antiproliferative effect.

Long-term effects on proliferation of human fetal lung fibroblast (HFLF) cells of repetitive exposure to a 1.5-T static magnetic field with exposure periods of 1 h/d for 3 weeks have been investigated (Wiskirchen et al., 1999). Results showed no changes in clonogenic activity, DNA synthesis, cell cycle, and proliferation kinetics. In a following paper, effects of static magnetic field levels of 0.2, 1.0, or 1.5 T on the cell cycle in both synchronized and nonsynchronized HFLF cells were evaluated (Wiskirchen et al., 2000). The exposure condition was 1 h/d for 5 consecutive days. Results showed no significant differences in cell cycle events between synchronized and nonsynchronized cells.

A series of research studies on growth enhancement by strong inhomogenous static magnetic fields have been reported. Tsuchiya et al. (1996) showed that the growth of *E. coli* was affected by both a strong homogenous static magnetic field strength of 7 T and inhomogenous field strengths of 5.2–6.1 or 3.2–6.7 T. In the stationary phase, the cell number under a high magnetic field was about two to three times higher than that of a control. The effect of the inhomogenous field was much stronger than that of the homogenous field. They also showed that the transcription activity of *E. coli* was enhanced by the strong inhomogenous static magnetic field levels of 5.2–6.1 T (Tsuchiya et al., 1999). The transcription levels of the *rpoS*, gene which encodes sigma factor of RNA polymerase, was increased in the stationary phase by the static magnetic field exposure. This transcription factor is specifically activated during the stationary phase and plays an important role in the transcription control of other genes in the stationary phase.

Horiuchi et al. (2001) showed that *E. coli* cell death in the stationary phase was suppressed by strong inhomogenous static magnetic field levels of 5.2–6.1 T with a gradient of 24 T/m and that the suppression was dependent on the addition of amino acids to the LB medium. The addition of glutamic acid enhanced cell death as pH increased in the stationary phase, and cell death was dramatically suppressed by the field exposure. At the same time, *rpoS* gene expression was increased 20% by the field exposure. They suggested that the increase of *rpoS* gene expression in the stationary phase by the field exposure might be related to the base resistance because the *rpoS*-disrupted strain showed a lower base resistance than the wild-type strain (Ishizaki et al., 2001). It has been shown that the medium supernatant, when used after the static magnetic field exposure, could enhance the suppression of cell death (Horiuchi et al., 2002). The pH of the medium after the static magnetic field exposure was only slightly different from that of the control (by a factor of 0.07 pH), and pH-adjusted medium from both exposed and control supernatants still had the suppression effect characteristics from the exposure to static magnetic fields. It has been suggested that other factors also were involved in the full suppression effect of cell death by the field exposure.

The growth advantage in stationary phase (GASP) phenomenon is described as follows. When *E. coli* cells grown for 10 d (aged culture) and *E. coli* cells grown only for 1 d (young culture) were mixed, the cell number of the young culture decreased, and the population of the young culture was taken over by the aged culture, and eventually only aged cells predominantly survived in the system. It has been found that the GASP phenomenon disappeared with the exposure to strong inhomogenous static magnetic field levels of 5.2–6.1 T (Okuno et al., 2001). They suggested that the disruption of the GASP phenomenon might be related to an effect on the *rpoS* gene by the static magnetic field exposure.

Gray et al. (2000) evaluated static electric and magnetic field effects on the action enhancement of the chemotherapeutic agent adriamycin in transplanted mammary adenocarcinoma in female B6C3F1 mice. Treatment consisted of using 10 mg/kg of adriamycin in combination with a 4-h exposure to a 110-mT field. Tumor regression in the groups exposed to a static magnetic field was greater than in the group treated with adriamycin only.

Tanioka et al. (1996) evaluated the effects of a 6.34-T static magnetic field on proliferation and metastatic activity in the B16 melanoma and EL-4 T-lymphoma cell lines. Cell cultures were incubated in the presence of magnetic fields for 12, 24, 36, or 48 h at 37°C. It was found that the proliferative and the metastatic activities of both cell lines were promoted under certain conditions.

Tofani et al. (2003) exposed immunocompetent mice bearing either the murine Lewis lung carcinoma or the B16 melanotic melanomas to static field levels of 3 and 4 mT and treated them with two commonly used anticancer drugs, cisplatin and cyclophosphamide, respectively. The survival time of mice treated with cisplatin and exposure to the magnetic fields was significantly longer than that of mice treated only with cisplatin or only exposed to the magnetic fields, surpassing that of mice treated with 10 mg/kg i.p. of the drug and showing that the magnetic field acts synergically with the pharmacological treatment. When mice treated with cyclophosphamide were exposed to the magnetic field, no synergic effects were observed. No clinical signs or toxicity were seen in any of the mice exposed to the magnetic field alone or along with cisplatin or cyclophosphamide treatment.

Raylman et al. (1996) studied the effects of exposure to a static magnetic field of 7 T for 64 h on cell viability in three malignant human cell lines, melanoma (HTB 63), ovarian carcinoma (HTB 77IP3), and lymphoma (Raji; CCL86). It has been reported that the static field exposure reduced the viable cell count and appeared to inhibit cell growth.

Using two types of mammalian cells, mouse leukemia cells P388 and Chinese hamster fibroblast cells V79, Sakurai et al. (1999) tested the effects on cell growth patterns of exposure to a 7-T static magnetic field for up to 5 d. No significant magnetic field effects on cells were found.

In a series of papers (Kula, 1996; Kula and Drozbz, 1996a,b), the magnetic field effects on cultured fibroblasts isolated from the BALB/c mouse have been investigated. The fibroblast cultures were exposed to a static magnetic field of 0.49 T and a 50-Hz AC magnetic field of 0.02 T for a time period of 2–64 min/d over four consecutive days. The following parameters were studied: the dynamics of culture growth; protein content;

thymidine incorporation; Zn, Fe, and Cu ion content; the activity of superoxide dismutase (SOD) and catalase (CT); and glycosaminoglycan metabolism. The static magnetic field exposure had no effect on both the vital functions and glycosaminoglycan metabolism and did not show any changes in the free-radical process in fibroblasts. Kula et al. (2000) further evaluated the activities of SOD, CT, glutathione peroxidase, and malondialdehyde (MDA) in the livers and kidneys of rats exposed to a static magnetic field of 0.49 T and a 50-Hz AC field of 0.018 T. While the 50-Hz magnetic field was found to influence free-radical processes in both liver and kidney tissue, the static magnetic field showed no effects. Magnetic field effects on the lipid peroxidation product, MDA, in mouse subcellular fibroblast have been evaluated by the same protocol in a previous report (Kula et al., 2002). It was found that exposure to a static magnetic field caused no changes in peroxidation of membrane structures.

Effects with a combination of static and alternating magnetic fields on cell attachment and induction of apoptosis in rat tendon fibroblast and rat bone marrow (RBM) osteoprogenitor cells have been reported (Blumenthal et al., 1997). Experiments utilized 60-and 1000-Hz AC magnetic fields of up to 0.25 $mT_{p\text{-}p}$ and static magnetic fields of up to 0.25 mT. It was found that AC fields and static magnetic fields tested with various combinations of field strengths and frequencies resulted in extensive detachment of preattached cells and prevented the normal attachment of cells not previously attached to substrates. Results suggested significant alterations in cell metabolism and cytoskeleton structure after the exposure.

Blanchard and Blackman (1994) proposed the ion parametric resonance (IPR) model for the prediction of the interaction between magnetic fields and biological systems (see also Chapter 9 on resonance phenomena by Liboff). According to the IPR model, the relationships among the strength of a static magnetic field, the AC magnetic field frequency, and the charge-to-mass ratio of ions of biological relevance were important key factors. Blackman et al. (1996) tested the influence of magnetic fields on neurite outgrowth in PC-12 cells and showed that the PC-12 cell response to perpendicular AC and static magnetic fields was distinct and predictably different from that found for parallel AC and static magnetic fields. It has been reported that the response to perpendicular fields was dominant in an intensity-dependent nonlinear manner.

The effects of the combination of AC and static magnetic fields on the behavior of Friend erythroleukemia cells have been studied (Eremenko et al., 1996). The combined fields were a geomagnetic field of 45 μT, together with a 70-μT field at 50 Hz which was produced in a solenoid coil, and 20-nT DC and 2.5-pT AC fields in a magnetically shielded room. It was found that the culture growth cycle of cells was slightly accelerated inside the solenoid, and the degree of acceleration appeared to depend on sensitivity of the cell cycle to the magnetic field. On the other hand, it was found that the culture growth cycle of cells inside the magnetically shielded room was slightly decreased.

Effects of a static magnetic field exposure of 10 T for up to 4 d on the rate of cell growth or cell cycle distribution in Chinese hamster ovary (CHO-K1) cells have been investigated (Nakahara et al., 2002). The exposure to the static magnetic field alone did not affect micronucleus formation. In x-ray irradiated cells, exposure to the 10-T static magnetic field resulted in a significant increase in micronucleus formation. Buemi et al. (2001) examined the effects of a 0.5-mT static magnetic field on the cell proliferation and cell death balance in monkey renal cells (VERO) and in rat cortical astrocyte cells. After 6 d of exposure to the magnetic field, they observed the effects on cell proliferation and cell death balance and suggested that the effects might vary depending on the cell type. Magnetic fields may also have a nephropathogenic effect.

Tofani et al. (2001) investigated the role of magnetic field characteristics on the growth of WiDr human colon adenocarcinoma, MCF-7 human breast adenocarcinoma, and MRC-5 embryonal lung fibroblast. Cell death induction was observed with a magnetic field exposure of greater than 1 mT when the combined static magnetic field and 50-Hz magnetic field was applied. The report showed that significant tumor growth inhibition appeared when the total field strength was greater than 3.59 mT.

Schiffer et al. (2003) studied the effects of four different types of magnetic field applications on the cell cycle progression in two different tumor cell lines, the human acute myeloid leukemia cell HL-60 and the mouse lymphoma cell EA2. The four types of magnetic field applications used were (1) the static magnetic field of 1.5 and 7.05 T, (2) the magnetic gradient field with ± 10 and ± 100 mT/m at 100 Hz; (3) the pulsed high frequency magnetic field (5.8 μT at 63.6 MHz); and (4) the combination of (1)–(3). The exposure duration ranged from 1 to 24 h. Cell cycle fractions at G_0/G_1, S, and G_2/M phases were analyzed by flow cytometry. Cell cycle analysis did not show differences between the exposed and control cells. In conclusion, during MRI, no influence of magnetic field on cell cycle progression was observed in these cell lines.

8.3.2.3 Cell Membrane and Cell Metabolic Activity

Chignell et al. (1998) studied the effects of static magnetic fields of 25–150 mT on the photohemolysis of human erythrocytes by ketoprofen. An application of a static magnetic field during UV (>300 nm) irradiation of ketoprofen and erythrocytes significantly decreased the time required for photohemolysis. It has been suggested that the magnetic field increases the concentration and lifetime of free radicals that escape from the radical pair.

Chionna et al. (2005) investigated the effects of a 6-mT static magnetic field, applied for 24 h, on cell shape, cell surface sugar residue, cytoskeleton, and apoptosis in the hepatic transformed cell line Hep G2. Significant modifications of cell shape and surface by the field exposure have been observed. The exposed cells were found to be elongated, with many irregular microvilli randomly distributed on the cell surface. The shape of the cells was found to be less flat at the end of the exposure, although the morphology of the organelles remained unmodified throughout the exposure period. It has been reported that cell proliferation was partially affected. Results suggested that the static magnetic field caused a time-dependent biological effect on Hep G2 cells.

Sonnier et al. (2000) found that there were no effects from the exposure of static magnetic field levels of 0.1, 0.5, 5, or 7.5 mT, applied for 5 sec, on resting potential in cultured neuroblastoma cells. They also used the patch-clamp technique to measure transmembrane Na^+, K^+ currents in neuroblastoma cells SH-Sy5Y exposed to static magnetic fields of up to 7.5 mT (Sonnier et al., 2003). The magnetic field exposure did not result in detectable changes in any of the action potential parameters.

Trabulsi et al. (1996) measured the excitatory postsynaptic potential (EPSP) after the exposure of a mouse hippocampal slice to static magnetic field levels of 2–3 mT and 8–10 mT for a period of 20 min. They observed biphasic effects at 2–3 mT and depression of EPSP at 8–10 mT. It has been suggested that changes in intracellular Ca^{2+} concentration were responsible for these effects. Isolated *Helix aspersa* neurons were exposed to static magnetic field levels of 0.07–0.7 T, and their action potential was measured (Azanza and del Moral, 1996). A decrease in the spike depolarization voltage has been observed, and it has been attributed to desensitization of the membrane Na^+-K^+-ATP-ase pumps through an anisotropic diamagnetism reorientation.

Wieraszko (2000) studied the effect of 2–3-mT static magnetic fields applied for 20 min on the evoked potential response in B57/J56 mice hippocampal slices. Results, which were based on measurements of hippocampal function, showed both an alteration of the evoked potential and an effect on the influence of dantrolene, an inhibitor of intracellular Ca^{2+} channels.

The voltage-activated calcium channel function in cultured GH3 cells has been investigated (Rosen, 1996). A static magnetic field of 120 mT was applied for 150 sec. Reversible changes in calcium channel function were observed and were found to be temperature dependent. Results indicated that these changes were a result of alterations in the membrane proper because of the magnetically induced deformation.

Using the whole-cell patch-clamp technique, voltage-activated Na^+ channels in GH3 cells were examined (Rosen, 2003). The effects of exposure to a static magnetic field of 125 mT for 150 sec on voltage-gated Na^+ channel kinetics included a slight shift in the current–voltage relationship, a 5% reduction in peak current, and an increase in the activation time constant, τ_m, during and at least 100 sec after the exposure to the field. Significant changes were only observed at 35°C and 37°C. It was suggested that the temperature dependence factor that affected this process was probably due to the greater ease with which the liquid crystal membrane was deformed. Results suggested that the changes might be due to the reorientation of diamagnetic anisotropic molecules in the membrane. Hinch et al. (2005) showed the effects of static magnetic fields on action potential propagation and excitation recovery in nerve. At a field level of 125 mT, which was the same condition previously used by Rosen, they did not observe major changes in the electrical functioning of neurons.

Aldinucci et al. (2003a,b) investigated the effects of a 4.75-T static magnetic field exposure applied for 1 h, and also a 1-h exposure using combined fields of 4.75 T with a pulsed field of 0.7 mT, on proinflammatory cytokines, in human peripheral blood mononuclear cells (PBMCs) and Jurkat cells. They measured Ca^{2+}, proliferation, and the eventual production of proinflammatory cytokines. The static magnetic field exposure alone did not show any effects on the physiologic behavior of normal lymphocytes; however, the combined static and alternating magnetic field exposure contributed synergistically to the increase of $[Ca^{2+}]_i$. The exposure of PBMCs was carried out in a static magnetic field of 10 T (Onodera et al., 2003). It was reported that the magnetic field exposure reduced the viability of phytohemagglutinin (PHA)-activated T cells in both CD4+ and CD8+ subclasses. Sabo et al. (2002) observed a decrease in the metabolic activity of human promyelocytic leukemic cells HL-60 when exposed to a field of 1 T for 72 h. The decrease was also observed in the presence of antineoplastic drugs, which included 5-fluorouracil, cisplatin, doxorubicin, and vincristine.

Miyamoto et al. (1996) studied the effects of strong 6-T homogenous magnetic fields on both active and passive Rb^+ influx into HeLa cells. Using field exposures of 1.6 T and lower, and of 2.0 T, at various temperatures did not cause any changes in active or passive Rb^+ influxes.

Mouse islet of Langerhans cells have a very regular oscillation of calcium concentration. Madec et al. (2003) showed no effects of combined AC and static magnetic fields on these calcium oscillations in mouse islet of Langerhans.

8.3.2.4 Gene Expression and Signal Transduction

Fanelli et al. (1999) found a decrease of apoptosis in the human cell lines U937 and CEM, following exposure to a static magnetic field of 600 μT. It was suggested that the protective antiapoptotic effect was due to cellular modifications from the static field exposure, which affected the ability of the cell to enhance Ca^{2+} influx from the extracellular medium. Cohly et al. (2003) examined the effects of a 0.618-mT static magnetic field on a human osteoblast cell line (MG-63) culture, in terms of proliferation, proline uptake, and gene expression. Results showed that the exposure might be detrimental to bone formation.

Mnaimneh et al. (1996) investigated the effects of static magnetic field levels from 1 to 100 mT and also an AC field of 1.6 mT delivered at 1 Hz on NO production by murine

BCG-activated macrophages. No significant differences were observed in NO levels after a 14-h exposure.

Brief exposure effects from a static magnetic field of 100 mT for 15 min on protein expression in cultured rat primary hippocampal cells have been reported (Hirai et al., 2002). Expression of DNA binding activator protein-1 (AP-1), neural marker protein (MAP2), and neural differentiation marker protein (GAP-43, c-Fos, Fos-B, Fra-2, c-Jun, Jun-B, and Jun-D) were examined. Cytoplasmic Ca^{2+} and lactate dehydrogenase activities were also analyzed. It was found that exposure to the static field increased AP-1 DNA binding through expression of Fra-2, c-Jun, and Jun-D in immature cultured hippocampal neurons.

Flipo et al. (1998) examined *in vitro* effects of the static magnetic field levels of 2.5–150 mT applied for 24 h on the mitogen response to concanavalin A, phagocytosis, apoptosis, and Ca^{2+} influx in C57Bl/6 murine macrophages, spleen lymphocytes, and thymic cells. The exposure resulted in a decrease of phagocytosis, an inhibition of mitogenic response in lymphocytes, and a marked increase of apoptosis in thymic cells.

Salerno et al. (1999) measured *in vitro* expressions of activation markers and interleukin release in human PBMCs after exposure to a static magnetic field of 0.5 T for 24 h. They observed that the expression of CD69 at 0.5 T was reduced after PHA stimulation. Increases in interferon-γ and interleukin 4L (IL-4L) releases were observed; however, no changes in tumor necrosis factor α (TNF-α), IL-6, and IL-10 releases were observed.

Effects of a static magnetic field of 1.5 T applied for 240 min on human L-132 cells have been investigated (Guisasola et al., 2002). Heat shock proteins hsp70, hsp27, and their corresponding messenger RNAs (mRNAs), along with cyclic AMP and Ca^{2+} ions were analyzed. No field exposure effects were observed.

Effects of the exposure of HL-60 cells to a 6-T spatially inhomogenous magnetic field with a strong gradient of 41.7 T/m and to a spatially homogenous magnetic field of 10 T have been studied (Hirose et al., 2003b). The expression of c-Jun protein increased in HL-60 cells after exposure to the 6-T static magnetic field for 24, 36, 48, and 72 h.

Using budding yeast (*S. cerevisiae*) as a model for an *in vitro* biological test system, Ikehata et al. (2003) examined the genome-wide gene expression profile of yeast cells after exposure to 5- and 10-T fields for periods of 2 and 24 h. Exposure to static magnetic fields did not affect gene expression. Slight changes in the expression of several genes were observed after exposure to 14 T for 24 h.

8.3.2.5 Genotoxicity

Previous studies have shown that static magnetic fields alone did not have a lethal effect on cell growth and survival under normal culture conditions, regardless of the strength of the magnetic field applied. Effects of 5-h exposures of HL-60 cells to 6-mT static magnetic fields, with or without camptothecin, which is a DNA topoisomerase I inhibitor, have been investigated (Teodori et al., 2002). Results indicated that the field exposure did not cause apoptogenic or necrogenic effects. It was reported that exposure to the static magnetic field alone or with camptothecin did not affect cell viability.

Potenza et al. (2004a) showed that the conformation of plasmid DNA was altered by exposure to static magnetic fields of 250 mT. Various DNA point mutations were found, while no DNA degradation was observed. It was shown that the DNA degradation from H_2O_2 was accelerated by simultaneous field exposure; however, the plasmid DNA in *E. coli* cells exposed to the same static magnetic field did not show any alteration. They suggested that the magnetic field could change DNA stability directly or by activating the reactivity of oxidant radicals. It has also been suggested that the genotoxic effect could be minimized in living organisms by the presence of protective cellular responses, such as the DNA repair system and the buffering action of heat shock proteins.

In order to reveal the genetic effects of a 0.6-T static magnetic field, mutagen-sensitive mutants of the fruit fly (*D. melanogaster*) were used for the somatic cell test (Koana et al., 1995). It was shown that the exposure resulted in damaging effects in larval cellular DNA, and somatic cells without normal DNA repair functions failed to continue cell division, which resulted in developmental lethality of mutant larvae. The genotoxic activity of the field exposure was estimated to be the same as that of UV irradiation with 0.14 mJ/m^2/sec. Further study has been conducted with *D. melanogaster* using a wing spot test to estimate possible mutagenic or carcinogenic activity of the static magnetic field (Koana et al., 1997). A DNA repair defective mutation *mei-41*^{D5} was introduced into the conventional *mwh/flr* test system to enhance mutant spot frequency. Third-instar larvae were exposed to a field of 5 T for 24 h. It was shown that the exposure significantly enhanced the somatic recombination, and the recombination was found to be suppressed by supplementation of vitamin E. Results indicated that the magnetic field enhanced the genotoxic effects of spontaneously produced free radicals (Takashima et al., 2004).

In a report describing the investigations of whether static magnetic fields have cytogenetic effects in BALB/c AnNCrj male mouse bone marrow cells, Suzuki et al. (2001) indicated that the frequency of micronuclei was significantly increased by exposure to a 3-T field for 48 and 72 or a 4.7-T field for 24, 48, and 72 h. The increase in micronucleus frequency was shown to be dose dependent.

Micronuclei in cells have been used as an indicator of DNA damage. A study for *in vitro* assessment of the effects of a 4.7-T static magnetic field on the frequency of micronucleated cells in the Chinese hamster CHL/IU cell line with preexisting damage induced by exposure to mitomycin C (MMC) has been carried out (Okonogi et al., 1996). Results indicated a decrease in the frequency of micronuclei formation after 6 h of exposure and also the influence of the static magnetic field on the DNA damage stage produced by MMC.

An *E. coli* mutation assay has been carried out to assess the mutagenic effects of strong static magnetic fields (Zhang et al., 2003). Results obtained with a wild-type *E. coli* strain GC4468 and several derivatives, which were defective in DNA repair enzymes or redox-regulating enzymes, showed no effects of the exposure in terms of the survival rate of cells. On the other hand, the mutation frequency was significantly increased by exposure to the 9-T static magnetic field for 24 h in *soxR* and *sodAsodB* mutants, which were defective in their defense mechanism against oxidative stress. Results indicated that static magnetic fields induced mutations by increasing the production of intracellular superoxide radicals.

Ikehata et al. (1999) reported that 2- and 5-T static magnetic fields did not have mutagenic potential in a bacterial mutation test using *Salmonella typhimurium* (TA98, TA100, TA1535, and TA1537) and *E. coli* (WP2 *uvrA*) strains. They also reported that the exposure resulted in an increased mutation rate of the WP2 *uvrA* strain when induced by the agents *N*-ethyl-*N*′- nitro-*N*-nitrosoguanidine (ENNG), *N*-methyl-*N*′-nitro-*N*-nitrosoguanidine (MNNG), ethylmethanesulfonate (EMS), 4-nitroquinoline-*N*-oxide (4NQO), 2-amino-3-methyl-3*H*-imidazo-[4,5-*f*]-quinoline (IQ), and 2-(2-furyl)-3-(5-nitro-2-fulyl) acrylamide (AF-2). The mutagenicities of 2-aminoanthracene (2-AA), 9-aminoacridine (9-AA), *N*-4-aminocytidine, and 2-acetoamidofluorene (2-AAF) were not affected by the exposure. The bacterial growth did not change after the exposure. They suggested that the mechanism of these effects might be related to *in vitro* interactions between the chemicals and DNA and to repair systems in the test strains.

8.3.2.6 Cell-Free System, Free Radical, Enzyme Activity

Markov and Pilla (1994, 1997) studied the Ca^{2+}/calmodulin-dependent myosin phosphorylation and observed magnetic field effects of 44-μT, ambient, and 200-μT vertical fields for 5 min on the Ca^{2+} binding property. Phosphorylation increased up to at 200 μT depending on the [Ca^{2+}] concentration. The magnetic field effect disappeared as [Ca^{2+}] approached saturation for calmodulin. They emphasized that very small alterations in ambient level static magnetic fields are sufficient to have a profound effect on a cell-free enzyme system. In an attempt to replicate the results of Markov and Pilla, Coulton et al. (2000) saw no effects on myosin phosphoration in a cell-free system in vertical static magnetic fields up to 400 μT. As a result, in this experiment, no effects of static magnetic fields on the calcium/calmodulin binding property were observed. Engstrom et al. (2002) investigated the effects of nonuniform static magnetic fields of 0.7–87 mT with a gradient of 0.4–20 T/m for a period of 5 min on myosin phosphorylation and reported that the magnetic field exerted an influence on the rate of myosin phosphorylation. Increased phosphorylation was observed. It can be seen that the magnetic field gradients played a specific role in this experiment. Liboff et al. (2003) investigated the effect of a 30-min exposure to a static magnetic field of 20 μT on calmodulin-dependent cyclic nucleotide phosphodiesterase activity in cell-free systems and reported that the activity was altered in comparison to zero magnetic field exposures.

After studying the theoretical background of a cell-free system—Ca^{2+}/calmodulin-dependent myosin light chain phosphorylation reaction—Markov (2004a,b) designed experiments to test the effects of a pulsed radio frequency (RF) field, pulsating magnetic fields, gradient magnetic fields, and homogenous static magnetic fields on this cell-free system. He suggested that the magnetic fields affect the cell-free enzyme system by modulating ion–protein interactions.

Watanabe et al. (1997) measured and evaluated lipid peroxidation in the liver, kidneys, heart, lung, and brain of 8-week-old male BALB/c mice. The mice were exposed to 3.0- and 4.7-T fields for 3–48 h. The lipid peroxidation level in the liver was increased after exposure to the 4.7-T field. In kidney, heart, lung, and brain, no changes in the level of lipid peroxidation were observed compared to the control. The exposure to the 3.0-T field showed no alteration of the lipid peroxide level in all the tissues. The combination of CCl_4 administration and 4.7-T field exposure increased the lipid peroxidation level in the liver. It was concluded that the exposure to high static magnetic fields could induce the increase of lipid peroxidation levels in the liver of mice and could enhance the hepatotoxicity caused by CCl_4 injection.

Using fireflies, *Hotaria parvula* and *Luciola cruciata*, as bioluminescence systems, Iwasaka and Ueno (1998a) studied the effects of 8- and 14-T static magnetic fields on the emission of light. They showed that changes in the emission intensities under a magnetic field were related to the change in certain biochemical systems of the firefly, systems such as the enzymatic process of luciferase and the excited singlet state responsible for subsequent light emission.

Zhadin et al. (1998) have undertaken experiments that investigated the combined action of static and AC magnetic fields on ionic current in aqueous glutamic acid solution. Results showed that the combined parallel static and AC magnetic field causes a rapid change in the ionic current flow when the AC frequency is equal to the cyclotron frequency.

During the last few years, Brocklehurst and McLauchlan (1996) discussed the free radical mechanism involved in the observed effects of environmental electromagnetic fields on biological systems. Grissom and Natarajan (1997) summarized the theory of magnetic field effects on chemical and enzymatic reactions. Magnetic field effects have been used as a powerful technique to study enzymatic and chemical reactions with radical pair intermediates. They suggested that the coenzyme B_{12}-dependent enzymes with radical pair intermediates are well suited for the study of this effect. Taoka et al. (1997) tested the magnetic field effects on coenzyme B_{12}-dependent enzymes. The end point was that ethanolamine ammonia lyase and human enzyme methylmalonyl–coenzyme A mutase catalyze coenzyme B_{12}-dependent rearrangement reactions. While the end point was affected, the authors speculate that the change would have little physiological significance.

Eichwald and Walleczek (1996) showed a model for magnetic field effects on radical pair recombination in enzyme kinetics. The magnetic field effects in radical pair chemistry have been reviewed (Grissom, 1995; Brocklehurst, 2002; see also Chapter 6 on free radicals by Engström).

8.4 Miscellaneous

8.4.1 Biological Sensing and Magnetite

Many studies have suggested that the magnetic field is an important marker for animal navigation and spatial discrimination (Wiltschko and Wiltschko, 1995). The blind mole-rat (*Spalax ehrenbergi*) was used as a model to examine the possibility of the perception and use of magnetic fields in their orientation in space (Kimchi and Terkel, 2001). Experiments were performed in an eight-arm maze under Earth's natural and artificial magnetic fields. Results showed that the blind mole-rat was able to perceive and use Earth's magnetic field to orient in space. Kimchi and Terkel showed that blind mole-rats spontaneously preferred to place their nests toward the south of the magnetic North. Deutchlander et al. (2003) showed that Siberian hamsters (*Phodopus sungorus*) used directional information from the magnetic field to set a position for their nests. In contrast to blind mole-rats, the directional preference for nest position demonstrated by Siberian hamsters appeared to be a learned response.

Since the neural substrate subserving magnetic orientation is not known, the combination of two techniques, a behavioral test for magnetic compass orientation and an immunocytochemical visualization of the transcription factor c-Fos as a neuronal activity marker, has been used to investigate magnetoreception in the mole rat (*Crytomys anselli*) (Nemec et al., 2001). Nemec et al. found that the superior colliculus of the hypothalamus contained neurons that would respond to magnetic stimuli, and thus determined the involvement of a specific mammalian brain structure in magnetoreception.

Edmond (1996) showed that a very sensitive magnetic compass is formed by the incorporation of a small quantity of ferromagnetic, single-domain crystals, such as magnetite, within a nematic liquid crystal. Winklhofer et al. (2001) localized high concentrations of Fe^{3+} in the upper-beak skin of homing pigeons (*Columba livia*), and identified the materials of magnetite nanocrystals as the core of a magnetic field receptor.

Lohman et al. (2001) found that hatchling loggerhead sea turtles (*Caretta caretta*), when they were exposed to magnetic fields found in three widely separated oceanic regions, swam in the direction that would help to keep them within the currents of the North Atlantic gyre and facilitate their migratory pathway. It was found that young loggerheads used a guidance system of magnetic fields to assist in their navigation.

Although the mechanism of magnetoreception has not been clearly identified, geomagnetic orientation has been well recognized. A biophysical model has shown that changes in the wavelength of light can influence magnetic field orientation through the interaction between the geomagnetic field and photoreceptors. Deutschlander et al. (1999a) found that light-dependent orientation in the newt (*Notophthalmus viridescens*), was mediated by extraocular photoreceptors located in the pineal complex or deeper in the brain. Using newts, Phillips et al. (2001, 2002b) showed the role of photoreceptors in magnetic compass orientation and the magnetic inclination for deriving map information. They investigated the possibility that the fixed-axis response of the newts was mediated by a magnetoreception mechanism involving single-domain particles of magnetite (Phillips et al., 2002a).

There are several explanations for the magnetic sensitivity in fish. The aquatic animal might perceive an electric voltage induced by the water current or by its own movement in the geomagnetic fields. Elasmobranch fish such as sharks, skates, and rays are known to possess a sensitivity to the induced electric field through the sensory organs called the ampullae of Lorenzini.

Yano et al. (1997) studied the migrating behavior of the chum salmon (*Oncorhynchus keta*) fitted with a magnet to investigate the role of magnetic compass orientation in the North Pacific off the coast of Kushiro, Hokkaido. The magnetic field strength was about 0.6 mT around the head area, with polarity changes every 11.25 min. There were no effects observed on the movement of salmon.

Effects of electric and magnetic fields have been observed in the behavior of marine animals and freshwater and terrestrial species. In addition, there are a growing number of questions concerning the effect on aquatic ecosystems of the growing spread of artificial techniques such as underwater sea DC cables. There are many underwater DC cables under various seas all over the world, which carry electrical currents (see also Chapter 1 on fields in the environment by Mild and Greenebaum). These electric currents induce static magnetic fields with intensities up to 3.5 mT around cables on the sea bottom, where there are many invertebrate and vertebrate species. Research has been carried out to examine whether the exposure to magnetic fields of 3.7 mT for several weeks could influence the survival rate and fitness of common benthic animals of the Baltic Sea (Bochert and Zettler, 2004). The investigation was carried out on the crustacean (*Crangon crangon*, *Rhithropanopeus harrisii*, and *Saduria entomon*); the mussel (*Mytilus edulis*); and the flounder (*Planthichthys flesus*). Results showed no differences between experimental and control animals. Since this is the first study for investigating the effects of static magnetic fields generated by sea-positioned DC cables, on aquatic organisms and marine benthic animals, further studies are required.

In a study to confirm the magnetite-based detection mechanism in rainbow trout (*O. mykiss*), magnetic crystals in the area of olfactory lamellae were found, and the arrangement of several magnetic crystals in a chain of about 1 μm has been confirmed (Diebel et al., 2000).

It was shown that magnetizable material abolishes the behavior of bobolink (*Dolichonyx oryzivorus*) by blocking the ophthalmic branch of the trigeminal nerve (Beason and Semm, 1996). The result was consistent with the hypothesis that magnetite is a constituent of the magnetoreceptors associated with the ophthalmic nerve.

It is suggested that migratory birds, amphibians, and reptiles may have the ability to sense the geomagnetic field and use it as a source of compass information. Phillips (1996) presented a graphical model that predicts qualitatively the changes in the direction of homing orientation. Munro et al. (1997) investigated the effect of pulse remagnetization on the orientation of inexperienced, juvenile migrant birds, such as the Australian silvereye (*Zosterops l. lateralis*). The ability of juvenile birds to maintain their normal magnetic orientation after pulse application indicated that the pulse does not impair the magnetic compass. On the other hand, the deflection observed in adult birds after pulse treatment appeared to reflect "false" map information, which leads to a change in course. This is consistent with evidence that the magnetic compass involves light-dependent magnetoreception mechanisms.

Wiltschko and Wiltschko (2001) studied the behavior of European robins (*Erithacus rubecula*), under monochromatic light of various wavelengths and intensities to investigate magnetoreception. At a quantal flux of 7×10^{15} quanta/sec/m^2, the birds were well oriented in their migratory direction under 424 nm blue, 510 nm turquoise, and 565 nm green, whereas they were disoriented under 590 nm yellow. Changes in behavior depended on increasing the light intensity. This finding suggested that light-dependent magnetoreception may involve receptors and a neuronal pathway of its own.

Ritz et al. (2000) postulated the possibility that magnetoreception involves radical pair processes as a biophysical mechanism. They first considered a system of radical pairs as a model for the magnetic sensory organ and evaluated the influence of the geomagnetic field on radical pair systems. European robins (*E. rubecula*), were used in this study, and the results showed the disruption of magnetic orientation behavior of robins when exposed to a vertically aligned broadband field of 0.1–10 MHz and 0.085 μT or the single frequency of 7 MHz and 0.47 μT, together with the geomagnetic field (Ritz et al., 2004). The disorientation observed was found to depend on the angle between the 7-MHz oscillating field and the geomagnetic field. The robins oriented in the migratory direction when the oscillating field was parallel to the geomagnetic field. The author suggested a magnetic compass based on a radical pair mechanism, due to the resonance effect on singlet–triplet transitions in the oscillating fields.

Fuller suggested the significance of the time constants of magnetic field sensitivity in animals (Fuller and Dobson, 2005). Conditioning experiments have had great success in the analysis of animal sensory physiology. Wiltschko and Wiltschko (1996) commented that the conditioning technique did not appear to be suitable for testing magnetic sensitivity.

Some insects are able to respond to magnetic fields, especially the geomagnetic field. Mosquitoes were tested for the presence of remanent ferromagnetic material and their behavioral response to magnetic fields. Most mosquitoes, when placed in a uniform static magnetic field of 0.1 mT, moved around until they were oriented parallel to the field. It was reported that a significant remanence found on the surface of both living and dead mosquitoes might be due to attraction of ferromagnetic dust onto the body (Strickman et al., 2000). It is well-known that magnetic fields influence honey bee behavior, moth navigation, beetle larvae, the behavior of hatchling loggerhead sea turtles, migration of birds, etc. Slowik et al. (1997b) speculated that the red imported fire ant (*Solenopsis invicta*) might use magnetic field information in their nesting activities and in orientation, since their first observation suggested that fire ant workers moved as a colony toward the magnetic field. In a second paper, Slowik et al. (1997a) suggested the presence of small amounts of ferromagnetic material in fire ants.

Many review papers have been published during the last few years. Deuschlander et al. (1999b) and Wiltschko and Wiltschko (2002) reviewed the light-dependent magnetoreception in animals. Lohmann and Johnsen (2000) described the difference between a magnetic directional sense and a magnetic map sense and reviewed the three hypotheses of vertebrate magnetoreception.

8.4.2 Plant Growth, Response, and Magnetotropism

The enhancement of plant growth using various magnetic field applications has been reported by many researchers (Phirke et al., 1996). Effects of magnetic fields on seed germination, crop growth, physiological response, sporulation, water uptake, and rate of seed have been studied; however, there has been no consistency in results among the various reports.

Studies directed toward investigating the effects of magnetic field treatment on seeds and water in terms of the rate and percentage of germination of rice (*Oryza sativa* L.) and on the length and weight of germinating barley (*Hordeum vulgare* L.) and wheat (*Triticum aestivum* L.) have been carried out (Carbonell et al., 2000; Martinez et al., 2000, 2002). The field strength ranged from 125 to 250 mT depending on the research strategies. Results by Carbonell et al. showed that the rate and percentage of germination increased after chronic exposure to 150-mT fields. Magnetically treated water was found to improve the germination of rice seeds. Martinez et al. (2000) showed that the magnetic field increased the length and weight of barley seed, and the degree of this effect depended on the duration of exposure. Effects of magnetic biostimulation on the initial growth stages of wheat, with an exposure duration of 0, 1, 10, 20 min, 1 h, 24 h, and chronic exposure were investigated (Martinez et al., 2002). In the report, they defined the magnetic doses in terms of magnetic field energy density (J/m^3). An increase in plant height had been observed as the magnetic dose increased; thus, it was suggested that the stimulatory effects might be related to the amount of magnetic field energy. Florez et al. (2004) investigated the effects of 125- and 250-mT static magnetic fields on the germination and the initial growth stages of rice seeds (*O. sativa* L.). The seeds were exposed to the magnetic fields for various time durations, and the germination time was found to be shortened when the seeds were exposed to these fields. The seeds' maximum length and weight were obtained for the chronic exposure. It was shown that magnetic treatments, when applied under specific conditions, affected germination and the first stage of growth.

Piatti et al. (2002) investigated the effects of inhomogenous static magnetic fields ranging between 6 and 10 mT on the growth and viability of the plant-growing bacteria *Serratia marcescens*, barley callus cells (*Hordeum vulgare*), and blackberry (*Rubus fructiosus*). While there was no field effect observed on blackberry cells, it was found that the exposure reduced the number of bacterial cells and lowered both the number and the viability of barley cells. Diamagnetic susceptibility and root growth response to magnetic field exposure on three plant species, *Lens culinaris*, *Glycine soja*, and *T. aestivum* were investigated (Penuelas et al., 2004). Magnetic fields of 17.6 mT reduced root growth in all three plants. The field strength of 2.1 mT had no significant effect on reduction in the cereal *T. aestivum*.

Among the many studies that examined gradient magnetic field effects on various engineering, biological, and physicochemical phenomena, a study of effects of gradient magnetic fields of up to 10 T on the germination and growing process of cucumber (*Cucumis sativus* L.) was carried out (Hirota et al., 1999). It was found that the shoot germinated toward the field center, whereas the root grew in the opposite direction of the shoot. This observation seemed to be a result of the magnetic force influencing the geotaxis of the cucumber.

After calculating the magnetic field dependence of the ionic current density across the cellular membrane, Reina and colleagues examined the effects of 0–10 mT static magnetic fields on the amount and rate of water absorption in the lettuce seed cell membrane (*Lactuca sativa*) in order to compare the calculated and experimental results (Reina and Pascual, 2001; Reina et al., 2001). Theoretical calculations showed that the static magnetic field induced changes in the ionic concentration and in the osmotic pressure, which regulates the entrance of water into the seeds. The magnetic field exposure was 10 min in a Helmholtz coil, and this was carried out immediately before placing the seeds in water inside a climatized room. The investigators demonstrated a close correlation between their theoretical calculations and the actual experimental results. It was shown that exposure to the static magnetic fields altered the water absorption in seeds, which may possibly explain the change of germination rate. Adair (2002) discussed and questioned these results. Amyan and Ayrapetyan (2004) investigated changes in the wet and dry weight of barley seed after treatment by static magnetic field levels of 1.25, 2.50, and 3.75 mT. Seed treatment was carried out in cold (4°C) and warm (20°C) distilled water. After pretreatment by the fields, the seeds were incubated for 72 h. Results suggested that effects depend on not only the field strength but also the incubation time period.

Growth and sporulation of phytopathogenic microscopic fungi have been investigated under exposure to a static magnetic field that ranged from 0.1 to 1 mT (Nagy, 2004). It was shown that the growth was decreased by the magnetic field exposures. Increases in the number of developed conidia of *Alternaria alternata* and *Curvularia inaequalis* and a decrease in the number of *Fusarium oxysporum* conidia have been observed as well.

Effects of combined AC and DC magnetic fields on the germination of hornwort seed (*Cryptotaenia japonica* Hassk) have been reported by Kobayashi et al. (2004). They tested three directions of the AC magnetic field, which were vertical, parallel, and perpendicular to the direction of total geomagnetic field (DC). The frequency and strength of the AC magnetic fields ranged between 3.5 and 14 Hz and 500 and 750 μT. The total geomagnetic field was 50 μT. The seeds were exposed to the fields for 16 d at 24 h/d. The vertical AC magnetic field applied simultaneously with the DC field was found to promote the germination of seeds. Field level applications of 7 Hz, 750 μT and 14 Hz, 500 μT showed the maximum effects.

Effects of static magnetic fields ranging from 0 to 350 μT on gravitropic bending in the apical stem segments of flax seedlings (*Linum bienne*) have been investigated (Belova and Lednev, 2001). In comparison with the control group kept in a geomagnetic field of 46.5 μT, stimulation of the gravitropic bending was observed at $0 \leq B_{DC} \leq 2\,\mu T$ and $200 \leq B_{DC} \leq 350\,\mu T$, and inhibition was observed at $100 \leq B_{DC} \leq 170\,\mu T$.

Investigations of static magnetic field effects on the curvature of primary roots of radish seedling (*Raphanus sativus* L.) have been carried out (Yano et al., 2001). When radish roots were exposed to an inhomogenous static magnetic field, they responded to the south pole of the magnet. Trophic response was found at a field level of 13–68 mT with a gradient of 1.8–14.7 T/m. A small response to the north pole of the magnet was found as well.

Jovanic and Sarvan (2004) studied the effects of static magnetic fields on fluorescence spectra and leaf temperature in intact plant bean (*Phaseolus vulgaris* L.) The field strengths were as high as 160 mT, and the plant was grown for 3 weeks. It was reported that significant changes in fluorescence spectra and leaf temperature were induced after the field exposure. An increase of fluorescence intensity ratio and changes in leaf temperature ΔT were observed in parallel with increase in field intensity.

8.4.3 Magnetotaxis

There have been many studies reporting that magnetic fields affect the swimming behavior of *Paramecium*. A decrease in swimming velocity and an increase in the frequency of directional changes were observed after exposure to a magnetic field of 0.126 T during a motility study of *Paramecium* (Rosen and Rosen, 1990). Nakaoka et al. (2002) found that a typical ciliated protozoan, *Paramecium*, swam perpendicular to a static magnetic field of 0.68 T. It was suggested that the diamagnetic anisotropy of cellular components cilia and trichocysts was important for the magnetic orientation of their swimming.

Effects of horizontal magnetic fields on the movement of *Euglena gracilis* (ca. 50 μm in length) have been reported (Tanimoto et al., 2001). When the horizontal magnetic field with a gradient of ca. 400 T^2/m was applied, living *E. gracilis* moved to the higher field (positive magnetotaxis), whereas dead *E. gracilis* moved to the opposite, lower field. *E. gracilis* was found to be oriented perpendicularly to the magnetic field regardless of whether they were alive or dead. Thus, magnetotaxis of living *E. gracilis* may be explained by taking into account both the environmental inhomogenous magnetic forces and the magnetic orientation of *E. gracilis*. In contrast, magnetotaxis was not observed in a uniform magnetic field of 8 T. Effects of strong magnetic field gradient (max. 8 T, ca. 400 T^2/m) on the movement of *E. coli* have been investigated (Tanimoto et al., 2005). *E. coli* cells were placed in a 5 mm (diameter) × 150 mm (length) glass tube containing viscous media that flowed in the tube. The speed at zero field was 0.65 cm/h. The observed velocities of the movement from a high field (8 T) to a low field (1.5 T) and the movement in the opposite direction around were 1.35 and 0.49 cm/h, respectively. Diamagnetic *E. coli* experienced a repulsive behavior to magnetic forces of increased magnetic gradient. Therefore, it has been speculated that the velocity of *E. coli* would be accelerated toward the direction of the lower-strength field, while it would be decelerated in the direction of the stronger fields. Results suggested the magnetic force specifically, could be an important mechanism of magnetic field effects when a low-frequency high magnetic field was applied, since the microorganism might respond to mechanical stress due to alternating magnetic forces.

8.4.4 Others

As an initial study for investigating the relationship between magnetic fields and amoebae, Berk et al. (1997) examined the inhibitory effects of static magnetic fields on the population growth of amoebae. They tested three species, *Acanthamoeba hatcheii*, *Acanthamoeba castellanii*, and *Acanthamoeba polyphaga*. Amoebae were exposed to magnetic field strengths of 71 and 106.5 mT with an exposure duration up to 72 h. Results showed that magnetic fields decreased the growth of all three potentially pathogenic ameba populations significantly within 72 h. It was reported that the inhibitory effect did not depend on the field strength, and it was shown that this research would be important and advantageous in the development of disinfection strategies for surface material, such as the surface of contact lenses.

Rai et al. (1997) investigated the effects of a 0.1-T static magnetic field on the electrical parameters of goat eye lens. Under magnetic field application, the complex impedance between real and imaginary parts was obtained in the form of a Cole–Cole plot. It was reported that the static magnetic field altered the current flow in the tissue.

Iwasaka and Ueno (1998b) investigated the effects of a static magnetic field of up to 14 T on the near-infrared spectrum of water molecules and glucose solutions. They demonstrated the possibility that the static magnetic field affected the formation of hydrogen bonds of water molecules and the hydration of glucose molecules.

Morariu et al. (2000) exposed human blood samples to zero magnetic fields for 72 h in order to observe the aging process of erythrocytes. The control samples were kept in a normal geomagnetic field. In a zero magnetic field, increases in the rate of Na^+ and Ca^{2+} influx, in the rate of K^+ outflow, and in homolysis were observed. Reduction in Na^+-K^+-ATPase and Ca^{2+}-ATPase activities has been observed in a zero magnetic field; thus, zero fields significantly accelerated the aging of erythocytes. Effects of zero magnetic fields were further investigated on Zn and Cu concentrations in the human blood serum during *in vitro* aging of blood with a 48-h exposure (Ciortea et al., 2001). Blood samples were collected from both healthy donors and chronic bronchial asthma (BA) patients. While the Zn concentration was not found to be affected by the zero magnetic field exposure, Cu concentration was found to be sensitive to this field. It was also reported that the aging effect appeared to be decelerated for most BA types.

8.5 Medical Applications

8.5.1 Biomagnetic Phenomena

Biomagnetic phenomena for different intensities of magnetic fields and their frequency are shown in Figure 8.1. It is important to know the intensities and frequencies of magnetic fields involved in biomagnetic phenomena while discussing the relationship

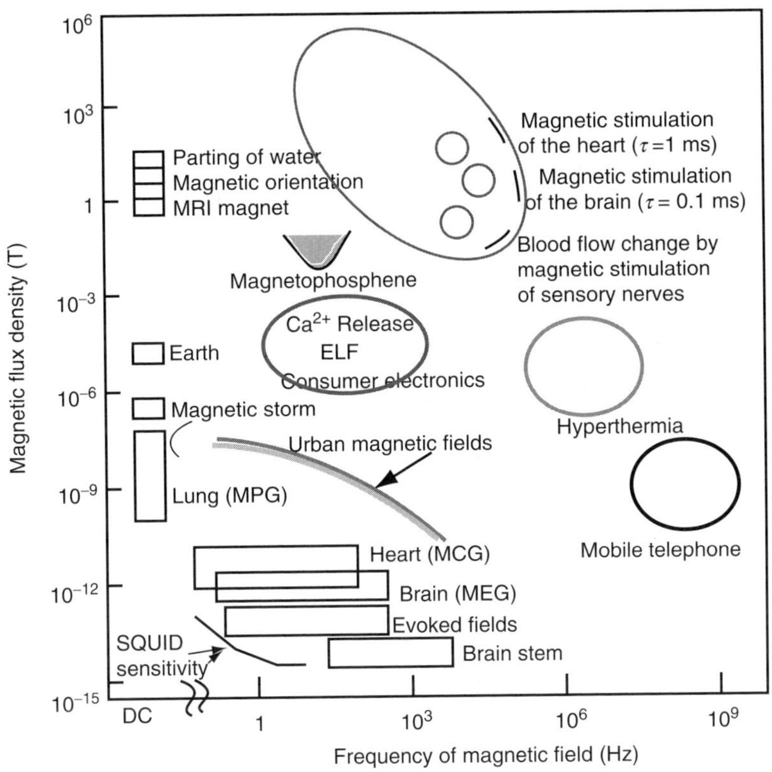

FIGURE 8.1
Biomagnetic phenomena for medical and therapeutic application.

between magnetism and living organisms. Regarding the effects of magnetism on living organisms, it should be realized that the actions of static and variable magnetic fields differ from each other in terms of the fundamental mechanism.

Studies on the biological effects of electromagnetic fields have resulted in significant developments in medical applications for electromagnetic fields, after the development of high-strength superconducting magnets. TMS, measurement of biomagnetic fields with the superconducting quantum interference device (SQUID), and MRI are the three mainstays of these medical applications. These techniques have also been leading the amazing progress in the understanding of the brain function. TMS locally stimulates the human cerebral cortex with millimeter-order spatial resolution from a figure-eight coil placed on the skull. A three-dimensional imaging of the brain neuron function has been enabled by utilization of SQUID in magnetoencephalography (MEG), functional magnetic resonance imaging (fMRI), and current-distribution MRI. Results from TMS and imaging studies indicate potential applications of biomagnetics in brain science and clinical neuropsychiatry.

In TMS, when a strong electric current is applied to a figure-eight coil positioned over the head for 0.1–0.2 ms, a pulsed magnetic field of 1 T is produced. This pulsed magnetic field generates eddy currents in the brain, which excite the targeted area of the nervous system.

Incidentally, unconscious and uncontrolled exposure of the brain to high-frequency electromagnetic waves has been increasing with the recent, rapid widespread use of cellular phones by the general public. Cellular phones in Japan are designed for operation with a frequency of 800 MHz and a microwave of 1.5 GHz (see Chapter 1 on environmental exposures by Mild and Greenebaum).

In biomagnetic measurements, MEG associated with auditory brain stem response is of the order of 10^{-15} T (1 fT), and it can be measured on the extremum of the sensitivity limitation of the SQUID gradiometer (Erne et al., 1988; Iramina and Ueno, 1995). In 10^{-12} T (1 pT)-order measurements, while α-wave spontaneous MEG can be detected without a signal-averaging technique, the technique is required to increase a signal-to-noise ratio (SNR) in order to detect various evoked responses and brain stem responses.

The recent development of noninvasive brain function measurement technologies, such as MEG and fMRI, has been contributing to the rapid progress in brain science research. Scientific discussions of mental problems such as thinking and psychomotor activities (e.g., joy, anger, sadness, and happiness) in terms of brain function became possible with the development of these new technologies.

In static magnetic fields of a few tesla, fibrin polymers, which are involved in blood coagulation, orient parallel to the magnetic fields in the course of polymerization (Yamagishi et al., 1989; Ueno et al., 1993). Furthermore, magnetic alteration of blood coagulation and dissolution processes by magnetic fields and magnetic orientation of biopolymers, such as fibrin and collagen, have been observed. These findings introduce a new aspect of biomagnetic applications in the regulation of living systems and biological materials.

8.5.2 Transcranial Magnetic Stimulation

TMS is the technique of locally applying magnetic stimulation by a strong pulsed magnetic field on the order of 1 T transcranially to the brain. When a strong electric current is applied to a figure-eight coil placed over the head for 150 μs, a pulsed magnetic field on the order of 1 T is produced that generates eddy currents in the brain, which excite the nervous system. The first study of magnetic stimulation in a human brain by Barker et al. (1985) utilized single coils; thus, localized magnetic stimulation of a targeted portion of the human brain was impossible. The localized vectorial magnetic stimulation of a human cortex using a figure-eight coil was developed by Ueno et al. (1978, 1986b, 1988, 1989, 1990a,b, 1991), which enabled stimulation of the motor cortex of a human brain at 5-mm resolution. Localized magnetic stimulation contributed to the creation of functional maps of the motor cortex related to hand and foot areas. An optimal direction of probe placement for the targeting of stimulating currents, which induce neural excitation in each functional area of the cortex, based on functional maps was observed, the so-called vectorial feature. Variations in the functional maps of the cortex with changes in orientation of the stimulating current were observed as well. It is proven that the vectorial feature allows for studies that reflect both functional and anatomical organizations of neural fibers in the brain. Localized magnetic nerve stimulation of the brain is suitable for investigations of brain function and construction without damaging any tissues.

Applications of TMS temporarily disturb brain function, which results in a virtual lesion in the brain. Zangaladze et al. (1999) showed that the disruption of the function of the occipital cortex with the use of focal TMS interferes with the tactile discrimination of grating orientation. Epstein et al. (2002) used TMS to investigate memory encoding and retrieval, particularly the role of the dorsolateral prefrontal cortex in associative memory for visual patterns. TMS disrupted associative learning of abstract patterns over the right frontal area, which suggests that the participating cortical networks may be lateralized in accordance with classic concepts of hemispheric specialization.

Traditionally, stimuli are applied at various scalp positions using a latitude- and longitude-based coordinate system referenced to Cz in the 10–20 international system at the vertex, while simultaneously, the amplitude of the motor evoked potentials generated in contralateral muscles is also measured (Ueno et al., 1989, 1990a). This gives a "map" of sites on the scalp from which responses can be obtained by each reference muscle.

Rothwell et al. (1987) revealed the enormous clinical importance of TMS, namely, for motor functional evaluation.

Recent developments in the navigated brain stimulation (NBS) stereotactic TMS devices allow noninvasive mapping of the spatial and temporal representation of any brain activity that reacts to magnetic stimuli (Krings et al., 2001), such as sensory, motor, language, and cognitive functions. Stereotactic TMS coil positioning and real-time visualization of the stimulating electromagnetic field effect using MRI allow precise replicability of stimulation parameters as well as accurate dose definition. Frameless NBS allows precise localization of a stimulation target in combination with other imaging modalities or by the use of anatomical landmarks. In a case where a brain tumor was resting adjacent to the precentral gyrus, the motor strip identified by TMS compared preoperative MRI and fMRI and revealed fine functional differences between results that were integrated on the navigation system. Distribution of the tumor margin and the motor cortex (both fMRI and TMS assisted) can be drawn on the patient's scalp using the navigation system. Skin incision, craniotomy, and operative approaches were considered from these results so as to avoid motor deterioration.

There has been no verification of which nerve cells are actually stimulated by TMS. There is one subject under discussion: whether a target neuron cell is directly stimulated by TMS (direct stimulation) or whether an interneuron is first stimulated and then a target neuron cell is stimulated indirectly (indirect stimulation). It is possible that alteration of eddy currents by heterogeneity of conductivity in the brain may affect the path of these currents and result in both neuronal excitation and excitatory directional changes at sites other than those targeted by the original intended direction of stimulation. Further investigation and analysis of TMS and construction of models using magnetic nerve stimulation are required to clarify how the relationship between a position of a coil and a site of stimulation can be affected by strength of stimulation, arrangement of neurons, heterogeneity of conductivity, and interneuronal participation. While a figure-eight coil is suitable for local stimulations at the surface of the cortex along the surface of the head, tridimensional localized stimulations are not possible with the coil at present.

Despite the problems described above related to TMS, there are high expectations for magnetic stimulation to contribute to a new era of brain science. A major and possibly very important future field of study is the application of TMS for obtaining therapeutic effects in neurological disorders. A number of animal studies testing the basic mechanisms of TMS-induced alterations of neurotrophic factors, gene expression, and changes in plasticity have been conducted (Fujiki and Stewart, 1997; Keck et al., 2000; Fujiki et al., 2003; Ogiue-Ikeda et al., 2003a).

There is strong evidence that the expression of certain genes such as the immediate early gene, astrocyte-specific glial fibrillary acidic protein mRNA (Fujiki and Stewart, 1997), and brain-derived neurotrophic factor is altered in response to repetitive TMS (rTMS). This indicates that the measurable effects of TMS reach the molecular and signaling levels. The most promising hypothesis is that magnetic field-induced neuroprotective or trophic factors may protect neurons from hypoxic insult (Fujiki et al., 2003). Long-duration rTMS modulates the monoamine neurotransmitter system in content and turnover and may also induce sprouting of mossy fibers in the hippocampus (Keck et al., 2000). Increased dopaminergic neurotransmission may contribute to the beneficial effects of rTMS in the treatment of affective disorders and Parkinson's disease.

The results of these studies provide strong evidence that noninvasive TMS can strongly modulate gene expression in neurons and astrocytes. Thus, TMS, originally used simply as a way to assess the function of descending motor tracts noninvasively, may in the end be used as a means to modulate gene expression and to induce restorative plasticity or tolerance against injury in the brain.

TMS does not cause any pain and requires no physical invasion of the body; therefore, it should become more important in functional, diagnostic, and therapeutic research of the brain. In brain functional research, application of magnetic stimulation for the temporary blockage or modification of the facultative information process and cognitive process of various sensory systems may be used to identify localization and connecting pathways of brain function. If a magnetic stimulation can effectively block and modify various sensory systems, it should be advantageous for pain treatment. Elucidation of the effects of magnetic stimulation on synaptic functions may lead to further research associated with brain plasticity. Further research for investigation of magnetic compensation and reconstruction of neuronal functions around damaged neurons may lead to the development of various magnetic field-based stimulation applications, including the treatment of depression, the prevention of dementia, and a safer and more effective magnetic pulse treatment, which may replace the current electroconvulsive therapy (ECT).

8.5.3 Magnetoencephalography

MEG measures the very weak magnetic fields of the order or 10^{-13} T (100 fT) generated by neuronal current flow, by the detection of magnetic signals measured by SQUID arrays. MEG can detect brain functions with high millisecond-order temporal resolution and high millimeter-order spatial resolution noninvasively; thus, it is useful for investigation of brain functions in humans, including higher brain functions such as memory and cognition. Since Cohen obtained a magnetoencephalogram for human α-waves with the use of a SQUID, a prototype developed by Zimmerman and Colleagues (1972), it was only until recently that the use of a whole-head MEG system that is able to carry out spontaneous measurement at multiple points has become practical (Squires, 1991; Ahonen et al., 1993; Vrva et al., 1993). In recent years, the whole-head MEG system has been incorporated into brain functional research all over the world and has accelerated progress in research.

Application of forward and inverse problems in MEG analysis is critical to estimate a localization of brain function. Ueno and Iramina (1991) measured MEG associated with short memory, cognition, and mental rotation in humans, constructed current-dipole and distributed intracerebral electrical source models, and carried out estimations for the localization of various brain functions during the processing of information. The electrical source of a visually evoked reaction with approximately 150 ms at latency localized in the primary visual cortex was described in a current-dipole model relatively well, while a distributed intracerebral electrical source model was more useful in estimation of the electrical source incident to a mental rotation with approximately 180 ms or higher at latency. In the distributed electrical source model, a chronological transition of electrical source groups from the occipital lobe area to the posterior temporal lobe area was captured. MEG is not only a tool for basic brain functional research, but is also applicable to medical research. Clinical applications of MEG include detection of epileptic spikes, measurements of slow waves associated with brain tumors and cerebrovascular diseases, and cerebroelectric activity of ELF induced by event-related potentials.

It is necessary to construct experimental paradigms that are able to perform more precise extraction of a specific brain function, allow understanding of brain function dynamics, and provide measuring techniques to assess the acquired information (Yoshida et al., 1995; Iwaki et al., 1999). It is also important to develop signal processing techniques for source determination of signals with very small SNRs and with distributed electrical sources, improve inverse problem approach methods, and construct suitable current source models (Ueno and Iramina, 1991; Iramina et al., 1994, 1995b). There are several factors to be taken into account in MEG inverse problem analysis: the shape of the head, heterogeneity of conductivity, alignment of neuron cells, interneurons, and thalamocortical specific projection system. An ideal electrical source model possesses electrophysiological features of complex cranial nerve systems with consideration of these factors.

In a study on language-related brain activities, Kuriki et al. (1995, 1998) used MEG imaging to examine the temporal and topographical characteristics of neural activities in the comprehension of Japanese complex sentences with a clause structure. The Korean language was also used as an experimental language (Kwon et al., 2005). The Korean language has a subject–object–verb order structure, ending with a verb. Semantic and syntactic violations, that is, errors introduced in a sentence, can be made by altering a single word, that is, a verb, in an inappropriate manner. Neural activities in response to such a violation are measured as the response elicited by the final verb in the verb-ending sentence. This study is aimed to identify neural activities in the cerebral cortex that occur during a latency course, processing syntactic and semantic aspects of spoken sentences.

In another MEG study on music, Kuriki and colleagues used melodies to measure the responses that are elicited by an out-of-key tone in musical phrases (Hirata et al., 1999; Kuriki et al., 2005). The musical context is established by the sense of a key and melodic pitch sequence. The responses would reflect the perception of these restricted aspects of melody. The results of the present study should provide an understanding of the spatiotemporal characteristics of cortical activities involved in melody perception. MEG measurements were also performed for musical tones and chord stimuli for well-experienced musicians and nonmusicians. The principal purpose of this study was to explore how the brain activity reflected in late auditory evoked responses would behave when exposed to the successively presented tones and chords stimuli and also how the activity would vary according to experience of musical training.

Although there is still ambiguity in the analytical technique, MEG still attracts medical researchers because it can reflect the chronological change of source signals to that of magnetic fields. It is obvious that MEG will become an essential technique in human brain function research, since there is only one technique that is noninvasive with millisecond-order high temporal resolution, electroencephalography, available at this time for estimation of brain function localization,. Development of MEG with higher sensitivity and operativity, construction of an intracerebral electric source model, and improvement of inverse problem analysis may become more important.

8.5.4 Magnetic Resonance Imaging

Since Lauterber suggested a linear magnetic field gradient in 1973, MRI has been rapidly developed (Lauterbur, 1973). MRI utilizes fusion techniques of spatially uniform direct current magnetic fields, spatially gradient direct current magnetic fields, and RF electromagnetic fields. A guideline of static magnetic field exposure to a human body by the International Commission on Non-Ionizing Radiation Protection (ICNIRP) suggests 2 T as the ceiling value for body parts, except for arms and legs, in occupational exposure. In the application of clinical MRI, although the exposure is carried out under supervision of doctors, the current exposure level is confirmed to be 2 T or less. It is not feasible to obtain resonance images, except from hydrogen atoms, in static magnetic fields at this strength.

The use of MRI conducted at high static magnetic field levels is fast growing (Robitaille et al., 1998). With the advent of the 8-T/80-cm MRI scanner (Schenck et al., 1992; Kangarlu et al., 1999) the safety of the static magnetic field became a paramount issue for MRI researchers. While the primary concern in high magnetic field MRI has been excessive RF deposition in human subjects (Kangarlu et al., 2003, 2005), the static magnetic field could equally cause alarm for its potential for interaction with biological cells and molecules. One concern regarding human exposure to a high static magnetic field is the orientation of the molecules. Magnetohydrodynamics, which describes the interaction of moving charged particles with the magnetic field, has also raised concern with high magnetic field applications. Investigation of this effect with its possible consequences on human cognition has not received enough attention. As such, Chakeres et al. (2003) have recently conducted a series of studies on the effect of high static magnetic fields on human cognitive function. In spite of the availability and use of high magnetic field instrumentation for three decades, high magnetic field exposures of human subjects for extended periods of time have not been conducted. Such exposures did not pose a significant biological hazard at field strengths of up to 8 T as measured within the capability of their experimental design. In addition, human neuropsychological performance as a measure of any possible static magnetic field modification of cognition was not detected (Chakeres et al., 2003). To our knowledge, such a study has not been performed in the past at a field strength of 8 T for such an extended period of exposure time (Schenck et al., 1992; Kangarlu et al., 1999). In spite of the complex nature of an investigation of such effects, our present lack of observation of any detectable change in human cognition as related to high magnetic field exposure is an important reason for further research. In this regard, studies such as this could serve as a good starting point for *in vivo* characterization of static field effects in humans, in light of the rapidly expanding applications of high magnetic field MRI.

A time-varying magnetic field with sufficient intensity may be excitatory to peripheral nerve stimulation (PNS) (Ueno et al., 1986a; Barker, 1991; Sandrey et al., 2002; So et al., 2004). The physiologic mechanism is presumably due to interaction of neuronal structures with the induced electric field, rather than a direct physiologic effect of the magnetic field (Reilly, 1989).

Relatively early in the development of MRI, it was recognized that the pulsed gradient magnetic fields might induce PNS or cardiac stimulation (Reilly, 1991; Ueno et al., 1992). The pulsed gradients are of audio frequency, and they modulate the frequency and phase of the signal from the precessing magnetization as part of the MRI image reconstruction process. Reilly (1989) projected that a long-duration (>1 ms) pulse of induced electric field of 6.2 V would be sufficient to stimulate a 20-μm-diameter nerve fiber. The same amplitude of electric field was estimated to be the 1-percentile rank for stimulation of the human heart. Because of the much longer chronaxie for cardiac muscle, about 3 vs. 0.38 ms, the gradient field intensity, expressed in terms of the time derivative $\mathrm{d}B/\mathrm{d}t$ of the magnetic field, required to achieve cardiac stimulation is far greater than that for PNS. The large projected values of $\mathrm{d}B/\mathrm{d}t$ required for cardiac stimulation were confirmed in measurements in dogs by Mouchawar et al. (1992) and by Bourland et al. (1999), who reported that cardiac stimulation by pulsed gradient fields requires a $\mathrm{d}B/\mathrm{d}t$ amplitude in excess of 2000 T/s for a 530-μs period. These values compare with representative $\mathrm{d}B/\mathrm{d}t$ intensities of less than 100 T/s in an MRI system. For healthy patients, cardiac stimulation in MRI is avoided by a wide margin.

Mild PNS in MRI is not thought to be harmful, but painful stimulation should be avoided. To determine the population distribution for physiologic response to the time-varying gradient fields, the MRI safety group at Purdue University undertook a study with 84 human volunteers (Bourland et al., 1999; Nyenhuis et al., 2001). The volunteers were exposed to magnetic field patterns similar to those that would be experienced in a cylindrical bore MRI system. (Nyenhuis et al., 1997). The volunteers were asked to rate their responses to a gradient pulse sequence on a scale covering the range of 1 = onset of PNS, 5 = uncomfortable but acceptable for the duration of a scan, and 10 = intolerable. The duration of the $\mathrm{d}B/\mathrm{d}t$ pulses ranged from 50 to 1000 μs, in order to determine parameters for the strength duration given by:

$$\frac{\mathrm{d}B}{\mathrm{d}t} = b\left(1 + \frac{c}{d}\right)$$

where b is rheobase for a long-duration pulse, c is chronaxie, and d is the pulse duration. The measured responses were well fit by a chronaxie of 380 μs. For onset of PNS (score = 1), the median rheobase b was found to be 18.8 T/s for the y-coil and 28.8 T/s for the z-coil. Median dB/dt intensities for scores of 5 and 10 were approximately 50% and 100% greater than the score = 1 values, respectively. From the population distribution, the lowest 1-percentile value for PNS (score = 1) was about half the median value, and the lowest 1-percentile for uncomfortable (score = 5) was approximately equal to the median of the PNS threshold.

den Boer et al. (2002) found good agreement among the Purdue and other studies for the PNS thresholds by the switched gradients. Accordingly, the results of these studies were used for determination of the allowable gradient field intensities in MRI, which were set to be 80% of the mean PNS threshold (IEC, 2002).

The values of chronaxie and rheobase in the Purdue study were determined for rectangular waveforms. Models based on the physiologic response to rectangular pulses can be used to predict the threshold intensities for nonrectangular waveforms (Havel et al., 1997; den Boer et al., 2002).

So et al. (2004) recently reported results of calculations incorporating a realistic human model of the rheobase electric field intensity for PNS in the Purdue study. The rheobase electric field intensities in subcutaneous fat ranged from 3.3 to 4.4 V/m for the different body models and coil configurations. These values are in reasonable agreement with a rheobase electric field of 5.36 V/m for PNS with a solenoidal coil enclosing the arm (Havel et al., 1997).

MRI of electrical phenomena in living bodies is potentially useful for quantitative evaluations of the biological effects of electromagnetic fields and for direct detection of neuronal electrical activities in the brain. Magnetic fields in an object cause a shift in the resonant frequency (Manassen et al., 1988; Sekino et al., 2004b) and a change in the phase of magnetic resonance signals (Joy et al., 1989). Spatial distributions of an externally applied magnetic field and electrical current can be estimated from these changes in magnetic resonance signals. These methods have use in certain medical applications, such as the imaging of current distributions in electrical defibrillation (Yoon et al., 2003).

The fMRI developed by Ogawa et al. (1992) utilizes a technology that reflects various magnetic features of hemoglobin in blood on magnetic resonance signal patterns. Tomograms of brain function can be obtained from information on localized blood flow in the brain. fMRI utilizes a blood oxygenation level dependent (BOLD) effect of localized blood flow on brain activation for indirect imaging of brain activities. However, no information on electrical conditions *in vivo* can be obtained with current MRI and fMRI systems.

Detection of electrical currents associated with neuronal or muscular electrical activities requires extremely high measurement sensitivity. The sensitivity for detecting weak magnetic fields in the human brain was estimated using numerical simulations (Hatada et al., 2005). The theoretical limit of sensitivity was approximately 10^{-8} T. The effect of neuronal electrical activities on magnetic resonance signals was investigated in several experimental studies (Kamei et al., 1999; Xiong et al., 2003). These studies potentially lead to a new method for visualizing brain function with a spatial resolution of millimeters and a temporal resolution of milliseconds.

Impedance-weighted magnetic resonance images were obtained during applications of external oscillating magnetic fields, which induce impedance-dependent eddy currents in a sample (Ueno and Iriguchi, 1998). In another study, spatial distribution of electrical impedance was obtained from the electrical current distributions by using an iterative algorithm (Khang et al., 2002). The apparent diffusion coefficient reflects electrical conductivity of a tissue, which enables an estimation of anisotropic conductivity of that tissue (Tuch et al., 2001; Sekino et al., 2004a). This method was applied to imaging of electrical conductivity in the human brain. Several regions in the white matter, such as the corpus callosum and the internal capsule, exhibited high anisotropy in conductivity. The magnitude and phase of magnetic resonance signals are affected by permittivity (Sekino et al., 2005). A distinctive signal inhomogeneity arises in images of objects whose dimension is comparable to the wavelength of the electromagnetic fields at the resonant frequency. This phenomenon, dielectric resonance, particularly appears in scanners with high static fields.

Once high-quality current distribution MRI of the detailed distribution of electric source incident to brain neural activities becomes available, comparison of results of MRI and fMRI will show the relationship between brain neural activities associated with BOLD effects and neural current distributions, which may lead to various new observations of dynamics in brain function localizations.

Impedance MRI may not be applied widely in brain function research; however, high-quality impedance MRI for impedance and admittance *in vivo* may lead to development of a new research field of impedance physiology. It is obvious that the information of impedance distributions is important for studying magnetic stimulation and MEG inverse problems.

8.5.5 Magnetic Orientation for Tissue Engineering

In the last decade, it has become possible to create static magnetic fields of 10 T and higher. With this development, studies regarding magnetic effects on macromolecules have increased. These studies include investigations into the magnetic effects on fibrin, collagen, erythocytes, and platelets (Higashi et al., 1993a,b; Iwasaka and Ueno, 1994; Iwasaka et al., 1998; Iino and Okuda, 2001). Recent research on effects of strong static magnetic fields includes their impact on morphogenesis, cell adhesion, and apoptosis (Tofani et al., 2001).

When technology for generation of stronger magnetic fields becomes available in the future, magnetic orientation research will be subdivided into several areas. Effects of magnetic orientation on cell functions such as morphogenesis, adhesion, motility, proliferation, differentiation, and apoptosis may become one of the important areas of research.

Macromolecules such as fibrin and collagen are oriented by static magnetic fields of several tesla. Fibrin polymers are diamagnetic materials that are oriented in a magnetic field. Collagen fibers orient perpendicular to the magnetic field orientation (Torbet and Ronziere, 1984). Polymerization and dissolution of fibrin in homogenous magnetic fields of up to 14 T have been investigated (Iwasaka et al., 1998). It was shown that the magnetic orientation of fibrin fibers accelerated both the polymerization and the dissolution of fibrin fibers.

Magnetic orientation of cells is associated with magnetic anisotropy of proteins and lipids. Erythrocytes orient the disk surface parallel to magnetic fields because of magnetic anisotropy of the biomembrane lipid bilayer. However, halophillic bacteria orient their membrane plane vertical to magnetic fields even though the purple membrane has a similar membrane structure as the erythrocyte membrane (Neugebauer et al., 1977). The purple membrane contains a membrane-bound protein (bacteriorhodopsin) that contributes 75% to the membrane weight. Since the magnetic anisotropy of bacteriorhodopsin is larger than that of the lipid bilayer, halophillic bacteria posses a different magnetic orientation from erythrocytes. Therefore, magnetic orientation is determined by the quantity and the alignment of cell components that possess magnetic anisotropy.

Higashi et al. (1996) found that an orientation of glutaraldehyde-fixed erythrocytes in strong static magnetic fields up to 8 T was perpendicular to the field. The effect was attributed to the paramagnetism of membrane-bound hemoglobin. The rates of sedimentation and aggregation of human erythrocytes in a homogenous magnetic field of 6.3 T have been studied (Iino, 1997; Iino and Okuda, 2001). It was reported that the cell aggregation accelerated the sedimentation rate. Results have suggested that the enhancement was especially significant in anisotropic erythocytes, and the increase in an intermembrane adhesive area might be due to the magnetic orientation of anisotropic erythrocytes.

DNA, which occupies most of the head portion in bovine sperm, may be involved in magnetic anisotropy and orientation determination of sperm. Cricket sperm with an acicular head part show the same magnetic orientation (vertical) as bovine sperm in magnetic fields of 0.09 T (Suzuki et al., 1995); this is because DNA, which is folded lengthwise, possesses large diamagnetic anisotropy. If DNA in the head part of bovine sperm is orderly aligned as in cricket sperm, the magnetic anisotropy may contribute to the magnetic orientation of sperm.

A significant feature of magnetic orientation of bovine sperm is the direction of orientation. Platelets orient parallel to magnetic fields because of the microtubules inside, which have a magnetic orientation parallel to the magnetic fields. Thus, bovine sperm without motility are assumed to orient parallel to magnetic fields as platelets do, since the tail (flagellum) consists of microtubules. On the contrary, the whole body of bovine sperm shows magnetic orientation vertical to magnetic fields, and the flat surface of the head also orients vertically to magnetic fields. A sperm with the tail removed shows the same orientation. Since it is impossible to obtain a tail without damaging flagellum, the magnetic orientation of a tail alone cannot be observed. In two separate experiments, Emura et al. (2001, 2003) studied the orientation of bull sperm cells and *Paramecium* cilia in static magnetic fields and measured their anisotropic diamagnetic susceptibility ($\Delta\chi$). Bovine sperm consists of a very flat head part (5 μm) and a long tail part (flagellum, 50 μm), which consists of microtubules. Compared to sperm of other species, the head, which contains DNA, is notably larger. The sperm showed an orientation perpendicular to the field of 1 T or lower. The diamagnetic cell components, such as cell membrane, DNA in the head, and microtubule in the tail, were thought to contribute to this orientation. It was observed that *Paramedium* cilia became oriented in parallel to the magnetic field at the strength of 8 T. The author suggested that $\Delta\chi$ for each was the quantitative index of the effect.

Iwasaka et al. (2003a) reported the effects of 14-T fields on assemblies of A7r5 smooth muscle cells. It was shown that the field affected the morphology of smooth muscle cell assemblies and the shapes of the cell colonies extended along the direction of the magnetic flux. They speculated that the mechanism was a diamagnetic torque force acting on cytoskeleton fibers, which are dynamically polymerizing and depolymerizing during cell division and cell migration. They also investigated the effects of the static magnetic field on the convection flow in a cell culture medium and on cell adhesion patterns (Iwasaka et al., 2003b). The mouse osteoblast cell line MC3T3-E1 and HeLa cell line were used in this study. The magnetic field of 6 T with a gradient of 60 T/m affected the convection of floating cell aggregations in a cell culture flask and reversibly changed the direction of convectional flow. After the exposure of MC3T3-E1 cells to the magnetic field of 8 T for 1 d, the thermal convectional flow in the medium was found to promote the cell orientation.

Iwasaka and Ueno (2003) examined the displacement of intracellular macromolecules under a static magnetic field of 14 T using linearly polarized light. The changes in polarized light intensity through the lamellar cell assembly under magnetic fields corresponded to the behavioral changes in cell components. They speculated that intracellular macromolecules rotated and showed a displacement due to diamagnetic torque forces during the exposure to the 14-T magnetic field for 2–3 h.

Matrix proteins provide a permissive environment for the orientation of cells, as demonstrated, for example, with the testing of smooth muscle cells and endothelial cells in collagen fibers under strong magnetic fields (Stefano and Tranquillo, 1993; Tranquillo et al., 1996).

Eguchi et al. (2003) observed the effect of a static magnetic field on orientation of Schwann cells. After a 60-h exposure, cultured Schwann cells from dissected sciatic nerves of neonatal rats oriented parallel to the field of 8 T, whereas Schwann cells suspended in a medium with collagen oriented perpendicular to the field after a 2-h exposure. It was suggested that magnetic field-oriented collagen fibers were the key factor in the orientation of Schwann cells.

Kotani et al. (2000) studied the effect of an 8-T magnetic field generated by a superconducting magnet on the orientation of osteoblasts alone and a mixture of osteoblasts and collagen. It was found that osteoblast cells oriented parallel to the magnetic field, but a mixture oriented perpendicular to the field.

Hirose et al. (2003a) investigated the preferred orientation of human glioblastoma cells A172 after exposure to a 10-T static magnetic field, in the presence or absence of collagen. It was found that A172 cells embedded in collagen gel oriented perpendicular to the direction of the static magnetic field.

By placing dorsal root ganglia (DRG) explants onto one end of magnetically aligned collagen gel formed into 4-mm-diameter rods, Dubey et al. (1999) developed an *in vitro* assay to study neurite elongation. The depth of neurite elongation from chick embryo DRG neurons into these aligned rods was found to be substantially greater than that under the control condition. The depth increased as the magnetic field strength increased, as did the collagen gel rod birefringence; collagen fibril aligned along the rod axis. These results may translate into an improved method of entubulation repair of transected peripheral nerves by directing and stimulating axonal growth through a tube filled with magnetically aligned collagen gel. The same research group later reported the improvement of peripheral nerve regeneration in mice after the treatment of magnetically aligned collagen gel filling of a collagen nerve guide (Ceballos et al., 1999). The hypothesis of this study was that contact guidance of regenerating axons or invading nonneuronal cells to the longitudinally aligned collagen fibrils would improve nerve regeneration. It was reported that mice exhibited regeneration with magnetically aligned collagen gel, including the appearance of nerve fascicle formation.

Application of magnetic orientation in the production of biologically functional materials and artificial organs has been started. By attaching aligned vascular smooth muscle cells and endothelial cells to artificial vascular walls in an orderly fashion, rheologically rational biological functions can be obtained. As techniques in bionics and biomaterials improve, application of magnetic orientation should expand.

8.5.6 Treatments of Pain, Cancer, and Other Diseases

Static magnetic fields or ELF-modulated static magnetic fields potentially have therapeutic effects on several diseases (see also Chapter 11 on medical applications of pulsed fields by Pilla). A static magnetic field in the 10-mT range blocks sensory neuron action potentials, which suggests that the magnetic field alleviates pain (Cavopol et al., 1995; McLean et al., 1995). To characterize the inhibitory effect of a static magnetic field, action potentials were elicited by intracellular application of 1-ms pulses of depolarizing current to the somata of mouse DRG neurons. During the control period, less than 5% of stimuli failed to elicit action potentials. During exposure to an approximate 11-mT static magnetic field produced by an array of four permanent center-charged magnets of alternating polarity, 66% of stimuli failed to elicit action potentials.

The efficacy of a nonpharmacologic, noninvasive static magnetic device was assessed for knee pain in patients with rheumatoid arthritis (Segal et al., 2001). Magnetic devices with four steep field gradients or one steep field gradient were taped to the knee of each subject for 1 week. Both devices demonstrated statistically significant pain reduction in comparison to baseline. Comparison between the two groups demonstrated a statistically insignificant difference.

Prato et al. (2005) reported effect of a magnetically shielded environment on opioid-induced analgesia. Mice were placed in a Mu-metal-lined box or an opaque Plexiglas box (sham condition) for 1 h/d for 10 consecutive days. Nociception was measured as the latency time to a foot lift/lick in response to an aversive thermal stimulus before and immediately after exposure. It was shown that mice can detect and will respond to the repeated absence of the ambient magnetic field, with the maximum analgesic response occurring over days 4–6 of exposure and returning to baseline thereafter. The effect was robust, independent of pre-exposure and intermittent testing, and seems to be opioid related, since the results obtained on day 5 were similar to those from a 5-mg/kg dose of morphine and were abolished with the opioid antagonist, naloxone.

Exposure to pulsed magnetic fields has been shown to have a therapeutic benefit in both animals (e.g., mice and snails) and humans. Shupak et al. (2004b) investigated the potential analgesic benefit of magnetic field exposure on sensory and pain thresholds following experimentally induced warm and hot sensations. Subjects were assigned to 30 min of magnetic field or sham exposure, between two sets of tests of sensory and pain thresholds and latencies at 1°C above and 2°C above pain thresholds. Results indicated that magnetic field exposure does not affect sensory thresholds. Pain thresholds were significantly increased following magnetic field exposure but not following sham exposure. A significant condition by gender interaction existed for postexposure pain thresholds. Taken together, these results indicate that magnetic field exposure does not affect basic human perception, but can increase pain thresholds in a manner indicative of an analgesic response.

Shupak et al. (2004a) showed an induction of analgesia in mice equivalent to a moderate dose of morphine (5 mg/kg) and the effect of both pulsed magnetic field (complex neuroelectromagnetic pulse, Cnp) exposure and morphine injection on some open-field activity. Cnp exposure was found to prolong the response latency to a nociceptive thermal stimulus (hot plate). Cnp plus morphine offset the increased movement activity found with morphine alone. These results suggest that pulsed magnetic fields can induce analgesic behavior in mice without the side effects often associated with opiates like morphine.

The effects of static and sinusoidal (AC) magnetic fields on myosin light chain phosphorylation were studied (Markov et al., 1993). In a cell-free preparation, exposure to DC (0–200 μT, vertically or horizontally controlled) or AC (16 Hz, 20.9 μT) magnetic fields significantly influenced myosin phosphorylation. Variations of the DC magnetic field (in the absence of AC components) were not only sufficient to alter the rate of phosphorylation but also gave the maximum effect.

The possibilities that magnetic fields cause antitumor activities *in vitro* (Tofani et al., 2001), *in vivo* (Tofani et al., 2002), and in human subjects (Ronchetto et al., 2004) have been investigated. *In vitro* experiments were carried out to study the role of magnetic field characteristics (intensity, frequency, and modulation) on two transformed cell lines (WiDr human colon adenocarcinoma and MCF-7 human breast adenocarcinoma) and one non-transformed cell line (MRC-5 embryonal lung fibroblast). Increase in cell death morphologically consistent with apoptosis was reported exclusively in the two transformed cell lines. Cell-death induction was observed with magnetic fields of more than 1 mT. Two different *in vivo* experiments were carried out on nude mice bearing a subcutaneous human colon adenocarcinoma (WiDr). In the first experiment, a significant increase in survival time (31%) was obtained in mice exposed daily to 70 min of modulated magnetic fields (static with a superimposition of 50 Hz) having a time average total intensity of 5.5 mT. In the second independent experiment, when mice bearing tumors were exposed to the same treatment for four consecutive weeks, significant inhibition of tumor growth (40%) was reported, together with a decrement in tumor cell mitotic index and proliferative activity. Human patients with heavily pretreated advanced cancer were enrolled in a pilot study, in which they were exposed to static magnetic fields that were amplitude modulated by ELF. Toxicity was assessed according to WHO criteria. ECG, chest x-ray, physical examination, blood cell count, and complete blood chemistry were performed before and at the end of the treatment. The results indicated that magnetic fields can be safely administrated according to the magnetic field exposure schedules.

Recently, several studies tested the application of pulsed magnetic stimulation as a form of cancer therapy. In one case, use of magnetizable beads and pulsed magnetic stimulation enabled targeted-cell destruction *in vitro* (Ogiue-Ikeda et al., 2003b). The cells were combined with the beads by an antigen–antibody reaction (cell–bead–antibody complex), aggregated by a magnet, and stimulated by a magnetic stimulator. The viability of the aggregated and stimulated cell–bead–antibody complexes was significantly decreased, and the cells were destroyed by the penetration of the beads into the cells or by rupturing of the cells by the beads. In another study, exposure to a pulsed magnetic stimulation caused a decrease of tumor weight in mice B16-BL6 melanoma models and induced the increase of cytokine (TNF-α and IL-2) production (Yamaguchi et al., 2006). These studies show the potential therapeutic possibilities of pulsed magnetic stimulation in cancer treatment.

Basic studies for magnetic stimulation treatment of depression, which has the potential to replace ECT, and also, magnetic treatment for a wide range of clinical problems, such as Parkinson's disease and various kinds of pain, are in progress. It is important to recognize the safety of magnetic stimulation and the limitations of its usefulness.

8.6 Conclusion

Over the last two decades, various studies have been carried out to examine the effects of static magnetic fields, including MRI fields, on biological systems. This chapter consisted of two parts. The first part focused on recent experiments covering behavior, cardiovascular system responses, reproduction and development, genotoxicity, molecular and cellular systems, cell-free systems, free radical and enzyme activity, etc. The second part concentrated on the recent development of medical and therapeutic applications of static magnetic fields.

There are many studies that have been mentioned in this chapter. With exposure to about 1 T and above, there are no adverse effects on reproduction and development, genotoxicity, and molecular and cellular systems, and no consistent evidence on behavioral effects. However, several studies suggest that static magnetic fields in millitesla ranges may affect microcirculation and blood pressure, and furthermore, higher-strength static magnetic fields at levels up to 10 T may reduce skin blood flow and lead to change in skin temperature. These findings need to be confirmed in further studies.

Although there are so many experiments to test the effects of static magnetic fields on the biology of living systems, using *in vivo* and *in vitro* techniques, the International Agency for Research on Cancer (IARC) has stated that static magnetic fields are *not classifiable as to their carcinogenicity to humans* by inconclusive carcinogenic evidence (IARC, 2002).

There are many experimental findings that suggest that animals use the static magnetic field, that is, the geomagnetic field for orientation, navigation, and migration. In order to establish the existence of a magnetoreception system in animals, Phillips argued that there are several key points that require further investigation: (1) establishing the lower limits of sensitivity to static magnetic, ELF, and RF fields in biological systems; (2) localizing specialized receptors responsible for sensing the geomagnetic field; (3) characterizing the underlying molecular and biophysical mechanisms; (4) identifying the regions of the brain involved in processing magnetic stimuli; and (5) understanding how the animal's perception of the magnetic field is physiologically processed for determining compass direction and spatial positioning (Phillips, 2005).

With the increasing exposure of humans to environmentally higher static magnetic fields generated from magnetic field equipment of higher capacity, it is necessary to investigate the possibilities of high static field effects on human biological and physiological processes.

There are an abundance of review papers and books published in recent years describing the possible physical and biological interactions of electromagnetic fields (Polk and Postow, 1986, 1997; Ueno, 1996; Andra and Nowak, 1998; Jin, 1999; Takebe et al., 1999; Lin, 2000; Shellock, 2001; Binhi, 2002; McLean et al., 2003a; Stavroulakis, 2003; Rosch and Markov, 2004). In addition, there have been many short reviews on the biological effects of static magnetic fields (Holden, 2005; Miyakoshi, 2005), since the physical interactions of static magnetic fields with living tissues were described (Schenck, 2005). In a report on the biological effects of exposure to MRI, an overview of the safety concerns regarding exposure to static magnetic fields, RF fields, and time-varying magnetic field gradients has been discussed (Formica and Silverstri, 2004). Application of novel high-throughput screening techniques for transcriptomics, proteomics, and metabolomics to determine *in vitro* effects of static magnetic fields have been suggested (Leszczynski, 2005). This report emphasized the research beyond screening that is required for the assessment of any possible health consequences. Possible physical mechanisms underlying the biological effects and interactions of zero-frequency (DC) and oscillating (AC) magnetic fields with biological matter have been reviewed (Binhi, 2001; Volpe, 2003). Effects of static and ELF electric and magnetic fields on human health have also been discussed (Repacholi and Greenebaum, 1999; McKinlay and Repacholi, 2005). Zhadin (2001) has introduced the Russian literature on the biological effects of DC and LF AC magnetic fields. These articles offer multidisciplinary information and knowledge for the understanding of magnetic field effects within living systems.

Acknowledgments

The authors wish to thank Drs. Ben Greenebaum and Frank Barnes for their suggestions and comments for organizing this chapter. The authors also wish to thank Drs. Minoru Fujiki, Yngve Hamnerius, Masateru Ikehata, Masakazu Iwasaka, Saeko Kanagawa, Alayar Kangarlu, Yohsuke Kinouchi, Shinya Kuriki, Tatsuki Matsumoto, Junji Miyakoshi, John Nyenhuis, Chiyoji Ohkubo, Hideyuki Okano, Frank Prato, Masaki Sekino, Masao Taki, Yoshifumi Tanimoto, Sotoshi Yamada, and Sachiko Yamaguchi for their valuable discussions.

References

Adair, R.K., Comments: influence of stationary magnetic fields on water relations in lettuce seeds, *Bioelectromagnetics*, 23, 550, 2002.

Ahonen, A.I., Hämäläinen, M.S., Knnutila, J.E.T., Kajola, M.J., Laine, P.P., Lounasmaa, O.V., Parkkonen, L.T., Simola, J.T., and Tesch, C.D., 122-Channel SQUID instrument for investigating the magnetic signals from the human brain, *Phys Scripta*, T49, 198, 1993.

Aldinucci, C., Garcia, J.B., Palmi, M., Sgaragli, G., Benocci, A., Meini, A., Pessina, F., Rossi, C., Bonechi, C., and Pessina, G.P., The effect of exposure to high flux density static and pulsed magnetic fields on lymphocyte function, *Bioelectromagnetics*, 24, 373, 2003a.

Aldinucci, C., Garcia, J.B., Palmi, M., Sgaragli, G., Benocci, A., Meini, A., Pessina, F., Rossi, C., Bonechi, C., and Pessina, G.P., The effect of strong static magnetic field on lymphocytes, *Bioelectromagnetics*, 24, 109, 2003b.

Amyan, A. and Ayrapetyan, S., On the modulation effect of pulsing and static magnetic field and mechanical vibrations on barley seed hydration, *Physiol Chem Phys Med NMR*, 36, 69, 2004.

Andra, W. and Nowak, H. (Eds.), *Magnetism in Medicine*, Wiley, New York, 1998.

Azanza, M.J. and del Moral, A., Isolated neuron amplitude spike decrease under static magnetic fields, *J Magn Magn Mater*, 157/158, 593, 1996.

Barker, A.T., An introduction to the basic principles of magnetic nerve stimulation, *J Clin Neurophysiol*, 8, 26, 1991.

Barker, A.T., Jalionus, R., and Freeston, I.L., Noninvasive magnetic stimulation of the human motor cortex, *Lancet*, 1, 1106, 1985.

Beason, R.C. and Semm, P., Does the avain ophthalmic nerve carry magnetic navigational information? *J Exp Biol*, 199, 1241, 1996.

Beaugnon, E. and Tournier, R., Levitation of organic materials, *Nature*, 349, 470, 1991.

Belova, N.A. and Lendnev, V.V., Activation and inhibition of gravitropic response in segments of flax stems exposed to static magnetic field with flux density ranging from 0 to 350 microtesla, *Biophysics*, 46, 117, 2001.

Belyaev, I.Y., Alipov, Y.D., and Harms-Ringdahl, M., Effects of zero magnetic field on the conformation of chromatin in human cells, *Biochim Biophys Acta*, 1336, 465, 1997.

Berk, S.G., Srikanth, S., Mahajan, S.M., and Ventrice, C.A., Static uniform magnetic fields and amoebae, *Bioelectromagnetics*, 18, 81, 1997.

Binhi, V.N., Theoretical concepts in magnetobiology, *Electr. Magnetobiol*, 20, 43, 2001.

Binhi, V.N., *Magnetobiology—Underlying Physical Problems*, Academic Press, New York, 2002.

Binhi, V.N., Alipov, Y.D., and Belyaev, I.Y., Effect of static field on *E. coli* cells and individual rotations of ion–protein complexes, *Bioelectromagnetics*, 22, 79, 2001.

Blackman, C.F., Blanchard, J.P., Benane, S.G., and House, D.E., Effect of AC and DC magnetic field orientation on nerve cells, *Biochem Biophys Res Commun*, 220, 807, 1996.

Blanchard, J.P. and Blackman, C.F., Clarification and application of an ion parametric resonance model for magnetic field interactions with biological systems, *Bioelectromagnetics*, 15, 217, 1994.

Blumenthal, N.C., Ricci, J., Breger, L., Zychlinsky, A., Solomon, H., Chen, G.G., Kuznetsov, D., and Dorfman, R., Effects of low-intensity AC and/or DC electromagnetic fields on cell attachment and induction of apoptosis, *Bioelectromagnetics*, 18, 264, 1997.

Bochert, R. and Zettler, M.L., Long-term exposure of several marine benthic animals to static magnetic fields, *Bioelectromagnetics*, 25, 498, 2004.

Bourland, J.D., Nyenhuis, J.A., and Schaefer, D.J., Physiologic effects of intense MRI gradient fields, *Neuroimaging Clin North Am* 9, 363, 1999.

Brocklehurst, B., Magnetic fields and radical reactions: recent development and their role in nature, *Chem Soc Rev*, 31, 301, 2002.

Brocklehurst, B. and McLauchlan, K.A., Free radical mechanism for the effects of environmental electromagnetic fields on biological systems, *Int J Radiat Biol*, 69(1), 3, 1996.

Buemi, M., Marino, D., Di Pasquale, G., Floccari, F., Senatore, M., Aloisi, C., Grasso, F., Mondio, G., Perillo, P., Frisina, N., and Corica, F., Cell proliferation/cell death balance in renal cell cultures after exposure to a static magnetic field, *Nephron*, 87, 269, 2001.

Carbonell, M.V., Martinez, E., and Amaya, J.M., Stimulation of germination in rice (*Oryza sativa* L.) by a static magnetic field, *Electr Magnetobiol*, 19, 121, 2000.

Carnes, K.I. and Magin, R.L., Effects of in utero exposure to 4.7 T MR imaging conditions on fetal growth and testicular development in the mouse, *Magn Reson Imaging*, 14, 263, 1996.

Cavopol, A.V., Wamil, A.W., Holcomb, R.R., and McLean, M.J., Measurement and analysis of static magnetic fields that block action potentials in cultured neurons, *Bioelectromagnetics*, 16, 197, 1995.

Ceballos, D., Navarro, X., Dubey, N., Wendelschafer-Crabb, G., Kennedy, W.R., and Tranquillo, R.T., Magnetically aligned collagen gel filling a collagen nerve guide improves peripheral nerve regeneration, *Exp Neurol*, 158, 290, 1999.

Chakeres, D.W. and de Vocht, F., Static magnetic field effects on human subjects related to magnetic resonance imaging systems, *Prog Biophys Mol Biol*, 87, 255, 2005.

Chakeres, D.W., Kangarlu, A., Boudoulas, H., and Young, D.C., Effect of static magnetic field exposure of up to 8 tesla on sequential human vital sign measurements, *J Magn Reson Imaging*, 18(3), 346, 2003.

Chignell, C.F. and Sik, R.H., The effect of static magnetic fields on the photohemolysis of human erythrocytes by ketoprofen, *Photochem Photobiol*, 67, 591, 1998.

Chionna, A., Tenuzzo, B., Panzarini, E., Dwikat, M.B., Abbro, L., and Dini, L., Time dependent modifications of Hep G2 cells during exposure to static magnetic fields, *Bioelectromagnetics*, 26, 275, 2005.

Ciortea, L.I., Morariu, V.V., Todoran, A., and Popescu, S., Life in zero magnetic field. 3. Effect on zinc and copper in human blood serum during *in vitro* aging, *Electr Magnetobiol*, 20, 127, 2001.

Cohen, D., Edelsack, E.A., and Zimmerman, J.E., Magnetocardiograms taken inside a shielded room with a superconducting point contact magnetometer, *Appl Phys Lett*, 16, 278, 1972.

Cohly, H.H.P., Abraham, G.E. III, Ndebele, K., Jenkins, J.J., Thompson, J., and Angel, M.F., Effects of static electromagnetic fields on characteristics of MG-63 osteoblasts grown in culture, *Biomed Sci Instrum*, 39, 454, 2003.

Coulton, L.A., Barker, A.T., Van Lierop, J.E., and Walsh, M.P., The effect of static magnetic fields on the rate of calcium/calmodulin-dependent phosphorylation of myosin light chain, *Bioelectromagnetics*, 21, 189, 2000.

Creanga, D.E., Morariu, V.V., and Isac, R.M., Life in zero magnetic field. IV. Investigation of developmental effects on fruit fly vision, *Electromagn Biol Med*, 21, 31, 2002.

Creim, J.A., Lovely, R.H., Miller, D.L., and Anderson, L.E., Rats can discriminate illuminance, but not magnetic fields, as a stimulus for learning a two-choice discrimination, *Bioelectromagnetics*, 23, 545, 2002.

Crozier, S. and Liu, F., Numerical evaluation of the fields induced by body motion in or near high-field MRI scanners, *Prog Biophys Mol Biol*, 87, 267, 2005.

Danielyan, A.A. and Ayrapetyan, S.N., Changes of hydration of rats' tissues after *in vivo* exposure to 0.2 tesla steady magnetic field, *Bioelectromagnetics*, 20, 123, 1999.

Danielyan, A.A., Mirakyan, M.M., Grigoryan, G.Y., and Ayrapetyan, S.N., The static magnetic field effects on quabain H3 binding by cancer tissue, *Physiol Chem Phys Med NMR*, 31, 139, 1999.

Del Seppia, C., Luschi, P., Ghione, S., Crosio, E., Choleris, E., and Papi, E., Exposure to hypogeomagnetic field or to oscillating magnetic fields similarly reduce stress induced analgesia in C57 male mice, *Life Sci*, 66, 1299, 2000.

Del Seppia, C., Mezzasalma, L., Choleris, E., Luschi, P., and Ghione, S., Effects of magnetic field exposure on open field behaviour and nociceptive responses in mice, *Behav Brain Res*, 144, 1, 2003.

den Boer, J.A., Bourland, J.D., Nyenhuis, J.A., Ham, C.L.G., Engels, J.M.L., Hebrank, F.X., Frese, G., and Schaefer, D.J., Comparison of the threshold for peripheral nerve stimulation during gradient switching in whole body MR systems, *J Magn Reson Imaging*, 15, 520, 2002.

Denegre, J.M., Valles, J.M., Lin, K., Jordan, W.B., and Mowry, K.L., Cleavage planes in frog eggs are altered by strong magnetic fields, *Proc Natl Acad Sci USA*, 95, 14729, 1998.

Deutschlander, M.E., Borland, S.C., and Phillips, J.B., Extraocular magnetic compass in newts, *Nature*, 400, 324, 1999a.

Deutschlander, M.E., Phillips, J.B., and Borland, S.C., The case for light-dependent magnetic orientation in animals, *J Exp Biol*, 202, 891, 1999b.

Deutschlander, M.E., Freake, M.J., Borland, S.C., Phillips, J.B., Madden, R.C., Anderson, L.E., and Wilson, B.W., Learned magnetic compass orientation by the Siberian hamster, *Phodopus sungorus*, *Anim Behav*, 65, 779, 2003.

Diebel, C.E., Proksch, R., Green, C.R., Neilson, P., and Walker, M.M., Magnetite defines a vertebrate magnetoreceptor, *Nature*, 406, 299, 2000.

Dubey, N., Letourneau, P.C., and Tranquillo, R.T., Guided neurite elongation and Schwann cell invasion into magnetically aligned collagen in simulated peripheral nerve regeneration, *Exp Neurol*, 158, 338, 1999.

Edmond, D.T., A sensitive optically detected magnetic compass for animals, *Proc R Soc Lond B*, 263, 295, 1996.

Eguchi, Y., Ogiue-Ikeda, M., and Ueno, S., Control of orientation of rat Schwann cells using an 8-T static magnetic field, *Neurosci Lett*, 351, 130, 2003.

Eichwald, C. and Walleczek, J., Model for magnetic field effects on radical pair recombination in enzyme kinetics, *Biophys J*, 71, 623, 1996.

Emura, R., Ashida, N., Higashi, T., and Takeuchi, T., Orientation of bull sperms in static magnetic fields, *Bioelectromagnetics*, 22, 60, 2001.

Emura, R., Takeuchi, T., Nakaoka, Y., and Higashi, T., Analysis of anisotropic diamagnetic susceptibility of a bull sperm, *Bioelectromagnetics*, 24, 347, 2003.

Engstrom, S., Markov, M.S., McLean, M.J., Holcomb, R.R., and Markov, J.M., Effects of non-uniform static magnetic fields on the rate of myosin phosphorylation, *Bioelectromagnetics* 23, 475, 2002.

Epstein, C.M., Sekino, M., Yamaguchi, K., Kamiya, S., and Ueno, S., Asymmetries of prefrontal cortex in human episodic memory: effects of transcranial magnetic stimulation on learning abstract patterns, *Neurosci Lett*, 320, 5, 2002.

Eremenko, T., Esposito, C., Pasquarelli, A., Pasquali, E., and Volpe, P., Cell-cycle kinetics of friend erythroleukemia cells in a magnetically shielded room and in a low-frequency/low-intensity magnetic field, *Bioelectromagnetics*, 18, 58, 1996.

Erne, S.N., Hoke, M., Lutkenhoner, B., Pantev, C., and Scheer, H.J., Brainstem auditory evoked magnetic fields, in *Biomagnetism '87* (Atsumi, K., Kotani, M., Ueno, S., Katila, T., and Williamson, S.J., Eds.), Tokyo Denki University Press, Tokyo, 158–161, 1988.

Espinar, A., Piera, V., Carmona, A., and Guerrero, J.M., Histophysiological changes during development of the cerebellum in the chick embryo exposed a static magnetic field, *Bioelectromagnetics*, 18, 36, 1997.

Eveson, R.W., Timmel, C.R., Brocklehurst, B., Hore, P.J., and McLauchlan, K.A., The effects of weak magnetic fields on radical recombination reactions in micelles, *Int J Radiat Biol*, 76, 1509, 2000.

Fanelli, C., Coppola, S., Barone, R., Colussi, C., Gualandi, G., Volpe, P., and Ghibelli, L., Magnetic fields increase cell survival by inhibiting apoptosis via modulation of Ca2+ influx, *FASEB J*, 13, 95, 1999.

Flipo, D., Fournier, M., Benquet, C., Roux, P., Le Boulaire, C., Pinsky, C., Labella, F.S., and Krzystyniak, K., Increased apoptosis, changes in intracellular Ca^{2+}, and functional alterations in lymphocytes and macrophages after *in vitro* exposure to static magnetic field, *J Toxicol Environ Health A*, 54, 63, 1998.

Florez, M., Carbonell, M.V., and Martinez, E., Early sprouting and first stages of growth of rice seeds exposed to a magnetic field, *Electromagn Biol Med*, 23, 157, 2004.

Formica, D. and Silvestri, S., Biological effects of exposure to magnetic resonance imaging: an overview, *Biomed Eng Online*, 3, 11, 2004.

Formicki, K. and Perkowski, T., The effect of a magnetic field on the gas exchange in rainbow trout *Oncorhynchus mykiss* embryos (Salmonidae), *Ital J Zool*, 65 Suppl, 475, 1998.

Formicki, K., and Winnicki, A., Effects of constant magnetic field on cardiac muscle activity in fish, *Publ Espec Inst Esp Oceanogr*, 21, 287, 1996.

Formicki, K. and Winnicki, A., Reactions of fish embryos and larvae to constant magnetic fields, *Ital J Zool*, 65 Suppl, 479, 1998.

Formicki, K., Bonislawska, M., and Jasinski, M., Spatial orientation of trout (*Salmo trutta* L.) and rainbow trout (*Oncorhynchus mykiss* Walb.) embryos in natural and artificial magnetic fields, *Acta Ichtyol Piscat*, 27, 29, 1997.

Fujiki, M. and Steward, O., High frequency transcranial magnetic stimulation mimics the effects of ECS in upregulating astroglial gene expression in the murine CNS, *Brain Res Mol Brain Res*, 44, 301, 1997.

Fujiki, M., Kobayashi, H., Abe, T., and Kamida, T., Repetitive transcranial magnetic stimulation for protection against delayed neuronal death induced by transient ischemia, *J Neurosurg*, 99, 1063, 2003.

Fuller, M. and Dobson, J., On the significance of the time constants of magnetic field sensitivity in animals, *Bioelectromagnetics*, 26, 234, 2005.

Gmitrov, J. and Ohkubo, C., Geomagnetic field decreases cardiovascular variability, *Electr Magnetobiol*, 18, 291, 1999a.

Gmitrov, J. and Ohkubo, C., Static magnetic field and calcium channel blocking agent combined effect on baroreflex sensitivity in rabbits, *Electr Magnetobiol*, 18, 43, 1999b.

Gmitrov, J. and Ohkubo, C., Artificial static and geomagnetic field interrelated impact on cardiovascular regulation, *Bioelectromagnetics*, 23, 329, 2002.

Gmitrov, J., Ohkubo, C., and Okano, H., Effect of 0.25 static magnetic field on microcirculation in rabbits, *Bioelectromagnetics*, 23, 224, 2002.

Gray, J.R., Frith, C.H., and Parker, J.D., *In vivo* enhancement of chemotherapy with static electric or magnetic fields, *Bioelectromagnetics*, 21, 575, 2000.

Grissom, C.B., Magnetic field effects in biology: a survey of possible mechanisms with emphasis on radical-pair recombination, *Chem Rev*, 95, 3, 1995.

Grissom, C.B. and Natarajan, E., Use of magnetic field effects on study coenzyme B_{12}-dependent reactions, *Methods Enzymol*, 281, 235, 1997.

Guisasola, C., Desco, M., Millano, O., Villanueva, F.J., and Garcia-Barreno, P., Biological dosimetry of magnetic resonance imaging, *J Magn Reson Imaging*, 15, 584, 2002.

Hata, N., The effect of external magnetic field on the photochemical reaction of isoquinoline *N*-oxide, *Chem Lett*, 5, 547, 1976.

Hatada, T., Sekino, M., and Ueno, S., Finite element method-based calculation of the theoretical limit of sensitivity for detecting weak magnetic fields in the human brain using magnetic-resonance imaging, *J Appl Phys*, 97, 10E109, 2005.

Havel, W.J., Nyenhuis, J.A., Bourland, J.D., Foster, K.S., Geddes, L.A., Graber, G.P., Waninger, M.S., and Schaefer, D.J., Comparison of rectangular and damped sinusoidal dB/dt waveforms in magnetic stimulation, *IEEE Trans Magn*, 33, 4269, 1997.

Hayashi, H., *Introduction to Dynamic Spin Chemistry*, World Scientific Publishing Co, Singapore, 2004.

Higashi, T., Sagawa, S., Kawaguchi, N., and Yamagishi, A., Effects of a strong static magnetic field on blood platelets. *Platelet*, 4, 341, 1993a.

Higashi, T., Yamagishi, A., Takeuchi, T., Kawaguchi, N., Sagawa, S., Onishi, S., and Date, M., Orientation of erythrocytes in a strong static magnetic field, *Blood*, 82, 1328, 1993b.

Higashi, T., Sagawa, S., Ashida, N., and Takeuchi, T., Orientation of glutaraldehyde-fixed erythocytes in strong static magnetic fields, *Bioelectromagnetics*, 17, 335, 1996.

High, W.B., Sikora, J., Ugurbil, K., and Garwood, M., Subchronic *in vivo* effects of a high static magnetic field (9.4 T) in rats, *J Magn Reson Imaging*, 12, 122, 2000.

Hinch, R., Lindsay, K.A., Noble, D., and Rosenberg, J.R., The effects of static magnetic field on action potential propagation and excitation recovery in nerve, *Prog Biophys Mol Biol*, 87, 321, 2005.

Hirai, T., Nakamich, N., and Yoneda, Y., Activator protein 1 complex expressed by magnetism in cultured rat hippocampal neuron, *Biochem Biophys Res Commun*, 292, 200, 2002.

Hirata, Y., Kuriki, S., and Pantev, C., Musicians with absolute pitch show distinct neural activities in the auditory cortex, *NeuroReport*, 10, 999, 1999.

Hirose, H., Nakahara, T., and Miyakoshi, J., Orientation of human glioblastoma cells embedded in type I collagen, caused by exposure to a 10 T static magnetic field, *Neurosci Lett*, 338, 88, 2003a.

Hirose, H., Nakahara, T., Zhang, Q.-M., Yonei, S., and Miyakoshi, J., Static magnetic field with a strong magnetic field gradient (41.7 T/m) induces C-jun expression in HL-60 cells, *In Vitro Cell Dev Biol Anim*, 39, 348, 2003b.

Hirota, N., Nakagawa, J., and Kitazawa, K., Effects of magnetic field on the germination of plants, *J Appl Phys*, 85, 5717, 1999.

Hirota, N., Ikezoe, Y., Uetake, H., Nakagawa, J., and Kitazawa, K., Magnetic field effect on the kinetics of oxygen dissolution into water, materials transaction, *JIM*, 41, 976, 2000.

Holden, A.V., The sensitivity of the heart to static magnetic fields, *Prog Biophys Mol Biol*, 87, 289, 2005.

Horiuchi, S., Ishizaki, Y., Okuno, K., Ano, T., and Shoda, M., Drastic high magnetic field effect on suppression of *Escherichia coli* death, *Bioelectrochemistry*, 53, 149, 2001.

Horiuchi, S., Ishizaki, Y., Okuno, K., Ano, T., and Shoda, M., Change in broth culture is associated with significant suppression of *Escherichia coli* death under high magnetic field, *Bioelectrochemistry*, 57, 139, 2002.

Houpt, T.A., Pittman, D.W., Barranco, J.M., Brooks, E.H., and Smith, J.C., Behavioral effects of high-strength static magnetic fields on rats, *J Neurosci*, 23, 1498, 2003.

Ichioka, S., Iwasaka, M., Shibata, M., Harii, K., Kamiya, A., and Ueno, S., Biological effects of static magnetic fields on the microcirculatory blood flow *in vivo*: a preliminary report, *Med Biol Eng Comput*, 36, 91, 1998.

Ichioka, S., Minegishi, M., Iwasaka, M., Shibata, M., Nakatsuka, T., Harii, K., Kamiya, A., and Ueno, S., High-intensity static magnetic fields modulate skin microcirculation and temperature *in vivo*, *Bioelectromagnetics*, 21, 183, 2000.

Ichioka, S., Minegishi, M., Iwasawa, M., Shibata, M., Nakatsuka, T., Ando, J., and Ueno, S., Skin temperature changes induced static magnetic field exposure, *Bioelectromagnetics*, 24, 380, 2003.

Iino, M., Effects of a homogeneous field on erythrocyte sedimentation and aggregation, *Bioelectromagnetics*, 18, 215, 1997.

Iino, M. and Okuda, Y., Osmolality dependence of erythrocyte sedimentation and aggregation in a strong magnetic field, *Bioelectromagnetics*, 22, 46, 2001.

Ikehata, M., Koana, T., Suzuki, Y., Shimizu, H., and Nakagawa, M., Mutagenicity and co-mutagenicity of static fields detected by bacterial mutation assay, *Mutat Res*, 427, 147, 1999.

Ikehata, M., Iwasaka, M., Miyakoshi, J., Ueno, S., and Koana, T., Effects of intense magnetic fields on sedimentation pattern and gene expression profile in budding yeast. *J Appl Phys*, 93, 6724, 2003.

International Agency for Research on Cancer (IARC), *IARC Monographs on the Evaluation of Carcinogenic Risks to Humans, Non-Ionising Radiation. Part I: Static and Extremely Low Frequency (ELF) Electric and Magnetic Fields*, vol. 80, IARC, Lyon, France, 2002.

International Electrotechnolgy Commission (IEC), *Particular Requirements for the Safety of Magnetic Resonance Equipment for Medical Diagnosis*, Standard 60601-2-33, 2nd ed., Geneva, Switzerland, 2002.

Iramina, K. and Ueno, S., Measurement of brainstem auditory evoked magnetic fields using a highly sensitive SQUID magnetometer with a variable base line, *IEEE Trans Magn*, 31, 4271, 1995.

Iramina, K., Ueno, K., and Ueno, S., Influence of spreading neuronal electric sources on spatio-temporal neuromagnetic fields, *J Appl Phys*, 75, 7168, 1994.

Iramina, K., Ueno, K., and Ueno, S., Spatio-temporal MEG pattern produced by spreading multiple dipoles, *IEEE Trans Magn*, 31, 4265, 1995.

Ishizaki, Y., Horiuchi, S., Okuno, K., Ano, T., and Shoda, M., Twelve hours exposure to inhomogeneous high magnetic field after logarithmic growth is sufficient for drastic suppression of *Escherichia coli* death, *Bioelectrochemistry*, 54, 101, 2001.

Iwaki, S., Ueno, S., Imada, T., and Tonoike, M., Dynamic cortical activation in mental image processing revealed by biomagnetic measurement, *Neuroreport*, 10, 1793, 1999.

Iwasaka, M. and Ueno, S., Effects of magnetic fields on fibrinolysis, *J Appl Phys*, 75, 7162, 1994.

Iwasaka, M. and Ueno, S., Bioluminescence under static magnetic fields, *J Appl Phys*, 83, 6456, 1998a.

Iwasaka, M. and Ueno, S., Structure of water molecules under 14 T magnetic field, *J Appl Phys*, 83, 6459, 1998b.

Iwasaka, M. and Ueno, S., Detection of intracellular macromolecule behavior under strong magnetic fields by linearly polarized light, *Bioelectromagnetics*, 24, 564, 2003.

Iwasaka, M., Takeuchi, M., Ueno, S., and Tsuda, H., Polymerization and dissolution of fibrin under homogeneous magnetic fields, *J Appl Phys*, 83, 6453, 1998.

Iwasaka, M., Miyakoshi, J., and Ueno, S., Magnetic field effects on assembly pattern of smooth muscle cells, *In Vitro Cell Biol Anim*, 39, 2003a.

Iwasaka, M., Yamamoto, K., Ando, J., and Ueno, S., Verification of magnetic field gradient effects on medium convection and cell adhesion, *J Appl Phys*, 93, 6715, 2003b.

Jajte, J., Grzegorczyk, J., Zmyslony, M., and Rajkowska, E., Effect of 7 mT static magnetic field and iron ions on rat lymphocytes: apoptosis, necrosis and free radical processes, *Bioelectrochemistry*, 57, 107, 2002.

Jin, J., *Electromagnetic Analysis and Design in Magnetic Resonance Imaging*, CRC Press, Boca Raton, FL, 1999.

Johnston, S.A., D'Andrea, J.A., Behavioral and cognitive effects of electromagnetic field exposures, in *Biological and Medical Aspects* (Barnes, F.S. and Greenebaum, B., Eds.), Taylor & Francis, Boca Raton, FL, 2006, Chapter 4.

Jovanic, B.R. and Sarvan, M.Z., Permanent magnetic field and plant leaf temperature, *Electromagn Biol Med*, 23, 1, 2004.

Jove, M., Torrente, M., Gilabert, R., Espinar, A., Cobos, P., and Piera, V., Effects of static electromagnetic fields on chick embryo pineal gland development, *Cell Tissues Organs*, 165, 74, 1999.

Joy, M., Scott, G., and Henkelman, M., *In vivo* detection of applied electric currents by magnetic resonance imaging, *Magn Reson Imaging*, 7, 89, 1989.

Kamei, H., Iramina, K., Yoshikawa, K., and Ueno, S., Neuronal current distribution imaging using magnetic resonance, *IEEE Trans Magn*, 35, 4109, 1999.

Kangarlu, A., Burgess, R.E., Zhu, H., Nakayama, T., Hamlin, R.L., Schenck, J.F., Abduljalil, A.M., and Robitaille, P.M.L., Cognitive, cardiac and physiological safety studies in ultra high field magnetic resonance imaging, *J Magn Reson Imaging*, 17, 1407, 1999.

Kangarlu, A., Shellock, F.G., and Chakeres, D.W., 8.0-Tesla human MR system: temperature changes associated with radiofrequency-induced heating of a head phantom, *J Magn Reson Imaging*, 17(2), 220, 2003.

Kangarlu, A., Ibrahim, T.S., and Shellock, F.G., Effects of coil dimensions and field polarization on RF heating inside a head phantom, *Magn Reson Imaging*, 23(1), 53, 2005.

Keck, M.E., Sillaber, I., Ebner, K., Welt, T., Toschi, N., Kaehler, S.T., Singewald, N., Philippu, A., Elbel, G.K., Wotjak, C.T., Holsboer, F., Landgraf, R., and Engelmann, M., Acute transcranial magnetic stimulation of frontal brain regions selectively modulates the release of vasopressin, biogenic amines and amino acids in the rat brain, *Eur J Neurosci*, 12, 3713, 2000.

Kefuss, J., M'Diaye, K., Bounias, M., Vanpoucke, J., and Ecochard, J., Biochemical effects of high intensity constant magnetic fields on worker honey bees, *Bioelectromagnetics*, 20, 117, 1999.

Khang, H.S., Lee, B.I., Oh, S.H., Woo, E.J., Lee, S.Y., Cho, M.H., Kwon, O., Yoon, J.R., and Seo, J.K., J-substitution algorithm in magnetic resonance electrical impedance tomography (MREIT): phantom experiments for static resistivity images, *IEEE Trans Med Imaging*, 21, 695, 2002.

Kimchi, T. and Terkel, J., Magnetic compass orientation in the blind mole rat *Spalax ehrenbergi*, *J Exp Biol*, 204, 751, 2001.

Kishioka, S., Yamada, A., and Aogaki, R., Analysis of gas dissociation rate into liquid phase under magnetic–field gradient, *Phys Chem Chem Phys*, 2, 4179, 2000.

Koana, T., Ikehata, M., and Nakagawa, M., Estimation of genetics effects of a static magnetic field by a somatic cell test using mutagen-sensitive mutants of *Drosophila melanogaster*, *Bioelectrochem Bioenerg*, 36, 95, 1995.

Koana, T., Okada, M.O., Ikehata, M., and Nakagawa, M., Increase in the mitotic recombination frequency in *Drosophila melanogaster* by magnetic field exposure and its suppression by vitamin E supplement, *Mutat Res*, 373, 55, 1997.

Kobayashi, M., Soda, N., Miyo, T., and Ueda, Y., Effects of combined DC and AC magnetic fields on germination of hornwort seeds, *Bioelectromagnetics*, 25, 552, 2004.

Kotani, H., Iwasaka, M., and Ueno, S., Magnetic orientation of collagen and bone mixture, *J Appl Phys*, 87, 6191, 2000.

Krings, T., Chiappa, K.H., Foltys, H., Reinges, M.H., Cosgrove, G.R., and Thron, A., Introducing navigated transcranial magnetic stimulation as a refined brain mapping methodology, *Neurosurg Rev*, 24, 171, 2001.

Kroeker, G., Parkinson, D., Vriend, J., and Peeling, J., Neurochemical effects of static magnetic field exposure, *Surg Neurol*, 45, 62, 1996.

Kula, B., A study of magnetic field effects on fibroblast cultures Part 3. The evaluation of the effects of static and extremely low frequency (ELF) magnetic fields on glycosaminoglycan metabolism in fibroblasts, cell coats and culture medium, *Bioelectrochem Bioenerg*, 39, 31, 1996.

Kula, B. and Drozbz, M., A study of magnetic field effects on fibroblast cultures Part 1. The evaluation of the effects of static and extremely low frequency (ELF) magnetic fields on vital functions of fibroblasts, *Bioelectrochem Bioenerg*, 39, 21, 1996a.

Kula, B. and Drozbz, M., A study on magnetic field on fibroblast cultures. Part 2. The evaluation of the effects of static and extremely low frequency (ELF) magnetic fields on free-radical processes in fibroblast cultures, *Bioelectrochem Bioenerg*, 39, 27, 1996b.

Kula, B., Sobczak, A., and Kuska, R., Effects of static and ELF magnetic fields on free-radical processes in rat liver and kidney, *Electr. Magnetobiol*, 19, 99, 2000.

Kula, B., Sobczak, A., and Kuska, R., A study of the effects of static and extremely low frequency magnetic fields on lipid peroxidant products in subcellular fibroblast fractions, *Electromagn Biol Med*, 21, 161, 2002.

Kuriki, S., Okita, Y., and Hirata, Y., Source analysis of magnetic field responses from the human auditory cortex elicited by short speech sounds, *Exp Brain Res*, 104, 144, 1995.

Kuriki, S., Takeuchi, F., and Hirata, Y., Neural processing of words in the human extrastriate visual cortex, *Cogn Brain Res*, 6, 193, 1998.

Kuriki, S., Isahai, N., and Otsuka, A., Spatiotemporal characteristics of the neural activities processing consonant/dissonant tones in melody, *Exp Brain Res*, 162, 46, 2005.

Kwon, H., Kuriki, S., Kim, J., Lee, Y., Kim, K., Park, Y., and Nam, K., MEG study on neural activities associated with syntactic and semantic violations in spoken Korean sentences, *Neurosci Res*, 51, 349, 2005.

Lauterbur, P.C., Image formation by induced local interactions: examples employing nuclear magnetic resonance, *Nature*, 242, 190, 1973.

Leszczynski, D., Rapporteur report: cellular, animal and epidemiological studies of the effects of static magnetic fields relevant to human health, *Prog Biophys Mol Biol*, 87, 247, 2005.

Levin, M. and Ernst, S.G., Applied DC magnetic fields cause alterations in the time of cell divisions and developmental abnormalities in early sea urchin embryos, *Bioelectromagnetics*, 18, 255, 1997.

Liboff, A.R., Geomagnetic cyclotron resonance in living cells, *J Biol Phys*, 13, 99, 1985.

Liboff, A.R., Cherng, S., Jenrow, K.A., and Bull, A., Calmodulin-dependent cyclic nucleotide phosphodiesterase activity is altered by 20 μT magnetostatic fields, *Bioelectromagnetics*, 24, 32, 2003.

Lin, J.C. (Ed.), *Advances in Electromagnetics Fields in Living Systems*, vol. 3, Kluwer Academic/Plenum Publishers, Hingham, MA, 2000.

Lockwood, D.R., Kwon, B., Smith, J.C., and Houpt, T.A., Behavioral effects of static high magnetic fields on unrestrained and restrained mice, *Physiol Behav*, 78, 635, 2003.

Lohmann, K.J. and Johnsen, S., The neurobiology of magnetoreception in vertebrate animals, *Trends Neurosci*, 23, 153, 2000.

Lohmann, K.J., Cain, S.D., Dodge, S.A., and Lohmann, C.M.F., Regional magnetic fields as navigational markers for sea turtles, *Science*, 294, 364, 2001.

Madec, F., Billaudel, B., Charlet de Sauvage, R., Sartor, P., and Veyret, B., Effects of ELF and static magnetic fields on calcium oscillations in islets of Langerhans, *Bioelectrochemistry*, 60, 73, 2003.

Magin, R.L., Lee, J.K., Klintsova, A., Carnes, K.I., and Dunn, F., Biological effects of long-duration, high-field (4 T) MRI on growth and development in the mouse, *J Magn Reson Imaging*, 12, 140, 2000.

Manassen, Y., Shalev, E., and Navon, G., Mapping of electrical circuits using chemical-shift imaging, *J Magn Reson*, 76, 371, 1988.

Markov, M.S., Myosin light chain modification depending on magnetic fields. II. Experimental, *Electromagn Biol Med*, 23, 125, 2004a.

Markov, M.S., Myosin light chain phosphorylation modification depending on magnetic fields. I. Theoretical, *Electromagn Biol Med*, 23, 55, 2004b.

Markov, M.S. and Pilla, A.A., Static field modulation of myosin phosphorylation: calcium dependence in two enzyme preparations, *Bioelectrochem Bioenerg*, 35, 57, 1994.

Markov, M.S. and Pilla, A.A., Weak static magnetic field modulation of myosin phosphorylation in a cell-free preparation: calcium dependence, *Bioelectrochem Bioenerg*, 43, 233, 1997.

Markov, M.S., Wang, S., and Pilla, A.A., Effects of weak low-frequency sinusoidal and DC magnetic-fields on myosin phosphorylation in a cell-free preparation, *Bioelectrochem Bioenerg*, 30, 119, 1993.

Martinez, E., Carbonell, M.V., and Amaya, J.M., A static magnetic field of 125 mT stimulates the initial growth stages of barley (*Hordeum vulgare* L.), *Electr. Magnetobiol*, 19, 271, 2000.

Martinez, E., Carbonell, M.V., and Florez, M., Magnetic biostimulation of initial growth stages of wheat, *Electromagn Biol Med*, 21, 43, 2002.

Mayrovitz, H.N., Groseclose, E.E., Markov, M., and Pilla, A.A., Effect of permanent magnets on resting skin blood perfusion in healthy persons assessed by laser Doppler flowmetry and imaging, *Bioelectromagnetics*, 22, 494, 2001.

Mayrovitz, H.N., Groseclose, E.E., and King, D., No effect of 85 mT permanent magnets on laser-Doppler measured blood flow response to inspiratory gasps, *Bioelectromagnetics*, 26, 331, 2005.

McKinlay, A.F. and Repacholi, M.H., More research is needed to determine the safety of static magnetic fields, *Prog Biophys Mol Biol*, 87, 173, 2005.

McLean, M.J., Holcomb, R.R., Wamil, A.W., Pickett, J.D., and Cavopol, A.V., Blockade of sensory neuron action potentials by a static magnetic field in the 10 mT range, *Bioelectromagnetics*, 16, 20, 1995.

McLean, M.J., Engstrom, S., and Holcomb, R.R., *Magnetotherapy: Potential Therapeutic Benefits and Adverse Effects*, TFG Press, New York, 2003a.

McLean, M.J., Engstrom, S., Holcomb, R.R., and Sanchez, D., A static magnetic field modulates severity of audiogenic seizures and anticonvulsant effects of phenytoin in DBA/2 mice, *Epilepsy Res*, 55, 105, 2003b.

Miyakoshi, J., Effects of static magnetic fields at the cellular level, *Prog Biophys Mol Biol*, 87, 213, 2005.

Miyamoto, H., Yamaguchi, H., Ikehara, T., and Kinouchi, Y., Effects of electromagnetic fields on K+ (Rb+) uptake by HeLa cell, in *Biological Effects of Magnetic and Electromagnetic Fields* (Ueno, S., Ed.), Plenum Press, New York, 101, 1996.

Mnaimneh, S., Bizri, M., and Veyret, B., No effect of exposure to static and sinusoidal magnetic fields on nitric oxide production by macrophages, *Bioelectromagnetics*, 17, 519, 1996.

Mohtat, N., Cozens, F.L., Hancock-Chen, T., Scaiano, J.C., McLean, J., and Kim, J., Magnetic field effects on the behavior of radicals in protein and DNA environments, *Photochem Photobiol*, 67, 111, 1998.

Morariu, V.V., Ciorba, D., and Neamtu, S., Life in zero magnetic field. I. *In vitro* human blood aging, *Electr. Magnetobiol*, 19, 289, 2000.

Morris, C. and Skalak, T., Static magnetic fields alter arteriolar tone *in vivo*, *Bioelectromagnetics*, 26, 1, 2005.

Mouchawar, G.A., Bourland, J.D., Nyenhuis, J.A., Geddes, L.A., Foster, K.S., Jones, J.T., and Graver, G.P., Closed-chest cardiac stimulation with a pulsed magnetic field, *Med Biol Eng Comput*, 30, 162, 1992.

Muehsam, D.J. and Pilla, A.A., Lorentz approach to static magnetic field effects on bound-ion dynamics and binding kinetics: thermal noise considerations, *Bioelectromagnetics*, 17, 89, 1996.

Munro, U., Munro, J.A., Phillips, J.B., Wiitschko, R., and Wiltschko, W., Evidence for a magnetite-based navigational "map" in birds, *Naturwissenschaften*, 84, 26, 1997.

Nagakura, S. and Molin, Y. (guest eds.), Magnetic field effects upon photophysical and photochemical phenomena, *Chem Phys*, 162(1), 1, 1992 (special issue).

Nagy, P. and Fischl, G., Effect of static magnetic field on growth and sporulation of some plant pathogenic fungi, *Bioelectromagnetics*, 25, 316, 2004.

Nakagawa, J., Hirota, N., Kitazawa, K., and Shoda, M., Magnetic field enhancement of water vaporization, *J Appl Phys*, 86, 2923, 1999.

Nakahara, T., Yaguchi, H., Yoshida, M., and Miyakoshi, J., Effects of exposure to CHO-K1 cells to a 10-T static magnetic field, *Radiology*, 224, 817, 2002.

Nakaoka, Y., Takeda, R., and Shimizu, K., Orientation of *Paramecium* swimming in a DC magnetic field, *Bioelectromagnetics*, 23, 607, 2002.

Narra, V.R., Howell, R.W., Goddu, S.M., and Rao, D.V., Effects of a 1.5-tesla static magnetic field on spermatogenesis and embryogenesis in mice, *Investig Radiol*, 31, 586, 1996.

Natarajan, E. and Grissom, C.B., The origin of magnetic field dependent recombination in alkylcobalamin radical pairs, *Photochem Photobiol*, 64, 286, 1996.

Nemec, P., Altmann, J., Marhold, S., Burda, H., and Oelschlaeger, H.H.A., Neuroanatomy of magntoreception: the superior colliculus involved in magnetic orientation in a mammal, *Science*, 294, 366, 2001.

Neugebauer, D.C., Blauruck, A., and Worcester, D.L., Magnetic orientation of purple membranes demonstrated by optical measurements and neutron scattering, *FEBS Lett*, 78, 31, 1977.

Nikolskaya, K. and Echenko, O., Alcohol addiction as the result of cognitive activity in altered natural magnetic field, *Electromagn Biol Med*, 21, 1, 2002.

Nikolskaya, K., Shtemler, V., Yeschenko, O., Savonenko, A., Osipov, A., and Nickolsky, S., The sensitivity of cognitive processes to the inhomogeneity of natural magnetic fields, *Electr Magnetobiol*, 15, 163, 1996.

Nikolskaya, K.A., Yeshchenko, O.V., and Pratusevich, V., The opioid system and magnetic field perception, *Electr. Magnetobiol*, 18, 277, 1999.

Nolte, C.M., Pittman, D.W., Kalevitch, B., Henderson, R., and Smith, J.C., Magnetic field conditioned taste aversion in rats, *Physiol Behav*, 63, 683, 1998.

Nyenhuis, J.A., Bourland, J.D., and Schaefer, D.J., Analysis from a stimulation perspective of the field patterns of magnetic resonance imaging coils, *J Appl Phys*, 81, 4314, 1997.

Nyenhuis, J.A., Bourland, J.D., Kildishev, A.V.D.J., and Schaefer, D.J., Health effects and safety of intense MRI gradient fields, in *Magnetic Resonance Procedures: Health Effects and Safety* (Shellock, F.G., Ed.), CRC Press, Boca Raton, FL, 31–54, 2001.

Ogawa, S., Tank, D.W., and Menon, R., Intrinsic signal changes accompanying stimulation: functional brain mapping with magnetic resonance imaging, *Proc Natl Acad Sci USA*, 89, 5951, 1992.

Ogiue-Ikeda, M., Kawato, S., and Ueno, S., The effect of repetitive transcranial magnetic stimulation on long-term potentiation in rat hippocampus depends on stimulus intensity, *Brain Res*, 993, 222, 2003a.

Ogiue-Ikeda, M., Sato, Y., and Ueno, S., A method to destruct targeted cells using magnetizable beads and pulsed magnetic force, *IEEE Trans Magn* 39, 3390, 2003b.

Ohkubo, C. and Xu, S., Acute effects of static magnetic fields on cutaneous microcirculation in rabbits, *In Vivo*, 11, 221, 1997.

Okano, H. and Ohkubo, C., Modulatory effects of static magnetic fields on blood pressure in rabbits, *Bioelectromagnetics*, 22, 408, 2001.

Okano, H. and Ohkubo, C., Anti-pressor effects of whole body exposure to static magnetic field on pharmacologically induced hypertension in conscious rabbit, *Bioelectromagnetics*, 24, 139, 2003a.

Okano, H. and Ohkubo, C., Effects of static magnetic fields on plasma levels of angiotensin II and aldosterone associated with arterial blood pressure in genetically hypertensive rats, *Bioelectromagnetics*, 24, 403, 2003b.

Okano, H., Gmitrov, J., and Ohkubo, C., Biphasic effects of static magnetic field on cutaneous microcirculation in rabbits, *Bioelectromagnetics*, 20, 161, 1999.

Okano, H., Masuda, H., and Ohkubo, C., Decreased plasma levels of nitric oxide metabolites, angiotensin II and aldosterone in spontaneously hypertensive rats exposed to 5 mT static magnetic field, *Bioelectromagnetics*, 26, 161, 2005a.

Okano, H., Masuda, H., and Ohkubo, C., Effects of 25 mT static magnetic field on blood pressure in reserpine-induced hypotensive Wistar–Kyoto rats, *Bioelectromagnetics*, 26, 36, 2005b.

Okazaki, R., Ootsuyama, A., Uchida, S., and Norimura, T., Effects of a 4.7 T static magnetic field on fetal development in ICR mice, *J Radiat Res*, 42, 273, 2001.

Okonogi, H., Nakagawa, M., and Tsuji, Y., The effects of a 4.7 tesla static magnetic field on the frequency of micronucleated cells induced by mitomycin C, *Tohoku J Exp Med*, 180, 209, 1996.

Okuda, T., Nishizawa, K., Ejima, Y., Nakatsugawa, S., Ishigaki, T., and Ishigaki, K., The effects of static magnetic fields and x-rays on instability of microsatellite repetitive sequences, *J Radiat Res*, 39, 279, 1998.

Okuno, K., Fujinami, R., Ano, T., and Shoda, M., Disappearance of growth advantage in stationary phase (GASP) phenomenon under a high magnetic field, *Bioelectromagnetics*, 53, 165, 2001.

Onodera, H., Jin, Z., Chida, S., Suzuki, Y., Tago, H., and Itoyama, Y., Effects of 10-T static magnetic field on human peripheral blood immune cells, *Radiat Res*, 159, 775, 2003.

Osuga, T. and Tatsuoka, H., Effect of 1.5 T steady magnetic field on neuroconduction of a bullfrog sciatic nerve in a partially active state within several hours after extraction, *Magn Reson Imaging*, 17, 791, 1999.

Pacini, S., Aterini, S., Pacini, P., Ruggiero, C., Gulisano, M., and Ruggiero, M., Influence of static magnetic field on the antiproliferative effects of vitamin D on human breast cancer cells, *Oncol Res*, 11, 265, 1999a.

Pacini, S., Vannelli, G.B., Barni, T., Ruggiero, M., Sardi, I., Pacini, P., and Gulisano, M., Effect of 0.2 T static magnetic field on human neurons: remodeling and inhibition of signal transduction without genome instability, *Neurosci Lett*, 267, 185, 1999b.

Pagnac, C., Geneviere, A.-M., Moreau, J.-M., Picard, A., Joussot-Dubien, J., and Veyret, B., No effects of DC and 60-Hz AC magnetic fields on the first mitosis of two species of sea urchin embryos, *Bioelectromagnetics*, 19, 494, 1998.

Pan, H. and Liu, X., Apparent biological effect of strong magnetic field on mosquito egg hatching, *Bioelectromagnetics*, 25, 84, 2004.

Pate, K., Benghuzzi, H., Tucci, M., Puckett, A., and Cason, Z., Morphological evaluation of MRC-5 fibroblasts after stimulation with static magnetic field and pulsating electromagnetic field, *Biomed Sci Instrum*, 39, 460, 2003.

Penuelas, J., Llusia, J., Martinez, B., and Fontcuberta, J., Diamagnetic susceptibility and root growth responses to magnetic fields in *Les culinaris*, *Glycine soja*, and *Triticum aestivum*, *Electromagn Biol Med*, 23, 97, 2004.

Phillips, J.B., Magnetic navigation, *J Theor Biol*, 180, 309, 1996.

Phillips, J.B, Animal magnetoreception: future directions, *BioEM 2005*, 160, 2005.

Phillips, J.B., Deutschlander, M.E., Freake, M.J., and Borland, S.C., The role of extraocular photoreceptors in newt magnetic compass orientation: parallels between light-dependent magnetoreception and polarized light detection in vertebrates, *J Exp Biol*, 204, 2543, 2001.

Phillips, J.B., Borland, S.C., Freake, M.J., Brassart, J., and Kirschvink, J.L., 'Fixed-axis' magnetic orientation by an amphibian: non-shoreward-directed compass orientation, misdirected homing or positioning a magnetite-based map detector in a consistent alignment relative to the magnetic field, *J Exp Biol*, 205, 3903, 2002a.

Phillips, J.B., Freake, M.J., Fischer, J.H., and Vorland, S.C., Behavioral titration of a magnetic map coordinate, *J Comp Physiol A*, 188, 157, 2002b.

Phirke, P.S., Kubde, A.B., and Umbarkar, S.P., The influence of magnetic field on plant growth, *Seed Sci Technol*, 24, 375, 1996.

Piatti, E., Cristina, A.M., Baffone, W., Fraternale, D., Citterio, B., Piera, P.M., Dacha, M., Vetrano, F., and Accorsi, A., Antibacterial effect of a magnetic field on *Serratia marcescens* and related virulence to *Hordeum vulgare* and *Rubus fruticosus* callus cells, *Comp Biochem Physiol B Biochem Mol Biol*, 132, 359, 2002.

Pilla, A.A., Mechanisms and therapuetic applications of time-varing and static magnetic fields, in *Biological and Medical Aspects* (Barnes, F.S., and Greenebaum, B., Eds.), Taylor & Francis, Boca Raton, FL, 2006, Chapter 11.

Poiata, A., Creanga, D.E., and Morariu, V.V., Life in zero magnetic field. V. *E. coli* resistance to antibiotics, *Electromagn Biol Med*, 22, 171, 2003.

Polk, C. and Postow, E. (Eds.), *Handbook of Biological Effects of Electromagnetic Fields*, CRC Press, Boca Raton, FL, 1986.

Polk, C. and Postow, E. (Eds.), *Handbook of Biological Effects of Electromagnetic Fields*, 2nd ed., CRC Press, Boca Raton, FL, 1997.

Potenza, L., Cucchiarini, L., Piatti, E., Angelini, U., and Dacha, M., Effects of high static magnetic field exposure on different DNAs, *Bioelectromagnetics*, 25, 352, 2004a.

Potenza, L., Ubaldi, L., De Sanctis, R., De Bellis, R., Cucchiarini, L., and Dacha, M., Effects of a static magnetic field on cell growth and gene expression in *Escherichia coli*, *Mutat Res*, 561, 53, 2004b.

Prato, F.S., Kavaliers, M., and Carson, J.J.L., Behavioural evidence that magnetic field effects in the land snail, *Cepaea nemoralis*, might not depend on magnetite or induced electric currents, *Bioelectromagnetics*, 17, 123, 1996a.

Prato, F.S., Kavaliers, M., and Carson, J.J.L., Behavioural responses to magnetic fields by land snails are dependent on both magnetic field direction and light, *Proc R Soc Lond B*, 263, 1437, 1996b.

Prato, F.S., Robertson, J.A., Desjardins, D., Hensel, J., and Thomas, A.W., Daily repeated magnetic field shielding induces analgesia in CD-1 mice, *Bioelectromagnetics*, 26, 109, 2005.

Rai, D.V., Kohli, K.S., Goyal, N., and Jindal, V.K., The effect of static magnetic field on electrical properties of lens, *Electr. Magnetobiol*, 16, 293, 1997.

Raylman, R.R., Clavo, A.C., and Wahl, R.L., Exposure to strong static magnetic field slows the growth of human cancer cell *in vitro*, *Bioelectromagnetics*, 17, 358, 1996.

Reilly, J.P., Peripheral nerve stimulation in induced electric currents: exposure to time-varying magnetic fields, *Med Biol Eng Comput*, 27, 101, 1989.

Reilly, J.P., Magnetic field excitation of peripheral nerves and the heart: a comparison of thresholds, *Med Biol Eng Comp*, 29, 571, 1991.

Reina, F.G. and Pascual, L.A., Influence of a stationary magnetic field on water relations in lettuce seeds. Part 1: theoretical considerations, *Bioelectromagnetics*, 22, 589, 2001.

Reina, F.G., Pascual, L.A., and Fundora, I.A., Influence of a stationary magnetic field on water relations in lettuce seeds. Part 2: experimental results, *Bioelectromagnetics*, 22, 596, 2001.

Repacholi, M.H. and Greenebaum, B., Interaction of static and extremely low frequency electric and magnetic fields with living systems: health effects and research needs, *Bioelectromagnetics*, 20, 133, 1999.

Ritz, T., Adem, S., and Schulten, K., A model for photoreceptor-based magnetoreception in birds, *Biophys J*, 78, 707, 2000.

Ritz, T., Dommer, D.H., and Phillips, J.B., Shedding light on vertebrate magnetoreception, *Neuron*, 34, 503, 2002.

Ritz, T., Thalau, P., Phillips, J.B., Wiltschko, R., and Wiltschko, W., Resonance effects indicate a radical-pair mechanism for avian magnetic compass, *Nature*, 429, 177, 2004.

Robitaille, P.M.L., Abduljalil, A.M., Kangarlu, A., Zhang, X., Yu, Y., Burgess, R., Bair, E., Noa, P., Yang, L., Zhu, H., Palmer, B., Jiang, Z., Chakeres, D.W., and Spigos, D., Human magnetic resonance imaging at eight tesla, *NMR Biomed*, 11, 263, 1998.

Ronchetto, F., Barone, D., Cintorino, M., Berardelli, M., Lissola, S., Orlassino, R., Ossala, P., and Tofani, S., Extremely low frequency-modulated static magnetic fields to treat cancer: a pilot study on patients with advanced neoplasm to assess safety and acute toxicity, *Bioelectromagnetics*, 25, 563, 2004.

Rosch, P.J. and Markov, M.S., *Bioelectromagnetic Medicine*, Marcel Dekker, New York, 2004.

Rosen, A.D., Inhibition of calcium channel activation in GH3 cells by static magnetic fields, *Biochim Biophys Acta*, 1282, 149, 1996.

Rosen, A.D., Effect of a 125 mT static magnetic field on kinetics of voltage activated Na^+ channels in GH3 cells, *Bioelectromagnetics*, 24, 517, 2003.

Rosen, M. and Rosen, A., Magnetic influence on *Paramecium* motility, *Life Sci*, 46, 1509, 1990.

Rothwell, J.C., Day, B.L., Thompson, P.D., Dick, J.P.R., and Marsden, C.D., Some experiences of techniques for stimulation of the human cerebral motor cortex through the scalp, *Neurosurgery*, 20(1), 156, 1987.

Ruggiero, M., Bottaro, D.P., Liguri, G., Gulisano, M., Peruzzi, B., and Pacini, S., 0.2 T magnetic field inhibits angiogenesis in chick embryo chorioallantoic membrane, *Bioelectromagnetics*, 25, 390, 2004.

Ruiz-Gomez, M.J., Prieto-Barcia, M.I., Ristori-Bogajo, E., and Martinz-Morillo, M., Static and 50 Hz magnetic field of 0.35 and 2.45 mT have no effect on the growth of *Saccharomyces cerevisiae*, *Bioelectromagnetics*, 64, 151, 2004.

Sabo, J., Mirossay, L., Horovcak, L., Sarissky, M., Mirossay, A., and Mojzis, J., Effect of magnetic field on human leukemic cell line HL-60, *Bioelectrochemistry*, 56, 227, 2002.

Sakurai, H., Okuno, K., Kubo, A., Nakamura, K., and Shoda, M., Effect of 7-tesla homogeneous magnetic field on mammalian cells, *Bioelectrochem Bioenerg*, 49, 57, 1999.

Salerno, S., Lo Casto, A., Caccamo, N., D'Anna, C., de Maria, M., Lagalla, R., Scola, L., and Cardinale, A.E., Static magnetic fields generated by a 0.5 T MRI unit affects *in vitro* expression of activation markers and interleukin release in human peripheral blood mononuclear cells (PBMC), *Int J Radiat Biol*, 75, 457, 1999.

Sandrey, M.A., Vesper, D.N., Johnson, M.T., Nindl, G., Swez, J.A., and Chamberlain, J., Effect of short duration electromagnetic field exposure on rat mass, *Bioelectromagnetics*, 23, 37658, 2002.

Satow, Y., Matsunami, K., Kawashima, T., Satake, H., and Huda, K., A strong constant magnetic field affects muscle tension development in bullfrog neuromuscular preparations, *Bioelectromagnetics*, 22, 53, 2001.

Saunders, R., Static magnetic fields: animal studies, *Prog Biophys Mol Biol*, 87, 225, 2005.
Schenck, J.F., Physical interactions of static magnetic fields with living tissues, *Prog Biophys Mol Biol*, 87, 185, 2005.
Schenck, J.F., Dumoulin, C.L., Redington, R.W., Kressel, H.Y., Elliott, R.T., and McDougall, I.L., Human exposure to 4.0-tesla magnetic fields in a whole-body scanner, *Med Phys*, 19, 1089, 1992.
Schiffer, I.B., Schreiber, W.G., Graf, R., Schreiber, E.M., Jung, D., Rose, D.M., Hehn, M., Ebhard, S., Sagemuller, J., Spiess, H.W., Oesch, F., Thelen, M., and Hengstler, J.G., No influence of magnetic fields on cell cycle progression using conditions relevant for patients during MRI, *Bioelectromagnetics*, 24, 241, 2003.
Schulten, K., Staerk, H., Weller, A., Werner, H.J., and Nickel, B., Magnetic field dependence of the geminate recombination of radical ion pairs in polar solvents, *Z Phys Chem N F*, 101, 371, 1976.
Segal, N.A., Toda, Y., Huston, J., Saeki, Y., Shimizu, M., Fuchs, H., Shimaoka, Y., Holcomb, R., and McLean, M.J., Two configurations of static magnetic fields for treating rheumatoid arthritis of the knee: a double-blind clinical trial, *Arch Phys Med Rehabil*, 82, 1453, 2001.
Sekino, M., Inoue, Y., and Ueno, S., Magnetic resonance imaging of mean values and anisotropy of electrical conductivity in the human brain, *Neurol Clin Neurophysiol*, 55, 1, 2004a.
Sekino, M., Matsumoto, T., Yamaguchi, K., Iriguchi, N., and Ueno, S., A method for NMR imaging of a magnetic field generated by electric current, *IEEE Trans Magn*, 40, 2188, 2004b.
Sekino, M., Mihara, H., Iriguchi, N., and Ueno, S., Dielectric resonance in magnetic resonance imaging: signal inhomogeneities in samples of high permittivity, *J Appl Phys.*, 97, 10R303, 2005.
Shellock, F.G. (Ed.), *Magnetic Resonance Procedures: Health Effects and Safety*, CRC Press, Boca Raton, FL, 2001.
Shupak, N.M., Hensel, J.M., Cross-Mellor, S.K., Kavaliers, M., Prato, F.S., and Thomas, A.W., Analgesic and behavioral effects of a 100 microT specific pulsed extremely low frequency magnetic field on control and morphine treated CF-1 mice, *Neurosci Lett*, 354(1), 30, 2004a.
Shupak, N.M., Prato, F.S., and Thomas, A.W., Human exposure to a specific pulsed magnetic field: effects on thermal sensory and pain thresholds, *Neurosci Lett*, 363(2), 157, 2004b.
Slowik, T.J., Green, B.L., and Thorvilson, H.G., Detection of magnetism in the red imported fire ant (*Solenopsis invicta*) using magnetic resonance imaging, *Bioelectromagnetics*, 18, 396, 1997a.
Slowik, T.J., Green, B.L., and Thorvilson, H.G., Response of red imported fire ant to magnetic field in the nest environment, *South Entomol*, 22, 301, 1997b.
Snyder, D.J., Jahng, J.W., Smith, J.C., and Houpt, T.A., c-Fos induction in visceral and vestibular nuclei of the rat brain stem by a 9.4 T magnetic field, *Neuro Report*, 11, 2681, 2000.
So, P., Stuchly, M.A., and Nyenhuis, J.A., Peripheral nerve stimulation by gradient switching fields in MRI, *IEEE Trans BioMed Eng*, 51, 1907, 2004.
Sonnier, H., Kolomytkin, O., and Marino, A, Resting potential of excitable neuroblastoma cell in weak magnetic fields, *Cell Mol Life Sci*, 57, 514, 2000.
Sonnier, H., Kolomytkin, O., and Marino, A., Action potentials from human neuroblastoma cells in magnetic fields, *Neurosci Lett*, 337, 163, 2003.
Squires, K.C., Development of a 37-channel SQUID-based magnetometer for study of the brain and heart, *Med Biol Eng Comput*, 29(Suppl), 761, 1991.
Stamenkovic-Radak, M., Kitanovic, I., Prolic, Z., Tomisic, I., Stojkovic, B., and Andjelkovic, M., Effect of permanent magnetic field on wing size parameters in *Drosophila melanogaster*, *Bioelectromagnetics*, 22, 365, 2001.
Stansell, M.J., Winters, W.D., Doe, R.H., and Dart, B.K., Increased antibiotic resistance of *E. coli* exposed static magnetic fields, *Bioelectromagnetics*, 22, 129, 2001.
Stavroulakis, P. (ed.), *Biological Effects of Electromagnetic Fields*, Springer-Verlag, Berlin, 2003.
Stefano, G. and Tranquillo, R.T., A methodology for the systematic and quantitative study of cell contact guidance in orientated collagen gels, *J Cell Sci*, 105, 317, 1993.
Stern, S., Laties, V.G., Nguyen, Q.A., and Cox, C., Exposure to combined static and 60 Hz magnetic fields: failure to replicate a reported behavioral effect, *Bioelectromagnetics*, 17, 279, 1996.
Steyn, P.F., Ramey, D.W., Kirschvink, J., and Uhring, J., Effect of a static magnetic field on blood flow to the metacarpus in horses, *J Am Vet Med Assoc*, 217, 874, 2000.
Strickman, D., Timberlake, B., Estrada-Franco, J., Weissman, M., Fenimore, P.W., and Novak, R.J., Effects of magnetic fields on mosquitoes, *J Am Mosq Control Assoc*, 16, 131, 2000.

Suzuki, M., and Nakamura, H., Orientation of sperm DNA under a magnetic field, *Proc Jpn Acad*, 71(Ser B), 36, 1995.

Suzuki,Y., Ikehata, M., Nakamura, K., Nishioka, M., Asanuma, K., Koana, T., and Shimizu, H., Induction of micronuclei in mice exposed to static magnetic fields, *Mutagenesis*, 16, 499, 2001.

Tablado, L., Perez-Sanchez, F., and Soler, C., Is sperm motility maturation affected by static magnetic field? *Environ Health Perspect*, 104, 1212, 1996.

Tablado, L., Perez-Sanchez, F., Nunez, J., Nunez, M., and Soler, C., Effects of exposure to static magnetic fields on the morphology and morphometry of mouse epididymal sperm, *Bioelectromagnetics*, 19, 377, 1998.

Tablado, L., Soler, C., Nunez, M., Nunez, J., and Perez-Sanchez, F., Development of mouse testis and epididymis following intrauterine exposure to a static magnetic field, *Bioelectromagnetics*, 21, 19, 2000.

Takashima, Y., Miyakoshi, M., Ikehara, M., Iwasaka, M., Ueno, S., and Koana, T., Genotoxic effects of strong static magnetic fields in DNA-repair defective mutants of *Drosphila melanogaster*, *J Radiat Res*, 45, 393, 2004.

Takebe, H., Shiga, T., Kato, M., and Masada, E., *Biological and Health Effects from Exposure to Power-Line Frequency Electromagnetic Fields—Confirmation of Absence of Any Effects at Environmental Field Strengths*, IOS Press, Amsterdam, 1999.

Tanimoto, Y., Hayashi, H., Nagakura, S., Sakurai, H., and Tokumaru, K. The external magnetic field effect on the singlet sensitized photolysis of dibenzoyl peroxide, *Chem Phys Lett*, 41, 267, 1976.

Tanimoto, Y., Izumi, S., Furuta, K., Suzuki, T., Fujiwara, Y., Fujiwara, M., Hirata, T., and Yamada, S., Effects of high magnetic field on *Euglena gracilis*, *Int J Appl Electromagn Mech*, 14, 311, 2001.

Tanimoto, Y., Ogawa, S., Fujitani, K., Fujiwara, Y., Izumi, S., and Hirata, T., Effects of a high magnetic field on *E. coli* movement, *Environ Sci*, 18, 53, 2005.

Tanioka, N., Sawada, S., and Hosokawa, T., Proliferative and metastatic activities of mouse tumor cells exposed to a strong static magnetic field, *Bioelectrochem Bioenerg*, 40, 29, 1996.

Taoka, S., Padmakumar, R., Grissom, C.B., and Banerjee, R., Magnetic field effects on coenzyme B12-dependent enzymes: validation of ethanolamine ammonia lyase results and extension to human methylmalonyl CoA mutase, *Bioelectromagnetics*, 18, 506, 1997.

Tenforde, T.S., Magnetcally induced electric fields and currents in the circulatory systems, *Prog Biophys Mol Biol*, 87, 279, 2005.

Teodori, L., Grabarek, J., Smolewski, P., Ghibelli, L., Bergamaschi, A., De Nocola, M., and Darzynkiewicz, Z., Exposure of cells to static magnetic field accelerated loss integrity of plasma membrane during apoptosis, *Cytometry*, 49, 113, 2002.

Testorf, M.F., Oberg, P.A., Iwasaka, M., and Ueno, S., Melanophore aggregation in strong static magnetic fields, *Bioelectromagnetics*, 23, 444, 2002.

Thomas, J.R., Schrot, J., and Liboff, A.R., Low-intensity magnetic fields alter operant behavior in rats, *Bioelectromagnetics*, 7, 349, 1986.

Till, U., Timmel, C.R., Brocklehurst, B., and Hore, P.J., The influence of very small magnetic field on radical recombination reactions in the limit of slow recombination, *Chem Phys Lett*, 298, 87, 1998.

Timmel, C.R., Till, U., Brocklehurst, B., McLauchlan, K.A., and Hore, P.J., Effects of weak magnetic fields on free radical recombination reactions, *Mol Phys*, 95, 71, 1998.

Tofani, S., Barone, D., Cintorino, M., de Santi, M.M., Ferrara, A., Orlassino, R., Ossola, P., Peroglio, F., Rolfo, K., and Ronchetto, F., Static and ELF magnetic fields induce tumor growth inhibition and apoptosis, *Bioelectromagnetics*, 22, 419, 2001.

Tofani, S., Cintorino, M. et al., Increased mouse survival, tumor growth inhibition and decreased immunoreactive p53 after exposure to magnetic fields, *Bioelectromagnetics*, 23, 230, 2002.

Tofani, S., Barone, D., Berardelli, M., Berno, E., Cintorino, M., Foglia, L., Ossola, P., Ronchetto, F., Toso, E., and Eandi, M., Static and ELF magnetic fields enhance the *in vivo* anti-tumor efficacy of cis-platin against Lewis lung carcinoma, but not of cyclophosphamie against B16 melanotic melanoma, *Pharmacol Res*, 48, 83, 2003.

Torbet, J. and Ronziere, M.C., Magnetic alignment of collagen during self-assembly, *Biochem J*, 219, 1057, 1984.

Torbet, J., Fryssinet, M., and Hudry-Clergeon, G., Oriented fibrin gels formed by polymerization in strong magnetic fields, *Nature*, 289, 91, 1981.

Trabulsi, R., Pawlowski, B., and Wieraszko, A., The influence of steady magnetic fields on the mouse hippocampal evoked potentials *in vitro*, *Brain Res*, 728, 135, 1996.

Tranquillo, R.T., Girton, T.S., Bromberek, B.A., Triebes, T.G., and Mooradian, D.L., Magnetically oriented tissue-equivalent tubes: application to a circumferentially oriented media-equivalent, *Biomaterials*, 17, 349, 1996.

Tsuchiya, K., Nakamura, K., Okuno, K., Ano, T., and Shoda, M., Effect of homogeneous and inhomogeneous high magnetic fields on the growth of *Escherichia coli*, *J Ferment Bioeng*, 81, 343, 1996.

Tsuchiya, K., Okuno, K., Ano, T., Tanaka, K., Takahashi, H., and Shoda, M., High magnetic field enhances stationary phase-specific transcription activity of *Escherichia coli*, *Bioelectrochem Bioenerg*, 48, 383, 1999.

Tsuji, Y., Nakagawa, M., and Suzuki, Y., Five-tesla static magnetic fields suppress food and water consumption and weight gain in mice, *Ind Health*, 34, 347, 1996.

Tuch, D.S., Wedeen, V.J., Dale, A.M., George, J.S., and Belliveau, J.W., Conductivity tensor mapping of the human brain using diffusion tensor MRI, *Proc Natl Acad Sci USA*, 98, 11697, 2001.

Ueno, S., Quenching of flames by magnetic fields, *J Appl Phys*, 65(3), 1243, 1989.

Ueno, S., *Biological Effects of Magnetic and Electromagnetic Fields*, Plenum Press, New York, 1996.

Ueno, S. and Harada, K., Redistribution of the dissolved oxygen concentration under strong DC magnetic field, *IEEE Trans Magn*, MAG-18, 1704, 1982.

Ueno, S. and Harada, K., Experimental difficulties in observing the effects of magnetic fields on biological and chemical processes, *IEEE Magn*, 22, 868, 1986.

Ueno, S. and Harada, K., Effects of magnetic fields on flames and gas flow, *IEEE Trans Magn*, MAG-23, 5, 2752, 1987.

Ueno, S. and Iramina, K., Modeling and source localization of MEG activities, *Brain Topogr*, 3, 151, 1991.

Ueno, S and Iriguchi, N., Impedance magnetic resonance imaging: a method for imaging of impedance distributions based on magnetic resonance imaging, *J Appl Phys*, 83, 6450, 1998.

Ueno, S. and Iwasaka, M., Parting of water by magnetic fields, *IEEE Trans Magn*, 30, 4698, 1994a.

Ueno, S. and Iwasaka, M., Properties of diamagnetic fluid in high gradient magnetic fields, *J Appl Phys*, 75, 7177, 1994b.

Ueno, S., Matsumoto, S., Harada, K., and Oomura, Y., Capacitative stimulatory effect in magnetic stimulation of nerve tissue, *IEEE Trans Magn*, MAG-14, 958, 1978.

Ueno, S., Lovsund, P., and Oberg, P.A., Effects of alternating magnetic fields and low-frequency electric currents on human skin blood flow, *Med Biol Eng Comput*, 24, 57, 1986a.

Ueno, S., Lovsund, P., and Oberg, P.A., Effects of time-varying magnetic fields on action potential in lobster giant axon, *Med Biol Eng Comput*, 24, 521, 1986b.

Ueno, S., Tashiro, T., and Harada, K., Localized stimulation of neural tissues in the brain by means of a paired configuration of time-varying magnetic fields, *J Appl Phys*, 64, 5862, 1988.

Ueno, S., Matsuda, T., and Fujiki, M., Localized stimulation of the human cortex by opposing magnetic fields, in *Advances in Biomagnetism* (Williamson, S.J., Hoke, M., Stroink, G., and Kotani, M., Eds.), Plenum Press, New York, 529, 1989.

Ueno, S., Matsuda, T., and Fujiki, M., Functional mapping of the human motor cortex obtained by focal and vectorial magnetic stimulation of the brain, *IEEE Trans Magn*, 26, 1539, 1990a.

Ueno, S., Matsuda, T., and Hiwaki, O., Estimation of structures of neural fibers in the human brain by vectorial magnetic stimulation, *IEEE Trans Magn*, 27, 5387, 1990b.

Ueno, S., Matsuda, T., and Hiwaki, O., Localized stimulation of the human brain and spinal cord by a pair of opposing pulsed magnetic fields, *J Appl Phys*, 66, 5838, 1991.

Ueno, S., Hiwaki, O., Matsuda, T. et al., Safety problems of dB/dt associated with echo planar imaging, *Ann NY Acad Sci*, 369, 96, 1992.

Ueno, S., Iwasaka, M., and Tsuda, H., Effects of magnetic fields on fibrin polymerization and fibrinolysis, *IEEE Trans Magn*, 29, 3352, 1993.

Valles, J.M. Jr., Model of magnetic field-induced mitotic apparatus reorientation in frog eggs, *Biophys J*, 82, 1260, 2002.

Valles, J.M. Jr., Lin, K., Denegre, J.M., and Mowry, K.L., Stable magnetic field gradient levitation of *Xenopus laevis*: toward low-gravity simulation, *Biophys J*, 73, 1130, 1997.

Valles, J.M. Jr., Wasserman, S.R.R.M., Schweidenback, C., Edwardson, J., Denegre, J.M., and Mowry, K.L., Processes that occur before second cleavage determine third cleavage orientation in *Xenopus*, *Exp Cell Res*, 274, 112, 2002.

van Rongen, E., International workshop ''effects of static magnetic fields relevant to human health'' Repporteurs report: dosimetry and volunteer studies, *Prog Biophys Mol Biol*, 87, 329, 2005.

Veliks, V., Ceihnere, E., Svikis, I., and Aivars, J., Static magnetic field influence on rat brain function detected by heart rate monitoring, *Bioelectromagnetics*, 25, 211, 2004.

Vink, C.B. and Woodward, J.R, Effect of a weak magnetic field on the reaction between neural free radicals in isotropic solution, *J Am Chem Soc*, 126, 16730, 2004.

Volpe, P., Interactions of zero-frequency and oscillating magnetic fields with biostructures and biosystems, *Photochem Photobiol Sci*, 2, 637, 2003.

Vrva, J., Bette, K., and Burband, M., Whole cortex 64 channel SQUID biomagnetometer system, *IEEE Trans*, AS 3, 1878, 1993.

Watanabe, Y., Nakagawa, M., and Miyakoshi, Y., Enhancement of lipid peroxidation in liver of mice exposed to magnetic fields, *Ind Health*, 35, 285, 1997.

Wieraszko, A., Dantrolene modulates the influence of steady magnetic fields on hippocampal evoked potentials *in vivo*, *Bioelectromagnetics*, 21, 175, 2000.

Wiltschko, R. and Wiltschko, W., *Magnetic Orientation in Animals*, *Zoophysiology*, vol. 33, Springer-Verlag, Berlin, 1995.

Wiltschko, R. and Wiltschko, W., Magnetoreception: why is conditioning so seldom successful? *Naturewissenschaften*, 83, 241, 1996.

Wiltschko, W. and Wiltschko, R., Light-dependent magnetoreception in birds: the behaviour of European robins, *Erithacus rubecula*, under monochromatic light of various wavelengths and intensities, *J Exp Biol*, 204, 3295, 2001.

Wiltschko, W. and Wiltschko, R., Magnetic compass orientation in birds and its physiological basis, *Naturwissenschaften*, 89, 445, 2002.

Winklhofer, M., Holtkamp-Roetzler, E., Hanzlik, M., Fleissner, G., and Petersen, N., Clusters of superparamagnetic magnetite particles in the upper-beak skin of homing pigeons: evidence of a magnetoreceptor? *Eur J Miner*, 13, 659, 2001.

Winnicki, A., Formicki, K., and Sobocinski, A., Application of constant magnetic field in transportation of gametes and fertilized eggs of salmonid fish, *Publ Espec Inst Esp Oceanogr*, 21, 301, 1996.

Wiskirchen, J., Groenwaeller, E.F., Kehlbach, R., Heinzelmann, F., Wittau, M., Rodemann, H.P., Claussen, C.D., and Duda, S.H., Long-term effects of repetitive exposure to a static magnetic field (1.5 T) on proliferation of human fetal lung fibroblasts, *Magn Reson Med*, 41, 464, 1999.

Wiskirchen, J., Groenwaller, E.F., Heinzelmann, F., Kehlbach, R., Rodegerdts, E., Wittau, M., Rodemann, H.P., Claussen, C.D., and Duda, S.H., Human fetal lung fibroblasts, *in vitro* study of repetitive magnetic field exposure at 0.2, 1.0, and 1.5 T, *Radiology*, 215, 858, 2000.

Xiong, J., Fox, P.T., and Gao, J.H., Directly mapping magnetic field effects of neuronal activity by magnetic resonance imaging, *Hum Brain Mapp*, 20, 41, 2003.

Xu, S., Okano, H., and Ohkubo, C., Subchronic effects of static magnetic fields on cutaneous microcirculation in rabbits, *In Vivo*, 12, 383, 1998.

Xu, S., Okano, H., and Ohkubo, C., Acute effects of whole-body exposure to static magnetic fields and 50-Hz electromagnetic fields on muscle microcirculation in anesthetized mice, *Bioelectromagnetics*, 53, 127, 2000.

Xu, S., Tomita, N., Ohata, R., Yan, Q., and Ikada, Y., Static magnetic field effects on bone formation of rats with an ischemic bone model, *BioMed Mater Eng*, 11, 257, 2001.

Yamagishi, A., Takeuchi, T., Higashi, T., and Date, M., Diamagnetic orientation of polymerized molecules under high magnetic field, *J Phys Soc Jpn*, 58, 2280, 1989.

Yamaguchi, S., Ogiue-Ikeda, M., Sekino, M., and Ueno, S., The effect of repetitive magnetic stimulation on the tumor development, *IEEE Trans Magn*, 40, 3021, 2004.

Yamaguchi, S., Ogiue-Ikeda, M., Sekino, M., and Ueno, S., Effects of pulsed magnetic stimulation on tumor development and immune functions in mice, *Bioelectromagnetics*, 27, 64, 2006.

Yan, Q.C., Tomita, N., and Ikada, Y., Effects of static magnetic field on bone formation of rat femurs, *Med Eng Phys*, 20, 397, 1998.

Yano, A., Ogura, M., Sato, A., Sakaki, Y., Shimizu, Y., Baba, N., and Nagasawa, K., Effect of modified magnetic field on the ocean migration of maturing chum salmon, *Oncorhynchus keta*, *Mar Biol*, 129, 523, 1997.

Yano, A., Hidaka, E., Fujiwara, K., and Iimoto, M., Induction of primary root curvature in radish seedlings in a static magnetic field, *Bioelectromagnetics*, 22, 194, 2001.

Yoon, R.S., DeMonte, T.P., Hasanov, K.F., Jorgenson, D.B., and Joy, M.L.G., Measurement of thoracic current flow in pigs for the study of defibrillation and cardioversion, *IEEE Trans Biomed Eng*, 50, 1167, 2003.

Yoshida, H., Ueno, S., Cheyne, D., and Weinberg, H., Measurements of low frequency brain magnetic fields associated with four-tone memory processes, *IEEE Trans Magn*, 31, 4268, 1995.

Zangaladze, A., Epstein, C.M., Grafton, S.T., and Sathian, K., Involvement of visual cortex in tactile discrimination of orientation, *Nature*, 401, 587, 1999.

Zhadin, M.N., Review of Russian literature on biological action of DC and low-frequency AC magnetic fields, *Bioelectromagnetics*, 22, 27, 2001.

Zhadin, M.N., Novikov, V.V., Barnes, F.S., and Pergola, N.F., Combined action of static and alternating magnetic fields on ionic current in aqueous glutamic acid solution, *Bioelectromagentics*, 19, 41, 1998.

Zhadin, M.N., Deryugina, O.N., and Pisachenko, T.M., Influence of combined DC and AC magnetic fields on rat behavior, *Bioelectromagnetics*, 20, 378, 1999.

Zhang, Q.-M., Tokiwa, M., Doi, T., Nakahara, T., Chang, P.-Q., Nakamura, N., Hori, M., Miyakoshi, J., and Yonei, S., Strong static magnetic field and the induction of mutations through elevated production of reactive oxygen species in *Escherichia coli* soxR, *Int J Radiat Biol*, 79, 281, 2003.

Zmyslony, M., Palus, J., Jajte, J., Dziubaltowska, E., and Rajkowska, E., DNA damage in rat lymphocytes treated *in vivo* with iron cations and exposed to 7 mT magnetic fields (static or 50 Hz), *Mutat Res*, 453, 89, 2000.

9

The Ion Cyclotron Resonance Hypothesis

A.R. Liboff

CONTENTS

9.1 Introduction

9.1.1 General Remarks

Ion cyclotron resonance (ICR) is one among a number of possible mechanisms that have been advanced to explain observed interactions between weak low-frequency electromagnetic fields and biological systems. Despite the failure to find a reasonable physical explanation, there remains an impressive body of experimental evidence that can be taken as an empirical basis for this hypothesis. The ICR suggestion has proven fruitful in framing both experimental and theoretical work, despite the biophysical situation being far from the literal cyclotron resonance model of an isolated classical charged particle moving in a vacuum under the influence of a magnetic field.

The properties of the applied fields that are used in ICR experiments include linear or circular polarization, the presence of a finite magnetostatic field, frequencies ranging from a few to several hundred hertz, magnetic intensities ranging from about 1 μT to 1 mT, and, most important, a directional constraint on the relative orientation of the time-varying electromagnetic field to the magnetostatic (DC) field. This orientation requires that time-varying magnetic fields be parallel to the DC field or, equivalently, that time-varying electric fields are perpendicular to the DC field.

The ICR hypothesis holds that the physiological activity of those ions implicated in cell signaling processes, including, among others, Ca^{2+}, Mg^{2+}, and K^+, can be altered when the ratio of applied signal frequency to the static magnetic field is equal to the ionic charge-to-mass ratio. This is expressed as

$$\omega/B = q/m \tag{9.1}$$

where the radial frequency $\omega = 2\pi f$, as measured in radians per second, is used instead of f, the frequency measured in hertz. In SI units, B is the DC field intensity measured in tesla, and q/m is the ratio of the ionic charge to mass, in coulombs per kilogram. For any given ionic species, the specific frequency that equals the product of B and q/m is called the cyclotronic frequency, ω_c.

The resonance concept is attractive for a number of reasons. There is a potential connection to interactions involving the Earth's magnetic field (geomagnetic field [GMF]). Further, the ICR mechanism may help provide the basis for at least some of the reports of low-frequency electromagnetic interactions that otherwise lack explanation. Finally, given the wide variety of biological systems in which ICR effects are observed, it is reasonable to ask if there are fundamental scientific questions connected to this phenomenon.

The ICR hypothesis has especial significance attached to magnetostatic fields whose intensity is of the order of the GMF (20–60 μT). This becomes apparent when the charge-to-mass ratios of key biological ions are substituted into Equation 9.1. These ratios range from about 2 to 8 $\times 10^6$ C/kg, implying that a static magnetic field of 50 μT corresponds to resonance frequencies of the order of 10–100 Hz (Figure 9.1). Such frequencies could

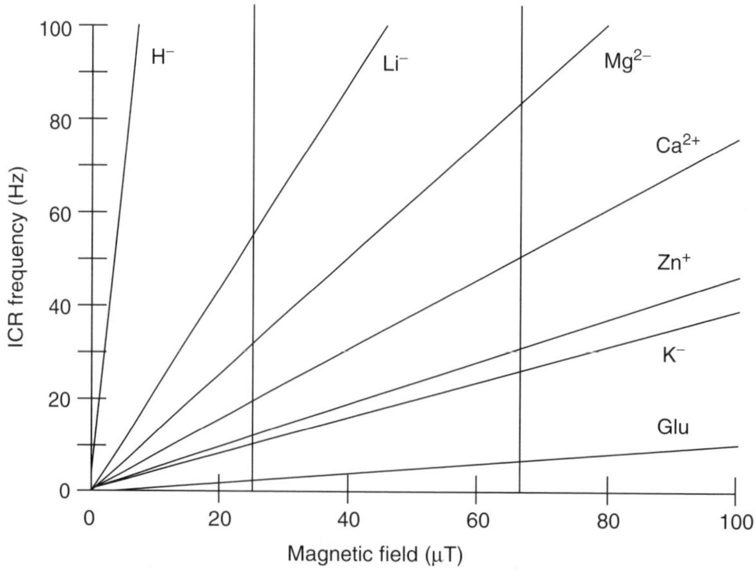

FIGURE 9.1
Ion cyclotron resonance frequencies for many biologically important ions in the Earth's magnetic field are in the ELF range.

TABLE 9.1

ICR Cation Possibilities

Ion	q/m (C/kg) $\times 10^{-6}$	f/B (Hz/μT)
H^+	95.76	15.241
Li^+	13.90	2.212
Mg^{2+}	7.937	1.263
H_3O^+	5.066	0.807
Ca^{2+}	4.814	0.766
Zn^{2+}	2.951	0.470
K^+	2.467	0.393
Arg^{2+}	1.235	0.197
Asn^+	0.838	0.133
Glu^+	0.747	0.119
Tyr^+	0.591	0.094

conceivably have physiological significance since they correspond approximately to the frequency range generated in the central nervous system [1]. This, coupled to the focus on the potential hazards attached to 50/60-Hz electromagnetic power delivery sources [2], has sparked study of the ICR hypothesis, in terms of both experiments specifically designed to test this hypothesis as well as theoretical models seeking an explanatory basis at the molecular level.

Some specific ions that have been implicated are listed in Table 9.1. Note that four polar amino acids and the hydronium ion are included. The ratios of frequency to DC magnetic field, as calculated from Equation 9.1, are shown in the right-hand column. This ratio can be regarded as an invariant characteristic for any given ion.

Although experimental evidence provides support for the ICR hypothesis [3], there is no widely accepted theoretical explanation. Indeed, because of constraints mainly arising from unfavorable damping conditions, there are strong arguments [4] against the occurrence in living tissue of any classical ICR mechanism [5], as occurs, say, for energetic charged particles moving in a vacuum under the influence of parallel static and AC magnetic fields. The circular and helical paths associated with such undamped motion are invariably the result of the Lorentz force, which imparts an acceleration **a** to a charged particle of mass m moving at velocity **v** in a magnetic field **B**:

$$\mathbf{a} = (q/m)(\mathbf{v} \times \mathbf{B}) \tag{9.2}$$

Nevertheless, arguments have been raised [6–11] that although the biological response may not correspond to the effects resulting from ICR-specific helical pathways of charged particles [4], the coupling is nevertheless a function of the ICR frequency as predicted by Equation 9.1. Although there has been no consistent experimental verification for any of these models, there is little question concerning the observed dependence on the cyclotron resonance frequency. Because the cyclotronic frequency is the common denominator in all these models, it is preferable to subsume all of them under the umbrella term ICR hypothesis.

The great variety of biosystems in which ICR effects have been observed implies a ubiquitous response that may have fundamental physiological significance. One can generalize this response R in terms of its functional dependence. From Equation 9.1, we can write

$$R = R(\omega, B, q/m) \tag{9.3}$$

Lednev [7] added a fourth variable, namely the intensity of the AC magnetic field, B_{AC}. Thus, the expanded expression for the response $R = R(\omega, B, B_{AC}, q/m)$, or, in terms of the two key variables,

$$R = R(\omega_c, B_{AC}) \tag{9.4}$$

There is no question as to the relevance of B_{AC} in studying the interactions between ICR field combinations and biological systems. However, it is not clear if the experimentally observed dependences on B_{AC} are a direct result of the underlying resonance mechanism, as has been suggested [7,9], or if there are other separate physiological factors that limit the levels of the AC field under which an ICR mechanism may be operative.

9.1.2 Background History

ICR was originally invoked [4] to explain an extraordinary set of observations by Blackman's group [12] indicating a strong dependence on the orientation of the magnetostatic field when studying the Ca-efflux model system [13]. The original discovery of the Ca-efflux effect [13] and subsequent studies [14–17] showed conclusively that the level of $_{45}Ca^{2+}$ efflux from preloaded chick brain was a nonlinear function of low-frequency (ca. 15 Hz) modulation signals when these brains were exposed to high-frequency carrier electric fields. Typically, this nonlinear signature (Figure 9.2), at first referred to as a "window," has the appearance of a resonance curve. The Blackman experiment [12] discovered that this resonance signature appeared only when certain specific values of the vertical DC magnetic field were superposed on the system. In Table 9.2, "Yes" indicates the appearance of a resonance signature for a given combination of f and B.

In addition to Blackman's original set of results, a fourth column has been added in Table 9.2 to show the putative charge-to-mass ratio as determined from Equation 9.1. One sees that the sign of the magnetic field direction, either pointing up or down, does not affect the outcome. The specific combination of 15 Hz and 38 μT is positive, as is the combination of 30 Hz and 76 μT, suggesting that the ratio of frequency to field is involved as a key factor.

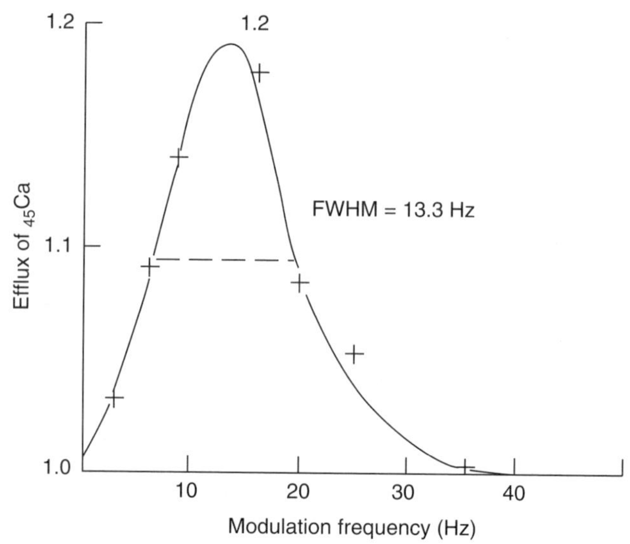

FIGURE 9.2
Comparison of shape of "window" data for Ca-efflux results [13] (seven points) with predicted resonance curve [18] (smooth curve). The best fit is for the charge-to-mass ratio for K^+, a magnetostatic field of 35.0 μT, and a collision time of 0.026 sec.

TABLE 9.2

Analysis of Blackman [12] Data

f (Hz)	*B* (μT)	Outcome	*q*/*m* (C/kg)
15	38	Yes	2.48×10^6
15	19	No	4.96×10^6
30	38	No	4.96×10^6
30	76	Yes	2.48×10^6
30	−76	Yes	2.48×10^6
30	50	No	3.77×10^6
30	25	Yes	7.54×10^6
30	−25	Yes	7.54×10^6
30	−83	No	2.27×10^6

Despite the fact that calcium was explicitly measured in this and the earlier Ca-efflux experiments, the data in Table 9.2 give reason to believe that the potassium ion was the primary target for the electromagnetic interactions. First, note that the charge-to-mass ratio of 2.48×10^6 C/kg is associated with positive outcomes. This ratio is less than 0.5% different from the q/m ratio for the potassium ion as shown in Table 9.1. The evidence linking K^+ to a positive outcome is further strengthened by examining the results obtained for the combination of 30 Hz and 25 μT. The positive outcome in this case suggests a q/m value of 7.54×10^6 C/kg, three times larger than the q/m ratio for K^+. In cyclotron resonance, one typically observes a set of resonance frequencies ω_n, where the fundamental at $n = 1$ is given in Equation 9.1, and the higher harmonic frequencies are restricted to the odd [19–21] harmonics $n = 3, 5, 7, \ldots$. The f/B ratio of 30/25 Hz/μT is nearly three times larger than the ratio 15/38 Hz/μT, again implying that the K^+ ion is interacting with the magnetic field, this time as a result of an excitation at the third harmonic.

There is further evidence that the K^+ ion is an important interactive factor in the nonlinear effects observed in the Ca-efflux experiments. McLeod and Liboff [18,22] derived the resonance signature for a charged particle as a function of frequency, showing that the relative conductivity, with and without the presence of an ion resonance field combination, is

$$\frac{\sigma_x}{\sigma_0} = \frac{(1 + (\omega_c + \omega)^2 \tau^2)}{(1 + [(\omega_c^2 - \omega^2)\tau^2]^2 + 4\omega^2\tau^2)} \tag{9.5}$$

This is a typical resonance expression that includes the effects of damping, expressed in terms of the collision time τ. By varying the choices of q/m ratios and collision times this expression can be directly compared to the results of Bawin and Adey [13], as shown in Figure 9.2. The smooth curve that best fits Equation 9.5 to the experimental points is also shown in Figure 9.2. This fitting procedure reveals that the most likely explanation for the data involves charged particles in cyclotron resonance with a q/m ratio equal to that of the potassium ion.

Thus, two independent sets of Ca-efflux data, one with DC magnetic fields applied as part of the experiment [12] and the other with the ambient magnetic field in the laboratory playing an unsuspected role [14], yield the same conclusion, that ICR stimulation of the K^+ ion results in the nonlinear resonance response.

9.2 Experimental Evidence

There is a surprisingly wide variety of biological systems in which ICR effects are observed. This suggests a heretofore unknown electromagnetic biological interaction. The model systems that have been examined in the literature can be conveniently divided into the categories of bone, cell culture, rat behavior, neural cell culture, diatom motility, complex biological systems, plants, and cell-free systems. These eight separate broad categories are listed respectively in Table 9.3 through Table 9.10. There is some

TABLE 9.3

ICR Effects in Skeletal Tissues

Frequency (Hz)	B_0 (μT)	Tuning	Ratio B/B_0	Comments	Reference
100	130	Ca^{2+}	1.0	Enhanced cell (fibroblast) proliferation; B varied between 50 and 500 μT	23
75	98	Ca^{2+}	1.0	Enhanced proliferation	23
16	42	Ca^{2+}	1.0	Enhanced proliferation	23
16	20.9	Ca^{2+}	1.0	Increases in rudiment length and midshaft diameter in embryonic chick femur	24
16	12.7	Mg^{2+}	1.0	Similar results to Ca^{2+} tuning	24
16	40.9	K^{+}	1.0	Results opposite to Ca^{2+} and Mg^{2+} cases: bone growth inhibited	24
80	20	Ca^{2+}/Mg^{2+}	1.0	Both fifth Ca^{2+} to third Mg^{2+} harmonics; enhanced collar thickness and length	24
72.6–80.6	20	Ca^{2+}	1.0	Resonance in IGF-II concentration at 76.6 Hz in osteosarcoma cell line	25
76.6	20	Ca^{2+}	1.0	Fifth harmonic: enhanced proliferation in osteosarcoma and human bone cells	25
15.3	20	Ca^{2+}	1.0	Reduction in tissue growth factor (TGF) β-1 inhibition in chondrocyte culture	26
25.4	20	Mg^{2+}	1.0	Reduction of TGF β-1 inhibition in chondrocyte culture	26
76.6	20	Ca^{2+}/Mg^{2+}	1.0	Reduction of TGF β-1 inhibition in chondrocyte culture	26
15.3	20	Ca^{2+}	1.0	Enhanced protoglycans synthesis in bovine cartilage	26
76.6	20	Ca^{2+}/Mg^{2+}	1.0	Mixed third and fifth harmonics reduce bone loss related to castration in rats	27
14.3–18.3	20		1.0	Increase in $_{45}Ca$ maximized at 16.3 Hz in osteosarcoma cell line	28
14.3–18.3	20		1.0	Increase in $_{45}Ca$ maximized at 15.3 Hz in a different osteosarcoma cell line	28
15.3	20	Ca^{2+}	1.0	370% increase in stiffness in ostectomized rabbit fibula after 24 h/28 d exposure	29
25.4	20	Mg^{2+}	1.0	137% increase in stiffness in ostectomized rabbit fibula after 24 h/28 d exposure	29
15.3	20	Ca^{2+}	1.0	Enhanced DNA synthesis and IGF-II levels in osteosarcoma cell line	30
13.3–17.3	20		1.0	Resonance maximum in IGF-II receptor number and affinity at 15.3 Hz	31
16	20.9	Ca^{2+}	1.0	Enhanced chick femoral diameter and glycosaminoglycans (GAGS) content	32
16	12.7	Mg^{2+}	1.0	Large (90%) GAGS enhancement	32
16	40.7	K^{+}	1.0	Opposite effects for Ca^{2+} and Mg^{2+} tuning, replicating Smith et al. [24]	32

TABLE 9.4

ICR Effects in Cell Culture

Frequency (Hz)	B_0 (μT)	Tuning	Ratio B/B_0	Comments	Reference
14.3	21	$_{45}Ca^{2+}$	1.0	Isotopic shift in q/m resonance confirmed, $_{40}$Ca to $_{45}$Ca	33
14.3	21	$_{45}Ca^{2+}$	1.0	Threefold incorporation of $_{45}$Ca into human lymphocytes	33
14.3	21	$_{45}Ca^{2+}$		Effect on human lymphocytes disappears at larger AC intensity	33
14.3	20.9	$_{45}Ca^{2+}$	1.0	2.3-fold uptake in $_{45}$Ca disappears with addition of calcium blocker nifedipine	34
38.15	50	Ca^{2+}	1.0	No effect on Ca^{2+} in four different cell lines as observed using calcium fluorochrome fura-2	35
16	20.9	Ca^{2+}	1.0	Enhanced proliferation (46%) for fibroblast culture at Ca^{2+} ICR tuning	36
?	20.9	K^+	1.0	Reduced proliferation (18%) for Raji cells exposed to K^+ ICR; frequency not provided	36
13.6	16.5	$_{45}Ca^{2+}$	1.2	ICR frequency off by 3.5 Hz; enhanced $_{45}Ca^{2+}$ levels (75–126%) in three cell lines	37
60	20	$_{45}Ca^{2+}$	1.0	Fifth harmonic for $_{45}Ca^{2+}$ uptake is enhanced by 37%	37
16	23.4	$_{45}Ca^{2+}$	1.8	Decreased Ca^{2+} influx in mitogen-activated lymphocytes but no effect on resting cells	38
16	51.1	K^+	1.0	Third harmonic: enhanced proliferation of lymphoma cells; very narrow FWHM[a]	39
16	40.9	K^+	1.0	Enhanced proliferation of human lymphoma cells	39
16	23.4	Ca^{2+}	3.8, 5.3	No change in Ca^{2+} influx at AC/DC ratio of 3.8 but enhanced influx at ratio of 5.3	40
16	20.9	Ca^{2+}	1.4	No effect on mouse lymphocytes as observed using Ca fluorochrome Quin-2	41
50	65.3	Ca^{2+}	1.4	No effect on mouse lymphocytes with and without mitogenic stimulation	41
5–100	50–60	?Ca^{2+}	2.3–3.0	Enhanced calcium oscillations over broad frequency range, maximized at 50 Hz	42
32	42	Ca^{2+}	2.5, 5.0	Increased micronuclei formation in human lymphocytes at Ca^{2+} ICR tuning	43
32.50	0			No change in micronuclei formation when DC field is zero	43
15.3	20	Ca^{2+}	1.0	ICR effect on fura-2 calcium activity only found for added serum in cell medium	44
76.6	20	Ca^{2+}	1.0	Fifth harmonic is also successful	44
100	130	Ca^{2+}	1.9	Another ICR fundamental successful	44
100	130	Ca^{2+}	2.8	Repeating ICR application to primary bone cell culture at higher AC intensity	44

[a]FWHM, full width at half maximum.

unavoidable overlap among these, particularly in Table 9.3 (skeletal systems), where references to bone research in cell cultures and animals are grouped together.

Although most of the reports summarized in Table 9.3 through Table 9.10 lend considerable weight to the hypothesis that ICR magnetic stimulation can affect biological systems, the effects on diatom motility (Table 9.7) are not as clear-cut, in that a number of observers [62–65] failed to find any effects whatsoever. The explanation for the poor reproducibility in this case may rest with difficulties in handling the diatom model system, one that is especially sensitive to sample preparation.

TABLE 9.5

ICR Effects on Rat Behavior

Frequency (Hz)	B_0 (μT)	Tuning	Ratio B/B_0	Comments	Reference
60	26	Ca^{2+}	1.9	Third harmonic: loss of short-term (temporal) memory in rats	45
60	26	Ca^{2+}	1.9	Third harmonic: AC threshold observed (27 μT) for above results	46
60	27	Ca^{2+}	1.9	Third harmonic: learning inhibited relative to controls	47
60	48	Mg^{2+}	1.0	Learning enhanced relative to controls	47
60	26	Ca^{2+}	1.9	Third harmonic: no effect	48
50	65	Ca^{2+}		Reduced short-term memory and aggressiveness	49
630	500	Mg^{2+}	0.5	Enhanced exploratory activity	50
380	500	Ca^{2+}	0.5	Reduced exploratory activity	50
63	50	Mg^{2+}	0.7	Enhanced locomotor and exploratory activity	51
38	50	Ca^{2+}	0.7	Reduced locomotor and exploratory activity	51
630	500	Mg^{2+}	0.7	Enhanced locomotor and exploratory activity	51
380	500	Ca^{2+}	0.7	Reduced locomotor and exploratory activity	51

TABLE 9.6

ICR Effects on Neural Cell Culture

Frequency (Hz)	B_0 (μT)	Tuning	Ratio B/B_0	Comments	Reference
16	15–40.8	? Co^{2+} or Fe^{2+}?	0.5–1.3	Enhanced proliferation over controls (60%) in neuroblastoma cell culture	52
16	15–40.8		0.5–1.3	Decreased neurite outgrowth for ?Co^{2+}/Fe^{2+} ICR stimulation; possible Na^+ ICR effect?	52
15.3	20	Ca^{2+}	1.0	Ca^{2+} tuning increases rate of neuronal differentiation in PC-12 cells	53
45	36.6	Mg^{2+}	0.03–1.81	Changes in neurite outgrowth for Mg^{2+} ICR at different AC intensities in PC-12 cells	54
25	20.3	Mg^{2+}	0.54–1.26	Similar Mg^{2+} ICR effect on PC-12 neurite outgrowth at another frequency	54
45	2.96	H^+	0.14–2.0	H^+ ICR alters PC-12 neurite outgrowth	55
30	1.97	H^+	0.57–1.4	Similar effects at different ICR combinations	55
45	59	Ca^{2+}	0.26–1.49	PC-12 cells at Ca^{2+} ICR exhibit changes in neurite outgrowth at different AC intensities	56
42.5–47.5	2.97	H^+	0.56–1.5	Bandwidth for PC-12 neurite outgrowth due to H^+ ICR is ±10%	57
40, 50	2.97		0.56–1.5	No effect	57

TABLE 9.7

ICR Effects on Diatom Motility

Frequency (Hz)	B_0 (μT)	Tuning	Ratio B/B_0	Comments	Reference
5–32	20.9		1.0	Maximum motility occurs at 16 Hz when Ca^{2+} concentration is 0.25 nM; no effect when fields are at 90°	58
16	20.9	Ca^{2+}	0.0–3.0	Motility maximized when B/B_0 ratio is 1	58
16	20.9	Ca^{2+}	0.7	Enhanced motility	59
32	20.9	Ca^{2+}	0.7	Even harmonic: no effect	59
48	20.9	Ca^{2+}	0.7	Third harmonic: enhanced motility	59
64	20.9	Ca^{2+}	0.7	Even harmonic: no effect	59
8	10.45	Ca^{2+}	1.4	Enhanced motility	60
12–64	15.7–83.6	Ca^{2+}		Additional ICR frequencies at 12, 16, 23, 31, 32, 46, 64 Hz also enhance motility	60
24, 40, 120	10.45	Ca^{2+}	1.4	Three ICR harmonics for 10.45 μT ($n = 3, 5, 15$) enhance motility	60
16–136	10.45	Ca^{2+}	1.40.73	Thirteen other frequencies ($n = 2, 4, 6, 7, 8, 9, 10, 11, 12, 13, 14, 16, 17$) fail to show effect	60
8	20.45	K^+	0.73	Motility inhibited	60
16	41	K^+	0.37	Motility inhibited	60
24, 40, 120	20.45	K^+	0.37	Three ICR harmonic frequencies for $B_0 = 20.45$ μT ($n = 3, 5, 15$) also inhibit motility	60
16–136	20.45	K^+	0.37	Thirteen other frequencies ($n = 2, 4, 6, 7, 8, 9, 10, 11, 12, 13, 14, 16, 17$) fail to show effect	60
16	21	Ca^{2+}	1.0	Enhanced motility	61
16	20.9	Ca^{2+}	1.0	No effect on motility	62
30	39.2	Ca^{2+}	1.0	No effect on motility	62
60	78.4	Ca^{2+}	1.0	No effect on motility	62
16	20.9	Ca^{2+}	1.0	No effect on motility	63
16	21	Ca^{2+}	1.0	No effect on motility	64
16	20.9	Ca^{2+}	1.0	No effect on motility as viewed with real-time video system	65

9.2.1 Rat Behavior

On the other hand, some of the experimental evidence merits special emphasis because of the way the results have been positively replicated and reinforced. This is particularly true of the work on rat behavior (Table 9.5), in which four independent groups [45,47,49,51] observed significant changes in behavior for ICR exposures that were tuned to either the calcium or the magnesium ion. The end points of these experiments included changes in short-term memory [45,51], learning capacity [47], and aggressiveness [49], behavioral factors that are conceivably interconnected. Most important, these were undoubtedly resonance effects, since changes were not observed when separate runs were made for exposures to either the AC magnetic field alone or the DC magnetic field alone. Only when the AC and DC magnetic fields were jointly applied and parallel and, moreover, when the combined field characteristics conformed to Equation 9.1 were the altered behavioral responses observed.

An interesting possible explanation for the neural interaction site in these experiments has been proposed by Lovely et al. [47,92]. Using a Y-maze setup this group observed precisely opposite learning abilities for Ca^{2+} tuning and for Mg^{2+} tuning. Since the ICR combined field affects learning capacity oppositely for Ca^{2+} tuning and Mg^{2+} tuning, it may be reasonable to assume that the glutamate receptor *N*-methyl D-aspartate (NMDA) is

TABLE 9.8

ICR Effects on Complex Biological Systems

Frequency (Hz)	B_0 (μT)	Tuning	Ratio B/B_0	Comments	Reference
3–770	10–220		0064–2.8	No effect on turtle colon transepithelial current as measured in Ussing chamber	66
33.7	44	Ca^{2+}	1.41	Synthesis and release of rat pineal melatonin is reduced by Ca^{2+} ICR tuning	67
15	21	Ca^{2+}	6.7	Fluctuations in heart rate in Daphnia are maximized at Ca^{2+} ICR frequency	68
60	78.4	Ca^{2+}	0.13	Cephalic regeneration in planaria is delayed by 48 h	69
60	51.1	K^+	1.0	Regeneration rate unchanged when K^+ tuning is used instead of Ca^{2+} tuning	69
60	78.4	Ca^{2+}	0.51	Regeneration anomalies occur at Ca^{2+} tuning when larger AC intensity is used	70
16	20.9	Ca^{2+}	1.8	Enhanced rate of blastema growth during cephalic regeneration in planaria	71
16	20.9	Ca^{2+}	0.24–9.6	Evidence that ICR effect on planaria regeneration has an intensity window	71
30	39.1	Ca^{2+}	1.8	Evidence corroborating Lednev [7] prediction: ICR effects are maximized at AC/DC ratio of 1.8	72
60	78.1	Ca^{2+}	1.8	Effect of light on ICR modulation of snail opioid analgesia is independent of frequency	72
120	156.2	Ca^{2+}	3.6	Effect due to light appears to scale with DC intensity	72
60	78	Ca^{2+}	0–5.3	Ca^{2+} ICR variations with AC intensity support Lednev [7] model	73
30	76	K^+	0–2.8	K^+ ICR effects on snail opioid analgesia are reversed with K^+ channel blocker	73
35	45	Ca^{2+}	1.8	Maximum influence on bioluminescence of dinoflagellate, agreement with PRM model	74
35	45	Ca^{2+}	5.3	Influence reversed, again in agreement with PRM model	74
35	45	Ca^{2+}	3.8	No effect at ratio of 3.8, again in agreement with PRM model	74

involved. NMDA receptors act as a graded switch for memory formation, to enhance learning and memory [93], and it is well established that NMDA activity is differently sensitive to calcium and magnesium concentrations [94,95]. Similar reversals of behavioral outcome depending on which ions are tuned have been observed by Zhadin et al. [51].

This explanation also serves to reinforce the original suggestion [4] concerning the molecular explanation for ICR stimulation, namely, in terms of enhanced ionic permeability within ion channels. Further support for locating the ion channel as the site of magnetic interaction is the fact that the changes in Ca^{2+} concentration within the cell that result from ICR stimulation tuned to the Ca^{2+} ion are not observed with the addition of nifedipine [34], a well-known calcium ion channel blocker (Figure 9.3).

9.2.2 Plants

Highly consistent results have also been independently obtained in studying the effects of ICR stimulation on plant growth [75–77,96] and seed germination [20,78] (Table 9.9,

TABLE 9.9

ICR Effects on Plants

Frequency (Hz)	B_0(μT)	Tuning	Ratio B/B_0	Comments	Reference
60	78.3	Ca^{2+}	0.26	Ca^{2+} ICR field combination stimulates radish growth after delaying germination	75
60	153.3	K^+	0.13	K^+ ICR field enhances germination while reducing growth	75
60	0			No effect	75
60	78.4	Ca^{2+}	0.26	Ca^{2+} fundamental stimulates growth but slows down germination	20
60	39.2	Ca^{2+}	0.51	Ca^{2+} second harmonic: no effect	20
60	26.1	Ca^{2+}	0.77	Ca^{2+} third harmonic: same result as Ca^{2+} fundamental	20
60	153.3	K^+	0.13	K^+ fundamental results opposite to those of Ca^{2+}: growth inhibited, germination enhanced	20
60	76.6	K^+	0.26	K^+ second harmonic results weakly opposite to fundamental and third harmonics	20
60	51.1	K^+	0.39	K^+ third harmonic: effect same as K^+ fundamental	20
60	47.5	Mg^{2+}	0.42	Mg^{2+} fundamental stimulates growth	20
60	9.5	Mg^{2+}	2.11	Mg^{2+} fifth harmonic stimulates growth	20
60	0			AC only: no effect	20
60	78.3	Ca^{2+}	0.26	Replication of Smith et al.'s [75] work	76
60	78.3	Ca^{2+}	0.26	No effect on mustard plant; possible effect on barley plant	76
50	65.3	Ca^{2+}	0.61	Replication of Smith et al.'s [75] work on radish, for the 50-Hz Ca^{2+} ICR condition	77
50	39.6	Mg^{2+}	0.60	Replication of Smith et al. [75] for 50-Hz Mg^{2+} condition	77
60	76.3	Ca^{2+}	0.26	Germination weakly enhanced following Ca^{2+} ICR exposure of dry radish seeds	78
60	153.5	Mg^{2+}	0.13	No effect on germination	78
60	47.6	K^+	0.42	Significantly greater (earlier) germination of dry seeds following K^+ ICR exposure	78
35.8	46.5	Ca^{2+}	1.84	Gravitropic response in millet, flax, and clover seedlings enhanced by Ca^{2+} ICR	79
58.7	46.5	Mg^{2+}	1.84	Gravitropic response unaffected by Mg^{2+} tuning	79
54.7	46.5	K^+	1.84	Gravitropic response inhibited by K^+ ICR	79
33.8–37.8	46.5	Ca^{2+}	1.84	Frequency-dependent gravitropic response exhibits Ca^{2+} peak: FWHM[a] = 1.6 Hz	80
60	48	Mg^{2+}	1.48	CO_2 uptake significantly below control in radish, replication of Smith et al. [75]	81

[a]FWHM, full width at half maximum.

Figure 9.4). The approach in the earlier reports [75] involved direct observations of aspects of plants that are readily measurable: plant height, aboveground height, root mass, stem diameter, leaf length, and width. Remarkably, all aspects related to growth are significantly affected, suggesting that magnetic fields play some unknown role in plant physiology. As observed in other systems (see Table 9.3), Ca^{2+} and Mg^{2+} tuning tends to enhance growth while tuning to the potassium ion acts as an inhibitor. Radish (*Raphanus sativus*) was used because of its rapid growth cycle (21 d), ease of handling, and seed availability. Davies [76] observed positive results when stimulating radish with ICR

TABLE 9.10

ICR Effects in Cell-Free Systems

Frequency (Hz)	B_0 (μT)	Tuning	Ratio B/B_0	Comments	Reference
8–20	20.9		1.0	Three frequencies (13.0, 14.0, and 16.0) affect calmodulin-dependent phosphorylation	82
100	0–260		>0.45	No Ca^{2+} ICR effect on conductance in pure bilipid layer	83
50–120	0–299		>0.31	Binding of Ca^{2+} to calmodulin is not enhanced using Ca^{2+} ICR fields	84
20	50	K^+	1.4	No changes from ICR tuning in gram A channel conductance in lipid bilayer	85
760	50	H^+	15.2	No changes from ICR tuning in gram A channel conductance in lipid bilayer	85
0.1–40	25	Asn^+	0.002	Enhanced aqueous conductivity in asparagine solution at 2.9 Hz; ICR prediction 2.9 Hz	86
0.1–40	25	Arg^{2+}	0.002	Enhanced aqueous conductivity in arginine solution at 4.4 Hz; ICR prediction 4.4 Hz	86
0.1–40	25	Glu^+	0.002	Enhanced aqueous conductivity in glutamic acid solution at 2.5 Hz; ICR prediction 2.6 Hz	86
0.1–40	25	Tyr^+	0.002	Enhanced aqueous conductivity in glutamic acid solution at 1.9 Hz; ICR prediction 2.1 Hz	86
0.1–40	25		0.2	ICR effect disappears at higher ratio of AC to DC intensities	86
0.1–40	0		0.002	ICR effect disappears for very small DC fields	86
0.1–40	25		0.002	ICR effects disappear when AC magnetic field is at 90° to DC field	86
12–60	20.9		1.0	No change in Ca^{2+} transport through patch-clamped cell membrane	87
10–22	20.9		1.0	No change in Ca^{2+} transport through patch-clamped system, measured over loner times	87
1–10	20–40	Glu^+	625–1.25 (×.001)	Changes in glutamic acid conductivity in solution; good agreement with Glu^+ ICR q/m prediction	88
1–10	40		0.25–2.0 (×.001)	Confirmation of earlier work; amino acid response at AC levels of 0.02–0.04 μT	88
0–10	40	Arg^{2+}	0.001	Sharp change in conductivity observed at 7.1 Hz, the ICR tuning point for Arg^{2+}	11
20.6	48	$H_3O^+(H_2O)$	0.02	ICR fields trigger long-term increases in electrical conductivity in pure water	89
40.1	48	H_3O^+	0.02	Data in agreement with ICR effect in hydronium ion	89
530	35	H^+	0.03	Data in agreement with ICR effect in proton	89

TABLE 9.10 (continued)

ICR Effects in Cell-Free Systems

Frequency (Hz)	B_0 (μT)	Tuning	Ratio B/B_0	Comments	Reference
16	20.9	Ca^{2+}	1.0	Binding of Ca^{2+} to calmodulin is not enhanced using Ca^{2+} ICR fields	90
25.4	37	$_{45}Ca^{2+}$	0.70	$_{45}Ca^{2+}$ efflux in plasma membrane vesicles: ICR peak observed	91
24	37		0–3.2	Agreement with Blanchard and Blackman's [9] IPR model	91

magnetic fields but reported observing no similar effect in mustard plants, implying that the influence on growth may be species specific. ICR effects on radish metabolism were also reported by Yano et al. [81] using a distinctly different assay, the rate of uptake of CO_2 as a surrogate for photosynthesis activity. Still another assay [97] that responds to ICR magnetic stimulation in radish is the optical transmittivity in leaf.

The work in radish was extended [96] to four species of orchid (*Brassavola*, *Encyclium*, *Phalaenopsis*, and *Bulbophuyllum*) (Figure 9.4 and Figure 9.5), with the magnetic exposures tuned to Ca^{2+} ICR lasting months instead of days. In all treated cases, plant heights were significantly higher compared to controls.

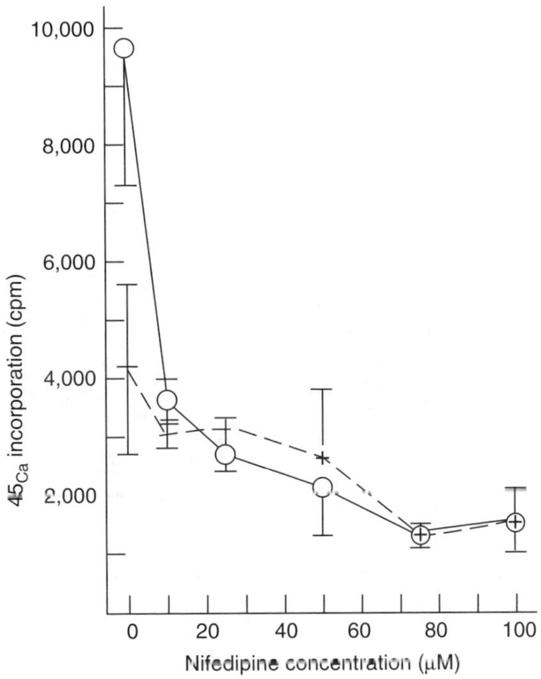

FIGURE 9.3

Incorporation of $_{45}Ca$ in human lymphocytes for unexposed cells (dashed line) and for ICR exposure tuned to the isotopic mass of $_{45}Ca$ (solid line) after 1 h. In the absence of calcium blocker nifedipine, there is a greater than twofold increase in calcium concentration over controls. The addition of nifedipine completely blocks the calcium uptake resulting from ICR stimulation, providing evidence that the ICR mechanism is related to ion channel transport. (From Rozek, R.J., Sherman, M.L., Liboff, A.R., McLeod, B.R., and Smith, S.D., Nifedipene is an antagonist to cyclotron resonance enhancement of $_{45}Ca$ incorporation in human lymphocytes, *Cell Calcium*, 8, 413, 1987.)

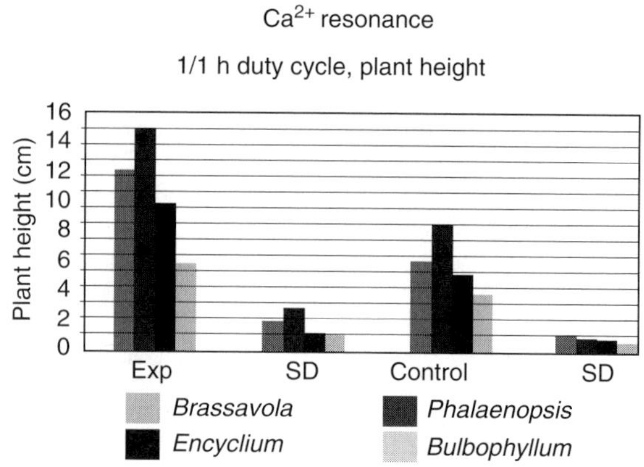

FIGURE 9.4
Comparisons of mean plant heights for four orchid varieties between unexposed controls and plants subjected to Ca^{2+} ICR stimulation. SD, standard deviation. (From Smith, S.D., Liboff, A.R., and McLeod, B.R., Calcium ICR and seedling growth in orchids (abstract), 20th Annual Meeting, Bioelectromagnetics Society, St. Petersburg Beach, FL, 1998.)

A distinctly different type of plant experiment has examined the rate of seed germination [21,78] instead of growth. In this case the assay simply involves a comparison of the time it takes for the seedling to be observed after the exposed seed is planted to the time it takes for a nonexposed seed to emerge. Unlike what is observed when examining growth rates, germination rates are significantly enhanced under K^+ stimulation and inhibited for Ca^{2+} tuning.

9.2.3 Bone

Elaborating on earlier work that used high-intensity pulsed magnetic fields [98] to treat bone disorders, ICR stimulation, operating at a much lower intensity, has proven very useful in repairing bone nonunions [29,99] and as an adjunct in enhancing spinal fusion following surgery [99]. Both these medical applications are approved by the U.S. Food and Drug Administration (FDA) and have been used to treat more than 100,000 patients in the United States [99]. There are two great advantages in these applications compared

FIGURE 9.5
Typical example of difference between treated (right) and unexposed (left) orchid (*Phalaenopsis*) plants. (From Smith, S.D., Liboff, A.R., and McLeod, B.R., Calcium ICR and seedling growth in orchids (abstract), 20th Annual Meeting, Bioelectromagnetics Society, St. Petersburg Beach, FL, 1998.)

to pulsed magnetic fields therapy. Much weaker AC magnetic field intensities, less than 0.1 mT in amplitude, are employed, requiring very small power and providing the patient with greater portability. The treatment is also more efficient, requiring only 30 min/d during the therapeutic regimen.

This FDA-approved ICR device (Figure 9.6) also makes use of harmonic frequencies. The orientation of the local GMF relative to the plane of the AC coil is sensed, and the frequency of the applied sinusoidal magnetic signal generated by the coil is automatically adjusted to be in cyclotron resonance with the GMF following the relation $\omega_n = nqB/m$, where n is an integer representing either the third or the fifth ICR harmonic. Table 9.11 lists the ICR harmonics for a number of ions. In the case of bone, tuning to either Ca^{2+} or Mg^{2+} tends to stimulate bone growth [24] (Figure 9.7 and Figure 9.8). The FDA-approved device makes use of this fact by employing one resonance condition (3.80 Hz/μT) that fits both types of stimulation, namely the third harmonic for Mg^{2+} and the fifth for Ca^{2+}.

It is likely that the level of efficacy of ICR magnetic treatment in repairing bony nonunions is far from optimal. There is good evidence [9,73,82] that the ICR response may depend on the ratio of the AC to DC magnetic fields that are used in combination. As such, one can expect the search for improvements in therapeutic signals for bone repair to continue.

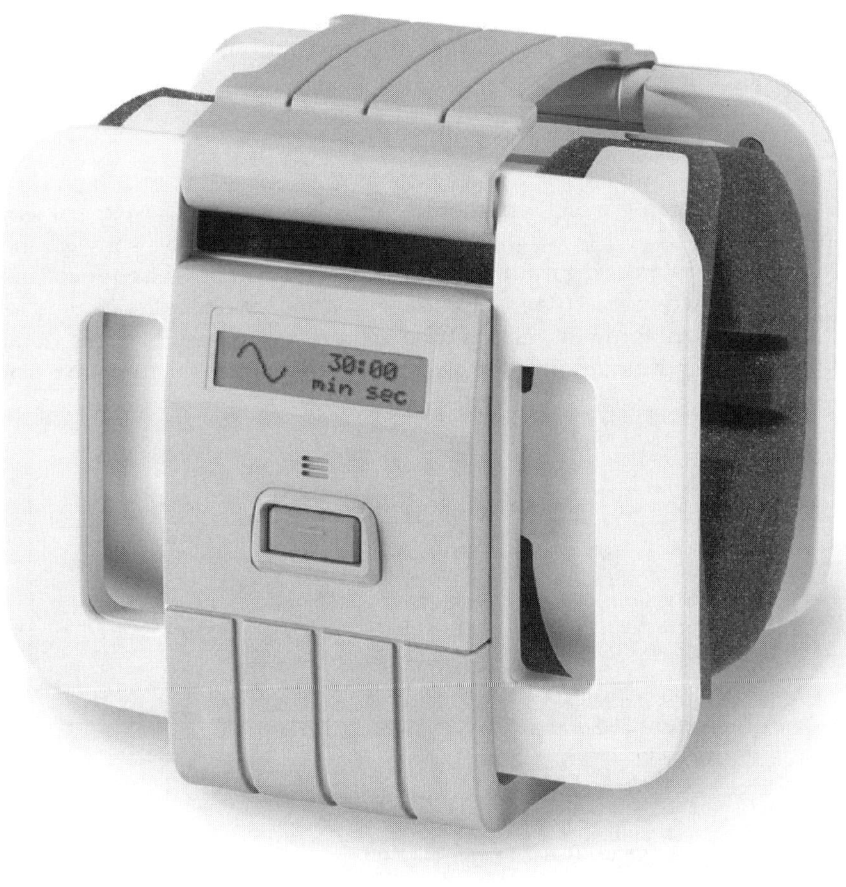

FIGURE 9.6
Device used to assist repair of bony nonunions. The two rectangular parallel coils are clamped over the defect, and the GMF component normal to the plane of these coils is automatically determined regardless of limb orientation. An AC magnetic field is applied parallel to the GMF field that is in ion resonance tuned to harmonics of calcium and magnesium. (Courtesy of OrthoLogic Corp., Tempe, AZ.)

TABLE 9.11

ICR Harmonics for Selected Ions

Ion	Fundamental (Hz/μT) f/B	Higher Harmonics (Hz/μT)		Subharmonics (Hz/μT)	
		$3f/B$	$5f/B$	$f/3B$	$f/5B$
Mg^{2+}	1.26	3.79	6.31	0.421	0.253
Ca^{2+}	0.77	2.30	3.83	0.255	0.153
Zn^{2+}	0.47	1.41	2.35	0.157	0.094
K^{+}	0.39	1.18	1.97	0.131	0.026

In attempting to further probe the response of skeletal tissues to resonant magnetic fields, Ryaby and colleagues [25–28,30,31] reported a number of associated metabolic changes in bone cells, most notably an increase in insulin-like growth factor-II (IGF-II) expression (Figure 9.9 and Figure 9.10) under ICR tuning for Ca^{2+}. This body of work was rather complete in that separate experiments were carried out to justify Equation 9.1. With the magnetostatic field held at one value the frequency was varied as shown in Figure 9.10. In addition, separate runs were made with a fixed frequency and different values of the DC magnetic field. Resonance peaks were observed for both arrangements in accordance with Equation 9.1.

9.2.4 Harmonics

In general, the question of which harmonics are observed in the experimental data remains unresolved. McLeod et al. [59], studying diatom motility, showed that when the DC field is kept constant, odd multiples of the ICR fundamental frequency also result in enhanced motility. The same type of frequency harmonic dependence was found for ICR stimulation of plant growth [21]. Similar experimental results were obtained by Blackman et al. [100] in determining the degree of radioactive calcium flux from chick

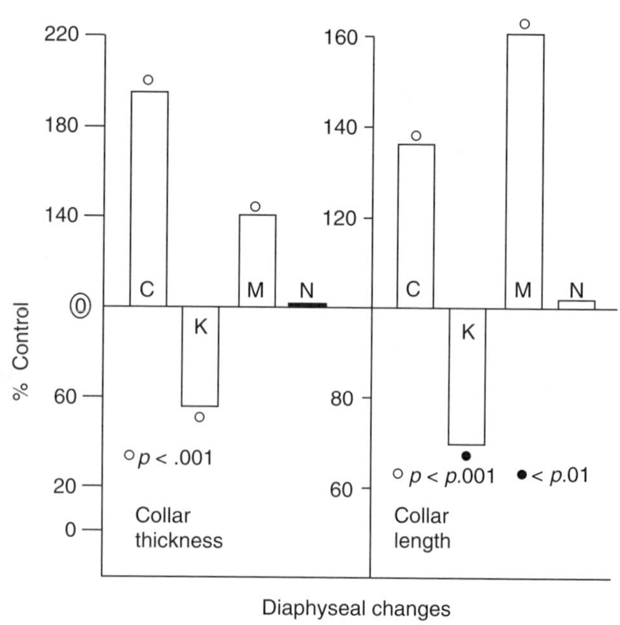

FIGURE 9.7
Ca^{2+} and Mg^{2+} ICR exposures aid bone growth. Changes in diaphysis of embryonic chick femur due to ICR stimulation are shown as percentage of controls. C and M correspond to Ca^{2+} and Mg^{2+} tuning, and K represents the effects of tuning to the q/m ratio for K^{+}. Note the reversal of effect following exposure to potassium-tuned magnetic fields. (From Smith, S.D., Liboff, A.R., and McLeod, B.R., Effects of resonant magnetic fields on chick femoral development *in vitro*, *J. Bioelectr.*, 10, 81, 1991.)

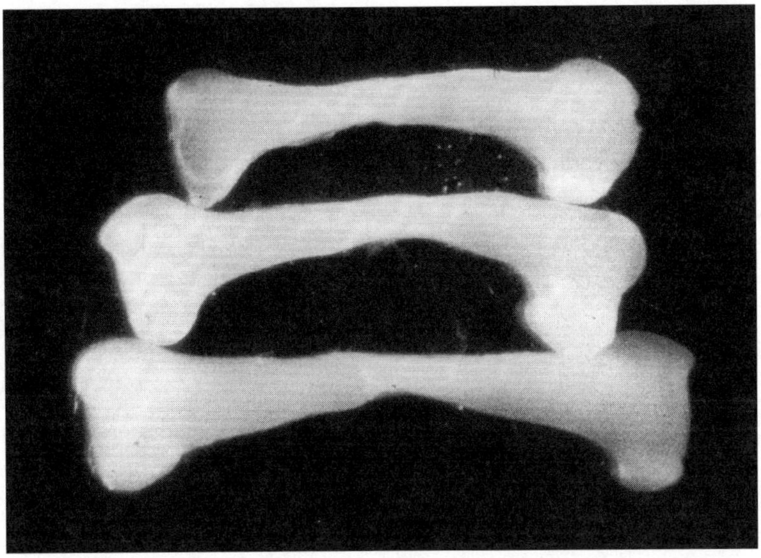

FIGURE 9.8
Typical examples of changes in embryonic chick femora with different ICR exposures. The longest femur corresponds to calcium ion stimulation, the shortest femur to potassium tuning, and the in-between size is the control. (From Smith, S.D., Liboff, A.R., and McLeod, B.R., Effects of resonant magnetic fields on chick femoral development *in vitro*, *J. Bioelectr.*, 10, 81, 1991.)

brain as a function of electric-field modulation frequency. These various findings are in reasonable agreement with the predicted higher harmonic ICR signatures listed in Table 9.11.

Because subharmonics are predicted [7,101] in several theoretical models, the first two predicted odd subharmonic characteristics are also listed for convenience in Table 9.11. It

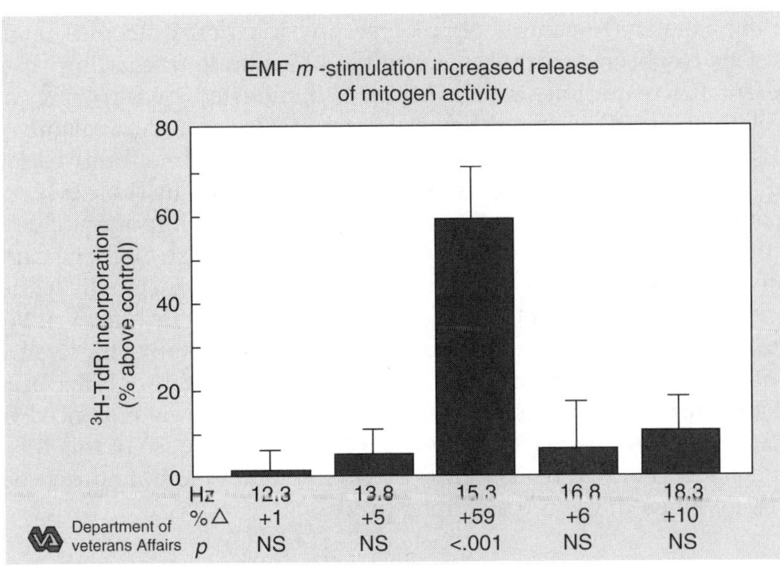

FIGURE 9.9
A sharp increase in mitogen expression in osteosarcoma cells exposed to a 20-μT DC magnetic field occurs when the AC magnetic field frequency is 15.3 Hz, corresponding to Ca^{2+} resonance. (Courtesy of Veterans Administration Hospital, Phoenix, AZ.)

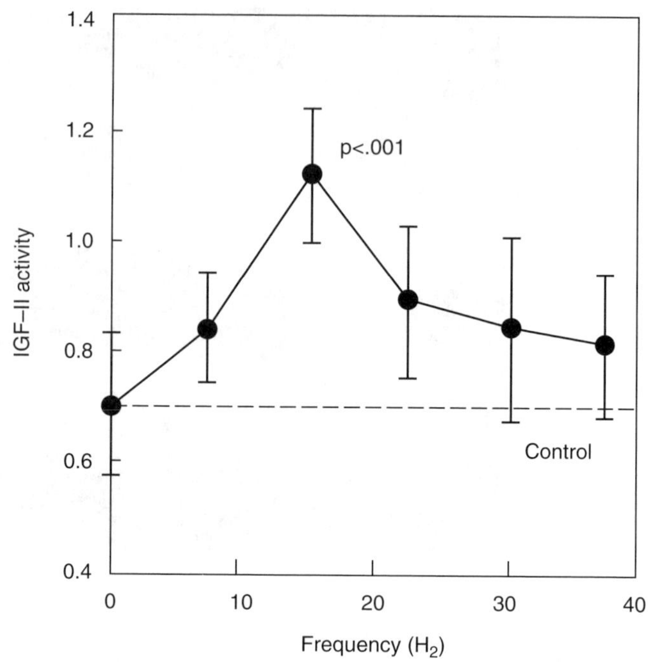

FIGURE 9.10
Peak in IGF-II expression at the fifth harmonic for Ca^{2+}. (Courtesy of J. Ryaby.) Fitzsimmons, R.J., Ryaby, J.T., Magee, F.P., and Baylink, D.J., Combined magnetic fields increase insulin-like growth factor-II in TE-85 human octeosarcoma bone cell cultures, *Endocrinology*, 136, 3100, 1995.)

is clear that a better fit between the theoretical predictions of harmonics and the nature of their experimental observations could be an important factor in further understanding of ICR phenomena in living things.

9.2.5 Physiological Reversals

There is substantial evidence [102] that changing the ICR tuning from one ionic species to another has the capacity to completely reverse physiological outcomes (Figure 9.8 and Figure 9.11). This has been reported in a number of separate fascinating observations in widely different ICR experiments. The effect is manifested by applying, at the same frequency, relatively small shifts in the local magnetic field or, equivalently, applying a changed frequency with the local magnetic field kept the same, both variations made according to Equation 9.1. Eight such sets of results are given in Table 9.12.

This type of opposite response in the data may provide an important clue in trying to understand the ICR effect at the molecular level. Such a response might be expected when looking at the effects of competitive ion concentrations on cell metabolism and signaling. One such example, mentioned above [47], is the difference in the effects of binding Ca^{2+} and Mg^{2+} to NMDA. The concentration of Ca^{2+} in the cytoplasm, itself an implicit function of other ionic concentrations, is carefully regulated through the use of calcium pumps. The production of key enzymes such as cyclic AMP is enhanced or inhibited depending on the cytoplasmic calcium ion concentration. Because of this it is reasonable to think that the action of ICR fields has its origin in membrane-bound ion channels or in those proteins involved in the cell-signaling process.

9.2.6 Water

There have been numerous puzzling reports [103–105] claiming that the physical properties of water (e.g., conductivity) can be sufficiently altered by exposure to magnetic

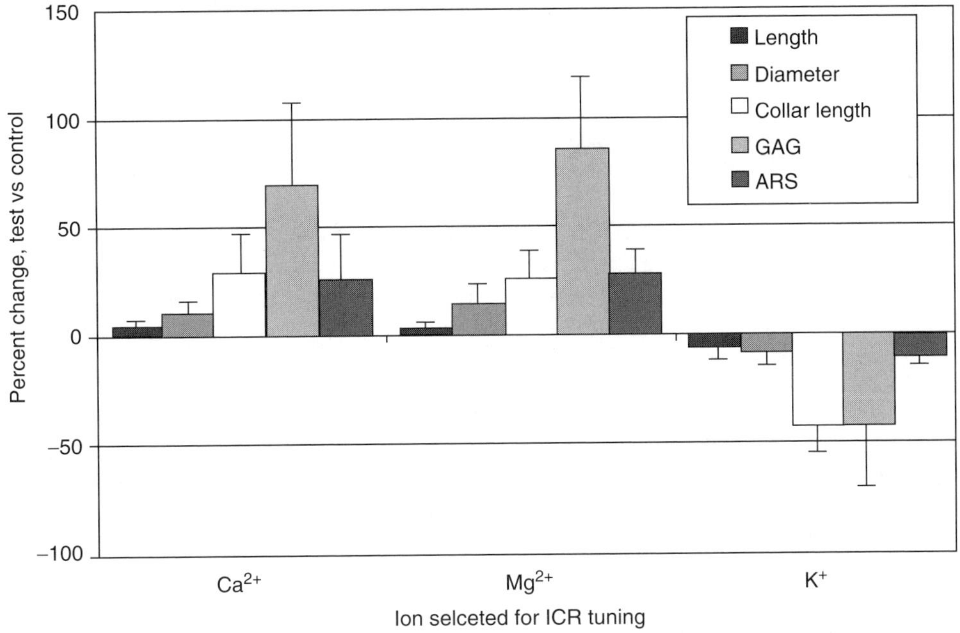

FIGURE 9.11
Independent confirmation of the results shown in Figure 9.8 for embryonic chick femur [32]. The additional assay for glycosaminoglycans (GAGS) reveals a greatly increased production of cartilage with Ca^{2+} or Mg^{2+} ICR stimulation and reduced cartilage production with K^+ tuning.

fields to alter biological properties. Zhadin [86,88] explored this question using combined AC and DC magnetic fields, to see if there might be an ICR response. The conductivity of polar amino acids in solution (Table 9.1 and Table 9.10) was chosen as an assay. The frequency of the magnetic field was swept slowly, and a single sharp current peak was observed at the cyclotronic frequency, determined from Equation 9.1 using the naked (unhydrated) molecular mass. Originally made in glutamic acid solutions [86], these observations were subsequently extended to additional amino acids, even to the case

TABLE 9.12
Opposite Responses Resulting from Shifts in Frequency or Field

Model System	Frequency (Hz)	B_0 (μT)	Ion	Response	Reference
Diatom motility	16	20.9	Ca^{2+}	Motility ↑	58
	16	41.0	K^+	Motility ↓	
Embryonic bone	16	20.9	Ca^{2+}	Growth ↑	24
	16	40.7	K^+	Growth ↓	32
Plant growth	60	78.3	Ca^{2+}	Growth ↑	75
	60	153.3	K^+	Growth ↓	
Rat behavior	60	48	Mg^{2+}	Learning ↑	47
	60	27	Ca^{2+}	Learning ↓	
Rat behavior	63	50	Mg^{2+}	Activity ↑	51
	38	50	Ca^{2+}	Activity ↓	
Gravitropic response	35.8	46.5	Ca^{2+}	Response ↑	79
	54.7	46.5	K^+	Response ↓	
Glycosaminoglycans (GAGs) concentration	16	20.9	Ca^{2+}	GAGs ↑	32
	16	40.7	K^+	GAGs ↓	

where it was possible to shift the ICR peak by changing the pH of the solution, and therefore the valence of the ion. This work has been independently confirmed in two-other laboratories [11,105]. One interesting aspect of these observations is the remarkably small AC magnetic field intensity required to see this effect, on the order of 0.05 μT.

It has also been shown [89] that very short (minutes) applications of ICR-tuned magnetic fields act to trigger large (days) continuing increases in conductivity in highly purified water. The required AC intensities (1 μT) are greater than that was found necessary in the Zhadin experiments, but they are still far less than reported elsewhere as interactive. The ions that are presumably affected are hydronium (H_3O^+) and its water clusters. One intriguing aspect of this work are the long-term changes in conductivity that are observed, even after removal of the magnetic field, a phenomenon somewhat akin to that of water "memory" [106,107].

9.3 Theoretical Approaches

9.3.1 Physical Constraints

The conditions in living things are such that charged particles are not free but subject to damping forces. If a particle such as an ion carries charge q and mass m and is moving with velocity $\mathbf{v}$, damping is manifested in the form of a retarding force, above and beyond the Lorentz force $q(\mathbf{v} \times \mathbf{B})$ and the Coulomb force $q\mathbf{E}$:

$$\mathbf{F} = q(\mathbf{v} \times \mathbf{B} + \mathbf{E}) - m\mathbf{v}/\tau \tag{9.6}$$

The last term in Equation 9.6 is the mean damping force acting on the particle expressed in terms of the collision rate τ^{-1}.

The possibility of ICR occurring in living systems is counterintuitive, mainly because of damping. Cyclotron resonance, whether involving ions [108] or electrons [109], is usually associated with motion in a vacuum, or at least a low-pressure gas. In sharp contrast, the biological milieu is generally approximated as a liquid. Electron cyclotron resonance is observed in metals [109], but only at extremely high frequencies ranging from megahertz to gigahertz, many orders of magnitude greater than the typically observed frequencies listed in Table 9.3 through Table 9.10. At very high frequencies, the periodic motion of charged particles in cyclotron resonance can take place because the periods are shorter than the collision times. At 15 Hz, however, one loop is executed in 70 ms, a time over which one might expect as many as 10^{12} collisions because of thermal scattering [110]. The problem of explaining ICR effects in living systems is indeed challenging.

A second argument against the ICR hypothesis is that ions in solution are never totally free but are surrounded by layers of water. Therefore, one would think that the values of the q/m ratio indicated in Equation 9.1 would have to be lowered to reflect the extra mass that is part of the overall ionic package.

One further criticism lies in estimates of the path radius for (free) charged particles in cyclotron resonance. This is based on equating the kinetic energy $mv^2/2$ of the charged particle to the thermal energy kT, and then arguing that the radius of the particle path must necessarily reflect the resulting velocity, because this radius is given as $\rho = v(m/qB)$. At a temperature of 37°C this results in radii that are of the order of meters, much larger than the extent of the system itself. However, it is hardly clear that one can use classical kinetics to discuss low energy charged particle interactions with molecular structures. Further, it must be noted that the cyclotronic frequency ω_c, as given in Equation 9.1 under the condition of vanishing damping, is entirely independent of particle radius and energy.

ICR tends to somewhat mute one of the strongest arguments against the likelihood of weak extremely low-frequency (ELF) magnetic interactions with biosystems. For an electric signal to initiate a biological effect, this signal must be greater than the electric potential generated by thermal noise [111]. The mean square thermal noise voltage $(\delta V)^2$ generated is proportional to Boltzmann's constant k, the temperature T, the tissue resistance R, and the bandwidth $\Delta\nu$ as follows:

$$(\delta V)^2 = 4RKT\Delta\nu \tag{9.7}$$

The original interaction site suggested by Weaver and Astumian [111] was the plasma membrane of the cell. The lipid membrane, however, is a highly electrically insulating material, with physical properties that are not conducive to any low-frequency interaction mechanism [112] except at high voltages. When more conductive substances such as proteins are considered as weak-field interaction sites, the effective resistance R can be many orders of magnitude lower than that found in lipids. Further, experimental evidence indicates that much narrower bandwidths, $\Delta\nu$, are encountered in ICR applications [39,106] than was previously [111,113] assumed. If the product $R\Delta\nu$ is reduced in Equation 9.7 by a factor of 10^{-4}, this lowers δV by a factor of 10^{-2} below the original estimates. The "kT" question becomes even less of a problem if one considers electric-field ICR [114], where the frequency of oscillation in Equation 9.1 is derived from endogenous sources within the living system.

9.3.2 Ion Channels

Placing the site of the ICR interaction within the lumen of membrane-bound ion channels deals with the problems of damping and hydration layers but leaves other questions unresolved. It is clear that damping within channels is quite different from that in liquids. The very function of ion channels helps define their intrinsically small effective resistivity, often less than 0.1 Ωm [115], and therefore their vanishing damping. Channels function as shunts across the insulating cell membrane, allowing ions to percolate through, often in single file [116]. The passage of ions is determined less by collision rate, as in the solution-like regions external to channels, but is more dependent on the molecular architecture [117] of the channel proteins: the gating mechanisms that permit entry into the channel lumen and the electrical potential of the atoms lining the lumen walls.

Further, this passage of ions is unencumbered by the layers of hydration that surround all ions in solution. These waters are replaced upon entry into the channel by an equivalent fixed cage lining the lumen (Figure 9.12). Although the energetics of this water replacement is still puzzling [118], there is little question that ions can pass through ion channels in a naked fashion, with no attached waters [119]. The path of ions in a channel is clearly determined in a manner that does not permit one to apply kT arguments in trivial ways.

Although one can readily deal with the two problems associated with first, damping and second, ionic hydration layers, it is not as easy to dismiss the disparity in relaxation times when one compares the very rapid passage of ions through channels to the far slower movements associated with resonance frequencies. The ratio of these times can be well in excess of 10^{10}. To illustrate the problems raised by this disparity, consider the argument that the helicity of ionic paths occurring under ICR might be such as to match the helicity of the electric potential in the lumen [120].

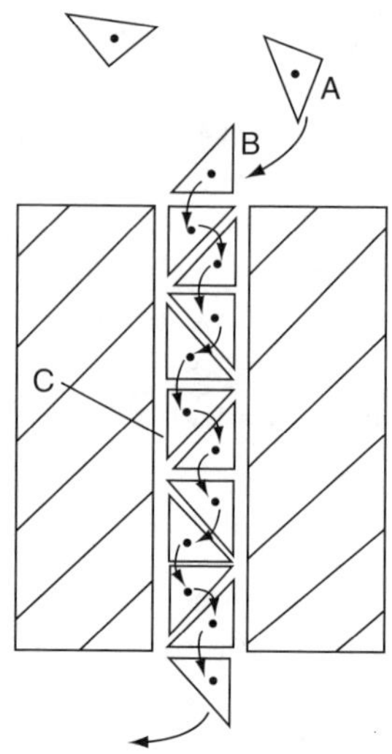

FIGURE 9.12
Hydrated ions near the mouth of an ion channel exchange their layers of water for an equivalent potential distribution within the lumen of the channel.

The electric field in Equation 9.6 can have a number of distinctly different origins. In living tissue, the electric field can be the result of endogenous sources independent of **B**. If **B** is time varying, and of sufficiently large intensity or frequency, an electric field will in addition be induced by Faraday's law. There are also experimental situations in which an electric field may be applied independently of the first two sources. When an electric field **E** is present, whatever the source, and it is at right angles to both **v** and **B**, then the circular motion becomes helical (Figure 9.13). It is then convenient to treat the velocity of the particle as composed of two components: $v\|$, directed parallel to the direction of **E**, and

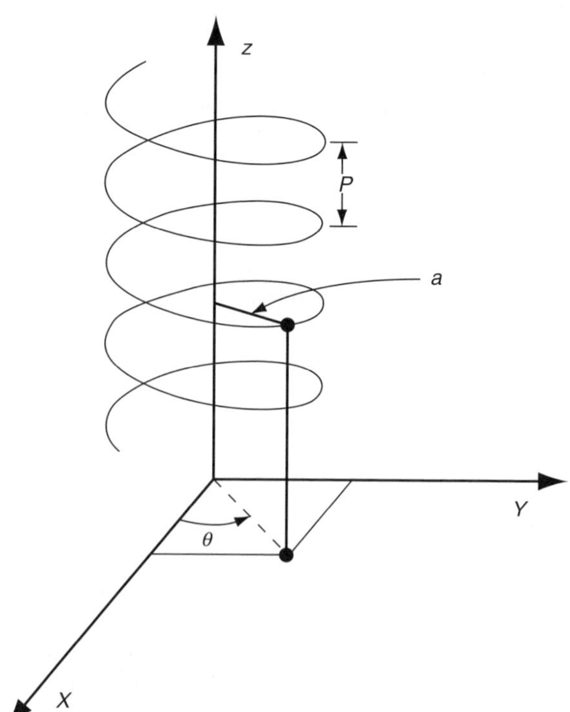

FIGURE 9.13
The Lorentz force acting on a charged particle often results in helical motion.

$v_\perp$, directed within the plane perpendicular to **E**. Then the radius of the helix is $a = mv_\perp/qB$, and its pitch is $p = 2\pi mv\|/qB$. It is not difficult to show that the total velocity v is

$$v = f_c(4\pi^2a^2 + p^2)^{1/2} \tag{9.8}$$

The ICR frequency f_c is often only 15 Hz, and for ion channels the dimensions a and p are of the order of ångstroms (10^{-10} m). This implies a velocity of about 10^{-9} m/sec, a value that is totally inconsistent with the much greater experimentally observed speeds with which ions are transported across membranes.

Even though transit times directly through the channel lumen are very rapid, much longer relaxation times are found in other compartments of the larger channel structure. Times as long as, if not longer than, 10 ms are associated with voltage gating [121], making it conceivable that ICR could play an interactive role in channel-gating mechanisms.

9.3.3 Dependence on AC Magnetic Field

Borrowing from the theory of atomic spectroscopy [122], Lednev [7] hypothesized a functional relationship between the strength of the ICR signature and B_{AC}. When the calcium ion is bound to a calcium-binding protein such as calmodulin, the energy levels for the ion protein complex will be split, Zeeman-like, by the application of a magnetostatic field B, such that the difference in frequency between the two new levels is equal to the cyclotron resonance frequency ω_c.

Application of an AC magnetic field at this frequency then leads to a resonant condition that can alter the transition to the ground state. The ICR response itself then becomes a function of the peak value of B_{AC} expressed in terms of the probability $p(B_{AC})$ for Ca^{2+} transitions to the ground state. The field-dependent part of this probability is

$$p(B_{AC}) = (-1)^n KJ_n\left(n\frac{B_{AC}}{B_0}\right) \tag{9.9}$$

where K is a constant, B_0 is the DC magnetic intensity parallel to B_{AC}, and J_n is the nth-order Bessel function having the argument nB_{AC}/B_0.

One great advantage of this expression for the probability is that one can readily design experiments that can test the predicted dependence on the AC magnetic field. Thus, the first extremum of $J_1(B_{AC}/B_0)$ occurs when the ratio B_{AC}/B_0 is equal to 1.84 (Figure 9.14). By varying this ratio while maintaining the same ICR condition, Prato et al. [72] obtained data tending to confirm this prediction. The effectiveness of this numerical ratio was also reported [74] in connection with studies on changes in calcium concentrations under ICR stimulation.

An alternative formulation to Equation 9.9 has been proposed by Blanchard and Blackman [9]. Termed ion parametric resonance (IPR) this approach considers a wider range of potential ELF interactions with metallic ions, beyond merely those processes associated with the activation of calcium-binding proteins. In the IPR formulation the probability $p(B_{AC})$ has the dependence $J_n(2nB_{AC}/B_0)$. The factor of 2 in the argument of the Bessel function has the effect of altering where the first extremum occurs, in this case at the B_{AC}/B_0 ratio of 0.92 (Figure 9.14). This ratio is close to the value of 1.0 successfully used by Smith [21,24,58,75] in all his experiments. However it is not clear in Smith's body of work whether further manipulation of the AC magnetic intensity away from this ratio of unity might have enhanced the responses that were observed. Koch et al. [91] were also able to obtain good agreement with the IPR predictions. One set of experimental results in sharp

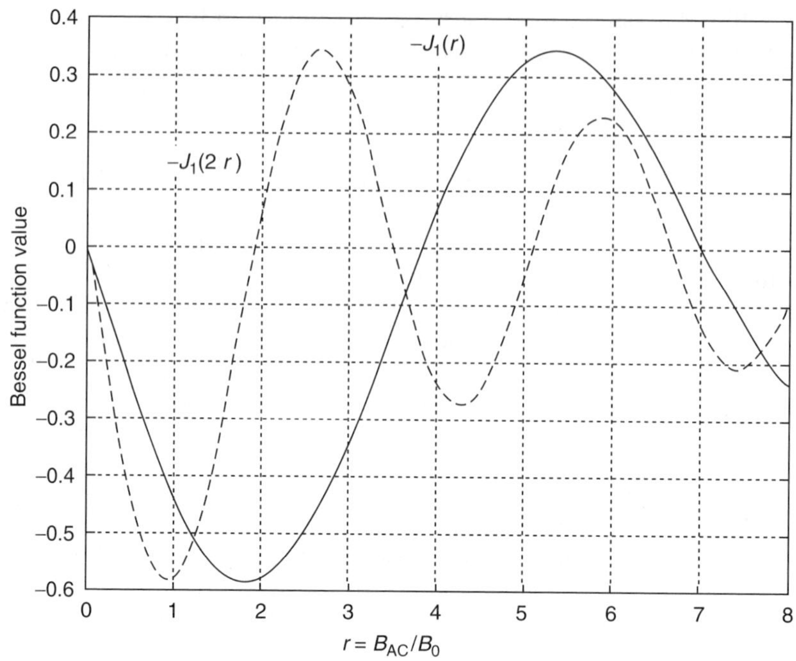

FIGURE 9.14
Comparing first-order Bessel functions with arguments B_{AC}/B_0 (solid line) and $2B_{AC}/B_0$ (dotted line). The first extremum is of experimental interest. By doubling the argument, the ratio where this extremum occurs is shifted from 1.84 to 0.92. In principle, one holds B_0 and the ICR frequency fixed and varies B_{AC} to determine where the maximum biological response occurs.

contrast with the predictions of parametric resonance are the consistent reports by Zhadin's group [86,88] and others [11], in which ICR exposures of polar amino acids in solution lead to narrow resonances, but only for B_{AC}/B_0 ratios of approximately 0.002.

The question of the dependence of the ICR interactive response on AC magnetic intensity is clouded by two additional possibilities. First, there is the problem of experimentally sorting out effects due to Faraday induction with larger AC fields [70]. Second, it is conceivable that biological responses to different AC fields may have little to do with the ICR response but rather with poorly understood physiological and energetic constraints. In any event, there is a clear record in the literature (e.g., Figure 9.15) of ICR threshold effects that do not appear to be related to functional variations such as Equation 9.9.

Adair [123,124] has criticized the parametric resonance approaches. Similar to earlier arguments [111,113] related to energetic constraints imposed by thermal noise, it was pointed out that the transition energy is only 10^{-11} of the thermal energy, kT, which would greatly suppress any meaningful effects due to Zeeman splitting. For this and other reasons [123], he concludes that application of the cyclotronic frequency ω_c cannot affect the transition rate of Ca^{2+}.

A very different approach to the problem of whether resonant magnetic fields can influence the binding of calcium has been suggested by Binhi [101]. Using Schrodinger's equation, it has been shown that ICR magnetic field combinations may redistribute the ionic probability density when one takes into account interference effects between quantum states. However, there is some question as to how much the wave function must shrink to allow ions as large as Ca^{2+} to escape.

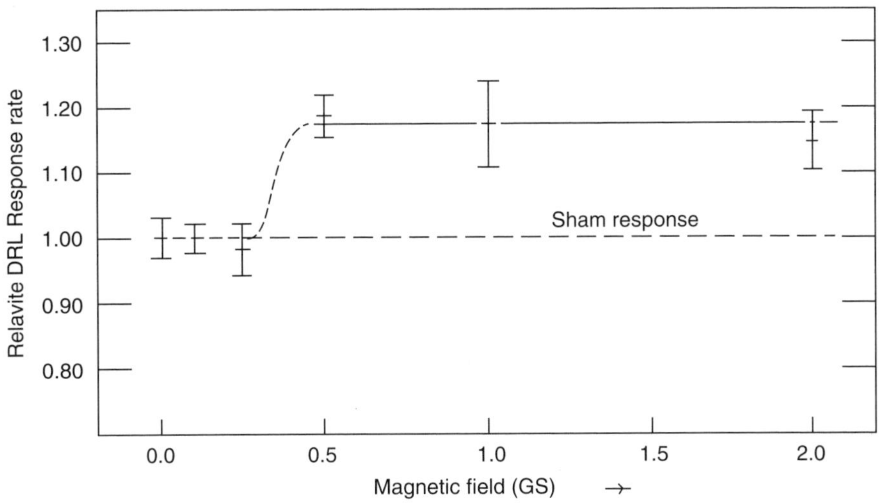

FIGURE 9.15
ICR-induced changes in rat behavior as a function of the AC magnetic field intensity. A clear threshold is evident, below which there is no effect on rat behavior. (From Liboff, A.R., Thomas, J.R., and Schrot, J., Intensity threshold for 60-Hz magnetically induced behavioral changes in rats, *Bioelectromagnetics*, 10, 111, 1989.)

9.3.4 Precessional Effects

It has been shown [8,10] that Larmor precessional effects can result from the application of weak ICR magnetic field combinations. In principle, these effects might carry physiological consequences, although no evidence along these lines has been reported. One interesting aspect of this approach [10] suggests the possibility of magnetically induced changes in the thermal energy distribution. The altered biological response in this model would not be the result of dissociative processes between ion and protein but would rather follow transfers of energy to the protein by the ion, resulting in appropriate conformational changes.

9.3.5 Coherence Domains

Del Giudice [11] has employed the principles of quantum electrodynamics (QED) in attempting to explain biological ICR, especially those results obtained in connection with the "Zhadin Experiment" [11,86,88]. The great advantage in this approach is that it finesses the limitations caused by Maxwell–Boltzmann statistics predictions [111,113], which require kT as a minimal threshold informational energy for electromagnetic interactions. In the QED model it is argued that in the case of electromagnetic interactions with water, coherent, highly ordered spherical domains are formed, with diameters close to 100 nm and embedded in a surrounding noncoherent water phase. An applied ICR frequency in this case couples to ions on the perimeter of the coherent spheres stimulating circulatory motion, allowing ions to escape into the noncoherent phase, where they act to enhance the measured conductivity.

One interesting aspect of this model is that because phase transitions are involved, temperature is expected to play a prominent role in the experimental outcome. The changes in amino acid conductivity under ICR conditions first reported by Novikoff and Zhadin [86] were temperature sensitive. Similar observations regarding the experimental relevance of temperature were reported in earlier [125,126] Ca-efflux studies.

9.4 Discussion

Accepting the abundant evidence that biological systems are sensitive to ICR-tuned field combinations, the question remains as to the molecular explanation. There is also the more subtle question dealing with the implications of this effect. From the physical standpoint, the seeming inability of theorists to come to grips with this phenomenon may not mean that a new physics is called for. But there is certainly good reason to believe that something is missing in the way we categorize biological function.

The fact that a low-frequency ion resonance interaction is found in an extraordinarily wide variety of organisms argues against the commonly held notion that this phenomenon merely represents one fortuitous example of the Lorentz force acting on materials that happen to occur within a biological context.

Indeed, the reports listed in Table 9.10 lend strength to another notion, that the weak-field ICR phenomenon is not limited to biological systems, but is actually a subtle physical effect that has not heretofore been recognized. For example, one can argue that this effect is particularly evident in water, which in turn results in the biological effects that are observed. In other words, because of the ubiquity of water in biological systems, any ELF-induced change in the properties of water might help explain the variety of biological effects that have been reported.

Whatever its molecular origin, it is likely that the ICR-related biophenomenon is an evolutionarily well-conserved property of the physiological armory, one that is enabled by the interaction of the ubiquitous Earth's magnetic field with the equally ubiquitous electric fields that are found throughout living things. This concept is embodied in the reasonable likelihood [114] that electric-field ICR processes were incorporated into living things over evolutionary times. Unlike the laboratory experiments involving the applications of parallel combined AC and DC magnetic fields, *E*-field resonance occurs maximally for time-varying electric and magnetostatic fields at right angles. This naturally occurring interaction therefore relates the GMF and the AC electric fields found within the biosystem. It has been hypothesized [127], for example, that this putative interaction mechanism can be used to explain the problem of bird navigation.

One can speculate that this ICR phenomenon conserved very early in living things has evolved into a number of separate biological expressions, where the physics is always the same but the physiological pathway leads to different end points. For example, two sharply different possible end points, both apparently relying on ICR, are the magnetic compasses in animals [124] and growth and repair mechanisms [24,37,75]. A third intriguing possibility is found in the ICR interactive changes in the central nervous system [45,47,49,51], an area where there are obvious effects but where the end points are still unresolved.

It is also tempting to speculate on the larger implications of the work that is detailed in Table 9.3 and Table 9.9, on bone and plants, respectively. In both cases there is evidence that ICR signals affect growth and repair responses more than they affect stasis. Drawing on possible endogenous mechanisms related to *E*-field ICR, it is conceivable that the interactive relation of the GMF to living systems is especially important as regards growth and repair during development.

Whatever be the explanation for the ICR effect, it is clear that while nothing new may be required in our understanding of physical law, new approaches are necessary in comprehending how physics has been incorporated in biology, particularly with regard to cell signaling and regulation. This is strongly implied by the remarkable data indicating reversed physiological outcomes with minor changes in electromagnetic field conditions. There are protein systems in physiology that are switch-like in their function. Two such examples are the calcium-binding proteins and NMDA. Another example is found in synaptic transmission. The electromagnetic reversal effects observed in ICR experiments are evocative of the way such switch-like proteins function.

This begs the fascinating question as to whether the ICR stimulus merely serves as an adjunct to an existing biochemical process, perhaps acting to sharpen enhancement or inhibition, or whether it acts in a totally different manner, as a separate and distinct class of interactions, one that is completely electromagnetic in its operation. In this regard, note that the GMF preceded by far the first appearance of living things on Earth and would therefore have been part of the original physical template that determined the course of early evolution.

References

1. Jenrow, K.A. and Liboff, A.R., Electromagnetic techniques in neural therapy, in *Bioelectromagnetic Medicine*, Rosch, P.J. and Markov, M.S., Eds., Marcel Dekker, New York, 2004, Chapter 14.
2. Wertheimer, N. and Leeper, E., Electrical wiring configurations and childhood cancer, *Am. J. Epidemiol.*, 109, 273–284, 1979.
3. Liboff, A.R., The charge-to-mass ICR signature in weak ELF bioelectromagnetic effects, in *Advances Electromagnetic Fields in Living Systems*, Vol. 4, Lin, J., Ed. Kluwer/Plenum, New York (in press).
4. Liboff, A.R., Geomagnetic cyclotron resonance in living cells, *J. Biol. Phys.*, 13, 99–102, 1985.
5. Tipler, P., *Physics*, 2nd ed., Vol. 2, Worth Publishers, New York, 1982, 729.
6. Zhadin, M.N. and Fesenko, E.E., Ionic cyclotron resonance in biomolecules, *Biomed. Sci.*, 1, 245–250, 1990.
7. Lednev, V.V., Possible mechanism for the influence of weak magnetic fields on biological systems, *Bioelectromagnetics*, 12, 71–75, 1991.
8. Edmonds, D.T., Larmor precession as a mechanism for the detection of static and alternating magnetic fields, *Bioelectrochem. Bioenerg.*, 30, 3–12, 1993.
9. Blanchard, J.P. and Blackman, C.F., Clarification and amplification of an ion parametric resonance model for magnetic field interactions with biological systems, *Bioelectromagnetics*, 15, 217–238, 1994.
10. Zhadin, M.N., Combined action of static and alternating magnetic fields on ion motion in a macromolecule: theoretical aspects, *Bioelectromagnetics*, 19, 279–292, 1998.
11. Del Giudice, E., Fleischmann, M., Preperata, G., and Talpo, G., On the "unreasonable" effects of ELF magnetic fields upon a system of ions *Bioelectromagnetics*, 23, 522–530, 2002.
12. Blackman, C.F., Benane, S.G., Rabinowitz, J.R., House, D.E., and Joines, W.T., A role for the magnetic field in the radiation-induced efflux of calcium ions from brain tissue *in vitro*, *Bioelectromagnetics*, 6, 327–337, 1985.
13. Bawin, S.M., Kazmarek, K.L., and Adey, W.R., Effects of modulated VHF fields on the central nervous system, *Ann. NY Acad. Sci.*, 247, 74–81, 1975.
14. Blackman, C.F., Elder, J.A., Weil, C.M., Benane, S.G., Eichinger, D.C., and House, D.E., Induction of calcium-ion efflux from brain tissue by radiofrequency radiation and field strength, *Radio Sci.*, 14, 93–98, 1979.
15. Blackman, C.F., Benane, S.G., Kinney, L.S., Joines, W.T., and House, D.E., Effects of ELF fields on calcium-ion efflux from brain tissue *in vivo*, *Rad. Res.*, 92, 510–520, 1982.
16. Dutta, S.K., Subramaniam, A., Ghosh, B., and Parshad, R., Microwave radiation-induced calcium ion efflux from human neuroblastoma cells in culture, *Bioelectromagnetics*, 5, 71–78, 1984.
17. Blackman, C.F., Benane, S.G., House, D.E., and Joines, W.T., Effects of ELF (1–120 Hz) and modulated (50 Hz) RF fields on the efflux of calcium ions from brain tissue, *in vitro*, *Bioelectromagnetics*, 6, 1–11, 1985.

18. McLeod, B.R. and Liboff, A.R., Dynamic characteristics of membrane ions in multifield configurations of low-frequency electromagnetic radiation, *Bioelectromagnetics*, 7, 177–189, 1986.
19. McLeod, B.R. and Liboff, A.R.. Cyclotron resonance in cell membranes: the theory of the mechanism, in *Mechanistic Approaches to Interactions of Electric and Electromagnetic Fields with Living Systems*, Blank, M. and Findl, E., Eds., Plenum Press, New York, 1987, 97–108.
20. McLeod, B.R., Smith, S.D., and Liboff, A.R., Calcium and potassium cyclotron resonance curves and harmonics in diatoms, *J. Bioelectr.*, 6, 153–168, 1987.
21. Smith, S.D., McLeod, B.R., and Liboff, A.R., Testing the ion cyclotron resonance theory of electromagnetic field interaction with odd and even harmonic tuning for cations, *Bioelectrochem. Boenerg.*, 38, 161–167, 1995.
22. Liboff, A.R. and McLeod, B.R., Kinetics of channelized membrane ions in magnetic fields, *Bioelectromagnetics*, 9, 39–51, 1988.
23. Ross, S.M., Combined DC and ELF magnetic fields can alter cell proliferation, *Bioelectromagnetics*, 11, 27–36, 1990.
24. Smith, S.D., Liboff, A.R., and McLeod, B.R., Effects of resonant mcLeod fields on chick femoral development *in vitro, J. Bioelectr.*, 10, 81–99, 1991.
25. Fitzsimmons, R.J., Baylink, D.J., Ryaby, J.T., and Magee, F., EMF-stimulated bone-cell proliferation, in *Electricity and Magnetism in Biology and Medicine*, Blank, M., Ed., San Francisco Press, San Francisco, 1993, 899–901.
26. Ryaby, J.T., Grande, D.A., Magee, F.P., and Weinstein, A.M., The effects of combined AC/DC magnetic fields on resting articular cartilage metabolism, in *Electricity and Magnetism in Biology and Medicine*, Blank, M., Ed., San Francisco Press, San Francisco, 1993, 371–374.
27. Ryaby, J.T., Magee, F.P., Weinstein, A.M., Fitzsimmons, R.J., and Baylink, D.J., Prevention of experimental osteopenia by use of combined magnetic fields, in *Electricity and Magnetism in Biology and Medicine*, Blank, M., Ed., San Francisco Press, San Francisco, 1993, 807–810.
28. Fitzsimmons, R.J., Ryaby, J.T., Magee, F.P., and Baylink, D.J., Combined magnetic fields increased net calcium flux in bone cells. *Calcif. Tissue Int.*, 55, 376–380, 1994.
29. Diebert, M.C., McLeod, B.R., Smith, S.D., and Liboff, A.R., Ion resonance electromagnetic field stimulation of fracture healing in rabbits with a fibular ostectomy, *J. Orthop. Res.*, 12, 878–885, 1994.
30. Fitzsimmons, R.J., Ryaby, J.T., Magee, F.P., and Baylink, D.J., Combined magnetic fields increase insulin-like growth factor-II in TE-85 human osteosarcoma bone cell cultures, *Endocrinology*, 136, 3100–3106, 1995.
31. Fitzsimmons, R.J., Ryaby, J.T., Magee, F.P., and Baylink, D.J., IGF-II receptor number is increased in TE-85 human osteosarcoma cells by combined magnetic fields, *J. Bone Miner. Res.*, 10, 812–819, 1995.
32. Regling, C., Brueckner, C., Liboff, A.R., and Kimura, J.H., Evidence for ICR magnetic field effects on cartilage and bone development in embryonic chick bone explants (abstract), 48th Annual Meeting, Orthopedic Research Society, Dallas, 2002.
33. Liboff, A.R., Rozek, R.J., Sherman, M.L., McLeod, B.R., and Smith, S.D., $_{45}Ca^{2+}$ resonance in human lymphocytes. *J. Bioelectr.*, 6, 99–102, 1987.
34. Rozek, R.J., Sherman, M.L., Liboff, A.R., McLeod, B.R., and Smith, S.D., Nifedipene is an antagonist to cyclotron resonance enhancement of 45Ca incorporation in human lymphocytes, *Cell Calcium*, 8, 413–427, 1987.
35. Parkinson, W.C. and Hanks, C.T., Search for cyclotron resonance in cells *in vitro, Bioelectromagnetics*, 10, 129–145, 1989.
36. Rochev, Y.A., Narimanov, A.A., Sosunov, E.A., Kozlov, A.N., and Lednev, V.V., Effects of weak magnetic field on the rate of cell proliferation in cell culture, *Stud. Biophys.*, 135, 93–98, 1990.
37. Lyle, D.B., Wang, X., Ayotte, R.D., Sheppard, A.R., and Adey, W.R., Calcium uptake by leukemic and normal T-lymphocytes exposed to low frequency magnetic fields, *Bioelectromagnetics*, 12, 145–156, 1991.
38. Yost, M.G. and Liburdy, R.P., Time-varying and static magnetic fields act in combination to alter calcium signal transduction in the lymphocyte, *FEBS Lett.*, 296, 117–122, 1992.
39. Liboff, A.R., Jenrow, K.A., and McLeod, B.R., ELF-induced proliferation at 511 mG in HSB-2 cell culture as a function of 60 Hz field intensity, in *Electricity and Magnetism in Biology and Medicine*, Blank, M., Ed., San Francisco Press, San Francisco, 1993.

40. Liburdy, R.P. and Yost, M.G., Time-varying and static magnetic fields act in combination to alter calcium signal transduction in the lymphocyte, in *Electricity and Magnetism in Biology and Medicine*, Blank, M., Ed., San Francisco Press, San Francisco, 1993.
41. Coulton, L.A. and Barker, A.T., Magnetic fields and intracellular calcium: effects on lymphocytes exposed to conditions for "cyclotron resonance", *Phys. Med. Biol.*, 38, 347–360, 1993.
42. Lindstrom, E., Lindstrom, P., Berglund, A., Lundgren, E., and Mild, K.H., Intracellular calcium oscillations in a T-cell line after exposure to extremely-low-frequency magnetic fields with variable frequencies and flux densities, *Bioelectromagnetics*, 16, 41–47, 1995.
43. Tofani, S., Ferrara, A., Anglesio, L., and Gilli, G., Evidence for genotoxic effects of resonant ELF magnetic fields, *Bioelectrochem. Bioenerg.*, 36, 9–13, 1995.
44. Reinbold, K.A. and Pollack, S.R., Serum plays a critical role in modulating $[Ca^{2+}]_c$ of primary cell culture bone cells exposed to weak ion-resonance magnetic fields, *Bioelectromagnetics*, 18, 203–214, 1997.
45. Thomas, J.R., Schrot, J., and Liboff, A.R., Low-intensity magnetic fields alter operant behavior in rats, *Bioelectromagnetics*, 7, 349–357, 1986.
46. Liboff, A.R., Thomas, J.R., and Schrot, J., Intensity threshold for 60-Hz magnetically induced behavioral changes in rats, *Bioelectromagnetics*, 10, 111–113, 1989.
47. Lovely, R.H., Creim, J.A., Miller, D.L., and Anderson, L.E., Behavior of rats in a radial arm maze during exposure to magnetic fields: evidence for effects of magnesium ion resonance (abstract), 15th Annual Meeting, Bioelectromagnetics Society, Los Angeles, 1993.
48. Stern, S., Laties, V.G., Nguyen, Q.A., and Cox, C., Exposure to combined static and 60Hz magnetic fields: failure to replicate a reported behavioral effect, *Bioelectromagnetics*, 17, 279–292, 1996.
49. Lyskov, Y.B., Chernysev, M.V., Michailov, V.O., Kozlov, A.P., Makarova, T.M., Vasilyeva, Y.V., Druzin, M.Y., Sokolov, G.V., and Vishnevski, A.M., The effect of a magnetic field with the frequency of 50 Hz on behavior in rats depends on the value of the constant magnetic field, *Biophysics*, 41, 881–886, 1996.
50. Derjugina, O.N., Pisachenko, T.M., and Zhadin, M.N., Combined action of alternating and static magnetic fields on behavior of rats in the "Open-field" test, *Biophysics*, 41, 762–764, 1996.
51. Zhadin, M.N., Deryugina, O.N., and Pisachenko, T.M., Influence of combined DC and AC magnetic fields on rat behavior, *Bioelectromagnetics*, 20, 378–386, 1999.
52. Smith, S.D., Liboff, A.R., McLeod, B.R., and Barr, E.J., Effects of ion resonance tuned magnetic fields on N-18 neuroblastoma cells, in *Charge and Field Effects in Biosystems-3*, Allen, M.J., Cleary, A.F., Sowers, A.E., and Shillady, D.D., Eds., Birkhauser, Boston, 1992, 263–271.
53. Horton, P., Ryaby, J.T., Magee, F.P., and Weinstein, A.M., Stimulation of specific neuronal differentiation proteins in PC-12 cells by combined AC/DC magnetic fields, in *Electricity and Magnetism in Biology and Medicine*, Blank, M., Ed., San Francisco Press, San Francisco, 1993, 619–622.
54. Blackman, C.F., Blanchard, J.P., Benane, S.G., and House, D.E., Empirical test of an ion parametric resonance model for magnetic field interactions with PC-12 cells, *Bioelectromagnetics*, 15, 239–250, 1994.
55. Trillo, M.A., Ubeda, A., Blanchard, J.P., House, D.E., and Blackman, C.F., Magnetic fields at resonant conditions for the hydrogen atom can affect neurite outgrowth in PC-12 cells, *Bioelectromagnetics*, 17, 10–20, 1996.
56. Blanchard, J.P., Blackman, C.F, Benane, S.G., and House, D.E., IPR response of PC-12 cells exposed to magnetic fields tuned for calcium ions, The Annual Review of Research on Biological Effects of Electric and Magnetic Fields from the Generation, Delivery and Use of Electricity, San Diego, CA, W/L Associates, Ltd, Frederick, MD, 1997, 4.
57. Blackman, C.F., Blanchard, J.P., Benane, S.G., and House, D.E., Experimental determination of hydrogen bandwidth for the Ion Parametric Resonance model, *Bioelectromagnetics*, 20, 5–12, 1999.
58. Smith, S.D., McLeod, B.R., Liboff, A.R., and Cooksey, K., Calcium cyclotron resonance and diatom motility, *Bioelectromagnetics*, 8, 215–227, 1987.
59. McLeod, B.R., Smith, S.D., Cooksey, K.E., and Liboff, A.R., Ion cyclotron resonance frequencies enhance Ca^{2+}-dependent motility in diatoms, *J. Bioelectr.*, 6, 1–12, 1987.
60. McLeod, B.R., Smith, S.D., and Liboff, A.R., Calcium and potassium cyclotron resonance curves and harmonics in diatoms, *J. Bioelectr.*, 6, 153–168, 1987.

61. Reese, J.A, Frazier, M.E., Morris, J.E., Buschbom, R.L., and Miller, D.L., Evaluation of changes in diatom motility after exposure to 16-Hz electromagnetic fields, *Bioelectromagnetics*, 12, 21–25, 1991.
62. Parkinson, W.C. and Sulik, G.L., Diatom response to extremely low frequency magnetic fields, *Rad. Res.*, 130, 319–330, 1992.
63. Saalman, E., Galt, S., Hamnerius, Y., and Norden, B., Diatom motility: replication study in search of cyclotron resonance effects, in *Interaction Mechanisms of Low-Level Electromagnetic Fields in Living Systems*, Norden, B. and Ramel, C., Eds., Oxford University Press, Oxford, 1992, 280–292.
64. Prasad, A.V., Miller, M.W., Cox, C., Carstensen, E.L., Hops, H., and Brayman, A.A., A test of the influence of cyclotron resonance exposures on diatom motility, *Health Phys.*, 66, 305–312, 1994.
65. Clarkson, H., Davies, M.S., and Dixey, R., Diatom motility: the search for independent replication of biological effects of extremely low-frequency electromagnetic fields, *Int. J. Rad. Biol.*, 75, 387–392, 1999.
66. Liboff, A.R. and Parkinson, W.C., Search for ion-cyclotron resonance in an Na^+ transport system, *Bioelectromagnetics*, 12, 77–83, 1991.
67. Lerchl, A., Reiter, R.J., Nonaka, K.O., and Stokken, K.-A., Evidence that extremely low frequency Ca^{2+}-cyclotron resonance depresses pineal melatonin synthesis *in vitro*, *Neurosci. Lett.*, 124, 213–215, 1991.
68. Chemeris, N.K. and Safranova, V.G., Weak low-frequency magnetic field initiates frequency-dependent fluctuations of period of Daphnia magna's heart beatings, *Biophysics*, 38, 511–519, 1993.
69. Jenrow, K.A., Smith, C.H., and Liboff, A.R., Weak extremely-low-frequency magnetic fields and regeneration in the planarian Dugesia tigrina, *Bioelectromagnetics*, 16, 106–112, 1995.
70. Jenrow, K.A., Smith, C.H., and Liboff, A.R., Weak, extremely-low-frequency magnetic field-induced regeneration anomalies in the planarian Dugesia tigrina, *Bioelectromagnetics*, 17, 467–474, 1996.
71. Tiras, K.P., Srebnitskaya, L.K., Il'jasova, A.A., and Lednev, V.V., The influence of weak combined magnetic fields on the rate of regeneration in planarian Dugesia tigrina, *Biophysics*, 42, 826–831, 1996.
72. Prato, F.S., Kavaliers, M., Cullen, A.P., and Thomas, A.W., Light-dependent and -independent behavioral effects of extremely low frequency magnetic fields, *Bioelectromagnetics*, 18, 284–291, 1997.
73. Prato, F.S., Kavaliers, M., and Thomas, A.W., Extremely low frequency magnetic fields can either increase or decrease analgesia in the land snail depending on field and light conditions, *Bioelectromagnetics*, 21, 287–301, 2000.
74. Berden, M., Zrimec, A., and Jerman, I., New biological detection system for weak ELF magnetic fields and testing of the parametric resonance model, *Electro. Magnetobiol.*, 20, 27–41, 2001.
75. Smith, S.D., McLeod, B.R., and Liboff, A.R., Effects of CR-tuned 60 Hz magnetic fields on sprouting and early development of *Raphanus sativus*, *Bioelectr. Bioenerg.*, 32, 67–76, 1993.
76. Davies, M.S., Effects of 60 Hz electromagnetic fields on early growth in three plant species and a replication of previous results, *Bioelectromagnetics*, 17, 154–161, 1996.
77. Tohtz, S.W., The Influence of Ultra Weak Magnetic Fields on Growth and Differentiation Using the ICR Hypothesis: An Experimental Study on Radish Seedlings and a Theoretical Discussion on Relevant Aspects of Cartilage, Bone, and Other Connective Tissues, Ph.D. Dissertation, Medical Faculty of Humboldt University, Berlin, 1996 (in German).
78. Smith, S.D., Liboff, A.R., and McLeod, B.R., Potassium ion cyclotron resonance magnetic fields stimulate germination, Proceedings of Millennium International Workshop on Biological Effects of Electromagnetic Fields, Heraclion, Crete, Greece, 2000, 347–351.
79. Belova, N.A. and Lednev, V.V., Activation and inhibition of gravitropic response in plants by weak combined magnetic fields, *Biophysics*, 45, 1069–1074, 2000.
80. Belova, N.A. and Lednev, V.V., Dependence of gravitropic response in the segments of flax stems on the frequency and amplitude of weak combined magnetic fields, *Biophysics*, 45, 1075–1078, 2000.
81. Yano, A., Ohashi, Y., Hirasaki, T., and Fujiwara, K., Effects of a 60 Hz magnetic field on photosynthetic CO_2 uptake and early growth of radish seedlings, *Bioelectromagnetics*, 25, 572–581, 2004.

82. Shuvalova, L.A., Ostrovskaja, M.V., Sosunov, E.A., and Lednev, V.V., Effect of weak magnetic field in the parametric resonance mode on the rate of calmodulin-dependent phosphorylation of myosin in the solution, Doklady Akademii Nauk SSSR (Reports of the Academy of Science of the USSR), Vol. 317, 1991, 227–230 (in Russian).
83. Durney, C.H., Kaminski, M., Anderson, A.A., Bruckner-Lea, C., and Rappaport, C., Investigation of AC–DC magnetic field effects in planar phospholipid layers, *Bioelectromagnetics*, 13, 19–33, 1992.
84. Bruckner-Lea, C., Durney, C.H., Janata, J., Rappaport, C., and Kaminski, M., Calcium binding to metallochromic dyes and calmodulin in the presence of combined AC–DC magnetic fields, *Bioelectromagnetics*, 13, 147–162, 1992.
85. Galt, S., Sandblom, A., Hamnerius, Y., Hojevic, P., Saalman, E., and Norden, B., Experimental search for combined AC and DC magnetic field effects on ion channels, *Bioelectromagnetics*, 14, 315–327, 1993.
86. Novikoff, V.V. and Zhadin, M.N., Combined action of weak constant and variable low-frequency magnetic fields on ionic currents in aqueous solutions of amino acids, *Biophysics*, 39, 41–45, 1994.
87. Hojevic, P., Sandblom, J., Galt, S., and Hamnerius, Y., Ca^{2+} ion transport through patch-clamped cells exposed to magnetic fields, *Bioelectromagnetics*, 16, 33–40, 1995.
88. Zhadin, M.N., Novikoff, V.V., Barnes, F.S., and Pergola, M.F., Combined action of static and alternating magnetic fields on ionic current in aqueous glutamic acid solution, *Bioelectromagnetics*, 19, 41–45, 1998.
89. Mohri, K., and Fukushima, M., Gradual decreasing characteristics and temperature stability of electric resistivity in water triggered with milligauss AC field, *IEEE Trans. Magn.*, 38, 3353–3355, 2002.
90. Liboff, A.R., Cherng, S., Jenrow, K.A., and Bull, A., Calmodulin dependent cyclic nucleotide phosphodiesterase activity is altered by 20 μT magnetostatic fields, *Bioelectromagnetics*, 24, 32–38, 2002.
91. Koch, C.L.M.B., Sommarin, M., Perrson, B.R.R., Salford, L.G., and Eberhardt, J.L., Interaction between weak low frequency magnetic fields and cell membranes, *Bioelectromagnetics*, 24, 395–402, 2003.
92. Creim, J.A., Lovely, R.H., Miller, D.L., and Anderson, L.E., Rats can discriminate illuminance, but not magnetic fields, as a stimulus for learning a two-choice discrimination, *Bioelectromagnetics*, 23, 545–549, 2002.
93. Tang, Y-P., Shimizu, E., Dube, G.R., Rampon, C., Kerchner, G.A., Zhuo, M., Liu, G., and Tsien, J.Z., Genetic enhancement of learning and memory in mice, *Nature*, 401, 63–69, 1999.
94. Muller, D., and Lynch, G., Synaptic modulation of *N*-methyl D-aspartate receptor mediated responses in hippocampus, *Synapse*, 5, 94–103, 1990.
95. Shiekhattar, R. and Aston-Jones, G., NMDA-receptor-mediated sensory responses of brain noradrenergic neurons are suppressed by *in-vivo* concentrations of extracellular magnesium, *Synapse*, 10, 103–109, 1992.
96. Smith, S.D., Liboff, A.R., and McLeod, B.R., Calcium ICR and seedling growth in orchids (abstract), 20th Annual Meeting, Bioelectromagnetics Society, St. Petersburg Beach, FL, 1998.
97. Poggi, C., private communication, 2004.
98. Bassett, C.A.L., Pawluk, R.J., and Pilla, A.A., Acceleration of fracture repair by electromagnetic fields: a surgically non-invasive method, in *Electrically Mediated Growth Mechanisms in Living Systems*, Liboff, A.R. and Rinaldi, R.A., Eds., *Ann. NY Acad. Sci.*, 238, 242–262, 1974.
99. Liboff, A.R., Electrical treatment of ununited bone fracture and spinal fusion, in *Encyclopedia of Medical Devices*, 2nd ed., Webster, Ed., John Wiley, New York, 2005.
100. Blackman, C.F., Benane, S.G., Elliott, D.J., House, D.E., and Pollock, M.M., Influence of electromagnetic fields on the efflux of calcium ions from brain tissue *in vitro*: a three-model analysis consistent with the frequency response up to 510 Hz, *Bioelectromagnetics*, 9, 215–227, 1988.
101. Binhi, V.N., Amplitude and frequency dissociation spectra of ion-protein complexes rotating in magnetic fields, *Bioelectromagnetics*, 21, 34–45, 2000.
102. Liboff, A.R., Comment on "Extremely low frequency magnetic fields can either increase or decrease analgesia in the Land Snail depending on field and light conditions", *Bioelectromagnetics*, 23, 406–407, 2002.

103. Ayrapetyan, S., Grigorian, K., Avanesyan, A., and Stamboltsian, K., Magnetic fields alter electrical properties of solutions and their physiological effects, *Bioelectromagnetics*, 15, 133–142, 1994.
104. Rai, S., Singh, U., Singh, K., and Singh, A., Germination responses of fungal spores to magnetically restructured water, *Electro. Magnetobiol.*, 13, 237–246, 1994.
105. Ruzic, R. and Jerman, I., Influence of Ca^{2+} in biological effects of direct and indirect ELF magnetic field stimulation, *Electro. Magnetobiol.*, 17, 205–216, 1998.
106. Pazur, A., Characterization of weak magnetic field effects in an aqueous glutamic acid solution by nonlinear dielectric spectroscopy and voltammetry, *Biomagn. Res. Technol.*, 2, 1–18, 2004.
107. Binhi, V.N., *Magnetobiology*, Academic Press, London, 2002, 333–343.
108. Beauchamp, J.L., Ion cyclotron resonance spectroscopy, *Ann. Rev. Phys. Chem.*, 22, 527–561, 1971.
109. Azbel, M.Ya. and Kaner, E.A., Cyclotron resonance in metals, *J. Phys. Chem. Solids*, 6, 113–135, 1958.
110. Hobbie, R.K., *Intermediate Physics for Medicine and Biology*, 3rd ed., Springer, New York, 1997, 83–84.
111. Weaver, J.C. and Astumian, R.D., The response of living cells to very weak magnetic fields: the thermal noise limit, *Science*, 247, 459–462, 1990.
112. Liboff, A.R. and Jenrow, K.A., Cell sensitivity to magnetic fields, *Electro. Magnetobiol.*, 19, 223–236, 2000.
113. Adair, R.K., Constraints on biological effects of weak extremely-low-frequency electromagnetic fields, *Phys. Rev. A*, 43, 1039–1048, 1991.
114. Liboff, A.R., Electric-field ion cyclotron resonance, *Bioelectromagnetics*, 18, 85–87, 1997.
115. Aidley, D.J. and Stanfield, P.B., *Ion Channels*, Cambridge University Press, Cambridge, 1996, 146.
116. Nelson, P.H., A permeation theory for single-file ion channels: concerted-association/dissociation, *J. Chem. Phys.*, 119, 6981–6982, 2003.
117. Yellen, G., Permeation in potassium channels: implications for channel structure, *Ann. Rev. Biophys. Biophys. Chem.*, 16, 227–246, 1987.
118. Liboff, A.R., Ion cyclotron resonance in biological systems: experimental evidence, in *Biological Effect of Electromagnetic Fields*, Stavroulakis, P., Ed., Springer, Berlin, 2003, 76–113.
119. Hille, B., *Ionic Channels of Excitable Membranes*, Sinauer Associates, Sunderland, MA, 183, 1984.
120. Li-Smerin, Y., Hackos, D.H., and Swartz, K.J., α-Helical structural elements within the voltage-sensing domain of a K^+ channel, *J. Gen. Physiol.*, 115, 33–49, 2000.
121. Kuyucak, S., Andersen, O.S., and Chung, S.-H., Models of permeation in ion channels, *Rep. Prog. Phys.*, 64, 1427–1472, 2001.
122. Lednev, V., Interference with the vibrational energy sublevels of ions bound in calcium-binding proteins as the basis for the interaction of weak magnetic fields with biological systems, in *On the Nature of Weak Magnetic Fields with Biological Systems*, Frey, A.H., Ed., Landes, Austin, TX, 1994, Chapter 5.
123. Adair, R.K., Criticism of Lednev's mechanism for the influence of weak magnetic field on biological systems, *Bioelectromagnetics*, 13, 231–235, 1992.
124. Adair, R.K., A physical analysis of the ion parametric resonance mode, *Bioelectromagnetics*, 19, 181–191, 1998.
125. Lee, Q.P., Guy, A.W., Lai, H., and Horita, A., The effects of modulated radiofrequency radiation on calcium efflux from chick brains *in vitro* (abstract), 9th Annual Meeting, Bioelectromagnetics Society, Portland, OR, 1987.
126. Blackman, C.F., Benane, S.G., and House, D.G., The influence of temperature during electric- and magnetic-field induced alteration of calcium-ion release from *in vitro* brain tissue, *Bioelectromagnetics*, 12, 173–182, 1991.
127. Liboff, A.R. and Jenrow, K.A., New model for the avian magnetic compass, *Bioelectromagnetics*, 21, 555–565, 2000.

10

Computational Methods for Predicting Field Intensity and Temperature Change

James C. Lin and Paolo Bernardi

CONTENTS

10.1 Introduction

Electromagnetic energy at both high and low-frequencies can be transmitted into biological materials through the use of antennas or applicators. Antennas launch the electromagnetic energy into the medium. They serve to couple the generating source of electromagnetic energy into the medium, which surrounds it. The spatial distribution of electromagnetic energy from an antenna is directional and varies with distance from the antenna. At distances sufficiently far from an antenna, so that local field distribution changes predictably and varies mostly with distance, the region is called a far field or radiation zone. In the near field or near zone close to the antenna, the electromagnetic energy distribution varies as a function of both angle and distance. Moreover, the behavior of electromagnetic fields (EMFs) and their coupling and interaction with biological systems are very different, depending on whether they are in the near or far zone. In fact, these differences constitute the major variances between radio frequency (RF) and low-frequency energy deposition into biological systems. As shown in subsequent sections, the induction of electric and magnetic fields, deposition of electromagnetic power, absorption of electromagnetic energy, and their penetration into tissue, all are functions of the source and its frequency or wavelength. In general, when considering the interaction of EMFs with biological systems, it is necessary to account for the frequency or wavelength and its relationship to the physical dimensions of the body.

In addition, the interaction of EMFs with biological systems is characterized by the electromagnetic properties of tissue media, specifically, dielectric permittivity. Biological materials have magnetic permeability values close to that of free space and are independent of frequency. In a medium such as biological tissue with a finite electrical conductivity σ, a conduction current, $\mathbf{J} = \sigma\mathbf{E}$, can be induced to flow, giving rise to energy loss by joule heating. Clearly, fields must be coupled into tissues, and energy must be deposited or absorbed in the biological systems, regardless of the mechanism that is accountable for an effect, for the system to respond in some manner. Thus, to achieve any biological response, the electric field, magnetic field, or EMF that is exerting its influence must be quantified and correlated with the observed phenomenon.

The purpose of this chapter is to present an account of electromagnetic interactions in biological media, with special emphasis on the energy coupling and distribution characteristics in models of biological structures. Such information is essential for analyzing the interrelationships among various observed biological effects, for separating known and substantiated effects from those that are speculative and unsubstantiated, for assessing the therapeutic effectiveness of electromagnetic waves, and for extracting diagnostic information from field effects.

There exist a wide variety of methods for quantifying fields in biological bodies. The extent of computer usage varies, depending on the specific information sought and the complexity of tissue geometry. This chapter outlines a number of techniques that have been successfully employed to analyze the propagation and absorption characteristics of electromagnetic energy in tissue structures. There are two general approaches: one involves extensive use of analytical development and the other relies more heavily on numerical formulation. Analytical computations are most suited for calculation of the distribution of absorbed energy in simplified tissue geometries such as plane slabs, cylinders, and spheroids, whereas numerical methods offer the opportunity of analyzing the coupling of electromagnetic energy to animal and human bodies, which is difficult, if not impossible, to approach analytically. The advantages and limitations of various methods for field computations, along with representative results, are provided in this chapter. In some cases, for additional details, the reader is referred to previous editions of this handbook [1,2]. This chapter will begin with a brief introduction to the concepts of induced field and power deposition and the characteristics of field intensities and dosimetric quantities.

10.1.1 Induced Field Intensity and Dosimetric Quantities

The quantities of import to characterize coupling of electromagnetic energy into biological systems include the incident field, induced field, power deposition, and absorbed energy. The metrics of specific absorption rate (SAR) and specific absorption (SA) in biological systems or tissue models have been adopted as the dosimetric quantities, especially at RF frequencies. The metric SAR (in W/kg) is defined as the time derivative of the incremental energy absorbed by (or dissipated in) an incremental mass contained in a volume of a given density. SA (in J/kg) is the total amount of energy deposited or absorbed and is given by the integral of SAR over a finite interval of time. Information on SA and SAR is of interest because it may serve as an index for comparison and extrapolation of experimental results from tissue to tissue, from animal to animal, from animal to human, and from human to human exposures. It is also useful in analyzing the relationships among various observed biological effects in different experimental models and subjects. This is in clear contrast to incident field or any other external measures of exposure, which often do not provide the same field inside biological systems of different sizes, species, or constitutions.

Moreover, determination of the induced field would be preferred because it (1) relates the field to specific responses of the body, (2) facilitates understanding of biological phenomena, and (3) is independent of mechanisms of interaction. Once the induced field is known, quantities such as SAR (in W/kg) can be derived from it by a simple conversion formula. For example, from an induced electric field E (in V/m), the SAR can be derived as

$$\mathrm{SAR} = \frac{\sigma E^2}{\rho_\mathrm{m}} \tag{10.1}$$

where σ is the bulk electrical conductivity (S/m) and ρ_m is the mass density (kg/m^3) of tissue. However, at present, a small, isotropic, implantable electric field probe has yet to be developed with sufficient sensitivity for practical use. Consequently, a common practice in experimental dosimetry relies on the use of temperature elevation produced under a short-duration (<30 sec), high-intensity exposure condition. The short duration is not enough for significant convective or conductive heat contribution to tissue temperature rises. In this case, the time rate of initial rises in temperature (slope of transient temperature response curve) can be related to SAR through a secondary procedure, that is,

$$\mathrm{SAR} = \frac{c\Delta T}{\Delta t} \tag{10.2}$$

where ΔT is the temperature increment (°C), c is the specific heat capacity of tissue (J/kg °C), and Δt is the duration (sec) over which ΔT is measured. Thus, the rise in tissue temperature during the initial or a transient period of RF energy absorption is linearly proportional to the value of SAR. It is important to distinguish the use of SAR and its derivation from temperature measurement. The quantity of SAR is merely a metric for energy deposition or absorption, and it should not be construed to imply any mechanism of interaction, thermal or otherwise. However, it is a quantity that pertains to a macroscopic phenomenon by virtue of the use of bulk electrical conductivity in its derivation (Equation 10.1).

It is of particular significance to emphasize the use of bulk electrical conductivity, the specific heat capacity, and the mass density (kg/m^3) of tissue in the derivation of SAR from electric field strength or temperature elevation. Their use in the definition means that a volume of tissue mass must be selected over which SAR is determined. It is self-evident that the numerical value of SAR would be the same, regardless of what volume is chosen, if the induced field or power deposition is uniform in a tissue medium. A difficulty arises when the absorption is not uniform or when tissues with differing properties and conductivities are within the same volume. Thus, in general, a smaller averaging mass or volume would allow SAR—as a metric—to provide a closer representation of its variation inside the body or tissue medium.

10.1.2 Characterizing EMFs

The space surrounding a source antenna can be divided into near and far zones as a function of distance from the antenna [3]. The demarcating boundary occurs at a conservative distance of $R = 2D^2/\lambda$, where D is the largest dimension of the antenna. Furthermore, the near zone can be divided into two subregions: the radiative region and the reactive region. In *the radiative region*, the region closer than $2D^2/\lambda$, the radiated power varies with distance from the antenna. The vicinity of the antenna where the reactive components predominate is known as *the reactive region*. The precise extent of these regions varies for different antennas. For most antennas, the transition point between reactive and radiative regions occurs from 0.2 to $0.4D^2/\lambda$. For a short dipole antenna, the reactive component predominates to a distance of approximately $\lambda/2\pi$, where the radiative and reactive components are equal to each other. However, the outer limit is on the order of a few wavelengths or less in most cases.

A typical wavelength in free space at extremely low-frequencies (ELF between 3 Hz and 3 kHz) is ~5000 km. The $\lambda/2\pi$ distance is about 800 km for the induction and radiation fields to have equal amplitudes. Therefore, for most purposes, ELF transmission line fields are not radiative but are inductive and quasi-static in nature. This fact governs the coupling and induced field characteristics of ELF and other low-frequency electric and magnetic fields in biological tissue. In particular, (1) the electric field is enhanced at the surface of the biological body and is nearly perpendicular to the surface of the body, (2) electric and magnetic fields are decoupled inside a biological body, (3) the electric field applied through air is weakened by a large dielectric permittivity (by about 10^{-6}) upon penetration into biological tissues, (4) the magnetically induced electric field encircles the magnetic field and produces an eddy current whose magnitude increases with distance from the center of the body, and (5) an eddy current appears in each region inside the body with a different conductivity and behaves as a unit with its own body center and radius or an equivalent radius. These observations apply to all frequencies where the wavelength is high or the largest dimension of the body is small compared with the wavelength [2]

In contrast, the $2D^2/\lambda$ distance is approximately 6 cm for a 10-cm RF antenna operating at 900 MHz in free space. Clearly, both near-zone reactive and far-zone radiative interactions are encountered in the vicinity of wireless RF telecommunication systems. In the near zone, the coupling of RF energy into the human head is substantial. As much as 40% to 50% of the radiated RF power is transferred back and forth between the radiating antenna and the head. SAR will vary with specific antenna configuration and its placement next to the head. The bulk of power deposition is on the side of the head nearest to the radiating structure of the cellular mobile telephone and follows an exponential trend away from the antenna side. The anatomy of the head and tissue inhomogeneity can influence the maximum value and distribution of SAR in the head of a mobile telephone user. However, the integrated SAR in the head is similar for a homogenous or inhomogenous model. Some of the major features of near-zone field are that (1) RF electric and magnetic fields are decoupled and are not uniform, (2) wave impedance varies from point to point, (3) beam width from the antenna is divergent and is small compared with the head or human body, (4) electric field ffect is weaker since dielectric permittivity of tissue is relatively high, and (5) inductive coupling of antenna-current-generated magnetic field dominates power deposition.

In the far field, coupling is characterized by plane wave RF field interaction and is independent of source configurations. Electric and magnetic fields are uniquely defined through the intrinsic impedance of the medium. Thus, determination of the electric field behavior is sufficient to characterize the interaction. The coupling of RF power from air into planar tissue ranges from 20% to 60% at wireless communication frequencies. However, enhanced coupling can occur at a greater depth in bodies with curved surfaces. In fact, RF energy is resonantly absorbed by the head at 400 to 1500 MHz, and SAR peaks or hot spots may occur near the center of the head. The interaction of RF energy with biological systems depends on electric field polarization for elongated bodies whose height-to-width ratio is large. It is significant to observe that the integrated SAR or total absorption in the biological body is similar for a homogenous and inhomogenous model.

10.2 Planar Tissue Models

When the radius of curvature of the body surface is large compared to the wavelength and beam width of the impinging radiation, planar tissue models may be used to estimate the absorbed energy and its distribution inside the body. As a first-order approximation, the plane wave configuration is often used for its simplicity to assess EMF interaction with planar biological tissues. This section presents a summary of specific results that have been obtained, using analytical approaches, for thick or semi-infinite layers and for multilayered planar models of biological tissue structures.

10.2.1 Thick or Semi-Infinite Layers

The reflection and transmission of a plane wave at a planar tissue interface depend on the frequency, polarization, and angle of incidence of the wave and on the dielectric constant and conductivity of the tissue. For a linearly polarized plane wave impinging normally on a boundary separating two semi-infinite media, the reflection and transmission coefficients are given by

$$\Gamma = \frac{\eta_2 - \eta_1}{\eta_2 + \eta_1} \tag{10.3}$$

and

$$T = \frac{2\eta_2}{\eta_2 + \eta_1} \tag{10.4}$$

respectively, where η_1 and η_2 are the intrinsic impedances ($= \sqrt{(\mu/\varepsilon)}$) of media 1 and 2. If intrinsic impedances of the two media are approximately equal or if the dielectric permittivities are comparable, most of the energy is transmitted into the second medium, and the reflected field is relatively small. Conversely, if intrinsic impedances differ greatly, or if the dielectric permittivities are very different, the transmitted field is small, and the quantity of reflected energy is large.

Table 10.1 summarizes the magnitude of the reflection coefficient at the boundary separating various tissues. The fraction of normally incident power reflected by the

TABLE 10.1

Reflection Coefficient (Magnitude in %) between Biological Tissues at 37°C

	Frequency (MHz)	Air	Fat (Bone)	Lung	Muscle (Skin)	Blood	Saline
Air	433	0	46	76	82	81	83
	915	0	43	73	78	79	80
	2,450	0	41	71	76	77	79
	5,800	0	39	70	75	76	78
	10,000	0	37	70	74	76	78
Fat (bone)	433		0	46	56	56	60
	915		0	43	52	54	57
	2,450		0	42	50	53	57
	5,800		0	42	50	53	56
	10,000		0	45	52	54	58
Lung	433			0	14	13	19
	915			0	12	14	18
	2,450			0	10	15	19
	5,800			0	10	14	19
	10,000			0	10	13	18
Muscle (skin)	433				0	4	6
	915				0	4	7
	2,450				0	5	10
	5,800				0	4	9
	10,000				0	3	9
Blood	433					0	6
	915					0	4
	2,450					0	5
	5,800					0	5
	10,000					0	6
Saline	433						0
	915						0
	2,450						0
	5,800						0
	10,000						0

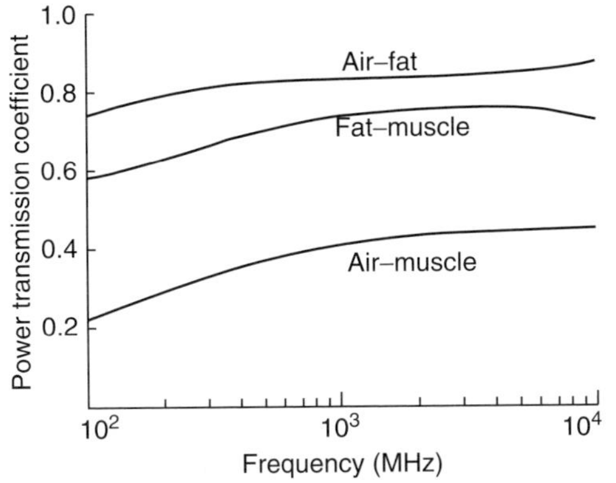

FIGURE 10.1
Power transmission coefficients at three tissue interfaces as functions of frequency.

discontinuity is given by Γ^2. Clearly, about one half of the incident power is reflected at these boundaries. Further, the reflection coefficient for tissue–tissue interfaces generally is smaller than for air–tissue interfaces. The percent reflected power for tissue–tissue interfaces ranges from a low of 5 for muscle–blood to a high of 50 for bone–biological fluid interfaces. This suggests that the closer are the dielectric properties on both sides of the interface, the smaller is the power reflection.

The fraction of power transmitted is related to the power transmission coefficient, T^2. It is readily apparent from Table 10.1 that the power transmitted at air–tissue interfaces is quite substantial at RF and microwave frequencies. Moreover, Figure 10.1 shows that the power transmission coefficient is highly frequency dependent, especially at lower frequencies.

As the transmitted wave propagates in the tissue medium, energy is extracted from the wave and absorbed by the medium. This absorption will result in a progressive reduction of the power density of the wave as it advances in the tissue. This reduction is quantified by the depth of penetration, which is the distance in which the power density decreases by a factor of e^{-2}. Table 10.2 presents the calculated depth of penetration in selected tissues using typical dielectric constants and conductivities. A graphical representation of penetration depth vs. frequency for blood, muscle, and fat is given in Figure 10.2. It is seen that the penetration depth is frequency dependent and takes on different values for different tissues. In particular, the penetration depth for fat and bone is nearly five times greater than for higher-water-content tissues.

TABLE 10.2
Depth of Penetration of an EMF in Biological Tissues as a Function of Frequency

	Tissue				
Frequency (MHz)	**Saline**	**Blood**	**Muscle (Skin)**	**Lung**	**Fat (Bone)**
Depth of Penetration (cm)					
433	2.8	3.7	3.0	4.7	16.3
915	2.5	3.0	2.5	4.5	12.8
2,450	1.3	1.9	1.7	2.3	7.9
5,800	0.7	0.7	0.8	0.7	4.7
10,000	0.2	0.3	0.3	0.3	2.5

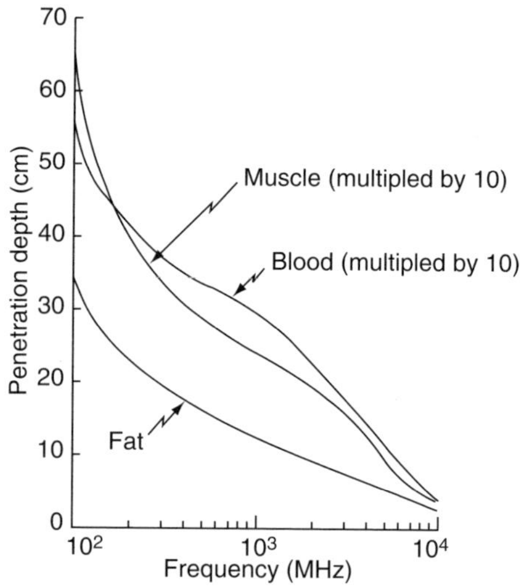

FIGURE 10.2
Depth of penetration for blood, muscle, and fat as functions of frequency.

A wave of general polarization usually is decomposed into its orthogonal linearly polarized components whose electric or magnetic field parallels the interface. These components can be treated separately and combined afterward. Figure 10.3 and Figure 10.4 illustrate the magnitude and phase of the reflection coefficients of representative tissue interfaces at a temperature of 37°C for irradiation at 2450 MHz. The figures clearly show the difference between *E* and *H* polarization. *E* polarization, also called perpendicular polarization, and *H* polarization, also referred to as parallel polarization, are defined in Chapter 1. For *E* polarization, there is only a slight variation in magnitude and phase of the reflection coefficient with incidence angle. For *H* polarization, however, there is a

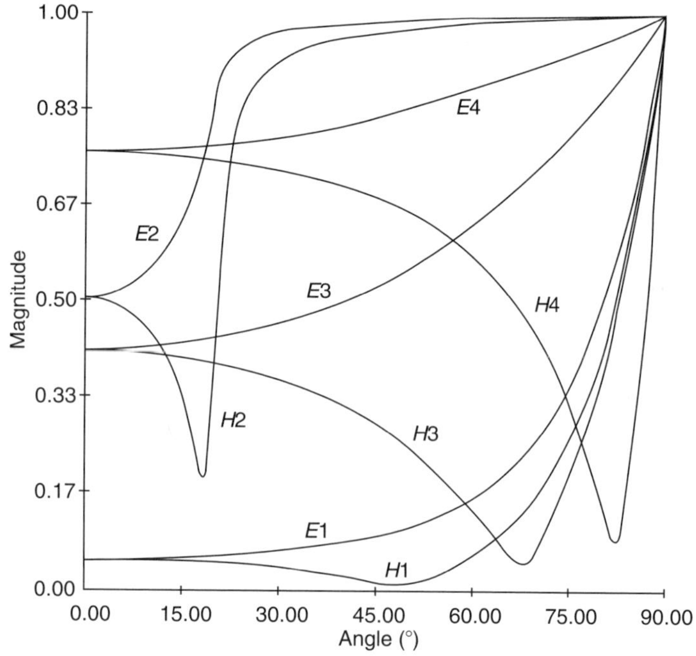

FIGURE 10.3
Magnitudes of reflection coefficients for *E*- and *H*-polarized plane waves at 2450 MHz.

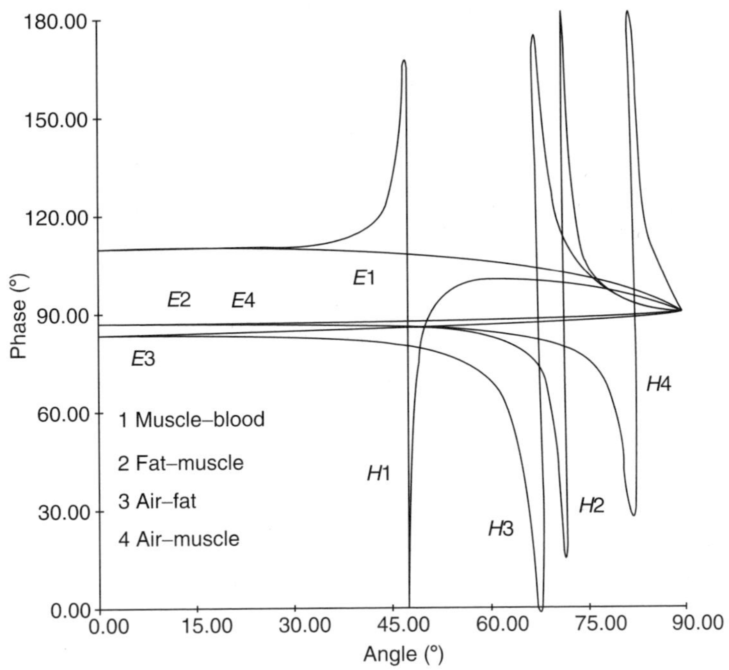

FIGURE 10.4
Phase of reflection coefficients for *E*- and *H*-polarized 2450 MHz plane waves.

pronounced dependence on incidence angle. The reflection coefficient reaches a minimum magnitude and has a phase angle of 90° at Brewster's angle. Thus, the *H* polarized wave is totally transmitted into the muscle medium at Brewster's angle.

10.2.2 Multiple Layers

When there are several layers of different tissues, the reflection and transmission characteristics become more complicated. Multiple reflections can occur between the skin and subcutaneous tissue boundaries, with a resulting modification of the reflection and transmission coefficients [4–7]. In general, the transmitted wave will combine with the reflected wave to form standing waves in each layer. This phenomenon becomes especially pronounced if the thickness of each layer is less than the penetration depth for that tissue. Plane waves impinging on the human body, considered as consisting of parallel layers of subcutaneous fat and more deeply lying muscle, have been studied in detail by Schwan and Li [5,6].

For the tissue model depicted in Figure 10.5, the electric field strength in the fat layer is given by

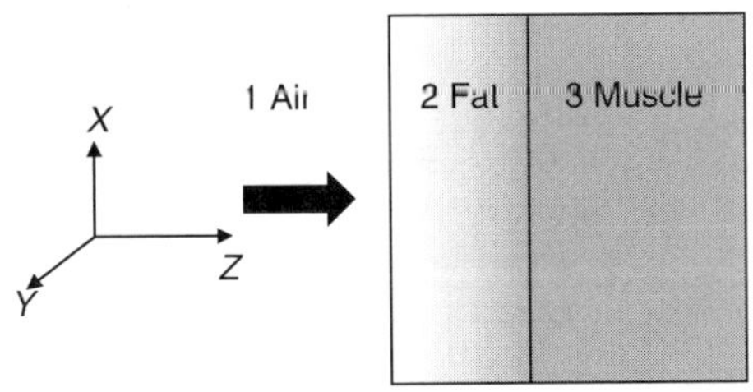

FIGURE 10.5
Plane wave impinging on a composite fat–muscle layer.

$$E_{\rm f} = F_{\rm l} E_0 \left[{\rm e}^{-(\alpha_2 + {\rm j}\beta_2)z} + \Gamma_{32} {\rm e}^{(\alpha_2 + {\rm j}\beta_2)z} \right] \tag{10.5}$$

and the electric field in the underlying muscular tissue is given by

$$E_{\rm m} = F_{\rm t} E_0 {\rm e}^{-(\alpha_3 + {\rm j}\beta_3)z} \tag{10.6}$$

where α_2, β_2 and α_3, β_3 are the attenuation and propagation coefficients in fat and muscle, respectively. The layer function $F_{\rm l}$ and the transmission function $F_{\rm t}$ are given by

$$F_{\rm l} = \frac{T_{12}}{{\rm e}^{(\alpha_2 + {\rm j}\beta_2)l} + \Gamma_{21}\Gamma_{32}{\rm e}^{-(\alpha_2 + {\rm j}\beta_2)l}} \tag{10.7}$$

$$F_{\rm t} = \frac{T_{12}T_{23}}{{\rm e}^{(\alpha_2 + {\rm j}\beta_2)l} + \Gamma_{21}\Gamma_{32}{\rm e}^{-(\alpha_2 + {\rm j}\beta_2)l}} \tag{10.8}$$

where T_{12} and T_{23} are the transmission coefficients at the air–fat and fat–muscle boundaries, respectively. Γ_{21} and Γ_{32} denote the reflection coefficients at these boundaries, respectively; l is the thickness of the fat layer. The power deposition in a given layer can be obtained from Equation 10.1.

Figure 10.6 shows the results of SAR distribution obtained using the dielectric data given in part 1. The values are normalized to the SAR in muscle at the fat–muscle boundary. Note that the absorbed energy is much lower in fat than in muscle. The standing-wave maximum becomes bigger in fat, and the penetration into muscle becomes less as the frequency increases.

The electromagnetic energy absorbed in models composed of planar layers of skin, fat, and muscle can be analyzed in a similar manner [5–8], except that the distribution of

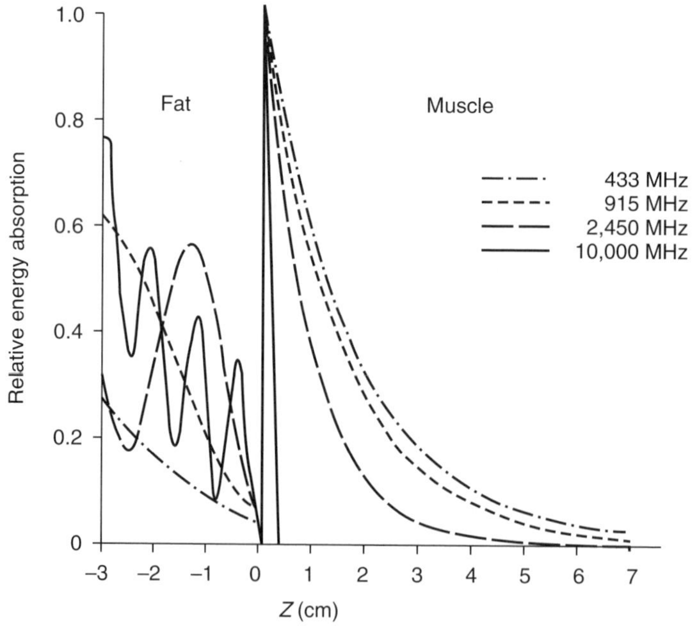

FIGURE 10.6
SAR (absorbed power density) in plane fat–muscle layers.

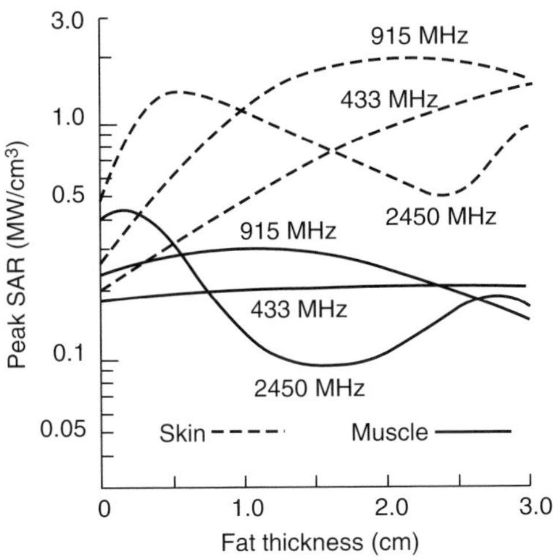

FIGURE 10.7
Peak SAR (absorbed power density) in models composed of fat–skin–muscle layers.

absorbed energy becomes more complex. Figure 10.7 shows that in addition to frequency dependence, the peak SAR exhibits considerable fluctuation with thickness of the subcutaneous fatty layer. The incident power density is, in this case, 10 W/m^2, and the skin layer is 0.2 cm thick. Note that the peak SAR is always higher in the skin layer for planar models at microwave frequencies. The depth of penetration for 10 GHz radiation in skin is less than 0.5 mm—the transmitted energy is almost completely absorbed in the skin, and the SAR is rather unaffected by changing fatty layer thickness. The fact that SAR is highest in the skin is significant, since skin is populated with thermosensitive free nerve endings, which may be excited along with cutaneous pain receptors when the absorbed energy exceeds the normal range that can be handled by thermoregulation.

Figure 10.8 shows the distribution of induced electric field strength in a layer of muscle beneath layers of fat, muscle, and bone for two frequencies [5–8]. It is seen that in addition

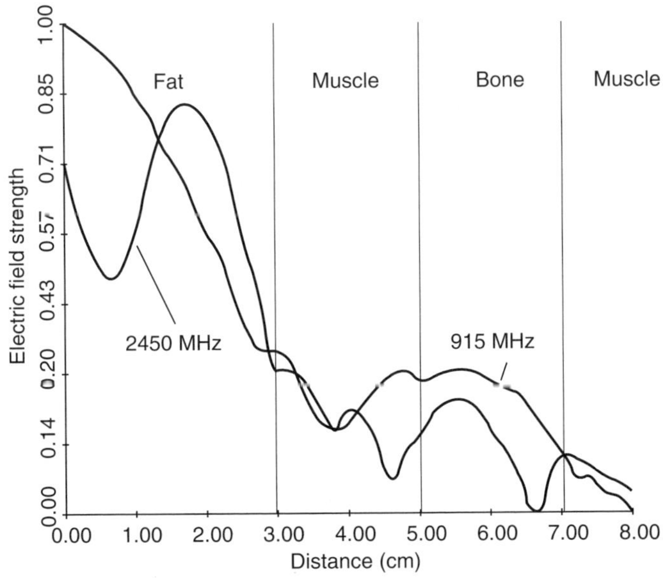

FIGURE 10.8
Distribution of electric field strength in planar layer of fat–muscle–bone–muscle tissue model.

to frequency dependence, the electric fields exhibit considerable fluctuation within each tissue layer. While the standing-wave oscillations are larger at 2450 than at 915 MHz, microwave energy at both frequencies penetrates to deeper tissues. This result, together with Figure 10.6 and Figure 10.7, implies that at frequencies between 300 and 3000 MHz, sufficient energy may be transmitted and reflected to allow examination of organs within the body. Furthermore, at these frequencies, electromagnetic energy can penetrate into more deeply situated tissues, making it especially desirable for therapeutic applications. They also call for special attention for safety considerations since electromagnetic energy in this frequency range can produce higher SAR at greater depth compared to superficial tissues.

10.3 Bodies of Revolution

Although depth of penetration and reflection and transmission characteristics in planar tissue provide considerable physical insight into the coupling and distribution of electromagnetic energy, biological structures generally are more complex in form and exhibit substantial curvature that can modify electromagnetic energy transmission and reflection. For bodies with complex shapes, the propagation characteristics depend critically on polarization and on orientation of the incident wave with respect to the body, as well as on the ratio of body size to wavelength. These complications place severe limitations on calculations of reflected and transmitted energy for bodies of arbitrary shape and complex permittivity. This section presents a summary of results for homogenous and multilayered models based on bodies of revolution that approximate certain mammalian tissue structures.

10.3.1 Spherical Models

Some representative calculations of the SAR are shown in Figure 10.9 for four different-sized models at 918 and 2450 MHz [7,8]. The 6-cm diameter sphere approximates a cat or rhesus monkey brain, and the 10-cm diameter sphere approximates the head of a child, while the 14- and 18-cm diameter spheres are more typical of human adult heads. The figures illustrate the SAR distributions along the three perpendicular axes whose origin coincides with the center of the sphere. An incident plane wave power density of $10\ \mathrm{W/m^2}$ is assumed. The plane wave is propagating in the positive z direction and is polarized along the x axis. It is seen that for 918 MHz, maximum absorption occurs near the center or inside of all the brain spheres. When the frequency is increased to 2450 MHz, the location of peak SAR for the cat-size brain sphere remains near the center, whereas that for a human-size brain sphere is moved to an anterior location.

In general, standing-wave patterns with many oscillations are observed. Note that while peak and average SARs in the cat brain are larger by a factor of 2 than in the human brain at 918 MHz, at 2450 MHz the peak absorption is four times and the average absorption is three times greater in the cat brain than in the human brain. Other studies [9–12] indicate that the peak absorption may be as much as five times greater than the average, and the enhanced absorptions near the center of these brain models may be two to three orders of magnitude greater than that expected from the planar tissue models. The increased absorptions are due to a combination of high dielectric constant and curvature of the model, which produces a strong focusing of energy toward the interior of the sphere that more than compensates for the transmission losses through the tissue.

The peak absorption per unit volume, average absorption per unit volume, and average absorption per unit surface area as functions of frequency and radius of the spherical brain model are illustrated in Figure 10.10. It can be seen that the absorbed energy varies widely with sphere size and frequency. In general, the absorption increases rapidly with increasing radius and is then followed by some resonant behavior. The peaks of these resonant oscillations are related to the maxima, or hot spots, in the distribution of absorbed energy inside the head model, as shown in Figure 10.9. Therefore, for $(2\pi a/\lambda_0) < 0.4$, where a is the sphere radius and λ_0 is the wavelength in vacuum, hot spots do not occur inside the sphere. However, for some combinations of irradiation frequency and radius, hot spots will occur, for example, in spheres with radii between 2 and 8 cm at 918 MHz and between 0.9 and 5 cm at 2450 MHz. For spheres whose radii exceed the size ranges mentioned above, the maximum absorption appears at the anterior portion (exposed surface) of the sphere, and the penetration depth at the surface becomes a dominating factor for exposures at frequencies in this range. The planar model discussed previously may be applied to obtain a theoretical estimation of the absorbed energy in this case.

The frequency dependence of energy absorption is illustrated in the upper graphs in Figure 10.10 for the head of a small animal, such as a cat or rhesus monkey, and a sphere the

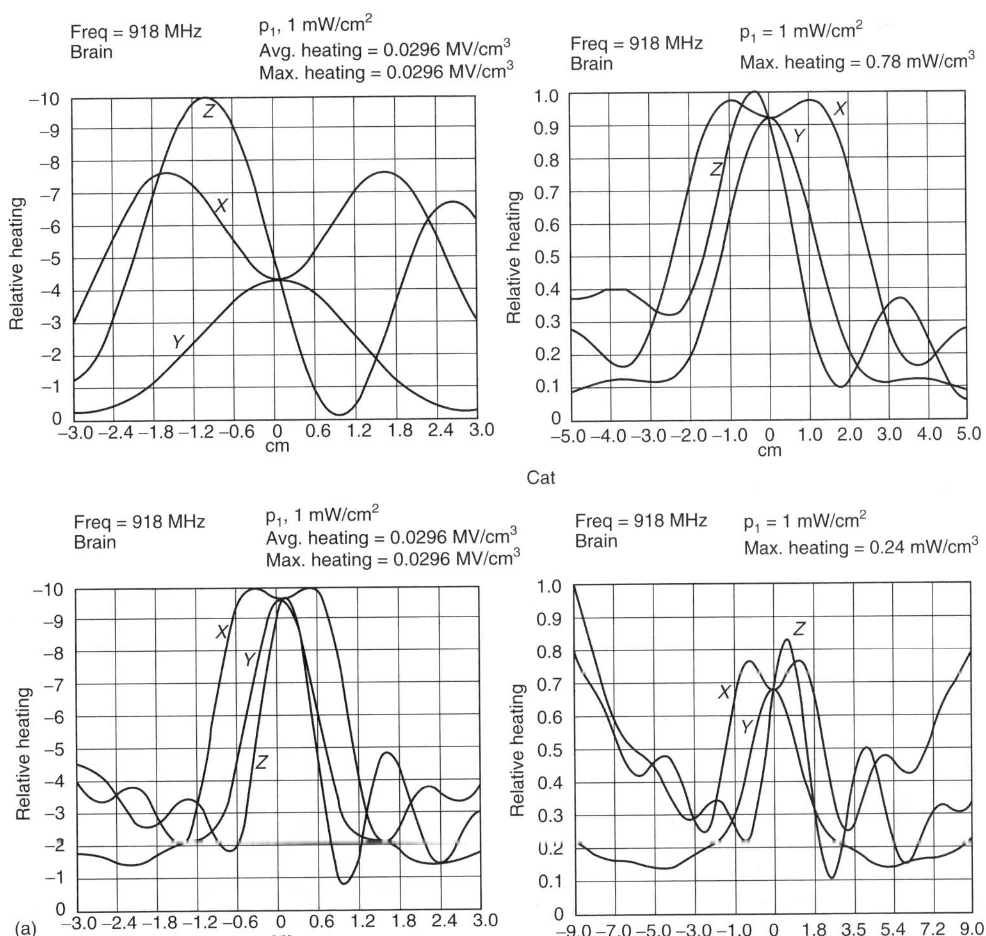

FIGURE 10.9
(a) Predicted SAR distribution (heating pattern) along the three rectangular axes of spherical models of brain exposed to 918-MHz uniform plane waves.

continued

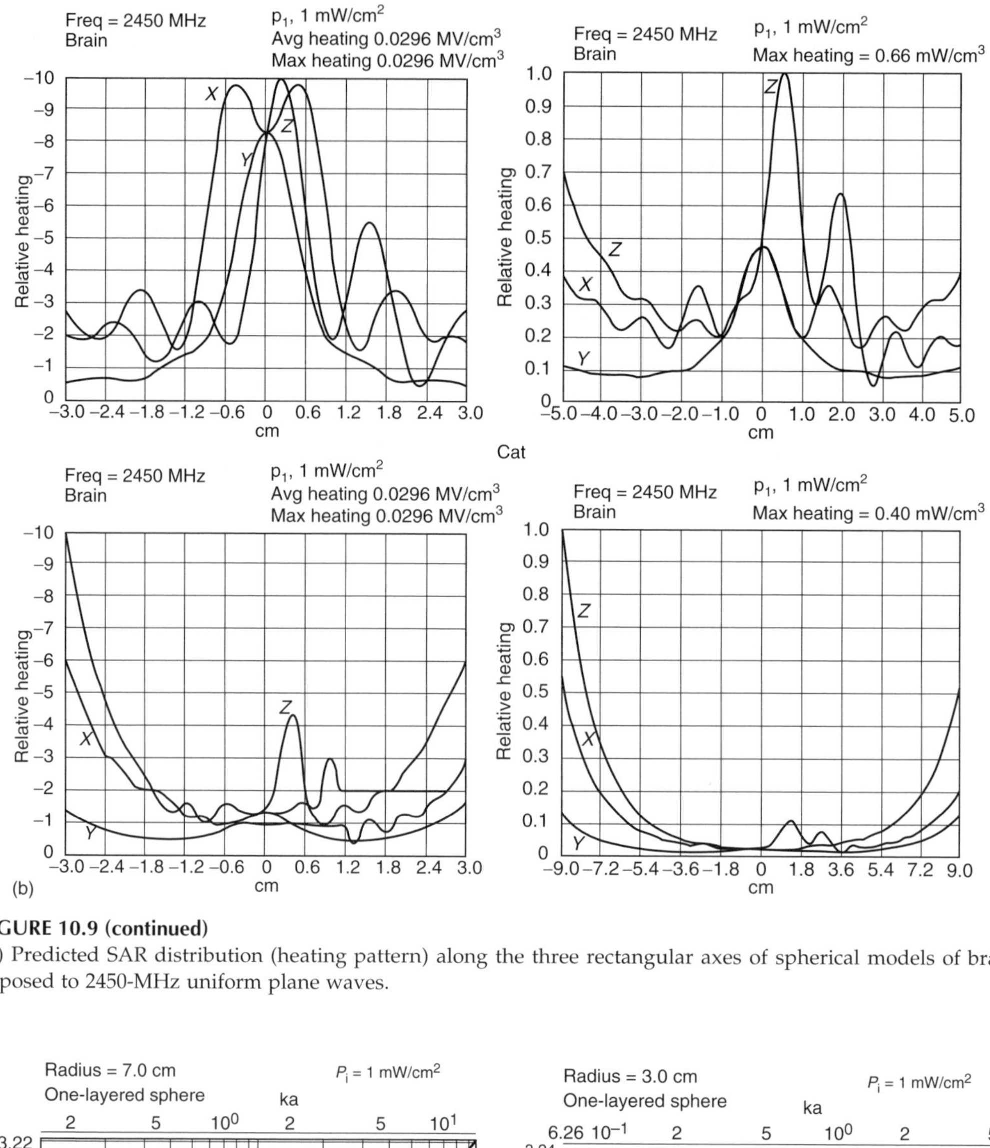

FIGURE 10.9 (continued)
(b) Predicted SAR distribution (heating pattern) along the three rectangular axes of spherical models of brain exposed to 2450-MHz uniform plane waves.

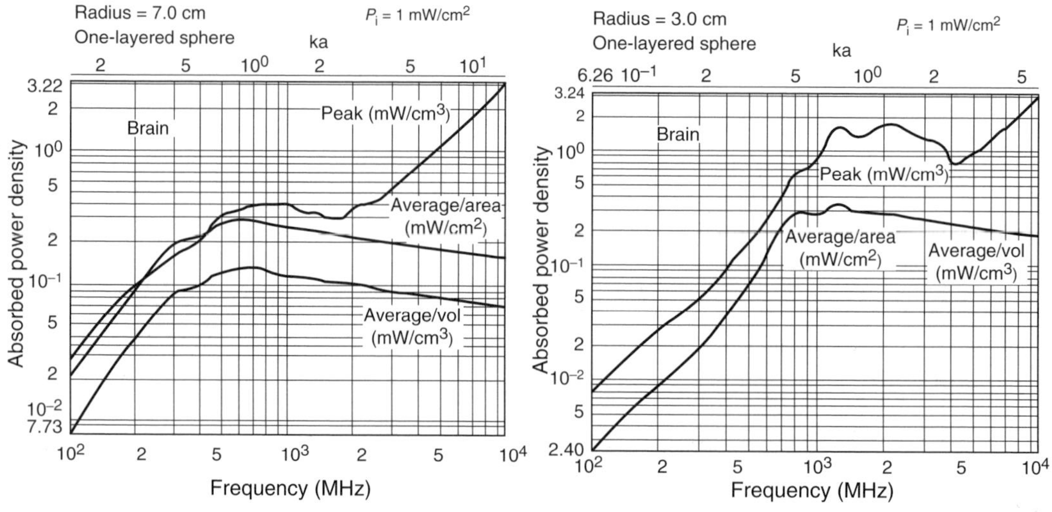

FIGURE 10.10
Electromagnetic energy absorption characteristics for spherical models of brain.

size of a human head. Besides the occurrence of resonant peaks, with increasing frequency, energy is absorbed in a decreasing volume as a result of shortened penetration depth.

The effects of skin, fat, bone, dura, and cerebrospinal fluid on the absorption of RF radiation by the brain have been investigated in several laboratories [9,13–15], using more complex spherical structures where the spherical core of brain is surrounded by five concentric shells of tissues. It is interesting to note that if brain sizes remain unchanged, but the overall sphere diameter is increased to account for the outer tissue layers, absorption in brain tissues may be increased by 25% for human- and cat-size heads at 918 MHz or decreased by 70% or more in the case of 2450 MHz. Moreover, surface absorption is greatly increased in the case of layered models, while fat and bone always absorb the least amount of energy.

If the outer diameter of the sphere remains the same, while the tissue layers are allowed to be either layered or homogenous, the peak and average SARs show very little change except when the radius of the spherical head is between 0.1 and 1.0 times the wavelength in air. The peak and average SARs for layered models may be several times greater than for homogenous models. Enhancement is apparently the result of resonant coupling of energy into the sphere by the outer tissue layers.

A study also has been made of the interaction of circularly polarized plane electromagnetic waves with six-layered spherical models of the mammalian head [15]. The approach is a classic one; Mie equations were modified to account for the two polarizations that are orthogonal in space. For example, calculations at 918 and 2450 MHz for a 7-cm diameter sphere representing a cat or monkey head and a 20-cm diameter sphere typical of a human adult head indicated that the maximum absorption for 918 MHz occurs near the center of a cat-size head, whereas the maximum absorption for a human-size head is at the surface, as in the case of linearly polarized plane waves. However, at 2450 MHz the location of maximum absorption for both the smaller and the larger spheres shifts to the leading surface. The distribution of absorbed energy for circularly polarized waves is more uniform compared with the linearly polarized case. In fact, the absorbed energy distribution in the planes transverse to the direction of propagation is rotationally symmetric, that is, it is independent of angular variation. Note also that the maximum energy absorbed in the spherical head models varies only slightly between these two frequencies. However, a greater quantity of energy is deposited in the inner sphere (representing the brain of a human head) for 918- than for 2450-MHz radiation.

Spherical models of muscle [16,17] have been used as a first-order approximation for the extrapolation to human beings of results obtained from laboratory animals and as an index of whole-body absorption of electromagnetic energy as a function of frequency. The spherical model is attractive since exact solutions for absorbed energy can be obtained for all frequencies and body sizes. While in this case the peak absorption is of very limited utility, the average absorption per unit surface area is related to the time and power required to overload the thermoregulatory capacity of an exposed subject. The absorptions for homo genous muscle spheres, whose volumes correspond to small animals, such as a rat, and standard man, computed as functions of frequency, showed that the average absorption for the rat model is at least ten times higher than for a muscle sphere representing a human body at frequencies greater than 500 MHz. The absorption increases rapidly with frequency until the free space wavelength of the impinging radiation approaches the diameter of the sphere. A number of resonant oscillations appear that tend to increase the amount and nonuniformity of absorbed energy. Above this range the absorption falls off slowly, indicating that details of body surface curvature are of little significance.

We have, thus far, dealt mainly with situations where the diameter of the sphere is comparable to or larger than the wavelength in air. It is interesting to note that when the sphere is small compared with the wavelength, the absorbed energy distribution varies almost as the square of the radius or distance from the axis parallel to the direction of the magnetic field vector. If the sphere is extremely small compared with the wavelength, the absorbed energy distribution becomes nearly uniform in the transverse directions but decreases continuously with distance from the exposed surface. This behavior can be explained by a quasi-static field theory [16]. The electric component of the incident field couples to the object in the same fashion as an electrostatic field. This gives rise to a constant induced electric field inside the sphere that has the same direction but is reduced by $3/\varepsilon$ from the applied electric field for biological materials and is independent of sphere size. Similarly, the magnetically induced electric field inside the body is identical to the quasi-static solution whose magnitude is given by $E = \pi f \mu r H$, where f is the frequency, μ is the permeability, r is the radius, and H is the magnetic field component. Thus, the magnetic component of the incident field produces an internal electric field that varies directly with distance from the axis and in proportion to the frequency. This magnetically induced electric field encircles the magnetic axis and gives rise to an eddy current whose magnitude increases with distance from the y axis. It indicates that while the H-induced energy absorption in a mouse or larger animal is much greater than the E-induced component, electrically and magnetically induced absorption may be equally significant in smaller animals at lower frequencies (below 30 to 40 MHz). Moreover, for a small insect or pupae the electric field will be the predominant factor.

The variation of average and maximum energy absorption with frequency for a human-size sphere is illustrated in Figure 10.11. In the frequency range from 1 to 20 MHz, the maximum absorption rate is only 10^{-6} to 10^{-3} (W/kg)/(W/m^2) of incident power. Inspection of the maximum absorption rate induced by a plane wave, a quasi-static electric field, and a quasi-static magnetic field shows that absorption at frequencies below 20 MHz is primarily due to the magnetically induced eddy current and is characterized by a square-of-frequency dependence. The approximate frequency dependence of average or total energy absorption throughout the frequency range from 1 MHz to 10 GHz is indicated by the dashed line. For frequencies below 20 MHz the average absorption varies as the square of the frequency. In the frequency range of 20 to 200 MHz, the average absorption increases directly in proportion to frequency and attains a maximum of about 2×10^{-3} (W/kg)/(W/m^2) of incident power at 200 MHz. The average absorption rate remains fairly constant with increasing frequency. (Its slow variation is inversely proportional to frequency for higher frequencies.) There is thus little doubt that electromagnetic energy absorption varies both with frequency and with body size, and in a predictable manner.

10.3.2 Prolate Spheroidal Models

Since the bodies of humans and experimental animals are seldom spherical in shape, a better geometric model is needed to analytically and numerically describe the induced fields and absorbed energy inside experimental subjects. A prolate spheroid emulates more closely the shape of mammalian bodies, but most analyses have been restricted to homogenous models for humans and experimental animals [18–22]. As in the case of spherical models, for frequencies below resonance, long-wavelength formulations [19,20] and quasi-static approximations [22] have been used to obtain absorption information. Geometric optics approximations also have been developed for computation of absorption characteristics of prolate spheroidal models of humans at frequencies whose wavelengths are short compared with body size [21].

Three orientations of the impinging plane wave with respect to the body must be distinguished: E-polarization, in which the electric field is parallel to the major axis of

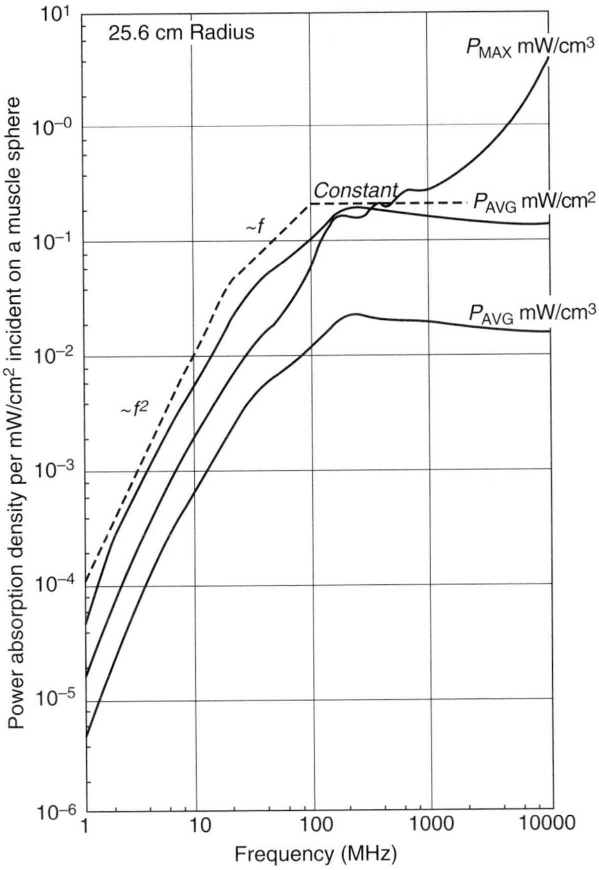

FIGURE 10.11
Frequency dependence of absorption in a spherical model of the human body.

the spheroid; *H*-polarization, in which the magnetic field vector is parallel to the major axis; and *K*-polarization, in which both electric and magnetic field vectors are perpendicular to the major axis of the spheroid. In general, *E*-polarization produces the highest energy absorption for frequencies up to and slightly beyond the resonance region.

For a plane wave with long wavelength, that is, $\lambda > a$, where a is the semimajor axis of the prolate spheroid, the induced fields within the spheroid are uniform and independent of size when the external field is uniform. For $\varepsilon_r > 1$, the field inside the spheroid is weaker than the applied field. Moreover, the whole-body energy absorption depends not only on the strength of impressed fields but also on the orientation of the field with respect to the major axis of the body. As in the case of spherical models, the absorption is produced by an electrically induced current in the direction of the applied *E*-field vector, combined with a circulating eddy current induced by the incident magnetic field [16]. One would therefore expect the electrically induced absorption to be uniform, whereas the absorption due to the circulating eddy current would be zero at the center of the body and increase as the square of the distance from the center.

Note that for a given incident field orientation, the average SAR for humans may be either higher or lower than for rats, depending on the frequency. For example, at 70 MHz, the average SAR is the highest for humans, having a value of 0.25 W/kg for an incident power density of 10 W/m^2; the average SAR for a rat is only 0.0125 W/kg. In contrast, the average SAR of 0.8 W/kg at 700 MHz is the highest for rats; the corresponding value for humans is less than 1/25. It is thus extremely important to take into account the body size and operating frequency to draw any relationship between the biological effects that arise in the laboratory and the corresponding effects that might occur in humans at a given incident power density [1,2].

The frequency for maximal absorption (resonance frequency) depends on the subject and its orientation with respect to the incident field. In general, the shorter the subject, the higher the resonance frequency and vice versa. Further, the frequency dependence of whole-body or average absorption may be partitioned into three regions. This may be illustrated using the orientation that is most efficient in energy coupling, *E*-polarization. For frequencies well below resonance such that the ratio of the longest body dimension (L) to free space wavelength (λ) is less than 0.2, the average SAR is characterized by an f^2 dependence. The average absorption goes through a resonance in the region where $0.2 < L/\lambda < 1.0$. In this case, the average SAR rapidly increases to a maximum near $L/\lambda = 0.4$ and then falls off as $1/f$. At frequencies for which $L/\lambda > 1.0$, the whole-body absorption decreases slightly but approaches asymptotically the geometrical optics limit of about one half of the incident power (1 − power reflection coefficient).

It should be noted that the resonant absorption length of 0.4λ is in good agreement with results from antenna analysis. In addition, whole-body absorptions for *H*- and *K*-polarizations are totally different. The resonances are not nearly as well-defined as for *E*-polarization. In fact, the whole-body absorption curve for *H*-polarization gradually reaches a plateau and stays at that plateau for higher frequencies.

At 10 MHz, the size of the spheroidal model approximating an average human body—height equals 1.75 m with a major-to-minor axis ratio of 6.34 and a 70-kg mass—is small compared with the wavelength. The distribution of absorbed energy in the spheroidal model is qualitatively similar to that for spherical models. But quantitatively, the difference could be as much as one order of magnitude. As expected, the absorbed energy is highest for *E*-polarization. There is a strong coupling of the applied electric field into the interior of the prolate spheroid, and a relatively weak eddy current contribution due to a smaller cross section for intercepting the magnetic flux. The current distribution along the direction of incident field indicates that the electrically and magnetically induced field components are nearly equal. The electric polarization field and the circulating eddy current add at the front side and subtract on the back side of the spheroid to render an absorption pattern that peaks at the front surface and is reduced to almost zero deeper inside the spheroid.

For *H*-polarization, the electrically induced current flows along the x axial direction of incident **E** field, and the eddy current field encircles the z axial direction of incident **H** field. The relatively low power on the z axis comes solely from the incident electric field. The combination of *E*- and *H*-induced components generates a displaced parabolic energy absorption pattern along both the x and the y axes. Clearly, the magnetically induced eddy current predominates in this case, and the absorption is highest along the transverse circumference at the middle of the prolate spheroid. For *K*-polarization, both the electric and the magnetic components of the incident field are along the minor axes of the spheroid: the electrically induced current flows along one axis, and the incident magnetic field induces an eddy current electric field that encircles another axis. The absorption is lowest at the center. Whereas in both *E*- and *H*-polarization cases, the peak absorption occurs at the front surface of the spheroid irradiated by the incident field, this is not the case for *K*-polarization. Maximum absorption appears at the surface of the narrow cross section, and the absorbed energy varies parabolically. This is the result of the large quantity of magnetic flux intercepted by the broad cross section (and the resulting concentration of eddy current). It should be noted that the results match very well with experimental measurements [23]. Moreover, the peak absorptions may be two orders of magnitude higher than those for dielectric spheres of equal mass.

10.4 Anatomically Based Models

We have summarized above some of the computational approaches to quantify the absorbed energy in simple models of biological objects. It should be recognized that while spheres and spheroids are good models of some animal bodies and certain body parts, they may not always be adequate for humans and experimental animals under a variety of exposure situations. More realistic models, such as models of human bodies formed from small-sized, computational-cell volumes have been developed to account for the irregular shapes [24–28]. These models, based on numerical techniques have been a great asset in efforts to accurately predict energy absorption and its distribution in biological objects exposed to EMFs and RF radiation. In what follows we shall summarize a number of computer techniques that have been applied with some success in solving electromagnetic energy absorption and SAR distribution problems. We shall also describe some results obtained using these methods.

10.4.1 Brief Survey of Numerical Methods

The numerical methods used to predict induced fields in biological bodies of realistic shape and composition include the quasi-static impedance method, the method of moments (MoM), the finite element method (FEM), and the finite difference time domain (FDTD) method. Note that the quasi-static impedance method is restricted to lower frequencies (<30 to 40 MHz for the human body), but the MoM, the FEM, and the FDTD method may be used for any frequency of interest. In addition, both the finite element and the FDTD methods involve solving Maxwell's equations in the differential form for the computation of induced fields.

10.4.1.1 Quasi-Static Impedance Method

For low-frequency situations, where the dimensions of the biological body are small compared to the wavelength, the impedance method has been found to be highly efficient as a numerical procedure for calculating internal current densities and induced electric fields [29–33]. In this method, the biological body or the exposed part thereof is represented by a three-dimensional (3-D) network of impedances whose individual values are obtained from the complex conductivities $\sigma + j\omega\varepsilon$ for the various locations of the body. The impedances for various directions for the 3-D network can be written as

$$Z_m^{i,j,k} = \frac{\delta_m}{\delta_\mathrm{n}\delta_\mathrm{p}(\sigma_m^{i,j,k} + j\omega\varepsilon_m^{i,j,k})} \tag{10.9}$$

where i, j, k indicate the cell index; m is the direction in x, y, or z for which the impedance is calculated; and σ_m and $j\omega\varepsilon_m$ are the conductivities and the dielectric permittivities for the cell (i,j,k). δ_m is the thickness of the cell in the mth direction, and δ_n and δ_p are the widths of the cell in directions at right angles to the mth direction.

In the impedance method formulation, it can be seen that the cells need not be identical so that fairly thin features of the body can be modeled as well as the interfaces between the various tissues and organs. Also, the conductivity for a given cell can be directionally dependent. This feature will be useful in allowing for the highly anisotropic conductivities of the tissues that have been reported for low-frequencies including the power-line frequencies [34,35].

Employing anatomically based models of the human body, the impedance method has been used for the following applications:

1. Calculation of SAR distributions for operator exposure to spatially variable fields of induction heaters [31]
2. SAR distributions for linearly or circularly polarized RF magnetic fields representative of magnetic resonance imagers [32]
3. SAR distributions due to capacitive-type electrodes used for hyperthermia [33]
4. SAR distributions for interstitial RF needle applicators for hyperthermia [36]

Some calculations using the impedance method are listed below:

1. Internal electric fields and current densities induced in the human body by exposure to magnetic fields of high-voltage power transmission lines [37]
2. Electric fields and current densities induced in the human head by magnetic fields of a hair dryer [38]
3. Current densities induced in the arm and the body by magnetic fields of an electric hand drill [39]
4. Currents induced in the anatomically based model of the human body by the electric and magnetic fields of electric blankets [40].

In the section that follows, the use of the impedance method is described for calculating currents in models of the human body exposed to electric and magnetic fields of both the conventional (pre-1990) electric blanket and the new low-magnetic-field electric blanket.

10.4.1.2 Volume Integral Equation MoM

The MoM [41] is used in conjunction with either the volume integral equation method or the surface integral equation method for finding solutions to the unknown fields inside the body. The approaches differ, however, in specifics, in that the surface integral equation MoM (SMoM) finds the unknown currents on the body surface and calculates the interior fields from the surface currents, the reciprocity theorem, and a ''measurement matrix.'' In contrast, the volume integral equation MoM (VMoM) requires determination of unknown fields throughout the volume of the body using the volume equivalence principle and the MoM.

The numerical technique that has been adopted for most of the early field intensity computations is the VMoM, employing the volume equivalence principle [26,27,42,43]. The MoM is used to transform the integral equation into a matrix equation by subdividing the body into N simply shaped cells. This is accomplished with the aid of an appropriate set of expansion functions, chosen to satisfy the boundary conditions, and a set of weighting (testing) functions to reduce the matrix fill-in time. The total electric field in each of the N cells is given by matrix inversion. A more detailed description of the volume integral equation method is included in the next section.

However, it should be noted that a fundamental limitation of this method is the use of full or nearly full matrices and, therefore, the requirement of extensive computer storage and long running time. Even with the availability of larger and faster computers, this difficulty is not completely resolved. The need for excessively large numbers of mathematical cells to render a more accurate representation of the body will give rise to an equally large and full matrix. The inversion of large, full matrices often leads to numerical instabilities in the solution. Nevertheless, the method does allow the use of inhomogenous models with up to 1000 cells. In fact, this method has been employed, successfully, to calculate whole-body averaged absorption and to obtain regional distribution of absorbed RF energy using inhomogenous block models composed of rectangular cells [29–28,44]. This method also has been used to study the interaction of the near-zone field of an antenna with biological bodies [45,46].

10.4.1.3 SMoM

Another approach for predicting the distribution of absorbed electromagnetic energy is the SMoM [47–50]. This method makes use of two coupled integral equations, that is, the electric field and magnetic field integral equations for the tangential components of the field on the surface separating the biological body from air. The unknown surface currents are found by Fourier decomposition and the moment method. The fields inside the biological body are calculated using the previously computed surface currents, the reciprocity theorem, and the concept of measurement matrix [48,51–53].

The method begins with the matrix representation of the coupled integral equations. If the body is assumed to be rotationally symmetric, the incident wave and the induced current could then be expanded in a Fourier series expansion in the angle of rotation. This reduces the problem to that of solving a system of orthogonal modes. The method further expands the surface components in terms of triangular expansion or basis functions and allows the testing functions to be the complex conjugate of the basis functions taking advantage of the orthogonality property. Thus, the major advantage of introducing the Fourier series is to enable each mode to be treated completely independently of all other modes. This results in a much smaller-size, manageable matrix equation to be evaluated for the unknown expansion coefficients that determine the surface currents. It should be noted that for biological bodies, triangular expansion and testing functions are preferred over flat pulse expansion functions [47,48]. In fact, an expansion function with a continuous first derivation may constitute an even better choice for the expansion basis function. In any event, once the surface currents are obtained, the fields everywhere, or SAR at each point inside the body, can be calculated using the reciprocity theorem [52,53]. The total absorption can be found by integrating the surface Poynting vector.

The validity of SMoM has been substantiated by using a dielectric sphere [47]. Calculations for a human torso modeled by a homogenous muscle body of revolution with a height of 1.78 m at 30, 80, and 300 MHz showed enhanced absorption in the neck region for all three frequencies and both vertical and horizontal polarizations [48]. Note that the vertical direction is aligned with the long dimension of the torso and serves as the axis of symmetry. The strongest absorption in the torso model was found to occur with vertical polarization and near the first resonance frequency of the torso (80 MHz). In general, the surface integral equation method is applicable to any arbitrarily shaped homogenous body of revolution. The method can be used not only with incident plane waves but also with a wide variety of other field exposure conditions, including direct contact situations and near-zone sources.

Since both the surface and volume integral equation methods for field intensity prediction rely on the MoM for implementation, it is instructive to compare the relative advantages of these two techniques. For simplicity, consider a homogenous cube with N samples on each side: the computer storage requirements are N^2 and N^3 for the surface and volume integral methods, respectively [48]. For sufficient sampling to ensure accurate description of field variations, N is usually a large number. Thus, the surface integral equation method requires significantly fewer unknowns for homogenous models. Moreover, in cases where permittivity and conductivity values are large, such as in biological bodies, the wavelength becomes contracted inside the body, and a much larger number of cells than that indicated above may actually be needed. If the model is inhomogenous, then the volume integral equation would prove to be more suitable. It is possible, however, to generalize the surface integral equation technique to account for inhomogeneities by employing the invariant imbedding procedure [54].

10.4.1.4 FEM

The FEM has been a preferred numerical algorithm in many fields of application. However, its use and popularity in predicting field intensities in biological systems have been modest until recent progress in mesh generation, boundary conditioning, and large matrix solvers. The FEM method is a near-neighbor, volume method for solving Maxwell's differential equations and is associated with a sparse system of equations [55,56]. Aside from the low memory requirement (on the order of N), an inherent attraction of FEM is its adaptability in modeling inhomogeneities and complex geometries. The feature of conforming and the variable-sized cell elements of the computational volume are extremely important in bioelectromagnetics.

The basic approach of the FEM method for predicting EMF distributions inside biological bodies starts by subdividing the physical space and biological body of interest into meshes of small volumes or cells of tetrahedral elements. This step is very important since the manner in which the volume is subdivided will dictate the computational resources required and the speed of the computation and accuracy of the results. Each cell element and node location will have to be systematically numbered and described. Once the volume has been subdivided, labeled, and the appropriate property values ascribed, the unknown field within each element is then approximated using linear extrapolation. A major step in FEM is the formulation of the system of linear equations using either the Ritz or the Galerkin algorithm with proper boundary conditions. There are two approaches to solve the system of linear algebraic equations: the direct method of Gaussian elimination or the iterative method that starts with an initial guess. In practice, either method can produce an approximate solution to the unknown field intensity with a prescribed accuracy.

It should be noted that a large region exterior to the biological body is often encountered in bioelectromagnetic situations, where the biological body, or portion of it, is part of a region into which electromagnetic energy is radiated and scattered. The region of space exterior to the biological body and applicator must be truncated with an artificial boundary to limit the volume elements and the number of unknowns. Consequently, an appropriate boundary condition needs to be established at this artificial boundary for a unique finite element determination of the induced fields inside the body. The most common boundary conditions selected for this purpose are the absorbing boundary conditions that minimize the nonphysical reflections from the artificial boundaries by making boundaries transparent to the scattered field.

Fairly large-scale calculations, on the order of 200,000 elements, have been conducted effectively in the workstation-computing environment. Specifically, detailed power deposition patterns have been simulated in full and partial models of the human body undergoing electromagnetic hyperthermia treatment for cancer [57]. In this case, the cell elements were generated from computerized tomographic data obtained on human patients.

10.4.1.5 FDTD Method

The FDTD approach is an attempt to solve Maxwell's curl equations by directly modeling propagation of waves into a volume of space containing the biological body. By repeatedly implementing a finite difference representation of the curl equations at each cell of the corresponding space lattice, the incident wave is tracked as it first propagates to the body and then interacts with it through surface current excitation, transmission, and diffraction. This wave tracking process is completed when the steady-state behavior is observed at each lattice cell. Considerable simplification is achieved by analyzing the interaction of the wavefront with a part of the body surface at a time, rather than attempting a simultaneous solution of the entire problem.

The FDTD method has become one of the most successful methods for SAR calculations. The method was first proposed by Yee [58] and later developed by Taflove [59–61], Holland [62], and Kunz and Lee [63]. Several books [64–66] are devoted to the FDTD method and some of its applications. For bioelectromagnetic applications the FDTD method has been found to be extremely versatile and has been used for whole-body or partial-body exposures due to spatially uniform or nonuniform fields (far or near fields), sinusoidally varying EMFs, and transient fields such as ultra-wide-band (UWB) and electromagnetic pulses (EMPs) [67–71]. Accordingly, some details of FDTD are included in this section.

10.4.1.5.1 The Traditional FDTD Method

In this method, the time-dependent Maxwell's curl equations

$$\nabla \times \mathbf{E} = -\mu \frac{\partial \mathbf{H}}{\partial t} \tag{10.10}$$

and

$$\nabla \times \mathbf{H} = \sigma \mathbf{E} + \varepsilon \frac{\partial \mathbf{E}}{\partial t} \tag{10.11}$$

are implemented for a lattice of subvolumes or Yee "cells" that may be cubical or parallelepiped with different dimensions, δ_x, δ_y, and δ_z in the x-, y-, or z-directions, respectively. The components of **E** and **H** are positioned about each of the cells as shown in Figure 10.12 and calculated alternately with half-time steps where the time

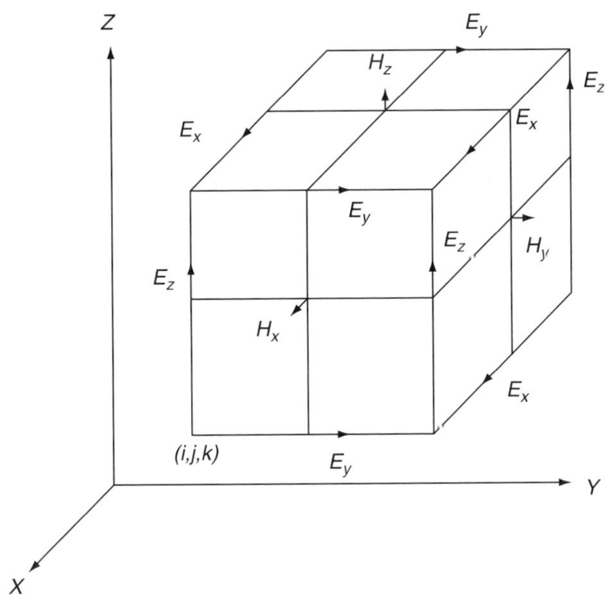

FIGURE 10.12
Unit cell of Yee lattice indicating positions for various field components.

step $\delta t = \delta/2c$. Here δ is the smallest of the dimensions used for any of the cells, and c is the maximum phase velocity of the fields in the modeled space. Since some of the modeled volume is air, c corresponds to the velocity of EM waves in air.

In the FDTD method, it is necessary to represent not only the scatterer or absorber, such as the human body or a part thereof, but also the EM sources, including their shapes, excitations, etc., if these sources are in the near-field region [70,71]. The far-field sources, on the other hand, are described by means of incident plane wave fields prescribed for a "source" plane [68,69], typically six to ten cells away from the exposed body. The source–body interaction volume is subdivided into Yee cells of the type shown in Figure 10.12. The interaction space consisting of several hundred thousand to a few million cells is truncated by means of absorbing boundaries. The prescribed incident fields are tracked in time for all cells of the interaction space. The solution is considered completed when either the fields have died off or, for sinusoidal excitation, when a sinusoidal steady-state behavior for **E** and **H** is observed for the interaction space.

The body of interest is mapped into the lattice space by first choosing the lattice increment and then assigning values of permittivity and conductivity to each cell. The boundary conditions at media interfaces are naturally generated by the curl equations. Thus, once a computer program is developed, the basic routines need not be changed for different model geometries. In fact, inhomogeneities and fine structural details could be modeled with a maximum resolution of one unit cell.

Time stepping for the FDTD method is accomplished by an explicit finite difference procedure [58,64]. For a cubic-cell lattice space, this procedure involves positioning the electric and magnetic field components about a unit cell of the lattice and then evaluating the components at alternate half-time steps. In this manner, centered difference expression can be used for both the space and the time increments without solving simultaneous equations to compute the fields at the latest time step.

The explicit formulation of the FDTD method is particularly suited for execution with minimum computer storage and run time using current array-processing computers. The required computer storage and run time increase only linearly with N, the number of cells. In fact, it has been shown that the FDTD method is capable of solving for more than 1 million unknown field components within a few minutes on an array-processing computer. Field intensities have been predicted to within 2.5% accuracy relative to known analytical and experimental bench marks. Recently, this FDTD technique has been improved to allow solutions for field penetration and absorption in large, complex, inhomogenous, and irregularly shaped biological bodies in three dimensions, with millimeter range spatial resolution. With the exception of a few early attempts with lossy biological objects [59,67], a majority of early efforts have been directed toward application of the FDTD method to electromagnetic interaction in time-varying inhomogenous media [60], and with metallic bodies of revolution [61]. However, during the last decade, the FDTD method has become the most extensively used numerical procedure for bioelectromagnetic computations.

10.4.1.5.2 Frequency-Dependent FDTD Formulation

For short pulses where wider bandwidths are generally involved, a frequency-dependent FDTD or $(FD)^2TD$ method is needed. Two general approaches have been used for the $(FD)^2TD$ method. One approach is to convert the complex permittivity from the frequency domain to the time domain and convolve this with the time domain electric fields to obtain time domain fields for the dispersive material. This discrete time domain method may be updated recursively for some rational forms of complex permittivity, which removes the need to store the time history of the fields and makes the method feasible. This method has been applied to materials such as water, for which the permittivity may be described by a first-order Debye relaxation equation [72–74] or more complex materials with dielectric properties given by a second-order Lorentz equation with multiple poles [75].

A second approach is to add a differential equation relating the electric flux density **D** to the electric field **E** and solve this new equation simultaneously with the standard FDTD equations. This method has been applied to 1-D and 2-D examples with materials described by a first-order Debye equation or second-order single-pole Lorentz equation [76], and to a 3-D sphere and homogenous two-thirds muscle-equivalent man model with properties described by a second-order Debye equation [77,78]. In the following we describe this differential equation approach, which has now been used for induced current and SAR calculations for a heterogenous model of the human body [79].

The time-dependent Maxwell's curl equations used for the FDTD method have already been given as Equation 10.10 and Equation 10.11. The curl $\boldsymbol{H}$ can also be written as follows:

$$\nabla \times \mathbf{H} = \frac{\partial \mathbf{D}}{\partial t} \tag{10.12}$$

where the flux density vector **D** is related to the electric field through the complex permittivity $\varepsilon^*(\omega)$ of the local tissue by the following equation:

$$\mathbf{D} = \varepsilon^*(\omega)\mathbf{E} \tag{10.13}$$

Since Equation 10.10 and Equation 10.12 are to be solved iteratively in the time domain, Equation 10.13 must also be expressed in the time domain. This may be done by choosing a rational function for $\varepsilon^*(\omega)$, such as the Debye equation with two relaxation constants (see Chapter 1):

$$\varepsilon^*(\omega) = \varepsilon_0 \left[\varepsilon_\infty + \frac{\varepsilon_{s1} - \varepsilon_\infty}{1 + j\omega\tau_1} + \frac{\varepsilon_{s2} - \varepsilon_\infty}{1 + j\omega\tau_2} \right] \tag{10.14}$$

Rearranging Equation 10.14 and substituting in Equation 10.13 gives

$$\mathbf{D}(\omega) = \varepsilon^*(\omega)\mathbf{E}(\omega) = \varepsilon_0 \frac{\varepsilon_s + j\omega(\varepsilon_{s1}\tau_2 + \varepsilon_{s2}\tau_1) - \omega^2\tau_1\tau_2\varepsilon_\infty}{1 + j\omega(\tau_1 + \tau_2) - \omega^2\tau_1\tau_2} \mathbf{E}(\omega) \tag{10.15}$$

where the static (zero frequency) dielectric constant is given by

$$\varepsilon_s = \varepsilon_{s1} + \varepsilon_{s2} - \varepsilon_\infty \tag{10.16}$$

Assuming $e^{j\omega t}$ time dependence, Equation 10.15 can be written as a differential equation in the time domain:

$$\tau_1\tau_2 \frac{\partial^2 \mathbf{D}}{\partial t^2} + (\tau_1 + \tau_2)\frac{\partial \mathbf{D}}{\partial t} + \mathbf{D} = \varepsilon_0 \left[\varepsilon_s \mathbf{E} + (\varepsilon_{s1}\tau_2 + \varepsilon_{s2}\tau_1)\frac{\partial \mathbf{E}}{\partial t} + \varepsilon_\infty\tau_1\tau_2 \frac{\partial^2 \mathbf{E}}{\partial t^2} \right] \tag{10.17}$$

For the $(FD)^2TD$ method, Equation 10.10 and Equation 10.12 need to be solved subject to Equation 10.17. These equations can be written in the difference form [77,78] and solved to find **E**, **H**, and **D** at each cell location. The **E** → **H** → **D** loop is then repeated until the pulse has died off.

10.4.2 Human Bodies Exposed to EMFs

Computational algorithms based on numerical techniques described in the previous section have been applied to predict field intensities and SAR distributions in anatomically realistic models of human bodies. However, the use of a given numerical method and the number of computational cells involved in the models typically dictate the applicability of the techniques, how real the model is, the degree of accuracy attainable, and its domain of applicability. In the following sections, a few of the models will be described, and they will be followed by some representative results obtained using the models.

10.4.2.1 Realistic Models of the Human Body

Models proposed as better representations of the complex geometry and composition of the human body include constructions using small-volume cubic cells or cell meshes and anatomically based models generated from computerized tomographic and magnetic resonance image data.

10.4.2.1.1 Cubic-Cell Models

Models of the human body consisting of 200 to 1000 cubic cells that account more realistically for the gross anatomic and biometric characteristics of human bodies have been used by several investigators [24–28]. The models are 1.75 m tall and can be made either homogenous or inhomogenous by choosing an equivalent or a volume-weighted complex permittivity for each cell. The cubic-cell model has been employed successfully to calculate whole-body averaged absorption. It is important to note that for subdivision with less than three cells per wavelength, the magnitude and phase resolutions would be such that even with convergence the reliability of the MoM-computed SAR would be questionable. Therefore, if the interest is primarily in whole-body SAR, this model may provide quite adequate results for frequencies lower than 30 MHz. To achieve more accurate structural representation of the human body, anatomically based models have been offered in recent years.

10.4.2.1.2 Millimeter-Resolution Model Based on Magnetic Resonance Imaging Scans of the Human Body

A new millimeter-resolution model of the human body has been developed from the magnetic resonance imaging (MRI) scans of a male volunteer of 176.4-cm height and 64-kg weight [80,81]. The MRI scans were taken with a resolution of 3 mm along the height of the body and 1.875 mm for the orthogonal axes in the cross-sectional planes. Even though the height of the volunteer was quite appropriate for an average adult male, the weight was somewhat lower than an average of 71 kg, which is generally assumed for an average male. This problem can, to some extent, be ameliorated by assuming that the cell dimensions for the cross sections are larger than 1.875 mm by the ratio of $(71/64)^{1/2} = 1.053$. By taking the larger cell dimensions of $1.053 \times 1.875 = 1.974$ mm for the cross-sectional axes, the volume of the model can be increased by $(1.053)^2 = 1.109$, that is, by about 10.9%, which results in an increase of its weight by approximately the same percentage, that is, to a new weight of approximately 71 kg. The MRI sections were converted into images involving 29 tissue types whose electrical properties can then be prescribed at the exposure frequency. The tissue types are fat, muscle, bone, cartilage, skin, brain, nerve, cerebrospinal fluid (CSF), intestine, spleen, pancreas, heart, blood, eye, eye humor, eye sclera, eye lens, ear, liver, kidney, lung, bladder, stomach, ligament, compact bone, testicle, spermatic cord, prostate gland, and erectile tissue. As described above, this model has been used to calculate the electromagnetic absorption in the human head, neck, and shoulders for cellular telephones operating at frequencies of 800 to 900 MHz. Because of the localized nature of EMFs, it was possible to use the model corresponding to the top 42 cm of the body for SAR calculations.

10.4.2.1.3 The ''Visible Human'' Model

The Visible Human (VH) Project, developed by the National Library of Medicine, is a 3-D digital image library representing an adult human male and female [82]. The dataset for both male and female includes photographic images obtained through cryosectioning of human cadavers and digital images obtained by computer tomography and MRI of the same cadavers. In particular, the photographic images represent a highly accurate and realistic counterpart of the anatomical cross sections contained in human anatomy atlases. The male dataset, the first to be constructed, consists of 1871 digital axial images obtained at 1.0-mm intervals, with a pixel resolution of 1 mm, while the female one contains 5189 digital axial images, obtained with a finer spatial step of 0.33 mm. While these digital datasets represent a unique tool to explore human anatomy, their direct use for computational electromagnetic dosimetry is limited by the fact that images cannot be directly used as an input for a numerical electromagnetic tool but must be converted to a so-called ''segmented'' version. A segmented model is a model where every pixel, usually called in such models ''voxel,'' does not contain information about the color (like in digital images) but rather a label that is uniquely associated to a given tissue. In such a way, it is possible to know which tissue fills each of the model voxels and hence assign the correct complex permittivity values to be used in numerical simulations.

Segmentation of the original image sets is a complex and time-consuming activity, which is difficult to carry out making exclusive use of automatic procedures, such as contour recognition algorithms, but inevitably requires intervention by experts in human anatomy. The segmentation procedure has been carried out for the male model by researchers at the Air Force Research Laboratory, Brooks Air Force Base, TX [83]. The final segmented model, made freely available to the scientific community, comprises $586 \times 340 \times 1878$ voxels with a resolution of $1 \times 1 \times 1$ mm, and is segmented in about 40 different tissue types [84]. The model has been widely used to study both whole-body and localized human exposure to EMFs radiated by different types of sources and is now being included in many commercially available electromagnetic simulation tools with capabilities for dosimetric evaluation.

10.4.2.2 Currents Induced in the Human Body by Low-Frequency EMFs

This section reports the results obtained for low-frequency EMF. Specifically, it includes the use of the impedance method to calculate currents induced in the human body by the EMFs of electric blankets. It also includes the use of the FDTD method for calculations of internal **E** and **H** fields and induced current densities for exposure to electric, magnetic, or combined electric and magnetic fields at power-line frequencies. The results given below were obtained using a 1.31-cm resolution, anatomically based model of the human body. Since the term $j\omega\varepsilon$ can be neglected as compared to σ for the various tissues at ELF including electric power frequencies (50/60 Hz), the impedances for the various cells of the model given by Equation 10.9 can be replaced by resistances. It is recognized that the conductivities of various tissues, for example, skeletal muscle, heart, and bone, are anisotropic for power-line frequencies [34,35,85]. This has been neglected in this case, however, and average values of conductivities given in Table 10.3 have been taken for the various tissues for the calculations.

TABLE 10.3

Tissue Conductivities Used for Calculations at the Power-Line Frequency of 60 Hz

Tissue Type	σ (S/m)
Air	0
Muscle	0.52 or 0.11
Fat, bone	0.04
Blood	0.6
Intestine	0.11
Cartilage	0.04
Liver	0.13
Kidney	0.16
Pancreas	0.11
Spleen	0.18
Lung[a]	0.04
Heart	0.11
Nerve, brain	0.12
Skin	0.11
Eye	0.11

[a]The dielectric properties of the lung consist of 33% lung tissue and 67% air.

10.4.2.2.1 Electric Blankets

To illustrate the use of the impedance method, currents induced in the human body by the EMFs of two types of electric blankets have been calculated [1,2]. The two models used for the blanket are (a) a low-magnetic-field blanket and (b) a conventional (pre-1990) electric blanket. The low-magnetic-field blanket uses two parallel leads carrying equal and opposite currents to reduce the net magnetic field around the conductors. The two leads are separated typically by 1.5 mm and are embedded in a positive temperature coefficient (PTC) conductive polymer and insulated by polyvinyl chloride (PVC). The PTC conductive polymer surrounding the two leads may be represented by a set of distributed resistors, which would result in linearly decreasing equal but opposite currents flowing through the two leads over the length of the wiring used for the blanket. By comparison, the conventional electric blanket uses a resistive alloy wire wrapped on a nylon cord and insulated with PVC. Because of the distributed resistance of the wire, this blanket would therefore have a linearly diminishing voltage and identical magnitude of current over the length of the wiring used for the blanket.

The validity of the calculated results has been established by comparing the results obtained using the impedance algorithm and those reported by others. The calculated fields are in excellent agreement with the data given by Florig et al. [86] and Hayashi et al. [87].

Currents are induced in the body by the following sources:

1. Time- and spatially varying magnetic fields of the blanket induced voltages in the various resistance loops of the body.
2. Currents launched into different subareas at the body surface by means of the capacitively coupled currents from the various conductors of the blanket.

The spatial variations of the magnetic fields were calculated from Biot-Savart's law for a short current-carrying conductor [2]. By integrating it over the entire length of the current-carrying conductors, one can obtain the vector magnetic fields at the centers of the cells representing the model of the human body or any of the other points in space. From the vector magnetic fields thus calculated for each of the cell centers for the impedance model of the human body, the induced voltages for each of the faces of the cells can be written [32]. This information is then used to calculate the induced currents for the various impedances, that is, resistances. The average current densities J_x, J_y, and J_z for each of the cell centers can be obtained by taking the average of the currents through the resistances representing each of the four edges of the cell in the respective directions and dividing the same by the cross-sectional area δ^2 ($= 1.31 \times 1.31$ cm).

To calculate the electric field distribution in air, a 3-D impedance model consisting of capacitors representing the space between the various faces of the cells was used. For cubic cells of dimension $\delta = 1.31$ cm, the capacitances used are $\varepsilon_0\, \delta^2/\delta = 0.116$ pF.

For currents induced in the human body due to electric fields, it should be recognized that the energized conductors of the blanket are capacitively coupled to the body. The capacitance between a given conductor and the highly conducting human body can be obtained by using an expression similar to that for a conductor at a distance S from the ground plane. Capacitance per unit length C of a wire of diameter d parallel to but separated at a distance S from the ground plane is given by

$$C = (2.73\varepsilon_{\text{eff}})/[\log_{10}(4S/d)], \quad \text{pF/m} \tag{10.18}$$

For a spacing $S = 5$ mm and a wire diameter $d = 0.8$ mm, and for $\varepsilon_{\text{eff}} = 2.5\varepsilon_0$, which is a value intermediate between the permittivity ε_0 for air and $4\varepsilon_0$ for the material of the blanket, we can calculate $C = 43.1$ pF/m. For a cell length $d = 1.31 \times 10^{-2}$ m, the coupling capacitance C_c between the wire and the cell can be calculated to be 0.565 pF. Since the interconductor spacing of 1.5 mm for a PTC blanket is fairly small as compared to the cell size, a proportionately smaller resistance is taken for the tissue-equivalent cells immediately underneath the conductors for the direction parallel to the interconductor spacing. Capacitances of 0.565 pF are taken from each of the conductors of the PTC blanket to the appropriate points on the impedance model of the human body. In the presence of an electrical grounding surface, the space underneath the model is represented by a 3-D network of capacitors, each of value 0.116 pF, representing the air space between the various faces of the cubic cells of dimension $d = 1.31$ cm for each of the sides.

For the PTC low-magnetic-field blanket, a constant voltage of 110 V AC is taken between the conductors of the twin-lead wiring for calculation of currents induced or injected into the human body as a result of electric fields. For calculating the magnetic fields, an input current of 1 A is taken. On account of the conductive polymer surrounding the parallel wires, this current diminishes linearly to zero at the end of the PTC wiring. This assumes a blanket input power of 110 W under normal operating conditions. If magnetic fields or induced current densities due to higher input powers are desired, the numbers calculated for 1 A input current may then be multiplied by the appropriate factor.

The conventional blanket, on the other hand, uses a resistive conductor for which the voltage diminishes linearly from 110 to 0 V over the length of the wiring. For this blanket the current throughout the length of the wiring is the same as at the input, that is, 1 A, which is assumed for the calculation of magnetic fields.

The magnetically induced, section-averaged magnitudes of the total current densities from head to feet for the two types of blankets are shown in Figure 10.13a and b, respectively. For these calculations, the wiring of the blanket was taken to be 0.5 cm from the surface of the body. Nearly identical current densities were also obtained for a grounding plate underneath the body at distances of 0.25, 0.5, and 1.0 m. It is interesting to

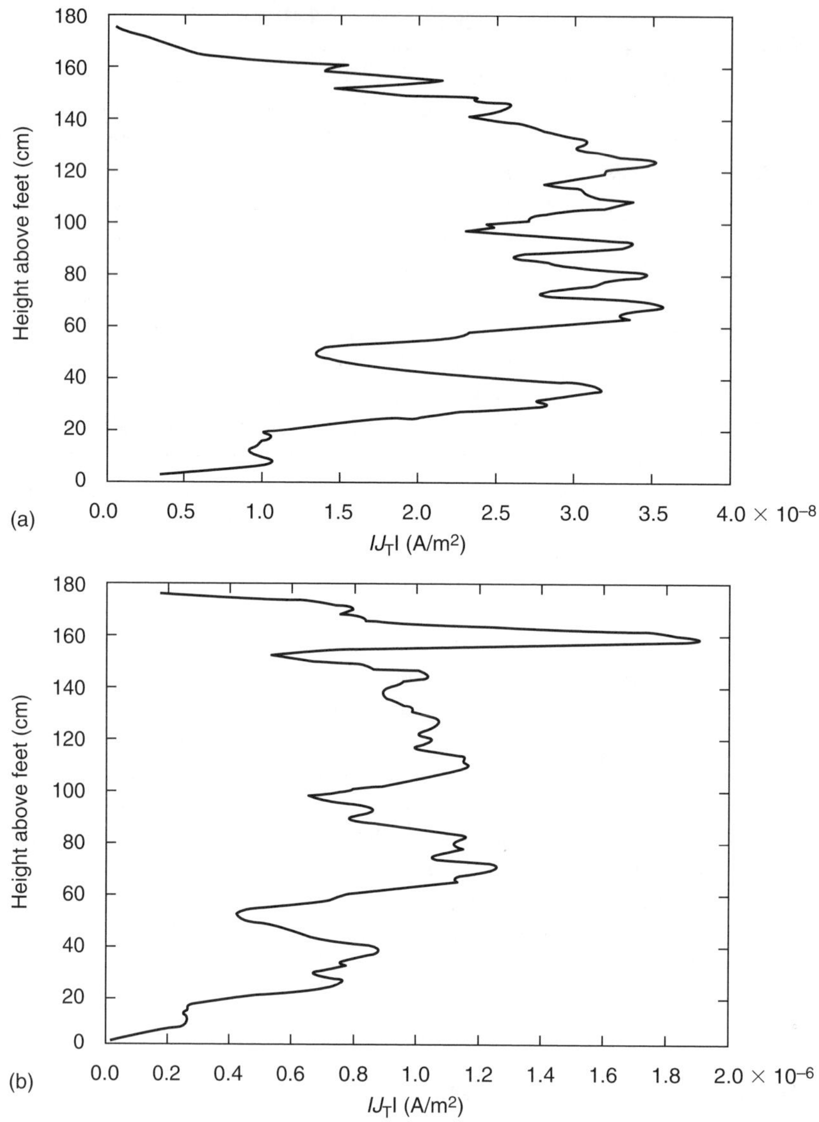

FIGURE 10.13
Section-averaged magnitudes of total current densities for the various sections of the body for magnetic fields of the blankets. Nearly identical current densities were obtained also for a grounding plate at distances of 0.25, 0.5, and 1.0 m underneath the body. Input current = 1 A.

note that the induced current densities are larger by a factor of about 500 for the conventional blanket vis-a-vis those for the low-magnetic-field blanket.

The calculated section-averaged magnitudes of the total current densities due to electric fields of both the blankets in the absence of a grounded plane are shown in Figure 10.14a and b, respectively. It should be noted that while the current densities induced by the electric fields of a low-magnetic-field blanket (Figure 10.14a) are considerably higher than those due to magnetic fields (Figure 10.25a), the converse is true for a conventional blanket. For this blanket, the current densities induced by the magnetic fields (Figure 10.13b) are higher than those due to electric fields (Figure 10.14b). In fact, while the current densities due to magnetic fields are fairly small for a low-magnetic-field blanket

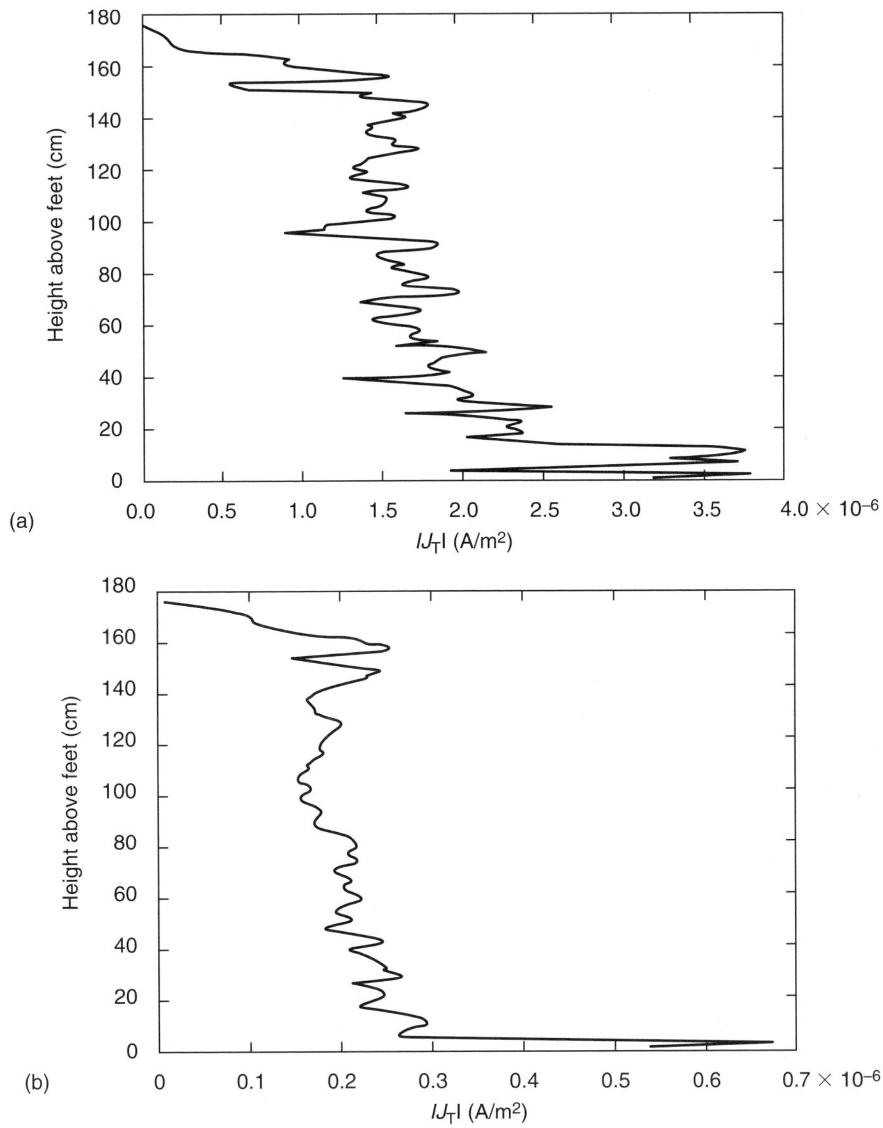

FIGURE 10.14
Section-averaged magnitudes of total current densities for the various sections of the body for electric fields of the blankets. No ground plane underneath the body.

as compared to those for a conventional blanket, the current densities due to electric fields of a low-magnetic-field blanket are even higher than those for a conventional blanket (see Figure 10.14a and b). The reasons for these observations can be seen from the values of magnetic and electric fields given in Table 10.4 for the two types of blankets, respectively. While fairly small magnetic fields are calculated for the low-magnetic-field blanket as compared to those for the conventional blanket, the converse is true for the electric fields created by these blankets. As seen in Table 10.5, somewhat higher electric fields are created by the low-magnetic-field blanket as compared to those for the conventional blanket. This is likely due to the higher potential difference between the twin-lead conductors that are used for the low-magnetic-field blankets.

TABLE 10.4

Comparison of the Calculated and Measured Magnetic Field (μT) and Electric Field (V/m) Close to a Flat Blanket [2]

	Grid Size (cm)	Calculated	Measured
Low-magnetic-field blanket[a]			
Magnetic field			
Average	1.31×1.31	0.056	—
	10.5×10.5	0.072	0.09
Peak	—	0.20	0.26
Electric field			
Average	1.31×1.31	103.7	—
	10.5×10.5	144.6	111.2
Peak	—	159.7	176.0
Conventional electric blanket[b]			
Magnetic field			
Average	1.31×1.31	2.16	—
	10.5×10.5	2.45	2.18
Peak	—	3.52	3.94
Electric field			
Average	1.31×1.31	57.3	—
	10.5×10.5	70.1	95.4
Peak	—	176.1	167.2

[a]For this blanket, the magnetic-field results are normalized for a blanket input current of 1.227 A, that is, a power input of 135 W.
[b]For this blanket, an input current of 1 A is assumed.

10.4.2.2.2 *Power Transmission Lines*

The FDTD method has been used for calculations of internal **E** and **H** fields and induced current densities for exposure to electric, magnetic, or combined electric and magnetic fields at power-line frequencies [37]. While recognizing that the conductivities of many biological tissues (skeletal muscle, bone, etc.) are highly anisotropic for power-line frequencies, however, the effect of anisotropy is neglected for the sake of simplicity. They could be included in more complex models by separately identifying these tissues.

Both sinusoidal and prescribed time-varying incident fields can be used with the FDTD procedure—hence the method is well suited also for transient exposures that are often of interest at power-line-related frequencies. For sinusoidally varying fields, the solution is completed when a sinusoidal steady-state behavior for **E** and **H** fields is observed for each of the cells. For lossy biological bodies this typically takes a time step on the order of three to four time periods of oscillation. Since Δt is fixed for a given cell size, a larger number of iterations are therefore needed at lower frequencies. Because of the large number of iterations, the FDTD procedure would be clearly inapplicable for calculations at power-line frequencies were it not for the quasi-static nature of the coupling at low-frequencies [16,88,89]. Thus, the field outside the body does not depend on the internal tissue properties, but it depends only on the shape of the body so long as the quasi-static approximation holds, that is, the size of the body is a factor of 10 or more smaller than the wavelength, and $| \sigma + j\omega\varepsilon | \gg \varepsilon_0$, where σ and ε are the conductivity and the permittivity of the tissues, respectively; $\omega = 2\pi f$ is the radian frequency; and ε_0 is the permittivity of the free space outside the body. Under these conditions, the electric fields in air are normal to the body surface, and the internal tissue electric fields can be obtained from the boundary conditions in terms of fields outside:

TABLE 10.5

Reported SAR Values, Averaged over the Whole Body (SAR_{WB}), for Plane Wave Exposures

	SAR (W/kg)					
Frequency (MHz)	**Grounded Shoes [98]**	**Grounded [102]**	**Isolated [102]**	**Isolated (resolution 3 mm) [103]**	**Isolated (resolution 5 mm) [103]**	**Grounded [105]**
10	0.027	0.045				
20	0.102	0.182	0.021			
30	0.180	0.313	0.054			
40	0.291	0.348	0.114			
50	0.230	0.293	0.199			
60	0.177	0.231	0.288			
70	0.152	0.188	0.302	0.270	0.290	
80	0.130	0.162	0.251			
90	0.107		0.195			
100	0.092	0.118	0.155			0.123[a]
200	0.062	0.081	0.080	0.048	0.051	0.078[b]
300	0.054					
400	0.060	0.063	0.063	0.064	0.060	
500	0.058					0.060
600	0.057		0.063	0.067	0.066	
700	0.059					
800	0.061			0.064	0.063	
900	0.061	0.062	0.064			
1000				0.063	0.061	0.057
1400			0.063			
1800		0.057	0.058	0.056	0.060	
2000				0.055	0.060	

[a]The frequency considered is 120 MHz.
[b]The frequency considered is 210 MHz.

$$j\omega\varepsilon_0 E_0 = (\sigma + j\omega\varepsilon)E_{\text{tissue}} \qquad (10.19)$$

A higher quasi-static frequency f', at 5 to 20 MHz, may therefore be used for irradiation of the E model, and the internal fields E thus calculated may be scaled back to frequency f of interest, for example, 60 Hz. Since in the FDTD method, one needs to calculate in the time domain until convergence is obtained, this frequency scaling to 5 to 20 MHz for f reduces the required number of iterations by over five orders of magnitude. From Equation 10.19 we can write

$$\omega'(\sigma + j\omega\varepsilon)E_{\text{tissue}}(f) = \omega(\sigma' + j\omega\varepsilon')E'_{\text{tissue}}(f') \qquad (10.20)$$

or

$$E_{\text{tissue}}(f) = (f\sigma'/f'\sigma)E'_{\text{tissue}}(f') \qquad (10.21)$$

assuming that $\sigma + j\omega\varepsilon \sim \sigma'$ at both f' and f [2,90]. To validate the use of a higher frequency f to obtain induced E fields at ELF frequencies, test cases involving homogenous and layered spheres have been used. Excellent agreement between the numerical and analytical results lends support to the validity of the FDTD method for calculating internal E fields and current densities at power-line-related frequencies. It should be noted that incident E and H fields of any orientation and relative magnitudes can be prescribed in the FDTD method, allowing the possibility of calculations for realistic exposure conditions. Also, the choice of a considerably higher frequency such as 5 to 20 MHz reduces the number of iterations needed to obtain converged results by five to six orders of magnitude as compared to those that would be needed at ELF frequencies of 10 Hz to 1 kHz.

Some calculated results using a 16-tissue, 1.31-cm resolution, anatomically based model of the human body are given in Figure 10.15. A frequency f' of 5 to 10 MHz was used to reduce the computation time. At the higher irradiation frequency f', $\sigma' = \sigma$ was assumed, that is, conductivities of the various tissues at 60 Hz. Furthermore, the incident **E** field, $\mathbf{E}_i(f') = 60\mathbf{E}_i(f)/f'$, was used to obtain $\mathbf{E}_{tissue}(f)$ at, say, $\mathbf{E}_i(f) = 10$ kV/m. The incident magnetic field $\mathbf{H}_i(f')$ has similarly been taken to be considerably lower ($=60\mathbf{H}_i(f)/f'$) to account for the fact that the induced current densities and internal electric fields are proportional to the frequency of the incident fields and would therefore be higher at the assumed frequency f'. Recognizing the anisotropy in the conductivity of skeletal muscles, two different values of muscle conductivities are taken for curves (1) and (2). For these curves a higher conductivity of 0.52 S/m is taken for the skeletal muscle, and an average value of 0.11 S/m is taken for the muscle in the interior of the body. For curves (3) and (4), however, a lower conductivity of 0.11 S/m is taken for all of the muscle, interior or skeletal. The results shown in Figure 10.15, curves (1), (3), and (4) are for $E_{inc} = 10$ kV/m (vertical) and $H_{inc} = 26.5$ A/m ($B_{inc} = 33.3\ \mu$T) from side to side of the model. To point out the preponderance of the induced currents due to incident electric field, $H_{inc} = 0$ is assumed for the calculations shown in curve (2). It is interesting to note that the layer currents due to E-field exposure alone are almost 98% to 99% of the currents calculated for the combined electric and magnetic fields. It is also interesting to note that the calculated foot currents of 155 to 160 μA are in excellent agreement with 165 μA that would be projected from the measurements of Deno [91] for the human body. The variations of the induced currents calculated along the height of the body have been checked against the results by DiPlacido et al. [92]. The agreement with the results of these two authors who had

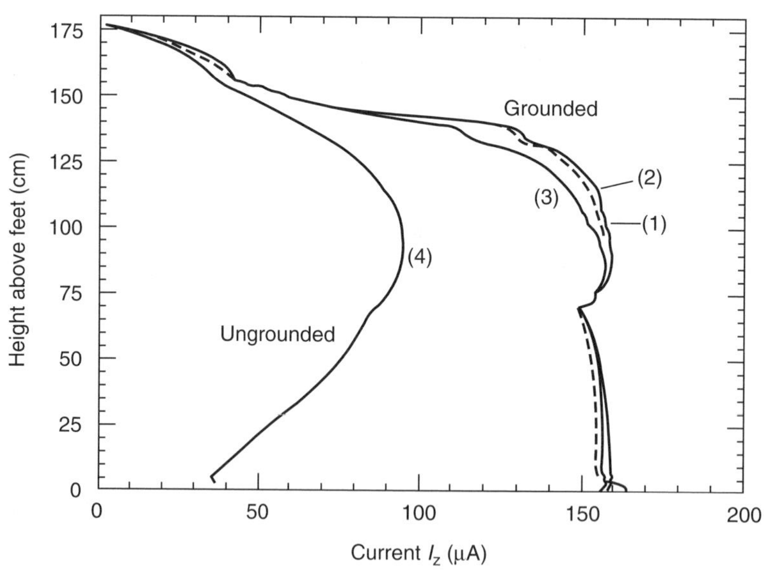

FIGURE 10.15
Calculated layer currents for anatomically based grounded and ungrounded models exposed to EMFs at 60 Hz. For curves (1) and (2), $s = 0.52$ S/m for skeletal muscle, and $s = 0.11$ S/m for the interior muscle. For curves (3) and (4), $s = 0.11$ S/m for all of the muscle. $E = 10$ kV/m (vertical), $H = 26.5$ A/m from side to side for all of the curves except for curve (2), for which only **E**-field exposure is assumed.

used a vertical electric field such as that under a high-voltage power line was found to be very good [37].

10.4.2.3 Absorption in Human Bodies Exposed to Far Field of RF Sources

As noted, the configuration and frequency of the electromagnetic source and the geometry and composition of the biological body will influence the induced field and power absorption and distribution inside the body. Moreover, the field emitted from a source is dictated by the frequency, size, and configuration of the source. Near an antenna, the radiated energy is in the form of a spherical wave in which the wave fronts are concentric shells. The spheroidal wave front expands as the wave propagates outward from the source. At distances far from the source, the radius of curvature of the spherical shells becomes so large that the wave front would essentially appear as a plane. They are therefore referred to as plane waves. Plane waves are important since their behavior is well quantified; the fields are uniform in planes normal to the direction of propagation, and the power density varies only in the direction of propagation. In this case, both electric and magnetic fields of the propagating wave are orthogonal in space and lie in the plane of the wave front, and are related through the intrinsic impedance of the medium. In other words, in the far or radiation zone, the electric and magnetic fields have only transverse components.

In this section, we shall briefly summarize some of the efforts devoted to field computation using various models of the human body, which consists of large quantities of numerical cells, and present some results obtained for plane wave exposures. Note that some, especially the simpler, models are of interest primarily for whole-body SAR and can provide quite adequate results for frequencies lower than 30 MHz. To achieve more accurate structural representation of the human body, anatomically based models are needed.

10.4.2.3.1 SAR Induced in Cubic-Cell Models

The VMoM for field computation has been used for models of the human body consisting of 200 to 1000 cubic cells. These models account for the gross anatomic and biometric characteristics of human bodies and have been used by several investigators [24–28]. The models are 1.75 m tall and can be made either homogenous or inhomogenous by choosing an equivalent or a volume-weighted complex permittivity for each cell. The cubic-cell model has been employed, successfully, to calculate whole-body averaged absorption. It is important to note that for subdivision with less than three cells per wavelength, the magnitude and phase resolutions would be such that even with convergence the reliability of the MoM computed SAR would be questionable.

According to the MoM, the body may be partitioned into N cubic subvolumes or cells that are sufficiently small for the electric field and dielectric permittivity to be constant within each cell. The integral equation is then transformed into a system of $3N$ simultaneous linear equations for the three orthogonal components of the electric field at the center of each cell. The simultaneous equations may be written in matrix form as

$$[G][E] = -[E^i] \tag{10.22}$$

where $[G]$ is a $3N \times 3N$ matrix and $[E^i]$ and $[E]$ are column matrices representing incident and induced electric fields at the center of each cell. The elements of $[G]$ can be evaluated as shown in Liversy and Chen [24]. In particular, the diagonal elements of the $[G]$ matrix may be evaluated exactly by approximating each subvolume with a sphere of equal volume centered at the position of an interior point. If the actual shape of the cell differs appreciably from that of a sphere, this approximation may lead to unsatisfactory numerical results [36]. In such cases, a small cylindrical volume may be created around an interior point. It may also be necessary to evaluate these terms by numerical integration throughout the cubic subvolume for increased accuracy. The evaluation of off-diagonal elements of the $[G]$ matrix is considerably simplified since it does not involve principal value operations. Therefore, for a given applied field configuration, the induced electric fields inside the body are obtained by matrix inversion. That is,

$$[E] - [G]^{-1}[E^i] \tag{10.23}$$

Factors that influence the computational accuracy include frequency, body size, cell dimensions, and computer memory. It has been found that reliable numerical results can be obtained if the linear dimensions of the cell do not exceed a quarter free-space wavelength [24]. For a computer with sufficient capacity to invert a 120×120 matrix, the maximum number of cells is limited to 40. If we assume, for simplicity, symmetries between the right and the left half and the front and the back of a 1.7-m-tall adult human body, this computer would handle approximately a cell size around $10^{-5}\,m^3$. Once the 10^{-5}-m^3 cell size is adopted, 750 MHz would be the highest frequency that can be considered for field intensity calculation without violating the criterion that the linear dimension of the cell not exceed a quarter free-space wavelength.

The computational resources necessary to obtain even a regional SAR using this MoM approach are quite extensive. A relatively full complex matrix, $3N \times 3N$ in dimensions, is required for a model with N cells. The computation time required for a noniterative solution of the matrix equation is therefore proportional to a value between N^2 and N^3, which increases rapidly as N increases. The faithfulness with which a cubic-cell model approximates the detailed structure of a biological body and the maximum usable frequency increases with the number of cells. In fact, substantial errors will occur if

$$N \leq (2\pi L)/(\lambda' 6^{1/2}) \tag{10.24}$$

where L/λ' is the ratio between the linear dimension of the body and the wavelength in the body.

The accuracy of the numerical method can be verified by comparison with known results from exact analytic solutions based on well-characterized geometric bodies, such as spheres. It should be noted that perfect agreement between the exact solution, based on Mie theory, and the numerical method, based on the volume integral equation, is not expected unless a large number of cubic cells are used to simulate the sphere. Figure 10.16 shows one eighth of a sphere approximated by one eighth of a "cubic model of a sphere," which is constructed from 73 cubic cells. Clearly, a better approximation can be achieved by a larger number of smaller cubic cells. Nevertheless, for a brain sphere constructed from 40 cubic cells at a frequency of 918 MHz, the computed maximum field intensity deviated from the exact solution by less than 9% [93].

A model of a human body consisting of 180 cubic cells that accounts for the anatomic and biometric characteristics of human beings is shown in Figure 10.17. The model is 1.75 m tall and can be made either homogenous or inhomogenous by using an equivalent or a volume-weighted complex permittivity for each cell [94]. The average absorption or whole-body SAR for the model of the human body shown in Figure 10.17 as a function of frequency is illustrated in Figure 10.18. The electric field vector is along the height of the body, and the plane wave propagates from front to back of the model with an incident power density of $10\,W/m^2$. A homogenous complex permittivity approximately two thirds of that for muscle is used in the calculations. Note that the whole-body SAR

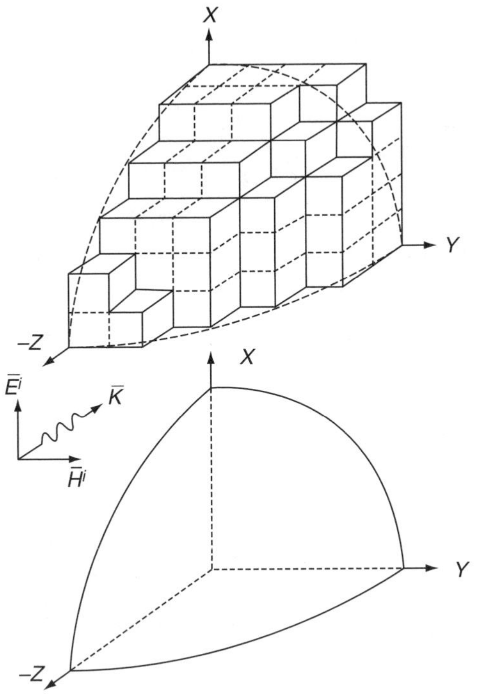

FIGURE 10.16
Approximation of one eighth of a sphere by an equivalent cubic-cell-formed structure.

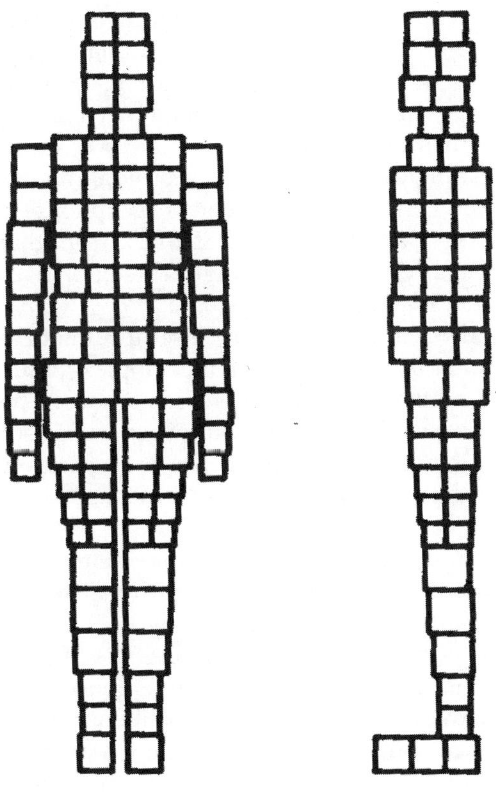

FIGURE 10.17
A cubic-cell representation of the human body.

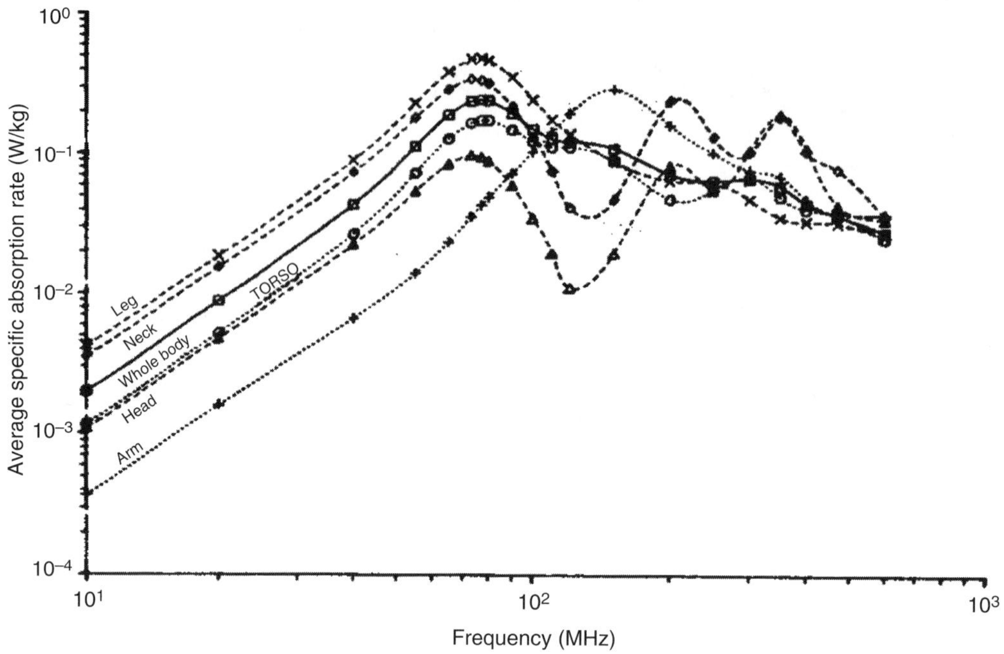

FIGURE 10.18
Average SARs for a homogenous 180-cell model of the human body exposed to a vertically polarized, 80-MHz plane wave. The incident power density is 10 W/m^2.

increases with frequency until it reaches a maximum of about 0.23 W/kg at 77 MHz (resonance frequency), and it then decreases as 1/frequency. The experimental data shown in Figure 10.19 are obtained from a saline-filled scale model of the human body. It can be seen that the calculated absorption is in good agreement with that found experimentally [27,95], except for the resonant frequency, which is somewhat lower (70 MHz) in the experimental case. It should be mentioned that whole-body SAR, given in Figure 10.18, is typically within 10% of that estimated from prolate spheroidal models of the same height and dielectric property. Further, when inhomogenous complex permittivities are used with the model, the whole-body SAR changes less than 2% from that depicted in Figure 10.18. Thus, if one is primarily concerned with average absorption over the body, a homogenous prolate spheroidal model may be quite adequate.

While the MoM based on the volume integral equation has been a useful numerical procedure for computation of average SAR and SAR distribution in complex tissue geometries, the requirement of a full $3N \times 3N$ matrix presents severe limitations. The computation times required to provide even regional SAR distribution of sufficient resolution to delineate the resonant frequency for the head region are enormous. A minimum of 340 cells was needed, increasing the computation time by a factor of 4 over the 180-cell models [95,96].

Matrix inversion operations consume the largest block of time in moment method solutions for the cubic-cell models. The computer time required is proportional to the cube of the number of cells. However, the matrix generated is usually diagonally dominant and well-conditioned. For a human-size body, iterative procedures for matrix inversion are practical at frequencies below about 60 MHz. The convergence rate decreases with increasing frequency and fails above 90 MHz. This is most likely caused by the decrease in the degree of diagonal dominance with increasing frequency. A number of approaches have been investigated to alleviate this difficulty. A semi-iterative procedure

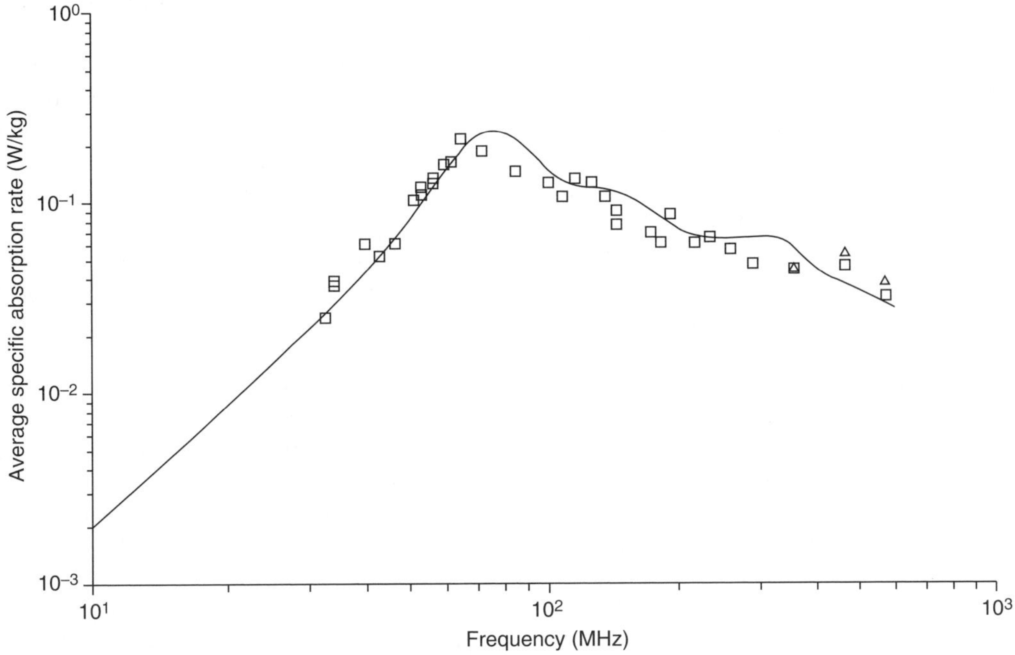

FIGURE 10.19
Whole-body averaged absorption for a homogenous cubic-cell model of humans exposed to vertically polarized plane in free space. The incident power density is 10 W/m^2.

called band approximation method appears to be an efficient algorithm that can be profitably used to invert large matrices generated by the number of cells for human models at arbitrarily high frequencies, and converges significantly faster than standard iterative algorithms [96].

10.4.2.3.2 SAR Induced in Fine Resolution Anatomical Models

To accurately evaluate the SAR induced in the human body, the FDTD method was introduced during the late 1980s, when the limitations of the MoM due to its memory requirements were reached. The capability of FDTD to take into account heterogeneities in models of the human body was first demonstrated using a model of the isolated human torso [69]. Later, a complete model of the human body was considered [97], and results for an isolated homogenous man model standing in free space were compared with results for an inhomogenous man model, under both isolated and grounded conditions [97]. The incident field was a plane wave propagating parallel to the ground plane and with the electric field vertically polarized (parallel to the long axis of the human body), at frequencies of 100 and 350 MHz. The human body model was obtained from cross-sectional diagrams and had a resolution of 2.62 cm at 100 MHz. The total occupied volume was $23 \times 12 \times 68$ cubic cells. At 350 MHz, the resolution was 1.31 cm for a total volume of $45 \times 24 \times 135$ cubic cells. The result, depicted as layer-averaged SAR or organ averaged SAR, demonstrated the importance of considering inhomogenous models of the human body. For example, the homogenous model was not able to predict the peak SAR obtained in the eyes at 350 MHz, or the difference in absorption among the different organs.

Since these first works on power absorption, several papers have been published using anatomical models of the human body with finer resolutions.

The VH body model has been used to evaluate power absorption and temperature increase as a function of frequency of the incident plane wave, by considering a grounded male, either barefoot or with shoes [98]. The model had a resolution of 5 mm, a total height of 180 cm, and weight of about 103 kg. The large mass was due to the use of the VH model, which is far from the so-called "reference man," as defined by International Commission on Radiological Protection (ICRP). The reference man weighs 73 kg and has a height of 176 cm [99]. In the referenced paper [98], for the frequency range between 10 and 900 MHz, SARs averaged over the whole body (SAR_{WB}) and locally, that is, averaged over 1.0 g (SAR_{1g}) and 10 g (SAR_{10g}), were evaluated. In particular, it has been found [98] that when the incident power density is equal to the reference levels set in the exposure standards, the basic restrictions on SAR_{WB} and on local SARs are never exceeded. Moreover, it has been shown [98] that the ratio, SAR_{10g}/SAR_{WB} was about the same (either 25 or 50 according to the body part considered) as the value used in the safety guidelines to convert basic restrictions on SAR_{WB} to basic restriction on local peak SAR [100]. On the other hand, the SAR_{1g}/SAR_{WB} ratio was found to be always higher than the value of 20 adopted in the safety guideline [101].

Figure 10.20 gives the SAR_{WB} as a function of frequency in a grounded male for an incident plane wave with a power density of 10 W/m^2 for two different human body models [98,102]. The SARs are slightly different because the two body models were different in height, weight, and tissue composition. The influence of the human body model on electromagnetic power absorption is further illustrated in Figure 10.21, where a comparison among the SAR_{WB} values obtained with the heterogenous VH model and a homogenous VH model consisting either of muscle or fat is reported. The frequencies considered are from 10 to 200 MHz to highlight the differences in power absorption at resonance.

It can be seen from the figures that a higher peak appeared at resonance in the homogenous muscle model, and lower absorptions were obtained in the homogenous fat model. Moreover, data from a lighter model (65.8 kg) obtained by reducing the cell dimension on the horizontal plane suggested that the lighter body absorbed more elec-

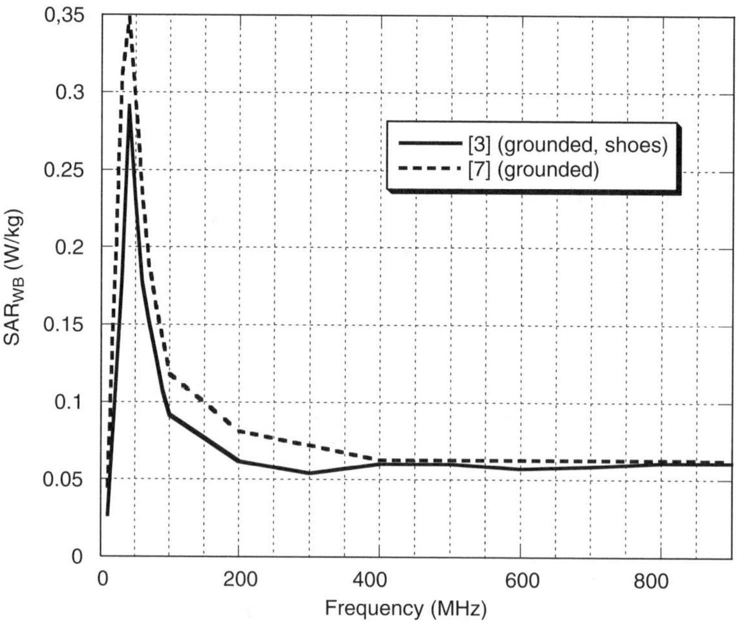

FIGURE 10.20
SAR as averaged over the whole body as a function of the frequency.

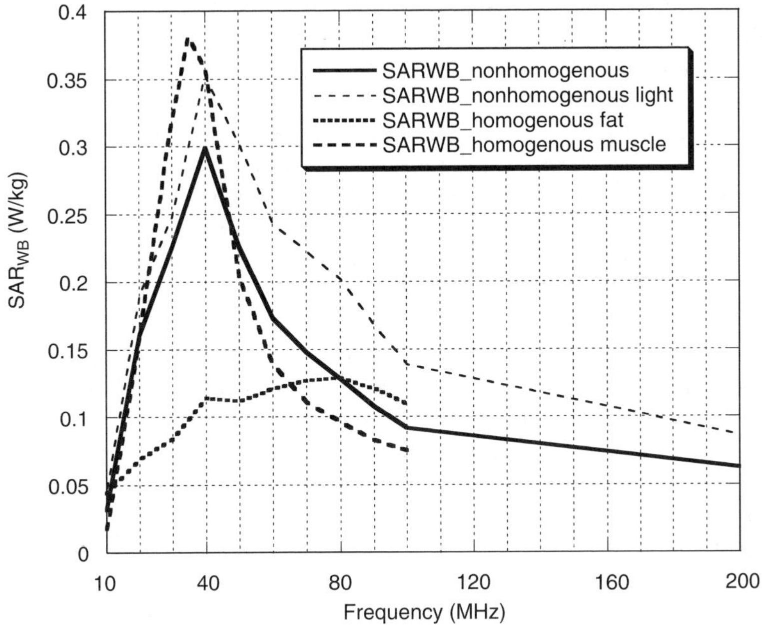

FIGURE 10.21
SAR as averaged over the whole body as a function of the frequency for different VH body models: nonhomogenous, nonhomogenous with a lighter weight (65.8 vs. 103 kg), homogenous fat, and homogenous muscle.

tromagnetic power than the heavier one. The frequency dependence of power absorption is clearly evidenced by Figure 10.20 and Figure 10.21. Indeed, this frequency dependence was the basis of the different limits imposed on the reference levels for different frequencies in the safety guidelines [100,101].

In Table 10.5 some literature data are summarized for the SAR_{WB} as a function of frequency (10 MHz to 2 GHz) for a grounded and isolated man model exposed to an incident power density of 1.0 mW/cm^2.

The first two columns refer to data in Figure 10.20, while the successive columns report data from published literature. Note the resonant frequencies of 40 vs. 70 MHz for grounded or isolated bodies. Moreover, the data in Table 10.5 show nonsignificant differences in SAR averaged over the whole body by changing the model resolution from 3 to 5 mm [103]. For the same reason the cubic-cell model described in the previous section had been employed, successfully, to calculate whole-body averaged absorption. The SAR_{WB} has a weak dependence on model resolution.

Likewise, a study on the SAR dependence on permittivity values [104] showed that uncertainty in permittivity values does not substantially affect the SAR as averaged over the whole body, while the same uncertainties have a greater effect on local SAR. In particular, considerations of different frequencies and orientations of the incident plane wave, or higher or lower permittivity values, showed that the maximum difference in SAR_{WB} was within $\pm 20\%$ [103]. Larger differences were found in local SAR, particularly when the permittivity of muscle, representing about 42% of the whole-body mass, was changed [103].

The data in the last column of Table 10.5 were obtained from a human body model [105] that was developed by using a new, semiautomatic procedure to construct numerically a frequency-dependent, dielectric anatomy model, starting from MRI images. The main difference between this human body model and the models usually considered in FDTD calculations is that in this model permittivity and conductivity can vary, even for the same tissue, thus reflecting the realistic spatial inhomogeneity of such parameters. The semiautomatic procedure requires a short time to construct the model; thus, it could be used for dosimetry studies based on the model of specific persons.

A high-resolution human body model ($1.974 \times 1.974 \times 3.0\,mm$) and a coarser one ($5.922 \times 5.922 \times 6.0\,mm$), both for isolated and grounded conditions, were employed to determine the power absorption in the head and neck region and to evaluate the frequencies at which the absorption may be maximized [106]. It was observed that under isolated conditions two resonant frequencies occurred for the head and neck, one associated with the whole-body resonance and the other with a local resonance of the head and neck. Under grounded conditions, three resonances were observed; the additional resonance was attributed to a torso resonance.

Some studies were conducted to evaluate power absorption in models of women and children. Specifically, a 10-year-old child and a 5-year-old child were considered by scaling the adult human body model [106]. It should be noted that simply scaling the adult human body model to the children's dimensions does not produce an accurate model since the different organs scale differently; however, the general features in terms of height and weight are fulfilled, thus allowing for the determination of general properties of electromagnetic power absorption. In this way, resonant frequencies of 104 MHz for the isolated model and 65 MHz for the grounded 10-year-old child were obtained, while for the 5-year-old model, they were 126 and 73 MHz, respectively.

In a different study, power absorption in scaled versions of the adult human body model representing 10-, 5-, and 1-year-old children was evaluated for both grounded and isolated conditions [102]. Figure 10.22 gives the SAR_{WB} obtained for the three child models under isolated conditions and an incident power density of $1\,mW/cm^2$. A shift in the resonant frequency with the height of the model—the taller the model, the lower

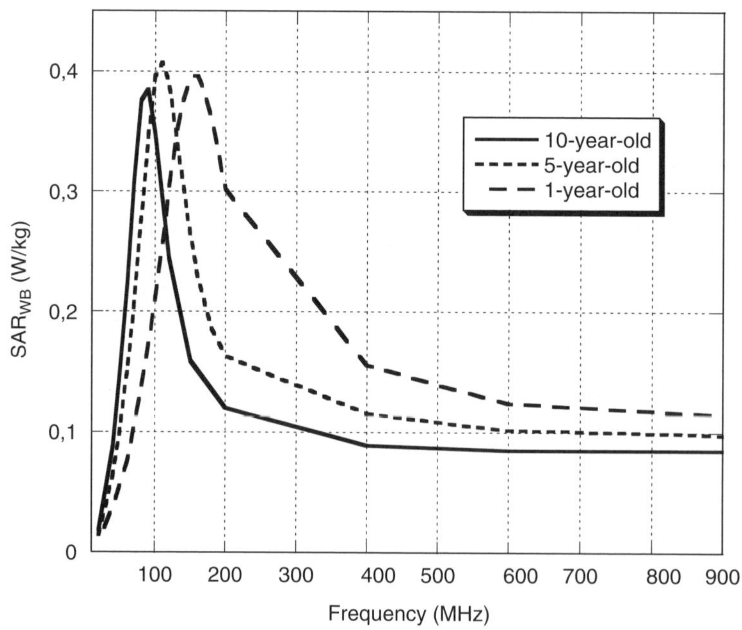

FIGURE 10.22
SAR as averaged over the whole body as a function of the frequency for different child body models: 10, 5, and 1 year old. Incident power: $1\,mW/cm^2$. (Data from Dimbylow, P.J., *Phys. Med. Biol.*, 47, 2835, 2002.)

the resonance frequency—can be observed from the figure; note also the higher absorption in the smaller child body model [102].

The potential differences between SAR induced in man and woman have been explored using slightly different models [107,108]. In one study, 2-mm resolution models were developed for evaluating power absorption in Japanese males and females. Note that the use of the 2-mm resolution body models had led to an overestimation of the skin weight by 50% or more than the average value for the Japanese reference body [107]. The computed results showed that the difference in SAR_{WB} between male and female models was small, within 1.1 dB. The authors concluded that gender does not affect SAR_{WB}. Similarly, they obtained no significant differences between the male and female models with regard to local SARs. However, the overestimation of skin weight, and perhaps other tissues, in the 2-mm resolution body models could have influenced their results and conclusions.

In contrast, a clear difference in power absorption was reported [108] between Caucasian male and female models. The two Caucasian body models were developed using the semiautomated procedure previously cited [105]. In particular, considerably greater SAR_{WB} was obtained in the female model than in the male one (about 40% higher in the frequency range between 500 MHz and 2.0 GHz and 25% higher in the frequency range between 2.0 and 4.0 GHz). The difference in local SARs (both SAR_{1g} and SAR_{10g}) were insignificant between genders for up to 3.0 GHz, while above this frequency, say up to 4 GHz, the SAR_{1g} and SAR_{10g} in the female model became larger than those in the male model. The authors observed that this result could be explained by the difference in subcutaneous fat between man and woman. It was noted that a better identification and modeling of the skin layer could influence the results. Clearly, further studies are needed to assess the similarities and differences in power absorption between male and female body models exposed to EMFs of radiating sources.

In summary, available data on electromagnetic power absorption in human body models exposed to plane wave fields show that the choice of the human body model affects the results obtained. The observed differences among the published data are usually more pronounced in local SARs than in SARs averaged over the whole body. The body height, mass, tissue distribution and composition, including fat and muscle, are important factors in power absorption and can explain some differences among the published data. Another fundamental aspect is the value assigned to the dielectric properties of different tissues or organs identified in the body model. In the case of children, the variation in tissue dielectric properties with age also may influence the computed results. It is noteworthy that the above publications have used dielectric properties from the same sources [109,110].

10.4.2.4 Human Exposure to the Field Radiated by Transceiver Base-Station Antennas

The enormous growth in the number of subscribers of mobile telecommunication systems during the past few years has pushed upward the system's capacity. As a result, more and more base stations have been installed on the rooftop of existing buildings in densely populated areas, and many more are expected to be set up as the next-generation mobile networks (UMTS, IMT2000, etc.) are deployed. These installations are giving rise to widespread concerns among the population about possible deleterious effects on human health from exposure to the EMFs radiated by the base-station antennas. Recently, increasing attention has been paid to the topic of numerical exposure and compliance assessment for base-station installations.

A great deal of work has been done in the area of field intensity prediction in the vicinity of base-station antennas to determine the so-called free-space compliance boundary. The studies were aimed toward a direct comparison with reference levels suggested by international exposure guidelines [100,101]. They generally have neglected both the influence of the environment in which the antennas operate and the dosimetric problem of SAR evaluation inside an exposed subject. In particular, considerable efforts have been spent on determining simplified and efficient analytical [111,112] and numerical [113,114] models to evaluate field levels near base-station antennas. Starting from the theory of collinear dipole arrays, typically employed in base transceiver station (BTS) antennas, practical analytical formulas have been derived to predict average power density falloff as a function of distance from the antenna. The space surrounding the antenna is often divided into a cylindrical-wave region, closest to the antenna, and a spherical-wave region, further away [111]. A complementary formulation also has been proposed, on the basis of an exact asymptotic solution for the radiated field, to derive approximate analytical formulas that allow a conservative prediction of equivalent peak power density as a function of the distance from the antenna [112].

Besides the aforementioned practical analytical formulations, simplified numerical models are often used, by subdividing the antenna into elementary radiators [113,114]. Under the hypothesis of weak coupling between the subelements, the near field can be quickly computed through a superposition of the fields independently radiated by the different elements. Once the radiation pattern of the subelements is known, the field is derived using the antenna gain-based formula [113,114]. For better accuracies in the vicinity of the antenna, an MoM simulation of the subelement can be invoked [114]. These simplified approaches represent extremely fast tools for field intensity prediction, but they are limited by a minimum distance from the antenna where they can be applied in order to maintain an acceptable computational accuracy. On the other hand, when field computations within a distance of a few wavelengths from the antenna have to be done, full-wave numerical techniques, such as FDTD, must be adopted. The accuracy of FDTD models for evaluating the near field of base-station antennas has been investigated and validated through a comparison with measurements carried out in a fully anechoic chamber [115].

Full-wave approaches require knowledge of the internal structure of the antenna, which is not always available. Cylindrical- and spherical-wave expansion techniques have been proposed to evaluate the near field, starting from measurements performed on a surface enclosing the antenna [116–118]. The basic approach consists of performing field measurements on a spherical surface surrounding the antenna and describing the measured field as a superposition of spherical modes [115]. Once the spherical-wave expansion coefficients have been determined, the near field can be extrapolated for all points lying outside the minimum sphere enclosing the antenna. This technique also has been improved to allow extrapolation of the field inside the minimum sphere [117]. To this end, the spherical-wave expansion coefficients are derived for each of the antenna subelements. This approach extends the range of applicability of the formulation to all points outside the minimum sphere of a single subelement, which is much closer to the antenna than the minimum sphere of the overall array. Recently, an alternative solution to extend field extrapolation close to the antenna was proposed, which consists of spherical-wave expansion outside the minimum sphere of the antenna and cylindrical-wave expansion for the region close to the antenna, with an appropriate matching of the two expansions [118].

The effect of the surrounding environment must be taken into account to some extent in dealing with the problem of evaluating induced SAR in a subject exposed to the field radiated by BTS antennas, since the antenna is not operating in a free-space condition. A very interesting approach, applicable to on-site evaluations, consists of using mixed experimental and numerical procedures [119–121]. One procedure is based on on-site measurement of the amplitude and phase of the exposure field distribution over a surface surrounding the antenna. The measurement is then used to numerically evaluate induced SAR distributions inside a phantom. The measured fields are used to excite the FDTD domain via the equivalence principle [119]. A much faster and efficient procedure uses previously stored FDTD-computed **E**-field distributions inside a phantom exposed to spatially impulsive electric fields, the equivalent spatial impulsive responses of the phantom or Green's functions. The on-site measurement of amplitude and phase of the exposure field is made over an equally spaced grid of points placed on an appropriately chosen surface [120]. The procedure has recently been enhanced and made faster by substituting the spatial impulse response with responses to spatial harmonic components [121]. In this way, good accuracy is achieved using only six to ten spatial harmonic components, as opposed to the 54 spatial impulse responses needed previously.

The aforementioned hybrid experimental–numerical procedures have the great advantage of allowing easy characterization of environmental perturbations to the exposure field by directly including them in the measured field. On the other hand, they require the antenna to be already installed and operating at the time of measurement. The last point makes such procedures not suitable for *a priori* compliance assessment evaluations during the planning stage of a cellular network. For such evaluations, a thorough numerical dosimetric analysis is required. Once again, a possible approach, when the environment can be neglected and the antenna can be supposed to operate under free-space conditions, consists of performing full-wave FDTD simulations and modeling both the BTS antenna and a numerical phantom of the exposed subject. The applicability of such an approach has also been demonstrated, through a comparison with SAR measurements, for exposure locations in close proximity to the antenna [122]. The main drawback of full-wave FDTD analysis is the large amount of memory required to discretize the simulation space for phantom locations not in the close proximity of the antenna. This problem can be faced by exploiting parallel computer architectures with parallelized versions of the FDTD code [123]. Parallel FDTD also has the potential to allow SAR computations for large antenna–phantom distances.

More efficient techniques have been developed that combine two different techniques, one to model BTS antenna and propagation in free space and the other, SAR inside the phantom [124,125]. In particular, if the antenna–phantom distance is such that mutual coupling can be neglected, a hybrid ray-tracing (RT)–FDTD approach can be used [124]. RT is used to model field propagation from the BTS antenna to an equivalent surface surrounding the phantom, and FDTD is employed to study absorption inside the phantom, using RT-derived exposure fields for excitation. For closer antenna–phantom distances, where the mutual coupling cannot be neglected, a hybrid FEM–MoM technique has been proposed [125]. In this case, MoM is used to model the BTS antenna, while FEM is used to study absorption inside the phantom. The MoM and FEM formulations are coupled together and are solved iteratively. These hybrid approaches allow very efficient SAR computation for different antenna–phantom distances and are well suited for evaluating free-space compliance distances, on the basis of SAR restrictions. They do not require the use of derived exposure field reference levels. For example, the RT–FDTD technique has been applied to a common 14-dBi-gain GSM900 antenna using the VH phantom. It was shown that for a total radiated power of 30 W, typical for urban area installations, SAR basic restrictions for the general population may be exceeded at distances of 2 m or less. Note that, at these distances, only occupational personnel are allowed [124].

The RT–FDTD hybrid technique also has been successfully employed to study human exposure to the field radiated by a BTS antenna in an urban scenario, including the effect of environmental perturbations to the exposure field [126]. In this case, image sources have been introduced to represent corner-reflector-like urban scenarios. Three different exposure conditions have been considered for a rooftop-mounted 14-dBi-gain BTS

TABLE 10.6

Spatial Maximum (E_{iMAX}) and Spatial Average (E_{iAVE}) of the Incident Field (rms Value); Maximum SAR Values Averaged over 1 g (SAR_{1g}) and over 10 g (SAR_{10g}), and SAR Value Averaged over the Whole Body (SAR_{WB}) for Three Exposure Conditions

	E_{iMAX} (V/m)	E_{iAVE} (V/m)	SAR_{1g} (mW/kg)	SAR_{10g} (mW/kg)	SAR_{WB} (mW/kg)
Rooftop	4.2	2.8	5.3	3.0	0.12
Balcony	8.1	5.5	13.2	8.5	0.46
Street	1.3	1.1	0.26	0.17	0.01

antenna, radiating 30 W in the GSM900 frequency band: (1) a subject standing on the rooftop, near the antenna mast; (2) a subject standing on a balcony of a building facing the antenna at a distance of 30 m, within the antenna main beam; and (3) a subject standing in the street below the 30-m tall building on which the BTS antenna was mounted. The computed results for the incident electric field and SARs, under these exposure conditions, are given in Table 10.6.

From Table 10.6, because of the high directivity over the vertical plane of base-station antennas, it appears that the highest field levels are not obtained on the rooftop of the building where the antenna is located. Instead, they are on the nearby building, in the direction of the maximum antenna radiation. As expected, the lowest field levels are experienced by a subject standing in the street, as a result of the large distance from the antenna and the off-axis position with respect to the antenna pointing direction. In all cases, the computed SARs are at least two orders of magnitude lower than the basic restrictions, confirming the expected low exposure levels for people living near a BTS installation in urban areas.

More recently, some hybrid techniques have been developed, with enhanced capabilities in modeling complex urban environments, by taking into account diffraction phenomena [127–129]. One such technique uses FEM to model the BTS antenna, the uniform theory of diffraction (UTD) to model the effects of the environment on field propagation, and FDTD to study power absorption in the exposed subject [127]. The technique has been employed to study exposure of a subject standing inside a room with a microcell BTS antenna mounted on the external wall. Another possible hybrid solution exploits time domain physical optics, instead of UTD, to model field scattering from the environment and FDTD to study absorption inside the exposed subject [128].

Finally, a hybrid UTD–FDTD technique has been developed to highlight some key points related to compliance assessment procedures for cellular base-station antennas, in a realistic urban environment [129]. The scenario analyzed consists of a room in one building where the field, radiated by a GSM900 or a UMTS BTS antenna installed on a facing building, penetrates through the room's external wall and window. The relation between SAR in an exposed subject and ambient field in the absence of the subject has been investigated for the complex scenario. As expected, the ambient field showed a highly nonuniform distribution resulting from the many reflections and diffractions that took place. The results showed that whole-body averaged SARs (SAR_{WB}) are closely correlated with the exposure field value averaged over the volume that would be occupied by the exposed subject. In particular, it has been estimated that assessing SAR_{WB} on the basis of volume-averaged field values yields an average error of approximately 6%. Peak 1-g and 10-g averaged SARs, instead, show a rather complex and difficult-to-predict relation with reference to the exposure field. Analysis of the results has revealed that the use of the volume-averaged exposure field value, in the absence of the subject, can lead to an underestimation of the peak local SARs, up to 36%. On the contrary, using the maximum volumetric value yields an overestimation of peak local SAR (up to approximately four times). The conclusion was that peak local SAR showed a good correlation (15% average error) with the maximum average exposure field value obtained by varying the position of a vertical averaging plane, having a surface equivalent to the projected human body area, inside the volume occupied by the subject.

10.4.2.4.1 Human Exposure to the Fields Produced by Coexisting Wireless Communication Systems

The discussions thus far have dealt with the problem of human exposure to fields radiated by a single base-station antenna from the cellular mobile communication systems (i.e., GSM, UMTS, etc.). However, as the development of communication systems making use of wireless technology expands, new exposure scenarios are encountered in everyday life. Following the enormous growth in the number of base stations in densely populated areas, one of the most promising systems in the near future may be the so-called Wi-Fi system, namely, wireless LAN adopting the IEEE 802.11b communication standard. Wi-Fi is characterized by completely different coverage ranges. Unlike base-station antennas of cellular systems, which are installed almost entirely in outdoor locations, access points (APs) of Wi-Fi systems would operate essentially inside buildings. Nonetheless, the EMFs radiated by the two systems will coexist in indoor environments, particularly if buildings located in front of a rooftop-mounted base-station antenna are considered. This poses new questions about human exposure in such environments. It becomes important to assess typical exposure levels attributable to each system.

The problem has been recently addressed by considering exposure of a subject standing inside a room with a Wi-Fi AP and facing a dual-band GSM900/GSM1800 BTS antenna mounted on the rooftop of a nearby building [130]. The AP radiates a power of 100 mW at 2.44 GHz, while the GSM BTS employs an antenna with a 18-dBi gain, radiating a total power of 30 and 20 W in the GSM900 and GSM1800 frequency bands, respectively. The computed results for exposure field values and SAR levels are summarized in Table 10.7 and Table 10.8. Specifically, the first two columns of Table 10.7 show the peak ($E_{vol\ peak}$) and average ($E_{vol\ ave}$) root mean square (rms) exposure field values over the entire parallelepiped volume where the subject will be placed, while the third column reports the average ($E_{sup\ ave}$) rms exposure field values over vertical sections of the parallelepiped volume. In particular, the minimum and maximum field values are given because the averages depend on where exactly the surface is placed. Table 10.8 presents whole-body, peak 1-g and 10-g averaged SARs inside the exposed subject.

It can be seen from Table 10.7 that the highest contribution to the total field level inside the room is not due to the indoor source but to the outdoor one. In particular, the average **E**-field

TABLE 10.7

Exposure Field (rms Values) for the Indoor Scenario (Coexisting Outdoor GSM BTS and Indoor Wi-Fi AP)

	$E_{vol\ peak}$ (V/m)	$E_{vol\ ave}$ (V/m)	$E_{sup\ ave}$, min.–max. (V/m)
GSM900	5.57	3.13	2.73–3.44
GSM1800	3.61	1.70	1.62–1.91
Total GSM	6.18	3.56	3.19–3.93
Wi-Fi	2.51	1.13	1.05–1.19
Total	6.30	3.74	3.40–4.09

TABLE 10.8

SAR Values for the Indoor Scenario (Coexisting Outdoor GSM BTS and Indoor Wi-Fi AP)

	SAR_{WB} (mW/kg)	SAR_{1g} (mW/kg)	SAR_{10g} (mW/kg)
GSM900	0.109	2.41	1.08
GSM1800	0.027	1.31	0.58
Total GSM	0.136	3.07	1.46
Wi-Fi	0.014	0.79	0.35
Total	0.150	3.66	1.60

value attributable to the Wi-Fi system is as low as 1 V/m. The computed data also demonstrate that coexistence of the two systems (GSM and Wi-Fi) is possible without exceeding the reference levels for the exposure field, as averaged over the volume occupied by the body or over an equivalent surface, even if the particularly stringent limits issued by some national regulations (e.g., 6 V/m) are considered. Finally, the SARs presented in Table 10.8 suggest that a typical exposure scenario results in RF absorption that is two orders of magnitude below the basic restrictions, both for whole-body and for locally averaged SAR.

10.4.2.5 Coupling of Transient EM Pulses into the Human Body

Electromagnetic transient radiations are widely used for studying the susceptibility of test objects to broadband EMPs, and increasingly, pulsed fields are being explored for telecommunication purposes. The main characteristics of these pulse fields are waveforms that include high peak powers, fast rise times, and a narrow pulse width. Earlier investigations on their interaction with biological systems relied on mathematical analyses of canonical shapes of dielectric equivalent bodies, such models as planar tissue layers and bodies of revolution [131–134]. The well-known effect of microwave hearing from pulse-induced thermoelastic pressure in the human head have been investigated both analytically [135–140] and numerically [141–143]. More recently, major strides have been made in the development of UWB systems for wireless telecommunications [144,145]. It promises a powerful combination of low power, high throughput, greater range, and better inherent security, using nanosecond pulses. This section presents predictions of fields and power depositions, which have been obtained from the frequency-dependent FDTD formulations described in Section 10.4.1.5, for models of the biological body.

10.4.2.5.1 Modeling of Tissue Properties with the Debye Equation

For UWB calculations using the $(FD)^2TD$ method, the measured properties for the various tissues may be fitted to the Debye equation (Equation 10.14) with two relaxation constants [77–79]. For the results shown here, the measured properties of biological tissues (muscle, fat, bone, blood, intestine, cartilage, lung, kidney, pancreas, spleen, lung, heart, brain/nerve, skin, and eye) were obtained from the literature. Optimized values for ε_{s1}, ε_{s2}, ε_{∞}, τ_1, and τ_2 in Equation 10.14 were obtained by nonlinear least squares matching to the measured data for fat and muscle (Table 10.9), with τ_1 and τ_2 being the average of the optimized values for fat and muscle. All other tissues have properties falling roughly between these two types of tissues. This was done to facilitate volume averaging of the tissue properties in cells of the heterogenous human model. Having τ_1 and τ_2 constant for all tissues, allowed linear (volume) averaging of the ε values for each tissue in a given cell to calculate ε values for that cell.

TABLE 10.9

Debye Constants for Tissues, $\tau_1 = 46.2 \times 10^{-9}$ s and $\tau_2 = 0.91 \times 10^{-10}$ s (Average of Optimum for Fat And Muscle)

Tissue	ε_∞	ε_{s1}	ε_{s2}
Muscle	40.0	3948	59.09
Bone/cartilage	3.4	312.8	7.11
Blood	35.0	3563	66.43
Intestine	39.0	4724	66.09
Liver	36.3	2864	57.12
Kidney	35.0	3332	67.12
Pancreas/spleen	10.0	3793	73.91
One-third lung	10.0	1224	13.06
Heart	38.5	4309	54.58
Brain/nerve	32.5	2064	56.86
Skin	23.0	3399	55.59
Eye	40.0	2191	56.99

10.4.2.5.2 Induced Currents and SAs

The $(FD)^2TD$ formulation has been used to calculate coupling of an ultrashort pulse to the heterogenous model of the human body. From the calculated internal fields, the vertical currents passing through the various layers of the body are calculated by using the following equation:

$$I_z(t) = \delta^2 \sum_{i,j} \frac{\partial D_z}{\partial t} \tag{10.25}$$

where δ is the cell size ($=1.31$ cm), and the summation is carried out for all cells in a given layer. The layer-averaged absorbed energy density or SA and the total energy W absorbed by the whole body can be calculated using the following relationships:

$$\mathrm{SA}|_{\text{layer } k} = \frac{\delta t}{N_k} \sum_{i,j,t} \frac{E(i,j,k,t)}{\rho(i,j,k)} \frac{\partial D(i,j,k,t)}{\partial t} \tag{10.26}$$

$$W = \delta t \delta^3 \sum_{i,j,k,t} E(i,j,k,t) \frac{\partial D(i,j,k,t)}{\partial t} \tag{10.27}$$

In Equation 10.26 and Equation 10.27, δt is the time step ($=\delta/2c = 0.02813$ nsec) used for the time domain calculations, N_k is the number of cells in layer k of the body, and $\rho\,(i,j,k)$ is the mass density (in kg/m^3) for each of the cells in the corresponding layers.

A typical time domain, UWB pulse with a peak amplitude of 1.1 V/m is shown in Figure 10.23. It is interesting to note that the pulse has a rise time of about 0.2 nsec and a total time duration of about 7 to 8 nsec. The Fourier spectrum of the pulse is shown in Figure 10.24. Most of the energy in the pulse is concentrated in the 200- to 900-MHz band with the peak of the energy being at about 500 MHz.

For purposes of illustration, the results that follow assume the incident fields to be vertically polarized, since this polarization is known to result in the strongest coupling for standing individuals [146]. Also, a uniform plane wave illumination of the whole body is assumed by the incident fields. The $(FD)^2TD$ procedure is used to calculate the temporal

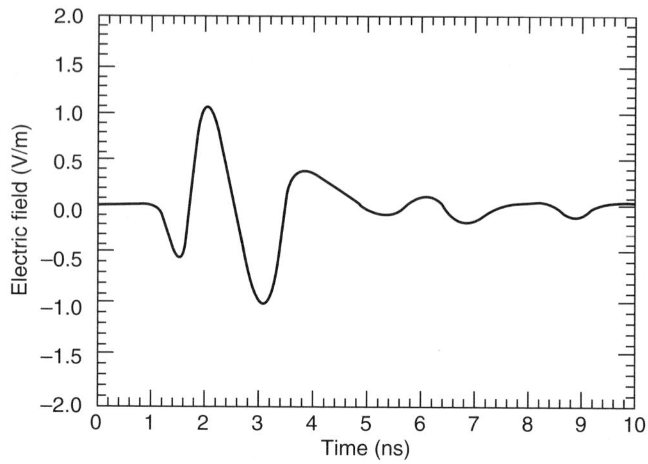

FIGURE 10.23
A representative UWB EMP. Peak incident field = 1.1 V/m.

variations of total vertical currents for the various sections of the body for both the shoe-wearing grounded and ungrounded exposure conditions of the model. The current variations for a couple of representative sections such as those through the eyes and the bladder are given in Figure 10.25a and b, respectively. The calculated peak currents for the various sections are on the order of 1.1 to 3.2 mA/(V/m). It is interesting to note that there is very little difference in the induced currents, whether the model is grounded or not. This is due to the fact that most of the energy in the pulse is at frequencies in excess of 300 MHz, where the effect of the ground plane on the induced currents or the SARs is minimal.

In Figure 10.26, the peak current for each section of the body is plotted with a section resolution of 1.31 cm. The maximum peak sectional current of 3.5 mA, which is equal to 3.2 mA/(V/m), occurs at a height of 96.3 cm above the bottom of the feet. A very similar result also had been observed for calculations using isolated and grounded models of the human body for plane wave exposures at frequencies of 350 to 700 MHz, where the highest induced currents on the order of 3.0 to 3.2 mA/(V/m) were calculated for sections of the body that are at heights of 85 to 100 cm relative to the feet.

The SA and the total absorbed energy for exposure to the UWB pulse can be calculated using Equation 10.26 and Equation 10.27. The total energy absorbed by the body exposed

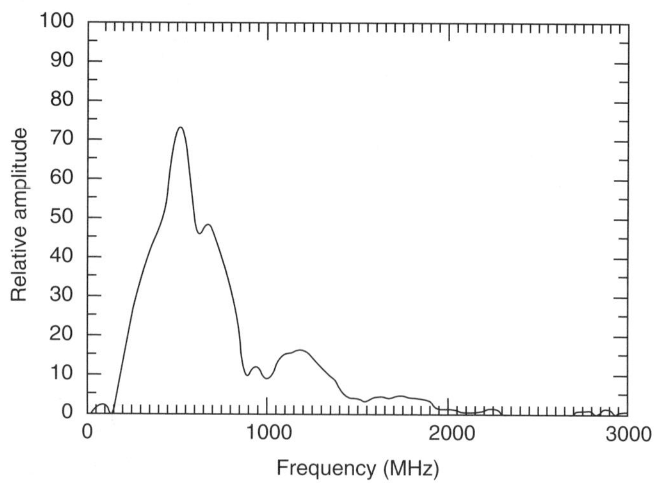

FIGURE 10.24
Fourier spectrum of the EMP of Figure 10.23.

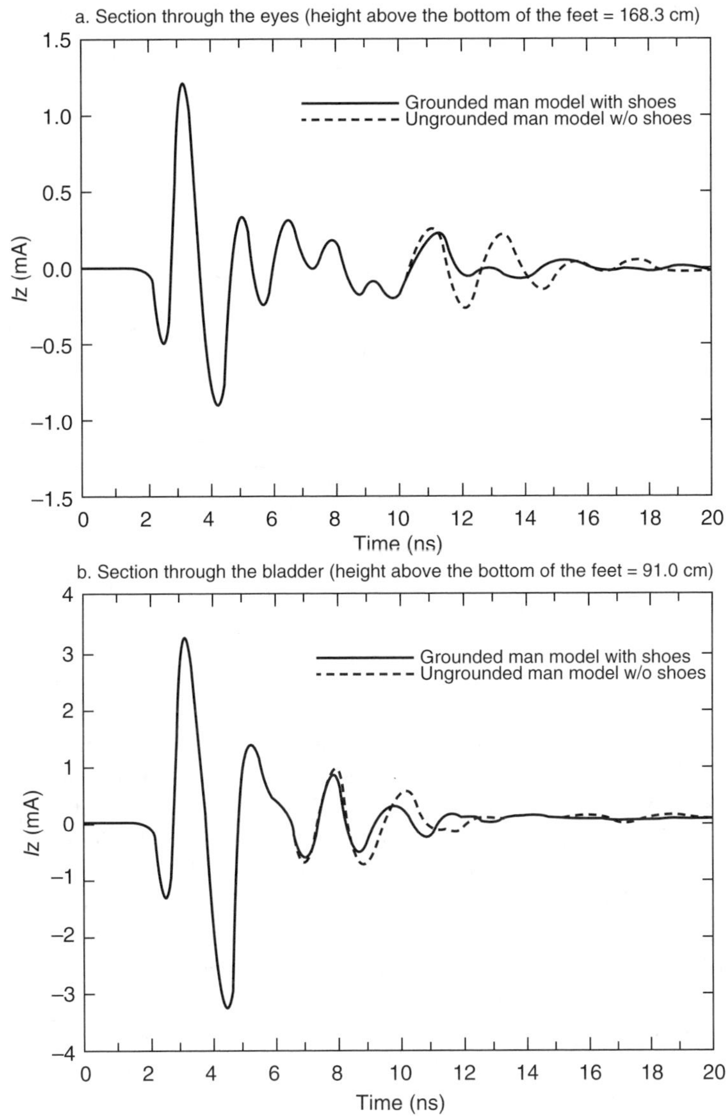

FIGURE 10.25
Currents induced for the various sections of the body for shoe-wearing grounded and ungrounded conditions of exposure. $E_{peak} = 1.1$ V/m.

to a single pulse of the type shown in Figure 10.23 is 2.0 and 1.91 pJ for isolated and shoe-wearing grounded conditions, respectively.

10.4.2.6 Absorption in the Head of Cellular Phone Users

The widespread use of cellular mobile telephone systems has brought about an increased concern for possible adverse health effects from the RF field emitted by the handset. Indeed, exposure standards have been mandated by various national bodies to limit human exposure to cell phone radiations. These RF exposure standards provide specifications in terms of power deposition per unit mass, that is, SAR induced in the user's head, with which cell phones must comply [100,101].

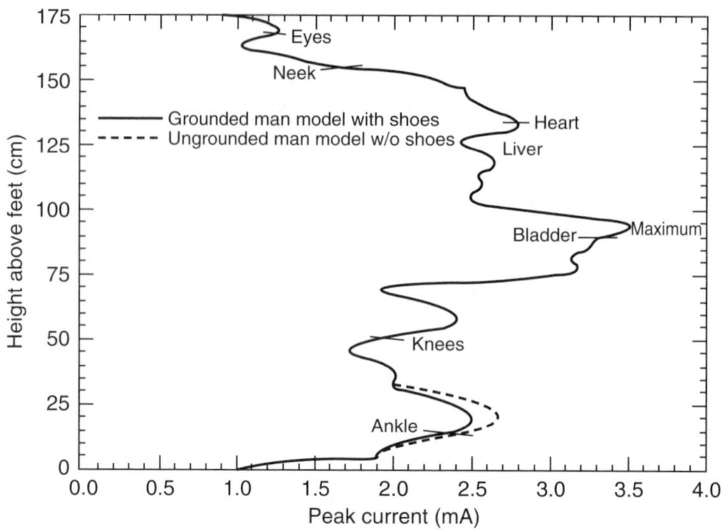

FIGURE 10.26
Peak currents induced for the various sections of the body for shoe-wearing grounded and ungrounded conditions of the model. $E_{peak} = 1.1$ V/m.

Laboratory procedures for compliance testing of mobile phones are based on experimental measurements, performed according to published protocols [147,148]. In these tests, real phones and phantom head models, shells filled with a material with dielectric properties equivalent to those of the brain tissue at the frequencies of interest, are used. Clearly, the phantom is a simplified model of the human head and is specifically designed for compliance testing. Consequently, it is not suited for an accurate analysis of the SAR distribution in various tissues and organs of the head. A detailed analysis of the distribution of the power absorption would be required to obtain a better understanding of SARs inside the head. Such information is needed for the necessary extrapolation of results from *in vivo* and *in vitro* experimental studies devoted to investigating the effects of RF radiation on humans. They would provide the exposure data needed in epidemiological studies aimed at evaluating any possible dose–effect relations.

Increasingly, advances in computational bioelectromagnetics have made detailed evaluation of SAR distribution inside the human head possible through the use of accurate and realistic models of the human head and the source, that is, the mobile phone, and the use of suitable numerical methods such as the FDTD technique.

The first numerical studies were performed by simulating the phone radiating element as a half-wavelength dipole or a quarter-wavelength monopole mounted on a box [149–154]. These antenna models, and the last in particular, can only be used as a rough model of the retractable antenna, which at the beginning was in nearly all cell phone handsets. However, at present, the need for more and more compact terminals and for dual-band operation has given rise to new antenna types. In particular, two types of antennas have been developed: planar integrated antennas and helical antennas. While half-wavelength dipoles and monopoles can be easily implemented inside an FDTD code, modeling of helix or planar antennas can become a rather difficult task.

The difficulties in modeling helical structures with the FDTD method were revealed in some recent studies. For example, only rather large structures have been studied employing a pure FDTD scheme [155,156]. For smaller structures, published reports had either employed equivalent sources [157] or a hybrid MoM–FDTD technique [158,159]. While these reports show some problems and drawbacks, investigations using FDTD, properly modified through the use of a graded mesh, have obtained good agreement with MoM and experimental results [160,161]. In these studies, both near-field and radiation patterns of dual-band cell phones equipped with a helical antenna have been reproduced. Moreover, the SAR distributions inside the VH model of the head have been computed, showing a higher penetration depth at 900 MHz and higher superficial SARs at 1800 MHz. Moreover, approximately 80% and 50% of the radiated power was absorbed inside the head at 900 and 1800 MHz, respectively. For a phone in contact with the ear and tilted to bring its axis along the ear–mouth direction (the so-called cheek position), radiating an average power of 250 mW at 900 MHz and 125 mW at 1800 MHz, peak 1-g SARs of 1.65 and 1.08 W/kg were obtained in the head at 900 and 1800 MHz, respectively. In the same examples, the peak 1-g SARs in the brain were 0.13 and 0.06 W/kg at the two considered frequencies, respectively.

Planar antennas can be mounted on the top, lateral, or back sides of the phone [162–165]. Shorted patch antennas typically have a 10-dB bandwidth between 5% and 10% that can be increased to about 12% by parasitically coupling another printed radiator in the vertical direction (stacked patch). For comparison, the bandwidth is about 30% for the monopole antenna [163]. In this case, an important consideration is the influence of the hand wrapped around the handset. For a cell phone equipped with a planar antenna, the hand has a detuning effect on the antenna resonant frequency and causes a reduction of the bandwidth; both are evident where the hand masks the antenna. About 30% of the radiated power was absorbed by the hand, while for the monopole the hand absorption was only 15% [162]. For a radiated power of 250 mW at the frequency of 900 MHz and a head–handset separation of 2 cm, the computed peak 1-g SAR was 0.95 W/kg for a laterally mounted planar inverted F antenna (PIFA), and the result was 0.49 W/kg for a monopole antenna [162]. When a phone equipped with a side-mounted PIFA was kept in contact with the ear, the peak 1-g SAR increased to 1.4 W/kg [164]. Note that the back-mounting configuration gave rise to a substantial (up to three times) reduction in the peak SAR [162].

Another important task for an accurate evaluation of the power absorption in the head is the model adopted for the phone case. The typical approach, followed in the literature, consists of representing the case as a box, that is, a plastic-coated metal parallelepiped [149–165]. In order to model the correct shape of cell phones, both CAD files [166] and topometric sensors [167] have been used. However, in most previous studies, the internal structures of the phone have been modeled simply as a homogenous perfect conductor. Recently, CAD files have been used also to model the internal structures (printed circuit board, battery, keypad and buttons, etc.) of the phone [168]. An alternate approach to a suitable numerical model of the mobile phone was proposed by Pisa et al. [169]. It starts with a simplified model, which includes only the main phone parts (antenna, keyboard, internal box, plastic coating, etc.) having "realistic" dimensions and electric properties. The realistic parameters are then tuned by using an optimization procedure, which minimizes a functional that depends on the differences between the measured and simulated electric and magnetic fields in front of the phone and on the SAR inside a cubic phantom. As an example for the applicability of the proposed optimization method, a numerical model of a commercial phone, operating at 900 MHz, was implemented, and the power deposition in the VH model of the human head was computed for various phone–head distances. The results in terms of peak SAR in various head organs and tissues and for various phone–head distances are presented in Table 10.10. The SAR_{1g} and SAR_{10g} show a monotonic decrease when the phone–head distance increases, while the peak SARs inside the head tissues reach their maxima when the phone is kept at specific distances from the head. This behavior is due to the fact that with the telephone pressed against the head, the power absorption is confined to a limited region in front of the

TABLE 10.10

Peak SAR Averaged over 1 g (SAR_{1g}) and 10 g (SAR_{10g}) and Peak SAR as Averaged over 1 g in Various Organs and Tissues, for Different Distances between the Phone and the Visible Human Phantom for a Radiated Power of 250 mW at 900 MHz

	SAR_{1g} (W/kg)	SAR_{10g} (W/kg)	$SAR_{1\ brain}$ (W/kg)	$SAR_{1\ eye}$ (W/kg)	$SAR_{1\ skin}$ (W/kg)	$SAR_{1\ muscle}$ (W/kg)	$SAR_{1\ fat}$ (W/kg)	$SAR_{1\ gland}$ (W/kg)
$d = 0$ mm	1.450	0.600	0.125	0.0102	0.504	0.223	0.142	0.448
$d = 2$ mm	0.740	0.500	0.123	0.0101	0.504	0.244	0.162	0.493
$d = 6$ mm	0.670	0.470	0.174	0.0100	0.482	0.278	0.165	0.471
$d = 8$ mm	0.630	0.450	0.109	0.0103	0.460	0.280	0.160	0.445
Cheek position ($d = 0$ mm) optimized	0.810	0.410	0.088	0.0360	0.171	0.447	0.301	0.420
Cheek position ($d = 0$ mm) nonoptimized	2.568	0.888	0.177	0.0578	0.315	0.721	0.400	0.664

antenna feed point, whereas by moving the handset away from the head, a greater portion of the head is exposed. When the distance was further increased, the SAR decreased monotonically as a result of the decay in field intensity. This study also showed that the use of inaccurate phone models (last row in Table 10.10) could give rise to SARs, averaged over 1 g, up to three times higher than those computed for the optimized model [169].

Faced with a rapid saturation of the cellular phone market, many cell phone manufacturers and service providers are turning their attention toward youths in promoting handsets that are cheap, with inexpensive service plans, or both [170]. An issue of particular interest is the possible difference in power absorption between children and adults. To answer this question, the first problem to be addressed is the realization of an accurate numerical model of a child's head. Because of ethical concerns, the availability of anatomical models of children has been limited. The common approach for obtaining a model of the child has been the reduction of the dimensions of the voxel size of adult models. Sometimes, this reduction was performed by employing different scaling factors for the different parts of the head. It appears that dielectric constants for children may be considerably higher than adults [171]. However, since the detailed data for the dielectric properties of tissues in children are scarce, they are usually assumed to be equal to those of adults or generically increased by a constant factor in most models. Nevertheless, using these models, some papers have reported increases of up to 50% in the peak 1-g SARs in the child head, compared to an adult head, for exposure to cell phones operating at frequencies around 835 and 1900 MHz [151,172]. A similar increase has been observed in the peak 1-g SARs obtained in the brain. A possible explanation of these results is the larger depth of penetration of power in the child models as compared to the adult one [151,172]. Other papers devoted to the investigation of differences between child and adult exposure to cellular phones have shown no significant difference in peak 1-g SAR between adults and children [173,174]. There are several possible explanations for the discrepancy. It has been suggested that the contradictory results may be due to the different phone excitation schemes used by different authors [175]. The disparity in distances of separation between the antenna and the head also was suggested as a pivotal factor in determining the reported discrepancies [170].

In an effort to help resolve the discrepancies, the SAR distributions induced in two child's head models, an isotropic scaling of the VH head (child size, CS) and an

TABLE 10.11

Peak SAR Averaged over 10 g (SAR_{10g}) and Peak SAR as Averaged over 1 g in Various Organs and Tissues, for Radiated Powers of 250 mW at 900 MHz and 125 mW at 1800 MHz in Children and Adults

		SAR_{10g} (W/kg)	$SAR_{1\ skin}$ (W/kg)	$SAR_{1\ muscle}$ (W/kg)	$SAR_{1\ bone}$ (W/kg)	$SAR_{1\ csf}$ (W/kg)	$SAR_{1\ brain}$ (W/kg)
900 MHz	VH	0.67	2.00	0.67	0.20	0.34	0.16
	CS	1.18	4.02	1.03	0.31	0.45	0.25
	CL	1.03	1.15	0.41	0.14	0.23	0.20
1800 MHz	VH	0.39	0.99	0.30	0.08	0.12	0.08
	CS	0.29	0.87	0.30	0.08	0.13	0.09
	CL	0.27	0.91	0.30	0.06	0.15	0.10

Note: Results are for visible human (VH) adult, isotropically reduced child size (CS), and anisotropically reduced child-like (CL) numerical phantoms.

tanisotropic scaling of the VH head (childlike, CL), have been computed and compared with SAR distributions induced in the VH by a mobile phone equipped with a back-mounted dual-band patch antenna [176]. Some of the results are presented in Table 10.11. It can be seen that the peak SAR_{10g} showed an increase of about 50% and a reduction of about 25% for the child models compared to the adult model at 900 and 1800 MHz, respectively. Moreover, as the brain is closer to the mobile phone in the case of CS and CL heads, the SAR_{1g} in the brain of children is slightly more significant than that for the adult.

Other exposure scenarios also have been investigated, including head exposure inside a car and cell phones not placed in contact with the ear. In some studies, the influence of the metallic and dielectric structures of a car on SAR induced by a cellular phone inside an adult head was analyzed [155,177–179]. Studies performed by modeling the whole car have shown that the main influence on SAR distribution was due to structures that were very close to the head [178,179]. In particular, the presence of a vertical glass wall in parallel with the antenna axis did not significantly influence the SAR distribution, while a metallic wall can cause up to an 80% increase in the peak 1-g SAR [155,177]. The presence of a reflecting wall placed horizontally over the head, simulating the roof of a car, rendered the SAR distribution more uniform, increasing the lower values and reducing the higher ones [155].

Cellular phones commonly are used in contact with the ear. However, when they are used with a headset, they can be positioned at different body locations. Moreover, unintentional exposures also might occur for a subject standing close to someone using a cellular phone, for example, in a crowded environment. The peak SAR produced by a phone placed slightly above the navel of the VH model has been calculated by using the FDTD method, and the results have been compared to those of a flat phantom—modeled as a multilayered transmission line—whose thickness was varied statistically [180]. The exposure of a human subject to a half-wavelength dipole placed at various positions at a distance of 9 mm from the body surface also has been investigated [181]. In particular, 11 different locations have been considered, that is, in front of the right ear and left ear, the nose, eye, heart, lung, shoulder, stomach, hip, lower back, and the groin. The computed results showed that the maximum values of SAR_{1g} and SAR_{10g} occurred, in all cases, when the phone is close to the ear. For all other positions, corresponding to possible locations of the phone when used in conjunction with a headset, peak 1-g SAR values were at least 30% lower than those obtained in the common ear position.

10.5 Temperature Elevations Induced in Biological Tissues by EM Power Absorption

10.5.1 Introduction

EM energy impinging on the human body induces currents and fields inside the body. A major biological response from absorption of EM energy in the RF and microwave frequency range is the elevation of tissue temperature. Consequently, most internationally recognized guidelines for limiting human exposure to EMFs in the RF and microwave range use SAR as the basic dosimetric metric [100,101]. Likewise, the vast majority of studies available in the literature, addressing the topic of human exposure to EMFs, focus their attention on the dosimetric problem of quantifying induced power absorption inside the exposed subject. A central premise of these exposure guidelines is to protect exposed subjects against temperature increases exceeding the threshold for induction of adverse thermal effects. Therefore, an increasing number of investigators are beginning to address the problem of human exposure to EMFs with a thermal analysis to estimate the temperature increment induced inside the exposed subject.

Another domain in which a thermal analysis can be very useful, or even essential, is that of therapeutic applications where EMFs are deliberately used to cause predefined temperature increases in specific target tissues in the body. Some of the applications include hyperthermia cancer treatment and microwave tissue ablation. In such cases, performing a numerical electromagnetic and thermal study of the applicator in its intended operating environment, inside the body, can be a valuable aid in designing the applicator, in establishing the clinical protocol (i.e., power to be delivered, time of application, etc.), and for treatment planning purposes.

In the following, an overview of the available analytical formulations to characterize heating induced by EMFs is presented. Some numerical implementations, suitable to study the thermal problem in realistic situations, are summarized with specific examples.

10.5.2 Bio-Heat Equation

The bio-heat equation (BHE) was originally proposed by Pennes in 1948 to analyze temperature distributions in a resting forearm [182]. Subsequently, it has been modified to study phenomena of heat transport and exchange for the whole body [183,184]. It is an analytical model that describes the temperature distribution $T = T(\boldsymbol{r},t)$ inside the body. One of the more general formulations of the BHE is given here for temperature rises associated with exposures to EMFs:

$$\nabla \cdot (K(\mathbf{r})\nabla T) + A(\mathbf{r},T) + Q_{\mathrm{v}}(\mathbf{r}) - R_{\mathrm{L}}(\mathbf{r}) - B(\mathbf{r},T)(T - T_{\mathrm{B}}) = C(\mathbf{r})\rho(\mathbf{r})\frac{\partial T}{\partial t}\,(\mathrm{W/m^3}) \qquad (10.28)$$

The five terms on the left side of Equation 10.28 represent heat accumulation (or loss) per unit time and per unit volume at a point inside the body. Specifically, the various ways through which heat is transferred, produced, or removed from the tissue are:

- Heat transfer through internal conduction, where K (W/(m°C)) is the tissue thermal conductivity
- Metabolic heat production (A [W/m^3])
- Electromagnetic power deposition (Q_{v} [W/m^3])

- Respiratory heat losses in the lungs (R_L [W/m^3])
- Heat exchange due to capillary blood perfusion, which is proportional to blood flow and is represented by the parameter B (W/(°C m^3)), and the difference between blood and tissue temperature ($T_B - T$); note that T_B is a function of time (i.e., $T_B = T_B(t)$)

The right side of Equation 10.28 denotes the temperature increase (or decrease) per unit time. The thermal capacitance per unit volume is given by the product between the tissue specific heat, C (J/(kg°C)) and density, ρ (kg/m^3).

It should be mentioned that the BHE assumes that heat exchange with blood takes place exclusively via capillary perfusion. In reality, heat exchange also occurs with large blood vessels. This mechanism does not take the form of a distributed exchange throughout the tissue volume, like the $B(T - T_B)$ term in the BHE, but instead, the form of a localized exchange at the blood vessel walls. To account for it would require the introduction of an additional term in the BHE [185,186]. Moreover, it would require precise knowledge of the structure of the vasculature inside the biological body, which is not always available. However, this mechanism only alters temperature distribution near large blood vessels and does not significantly affect the overall temperature distribution, especially the maximum temperature increases elsewhere in an exposed body [187]. Therefore, this mechanism can generally be neglected without significant loss of accuracy, if the principal purpose is to assess safety compliance of a given exposure situation, from the thermal point of view. On the other hand, proper inclusion of large blood vessels may be important when planning hyperthermia or ablation treatments. The presence of a large blood vessel in the target region may cause temperature elevations to remain below the minimum threshold required for effective treatment.

A first step in using the BHE to compute the temperature increases induced by exposure to EMFs is the evaluation of SAR or local power deposition. In fact, $Q_v = \rho$ SAR is the exogenous heat source responsible for the alteration in temperature profiles inside the exposed subject. Once the Q_v term is determined, the BHE would provide the time evolution of temperature, provided that appropriate initial and boundary conditions are imposed, as discussed later. In this manner, the BHE allows assessment of both the transient response and steady-state temperature increases.

An implicit assumption in the above discussion is that the electromagnetic and thermal problems are independent and can be investigated sequentially. This is equivalent to assuming that electromagnetic transients are irrelevant and therefore the steady-state SAR distributions can be used as the input for the thermal analysis, and that changes in tissue temperature do not alter the field distribution inside the tissue. Concerning the first assumption, the time constants of the electromagnetic and thermal processes are of orders of magnitude different, with the EMF reaching steady state at most after a few microseconds, while thermal constants inside living biological tissues are of the order of a few minutes under usual circumstances. This means that electromagnetic transients can indeed be neglected for the thermal analysis. The second assumption, instead, deserves some more attention. In fact, dielectric constants of biological tissues are temperature dependent, and therefore, the EMF distribution could change as heating proceeds and temperature increases. However, this effect may be neglected so long as the temperature elevation is small, that is, on the order of a few degree Celsius, as is expected for common EMF exposures. If temperature increase is large and the induced variation in dielectric constants is no longer negligible, for example, when heating food in a microwave oven, the electromagnetic and thermal problems must be solved in a coupled manner, iteratively updating the electromagnetic solution as heating progresses [188–190].

10.5.2.1 Initial Conditions

The BHE is a partial differential equation in time and space; its solution requires the specification of both initial and boundary conditions. For studies involving RF- and microwave-induced heating inside the human body, the initial temperature distribution typically is set to the physiological norm, computed as the steady-state solution of Equation 10.28 in the absence of external power deposition ($Q_v = 0$). Thus, the resulting equation is

$$\nabla \cdot (K(\mathbf{r})\nabla T) + A_0(\mathbf{r}) - R_L(\mathbf{r}) - B_0(\mathbf{r})(T - T_{B0}) = 0 \ (\mathrm{W/m^3}) \tag{10.29}$$

In this case, thermal parameters do not depend on temperature and are set to their physiological values at approximately 37°C. Similarly, the physiological value at rest is used for blood temperature. Note that Equation 10.29 does not contain time derivatives and, therefore, does not require any initial condition to be solved.

10.5.2.2 Boundary Conditions

Boundary conditions are needed to account for the heat exchange between the body surface, namely, the skin, and the external environment, in both the general and the steady-state formulations of the BHE as represented by Equation 10.28 and Equation 10.29, respectively. The simplest boundary conditions that can be applied are the adiabatic condition, that is, a thermally insulated surface, or the Dirichelet boundary condition, that is, an enforced surface temperature. Adiabatic conditions can be used to model tissue surfaces in close contact with highly insulating materials, such as the catheters used to insert antennas employed in hyperthermia or ablation treatment. In contrast, Dirichelet boundary conditions can be used for surfaces in close contact with a circulating fluid kept at a constant temperature (forced convection).

Adiabatic and Dirichelet boundary conditions are rather simple but they are not suitable for representing the general heat exchanges that take place at the skin. A general boundary condition, obtained by imposing the continuity of the heat flow perpendicular to the surface of the body, can be expressed as [191]:

$$-K(\mathbf{r})(\nabla T \cdot \mathbf{n}_0)_S = H(T_s - T_A) + \mathrm{SW}(T)(\mathrm{W/m^2}) \tag{10.30}$$

where S is the skin surface and $\mathbf{n}_0$ is the outward unit vector normal to S. The terms on the right side of Equation 10.30 represent the two ways in which heat is exchanged with the environment. In particular, the first term describes heat loss due to convection, and it is proportional to the difference between skin temperature (T_S) and ambient air temperature (T_A) through the convection coefficient H (W/(m^{2}°C)). The last term represents heat loss due to sweating (SW).

A few words are needed about radiative heat exchange. If one assumes the body surface is surrounded by objects that are all at the same ambient temperature T_A, the expression for heat exchange through radiation from the body surface to the environment, per unit area, is given by [192,193]:

$$Q_r = e\sigma(T_s^4 - T_A^4)(\mathrm{W/m^2}) \tag{10.31}$$

where e is the surface emissivity, σ is Stefan–Boltzmann's constant, and the temperatures are expressed in degrees Kelvin (K). Under normal conditions, T_S and T_A do not differ significantly from about 300 K, and Equation 10.31 can be approximated as [193]:

$$Q_r = H_r(T_s - T_A)(\mathrm{W/m^2}) \tag{10.32}$$

where H_r is an equivalent convection coefficient. Therefore, the convective term in Equation 10.32 can effectively model both convective and radiative heat exchanges, by using an overall convection coefficient that also takes into account the equivalent convection parameter H_r in Equation 10.32. There are cases, however, in which the surrounding objects are at different temperatures. In such circumstances, the problem becomes very complex and a possible solution, based on the use of an RT method to connect mutually visible surfaces and consider radiant heat transfer between them, has been proposed [194]. However, the simple approach of an equivalent convection is generally sufficient for an accurate analysis of the thermal problem.

It is worthy of mention that the convective boundary condition also can be used to represent an adiabatic condition by simply setting the convective coefficient to zero. It can also represent an approximate Dirichelet boundary condition, by using a very high convective coefficient and setting T_A equal to the imposed surface temperature.

10.5.3 Thermoregulatory Responses

The temperature of the human body is regulated to within a narrow range of about 37°C, in its core. Under normal circumstances, this is accomplished through an exquisite thermoregulatory mechanism involving sweating and vasodilatation (see Chapter 5 by Black for more details on thermoregulation [285]). The thermoregulatory mechanism is activated whenever the temperature $T(\mathbf{r})$ in specific parts of the body, where thermal sensors are placed, shifts from its basal value $T_0(\mathbf{r})$. In particular, the basal temperature distribution $T_0(\mathbf{r})$, which corresponds to a state of "thermal comfort," is the one obtained in a naked subject when the external air temperature T_A is about 30°C (in a dry environment) [184,195].

Since a part or all of the absorbed electromagnetic energy is converted into heat inside the human body, computations of tissue temperature must take into account the thermoregulatory mechanisms in response to the heat input from RF and microwave absorption, starting from a state of thermal comfort [195]. Specifically, the presence of thermoregulatory mechanisms causes some of the terms in Equation 10.28 and Equation 10.30 to vary with body temperature. The first term that shows a dependence on temperature in Equation 10.28 is metabolic heat production, which may be characterized by the following equation [196]:

$$A(\mathbf{r},T(\mathbf{r})) = A_0(\mathbf{r})(1.1)^{(T(\mathbf{r})-T_0(\mathbf{r}))} \tag{10.33}$$

where A_0 is the basal metabolic rate in the tissue. Equation 10.33 shows that metabolic heat production depends only on local tissue temperature. It must be noted that this dependence is not related to thermoregulation but rather to the fact that metabolic processes are slightly accelerated when the temperature increases.

With regard to the variation of blood flow, there are two different, but essential, phenomena: one for internal tissue perfusion and the other for peripheral (skin) perfusion. For blood perfusion to the internal tissues, the regulation depends only on local tissue temperature [196,197]. Thus, in a simple model, blood perfusion could be assumed to be at its basal value B_0 until the local temperature reaches 39°C. When the local temperature exceeds 39°C, the blood perfusion starts to increase linearly with temperature in order to enhance the local heat removal process, until the local temperature rises to above a value of about 44°C. At this point, the increasing rate of blood perfusion arrives at a maximum. Accordingly, the internal blood perfusions are modeled by the following expressions:

$$B(\mathbf{r},T(\mathbf{r})) = B_0(\mathbf{r}) \; T(\mathbf{r}) \leq 39°\text{C} \tag{10.34}$$

$$B(\mathbf{r},T(\mathbf{r})) = B_0(\mathbf{r})[1 + S_B(T(\mathbf{r}) - 39)] \; 39°\text{C} < T(\mathbf{r}) < 44°\text{C} \tag{10.35}$$

$$B(\mathbf{r},T(\mathbf{r})) = B_0(\mathbf{r})(1 + 5S_B) \; T(\mathbf{r}) \geq 44°\text{C} \tag{10.36}$$

The above model of the internal temperature regulation mechanism is rather simple. A more complex temperature control model, instead, would include regulation of blood perfusion in the skin through vasodilatation. In particular, two different signals are used as feedback to regulate vasodilatation: one is the hypothalamic temperature increase ($T_H - T_{H0}$), used as an indicator of the elevation of the body core temperature, and the other is the average skin temperature increase $\overline{\Delta T}_s$, defined as follows:

$$\overline{\Delta T}_s = \frac{\int_S (T(\mathbf{r}) - T_0(\mathbf{r}))\,dS}{S} \tag{10.37}$$

where S is the skin surface of the body. The two feedback signals are assigned different weights, with a greater importance given to the hypothalamic temperature, and then used, together with local skin temperature, to regulate skin blood flow according to [183,184]:

$$B(\mathbf{r},T(\mathbf{r})) = \left[B_0(\mathbf{r}) + F_{HB}(T_H - T_{H0}) + F_{SB}\overline{\Delta T_S}\right] 2^{(T(\mathbf{r})-T_0(\mathbf{r}))/6} \tag{10.38}$$

where F_{HB} and F_{SB} are the weights of the hypothalamic and skin temperature signals, respectively.

From the above discussion, it can be noted that blood acts as a heat transfer agent, taking heat away from the inner body parts, whose temperature is higher than that of the blood, and bringing this heat to the body periphery. There, heat is passed to the skin layers, whose temperature is lower than that of the blood, and dissipated through sweating and evaporation. During microwave irradiation, the net heat exchange, between blood and the various body tissues, is different from zero, and consequently the blood temperature T_B varies according to the following equation:

$$Q_{BTOT} = C_B \rho_B V_B \frac{\partial T_B}{\partial t} \text{ (W)} \tag{10.39}$$

where Q_{BTOT} is the net rate of heat acquisition of the blood from the body tissues, C_B and ρ_B are the blood specific heat and mass density, respectively, and V_B is the total blood volume, assumed equal to about 5 L [198]. When the thermal equilibrium is reached, the net heat exchange is null, and therefore blood temperature stays at a constant value, slightly higher than the basal one.

The feedback mechanism that regulates sweating (SW in Equation 10.30) is very similar to that regulating peripheral blood flow and can be described as follows [183,184]:

$$\text{SW}(\mathbf{r},T(\mathbf{r})) = \left[\text{PI} + F_{HS}(T_H - T_{H0}) + F_{SS}\overline{\Delta T_S}\right] 2^{(T(\mathbf{r})-T_0(\mathbf{r}))/10} \tag{10.40}$$

where PI represents *perspiratio insensibilis* (insensible perspiration), that is the basal evaporative heat loss from the skin.

In fact, this model still represents a simplification of the thermoregulatory system of the human body, which is very complex. For example, the skin from different parts of the body does not have the same sweating behavior, as implied by Equation 10.40, and there exist other internal temperature sensors, besides the hypothalamus. Notwithstanding these limitations and the great variability in thermoregulatory behavior among different subjects, the model can be considered a good starting point to assess thermal responses in a human subject exposed to an EMF. It must also be observed that in most practical situations, induced thermal elevations are very small and thermoregulatory responses may not be invoked, so that basal physiological values for thermal parameters may be assumed for the BHE.

10.5.4 Numerical Methods for Solving the Thermal Problem

The combination of the BHE and the thermal boundary condition represent a complicated problem, which can become nonlinear if thermoregulatory mechanisms are considered. Analytical solutions of this problem, neglecting thermoregulation, can be obtained only for simplified body geometries and exposure conditions, which allow an analytical determination of the SAR distribution. For example, an analytical solution in stratified media may be obtained as an expansion in eigenfunctions, which are applicable to planar, cylindrical, and spherical multilayer geometries [199,200]. Also, an analytical solution for the case of a multilayer slab has been investigated in terms of Green's functions [201]. More complicated solutions, able to take into account thermoregulatory mechanisms, have also been developed, based on a simplified cylindrical segment approximation of the human body [183,184,202,203]. In particular, the thermal behavior of the human body is simulated by means of two systems: a controlling system and a controlled one. The controlled system, modeled by the BHE, determines the temperature distribution inside the body, while the controlling system provides feedback signals able to modify the thermal parameters of the controlled one in order to maintain a constant body core temperature (thermoregulation). The principal limitation of this approach stems from modeling the body as a few homogenous cylindrical segments, each having a uniform SAR and temperature distribution.

When studying more realistic and detailed geometries, like an anatomically based body model, analytical or cylindrical segment solutions are no longer feasible, and a numerical approach becomes necessary. One possibility is the development of a finite element solution of the BHE [204,205]. However, the most common approach is to use a finite difference scheme, which will be discussed in some detail in the following. One of the main reasons for preferring a finite difference solution, besides its computational efficiency, is that it allows a very simple link with the FDTD method, which is the most popular numerical method for SAR computations. Earlier finite difference solutions of the BHE, used in conjunction with cubic cell models of the human body, comprised only a few hundred voxels. They used an implicit formulation to avoid the restrictions on the size of the time step. Since these solutions required matrix inversions, they were computationally intensive [16]. As more detailed body models, comprising thousands of voxels, became available, new finite difference solutions were developed. These techniques, based either on explicit or on alternate direction implicit (ADI) formulations, are discussed in the following.

10.5.4.1 Explicit Finite Difference Formulation

One approach to obtain a finite difference explicit formulation of the BHE is based on the thermal balance approach [98]. The body under consideration is divided into cubic cells of side δ, and the temperature is evaluated at the center of each cell. Temperatures are computed at equal-time steps, δt. In the following, the expression $T^n(i,j,k)$ represents temperature computed at time $n\delta t$ in the (i,j,k) cell. The finite difference formulation is derived by imposing the thermal balance to each cell [192], such that

$$Q_{\mathrm{tot}}^{n,n+1}(i,j,k) = \Gamma(T^{n+1}(i,j,k) - T^n(i,j,k))(\mathrm{J}) \tag{10.41}$$

where $Q_{\mathrm{tot}}^{n,n+1}$ is the total heat accumulated (positive) or lost (negative) in the cell during the time interval $[n\delta t - (n+1)\delta t]$, $(T^{n+1} - T^n)$ is the variation of the cell temperature in the same time interval, and $\Gamma = C\rho\delta^3$ is the thermal capacitance of the cell. The heat $Q_{\mathrm{tot}}^{n,n+1}$ is accumulated (or lost) in the cell through the five mechanisms present in the BHE, and, for boundary cells, also through convection to external air and sweating (Equation 10.30).

Heat transfer through internal conduction is governed by Fourier's law. For the (i,j,k) cell, the heat $Q_{\mathrm{K}}^{n,n+1}$ entering the cell through conduction from the $(i-1,j,k)$ cell in the time interval $[n\delta t - (n+1)\delta t]$ can be derived by exploiting the well-known analogy between heat conduction and electrical current conduction [185,192]. In particular, if K_1 and K_2 are the thermal conductivities of the two cells under examination, heat flows through the series connection of two thermal conductances, equaling $K_1\delta^2/(\delta/2)$ and $K_2\delta^2/(\delta/2)$, respectively. Therefore, heat flowing from the center of the $(i-1,j,k)$ cell to the center of the (i,j,k) cell, through the boundary face, experiences an overall thermal conductance equal to $2K_1K_2\delta^2/((K_1 + K_2)\delta)$. As a result, we obtain:

$$Q_K^{n,n+1}(i,j,k) = \frac{2K_1K_2}{K_1 + K_2}\delta^2\frac{T^n(i-1,j,k) - T^n(i,j,k)}{\delta}\delta t \tag{10.42}$$

An expression similar to Equation 10.42 holds for heat exchanged with the other neighboring cells.

The contributions of metabolic heat production and electromagnetic power deposition are volumetric heat sources, and therefore the contribution they give to $Q_{\mathrm{tot}}^{n,n+1}$ can be immediately derived and is expressed by Equation 10.43 and Equation 10.44, respectively:

$$Q_A^{n,n+1}(i,j,k) = A(i,j,k)\delta^3\delta t \tag{10.43}$$

$$Q_{Q_v}^{n,n+1}(i,j,k) = Q_v(i,j,k)\delta^3\delta t \tag{10.44}$$

Respiratory losses in the lungs are taken into account by subtracting the volumetric loss R_L from the metabolic heat production A in the corresponding cells for the lung.

Finally, the contribution to $Q_{\mathrm{tot}}^{n,n+1}$ due to capillary blood perfusion takes the form of a volumetric term and can be represented, as in metabolic heat production and exogenous heat deposition, as:

$$Q_B^{n,n+1}(i,j,k) = B(i,j,k)((T_B - T^n(i,j,k))\delta^3\delta t \tag{10.45}$$

For boundary cells in contact with air, heat flow through the face is governed by convection rather than conduction. For a generic cell (i,j,k), considering a face in direct contact with air, the heat $Q_{\mathrm{H}}^{n,n+1}$ entering the cell through convection in the time interval $[n\delta t - (n+1)\delta t]$ is:

$$Q_H^{n,n+1}(i,j,k) = H(T_A - T^n(i,j,k))\delta^2\delta t \quad (10.46)$$

Heat losses due to sweating at the skin surface (SW) are converted to a volumetric heat loss term and are directly subtracted from the metabolic heat production term A in the skin cells.

Note that in Equation 10.42, Equation 10.45, and Equation 10.46, the temperature has been referred to the time instant $n\delta t$ in order to obtain, at the end, an explicit formulation. Starting from the equations given above, general explicit formulations that hold for each cell can be derived. (See the paper by Bernardi et al. [98] for additional details)

10.5.4.1.1 Stability Criterion

Explicit finite difference formulations are straightforward to implement and are computationally efficient, but they have a limitation on the maximum time step δt that can be used without incurring numerical instability.

The stability criterion for the previously mentioned finite difference scheme can be obtained through Fourier's analysis, which yields the following restriction on δt, derived for internal cells (no convective contributions) [206]:

$$\delta t \leq \frac{1}{6(K/C\rho\delta^2) + (B_0/2C\rho)} \quad (10.47)$$

Because of the typical values of thermal parameters and convection coefficients for the human body, the stability criterion for the peripheral voxels, where some of the faces exchange heat through convection rather than conduction, is less stringent than that of Equation 10.47. Consequently, Equation 10.47 may be safely assumed as the stability criterion for the overall scheme.

10.5.4.2 ADI Formulation

While the explicit finite difference formulation of the BHE has been applied to many simulations, it becomes computationally unaffordable when very small cell sizes are used or when high thermal conductivity materials are present in the domain under study. This stems from the extremely small time steps δt that would be required according to Equation 10.47. In such cases, the ADI formulations can be used [207]. ADI is a general method, developed for numerical solution of parabolic equations. It combines unconditional stability, typical of implicit methods, with the computational efficiency due to the tri-diagonal nature of the resulting matrices [208–210]. Note that the Fourier heat conduction equation is a typical parabolic equation. While the BHE is no longer parabolic, because of the presence of the term related to blood flow, ADI can still be applied, but it loses its unconditional stability. In any case, for time steps on the order of a few seconds, the scheme has proved to be stable in typical applications. It can yield reductions on the order of tenfold or more [207] in execution time, over the classical explicit formulation.

The basic idea behind the ADI technique is to extend to the 3-D case the 1-D Crank–Nicolson's scheme, which averages the outcome from the explicit and the implicit formulations to obtain second-order accuracy both in space and in time variables [192]. In particular, Crank–Nicolson's scheme can be extended to the full 3-D case, by using a sequence of approximate Crank–Nicolson solutions along the three axes, indicated as $T^*(i,j,k)$, $T^{**}(i,j,k)$, and $T^{***}(i,j,k)$, the last one being used as the final estimate for $T^{n+1}(i,j,k)$.

The first approximate solution is obtained using Crank–Nicolson's scheme along the x axis only, while backward differencing is used along y and z:

$$
\begin{aligned}
&\frac{K}{\rho C}\frac{1}{2}\left(\frac{T^*(i-1,j,k)-2T^*(i,j,k)+T^*(i+1,j,k)}{(\delta x)^2}+\frac{T^n(i-1,j,k)-2T^n(i,j,k)+T^n(i+1,j,k)}{(\delta x)^2}\right)\\
&\quad+\frac{K}{\rho C}\frac{T^n(i,j-1,k)-2T^n(i,j,k)+T^n(i,j+1,k)}{(\delta y)^2}\\
&\quad+\frac{K}{\rho C}\frac{T^n(i,j,k-1)-2T^n(i,j,k)+T^n(i,j,k+1)}{(\delta z)^2}\\
&=\frac{T^*(i,j,k)-T^n(i,j,k)}{\delta t}-\frac{A_0+Q_v+B_0T_B}{\rho C}\\
&\quad+\frac{B_0}{\rho C}\frac{T^*(i,j,k)+T^n(i,j,k)}{2}
\end{aligned}
\tag{10.48}
$$

Extending subsequently Crank–Nicolson's solution along the y axis and the z axis, the following expressions are obtained for the second and third estimates:

$$
\begin{aligned}
&\frac{K}{\rho C}\frac{1}{2}\left(\frac{T^*(i-1,j,k)-2T^*(i,j,k)+T^*(i+1,j,k)}{(\delta x)^2}+\frac{T^n(i-1,j,k)-2T^n(i,j,k)+T^n(i+1,j,k)}{(\delta x)^2}\right)\\
&\quad+\frac{K}{\rho C}\frac{1}{2}\left(\frac{T^{**}(i,j-1,k)-2T^{**}(i,j,k)+T^{**}(i,j+1,k)}{(\delta y)^2}+\frac{T^n(i,j-1,k)-2T^n(i,j,k)+T^n(i,j+1,k)}{(\delta y)^2}\right)\\
&\quad+\frac{K}{\rho C}\frac{T^n(i,j,k-1)-2T^n(i,j,k)+T^n(i,j,k+1)}{(\delta z)^2}\\
&=\frac{T^{**}(i,j,k)-T^n(i,j,k)}{\delta t}-\frac{A_0+Q_v+B_0T_B}{\rho C}\\
&\quad+\frac{B}{\rho C}\frac{T^{**}(i,j,k)+T^n(i,j,k)}{2}
\end{aligned}
\tag{10.49}
$$

$$
\begin{aligned}
&\frac{K}{\rho C}\frac{1}{2}\left(\frac{T^*(i-1,j,k)-2T^*(i,j,k)+T^*(i+1,j,k)}{(\delta x)^2}+\frac{T^n(i-1,j,k)-2T^n(i,j,k)+T^n(i+1,j,k)}{(\delta x)^2}\right)\\
&\quad+\frac{K}{\rho C}\frac{1}{2}\left(\frac{T^{**}(i,j-1,k)-2T^{**}(i,j,k)+T^{**}(i,j+1,k)}{(\delta y)^2}+\frac{T^n(i,j-1,k)-2T^n(i,j,k)+T^n(i,j+1,k)}{(\delta y)^2}\right)\\
&\quad+\frac{K}{\rho C}\frac{1}{2}\left(\frac{T^{n+1}(i,j,k-1)-2T^{n+1}(i,j,k)+T^{n+1}(i,j,k+1)}{(\delta z)^2}+\frac{T^n(i,j,k-1)-2T^n(i,j,k)+T^n(i,j,k+1)}{(\delta z)^2}\right)\\
&=\frac{T^{n+1}(i,j,k)-T^n(i,j,k)}{\delta t}-\frac{A_0+Q_v+B_0T_B}{\rho C}\\
&\quad+\frac{B_0}{\rho C}\frac{T^{n+1}(i,j,k)+T^n(i,j,k)}{2}
\end{aligned}
\tag{10.50}
$$

Starting from the above expressions, it is possible to derive the general ADI formulation that holds for each internal cell. (See the paper by Pisa et al. [207] for additional details.) The formulation can also be adapted to boundary cells, where convective heat exchange must be considered, by introducing a fictitious external node.

10.5.5 Temperature Elevations in Subjects Exposed to EM Fields

When an EM field impinges on the biological body, a fraction of the incident power is absorbed by the body and is converted into heat in the body tissue. Thus, the absorbed energy can cause temperature increases in various body organs and tissues. If the temperature increase is small, it has little effect and is controlled by the thermoregulatory mechanisms of the body. However, if the temperature increment is large, it can produce irreversible biological damage. For example, a temperature increase of about 4.5°C for more than 30 min would produce neuronal damage [211]. Experiments performed on the rabbit eye indicated that a threshold increase of 3°C to 5°C in the lens can induce cataract formation [212,213]. The temperature increase necessary to induce thermal damage to the skin is about 10°C [214,215], while it is 8°C for muscle tissues [216]. Also, experiments performed using laboratory animals have shown various physiological and behavioral effects when the body core temperature rises more than 1°C to 2°C [100]. (See also Chapter 5 by Black [285].)

Moreover, most RF protection standards have adopted basic restrictions in order to keep the thermal increments below some agreed upon level. In particular, the value of 4 W/kg averaged over the whole body (SAR_{WB}) had been adopted by exposure guidelines as the threshold for the induction of adverse thermal effects associated with an increase of the body core temperature of about 1°C in animal experiments. Restrictions on local SAR were introduced in order to limit local temperature increments, since the ratio of the local peak to whole-body averaged SAR can be as high as 20:1 for exposure to a uniform plane wave [217]. However, it is important to note that tissue heating during EM exposure is strongly influenced not only by the power dissipated in the local tissue volume, but also by the way in which absorption is distributed in the surrounding area, by the thermal characteristics of the tissue and its neighbors, and, finally, by the heat exchange with the external environment. The correlation between local SAR values and temperature increases and between SAR and temperature distributions is not straightforward. There have been many studies dealing with thermal increments due to exposures both to the far field of radiating sources and to the near field of cellular phones. In the following some of these studies will be summarized and discussed.

10.5.5.1 Temperature Increments in the Human Body Exposed to the Far Field of Radiating RF Sources

In the past, temperature elevations due to EM power absorption in the human body have been evaluated using several models. The earlier studies were based on a simplified cylindrical segment approximation of the human body [183,184,202,203]. In these studies, the thermal behavior of the human body was simulated by means of two systems: a controlling system and a controlled one. The controlled system, modeled by the BHE, determines the temperature distribution inside the body, while the controlling system provides feedback signals able to modify the thermal parameters of the controlled one in order to maintain a constant body core temperature (thermoregulation). These studies are limited to modeling the body as few homogenous cylindrical segments, each one having a uniform SAR and temperature distribution. A subsequent study [197] used a cubic-cell model with tissue inhomogeneity and took into account the thermoregulatory mechanisms in the BHE. In this case, the BHE was solved using an implicit formulation, which

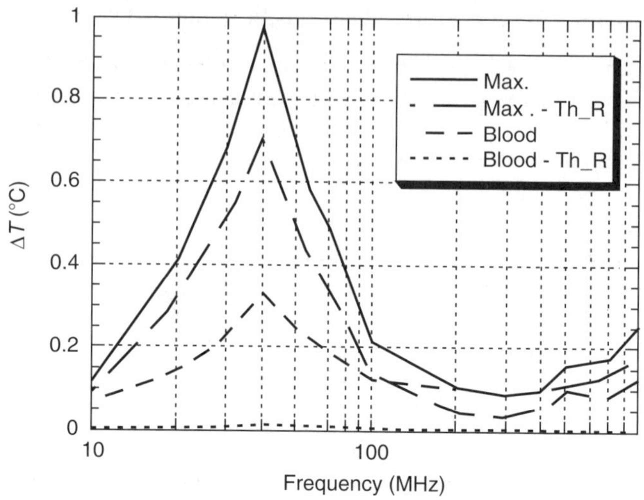

FIGURE 10.27
Maximum temperature increases (ΔT) in the body and blood of a subject wearing shoes, as a function of frequency, with or without thermoregulation (Th_R) for a P_{inc} equal to the limits set by ICNIRP [101].

avoided the restriction in the size of the time step. The explicit formulation requires a computationally intensive matrix inversion. Lately, explicit formulations of the BHE have been developed to study thermal responses in anatomically accurate body models [196,218,219]. In most of these studies, only limited body regions were considered, and thermoregulation mechanisms were neglected.

More recently, the electromagnetic and thermal problems have been combined in a detailed anatomical model of the human body (5-mm resolution) [98]. The FDTD method was used to compute the EMF distribution inside the exposed body, while an explicit finite difference formulation of the BHE, together with an accurate model of the human thermoregulatory system, were developed to compute the corresponding temperature increase.

For an incident power density equal to the maximum permissible value in the International Commission on Non-Ionizing Radiation Protection (ICNIRP) exposure guidelines for the general public [101], Figure 10.27 shows the maximum temperature increase ΔT_{max} obtained inside the body and in the blood as a function of frequency for a subject wearing shoes, both with and without thermoregulation [98]. It can be seen that the highest ΔT values are obtained at 40 MHz. It is also interesting to note that when thermoregulation is considered, the increase in blood temperature is practically zero, while the maximum temperature rise in the body can reach 0.72°C. This happens since blood temperature is very close to the body core temperature, which thermoregulation tends to keep as constant, while the maximum temperature increments are usually found in the superficial tissue layers.

It is worth noting that at 40 MHz, the maximum temperature increase (i.e., 0.72°C in the presence of thermoregulation) is found in the muscle tissues of the ankle. The threshold for the induction of thermal damage in muscle is about 45°C [216], which corresponds to a temperature increase of about 8°C. This temperature differential coincides with the safety factor of 10 promulgated by ICNIRP [100] for occupational exposure in limiting the power density from inducing a thermal effect. Moreover, a safety factor of 50 can be deduced from the temperature increase obtained at 10 MHz and in the frequency range between 100 and 900 MHz, where the maximum temperature increases are less than 0.16°C.

Figure 10.28 shows the distribution of the maximum temperature increase for each horizontal layer of the body at 40 MHz (a) and 900 MHz (b), for an incident power density of 1 mW/cm.^{2} At 40 MHz, the maximum temperature increases are obtained at the level

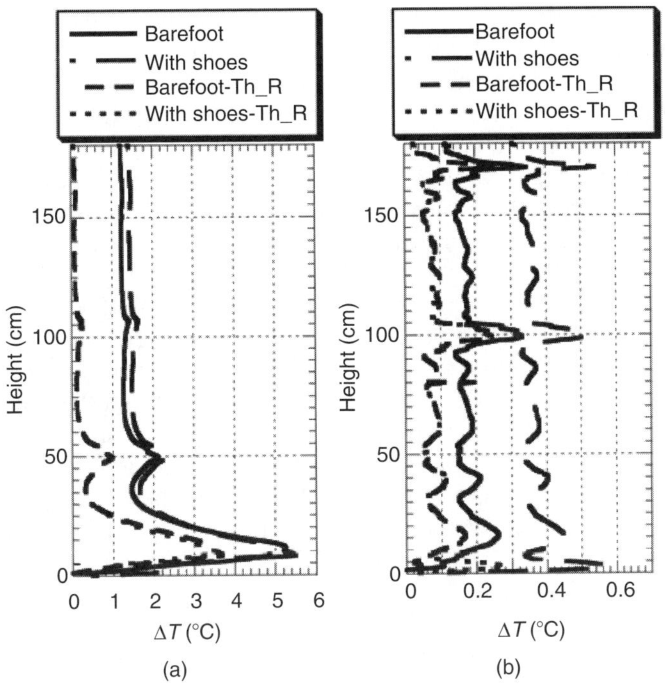

FIGURE 10.28
Layer peak temperature increase (ΔT) in the absence or presence of thermoregulation (Th_R) for a grounded subject barefoot or with shoes (P_{inc} = 1 mW/cm^2). (a) f = 40 MHz and (b) f = 900 MHz.

of the ankles, where the maximum local SARs also are located. The result showed that the presence of the shoes could influence temperature change. Since shoes are isolators from the thermal point of view, they prevent heat flow from going to the ground through them and provoking higher heating within the body. The only exception is in the ankle section, where heating was higher in the barefoot model because of the higher SAR values, at 40 MHz.

Also, the presence of the thermoregulatory mechanisms reduced the temperature increase almost uniformly along the body. Figure 10.28 illustrates clearly the role of blood convection in heat exchanges. In fact, even without active thermoregulation, heat could be spread from the point where its deposition is relatively high (e.g., the ankle at 40 MHz) to the rest of the body, by an increase in the temperature of the circulating blood. Therefore, even if power absorption is limited to one body region, temperature elevations may occur throughout the body. This is clearly visible at 40 MHz, where, although power absorption is mainly confined to the ankle region, in the absence of thermoregulation significant temperature elevations (about 1.4°C, P_{inc} = 1.0 mW/cm^2—see Figure 10.28a) can be observed in the brain. The corresponding elevation in blood temperature elevation is about 1.46°C. It is interesting to note that at 900 MHz, in the absence of active thermoregulation, the brain temperature elevation (0.5°C, P_{inc} = 1.0 mW/cm^2—see Figure 10.28b) is higher than blood temperature elevation (0.35°C), as a result of power deposition inside the brain.

10.5.5.2 Temperature Increments in the Head of a Cellular Telephone User

One consequence of RF power absorption in the human head exposed to the field emitted by a cellular phone is temperature increase in the head. In practical situations, in addition to RF power deposition, there are two other causes for temperature increase. The first one is the contact between the phone case and the user's head (ear and cheek, in particular), which blocks the convective heat exchange between the skin layers and the air. It causes the temperature to rise in tissues around the contact zone. Obviously, this heating is independent of the radiated power, and indeed it was observed also for a wired phone. The second cause for temperature increase is the heating of the phone itself, because of the power dissipated in the internal circuitry, especially the power amplifier. This heating is transferred to the head tissues via thermal conduction.

Studies on temperature rises in the human head, associated with the field emitted by the cell phone, are usually conducted by using anatomically based head models, a numerical solution of the electromagnetic problem, and finite difference formulations of the BHE [161,164,172,187,220–223]. For example, the dissipation in the power amplifier was simulated by adding a power deposition of 250 mW at 900 MHz and 125 mW at 1800 MHz, uniformly distributed inside the upper part of the phone with a 50% efficiency [161]. The heating effects due to SAR, phone contact, and power dissipation in the amplifier were considered separately and were subsequently added together in order to obtain the temperature elevation. The maximum temperature elevations obtained in the ear and in the brain of a user's head are given in Table 10.12 for 900 and 1800 MHz. The ambient air temperature was assumed to be 24°C and a time interval of 15 min, approximating the duration of a long phone call. Although a steady state would not have been reached, the temperature elevation after 15 min is expected to be very close to the steady-state value [164].

The data shown in Table 10.12 reveal some interesting aspects. First, the temperature elevation induced inside the brain by SAR alone was less than 0.1°C, especially when the phone was kept in the "cheek" position. This configuration resulted in a marked reduction in power deposition inside the brain. Table 10.12 also shows that the mere contact of the cellular phone with the ear and cheek, even in a standby mode, in which no RF power is radiated, can cause a temperature elevation in the ear reaching as high as 1.5°C. This is due to the highly insulating properties assumed for the phone plastic shell. A negligible maximum temperature elevation of about 0.01°C, instead, is obtained in the external brain region. If the contribution from power dissipation in the amplifier is included, the induced temperature elevations are not significantly altered. It must be noted, however, that this result arises from consideration of power dissipation in the power amplifier alone. In the real situation, additional power dissipation is present in the internal circuitry of the handset, besides the power amplifier, and hence slightly higher temperature elevations may be observed [221].

TABLE 10.12

Temperature Elevations Induced in the User's Head, after 15 min, by a Phone Equipped with a Dual-Band Monopole-Helix Antenna (Average Radiated Power: 250 mW at 900 MHz and 125 mW at 1800 MHz)

Frequency (MHz)	Position	Heating Cause	ΔT_{max} (°C)	$\Delta T_{maxbrain}$ (°C)
900	Vertical	SAR	0.221	0.061
	Cheek	SAR	0.136	0.023[a]
		Contact	1.543	0.012[b]
		Contact + power dissipation	1.544	0.012[b]
		Contact + power dissipation + SAR	1.581	0.023[a]
1800	Vertical	SAR	0.155	0.036
	Cheek	SAR	0.085	0.011[a]
		Contact	1.543	0.012[b]
		Contact + power dissipation	1.543	0.012[b]
		Contact + power dissipation	1.549	0.012[b]

[a] $\Delta T_{maxbrain}$ located in the upper external brain region.
[b] $\Delta T_{maxBRAIN}$ located in the lower external brain region.

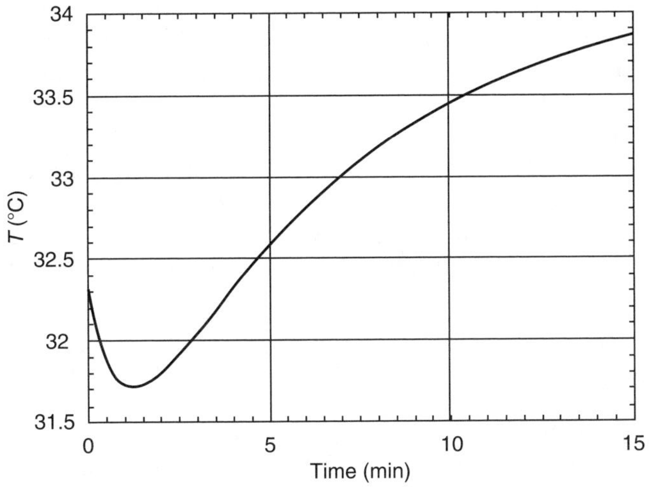

FIGURE 10.29
Time evolution of the temperature at a point on the ear in direct contact with a nonradiating phone.

The effect of the phone contact on the temperature evolution in the ear region is shown in Figure 10.29. It can be seen that when the phone was put in contact with the ear, the ear temperature experienced a quick decrease. This is because the phone was initially at ambient temperature (24°C), which was lower than the ear temperature. However, soon afterward, the heat supplied by the blood and conducted from the neighboring tissues stopped the decrease. Indeed, the temperature started to elevate, going beyond the initial value, because of the suppressed convective exchange with air.

When all heating sources are considered simultaneously, the results indicate that the maximum temperature elevation in the ear was almost entirely due to the contact effect, with only a very slight contribution due to SAR deposition in the ear region. However, the situation was completely different in the brain region. In fact, the heating effects due to SAR and phone contact tend to heat different parts of the brain. The contact-effect heating occurred in the lower external brain region, and the SAR-induced heating appeared mostly in the upper external brain region. Therefore, these two heating effects are not additive in the brain, and when they are simultaneously present, as opposed to the case when only one is considered, the result is that the portion of the brain affected by heating becomes larger. The peak temperature increase in the brain is therefore governed by the more significant of the two heating causes. At 900 MHz, SAR is more dominant, while at 1800 MHz, because of the lower radiated power, the two heating effects are comparable. Note that the ANSI/IEEE safety guidelines, which restrict the 1-g averaged spatial peak SAR to 1.6 W/kg, are associated with maximum temperature rises in the brain between 0.03°C and 0.09°C. These values are about 50 times lower than the threshold for thermal damage. The ICNIRP safety guideline of a 10-g averaged spatial peak SAR equal to 2 W/kg, results in maximum temperature rises in the brain between 0.1°C and 0.2°C, which is about 25 times lower than the threshold value.

10.6 Thermal Therapeutic Applications of Microwave Energy

Recent advances in technological development and in the understanding of EMF interaction with biological materials and systems have launched a broad range of biomedical applications. The following is a brief overview of computational techniques involved in the development of ablative and hyperthermia therapies, including microwave cardiac ablation, endometrial ablation, and hyperthermia treatment of cancer. All of these therapeutic applications involve the use of microwave technologies to produce localized depositions of electromagnetic energy and, as a result, a rise of the local temperature in biological tissues. The target tissue temperature for the ablative treatments is about 65°C [224,225], whereas for the hyperthermia treatments, it is about 43°C [226]. The ablation treatments include the treatment for cardiac arrhythmias and for endometrial disorders such as menorrhagia. In all cases, the distribution of temperature increase must be carefully controlled in order to avoid excessive temperature increase in the surrounding healthy tissue. (See also Chapter 12 by Chou [284].)

10.6.1 Ablation for Cardiac Arrhythmias

Cardiac arrhythmias are mainly due to the presence of abnormal electrical sources or current paths in the cardiac tissues and are usually treated by destroying the tissue substrate where the electrical anomaly is localized [227–230]. This was performed by surgical resection initially, and later on by delivering a high-intensity DC shock with a defibrillator [227,230]. However, the high voltages and currents associated with the DC shock resulted in uncontrollable cardiac damages and severe complications. Subsequently, the use of RF EMFs to heat and destroy the arrhythmic tissue was proposed [230], and the advantages of microwave technology were demonstrated [228,229].

The RF ablation of cardiac tissue is usually performed using a unipolar arrangement, in which a catheter electrode placed into the heart delivers RF current to a large dispersive ground electrode on the skin surface of the patient. The RF current that flows from the catheter to the ground electrode following radial paths generates heat. While flowing from the catheter, the RF current density decreases as the second power of distance; consequently, the heat generation decreases according to the fourth-power law with the distance from the electrode. With this arrangement, the lesion (ablated tissue) is restricted to a small region (usually about 2 to 3 mm in diameter) contiguous to the catheter electrode, and the lesion dimensions cannot be extended by increasing the delivered power [225,227,229–231]. In microwave ablation, a catheter antenna is inserted into the cardiac chamber and is used to deliver MW power [225,228]. Since MW power deposition inside tissue decays with distance by following a second-power law, deeper lesions can be obtained compared to RF ablation [229,232–234].

In cardiac ablation treatments the lesion is usually defined as the region where temperature exceeds 65°C, which represents the threshold for irreversible damage of this tissue. This region must be large enough to ablate all the abnormal tissue responsible for the arrhythmia. Catheter antennas used for MW ablation have included monopoles [235,236], dipoles [229,237,238], and helices [232,239]. The design of a MW catheter antenna is a crucial point in MW ablation treatments and requires a complete study of the performance of the antenna imbedded in blood and in contact with the cardiac tissue. Performance parameters usually analyzed include the radiation pattern in the tissue; antenna impedance, which governs the bandwidth of the antenna; surface current suppression, in order to prevent heating of tissue along the catheter feeding cable; and dimensions of the induced lesion [240,241].

The design and performance evaluation of the microwave ablation antennas can be conducted by using computational tools and tested through *in vitro* or *in vivo* experimental studies [229,233,238,242]. Numerical tools allow the analysis of antenna performances under different operating conditions, and the evaluation of both electromagnetic power deposition and temperature increase in the tissue. In particular, the numerical evaluation of temperature increase induced inside the cardiac tissue allows the evaluation of blood flow influence on the induced lesion. This evaluation is of fundamental importance since the

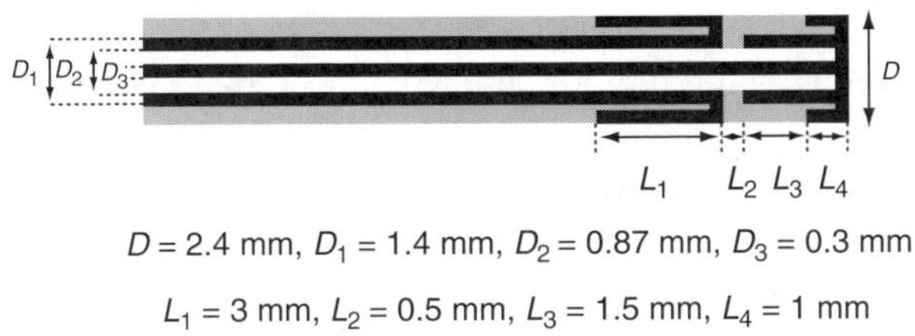

FIGURE 10.30
The cap-choke microwave catheter antenna.

presence of high blood flow rates inside the heart chambers provides a very effective heat removal process at the heart tissue surface so that the blood flow strongly limits the efficiency of both RF and MW ablation systems. Several numerical studies of SAR and temperature increase have been performed with the FEM [236,243–245] as well as the FDTD [246].

Figure 10.30 shows a cap-choke antenna mounted on an RG/178BU flexible coaxial cable operating at 2450 MHz [238]. The structure was realized by connecting an annular cap to the inner conductor of the cable, and a cylindrical coaxial choke to the outer conductor. A junction, filled with high-temperature epoxy resin, was used to separate the cap from the choke. Figure 10.31 shows the same antenna inserted in a two-layer cylindrical model of the heart. The heart model was used to study the influence of antenna position (touching or pressed into the cardiac muscle) and of blood flow (high or low) on the induced lesion. As an example, some computed results are presented in Table 10.13 for the amount of power to be delivered to the cardiac tissue to obtain lesions of different depths, under four operating conditions for an exposure duration of 60 sec [246].

10.6.2 Ablation for Endometrial Disorders

While still in the beginning stage, microwave and RF thermal ablation of the endometrium have been suggested recently as efficient and cost-effective alternatives to surgical resection as a treatment for dysfunctional uterine bleeding [247–251]. Techniques presently in

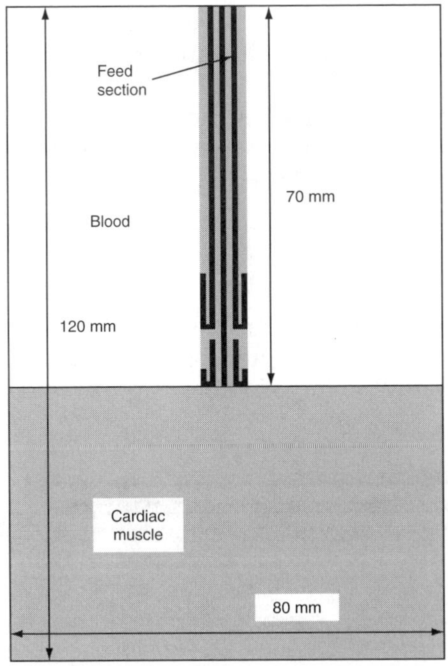

FIGURE 10.31
Cross section of the two-layer numerical model. The antenna's radial dimensions are not in scale.

TABLE 10.13

Microwave Power Required to Produce a Given Lesion Depth for Different Operating Conditions (Exposure Duration: 60 sec)

	Power (W)			
Lesion Depth (mm)	Touch Low Perfusion	Touch High Perfusion	Pressed Low Perfusion	Pressed High Perfusion
1	2.5	4.0	2.0	2.5
3	6.2	10	5.0	6.2
5	16	23	13	16
7	44	58	38	44

use include electrosurgery resection and Nd:YAG laser coagulation under direct hysteroscopic visualization [247,248,252–256]. The reported rates of complete amenorrhea are between 25% and 70% [253,256]. The variability arises principally from the unpredictable nature of induced thermal injury and perforation of the uterine wall [253]. Quantitative measures and patients' subjective reports of microwave endometrial ablation suggest a higher rate of satisfaction, acceptability, and life quality improvement [257,258].

In endometrial ablation, as in cardiac ablation, the design of the catheter antenna should be carefully tested to ensure the desired temperature increase (up to an increase of about 40°C in this case) in the tissue of interest. A sleeved-slot antenna designed for endometrial ablation operation at 915 MHz is depicted in Figure 10.32. The antenna configuration is similar to that of the catheter antenna for cardiac ablation, and is mounted on an RG8U coaxial cable. Figure 10.33 shows a comparison of the axial SAR distribution obtained from an FDTD analysis and from temperature measurements made in a cylindrical phantom filled with a muscle-equivalent material [259]. Two radial distances from the antenna axis, 2.5 and 7.5 mm, are shown. The SAR data obtained experimentally were normalized to the ones obtained from the FDTD simulation with 1.0 W of radiated power.

Figure 10.34 presents the lesion depth, defined as the region around the antenna where the temperature increase was greater or equal to 38°C (final temperature 75°C), as a function of the time of the exposure at different radiated powers. The curves were obtained by considering an antenna inserted in a cylindrical phantom filled with muscle-equivalent material. They simulate an antenna touching the endometrial tissue to be ablated. The lesion dimensions were evaluated as the depth from the antenna outer conductor. These curves can serve as guides in clinical trials where the antenna feeding power and time of exposure are key parameters to be chosen for a desired lesion dimension.

10.6.3 Microwave Interstitial Hyperthermia for Cancer Treatment

Hyperthermia cancer therapy is a treatment procedure in which tumor temperatures are elevated to above 43°C [226]. Investigations performed on cell cultures, tumor-bearing animals, and human patients have clearly shown that hyperthermia affords preferential

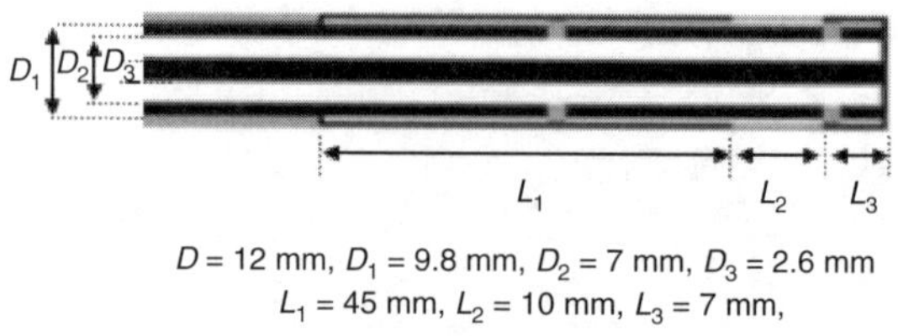

FIGURE 10.32
A microwave catheter antenna for endometrial ablation.

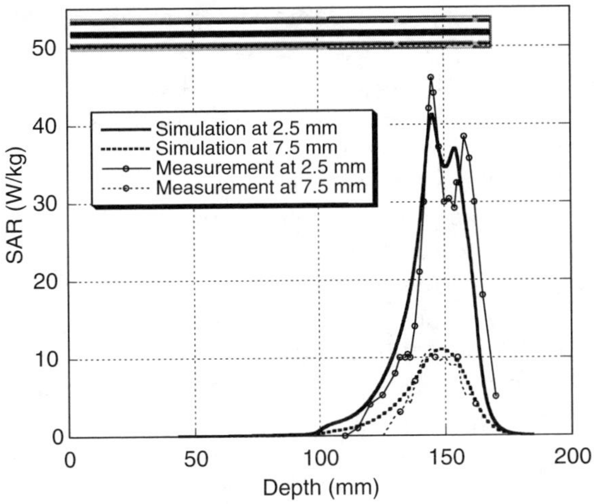

FIGURE 10.33
Comparison of axial SAR distributions obtained from FDTD and from temperature measurements made in tissue phantoms for the endometrial sleeved-slot antenna. The data are shown for radial distances of 2.5 and 7.5 mm.

killing of malignant cells, enhances the cytotoxic effects of many anticancer drugs, and potentiates the cell-killing ability of ionizing irradiation [260–266].

An important aspect of the development of applicators for microwave hyperthermia is the production of required temperature distributions in superficial and deep-seated tumors, that is, to produce temperatures in excess of 43°C in the tumor tissue, in order to guarantee the destruction of the malignant cells, while maintaining the temperature in the healthy tissue below 42°C to avoid thermal damage. In microwave hyperthermia, the antenna must provide efficient power delivery, good impedance matching at the frequency of operation, and a uniform SAR distribution in the tumor region. However, if the region to be treated is large compared to the field penetration depth, the required SAR uniformity cannot be achieved with a single microwave antenna, and the use of antenna

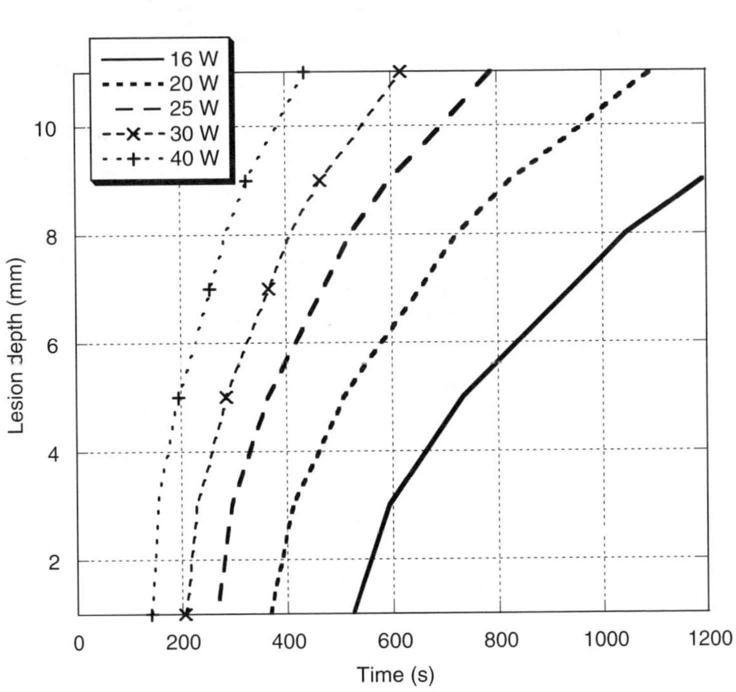

FIGURE 10.34
Endometrial lesion depth as a function of exposure duration for different radiated powers.

arrays becomes necessary. The following paragraphs present some computational results for thin catheter antennas designed for interstitial hyperthermia treatment of brain tumors.

Many investigations have been conducted on electromagnetic power deposition with regard to hyperthermia treatment for cancer. Most earlier studies were conducted analytically, modeling the catheter dipole antenna and surrounding tissue with a lossy transmission line [267–270]. Later, numerical and experimental studies were conducted on several antenna types, with the aim of optimizing antenna performances [270–274]. Considerable attention was devoted to arrays of antennas that have been studied both experimentally [275–277], by measuring the rate of temperature change caused by the radiated power, and theoretically, using antenna theory approaches [275,276,278] or approximate numerical procedures [277,279]. Indeed, antenna arrays have proved to be more suitable for increasing the region of uniform SAR deposition. More recent studies calculated, besides the electromagnetic power deposition, also the corresponding temperature increase. In fact, thermal analysis is a fundamental step in evaluating the effectiveness of the microwave applicator since it allows the assessment of the region where the temperature is above the threshold, and estimation of the required input power. Temperature distributions produced by interstitial antennas in tumor tissues have been experimentally evaluated by using microwave radiometry [277] and computed by using finite difference explicit solutions of the BHE [280,281]. A numerical solution of the BHE allows the analysis of the influence of different parameters such as blood flow on the temperature increase [280] and the development of clinical protocols [281].

Recently, an ADI solution of the BHE has been developed to study a 3-D array of catheter antennas inserted into a brain-equivalent phantom [282]. The antenna considered (Figure 10.35) was a sleeved-slot antenna mounted on a UT-34 coaxial cable [259,283]. An equilateral triangular array of the sleeved-slot antennas was studied in a phantom of brain-equivalent tissue in order to assess its capability in heating tumor regions of various dimensions [282]. Figure 10.36a shows the SAR distribution for 15-mm spacing among the antennas in the array; while in Figure 10.36b, contour plots at $\Delta T = 6°C$ for various radiated powers and for a 15-mm spacing are presented. The graphs shown in Figure 10.36a and b are for a horizontal plane passing through the antenna slot [282].

Results such as those in Figure 10.36 can provide guidance to clinical protocols regarding the input power and optimal geometry (array spacing) for achieving an efficient tumor heating. For example, there is a minimum input power below which three separated regions around the three interstitial antennas constitute the region with temperatures above 43°C. Once this power is exceeded, a simply connected domain is obtained.

It is interesting to note that blood flow is remarkably low in the necrotic core of tumor tissues compared to normal tissue. Moreover, the rate of blood flow in normal tissue increases during hyperthermia. These phenomena may be investigated by varying the blood perfusion parameter in the BHE. Computer simulations, performed by varying the blood perfusion parameter from 5% to 200% of the nominal value, showed that the heat removal mechanism becomes less efficient as the blood flow is decreased, and the effective region of hyperthermia therapy is enlarged [282]. Consequently, the reduced blood flow in the necrotic tumor core and the increase in blood flow in normal tissue can serve to facilitate hyperthermia therapy.

10.7 Concluding Remarks

Knowledge of internal electric and magnetic fields, induced current densities, and SARs is fundamental in the study of biological responses, health effects, and medical applications

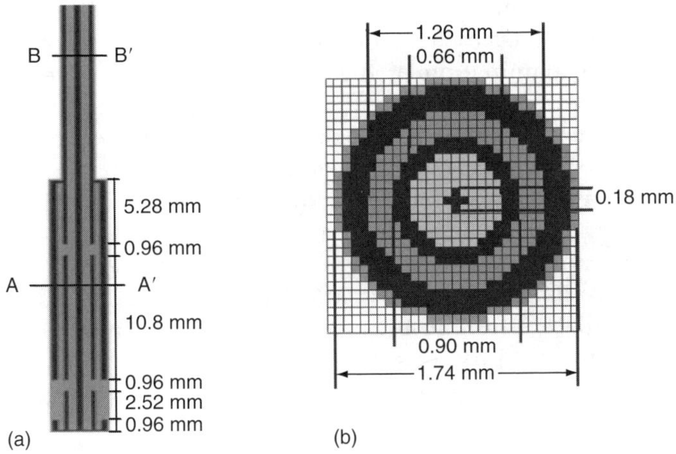

FIGURE 10.35
Longitudinal (a) and horizontal AA′ (b) sections of the sleeved-slot antenna.

of EMFs. Complexities of biological tissues and of the incident fields make closed-form analytical solutions impractical, and computer methods are needed to predict the internal fields and their distributions. Great strides have been made during the past decade in the area of numerical dosimetry using anatomically based models of the human body. Among the most valuable of the numerical methods for predicting field intensities and SAR calculations are the impedance method for use at lower frequencies, where quasi-static approximations may be made ($<\sim 40$ MHz for the human body), and the FEM and FDTD methods, which may be used at any frequency of interest. For numerical calculations, the FDTD method requires a computer memory and computation time that is proportional to N. This is a considerable advantage over the computing methods, such as MoM or MoM–FFT. This chapter described the salient features of these methods and the many bioelectromagnetic exposure conditions for which they have been applied. Because of the limitations on the length of the chapter, only a few of the important recent applications of some of these methods were presented in some detail.

Note that numerical methods have matured to a level that they are being increasingly used by researchers in many laboratories for dosimetric calculations for important and

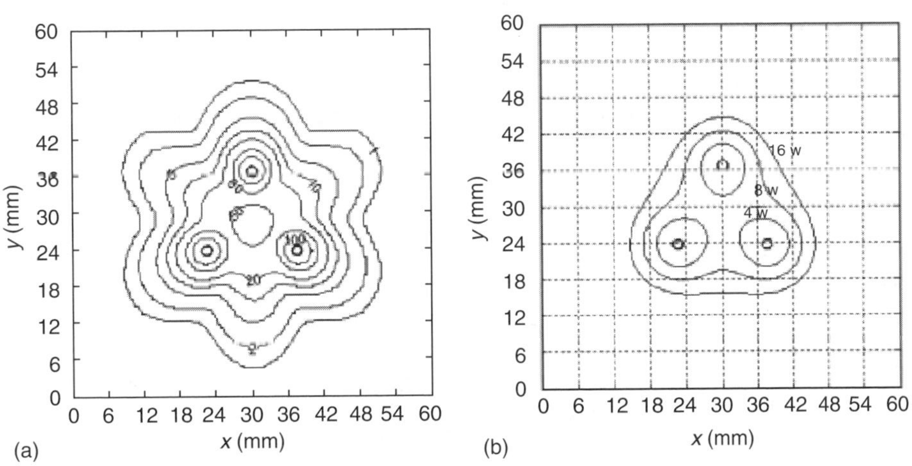

FIGURE 10.36
SAR distribution (a) and contour plots at $\Delta T = 6°C$ for various radiated powers (b), on a horizontal plane passing through the antenna slot, for 15-mm spacing among the antennas in the array.

meaningful bioelectromagnetic problems. Some of the future developments will involve improving the efficiencies of the various codes by techniques such as use of the expanding grid rather than the regular grid, elimination of the relatively shielded interior regions of the modeled space at higher frequencies, and use of truncated models of the body at microwave frequencies where there is a lack of coupling between the various regions of the body. Because of accurate modeling of the tissue heterogeneities and shapes, these models have and will continue to play an important role in emerging technologies with bioelectromagnetic concerns. Some of the likely applications are personal wireless and mobile communications systems, automotive devices such as electric automobiles and collision avoidance systems, medical devices such as MRIs, implantable pacemakers and defibrillators, applicators for hyperthermia, and other minimally invasive therapeutic and surgical procedures.

Acknowledgment

We wish to thank Stefano Pisa, Marta Cavagnaro, Emanuele Piuzzi, and Zhangwei Wang for their valuable assistance in preparing the manuscript.

References

1. Lin, J.C., Computer methods for field intensity predictions, in *CRC Handbook of Biological Effects of Electromagnetic Fields*, Polk, C., and Postow, E., Eds., CRC Press, Boca Raton, FL, 1986, 273–313.
2. Lin, J.C. and Gandhi, O.P., Computer methods for predicting field intensity, in *Handbook of Biological Effects of Electromagnetic Fields*, Polk, C., and Postow, E., Eds., CRC Press, Boca Raton, FL, 1996, 337–402.
3. Lin, J.C., Mechanisms of field coupling into biological systems at ELF and RF frequencies, in *Advances in Electromagnetic Fields in Living Systems*, vol. 3, Kluwer/Plenum, New York, 2000, 1–38.
4. Schwan, H.P. and Piersol, G.M., Absorption of electromagnetic energy in body tissues, *Am. J. Phys. Med.*, 33, 371, 1954.
5. Schwan, H.P. and Li, K., Hazards due to total body irradiation by radar, *Proc. IRE*, 44, 1572, 1956.
6. Schwan, H.P. and Li, K., The mechanism of absorption of ultrahigh frequency electromagnetic energyin tissue as related to the problem of tolerance dosage, *IRE Trans. Med. Electron.*, 4, 45, 1956.
7. Johnson, C.C. and Guy, A.W., Nonionizing electromagnetic wave effects in biological materials and systems, *Proc. IEEE*, 60, 692, 1972.
8. Michaelson, S.M. and Lin, J.C., *Biological Effects and Health Implications of Radiofrequency Radiation*, Plenum Press, New York, 1987.
9. Shapiro, A.R., Lutomirski, R.F., and Yura, H.T., Induced fields and heating within a cranial structure irradiated by an electromagnetic plane wave, *IEEE Trans. Microwave Theory Technol.*, 19, 187, 1971.
10. Kritikos, H.N. and Schwan, H.P., Hot spot generated in conduction spheres by EM waves and biological implications, *IEEE Trans. Biomed. Eng.*, 19, 53, 1972.
11. Lin, J.C., Guy, A.W., and Kraft, G.H., Microwave selective brain heating, *J. Microwave Power*, 8, 275, 1973.
12. Ho, H.S. and Guy, A.W., Development of dosimetry for RF and microwave radiation, *Health Phys.*, 29, 317, 1975.

13. Weil, C.M., Absorption characteristics of multi-layered sphere models exposed to UHF/microwave radiation, *IEEE Trans. Biomed. Eng.*, 22, 468, 1975.
14. Joines, W.T. and Spiegel, R.J., Resonance absorption of microwaves by the human skull, *IEEE Trans. Biomed. Eng.*, 21, 46, 1975.
15. Lin, J.C., Interaction of two cross-polarized electromagnetic waves with mammalian cranial structures, *IEEE Trans. Biomed. Eng.*, 23, 371, 1976.
16. Lin, J.C., Guy, A.W., and Johnson, C.C., Power deposition in a spherical model of man exposed to 1–20 MHz electromagnetic fields, *IEEE Trans. Microwave Theory Technol.*, 21, 791, 1973 (see also corrections *IEEE MTT*, 23, 265, 1975).
17. Kritikos, H.N. and Schwan, H.P., The distribution of heating potential inside lossy spheres, *IEEE Trans. Biomed. Eng.*, 22, 457, 1975.
18. Johnson, C.C., Durney, C.H., and Massoudi, H., Long-wavelength electromagnetic power absorption in prolate spheroidal models of man and animals, *IEEE Trans. Microwave Theory Technol.*, 23, 739, 1975.
19. Durney, C.H., Johnson, C.C., and Massaudi, A., Long wave-length analysis of plane wave irradiation of a prolate spheroidal model of man, *IEEE Trans. Microwave Theory Technol.*, 23, 246, 1975.
20. Massoudi, H., Durney, C.H., and Johnson, C.C., Long wavelength electromagnetic power absorption in ellipsoidal models of man and animals, *IEEE Trans. Microwave Theory Technol.*, 24, 41, 1977.
21. Rowlandson, G.I. and Barber, P.W., Absorption of high frequency RF energy by biological models: calculations based on geometrical optics, *Radio Sci.*, 14, 43S, 1979.
22. Lin, J.C., Whole-body exposure in the near zone of HF electromagnetic fields, *URSI International Electromagnetic Waves and Biology Symposium*, Jouy en Josas, France, 1980.
23. Guy, A.W., Webb, M.D., and Sorenson, C.C., Determination of power absorption in man exposed to HF electromagnetic fields by thermographic measurements on scale models, *IEEE Trans. Biomed. Eng.*, 23, 361, 1976.
24. Liversy, D.E. and Chen, K.M., Electromagnetic fields induced inside arbitrary shaped biological bodies, *IEEE Trans. Microwave Theory Technol.*, 22, 1273, 1974.
25. Guru, B.S. and Chen, K.M., Experimental and theoretical studies in electromagnetic field induced inside finite biological bodies, *IEEE Trans. Microwave Theory Technol.*, 24, 433, 1976.
26. Chen, K.M. and Guru, B.S., Internal EM field and absorbed power density in human torsos induced by 1–500 MHz EM waves, *IEEE Trans. Microwave Theory Technol.*, 25, 746. 1977.
27. Hagman, M.J., Gandhi, O.P., and Durney, C.H., Numerical calculation of electromagnetic energy deposition for a realistic model of man, *IEEE Trans. Microwave Theory Technol.*, 27, 804, 1979.
28. Gandhi, O.P., Electromagnetic absorption in inhomogeneous model of man for realistic exposure conditions, *Bioelectromagnetics*, 3, 81, 1982.
29. Armitage, D.W., Leveen, H.H., and Pethig, R., Radio-frequency induced hyperthermia: computer simulation of specific absorption rate distributions using realistic anatomical models, *Phys. Med. Biol.*, 28, 31, 1983.
30. Gandhi, O.P., DeFord, J.F., and Kanai, H., Impedance method for calculation of power deposition patterns in magnetically induced hyperthermia, *IEEE Trans. Biomed. Eng.*, 31, 644, 1984.
31. Gandhi, O.P. and DeFord, J.F., Calculation of EM power deposition for operator exposure to RF induction heaters, *IEEE Trans. Electromagn. Compat.*, 30, 63, 1988.
32. Orcutt, N. and Gandhi, O.P., A 3-D impedance method to calculate power deposition in biological bodies subjected to time-varying magnetic fields, *IEEE Trans. Biomed. Eng.*, 35, 577, 1988.
33. Orcutt, N. and Gandhi, O.P., Use of the impedance method to calculate 3-D power deposition patterns for hyperthermia with capacitive plate electrodes, *IEEE Trans. Biomed. Eng.*, 37, 36, 1990.
34. Epstein, B.R. and Foster, K.R., Anisotropy in dielectric properties of skeletal muscle, *Med. Biol. Eng. Comput.*, 21, 51, 1983.
35. Zheng, E., Shao, S., and Webster, J.G., Impedance of skeletal muscle from 1 Hz to 1 MHz, *IEEE Trans. Biomed. Eng.*, 31, 477, 1984.

36. Zhu, X.L. and Gandhi, O.P., Design of RF needle applicators for optimum SAR distributions in irregularly-shaped tumors, *IEEE Trans. Biomed. Eng.*, 35, 382, 1988.
37. Gandhi, O.P. and Chen, J.Y., Numerical dosimetry at power-line frequencies using anatomically based models, *Bioelectromagnetics*, Supplement 1, 43, 1992.
38. Tofani, S., Ossola, P., d'Amore, G., and Gandhi, O.P., Electric fields and current density distributions induced in an anatomically based model of the human head by magnetic fields from a hair dryer, *Health Phys.*, 68, 71–79, 1995.
39. Wu, D., Ren, J., and Gandhi, O.P., Currents induced in the arm and the body by the magnetic fields of an electric hand drill, *Abstracts of the Fifteenth Annual Meeting of the Bioelectromagnetics Society*, Los Angeles, CA, 1993, 31.
40. Ren, J., Gandhi, O.P., Casamento, J.P., and Bassen, H.I., Currents induced in the human body by electric blankets, *Abstracts of the 15th Annual Meeting of the Bioelectromagnetics Society*, Los Angeles, CA, 1993, 32.
41. Harrington, R.F., *Field Computation by Moment Methods*, McGraw-Hill, New York, 1968.
42. Schelkunoff, S.A., Field equivalence theorems, *Comm. Pure Appl. Math.*, 4, 43, 1951.
43. Lin, J.C. and Wu, C.L., Scattering of microwaves by dielectric materials used in laboratory animal restrainers, *IEEE Trans. Microwave Theory Technol.*, 24, 219, 1976.
44. Gandhi, O.P., Hagman, M.J., and D'Andrea, J.A., Part-body and multi-body effects on absorption of radio frequency electromagnetic energy by animals and by models of man, *Radio Sci.*, 14, 155, 1979.
45. Karimullah, K., Chen, K.M., and Nyquist, D.P., Electromagnetic coupling between a thin-wire antenna and a neighboring biological body, *IEEE Trans. Microwave Theory Technol.*, 28, 1218, 1980.
46. Chatterjee, I., Hagman, M.J., and Gandhi, O.P., Electromagnetic energy deposition in an inhomogeneous block model for near-field irradiation conditions, *IEEE Trans. Microwave Theory Technol.*, 28, 1452, 1980.
47. Wu, T.K. and Tsai, L.L., Electromagnetic fields induced inside arbitrary cylinders of biological tissue, *IEEE Trans. Microwave Theory Technol.*, 25, 61, 1977.
48. Wu, T.K. and Tsai, L.L., Scattering from arbitrary-shaped lossy dielectric bodies of revolution, *Radio Sci.*, 12, 709, 1977.
49. Poggio, A.J. and Miller, E.K., Integral equation solutions of three-dimensional scattering problems, in *Computer Techniques for Electromagnetics*, Mittra, R., Ed., Pergamon Press, Elmsford, New York, 1973, 159.
50. Massoudi, H., Durney, C.H., Barber, P.W., and Iskander, M.F., Post resonance EM absorption by man and animals, *Bioelectromagnetics*, 3, 333, 1982.
51. Wu, T.K., Electromagnetic fields and power deposition in body of revolution models of man, *IEEE Trans. Microwave Theory Technol.*, 27, 279, 1979.
52. Mautz, J.R. and Harrington, R.F., Radiation and scattering from bodies of revolution, *Appl. Sci. Res.*, 20, 405, 1969.
53. Harrington, R.F. and Mautz, J.R., Green's functions for surfaces of revolution, *Radio Sci.*, 7, 603, 1972.
54. Pogorzelski, R.J. and Wu, T.K., Computations of scattering from inhomogeneous penetrable elliptic cylinders by means of invariant imbedding, in *URSI Symposium of Electromagnetic Wave Theory*, Stanford, CA, 1977, 323.
55. Jin, J., *The Finite Element Method in Electromagnetics*, Wiley, New York, 1993.
56. Boyes, W.E., Lynch, D.R., Paulsen, K.D., and Minerbo, G.N., Nodal-based finite element modelling of Maxwell's equations in three dimensions, *IEEE Trans. Antennas Propagat.*, 40, 642, 1992.
57. Paulsen, K.D., Jia, X., and Sullivan, J.M., Jr., Finite element computations of specific absorption rates in anatomically conforming full-body models for hyperthermia treatment analysis, *IEEE Trans. Biomed. Eng.*, 40, 933, 1993.
58. Yee, K.S., Numerical solutions of initial boundary value problems involving Maxwell's equations in isotropic media, *IEEE Trans. Antennas Propagat.*, 14, 303, 1966.
59. Taflove, A. and Brodwin, M.E., Computation of the electromagnetic fields and induced temperatures within a model of the microwave-irradiated human eye, *IEEE Trans. Microwave Theory Technol.*, 23, 888, 1975.

60. Taflove, A. and Brodwin, M.E., Numerical solution of steady-state EM scattering problems using the time dependent Maxwell's equation, *IEEE Trans. Microwave Theory Technol.*, 23, 623, 1975.
61. Taflove, A., Application of the finite-difference time domain method to sinusoidal steady-state electromagnetic-penetration problems, *IEEE Trans. Electromagn. Compat.*, 22, 191, 1980.
62. Holland, R., THREDE: a free field EMP coupling and scattering code, *IEEE Trans. Nuclear Sci.*, 24, 2416, 1977.
63. Kunz, K.S. and Lee, K.M., A three-dimensional finite-difference solution of the external response of an aircraft to a complex transient EM environment: part 1—the method and its implementation, *IEEE Trans. Electromagn. Compat.*, 20, 328, 1978.
64. Kunz, K.S. and Luebbers, R.J., *The Finite-Difference Time-Domain Method for Electromagnetics*, CRC Press, Boca Raton, FL, 1993.
65. Taflove, A., *Computational Electrodynamics: The Finite-Difference Time-Domain Method*, Artech House, London, 1995.
66. Taflove, A. and Hagness, S.C., Eds., *Computational Electrodynamics: The Finite-Difference Time-Domain Method*, Artech House, London, 2000.
67. Spiegel, R.J., Fatmi, M.B.A., Stuchly, S.S., and Stuchly, M.A., Comparison of finite-difference time-domain SAR calculations with measurements, in a heterogeneous model of man, *IEEE Trans. Biomed. Eng.*, 36, 849, 1989.
68. Chen, J.Y. and Gandhi, O.P., Currents induced in an anatomically based model of a human for exposure to vertically polarized electromagnetic pulses, *IEEE Trans. Microwave Theory Technol.*, 39, 31, 1991.
69. Sullivan, D.M., Borup, D.T., and Gandhi, O.P., Use of the finite-difference time-domain method in calculating EM absorption in human tissues, *IEEE Trans. Biomed. Eng.*, 34, 148, 1987.
70. Chen, J.Y. and Gandhi, O.P., Numerical simulation of annular-phased arrays of dipoles for hyperthermia of deep-seated tumors, *IEEE Trans. Biomed. Eng.*, 39, 209, 1992.
71. Chen, J.Y., Gandhi, O.P., and Conover, D.L., SAR and induced current distributions for operator exposure to RF dielectric sealers, *IEEE Trans. Electromagn. Compat.*, 33, 252, 1991.
72. Luebbers, R., Hunsberger, F.P., Kunz, K.S., Standler, R.B., and Schneider, M., A frequency-dependent finite-difference time-domain formulation for dispersive materials, *IEEE Trans. Electromagn. Compat.*, 32, 222, 1990.
73. Bui, M.D., Stuchly, S.S., and Costache, G.I., Propagation of transients in dispersive dielectric media, *IEEE Trans. Microwave Theory Technol.*, 39, 1165, 1991.
74. Sullivan, D.M., A frequency-dependent FDTD method for biological applications, *IEEE Trans. Microwave Theory Technol.*, 40, 532, 1992.
75. Luebbers, R.J., Hunsberger, F., and Kunz, K.S., FDTD for nth order dispersive media, *IEEE Trans. Antennas Propagat.*, 40, 1297, 1992.
76. Joseph, R.M., Hagness, S.C., and Taflove, A., Direct time integration of Maxwell's equations in linear dispersive media with absorption for scattering and propagation of femtosecond electromagnetic impulses, *Optics Lett.*, 16, 1412, 1991.
77. Gandhi, O.P., Gao, B.Q., and Chen, J.Y., A frequency-dependent finite-difference time-domain formulation for induced current calculations in human beings, *Bioelectromagnetics*, 13, 543, 1992.
78. Gandhi, O.P., Gao, B.Q., and Chen, J.Y., A frequency-dependent finite-difference time-domain formulation for general dispersive media, *IEEE Trans. Microwave Theory Technol.*, 41, 658, 1993.
79. Furse, C.M., Chen, J.Y., and Gandhi, O.P., A frequency-dependent finite-difference time-domain method for induced current and SAR calculations for a heterogeneous model of the human body, *IEEE Trans. Electromagn. Compat.*, 36, 128, 1994.
80. Gandhi, O.P. and Chen, J.Y., Electromagnetic absorption in the human head for a proposed 6 GHz mobile communication system, *IEEE Trans. Electromagn. Compat.*, 37, 547, 1995.
81. Gandhi, O.P., Some numerical methods for dosimetry: ELF to microwave frequencies, *Radio Sci.*, 30, 161–177, 1995.
82. Ackerman, M.J., The visible human project, *Proc. IEEE*, 86, 504, 1998.

83. Mason, P.A., Ziriax, J.M., Hurt, W.D., Walters, T.J., Ryan, K.L., Nelson, D.A., Smith, K.I., and D'Andrea, J.A., Recent advancements in dosimetry measurements and modeling, in *Radio Frequency Radiation Dosimetry*, Klauenberg, B.J. and Miklavcic, D., Eds., Kluwer, Newyork, 2000, 141.
84. fttp://starview.brooks. af.mil/EMF/dosimetry_models/
85. Rush, S., Abildskov, J.A., and McFee, R., Resistivity of body tissues at low frequencies, *Circ. Res.*, 12, 40, 1963.
86. Florig, H.K., Hoburg, J.F., and Morgan, M.G., Electric-field exposure from electric blankets, *IEEE Trans. Power Deliv.*, 2, 527, 1987.
87. Hayashi, N., Isaka, K., and Yokoi, Y., Analysis of magnetic-field profiles in electric blanket users, *IEEE Trans. Power Deliv.*, 4, 1897, 1989.
88. Kaune, W.T. and Gillis, M.F., General properties of the interaction between animals and ELF electric fields, *Bioelectromagnetics*, 2, 1, 1981.
89. Guy, A.W., Davidow, S., Yang, G.Y., and Chou, C.K., Determination of electric current distributions in animals and humans exposed to a uniform 60-Hz high-intensity electric field, *Bioelectromagnetics*, 3, 47, 1982.
90. Stratton, J.A., *Electromagnetic Theory*, McGraw-Hill, New York, 1941.
91. Deno, D.W., Currents induced in the human body by high voltage transmission line electric field—measurement and calculation of distribution and dose, *IEEE Trans. Power Apparatus Syst.*, 96, 1517, 1977.
92. DiPlacido, J., Shih, C.H., and Ware, B.J., Analysis of the proximity effects in electric field measurements, *IEEE Trans. Power Apparatus Syst.*, 97, 2167, 1978.
93. Rukspollmuang, S. and Chen, K.M., Heating of spherical vs. realistic models of human and infrahuman heads by electromagnetic waves, *Radio Sci.*, 14, 51, 1979.
94. Hagman, M.J., Gandhi, O.P., D'Andrea, J.A., and Chatterjee, I., Head resonance: numerical solutions and experimental results, *IEEE Trans. Microwave Theory Technol.*, 27, 809, 1979.
95. Gandhi, O.P., Hunt, E.L., and D'Andrea, J.A., Deposition of EM energy in animals and in models of man with and without grounding and reflector effects, *Radio Sci.*, 12, 39S, 1977.
96. Deford, J.F., Gandhi, O.P., and Hagman, M.J., Moment-method solutions and SAR calculations for inhomogeneous models of man with large number of cells, *IEEE Trans. Microwave Theory Technol.*, 31, 848, 1983.
97. Sullivan, D.M., Gandhi Om, P., and Taflove, A., Use of the finite-difference time-domain method for calculating EM absorption in man models, *IEEE Trans. Biomed. Eng.*, 35, 179, 1988.
98. Bernardi, P., Cavagnaro, M., Pisa, S., and Piuzzi, E., Specific absorption rate and temperature elevation in a subject exposed in the far-field of radio-frequency sources operating in the 10–900-MHz range, *IEEE Trans. Biomed. Eng.*, 50, 295, 2003.
99. Dimbylow, P.J., FDTD calculations of the whole-body averaged SAR in an anatomically realistic voxel model of the human body from 1 MHz to 1 GHz, *Phys. Med. Biol.*, 42, 479, 1997.
100. ICNIRP, Guidelines for limiting exposure to time-varying electric, magnetic, and electromagnetic fields (up to 300 GHz), *Health Phys.*, 74, 494, 1998.
101. IEEE, IEEE standard for safety levels with respect to human exposure to radio frequency electromagnetic fields, 3 kHz to 300 GHz, in *IEEE Standard C95.1*, 1999.
102. Dimbylow, P.J., Fine resolution calculations of SAR in the human body for frequencies up to 3 GHz, *Phys. Med. Biol.*, 47, 2835, 2002.
103. Mason, A.P., Hurt, W.D., Walters, T.J., D'Andrea, J.A., Gajšek, P., Ryan, K.L., Nelson, D.A., Smith, K.I., and Ziriax, J.M., Effects of frequency, permittivity, and voxel size on predicted specific absorption rate values in biological tissue during electromagnetic-field exposure, *IEEE Trans. Microwave Theory Technol.*, 48, 2050, 2000.
104. Gajšek, P., Hurt, W.D., Ziriax, J.M., and Mason, A.P., Parametric dependence of SAR on permittivity values in a man model, *IEEE Trans. Biomed. Eng.*, 48, 1169, 2001.
105. Mazzurana, M., Sandrini, L., Vaccari, A., Malacarne, C., Cristoforetti, L., and Pontalti, R., A semi-automatic method for developing an anthropomorphic numerical model of dielectric anatomy by MRI, *Phys. Med. Biol.*, 48, 3157, 2003.
106. Tinniswood, A.D., Furse, C.M., and Gandhi Om, P., Power deposition in the head and neck of an anatomically based human body model for plane wave exposures, *Phys. Med. Biol.*, 43, 2361, 1998.

107. Nagaoka, T., Watanabe, S., Sakurai, K., Kunieda, E., Watanabe, S., Taki, M., and Yamanaka, Y., Development of realistic high-resolution whole-body voxel models of japanese adult males and females of average height and weight, and application of models to radio-frequency electromagnetic-field dosimetry, *Phys. Med. Biol.*, 49, 1, 2004.
108. Sandrini, L., Vaccari, A., Malacarne, C., Cristoforetti, L., and Pontalti, R., RF dosimetry: a comparison between power absorption of female and male numerical models from 0.1 to 4 GHz, *Phys. Med. Biol.*, 49, 5185, 2004.
109. Gabriel, S., Lau, R.W., and Gabriel C., The dielectric properties of biological tissues: III. Parametric models for the dielectric spectrum of tissues, *Phys. Med. Biol.*, 41, 2271, 1996.
110. Gabriel, C., Compilation of the Dielectric Properties of Body Tissues at RF and Microwave Frequencies, Technical Report, AL/OE-TR-1996–0037, Brooks Air Force, Brooks AFB, TX, 1996.
111. Faraone, A., Tay, R.Y.-S., Joyner, K.H., and Balzano, Q., Estimation of the average power density in the vicinity of cellular base-station collinear array antennas, *IEEE Trans. Veh. Technol.*, 49, 984, 2000.
112. Cicchetti, R. and Faraone, A., Estimation of the peak power density in the vicinity of cellular and radio base station antennas, *IEEE Trans. Electromagn. Compat.*, 46, 275, 2004.
113. Bizzi, M. and Gianola, P., Electromagnetic fields radiated by GSM antennas, *Electron. Lett.*, 35, 855, 1999.
114. Altman, Z., Begasse, B., Dale, C., Karwowski, A., Wiart, J., Wong, M.F., and Gattoufi, L., Efficient models for base station antennas for human exposure assessment, *IEEE Trans. Electromagn. Compat.*, 44, 588, 2002.
115. Bernardi, P., Cavagnaro, M., Cristoforetti, L., Malacarne, C., Pisa, S., Piuzzi, E., Pontalti, R., and Vaccari, A., Modelling of BTS antennas: Dependence of the accuracy on FDTD mesh size and implementation criteria, in *Proceedings of the 2nd International Workshop on Biological Effects of Electromagnetic Fields*, Rhodes, Greece, Kostarakis, P., Ed., 2002, 74.
116. Blanch, S., Romeu, J., and Cardama, A., Near field in the vicinity of wireless base-station antennas: an exposure compliance approach, *IEEE Trans. Antennas Propagat.*, 50, 685, 2002.
117. Adane, Y., Gati, A., Wong, M.-F., Dale, C., Wiart, J., and Hanna V.F., Optimal modeling of real radio base station antennas for human exposure assessment using spherical-mode decomposition, *Antennas Wireless Propagat. Lett.*, 1, 215, 2002.
118. Fridén, J., RF exposure compliance boundary analysis of base station antennas using combined spherical–cylindrical near-field transformations, *Electron. Lett.*, 39, 1783, 2003.
119. Nicolas, E., Lautru, D., Jacquin, F., Wong, M.F., and Wiart, J., Specific absorption rate assessments based on a selective isotropic measuring system for electromagnetic fields, *IEEE Trans. Instrum. Meas.*, 50, 397, 2001.
120. Lazzi, G. and Gandhi, O.P., A mixed FDTD-integral equation approach for on-site safety assessment in complex electromagnetic environments, *IEEE Trans. Antennas Propagat.*, 48, 1830, 2000.
121. Gandhi, O.P. and Lam, M.S., An on-site dosimetry system for safety assessment of wireless base stations using spatial harmonic components, *IEEE Trans. Antennas Propagat.*, 51, 840, 2003.
122. Cooper, J., Marx, B., Buhl, J., and Hombach, V., Determination of safety distance limits for a human near a cellular base station antenna, adopting the IEEE standard or ICNIRP guidelines, *Bioelectromagnetics*, 23, 429, 2002.
123. Catarinucci, L., Palazzari, P., and Tarricone, L., Human exposure to the near field of radiobase antennas—a full-wave solution using parallel FDTD, *IEEE Trans. Microwave Theory Technol.*, 51, 935, 2003.
124. Bernardi, P., Cavagnaro, M., Pisa, S., and Piuzzi, E., Human exposure in the vicinity of radio base station antennas, in *Proceedings of EMC Europe 2000 (4th European Symposium on Electromagnetic Compatibility)*, Brugge, Belgium, 2000, 187.
125. Meyer, F.J.C., Davidson, D.B., Jakobus, U., and Stuchly M.A., Human exposure assessment in the near field of GSM base-station antennas using a hybrid finite element/method of moments technique, *IEEE Trans. Biomed. Eng.*, 50, 224, 2003.
126. Bernardi, P., Cavagnaro, M., Pisa, S., and Piuzzi, E., Human exposure to cellular base station antennas in urban environment, *IEEE Trans. Microwave Theory Technol.*, 48, 1996, 2000.

127. Alaydrus, M., Bitz, A., Streckert, J., and Hansen, V., Combination of the hybrid$^{(2)}$–method and the FDTD for safety assessment of human for base station antennas mounted in real environments, in XXVIIth General Assembly of International Union of Radio Science, Maastricht, The Netherlands, 2002.
128. Martinez-Burdalo, M., Nonidez, L., Martin, A., and Villar, R., Near-field time-domain physical-optics and FDTD method for safety assessment near a base-station antenna, *Microwave Opt. Technol. Lett.*, 39, 393, 2003.
129. Bernardi, P., Cavagnaro, M., Cicchetti, R., Pisa, S., Piuzzi, E., and Testa, O., A UTD/FDTD investigation on procedures to assess compliance of cellular base-station antennas with human-exposure limits in a realistic urban environment, *IEEE Trans. Microwave Theory Technol.*, 51, 2409, 2003.
130. Bernardi, P., Cavagnaro, M., Cicchetti, R., Pisa, S., Piuzzi, E., and Testa, O., Human exposure to the electromagnetic fields produced by coexisting wireless communication systems in a complex indoor environment, in *Proceedings of EMC Europe 2004 (International Symposium of Electromagnetic Compatibility)*, Eindhoven, The Netherlands, 2004, 79.
131. Lin, J.C., Interaction of electromagnetic transient radiation with biological materials, *IEEE Trans. Electromagn. Compat.*, 17, 93, 1975.
132. Lin, J.C., Wu, C.L., and Lam, C.K., Transmission of electromagnetic pulse into the head, *Proc. IEEE*, 63, 1726, 1975.
133. Lin, J.C., Electromagnetic pulse interaction with mammalian cranial structures, *IEEE Trans. Biomed. Eng.*, 23, 61, 1976.
134. Lin, J.C. and Lam, C.K., Coupling of Gaussian electromagnetic pulse into muscle-bone model of biological structure, *J. Microwave Power*, 11, 67, 1976.
135. Lin, J.C., The microwave auditory phenomenon, *Proc. IEEE*, 68, 67, 1980.
136. Lin, J.C., Microwave-induced hearing sensation: some preliminary theoretical observations, *J. Microwave Power*, 11, 295, 1976.
137. Lin, J.C., On Microwave-induced hearing sensation, *IEEE Trans. Microwave Theory Technol.*, 25, 605, 1977.
138. Lin, J.C., Further studies on the microwave auditory effects, *IEEE Trans. Microwave Theory Technol.*, 25, 936, 1977.
139. Lin, J.C., Calculations of frequencies and threshold of microwave-induced auditory signals, *Radio Sci.*, 12, SS-1, 237, 1977.
140. Lin, J.C., *Microwave Auditory Effects and Applications*, Charles C. Thomas, Publisher, Springfield, IL, 1978.
141. Watanabe, Y., Tanaka, T., Taki, M., and Watanabe, S., FDTD analysis of microwave hearing effect, *IEEE MTT*, 48, 2126, 2000.
142. Lin, J.C. and Wang, Z.W., RF thermoelastic pressures induced in a human head model in high-pass MRI birdcage coils, in ISMRM, Miami, 2005.
143. Wang, Z.W. and Lin, J.C., RF induced thermoelastic pressure waves in different size human head models in MRI birdcage coils, in 27th Annual Meeting of Bioelectromagnetics Society, Dublin, Ireland, June 2005.
144. Yang, L.Q. and Giannakis, G.B., Ultra-wideband communications, *IEEE Signal Process. Mag.*, 21, 26, 2004.
145. Qiu, R.C., Liu, H.P., and Shen, X.M., Ultra-wideband for multiple access communications, *IEEE Commun. Mag.*, 43, 80, 2005.
146. Gandhi, O.P., Gu, Y.G., Chen, J.Y., and Bassen, H.I., Specific absorption rates and induced current distributions in an anatomically based human model for plane-wave exposures, *Health Phys.*, 63, 281, 1992.
147. FCC, Evaluating compliance with FCC guidelines for human exposure to radiofrequency electromagnetic fields, in *OET Bulletin 65*, FCC, Washington, DC, Aug. 1997.
148. IEC, human exposure to radio frequency fields from hand-held and body-mounted wireless communication devices—human models, instrumentation, and procedures—part 1: procedure to determine the specific absorption rate (SAR) for hand-held devices used in close proximity to the ear (frequency range of 300 MHz to 3 GHz), in *IEC Standard, 62209-1*, 2005.

149. Toftgard, J., Hornsleth, S.N., and Andersen, J.B., Effects on portable antennas of the presence of a person, *IEEE Trans. Antennas Propagat.*, 41, 739, 1993.
150. Dimbylow, P.J. and Mann, S.M., SAR calculations in an anatomically realistic model of the head for mobile communication transceivers at 900 MHz and 1.8 GHz, *Phys. Med. Biol.*, 39, 1537, 1994.
151. Gandhi, O.P., Lazzi, G., and Furse, C.M., Electromagnetic absorption in the human head and neck for mobile telephones at 835 and 1900 MHz, *IEEE Trans. Microwave Theory Technol.*, 44, 1884, 1996.
152. Okoniewski, M., and Stuchly, M.A., A study of the handset antenna and human body interaction, *IEEE Trans. Microwave Theory Technol.*, 44, 1855, 1996.
153. Hombach, V., Meier, K., Burkhardt, M., Kuhn, E., and Kuster, N., The dependence of EM energy absorption upon human head modeling at 900 MHz, *IEEE Trans. Microwave Theory Technol.*, 44, 1865, 1996.
154. Watanabe, S., Taki, M., Nojima, T., and Fujiwara, O., Characteristics of the SAR distributions in a head exposed to electromagnetic fields radiated by a hand-held portable radio, *IEEE Trans. Microwave Theory Technol.*, 44, 1874, 1996.
155. Bernardi, P., Cavagnaro, M., and Pisa, S., Evaluation of the SAR distribution in the human head for cellular phones used in a partially closed environment, *IEEE Trans. Electromagn. Compat.*, 38, 357, 1996.
156. Colburn, J.S., and Rahmat-Samii, Y., Human proximity effects on circular polarized handset antennas in personal satellite communications, *IEEE Trans. Antennas Propagat.*, 46, 813, 1998.
157. Lazzi, G., and Gandhi, O.P., On modeling and personal dosimetry of cellular telephone helical antennas with the FDTD code, *IEEE Trans. Antennas Propagat.*, 46, 525, 1998.
158. Mangoud, M.A., Abd-Alhameed, R.A., and Excell, P.S., Simulation of human interaction with mobile telephones using hybrid techniques over coupled domains, *IEEE Trans. Microwave Theory Technol.*, 48, 2014, 2000.
159. Cerri, G., Russo, P., Schiavoni, A., Tribellini, G., and Bielli, P., A new MoM-FDTD hybrid technique for the analysis of scattering problems, *Electron. Lett.*, 34, 438, 1998.
160. Bernardi, P., Cavagnaro, M., Pisa, S., and Piuzzi, E., A graded-mesh FDTD code for the study of human exposure to cellular phones equipped with helical antennas, *Appl. Comput. Electromagn. Soc. J.*, 16, 90, 2001.
161. Bernardi, P., Cavagnaro, M., Pisa, S., and Piuzzi, E., Power absorption and temperature elevations induced in the human head by a dual-band monopole-helix antenna phone, *IEEE Trans. Microwave Theory Technol.*, 49, 2539, 2001.
162. Jensen, M.A. and Rahmat-Samii, Y., EM interaction of handset antennas and a human in personal communications, *Proc. IEEE*, 83, 7, 1995.
163. Rowley, J.T. and Waterhouse, R.B., Performance of shorted patch antennas for mobile communication handsets at 1800 MHz, *IEEE Trans. Microwave Theory Technol.*, 47, 815, 1999.
164. Bernardi, P., Cavagnaro, M., Pisa, S., and Piuzzi, E., Specific absorption rate and temperature increases in the head of a cellular-phone user, *IEEE Trans. Microwave Theory Technol.*, 48, 1118, 2000.
165. de Salles, A.A., Fernandez, C.R., and Bonadiman, M., FDTD simulations and measurements on planar antennas for mobile phones, *Proc. SBMO/IEEE MTT-S IMOC*, 1043, 2003.
166. Tinniswood, A.D., Furse, C.M., and Gandhi, O.P., Computations of SAR distributions for two anatomically based models of the human head using CAD files of commercial telephones and the parallelized FDTD code, *IEEE Trans. Antennas Propagat.*, 46, 829, 1998.
167. Schiavoni, A., Bertotto, P., Richiardi, G., and Bielli, P., SAR generated by commercial cellular phones—phone modeling, head modeling, and measurements, *IEEE Trans. Microwave Theory Technol.*, 48, 2064, 2000.
168. Chavannes, N., Tay, R., Nikoloski, N., and Kuster, N., Suitability of FDTD-based TCAD tools for RF design of mobile phones, *IEEE Antennas Propagat. Mag.*, 45, 52, 2003.
169. Pisa, S., Cavagnaro, M., Lopresto, V., Piuzzi, E., Lovisolo, G.A., and Bernardi, P., A procedure to develop realistic numerical models of cellular phones for an accurate evaluation of SAR distribution in the human head, *IEEE Trans. Microwave Theory Technol.*, 53, 2005.
170. Lin, J.C., Cellular mobile telephones and children, *IEEE Antennas Propagat. Mag.*, 44, 142, 2002.
171. Peyman, A., Rezazadeh, A.A., and Gabriel, C. Changes in the dielectric properties of rat tissue as a function of age at microwave frequencies, *Phys. Med. Biol.*, 46, 1617, 2001.

172. Gandhi, O.M. and Kang, G., Some present problems and a proposed experimental phantom for SAR compliance testing of cellular telephone at 835 and 1900 MHz, *Phys. Med. Biol*, 47, 1501, 2002.
173. Schoenborn, F., Burkhardt, M., and Kuster, N., Differences in energy absorption between heads of adults and children in the near field of sources, *Health Phys.*, 74, 160, 1998.
174. Guy, A.W., Chou, C.K., and Bit-Babik, G., FDTD derived SAR distributions in various size human head models exposed to simulated cellular telephone handset transmitting 600 mW at 835 MHz, in *24th Bioelectromagnetics Society Annual Meeting*, Quebec, Canada, 7, 2002.
175. Wang, J. and Fujiwara, O., Comparison and evaluation of electromagnetic absorption characteristic in realistic human head models of adult and children for 900-MHz mobile telephones, *IEEE Trans. Microwave Theory Technol.*, 51, 966, 2003.
176. Hadjem, A., Lautru, D., Dale, C., Wong, M.F., Hanna, V.H., and Wiart, J., Study of specific absorption rate (SAR) induced in two child head models and in adult heads using mobile phones, *IEEE Trans. Microwave Theory Technol.*, 53, 4, 2005.
177. Dominguez, H., Raizer, A., and Carpes, W.P., Jr., Electromagnetic fields radiated by cellular phone in close proximity to metallic walls, *IEEE Trans. Magn.*, 38, 793, 2002.
178. Anzaldi, G., Delgado, E.C., Riu, P.J., and Silva, F., FDTD analysis of SAR from a cell phone inside a vehicle, in *Proceedings of the 16th International Symposium on Electromagnetic Compatibility*, Zurich, Switzerland, 2005, 155.
179. Ruddle, A.R., Modelling electromagnetic field exposure and SAR in vehicles due to on-board transmitters, in *Proceedings of the 16th International Symposium on Electromagnetic Compatibility*, Zurich, Switzerland, 2005, 145.
180. Onishi, T., Iyama, T., and Uebayashi, S., The estimation of the maximum SAR with respect to various types of wireless device usage, in *Proceedings of 16th International Symposium on Electromagnetic Compatibility*, Zurich, Switzerland, 2005, 151.
181. Bernardi, P., Cavagnaro, M., Pisa, S., Piuzzi, E, and Lin, J., SAR distribution produced by cellular phones radiating close to the human body, in BioEM-2005 Meeting, Dublin, Ireland, 2005.
182. Pennes, H.H., Analysis of tissue and arterial blood temperatures in resting forearm, *J. Appl. Physiol.*, 1, 93, 1948.
183. Stolwijk, J.A.J. and Hardy, J.D., Control of body temperature, in *Handbook of Physiology—Reaction to Environmental Agents*, Lee, D.H.K., Ed., Am. Physiol. Soc. Bethesda, MD, 1977, 45.
184. Spiegel, R.J., A review of numerical models for predicting the energy deposition and resultant thermal responses of humans exposed to electromagnetic fields, *IEEE Trans. Microwave Theory Technol.*, 32, 730, 1984.
185. Mooibroek, J. and Lagendijk, J.J.W., A fast and simple algorithm for the calculation of convective heat transfer by large vessels in three-dimensional inhomogeneous tissues, *IEEE Trans. Biomed. Eng.*, 38, 490, 1991.
186. Kolios, M.C., Sherar, M.D., and Hunt, J.W., Large blood vessel cooling in heated tissues: a numerical study, *Phys. Med. Biol.*, 40, 477, 1995.
187. Van Leeuwen, G.M.J., Lagendijk, J.J.W., Van Leersum, B.J.A.M., Zwamborn, A.P.M., Hornsleth, S.N., and Kotte, A.N.T.J., Calculation of change in brain temperatures due to exposure to a mobile phone, *Phys. Med. Biol.*, 44, 2367, 1999.
188. Ma, L., Paul, D.-L., Pothecary, N., Railton, C., Bows, J., Barratt, L., Mullin, J., and Simons, D., Experimental validation of a combined electromagnetic and thermal FDTD model of a microwave heating process, *IEEE Trans. Microwave Theory Technol.*, 43, 2565, 1995.
189. Torres, F. and Jecko, B., Complete FDTD analysis of microwave heating process in frequency-dependent and temperature-dependent media, *IEEE Trans. Microwave Theory Technol.*, 45, 108, 1997.
190. Lu, C.-C., Li, H.-Z., and Gao, D., Combined electromagnetic and heat-conduction analysis of rapid rewarming of cryopreserved tissues, *IEEE Trans. Microwave Theory Technol.*, 48, 2185, 2000.
191. Gordon, R.G., Roemer, R.B., and Horvath, S.M., A mathematical model of the human temperature regulatory system—transient cold exposure response, *IEEE Trans. Biomed. Eng.*, 23, 434, 1976.

192. Ozisik, N., *Heat Transfer: A Basic Approach*, McGraw-Hill, New York, 1985.
193. Gagge, A.P. and Nishi, Y., Heat exchange between human skin surface and thermal environment, in *Handbook of Physiology—Reaction to Environmental Agents*, Lee, D.H.K., Ed., Am. Physiol. Soc. Bethesda, MD, 1977, 69.
194. Haala, J. and Wiesbeck, W., Modeling microwave and hybrid heating processes including heat radiation effects, *IEEE Trans. Microwave Theory Technol.*, 50, 1346, 2002.
195. Adair, E.R., Thermal physiology of radiofrequency radiation (RFR) interactions in animals and humans, in *Radiofrequency Radiation Standards*, Klanenberg, B.J., Grandolfo, M., and Erwin, D.N., Eds., Plenum Press, New York, 1994, 245.
196. Hoque, M. and Gandhi, O.P., Temperature distributions in the human leg for VLF-VHF exposures at the ANSI recommended safety levels, *IEEE Trans. Biomed. Eng.*, 35, 442, 1988.
197. Chatterjee, I. and Gandhi, O.P., An inhomogeneous thermal block model of man for the electromagnetic environment, *IEEE Trans. Biomed. Eng.*, 30, 707, 1983.
198. Guyton, A.C., *Textbook of Medical Physiology*, W.B. Saunders Company, Philadelphia, PA, 1991.
199. Bardati, F., Gerosa, G., and Lampariello, P., Temperature distribution in simulated living tissues irradiated electromagnetically, *Alta Freq.*, XLIX, 61, 1980.
200. Durkee, J.W., Antich, P.P., and Lee, C.E., Exact solutions to the multiregion time-dependent bioheat equation. I: solution development, *Phys. Med. Biol.*, 35, 847, 1990.
201. Vyas, R. and Rustgi, M.L., Green's function solution to the tissue bioheat equation, *Med. Phys.*, 19, 1319, 1992.
202. Way, W.I., Kritikos, H., and Schwan, H., Thermoregulatory physiological responses in the human body exposed to microwave radiation, *Bioelectromagnetics*, 2, 341, 1981.
203. Charny, C.K., Hagmann, M.J., and Levin, R.L., A whole body thermal model of man during hyperthermia, *IEEE Trans. Biomed. Eng.*, 34, 375, 1987.
204. Scott, J.A., A finite element model of heat transport in the human eye, *Phys. Med. Biol.*, 33, 227, 1988.
205. Labonté, S., Blais, A., Legault, S.R., Ali, H.O., and Roy, L., Monopole antennas for microwave catheter ablation, *IEEE Trans. Microwave Theory Technol.*, 44, 1832, 1996.
206. Wang, J. and Fujiwara, O., FDTD computation of temperature rise in the human head for portable telephones, *IEEE Trans. Microwave Theory Technol.*, 47, 1528, 1999.
207. Pisa, S., Cavagnaro, M., Piuzzi, E., Bernardi, P., and Lin, J.C., Power density and temperature distributions produced by interstitial arrays of sleeved-slot antennas for hyperthermic cancer therapy, *IEEE Trans. Microwave Theory Technol.*, 51, 2418, 2003.
208. Peaceman, D.W. and Rachford, H., The numerical solution of parabolic and elliptic differential equations, *J. Soc. Ind. Appl. Math.*, 3, 28, 1955.
209. Douglas, J., On the numerical integration of $u_{xx}+u_{yy} = u_t$ by implicit methods, *J. Soc. Ind. Appl. Math.*, 3, 42, 1955.
210. Douglas, J., Alternating direction methods for three space variables, *Numer. Math.*, 4, 41, 1962.
211. Guyton, A.C., *Textbook of Medical Physiology*, W.B. Saunders Company, Philadelphia, PA, 1991.
212. Guy, A.W., Lin, J.C., Kramar, P.O., and Emery, A.F., Effect of 2450 MHz radiation on the rabbit eye, *IEEE Trans. Microwave Theory Technol.*, 23, 492, 1975.
213. Appleton, B., Hirsch, S.E., and Brown, P.V.K., Investigation of single-exposure microwave ocular effects at 3000 MHz, *Ann. NY Acad. Sci.*, 247, 125, 1975.
214. Sliney, D.H. and Stuck, B.E. Microwave exposure limits for the eye: applying infrared laser threshold data, in *Radiofrequency Radiation Standards*, Klanenberg, B.J., Grandolfo, M., and Erwin, D.N., Eds., Plenum Press, New York, 1994, 79.
215. Hardy, J.D., Wolff, H.G., and Goodell, H., *Pain Sensations and Reactions*, Williams and Wilkins, Baltimore, MD, 1952.
216. Foster, K.R., and Erdreich, L.S., Thermal models for microwave hazards and their role in standards development, *Bioelectromagnetics*, 20, 52, 1999.
217. Riu, P.J. and Foster, K.R., Heating of tissue by near-field exposure to a dipole: a model analysis, *IEEE Trans. Biomed. Eng.*, 46, 911, 1999.
218. Cherry, P.C. and Iskander, M.F., Calculations of heating patterns of an array of microwave interstitial antennas, *IEEE Trans. Biomed. Eng.*, 40, 771, 1993.

219. Bernardi, P., Cavagnaro, M., Pisa, S., and Piuzzi, E., SAR distribution and temperature increase in an anatomical model of the human eye exposed to the field radiated by the user antenna in a wireless LAN, *IEEE Trans. Microwave Theory Technol.*, 46, 2074, 1998.
220. Wainwright, P., Thermal effects of radiation from cellular telephones, *Phys. Med. Biol.*, 45, 2363, 2000.
221. Gandhi, O.M., Li, Q.X., and Kang, G., Temperature rise for the human head for cellular telephones and for peak SARs prescribed in safety guidelines, *IEEE Trans. Microwave Theory Technol.*, 49, 1607, 2001.
222. Hirata, A., Morita, M., and Shiozawa, T., Temperature increase in the human head due to a dipole antenna at microwave frequencies, *IEEE Trans. Electromagn. Compat.*, 45, 109, 2003.
223. Hirata, A. and Shiozawa, T., Correlation of maximum temperature increase and peak SAR in the human head due to handset antennas, *IEEE Trans. Microwave Theory Technol.*, 51, 1834, 2003.
224. Beckman, K.J., Lin, J.C., Wang, Y., Illes, R.W., Papp, M.A., and Hariman, R.J., Production of reversible and irreversible atrio-ventricular block by microwave energy, in *The 60th Scientific Sessions, American Heart Association, Anaheim*, CA, 1987; also in *Circulation* 76, 1612, 1987.
225. Lin, J.C., Beckman, K.J., Hariman, R.J., Bharati, S., Lev, M., and Wang, Y.J., Microwave ablation of the atrioventricular junction in open heart dogs, *Bioelectromagnetics*, 16, 97, 1995.
226. Lin, J.C., Hyperthermia therapy, in *Encyclopedia of Electrical and Electronics Engineering*, vol. 9, Webster, J.G., Ed., Wiley, New York, 1999, 450–460.
227. McRury, I.D. and Haines, D.E., Ablation for the treatment of arrhythmias, *Proc. IEEE*, 84, 404, 1996.
228. Lin, J.C., Hariman, R.J., Wang, Y.G., and Wang, Y.J., Microwave catheter ablation of the atrioventricular junction in closed-chest dogs, *Med. Biol. Eng. Comput.*, 34, 295, 1996.
229. Lin, J.C., Catheter microwave ablation therapy for cardiac arrhythmias, *Bioelectromagnetics*, 20, S4, 120, 1999.
230. Huang, S.K.S. and Wilber, D.J., Eds., *Radiofrequency Catheter Ablation of Cardiac Arrhythmias: Basic Concepts and Clinical Applications*, 2nd ed., Futura, Armonk, New York, 2000.
231. Lin, J.C., Biophysics of radiofrequency ablation, in *Radiofrequency Catheter Ablation of Cardiac Arrhythmias: Basic Concepts and Clinical Applications*, 2nd ed, Huang, S.K.S. and Wilber, D.J., Eds., Futura, Armonk, New York, 2000,13–24.
232. Wonnel, T.L., Stauffer, P.R., and Langberg, J.J., Evaluation of microwave and radio-frequency catheter ablation in a myocardium-equivalent phantom model, *IEEE Trans. Biomed. Eng.*, 39, 10, 1086, 1992.
233. Lin, J.C., Wang, Y.L., and Heriman, R.J., Comparison of power deposition patterns produced by microwave and radio frequency cardiac ablation catheters, *Electron. Lett.*, 30, 12, 922, 1994.
234. Lin, J.C., Studies on microwaves in medicine and biology: from snails to humans, *Bioelectromagnetics*, 25, 146, 2004.
235. Labonté, S., Blais, A., Legault, S.R., Ali, H.Q., and Roy, L., Monopole antennas for microwave catheter ablation, *IEEE Trans. Microwave Theory Technol.*, 44, 1832, 1996.
236. Kaouk, Z., Khebir, A., and Savard, P., A finite element model of a microwave catheter for cardiac ablation, *IEEE Trans. Microwave Theory Technol.*, 44, 1848, 1996.
237. Lin, J.C. and Wang, Y.J., The cap-slot catheter antenna for microwave ablation therapy, *IEEE Trans. Biomed. Eng.*, 43, 657, 1996.
238. Lin, J.C. and Wang, Y.J., A catheter antenna for percutaneous microwave therapy, *Microwave Opt. Technol. Lett.*, 8, 70, 1995.
239. Gu, Z., Rappaport, C.M., Wang, P.J., and VanderBrink, B.A., Development and experimental verification of the wide-aperture catheter-based microwave cardiac ablation antenna, *IEEE Trans. Microwave Theory Technol.*, 48, 1892, 2000.
240. Nevels, R.D., Arndt, G.D., Raffoul, G.W., Carl, J.R., and Pacifico, A., Microwave catheter design, *IEEE Trans. Biomed. Eng.*, 45, 885, 1998.
241. Rosenbaum, R.M., Greenspon, A.J., Hsu, W.L., Walinsky, S.P., and Rosen, A., RF and microwave ablation for the treatment of ventricular tachycardia, *IEEE MTT-S Int. Microwave Symp. Digest*, 1993, 1155.
242. Greenspon, A.J., Advances in catheter ablation for the treatment of cardiac arrhythmias, *IEEE Trans. Microwave Theory Technol.*, 48, 2670, 2000.

243. Panescu, D., Whayne, J.G., Fleischman, S.D., Mirotznik, M.S., Swanson, D.K., and Webster, J.G., Three-dimensional finite element analysis of current density and temperature distributions during radio-frequency ablation, *IEEE Trans. Biomed. Eng.*, 42, 879, 1995.
244. Cao, H., Vorperian,, V.R., Tungjitkusolmun, S., Tsai, J.Z., Haemmerich, D., Choy, Y.B., and Webster, J.G., Flow effect on lesion formation in RF cardiac catheter ablation, *IEEE Trans. Biomed. Eng.*, 48, 425, 2001.
245. Tungjitkusolmun, S., Vorperian, V.R., Bhavaraju, N., Cao, H., Tsai, J.Z., and Webster, J.G., Guidelines for predicting lesion size at common endocardial locations during radio-frequency ablation, *IEEE Trans. Biomed. Eng.*, 48, 194, 2001.
246. Bernardi, P., Cavagnaro, M., Lin, J.C., Pisa, S., and Piuzzi, E., Distribution of SAR and temperature elevation induced in a phantom by a microwave cardiac ablation catheter, *IEEE Trans. Microwave Theory Technol.*, 52, 1978, 2004 (special issue on medical applications and biological effects of RF/microwaves).
247. McLucas, B., Management of uncontrolled menorrhagia with the resectoscope, *Minim. Invasive Ther.*, 1, 131, 1992.
248. Brooks, P.G., Clouse, J., and Morris, L.S., Hysterectomy vs resectoscopic endometrial ablation for the control of abnormal uterine bleeding—a cost-comparative study, *J. Reprod. Med.*, 39, 755, 1994.
249. Pinion, S.B., Parkin, D.E., Abramovich, D.R., Naji, A., Alexander, D.A., Russel, I.T., and Kitchener, H.C., Randomized trial of hysterectomy, endometrial laser ablation, and transcervical endometrial resection for dysfunctional uterine bleeding, *Br. Med.J.*, 309, 979, 1994.
250. Seymour, J., Wallage, S., Graham, W., Parkin, D., and Cooper, K., The cost of microwave endometrial ablation under different anaesthetic and clinical settings, *BJOG: Int. J. Obstet. Gynaecol.*, 110, 922, 2003.
251. Wallage, S., Cooper, K.G., Graham, W.J., and Parkin, D.E., A randomised trial comparing local versus general anaesthesia for microwave endometrial ablation, *BJOG: Int. J. Obstet. Gynaecol.*, 110, 799, 2003.
252. Lalonde, A., Evaluation of surgical options in menorrhagia, *Br.J. Obstet. Gynaecol.*, 101, 8, 1994.
253. Erian, J., Endometrial ablation in the treatment of menorrhagia, *Br.J. Obstet. Gynaecol*, 104, 19, 1994.
254. Daniell, J.F., Kurtz, B.R., and Ke, R.W., Hysteroscopic endometrial ablation using the rollball electrode, *Obstet. Gynecol.*, 80, 329, 1992.
255. Reid, P.C., Turrell, W., Smith, J.H.F., Kenndy, A., and Sharp, F., Nd-YAG laser endometrial ablation—histological aspects of uterine healing, *Int. J. Gynecol. Pathol.*, 11, 174, 1992.
256. Phillips, D.R., A comparison of endometrial ablation using the Nd:YAG laser or electrosurgical techniques, *J. Am. Assoc. Gynecol. Laparosc.*, 1, 235–239, 1993.
257. Sharp, N.C., Cronin, N., Feldberg, I., Evans, M., Hodgson, D., and Ellis, S., Microwaves for menorrhagia: s new fast technique for endometrial ablation, *Lancet*, 346, 1003, 1995.
258. Cooper, K.G., Bain, C., and Parkin, D.E., Comparison of microwave endometrial ablation and transcervical resection of the endometrium for treatment of heavy menstrual loss: a randomised trial, *Lancet*, 354, 1859, 1999.
259. Pisa, S., Cavagnaro, M., Bernardi, P., and Lin, J.C., A 915-MHz antenna for microwave thermal ablation treatment: physical design, computer modeling and experimental measurement, *IEEE Trans. Biomed. Eng.*, 48, 599, 2001.
260. Stea, B., Kittleson, J., and Cassady, J.R., Treatment of malignant glioma with interstitial irradiation and hyperthermia, *Int. J. Radiat. Oncol. Biol. Phys.*, 24, 657, 1992.
261. Overgaard, J., Gonzalez, D.G., Hulshof, M., Arcangeli, G., Dahl, O., Mella, O., and Bentzen, S.M., Randomised trial of hyperthermia as adjuvant to radiotherapy for recurrent or metastatic malignant melanoma, *Lancet*, 345, 540, 1995.
262. Kuwano, H., Sumiyoshi, K., Watanabe, M., Sadanaga, N., Nozoe, T., Yasuda, M., and Sugimachi, K., Preoperative hyperthermia combined with chemotherapy and irradiation for the treatment of patients with esophageal carcinoma, *Tumorigenesis*, 81, 18, 1995.
263. Nakajima, T., Roberts, D.W., Ryan, T.P., Hoopes, P.J., Coughlin, C.T., Trembly, B.S., and Strohbehn, J.W., Pattern of response to interstitial hyperthermia and brachytherapy for malignant intracranial tumour: a CT analysis, *Int. J. Hyperthermia*, 9, 491, 1993.
264. Sneed, P.K., Stauffer, P.R., McDermott, M.W., Diederich, C.J., Lamborn, K.R., Prados, M.D., Chang, S., Weaver, K.A., Spry, L., Malec, M.K., Lamb, S.A., Voss, B., Davis, R.L., Wara, W.M.,

Larson, D.A., Phillips, T.L., and Gutin, P.H., Survival benefit of hyperthermia in a prospective randomized trial of brachytherapy boost +/− hyperthermia for glioblastoma multiforme, *Int. J. Radiat. Oncol. Biol. Phys.*, 40, 287, 1998.
265. Coughlin, C.T., Douple, E.B., Strohbehn, J.W., Eaton, J.W., Jr., Trembly, B.S., and Wong, T.Z., Interstitial hyperthermia in combination with brachytherapy, *Radiobiology*, 148, 285, 1983.
266. Hiraoka, M., Mitsumori, M., Hiroi, N., Ohno, S., Tanaka, Y., Kotsuka, Y., and Sugimachi, K., Development of RF and microwave heating equipment and clinical applications to cancer treatment in Japan, *IEEE Trans. Microwave Theory Technol.*, 48, 1789, 2000.
267. King, R.W.P., Trembly, B.S., and Strohbehn, J.W., The electromagnetic field of an insulated antenna in a conducting or dielectric medium, *IEEE Trans. Microwave Theory Technol.*, 31, 574, 1983.
268. Jones, K.M., Mechling, J.A., Trembly, B.S., and Strohbehn, J.W., SAR distribution for 915 MHz interstitial microwave antennas used in hyperthermia for cancer therapy, *IEEE Trans. Biomed. Eng.*, 35, 851, 1988.
269. Iskander, M.F. and Tumeh, A.M., Design optimization of interstitial antennas, *IEEE Trans. Biomed. Eng.*, 36, 238, 1989.
270. Su, D.W.-F. and Wu, L.-K., Input impedance characteristics of coaxial slot antennas for interstitial microwave hyperthermia, *IEEE Trans. Microwave Theory Technol.*, 47, 302, 1999.
271. Camart, J.C., Despretz, D., Chive, M., and Pribetich, J., Modeling of various kinds of applicators used for microwave hyperthermia based on the FDTD method, *IEEE Trans. Microwave Theory Technol.*, 44, 1811, 1996.
272. Ahn, H.R. and Lee, K., Interstitial antennas tipped with reactive load, *IEEE Microwave Wireless Comp. Lett.*, 15, 83, 2005.
273. Longo, L., Gentili, G.B., Cerretelli, M., and Tosoratti, N., A coaxial antenna with miniaturized choke for minimally invasive interstitial heating *IEEE Trans. Biomed. Eng.*, 50, 82, 2003.
274. Brace, C.L., Laeseke, P.F., van der Weide, D.W., and Lee, F.T., Jr., Microwave ablation with a triaxial antenna: results in *ex vivo* bovine liver, *IEEE Trans. Microwave Theory Technol.*, 53, 215, 2005.
275. Jones, K.M., Mechling, J.A., Strohbehn, J.W., and Trembly, B.S., Theoretical and experimental SAR distributions for interstitial dipole antenna arrays used in hyperthermia, *IEEE Trans. Microwave Theory Technol.*, 37, 1200, 1989.
276. Lin, J.C., Hirai, S., Chiang, C.L., Hsu, W.L., and Wang, Y.J., Computer simulation and experimental studies of SAR distributions of interstitial arrays of sleeved-slot microwave antennas for hyperthermia treatment of brain tumors, *IEEE Trans. Microwave Theory Technol.*, 48, 2191, 2000.
277. Camart, J.C., Fabre, J.J., Prevost, B., Pribetich, J., and Chive, M., Coaxial antenna array for 915 interstitial hyperthermia: design and modelization—power deposition and heating pattern—phased array, *IEEE Trans. Microwave Theory Technol.*, 40, 2243, 1992.
278. Clibbon, K.L., McCowen, A., and Hand, J.W., SAR distribution in interstitial microwaveantenna arrays with a single dipole displacement, *IEEE Trans. Biomed. Eng.*, 40, 925, 1993.
279. Cherry, P.C. and Iskander, M.F., FDTD analysis of power deposition patterns of an array of interstitial antennas for use in microwave hyperthermia, *IEEE Trans. Microwave Theory Technol.*, 40, 1692, 1992.
280. Cherry, P.C. and Iskander, M.F., Calculations of heating patterns of an array of microwave interstitial antennas, *IEEE Trans. Biomed. Eng.*, 40, 771, 1993.
281. Gentili, G.B., Leoncini, M.L., Trembly, B.S., and Schweizer, S.F., FDTD electromagnetic and thermal analysis of interstitial hyperthermic applicators, *IEEE Trans. Biomed. Eng.*, 42, 973, 1995.
282. Pisa, S., Cavagnaro, M., Piuzzi, E., Bernardi, P., and Lin, J.C., Power density and temperature distributions produced by interstitial arrays of sleeved-slot antennas for hyperthermic cancer therapy, *IEEE Trans. Microwave Theory Technol.*, 5, 2418, 2003.
283. Lin, J.C. and Wang, Y.J., Interstitial microwave antennas for thermal therapy, *Int. J. Hyperthermia*, 3, 37, 1987.
284. Chou, C.-K., Therapeutic heating applications of radio frequency energy, *Biological and Medical Aspects*, Barnes, F.S. and Greenebaum, B., Eds., Taylor & Francis, Boca Raton, FL, 2006, Chapter 12.
285. Black, D., Thermoregulation in the presence of radiofrequency fields, in *Biological and Medical Aspects*, Barnes, F.S. and Greenebaum, B., Eds., Taylor & Francis, Boca Raton, FL, 2006, Chapter 5.

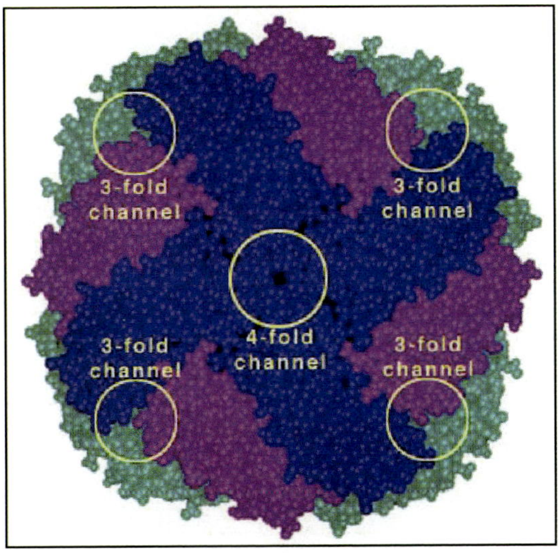

FIGURE 4.7
Model of the ferritin protein showing the peptide subunits and iron transport channels. (From www.chemistry.wustl.edu/~edudev/LabTutorials/Ferritin/FerritinTutorial.html. With permission.)

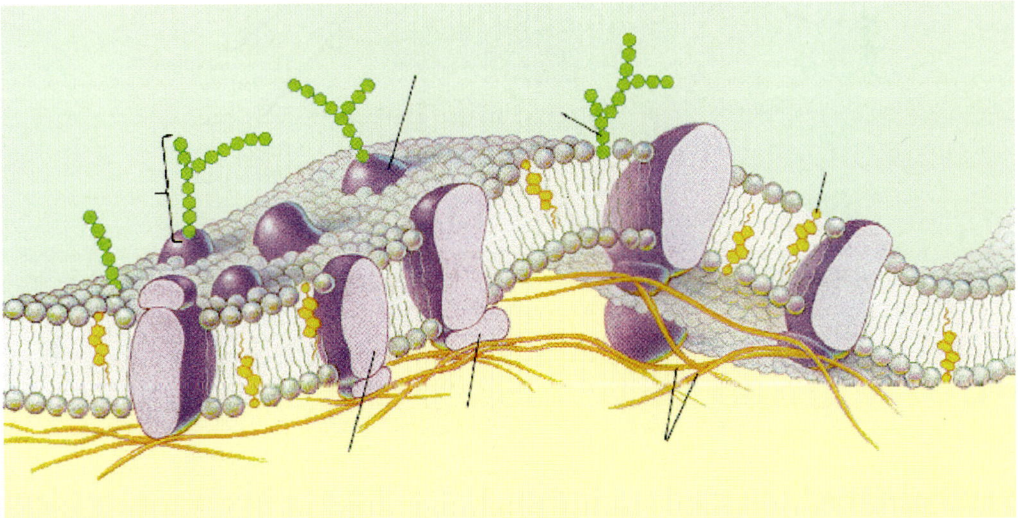

(a)

FIGURE 5.1
(a) Model of a cell membrane. (From Chiras, D., *Human Biology*, © Jones and Baretlett Publishers, Boston. With permission.)

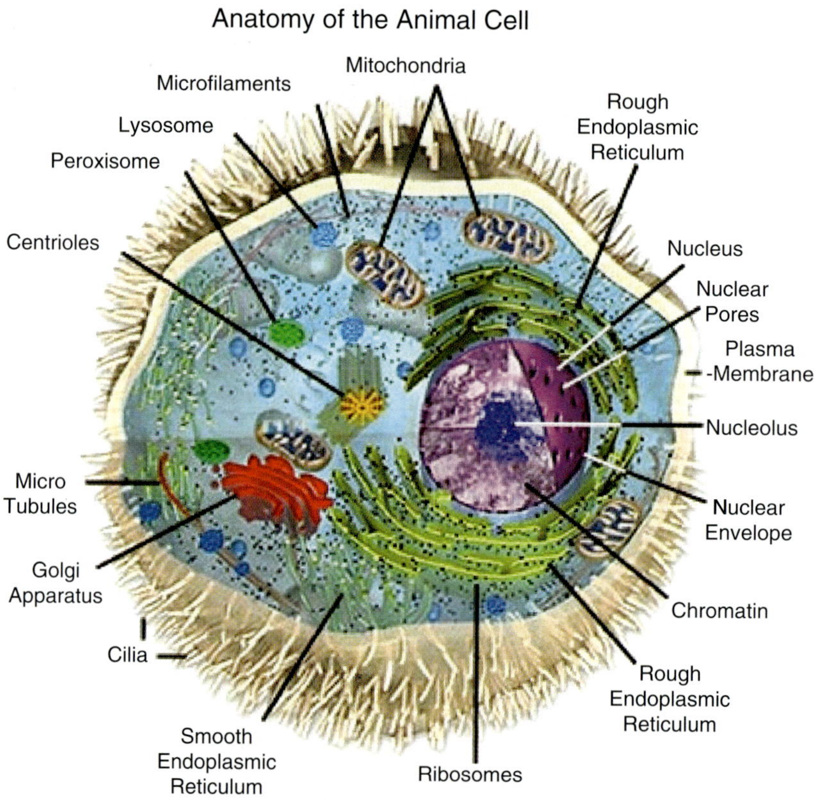

FIGURE 5.1 (continued)
(b) Anatomy of the animal cell. (From *Molecular Expression*, http://microscopy.fsu.edu, accessed Sep. 30, 2005; drawing © M.W. Davidson and Florida State University. With permission).

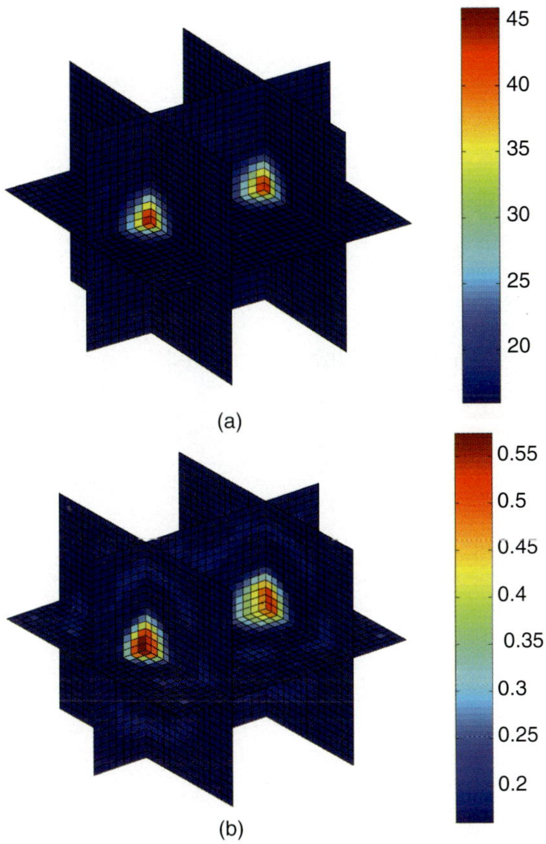

FIGURE 12.7
Inversion for Figure 12.6 on three orthogonal slices through the center of the anomalies. Inverted dielectric constant ε_r (a) and conductivity σ (b).

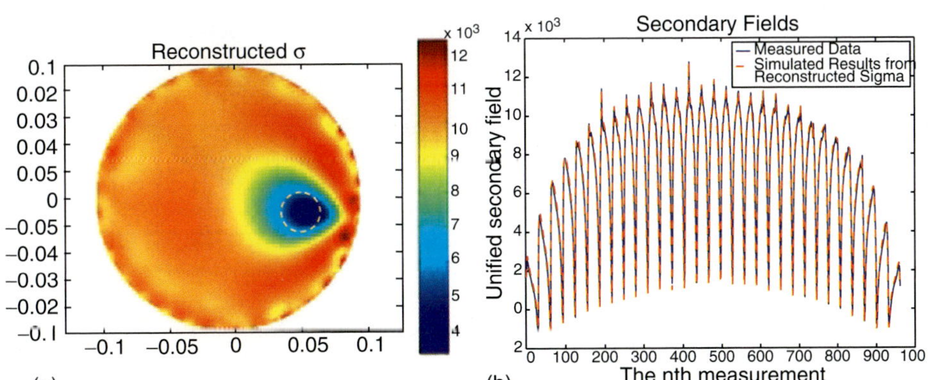

FIGURE 12.8
Left: Image reconstructed from the measured data in Case 1. The dashed circle indicates the ground truth. Right: Comparison of the secondary field between measurements and simulated by forward solver with the reconstructed σ in Case 1.

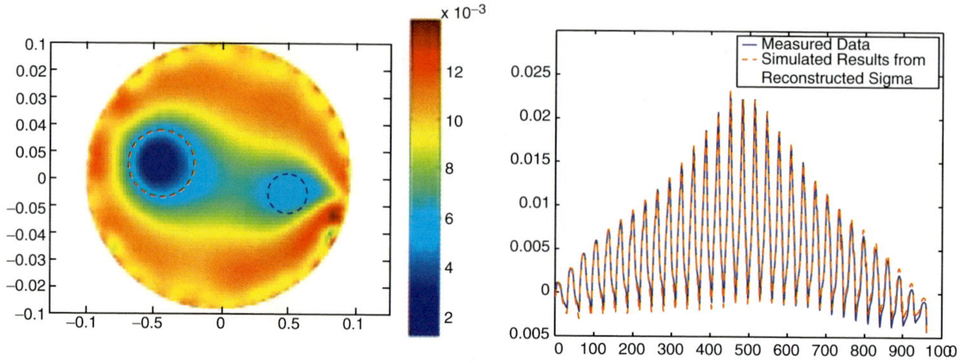

FIGURE 12.11
Left: Image reconstructed from the measured data in Case 2. The dashed circles indicate the ground truth. Right: Comparison of the secondary field between measurements and simulated by forward solver with the reconstructed σ in Case 2.

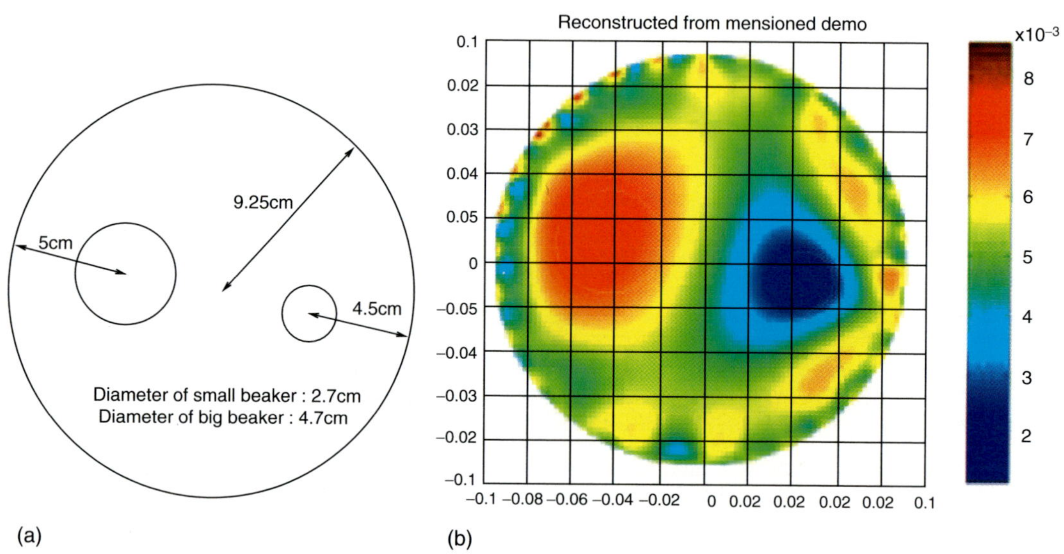

FIGURE 12.12
Left: The setup for Case 3 with one larger metal object and one smaller insulator object. Right: The reconstructed σ. The two circles denote the locations of the objects.

11

Experimental EMF Exposure Assessment

Sven Kühn and Niels Kuster

CONTENTS

11.1 Objectives and Limitations

Experimental electromagnetic field (EMF) exposure assessments can be divided into two categories: (1) densitometry, that is, quantification of the incident EMF or the field characteristics at the location of exposed bodies but without their presence, and (2) dosimetry, that is, quantification of the EMF induced in biological tissues or bodies. In most cases of compliance evaluation, verification that the actual exposure is below a certain value is paramount to accurate determination of the individual exposure. However, highly accurate predictions are needed for some medical applications, for instance, hyperthermia. The focus of this chapter is on dosimetry for frequencies above 10 MHz, often referred to as radio frequency (RF) dosimetry. At lower frequencies, experimental dosimetry is usually limited to measurement of the incident magnetic fields.

11.1.1 Dosimetry

Dosimetry is the science of quantifying the three-dimensional (3–D) distribution of EMF inside tissues and organs of biological bodies. The term is also applied with media having dielectric characteristics similar to biological bodies, for example, cell cultures, tissue-simulating liquids, etc. The induced field is the only field parameter that can interact with biological processes and therefore is referred to as the primary quantity.

Dosimetry usually refers to the assessment of the induced fields on a macro level, that is, the averaged induced fields across cells. Microdosimetry refers to the evaluation of fields across membranes, proteins, etc. This is a new field facing various basic problems such as material models and transitions between classical and quantum electrodynamics. Since the field of microdosimetry has not yet matured and since microdosimetry can be directly developed from locally averaged induced fields in the frequency range considered (<300 MHz), the term dosimetry in this chapter refers to marcodosimetry only.

The distribution of induced fields, particularly at RF, is a complex function of numerous parameters such as frequency, incident field strength, incident angle, field impedance, incident field distribution, polarization, size and shape (posture) of the biological body, tissue distribution, dielectric characteristics of the tissues, etc. In general, the dynamics of the induced field strength range over several orders of magnitude. In other words, the strength and distributions of the fields induced by the same incident exposure greatly vary with anatomy and body orientation with respect to the field and posture.

It is practically impossible to measure the fields noninvasively or *in vivo*; thus, measurements can only be obtained postmortem. The limitations [1] associated with postmortem evaluations include (1) accessibility to only certain tissues, (2) dielectric changes (e.g., lower tissue temperature and decreased blood content), (3) field distortions by the invasively introduced probe, and (4) large uncertainties associated with obtaining accurate measurements near and across tissue boundaries [2,3]. Only the integrated, totally absorbed power can be determined relatively easily by the calorimeter method (see Section 11.3.2.4).

Progress in computational electromagnetics along with the exponential growth of computational power and memory have facilitated determination of the field distributions in full anatomical human bodies with resolutions much smaller than 1 mm^3 (see Chapter 10 on RF modeling by Lin). The lossy and low-resonance properties of complex anatomical structures pose no special challenges for numerical methods (Figure 11.1). Limitations are only due to inadequate phantoms providing insufficient spatial resolution, for example, to accurately represent the scull [3].

Nevertheless, experimental dosimetry is often superior to numerical approaches for the compliance testing of commercial devices. Sources usually consist of highly resonant structures tightly assembled with other electronic and metallic structures. It is difficult to predict with reasonable and known uncertainty whether and how distortion of resonances and excitations of secondary structures might occur through simulations, especially when also considering the scattered fields of biological bodies. Small differences can

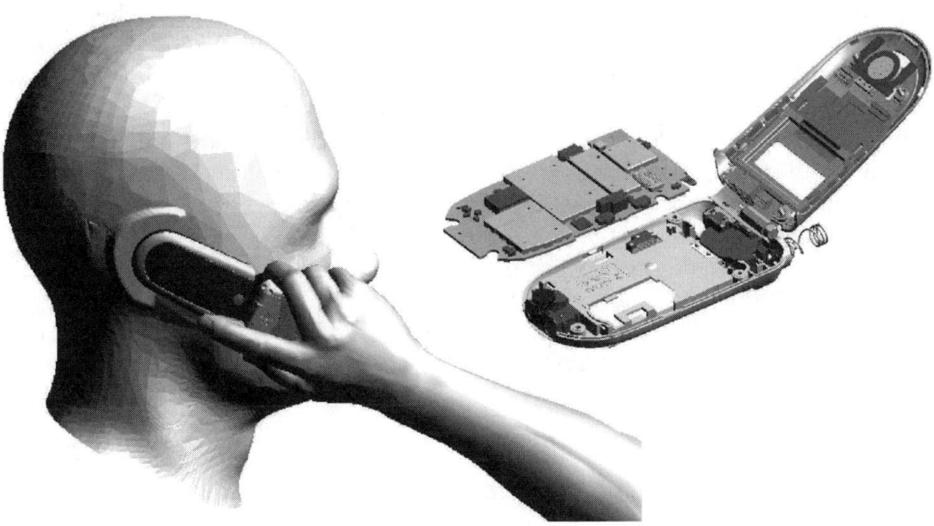

FIGURE 11.1
The latest advances in numerical techniques enable the computation of realistic scenarios without restrictions regarding spatial resolution and material description. The remaining difficulties are material parameters and manufacturing tolerances.

easily result in deviations of more than a factor of 2 from reality. Only when the structure is electromagnetically well-defined can good correspondence between simulations and measurement generally be achieved, that is, with an uncertainty smaller than 20% [4]. On the other hand, experimental techniques allow testing of the physical device under test (DUT) and do not require verification of the computational model.

The detailed information of the field distributions inside anatomical bodies obtainable by numerical methods is often irrelevant, since it cannot be generalized. For safety considerations, the upper boundary of exposure for the entire population rather than individual exposure is relevant. Hence, worst-case phantoms are often applied to assess the upper exposure boundary for specific exposure conditions, for example, from mobile transmitters. Since the degree of freedom for these phantoms is relatively large, they can be chosen homogenously (see Section 11.4.1). Recent evaluation of the procedures for mobile phone compliance testing has confirmed that conservative assessments are possible with a simplified experimental technique using homogenous phantoms [4–6].

During the past decade, considerable progress has been achieved in experimental dosimetric assessment. Since the late 1990s, experimental techniques have been routinely applied for evaluating transmitters operating in the closest vicinity of the body. They have proven to be more reliable, cheaper, and faster. A summary of these techniques and procedures is provided in this chapter. However, dosimetric assessments are limited to homogenous phantoms or to a few tissues in biological bodies. These assessments are normally applied for testing compliance with the basic restrictions or for the validation of results obtained by computational techniques.

11.1.2 Densitometry or Incident Exposure Assessment

Experimental dosimetry requires sophisticated instrumentation, significant expertise, and time. It is impractical for *in situ* exposure assessments. Therefore, easy to apply techniques and methods using a worst-case approach have been developed to determine compliance with potentially hazardous induced fields by determining the incident fields. The worst-case concept derives the conditions for maximum induced fields inside the human body for an incident field strength in terms of the polarization, field distribution (plane wave), field impedance (plane wave), size, and posture of the human body and its dielectric properties. The induced fields are lower than a value X if the local maximum incident field strength of the E field and H field is below the value Y. Attempts to alter this conservative approach have yielded alternatives with an insufficient scientific and engineering basis (see Section 11.5).

Assessment of the incident exposure is simple for plane wave or far-field conditions. Under these conditions, the E field vector is perpendicular to the H field vector, and both are orthogonal to the direction of propagation. The ratio of the E and H fields is equal to the wave impedance Z at any location:

$$Z = \frac{E}{H} \tag{11.1}$$

and in free space

$$Z_0 = 377\,\Omega \tag{11.2}$$

Furthermore, the amplitude is constant over the entire volume of the absent exposed body. In these cases, it is necessary to measure only one component (E or H field) at one location in space. Unfortunately, far-field conditions rarely occur. However, far-field conditions are approximately met locally by changing the amplitude in space at distances larger than the extension of the reactive near-field zone:

$$r > \frac{2D^2}{\lambda}, \quad D \sim \frac{\lambda}{2} \tag{11.3}$$

$$r > \frac{\lambda}{2\pi}, \quad D \ll \frac{\lambda}{2} \tag{11.4}$$

where r is the radius or distance from the transmitting antenna/structure, λ the wavelength, and D the largest antenna dimension.

In other words, for distances meeting the requirements of Equation 11.3 and Equation 11.4, only the maximum of the field components must be determined to demonstrate compliance.

For any distance smaller than the requirements of Equation 11.3 and Equation 11.4, the maximum of both components must be spatially scanned to reliably predict that the maximal induced fields are below a certain limit. Fine volume scanning of transmitter antennas in the very near field yields greater uncertainty and is more time-consuming than dosimetric measurement in homogenous phantoms. Since such near-field assessments are more conservative than specific absorption rate (SAR) evaluations, they are rarely conducted in the context of exposure assessments.

11.2 Fundamental Quantities of EMF

The basic quantities necessary for electromagnetic exposure assessment are summarized in Table 11.1.

TABLE 11.1

Quantities, Symbols, and Units Used in Experimental Exposure Assessment

Symbol	Quantity	Unit
$\vec{E}$	E field (vector)	V/m
$\vec{H}$	E field (vector)	A/m
S	Power density (scalar)	W/m^2
$\vec{S}$	Pointing vector, $\vec{S} = \vec{E} \times \vec{H}$	W/m^2
W	Energy density	W/m^3
SAR	Specific absorption rate	W/kg
ΔT	Temperature increase	K
J	Current density	A/m^2
$\bar{\varepsilon}$	Complex permittivity, $\bar{\varepsilon} = \varepsilon_0 \cdot \bar{\varepsilon}_r$	F/m
ε_0	Permittivity of free space, $\varepsilon_0 = 8.854 \times 10^{-12}$	F/m
$\overline{\varepsilon_r}$	Complex relative permittivity, $\overline{\varepsilon_r} = \varepsilon_r' - j\varepsilon_r''$	
σ	Conductivity	S/m
μ	Permeability, $\mu = \mu_r \cdot \mu_0$	H/m
μ_0	Permeability of free space, $\mu_0 = 4\pi \times 10^{-7}$	H/m
μ_r	Relative permeability	
c	Specific heat capacity	J/(kg K)
ρ	Mass density	kg/m^3

11.2.1 Primary Quantities: Basic Restrictions

Restrictions on exposure to time-varying electric fields, magnetic fields, and EMF that are based directly on established health effects are termed basic restrictions (see Chapter 8 on standards by Van Deventer et al. [68]). Depending on the frequency of the field, the physical quantities used to specify these restrictions are either the current density (J) or the SAR. The dosimetric quantities used in current guidelines [7–10] are J for frequencies up to 10 MHz and the SAR for the frequency range from 100 kHz to 10 GHz. J is related to the internal electric field by Ohm's law:

$$J = \sigma E \tag{11.5}$$

E is the internal electric field, and σ is the complex conductivity of the tissue.

SAR is the ratio of the average rate of the absorbed power to the absorbing mass. It is defined as follows:

$$\text{SAR} = \frac{d}{dt}\left(\frac{dW}{dm}\right) = \frac{d}{dt}\left(\frac{dW}{\rho dV}\right) \tag{11.6}$$

where dW is the incremental energy dissipated in an incremental mass dm included in an incremental volume dV and ρ is the mass density. SAR can also be calculated directly from the electrical loss, which is proportional to the mean square of the locally induced electric field strength E:

$$\text{SAR} = \frac{\sigma E^2}{\rho} = \frac{J^2}{\sigma\rho} \tag{11.7}$$

and to a temperature increase by:

$$\text{SAR} = c\frac{dT}{dt} \tag{11.8}$$

where c is the local specific heat capacity of the tissue. Equation 11.8 is valid only if the exposed body is in thermal equilibrium or in a steady thermal state at the beginning of the exposure, and either heat exchange processes can be neglected during the measurement interval or the processes are known to correct dT correspondingly. Current safety standards [7–10] for limiting EMF exposure provide maximum limits for basic restrictions for the uncontrolled/general public as well as for controlled/occupational exposure over the whole considered frequency range. The standards are ambivalent with respect to the quantity SAR. The debate among experts is whether SAR is a dosimetric quantity only relevant as a surrogate for thermally based models of EMF interaction or whether it can describe effects in addition to those related to temperature. SAR is directly related to the induced internal E fields as well as to the current density (Equation 11.7) but only directly related to the induced H fields for special cases.

11.2.2 Derived Quantities: Reference Levels or Maximum Permissible Exposure

The derived quantities are the electric field strength (E), magnetic field strength (H), magnetic flux density (B), and power density (S). As discussed in the previous sections, these quantities were derived from the basic restrictions using experimental or computational methods and represent conservative limits for worst-case exposure scenarios. Derived limits are called reference [7] or maximum permissible exposure levels [8,9]. Compliance with these quantities implies compliance with the basic restrictions. However, if the reference quantities exceed the derived limits, the relevant basic restrictions are not necessarily exceeded. In such cases, compliance can be demonstrated by dosimetric means.

In summary, reference levels are easy to assess if the plane wave or far-field conditions are met (see Section 11.1.2) and the resulting SAR and induced current densities are below the corresponding basic restrictions under all circumstances. Typical reference limits for occupational/controlled and general public/uncontrolled exposure are given in Refs. [7–10].

11.3 Experimental Techniques

11.3.1 Field Probes

E field probes have been used for microwave measurement since the early 1970s. The first power meters were developed by Aslan [11] and Rudge [12]. Aslan used a thermocoupling model consisting of two pairs of thin-film, vacuum-evaporated electrothermic elements that functioned as both an antenna and a detector. Rudge employed two small diode-loaded dipoles as sensor elements. The first prototype of an isotropic, miniature field probe was introduced by Bassen et al. [13] in 1975. Additionally, fiber optic field probes had already been proposed in the 1970s [14]. Comprehensive overviews of field probes are reported in Refs. [2,15].

11.3.1.1 Broadband E Field Probes

Diode-based field probes are well established and commonly used for dosimetric assessment. These probes consist of an appropriate field sensor, a detector, transmission lines, and read-out electronics (Figure 11.2). Three mutually orthogonal diode-loaded dipoles with an isotropic receiving pattern constitute the probe. To achieve good spatial resolution and broadband performance, electrically short dipoles are employed:

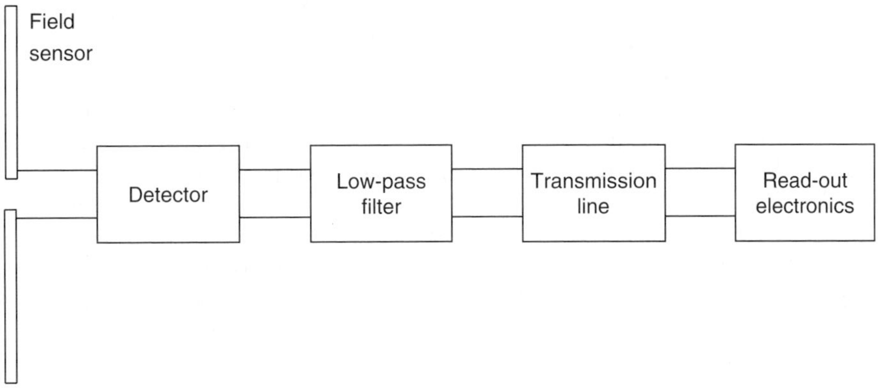

FIGURE 11.2
Simplified schematic of a broadband field probe.

$$\beta h = \frac{2\pi h}{\lambda} \ll 1 \tag{11.9}$$

where h is the length of a dipole arm, β the propagation constant, and λ the wavelength [15]. A flat frequency dependence cannot be achieved if the length of the probe is larger than a fraction of the wavelength ($<0.05\lambda$). However, this limitation can be overcome by gradually resistive dipoles [16].

Typical isotropic E field probe sensors are shown in Figure 11.3. An RF detector diode (usually Schottky type) is located in the center of the dipole arms. If the detector diode is operated in the square-root law region, the diode current is proportional to the RF power delivered over the dipole to the detector diode. The detector diode is connected with a highly resistive line to the data acquisition electronics. The transmission lines must be designed precisely to eliminate distortions, such as parasitic sensor elements and scattering, which cause degradation of the pattern and unwanted polarization characteristics of the actual receiving antenna. High-resistive transmission lines are ideal for minimizing field perturbation and pickup effects. A detailed investigation of transmission line design

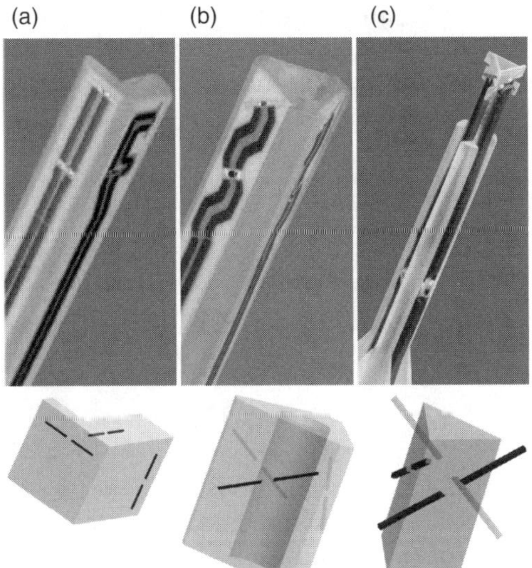

FIGURE 11.3
E field probe configurations. (a) E field probe for measurements in air with one sensor aligned to the probe axis and two orthogonally, (b) dosimetric probe for E field measurements in tissue-simulating liquids allowing integration of an optical proximity sensor in its center, and (c) miniature dosimetric probe with interleaved dipoles. (From IT'IS Foundation. With permission.)

can be found in Smith [17]. Current probe designs apply either thin-film ($R \sim 10\ \text{M}\Omega/\text{m}$) [14] or thick-film ($R \sim 500\ \text{M}\Omega/\text{m}$) [18] techniques.

The induced fields are recorded by the read-out electronics. The electronics typically consist of a measurement amplifier and an analog digital converter. The read-out field values are then optically forwarded. The optical transmission provides a galvanic decoupling and reduction of field-perturbing conductive parts near the DUT.

In addition to *E* field probes, *H* field probes are also available, the basic theory of which can be found in Whiteside and King [19]. *H* field probes and *E* field probes are basically similar, except for the field sensor element, that is, *H* field probes employ a small loop element instead of a dipole sensor. The disadvantages of loop-based sensors include a strong frequency dependence and currents induced by both *H* and *E* fields. Different methods for flattening the frequency dependence of loop probes have been proposed in Refs. [2,20]. The ratio of the voltage induced in a circular loop with diameter *d* by the *E* field and the *H* field is [19]:

$$\frac{V_E}{V_H} \simeq -\mathrm{j}2\pi\frac{d}{\pi} \tag{11.10}$$

Therefore, for electrically small loops only, that is, $d/\lambda \ll 0.01$, the current will mainly be determined by the magnetic field. In Pokovíć [2], lossy covers were proposed to suppress the *E* field sensitivity of the loop.

In contrast to dipole-based sensors, thermocouple probes are true square-root law [a1]detectors. Such sensors are, for example, used in free space field probes by Narda [21]. These sensors are, however, impractical for dosimetric and near-field applications because of their generally lower sensitivity and dynamic range.

Thermistors are also true square-root law detectors and also small. They can have higher resolution than thermocouples but need more frequent calibration.

Based on the setup of these isotropic field probes, the measured field magnitude is yielded from the root sum square of the three orthogonal components:

$$|X| = \sqrt{|X_1|^2 + |X_2|^2 + |X_3|^2}\ X \in \{E,H\} \tag{11.11}$$

This summation is carried out regardless of any phase differences between the respective components. Hence, if maximum hold peak detectors are applied, the field reading represents the upper bound of the field magnitude.

The output signal of the probes just described is dependent on the following parameters:

- Frequency, modulation, and field strength
- Polarization, direction, and field gradients
- Material boundaries near the probe sensors
- Sources of interference (noise, static and low-frequency fields, vibration, temperature, etc.).

Therefore, it is necessary to quantify the influence of these parameters. Calibration under well-defined conditions is carried out to characterize the most crucial parameters for each probe individually. A detailed summary of different calibration methods for field probes and characterization of the parameters contributing to the measurement uncertainty is given in Pokovíć [2]. All possible influences must be included in the resulting uncertainty assessment, since the conditions during the actual application of the probes differ

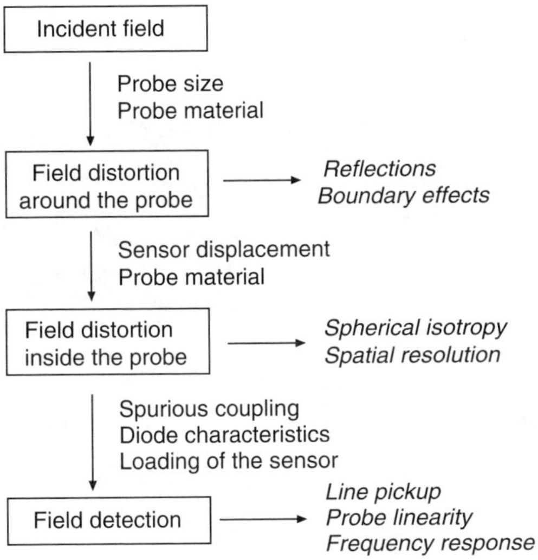

FIGURE 11.4
Impact of the constructional details on the probe characteristics.

considerably from the calibration scenario. Figure 11.4 provides an overview of different probe construction parameters that contribute to the probe characteristics.

Modern dosimetric field probes are available in the frequency range from 10 MHz up to 6 GHz. They have an isotropy error lower than ± 0.5 dB and sensitivities in the range of 1–10 μW/g [22]. Modern probes have very small sensor tip dimensions (2.5 mm) and high spatial resolution, allowing measurements very close to material boundaries. A probe with reduced size (tip diameter 1.0 mm) was introduced by Poković et al. [23] and enables accurate dosimetric measurements for frequencies exceeding 10 GHz. Although this probe consists of only a single sensor element, isotropy is obtained by 120° rotation around the probe axis and an appropriately aligned sensor element. Probes for determining both the electrical and the magnetic field vector information are presented by Poković et al. [24]. This technique is based on measurements of five or more field components in space and reconstruction of the ellipse parameters by a combination of a downhill simplex and a Givens updating algorithm.

11.3.1.2 *Electro-Optical Sensors*

In contrast to RMS sensors, it is possible to measure fields over a broad frequency band in the time domain by exploiting electro-optic methods. Recent advances in semiconductor and photonic crystal research have provided the foundation for the next generation of near-field measurement equipment based on electro-optical methods. The basic concept of electro-optic probes is shown in Figure 11.5.

The most important electro-optic effects for measuring EMF are the Pockels effect and the quantum confined Stark effect (QCSE). The Pockels effect describes the refraction index dependence on the induced *E* field of anisotropic dielectric crystals such as cadmium telluride (CdTe) [25] or lithium nobiat ($LiNbO_3$) [26,27]. *H* field-sensitive crystals have also been determined, for example, cadmium manganese telluride (CdMnTe) [25]. The QCSE is based on the change of the absorption spectrum of a semiconductor structure under the influence of the electric field. Sensors exploiting the Pockels effect are reported in Refs. [25–27]. A detailed summary of electro-optic phenomena, materials, and applications can be found in Agullo-Lopez et al. [28]. Alternatively, the signal from the field sensor can modulate a small laser diode rather than the external electro-optical modulation.

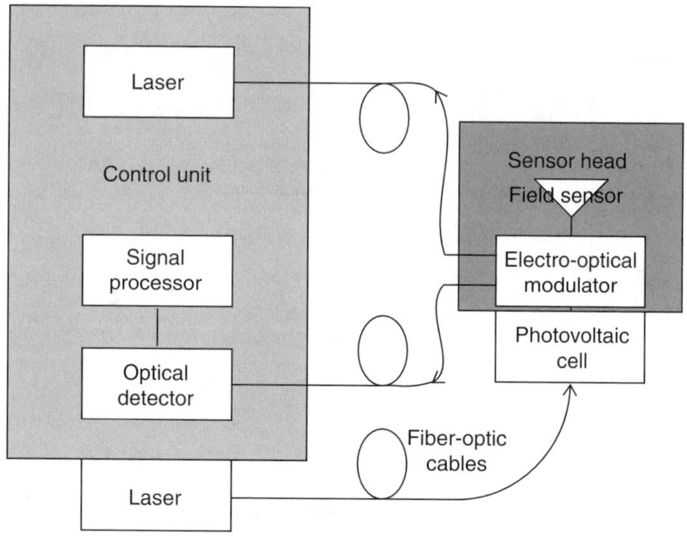

FIGURE 11.5
Principle of an electro-optic probe for EMF measurements.

The most common modulators are the Mach–Zehnder interferometer (MZI) as well as the already mentioned method of the direct modulation of the laser-diode current. The MZI (Figure 11.6) splits the light of a single-mode fiber into two parallel branches of $LiNbO_3$ waveguides. The feedpoints of the dipole are connected to electrodes that are located between the parallel branches, resulting in the antipodal modulation of the phase of the light. For recombined light, the phase modulation results in an intensity modulation. Sensitivities of 1–10 mV/(m Hz) can be reached. The upper cutoff frequency of these sensors is primarily determined by the length of the branches. For high sensitivity, the transit time through the interferometer must be much lower than the periodic time of the signal being measured. In Refs. [26,27], interferometers with a 3-dB bandwidth of 10 GHz were realized. A possibility to overcome this limitation could be the application of traveling wave modulation [29].

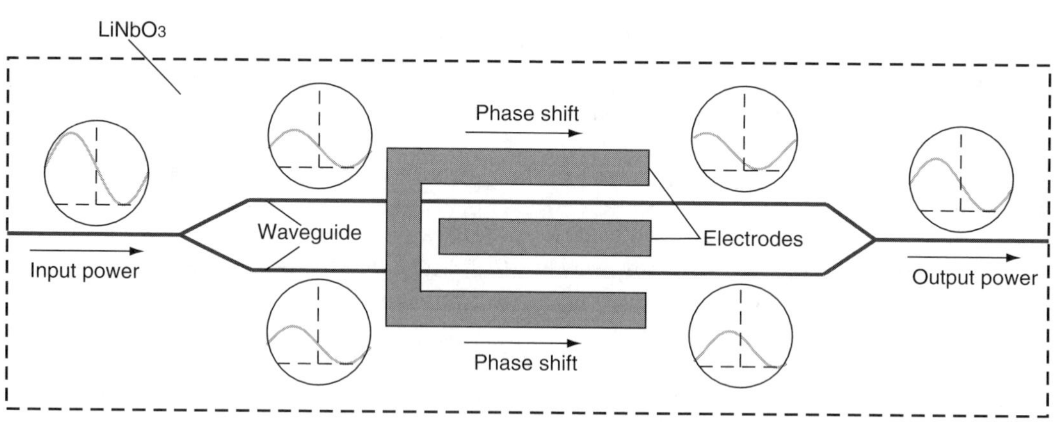

FIGURE 11.6
Schematic of an MZI. The light through the waveguide is modulated antipodally by the electrodes by changing the refractive index of the $LiNbO_3$ substrate.

Another approach reported in Ref. [25] uses only nonconductive parts in the sensor tip. Different types of crystals that are sensitive to electric and magnetic fields are used. Linear polarized light passing through the electro-optic crystal is reflected on a dielectric mirror and passes through the crystal again, where the polarization is changed. A beam-split polarizer transduces the ellipticity change into a change of the optical intensity. Sensors with a sensitivity of 10–400 V/m and 0.1–0.8 A/m in the frequency range from DC to 2 GHz were realized.

For modulation using QCSE, semiconductors with pn- or pin-junction structures are used. A varying reverse bias over the junction modulates the electric field and consequently the absorption and reflection spectra of the semiconductor as well. If the modulator is operated at a constant frequency, different reflection and transmission characteristics of the light and an intensity modulation result.

In addition to passive modulation (QCSE), direct modulation of a vertical cavity surface emitting laser (VCSEL) [30–32] has recently been applied. The current through the laser is directly modulated by the measurement signal. The VCSEL is especially suitable for modulation because of its high bandwidth, low threshold current, and low noise.

The MZI is the most popular modulation method among the three mentioned, and several commercial products are available based on this technology. Disadvantages include the limited ability to reduce its size as well as the bandwidth limitations. Passive modulators are relatively simple to manufacture as integrated circuits. Although passive modulators are inferior to MZI regarding sensitivity, the development of miniature field probes applying the VSCEL technology is currently the most promising approach.

11.3.2 Temperature Instrumentations

11.3.2.1 Temperature Probes

The locally induced SAR can also be assessed by temperature measurements, as summarized in Equation 11.8. A typical temperature rise curve as applied for SAR measurements is shown in Figure 11.7. Two types of temperature probes exist: those based on thermistors and those based on optical effects. The requirements for temperature probes for SAR assessments are:

- *Small size:* The probe must be small to resolve high temperature gradients, without disturbing the temperature distribution as well as the RF field.
- *Nonconductive materials:* Only electrically nonconductive materials prevent heating of the probe by induced currents and are transparent to EMF.
- *Low noise level:* Especially for dynamic temperature measurements, for example, SAR, small differences must be detected accurately. Therefore, the noise level should be much smaller than 10 mK.
- *Short reaction time:* A short reaction time is essential for SAR measurements, since the temperature rise ($\mathrm{d}T/\mathrm{d}t$) is only proportional to the SAR if heat diffusion does not occur. An appropriate probe must have reaction times much lower than 100 msec [33].

Temperature probes for SAR measurement using thermistors were first described by Bowman [34]. These probes utilize high-resistance thermistors that are connected through high-resistance lines to the read-out electronics. In Burkhardt et al. [35], a temperature probe based on a VITEK thermistor (BSD Medical Devices, U.S.A.) is presented. This probe has a noise level of 5 mK, a sensitivity of 5 mK/sec, and a tip diameter of

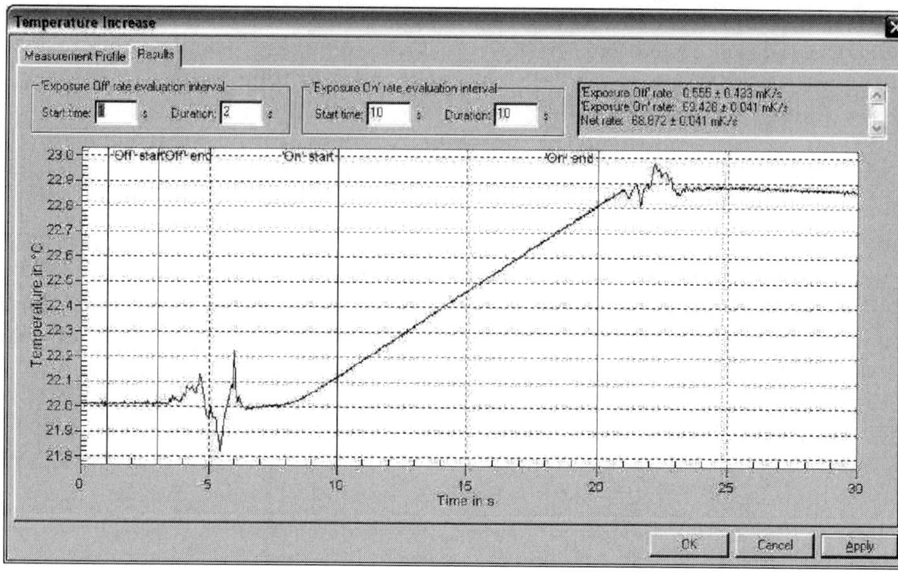

FIGURE 11.7
Typical temperature rise curve for SAR assessment. The linear part of the rise curve is applied for the calculation of the SAR. (Acquired using DASY4 from SPEAG, Switzerland.)

1 mm; however, it has a relatively slow response time of 240 msec. In Schuderer et al. [36], a novel temperature probe design for dosimetric assessments was introduced. The probe provides a spatial resolution of 0.02 mm^3, noise level of 4 mK, a sensitivity of 0.5 mK/sec, as well as a response time of only 10 msec.

In addition to the thermistor concept, temperature probes based on thermo-optical effects are also available. These probes are applied in high-voltage transformers, industrial microwave ovens, and hyperthermia treatment. One of the exploited effects is the decay rate of a phosphorescent layer at the tip of a fiber optic cable [37]. These modern commercially available probes have a noise level of 0.1 K and reaction times of 250 msec. Another exploited effect is the interferometric application of a cavity that is filled with materials that have highly temperature-dependent refraction indices. These probes reach sensitivities of 2–3 mK/sec [35].

11.3.2.2 Infrared Photography

The measurement of temperature by black body-equivalent radiation (infrared photography) is an alternative to invasive measurements using temperature probes. The resolution of infrared thermographs can be very high (<1 mm), and the sensitivity of affordable systems has continuously improved over the years. This was also one of the first methods to measure SAR [38], since the surface radiation can be recorded quickly using infrared cameras without perturbation of the incident field. Infrared cameras were also used to measure the temperature of Global System for Mobile Communications (GSM) mobile phones [39]. The technique has several disadvantages:

- It has limited sensitivity compared to temperature or dosimetric probes.
- It is limited to surfaces.

- The thermal radiation characteristics of the materials must be determined accurately.
- The background radiation must be homogenous.
- Evaporation and convection can cause substantial errors and must be controlled.
- Different aspect angles of the camera can cause different results.

11.3.2.3 Microcapsulated Thermo-Chromic Liquid Crystals

A novel idea to assess 3-D temperature distributions optically and in quasi real time was proposed by Suzuki et al. [40]. Microcapsulated thermo-chromic liquid crystals are suspended uniformly in a gel with the dielectric properties of human muscle tissue. The temperature of the gel is determined by measuring the light scattered from a laser beam that scans through the liquid. The technique has a limited dynamic range and sensitivity.

11.3.2.4 Calorimeters

Calorimetry specifies methods for measuring heat due to biological, chemical, or physical processes that are endothermic or exothermic. Calorimetric methods are suitable for determining average whole-body SAR, but they cannot reveal SAR distributions.

Calorimetry can be subdivided into two types:

- *Direct calorimetry:* The heat is determined directly using calorimeters.
- *Indirect calorimetry:* The amount of expressed heat is determined indirectly by measuring the amount of oxygen consumption and relating it to the oxicaloric equivalent of the reaction.

Basically, calorimetric dosimetry analyzes the heating and cooling processes of an exposed sample. Typical calorimeters used in microwave dosimetry are the Dewer flask calorimeter and the twin-well calorimeter [41].

11.3.3 Measurement Antennas

Different types of broadband matched antennas are usually applied for the frequency-selective exposure assessment of external fields. These measurement antennas are matched to 50 Ω to be compatible with standard RF measurement equipment. They are usually applied for far-field measurements of emissions by cellular telephony base stations, broadcast services, etc. Common broadband RF measurement antennas such as horn or logarithmic periodic antennas have a certain directivity. For example, for an antenna with a 45° beam width, more than 18 measurement directions are necessary for each polarization, since the receiving patterns do not have the shapes of square sectors. This reduces the applicability of these antennas for complex propagation scenarios, that is, locations where the received field is not dominated by a direct line of sight propagation path but by multipath reception. Tuned dipole antennas do not show a directivity in the radial direction, but they lack broadband operation. Conical [42] and biconical [43] antennas have the advantage of nondirectiveness in the radial direction and generally good broadband characteristics. If, for example, the ADD3D method [44] is applied in combination with a conical dipole antenna, the measurement orientations for a fully isotropic scan can be reduced to three different directions. In this case, the isotropy is obtained by an antenna alignment similar to near-field probes, such that the resulting field is:

$$|E| = \sqrt{|E_1|^2 + |E_2|^2 + |E_3|^2} = \sqrt{|U_1|^2 + |U_2|^2 + |U_3|^2} \cdot AF \tag{11.12}$$

where AF is the frequency-dependent antenna factor in linear quantities (1/m) and U_i is the antenna output voltage at the three different antenna orientations. To obtain the antenna output voltages, a measurement receiver (see Section 11.5.1) connected via a well-characterized cable is applied. The antenna output voltage is:

$$|U_{i_{\text{antenna}}}| = |U_{i_{\text{receiver}}}| \cdot \text{ATT}_{\text{cable}} \tag{11.13}$$

where $|U_{i_{\text{receiver}}}|$ is the voltage measured with the receiver and $\text{ATT}_{\text{cable}}$ the attenuation of the cable in linear units. Other sources of attenuation, for example, attenuators to reduce the Voltage standing wave ratio (VSWR) of the measurement antenna, must be similarly considered.

11.4 Near-Field Scanners

11.4.1 Scanners for Dosimetric Compliance Testing

In dosimetric compliance tests, exposures from RF transmitters operating in close proximity to the body, for example, mobile phones or body-worn wireless devices, are compared to the basic restrictions (see Section 11.2.1). This is a complex task, since the exposure greatly depends on the device design, the position of the device with respect to the body, the external and internal anatomy, as well as the effect of the backscattered field on the device (Figure 11.8). In order to obtain a reliable assessment with acceptable effort, the standards must determine the maximum or 90th percentile exposure for the entire

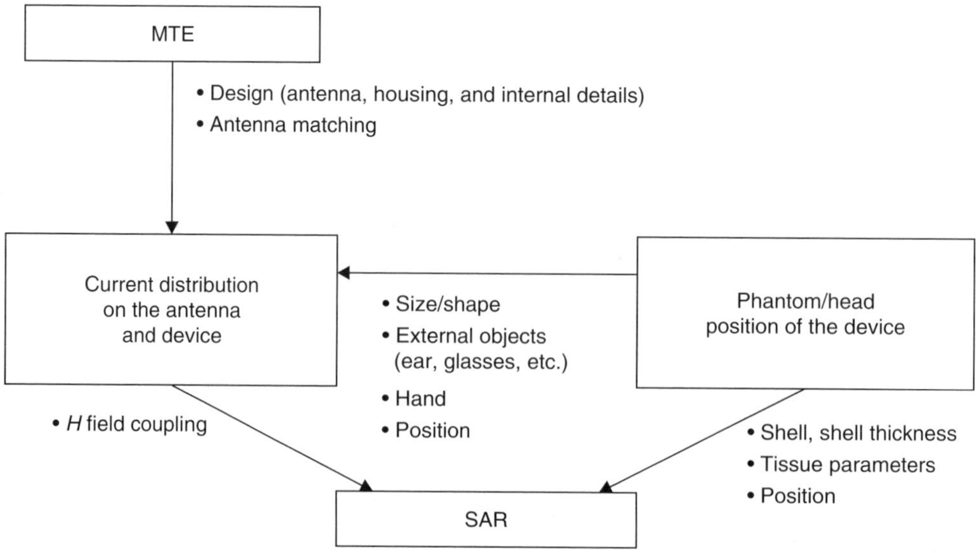

FIGURE 11.8
Overview of impact parameters for dosimetric compliance testing of mobile terminal equipment (MTE).

population that may be operating the DUT. Measurement standards have been developed by different organizations, for example, by the International Electrotechnical Commission (IEC), the Institute of Electrical and Electronics Engineers (IEEE), the Association of Radio Industries and Businesses (ARIB), Korea Electromagnetic Engineering Society (KEES), and others.

In general, a dosimetric evaluation requires the measurement of several hundreds of points distributed over a complex 3-D phantom. The task is divided into (1) searching for the location of the maximum absorption on a 2-D grid and (2) determining the peak spatial SAR value on a fine 3-D grid. Especially at high frequencies, these points must be determined with high accuracy to achieve low measurement uncertainty despite high attenuation and large spatial field variations. Automated dosimetric assessment systems are utilized to perform these compliance tests. A typical configuration of a dosimetric assessment system is shown in Figure 11.9 and Figure 11.10. A computer-controlled six-axis positioner with excellent positioning repeatability of ± 0.2 mm is used to move the dosimetric E field probe within the scanning grid, which can be adaptive, for example, along a surface that is being detected during the scanning job. The different field probe designs have already been discussed in Section 11.3.1. Phantoms, for example, the specific anthropomorphic mannequin (SAM) phantom, elliptical phantom, etc., and tissue-simulating liquids (see Table 11.2) have been developed and validated with respect to the 90th percentile requirements [45]. A detailed description is provided by Schmid et al. [18]. Because of the strong curvature of the SAM phantom, sufficient accuracy can be obtained only when the measurements are taken with the probes aligned normal to the phantom boundary. The most elaborated components, procedures, and algorithms are employed (Figure 11.10), and an expanded total uncertainty ($k = 2$) for the compliance test of less than 20% can be achieved with even better repeatability.

It should be noted that this approach provides reliable conservative estimates of the maximum peak spatial SAR that might occur in the user population but little information about the exposure of specific tissues.

11.4.2 Fast Dosimetric Scanners

Dosimetric compliance tests of near-field sources are extremely time-consuming, especially as all configurations with all accessories must be evaluated. SAR patterns provide

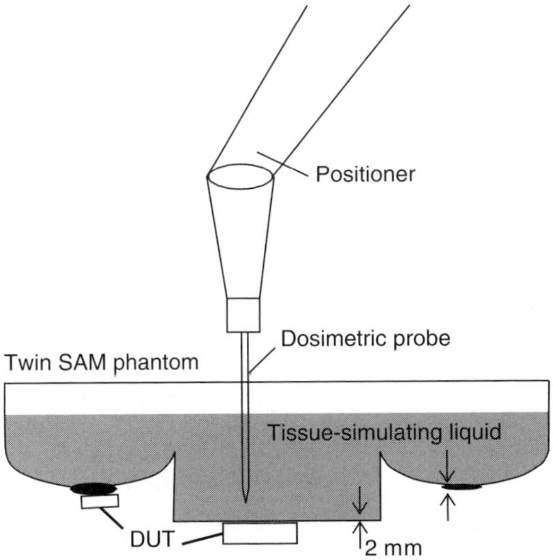

FIGURE 11.9
A dosimetric assessment systems consists of a computer-controlled positioner, a dosimetric field probe, and a phantom (left and right head, flat region for system validation) filled with tissue-simulating liquid.

FIGURE 11.10
DASY4 dosimetric assessment system. (From SPEAG, Switzerland.)

further valuable information such as (1) the output power under loaded conditions as indicated by a different peak spatial SAR if the absorption pattern is not altered and (2) changes in the internal RF path, for instance, by poor contacts, as indicated by a modified SAR pattern. Therefore, fast SAR scanners are desirable for precompliance testing as well as for research and development. Such scanners would also be greatly beneficial for

TABLE 11.2
Electric Parameters of Head and Body Tissue Equivalent Liquids at Various Frequencies

	Head		Body	
Frequency (MHz)	ε_r	σ (S/m)	ε_r	σ (S/m)
150	52.3	0.76	61.9	0.80
300	45.3	0.87	58.2	0.92
450	43.5	0.87	56.7	0.94
835	41.5	0.90	55.2	0.97
900	41.5	0.97	55.0	1.05
915	41.5	0.98	55.0	1.06
1450	40.5	1.20	54.0	1.30
1610	40.3	1.29	53.8	1.40
1800–2000	40.0	1.40	53.3	1.52
2450	39.2	1.80	52.7	1.95
3000	38.5	2.40	52.0	2.73
5800	35.3	5.27	48.2	6.00

quality assurance purposes, when integrated into production lines (every device tested for compliance, output power, and RF performance).

An algorithm to extrapolate peak spatial SAR values from the 2-D area scan performed at the phantom surface was presented by Kanda et al. [46]. Another algorithm reducing the measurement time of an area scan lasting 5–10 min was proposed by Merckel and Fleury [47].

Two additional approaches have been presented to further accelerate the assessment. The first approach, using a scanner based on incident *H* field measurements, is described in the next section. The second approach is based on a sensor array implanted in a solid flat phantom (Figure 11.11). The phantom is filled with a broadband tissue-simulating gel (300 MHz up to 6 GHz), with sensors located 4 mm below the surface. The density of the sensor array (15 mm) is sufficient to reliably assess the exposure. The measured SAR values of all sensors are acquired and integrated in parallel, such that the total assessment time is less than 3 sec, even for TDMA (Time Division Multiple Access) signals with complicated frame structures.

11.4.3 Incident Near-Field Scanners

In the reactive near field, field gradients are generally very high, and the field impedance differs greatly from the far-field impedance and rapidly changes over short distances. Hence, both *E* and *H* fields must be assessed, based on requirements regarding spatial resolution and isotropy that are similar to those of dosimetric scanners. In addition, distortions by reflections from instrumentation must also be carefully evaluated and included in the uncertainty budget.

General purpose scanner systems are equal or similar to those used for dosimetric evaluations, except that *E* and *H* field probes optimized for free space are used instead of

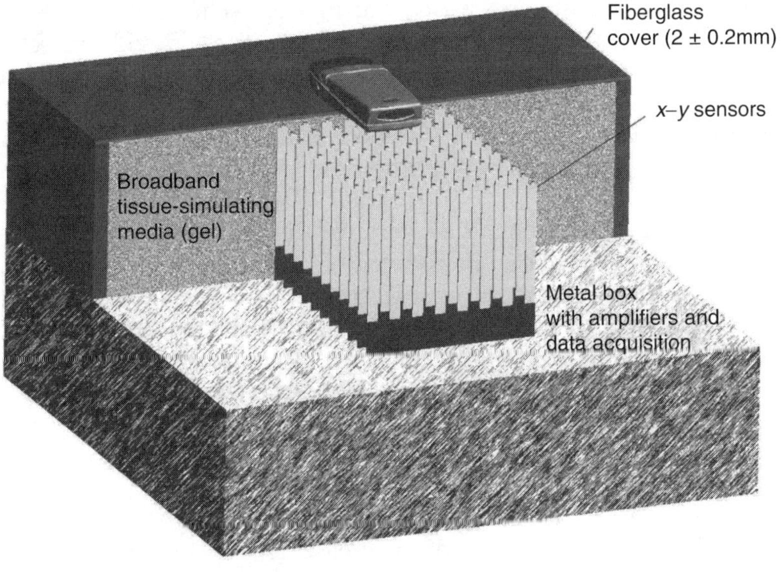

FIGURE 11.11
The concept of a fast dosimetric scanner based on a dense array of sensors completely immersed in a broadband tissue-simulating medium. In order to provide a quick response, the signals of all sensors must be amplified and integrated in parallel for the frame structure and to increase the signal-to-noise ratio. The high-resistive sensor leads are vertical to minimize field distortions.

dosimetric probes. Since the scan procedures do not need to follow complex surfaces, three-axis systems [48] are also suitable. However, such evaluations are rarely used to test compliance with safety limits, since the limits are significantly more conservative and the measurement resources, time required, and uncertainty budget are higher than for dosimetric assessments. Nevertheless, such measurements are valuable for the validation of numerical results [49] or to test compliance with interference limits in the close near field, for example, for hearing aids [50].

A specialized near-field scanner was proposed and developed to test the precompliance of mobile handsets with safety limits [51]. Its underlining concept is based on the primary interaction mechanism of near-field exposure [52], that is, the local SAR is approximately proportional to the square of the incident magnetic field at the surface of the phantom:

$$\mathrm{SAR} = \alpha H_s^2 \tag{11.14}$$

The magnetic field is scanned in a reference plane above the DUT using a loop antenna array. The factor α is determined by relating a traditional compliance measurement inside the phantom to the magnetic field scan in free space for a certain DUT. The system can predict the SAR within 30 sec. The main disadvantage of such a system is that it does not simulate a realistic load to the antenna of the DUT.

11.5 Incident Field Evaluations in the Far Field of Transmitters

Evaluation of the exposure in the far field of a transmitter is usually conducted for fixed transmitters such as radio and TV broadcast antennas, radar sites, or cellular base stations. Exposure assessments are carried out in areas that are generally accessible or restricted to qualified working personnel. Compliance is tested with the reference limits by assuming free space field impedance, that is, by E field evaluation. As described in Section 11.1.2, only one measurement point is required under real far-field conditions. However, actual environments usually involve nearby reflectors, that is, a scanning procedure or statistical knowledge about the field distribution is needed to determine the maximum exposure.

Broadband instant measurements are often insufficient, since the evaluated transmitters do not always operate at maximum power, for example, in the case of base stations, the transmitted power is dependent on the traffic. In such cases, information on the maximum exposure with respect to the measured exposure must be available and soundly applied to extrapolate the worst-case exposure. Table 11.3 lists the parameters that are necessary to extrapolate the worst-case exposure and to reduce the uncertainty of the actual measurement campaign. It is easier to determine the measurement methods when additional parameters are known. General sources of error are:

- Field perturbation by measurement personnel, for instance, reflection and absorption of EMF due to the body of the measurement engineer
- Application of the measurement antenna, for example, nonobservance of antenna directivity and polarization
- Application of ineffectively decoupled cables, for instance, acting as secondary antennas

TABLE 11.3

Important Parameters of RF Transmitter Sites Assessed in the Far Field

Site Parameter	Explanation
Location	The location of the transmitter with respect to the measurement point
Line of sight/nonline of sight	Determines if a prevalent propagation path may be expected
Type of site	Single- or multiple-antenna site
Antenna directivity	Antenna beam characteristics
Antenna radiation direction	The direction the transmitter radiates
Antenna power at measurement	The antenna input power at the time the measurement takes place
Maximum antenna input power	Maximum permissible antenna input power
Frequency	Frequencies at which the site transmits
Communication system	Communication system that is used, that is, which signal characteristics are expected
Other sources of radiation	The field at the measurement point if the assessed transmitter is switched off

- Application of the measurement receiver, for example, incorrect measurement settings
- Selection of the measurement point, for example, measurement points that are not feasible to give the maximum EMF exposure, measurement points close to bodies that influence the measurement antenna's calibration.

Different methods for assessing EMF exposure in the far field have been proposed. One approach is the scanning method. This method requires the engineer to slowly move the measurement antenna with varying polarizations and directions through the volume of interest [53]. The measurement receiver operates in maximum hold mode during the assessment, that is, the maximum field value is determined and compared to the reference values. Another method is based on the examination of several well-defined points in the area of interest. In this case, the antenna is mounted on a tripod, and the different directions and polarizations are examined at the considered points [54]. The first method is conservative but sensitive to the position of the measurement operator with respect to the antenna. The second method can be performed with the measurement engineer further away, but the number of measurements in the volume is small. A combination of both methods is presented in Coray et al. [55], that is, first the region is scanned for the field maximum in the area of interest, and then an isotropic and frequency-selective measurement is performed at the location of the maximum.

Far-field techniques are also often employed in the near field of transmitters, for example, on transmitter towers. Some standards allow a spatial averaging of *E* field evaluations [9], the rationale of which is based on the whole-body SAR limit. However, this constitutes a relaxation of the safety concept, since it does not consider *H* field coupling as the dominant mechanism in the near field nor the limits of peak spatial SAR. Based on current knowledge, such relaxations do not exclude violations of the basic restrictions.

The advantages and limitations of different measurement techniques to assess exposure from unknown transmitters are discussed below.

11.5.1 Broadband Measurements

Broadband measurements are especially applicable for survey measurements. The field is measured and automatically summed over a broad frequency range. No information on

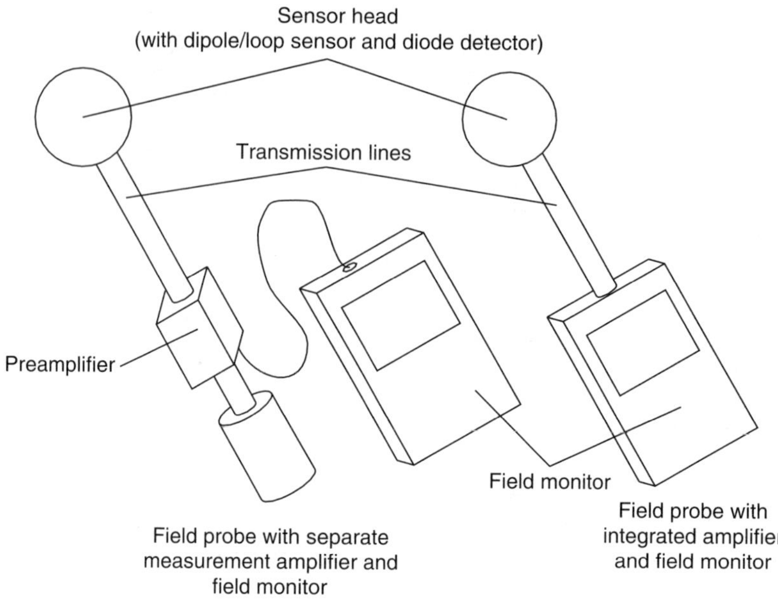

FIGURE 11.12
Schematic of the most common designs of broadband RF survey meters.

the spectral characteristics of the field is available. Therefore, if a broadband meter is used for compliance testing, the measured field value must comply with the lowest permissible limit defined in the measurement range of the meter. Broadband survey meters are relatively inexpensive and easy to apply. Hence, these probes are often used for field measurements.

Figure 11.12 displays typical broadband field survey meter designs. Figure 11.13 displays the frequency response of two broadband probes. The field value measured with probe 1 must comply with the lowest limit in the frequency range from 10 MHz to 1 GHz, whereas probe 2 must comply with the lowest limit between 100 MHz and 10 GHz. The overlapping frequency range is counted twice if the exposure values are added to cover the entire frequency range. Some broadband probes are designed to reflect the frequency dependence of the limits. In all cases, it is advised that the off-band response of

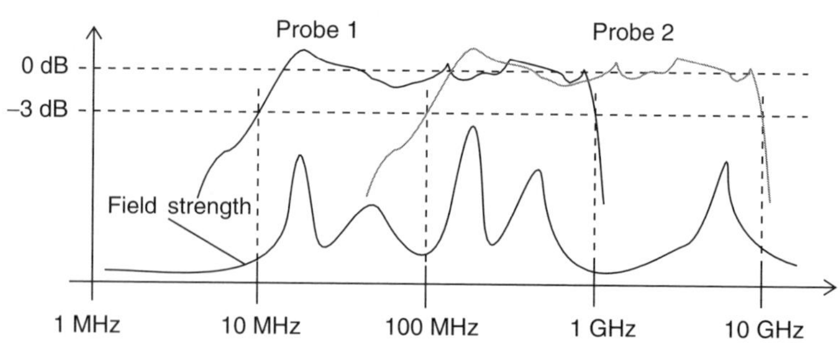

FIGURE 11.13
Frequency response of two broadband field survey meters. The fields in the frequency ranges are summed up. If the fields from probes 1 and 2 are summed again, then the fields in the overlapping frequency range are accounted for twice.

these probe systems be carefully evaluated. If a specific transmitter is the dominant source, compliance testing is greatly simplified [56].

In addition to the frequency response, broadband probes have a certain time-domain transfer function. When pulsed fields are measured, this response must be compensated. However, for this compensation, information on the time-domain characteristics of the measured field is necessary. In summary, the main sources of uncertainty regarding broadband survey meters are:

- Calibration
- Linearity
- Frequency response
- Isotropy
- Time-domain response
- Temperature response

In conclusion, the accuracy of broadband evaluations is significantly limited but generally conservative.

11.5.2 Frequency-Selective Measurements

Frequency-selective measurement techniques can overcome the issue of the unknown spectral composition of the field. However, the execution of the measurement is more complicated, such that insufficiently trained personnel are likely to produce erroneous results.

Measurements in the frequency domain are performed with a measurement antenna, as described in Section 11.3.3, that is connected to a spectrum analyzer (Figure 11.14). The spectrum analyzer mixes the received RF signal down to the base band. A filter is swept in frequency over the considered sweep bandwidth. The signal after the filter is detected using user-definable detectors. Most spectrum analyzers provide video filters for additional filtering (smoothing) of the spectral signal. Setting the sweep and filter parameters can significantly impact the measurement result. Optimal settings for GSM and Universal Mobile Telecommunications System (UMTS) based on a simulation approach were recently presented by Olivier and Martens [57]. The impact of summing up parts of the spectrum because of nonideal filters was investigated by Joseph et al. [58]. Modern analyzers also provide a zero-span mode. The RF signal is mixed down to the base band, and only the time-domain envelope of the signal is displayed afterward. This mode may be applied, for example, to investigate the time-domain characteristics of an unknown communication system. Additionally, it is especially useful to measure pulsed

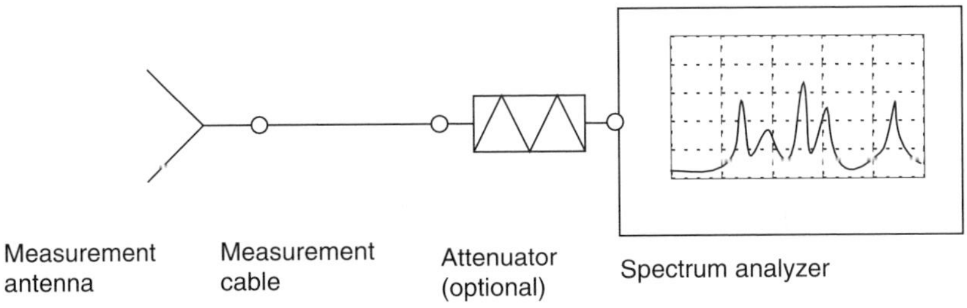

FIGURE 11.14
Schematic of a frequency-selective measurement of EMF.

signal forms, as this signal form is often used in communication systems. In summary, the application of spectrum analyzers is a complex subject. Measurement recommendations dealing with frequency-selective measurements should always describe the spectrum analyzer settings to produce feasible and comparable results. Nevertheless, the measurement engineer should still test the actual applicability of these settings for his particular measurement equipment.

The main sources of uncertainty regarding frequency-selective measurements are:

- Calibration of the spectrum analyzer, cable, and measurement antenna
- Linearity of the spectrum analyzer, cable, and measurement antenna
- Frequency response of the spectrum analyzer, cable, and measurement antenna
- Demodulation method of the spectrum analyzer (detector type)
- Temperature response of the spectrum analyzer, cable, and measurement antenna
- Mismatch between measurement equipment

Although the frequency-selective measurement method overcomes most of the issues regarding broadband survey meters, it is not sufficient to soundly identify the exposure from different transmitters at the same frequency. In this case, measurement receivers should be applied.

11.5.3 Code-Selective Measurements

Code-selective measurements are especially necessary if the exposure from a specific transmitter applies code division multiple access (CDMA), for example, if a UMTS is to be assessed. In the case of UMTS, all base stations usually transmit in a single-frequency band. With a frequency-selective receiver, it is not possible to discriminate between exposure from different base stations, since a single-frequency band is used and the channels are multiplexed in the code domain. Code-selective receivers decode the signal received from a base station, that is, the receiver is able to discriminate between a field strength received from the base station of interest and other noise-like sources. The receiver measures the field received from the base station of interest only if the particular scrambling code is used for decoding. Basically, the same sources of uncertainty must be considered for code- and frequency-selective measurements. However, in contrast to frequency-selective measurements, the modulation of the signal does not increase the uncertainty but rather the possible nonorthogonality of the respective scrambling codes. Many of the typical measurement receivers for UMTS base station measurements (e.g., Rohde and Schwarz TSMU [59], Anritsu ML8720B [60]) provide insufficient sampling rates for swept scanning. Others overcome this limitation, for example, Narda-STS SRM-3000 [61].

11.6 Typical Maximum Peak Spatial SAR Exposures

The daily local RF exposure of the general public has increased by several orders of magnitude with the introduction and proliferation of mobile handsets. This has triggered concern among health agencies and the public, since the highest exposed tissue is the brain. Figure 11.15 and Figure 11.16 display the frequency of worst-case SAR from mobile phones measured according to Refs. [62,63]. Figure 11.15 represents the typical SAR

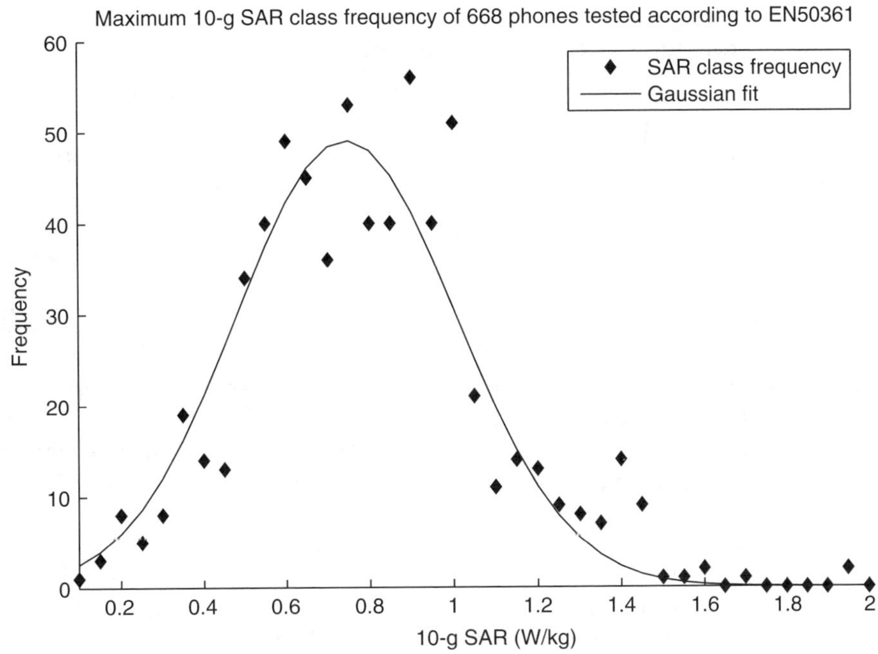

FIGURE 11.15
Statistical distribution of maximum 10-g SAR measured for 668 mobile phones according to EN50361. (From German Federal Office for Radiation Protection, http://www.bfs.de/elektro/hff/oekolabel.html, 2005.)

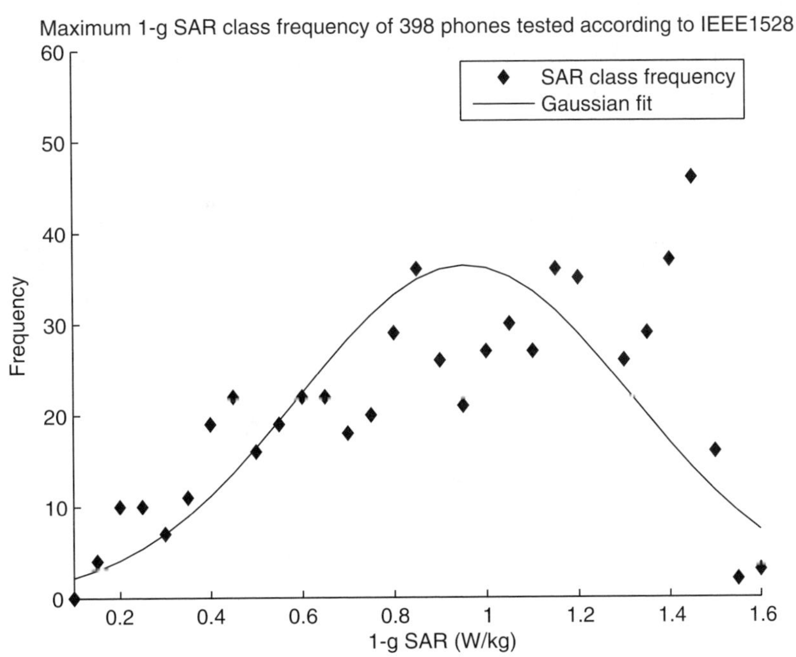

FIGURE 11.16
Statistical distribution of maximum 1-g SAR measured for 687 mobile phones according to IEEE-1528. (From Federal Communications Commission, http://www.fcc.gov/cgb/sar/, 2005.)

TABLE 11.4

Worst-Case Peak Spatial SAR of a Set of Wireless Indoor Devices

Device Class	Frequency Range (MHz)	Worst-Case 10-g SAR (W/kg)
Baby surveillance	40–863	0.077
DECT	1880–1900	0.055
WLAN	2400–2484	0.81
Bluetooth	2402–2480	0.49
PC peripherals	27–40	≤0.005

Note: An absolute worst case for all commercially available products cannot be estimated based on these data.

values for Europe (mean 10-g SAR: 0.74) [64] and Figure 11.16 for North America (mean 1-g SAR: 0.96) [65]. The different averaging masses are due to different legal regulations in Europe and the United States. These values are a considerable percentage of the limits. A recent statistical analysis of the Federal Communications Commission (FCC) SAR database found that the SAR values of newer phones are typically lower than of older phones despite their greatly reduced size.

SAR measurements of devices operated in home and office environments were reported by Kramer et al. [66]. A summary of the maximum SAR values determined in this study for several types of transmitters is given in Table 11.4. It should be noted that under worst-case scenarios, the SAR values measured for WLAN and Bluetooth communication systems are in the same range as those for mobile phones.

11.7 Typical Far-Field Exposures

A study regarding indoor incident field exposure from cellular base station sites was conducted by ARCS (Austria) [55] in the city of Salzburg. Table 11.5 shows two

TABLE 11.5

Results from Indoor Incident *E* Field Measurements Conducted in an Austrian City [55]

Base Station	Distance to Base Station (m)	Cumulative Incident *E* Field (V/m)	Distance to Base Station (m)	Cumulative Incident *E* Field (V/m)
1	196	0.37	374	0.35
2	88	0.51	108	0.89
3	9	0.034	15	0.037
4	16	0.62	8	1.00
5	85	0.94	152	0.75
6	81	1.8	85	1.71
7	4	3.7	25	1.02
8	93	0.19	208	0.19
9	34	0.40	55	0.63
10	39	1.9	76	2.8
11	174	0.59	220	0.45
12	41	0.70	107	0.67
13	2.5	0.67	5.5	0.15

Note: Two cumulative incident field exposure values (sum of incident field exposures from multiple transmitters at one site) at different distances are shown as exemplary for each base station site.

TABLE 11.6

Worst-Case *E* Field of Typical Wireless Indoor Devices at 20-cm and 1-m Distance

Device Class	Frequency Range (MHz)	Worst-Case *E* Field (V/m) (20 cm)	Worst-Case *E* Field (V/m) (1 m)
Baby surveillance	40–863	8.5	3.2
DECT	1880–1900	11.5	2.9
WLAN	2400–2484	3.9	1.1
Bluetooth	2402–2480	3.1	1.0
PC peripherals	27–40	≤1.5	≤1.5

cumulative incident field exposure values (sum of incident field exposures from multiple transmitters at one site) measured at different distances from the considered base station sites. The values are within 0.1 to 1 V/m for distances up to several hundreds of meters. These data also underline that the distance to the base station site has a poor correlation for the incident exposure.

Similar results were reported by Bornkessel and Schubert [67]. This study also included outdoor measurement points and addressed the time dependence, that is, traffic dependence of the exposure from cellular base stations. The results showed a significant time dependence for base stations with multiple traffic channels. In these cases, clearly lower exposure can be expected at night and on weekends.

The incident field exposures from typical devices used in home and office environments were assessed by Kramer et al. [66]. The maximum *E* field exposure values for different device categories are summarized in Table 11.6. The incident field exposure from cellular base stations may be exceeded by the exposure from these devices because of the generally closer distances.

Additionally, an incident exposure of 1 V/m translates to a peak spatial SAR value in the brain that is approximately a factor 10,000 times lower than the maximum exposure by a handset. Thus, exposure by handsets is by far the most dominant source of RF exposure for the general population.

Acknowledgment

The authors wish to thank Prof. Dr. Quirino Balzano, Dr. Urs Lott, and Dr. Katja Pokovic for their critical review of this manuscript. The dosimetric research at the IT'IS Foundation is supported by many institutions, the most significant of which are European and Swiss Funding Agencies, Motorola Inc. (USA), the Mobile Manufacturers Forum (Belgium), MTT (Japan), and Schmid & Partner Engineering AG (SPEAG, Switzerland).

References

1. Schmid, G., Neubauer, G., Illievich, U.M., and Alesch, F., "Dielectric properties of porcine brain tissue in the transition from life to death at frequencies from 800 to 1900 MHz," *Bioelectromagnetics*, Vol. 24, No. 6, Sept. 2003, pp. 413–422.

2. Poković, K., Advanced Electromagnetic Probes for Near-Field Evaluations, Ph.D. thesis, Diss. ETH Nr. 13334, Zurich, Switzerland, 1999.
3. Schönborn, F., Poković, K., and Kuster, N., "Dosimetric analysis of the carousel setup for the exposure of rats at 1.62 GHz," *Bioelectromagnetics*, Vol. 25, No. 1, Jan. 2004, pp. 16–26.
4. Christ, A., Chavannes, N., Nikoloski, N., Gerber, H.-U., Poković, K., and Kuster, N., "A numerical and experimental comparison of human head phantoms for compliance testing of mobile telephone equipment," *Bioelectromagnetics*, Vol. 26, No. 2, Feb. 2005, pp. 125–137.
5. Kainz, W., Christ, A., Kellom, T., Seidman, S., Nikoloski, N., Beard, B., and Kuster, N., "The specific anthropomorphinc mannequin (SAM) compared to 14 anatomical head models using a novel definition for the mobile phone positioning," *Physics in Medicine and Biology*, Vol. 50, No. 14, July. 2005, pp. 3423–3445.
6. Kuster, N., Schuderer, J., Christ, A., Futter, P., and Ebert, S., "Guidance for exposure design of human studies addressing health risk evaluations of mobile phones," *Bioelectromagnetics*, Vol. 25, No. 7, Oct. 2004, pp. 524–529.
7. ICNIRP, "Guidelines for limiting exposure to time-varying electric, magnetic, and electromagnetic fields (up to 300 GHz)," *Health Physics*, Vol. 74, No. 4, 1998, pp. 494–522.
8. ANSI/IEEE, *C95.6–2002, IEEE Standard for Safety Levels with Respect to Human Exposure to Electromagnetic fields, 0–3 kHz*, The Institute of Electrical and Electronics Engineers, Inc., New York, 2002.
9. ANSI/IEEE, *C95.1–2001, IEEE Standard for Safety Levels with Respect to Human Exposure to Radio Frequency Electromagnetic Fields, 3 kHz to 300 GHz*, The Institute of Electrical and Electronics Engineers, Inc., New York, 2001.
10. National Council on Radiation Protection and Measurements (NCRP), Ed., *Biological Effects and Exposure Criteria for Radiofrequency Electromagnetic Fields*, NCRP, 1986.
11. Aslan, E., "Electromagnetic radiation survey meter," *IEEE Transactions on Instrumentation and Measurement*, Vol. IM-19, No. 4, 1970, pp. 368–372.
12. Rudge, A.W., "An electromagnetic radiation probe for near-field measurements at microwave frequencies," *Journal of Microwave Power*, Vol. 5, No. 3, 1970, pp. 155–174.
13. Bassen, H., Swicord, M., and Abita, J., "A miniature broad-band electric field probe," *Annals of the New York Academy of Sciences, Biological Effects of Nonionizing Radiation*, Vol. 247, No. 2, Feb. 1975, pp. 481–493.
14. Bassen, H., Herman, W., and Hoss, R., "EM probe with fiber optic telemetry system," *Microwave Journal*, Vol. 20, No. 4, Apr. 1977, pp. 35–47.
15. Bassen, H.I. and Smith, G.S., "Electric field probes—a review," *IEEE Transactions on Antennas and Propagation*, Vol. 31, No. 5, 1983, pp. 710–718.
16. Hopfer, S., "The design of broad-band resistive radiation probes," *IEEE Transactions on Instrumentation and Measurements*, Vol. IM–21, No. 4, Nov. 1972, pp. 416–421.
17. Smith, G., "Analysis of miniature electric field probes with resistive transmission lines," *IEEE Transactions on Microwave Theory and Techniques*, Vol. MTT–29, No. 11, Nov. 1981, pp. 1213–1224.
18. Schmid, T., Egger, O., and Kuster, N., "Automated E-field scanning system for dosimetric assessments," *IEEE Transactions on Microwave Theory and Techniques*, Vol. 44, No. 1, Jan. 1996, pp. 105–113.
19. Whiteside, H. and King, R., "The loop antenna as a probe," *IEEE Transactions on Antennas and Propagation*, Vol. AP–12, No. 5, May 1964, pp. 291–297.
20. Kanda, M., "Standard probes for electromagnetic field measurement," *IEEE Transactions on Antennas and Propagation*, Vol. AP–41, No. 10, Oct. 1993, pp. 1349–1364.
21. Narda Safety Test Solutions, "Electric field probe for EMR-200 and EMR-300," http://www.narda-sts.com, 2005, accessed Dec. 16, 2005.
22. SPEAG Switzerland, "SPEAG dosimetric near-field probes (product flyer)," http://www.dasy4.com/flyers/probes.pdf, 2005, accessed Dec. 16, 2005.
23. Poković, K., Schmid, T., and Kuster, N., "Millimeter-resolution E-field probe for isotropic measurement in lossy media between 100 MHz and 20 GHz," *IEEE Transactions on Instrumentation and Measurements*, Vol. 49, No. 4, Aug. 2000, pp. 873–878.

24. Poković, K., Schmid, T., Fröhlich, J., and Kuster, N., "Novel probes and evaluation procedures to assess field magnitude and polarization," *IEEE Transactions on Electromagnetic Compatibility*, Vol. 42, No. 2, May 2000, pp. 240–245.
25. Cecelja, F. and Balachandran, W., "Optimized CdTe sensors for measurement of electric and magnetic fields in the near-field region," *IEEE Transactions on Instrumentation and Measurements*, Vol. 49, No. 3, June 2000, pp. 483–487.
26. Tokano, Y., Kobayashi, H., and Miyakawa, T., Houjyo, Y. "A gigahertz-range micro optical electric field sensor," in *Electromagnetic Compatibility 2001*, Zurich, Switzerland, 2001, p. 127.
27. Tajima, K., Kobayashi, R., Kuwabara, N., and Tokuda, M., "Development of optical isotropic E-field sensor operating more than 10 GHz using Mach–Zehnder interferometers," *IEICE Transactions on Electronics*, Vol. 85, 2002, p. 961.
28. Agullo-Lopez, F., Cabrera, J., and Agullo-Rueda, F., *Electrooptics*, Academic Press, New York, 1994.
29. Rigrod, W.W. and Kaminow, I.P., "Wide-band microwave light modulation," *IEEE*, Vol. 51, Jan. 1963, pp. 137–140.
30. Lei, C. and Kilcoyne, S., Eds., "Vertical-cavity surface-emitting lasers VI," in *Proceedings of SPIE*, SPIE—The international society for optical engineering, 2002.
31. Choquette, K. and Lei, C., Eds., "Vertical-cavity surface-emitting lasers V," in *Proceedings of SPIE*, SPIE—The international society for optical engineering, 2001.
32. Wilmsen, C., Temkin, H., and Coldren, L., *Vertical-Cavity Surface-Emitting Lasers*, Cambridge University Press, The Edinburgh Building, Cambridge CB2 2RU, U.K. 1999.
33. Schuderer, J., EMF Risk Assessment: In Vitro Research and Sleep Studies, Ph.D. thesis, Diss. ETH Nr. 15347, Zurich, Switzerland, 2003.
34. Bowman, R.-R., "A probe for measuring temperature in radio-frequency heated material," *IEEE Transactions on Microwave Theory and Techniques*, Vol. 24, 1976, pp. 43–45.
35. Burkhardt, M., Poković, K., Gnos, M., Schmid, T., and Kuster, N., "Numerical and experimental dosimetry of petri dish exposure setups," *Journal of the Bioelectromagnetic Society*, Vol. 17, No. 6, 1996, pp. 483–493.
36. Schuderer, J., Schmid, T., Urban, G., Samaras, T., and Kuster, N., "Novel high resolution temperature probe for RF dosimetry," *Physics in Medicine and Biology*, Vol. 49, No. 6, Mar. 2004, pp. N83–N92.
37. Wickersheim, K.-A. and Sun, M.-H., "Fiberoptic thermometry and its application," *Journal of Microwave Power*, Vol. 22, 1987, pp. 85–93.
38. Guy, A.W., "Analyses of electromaagnetic fields induced in biological tissue by thermographic studies on equivalent phantom models," *IEEE Transactions on Microwave Theory and Techniques*, Vol. 34, No. 6, June 1971, pp. 671–680.
39. Taurisano, M.D. and Vorst, A.V., "Experimental thermographic analysis of thermal effects induced on a human head exposed to 900 MHz fields of mobile phones," *IEEE Transactions on Microwave Theory and Techniques*, Vol. 48, No. 11, Nov. 2000, pp. 2022–2032.
40. Baba, M., Suzuki, Y., Taki, M., Fukunaga, K., and Watanabe, S., "Three dimensional visualization of the temperature distribution in a phantom for the assessment of localized exposure to microwaves," in *Biolelectromagnetic Society 2005 Annual Meeting*, Bioelectromagnetic Society, Dublin, Ireland, June 2005, pp. 250–251.
41. Durney, C.H., Massoudi, H., and Iskander, M.F., *Radiofrequency Radiation Dosimetry Handbook*, 4th ed. Brooks Air Force, Brooks Air Force Base, TX, 1997.
42. Seibersdorf Research, Austria, "Precision conical dipole PCD 8250," http://www.seibersdorf-rf.com/docs/leaflet/W_ Products_PCD8250_engl_.pdf, 2005, accessed Dec. 16, 2005.
43. Seibersdorf Research, Austria, "Precision biconical antenna PBA 320," http://www.seibersdorf-rf. com/docs/leaflet/W_ Products_PBA320_engl.pdf, 2005, accessed Dec. 16, 2005.
44. Müller, W. and Georg Neubauer, H.H., "ADD3D, a new technique for precise power flux density measurements at mobile communications base stations," http://www.seibersdorf-rf.com/docs/pubdl/Pub4_Add3Denglish.pdf, 2000, accessed Dec. 16, 2005.

45. Drossos, A., Santomaa, V., and Kuster, N., "The dependence of electromagnetic energy absorption upon human head tissue composition in the frequency range of 300–3000 MHz," *IEEE Transactions on Microwave Theory and Techniques*, Vol. 48, No. 11, Nov. 2000, pp. 1988–1995.
46. Kanda, M.Y., Douglas, M.G., Mendivil, E.D., Ballen, M., Gessner, A.V., and Chou, C.-K., "Faster determination of mass-averaged SAR from 2-D area scans," *IEEE Transactions on Microwave Theory and Techniques*, Vol. 52, No. 8, 2004, pp. 2013–2020.
47. Merckel, O. and Fleury, G., "Rapid SAR measurement of mobile phones," *International Symposium on International Journal of Applied Electromagnetics and Mechanics*, Vol. 19, 2004, pp. 183–186.
48. TEM Consulting, "Near field RF scanner," http://www.temconsulting. com/Products/Scanning Positioner/RF_Scanner.htm, 2005, accessed Dec. 16, 2005.
49. Chavannes, N., Tay, R., Nikoloski, N., and Kuster, N., "Suitability of FDTD based TCAD tools for RF design of mobile phones," *IEEE Antennas and Propagation Magazine*, Vol. 45, Dec. 2003, pp. 52–66.
50. ANSI C63.19, "American National Standard for Methods of Measurement of Compatibility between Wireless Communications Devices and Hearing Aids (Revision Draft 3.8)," Oct. 2005.
51. Ogawa, K., Ozaki, A., Kajiwara, S., Yamamoto, A., Koyanagi, Y., and Saito, Y., "High-speed SAR prediction for mass production stages in a factory by H-field measurements," *IEEE Antennas and Propagation Society Symposium, 2004*, Vol. 2, June 2004, pp. 1843–1846.
52. Kuster, N. and Balzano, Q., "Experimental and numerical dosimetry," in *Mobile Communications Safety*, Kuster, N., Balzano, Q., and Lin, J.C., Eds. Chapman & Hall, London, 1996, pp. 13–64.
53. Abteilung Nichtionisierende Strahlung, Mobilfunk Basisstationen (GSM) Messempfchlung, Technical report, BUWAL, Switzerland, 2002.
54. ANFR, Protocol de measure *in situ*, Technical Report 2.1, Agence National des Frequences, France, May 2004.
55. Coray, R., Krähenbühl, P., Riederer, M., Stoll, D., Neubauer, G., and Szentkuti, B., Immissionen in Salzburg, Technical Report, BAKOM, Switzerland, 2002.
56. CENELEC EN 50400, "Basic standard to demonstrate the compliance of fixed equipment for radio transmission (110 MHz–40 GHz) intended for use in wireless telecommunication networks with the basic restrictions or the reference levels related to general public exposure to radio frequency electromagnetic fields, when put into service," Sept. 2005.
57. Olivier, C. and Martens, L., "Optimal settings for frequency selective electromagnetic field measurements for exposure assessment around base stations," In *Biolelctromagnetic Society 2005 Annual Meeting*, Bioelectromagnetic Society, Dublin, Ireland, June 2005, pp. 466–468.
58. Joseph, W., Olivier, C., and Martens, L., "A robust, fast and accurate deconvolution algorithm for EM-field measurements around GSM and UMTS base stations with a spectrumanalyser," *IEEE Transactions on Instrumentation and Measurements*, Vol. 51, No. 6, Feb. 2003, pp. 1163–1169.
59. Rohde and Schwarz, Germany, "Radio network analyzer R&S TSMU," http://www.rohde-schwarz.com/www/download_files.nsf/file/TSMU_en.pdf/-$file/TSMU_en.pdf, 2004, accessed Dec. 16, 2005.
60. Anritsu, Japan, "ML8720B UMTS/W-CDMA field strength area tester," http://www.eu.anrit-www.eu.anritsu.com/files/ML8720B_E11000.pdf, 2005, accessed Dec. 16, 2005.
61. Narda Safety Test Solutions, "SRM-3000 frequency-selective field strength measuring system for rapid, reliable safety assessments," http://www.narda-sts.com, 2005, accessed Dec. 16, 2005.
62. CENELEC EN 50361, "Basic standard for the measurement of specific absorption rate related to human exposure to electromagnetic fields from mobile phones (300 MHz–3 GHz)," 2001.
63. IEEE SC34, Recommended Practice for Measurements and Computations with Respect to Human Exposure to Radio Frequency Electromagnetic Fields, 100kHz to 300 GHz (IEEE Standard C95.3), Technical report, IEEE, 2002.
64. German Federal Office for Radiation Protection, "Strahlenschutzkriterien für ein Mobiltelefon-Ökolabel," http://www.bfs.de/elektro/hff/oekolabel.html, 2005, accessed Dec. 16, 2005.
65. Federal Communications Commission, "FCC SAR database," http://www.fcc.gov/cgb/sar/, 2005, accessed Dec. 16, 2005.
66. Kramer, A., Kühn, S., Lott, U., and Kuster, N., Development of Procedures for the Assessment of Human Exposure to EMF from Wireless Devices in Home and Office Environments, Technical

report, IT'IS Foundation for Research on Information Technologies in Society, Zurich, Switzerland, Jan. 2005.
67. Bornkessel, C. and Schubert, M., Zwischenbericht: Analyse der Immissionsverteilung, Technical report, IMST GmbH Kamp-Lintfort, 2004.
68. van Deventer, E., Simunic, D., and Repacholi, M., EMF Standards for human health, in *Biological and Medical Aspects*, Barnes, F.S. and Greenebaum, B., Eds., Taylor & Francis, Boca Raton, FL, 2006, Chapter 8.

12

Electromagnetic Imaging of Biological Systems

William T. Joines, Qing H. Liu, and Gary Ybarra

CONTENTS

12.1 Introduction

Two emerging types of electromagnetic imaging are presented: electrical impedance imaging (or tomography, EIT) and microwave imaging (MWI). Both techniques rely on the contrast in electromagnetic properties (complex permittivity) of the tissues to be imaged with that of the neighboring tissue region. Both techniques use an array of sensors around the region to be imaged. For EIT, a current due to a known voltage is passed between all electrodes within an array to determine the electrical impedance (or admittance) of the imaged tissue. For MWI, a microwave signal is sequentially transmitted through the imaged tissue to all antennas within an array to determine the scattering parameters (ratio of reflected and transmitted signals to the incident signal) of the imaged region. EIT uses a lower frequency (in the kHz or MHz range) so that the imaged region is small compared with the signal wavelength. MWI uses a higher frequency (in the GHz range) so that the imaged region is comparable to the signal wavelength. The MWI frequency is chosen to be low enough to yield an adequate depth of penetration into the imaged tissue, but high enough to allow the use of a number of small, closely spaced antennas in the array. Since antenna size is also comparable to signal wavelength, microstrip patches or waveguide apertures on high-permittivity substrates are often used to reduce the wavelength at the antenna.

Once MWI and electrical impedance imaging are fully developed for clinical use, they have great potential for the early detection of breast cancer. Other imaging methods are now available, but they have certain disadvantages that limit their acceptance as the method of choice for breast cancer screening. Magnetic resonance imaging (MRI) relies on large electrical currents in cryogenically cooled conductors to produce a strong magnetic field. Hardware and safety requirements may continue to make MRI too expensive for widespread use in breast cancer screening. X-ray imaging is now used fairly extensively, but many women dread the discomfort or pain associated with having their breast sandwiched between two hard, flat surfaces in preparation for the screening, and their aversion to this test may prevent them from participating.

For testing patients in the clinic, both MWI and EIT will most likely use the same arrangement as far as the patient is concerned. The patient will lie face down in a comfortable position on a cot or gurney, as illustrated in Figure 12.1. The breast to be screened extends into an opening containing body-temperature liquid that has approximately the same electrical properties as the normal female breast. This liquid is held within a cylindrical or rectangular thin-walled plastic container that is attached beneath the cot. For MWI, numerous small antennas are attached to the entire outer surface of the plastic container. For EIT, numerous small metal electrodes are attached on the inside surface of the plastic container so that the electrodes can flow current through the liquid and the breast tissue under test. These are the techniques, along with numerical modeling,

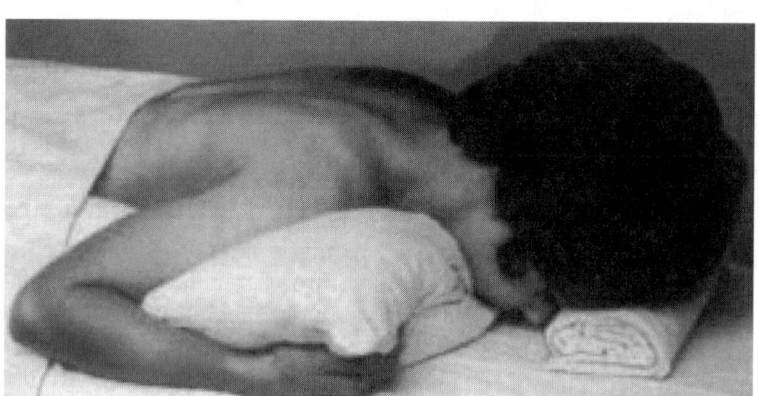

FIGURE 12.1
Breast imaging arrangement with patient lying comfortably face down.

that we are now using in the laboratory to obtain the results reported herein. Although we describe how we obtain the breast cancer images using phantom (artificial) tissues, we have not yet moved our imaging systems into the clinic to image breast cancer in real patients. The research and development we are now doing are necessary steps that must be taken before successful clinical applications can occur.

12.2 Development of MWI

Active MWI is an emerging technique for several biomedical imaging applications. Besides being economical and easily portable, MWI takes advantage of the high contrast in electrical properties that exists between anomalous and normal tissue over certain ranges of frequency [1–7]. Fortunately for breast cancer imaging, the contrast between normal and malignant tissue appears to be greatest over the approximate frequency range of 600 to 1000 MHz. This frequency range is both low enough and high enough to meet most of the requirements cited earlier.

While the high contrast is advantageous, it does make multiple scattering in the region more pronounced, and this adds to the difficulty of image formation. Although the specific contrasts vary with frequency, there is now a general belief that these contrasts are substantial, especially near 800 MHz (see the references cited in Refs. [8,9] and the review in Fear et al. [10]). For example, according to Joines et al. [6], at 800 MHz, the permittivity relative to air (ε_r) and the electrical conductivity (σ) for malignant breast tissue are approximately $\varepsilon_r \cong 57.2$ and $\sigma \cong 1.08$ S/m, respectively, while they are $\varepsilon_r \cong 16$ and $\sigma \cong 0.16$ S/m, respectively, for normal mammary tissue. The contrast is 3.75 for the relative permittivity and 6.75 for the electrical conductivity. Over the past few years, several research groups have been working on both hardware and software aspects of microwave breast imaging [8,9,11–15] to take advantage of the high contrast in applications of near-field MWI. The reason for the high contrast between malignant and normal tissue may be better understood through a brief examination of the frequency-dependent electrical properties of biological tissue.

12.2.1 Frequency Dependence

The frequency dependence of ε and σ is directly related to the polarization of molecules and structural interfaces caused by an applied electric field within the biological tissue. A specific polarization effect is important in determining ε and σ up to a relaxation frequency, f_r. Above this frequency the induced polarization can no longer change as fast as the applied field. Thus, above f_r the energy storage term (ε) is less, and the energy dissipation term (σ) is greater. Structural or Maxwell–Wagner relaxation, because of cellular membranes and other layered structures within the tissue, is of importance at frequencies below about 100 MHz [16]. Polar or Debye relaxation, because of the rotation of molecules or molecular groups by the applied field, is of importance in determining ε and σ at all frequencies above about 30 MHz. In the 30- to 2000-MHz range, the values of ε and σ versus frequency are highly dependent on the free-water content ($f_r \cong 25{,}000$ MHz) of the tissue. However, bound water (f_r in the 100- to 1000-MHz range) and protein molecules (f_r in the 40- to 300-MHz range) also make significant contributions to ε and σ.

Because most types of polarization can be described formally in the same qualitative manner, the Debye equations often apply very well, even though they were derived for the case of molecular rotation. For a given relaxation frequency, the Debye relations for ε and σ are [17]:

$$\varepsilon = \varepsilon_H + \frac{\varepsilon_L - \varepsilon_H}{1 + \left(\frac{f}{f_r}\right)^2} \tag{12.1}$$

and

$$\sigma = \sigma_L + \frac{(\varepsilon_L - \varepsilon_H)2\pi f_r}{1 + (f_r/f)^2} \tag{12.2}$$

where $\varepsilon = \varepsilon_H$ at f well above f_r and $\sigma = \sigma_L$, $\varepsilon = \varepsilon_L$ at f well below f_r.

Two tissues having different constituencies (e.g., normal and malignant tissues) most likely will have different relaxation frequencies (f_{r1} and f_{r2}). If the operating frequency of an applied electromagnetic wave is between f_{r1} and f_{r2}, then an examination of the Debye equations shows that this is the frequency where the greatest contrast occurs between ε and σ for the two tissues [5].

12.2.2 Scattering Parameters

As stated earlier, to achieve MWI an electromagnetic wave is sequentially transmitted through the tissue region to be imaged and to all antennas within an array surrounding the region in order to determine the scattering parameters (ratio of reflected and transmitted signals to the incident signal) of the imaged region. This is important because the scattering parameters may be measured as well as calculated.

Taking just two antennas in the array, one transmitting and one receiving, the scattering parameters may be simply expressed. The electric field intensity on the transmitter side (E_1) of the region is made up of incident and reflected (scattered) components and likewise, for the electric field intensity on the receiver side (E_2) of the imaged region. This relationship is expressed as

$$E_1 = E_{1i} + E_{1r} \tag{12.3}$$

and

$$E_2 = E_{2i} + E_{2r} \tag{12.4}$$

where E_{1i} is a wave incident from antenna 1 into the imaged region, E_{1r} is reflected into antenna 1 from the region, E_{2i} is incident from antenna 2 into the region, and E_{2r} passes from the region and into antenna 2. Since the incident and scattered fields completely determine the transmitted and received signals, it is convenient to express the scattered fields as functions of the incident fields, as

$$E_{1r} = S_{11}E_{1i} + S_{12}E_{2i} \tag{12.5}$$

and

$$E_{2r} = S_{21}E_{1i} + S_{22}E_{2i} \tag{12.6}$$

where the scattering parameters are defined as:

$$S_{11} = E_{1r}/E_{1i} \,|_{E_{2i}=0}, \; S_{12} = E_{1r}/E_{2i} \,|_{E_{1i}=0}, \; S_{21} = E_{2r}/E_{1i} \,|_{E_{2i}=0}, \; S_{22} = E_{2r}/E_{2i} \,|_{E_{1i}=0} \quad (12.7)$$

Thus, S_{11} is the reflection coefficient at antenna 1, under the condition that antenna 2 is terminated in the impedance of its connecting cable (usually 50 Ω), so that no signal enters the region from antenna 2. Under the same condition at antenna 2, S_{21} is the forward transmission coefficient of signals from antenna 1 to antenna 2. Likewise, S_{22} is the reflection coefficient at antenna 2, under the condition that antenna 1 is terminated in the impedance of its connecting cable (usually 50 Ω), so that no signal enters the region from antenna 1. Under the same condition at antenna 1, S_{12} is the reverse transmission coefficient of signals from antenna 2 to antenna 1.

Using the very same concepts and definitions, the scattering parameters for N antennas are expressed as

$$\begin{aligned} E_{1r} &= S_{11}E_{1i} + S_{12}E_{2i} + S_{13}E_{3i} + \cdots + S_{1N}E_{Ni} \\ E_{2r} &= S_{21}E_{1i} + S_{22}E_{2i} + S_{23}E_{3i} + \cdots + S_{2N}E_{Ni} \\ E_{3r} &- S_{31}E_{1i} + S_{32}E_{2i} + S_{33}E_{3i} + \cdots + S_{3N}E_{Ni} \\ &\vdots \\ E_{Nr} &= S_{N1}E_{1i} + S_{N2}E_{2i} + S_{N3}E_{3i} + \cdots + S_{NN}E_{Ni} \end{aligned} \quad (12.8)$$

12.2.3 Scattering-Parameter Imaging of Tissue Permittivity

This method of imaging relies on the propagation of electromagnetic waves into and through the region of interest as shown in Figure 12.2. The reflected and transmitted electric fields are referenced to the incident electric field to determine the magnitude and phase delay of the reflection coefficient at the nth port ($S_{nn} = | S_{nn} | \angle\phi_R = E_r/E_i$) and the transmission coefficient from port n to m ($S_{mn} = | S_{mn} | \angle\phi_T = E_t/E_i$) at multiple transmitter–receiver sites around the circumference of the region.

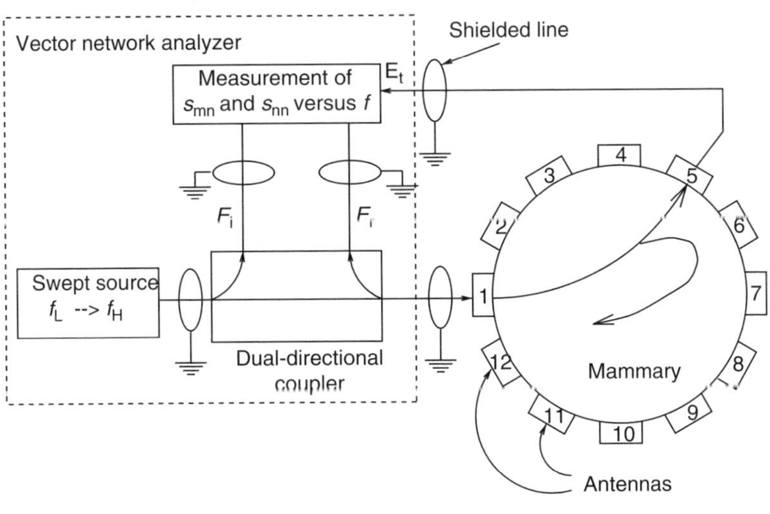

FIGURE 12.2
S-parameter measurement system.

A further visualization of the transmission–reflection method is shown in Figure 12.3, where each antenna on the circumference is a transmitter–receiver and the lines between antennas may be considered average ray paths for transmitted waves. Thus, position 1 in Figure 12.3 transmits to all other receivers at the same time and also receives reflected signals from various points within the target region. Next, position 2 transmits to all positions, and so on around the circumference. Both the amplitude and phase delay of transmitted and reflected signals are functions of the complex permittivity that the particular ray path encounters in traversing or partially traversing the normal and malignant breast tissue regions. The data collected with this method are processed by computer to produce images of subregions of differing permittivity and conductivity within the mammary tissue.

At present, we use a vector network analyzer (HP 8753A, 0.3 MHz to 3 GHz) to perform the measurement functions in Figure 12.2. In a fully developed MWI system, the network analyzer will be replaced with lower-cost, application-specific, individual components.

As an example to illustrate the transmission–reflection imaging method, we use a simplified multiport network theory; a more rigorous full-wave theory will be summarized in Section 12.4. Let d be the distance of an average ray path from a transmitter at port 1 to a receiver at port 7 in Figure 12.3, let Z_0 be the receiver or transmitter impedance, and let Z_M be the intrinsic impedance that the rays encounter in the bulk tissue between transmitter and receiver. From two-port network theory, the reflection coefficient (S_{11}) looking from the transmitter antenna into the bulk tissue is

$$S_{11} = \frac{(Z_M^2 - Z_0^2)\tanh \gamma d}{2Z_0 Z_M + (Z_M^2 - Z_0^2)\tanh \gamma d} = \frac{E_\mathrm{r}}{E_\mathrm{i}} \tag{12.9}$$

and the transmission coefficient (S_{71}) from port 1 to port 7 is

$$S_{71} = \frac{2Z_0 Z_M \sqrt{1 - \tanh^2 \gamma d}}{2Z_0 Z_M + (Z_M^2 - Z_0^2)\tanh \gamma d} = \frac{E_\mathrm{t}}{E_\mathrm{i}} \tag{12.10}$$

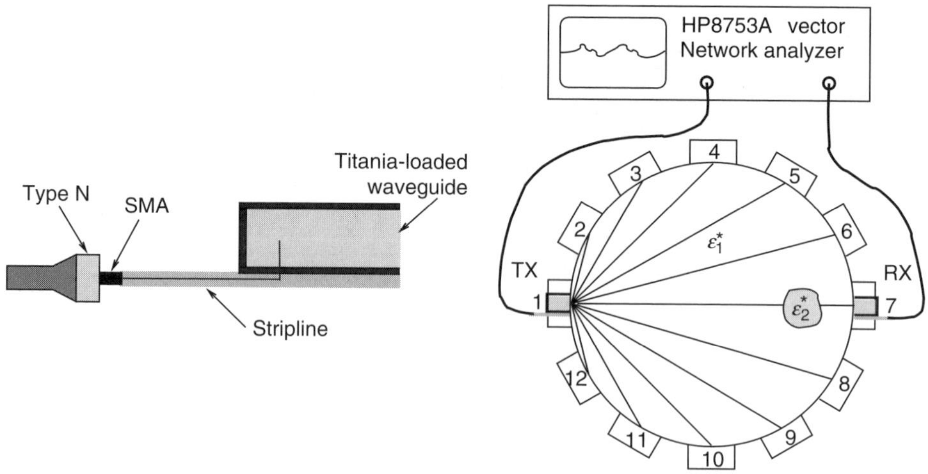

FIGURE 12.3
Top view of the cylindrical plastic container showing some of the antennas in the total array.

where $\gamma = j\omega\sqrt{\mu_0\varepsilon^*}$ is the propagation constant of the bulk tissue, and $\omega = 2\pi f$, where f is the frequency in hertz. Let $Z_M = \sqrt{\mu_0/\varepsilon^*} = 377/\sqrt{\varepsilon_r^*}$ and $Z_0 = 50\ \Omega$, then $Z_0/Z_M = \sqrt{\varepsilon_r^*}/7.54$, and S_{11} and S_{71} become

$$S_{11} = \frac{(56.85 - \varepsilon_r^*)\mathrm{j}\tan\left(2\pi f\sqrt{\varepsilon_r^*}d/c\right)}{15.08\varepsilon_r^* + (56.85 + \varepsilon_r^*)\mathrm{j}\tan\left(2\pi f\sqrt{\varepsilon_r^*}d/c\right)} = \frac{E_r}{E_i} \tag{12.11}$$

$$S_{71} = \frac{15.08\sqrt{\varepsilon_r^*}\sqrt{1 - \mathrm{j}\tan^2\left(2\pi f\sqrt{\varepsilon_r^*}d/c\right)}}{15.08\varepsilon_r^* + (56.85 + \varepsilon_r^*)\mathrm{j}\tan\left(2\pi f\sqrt{\varepsilon_r^*}d/c\right)} = \frac{E_t}{E_i} \tag{12.12}$$

where $c = 3 \times 10^8\,\mathrm{m/s}$. For example, if $d = 0.1$ m, $f = 800$ MHz, and the measured values are

$$S_{11} = 0.200\angle 81.76^\circ = -13.94\,\mathrm{dB}\angle 81.76^\circ \tag{12.13}$$

$$S_{71} = 0.011\angle 86.49^\circ = -39.17\,\mathrm{dB}\angle 86.49^\circ \tag{12.14}$$

then the complex permittivity of the bulk tissue medium at $f = 800$ MHz is determined as

$$\varepsilon_r^* = 36 - \mathrm{j}36 = \frac{\varepsilon}{\varepsilon_0} - \mathrm{j}\frac{\sigma}{\omega\varepsilon_0} \tag{12.15}$$

where $\varepsilon = 36\varepsilon_0$ and $\sigma = 1.60$ S/m. From our previous measurements [6], normal mammary tissue at 800 MHz would yield $\varepsilon_r^* = 17 - \mathrm{j}4$, or $\varepsilon = 17\varepsilon_0$ and $\sigma = 0.18$ S/m. Thus, $\varepsilon_r^* = 36 - \mathrm{j}36$ would represent a very large difference in expected tissue properties along the path from port 1 to port 7, which passes through the region occupied by ε_2^* in Figure 12.3. This example is intended to illustrate how the contrast in complex permittivity of a region can not only be measured but also located by coordinate position.

12.2.4 Power, Signal Attenuation, and Signal-to-Noise Ratio

At an operating frequency of 800 MHz, the free space wavelength is $\lambda_0 = 37.5$ cm. In normal breast tissue the wavelength is reduced to $\lambda_{\mathrm{normal}} = \frac{\lambda_0}{\sqrt{\varepsilon_r}} = 9.375$ cm, assuming that the dielectric constant is 16 for normal breast tissue. The distance between any pair of transmitting and receiving antennas is less than 20 cm. Therefore, all measurements are made within a region comparable to the wavelength. The propagation loss between transmitting and receiving antennas is typically in the range of 8 to 16 dB. A typical dielectrically loaded waveguide antenna used for MWI has a 3 × 3 cm aperture. If the transmitting antenna is operated at a transmitted power of 0 dB m or 1 mW, it results in a transmitted power density of 0.11 mW/cm^2, nine times below the ANSI safety level. The received power is −16 to −8 dB m or 25 to 158 μW. The scattered signal from the tumor is typically 0.01% of of the incident wave. Since the network analyzer's noise level is approximately −90 dB m, the signal-to-noise ratio (SNR) at the receiving antenna is on the order of 30 dB.

12.3 Three-Dimensional Formulation

Currently, there are few methods developed for MWI in three dimensions because of the large computational demand in both forward and inverse problems of inhomogenous media. While two-dimensional (2-D) and 3-D imaging techniques based on the scalar-wave approximation have been developed with some success, breast cancer imaging is inherently a 3-D problem and requires the full 3-D inverse scattering algorithms based on the vectorial Maxwell's equations. In our imaging research we have developed a fast inverse scattering method based on the combination of the contrast source inversion and the fast Fourier transform (FFT) algorithms.

Because of the volumetric inhomogeneities in biological tissues, surface integral equation methods become impractical. Methods based on volumetric techniques are more appealing. We focus on the frequency-domain solution of a volume integral equation (VIE) for inhomogenous media. In a typical inhomogenous tissue medium, both the forward and the inverse scattering problems can be formulated using VIEs. The conventional forward scattering method for VIEs is the method of moments (MoM) [18], but the computational cost is prohibitively high; several 3-D forward problems have been solved in a number of articles [19–22] using this approach. If the VIE involves N unknowns, the MoM has a memory requirement of $O(N^2)$, and a CPU time requirement of $O(N^2)$ or $O(N^3)$ depending on whether the resulting matrix equation is solved iteratively or by direct inversion.

An important improvement over the MoM is a variant of Bojarski's k-space method (see references in Ref. [23]), the so-called conjugate-gradient fast Fourier transform (CG-FFT) method proposed during the 1980s [24,25], which uses iterative the Krylov subspace method [26] combined with FFT or nonuniform FFT [27]. Zhang and Liu [28] recently developed a biconjugate-gradient FFT method based on the weak-form discretization of Zwamborn and van den Berg and showed a significant improvement over the CG-FFT method for wave scattering problems. This method has been further accelerated by the stabilized biconjugate-gradient fast Fourier transform (BCGS-FFT) method for wave scattering by Xu et al. [29]. Recently, the adaptive integral method developed for surface integral equations [30] has been further developed for VIEs to accelerate MoM by using two sets of basis functions to represent near-field and far-field interactions [31].

In the present work, we apply the BCGS-FFT method to MWI by solving the VIE. The problem under consideration is schematically shown in Figure 12.4, where an arbitrary

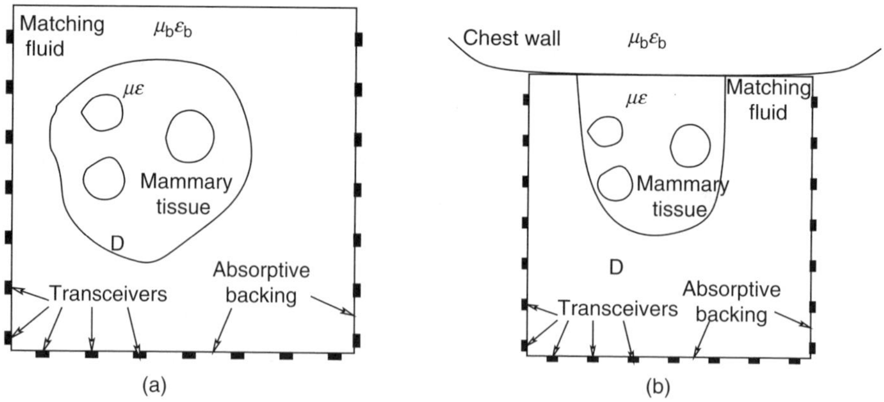

FIGURE 12.4
Imaging chamber from above (a) and from the side (b).

number of antennas are mounted on the plastic container surrounding the phantom model of breast tissue. Specific configurations for simulation will be described in a later section. We calculate the electromagnetic fields both inside the tissue medium and at the receiver array for any source locations. For MWI of breast tissue, the excitation sources (antennas) are in the near-field zone, and the incident field cannot be approximated as a plane wave. Therefore, we include the effects of the finite sources in the near-field zone.

In the following section, we derive from the vectorial Maxwell's equations the VIE for 3-D MWI in the discretized form that we use. Iterative methods for the solution of this discretized linear system are then described. Numerical results are shown to validate the method for MWI.

12.4 Wave Equation and VIE

From Maxwell's equations with an assumed time dependence $e^{j\omega t}$,

$$\nabla \times \mathbf{E} = -j\omega\mu\mathbf{H} \tag{12.16}$$

$$\nabla \times \mathbf{H} = (\sigma + j\omega\varepsilon)\mathbf{E} + \mathbf{J} = j\omega\left(\varepsilon - j\frac{\sigma}{\omega}\right)\mathbf{E} + \mathbf{J} = j\omega\varepsilon^*\mathbf{E} + \mathbf{J} \tag{12.17}$$

$$\nabla \cdot \mathbf{D} = \rho \tag{12.18}$$

$$\nabla \cdot \mathbf{B} = 0 \tag{12.19}$$

we may let

$$\mathbf{B} = \nabla \times \mathbf{A} \tag{12.20}$$

since $\nabla \cdot \nabla \times \mathbf{A} \triangleq 0$ maintains $\nabla \cdot \mathbf{B} = 0$, where $\mathbf{A}$ is the vector potential. Substituting Equation 12.20 into Equation 12.16 yields

$$\nabla \times \mathbf{E} = -j\omega(\nabla \times \mathbf{A}) \tag{12.21}$$

or

$$\nabla \times (\mathbf{E} + j\omega\mathbf{A}) = 0 \tag{12.22}$$

Since $\nabla \times (\nabla\phi) \triangleq 0$, we may let the term in parentheses in Equation 12.22 be the negative gradient of the scalar potential ϕ, and express $\mathbf{E}$ as

$$\mathbf{E} = -\nabla\phi - j\omega\mathbf{A} \tag{12.23}$$

Note that if $\omega = 0$, then $\mathbf{E} = -\nabla\phi$, as expected for static fields. We now invoke the Lorenz condition, which defines the divergence of $\mathbf{A}$ as

$$\nabla \cdot \mathbf{A} = -j\omega\mu\varepsilon^*\phi \tag{12.24}$$

Substituting this condition into Equation 12.23 yields

$$\mathbf{E} = -j\omega\left[1 + \frac{\nabla\nabla\cdot}{k^2}\right]\mathbf{A} \tag{12.25}$$

where $k^2 = \omega^2\mu\varepsilon^*$. Since **E** and **H** are propagating field intensities that satisfy a wave equation, then **A** must satisfy a wave equation obtained as follows. Substitute Equation 12.20 and Equation 12.23 into Equation 12.17 to yield

$$\nabla\times\nabla\times\mathbf{A} \stackrel{\Delta}{=} \nabla(\nabla\cdot\mathbf{A}) - \nabla^2\mathbf{A} = j\omega\mu\varepsilon^*(-\nabla\phi + j\omega\mathbf{A}) + \mu\mathbf{J} \tag{12.26}$$

or (again using the Lorenz condition),

$$\nabla^2\mathbf{A} = -\omega^2\mu\varepsilon^*(\mathbf{r})\mathbf{A} - \mu\mathbf{J} \tag{12.27}$$

A solution to Equation 12.27 is in general not available in a closed form because ε^* (**r**) is inhomogenous. However, for a homogenous medium with constant complex permittivity $\varepsilon_b{}^*$, one can find the solution in closed form

$$\mathbf{A}^{\text{inc}}(\mathbf{r}) = \mu\int_V \mathbf{J}(\mathbf{r}')g(\mathbf{r},\mathbf{r}')\,\mathrm{d}V' \tag{12.28}$$

where **r**, **r**′, and $\mathbf{R} = \mathbf{r} - \mathbf{r}'$ are vectors from the origin to the field point, from the origin to the source point, and from the source point to the field point, respectively. Green's function in Equation 12.28 for the homogenous medium is given by $g(\mathbf{r},\mathbf{r}') = \mathrm{e}^{-jk_bR}/4\pi R$, where $k_b = \omega\sqrt{\mu\varepsilon_b^*}$ is the complex wavenumber of the medium. We call this solution $\mathbf{A}^{\text{inc}}$, the vector potential for the incident field in a homogenous background medium. The corresponding incident electric field can be found from Equation 12.25 as

$$\mathbf{E}^{\text{inc}} = -j\omega\left[1 + \frac{\nabla\nabla\cdot}{k_b^2}\right]\mathbf{A}^{\text{inc}} \tag{12.29}$$

For an inhomogenous medium, even though a closed form solution is not available, one can express the solution in terms of an integral equation through the equivalence principle, as discussed below.

In order to introduce the equivalence principle, the key is to rewrite Maxwell's equations into a second-order partial differential equation for the electric field:

$$\nabla^2\mathbf{E} + k_b^2\mathbf{E} = j\omega\mu[\mathbf{J} + \mathbf{J}_{\text{eq}}] \tag{12.30}$$

where

$$\mathbf{J}_{\text{eq}} = j\omega[\varepsilon^*(\mathbf{r}) - \varepsilon_b]\mathbf{E} \tag{12.31}$$

is the volume equivalent electric current density induced in the inhomogenous medium.

Thus, from Equation 12.30, the total field **E** can be written as the superposition of the incident field $\mathbf{E}^{\text{inc}}$ due to the primary source **J** and the scattered field $\mathbf{E}^{\text{sct}}$ due to the induced source $\mathbf{J}_{\text{eq}}$. Similar to the incident vector potential and incident electric field, the scattered vector potential and scattered electric fields are

$$\mathbf{A}^{\mathrm{sct}}(\mathbf{r}) = \mu \int_V \mathbf{J}_{\mathrm{eq}}(\mathbf{r}')g(\mathbf{r}, \mathbf{r}')\,\mathrm{d}V' \tag{12.32}$$

$$\mathbf{E}^{\mathrm{sct}} = -\mathrm{j}\omega\left[1 + \frac{\nabla\nabla\cdot}{k_{\mathrm{b}}^2}\right]\mathbf{A}^{\mathrm{sct}} \tag{12.33}$$

The total electric field **E(r)** is composed of an incident field plus a scattered field, and it is the scattered field that we need to determine. Thus, combining Equation 12.31 through Equation 12.33, we have

$$\mathbf{E}(\mathbf{r}) - \mathbf{E}^{\mathrm{inc}} = \omega^2\mu\left(1 + \frac{\nabla\nabla\cdot}{k_b^2}\right)\int_V \mathbf{E}(\mathbf{r}')g(\mathbf{r}, \mathbf{r}')[\varepsilon^*(\mathbf{r}') - \varepsilon_b^*]\,\mathrm{d}V' \tag{12.34}$$

where the subscript b denotes a parameter of the background medium (normal tissue and liquid with the same properties). Equation 12.34 is the integral equation representation of the scattered electric field everywhere in space. In particular, for $\mathbf{r} \in V$, Equation 12.34 is a Fredholm integral equation of the second kind. This is the integral equation we solve for the internal electric field **E** for $\mathbf{r} \in V$, from which the field everywhere in the region can be obtained. This type of VIE has been solved by using the MoM [18,20,21]. In our work we will use an alternative discretization method coupled with the Krylov subspace iterative technique to significantly speed up the numerical solution of the problem.

12.4.1 Microwave Imaging

In MWI, scattering parameters are measured using the antennas mounted on the surface of the plastic container in Figure 12.4. The objective of MWI is to reconstruct the distribution of the complex permittivity inside the tissue given the measured scattering parameters.

First, for the MWI system design optimization, the forward problem must be solved. This is to calculate the electric field distribution, given a set of antennas and known distribution of the complex permittivity. In the general problem of microwave interaction with a tissue medium shown in Figure 12.4, an inhomogenous medium with a finite volume V is embedded in an isotropic, homogenous background medium with constant permittivity ε_b, electric conductivity σ_b, and permeability μ_b. This background medium may be air or a matching fluid that is designed to approximately match the electrical properties of the tissue to enhance the SNR in the measurement of the scattered field [9]. The inhomogenous volume V is characterized by nonuniform distributions of permittivity $\varepsilon(\mathbf{r})$ and conductivity $\sigma(\mathbf{r})$; and permeability is assumed constant, that is, $\mu = \mu_b$. The objective is to solve for the electric field everywhere in space due to a finite antenna (usually electrically small because a large array of antennas is needed in an array imaging system). For more details of the BCGS-FFT method, the reader is referred to Refs. [9,28,29,32].

The above discussion is for the forward problem where the distribution of the complex permittivity is known. In reality, for the clinic application of MWI and electrical impedance tomography, we need to solve the inverse scattering problem where the complex permittivity is an unknown distribution. From some limited measurement data collected on the surface of the container, we infer such unknown permittivity distribution by solving the inverse problem through Equation 12.34. In our work, we apply both the distorted Born iterative method and the contrast source inversion method to solve this inverse problem. For details of such inverse solvers, the reader is referred to Refs. [9,33–35].

12.4.2 Electrical Impedance Tomography

This imaging method uses multiple planar electrodes positioned around the region to be imaged, as in Figure 12.5. Impedance or admittance measurements are made between all electrodes, two at a time. Thus, in a parallel-plate capacitor sense, the admittance between any pair of electrodes is the ratio of current to voltage as:

$$Y = \frac{I}{V} = \frac{\int_S \mathbf{J} \cdot d\mathbf{S}}{\int_0^d \mathbf{E} \cdot d\mathbf{l}} = \frac{\int_S (\sigma + j\omega\varepsilon)\mathbf{E} \cdot d\mathbf{S}}{\int_0^d \mathbf{E} \cdot d\mathbf{l}} = (\sigma + j\omega\varepsilon)\frac{A}{d} = j\omega\varepsilon^* \frac{A}{d} \tag{12.35}$$

or $Y = G + j\,\omega C$, where $G = \sigma\,(A/d)$, $C = \varepsilon\,(A/d)$, and $\varepsilon^* = \varepsilon - j\,\frac{\sigma}{\omega}$ is the measured complex permittivity if A/d is known from calibration data. The real and imaginary parts of ε^*, the permittivity ε, and the conductivity σ are the electrical properties of the composite tissue between pairs of electrodes. The factor (A/d) is an effective area-to-distance ratio that may be different for each electrode pair, but this ratio is determined by measuring a material with known electrical properties (such as 0.15 M NaCl at 24°C).

In general, the electric field will be nonuniform but most intense in the region between the electrode pair selected for measurement. The fields between adjacent electrodes will measure properties near the tissue surface, while more diametrically opposed electrodes will measure the composite properties across the tissue region. Such a sequence of measurements is stepped around to include all electrode pairs surrounding the region in 3-D, so that the next set of measurements would be between electrode 2 and all the other electrodes, and so on. Thus, the mapping and imaging of a region are done based on the differing electrical properties within the region.

Impedance imaging is a subject that has been under continuing investigation [36–40]. Research into impedance imaging in our research group began in 1975. We used an open-ended coaxial probe that produced a nonradiating, fringing field to measure the admittance of the material against which the probe was held. This was essentially a two-electrode system that measured the admittance between the inner and outer conductors as given in Zhang and Liu [35]. Moving this noninvasive probe to different positions on the surface of the body, we were able to locate and map out the tumors lying near the skin surface on the bodies of two cancer patients. This work indicated that even

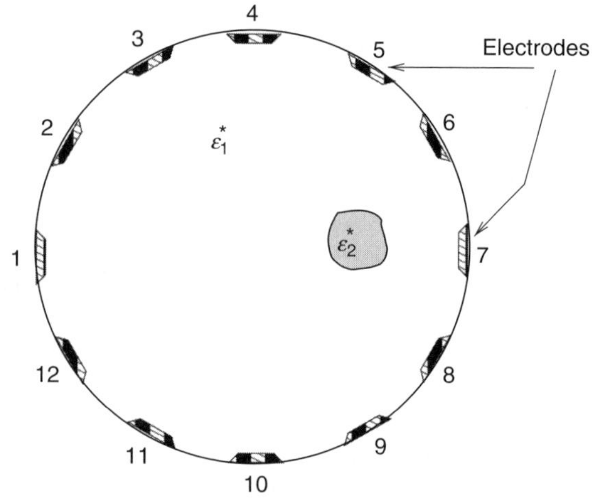

FIGURE 12.5
Configuration of electrical impedance tomography.

when measured through the intact skin, cancerous tissue generally has greater electrical conductivity and permittivity than normal surrounding tissue [41].

The forward and inverse problems of EIT can also be formulated in the same way as for the MWI, that is, using the volume integral Equation 12.34. This is especially convenient if the electrodes are small and can be considered as point electrodes. For finite-size electrodes, in order to apply the boundary conditions on the electrode surface, we use the high-order finite element and spectral element methods to solve the partial differential equations directly in the forward problem. For the inverse problem, we use the distorted Born iterative method [42,43].

12.5 Three-Dimensional Images Reconstructed from Simulated Three-Dimensional MWI Data

Figure 12.6 shows the measurement setup to image two identical spherical anomalies both with $\varepsilon_r = 48$ and $\sigma = 0.8$. The sources and receivers are evenly distributed over the six surfaces of the cuboid. The two spheres of radius 1.1 cm are located at (3.9,0,0) and (−3.9,0,0) cm, respectively. The imaged domain is discretized into 31 × 31 × 31 voxels. The reconstructed ε_r and σ are displayed on three orthogonal slices in Figure 12.7, showing a high fidelity to the ground truth.

12.6 Two-Dimensional Images Reconstructed from Measured Two-Dimensional EIT Data

This section presents three examples of images reconstructed from measured data obtained from a Two-Dimensional EIT system.

12.6.1 Case 1: One Insulator Object Inside the Container

The first example is Case 1, shown in the left panel of Figure 12.9. The difference between the total field and the background field in the right panel of Figure 12.9 is the measured

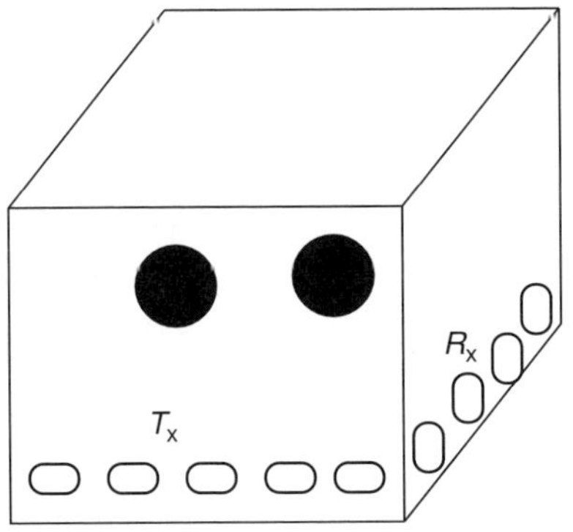

FIGURE 12.6
The setup of Three-Dimensional imaging of two spherical anomalies separated by 7 cm.

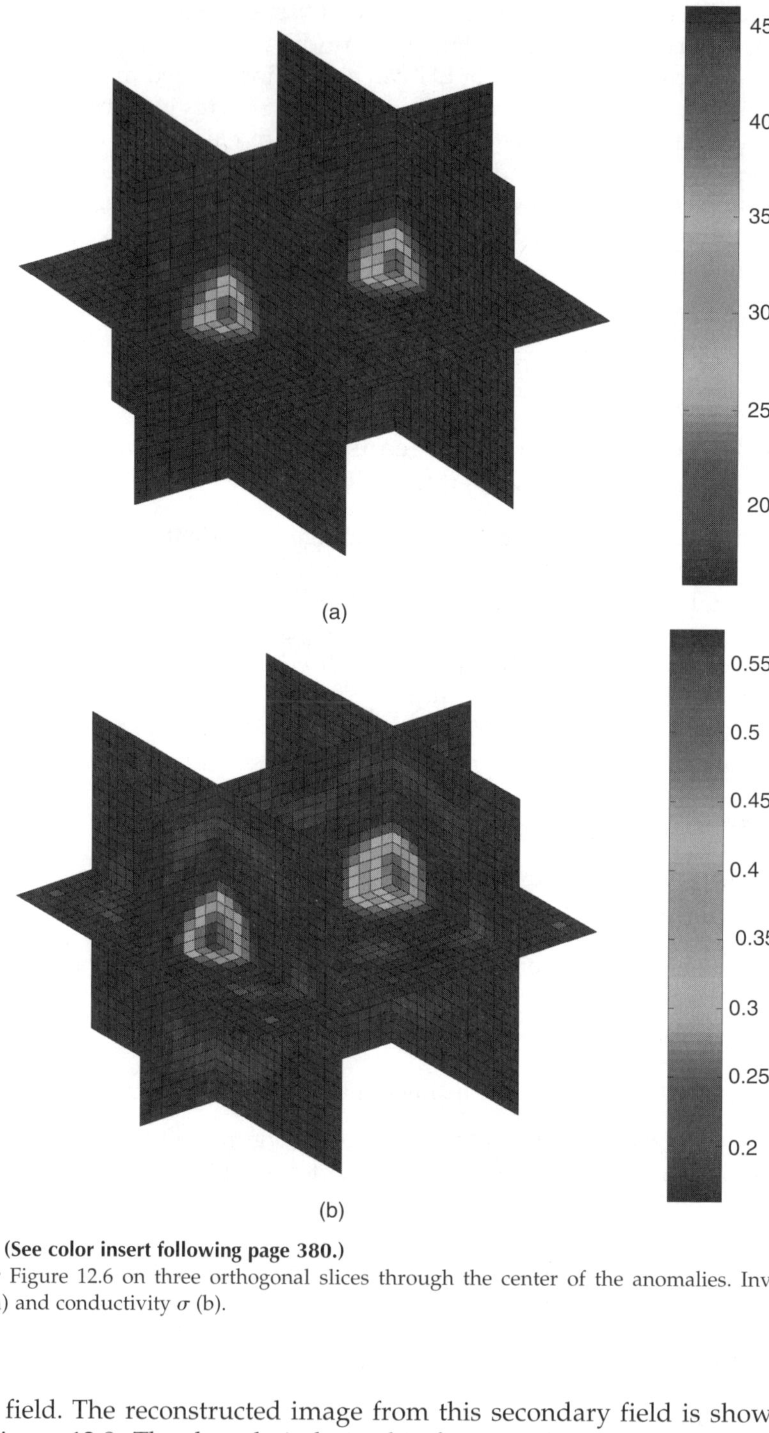

FIGURE 12.7 (See color insert following page 380.)
Inversion for Figure 12.6 on three orthogonal slices through the center of the anomalies. Inverted dielectric constant ε_r (a) and conductivity σ (b).

secondary field. The reconstructed image from this secondary field is shown in the left panel of Figure 12.8. The dotted circle in this figure indicates the ground truth of the object. It is observed that the reconstructed image matches well with the ground truth, although the absolute values of the conductivity of the highly resistive object are not well recovered because of the extreme contrast.

In order to show the misfit between the reconstructed data and the measured data, we take the reconstructed 2-D conductivity map in the left panel of Figure 12.8 and use the

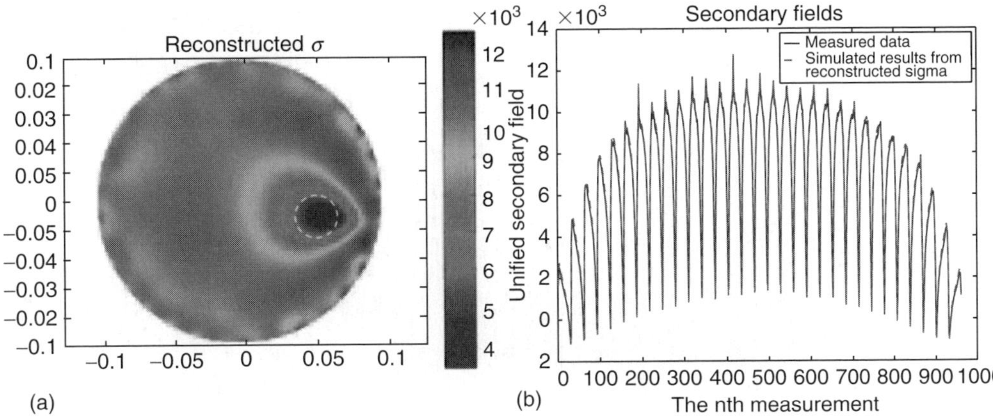

FIGURE 12.8 (See color insert following page 380.)
Left: Image reconstructed from the measured data in Case 1. The dashed circle indicates the ground truth. Right: Comparison of the secondary field between measurements and simulated by forward solver with the reconstructed σ in Case 1.

forward simulator to predict the data corresponding to this image. The comparison between the measured data (blue curve) and the simulated data (red curve) using the reconstructed image is shown in the right panel of Figure 12.8. We observe that these two sets of results have excellent agreement, indicating small data misfit from the reconstructed data.

12.6.2 Case 2: Two Insulator Objects Inside the Container

The setup of the second example (Case 2) is shown in the left panel of Figure 12.10. It is similar to Case 1, except that two insulators (beakers) are inserted into the container. The measured total field and background field are shown in the right panel of Figure 12.10. From the secondary field, the distorted Born iterative method (DBIM) reconstructs the

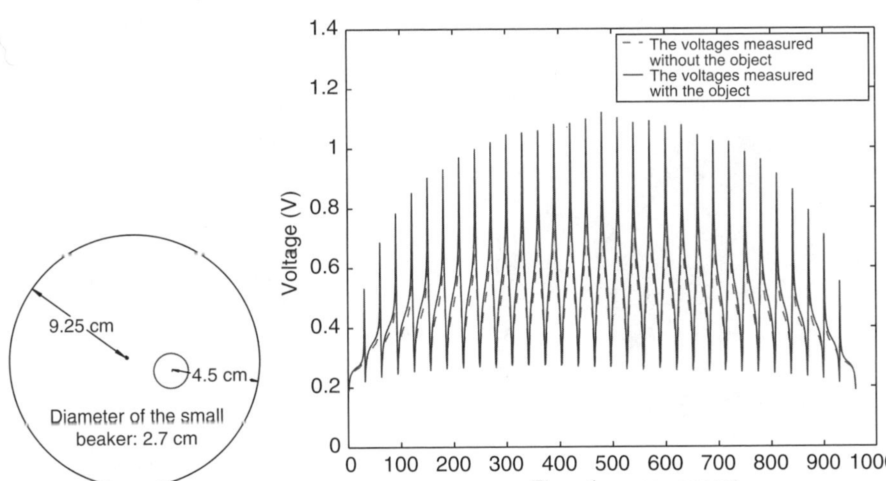

FIGURE 12.9
Left: The setup of Case 1 with one insulator object (beaker). Right: The measured voltage with and without the object in Case 1.

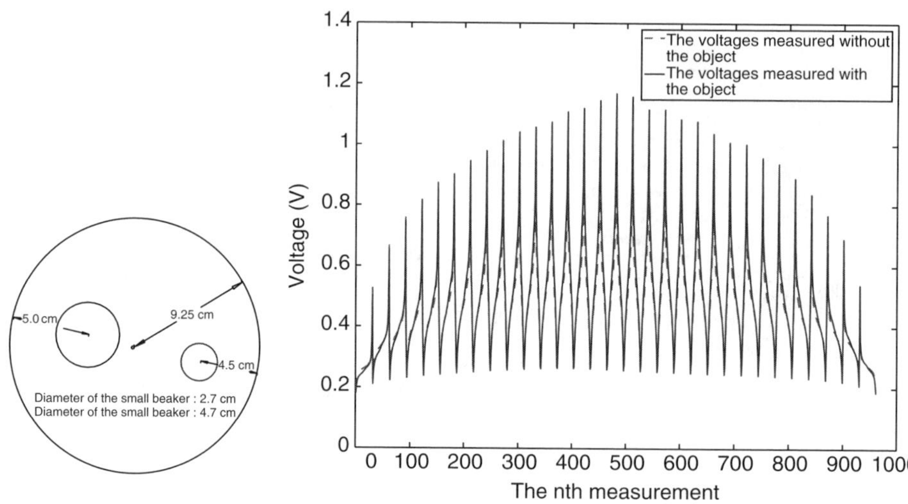

FIGURE 12.10
Left: The setup of Case 2 with two nonconducting circular objects (beakers). Right: The measured voltage with and without the objects in Case 2.

image shown in the left panel of Figure 12.11. The dotted circles in this figure indicate the ground truth of the objects. It is observed that the reconstructed image matches well with the ground truth, although the absolute values of the conductivity of the highly resistive objects are again not well recovered because of the extreme contrasts.

From the reconstructed conductivity image, we use the forward simulator to predict the secondary field data. This simulated result is then compared with the measured secondary field in the right panel of Figure 12.11. Again, we observe that these two sets of results have excellent agreement, indicating small data misfit from the reconstructed data.

12.6.3 Case 3: One Conductive and One Resistive Object Inside the Container

The third case, shown in the left panel of Figure 12.12, consists of two objects in the container. The bigger object is a conductive metal cylinder, while the smaller

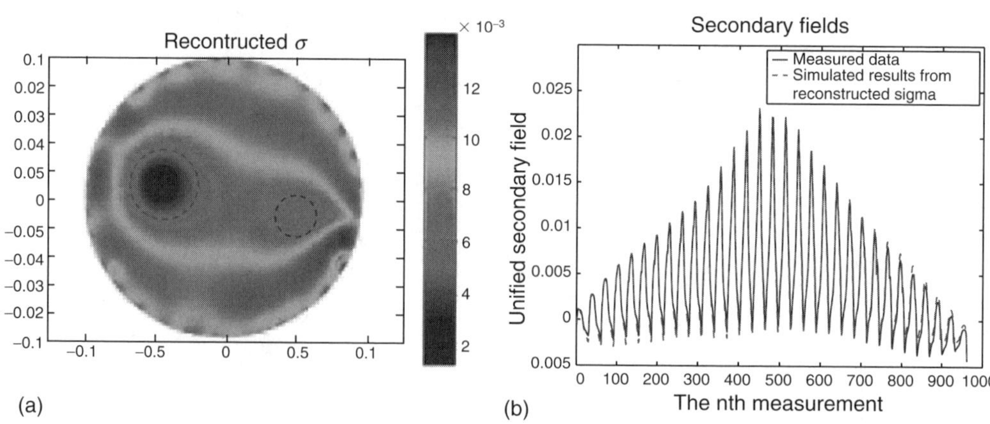

FIGURE 12.11 (See color insert following page 380.)
Left: Image reconstructed from the measured data in Case 2. The dashed circles indicate the ground truth. Right: Comparison of the secondary field between measurements and simulated by forward solver with the reconstructed σ in Case 2.

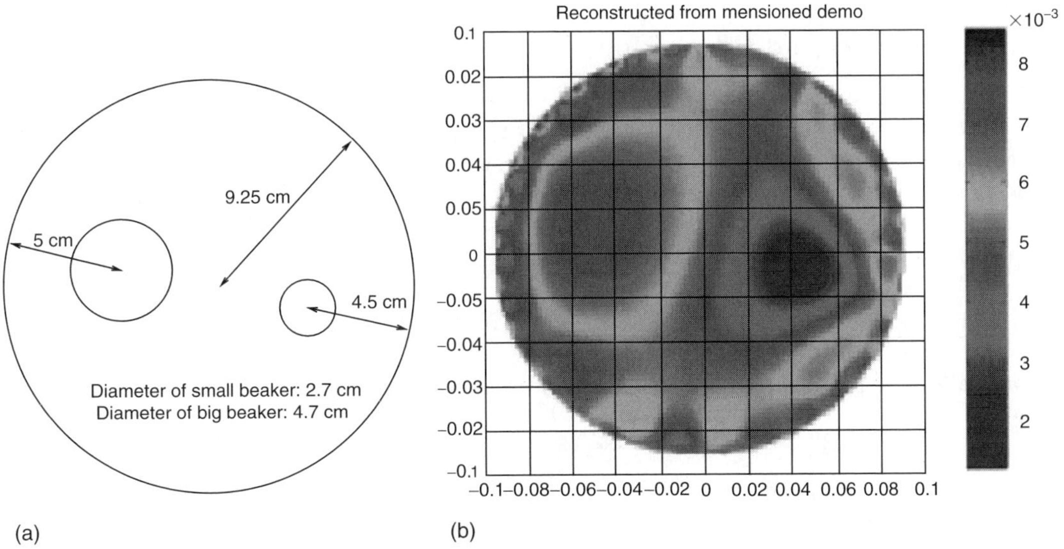

FIGURE 12.12 (See color insert following page 380.)
Left: The setup for Case 3 with one larger metal object and one smaller insulator object. Right: The reconstructed σ. The two circles denote the locations of the objects.

object is a resistive beaker (an insulator). This represents a more interesting case for reconstruction as the conductivity contrast is positive in one region and negative in another.

The reconstructed σ is shown in the right panel of Figure 12.12. The anomalies reconstructed match well with the original objects in location. However, the size of both objects has been somewhat overestimated. The conductivity value of the bigger object is indeed greater than the background, and the smaller object has conductivity values smaller than the background. These indeed match with the original setting that one object is a conductor and another is an insulator, although again the exact values of conductivity are not obtained because of the large contrasts. Nevertheless, the reconstructed images show high-quality reconstruction.

The CPU time for the above image reconstruction examples is less than 3 min on a Pentium IV computer. We emphasize that this speed is expected to be greatly reduced once the program is optimized. Furthermore, in clinical application, the image reconstruction will be performed offline, and thus the computation time of a few minutes is not a concern.

Observations—There are some artifacts in the areas close to the electrodes, perhaps caused by the surface resistance at the interface between the saline solution and the electrodes. However, the overall reconstructed images are excellent. The size and shape of the objects can be well predicted by the reconstructed images. Our future work is to extend the methodology reported here to a full 3-D EIT system. In such a system, we will further improve the sensitivity and resolution of the system by incorporating higher-precision multimeters and by placing a denser 3-D electrode array in the system. Furthermore, to match the higher sensitivity and resolution requirements, we will improve the accuracy of the forward and inverse solvers by incorporating higher-order and spectral methods. From the successful data acquisition and image reconstruction with our 2-D EIT system, it is believed that the 3-D EIT system is highly promising for breast cancer detection.

12.7 Summary and Conclusions

This chapter is a brief summary of MWI and electrical impedance tomography projects ongoing at Duke University. The modeled and measured results presented herein on artificial materials show the kinds of clear images one would fully expect to generate in the clinic using the same techniques. Other research groups are also developing improved MWI and impedance imaging systems, and some breast cancer images derived from clinical data have been published [14,44,45]. Because of 2-D artifacts or algorithm limitations, the earlier clinical images are not as clear as the ones that can be generated with the improved techniques now available. While improvements in current prototypes are encouraging, more work remains before MWI and EIT systems are integrated into clinical applications to produce the high-resolution breast cancer images that are now obtained in the laboratory.

Acknowledgment

We wish to thank the students and postdoctoral research associates involved in the work. Rebecca Willett, Adam Bryan, Erika Ward, Patrick Mathias, John Stang, and Rodger Dalton made significant contributions to the development of MWI hardware. Zhong Qing Zhang helped to develop the forward and inverse scattering algorithms for MWI. Jim Di Sarro, Jackie Hu, Guining Shi, Gang Ye, and Kyle McCarter made significant contributions to the development of electrical impedance tomography hardware. Kim Hwa Lim and Jun Ho Lee developed the forward and inverse solvers for electrical impedance tomography.

References

1. W.T. Joines, R.L. Jirtle, M.D. Rafal, and D.J. Schaefer, "Microwave power absorption differences in normal and malignant tissue," *Int. J. Radiat. Oncol. Biol. Phys.*, Vol. 6, No. 6, 681–687, June 1980.
2. W.T. Joines, S. Shrivastav, and R.L. Jirtle, "A comparison using tissue electrical properties and temperature rise to determine relative absorption of microwave power in malignant tissue," *Med. Phys. J.*, Vol. 16, No. 3, 840–844, Dec. 1989.
3. R. Pethig, "Dielectric properties of biological materials," *IEEE Trans. Elect. Insul.*, Vol. 19, 453–474, 1984.
4. S.S. Chaudhary, R.K. Mishra, A. Swarup, and J.M. Thomas, "Dielectric properties of normal and malignant human breast tissues at radiowave and microwave frequencies," *Indian J. Biochem. Biophys.*, Vol. 21, 76–79, 1984.
5. W.T. Joines, "Frequency dependent absorption of electromagnetic energy in biological tissue," *IEEE Trans. Biomed. Eng.*, Vol. 31, 215–222, Jan. 1984.
6. W.T. Joines, Y. Zhang, C. Li, and R.L. Jirtle, "The measured electrical properties of normal and malignant human tissues from 50 to 900 MHz," *Med. Phys. J.*, Vol. 21, No. 4, 547–550, Apr. 1994.
7. A.M. Campbell and D.V. Land, "Dielectric properties of female human breast tissue measured at 3.2 GHz," *Phys. Med. Biol.*, Vol. 37, No. 1, 193–210, June 1992.
8. L. Jofre, M.S. Hawley, A. Broquetas, E. de los Reyes, M. Ferrando, and A.R. Elias-Fuste, "Medical imaging with a microwave tomographic scanner," *IEEE Trans. Biomed. Eng.*, Vol. 37, No. 3, 303–312, 1990.

9. Q.H. Liu, Z.Q. Zhang, T. Wang, G. Ybarra, L.W. Nolte, J.A. Bryan, and W.T. Joines, "Active microwave imaging I: 2-D forward and inverse scattering methods," *IEEE Trans. Microwave Theory Technol.*, Vol. 50, No. 1, 123–133, Jan. 2002.
10. E.C. Fear, S.C. Hagness, P.M. Meaney, M. Okoniewski, and M.A. Stuchly, "Enhancing breast tumor detection with near-field imaging," *IEEE Microwave Mag.*, 48–56, Mar. 2002.
11. P.M. Meaney, K.D. Paulsen, A. Hartov, and R.K. Crane, "An active microwave imaging system for reconstruction of 2-D electrical property distributions," *IEEE Trans. Biomed. Eng.*, Vol. 42, No. 10, 1017–1025, 1995.
12. S.C. Hagness, A. Taflove, and J.E. Bridges, "Two-dimensional FDTD analysis of a pulsed microwave confocal system for breast cancer detection: fixed-focus and antenna-array sensors," *IEEE Trans. Biomed. Eng.*, Vol. 45, No. 12, 1470–1479, 1998.
13. S.C. Hagness, A. Taflove, and J.E. Bridges, "Three-dimensional FDTD analysis of a pulsed microwave confocal system for breast cancer detection: design of an antenna-array element," *IEEE Trans. Antennas Propagat.*, Vol. 47, No. 5, 783–791, 1999.
14. P.M. Meaney, M.W. Fanning, D. Li, P. Poplack, and K.D. Paulsen, "A clinical prototype for active microwave imaging of the breast," *IEEE Trans. Microwave Theory Technol.*, Vol. 48, No. 11, 1841–1853, 2000.
15. E.C. Fear and M.A. Stuchly, "Microwave detection of breast cancer," *IEEE Trans. Microwave Theory Tech.*, Vol. 48, No. 11, 1854–1863, 2000.
16. Pethig R., *Dielectric and Electronic Properties of Biological Materials*, Wiley, New York, 1979.
17. P. Debye, *Polar Molecules*, Chemical Catalog Co., New York, 1929.
18. R.F. Harrington, *Field Computation by Moment Methods*, Macmillan, New York, 1968.
19. D.E. Livesay and K.M. Chen, "Electromagnetic fields induced inside arbitrarily shaped biological bodies," *IEEE Trans. Microwave Theory Technol.*, Vol. 22, 1273–1280, Dec. 1974.
20. D.H. Schaubert, D.R. Wilton, and A.W. Glisson, "A tetrahedral modeling method for electromagnetic scattering by arbitrarily shaped inhomogeneous dielectric bodies," *IEEE Trans. Antennas Propagat.*, Vol. 32, 77–85, Jan. 1984.
21. C.T. Tsai, H. Massoudi, C.H. Durney, and M.F. Iskander, "A procedure for calculating fields inside arbitrarily shaped, inhomogeneous dielectric bodies using linear basis functions with the moment method," *IEEE Trans. Microwave Theory Technol.*, Vol. 34, 1131–1139, Nov. 1986.
22. H. Massoudi, C.H. Durney, and M.F. Iskander, "Limitations of the cubical block model of man in calculating SAR distributions," *IEEE Trans. Microwave Theory Technol.*, Vol. 32, 746–751, Aug. 1984.
23. N.N. Bojarski, "*k*-Space formulation of the scattering problem in the time domain," *J. Acous. Soc. Am.*, Vol. 72, 570–584, 1982.
24. T.K. Sarkar, E. Arvas, and S.M. Rao, "Application of fast Fourier transform and the conjugate gradient method for efficient solution of electromagnetic scattering from both electrically large and small conducting bodies," *Electromagnetics*, Vol. 5, 99–122, 1985.
25. H. Gan, and W.C. Chew, "A discrete BCG-FFT algorithm for solving 3D inhomogeneous scatterer problems," *J. Electromagn. Waves Appl.*, Vol. 9, 1339–1357, 1995.
26. C. Lanczos, "An iteration method for the solution of the eigenvalue problem of linear differential and integral operators," *J. Res. Nat. Bur. Stand.*, Vol. 45, 255–282, 1950.
27. Q.H. Liu, X.M. Xu, B. Tian, and Z.Q. Zhang, "Applications of nonuniform fast transform algorithms in numerical solutions of differential and integral equations," *IEEE Trans. Geosci. Remote Sensing*, Vol. 38, 1551–1560, 2000.
28. Z.Q. Zhang and Q.H. Liu, "Three-dimensional weak-form conjugate- and bi-conjugate FFT methods for volume integral equations," *Microwave Opt. Technol. Lett.*, Vol. 5, 350–356, 2001.
29. X. Xu, Q.H. Liu, and Z.Q. Zhang, "The stabilized biconjugate gradient fast Fourier transform method for electromagnetic scattering." *J. Appl. Computat. Electromagn. Soc.*, Vol. 17, No. 1, 97–103, Mar. 2002.
30. E. Bleszynski, M. Bleszynski, and J. Jaroszewicz, "Adaptive integral method for solving large-scale electromagnetic scattering and radiation problems," *Radio Sci.*, Vol. 31, 1225–1251, 1996.
31. Z.Q. Zhang and Q.H. Liu, "A volume adaptive integral method (VAIM) for 3D inhomogeneous objects," *IEEE Antennas Wireless Propagat. Lett.*, Vol. 1, No. 6, 102–105, 2002.
32. Z.Q. Zhang, Q.H. Liu, and X.M. Xu, "RCS computation of large inhomogeneous objects using a fast integral equation solver," *IEEE Trans. Antennas Propagat.*, Vol. 51, No. 3, 613–618, Mar. 2003.

33. Z.Q. Zhang, and Q.H. Liu, "Two nonlinear inverse methods for electromagnetic induction measurements," *IEEE Trans. Geosci. Remote Sensing*, Vol. 39, No. 6, 1331–1339, June 2001.
34. Z.Q. Zhang, Q.H. Liu, C. Xiao, E. Ward, G. Ybarra, and W.T. Joines, "Microwave breast imaging: 3-D forward scattering simulation," *IEEE Trans. Biomed. Eng.*, Vol. 50, No. 10, 1180–1189, Oct. 2003.
35. Z.Q. Zhang and Q.H. Liu, "3-D nonlinear image reconstruction for microwave biomedical imaging," *IEEE Trans. Biomed. Eng.*, Vol. 51, No. 3, 544–548, 2004.
36. L.E. Baker, "Principles of the impedance technique," *IEEE Eng. Med. Biol. Mag.*, 11–15, Mar. 1989.
37. S. Caorsi, G.L. Gragnani, and M. Pastorino, "Redundant electromagnetic data for microwave imaging of three-dimensional dielectric objects," *IEEE Trans. Antennas Propagat.*, Vol. 42, No. 5, 581–589, May 1994.
38. M. Cheney and D. Isaacson, "Distinguishability in impedance imaging," *IEEE Trans. Biomed. Eng.*, Vol. 39, No. 8, 852–860, Aug. 1992.
39. B.M. Eyuboglu, B.H. Brown, and D.C. Barber, "*In vivo* imaging of cardiac related impedances changes," *IEEE Eng. Med. Biol. Mag.*, 39–45, Mar. 1989.
40. A.D. Seagar, D.C. Barber, and B.H. Brown, "Electrical impedance imaging," *IEEE Proc.*, Vol. 134, Pt. A, 201–210, 1987.
41. W.T. Joines, E. Tanabe, and Raymond U, "Determining the electrical properties of biological tissue *in vivo*," *IEEE Proc. Southeast. Conf.*, 382–385, Apr. 1976.
42. W.C. Chew, *Waves and Fields in Inhomogeneous Media*, Van Nostrand Reinhold, New York, 1990.
43. Q.H. Liu, "Nonlinear inversion of electrode-type resistivity measurements," *IEEE Trans. Geosci. Remote Sensing*, Vol. 32, No. 3, 499–507, 1994.
44. T.E. Kerner, K.D. Paulsen, A. Hartov, S.K. Soho, and S.P. Poplack, "Electrical impedance spectroscopy of the breast: clinical imaging results in 26 subjects," *IEEE Trans. Med. Imaging*, Vol. 21, No. 6, 638–645, June 2002.
45. A. Malich, T. Fritsch, R. Anderson, T. Boehm, M.G. Freesmeyer, M. Fleck, and W.A. Kaiser, "Electrical impedance scanning for classifying suspicious breast lesions: first results," *Eur. Radiol.*, Vol. 10, No. 10, 1555–1561, 2000.

Index

F

HANDBOOK OF BIOLOGICAL EFFECTS OF ELECTROMAGNETIC FIELDS

THIRD EDITION

Biological and Medical Aspects of Electromagnetic Fields

HANDBOOK OF BIOLOGICAL EFFECTS OF ELECTROMAGNETIC FIELDS

THIRD EDITION

Biological and Medical Aspects of Electromagnetic Fields

EDITED BY

Frank S. Barnes
University of Colorado-Boulder
Boulder, CO, U.S.A.

Ben Greenebaum
University of Wisconsin-Parkside
Kenosha, WI, U.S.A.

CRC Taylor & Francis
Taylor & Francis Group
Boca Raton London New York

CRC is an imprint of the Taylor & Francis Group,
an informa business

CRC Press
Taylor & Francis Group
6000 Broken Sound Parkway NW, Suite 300
Boca Raton, FL 33487-2742

Printed in the United States of America on acid-free paper
10 9 8 7 6 5 4 3 2 1

International Standard Book Number-10: 0-8493-9538-0 (Hardcover)
International Standard Book Number-13: 978-0-8493-9538-3 (Hardcover)

Visit the Taylor & Francis Web site at
http://www.taylorandfrancis.com

and the CRC Press Web site at
http://www.crcpress.com

Preface

We are honored to have been asked to carry on the tradition established by Dr. Postow and the late Dr. Polk in the first two editions of the *Handbook of Biological Effects of Electromagnetic Fields*. Their editions of this handbook were each recognized as the authoritative standards of their time for scientists working in bioelectromagnetics, the science of electromagnetic field effects on biological systems, and for others seeking information about this field of research.

In revising and updating this edition of the *Handbook of Biological Effects of Electromagnetic Fields*, we have expanded the coverage to include more material on diagnostic and therapeutic applications. At the same time, in updating and expanding the previous editions' coverage of the basic science and studies related to the possible biological effects of the electromagnetic fields, we have added new material on the related physics and chemistry as well as reviews of the recent developments in the setting standards for exposure limits. Following the previous edition's lead, we have charged the authors of the individual chapters with providing the reader, whom we imagine is fairly well founded in one or more of the sciences underlying bioelectromagnetics but perhaps not in the others or in the interdisciplinary subject of bioelectromagnetics itself, with both an introduction to their topic and a basis for further reading. We asked the chapter authors to write what they would like to be the first thing they would ask a new graduate student in their laboratory to read. We hope that this edition, like its two predecessors, will be useful to many as a reference book and to others as a text for a graduate course that introduces bioelectromagnetics or some of its aspects.

As a "handbook" and not an encyclopedia, this work does not intend to cover all aspects of bioelectromagnetics. Nevertheless, taking into account the breadth of topics and growth of research in this field since the last edition, we have expanded the number of topics and the number of chapters. Unavoidably, some ideas are duplicated in chapters, sometimes from different viewpoints that could be instructive to the reader; and different aspects of others are presented in different chapters. The increased amount of material has led to the publication of the handbook as two separate, but inter-related volumes: *Biological and Medical Aspects of Electromagnetic Fields (BMA)* and *Bioengineering and Biophysical Aspects of Electromagnetic Fields (BBA)*. Because there is no sharp dividing line, some topics are dealt with in parts of both volumes. The reader should be particularly aware that various theoretical models, which are proposed for explaining how fields interact with biological systems at a biophysical level, are distributed among a number of chapters. No one model has become widely accepted, and it is quite possible that more than one will in fact be needed to explain all observed phenomena. Most of these discussions are in the *Biological and Medical* volume, but the *Bioengineering and Biophysics* volume's chapters on electroporation and on mechanisms and therapeutic applications, for example, also have relevant material. Similarly, the chapters on biological effects of static magnetic fields and on endogenous electric fields in animals could equally well have been in the *Biological and Medical* volume. We have tried to use the index and cross-references in the chapters to direct the reader to the most relevant linkages, and we apologize for those we have missed.

Research in bioelectromagnetics stems from three sources, all of which are important; and various chapters treat both basic physical science and engineering aspects and the biological and medical aspects of these three. Bioelectromagnetics first emerged as a separate scientific subject because of interest in studying possible hazards from exposure to electromagnetic fields and setting exposure limits. A second interest is in the beneficial use of fields to advance health, both in diagnostics and in treatment, an interest that is as old as the discovery of electricity itself. Finally, the interactions between electromagnetic fields and biological systems raise some fundamental, unanswered scientific questions and may also lead to fields being used as tools to probe basic biology and biophysics. Answering basic bioelectromagnetic questions will not only lead to answers about potential electromagnetic hazards and to better beneficial applications, but they should also contribute significantly to our basic understanding of biological processes. Both strong fields and those on the order of the fields generated within biological systems may become tools to perturb the systems, either for experiments seeking to understand how the systems operate or simply to change the systems, such as by injecting a plasmid containing genes whose effects are to be investigated. These three threads are intertwined throughout bioelectromagnetics. Although any specific chapter in this work will emphasize one or another of these threads, the reader should be aware that each aspect of the research is relevant to a greater or lesser extent to all three.

The reader should note that the chapter authors have a wide variety of interests and backgrounds and have concentrated their work in areas ranging from safety standards and possible health effects of low-level fields to therapy through biology and medicine to the fundamental physics and chemistry underlying the biology. It is therefore not surprising that they have different and sometimes conflicting points of view on the significance of various results and their potential applications. Thus authors should only be held responsible for the viewpoints expressed in their chapters and not in others. We have tried to select the authors and topics so as to cover the scientific results to date that are likely to serve as a starting point for future work that will lead to the further development of the field. Each chapter's extensive reference section should be helpful for those needing to obtain a more extensive background than is possible from a book of this type.

Some of the material, as well as various authors' viewpoints, are controversial, and their importance is likely to change as the field develops and our understanding of the underlying science improves. We hope that this volume will serve as a starting point for both students and practitioners to come up-to-date with the state of understanding of the various parts of the field as of late 2004 or mid-2005, when authors contributing to this volume finished their literature reviews.

The editors would like to express their appreciation to all the authors for the extensive time and effort they have put into preparing this edition, and it is our wish that it will prove to be of value to the readers and lead to advancing our understanding of this challenging field.

Frank S. Barnes
Ben Greenebaum

Editors

Frank Barnes received his B.S. in electrical engineering in 1954 from Princeton University and his M.S., engineering, and Ph.D. degrees from Stanford University in 1955, 1956, and 1958, respectively. He was a Fulbright scholar in Baghdad, Iraq, in 1958 and joined the University of Colorado in 1959, where he is currently a distinguished professor. He has served as chairman of the Department of Electrical Engineering, acting dean of the College of Engineering, and in 1971 as cofounder/director with Professor George Codding of the Political Science Department of the Interdisciplinary Telecommunications Program (ITP).

He has served as chair of the IEEE Electron Device Society, president of the Electrical Engineering Department Heads Association, vice president of IEEE for Publications, editor of the *IEEE Student Journal* and the *IEEE Transactions on Education*, as well as president of the Bioelectromagnetics Society and U.S. Chair of Commission K—International Union of Radio Science (URSI). He is a fellow of the AAAS, IEEE, International Engineering Consortium, and a member of the National Academy of Engineering.

Dr. Barnes has been awarded the Curtis McGraw Research Award from ASEE, the Leon Montgomery Award from the International Communications Association, the 2003 IEEE Education Society Achievement Award, Distinguished Lecturer for IEEE Electron Device Society, the 2002 ECE Distinguished Educator Award from ASEE, The Colorado Institute of Technology Catalyst Award 2004, and the Bernard M. Gordon Prize from National Academy of Engineering for Innovations in Engineering Education 2004. He was born in Pasadena, CA, in 1932 and attended numerous elementary schools throughout the country. He and his wife, Gay, have two children and two grandchildren.

Ben Greenebaum retired as professor of physics at the University of Wisconsin–Parkside, Kenosha, WI, in May 2001, but was appointed as emeritus professor and adjunct professor to continue research, journal editing, and university outreach projects. He received his Ph.D. in physics from Harvard University in 1965. He joined the faculty of UW–Parkside as assistant professor in 1970 following postdoctoral positions at Harvard and Princeton Universities. He was promoted to associate professor in 1972 and to professor in 1980. Greenebaum is author or coauthor of more than 50 scientific papers. Since 1992, he has been editor in chief of *Bioelectromagnetics*, an international peer-reviewed scientific journal and the most cited specialized journal in this field. He spent 1997–1998 as consultant in the World Health Organization's International EMF Project in Geneva, Switzerland. Between 1971 and 2000, he was part of an interdisciplinary research team investigating the biological effects of electromagnetic fields on biological cell cultures. From his graduate student days through 1975, his research studied the spins and moments of radioactive nuclei. In 1977 he became a special assistant to the chancellor and in 1978, associate dean of faculty (equivalent to the present associate vice chancellor position). He served 2 years as acting vice chancellor (1984–1985 and 1986–1987). In 1989, he was appointed as dean of the School of Science and Technology, serving until the school was abolished in 1996.

On the personal side, he was born in Chicago and has lived in Racine, WI, since 1970. Married since 1965, he and his wife have three adult sons.

Contributors

Larry E. Anderson Battelle Pacific Northwest Laboratories, Richland, Washington.

Pravin Betala Pritzker School of Medicine, The University of Chicago, Chicago, Illinois

David Black University of Auckland, Auckland, New Zealand

Sigrid Blom-Eberwein Lehigh Valley Hospital Burn Center, Allentown, Pennsylvania

Elena N. Bodnar Pritzker School of Medicine, The University of Chicago, Chicago, Illinois

Yuri Chizmadzhev The A.N. Frumkin Institute of Electrochemistry, Russian Academy of Sciences, Moscow, Russia

C-K. Chou Motorola Laboratories, Fort Lauderdale, Florida

John A. D'Andrea Naval Health Research Center, Detachment, Brooks City Base, Texas

Edward C. Elson Walter Reed Army Institute of Research, Washington, D.C.

Maria Feychting Institute of Environmental Medicine, Karolinska Institutet, Stockholm, Sweden

Sheila A. Johnston Independent Neuroscience Consultant, London, U.K.

Leeka Kheifets School of Public Health, University of California at Los Angeles, Los Angeles, California

Raphael C. Lee Pritzker School of Medicine, The University of Chicago, Chicago, Illinois

David L. McCormick IIT Research Institute, Chicago, Illinois

Tatjana Paunesku Robert H. Lurie Comprehensive Cancer Center, Northwestern University, Chicago, Illinois

Arthur A. Pilla Columbia University and Mount Sinai School of Medicine, New York, New York

Michael Repacholi Radiation and Environmental Health Unit, World Health Organization, Geneva, Switzerland

Riti Shimkhada School of Public Health, University of California at Los Angeles, Los Angeles, California

Dina Simunic University of Zagreb, Zagreb, Croatia

Emilie van Deventer Radiation and Environmental Health Unit, World Health Organization, Geneva, Switzerland

James C. Weaver Massachusetts Institute of Technology, Cambridge, Massachusetts

Gayle E. Woloschak Robert H. Lurie Comprehensive Cancer Center, Northwestern University, Chicago, Illinois

Table of Contents

Introduction

Charles Polk*

Revised for the 3rd Edition by Ben Greenebaum

Much has been learned since this handbook's first edition, but a full understanding of biological effects of electromagnetic fields has is to be achieved. The broad range of what must be studied has to be a factor in the apparent slow progress toward this ultimate end. The broad range of disciplines involved includes basic biology, medical science and clinical practice, biological and electrical engineering, basic chemistry and biochemistry, and fundamental physics and biophysics. The subject matter ranges over characteristic lengths and timescales from, at one extreme, direct current (dc) or $\sim 10^4$ km-wavelengths, multimillisecond ac fields and large, long-lived organisms to, at the other extreme, submillimeter wavelength fields with periods below 10^{-12} s and subcellular structures and molecules with subnanometer dimensions and characteristic times as short as the 10^{-15} s or less of biochemical reactions.

This chapter provides an introduction and overview of the research and the contents of this handbook.

0.1 Near Fields and Radiation Fields

In recent years it has become, unfortunately, a fairly common practice—particularly in nontechnical literature—to refer to the entire subject of interaction of electric (E) and magnetic (H) fields with organic matter as biological effects of nonionizing radiation, although fields that do not vary with time and, for most practical purposes, slowly time-varying fields do not involve radiation at all. The terminology had its origin in an effort to differentiate between relatively low-energy microwave radiation and high-energy radiation, such as UV and x-rays, capable of imparting enough energy to a molecule or an atom to disrupt its structure by removing one or more electron\s with a single photon. However, when applied to dc or extremely low-frequency (ELF), the term "nonionizing radiation" is inappropriate and misleading.

A structure is capable of efficiently radiating electromagnetic waves only when its dimensions are significant in comparison with the wavelength λ. But in free space $\lambda = c/f$, where c is the velocity of light in vacuum (3×10^8 m/s) and f is the frequency in hertz (cycles/s); therefore the wavelength at the power distribution frequency of 60 Hz, e.g., is 5000 km, guaranteeing that most available human-made structures are much smaller than one wavelength.

The poor radiation efficiency of electrically small structures (i.e., structures whose largest linear dimension $L \ll \lambda$ can be illustrated easily for linear antennas. In free space the radiation resistance, R_r of a current element, i.e., an electrically short wire of length ℓ carrying uniform current along its length [1], is

$$R_r = 80\pi^2 \left(\frac{\ell}{\lambda}\right)^2 \qquad (0.1)$$

*Deceased.

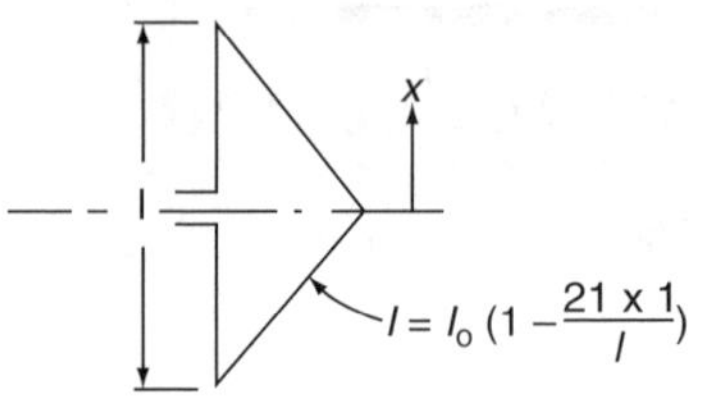

FIGURE 0.1
Current distribution on short, thin, center-fed antenna.

whereas the R_r of an actual center-fed radiator of total length ℓ with current going to zero at its ends, as illustrated in Figure 0.1, is

$$R_r = 20\pi^2 \left(\frac{\ell}{\lambda}\right)^2 \tag{0.2}$$

Thus, the R_r of a 0.01 λ antenna, 50 km long at 60 Hz, would be 0.0197 Ω. As the radiated power $P_r = I^2R_r$ where I is the antenna terminal current, whereas the power dissipated as heat in the antenna wire is I^2R_d; when I is uniform, the P_r will be very much less than the power used to heat the antenna, given that the ohmic resistance R_d of any practical wire at room temperature will be very much larger and R_r. For example, the resistance of a 50-km long, 1/2-in. diameter solid copper wire could be 6.65 Ω. At dc, of course, no radiation of any sort takes place, as acceleration of charges is a condition for radiation of electromagnetic waves.

The second set of circumstances, which guarantees that any object subjected to low-frequency E and H fields usually does not experience effects of radiation, is that any configuration that carries electric currents sets up E and H field components which store energy without contributing to radiation. A short, linear antenna in free space (short electric dipole) generates, in addition to the radiation field E_r, an electrostatic field E_s and an induction field E_i. Neither E_s nor E_i contribute to the P_r [2,3]. Whereas E_r varies as $1/r$, where r is the distance from the antenna, E_i varies as $1/r^2$, and E_s as $1/r^3$. At a distance from the antenna of approximately one sixth of the wavelength ($r = \lambda/2\pi$), the E_i equals the E_r, and when $r \ll \lambda/6$ the E_r quickly becomes negligible in comparison with E_i and E_s. Similar results are obtained for other antenna configurations [4]. At 60 Hz the distance $\lambda/2\pi$ corresponds to about 800 km and objects at distances of a few kilometers or less from a 60-Hz system are exposed to nonradiating field components, which are orders of magnitude larger than the part of the field that contributes to radiation.

A living organism exposed to a static (dc) field or to a nonradiating near field may extract energy from it, but the quantitative description of the mechanism by which this extraction takes place is very different than at higher frequencies, where energy is transferred by radiation:

1. In the near field the relative magnitudes of E and H are a function of the current or charge configuration and the distance from the electric system. The E field may be much larger than the H field or vice versa (see Figure 0.2).
2. In the radiation field the ratio the E to H is fixed and equal to 377 in free space, if E is given in volt per meter and H in ampere per meter.
3. In the vicinity of most presently available human-made devices or systems carrying static electric charges, dc, or low-frequency (<1000 Hz) currents, the E and H fields will only under very exceptional circumstances be large enough to produce heating effects inside a living object, as illustrated by Figure 0.3. (This statement assumes that the living object does not form part of a conducting path

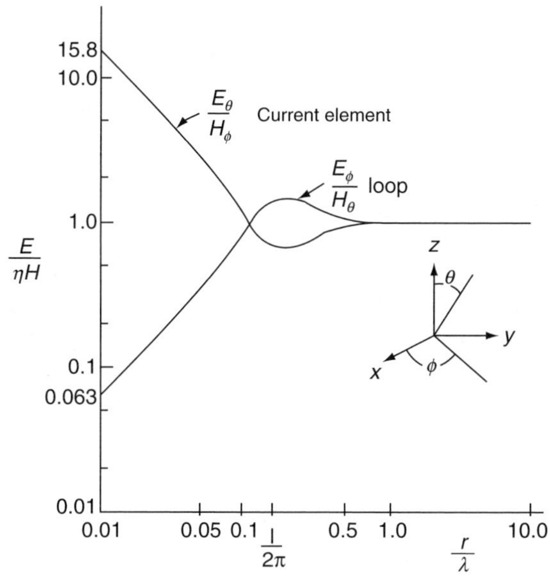

FIGURE 0.2
Ratio of E to H field (divided by wave impedance of free space $\eta = 377\ \Omega$) at $\theta = 90°$ for electric current element at origin along z-axis and for electrically small loop centered at the origin in x–y plane.

that permits direct entrance of current from a wire or conducting ground.) However, nonthermal effects are possible; thus an E field of sufficient magnitude may orient dipoles, or translate ions or polarizable neutral particles (see Chapter 3 and Chapter 4 in *BBA**).

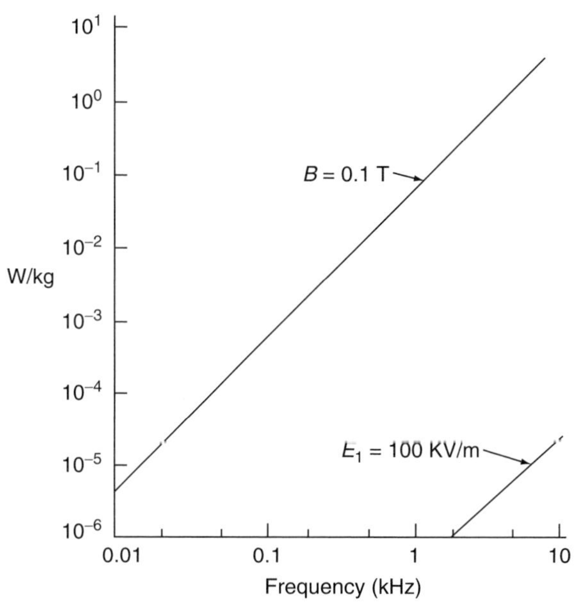

FIGURE 0.3
Top line: Eddy current loss produced in cylinder by sinusoidally time-varying axial H field. Cylinder parameters are conductivity $\sigma = 0.1\,\text{S/m}$, radius 0.1 m, density $D = 1100\,\text{kg/m}^3$, RMS magnetic flux density $0.1\,\text{T} = 1000\,\text{G}$. Watt per kilogram $= \sigma B^2 r^2 w^2/8D$; see Equation 0.15 and use power per volume $= J^2/\sigma$, *Lower line:* loss produced by 60-Hz E field in Watt per kilogram $= \sigma\, E_{\text{int}}{}^2/D$, where external field E_1 is related to E_{int} by Equation 0.9 with $\varepsilon_2 = \varepsilon_0 \times 10^5$ at 1 kHz and $\varepsilon_0 = 8 \times 10^4$ at 10 kHz.

**BBA*: *Bioengineering and Biophysical Aspects of Electromagnetic Fields* (ISBN 0-8493-9539-9); *BMA*: *Biological and Medical Aspects of Electromagnetic Fields* (ISBN 0-8493-9538-0).

4. With radiated power it is relatively easy to produce heating effects in living objects with presently available human-made devices (see Chapter 10 in *BBA* and Chapter 5 in *BMA*). This does not imply, of course, that all biological effects of radiated radio frequency (RF) power necessarily arise from temperature changes.

The results of experiments involving exposure of organic materials and entire living organisms to static *E* and ELF *E* fields are described in *BBA*, Chapter 3. Various mechanisms for the interaction of such fields with living tissue are also discussed there and in *BBA*, Chapter 5. In the present introduction, we shall only point out that one salient feature of static (dc) and ELF *E* field interaction with living organisms is that the external or applied *E* field is always larger by several orders of magnitude than the resultant average internal *E* field [5,6]. This is a direct consequence of boundary conditions derived from Maxwell's equations [1–3].

0.2 Penetration of Direct Current and Low-Frequency Electric Fields into Tissue

Assuming that the two materials illustrated schematically in Figure 0.4 are characterized, respectively, by conductivities σ_1 and σ_2 and dielectric permittivities ε_1 and ε_2, we write *E*-field components parallel to the boundary as E_P and components perpendicular to the boundary as $E_\perp$. For both static and time-varying fields

$$E_{P1} = E_{P2} \tag{0.3}$$

and for static (dc) fields

$$\sigma_1 E_{\perp 1} = \sigma_2 E_{\perp 2} \tag{0.4}$$

as a consequence of the continuity of current (or conservation of charge). The orientations of the total *E* fields in media 1 and 2 can be represented by the tangents of the angles between the total fields and the boundary line

$$\tan\theta_1 = \frac{E_{\perp 1}}{E_{P1}}, \quad \tan\theta_2 = \frac{E_{\perp 2}}{E_{P2}} \tag{0.5}$$

From these equations it follows that

$$\tan\theta_1 = \frac{\sigma_2}{\sigma_1}\frac{E_{\perp 1}}{E_{P1}} = \frac{\sigma_2}{\sigma_1}\frac{E_{\perp 2}}{E_{P2}} = \frac{\sigma_2}{\sigma_1}\tan\theta_2 \tag{0.6}$$

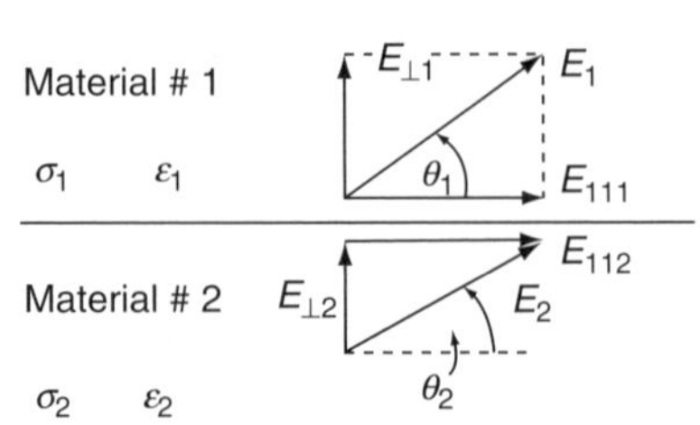

FIGURE 0.4
Symbols used in description of boundary conditions for *E*-field components.

If material 1 is air with conductivity [7] $\sigma_1 = 10^{-13}$ S/m and material 2 a typical living tissue with $\sigma_2 \approx 10^{-1}$ S/m (compare Chapter 3 in *BBA*), $\tan\theta_1 = 10^{12}\tan\theta_2$, and therefore even if the field in material 2 (the inside field) is almost parallel to the boundary so that $\theta_2 \cong 0.5°$ or $\tan\theta_2 \approx (1/100)$, $\tan\theta_1 = 10^{10}$ or $\theta_1 = (\pi/2 - 10)^{-10}$ radians. Thus an electrostatic field in air, at the boundary between air and living tissue, must be practically perpendicular to the boundary. The situation is virtually the same at ELF although Equation 0.4 must be replaced by

$$\sigma_1 E_{\perp 1} - \sigma_2 E_{\perp 2} = -j\omega\rho_s \tag{0.7}$$

and

$$\varepsilon_1 E_{\perp 1} - \varepsilon_2 E_{\perp 2} = \rho_s \tag{0.8}$$

where $j = \sqrt{-1}$, ω is the radian frequency ($= 2\pi \times$ frequency), and ρ_s is the surface charge density. In Chapter 3 in *BBA* it is shown that at ELF the relative dielectric permittivity of living tissue may be as high as 10^6 so that $\varepsilon_2 = 10^6\ \varepsilon_0$, where ε_0 is the dielectric permittivity of free space $(1/36\ \pi)\ 10^{-9}$ F/m; however, it is still valid to assume that $\varepsilon_2 \leq 0^{-5}$. Then from Equation 0.7 and Equation 0.8

$$E_{\perp 1} = \frac{\sigma_2 + j\omega\varepsilon_2}{\sigma_1 + j\omega\varepsilon_1} E_{\perp 2} \tag{0.9}$$

which gives at 60 Hz with $\sigma_2 = 10^1$ S/m, $\sigma_1 = 10^{-13}$ S/m, $\varepsilon_2 \approx 10^{-5}$ F/m, and $\varepsilon_1 \approx 10^{-11}$ F/m

$$E_{\perp 1} = \frac{10^{-1} + j_4 10^{-3}}{10^{-13} + j_4 10^{-9}} E_{\perp 2} \approx \frac{\sigma_2}{j\omega\varepsilon_1} = -j\,(2.5 \times 10^7) E_{\perp 2} \tag{0.10}$$

This result, together with Equation 0.3 and Equation 0.5, shows that for the given material properties, the field in air must still be practically perpendicular to the boundary of a living organism: $\tan\theta_1$: $2.5(10^7)\tan\theta_2$.

Knowing now that the living organism will distort the E field in its vicinity in such a way that the external field will be nearly perpendicular to the boundary surface, we can calculate the internal field by substituting the total field for the perpendicular field in Equation 0.4 (dc) and Equation 0.9 (ELF). For the assumed typical material parameters we find that in the static (dc) case

$$\frac{E_{\text{internal}}}{E_{\text{external}}} \approx 10^{-12} \tag{0.11}$$

$$\rho_f = \frac{3(\sigma_2\varepsilon_1 - \sigma_1\varepsilon_2)E_0}{2\sigma_1 + \sigma_2} \cos\vartheta \text{ C/m}^2$$

and for 60 Hz

$$\frac{E_{\text{internal}}}{E_{\text{external}}} \approx 4(10^{-8}) \tag{0.12}$$

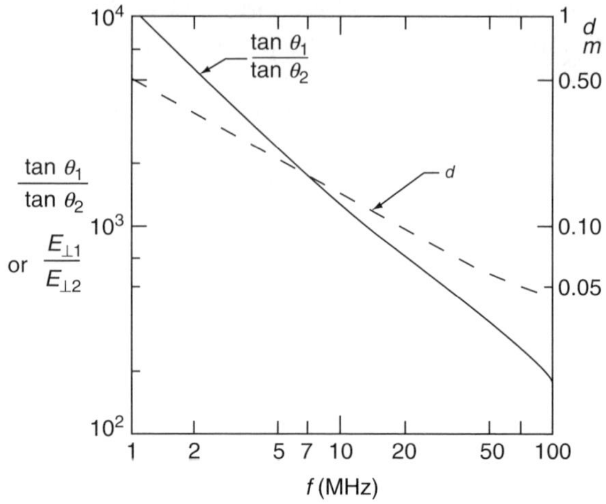

FIGURE 0.5
Orientation of *E*-field components at air–muscle boundary (or ratio of fields perpendicular to boundary); depth (*d*) at which field component parallel to boundary surface decreases by approximately 50% ($d = 0.693\delta$).

Thus, a 60-Hz external field of 100 kV/m will produce an average E_{internal} field of the order of 4 mV/m.

If the boundary between air and the organic material consists of curved surfaces instead of infinite planes, the results will be modified only slightly. Thus, for a finite sphere (with ε and σ as assumed here) embedded in air, the ratios of the internal field to the undisturbed external field will vary with the angle θ and distance r as indicated in Figure 0.5, but will not deviate from the results indicated by Equation 0.7 and Equation 0.8 by more than a factor of 3 [3,8]. Long cylinders ($L \ll r$) aligned parallel to the external field will have interior fields essentially equal to the unperturbed external field, except near the ends where the field component perpendicular to the membrane surface will be intensified approximately as above (see Chapter 9 and Chapter 10 in this volume).

0.3 Direct Current and Low-Frequency Magnetic Fields

Direct current *H* fields are considered in more detail in the Chapter 3, Chapter 5, and Chapter 8 in *BBA*. ELF *H* fields are considered in various places, including Chapter 5 and Chapter 7 in *BBA* and Chapter 2 and Chapter 11 in *BMA*. As the magnetic permeability μ of most biological materials is practically equal to the magnetic permeability μ_0 of free space, $4\pi(10^{-7})$ H/m, the dc or ELF *H* field "inside" will be practically equal to the *H* field "outside." The only exceptions are organisms such as the magnetotactic bacteria, which synthesize ferromagnetic material, discussed in Chapter 8 of *BBA*. The known and suggested mechanisms of interaction of dc *H* fields with living matter are:

1. Orientation of ferromagnetic particles, including biologically synthesized particles of magnetite.
2. Orientation of diamagnetically or paramagnetically anisotropic molecules and cellular elements [9].
3. Generation of potential differences at right angles to a stream of moving ions (Hall effect, also sometimes called a magnetohydrodynamic effect) as a result of the magnetic force $F_m = qvB \sin\theta$, where q is the the electric charge, v is the

velocity of the charge, B is the magnetic flux density, and $\sin\theta$ is the sine of the angle θ between the directions v and B. One well-documented result of this mechanism is a "spike" in the electrocardiograms of vertebrates subjected to large dc H fields.

4. Changes in intermediate products or structural arrangements in the course of light-induced chemical (electron transfer) reactions, brought about by Zeeman splitting of molecular energy levels or effects upon hyperfine structure. (The Zeeman effect is the splitting of spectral lines, characteristic of electronic transitions, under the influence of an external H field; hyperfine splitting of electronic transition lines in the absence of an external H field is due to the magnetic moment of the nucleus; such hyperfine splitting can be modified by an externally applied H field.) The magnetic flux densities involved not only depend upon the particular system and can be as high as 0.2 T (2000 G) but also <0.01 mT (100 G). Bacterial photosynthesis and effects upon the visual system are prime candidates for this mechanism [10,11].
5. Induction of E fields with resulting electrical potential differences and currents within an organism by rapid motion through a large static H field. Some magnetic phosphenes are due to such motion [12].

Relatively slow time-varying H fields, which are discussed in the basic mechanisms and therapeutic uses chapters (Chapter 5 of *BBA* and Chapter 11 in *BMA*), among others, may interact with living organisms through the same mechanisms that can be triggered by static H fields, provided the variation with time is slow enough to allow particles of finite size and mass, located in a viscous medium, to change orientation or position where required (mechanism 1 and 2) and provided the field intensity is sufficient to produce the particular effect. However, time-varying H fields, including ELF H fields, can also induce electric currents into stationary conducting objects. Thus, all modes of interaction of time-varying E fields with living matter may be triggered by time-varying, but not by static, H fields.

In view of Faraday's law, a time-varying magnetic flux will induce E fields with resulting electrical potential differences and "eddy" currents through available conducting paths. As very large external ELF E fields are required (as indicated by Equation 0.9 through Equation 0.12) to generate even small internal E fields, many human-made devices and systems generating both ELF E and H fields are more likely to produce physiologically significant internal E fields through the mechanism of *magnetic* induction.

The induced voltage V around some closed path is given by

$$V = \oint \mathrm{E} \cdot \mathrm{d}\ell = -\iint \frac{\partial B}{\partial t}\,\mathrm{ds} \tag{0.13}$$

where E is the induced E field. The integration $\oint E\,\mathrm{d}\ell$ is over the appropriate conducting path, $\partial B/\partial t$ is the time derivative of the magnetic flux density, and the "dot" product with the surface element, ds, indicates that only the component of $\partial B/\partial t$ perpendicular to the surface, i.e., parallel to the direction of the vector ds, enclosed by the conducting path, induces an E field. To obtain an order-of-magnitude indication of the induced current that can be expected as a result of an ELF H field, we consider the circular path of radius r, illustrated by Figure 0.6. Equation 0.13 then gives the magnitude of the E field as

$$E = \frac{\omega B r}{2} \tag{0.14}$$

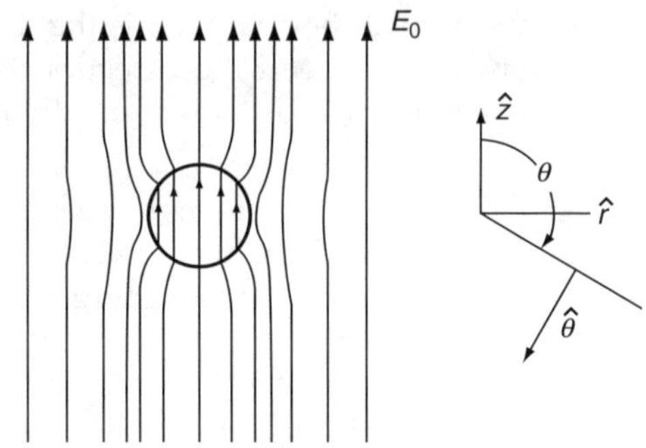

FIGURE 0.6
E field when sphere of radius R, conductivity σ_2, and dielectric permittivity ε_2 is placed into an initially uniform static field ($E = ?F_0$) within a medium with conductivity σ_1 and permittivity ε_1. The surface charge density is $\rho_r = \frac{3(\sigma_2\varepsilon_1 - \sigma_1\varepsilon_2)E_0}{2\sigma_1 + \sigma_2} \cos\theta\ \text{C/m}^2$.

$$r < R \quad \bar{E} = \frac{3\sigma_1 E_0}{2\sigma_1 + \sigma_2}\hat{z} \qquad \varepsilon_2, \sigma_2$$

$$r < R \quad \bar{E} - E_0 \cos\theta\left[1 + \frac{2R^3(\sigma_2 \propto \sigma_1)}{r^3(2\sigma_1+\sigma_2)}\right]\hat{r} \qquad \varepsilon_1, \sigma_1$$

$$- E \sin\theta\left[1 - \frac{R^3(\sigma_2 \propto \sigma_1)}{r^3(2\sigma_1+\sigma_2)}\right]\hat{\theta}$$

where ω is the $2\pi f$ and f is the frequency. The magnitude of the resulting electric current density J in ampere per square meter is*

$$J = \sigma E = \frac{\sigma\omega B r}{2} \tag{0.15}$$

where σ is the conductivity along the path in Siemens per meter. In the SI (Systeme Internationale) units used throughout this book, B is measured in tesla ($T = 10^4$ G) and r in meters. Choosing for illustration a circular path of 0.1 m radius, a frequency of 60 Hz, and a conductivity of 0.1 S/m, Equation 0.14 and Equation 0.15 give $E = 18.85$ B and $J = 1.885$ B. The magnetic flux density required to obtain a current density of 1 mA/m^2 is 0.53 mT or about 5 G. The E field induced by that flux density along the circular path is 10 mV/m. To produce this same 10 mV/m E_{internal} field by an external 60 Hz E_{external} field would require, by Equation 0.12, a field intensity of 250 kV/m.

As the induced voltage is proportional to the time rate of change of the H field (Equation 0.13), implying a linear increase with frequency (Equation 0.14), one would expect that the ability of a time-varying H field to induce currents deep inside a conductive object would increase indefinitely as the frequency increases; or conversely, that the magnetic flux density required to induce a specified E field would decrease linearly with frequency, as indicated in Figure 0.7. This is not true however, because the displacement current density $\partial D/\partial t$, where $D = \varepsilon E$, must also be considered as the frequency increases. This leads to the wave behavior discussed in Part III, implying that at sufficiently high frequencies the effects of both external E and H fields are limited

*Equation 0.15 neglects the H field generated by the induced eddy currents. If this field is taken into account, it can be shown that the induced current density in a cylindrical shell of radius r and thickness Δ is given by $\Delta r < 0.01$ m^2/[$1 + j\Delta r/\delta^2$], where $H_0 = B_0/\mu_0$ and δ is the skin depth defined by Equation 0.17 below. However, for conductivities of biological materials ($\sigma < 5$ s/m) one obtains at audio frequencies $\delta > 1$ m and as for most dimensions of interest $\Delta r < 0.01$ m^2 the term $j\Delta r/\delta^2$ becomes negligible. The result $-jrH_0/\delta^2$ is then identical with Equation 0.15.

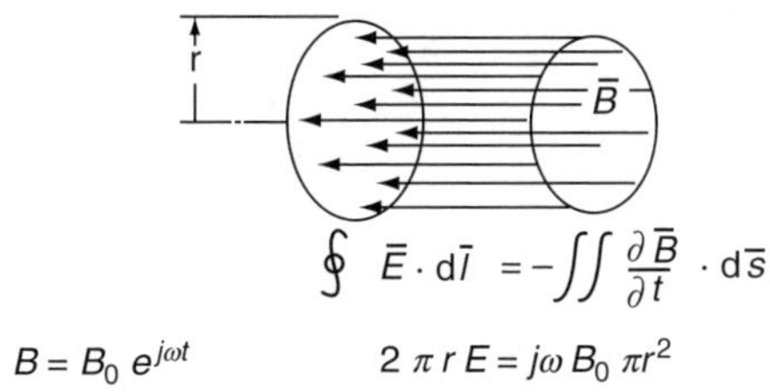

FIGURE 0.7
Circular path (loop) of radius r enclosing uniform magnetic flux density perpendicular to the plane of the loop. For sinusoidal time variation $B = B_0 e^{j\omega t}$.

by reflection losses (Figure 0.8 through Figure 0.10) as well as by skin effect [13], i.e., limited depth of penetration d in Figure 0.5.

0.4 RF Fields

At frequencies well below those where most animals and many field-generating systems have dimensions of the order of one free space wavelength, e.g., at 10 MHz where $\lambda = 30$ m, the skin effect limits penetration of the external field. This phenomenon is fundamentally different from the small ratio of internal to external E fields described in Equation 0.4 (applicable to dc) and Equation 0.9.

Equation 0.9 expresses a "boundary condition" applicable at all frequencies, but as the angular frequency ω increases (and in view of the rapid decrease with frequency of the dielectric permittivity ε_2 in biological materials—see Chapter 3 of *BBA*, the ratio of the normal component of the external to the internal E field at the boundary decreases

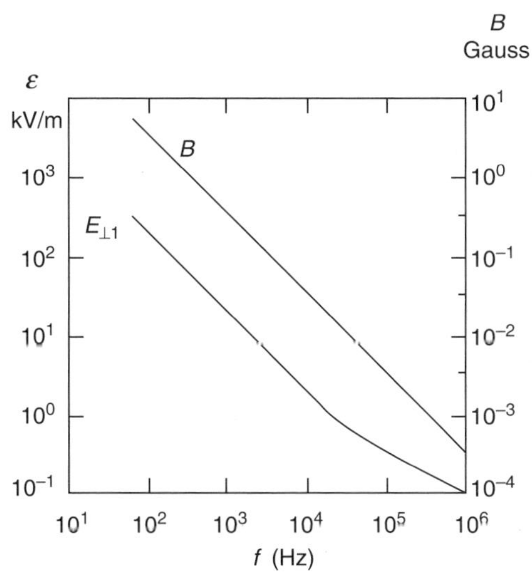

FIGURE 0.8
External E and H field required to obtain an internal E field of 10 mV/m (conductivity and dielectric permittivity for skeletal muscle from Foster, K.R., Schepps, J.L., and Schwan, H.P. 1980. *Biophys. J.*, 29:271–281. H-field calculation assumes a circular path of 0.1-m radius perpendicular to magnetic flux).

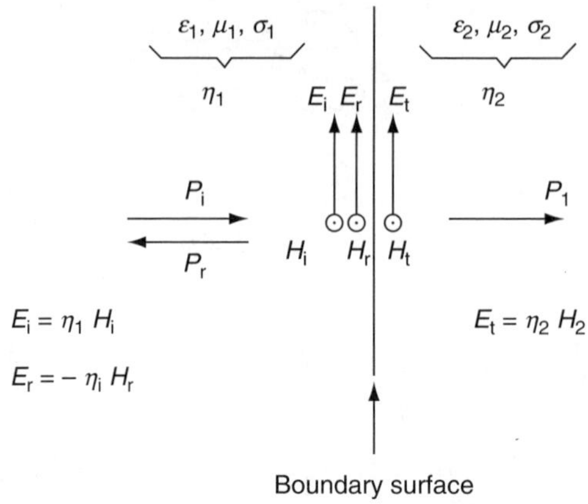

FIGURE 0.9
Reflection and transmission of an electromagnetic wave at the boundary between two different media, perpendicular incidence; P_i = incident power, P_r = reflected power, P_t = transmitted power.

with increasing frequency. This is illustrated by Figure 0.10 where $\tan\theta_1/\tan\theta_2$ is also equal to $E_{\perp 1}/E_{\perp 2}$ in view of Equation 0.3, Equation 0.5, and Equation 0.9. However, at low frequencies the total field inside the boundary can be somewhat larger than the perpendicular field at the boundary; and any field variation with distance from the boundary is not primarily due to energy dissipation, but in a homogeneous body is a consequence of shape. At RF, on the other hand, the *E* and *H* fields of the incoming

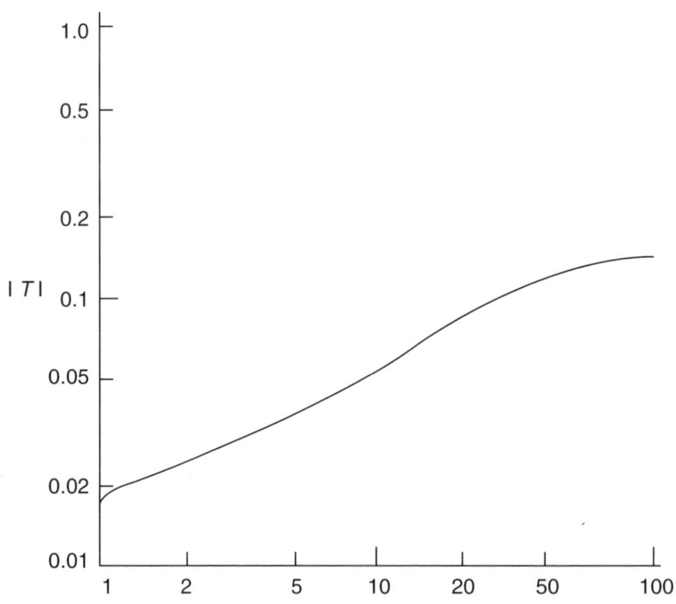

FIGURE 0.10
Magnitude of transmission coefficient *T* for incident *E* field parallel to boundary surface. $T = E_t/E_i$: reflection coefficient $r = E_r/E_i = T-1$. Γ and *T* are complex numbers; ε_r and σ for skeletal muscle from Chapter 3 in *BBA*.

electromagnetic wave, after reflection at the boundary, are further decreased due to energy dissipation. Both E and H fields decrease exponentially with distance from the boundary

$$g(z) = Ae^{-\frac{Z}{\delta}} \tag{0.16}$$

where $g(z)$ is the field at the distance z and A is the magnitude of the field just *inside* the boundary.

As defined by Equation 0.16 the skin depth δ is the distance over which the field decreases to $1/e$ (= 0.368) of its value just *inside* the boundary. (Due to reflection, the field A just inside the boundary can already be very much smaller than the incident external field; see Figure 0.8 and Figure 0.9.)

Expressions for δ given below were derived [2,3,13,14] for plane boundaries between infinite media. They are reasonable accurate for cylindrical structures if the ratio of radius of curvature to skin depth (r_0/δ) is larger than about five [13]. For a good conductor

$$\delta = \frac{1}{\sqrt{\pi f \mu \sigma}} \tag{0.17}$$

where a good conductor is one for which the ratio p of conduction current, $J = \sigma E$, to displacement current, $\partial D/\partial t = \varepsilon\,(\partial E/\partial t) = j\omega\varepsilon E$ is large:

$$p = \frac{\sigma}{\omega\varepsilon} \gg 1 \tag{0.18}$$

Since for most biological materials p is of the order of one ($0.1 < p < 10$) over a very wide frequency range (see Chapter 3 of *BBA*), it is frequently necessary to use the more general expression [13]

$$\delta = \frac{1}{\omega\left[\frac{\mu\varepsilon}{2}(\sqrt{1+p^2}-1\right]^{1/2}} \tag{0.19}$$

The decrease of field intensity with distance from the boundary surface indicated by Equation 0.16 becomes significant for many biological objects at frequencies where $r_0/\delta \geq 5$ is not satisfied. However, the error resulting from the use of Equation 0.16 and Equation 0.17 or Equation 0.19 with curved objects is less when $z < \delta$. Thus at $z = 0.693\,\delta$, where $g(z) = 0.5$ A from Equation 0.16 and Equation 0.17, the correct values of $g(z)$, obtained by solving the wave equation in cylindrical coordinates, differs only by 20% (it is 0.6 A) even when r_0/δ is as small as 2.39 [14]. Therefore, Figure 0.10 shows the distance $d = 0.693\,\delta$, at which the field decreases to half of its value just inside the boundary surface, using Equation 0.19 with typical values for σ and ε for muscle from Figure 0.11. It is apparent that the skin effect becomes significant for humans and larger vertebrates at frequencies >10 MHz.

Directly related to skin depth, which is defined for fields varying sinusoidally with time, is the fact that a rapid transient variation of an applied magnetic flux density constitutes an exception to the statement that the dc H field inside the boundary is equal to the H field outside. Thus, from one viewpoint one may consider the rapid application or removal of a dc H field as equivalent to applying a high-frequency field during the switching period, with the highest frequencies present of the order of $1/\tau$, where τ is the rise time of the applied step function. Thus, if $\tau < 10^{-8}$ s, the skin effect will be important during the transient period, as d in Figure 0.5 is <5 cm above 100 MHz. It is also possible to calculate directly the magnetic flux density inside a conducting cylinder as a function of radial position r and time t when a magnetic pulse is applied in the axial

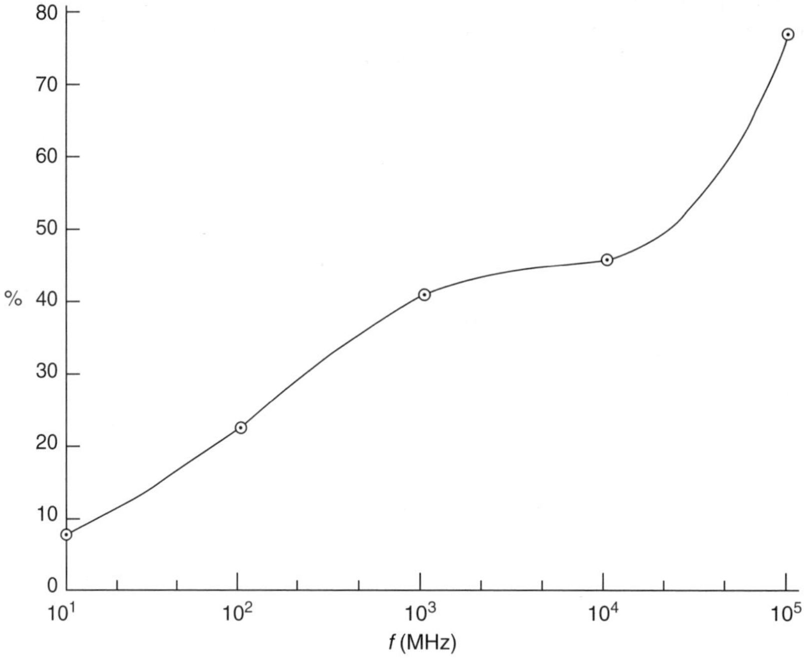

FIGURE 0.11
Ratio of transmitted to incident power expressed as percent of incident power. Air–muscle interface, perpendicular incidence (Equation 0.31, Table 0.1).

direction [15,16]. Assuming zero rise time of the applied field B_0, i.e., a true step function, one finds that the field inside a cylinder of radius a is

$$B = B_0\left[1 - \sum_{k=1}^{\infty} J_0\left(r\frac{v_k}{a}\right)e^{-t/T_k}\right] \tag{0.20}$$

where $J_0\ (r\ v_k/a)$ is the zero-order Bessel function of argument $r\ v_k/a$ and the summation is over the nulls of J_0 designated v_k (the first four values of v_k are 2.405, 5.520, 8.654, and 11.792).* T_k is the rise time of the kth term in the series and is given by

$$T_k = \frac{\mu_0\sigma a^2}{v_k} \tag{0.21}$$

As v_k increases, the rise time decreases and therefore the longest delay is due to the first term in the summation with $k = 1$

$$T_1 = \frac{\mu_0\sigma a^2}{2.405} \tag{0.22}$$

For a cylinder with 0.1 m radius and a conductivity $\sigma \approx 1$ S/m, which is a typical value for muscle between 100 and 1000 MHz, Equation 0.22 gives $T_1 = 2.6 \times 10^{-8}$ s. This finite rise time (or decay time in case of field removal) of the internal H field may be of some importance when pulsed H fields are used therapeutically [17]. It might also be used

*This result is based on solution of $\partial B/\partial t = (1/\mu_0)\nabla^2 B$, which is a consequence of Ampere's and Farraday's laws when displacement is disregarded. Equations 0.20 to 0.22 are therefore only correct when $p \gg 1$.

to measure noninvasively the conductivity of biological substances *in vivo* through determination of the final decay rate of the voltage induced into a probe coil by the slowly decaying internal field after the applied field is removed [16].

The properties of biological substances in the intermediate frequency range, above ELF (>300 Hz), and below the higher RFs, where wave behavior and skin effect begin to be important (~20 MHz), are discussed in Chapter 3 of *BBA*. However, many subsequent chapters are concerned with biological effects at dc and ELF frequencies below a few kilohertz, while others deal primarily with the higher RFs, >50 MHz. One reason for this limited treatment of the intermediate frequency range is that very little animal data are available for this spectral region in comparison with the large number of experiments performed at ELF and microwave frequencies in recent years.* Another reason is that most electrical processes known to occur naturally in biological systems—action potentials, EKG, EEG, ERG, etc.—occur at dc and ELF frequencies. Therefore, one might expect some physiological effects from external fields of appropriate intensity in the same frequency range, even if the magnitude of such fields is not large enough to produce thermal effects. As illustrated by Figure 0.3 and Figure 0.7, most *E* fields below 100 kHz set up by currently used human-made devices, and most *H* fields below 10 kHz except the very strongest, are incapable of producing thermal effects in living organisms, excluding, of course, fields accompanying currents directly introduced into the organism via electrodes. Thus, the frequencies between about 10 and 100 kHz have been of relatively little interest because they are not very likely to produce thermal or other biological effects. On the other hand, the higher RFs are frequently generated at power levels where enough energy may be introduced into living organisms to produce local or general heating. In addition, despite skin effect and the reflection loss to be discussed in more detail below, microwaves modulated at an ELF rate may serve as a vehicle for introducing ELF fields into a living organism of at least the same order of magnitude as would be introduced by direct exposure to ELF. Any effect of such ELF-modulated microwaves would, of course, require the existence of some amplitude-dependent demodulation mechanism to extract the ELF from the microwave carrier.

Among the chapters dealing with RF, Chapter 10 and Chapter 11 of *BBA* give the necessary information for establishing the magnitude of the fields present in biological objects: (1) experimental techniques and (2) analytical methods for predicting field intensities without construction of physical models made with "phantom" materials, i.e., dielectric materials with properties similar to those of living objects which are to be exposed. As thermal effects at microwave frequencies are certainly important, although one cannot assume *a priori* that they are the only biological effects of this part of the spectrum, and as some (but not all) thermal effects occur at levels where the thermoregulatory system of animals is activated, thermoregulation in the presence of microwave fields is discussed in Chapter 5 of *BMA*, as well as in Chapter 10 of *BBA*. Not only are the therapeutic applications of microwaves based upon their thermal effects, but also the experimental establishment of possible nonthermal effects at the threshold of large scale tissue heating in particular living systems and also requires thorough understanding of thermoregulatory mechanisms. The vast amount of experimental data obtained on animal systems exposed to microwave is discussed in Chapter 3 and Chapter 4 in *BMA*. Both nonmodulated fields and modulated fields, where the type of modulation had no apparent effect other than modification of the average power level, are considered. These chapters and the Chapter 9 in *BMA* are considered to be very new extension of experiments into exposures to ultra-short and to ultra-high power pulses.

*Though this statement was written in for the second edition in 1995, it continues to be true in 2005—Ben Greenebaum.

At the higher RFs, the external E field is not necessarily perpendicular to the boundary of biological materials (see Figure 0.4 and Figure 0.10), and the ratio of the total external E field to the total internal field is not given by Equation 0.9. However, the skin effect (Equation 0.16 through Equation 0.19) and reflection losses still reduce the E field within any biological object below the value of the external field. As pointed out in Chapter 3, dielectric permittivity and electrical conductivity of organic substances both vary with frequency. At RF, most biological substances are neither very good electrical conductors nor very good insulators, with the exception of cell membranes, which are good dielectrics at RF but at ELF can act as intermittent conductors or as dielectrics and are ion-selective [18–20]). The ratio p (Equation 0.18) is neither much smaller nor very much larger than values shown for typical muscle tissue [21,22] in Table 0.1.

Reflection loss at the surface of an organism is a consequence of the difference between its electrical properties and those of air. Whenever an electromagnetic wave travels, from one material to another with different electrical properties, the boundary conditions (Equation 0.3 and Equation 0.8) and similar relations for the H field require the existence of a reflected wave. The expressions for the reflection coefficient

$$\Gamma = \frac{E_r}{E_i} \tag{0.23}$$

and the transmission coefficient

$$T = \frac{E_t}{E_i} \tag{0.24}$$

become rather simple for loss-free dielectrics ($p \ll 1$) and for good conductors ($p \gg 1$). As biological substances are neither the most general expressions for Γ and T, applicable at plane boundaries, are needed [3,13]. For perpendicular incidence, illustrated by Figure 0.8,

$$\Gamma = \frac{\eta_2 - \eta_1}{\eta_2 + \eta_1} \tag{0.25}$$

$$T = \frac{2\eta_2}{\eta_2 + \eta_1} = 1 + \Gamma \tag{0.26}$$

TABLE 0.1

Ratio p of Conduction Current to Displacement as a Function of Frequency

f (MHz)	σ	ε_r	$p = \dfrac{\sigma}{\omega\varepsilon_0\varepsilon_r}$
1	0.40	2000	3.6
10	0.63	160	7.1
100	0.89	72	2.2
10^3	1.65	50	0.59
10^4	10.3	40	0.46
10^5	80	6	2.4

where η_1 and η_2 are the wave impedances, respectively, of mediums 1 and 2. The wave impedance of a medium is the ration of the E to the H field in a plane wave traveling through that medium; it is given by [13]

$$\eta = \left(\frac{j\omega\mu}{\sigma + j\omega\varepsilon}\right)^{1/2} \tag{0.27}$$

Clearly Γ and T are in general complex numbers, even when medium 1 is air for which Equation 0.27 reduces to the real quantity $\eta_0 = \sqrt{\mu_0/\varepsilon_0}$, because medium 2, which here is living matter, usually has a complex wave impedance at RFs.

The incident, reflected, and transmitted powers are given by [13]

$$P_i = R_1|E_i|^2 \frac{1}{\eta_1^*} = \frac{|E_i|^2}{|\eta_1|^2} R_1 \tag{0.28}$$

$$P_r = R_1|E_r|^2 \frac{1}{\eta_1^*} = \frac{|E_r|^2}{|\eta_1|^2} R_1 \tag{0.29}$$

$$P_t = R_1|E_t|^2 \frac{1}{\eta_2^*} = \frac{|E_t|^2}{|\eta_2|^2} R_2 \tag{0.30}$$

where the E fields are effective values ($E_{\text{eff}} = E_{\text{peak}}/\sqrt{2}$) of sinusoidal quantities, R_1 signifies "real part of," η^* is the complex conjugate of η, and R_1 and R_2 are the real parts of η_1 and η_2. If medium 1 is air, $\eta_1 = R_1 = 377\ \Omega$, it follows from Equation 0.23, Equation 0.24, and Equation 0.28 through Equation 0.30 and conservation of energy that the ratio of the transmitted to the incident real power is given by

$$\frac{P}{P_1} = |T|^2 \frac{\eta_1\eta_2^* + \eta_1^*\eta_2}{2|\eta_2|^2} = 1 - \frac{P_r}{P_i} = 1 - |\Gamma|^2 \tag{0.31}$$

The magnitude of the transmission coefficient T for the air–muscle interface over the 1- to 100-MHz frequency range is plotted in Figure 0.9, which shows that the magnitude of the transmitted E field in muscle tissue is considerably smaller than the E field in air. The fraction of the total incident power that is transmitted (Equation 0.31) is shown in Figure 0.11, indicating clearly that reflection loss at the interface decreases with frequency. However, for deeper lying tissue this effect is offset by the fact that the skin depth δ (Equation 0.19) also decreases with frequency (Figure 0.12) so that the total power penetrating beyond the surface decreases rapidly.

In addition to reflection at the air–tissue boundary, further reflections take place at each boundary between dissimilar materials. For example, the magnitude of the reflection coefficient at the boundary surface between muscle and organic materials with low-water content, such as fat or bone, is shown in Table 0.2.

The situation is actually more complicated than indicated by Figure 0.9 and Figure 0.11, because the wave front of the incident electromagnetic wave may not be parallel to the air–tissue boundary. Two situations are possible: the incident E field may be polarized perpendicular to the plane of incidence defined in Figure 0.13 (perpendicular polarization, Figure 0.13a) or parallel to the plane of incidence (parallel polarization, Figure 0.13b). The transmission and reflection coefficients [8] are different for the two types of polarization and also become functions of the angle of incidence α_1:

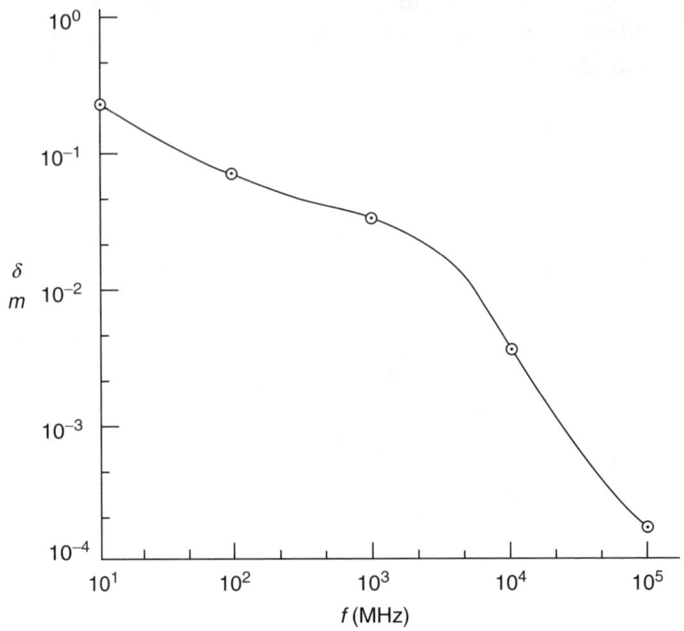

FIGURE 0.12
Electromagnetic skin depth in muscle tissue from plane wave expression (Equation 0.19, Table 0.1).

$$\text{Perpendicular polarization}\begin{cases} T_{\perp} = \dfrac{2\eta_2 \cos\alpha_1}{\eta_2\cos\alpha_1 + \eta_1\cos\alpha_2} \\[2ex] \Gamma_{\perp} = \dfrac{\eta_2\cos\alpha_1 - \eta_1\cos\alpha_2}{\eta_2\cos\alpha_1 + \eta_1\cos\alpha_2} \end{cases} \qquad (0.32),(0.33)$$

$$\text{Parallel polarization}\begin{cases} T_{\mathrm{p}} = \dfrac{2\eta_2 \cos\alpha_1}{\eta_2\cos\alpha_2 + \eta_1\cos\alpha_1} \\[2ex] \Gamma_{\mathrm{p}} = \dfrac{\eta_1\cos\alpha_1 - \eta_2\cos\alpha_2}{\eta_2\cos\alpha_2 + \eta_1\cos\alpha_1} \end{cases} \qquad (0.34),(0.35)$$

where α_2 is given by the generalized Snell's law (when both the media have the magnetic permeability of free space) by

TABLE 0.2

Reflection Coefficient "Capital Gamma" for Low–Water-Content Materials

	Fat or Bone		
f (MHz)	σ (S/m)	ε_r	Muscle[a]–Fat (Γ)
10^2	0.048	7.5	0.65
10^3	0.101	5.6	0.52
10^4	0.437	4.5	0.52

[a] σ and ε_r for muscle from Table 0.1.

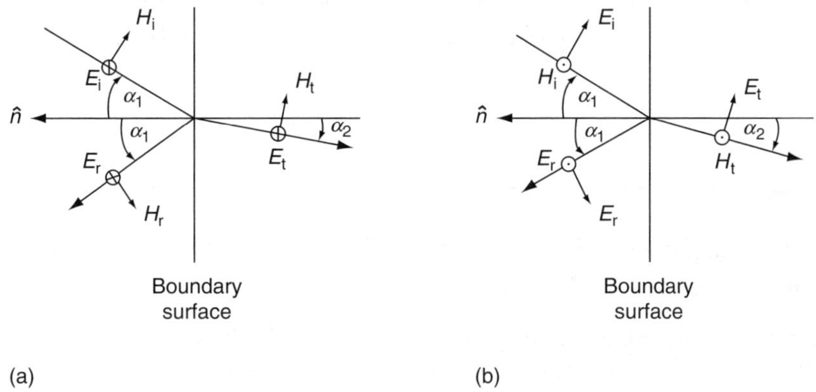

FIGURE 0.13
Oblique incidence of an electromagnetic wave at the boundary between two different media. (a) Perpendicular polarization (*E* vector perpendicular to plane of incidence); (b) parallel polarization (*E* vector parallel to plane of incidence). The plane of incidence is the plane formed by the surface normal (unit vector **n** and the direction of the incident wave); ⊗ indicates a vector into the plane of the paper; ⊙ indicates a vector out of the plane of the paper. The orientation of the field vectors in the transmitted field is shown for loss-free dielectrics. For illustration of the transmitted wave into a medium with finite conductivity, where the wave impedance η_2 becomes a complex number, see Stratton, J.A., *Electromagnetic Theory*, McGraw-Hill, New York, 1941, p. 435.

$$\sin\alpha_2 = \frac{\sqrt{\varepsilon_1}}{\sqrt{\varepsilon_2 - \mathrm{j}\frac{\sigma_2}{\omega}}} \tag{0.36}$$

so that $\cos\alpha_2 = \sqrt{1-\sin^2\alpha_2}$ is a complex number unless $\rho_2 = (\sigma_2/\omega\varepsilon_2) = 1$.

As illustration, the variation with angle of incidence of the transmission coefficient for parallel polarization at the air–muscle interface at 10 MHz, is shown in Figure 0.14. It is apparent that the transmitted field is not necessarily maximized by perpendicular incidence in the case of parallel polarization. Furthermore, whenever $p \approx 1$ or $p > 1$ (see Table 0.1, above), α_2 is complex, which causes the waves entering the tissue to be inhomogeneous—they are not simple plane waves, but waves where surfaces of constant phase and constant amplitude do not coincide [3,23]; only the planes of constant amplitude are parallel to the boundary surface.

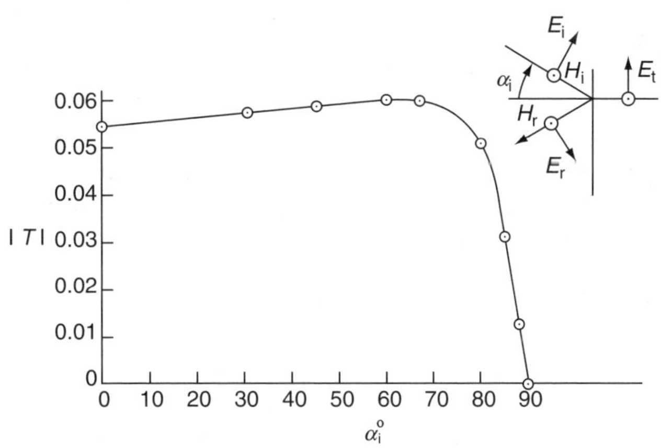

FIGURE 0.14
Magnitude of complex transmission coefficient for parallel polarization versus angle of incidence α_1 at 10 MHz (*E* field in plane of incidence, *H* field parallel to boundary plane; $\sigma_2 = 0.7\,\mathrm{S/m}$, $\varepsilon_{r2} = 150$, $T = E_t/E_r$).

Analytical solutions for nonplanar structures taking into account size and shape of entire animals have been given [24] and are also described in the RF modeling Chapter 10 of *BBA*.

0.5 Biophysical Interactions of Fields: Ionization, Ionizing Radiation, Chemical Bonds, and Excitation

RF fields can be characterized as nonionizing radiation. By this we mean that there is not enough energy in a single quantum of energy, hf, to ionize an atom or a molecule at RFs, where h is Planck's constant and f is the frequency. By comparison radiation in the UV or x-rays often lead to ionization. It is desirable to begin by reviewing the differences between ionizing and nonionizing radiations, to explain ionization phenomena and also to discuss related excitation phenomena, which require less energy than ionization. Then a number of the proposed models concerning atomic or molecular-level interactions of fields will be introduced. A number of these theories will be discussed and their predictions compared with experimental results in later chapters, including Chapter 5 through Chapter 7 and Chapter 9 in *BBA*; Chapter 9 and Chapter 11 in *BMA*. Heating, cell excitation, electroporation, and other results of high-intensity fields have been accepted as explanations for many bioelectromagnetic phenomena. For low-intensity exposure, however, no theory is widely accepted as a general explanation for bioelectromagnetic phenomena, and few specific phenomena have accepted explanations. It is quite possible that no general explanation exists and that more than one mechanism of interaction between fields will be found to be operating, depending on the situation. Binhi's book [25] contains a good summary of most recent theoretical proposals, including comparisons with data and critiques of their strong and weak points, as well as his own theory.

We note first that the energy of electromagnetic waves is quantized with the quantum of energy (in joules) being equal to Planck's constant ($h = 6.63 \times 10^{-34}$ J s) times the frequency. This energy can also be expressed in electron volts, i.e., in multiples of the kinetic energy acquired by an electron accelerated through a potential difference of 1 eV (1 eV $\approx 1.6 \times 10^{-19}$ J). Energy quanta for a few frequencies are listed in Table 0.3.

Quantized energy can "excite" molecules; appropriate frequencies can couple to vibrational and rotational oscillation; and if the incident energy quantum has sufficient magnitude it can excite other changes in the electron configuration, such as changing an electron to another (unoccupied) energy level or tearing an electron away from one of the constituent atoms, the latter process called as ionization. The energy required to remove one electron from the highest energy orbit of a particular chemical element is called its "ionization potential." Typical ionization potentials are of the order 10 eV; for example, for the hydrogen atom it is 13.6 eV and for gaseous sodium it is 5.1 eV. As chemical binding forces are essentially electrostatic, ionization implies profound chemical changes. Therefore ionization by any outside agent of the complex compounds that make up a living system leads to profound and often irreversible changes in the operation of that system.

Table 0.3 shows that even the highest RF (millimeter waves) has quantum energies well below the ionization potential of any known substance; thus one speaks of nonionizing radiation when referring to electromagnetic waves below UV light frequencies. Ionizing radiation includes UV and higher frequency electromagnetic waves (x-rays, γ-rays).

TABLE 0.3

Wave and Quantum Characteristics of Various Types of Radiation

Name of Radiation or Application	Frequency (Hz)	Wavelength (m)	Energy of 1 Quantum of Radiation (eV)
UHF TV	7×10^8	0.43	2.88×10^{-6}
Microwave radar	10^{10}	3×10^{-2}	4.12×10^{-5}
Millimeter wave	3×10^{11}	1×10^{-3}	1.24×10^{-3}
Visible light	6×10^{14}	5×10^{-7}	2.47
Ionizing UV	10^{16}	3×10^{-4}	41.2
Soft x-ray	10^{18}	3×10^{-10}	4120
Penetrating x-ray	10^{20}	3×10^{-12}	4.12×10^5

This explanation of the difference between ionizing and nonionizing radiation should not imply that nonionizing electromagnetic radiation cannot have profound effects upon inorganic and organic substances. As excitation of coherent vibrational and rotational modes requires considerably less energy than ionization, it could occur at RF; this will be discussed in later chapters. In addition, many other possible biological effects require energies well below the level of ionizing potentials. Examples are tissue heating, dielectrophoresis, depolarization of cell membranes, mechanical stress due to piezoelectric transduction, or dielectric saturation, resulting in the orientation of the polar side chains of macromolecules and leading to the breaking of hydrogen bonds. These and other mechanisms will be discussed by the authors of several chapters (see especially Chapter 5 through Chapter 7 of *BBA* and Chapter 9 of *BMA*), who will also give estimates of rates at which energy must be delivered to produce particular effects.

Returning to the discussion of ionization, it is important to note that ionization of a chemical element can be brought about not only by absorption of electromagnetic energy, but also by collision either with foreign (injected) atoms, molecules, or subatomic particles of the requisite energy, or by sufficiently violent collision among its own atoms. The latter process constitutes ionization by heating, or thermal breakdown of a substance, which will occur when the kinetic energy of the colliding particles exceeds the ionization potential. As the average thermal kinetic energy of particles is related to temperature [26] by $W = kT$ where k is Boltzmann's constant ($= 1.38 \times 10^{-23}$ J/K), we find that the required temperature is

$$1.38(10^{-23})T \approx 5 \ \text{eV} \approx (5)1.6(10^{-19})\,\text{J}$$

$$T \approx 5(10^4)\,\text{K}$$

which is about twice the temperature inside a lightning stroke [27] and orders of magnitude higher than any temperature obtainable from electromagnetic waves traveling through air.

Actually, initiation of lightning strokes is an example of ionization by collision with injected energetic particles. The few free electrons and ions always present in the air due to ionization by cosmic rays are accelerated by the E fields generated within clouds to velocities corresponding to the required ionization energy. Only when the field is large enough to impart this energy over distances shorter than the mean free path of the free electrons or ions at atmospheric pressure can an avalanche process take place: an accelerated electron separates a low-energy electron from the molecule with which it collides and in the process loses most of its own energy; thus, one high-energy free electron is exchanged for two free low-energy electrons and one positive ion. Both the electrons are in turn accelerated again by the field, giving them high kinetic energy before they collide with neutral molecules; their collision produces four free electrons and the multiplication process continues. The breakdown field strength for air at atmospheric pressure is approximately 3×10^6 V/m, implying a mean free path of electrons

$$\Delta\ell \approx [5\,\text{eV}/3 \times 10^6\,\text{V/m}] \approx 10^{-6}\,\text{m}$$

However, this model is not entirely accurate because the actual mean free path corresponds to energies of the order of 0.1 eV, which is only sufficient to excite vibrational modes in the target molecule. Apparently such excitation is sufficient to cause ionization if the collision process lasts long enough [28].

Except for some laboratory conditions where a sufficiently high potential difference can be applied directly across a biological membrane to bring about its destruction, collisional ionization is generally not a factor in the interaction of electromagnetic waves with tissue: The potential difference required for membrane destruction [29] is between 100 nV and 300 mV, corresponding to a field strength of the order of 2×10^7 V/m, assuming a membrane thickness ($d = 100$ Å; $E = V/d$). However, there is a third mechanism of ionization that is particularly important in biological systems. When a chemical compound of the type wherein positive and negative ions are held together by their electrostatic attraction, such as the ionic crystal NaCl, is placed in a suitable solvent, such as H_2O, it is separated into its ionic components. The resulting solution becomes an electrolyte, i.e., an electrically conducting medium in which the only charge carriers are ions.

In this process of chemical ionization, the Na^+ cations and Cl^- anions are separated from the original NaCl crystal lattice and individually surrounded by a sheet of solvent molecules, the "hydration sheath." If the solvent is H_2O, this process is called "hydration," or more generally, for any solvent, "solvation."

A dilute solution of NaCl crystals in H_2O is slightly cooler than the original constituents before the solvation process, indicating that some internal energy of the system was consumed. Actually energy is consumed in breaking up the original NaCl bonds and some, but less, is liberated in the interaction between the dipole moment of the solvent molecule (H_2O in our example) and the electric charges on the ions. Thus, solvents with higher relative dielectric constant ε_r, indicating higher inherent electric dipole moment per unit volume (P), solvate ions more strongly ($\varepsilon_r = 1 + P/[\varepsilon_o E]$, where E is the electric field applied during the measurement of ε_r). For example, H_2O with $\varepsilon_r \approx 80$ solvates more strongly than methanol with $\varepsilon_r \approx 33$. For biological applications it is worth noting that solvation may affect not only ionic substances, but also polar groups, i.e., molecular components which have an inherent dipole moment, such as—C=O, —NH, or —NO_2. Details of the process are discussed in texts on electrochemistry [30,31].

In biological processes not only chemical ionization and solvation of ionic compounds, but also all kinds of chemical reaction take place. One of the central questions in the study of biological effects of E and H fields is therefore not only whether they can cause or influence ionization, but also whether they can affect—speed up, slow down, or modify—any naturally occurring biologically important chemical reaction.

In Table 0.4 typical energies for various types of chemical bonds are listed. For comparison the thermal energy per elementary particle at 310 K is also shown. Complementing the numbers in Table 0.4 one should also point out that:

1. The large spread in the statistical distribution of energies of thermal motion guarantees that at physiological temperatures some molecules always have sufficient energy to break the strongest weak bonds [32].

TABLE 0.4

Bond and Thermal Energies

Type of Bond	Change in Free Energy (Binding Energy) kcal/mol	eV/Molecule
Covalent	50–100	2.2–4.8
Van der Waals	1–2	0.04–0.08
Hydrogen	3–7	0.13–0.30
Ionic[a]	5	0.2
Avg. thermal energy at 310 K	0.62	0.027

[a]For ionic groups of organic molecules such as COO^-, $NH3_3^-$ in aqueous solution.

2. The average lifetime of a weak bond is only a fraction of a second.
3. The weak binding forces are effective only between the surfaces in close proximity and usually require complementary structures such as a (microscopic) plug and hole, such as are thought to exist, for instance, between antigen and antibody [33].
4. Most molecules in aqueous solution form secondary bonds.
5. The metabolism of biological systems continuously transforms molecules and therefore also changes the secondary bonds that are formed.

Comparison of the last columns in Table 0.3 and Table 0.4 shows that millimeter waves have quantum energies, which are only about one order of magnitude below typical Van der Waals energies (waves at a frequency of 10^{12} Hz with a quantum energy of 0.004 eV have a wavelength of 0.3 mm and can still be classified as millimeter waves). One might expect therefore that such waves could initiate chemically important events, such as configurational changes, by e.g., multiple transitions between closely spaced vibrational states at successively high-energy levels [46].

Energies associated with transition from one to another mode of rotation of a diatomic molecule are given by $W = \ell(\ell+1)A$ [26,33], where $\ell = 0, 1, 2, 3 \ldots$ and $A = 6 \times 10^{-5}$ eV; thus an electromagnetic wave with a frequency as low as 29 GHz—still in the microwave region—can excite a rotational mode. Vibrational modes of diatomic molecules [26,33] correspond to energies of the order of 0.04 eV, requiring excitation in the IR region. Vibrational frequencies in a typical H-bonded system [34] are of the order of 3000 GHz; however, attenuation at this frequency by omnipresent free H_2O may prevent any substantial effect [34].

Kohli et al. [34] predict that longitudinal and torsional modes of double helical DNA should not be critically damped at frequencies >1 GHz, although relaxation times are of the order of picoseconds, and Kondepudi [36] suggests the possibility of an influence of millimeter waves at approximately 5×10^{11} Hz upon oxygen affinity of hemoglobin due to resonant excitation of heme plane oscillations. Although Furia et al. [37] did not find resonance absorption at millimeter waves in yeast, such was reported by Grundler et al. [38,47]. The latter experiment has been interpreted [39,40] as supporting Fröhlich's theory of cooperative phenomena in biological systems. That theory postulates "electric polarization waves" in biological membranes which are polarized by strong biologically generated [18] fields (10^7 V/m). Fröhlich [41,42] suggests that metabolically supplied energy initiates mechanical vibrations of cell membranes. The frequency of such vibrations is determined by the dimensions and the elastic constants of the membranes;

based on an estimate of the sound velocity in the membrane of 10^3 m/s and a membrane thickness of 100 Å (equal to one half wavelength) one obtains a frequency of $5(10^{10})$ Hz. Individual molecules within and outside the membrane may also oscillate, and frequency estimates vary between 10^9 Hz for helical RNA [43] and 5×10^{13} Hz for hydrogen-bonded amide structures [44]. As the membranes and molecules involved are strongly polarized, the mechanically oscillating dipole electromagnetic fields that are able to transmit energy, at least in some situations, over distances much larger than the distance to the next adjacent molecule.

Electromagnetic coupling of this type may produce long-range cooperative phenomena. In particular, Fröhlich [45] has shown that two molecular systems may exert strong forces upon each other when their respective oscillation frequencies are nearly equal, provided the dielectric permittivity of the medium between them is strongly dispersive or excitation is supplied by pumping, i.e., by excitation at the correct frequency from an external source. The mechanism is nonlinear in the sense that it displays a steplike dependence on excitation intensity. Possible long-range effects may be, for example, attraction between enzyme and substrate [42]. These and related topics have been discussed in detail by Illinger [34] and are reviewed in the present volume in Chapter 11 and Chapter 5 of *BBA*.

References

1. Jordan, E.C., *Electromagnetic Waves and Radiating Systems*. Prentice-Hall, Englewood Cliffs, NJ, 1950.
2. Schelkunoff, S.A., *Electromagnetic Waves*, D Van Nostrand, New York, 1943, p. 133.
3. Stratton, J.A., *Electromagnetic Theory*, McGraw-Hill, New York, 1941, p. 435.
4. Van Bladel, J., *Electromagnetic Fields*, McGraw-Hill, New York, 1964, p. 274.
5. Kaune, W.T. and Gillis, M.F., General properties of the interaction between animals and ELF electric fields, *Bioelectromagnetics*, 2, 1, 1981.
6. Bridges, J.E. and Preache, M., Biological influences of power frequency electric fields—a tutorial review from a physical and experimental viewpoint, *Proc. IEEE*, 69, 1092, 1981.
7. Iribarne, J.V. and Cho, H.R., *Atmospheric Physics*, D. Reidel, Boston, 1980, p. 134.
8. Zahn, M., *Electromagnetic Field Theory, A Problem Solving Approach*, John Wiley & Sons, New York, 1979.
9. Raybourn, M.S., The effects of direct-current magnetic fields on turtle retina in vitro, *Science*, 220, 715, 1983.
10. Schulten, K., Magnetic field effects in chemistry and biology, *Festkörperprobleme/Advances in Solid State Physics*, Vol. 22, Heyden, Philadelphia, 1982, p. 61.
11. Blankenship, R.E., Schaafsma, T.J., and Parson, W.W., Magnetic field effects on radical pair intermediates in bacterial photosynthesis, *Biochim. Biophys. Acta*, 461, 297, 1977.
12. Sheppard, A.R., Magnetic field interactions in man and other mammals: an overview, in *Magnetic Field Effect on Biological Systems*, Tenforde, T.S., Ed., Plenum Press, New York, 1979, p. 33.
13. Jordan, E.C., *Electromagnetic Waves and Radiating Systems*, Prentice-Hall, Englewood Cliffs, NJ, 1950, p. 132.
14. Ramo, S., Whinnery, J.R., and Van Duzer, T., *Fields and Waves in Communication Electronics*, John Wiley & Sons, New York, 1965, p. 293.
15. Smyth, C.P., *Static and Dynamic Electricity*, McGraw-Hill, New York, 1939.
16. Bean, C.P., DeBlois, R.W., and Nesbitt, L.B., Eddy-current method for measuring the resistivity of metals, *J. Appl. Phys.*, 30(12), 1959, 1976.
17. Bassett, C.A.L., Pawluk, R.J., and Pilla, A.A., Augmentation of bone repair by inductively coupled electromagnetic fields, *Science*, 184, 575, 1974.

18. Plonsey, R. and Fleming, D., *Bioelectric Phenomena*, McGraw-Hill, New York, 1969, p. 115.
19. Houslay, M.D. and Stanley, K.K., *Dynamics of Biological Membranes*, John Wiley & Sons, New York, 1982, p. 296.
20. Wilson, D.F., Energy transduction in biological membranes, in *Membrane Structure and Function*, Bittar, E.D., Ed., John Wiley & Sons, New York, 1980, p. 182.
21. Johnson, C.C. and Guy, A.W., Nonionizing electromagnetic wave effects in biological materials and systems, *Proc. IEEE*, 60, 692, 1972.
22. Schwan, H.P., Field interaction with biological matter, *Ann. NY Acad. Sci.*, 303, 198, 1977.
23. Kraichman, M.B., *Handbook of Electromagnetic Propagation in Conducting Media*, NAVMAT P-2302, U.S. Superintendent of Documents, U.S. Government Printing Office, Washington, D.C., 1970.
24. Massoudi, H., Durney, C.H., Barber, P.W., and Iskander, M.F., Postresonance electromagnetic absorption by man and animals, *Bioelectromagnetics*, 3, 333, 1982.
25. Binhi, V.N., *Magnetobiology: Understanding Physical Problems*, Academic Press, London, 473 pp.
26. Sears, F.W., Zemansky, M.W., and Young, H.D., *University Physics*, 5th ed., Addison-Wesley, Reading, MA, 1976, p. 360.
27. Uman, M.A., *Lightning*, McGraw-Hill, New York, 1969, p. 162.
28. Coelho, R., *Physics of Dielectrics for the Engineer*, Elsevier, Amsterdam, 1979, p. 155.
29. Schwan, H.P., Dielectric properties of biological tissue and biophysical mechanisms of electromagnetic field interaction, in *Biological Effects of Nonionizing Radiation*, Illinger, K.H., Ed., ACS Symposium Series 157, American Chemical Society, Washington, D.C., 1981, p. 121.
30. Koryta, J., *Ions, Electrodes and Membranes*, John Wiley & Sons, New York, 1982.
31. Rosenbaum, E.J., *Physical Chemistry*, Appleton-Century-Crofts, Education Division, Meredith Corporation, New York, 1970, p. 595.
32. Watson, J.D., *Molecular Biology of the Gene*, W.A. Benjamin, Menlo Park, CA, 1976, p. 91.
33. Rosenbaum, E.J., *Physical Chemistry*, Appleton-Century-Crofts, Education Division, Meredith Corporation, New York, 1970, p. 595.
34. Illinger, K.H., Electromagnetic-field interaction with biological systems in the microwave and far-infrared region, in *Biological Effects of Nonionizing Radiation*, Illinger, K.H., Ed., ACS Symposium Series 157, American Chemical Society, Washington, D.C., 1981, p. 1.
35. Kohli, M., Mei, W.N., Van Zandt, L.L., and Prohofsky, E.W., Calculated microwave absorption by double-helical DNA, in *Biological Effects of Nonionizing Radiation*, Illinger, K.H., Ed., ACS Symposium Series 157, American Chemical Society, Washington, D.C., 1981, p. 101.
36. Kondepudi, D.K., Possible effects of 10^{11} Hz radiation on the oxygen affinity of hemoglobin, *Bioelectromagnetics*, 3, 349, 1982.
37. Furia, L., Gandhi, O.P., and Hill, D.W., Further investigations on resonant effects of mm-waves on yeast, Abstr. 5th Annu. Sci. Session, Bioelectromagnetics Society, University of Colorado, Boulder, June 12 to 17, 1983, 13.
38. Grundler, W., Keilman, F., and Fröhlich, H., Resonant growth rate response of yeast cells irradiated by weak microwaves, *Phys. Lett.*, 62A, 463, 1977.
39. Fröhlich, H., Coherent processes in biological systems, in *Biological Effects of Nonionizing Radiation*, Illinger, K.H., Ed., ACS Symposium Series 157, American Chemical Society, Washington, D.C., 1981, p. 213.
40. Fröhlich, H., What are non-thermal electric biological effects? *Bioelectromagnetics*, 3, 45, 1982.
41. Fröhlich, H., Coherent electric vibrations in biological systems and the cancer problem, *IEEE Trans. Microwave Theory Tech.*, 26, 613, 1978.
42. Fröhlich, H., The biological effects of microwaves and related questions, in *Advances in Electronics and Electron Physics*, Marton, L. and Marton, C., Eds., Academic Press, New York, 1980, p. 85.
43. Prohofsky, E.W. and Eyster, J.M., Prediction of giant breathing and rocking modes in double helical RNA, *Phys. Lett.*, 50A, 329, 1974.
44. Careri, J., Search for cooperative phenomena in hydrogen-bonded amide structures, in *Cooperative Phenomena*, Haken, H. and Wagner, W., Eds., Springer-Verlag, Basel, 1973, p. 391.
45. Fröhlich, H., Selective long range dispersion forces between large systems, *Phys. Lett.*, 39A, 153, 1972.

46. Barnes, F.S. and Hu, C.-L.J., Nonlinear Interactions of electromagnetic waves with biological materials, in *Nonlinear Electromagnetics*, Uslenghi, P.L.E., Ed., Academic Press, New York, 1980, p. 391.
47. Grundler, W., Keilmann, F., Putterlik, V., Santo, L., Strube, D., and Zimmermann, I., Nonthermal resonant effects of 42 GHz microwaves on the growth of yeast cultures, in *Coherent Excitations of Biological Systems*, Frölich, H. and Kremer, F., Eds., Springer-Verlag, Basel, 1983, p. 21.

1

Effects of Radiofrequency and Extremely Low-Frequency Electromagnetic Field Radiation on Cells of the Immune System

Tatjana Paunesku and Gayle E. Woloschak

CONTENTS

1.1 Introduction

Establishing the biological effects of exposure to electromagnetic fields (EMF), particularly extremely low-frequency (ELF, 0 to ~3000 Hz) fields and radiofrequency (RF) radiation, remains an elusive goal to this day. Over the last 20 to 30 y, hundreds of scientists across the world have used different types of EMF and RF sources and exposure regimens on different biological experimental models, and different types of data analyses have been applied to both experimental and epidemiologic data. Despite the breadth of these studies (or perhaps because of it), no universally accepted hypotheses exist in this field of research. In this chapter, we attempt to present selected examples from the plethora of available material that best emphasize positive and negative findings from the literature pertinent to the immune system and discuss possible causes for the disparity of the published data.

Before going further into an exploration of the effects of RF and ELF EMF on the immune system, we emphasize some unique aspects of this view on the biological effects of EMF. First of all, many studies were done on the cells of the immune system because it is relatively simple to obtain these cells in a noninvasive way from humans and whole organisms; second, many chose to study effects of EMF on the immune system because this system is one of the first to interact with the environment and, it is generally believed, the most likely to be affected by stimuli from the environment. Clearly, cells of the immune system are well attuned to a broad variety of perturbations in the environment. Nevertheless, it is also true that immune cells are very difficult to study for prolonged periods since their lifetime in culture is relatively short; and any stimulus that these cells are exposed to is, of necessity, added to an already long list of stimuli to which the immune cells respond. Therefore, the immune system may be a difficult subject of study when one examines the effects of an agent without a direct target such as EMF exposure.

A broad range of different exposure conditions has been used for studies on the immune system, and often it is difficult to compare studies using such different exposure conditions. Some of the RF radiation qualities used in the studies reviewed here include continuous-wave (CW) or pulsed-wave RF radiation at 1.748, 1.9, or 2.45 GHz with power densities up to 5 mW/cm^2 and calculated specific absorption rates (SARs) in biological material between 0.1 and 12 W/kg. Some studies involved microwave (MW) with a frequency of 42.2 GHz, peak power density of 31 $\pm$ 5 mW/cm^2, and peak SARs of 622 $\pm$ 100 W/kg. Another large group of RF studies included variations of different possible mobile phone type RF—plane wave or CW with frequencies around values of 900 and 800 MHz, power densities up to 950 W/m^2, and SARs between 0.008 and 8.8 W/kg.

Electric and magnetic fields were, most often, tested with low-frequency (50 to 60 Hz) and magnetic field intensities between 0.2 and 500 μT. Static electric and magnetic fields were tested as well, often with magnetic intensities as high as 10 T. Furthermore, different investigators used continuous and/or intermittent exposures; according to some authors intermittent exposures are more efficient in causing biological effects, [1] while some case report studies concentrated on exposures that included peaked EMF exposures, such as when a surge occurs as a power line is severed. Homogeneous and gradient static MF exposures were compared, e.g., Hirose and others found the gradient MF to be more biologically effective [2]. EMF pre-, co-, or post-exposures were combined with exposures to chemical (mitomycin C [MMC], vinblastine, and others) and/or physical (gamma- and x-ray ionizing radiation, UVB, and UVA radiation) mutagens or substances that are responsive to magnetic fields but are not genotoxic on their own ($FeCl_2$). Experimental exposure lengths used varied between 5 min and 350 d, while case report studies included exposures that were years in duration.

In addition to case report studies and testing of blood samples from humans occupationally exposed to RF or ELF EMF, planned experiments were set up with biological targets that included human or rodent peripheral blood cells (PBC) exposed *ex vivo*, cultured cells of lymphoblast and monocyte lineages (Jurkat, HL-60, Mono Mac-6, K-562, U-937, and DG-75), or blood cells derived from exposed animals (Sprague-Dawley rats or mice: Swiss, C57BL/6, BALB/c, DDY, pKZ1, E mu-Pim1).

Different endpoints were tested by these studies to estimate the effects of EMF and RF radiation exposures on the immune system. We have broadly grouped the studies reviewed here into several groups based on the most significant endpoints of the study.

The effects of any type of stress on the immune system include multiple events in different cell types, and it is very rare that any study of a stress factor includes more than several endpoints. These always represent only a fraction of all the ongoing events. Cells of the immune system interact intensely with each other and their surroundings, and the more elaborate the conditions of exposure, e.g., *in vivo* exposure compared to less complex *in vitro* exposure, the more numerous these interactions will be. Therefore, it is not really possible to extrapolate results of studies involving PBC *in vitro* to the possible fate of the same population of cells exposed *in vivo*. In different studies presented here, the same endpoint was often pursued in systems with different complexities.

At present, many groups are using recently developed tools for studies of gene expression such as microarrays or different proteomics approaches; unfortunately, at the moment, these data are still largely unpublished. One rare example of this type is work by Olivares-Banuelos on chromaffin cells [3]. In these studies, chromaffin cells were allowed to differentiate in the EMF before RNA harvesting, which may limit the applicability of the data. It is clear that the wealth of information obtained from microarray and proteomics studies will have a great impact on the field.

1.2 Proliferation, Activity, and Function of Lymphocytes

There are several recent studies that have quantified the number of lymphocytes and their subtypes in humans exposed to whole-body RF or ELF EMF. In some cases, comparisons between exposed and control individuals were made, while in others, the results obtained on the same individuals were compared over time. Data obtained were not consistent from one study to another and often were not consistent even within the same study. While this may be explained by insufficiently controlled exposure conditions in some cases, it is far more likely that the differences in the results obtained are based on the different experimental conditions and individual differences among the subjects of these studies.

1.2.1 Human Studies

An interesting but small study compared six individuals, occupationally exposed for 8 h/d for 5 y to low-frequency (50 Hz) EMFs of 0.2 to 6.6 μT, with six control subjects [4]. Total lymphocytes, CD4+, and CD3+ cell counts decreased, while natural killer (NK) cells increased at statistically significant levels. Six months after cessation of exposure, to tal lymphocte counts had significantly increased, particularly CD4+, CD3+, and CD19+ counts (+28, +22, and +17%, respectively), in the same human subjects. The authors pro ceeded to expose 12 Swiss male mice to identical conditions for 109 d, comparing them to 12 additional mice as unexposed controls. Similar to the human data, total lymphocytes, leukocytes, polymorphonuclear neutrophils, and CD4+ cells were significantly lower in exposed animals. However, unlike the human studies, the NK counts were lower in the exposed mice at the time of completion of the exposure. In their earlier work, the same authors investigated 13 subjects occupationally exposed for 5 y to low-frequency (50 Hz) EMF of 0.2 to 6.6 μT [5]. In comparison to 13 controls matched for gender, age, and socioeco nomic status, the exposed group showed a significant decrease in total lymphocytes,

CD4+, CD3+, and CD2+ lymphocytes and a rise in NK cells. Contradictory results were published by others who counted lymphocyte subsets and found that the number of NK cells showed statistically significant differences compared to controls in workers occupationally exposed to magnetic fields generated at induction heaters [6]. Another similar study was done on subjects exposed to ELF EMF during electric arc welding (field intensities: 0.10 and 0.25 mT) [7]. Comparisons were done between 16 male welders and 14 healthy males who were 20 to 40 y of age, from the same geographic area, and with similar lifestyles. It was found that while the hematocrit levels of the welders had elevated, their CD4+ and CD8+ cells had decreased.

1.2.2 Peripheral Blood Cells Cultured *Ex Vivo*

PBC were obtained from a number of individuals and in some cases pooled together. This led to an increase in variability within individual exposure samples, while differences from one sample to another could be, in essence, disregarded. The effect of 60 Hz sinusoidal magnetic fields (MF) at magnetic flux densities of 1.0, 1.5, and 2.0 mT on human PBC cultures was studied at an exposure time of 72 h and, in some cases, with MMC treatment [8]. MF treatment was shown to increase cell proliferation; and even cells treated with MMC and MF showed a higher proliferation index—a reversal of the decrease in proliferation induced by MMC treatment alone. The same was true for the mitotic index as well.

Capri and others exposed peripheral blood mononuclear cells (PBMC) *ex vivo* to a 900 MHz global system of mobile communication (GSM) mobile telephone or CW RF field for 1 h/d for 3 d in a transverse electromagnetic mode (TEM) cell system (70–76 mW/kg of average SAR) [9]. A slight decrease in cell proliferation was found following stimulation of PBMC with the lowest mitogen concentration, while a slight increase in the number of cells with an altered distribution of phosphatidylserine across the membrane was observed following exposure to GSM, but not CW RF field.

1.2.3 Animal Studies

Many experiments have been done using animal systems. In these experiments, contradictory data were obtained. Often a decrease in the number of lymphocytes was noted, similar to the data obtained in human exposures; however, in some cases an increase in proliferation was observed as well, occasionally in the same experiments [10].

C57BL/6 mice, exposed for 2 h/d for 1, 2, or 4 weeks in a TEM cell of 900 MHz GSM-modulated radiation and a SAR of 1 or 2 W/kg, were monitored for effects on spleen cells. The authors tested splenic cellularity, the percentages of B and T cells, and the distribution of CD4+ and CD8+ populations and found that none of these parameters were altered by the exposure. Nevertheless, at 1 week of exposure to 1 or 2 W/kg, an increase in IFN-gamma production was observed, which was lost after prolonged exposure for 2 or 4 weeks [11].

Continuous exposure of twelve 6-week-old Swiss mice for 350 d to ELF EMF of 5.0 μT was generated by a transformer station and high current bus bars [10]. When compared on different days, the exposed animals sometimes showed a decrease in the numbers of leukocytes and/or neutrophils in comparison to controls (days 20, 90), while at other times they did not (days 43, 63, 350). Individual differences among the mice were also observed, and two of the exposed animals showed a significant increase in leukocyte and polymorphonuclear neutrophil counts compared to the unexposed mice.

Murabayashi and others investigated the effect of ELF-pulsed magnetic fields on proliferation (measured by bromo-2′-deoxy-uridine incorporation) of mitogen concanavalin

A (ConA)-stimulated spleen lymphocytes from DDY mice for 72 h [12]. Exposure times varied between 20 min and 2 h. The authors found that TVMF exposure augmented proliferation if the exposure took place within the first 40 min following ConA stimulation, but not afterward. They speculated that Ca^{2+} influx may be responsible for this early ConA response.

Another study concentrated on the effects of 25–150 mT (250–1500 Gauss) static magnetic fields' (SMF) exposure on the cellular immune functions of macrophages, splenic lymphocytes, and thymic cells isolated from the C57BI/6 mice and cultured *ex vivo* [13]. A decrease in phagocytic uptake of fluorescent latex microspheres and increased intracellular Ca^{2+} levels were found in macrophages; mitogenic responses in lymphocytes measured by [^{3}H]thymidine incorporation were decreased, while Con A-stimulated lymphocytes showed an increased Ca^{2+} influx. Thymocytes showed enhanced levels of apoptosis, as determined by flow cytometry.

Exposure of the nasal area of BALB/c mice for 30 min/d for 3 d to low-power electromagnetic millimeter waves (frequency: 42.2 GHz; peak incident power density: 31 +/− 5 mW/cm^2, and peak SAR at the skin surface: 622 $\pm$ 100 W/kg) combined with treatment with cyclophosphamide (CPA) at 100 mg/kg, acting as an immunosuppressant, was studied [14]. The mice exposed to combination treatment showed statistically significant proliferation recovery of splenocytes and altered activation of CD4+ T cells.

1.3 Cell Death

A large number of studies examined potential detrimental effects of EMF exposure on lymphocytes.

1.3.1 Cell Culture and Peripheral Blood Cells *Ex Vivo* Studies

PBMC from human volunteers were exposed to a 10-T SMF [15]. In the absence of lymphocyte stimulation, there were no significant differences in the viability of the exposed and unexposed T cells, B cells, and NK cells. However, when T cells were activated by mitogen phytohemagglutinin (PHA), both the CD4+ and CD8+ T cell populations expressed enhanced death due to apoptosis. Another report on the exposure of cultured human lymphoblast cancer cells lines predominantly of the myeloid lineage (K-562, U-937, DG-75, and HL-60) to a pulsating EMF (sinusoidal wave form, 35 mT peak, 50 Hz) for 4 h tested the cells for cell death at the conclusion of exposure or 24 h later [16]. Necrosis was induced in these experimental conditions, but the extent of the reaction depended also on the medium composition, its pH value, conductivity, and temperature.

Jajte and others exposed rat lymphocytes for 3 h to 7 mT SMF and 10 μg/ml of $FeCl_2$. This resulted in an increase in the percentage of apoptotic and necrotic cells [17].

Negative results were also obtained from these types of experiments. For example, Ikeda and others exposed human PBMC to three different magnetic field polarizations at 50 and 60 Hz: linearly polarized (vertical) with flux densities of 500, 100, 20, or 2 μT; circularly polarized; and elliptically polarized with flux density of 500 μT [18]. The authors used the cytotoxicity assay and found no effects of ELF MFs on cytotoxicity of human PBMC.

1.3.2 Studies in Animals

Cell death of thymocytes during thymus involution has been shown to be accelerated by continuous light exposure. Studies on the effects of long-term exposure to extremely low-frequency (50 Hz) sinusoidal electric and magnetic fields at two levels of field strength on thymocytes of rats housed in a regular dark/light cycle or under continuous light were done to test thymus involution-promoting effects of EMF [19]. The concurrent exposure to continuous light and ELF EMF significantly increased the number of cell deaths, in comparison with animals exposed to continuous light only or with animals exposed to EMF in a 12 h dark–light cycle. The authors concluded that the long-term exposure to ELF EMF in animals experiencing a photic stress leads to a more rapid thymus involution.

1.4 Micronucleus Formation

A large number but not all studies that examined micronucleus formation following RF or ELF EMF exposure resulted in acquisition of positive data.

1.4.1 Cell Culture and Peripheral Blood Cells *Ex Vivo* Studies

Human PBC were exposed *ex vivo* for 15, 30, or 60 min to different EMF frequencies (2.45 and 7.7 GHz) and power densities (10, 20, and 30 mW/cm^2) [20]. Exposures for 30 min or more to a power density of 30 mW/cm^2 led to formation of an increased number of micronuclei. On the other hand, another group of investigators studied the effect of 15 min *ex vivo* exposure to 1.748-GHz, continuous (CW) or phase-only modulated wave (GMSK) on human peripheral blood lymphocytes (PBL) [21]. The maximum SAR was 5 W/kg. No changes were found in cell proliferation kinetics after exposure to either field; but the micronucleus frequency showed a statistically significant increase following exposure to GMSK, but not CW. Tice and others studied the effects of RF signals emitted by four different types of cellular telephones: analog signal generator with voice modulated at 837 MHz, time division multiple access (TDMA) voice modulated at 837 MHz, 837 MHz code division multiple access (CDMA) without voice modulation, and 1909.8 MHz GSM-type personal communication systems (PCS) with voice modulation [22]. Exposure of PBL for 24 h at an average SAR of 5.0 or 10.0 W/kg to any one of the four RF sources produced a significant and reproducible increase of micronucleated lymphocytes.

Human PBC from 20 donors were exposed to EM field at a frequency used for mobile telephones (900 Hz) using CW-intermittent exposure (SAR: 1.6 W/kg) for 14 cycles with a 6 min exposure and a 3 h pause, a GSM signal and the same type of intermittent exposure, a GSM signal and the same type of intermittent exposure 24 h before stimulation with PHA (8 on/off cycles), and a GSM signal, with intermittent exposure (SAR: 0.2 W/kg) for 1 h/d for 3 d [23]. No statistically significant changes in micronucleus frequency or cell cycle kinetics under any of the experimental conditions were found.

Lymphocytes from 18 individuals were exposed to 50 Hz ELF EMF of 80 and 800 μT for a 72 h period and challenged with vinblastine in concentrations of 0, 5, 10, or 15 ng/ml during the last 48 h of this period [24]. Vinblastine- and ELF EMF-exposed groups expressed a higher proportion of micronuclei than the group with vinblastine treatment alone. On the other hand, an increase in the nuclear division index (NDI) was observed in the ELF EMF 800 μT treated group.

1.4.2 Studies in Animals

Rats exposed for 24 h to 2450 MHz CW RF radiation at an average whole-body SAR of 12 W/kg showed no increase of micronuclei in PB or bone marrow cells [25].

1.5 Chromosome Damage (Sister Chromatid Exchange, Aberrations, etc.)

Understanding whether EMF causes chromosomal damage would play a large role in determining any possible role exposure might play in the onset or progression of cancers, particularly those of the immune system. Several studies have examined the role of EMF exposure on the development of chromosomal abnormalities in lymphoid cells.

1.5.1 Studies on Human Subjects

Skyberg and others performed a study on 24 employees engaged in high-voltage testing and generator soldering and 24 controls from a Norwegian transformer factory [26]. Blood samples taken from the subjects were either analyzed immediately or the lymphocytes were briefly cultured and treated with hydroxyurea and caffeine to inhibit DNA synthesis and repair. In the latter type of cultures, samples from the high-voltage laboratory testers showed twice the median number of sister chromatid breaks (5 vs. 2.5 per 50 cells) compared to controls. The median number of aberrant cells was 5 vs. 3.5 for the high-voltage laboratory testers compared to controls.

1.5.2 Cells in Culture and Peripheral Blood Cells Cultured *Ex Vivo*

In another experiment, PBL were exposed to continuous 830 MHz EMFs at average SARs between 1.6 and 8.8 W/kg [27]. The exposure occurred for 72 h in a parallel plate resonator at temperatures between 34.5 and 37.5°C. The authors found an increase in chromosome 17 aneuploidy as a function of the increased SAR value.

Cho and Chung used human lymphocytes to investigate combined effects of ELF EMF and benzo(a)pyrene (BP) on the frequency of micronuclei and sister chromatid exchange (SCE) [28]. BP co-exposure with a 60 Hz field of 0.8 mT for 24 h, followed by BP exposure for 48 h led to significant increases in the frequencies of micronuclei and SCE compared to BP-alone treatment for 72 h. At the same time, no significant difference was observed between ELF EMF and sham-exposed control cells. The authors concluded from these findings that low-density ELF EMF is not an initiator, but a promoter of carcinogenesis.

Another group investigated 50 Hz magnetic fields alone or in combination with either MMC or x-rays, and found that magnetic fields up to 2500 μT did not change the frequencies of the chromosome aberrations and SCE, either alone or in combination with the mutagens [29]. Nevertheless, there were exceptions—504 μT MF treatment before and during cultivation showed a significant decrease in SCE frequency, while cells cultivated at 88.4 μT showed a significant increase in chromosome aberration frequency following x-ray exposure when compared to x-ray exposure alone.

Vijayalaxmi and others tested the effect of exposure to 835.62 MHz RF radiation, a frequency employed in customer-to-base station transmission of cellular telephone communications, on PB samples [30]. An analog signal generator and the frequency division multiple access (FDMA, CW) technology were used; a forward power of 68 W was used, with a power density of 860 W/m^2; and the mean SAR was 4.4 or 5.0 W/kg. Following a 24 h exposure, the lymphocytes were stimulated with the mitogen PHA and cultured for an additional 48 or 72 h to assess the frequencies of chromosomal aberrations and micronuclei. No evidence of either was found.

The same group also tested the effect of exposure to 847.74 MHz CW RF on PB samples from four individuals [31]. CDMA technology was used, with a nominal net forward power of 75 W, a nominal power density of 950 W/m^2, and a mean SAR of 4.9 or 5.5 W/kg. No differences between test and control samples were found for cellular mitotic indices; frequencies of exchange aberrations; or presence of excess fragments, binucleate cells, or micronuclei.

Another group exposed human blood lymphocytes simultaneously to 2.0 Gy of γ rays and a 50 Hz power frequency magnetic field of 0.23, 0.47, or 0.7 mT [32]. No differences in the frequency of chromosomal aberrations were observed between cells not exposed to or co-exposed to EMF.

1.5.3 Studies in Animals

In one of the recent articles dealing with chromosome aberrations in RF-exposed animals, the authors have concentrated on comparing 4 W/kg pulsed 900 MHz RF delivered daily for 30 min to plane-wave fields of 900 MHz with a pulse repetition frequency of 217 Hz and a pulse width of 0.6 ms (GSM pulsing pattern) for 1, 5, or 25 d. They studied somatic intrachromosomal recombination in the spleen in recombination mutagenesis model beta-galactosidase transgenic mouse pKZ1 [33]. While somatic intrachromosomal recombination inversion events are usually found in spleen cells of these transgenic mice, a decrease in the frequency of inversions was observed in the 25 d exposure group.

1.6 DNA Damage as Single-Strand and Double-Strand Breaks

A variety of studies have been done to examine the effects of EMF on single-strand (SSB) and double-strand breaks (DSB), predominantly as an assessment of the ability of the exposure to induce DNA damage and therefore have mutagenic potential. Most of the results demonstrated little effect of EMF exposure on DNA damage.

PBC were used to study the effect of a 2 h-long exposure to pulsed-wave 2450 MHz RF radiation with a net forward power of 21 W, transmitted from a standard gain rectangular antenna horn in a vertically downward direction [34]. The average power density at the position of the cells in the flask was 5 mW/cm^2, while the SAR was 2.135 $\pm$ 0.005 (SE) W/kg. To estimate primary DNA damage (SSB and alkali-labile lesions), the authors used the alkaline comet assay performed immediately after the exposures or 4 h post exposure—a sensitive method for monitoring DNA strand breaks. No RF exposure-related increase in DNA damage was observed.

Another group investigated the effect of a CW 1.9 GHz RF on PBC using a series of six circularly polarized, cylindrical waveguides [35]. Mean SARs of 0.0, 0.1, 0.26, 0.92, 2.4, and 10 W/kg were achieved, while the temperature was maintained steadily at 37.0°C. Exposures lasted 2 h, and no increase of primary DNA damage or micronuclei was found under any of the experimental conditions. The same investigators also studied the effect of a pulse-modulated 1.9 GHz RF field on human leukocytes [36]. Again, six circularly polarized cylindrical wave guides were used. Mean SARs ranged from 0 to

10 W/kg, and the temperature was maintained at 37°C. Exposures lasted 2 h, and no primary DNA damage was detected by the alkaline comet assay, nor were there any changes in the frequency of binucleated or micronucleated cells in comparison to sham-treated controls.

Maes and others studied effects of a 900 MHz RF radiation with different modes of exposure (continuous, pseudorandom, and dummy burst) and different power outputs (0, 2, 8, 15, 25, 50 W) on human blood cells collected from six donors [37]. The SAR varied between 0 and 10 W/kg. The authors investigated the effects of the 900 MHz radiation alone or in combination with a subsequent exposure to the chemical or physical mutagens, MMC (0.1 μg/ml) and x-rays (1 Gy exposure). A transverse electromagnetic transmission line (TEM) cell was used and PB cells were exposed for 2 h. No indication was found for mutagenic and/or co-mutagenic/synergistic effects between EMF and mutagen exposure under these experimental conditions.

Nevertheless, some reports suggest that EMF can serve as a co-mutagen and increase the amount of damage accumulated, most likely by slowing down repair processes. This was also reported recently in some nonlymphoid cell types, e.g., ELF EMF potentiates x-ray-induced DNA strand breaks in glioma MO54 cells [38]. Also, Nindl and others tested the combined effects of a low-frequency, low-intensity EMF of 100 Hz, 1 mT and UVB irradiation on Jurkat cell line [38]. When the UVB exposure was accompanied by EMF exposure, a decrease in DNA repair synthesis of 34 $\pm$ 13% was observed, compared to UVB treatment alone.

Zhang and others studied exposure to 2450 MHz MW radiation (5.0 mW/cm^2) concomitantly with MMC (0.0125, 0.025, or 0.1 μg/ml) and found synergistic genotoxic effects using single-cell gel electrophoresis comet assay and the cytokinesis-blocked micronucleus test *in vitro* [39]. Pooled whole blood cells from a male human donor and a female human donor were exposed to MW for 2 h, MMC only, or MMC for 24 h after exposure to MW. The last of these led to significant increase in DNA damage.

1.7 Leukemia and Lymphoma

Leukemia and lymphoma of the immune system, just as any other cancer, develop following accumulation of several mutations in the same cell clone. Therefore, a potential mutagenic influence of EMF on the immune system could result in the development of these types of cancer. For reasons listed in Section 1.1, and because of the intrinsic complications with evaluation of the epidemiologic data, we limit ourselves to work done in animal models. Recent results obtained in experimental systems targeted at studying links between RF radiation or EMF exposure and the development of leukemia and lymphoma have often been negative.

For example, in 1997 Repacholi and others found increased odds' ratio (OR) of 2.4, $P = 0.006$, 95% confidence interval (CI) of 1.3–4.5 for the development of lymphoma in lymphoma-prone E mu-Pim1 transgenic mice exposed to pulse-modulated RF fields twice for 30 min/d for up to 18 months (plane-wave fields of 900 MHz with a pulse repetition frequency of 217 Hz, pulse width of 0.6 ms; incident power densities were 2.6 to 13 W/m^2, and SAR of 0.008 to 4.2 W/kg, averaging 0.13–1.4 W/kg) [40]. However, no such increased risk of lymphomas was found when Utteridge and others exposed 120 E mu-Pim1 heterozygous mice and 120 wild-type mice to GSM-modulated 898.4 MHz radiation for 1 h/d, 5 d/week at SARs of 0.25, 1.0, 2.0, and 4.0 W/kg in "Ferris-wheel" exposure systems for up to 104 weeks (26 months) [41]. As controls, 120 heterozygous and

120 wild-type mice were sham exposed, and some additional animals served as an unrestrained negative control group.

Other recent animal studies have involved 4 groups of 50 female CBA/S mice, 3 to 5 weeks of age, to test the effects of combined exposure to ionizing radiation and RF radiation [42]. At the beginning of the study, the mice were irradiated and then exposed to RF radiation for 1.5 h/d, 5 d a week for 78 weeks. Two sources of RF were used: a continuous NMT (Nordic Mobile Telephones)-type frequency-modulated RF radiation at a frequency of 902.5 MHz and SAR of 1.5 W/kg, and a pulsed GSM-type RF radiation with a carrier wave frequency of 902.4 MHz, pulse frequency of 217 Hz, and SAR of 0.35 W/kg. No change in cancer incidence was found in association with RF exposure. Nevertheless, the same group found that co-exposure to low-level RF radiation and ultraviolet (UV) light slightly accelerated skin cancer development [43]. The same was true for co-exposure to UV and 50 Hz magnetic fields [44].

Therefore, for RF radiation exposures, the consensus at this moment is that they have no lymphoid cancer-initiating or -promoting capacity (see, e.g., the review by Elwood) [45].

1.8 Possible Mechanisms of Action of Radiofrequency Radiation and Electromagnetic Field

As evidence accumulates that direct DNA damage is not induced by either RF radiation or EMF exposure and yet at the same time cellular damage such as micronuclei formation frequently arises from these exposures, many different avenues are being pursued in the search for possible mechanisms of action of RF and ELF EMF. Some more recent ideas have proposed that RF and ELF EMF interact with electric charges in proteins and nucleic acids at the level of the molecules themselves, perturbing their electron clouds. Blank and Goodman found DNA sequences, which they call *electromagnetic field response elements* in the promoters of c-myc and HSP70 genes [46]. They maintain that EMF exposure-dependent activation of these genes relies on destabilization of hydrogen bonds in these DNA response elements under the influence of the field [47]. On the other hand experimental proofs of local changes in protein conformation, other than for *in vitro* experiments with isolated proteins, are still largely unavailable. For example, protein aggregation and the formation of amyloid fibers are increased by RF exposures to 1 GHz at SAR of 0.5 W/kg [48]. Several of the "older" ideas and some supporting, as well as contradictory findings, especially in the cells of the immune system, are presented below, including Ca^{2+} homeostasis, nonthermal heat shock effects, gene/protein activation, and induction of reactive oxygen species (ROS).

1.8.1 Ca^{2+} Homeostasis

Mixed results were obtained from studies investigating changes in Ca^{2+} as a potential biochemical pathway that mediates biological effects of EMF. The assumption was that the magnetic fields increase the cytosolic calcium concentration in lymphocytes in the same manner as a physiological stimulus, e.g., interaction with antibodies directed against the CD3 complex. Nevertheless, in some cases, nearly identical experimental conditions were used with different outcomes, e.g., the work by Lindstrom and others in 1993 and 1995 and that by Wey and others in 2000 using Jurkat T-cells exposed to 50 Hz MF. [49–51] Below are several such studies as an illustration of the breadth of the results obtained.

Studies showing no effect of RF exposure on Ca^{2+} flux were obtained by Cranfield and others who tested the effect of continuous-wave and pulsed-wave GSM mobile phone signals (915 MHz, 2 W/kg) on calcium concentration, using Fluo-3 fluorescence, in the Jurkat human lymphocyte cell line treated (or not) with PHA [52]. The only changes were found in PHA-activated cells treated by the pulsed wave.

Positive effects of exposure were obtained in another group of studies where a 2–150 mT SMF exposure of macrophages and ConA-stimulated lymphocytes from the C57BL/6 mice cultured *ex vivo* led to increased intracellular Ca^{2+} levels [13]. This was accompanied by a decrease in phagocytic uptake of fluorescent latex microspheres by macrophages. Aldinucci and others tested the effects of a combination of a static EMF on Jurkat cells and human lymphocytes at a flux density of 4.75 T and a pulsed EMF with a flux density of 0.7 mT generated by an NMR apparatus [53]. The authors monitored movements of Ca^{2+}, cell proliferation, and the production of cytokines following 1 h exposure in cells treated by the EMF fields alone or cells additionally activated by 5 mg/ml of PHA immediately before the exposure. EMF induced movements of Ca^{2+}, but had no other effect, be it proliferative, activating, or proinflammatory, on nonstimulated or on PHA-stimulated lymphocytes. In Jurkat cells (T-cell lineage), exposure to EMF led to a decreased intracellular Ca^{2+} concentration and proliferation.

Another group of investigators studied the effects of low-intensity 60-Hz electric fields (E: 0.5 V/m; current density: 0.8 A/m^2) on agonist-induced Ca^{2+} flux in HL 60 cells (myeloid lineage) [54]. The 60 min exposure to the E field was accompanied by the addition of 1 μM ATP or 100 μM histamine. ATP-activated and E field-exposed cells showed a 20–30% decrease in intracellular Ca^{2+} concentration compared to ATP alone, while histamine and E field-exposed cells showed a 20–40% increase of intracellular Ca^{2+}.

McCreary and others examined the influence of the cell cycle, pH of the medium, and response to a calcium agonist to changes in Ca^{2+} in Jurkat E6.1 cells treated by 60 Hz, 100 μT (peak) sinusoidal AC magnetic field (MF), 78 μT static MF (DC), and the combination of both MFs [55]. When the effect of the cell cycle and the quality of the alpha-CD3 monoclonal antibody response were taken into account, all exposure groups were significantly different from the control.

1.8.2 Nonthermal Heat Shock Effects

Some of the earliest literature on EMF-related biological effects implicated heat shock proteins in biological response to EMF; however, due to possible "contamination" of EMF effects with thermal effects, it took long to prove that the nonthermal effect of EMF also induced heat shock proteins and cell behaviors [56]. In recent years more articles have been published on the nonthermal EMF effects [57]; listed below are a few that focused predominantly on cells of the immune system.

Sarimov and others investigated GSM MWs at different frequencies ranging between 895 and 915 MHz in experiments with PBL from seven healthy persons [58]. Exposures were done in transverse electromagnetic transmission line cell (TEM cell) using a GSM test-mobile phone, 2 W output power in the pulses, and SAR of 5.4 mW/kg. The authors measured changes in chromatin conformation by anomalous viscosity time dependencies' method and found that 30 min exposure to all tested MW qualities led to chromosome condensation similar to the effects of heat shock in lymphocytes of at least some (1 in 5 to 4 in 5) of the tested individuals.

Belyaev and Alipov studied the effect of ELF EMF of 21 μT rms on human lymphocytes, using anomalous viscosity time dependencies to measure chromosome condensation [59]. The authors observed two resonance-type frequency windows with strong effects on human lymphocytes at 8 and 58 Hz and proceeded to compare these frequency windows with predictions at frequencies of harmonics and subharmonics for natural isotopes of Na, K, Ca, Mg, and Zn ions, based on the ion cyclotron resonance model and the magnetic parametric resonance model. The overall good overlap between these frequencies suggests that the effective resonance-type frequency windows may interfere with most abundant cellular ions.

On the other hand, human PBC exposed to 900-MHz CW and GSM fields at three average SARs of 0.4, 2.0, and 3.6 W/kg for 20 min, 1 h, or 4 h in a TEM cell at 37°C and 5% CO_2 showed no changes in heat shock proteins, hsp70 and hsp27, using flow cytometry approaches [60]. Similarly, exposure of PBMC from young and old donors to RF radiation (1800-MHz RF; three signal modulations of the GSM-type: GSM Basic, discontinuous transmission, and Talk; at SARs of 1.4 and 2.0 W/kg) does not affect apoptosis or hsp70 level [61].

1.8.3 Gene and Protein Activation

In a number of studies, different proteins have been reported as activated by EMF or RF radiation. Some of the more recent examples are listed below.

Cultured human monocytes (Mono Mac-6) were exposed to high peak power pulsed RF at a frequency of 8.2 GHz. The peak to average power ratio was 455:1; 2.2 μs pulse width pulse repetition rate of 1000 pulses/s; average power density was 50 W/m^2; SAR was 10.8 $\pm$ 7.1 W/kg at the position where the cells were located [62]. Following a 90 min exposure and 4 h incubation, the cells were harvested and an increase of 3.6-fold was detected in the DNA-binding activity of NF-κB transcription factor from nuclear extracts.

DNA-binding activity of the transcription factor cyclic-AMP responsive element-binding protein (CREB) changes following exposure of HL60 cells to a 0.1 mT 50 Hz ELF sinusoidal MF [63]. Magnetic field exposure induced increased binding for CREB *cis* element after 10 min, reached a peak at 1 h, and returned to basal level 4 h after exposure. Moreover, a new ATF2/ATF2 homodimer was formed after MF exposure for 30 min, 1, and 2 h. The authors argue that this activity was dependent on both extracellular and intracellular Ca^{2+}. Romano-Spica and others used an exposure system generating 50 MHz RF nonionizing radiation modulated with a 16 Hz frequency (field intensity of 0.2 μT), magnetic field parallel to the ground, and a 60 V/m electric field orthogonal to the earth's magnetic field and found overexpression of the ets1 mRNA in Jurkat cells [64]. This occurred only at 16 Hz modulation; according to the authors, this is due to correspondence with the resonance frequency for calcium ions with a DC magnetic field of 45.7 μT. A static MF exposure system with a spatially inhomogeneous 6 T strong MF having a gradient of 41.7 T/m or that with a spatially homogeneous 10 T static MF was compared as well [2]. Exposure to the gradient MF induced c-Jun protein expression in HL-60 cells exposed for 24 to 72 h.

1.8.4 Induction of Reactive Oxygen Species

Some investigators believe that EMF or RF exert its effects on the biological systems by enhancing (or creating) ROS. ROS are known to mediate a large variety of biological consequences, including DNA damage and possibly mutation induction. In a recent article, Simko and Mattsson have provided some lines of evidence suggesting that ELF EMF exposure increases the level of free radicals in cells through several routes mainly relying on the activation of macrophages (or other ROS-producing cell type) [65].

Zmyslony and others tested combined effects of 930 MHz CW RF, a carrier frequency for signals emitted by cellular phones, and $FeCl_2$ on creation of ROS in cells [66]. ROS were measured by the fluorescent probe—dichlorofluorescin diacetate. Exposure of rat lymphocytes was done in a GTEM cell, and the power density was 5 W/m^2 and the calculated SAR was 1.5 W/kg. RF radiation exposures alone of 5 and 15 min did not alter production of ROS, but combined with addition of 10 μg/ml $FeCl_2$ a significant increase of ROS occurred. The same authors also tested the effects of co-exposure to UVA (150 J/m^2) and EMF (50 Hz, 40 μT rms) on rat lymphocytes [67]. UVA exposure of 5 min accompanied by 5 or 60 min of exposure to EMF showed that 60, but not 5, minutes' exposure to EMF coupled with UVA increased damage to DNA, which the authors attributed to increased ROS presence in doubly exposed cells. Similarly, for lymphocytes exposed to 7 mT static SMF and $FeCl_2$, a combined 3 h exposure to SMF and 10 μg/ml $FeCl_2$ led to an increase in the percentage of apoptotic and necrotic cells [17]. This was accompanied by a significant increase in the amount of lipid peroxidation end products, MDA+4 HNE.

Moreover, Ding and others exposed HL-60 cells simultaneously to an ELF magnetic field (60 Hz, 5 mT) and H_2O_2 (85 or 100 μM) [68]. A 24 h co-exposure significantly decreased the number of viable cells in comparison to cells treated with H_2O_2 alone, thereby establishing ELF EMF as an apoptosis promoter. Interestingly, while no changes in Bax and Bcl-2 protein levels or caspase-3 activation were observed, activation of caspase-7 as well as poly (ADP-ribose) polymerase cleavage increased.

1.9 Conclusions—Possible Explanations of Contradictory Results

Brusick and others reviewed 100 different studies and concluded that the vast majority of adverse effects from exposure to high frequencies and high-power intensities of RF radiation are the result of hyperthermia; nevertheless, the authors indicate that there may be some subtle and indirect effects on the DNA replication and/or gene regulation under restricted exposure conditions [69]. Black and Heynick reviewed more recent literature on RF EMF and concluded that leukocytes are much more sensitive to RF EMF than erythrocytes, but that these effects are consistent with systemic temperature fluctuation [70]. Moreover, lifetime studies of RF EMF-exposed animals did not show cumulative adverse effects on their immune systems. Meltz emphasized similar points about RF, claiming that for all tested frequencies and modulations regardless of length of exposure, there is no induction of DNA damage on any scale (from mutations to chromosome aberrations) or promotion of DNA damage caused by chemical or physical mutagens; in other words no evidence that RF is either a carcinogen, cocarcinogen, tumor inducer, or tumor promoter [71]. The only biological effect of RF with acknowledged mixed evidence is that certain RF exposure conditions cause micronuclei formation. Reviewing the epidemiologic data, Elwood similarly emphasized on the inconsistency of data obtained from the effects of RF radiation exposures [45].

Perhaps the best summary of the controversy in the field of RF-related biological research is the review done by Vijayalaxmi [72]. She listed findings from the period 1990–2003 and calculated percentages of studies that showed or did not show biological effects of exposure to RF radiation. Negative reports comprised 58% of RF-related literature, positive 23%, and inconclusive 19%.

Research related to biological effects of ELF EMF is even more riddled by conflicting findings. Some reviews attempted to deal with it by limiting the literature available to that describing a specific experimental setup. McCann and others reviewed studies concerning potential genotoxicity of electric and magnetic fields and ascribed the highest merit to studies fulfilling one of the following criteria: ELF magnetic fields, 150 μT to 5 mT; combined ELF electric and ELF magnetic fields, 0.2 mT and 240 mV/m; and SMFs, 1–3.7 T [73]. The authors believed that some of the studies showing positive data met these criteria, but nevertheless wanted independent repetition before they could be admitted as truly positive data [74–80].

Miyakoshi reviewed findings on the effects of static MF at the cellular level and summarized that most but not all studies suggested that this type of MF has no effect on growth rate or cell cycle distribution [81]. On the other hand, (co-)genotoxicity of static MF following exposure to known mutagens, e.g., X-irradiation, or coupled with treatment with trace amounts of ferrous ions, appears to be firmly established.

At this juncture, it is also relevant to mention studies of Marino and others who developed a novel statistical procedure that can be adjusted to detect both nonlinear and linear effects of treatments of biological samples [82,83]. In their experiments (exposure of male and female mice to 100 or 500 μT at 60 Hz for 1 to 105 d or continuously for 175 d), the authors found statistically significant repression of lymphocytes that could, nevertheless, be scored only when the data were analyzed as if they were governed by nonlinear laws. When the relationship of the immune variables and the EMF was modeled as nonlinear, the results started to show a pattern, which suggested that lymphocytes respond to and are vulnerable to EMF transduction, but that it may be impossible to predict specific changes of immune-system variables, due to the innate nonlinearity of the regulatory mechanisms of the immune system. Moreover, the authors hypothesize that the inconsistencies in EMF research stem from the attempt to use linear methods and models to study inherently nonlinear input–output relationships and believe that the underlying behavior of the immune system may be extremely sensitive to its prior states.

In conclusion, the final pattern emerging from all of these studies suggests that while exposure to EMF and RF radiation has no direct effect on DNA, it does under some circumstances exert a biological effect reminiscent of heat shock and/or stress. This effect is mild and heavily dependent on the state of cell homeostasis prior to exposure. When experiments were done on animals, individual variation was often more pronounced than the results of the exposures. Moreover, most protracted exposure experiments on animals found that the dysbalance in the cell homeostasis introduced at the beginning of exposure was most often overcome before the end of the experiment.

Considering the variations in responses to EMF and RF, it is very difficult to either outrightly accept or reject any study that was controlled for "contamination" with thermal stress and included reasonable biological negative controls. In other words, the fact that the majority of studies failed to produce reproducible results depends on the innate complexity of the experimental design rather than any intrinsic "falseness" of the tests conducted. The immune system is one of the most important lines of defense of an organism against the hostile environment. The provocations coming from the environment can be met by the immune system with vehemence or almost passively, and two organisms of the same species do not always react the same way to the same stimuli. Furthermore, even a single individual does not respond the same way to the same stimulus at different times; as the immune system adjusts to the environment reactions to certain stimuli become intensified, while others gradually decrease.

Taking into consideration that many studies of the effects of EMF exposure showed opposite results for both EMF and RF exposures, we can either concern ourselves with trying to "prune" these data until they produce a coherent picture, or we can approach it in an all-encompassing manner. And while the first approach inevitably leads to the conclusion that the data "average" into a negative result, the second approach brings several tentative views on possible effects of EMF and RF exposure(s). If we consider the totality of the data, we can claim that the effects of these exposures are mild in comparison to other known stressors, such as heat shock, ionizing radiation, nutrient deprivation, etc. and unpredictable based on our current modeling capabilities. Therefore, the most significant role of these exposures is likely going to be to intensify the effects of exposures to other physical or chemical stresses. Any future studies in this field would probably benefit the most from work *in vivo*, using protracted exposures and combining EMF and RF exposures with exposures to known stress factors. Needless to say, a very rigorous control of the exposure conditions and use of multiple biological "negative controls" would be a necessary condition of any such experimentation.

References

1. Rannug, A., Holmberg, B., Ekstrom, T., Mild, K.H., Gimenez-Conti, I., and Slaga, T.J., Intermittent 50 Hz magnetic field and skin tumor promotion in SENCAR mice, *Carcinogenesis* 15 (2), 153–7, 1994.
2. Hirose, H., Nakahara, T., Zhang, Q.M., Yonei, S., and Miyakoshi, J., Static magnetic field with a strong magnetic field gradient (41.7 T/m) induces c-Jun expression in HL-60 cells, *In Vitro Cell Dev. Biol. Anim.* 39 (8–9), 348–52, 2003.
3. Olivares-Banuelos, T., Navarro, L., Gonzalez, A., and Drucker-Colin, R., Differentiation of chromaffin cells elicited by ELF MF modifies gene expression pattern, *Cell Biol. Int.* 28 (4), 273–9, 2004.
4. Bonhomme-Faivre, L., Marion, S., Forestier, F., Santini, R., Auclair, H., Bezie, Y., Fredj, G., and Hommeau, C., Effects of electromagnetic fields on the immune systems of occupationally exposed humans and mice. *Arch. Environ. Health* 58 (11), 712–17, 2003.
5. Bonhomme-Faivre, L., Marion, S., Bezie, Y., Auclair, H., Fredj, G., and Hommeau, C., Study of human neurovegetative and hematologic effects of environmental low-frequency (50-Hz) electromagnetic fields produced by transformers, *Arch. Environ. Health* 53 (2), 87–92, 1998.
6. Tuschl, H., Neubauer, G., Schmid, G., Weber, E., and Winker, N., Occupational exposure to static, ELF, VF and VLF magnetic fields and immune parameters, *Int. J. Occup. Med. Environ. Health* 13 (1), 39–50, 2000.
7. Dasdag, S., Sert, C., Akdag, Z., and Batun, S., Effects of extremely low-frequency electromagnetic fields on hematologic and immunologic parameters in welders, *Arch. Med. Res.* 33 (1), 29–32, 2002.
8. Heredia-Rojas, J.A., Rodriguez-De La Fuente, A.O., del Roble Velazco-Campos, M., Leal-Garza, C.H., Rodriguez-Flores, L.E., and de La Fuente-Cortez, B., Cytological effects of 60 Hz magnetic fields on human lymphocytes *in vitro*: sister-chromatid exchanges, cell kinetics and mitotic rate, *Bioelectromagnetics* 22 (3), 145–9, 2001.
9. Capri, M., Scarcella, E., Fumelli, C., Bianchi, E., Salvioli, S., Mesirca, P., Agostini, C., Antolini, A., Schiavoni, A., Castellani, G., Bersani, F., and Franceschi, C., *In vitro* exposure of human lymphocytes to 900 MHz CW and GSM modulated radiofrequency: studies of proliferation, apoptosis and mitochondrial membrane potential, *Radiat. Res.* 162 (2), 211–18, 2004.
10. Bonhomme-Faivre, L., Mace, A., Bezie, Y., Marion, S., Bindoula, G., Szekely, A.M., Frenois, N., Auclair, H., Orbach-Arbouys, S., and Bizi, E., Alterations of biological parameters in mice chronically exposed to low-frequency (50 Hz) electromagnetic fields, *Life Sci.* 62 (14), 1271–80, 1998.
11. Gatta, L., Pinto, R., Ubaldi, V., Pace, L., Galloni, P., Lovisolo, G.A., Marino, C., and Pioli, C., Effects of *in vivo* exposure to GSM-modulated 900 MHz radiation on mouse peripheral lymphocytes, *Radiat. Res.* 160 (5), 600–5, 2003.
12. Murabayashi, S., Yoshikawa, A., and Mitamura, Y., Functional modulation of activated lymphocytes by time-varying magnetic fields, *Ther. Apher. Dial.* 8 (3), 206–11, 2004.
13. Flipo, D., Fournier, M., Benquet, C., Roux, P., Le Boulaire, C., Pinsky, C., LaBella, F.S., and Krzystyniak, K., Increased apoptosis, changes in intracellular Ca^{2+}, and functional alterations in

lymphocytes and macrophages after *in vitro* exposure to static magnetic field, *J. Toxicol. Environ. Health A* 54 (1), 63–76, 1998.
14. Makar, V., Logani, M., Szabo, I., and Ziskin, M., Effect of millimeter waves on cyclophosphamide induced suppression of T cell functions, *Bioelectromagnetics* 24 (5), 356–65, 2003.
15. Onodera, H., Jin, Z., Chida, S., Suzuki, Y., Tago, H., and Itoyama, Y., Effects of 10-T static magnetic field on human peripheral blood immune cells, *Radiat. Res.* 159 (6), 775–9, 2003.
16. Radeva, M. and Berg, H., Differences in lethality between cancer cells and human lymphocytes caused by LF-electromagnetic fields, *Bioelectromagnetics* 25 (7), 503–7, 2004.
17. Jajte, J., Grzegorczyk, J., Zmyslony, M., and Rajkowska, E., Effect of 7 mT static magnetic field and iron ions on rat lymphocytes: apoptosis, necrosis and free radical processes, *Bioelectrochemistry* 57 (2), 107–11, 2002.
18. Ikeda, K., No effects of extremely low frequency magnetic fields found on cytotoxic activities and cytokine production of human peripheral blood mononuclear cells *in vitro*, *Bioelectromagnetics* 24 (1), 21–31, 2003.
19. Quaglino, D., Capri, M., Zecca, L., Franceschi, C., and Ronchetti, I.P., The effect on rat thymocytes of the simultaneous *in vivo* exposure to 50-Hz electric and magnetic field and to continuous light, *Sci. World J.* 4 (Suppl. 2), 91–9, 2004.
20. Zotti-Martelli, L., Peccatori, M., Scarpato, R., and Migliore, L., Induction of micronuclei in human lymphocytes exposed *in vitro* to microwave radiation, *Mutat. Res.* 472 (1–2), 51–8, 2000.
21. d'Ambrosio, G., Cytogenetic damage in human lymphocytes following GMSK phase modulated microwave exposure, *Bioelectromagnetics* 23 (1), 7–13, 2002.
22. Tice, R.R., Hook, G.G., Donner, M., McRee, D.I., and Guy, A.W., Genotoxicity of radiofrequency signals. I. Investigation of DNA damage and micronuclei induction in cultured human blood cells, *Bioelectromagnetics* 23 (2), 113–26, 2002.
23. Zeni, O., Chiavoni, A.S., Sannino, A., Antolini, A., Forigo, D., Bersani, F., and Scarfi, M.R., Lack of genotoxic effects (micronucleus induction) in human lymphocytes exposed *in vitro* to 900 MHz electromagnetic fields, *Radiat. Res.* 160 (2), 152–8, 2003.
24. Verheyen, G.R., Pauwels, G., Verschaeve, L., and Schoeters, G., Effect of coexposure to 50 Hz magnetic fields and an aneugen on human lymphocytes, determined by the cytokinesis block micronucleus assay, *Bioelectromagnetics* 24 (3), 160–4, 2003.
25. Vijayalaxmi, Pickard, W.F., Bisht, K.S., Prihoda, T.J., Meltz, M.L., LaRegina, M.C., Roti Roti, J.L., Straube, W.L., and Moros, E.G., Micronuclei in the peripheral blood and bone marrow cells of rats exposed to 2450 MHz radiofrequency radiation, *Int. J. Radiat. Biol.* 77 (11), 1109–15, 2001.
26. Skyberg, K., Hansteen, I.L., and Vistnes, A.I., Chromosomal aberrations in lymphocytes of employees in transformer and generator production exposed to electromagnetic fields and mineral oil, *Bioelectromagnetics* 22 (3), 150–60, 2001.
27. Mashevich, M., Folkman, D., Kesar, A., Barbul, A., Korenstein, R., Jerby, E., and Avivi, L., Exposure of human peripheral blood lymphocytes to electromagnetic fields associated with cellular phones leads to chromosomal instability, *Bioelectromagnetics* 24 (2), 82–90, 2003.
28. Cho, Y.H. and Chung, H.W., The effect of extremely low frequency electromagnetic fields (ELF-EMF) on the frequency of micronuclei and sister chromatid exchange in human lymphocytes induced by benzo(a)pyrene, *Toxicol. Lett.* 143 (1), 37–44, 2003.
29. Maes, A., Collier, M., Vandoninck, S., Scarpa, P., and Verschaeve, L., Cytogenetic effects of 50 Hz magnetic fields of different magnetic flux densities, *Bioelectromagnetics* 21 (8), 589–96, 2000.
30. Vijayalaxmi, Leal, B.Z., Meltz, M.L., Pickard, W.F., Bisht, K.S., Roti Roti, J.L., Straube, W.L., and Moros, E.G., Cytogenetic studies in human blood lymphocytes exposed *in vitro* to radiofrequency radiation at a cellular telephone frequency (835.62 MHz, FDMA), *Radiat. Res.* 155 (1 Pt 1), 113–21, 2001.
31. Vijayalaxmi, Bisht, K.S., Pickard, W.F., Meltz, M.L., Roti Roti, J.L., and Moros, E.G., Chromosome damage and micronucleus formation in human blood lymphocytes exposed *in vitro* to radiofrequency radiation at a cellular telephone frequency (847.74 MHz, CDMA), *Radiat. Res.* 156 (4), 430–2, 2001.
32. Lloyd, D., Hone, P., Edwards, A., Cox, R., and Halls, J., The repair of gamma-ray-induced chromosomal damage in human lymphocytes after exposure to extremely low frequency electromagnetic fields, *Cytogenet. Genome Res.* 104 (1–4), 188–92, 2004.

33. Sykes, P.J., McCallum, B.D., Bangay, M.J., Hooker, A.M., and Morley, A.A., Effect of exposure to 900 MHz radiofrequency radiation on intrachromosomal recombination in pKZ1 mice, *Radiat. Res.* 156 (5 Pt 1), 495–502, 2001.
34. Vijayalaxmi, Leal, B.Z., Szilagyi, M., Prihoda, T.J., and Meltz, M.L., Primary DNA damage in human blood lymphocytes exposed *in vitro* to 2450 MHz radiofrequency radiation, *Radiat. Res.* 153 (4), 479–86, 2000.
35. McNamee, J.P., Bellier, P.V., Gajda, G.B., Miller, S.M., Lemay, E.P., Lavallee, B.F., Marro, L., and Thansandote, A., DNA damage and micronucleus induction in human leukocytes after acute *in vitro* exposure to a 1.9 GHz continuous-wave radiofrequency field, *Radiat. Res.* 158 (4), 523–33, 2002.
36. McNamee, J.P., Bellier, P.V., Gajda, G.B., Lavallee, B.F., Lemay, E.P., Marro, L., and Thansandote, A., DNA damage in human leukocytes after acute *in vitro* exposure to a 1.9 GHz pulse-modulated radiofrequency field, *Radiat. Res.* 158 (4), 534–7, 2002.
37. Maes, A., Collier, M., and Verschaeve, L., Cytogenetic effects of 900 MHz (GSM) microwaves on human lymphocytes, *Bioelectromagnetics* 22 (2), 91–6, 2001.
38. Miyakoshi, J., Yoshida, M., Shibuya, K., and Hiraoka, M., Exposure to strong magnetic fields at power frequency potentiates X-ray-induced DNA strand breaks, *J. Radiat. Res. (Tokyo)* 41 (3), 293–302, 2000.
39. Zhang, M.B., He, J. L., Jin, L.F., and Lu, D.Q., Study of low-intensity 2450-MHz microwave exposure enhancing the genotoxic effects of mitomycin C using micronucleus test and comet assay *in vitro*, *Biomed. Environ. Sci.* 15 (4), 283–90, 2002.
40. Repacholi, M.H., Basten, A., Gebski, V., Noonan, D., Finnie, J., and Harris, A.W., Lymphomas in E mu-Pim1 transgenic mice exposed to pulsed 900 MHz electromagnetic fields, *Radiat. Res.* 147 (5), 631–40, 1997.
41. Utteridge, T.D., Gebski, V., Finnie, J.W., Vernon-Roberts, B., and Kuchel, T.R., Long-term exposure of E-mu-Pim1 transgenic mice to 898.4 MHz microwaves does not increase lymphoma incidence, *Radiat. Res.* 158 (3), 357–64, 2002.
42. Heikkinen, P., Kosma, V.M., Hongisto, T., Huuskonen, H., Hyysalo, P., Komulainen, H., Kumlin, T., Lahtinen, T., Lang, S., Puranen, L., and Juutilainen, J., Effects of mobile phone radiation on X-ray-induced tumorigenesis in mice, *Radiat. Res.* 156 (6), 775–85, 2001.
43. Heikkinen, P., Kosma, V.M., Alhonen, L., Huuskonen, H., Komulainen, H., Kumlin, T., Laitinen, J.T., Lang, S., Puranen, L., and Juutilainen, J., Effects of mobile phone radiation on UV-induced skin tumourigenesis in ornithine decarboxylase transgenic and non-transgenic mice, *Int. J. Radiat. Biol.* 79 (4), 221–33, 2003.
44. Kumlin, T., Kosma, V.M., Alhonen, L., Janne, J., Komulainen, H., Lang, S., Rytomaa, T., Servomaa, K., and Juutilainen, J., Effects of 50 Hz magnetic fields on UV-induced skin tumourigenesis in ODC-transgenic and non-transgenic mice, *Int. J. Radiat. Biol.* 73 (1), 113–21, 1998.
45. Elwood, J.M., Epidemiological studies of radio frequency exposures and human cancer, *Bioelectromagnetics* Suppl. 6, S63–73, 2003.
46. Lin, H., Blank, M., Rossol-Haseroth, K., and Goodman, R., Regulating genes with electromagnetic response elements, *J. Cell. Biochem.* 81 (1), 143–8, 2001.
47. Blank, M. and Goodman, R., Initial interactions in electromagnetic field-induced biosynthesis, *J. Cell. Physiol.* 199 (3), 359–63, 2004.
48. de Pomerai, D.I., Smith, B., Dawe, A., North, K., Smith, T., Archer, D.B., Duce, I.R., Jones, D., and Candido, E.P., Microwave radiation can alter protein conformation without bulk heating, *FEBS Lett.* 543 (1–3), 93–7, 2003.
49. Lindstrom, E., Lindstrom, P., Berglund, A., Mild, K.H., and Lundgren, E., Intracellular calcium oscillations induced in a T-cell line by a weak 50 Hz magnetic field, *J. Cell. Physiol.* 156 (2), 395–8, 1993.
50. Lindstrom, E., Lindstrom, P., Berglund, A., Lundgren, E., and Mild, K.H., Intracellular calcium oscillations in a T-cell line after exposure to extremely-low-frequency magnetic fields with variable frequencies and flux densities, *Bioelectromagnetics* 16 (1), 41–7, 1995.
51. Wey, H.E., Conover, D.P., Mathias, P., Toraason, M., and Lotz, W.G., 50-Hertz magnetic field and calcium transients in Jurkat cells: results of a research and public information dissemination (RAPID) program study, *Environ. Health Perspect.* 108 (2), 135–40, 2000.

52. Cranfield, C.G., Wood, A.W., Anderson, V., and Menezes, K.G., Effects of mobile phone type signals on calcium levels within human leukaemic T-cells (Jurkat cells), *Int. J. Radiat. Biol.* 77 (12), 1207–17, 2001.
53. Aldinucci, C., Garcia, J. B., Palmi, M., Sgaragli, G., Benocci, A., Meini, A., Pessina, F., Rossi, C., Bonechi, C., and Pessina, G.P., The effect of exposure to high flux density static and pulsed magnetic fields on lymphocyte function, *Bioelectromagnetics* 24 (6), 373–9, 2003.
54. Kim, Y.V., Conover, D.L., Lotz, W.G., and Cleary, S.F., Electric field-induced changes in agonist-stimulated calcium fluxes of human HL-60 leukemia cells, *Bioelectromagnetics* 19 (6), 366–76, 1998.
55. McCreary, C.R., Thomas, A.W., and Prato, F.S., Factors confounding cytosolic calcium measurements in Jurkat E6.1 cells during exposure to ELF magnetic fields, *Bioelectromagnetics* 23 (4), 315–28, 2002.
56. Lin, H., Head, M., Blank, M., Han, L., Jin, M., and Goodman, R., Myc-mediated transactivation of HSP70 expression following exposure to magnetic fields, *J. Cell. Biochem.* 69 (2), 181–8, 1998.
57. de Pomerai, D., Daniells, C., David, H., Allan, J., Duce, I., Mutwakil, M., Thomas, D., Sewell, P., Tattersall, J., Jones, D., and Candido, P., Non-thermal heat-shock response to microwaves, *Nature* 405 (6785), 417–18, 2000.
58. Sarimov, R., Nonthermal GSM microwaves affect chromatin conformation in human lymphocytes similar to heat shock, *IEEE Trans. Plasma Sci.* 32 (4), 1600–1608, 2004.
59. Belyaev, I. Y. and Alipov, E.D., Frequency-dependent effects of ELF magnetic field on chromatin conformation in *Escherichia coli* cells and human lymphocytes, *Biochim. Biophys. Acta* 1526 (3), 269–76, 2001.
60. Lim, H.B., Cook, G.G., Barker, A.T., and Coulton, L.A., Effect of 900 MHz electromagnetic fields on nonthermal induction of heat-shock proteins in human leukocytes, *Radiat. Res.* 163 (1), 45–52, 2005.
61. Capri, M., Scarcella, E., Bianchi, E., Fumelli, C., Mesirca, P., Agostini, C., Remondini, D., Schuderer, J., Kuster, N., Franceschi, C., and Bersani, F., 1800 MHz radiofrequency (mobile phones, different Global System for Mobile communication modulations) does not affect apoptosis and heat shock protein 70 level in peripheral blood mononuclear cells from young and old donors, *Int. J. Radiat. Biol.* 80 (6), 389–97, 2004.
62. Natarajan, M., Vijayalaxmi, Szzliagyl, M., Roldan, F.N., and Meltz, M.L., NF-kappaB DNA-binding activity after high peak power pulsed microwave (8.2 GHz) exposure of normal human monocytes, *Bioelectromagnetics* 23 (4), 271–7, 2002.
63. Zhou, J., Yao, G., Zhang, J., and Chang, Z., CREB DNA binding activation by a 50-Hz magnetic field in HL60 cells is dependent on extra- and intracellular Ca(2+) but not PKA, PKC, ERK, or p38 MAPK, *Biochem. Biophys. Res. Commun.* 296 (4), 1013–18, 2002.
64. Romano-Spica, V., Mucci, N., Ursini, C.L., Ianni, A., and Bhat, N.K., Ets1 oncogene induction by ELF-modulated 50 MHz radiofrequency electromagnetic field, *Bioelectromagnetics* 21 (1), 8–18, 2000.
65. Simko, M. and Mattsson, M.O., Extremely low frequency electromagnetic fields as effectors of cellular responses *in vitro*: possible immune cell activation, *J. Cell. Biochem.* 93 (1), 83–92, 2004.
66. Zmyslony, M., Politanski, P., Rajkowska, E., Szymczak, W., and Jajte, J., Acute exposure to 930 MHz CW electromagnetic radiation *in vitro* affects reactive oxygen species level in rat lymphocytes treated by iron ions, *Bioelectromagnetics* 25 (5), 324–8, 2004.
67. Zmyslony, M., Palus, J., Dziubaltowska, E., Politanski, P., Mamrot, P., Rajkowska, E., and Kamedula, M., Effects of *in vitro* exposure to power frequency magnetic fields on UV-induced DNA damage of rat lymphocytes, *Bioelectromagnetics* 25 (7), 560–2, 2004.
68. Ding, G.R., Nakahara, T., Hirose, H., Koyama, S., Takashima, Y., and Miyakoshi, J., Extremely low frequency magnetic fields and the promotion of H_2O_2-induced cell death in HL-60 cells, *Int. J. Radiat. Biol.* 80 (4), 317–24, 2004.
69. Brusick, D., Albertini, R., McRee, D., Peterson, D., Williams, G., Hanawalt, P., and Preston, J., Genotoxicity of radiofrequency radiation. DNA/Genetox Expert Panel, *Environ. Mol. Mutagen* 32 (1), 1–16, 1998.
70. Black, D. R. and Heynick, L.N., Radiofrequency (RF) effects on blood cells, cardiac, endocrine, and immunological functions, *Bioelectromagnetics* Suppl. 6, S187–95, 2003.

71. Meltz, M.L., Radiofrequency exposure and mammalian cell toxicity, genotoxicity, and transformation, *Bioelectromagnetics* Suppl. 6, S196–213, 2003.
72. Vijayalaxmi and Obe, G., Controversial cytogenetic observations in mammalian somatic cells exposed to radiofrequency radiation, *Radiat. Res.* 162 (5), 481–96, 2004.
73. McCann, J., The genotoxic potential of electric and magnetic fields: an update, *Mut. Res. – Rev. Mut. Res.* 411 (1), 45–86, 1998.
74. Miyakoshi, J., Kitagawa, K., and Takebe, H., Mutation induction by high-density, 50-Hz magnetic fields in human MeWo cells exposed in the DNA synthesis phase, *Int. J. Radiat. Biol.* 71 (1), 75–9, 1997.
75. Miyakoshi, J., Yamagishi, N., Ohtsu, S., Mohri, K., and Takebe, H., Increase in hypoxanthine–guanine phosphoribosyl transferase gene mutations by exposure to high-density 50-Hz magnetic fields, *Mutat. Res.* 349 (1), 109–14, 1996.
76. Lai, H. and Singh, N.P., Melatonin and *N-tert*-butyl-alpha-phenylnitrone block 60-Hz magnetic field-induced DNA single and double strand breaks in rat brain cells, *J. Pineal Res.* 22 (3), 152–62, 1997.
77. Lai, H. and Singh, N.P., Acute exposure to a 60 Hz magnetic field increases DNA strand breaks in rat brain cells, *Bioelectromagnetics* 18 (2), 156–65, 1997.
78. Koana, T., Estimation of genetic effects of a static magnetic field by a somatic cell test using mutagen-sensitive mutants of *Drosophila melanogaster*, *Bioelectrochem. Bioenerget.* 36 (2), 95–100, 1995.
79. Tabrah, F., Enhanced mutagenic effect of a 60 Hz time-varying magnetic-field on numbers of azide-induced TA100 revertant colonies, *Bioelectromagnetics* 15 (1), 85–93, 1994.
80. Tofani, S., Evidence for genotoxic effects of resonant ELF magnetic fields, *Bioelectrochem. Bioenerget.* 36 (1), 9–13, 1995.
81. Miyakoshi, J., Effects of static magnetic fields at the cellular level, *Prog. Biophys. Mol. Biol.* 87 (2–3), 213–223, 2005.
82. Marino, A.A., Wolcott, R.M., Chervenak, R., Jourd'heuil, F., Nilsen, E., and Frilot, C., II, Nonlinear determinism in the immune system. *In vivo* influence of electromagnetic fields on different functions of murine lymphocyte subpopulations, *Immunol. Invest.* 30 (4), 313–34, 2001.
83. Marino, A.A., Wolcott, R.M., Chervenak, R., Jourd'heuil, F., Nilsen, E., and Frilot, C., II, Nonlinear dynamical law governs magnetic field induced changes in lymphoid phenotype, *Bioelectromagnetics* 22 (8), 529–46, 2001.

2

Evaluation of the Toxicity and Potential Oncogenicity of Extremely Low-Frequency Magnetic Fields in Experimental Animal Model Systems

David L. McCormick

CONTENTS

2.1 Introduction

Over the past two decades, the possible relationship between exposure to power frequency (50 and 60 Hz) electromagnetic fields (EMF) and adverse human health outcomes has received significant attention in both the scientific community and the general population. Based on widely circulated accounts in the popular press [1], and on mass media reports of the results of selected epidemiologic investigations [2,3], a public perception has developed that human exposure to EMF may be associated with a range of adverse health effects, including reproductive dysfunction, developmental abnormalities, and cancer. The results of a number of epidemiologic studies provide limited support for the hypothesis that EMF exposure may be associated with an increased risk of neoplasia in several organ sites in humans. Among these sites, the hematopoietic system, breast, and brain have been identified most commonly as possibly sensitive targets for EMF action (reviewed in [4]).

Although individual epidemiologic studies provide suggestive evidence of the potential oncogenicity of EMF, the total body of epidemiologic data linking EMF exposure and cancer risk is by no means conclusive. Since 1990, more than 50 epidemiologic studies designed to investigate the possible association between occupational or residential exposure to magnetic fields and cancer risk have been published (reviewed in [5]). The methods used in these more recent studies are often substantially improved in comparison to methods used in earlier investigations; specific areas of improvement include the use of larger sample sizes and better exposure assessment. However, despite these improvements in epidemiologic methods, EMF cancer epidemiology studies continue to generate both positive and negative results. When considered together, the results of these studies are insufficient to either support or refute the hypothesis that exposure to EMF is a significant risk factor for human cancer.

In situations where epidemiology does not support the conclusive identification and quantitation of the potential risks associated with exposure to an environmental agent, laboratory studies conducted in appropriate experimental model systems increase in importance. Well-designed and controlled animal studies permit evaluation of biological effects *in vivo* under tightly controlled exposure and environmental conditions, and in the absence of potential confounding variables. In consideration of the conflicting results of EMF epidemiologic studies, and difficulties associated with exposure assessment in such studies, animal studies may provide the best opportunity to identify effects of EMF exposure that could translate into human health hazards.

The body of published scientific literature addressing the potential toxicity and oncogenicity of EMF in animal model systems has expanded greatly in the past 20 y. This chapter provides a brief review and analysis of this literature. For the purposes of the chapter, published toxicologic and carcinogenic bioassays of EMF in animal model systems are divided into five sections.

- In Section 2.2, studies designed to investigate the general and organ-specific toxicity of exposure to power frequency EMF are summarized. These studies include both general toxicity bioassays that have been conducted using standardized rodent model systems, and specialized toxicologic studies that are designed to identify possible specific toxic effects of EMF such as developmental toxicity (teratology), reproductive toxicity, or immunotoxicity.
- In Section 2.3, investigations into the possible oncogenicity of EMF as a single agent are reviewed. The most comprehensive studies reviewed in this section are chronic oncogenicity bioassays that are performed in rodents using standardized toxicity testing protocols. The designs of these studies include evaluations of the risk of oncogenesis in essentially all major organ systems; the experimental designs have been developed to satisfy the safety assessment requirements of regulatory agencies such as the United States Environmental Protection Agency and the United States Food and Drug Administration. Over several decades, an extensive database has been developed that supports the utility of the chronic oncogenicity bioassay in rodents as a predictor of oncogenicity in humans.
- In Section 2.4, studies designed to evaluate the activity of EMF exposure as a cocarcinogen or tumor promoter will be reviewed. These studies involve simultaneous or sequential exposures to EMF in combination with another agent in order to identify possible activity as a tumor initiator, tumor promoter, or cocarcinogen in specific organ sites. Most such studies have been conducted using animal models that were designed primarily for use as research tools to study site-specific carcinogenesis. As a result, the utility of these assays as

predictors of human oncogenicity is less completely understood than the results of the 2 y bioassay.

- Section 2.5 encompasses a discussion of studies in which the potential oncogenicity of EMF exposure has been evaluated using genetically modified (transgenic or gene knockout) animals. Via insertion of cancer-associated genes (oncogenes) or deletion of tumor suppressor genes from the germ line, transgenic and knockout animal strains have been developed that demonstrate a genetic predisposition to neoplasia. The use of these models to evaluate the potential oncogenicity of EMF exposure may identify biological effects that occur in sensitive subpopulations, but which may not occur in the general population. Given the relative novelty of the genetically engineered model systems used in these studies, however, little information exists on which their utility as predictors of human oncogenicity can be evaluated.
- Section 2.6 is a brief review of potential biological mechanisms through which EMF exposure has been proposed to induce or stimulate carcinogenesis. An exhaustive compilation and discussion of the literature related to all possibly relevant biological activities of EMF is well beyond the scope of this chapter. For this reason, this section is focused specifically on the effects of EMF exposure on biochemical and molecular markers that may be mechanistically linked to the process of neoplastic development, with specific emphasis on studies performed in whole animal model systems.

2.2 Animal Toxicity Bioassays of Electromagnetic Fields

2.2.1 General Toxicologic Bioassays of Electromagnetic Fields

Although many thousands of humans have received occupational exposure to EMF for extended periods, no documented case reports of generalized human toxicity resulting from EMF exposure have been published in the peer-reviewed literature. On this basis, it is generally agreed that acute, subchronic, or chronic exposure to EMF is not a significant risk factor for systemic or generalized toxicity in humans. Because certain occupational environments contain high-field areas in which EMF flux densities often exceed residential EMF levels by factors of tens to hundreds, it is logical to expect that populations working in such high-field environments would be the primary groups in which EMF toxicities would be observed. Because no case reports of EMF-associated generalized toxicity or malaise in any exposed worker have been published, it is logical to conclude that EMF exposure is not a significant risk factor for the induction of such generalized toxicity.

The lack of systemic toxicity of EMF exposure in humans is clearly supported by the results of acute and subchronic bioassays conducted in animal model systems. In these studies, rats or mice have been exposed to EMF flux densities ranging from low milli-Gauss (mG) levels up to maximum flux densities as high as 20 G (Gauss; equivalent to 2000 μT; bioassays conducted in different laboratories have studied the effects of both linearly polarized and circularly polarized magnetic fields.

Although the experimental end points evaluated in these bioassays vary by study, most include evaluations of the possible effects of EMF exposure on survival, overall health as monitored by clinical and physical observations, body weight, food consumption, hematology, clinical chemistry, and histopathologic evaluation of tissues. Although occasional instances of statistically significant differences in EMF-exposed animals from sham-exposed controls have been observed in one or more end points, no reproducible pattern of EMF toxicity has been identified in any bioassay with known predictability for human responses (see for example [6–8]; other studies reviewed in [4]). The lack of hematologic toxicity of EMF identified in these animal studies is consistent with the data from an experimental exposure study conducted in humans [9].

Although long-term studies have generally been focused on the possible activity of EMF as a carcinogen or tumor promoter, many of the chronic EMF exposure studies (discussed in succeeding sections of this chapter) also included evaluations of toxicologic end points such as body weight, clinical and physical observations, and hematology. As was the case with the acute and subchronic exposure studies, studies involving chronic EMF exposures have failed to identify any evidence of systemic toxicity (see for example [10–12]).

Summarizing the available data, no evidence of generalized toxicity associated with EMF exposure has been identified in epidemiologic studies of human populations receiving occupational exposure, or in experimental exposure studies in either animals or humans. On this basis, it can be concluded that the risks of generalized or systemic toxicity resulting from acute, subchronic, or chronic exposure to EMF are very small.

2.2.2 Developmental and Reproductive Toxicity Studies of Electromagnetic Fields

Investigations into the possible developmental and reproductive toxicity of EMF were stimulated by early reports of teratogenesis in chick embryos exposed to high-EMF flux densities (reviewed in [13]). Although the processes of prenatal development in chickens differ substantially from those in humans, the induction of terata in early chick embryos has been interpreted by some investigators to support the hypothesis that EMF is a teratogen. By contrast, however, teratologists who use animal models to develop human hazard assessments for chemical and physical agents have questioned the value of the chick embryo model for human hazard identification. In consideration of both the positive and negative results of studies designed to examine the effects of EMF exposure on chick embryo development, and the unknown ability of studies in chickens and other avian species to predict human developmental toxicity, data from developmental toxicity studies of EMF in chickens are considered to offer relatively little insight into human hazard assessment [13].

It should be noted, however, that the early teratologic data generated in chicken models were extended by reports of small increases in the incidence of birth defects and spontaneous abortions in women who used electric blankets and electric water bed heaters during pregnancy (reviewed in [4]). These early findings were hypothesis generating, but were limited in that they were retrospective studies that included no measurements of EMF exposure. Furthermore, the design of these studies essentially precluded the separation of possible EMF effects from thermal effects that could result from the use of an electric blanket or bed heater.

Since the publication of these early findings, a large number of epidemiologic studies have been performed to investigate the hypothesis that EMF exposure is a risk factor for teratogenesis. These studies have examined the incidence of birth defects, growth retardation, and premature delivery, among other end points, in the offspring of women who used electric blankets or electric bed heaters during pregnancy. Other studies have examined rates of these abnormalities in pregnant women with different levels of residential EMF exposure and in pregnant women with different measured or imputed levels of occupational EMF exposure. In general, although occasional EMF-associated differences have been reported, the results of these studies failed to confirm the hypothesis that

EMF exposure during pregnancy is a significant risk to the developing fetus (reviewed in [4]). In most studies, no statistically significant relationships have been found between EMF exposure and the risk of birth defects or other developmental abnormalities. The source of the EMF exposure (use of an electric blanket or bed heater, overall higher level of residential EMF, or occupational EMF exposure) was inconsequential with respect to possible effects on the risk of teratogenesis.

Supporting these epidemiologic data are the negative results of more than a dozen teratology studies in which pregnant rodents were exposed to EMF flux densities of up to 20 G (2000 μT; cf. [14–18]; others reviewed in [13]). It is important to note that the rodent models, study designs, and periods of exposure used in several EMF developmental toxicity studies have been used for many years to evaluate the possible teratogenicity of pharmaceuticals and environmental chemicals; these models have been validated extensively, and have been shown to be generally predictive of human responses. The results of these EMF teratology studies, which have been conducted by several different teams of investigators in several different countries, have been uniformly negative.

When the negative results generated in these experimental evaluations of EMF teratogenesis are considered together with the negative results of most epidemiologic studies, the overwhelming majority of available teratology data do not support the hypothesis that exposure to power frequency EMF during pregnancy constitutes an important hazard to the developing fetus. Interestingly, although Mevissen et al. [19] found no significant adverse effects of fetal exposure to 50 Hz EMF in rats, they reported a significant reduction in fetal survival in rats receiving gestational exposure to 30 mT static magnetic fields.

As is the case with developmental toxicity (teratology) studies of EMF, no compelling body of evidence from animal studies implicates EMF as a significant risk for either male or female fertility. The total number of relevant studies is small, and includes assessments of the potential reproductive toxicity of both power frequency electric and magnetic fields ([20,21]; other studies reviewed in [22]). However, the results of these studies, which have commonly involved exposure of several generations of rodents to EMF, have failed to identify any decreases in fertility in either male or female animals receiving such exposures.

2.2.3 Immunotoxicity Studies of Electromagnetic Fields

Evaluation of the possible effects of EMF exposure on immune function is important to support our understanding of the possible role of EMF in carcinogenesis and as a modifier of host resistance to infection. Any reproducible downregulation of immune function in EMF-exposed animals could have important consequences in terms of the host's ability to resist infectious disease challenge and to provide effective immune surveillance against neoplastic cells.

Over the past 20 y, an increasing database has developed in which immunotoxic effects observed in rodent models have been linked to similar effects in humans. On this basis, the results of immune function studies in established rodent models appear to provide the best experimental tool to predict possible effects in humans. Studies have been reported in which the effects of EMF on immune-related parameters have been evaluated in non-rodent species such as sheep [23]; however, the utility of these species as predictors of human responses is unknown. For this reason, the most relevant data set for the possible effects of EMF exposure on immune function is that which has been generated in rodents.

In the rodent studies, a battery of immune function assays (lymphoid organ weight, lymphoid organ cellularity, lymphoid organ histology, B- and T-cell-mediated immune responses, natural killer (NK) cell function, lymphocyte subset analysis, host resistance assays) was evaluated in mice or rats exposed to magnetic fields flux densities of up to 20 G for period ranging from 4 weeks to 3 months. The results of these studies demonstrated that EMF-exposed animals demonstrated no deficits in lymphoid organ weights, lymphoid organ cellularity, or lymphoid organ histology; similarly, no deficits in T- or B-cell function were identified, and lymphocyte subset analysis did not reveal any effects of EMF exposure [24,25]. Interestingly, however, a statistically significant suppression of NK cell function was seen in two studies performed in female mice; this alteration was not seen in male mice of the same strain or in either sex of rat evaluated in a parallel study [24,26].

Although the authors conclude that this suppression of NK cell function is real [26], it appears to have little or no biological significance. The primary role of NK cells appears to be in immune surveillance against cancer cells. However, completely negative results were obtained in a 2-y carcinogenesis bioassay of EMF performed in the same strain of mice that demonstrated the reduction in NK cell function [12]. Because the deficit in NK cell function was not associated with an increase in cancer risk, it is concluded that either (a) the quantitative reduction (~30%) in NK cell function induced by EMF is too small to be biologically significant, or (b) other arms of the host immune system were able to perform required immune surveillance in these animals.

2.3 Assessment of the Possible Oncogenic Activity of Electromagnetic Field Exposure Using Chronic Exposure Studies in Rodents

2.3.1 Design of Chronic Exposure Studies

Several teams of investigators have published the results of studies that are designed to investigate the possible carcinogenic activity of chronic exposure to EMF alone (as opposed to EMF exposure in combination with simultaneous or sequential exposure to other agents). These studies have generally been conducted using a standardized study design that is commonly referred to as the chronic oncogenicity bioassay in rodents. This study design has been used widely as the basis to evaluate the potential carcinogenicity of new drugs, agricultural chemicals, occupational chemicals, and a wide range of environmental agents, and is considered to be a useful predictor of human oncogenicity. Both the International Agency for Research on Cancer (IARC) and United States National Toxicology Program (NTP) have performed analyses demonstrating the utility of the rodent chronic bioassay as a predictor of predict human carcinogenicity. As a result, it is generally accepted among toxicologists that the chronic rodent bioassay currently provides the best available experimental approach to identify agents that may be carcinogenic in humans.

In the chronic rodent oncogenicity bioassay, groups of rats or mice are exposed to the test agent for a period than encompasses the majority of their normal life span (2 y in most studies; some studies in mice are conducted for 18 months). The bioassay includes a standard series of in-life toxicologic observations and evaluations; however, its key experimental end point is histopathology.

Histopathologic evaluations performed at the conclusion of chronic rodent bioassays generally include classification of all gross lesions identified at necropsy, and examination of 45 to 50 tissues from each animal in the high dose and control groups. In addition to the complete histopathologic evaluations of high dose and control animals, histopathology is generally performed on gross lesions and identified target tissues in animals in the middle and low dose groups.

A significant strength of the use of the chronic rodent bioassay to identify potentially carcinogenic agents is the ability to compare tumor incidence patterns of exposed groups to both contemporaneous controls, i.e., the control group from that specific study, as well as to tumor incidence patterns in historical control animals. Historical control databases for tumor incidence are commonly maintained both within conducting laboratories and by animal suppliers. The historical control database is particularly large for studies conducted for the NTP. Because all rats and mice used in NTP chronic oncogenicity studies are bred in the same animal colonies, the NTP database includes the results of tumor incidence data generated during histopathologic evaluations of many thousands of control animals.

2.3.2 Results of Chronic Rodent Oncogenicity Bioassays

2.3.2.1 IIT Research Institute (United States): Chronic Oncogenicity Bioassays in Rats and Mice

The largest and most comprehensive evaluations of the potential carcinogenicity of EMF exposure in rodents were conducted at IIT Research Institute (IITRI), under the sponsorship of the NTP. The quality and comprehensive nature of these studies was specifically highlighted in a commentary written by the Director of the Food and Drug Administration's National Center for Toxicological Research; this commentary was published in *Toxicologic Pathology* [27].

In this evaluation program, parallel studies were conducted in F344 rats [11] and in B6C3F1 mice [12]. In each study, groups of 100 animals per sex were exposed for 18.5 h/d for 2 y to pure, linearly polarized 60 Hz magnetic fields. It is useful to note that this group size is twice that commonly used in chronic rodent bioassays; this increase in group size confers increased statistical power to the experimental design, and thereby increases its ability to identify weak effects. Experimental groups included animals receiving continuous exposure to field strengths of 20 mG (equivalent to 2 μT), 2 G (200 μT), or 10 G (1000 μT), or intermittent exposures (1 h on/1 h off) to 10 G (1000 μT). Parallel sham control groups were housed within an identical exposure array, but were exposed to ambient magnetic fields only.

The design of these studies included continuous monitoring of magnetic field strength and waveform throughout the 2 y exposure period. At termination, all animals received a complete necropsy, and complete histopathologic evaluations were performed on all gross lesions collected from all study animals, in addition to approximately 50 tissues per animal in all study groups.

The results from these studies provided no evidence of increased cancer incidence in any of the putative target tissues (hematopoietic system, brain, breast) for EMF oncogenicity, and do not support the hypothesis that EMF is a significant risk factor for human neoplasia in any site:

- The results of the study conducted in B6C3F1 mice demonstrated no increases from sham control in the incidence of neoplasia in any tissue. When compared to sham controls, the only statistically significant differences in tumor incidence in groups exposed to EMF were:

  (1) A statistically significant *decrease* in the incidence of malignant lymphoma in female mice exposed intermittently to 10 G EMF

  (2) Significant *decreases* in the incidence of lung tumors in both sexes of mice exposed to EMF at 20 mG or 2 G (but not at 10 G)

- The results of the oncogenicity study conducted in F344 rats also failed to provide evidence to support the hypothesis that EMF is a significant risk factor for human cancer. Similar to the results of the study in mice, no significant increases in the incidence of neoplasia in the hematopoietic system, breast, or brain were seen in any study group. Significant differences from control tumor incidences included:

  (1) A statistically significant *decrease* in the incidence of leukemia in male rats exposed intermittently to 10 G EMF

  (2) A statistically significant *decrease* in the incidence of preputial gland carcinomas in male rats exposed to 2 G (but not to 10 G) EMF

  (3) A statistically significant *decrease* in the incidence of trichoepitheliomas of the skin in male rats exposed continuously to 10 G EMF

  (4) A statistically significant *decrease* in the incidence of adrenal cortical adenomas in female rats exposed intermittently to 10 G EMF

  (5) A statistically significant *increase* in incidence of thyroid C-cell tumors in male rats exposed to EMF at field strengths of 20 mG or 2 G (but not at 10 G)

When considered together, the results of these studies provide no evidence that EMF exposure is a significant risk factor for human cancer. No increases in the incidence of neoplasia in the brain, breast, or hematopoietic system were observed in either sex of either species in any group receiving chronic EMF exposure. Similarly, although a number of statistically significant differences from control tumor incidence were observed in the two studies, seven of the nine differences were *reductions*, rather than increases in tumor incidence.

2.3.2.2 Institut Armand Frappier (Canada): Chronic Oncogenicity Bioassay in Rats

Scientists at the Institut Armand Frappier (IAF) in Laval, Quebec conducted a chronic bioassay in which F344 rats were exposed to pure, linearly polarized 60 Hz magnetic fields for 20 h/d for 2 y [25]. Although group sizes used in the IAF study were smaller than those used in the chronic rat oncogenicity study conducted at IITRI, the experimental designs of the two rat oncogenicity bioassays were generally comparable. Two factors did distinguish the experimental design of the IAF study:

- EMF exposures were initiated 2 d before animal birth, rather than in young adult animals
- Study participants were "blinded" to identify study groups until after study completion

The most common approach to the conduct of rodent oncogenicity bioassays is to begin exposure to the test agent when animals are young adults (6 to 8 weeks old), and to identify the exposure status of each experimental group so that all study participants are aware of group identities. Because the IAF study included exposures during the perinatal and juvenile periods, its design addresses the possibly enhanced sensitivity of younger animals to EMF bioeffects. Furthermore, the conduct of the bioassay using a blinded design precludes any possible influence of investigator bias on study results.

The design of this study included 50 rats per sex per group. Experimental groups included cage controls, sham controls, and groups exposed to EMF at field strengths of 20 mG (2 μT), 200 mG (20 μT), 2 G (200 μT), and 20 G (2000 μT). Toxicologic end points included a standard battery of in-life evaluations and histopathologic evaluation of gross lesions and approximately 50 tissues per animal.

Similar to the data generated in the IITRI studies, the results of the IAF study failed to support an association between EMF exposure and increased risk of neoplasia in the brain, breast, hematopoietic system, or any other tissue examined. The study authors summarize their data by stating that "no statistically significant, consistent, positive dose-related trends with the number of tumor-bearing animals per study group could be attributed to EMF exposure." Although not statistically significant at the 5% level of confidence, total tumor incidence in all groups exposed to magnetic fields was *decreased* in comparison to tumor incidences observed in the cage control and sham control groups.

In the IAF study, the incidences of brain tumors (astrocytomas, gliomas, etc.), malignant mammary tumors (adenocarcinomas), and leukemias were low in all groups, and demonstrated no pattern of differences between control groups and groups exposed to EMF. Benign mammary tumors (fibroadenomas) were common in all groups, and were present in similar incidences, regardless of EMF exposure. The most common tumor identified in the study was a benign pituitary tumor (adenoma); although differences did not always attain statistical significance, the incidences of pituitary adenomas in all groups exposed to EMF were *decreased* from those seen in both the sham control group and the cage control group. *In toto*, the results of the IAF study are comparable to those from the IITRI study, and extend these results to include EMF exposures that were initiated during the perinatal period.

2.3.2.3 Mitsubishi Kasei Institute (Japan): Chronic Oncogenicity Bioassay in Rats

Scientists at the Mitsubishi Kasei Institute (MKI) of Toxicological and Environmental Sciences, Ibaraki, Japan conducted a 2 y bioassay in F344 rats to evaluate the potential oncogenicity of chronic exposure to linearly polarized, 50 Hz EMF [28]. The 50 Hz sine wave is the primary component of power frequency EMF associated with the generation and distribution of electricity in Europe and Asia. In this study, groups of 48 rats per sex were exposed for >21 h/d for 2 y to 50 Hz EMF at field strengths of 5 G (500 μT) or 50 G (5000 μT); an identically sized control group received sham exposure for the same period. At study termination, all animals received a complete necropsy. Complete histopathologic evaluations were performed on all gross lesions, and approximately 45 additional tissues per animal from all animals in all study groups.

Survival was comparable in all study groups, and differential white blood cell counts performed after 52, 78, and 104 weeks of exposures failed to identify any effects of EMF exposure. Consistent with the results of studies conducted at IITRI and IAF, the MKI study found no increased risk of neoplasia in animals receiving chronic exposure to EMF. The MKI team reported that EMF exposure had no effect on animal survival in either sex, and had no effect on the total incidence or number of neoplasms. Within a sex, incidences of leukemia, brain tumors, and mammary tumors in groups exposed to EMF were comparable to those of sex-matched sham controls.

The only statistically significant histopathologic findings from the study were an *increase* in the incidence of a benign lesion (fibroma) in the subcutis of male rats exposed to 50 G EMF, and a *decrease* in the total incidence of invasive neoplasms in male rats in the 5 G EMF group. Although the increased incidence of fibromas in male rats exposed to 50 G EMF was statistically significant when compared to concurrent male sham controls, lesion incidence was comparable to that observed in historical controls in the laboratory. On this basis, the authors concluded that this increase in fibroma incidence in rats exposed to 50 G EMF was not biologically significant. Similarly, because no differences in the incidence of metastatic neoplasms were seen in the study, the observed decrease in the incidence of invasive malignancy appears to be without biological significance.

2.3.2.4 Overview of Chronic Rodent Oncogenicity Studies

Although minor differences can be identified, the overall experimental designs used in the bioassays performed at IITRI (United States), IAF (Canada), and MKI (Japan) were very similar; all study designs conform with harmonized regulatory requirements for chronic oncogenicity testing in rodents. The results of the four studies conducted in these three laboratories are also in close agreement. In all studies, chronic exposure to power frequency EMF had no effect on the incidence of neoplasia in any of the three putative targets for EMF action (brain, breast, and hematopoietic system). Furthermore, chronic exposure to EMF did not induce any pattern of biologically significant increases in cancer incidence in other tissues. The very close comparability of the results of four chronic rodent oncogenicity bioassays conducted independently in three laboratories strengthens the conclusion that exposure to magnetic fields does not induce cancer in rodents. On the basis of the demonstrated predictive nature of the rodent 2 y bioassay, these findings do not support the hypothesis that EMF exposure is a significant risk factor in the etiology of human neoplasia.

2.3.3 Results of Site-Specific Oncogenicity Bioassays in Rodents

2.3.3.1 University of California at Los Angeles (United States): Lymphoma and Brain Tumor Bioassay in Mice

A chronic murine bioassay was conducted at the University of California at Los Angeles (UCLA) and Veterans Administration Hospital to evaluate the potential interaction between EMF and ionizing radiation in the induction of hematopoietic neoplasia in mice [29]. Although the primary focus of this study was the interaction between EMF and ionizing radiation, the study design did include several experimental groups that were exposed to EMF alone. Neoplastic responses in EMF-exposed and sham control groups that received no exposure to ionizing radiation are suitable for use in assessing the potential oncogenicity of exposure to EMF.

In this study, one group of 380 female C57BL/6 mice was exposed to circularly polarized 60 Hz EMF at a field strength of 14.2 G (1.42 mT) for up to 852 d. The incidence of hematopoietic neoplasia in these EMF-exposed mice was compared with that observed in a negative (untreated) control group consisting of 380 female C57BL/6 mice, and in a sham control group consisting of 190 female C57BL/6 mice. In addition to the investigation of EMF effects on hematopoietic neoplasia, a *post hoc* histopathologic analysis of brain tissues from this study was performed to investigate the possible activity of EMF as a causative agent for primary brain tumors [30].

Chronic exposure to EMF had no statistically significant effects on animal survival or on the incidence or latency of hematopoietic malignancy in this study [29]. At study termination, the final incidence of lymphoma in mice exposed to circularly polarized EMF was 36.8%, as compared to a lymphoma incidence of 34.7% in sham controls. The incidence of histiocytic sarcomas was 23.7% in mice receiving EMF exposure versus 22.1% in sham controls, yielding a total incidence of hematopoietic neoplasia of 56.3% in sham controls versus a total incidence of 59.2% in the group exposed to circularly polarized EMF. Relative risks for various types of hematopoietic neoplasia in the EMF group ranged from 0.94 to 1.03; all 95% confidence intervals included 1.00, and none differed from sham control at the 5% level of significance.

Consistent with the results for hematopoietic neoplasia, histopathologic evaluation of brains from this study provided no support for the hypothesis that EMF exposure is a significant risk factor for brain tumor induction. In this study, brains from a total of 950 nonirradiated mice were evaluated (total of control and EMF-exposed groups that received no exposure to ionizing radiation); only two benign brain tumors were identified [30]. Because the incidence of brain cancer in the negative control group, sham control group, and the EMF-exposed group were all below 0.25%, these data provide no evidence that exposure to circularly polarized EMF is oncogenic in the brain of C57BL/6 mice.

Although lack of proximity to power transmission and distribution lines ensures that most humans are not exposed to circularly polarized EMF, the possible carcinogenicity of circularly polarized fields has received only very limited study. The primary strengths of this study are that it evaluated the potential oncogenicity of a previously unstudied exposure metric (circularly polarized rather than linearly polarized EMF), and used very large experimental groups. The large group sizes used in the study greatly increase its statistical power, and therefore improve its ability to identify effects of modest magnitude.

It must be noted, however, that the study design was quite limited in that it included histopathologic evaluations of only the hematopoietic system, lymphoid system, and brain. As such, while the study data provide no support for the hypothesis that EMF exposure is causally associated with the induction of neoplasia in the hematopoietic system and brain, the lack of systematic evaluation of other tissues precludes any conclusions about possible influences on neoplasia in other sites.

2.3.3.2 Technical University of Nova Scotia (Canada): Lymphoma Bioassay in Mice

Using a unique study design in which three consecutive generations of CFW mice were exposed to extremely high flux densities (250 G [25 mT] of 60 Hz EMF [31]), data were presented that suggested an increased incidence of malignant lymphoma in the second and third generations of exposed animals. The strength of this study lies in its innovative design involving exposure of multiple generations, and its examination of flux densities that greatly exceed those that have been used in other studies. In consideration of the exposure regimen used in this study, it can be argued that this investigation provided the maximum likelihood of identifying a positive oncogenic activity of EMF. Indeed, this study is the only report in the peer-reviewed literature in which long-term EMF exposure in otherwise untreated animals has been associated with an increased incidence of malignancy.

Although the results of this study were clearly a novel finding, it must be noted that the experimental design and interpretation of this study has been strongly criticized for several reasons [32,33]. These reasons include:

- Group sizes are small and variable, and the number of observed malignant lesions is very small.
- Exposure assessment, environmental control, and control of possible confounding of study results by the heat and vibration generated by the EMF exposure system present serious questions as to the validity of results. Indeed noise and vibration associated with ventilation equipment needed to dissipate the heat generated by the EMF generation and exposure system as well as temperature control for exposed animals were very different than control animals that were housed in a different room, and were not exposed to these environmental factors.
- The onset of lymphoma in mice is age related, yet the study design included large discrepancies between the ages of EMF-exposed mice and the control groups to which they were compared.
- Peer review of photomicrographs indicates that some lesions identified by the investigators as neoplastic may, in fact, represent hyperplastic rather than malignant lesions.

2.4 Assessment of the Possible Cocarcinogenic or Tumor-Promoting Activity of Electromagnetic Field Exposure Using Multistage Rodent Models

It is generally accepted that exposure to power frequency EMF does not induce mutations or other genetic damage that is identifiable using standard genetic toxicology model systems (reviewed by McCann et al. [34]). Because EMF is apparently not genotoxic, it is logical to conclude that any oncogenic effects of exposure are mediated through cocarcinogenic or tumor promotion mechanisms. To address this possibility, a relatively large number of *in vivo* studies have been conducted to identify possible cocarcinogenic or tumor-promoting effects of EMF in various target tissues. The design of these studies involves simultaneous or sequential exposure to EMF in combination with another agent, and generally focuses on neoplastic development in a specific organ. It should be noted, however, that most multistage tumorigenesis studies of EMF have been conducted using animal models that were designed as research tools to study site-specific carcinogenesis. As a result, the utility of these assays as predictors of human oncogenicity is less completely understood than are the results of the chronic oncogenicity bioassay.

The design of most multistage tumorigenesis studies is based on the initiation–promotion paradigm that was originally developed in mouse skin. In the initiation–promotion model in mouse skin, mice are exposed to a single subthreshold dose of a genotoxic agent (e.g., benzo[*a*]pyrene or 7,12-dimethylbenz[*a*]anthracene [DMBA]), followed by repetitive administration of a nongenotoxic agent (e.g., 12-*O*-tetradecanoylphorbol-13-acetate). Over the past 50 y, this paradigm has been extended to include tumor induction models in more than a dozen organ sites, several of which have been used in EMF studies. Most multistage evaluations of the potential oncogenicity of EMF have been performed in animal models for neoplastic development in the breast, brain, skin, liver, and hematopoietic system.

In view of the relative large literature related to possible tumor promotion by EMF, analysis and discussion of all possibly relevant multistage oncogenicity evaluations is beyond the scope of this chapter. For this reason, discussion in this section is limited to a brief review of the studies that have been conducted in organ systems that have been suggested by epidemiologic data as likely targets for EMF oncogenicity (brain, breast, and hematopoietic system).

2.4.1 Design of Multistage Oncogenicity Studies

The design of multistage oncogenicity studies is site specific: the selection of animal species and strain, carcinogen exposure regimen, observation period, and end point analyses vary by animal model system. As a general rule, animals are exposed to one or a few doses of a genotoxic agent (the initiator), which is then followed by subchronic or chronic exposure to EMF (the putative promoter). Many, but not all, multistage studies that are designed to evaluate the activity of EMF as a promoter include a positive control group that is exposed to the same dose of initiating agent, followed by exposure to an agent with known promoting activity.

In consideration of their site specificity, multistage tumorigenesis studies are often conducted in a single sex, and employ group sizes that are smaller than those used in chronic rodent bioassays. The species and strain of animals used in these organ-specific tumorigenesis studies are selected on the basis of previous literature demonstrating appropriate sensitivity to neoplastic development in the organ site of interest. Under optimal conditions, the carcinogen-dosing regimen (initiating regimen) is selected to be a threshold dose that will induce a low incidence (<10%) of neoplasia in otherwise untreated animals; unfortunately, the value of a several multistage tumorigenicity evaluations of EMF is undermined by inappropriate selection of initiator doses. The period of observation and type of end point observations are also model specific; optimal study design in these areas is based on previous studies using the same model system.

2.4.2 Multistage Oncogenicity Studies in the Brain

The peer-reviewed literature contains two reports of experimental studies in which the activity of 60 Hz EMF as a promoter of brain tumor induction has been evaluated. These reports include an evaluation of the activity of EMF as a promoter of neurogenic tumors induced in F344 rats by transplacental exposure to *N*-ethyl-*N*-nitrosourea (ENU) [35], and a study to evaluate the activity of EMF as a promoter of brain tumors induced in C57BL/6 mice by ionizing radiation [30]. Neither study demonstrated any significant activity of EMF as a promoter of brain tumor induction.

2.4.2.1 Institut Armand Frappier (Canada): Brain Tumor Promotion Study in Rats

Mandeville et al. [35] have reported a study to evaluate the influence of EMF exposure on the induction of brain and spinal tumors by transplacental exposure to the nitrosamide, ENU. The transplacental ENU model in rats has been used widely in studies of neurogenic carcinogenesis, and is considered to be an appropriate model system for such evaluations.

In this study, groups of 50 pregnant female F344 rats received a single intraperitoneal dose of 5 mg ENU/kg body weight on day 18 of gestation. The F1 generation was exposed to EMF beginning on day 20 of gestation and continuing until study termination; flux densities (20 mG [2 μT], 200 mG [20 μT], 2 G [200 μT], and 20 G [2000 μT]) and environmental conditions in the laboratory were identical to those used in the chronic oncogenicity bioassay conducted at IAF.

EMF exposure had no effect on animal survival in the study, and had no influence on the incidence of neurogenic tumors. The results of this study do not support the hypothesis that EMF is a potent promoter of neural tumorigenesis.

2.4.2.2 University of California at Los Angeles (United States): Brain Tumor Promotion Study in Mice

A chronic murine bioassay was conducted at the UCLA and Veterans Administration Hospital to evaluate the potential interaction between EMF and ionizing radiation in the induction of hematopoietic neoplasia in mice [29]. As an adjunct to this study, the possible activity of EMF as a promoter of brain tumor induction was evaluated via histopathologic evaluation of brains from mice exposed to graded doses of ionizing radiation followed by exposure to either ambient magnetic fields (sham controls) or circularly polarized EMF (14.2 G [1.4 mT]) for up to 852 d [30].

When compared to matched sham controls, EMF exposure had no statistically significant effect on animal survival in any group. Total brain tumor incidence in the study was <0.5% (a total of seven brain tumors in 1707 irradiated mice), and no evidence of increased brain tumor incidence was observed in any group exposed to EMF. These data provide no evidence that exposure to circularly polarized EMF has tumor-promoting activity in the brain of C57BL/6 mice.

In spite of its large group sizes, one issue in the interpretation of the results of this study is its sensitivity and ability to identify any neurocarcinogenic effects of EMF exposure. The small number of brain cancers seen in all experimental groups could be interpreted as evidence that the mouse brain is relatively insensitive to neural tumor induction. Alternatively, however, the low incidence of brain cancers observed in otherwise unexposed animals could suggest that brain cancer incidence in study animals is at a response threshold and that exposure to any agent that stimulates cancer induction should increase brain cancer response. Regardless of the strength of the finding, the data from the present study provide no support for the hypothesis that EMF exposure is a significant risk factor for cancer induction in the brain.

2.4.3 Multistage Oncogenicity Studies in the Mammary Gland

The carcinogen-induced mammary adenocarcinoma in female rats is an organ-specific cancer model that is highly suitable for use as a test system in which to identify and quantitate EMF bioeffects. Epidemiologic evidence suggests that the breast may be a sensitive target tissue for the biological activity of EMF, and a well-defined and extensively studied multistage model system that is suitable for the identification of possible mammary gland carcinogens and tumor promoters has been established [36]. Numerous studies have been conducted to investigate the possible activity of EMF as a promoter of carcinogenesis in the rat mammary gland, and considerable controversy has developed concerning whether EMF exposure does indeed stimulate neoplastic development in this model system.

The general design of studies to evaluate the possible activity of EMF as a promoter of tumor induction in the rat mammary gland involves administration of one to four doses of a chemical carcinogen (DMBA or *N*-methyl-*N*-nitrosourea [MNU]) to young adult rats, followed by whole body exposure to exposure to either EMF or sham fields. Tumor development is monitored by palpation at regular intervals throughout the in-life period; at study termination, tumors are excised and evaluated histopathologically to confirm their malignant phenotype.

Several critical factors must be considered in evaluating the quality of studies reporting evaluations of the potential oncogenicity of EMF in the rat mammary gland.

- It is essential that study designs include histopathological evaluation of induced tumors. In the absence of such histologic confirmation, it is impossible to demonstrate that masses palpated during in-life observations are indeed malignant, and do not represent either benign tumors (adenomas or fibroadenomas) or nonmalignant lesions (e.g., mammary cysts).
- Dose selection for the chemical carcinogen is also a critical factor. To demonstrate promotional activity, the dose of carcinogen that is administered to test animals must induce a relatively low incidence of lesions, thereby providing a baseline cancer incidence against which a statistically significant increase in incidence can be achieved. If the cancer incidence induced by the carcinogen dose is too high, demonstration of promotional activity will be impossible.
- The mammary tumor response to the carcinogen must fall within accepted incidence and multiplicity ranges; these ranges have been established over more than 30 y through the work of many independent researchers. It is critical to note that this model has been used for several decades in studies of hormonal carcinogenesis and in efficacy evaluations of cancer chemopreventive agents. As such, dose–response parameters for mammary tumor induction are well established.

In instances, where reported dose–response relationships fall outside of the established dose–response parameters, it becomes impossible to exclude technical deficiencies in study conduct.

2.4.3.1 Oncology Research Center (Georgia): Mammary Tumor Promotion Studies in Rats

In 1991, Beniashvili et al. [37] published the first report of the enhancement of rat mammary carcinogenesis by EMF exposure and have subsequently published a confirming study [38]. In this work, exposure of rats to either static or 50 Hz magnetic fields alone did not induce mammary cancer. However, exposure of rats to static or 50 Hz EMF following administration of a chemical carcinogen (MNU) increased the incidence and reduced the latency of mammary tumors in comparison to controls exposed to MNU only. Tumor incidence was higher in animals exposed to MNU + 50 Hz EMF than to either MNU + static fields or MNU only. The authors interpret these findings to suggest that EMF is not oncogenic by itself, but may present a cancer hazard for both humans and animals.

The value of the data from these papers is compromised by the lack of information regarding the exposure system used and limited detail regarding experimental methods. However, these data were clearly suggestive and stimulated the conduct of a number of similar studies in other laboratories.

2.4.3.2 Hannover School of Veterinary Medicine (Germany): Mammary Tumor Promotion Studies in Rats

The largest number of studies investigating the influence of EMF on rat mammary carcinogenesis have been published by Loscher and colleagues [39–46] from the School of Veterinary Medicine at Hannover in the Federal Republic of Germany. In these studies, rats were exposed to DMBA followed by exposure to 50 Hz EMF or sham fields for periods ranging from 13 to 27 weeks.

The initial study from this group [45] reported an increase in mammary tumor incidence in rats exposed to 50 Hz EMF at a flux density of 1 G (100 μT). However, no histopathologic evaluations were performed in this study, and tumor size measurements were based on estimates performed via palpation. In a follow-up study conducted at lower doses that are more relevant to residential environments, Loscher et al. [39] reported that exposure to a 50 Hz EMF gradient of field strengths ranging from 3 to 10 mG (0.3 to 1.0 μT) reduced nocturnal melatonin levels, but had no effect on mammary carcinogenesis. In this study, EMF exposure for 13 weeks had no effect on mammary cancer incidence, mammary tumor size, or the incidence of preneoplastic lesions. The data from these studies were interpreted to suggest that exposure to EMF at high flux densities stimulates mammary tumor induction in rats, but that such effects do not occur as a result of exposure to EMF at low field strengths, such as those encountered in residential environments.

In a subsequent study, Loscher and Mevissen [40] reported a linear increase in mammary tumor incidence at 13 weeks in rats receiving DMBA and exposed to 50 Hz EMF at field strengths of 100 mG, 500 mG, and 1 G (10, 50, and 100 μT). Confirming these data, the same group of investigators reported that exposure to 500 mG (50 μT) EMF increased both the number and growth rate of mammary tumors [41] and the total incidence of mammary tumors [46]. By contrast, however, another confirming study [42] published the same year as the initial investigation of this series reported no increase in mammary tumor incidence, number, or size in rats exposed to 50 Hz EMF at 100 mG (10 μT). A fourth confirming study conducted for 26 weeks [43] found a much smaller increase from sham control in the incidence of mammary tumors in groups exposed to the 500 mG magnetic field; although the authors state statistical significance for the finding, the reported difference appears to be within the intrinsic variability of the DMBA mammary cancer system.

2.4.3.3 Battelle-Pacific Northwest Laboratories (United States): Mammary Tumor Promotion Studies in Rats

A series of studies supported by the NTP and conducted at Battelle-Pacific Northwest Laboratories (PNL) was designed to replicate and extend the mammary carcinogenesis studies reported by Loscher and colleagues. In these studies, groups of 100 female rats received sham EMF exposure or were exposed to either 50 Hz or 60 Hz EMF at field strengths of 1 G or 5 G (100 or 500 μT).

The initial study conducted was a 13 week bioassay that was designed to replicate the findings of Loscher and Mevissen [40] and Loscher et al. [45]. The results of this replication study demonstrated no differences from DMBA-treated sham controls in mammary cancer incidence, mammary cancer number, or mammary cancer size in DMBA-treated rats that were exposed to either 50 Hz or 60 Hz EMF [47]. Therefore, the replication study conducted at PNL failed to confirm the findings of the Hannover group.

The second study conducted at PNL used the same exposure metrics, and was designed to replicate the results of the 26 week mammary cancer bioassays reported by Loscher et al. [45]. Similar to the results of the 13 week study conducted at PNL, the results of the 26 week study demonstrated no significant increases in mammary cancer incidence in any group exposed to EMF. Furthermore, statistically significant decreases in the total number of mammary cancers was seen in groups exposed to 50 Hz EMF or 60 Hz EMF field strengths of 1 G [10].

Investigators at Battelle and in Hannover have collaborated in the attempt to identify reasons underlying the inability of Battelle investigators to reproduce the results reported by Loscher and colleagues [48]. A recent paper published by the Hannover group has attributed the inability of the Battelle group to reproduce their findings to differences in EMF responsiveness of different substrains of Sprague-Dawley rats [44].

2.4.3.4 National Institute for Working Life (Sweden): Mammary Tumor Promotion Studies in Rats

One study has been published by investigators from the National Institute for Working Life in Sweden in which the possible promoting activity of 50 Hz EMF was evaluated in the DMBA rat mammary carcinogenesis model system [49]. In this study, groups of female rats were exposed to DMBA followed by exposure to a transient-producing magnetic field at flux densities of 2.5 or 5 G (250 or 500 μT) for 25 weeks. No statistically significant differences from sham control were observed in mammary tumor incidence, mammary tumor number, total tumor weight, or tumor volume in either group exposed to EMF. The results of this study do not support the hypothesis that EMF exposure is an important risk factor for breast cancer.

2.4.4 Multistage Oncogenicity Studies in the Hematopoietic System

Two studies have been reported in which the activity of EMF as an inducer of lymphoma in standard (nontransgenic) animals has been evaluated. Several additional studies using genetically modified animals (transgenic and knockout strains) have also been published; the results of these studies are summarized in Section 2.5.

2.4.4.1 Zhejiang Medical University: Lymphoma Promotion Study in Mice

In 1997, Shen et al. [50] presented the results of a study designed to evaluate the possible activity of a 1 mT 50 Hz magnetic field in the promotion of lymphoma induction in newborn mice treated with DMBA. In this study, 320 newborn mice received a single injection of carcinogen within 24 h of birth; sham exposure (155 mice) or EMF exposure (165 mice) was initiated at 2 weeks of age, and was continued until study termination at 16 weeks. The incidence of lymphoma was histologically confirmed.

At the termination of the study, lymphoma incidence was 22% in sham controls, and 24% in the group receiving EMF exposure, demonstrating that EMF exposure had no promoting activity under the conditions of this study. However, the incidence of metastatic infiltration into the liver was higher in EMF-exposed mice than in controls; the significance of this finding, if any, is unclear.

2.4.4.2 University of California at Los Angeles (United States): Lymphoma Promotion Study in Mice

A chronic murine bioassay was conducted at the UCLA and Veterans Administration Hospital to evaluate the potential interaction between EMF and ionizing radiation in the induction of hematopoietic neoplasia in mice [29]. In this study, groups of mice were exposed to graded doses of ionizing radiation followed by exposure to either ambient magnetic fields (sham controls) or circularly polarized EMF (14.2 G [1.4 mT]) for up to 852 d. Details of the design of this study were provided in Section 2.3.

When compared to matched sham controls, EMF exposure had no statistically significant effect on animal survival in any group, and did not increase the incidence of lymphoma in any group. In fact, the only statistically significant difference from matched sham controls was a reduction in lymphoma risk in mice exposed to EMF after exposure to an ionizing radiation dose of 5.1 Gy. Extending the results of the complete carcinogenesis component of this study, the results of the lymphoma promotion cohorts of the study provide no evidence that exposure to circularly polarized EMF promotes the induction of lymphoma in C57BL/6 mice.

2.4.4.3 Overview of Multistage Oncogenicity Studies

Overall, the peer-reviewed literature provides little compelling evidence that exposure to EMF has cocarcinogenic or tumor-promoting activity in any organ site in experimental animals. Although the results of epidemiologic studies have identified the brain, breast, and hematopoietic system as possible targets of EMF action, no consistent pattern of EMF effects has been seen in any of these sites in experimental carcinogenesis models. Similarly, carcinogenesis studies conducted in animal model systems have not identified any other organ as a potential target of EMF oncogenicity.

Studies conducted in animal models for lymphoma and brain cancer provide no evidence of tumor-promoting or cocarcinogenic activity associated with exposure to EMF. The observed lack of activity in the promotion of hematopoietic neoplasia in multistage model systems is supported by similar results from studies conducted in transgenic and knockout models of lymphoma induction.

Several reports of tumor-promoting activity in the rat mammary gland have been presented in the peer-reviewed literature. However, these reports come from only two laboratories, and tumor-promoting activity is identified on the basis of different end points in different studies. More importantly, however, independent efforts to replicate these studies in other laboratories have failed. On this basis, the body of evidence supporting tumor-promoting activity for EMF in the rat mammary gland must be considered to be weak, at best.

Although not reviewed in detail in this chapter, similar negative results have been obtained in experimental studies conducted in animal models for neoplasia of the skin and liver. Although one report of skin tumor promotion by EMF has been presented [51], subsequent studies conducted in several laboratories have failed to identify either any tumor-promoting effects of EMF in mouse skin [52–55] or any effects of EMF exposure on biomarkers of tumor promotion in the skin [56]. Studies designed to determine the effects of EMF exposure on the induction of altered liver foci (a preneoplastic lesion in rat liver) also fail to provide clear evidence of tumor promotion by EMF [57,58].

2.5 Assessment of the Possible Oncogenic Activity of Electromagnetic Field Exposure Using Genetically Modified (Transgenic and Gene Knockout) Rodent Models

Recent insights into the molecular mechanisms of disease provide the scientific basis for the development and use of genetically modified animals to study disease processes, and the application of these models in hazard identification and toxicity evaluation. The use of genetically modified animals in safety assessments is an emerging technology that lacks the long historical precedent of the 2 y bioassay and other standardized approaches to toxicity testing. However, our understanding of the utility of these model systems in safety assessment is evolving rapidly, and their use in toxicology is expanding. In this regard, it should be noted that oncogenicity studies in one genetically modified model, the p53 knockout mouse, are routinely required by the FDA to support safety assessments of novel pharmaceuticals and natural products that induce chromosomal damage or other genetic effects in *in vitro* test systems.

The science supporting the use of genetically modified models in safety assessment stems from initial observations that the incidence of malignancy is increased and its latency is decreased in animal strains into whose germ line a tumor-associated gene (oncogene) has been inserted. Subsequent work demonstrated similar changes in patterns of neoplasia in animal strains from whose germ line a tumor suppressor gene has been deleted. On this basis, transgenic animals (into whose germ line a tumor-associated gene has been inserted) and knockout animals (from whose genome a tumor suppressor gene has been deleted) demonstrate a genetic predisposition to neoplasia; this predisposition may be exploited as an approach to the use of animal models whose sensitivity to oncogenesis is increased.

These models have several desirable attributes that support their use in the study of EMF bioeffects. First, oncogenicity studies in transgenic or knockout models involve evaluations of biological activity in animals in which exposure to a test agent is superimposed upon a genetic predisposition to disease. Such evaluations may identify oncogenic effects that occur in sensitive subpopulations, and rarely or not at all in the general population. As a result, genetically modified models may demonstrate sensitivity to EMF bioeffects that do not occur in standard animal strains. Secondly, transgenic or knockout animals provide experimental models that may be considered as analogous to cocarcinogenesis and initiation–promotion models; tumorigenesis in some transgenic and knockout model systems does not require exposure to a chemical carcinogen. Finally, the conduct of parallel assessments of the possible oncogenicity of EMF in animals with different genetic alterations can generate information regarding (a) the generality of any observed effects, and (b) possible molecular mechanisms underlying such effects.

Because genetically modified animals have a limited history of use in safety assessment, few data exist concerning their utility in the prediction of human oncogenicity. When considered alone, oncogenicity assessments in genetically modified animals may be of limited value in predicting human responses. However, when considered in combination with studies performed using standardized model systems, the increased sensitivity of transgenic and knockout models to disease processes may support the identification of low dose effects and effects in sensitive subpopulations. Given their rapidly expanding use in submissions to the FDA, studies in p53 knockout mice may become particularly valuable in assessing possible oncogenic hazards.

2.5.1 Design of Oncogenicity Studies in Genetically Modified Animals

The design of oncogenicity studies in transgenic and knockout animals generally parallels the designs used for oncogenicity bioassays in standard strains of rodents. However, these bioassays differ from the design of chronic oncogenicity bioassays in duration, group size, and experimental end points. Primary differences from the 2 y bioassay are as follows:

- The duration of oncogenicity studies in genetically modified animals varies by animal strain. The primary determinant of study duration is the kinetics of spontaneous oncogenesis in the animal model used in the bioassay. Most commonly, oncogenicity assessments are designed for completion using an in-life period of 6 months; however, the in-life period in certain models can be extended to include much longer periods of exposure. For example, the protocol developed by the NTP for oncogenicity bioassays in heterozygous p53 knockout mice involves 26 weeks of agent exposure; this protocol is now used widely in safety assessments of novel drugs and natural products, being submitted for FDA approval. By contrast, studies in PIM transgenic mice (as cited below) have ranged in duration from 6 to 18 months.
- Group size in these studies rarely exceeds 30 per sex per group versus 50 to 100 in standard oncogenicity bioassays. A group size of 25 per sex per group is routinely required by the FDA for studies in the p53 knockout mouse.
- In most cases, studies in genetically modified animals are designed to examine effects of a test agent on neoplastic development in a specific site, and histopathologic evaluations are limited to that site. In EMF studies, the primary focus has been on the possible influence of EMF exposure on the incidence of lymphoma. As such, histopathologic evaluation of tissues in these studies has been limited to lymphoid organs.

2.5.2 Results of Oncogenicity Studies in Genetically Modified Animals

2.5.2.1 IIT Research Institute (United States): Lymphoma Bioassays in PIM Transgenic and Heterozygous p53 Knockout Mice

As part of a larger program supported by the NTP and the United States National Institute of Environmental Health Sciences (NIEHS), scientists at IITRI conducted 6 month studies to determine the influence of EMF exposure on lymphoma induction in PIM transgenic mice and heterozygous p53 knockout mice [59]. Both genetically modified mouse strains were used as models for lymphoma; the PIM mouse receives a single dose of carcinogen (ENU) prior to EMF exposure, and is considered to be a high-incidence lymphoma model over a 6 month observation period. The heterozygous p53 knockout mouse receives no carcinogen exposure, and is a low incidence lymphoma model system over this same observation period.

2.5.2.1.1 *Results of IITRI (United States) Lymphoma Bioassay in PIM Transgenic Mice*

The PIM transgenic mouse carries the *pim-1* oncogene [60] and is a rapid model for the study of lymphoma. We have previously demonstrated that this model system is useful for studies of the modulation of lymphoma promotion by cancer chemopreventive agents [61]. It should be noted, however, that the *pim-1* oncogene appears to be specific for the induction of murine lymphoma and has no known homology to any human oncogenes. Furthermore, there are few data to correlate neoplastic responses in other organs in the PIM mouse model with responses to chemicals in 2-y bioassays for oncogenicity. As such, the PIM mouse must be considered to be at an early stage of development for use in the identification of agents that may stimulate the induction of leukemia and lymphoma.

EMF exposure metrics used in the PIM mouse oncogenicity bioassay [59] were identical to those used in the chronic animal bioassays of EMF that were conducted at IITRI. Groups of 30 mice per sex received a single intraperitoneal dose of 25 mg ENU and were then exposed for 18.5 h/d for 23 weeks to pure, linearly polarized 60 Hz magnetic fields. Experimental groups included animals receiving continuous exposure to field strengths of 20 mG (2 μT), 2 G (200 μT), or 10 G (1000 μT), or intermittent exposures (1 h on/1 h off) to 10 G (1000 μT). Parallel sham control groups were housed within an identical exposure array, but were exposed only to ambient magnetic fields. At the conclusion of the exposures, animals underwent a limited gross necropsy focused on the lymphoid system, and histopathologic evaluations were performed on the spleen, thymus, mandibular lymph node, liver, kidney, and lung.

When compared to sham controls, exposure to 60 Hz EMF did not increase the incidence of lymphoma in any experimental group. In fact, the only statistically significant difference from sham control was a significant *decrease* in the incidence of lymphoma in male mice exposed continuously to EMF (23% incidence versus 49% incidence in sham controls; $p = 0.041$ via Fisher's exact test). On this basis, the results of this study do not support the hypothesis that exposure to EMF is a significant risk factor for hematopoietic neoplasia.

2.5.2.1.2 *Results of IITRI (United States) Lymphoma Bioassay in p53 Knockout Mice*

As a result of deletion of one allele of the p53 tumor suppressor gene, heterozygous p53 knockout mice are highly sensitive to spontaneous neoplasia in several sites, as well as to the induction of neoplasia by carcinogens. By 18 months of age, approximately 50% of otherwise untreated heterozygous p53 knockout mice will develop malignant lesions, with particular predisposition to tumors of the bone and hematopoietic system [62]. Carcinogen exposure greatly accelerates neoplastic development in this model; in this context, p53 knockout mice were used as a model system for the study of EMF effects on lymphoma induction.

In this study, groups of 30 mice per sex were exposed for 18.5 h/d for 23 weeks to either pure, linearly polarized 60 Hz EMF at 10 G (1000 μT), or to sham fields. At the conclusion of the exposures, animals underwent a limited gross necropsy focused on the lymphoid system, and histopathologic evaluations were performed on the spleen, thymus, mandibular lymph node, liver, kidney, and lung.

Consistent with the results of the lymphoma bioassay in PIM mice, exposure to 60 Hz EMF at a flux density of 10 G had no effect on lymphoma induction in heterozygous p53 knockout mice. Lymphoma incidence in both sexes in the sham control group was 3% (1/30). By comparison, lymphoma incidence in male mice in the 10 G group was 0%

(0/30), while female mice exposed to 10 G had a 7% incidence of lymphoma (2/30). Although this study was of limited statistical power, it provides no evidence to support the hypothesis that EMF exposure is a risk factor for lymphoma induction.

2.5.2.2 Walter and Eliza Hall Institute of Medical Research (Australia): Lymphoma Bioassay in PIM Transgenic Mice

A very large lymphoma induction study using the PIM mouse model was conducted at the Walter and Eliza Hall Institute of Medical Research in Melbourne, Australia [63]. In this study, groups of approximately 100 female PIM mice were continuously exposed for 18 months to linearly polarized 50 Hz EMF at flux densities of 10 mG (1 μT), 1 G (100 μT), or 10 G (1000 μT), or intermittently exposed (15 min on/15 min off) to 10 G (1000 μT) for the same period. In contrast to the design of the PIM mouse study conducted at IITRI, EMF and sham-exposed mice in the present study received no carcinogen.

Although the study designs were very different, the conclusions drawn from the results of this study were quite similar to those of the PIM mouse study conducted at IITRI. When compared to sham controls, no statistically significant increase in the incidence or latency of lymphoma was seen in any group of mice exposed to EMF; in comparison to a lymphoma incidence of 29% in the sham control group, the incidence of lymphoma in groups exposed to EMF ranged from 26% to 35% ($p > 0.05$ for all comparisons). By contrast, mice exposed to a positive control material (ENU) demonstrated a lymphoma incidence of 60% at 9 months.

2.5.2.3 Overview of Oncogenicity Studies in Genetically Modified Animals

Although the total number of completed oncogenicity evaluations of EMF in genetically modified animals is small, both the goals and results of these studies are very consistent. The peer-reviewed literature contains reports of three studies in which the influence of EMF exposure on lymphoma induction has been assessed using genetically modified animals. The three studies were conducted using vastly different experimental models and designs (carcinogen-treated PIM mouse, noncarcinogen-treated PIM mouse, and noncarcinogen-treated p53 knockout mouse). Furthermore, the three models used demonstrated large differences in the incidence of lymphoma in sham control groups. However, the results of all three studies were very comparable; in all studies, exposure to EMF was not associated with a significant increase in any experimental group.

The relatively recent development of genetically modified animals precludes a long history of their use in safety assessment. As a result, data are insufficient to support a broadly based assessment of the predictive nature of these models for human oncogenesis. However, the results of oncogenicity evaluations of EMF exposure that has been conducted in standard animal strains and in genetically modified animal models are quite comparable. On this basis, the results of EMF lymphoma studies in the PIM transgenic and p53 knockout model systems can be considered to confirm and extend the results obtained in 2 y bioassays of EMF.

2.6 Proposed Mechanisms of Electromagnetic Field Action in Oncogenesis

A number of mechanisms have been proposed through which exposure to EMF may induce or stimulate neoplastic development; these possible mechanisms include a wide range of effects at the molecular, cellular, and organismal level. An exhaustive presentation and evaluation of the literature relevant to possible mechanisms of EMF action is beyond the scope of this chapter. However, a brief discussion of the more commonly proposed mechanisms for EMF action is relevant to an evaluation of exposure to EMF as a risk factor for cancer and other diseases. Detailed reviews of possible mechanisms of EMF action have been presented in the assessment of EMF health effects prepared by the U.S. NIEHS [5] and in the IARC evaluation of the potential carcinogenicity of EMF [6].

Review of the published scientific literature provides no clear support for any specific mechanism of EMF action in carcinogenesis. Furthermore, no reproducible body of scientific evidence exists to link EMF exposure to alterations in any biological end point that is mechanistically related to neoplastic development in any tissue.

Most hypothesized mechanisms of EMF action in carcinogenesis can be grouped into five types of effects:

- Genetic toxicity (DNA damage, clastogenesis, or mutagenesis)
- Alterations in gene expression
- Biochemical changes associated with regulation of cell proliferation
- Alterations in immune function
- Alterations in melatonin levels or action

These general classifications are necessarily arbitrary and were developed to simplify the purposes of this chapter. It is clear that several of these mechanisms overlap and that all possible mechanisms of EMF action are not included in these five groups. However, these groups encompass the vast majority of mechanisms proposed for EMF bioeffects, and include mechanisms that have received the most extensive study to identify possible linkages between EMF exposure and carcinogenesis.

2.6.1 Genetic Toxicology Studies of Electromagnetic Field Exposure

The scientific literature includes nearly 100 reports of studies that were conducted to determine if exposure to EMF induces clastogenesis, mutagenesis, or other types of DNA damage. Many of these studies were conducted using standardized experimental model systems (Ames test, micronucleus test) that are widely used in safety assessments of new drugs, industrial and agricultural chemicals, environmental contaminants, and other agents. As such, the predictive nature of several of the assay systems used to study the genetic toxicology of EMF is well known.

The authors of a comprehensive review of the genetic toxicology literature related to EMF concluded "the preponderance of evidence suggests that extremely low-frequency (ELF) electric or magnetic fields do not have genotoxic potential" [34]. Several investigators have presented study data that indicate positive mutagenic or clastogenic activity of EMF exposure; however, an independent scientific review of these studies suggests that none of these results satisfy basic quality criteria for data reproducibility, consistency, and completeness. By contrast, a number of well-designed, complete, reproducible, and apparently well-conducted studies have failed to demonstrate any genotoxic effects of EMF [34].

It is generally agreed by investigators in the field that ELF magnetic fields do not possess sufficient energy to damage DNA via direct interaction (reviewed in [4]). Although it is possible that power frequency EMF could damage DNA through an indirect mechanism (i.e., via generation of free radicals), the vast majority of well-designed genetic toxicology studies conducted in models with demonstrated predictiveness for DNA-damaging activity have failed to identify any significant mutagenic or clastogenic effects of EMF exposure. As such, it is the consensus among toxicologists that exposure to power frequency EMF presents little or no risk of genetic toxicity. On this basis, the induction of DNA damage by EMF exposure appears unlikely to provide a mechanism for cancer induction.

2.6.2 Alterations in Gene Expression

The possible influence of EMF exposure on the expression of cancer-related genes is one of the most controversial areas of research in bioelectromagnetics. Several investigators have reported increases in the expression of one or more protooncogenes (generally c-*myc* or c-*fos*) in cell cultures exposed to 60 Hz EMF [64–66]. However, replication studies conducted by numerous other investigators have not confirmed these findings [67–71], and other, more broadly based screening efforts to identify genes that are up- or down regulated by EMF exposure have generally yielded negative results [72,73]. On this basis, the robustness of the reports of altered gene expression, as well as the generality of those findings, must be questioned.

Given the huge universe of genetic targets in the mammalian cell, it is not possible to exclude the possibility that some condition of exposure to 60 Hz EMF may alter the expression of one or more genes. It is important to note, however, that a finding of enhanced expression of an oncogene or any other gene is not, by itself, sufficient evidence to suggest physiological or pathophysiological activity. Because many cellular pathways exhibit biological redundancy, alterations in one pathway may be compensated for by an antagonistic change in another pathway. For this reason, although findings of altered gene expression may be considered as generally suggestive of a biologically important effect of EMF exposure, demonstration of up- or downregulation of some effector function is necessary to establish the biological relevance of the observed alteration in gene expression. Lacking the demonstration of any physiological and pathophysiological effect, an isolated finding of altered gene expression cannot be construed to be biologically significant. Restated with emphasis on the process of carcinogenesis, a finding of altered expression of an oncogene or a tumor suppressor gene as a result of EMF exposure can by no means be equated with any activity of EMF as a carcinogen or tumor promoter.

2.6.3 Biochemical Changes Associated with Regulation of Cell Proliferation

Changes in the expression or activity of enzymes and protein products involved in cell proliferation, cell cycle regulation, and apoptosis would be expected if EMF exposure is significantly associated with either complete carcinogenesis or tumor promotion. For example, several reports of EMF-associated alterations in the activity of ornithine decarboxylase, the rate-limiting step in polyamine biosynthesis, have been presented in the literature [74,75]. However, other investigators have either failed to confirm these findings in replication studies [76] or have found no effects of EMF on ornithine decarboxylase activity in other *in vitro* or *in vivo* model systems [56,77,78]. A similar lack of consistency has been found in studies of biochemical end points, such as induction of heat shock proteins [79–82].

Clearly, a single finding of an alteration in a specific biochemical parameter is insufficient to support its alteration as a general mechanism for EMF activity. As discussed above with regard to possible alterations in gene expression, an isolated finding of changes in enzyme activity or other pathway in cells that are exposed to EMF can be considered to be no more than suggestive of a possible adverse outcome. Linkage of these biochemical or cell-based effects to the physiological or pathophysiological effect is essential to demonstrate their biological significance.

2.6.4 Alterations in Immune Function

Downregulation of the host immune response by EMF is a tempting hypothesis through which EMF exposure could be associated with risk of carcinogenesis; however, this hypothesis is supported by very little experimental evidence. In fact, several studies conducted in standard rodent models for immunotoxicology testing have failed to demonstrate any significant adverse effects of EMF exposure on either cell-mediated or humoral immune responses [24,25,83].

The only exception to these negative studies is an apparently reproducible suppression of NK cell function that was seen in B6C3F1 female mice in two studies [24,26]. Interestingly, this suppression was not observed in male B6C3F1 mice, or male and female F344 rats used in these same studies. Furthermore, the reduction in NK cell function does not appear to result in a functional deficit, since negative results were obtained in a 2 y carcinogenesis bioassay of EMF in the same animal species and sex (female B6C3F1 mouse) in which the NK cell functional deficits were observed [12]. On this basis, although the deficit in NK cell function was statistically significant, it appears to be without biological significance in terms of host resistance to neoplasia.

As such, although the hypothesis that EMF exposure may stimulate tumorigenesis through an indirect mechanism involving suppression of host immunity may be intellectually appealing, it is unsupported by any credible body of experimental evidence.

2.6.5 Alterations in Melatonin Levels or Action

The melatonin hypothesis is perhaps the most widely accepted potential mechanism for the biological effects of EMF exposure. A simplified version of this hypothesis states that EMF bioeffects are mediated through alterations in the function of the pineal gland; these changes in pineal function are then responsible for alterations in patterns of melatonin synthesis, kinetics, or action [84]. Ultimately, reductions in melatonin synthesis may have a number of potential adverse effects on the host, including reduced activity as an endogenous free radical scavenger, and influences on the hypothalamic–pituitary–gonadal axis [85].

A considerable body of evidence to support the melatonin hypothesis was generated in bioelectromagnetics studies performed during the 1980s and early 1990s (reviewed in [86,87]). In these studies, reductions in circulating melatonin levels and time shifts in the circadian rhythm of melatonin biosynthesis were reported in both experimental animals and humans (reviewed in [4]). More recently, however, several investigators who had previously reported significant effects of EMF exposure on pineal function have been unable to replicate their own data [88–91]. This failure of replication was often associated with the acquisition of new, more sophisticated EMF exposure and monitoring equipment. Furthermore, the majority of recent papers from other laboratories have failed to demonstrate any consistent pattern of effects of EMF exposure on melatonin synthesis or kinetics in humans or in rodent model systems.

One notable exception to this trend is a series of studies that have been conducted using the Siberian hamster as a model system; EMF effects on melatonin secretion in this species appear to be a generally reproducible finding [92,93]. However, this animal demonstrates a seasonal circadian rhythm that is considerably different from that seen in either humans or rodent species that are commonly used for safety assessments; as such, a relationship between this finding and possible human health effects is difficult to establish.

Concomitant with the weakening of the data set relating EMF exposure to alterations in pineal function, the melatonin hypothesis has been revised to suggest that EMF influences on melatonin may be affected through alteration in melatonin effects at the level of the target cell, rather than through effects on its synthesis or secretion. The database supporting this hypothesis is very limited and is considered insufficient to provide strong support for this potential mechanism of EMF action.

2.7 Conclusion

The results of acute, subchronic, and chronic toxicity, and oncogenicity studies conducted in experimental animal model systems provide little support for the hypothesis that exposure to power frequency magnetic fields is a significant risk factor for human disease. Although positive results have been reported in a small number of animal studies, essentially all of those results either (a) have not been replicated in subsequent studies conducted in other laboratories or (b) were generated using nontraditional experimental models whose value as predictors of human responses is largely or completely unknown.

Clearly, the safety of human exposure to EMF (or any other chemical or physical agent) cannot be conclusively determined on the basis of studies in animal models alone. Interspecies extrapolation of results, and requirements for extrapolation of high dose exposure data in animals to low dose exposures in humans both require the acceptance of mechanistic assumptions that may ultimately prove to be inaccurate. For this reason, the most effective hazard assessments are based on the integration of results from appropriate animal test systems with available data from studies of exposed human populations.

In the case of EMF hazard assessment, evaluation of data from specific toxicologic end points that have been examined both in experimental models and in epidemiologic studies demonstrates that the results of animal studies agree quite closely with data from studies in human populations. This relationship suggests that, where appropriate comparative data exist, the results of well-designed studies in appropriate animal model systems are largely predictive of human responses to EMF.

Ultimately, the key issue in EMF hazard assessment is whether animal data can be used to predict human risk for cancer and other disease processes for which the epidemiologic data are inconsistent or unavailable. On the basis of the close agreement between animal and human data in both general toxicology and organ site-specific toxicity studies, and the negative results of the overwhelming majority of oncogenicity evaluations of EMF in animal model systems, it is concluded that exposure to EMF is unlikely to be a significant risk factor for human oncogenesis.

References

1. Brodeur, P., *The Great Power-Line Cover-Up*, New York: Little, Brown, and Company, 1993.
2. Wertheimer, N. and Leeper, E., Electrical wiring configurations and childhood cancer, *Am. J. Epidemiol.*, 190, 273–284, 1979.
3. Savitz, D., Wachtel, H., Barnes, F.A., John, E.M., and Tvrdik, J.G., Case–control study of childhood cancer and exposure to 60-Hz magnetic fields, *Am. J. Epidemiol.*, 128, 21–38, 1998.

4. Portier, C.J., and Wolfe, M.S., Eds., *Assessment of Health Effects from Exposure to Power-Line Frequency Electric and Magnetic Fields*, NIEHS Working Group Report. NIH Publication No. 98-3981, Research Triangle Park, North Carolina, 1998, pp. 85–208.
5. International Agency for Research on Cancer, *IARC Monographs on the Evaluation of Carcinogenic Risks to Humans*, Non-ionizing radiation, Part 1: Static and extremely low-frequency (ELF) electric and magnetic fields, Vol. 80, IARC Press, Lyon, France, 2002.
6. Margonato, V., Veicsteinas, A., Conti, R., Nicolini, P., and Cerretelli, P., Biologic effects of prolonged exposure to ELF electromagnetic fields in rats: I. 50 Hz electric fields, *Bioelectromagnetics*, 14, 479–493, 1993.
7. Margonato, V., Nicolini, P., Conti, R., Zecca, L., Veicsteinas, A., and Cerretelli, P., Biologic effects of prolonged exposure to ELF electromagnetic fields in rats: II. 50 Hz magnetic fields, *Bioelectromagnetics*, 16, 343–355, 1995.
8. Boorman, G.A., Gauger, J.R., Johnson, T.R., Tomlinson, M.J., Findlay, J.C., Travlos, G.S., and McCormick, D.L., Eight-week toxicity study of 60 Hz magnetic fields in F344 rats and B6C3F1 mice, *Fundam. Appl. Toxicol.*, 35, 55–63, 1997.
9. Selmaoui, B., Bogdan, A., Auzeby, A., Lambrozo, J., and Touitou, Y., Acute exposure to 50 Hz magnetic field does not alter hematologic or immunologic functions in healthy young men: a circadian study, *Bioelectromagnetics*, 17, 364–372, 1996.
10. Boorman, G.A., Anderson, L.E., Morris, J.E., Sasser, L.B., Mann, P.C., Grumbein, S.L., Hailey, J.R., McNally, A., Sills, R.C., and Haseman, J.K., Effect of 26-week magnetic field exposures in a DMBA initiation–promotion mammary gland model in Sprague-Dawley rats, *Carcinogenesis*, 20, 899–904, 1999.
11. Boorman, G.A., McCormick, D.L., Findlay, J.C., Hailey, J.R., Gauger, J.R., Johnson, T.R., Kovatch, R.M., Sills, R.C., and Haseman, J.K., Chronic toxicity/oncogenicity evaluation of 60 Hz (power frequency) magnetic fields in F344/N rats, *Toxicol. Pathol.*, 27, 267–278, 1999.
12. McCormick, D.L., Boorman, G.A., Findlay, J.C., Hailey, J.R., Johnson, T.R., Gauger, J.R., Pletcher, J.M., Sills, R.C., and Haseman, J.K., Chronic toxicity/oncogenicity evaluation of 60 Hz (power frequency) magnetic fields in B6C3F1 mice, *Toxicol. Pathol.*, 27, 279–285, 1999.
13. Brent, R.L., Reproductive and teratologic effects of low-frequency electromagnetic fields: a review of *in vivo* and *in vitro* studies using animal models, *Teratology*, 59, 261–286, 1999.
14. Ryan, B.M., Mallett, E., Jr., Johnson, T.R., Gauger, J.R., and McCormick, D.L., Developmental toxicity study of 60 Hz (power frequency) magnetic fields in rats, *Teratology*, 54, 73–83, 1996.
15. Ryan, B.M., Polen, M., Gauger, J.R., Mallett, E., Jr., Kearns, M.B., Bryan, T.L., and McCormick, D.L., Evaluation of the developmental toxicity of 60 Hz magnetic fields and harmonic frequencies in Sprague-Dawley rats, *Radiat. Res.*, 153, 637–641, 2000.
16. Rommereim, D.N., Rommereim, R.L., Miller, D.L., Buschbom, R.L., and Anderson, L.E., Developmental toxicology evaluation of 60-Hz horizontal magnetic fields in rats, *Appl. Occup. Environ. Hyg.*, 11, 307–312, 1996.
17. Negishi, T., Imai, S., Itabashi, M., Nishimura, I., and Sasano, T., Studies of 50 Hz circularly polarized magnetic fields of up to 350 microT on reproduction and embryo–fetal development in rats: exposure during organogenesis or during preimplantation, *Bioelectromagnetics*, 23, 369–389, 2002.
18. Chung, M.K., Kim, J.C., Myung, S.H., and Lee, D.I., Developmental toxicity evaluation of ELF magnetic fields in Sprague-Dawley rats, *Bioelectromagnetics*, 24, 231–240, 2003.
19. Mevissen, M., Buntenkotter, S., and Loscher, W., Effects of static and time-varying (50-Hz) magnetic fields on reproduction and fetal development in rats, *Teratology*, 50, 229–237, 1994.
20. Rommereim, D.N., Rommereim, R.L., Miller, D.L., Buschbom, R.L., and Anderson, L.E., Developmental toxicology evaluation of 60-Hz horizontal magnetic fields in rats, *Appl. Occup. Environ. Hyg.*, 11, 307–312, 1996.
21. Ryan, B.M., Symanski, R.R., Pomeranz, L.E., Johnson, T.R., Gauger, J.R., and McCormick, D.L., Multigeneration reproductive toxicity assessment of 60-Hz magnetic fields using a continuous breeding protocol in rats, *Teratology*, 59, 156–162, 1999.
22. Chernoff, N., Rogers, J.M., and Kavet, R., A review of the literature on potential reproductive and developmental toxicity of electric and magnetic fields, *Toxicology*, 74, 91–126, 1992.

23. Hefeneider, S.H., McCoy, S.L., Hausman, F.A., Christensen, H.L., Takahashi, D., Perrin, N., Bracken, T.D., Shin, K.Y., and Hall, A.S., Long-term effects of 60-Hz electric vs. magnetic fields on IL-1 and IL-2 activity in sheep, *Bioelectromagnetics*, 22, 170–177, 2001.
24. House, R.V., Ratajczak, H.V., Gauger, J.R., Johnson, T.R., Thomas, P.T., and McCormick, D.L., Immune function and host defense in rodents exposed to 60-Hz magnetic fields, *Fundam. Appl. Toxicol.*, 34, 228–239, 1996.
25. Mandeville, R., Franco, E., Sidrac-Ghali, S., Paris-Nadon, L., Rocheleau, N., Mercier, G., Desy, M., and Gaboury, L., Evaluation of the potential carcinogenicity of 60 Hz linear sinusoidal continuous-wave magnetic fields in Fischer F344 rats, *FASEB J.*, 11, 1127–1136, 1997.
26. House, R.V. and McCormick, D.L., Modulation of natural killer cell function after exposure to 60 Hz magnetic fields: confirmation of the effect in mature B6C3F1 mice, *Radiat. Res.*, 153, 722–724, 2000.
27. Schwetz, B., Commentary: rodent carcinogenicity studies on magnetic fields, *Toxicol. Pathol.*, 27, 286, 1999.
28. Yasui, M., Kikuchi, T., Ogawa, M., Otaka, Y., Tsuchitani, M., and Iwata, H., Carcinogenicity test of 50 Hz sinusoidal magnetic fields in rats, *Bioelectromagnetics*, 18, 531–540, 1997.
29. Babbitt, J.T., Kharazi, A.I., Taylor, J.M., Bonds, C.B., Mirell, S.G., Frumkin, E., Zhuang, D., and Hahn, T.J., Hematopoietic neoplasia in C57BL/6 mice exposed to split-dose ionizing radiation and circularly polarized 60 Hz magnetic fields, *Carcinogenesis*, 21, 1379–1389, 2000.
30. Kharazi, A.I., Babbitt J.T., and Hahn, T.J., Primary brain tumor incidence in mice exposed to split-dose ionizing radiation and circularly polarized 60 Hz magnetic fields, *Cancer Lett.*, 147, 149–156, 1999.
31. Fam, W.Z. and Mikhail, E.L., Lymphoma induced in mice chronically exposed to very strong low-frequency electromagnetic field, *Cancer Lett.*, 105, 257–269, 1996.
32. Boorman, G.A., McCormick, D.L., Ward, J.M., Haseman, J.K., and Sills, R.C., Magnetic fields and mammary cancer in rodents: a critical review and evaluation of published literature, *Radiat. Res.*, 153, 617–626, 2000.
33. Boorman, G.A., Rafferty, C.N., Ward, J.M., and Sills, R.C., Leukemia and lymphoma incidence in rodents exposed to low-frequency magnetic fields, *Radiat. Res.*, 153, 627–636, 2000.
34. McCann, J., Dietrich, F., and Rafferty, C., The genotoxic potential of electric and magnetic fields: an update, *Mutat. Res.*, 411, 45–86, 1998.
35. Mandeville, R., Franco, E., Sidrac-Ghali, S., Paris-Nadon, L., Rocheleau, N., Mercier, G., Desy, M., Devaux, C., and Gaboury, L., Evaluation of the potential promoting effect of 60 Hz magnetic fields on *N*-ethyl-*N*-nitrosourea induced neurogenic tumors in female F344 rats, *Bioelectromagnetics*, 21, 84–93, 2000.
36. McCormick, D.L. and Moon, R.C., Vitamin A deficiency and cancer, in *Vitamin A Deficiency and Its Control*, Bauernfeind, J.C., Ed., Academic Press, New York, 1985, pp. 245–284.
37. Beniashvili, D.S., Bilanishvili, V.G., and Menabde, M.Z., Low-frequency electromagnetic radiation enhances the induction of rat mammary tumors by nitrosomethyl urea, *Cancer Lett.*, 61, 75–79, 1991.
38. Anisimov, V.N., Zhukova, O.V., Beniashvili, D.S., Menabde, M.Z., and Gupta, D., Effect of light/dark regimen and electromagnetic fields on mammary carcinogenesis in female rats, *Biofizika*, 41, 807–814, 1996.
39. Loscher, W., Wahnschaffe, U., Mevissen, M., Lerchl, A., and Stamm, A., Effects of weak alternating magnetic fields on nocturnal melatonin production and mammary carcinogenesis in rats, *Oncology*, 51, 288–295, 1994.
40. Loscher, W. and Mevissen, M., Linear relationship between flux density and tumor co-promoting effect of prolonged magnetic field exposure in a breast cancer model, *Cancer Lett.*, 96, 175–180, 1995.
41. Mevissen, M., Lerchl, A., and Loscher, W., Study on pineal function and DMBA-induced breast cancer formation in rats during exposure to a 100-mG, 50 Hz magnetic field, *J. Toxicol. Environ. Health*, 48, 169–185, 1996.
42. Mevissen, M., Lerchl, A., Szamel, M., and Loscher, W., Exposure of DMBA-treated female rats in a 50-Hz, 50 microTesla magnetic field: effects on mammary tumor growth, melatonin levels, and T lymphocyte activation, *Carcinogenesis*, 17, 903–910, 1996.

43. Thun-Battersby, S., Mevissen, M., and Loscher, W., Exposure of Sprague-Dawley rats to a 50-Hertz, 100-microTesla magnetic field for 27 weeks facilitates mammary tumorigenesis in the 7,12-dimethylbenz[*a*]anthracene model of breast cancer, *Cancer Res.*, 59, 3627–3633, 1999.
44. Fedrowitz, M., Kamino, K., and Loscher, W., Significant differences in the effects of magnetic field exposure on 7,12-dimethylbenz[*a*]anthracene-induced mammary carcinogenesis in two substrains of Sprague-Dawley rats, *Cancer Res.*, *64*, 243–251, 2004.
45. Loscher, W., Mevissen, M., Lehmacher, W., and Stamm, A., Tumor promotion in a breast cancer model by exposure to a weak alternating magnetic field, *Cancer Lett.*, 71, 75–81, 1993.
46. Mevissen, M., Haussler, M., Lerchl, A., and Loscher, W., Acceleration of mammary tumorigenesis by exposure of 7,12-dimethylbenz[*a*]anthracene-treated female rats in a 50-Hz, 100-microT magnetic field: replication study, *J. Toxicol. Environ. Health*, 53, 401–418, 1998.
47. Anderson, L.E., Boorman, G.A., Morris, J.E., Sassar, L.B., Mann, P.C., Grumbein, S.L., Hailey, J.R., McNally, A., Sills, R.C., and Haseman, J.K., Effect of 13-week magnetic field exposures on DMBA-initiated mammary gland carcinomas in female Sprague-Dawley rats, *Carcinogenesis*, 20, 1615–1620, 1999.
48. Anderson, L.E., Morris, J.E., Sasser, L.B., and Loscher, W., Effects of 50- or 60-hertz, 100 microT magnetic field exposure in the DMBA mammary cancer model in Sprague-Dawley rats: possible explanations for different results from two laboratories, *Environ. Health Perspect.*, 108, 797–802, 2000.
49. Ekstrom, T., Mild, K.H., and Holmberg, B., Mammary tumours in Sprague-Dawley rats after initiation with DMBA followed by exposure to 50 Hz electromagnetic fields in a promotional scheme, *Cancer Lett.*, 123, 107–111, 1998.
50. Shen, Y.H., Shao, B.J., Chiang, H., Fu, Y.D., and Yu, M., The effects of 50 Hz magnetic field exposure on dimethylbenz(alpha)anthracene induced thymic lymphoma/leukemia in mice, *Bioelectromagnetics*, 18, 360–364, 1997.
51. Stuchly, M.A., McLean, J.R., Burnett, R., Goddard, M., Lecuyer, D.W., and Mitchel, R.E., Modification of tumor promotion in the mouse skin by exposure to an alternating magnetic field, *Cancer Lett.*, 65, 1–7, 1992.
52. McLean, J.R., Thansandote, A., McNamee, J.P., Tryphonas, L., Lecuyer, D., and Gajda, G., A 60 Hz magnetic field does not affect the incidence of squamous cell carcinomas in SENCAR mice, *Bioelectromagnetics*, 24, 75–81, 2003.
53. Sasser, L.B., Anderson, L.E., Morris, J.E., Miller, D.L., Walborg, E.F., Jr., Kavet, R., Johnston D.A., and DiGiovanni, J., Lack of a co-promoting effect of a 60 Hz magnetic field on skin tumorigenesis in SENCAR mice, *Carcinogenesis*, 19, 1617–1621, 1999.
54. McLean, J.R., Thansandote, A., Lecuyer, D., and Goddard, M., The effect of 60-Hz magnetic fields on co-promotion of chemically induced skin tumors on SENCAR mice: a discussion of three studies, *Environ. Health Perspect.*, 105, 94–96, 1997.
55. Rannug, A., Ekström, T., Mild, K.H., Holmberg, B., Gimenez-Conti, I., and Slaga, T.J., A study on skin tumour formation in mice with 50 Hz magnetic field exposure, *Carcinogenesis*, 14, 573–578, 1993.
56. DiGiovanni, J., Johnston, D.A., Rupp, T., Sasser, L.B., Anderson, L.E., Morris, J.E., Miller, D.L., Kavet, R., and Walborg, E.F., Jr., Lack of effect of a 60 Hz magnetic field on biomarkers of tumor promotion in the skin of SENCAR mice, *Carcinogenesis*, 20, 685–689, 1999.
57. Rannug, A., Holmberg, B., and Mild, K.H., A rat liver foci promotion study with 50-Hz magnetic fields, *Environ. Res.*, 62, 223–229, 1993.
58. Rannug, A., Holmberg, B., Ekström, T., and Mild, K.H., Rat liver foci study on coexposure with 50 Hz magnetic fields and known carcinogens, *Bioelectromagnetics*, 14, 17–27, 1993.
59. McCormick, D.L., Ryan, B.M., Findlay, J.C., Gauger, J.R., Johnson, T.R., Morrissey, R.L., and Boorman, G.A., Exposure to 60 Hz magnetic fields and risk of lymphoma in PIM transgenic and TSG p53 (p53 knockout) mice, *Carcinogenesis*, 19, 1649–1653, 1998.
60. Breuer, M.L., Slebos, R., Verbeek, S., van Lohuizen, M., Wientjens, E., and Berns, A., Very high frequency of lymphoma induction by a chemical carcinogen in *pim-1* transgenic mice, *Nature*, 340, 61–63, 1989.
61. McCormick, D.L., Johnson, W.D., Rao, K.V.N., Bowman-Gram, T.A., Steele, V.E., Lubet, R.A., and Kelloff, G.J., Comparative activity of *N*-(4-hydroxyphenyl)- all-*trans*-retinamide and

α-difluoro-methyl-ornithine as inhibitors of lymphoma induction in PIM transgenic mice, *Carcinogenesis*, 17, 2513–2517, 1996.
62. Donehower, L.A., Harvey, M., Slagle, B.L., McArthur, M.J., Montgomery, C.A., Jr., Butel, J.S., and Bradley, A., Mice deficient for p53 are developmentally normal but susceptible to spontaneous tumours, *Nature*, 356, 215–221, 1992.
63. Harris, A.W., Basten, A., Gebski, V., Noonan, D., Finnie, J., Bath, M.L., Bangay, M.J., and Repacholi, M.H., A test of lymphoma induction by long-term exposure of E mu-Pim1 transgenic mice to 50 Hz magnetic fields, *Radiat. Res.*, 149, 300–307, 1998.
64. Wei, L.X., Goodman, R., and Henderson, A., Changes in levels of c-*myc* and histone H2B following exposure of cells to low-frequency sinusoidal electromagnetic fields: evidence for a window effect, *Bioelectromagnetics*, 11, 269–272, 1990.
65. Phillips, J.L., Effects of electromagnetic field exposure on gene transcription, *J. Cell Biochem.*, 51, 381–386, 1993.
66. Karabakhtsian, R., Broude, N., Shalts, N., Kochlatyi, S., Goodman, R., and Henderson, A.S., Calcium is necessary in the cell response to EM fields, *FEBS Lett.*, 349, 1–6, 1994.
67. Saffer, J.D. and Thurston, S.J., Short exposures to 60 Hz magnetic fields do not alter MYC expression in HL60 or Daudi cells, *Radiat. Res.*, 144, 18–25, 1995.
68. Lacy-Hulbert, A., Wilkins, R.C., Hesketh, T.R., and Metcalfe, J.C., No effect of 60 Hz electromagnetic fields on MYC or beta-actin expression in human leukemic cells, *Radiat. Res.*, 144, 9–17, 1995.
69. Owen, R.D., MYC mRNA abundance is unchanged in subcultures of HL60 cells exposed to power-line frequency magnetic fields, *Radiat. Res.*, 150, 23–30, 1998.
70. Jahreis, G.P., Johnson, P.G., Zhao, Y.L., and Hui, S.W., Absence of 60-Hz, 0.1-mT magnetic field-induced changes in oncogene transcription rates or levels in CEM-CM3 cells, *Biochim. Biophys. Acta*, 1443, 334–342, 1998.
71. Yomori, H., Yasunaga, K., Takahashi, C., Tanaka, A., Takashima, S., and Sekijima, M., Elliptically polarized magnetic fields do not alter immediate early response genes expression levels in human glioblastoma cells, *Bioelectromagnetics*, 23, 89–96, 2002.
72. Loberg, L.I., Gauger, J.R., Buthod, J.L., Engdahl, W.R., and McCormick, D.L., Gene expression in human breast epithelial cells exposed to 60 Hz magnetic fields, *Carcinogenesis*, 20, 1633–1636, 1999.
73. Balcer-Kubiczek, E.K., Harrison, G.H., Davis, C.C., Haas, M.L., and Koffman, B.H., Expression analysis of human HL60 cells exposed to 60 Hz square- or sine-wave magnetic fields, *Radiat. Res.*, 153, 670–678, 2000.
74. Byus, C.V., Pieper, S.E., and Adey, W.R., The effects of low-energy 60-Hz environmental electromagnetic fields upon the growth-related enzyme ornithine decarboxylase, *Carcinogenesis*, 8, 1385–1389, 1987.
75. Mullins, J.M., Penafiel, L.M., Juutilainen, J., and Litovitz, T.A., Dose–response of electromagnetic field-enhanced ornithine decarboxylase activity, *Bioelectrochem. Bioenerg.*, 48, 193–199, 1999.
76. Cress, L.W., Owen, R.D., and Desta, A.B., Ornithine decarboxylase activity in L929 cells following exposure to 60 Hz magnetic fields, *Carcinogenesis*, 20, 1025–1030, 1999.
77. Kumlin, T., Alhonen, L., Jänne, J., Lang, S., Kosma, V.M., and Juutilainen, J., Epidermal ornithine decarboxylase and polyamines in mice exposed to 50 Hz magnetic fields and UV radiation, *Bioelectromagnetics*, 19, 388–391, 1998.
78. McDonald, L.J., Loberg, L.I., Savage, R.E., Jr., Zhu, H., Lotz, W.G., Mandeville, R., Owen, R.D., Cress, L.W., Desta, A.B., Gauger, J.R., and McCormick, D.L., Ornithine decarboxylase activity in tissues from rats exposed to 60 Hz magnetic fields including harmonic and transient field characteristics, *Toxicol. Mech. Meth.*, 13, 31–38, 2003.
79. Lin, H., Head, M., Blank, M., Han, L., Jin, M., and Goodman, R., Myc-mediated transactivation of HSP70 expression following exposure to magnetic fields, *J. Cell Biochem.*, 69, 181–188, 1998.
80. Malagoli, D., Lusvardi, M., Gobba, F., and Ottaviani, E., 50 Hz magnetic fields activate mussel immunocyte p38 MAP kinase and induce HSP70 and 90, *Comp. Biochem. Physiol. C Toxicol. Pharmacol.*, 137, 75–79, 2004.
81. Kang, K.I., Bouhouche, I., Fortin, D., Baulieu, E.E., and Catelli, M.G., Luciferase activity and synthesis of Hsp70 and Hsp90 are insensitive to 50 Hz electromagnetic fields, *Life Sci.*, 63, 489–497, 1998.

82. Morehouse, C.A. and Owen, R.D., Exposure to low-frequency electromagnetic fields does not alter HSP70 expression or HSF-HSE binding in HL60 cells, *Radiat. Res.*, 153, 658–662, 2000.
83. Thun-Battersby, S., Westermann, J., and Löscher, W., Lymphocyte subset analyses in blood, spleen and lymph nodes of female Sprague-Dawley rats after short or prolonged exposure to a 50 Hz 100-microT magnetic field, *Radiat. Res.*, 152, 436–443, 1999.
84. Stevens, R.G. and Davis, S., The melatonin hypothesis: electric power and breast cancer, *Environ. Health Perspect.*, 104 (Suppl. 1), 135–140, 1996.
85. Reiter, R.J., Tan, D.X., Manchester, L.C., Lopez-Burillo, S., Sainz, R.M., and Mayo, J.C., Melatonin: detoxification of oxygen and nitrogen-based toxic reactants, *Adv. Exp. Med. Biol.*, 527, 539–548, 2003.
86. Reiter, R.J., Alterations of the circadian melatonin rhythm by the electromagnetic spectrum: a study in environmental toxicology, *Regul. Toxicol. Pharmacol.*, 15, 226–244, 1992.
87. Brainard, G.C., Kavet, R., and Kheifets, L.I., The relationship between electromagnetic field and light exposures to melatonin and breast cancer risk: a review of the relevant literature, *J. Pineal Res.*, 26, 65–100, 1999.
88. Graham, C., Cook, M.R., Riffle, D.W., Gerkovich, M.M., and Cohen, H.D., Nocturnal melatonin levels in human volunteers exposed to intermittent 60 Hz magnetic fields, *Bioelectromagnetics*, 17, 263–273, 1996.
89. Graham, C., Cook, M.R., Sastre, A., Riffle, D.W., and Gerkovich, M.M., Multi-night exposure to 60 Hz magnetic fields: effects on melatonin and its enzymatic metabolite, *J. Pineal Res.*, 28, 1–8, 2000.
90. Graham, C., Cook, M.R., Gerkovich, M.M., and Sastre, A., Examination of the melatonin hypothesis in women exposed at night to EMF or bright light, *Environ. Health Perspect.*, 109, 501–507, 2001.
91. Yellon, S.M., Acute 60 Hz magnetic field exposure effects on the melatonin rhythm in the pineal gland and circulation of the adult Djungarian hamster, *J. Pineal Res.*, 16, 136–144, 1994.
92. Löscher, W., Mevissen, M., and Lerchl, A., Exposure of female rats to a 100-microT 50 Hz magnetic field does not induce consistent changes in nocturnal levels of melatonin, *Radiat. Res.*, 150, 557–567, 1998.
93. Brendel, H., Niehaus, M., and Lerchl, A., Direct suppressive effects of weak magnetic fields (50 Hz and 16 2/3 Hz) on melatonin synthesis in the pineal gland of Djungarian hamsters (*Phodopus sungorus*), *J. Pineal Res.*, 29, 228–233, 2000.

3

Interaction of Nonmodulated and Pulse-Modulated Radio Frequency Fields with Living Matter: Experimental Results

Sol M. Michaelson,* Edward C. Elson, and Larry E. Anderson†

CONTENTS

*Deceased.
†Author of revisions for the third edition.

3.1 Introduction

Over the past several decades, research on electromagnetic fields (EMF) has been motivated primarily by public health considerations. Through the 1980s and early 1990s, investigations of extremely low-frequency (ELF) were preeminent in EMF research due to increased societal concerns in the area of power frequency. This work focused on biological effects of power transmission lines and appliances. Since the mid-1990s, however, as human-made sources of radio frequency (RF) energy continued to increase in the environment, so also did environmental and health concerns. Due to elevated levels of RF exposure, specifically the exponential increase in exposure from cellular telephony, the output from RF effects research has increased significantly since the publication of the last edition of the handbook, as judged from a survey of leading research journals from 1994 to 2004.

Other factors have also influenced the enthusiasm with which new investigators have ventured into EMF research and have affected the motivation of research administrators who influence public policy on resource allocation. Concerns over nonreproducibility of results and nonrobustness of effects have produced a voluble exchange of views on direction and magnitude of support for EMF research (Pickard and Foster 1987). Such problems are brought into focus by the fact that the database of this research typically enters directly into the public health policy forum and is used to establish permissible exposure limits, directly affecting diverse economic interests (IEEE 1991). EMF research is therefore subjected to a second, critical review, following scholarly peer review but in a more pragmatic context. The veracity of the work is again analyzed and, in addition, its relevance for formulating safe limits for human exposure is scrutinized. This exercise requires extrapolation from conditions of experiments, often with animals, to human exposure. Inevitably the process requires the imposition of interpretations and judgments as to whether an effect might or might not constitute a hazard to humans and the database is seldom clear on that point. Utilization of the database in standards setting has been described by Petersen (1991). The role of culture and philosophy in the divergence of Eastern European and Western safety standards has been the subject of commentary by Sliney et al. (1985).

The fundamental interaction of RF radiation (3 kHz to 300 GHz (IEEE 1991)) with biological systems is set forth in detail in other chapters of this book. At sufficiently high RF intensities bulk thermal energy is imparted to the system, causing an increase in random kinetic energy and temperature. The thermal gradients differ from those of ambient heating, a fact that is marshaled occasionally to explain differences in effects between RF and ambient heating. Above certain intensities RF energy can quickly produce morbidity, and, after thermoregulatory mechanisms are overwhelmed, mortality. In recent years, work has focused on possible effects at nonthermal levels or under conditions in which physiologic temperature can be maintained in the presence of normally thermalizing specific absorption rates (SAR).

Studies of human perception indicate that the greatest cutaneous sensitivity to microwave (MW) heating is at frequencies with wavelengths comparable to or smaller than the thickness of skin, the millimeter wave range (EPA 1984). In this range, most of the energy is absorbed in the superficial dermis containing thermal sensors. At lower frequencies, 1 to 10 GHz, with wavelengths equal to or longer than the human body, much of the energy is absorbed below the superficial dermis. In this range, the threshold temperature for cellular injury about 42°C is below the threshold for pain (about 45°C). Consequently cutaneous perception of RF energy may not be a reliable response, which protects against potentially harmful levels of RF radiation at the lower microwave frequencies (Hardy, 1978).

3.1.1 Thermal versus Athermal or Nonthermal Effects

With exposure to an energy source such as EMF, there will always be some temperature change because of the finite heat capacity of any material system. From an experimental point of view, however, one can specify an effect as a thermal or nonthermal if during exposure of a system, using a nonperturbing temperature probe, the real-time temperature excursion is below an arbitrarily finite ΔT, deemed nonthermal. A typical excursion could be $\Delta T = 1°C$, but, whatever ΔT is chosen as the nonthermal limit, it represents a judgment by an investigator carrying a set of presumptions into the experiment which he may or may not have made explicit. An effect can also be identified as nonthermal, even in the presence of a temperature excursion exceeding an arbitrary threshold if an RF-induced effect is found and the same effect is not elicited with ambient heating producing an identical ΔT.

Theoretical definitions are also arbitrary. The introduction of RF energy into a system is inherently diffuse on a large scale compared with macromolecules or whole cells. In thermodynamic terms, one can posit a target structure, possessing a mode (e.g., vibration, rotational degree of freedom) in equilibrium with a surrounding heat bath (tissue). RF energy couples into the structure and heat bath simultaneously but unequally; perturbed from equilibrium, they proceed to interact with each other to achieve equilibrium. In this construct, a rigorously nonthermal interaction would be one resulting from direct electromagnetic coupling with the target structure independent of thermal interactions with the heat bath. In any experimental setting such a distinction would be difficult to make.

The displacive phase transition in a macromolecule produced by pumping RF energy into low-frequency vibrational modes of DNA, leading to unwinding and strand separation, was postulated by Girirajan et al. (1989). This process, if indeed it happens, could be considered nonthermal. One also finds in the literature the term "field" effect, suggesting again the concept of direct transduction of RF energy into a specific kinetic or metabolic change mediating the effect unrelated to a change in random kinetic energy.

A more empirical way of approaching the thermal versus nonthermal argument would be to identify an experimental effect that occurs or is observed at low RF energy and look for the presence or absence of the effect following the insertion of energy by other means such as conductive heating from a water bath. Changing the ambient temperature of a biological system, by whatever means, produces thermal gradients, which are different from those produced by RF energy. If such differences can be assumed not significant (an analysis is required for each experiment), then one may look for the presence or absence of a biological effect under each exposure condition and identify the effect as nonthermal if it occurs following RF exposure but not after ambient heating to the equivalent small ΔT.

In evaluating the claims for thermal versus nonthermal effects, one must remember how elusive the distinction can be. Ease of application varies, depending on the experimental problem. Early claims that RF energy could produce a change in the intrinsic permeability of the blood–brain barrier (BBB) were later challenged, both as to the reality of the altered permeability and also as to whether generalized heating was the possible effector. "Window" effects with regard to calcium exchanges between compartments in the brain are viewed as inherently nonthermal effects by many investigators, but the validity of such effects has been challenged.

3.1.2 Vagaries and Pitfalls in the Literature

A critical analysis of the literature is a daunting task. A majority of interesting reports have not been replicated and often efforts to replicate studies have shown antithetical results. Discussions then arise over whether an attempt to reproduce a result was the same or, in some way, a significantly different experiment. The literature reports presented in this chapter will identify still unresolved issues and, where possible, describe a prevailing concensus. In addition, rather than presenting all publications available for specific subjects, this chapter will present a broad representative sample of the literature through 2004.

Table 3.1 and Table 3.2 identify factors which must be addressed in performing research in the field. In fact, since some of the parameters are not addressed in research reports appearing in journals known for meticulous peer review, perhaps because of practical limitations of space, the reader does not know if they were considered. It is even harder to establish whether a particular laboratory is imbued with a culture of rigorous science. Does it have a systematic approach to measurement assurance and quality control? The literature alone does not answer such questions. Because of the large number of potential sources of error and sources of variation from one laboratory to another, it is not surprising that the literature contains contradictions and produces frustration, especially in the risk assessment community. Conclusions that can be drawn therefore depend to a significant degree on the weight of the evidence instead of specific individual reports.

3.1.3 Physical and Physiologic Scaling

The cost–benefit analyses used to justify exposure of human beings to drugs of unknown hazard but potential benefit do not in general allow systematic RF experimentation on human beings for RF effects. A noteworthy exception has existed in medical oncology and

TABLE 3.1

Factors That Affect Microwave and RF Absorption

Physical Parameters	Biological Parameters	Artifacts	Environmental Factors
Frequency	Tissue dielectric properties	Ground or conductor plane	Temperature
Polarization	Size; geometry	Container (material, size)	Humidity
Modulation (AM, FM, Pulse, CW)	Animal orientation relative to polarization	Metal implants	
Power density (peak and average)	Spatial relations among animals	Shielding implants	
Field pattern (near or far)		Metal or nonmetallic objects in the field	
Field uniformity			
Type of transmitting and radiating equipment			
Chamber material			
Chamber dimensions			

more recently in the treatment of musculoskeletal, degenerative disorders where treatment efficacy has been established. Research in these areas focuses on maximizing benefit while minimizing adverse consequences.

Although exposure of humans is of primary importance and concern, the bulk of EMF research has been carried out using animal models. This requires an understanding and appreciation of biophysical principles and "comparative medicine." Such studies require interspecies scaling (Guy et al. 1976; Gandhi et al. 1977; Massoudi et al. 1977; Gandhi 1980; Durney et al. 1986), the selection of biomedical parameters that reflect basic physiological functions, and differentiation between adaptational or compensatory changes and pathological effects. In comparing results of experiments performed in the same or different laboratories, standardization of conditions is important and, unfortunately, all too often not attained.

TABLE 3.2

Factors That Influence Biological Responses to the Same SAR

Subject Variables	Concomitant Variables	Environmental Variables	Experimental Variables
Species	Genetic predisposition	Temperature	Acclimation procedures
Sex	Baseline of the response	Humidity	Duration of exposure
Age	Functional and metabolic disorders	Air flow	Number and schedule of exposures
Weight		Lighting	Mode of exposure (partial or whole body)
Sensitivity		Noise	Time between exposure and sampling
Number of subjects		Odor	Time of day of exposure
Interventions (anesthetics, drugs, electrodes, lesions)			Restraint devices
Animal husbandry			Investigator—animal interaction

Even by using approaches where absorbed energy patterns in a test animal are set to approximate as closely as possible the patterns that may exist in man under certain exposure conditions, the intrinsic physical and physiological dissimilarities between species further confound the problem of extrapolating from animals to man. In addition to the obvious external geometric differences, the differences in internal vascular anatomy and mechanisms of heat dissipation in fur-bearing animals compared to man must be taken into consideration.

Because of the use of animals as a surrogate for humans in hazard analysis, one must create a set of experimental conditions, which are as relevant as possible for the purpose of the study. Many factors, such as methods of animal care, the role of seasonal and circadian rhythms, temperature and humidity, etc., as well as psychosocial interactions, must be considered in the experimental design as well as in the analysis of results. Additionally, one should not extrapolate results obtained in small laboratory animals to larger animals or man without consideration of size and energy distributions, as well as metabolic and physiological differences.

The possible difference between a biological response (effect) and a deleterious change in function is an important question in hazard assessment. Remarkable changes following exposure of cells or tissues *in vitro* do not necessarily have any significance *in vivo.* Most exposures are necessarily short term and do not provide information on possible long-term consequences. Even short-term exposures can produce interpretive problems. For example, could the human central nervous system (CNS) tolerate specific absorptions up to 30 kJ/kg delivered in periods from 360 ms to 1 μs with no long-term sequelae, as do rats (Guy and Chou 1982). Although rats are stunned in such an exposure, they appear to make a complete recovery within minutes. No responsible investigator would be willing to assert that a human would make a similar recovery.

Johnson (1975) has described factors that affect absorption of electromagnetic energy in animals. This absorption is dependent on the size and geometry of the animal relative to wavelength and polarization. The wavelength-to-animal size relationship (λ-to-a, where a is the longest axis dimension of the body and the electric field vector is parallel to the longest axis) is a critical factor in the relative absorption cross section, the ratio of the absorbed energy per second to the power incident on the geometrical cross-sectional area of the animal (Anne et al. 1962). This produces the immediate result that, at a given frequency and power density, the SAR is vastly different in animals depending on size. To produce an identical whole body average SAR, one must scale from one frequency to another. It is of practical importance to realize that experiments on biological effects at 2.45 GHz on small animals, such as mice and rats, do not scale to humans at 2.45 GHz, but approximately to effects on humans at 100 MHz. Much information on absorption of RF energy has been collected in the *Air Force RF Handbook* (Durney et al. 1986). Subsequently, considerable research has been done on the variation of localized absorption in animals and man, through both experimental measurements (see Adair 1995) and modeling (Gandhi 1990; see also Chapter 10 of *BBA*).

The process of scaling in RF bioeffects research is not restricted to extrapolation of the physical absorption of energy, but includes the problems of quantitatively comparing the relative efficiency of various biological processes in different animals to each other and to those processes in humans. The same biological processes in different animals are not necessarily affected by the same SAR. It is also possible that the absorbed energy in a particular area or tissue may not be the most useful indicator of disturbances of some biological functions, if those functions are perturbed by a systemic stimulus instead of a localized stimulus. The need for proper dosimetry in experimental procedures and the importance of using appropriate scaling factors for extrapolation of data obtained with small laboratory animals to man are thus clearly indicated.

3.1.4 Specific Absorption Rate

RF energy absorption is converted to thermal energy. This thermal energy is then rapidly redistributed by conduction, convection (blood flow) and to a lesser extent, radiation from the biological target. The tissue heat capacity and heat transfer processes influence the dose and dose distribution within the body. Present modeling techniques use the internal geometry and dielectric properties of different tissues to develop a profile of energy absorption and current flow as a function of position as described in Chapter 10 of *BBA*. These calculations produce excellent agreement with phantoms (anatomically realistic models of the body with similar dielectric properties), but are not yet capable of simulating the complex redistribution of absorbed energy in living systems. Of course, they have only limited predictive value for actual physiologic or pathologic response. There is no substitute for animal experimentation.

There is a consensus that an important dosimetric measure of RF exposure is the SAR. This is the unit-mass, time-average rate of RF energy absorption specified in SI units of watts per kilogram (W/kg) (NCRP 1981). The amount of energy absorbed by a given mass of material, which is termed specific absorption (SA) is given in joule per kilogram (J/kg) and is the integral of the instantaneous SAR over the duration of exposure in seconds. Thus the SAR is the rate at which RF electromagnetic energy is imparted to unit mass of a biological body. The SAR is applicable to any tissue or organ of interest, or is expressed as a whole body average. As the science of dosimetry has evolved, localized SARs are more frequently reported, i.e., specified for a particular location or tissue, such as the eye.

Whole body SAR is maximal when the long axis of a body is parallel to the E field vector and is four tenths the wavelength of the incident field. At 2.45 GHz ($\lambda = 12.5$ cm), for example, a standard human (long axis is 175 cm) will absorb about half of the incident energy. If the human whole body SAR is divided by the basal metabolic rate (BMR) for humans, a ratio is obtained that provides a measure of the thermal load incurred due to a known incident power density (Stuchly 1978). Table 3.3 illustrates the variation of this ratio with frequency at two incident power densities. In the region of human whole body

TABLE 3.3

Ratio of SAR to Basal Metabolic Rate for an Average Man Exposed to Far Field Incident Power Densities of 10 and 50 W/m^2

	Average SAR/BMR (%)	
Frequency (GHz)	**10 W/m^2**	**50 W/m^2**
0.01	0.13	0.65
0.02	0.60	3.00
0.05	5.80	29.00
0.06	10.00	50.00
0.08	16.00	80.00
0.10	12.00	60.00
0.20	5.20	26.00
0.50	3.70	18.00
1.00	2.90	14.50
2.00	2.50	12.50
5.00	2.50	12.50
10.00	2.50	12.50
20.00	2.50	12.50

Source: From Stuchly, M.Q., *Health Aspects of Radio Frequency and Microwave Radiation Exposure*, Part 2, Department of National Health and Welfare, Ottawa, 1978.

resonance (60 to 80 MHz), this ratio reaches a maximum value (about 16 for an incident far field power density of 10 W/m^2). The ratio drops off rapidly on either side of this peak.

3.2 Response to Local Radio Frequency Exposure

Considerable attention has been paid in animal and human experiments to the temperature rise in various organs and tissues as functions of incident energy wavelength, thickness of the subcutaneous fatty layer, blood circulation rate, and other factors. The increase in temperature of tissues during local exposure is linear for short periods (1 to 3 min) and proportional to the magnitude of the MW energy absorbed. With exposures in excess of 3 min, the extent of the thermal effect and distribution of heat in tissues is determined by heat-regulating mechanisms. The thermal effect depends on exposure duration (Adair and Black 2003). Deep-lying muscles are heated to a greater extent only during the first 20 min of exposure. When the thigh region is exposed to microwaves, there is a greater temperature rise in the muscles than in the skin and subcutaneous (sc) fatty layer (Cook 1952; Herrick and Krusen 1953). For a full discussion and review of thermal interactions with animals and humans, including specific tissues (see Chapter 10 of *BBA* and Chapter 5 of this volume).

3.3 Biochemical Changes

Biochemical alterations have been reported to result from exposure to RF energies. Such effects generally appear to be reversible and no well-defined characteristic response pattern has been determined, nor is it known whether the changes are direct or indirect effects of exposure. This is an area of research in which considerable work was conducted in the 1970s and 1980s. However, very few results have been reported in publications over the past decade.

Effects on mitochondria isolated from exposed animals have been reported (Dumansky and Rudichenko 1976) but there was no effect on rat liver mitochondria exposed *in vitro* to 2.4 GHz, 1 to 4 W/kg, or 10 to 12 GHz, at a maximum of 20 W/m^2. No effect of microwave exposure has been found on a number of enzymes and proteins irradiated *in vitro* (Belkhode et al. 1974; Allis 1975; Ward et al. 1975; Bini et al. 1978). Albert et al. (1974) exposed Chinese hamsters at 2.45 GHz, 500 W/m^2 for 0.5 to 4.5 h over a period of 1 to 21 d and found no change in liver ATP.

The limited number of studies on oxidative enzyme systems has yielded mixed results. Exposure of suspension of the membrane-bound enzyme cytochrome oxidase to sinusoidally modulated 2.45 GHz MW energy at an SAR of 26 W/kg did not significantly affect its activity during exposure. On the other hand, Sanders et al. (1980), exposing the rat brain surface *in vivo* to 591 MHz continuous wave (CW) RF at 50 or 138 W/m^2 under conditions of negligible systemic temperature change, reported that increases in NADH and decreases in ATP concentration occurred supporting an hypothesis of radio frequency radiation (RFR) inhibition of electron transport in brain mitochondria. These observations were later extended to effects of pulsed, and sinusoidal amplitude-modulated microwaves on brain energy metabolism where similar changes were found (Sanders et al. 1985).

Dumansky et al. (1972) reported a decrease in liver glycogen content together with increased lactic acid levels and phosphorylase activity, which was interpreted as evidence of impairment of glycogen synthesis in the livers of rats chronically exposed to 2.45 or 10 GHz MWs at 500 to 2000 W/m^2. Exposure to 2.45 or 10 GHz MWs at 0.25 to 10 W/m^2 decreased the proteolytic activity of the mucous membrane of the small intestines of experimental animals, where the invertase and adenosine triphosphatase activity increased. Altered proteolytic activity was suggested to be due to alteration in the structure and physicochemical characteristics of the mucous membrane. Reduced synthesis of macroglobulin and macroglobulin antibodies resulted from exposure of experimental animals to 0.5 W/m^2 (Malyshev and Tkachenko 1972).

Dose-dependent transient elevations in serum glucose, blood urea nitrogen, and uric acid were noted following far field exposure of rabbits to 2.45 GHz for 2 h at intensities of 50, 100, and 250 W/m^2 (Wangemann and Cleary 1976). There were detectable differences between CW and PW exposures of equivalent average power density. There was an increase in colonic temperature of 1.7 to 3.0°C at 100 and 250 W/m^2. Exposure of rats for 15 min to pulsed 2.86 GHz at 50, 100, 500, or 1000 W/m^2 resulted in statistically significant changes in serum albumin and phosphorous levels only at 1000 W/m^2, and there was no change in serum glucose levels in rats exposed to 2.86 and 0.43 GHz pulse-modulated fields at an average power density of 50 W/m^2. Single or repeated exposures of rabbits to 3 or 10 GHz at 500 to 2500 W/m^2 resulted in alterations in serum albumin–globulin ratio, which was attributed to effects on the liver or adrenals (Swiecicki and Edelwejn 1963).

Baranski (1972a) found inconsistent changes in cholinesterase activity in both rabbit and guinea pig brains following 3 months of exposure, 1 h/d, with pulsed fields of 250 W/m^2. He also found a decrease in cholinesterase activity in the brains of guinea pigs after a single 3-h exposure to 35 W/m^2 of pulsed 2.45 GHz MWs. Increase to 250 W/m^2 caused a further decrease in activity. Pulsed energy was found to produce a more severe effect than CW exposures of the same average power density, suggesting that these effects are due to peak fields. Nikogosyan (1962) found an increase in blood cholinesterase activity after a single 90-min exposure to 3 GHz waves at 400 W/m^2. Revutsky and Edelman (1964) also reported an increase in specific cholinesterase activity in rabbit blood exposed *in vitro* to 2.45 GHz MWs. It should be noted that what is measured in blood is pseudocholinesterase, which has no neural-related activity. On the other hand, Olcerst and Rabinowitz (1978) found no effect on aqueous cholinesterase exposed to 2.45 GHz CW up to 1.25 kW/m^2 for half an hour or 250 W/m^2 for 3 h. No effect was found on cholinesterase activity in defibrinated rabbit blood exposed for 3 h to 210, 350, or 640 W/m^2, 2.45 GHz, CW or pulsed. Under similar exposure conditions, there was no effect on release of bound calcium or magnesium from rabbit red blood cells (RBCs).

Mitchell et al. (1989) exposed or sham-exposed rats to 100 W/m^2 CW radiation at 2.45 GHz for a period of 7 h. Animals were subjected to behavioral, biochemical, and electrophysiological measures during and immediately after exposure. In this joint U.S. and former USSR study, the U.S. group found Na^+, K^+, and ATPase activity to be significantly lower in microwave exposed animals than in the shams. No significant differences were found by the former Soviet Union group. On review, the authors concluded that the U.S. result was spurious. This and other studies done emphasized the value of multilaboratory collaboration in promoting rigor in scientific design.

Ho and Edwards (1977, 1979) studied oxygen consumption in mice exposed to 2.45 GHz radiation at various SARs in a waveguide exposure system and found that the animals compensated homeostatically to SARs of 10.4 W/kg or greater by a decrease in metabolic rate to compensate for thermal loading. Normal metabolic activity was resumed following cessation of exposure.

Oxidative stress has been proposed as a possible mechanism for the biological effects of MW exposure (Lai and Singh 1995–1997) and some experimental support for such a mechanism has been reported. Other experimental work, however, has not fully supported this proposal. Induction of stress proteins was investigated in HeLa and CHO cells following 2 h of RF exposure at 2450 or 2700 MHz, respectively (Cleary et al. 1997). The study was conducted under isothermal conditions at an SAR or 25 W/kg (HeLa cells) or 27 W/kg (CHO cells). Exposures had no detectable effect on stress protein induction with comparisons made to both sham and positive control cells. Reports of decreased camphor binding in rat olfactory tissue that potentially could indicate oxidative stress effects in exposed samples did not show correlation with SAR (Bruce-Wolfe and Adair 1985; Philippova et al. 1988, 1994). High levels of MW, which generated elevated temperatures in liposomes (*in vitro*), showed no effect on peroxidation (Logani and Ziskin 1996). However, in a series of studies in which the RF levels resulted in considerable temperature elevations, changes could be demonstrated in membrane fluidity, permeability, and protein shedding in a way that might be related to oxidative stress (Liburdy and Penn 1984; Liburdy and Magin 1985; Liburdy and Vanek 1985, 1987; Liburdy et al. 1988). Further work, suggesting that these observations were indeed due to temperature elevation, showed that even at high SAR RF exposure of sheep RBCs, no effect on NADH oxidase or glucose oxidase activity was observed when the temperature of the preparation was controlled (Kiel and Erwin 1986; Kiel et al. 1988). Using lower levels of RF, there was an apparent protection of temperature-induced oxidative hemolysis in RBCs (Kiel and Erwin 1984).

3.4 Cellular and Molecular Biology

Studies to determine whether RF exposure can produce effects in biological systems can be approached on several levels. A primary goal in such experimental studies would be to establish reproducible effects resulting from exposure under well-controlled conditions. One such approach is to utilize cultures of either animal or plant cells under *in vitro* conditions. Cell culture experiments are relatively straightforward to reproduce; however, the behavior of cells in culture is never precisely the same as within an organism (*in vivo*) due to the isolation of cells from their usual environment, including contact with other cell types, missing intercellular factors, and lack of normal tissue architecture. *In vitro* experiments, therefore, can sometimes be more and sometimes be less prone to evidence responsiveness to an environmental exposure. The advantages of cellular experiments accrue to precise control of conditions, good dosimetry, and strong statistical power due to sample size and replicate designs. Cell culture studies, however, also have significant limitations in long-term exposure experiments and are often only remotely connected to potential human health issues.

Significant advances in our understanding of biological systems have occurred over the 10 y since the previous edition of this handbook. These advances have had demonstrable effects on the understanding of biology, which in turn have facilitated efforts to determine if and how RF could influence biological systems. The design of experiments and collection of data from culture studies has also improved from technical advances in modeling the inherent complexity of biological systems, thus providing a theoretical foundation for interpreting experimental data. Cell and tissue culture techniques in the last decade or so have moved from simple biochemical assays and single-molecule detection methods to investigations using identified signal transduction pathways, polymerase chain reaction technology permitting DNA amplification from even single cells, and differential display techniques and cDNA microarray approaches that permit studies of gene expression of large numbers of genes at one time. Other new techniques include high performance liquid chromatography, mass spectroscopy of protein chips, and studies of protein distributions, protein dynamics, and protein–protein interactions. In addition, developments in microscopic techniques provide a basis for advanced studies of processes such as cell dynamics and inter- and intracellular interactions. These advanced techniques are not fully developed in MW research but are just recently beginning to be employed in the investigation of biological preparation response to MW exposure. Some aspects of MW effects on cellular biology have recently been reviewed by Heynick et al. (2003).

3.4.1 Chromosome and Genetic Effects

Early on in MW bioeffects research, investigators have focused on chromosome changes, including translocations and structural aberrations (Heller 1970; Chen et al. 1974; Yao 1978, 1982), polyploidy (Yao 1976), and stickiness (Manikowska et al. 1979). Exposures ranged from 70 W/m^2 to more than 2 kW/m^2. In a study of meiotic cells removed from exposed male mice (Manikowska et al. 1979), translocations were observed at meiosis I at 10 and 50 W/m^2. Other investigators reported no exposure-related effects (Huang et al. 1977; Lloyd et al. 1984, 1986). An extensive study conducted by Kerbacher et al. (1990) examined Chinese hamster ovary (CHO) cells exposed to 2450 MHz at high SAR levels (33.8 W/kg). Even though the temperature of the cultures was increased to 40°C by the exposure, an assessment of a large number of indicators of chromosomal alteration showed no changes due to exposure. This study included many experiments and independent replication for each exposure condition. Following this work, a number of more recent studies have reported results in agreement with these findings that induction of chromosomal aberrations is not a result of RF exposure (Maes et al. 1997, 2000, 2001; Vijayalaxmi et al. 1997a,b, 1998, 2001a,b, 2003). Induction of micronuclei (MN) as an indicator of chromosomal damage has also been thoroughly investigated following MW exposure. Although there are reports of MN induction in mammalian cells exposed to RF *in vitro* (D'Ambrosio et al. 2002; Tice et al. 2002), there are a far greater number of reports suggesting the lack of MW involvement in MN formation (Vijayalaxmi et al. 2001a,b, 2003; Bisht et al. 2002; McNamee et al. 2002a,b). Even studies that are positive for MN production on exposure failed to show a corresponding evidence of DNA breakage (Tice et al. 2002). The problems inherent in the use of chromosomal aberrations as indicators of genetic damage have been noted by Savage (1977).

Ciaravino et al. (1991) exposed CHO cells simultaneously to adriamycin and pulsed 2.45 GHz microwaves at a SAR of 33.8 W/kg for 2 h. Temperatures did not exceed 39.7°C. RF radiation did not affect changes in cell progression produced by adriamycin and did not change the number of sister chromatid exchanges induced by adriamycin, an agent which is known to damage DNA. Similar results were obtained for mitomycin C (Ciaravino et al. 1987). Lloyd et al. (1986) examined human peripheral blood lymphocytes *in vitro* exposed up to 200 W/kg of 2.45 GHz radiation for 20 min with temperatures ranging from 37 to 40°C. Using standard techniques for evaluating unstable chromosome and chromatid-type breaks, exchange aberrations or sister chromatid exchanges, no significant variations from appropriate control cultures were found. In a series of studies, Maes et al. investigating sister chromatid exchange (SCE) in exposed cultures initially reported an increase in SCE (Maes et al. 1996), then inconsistent changes

(Maes et al. 1997), and finally two studies with no evidence of an effect (Maes et al. 2000, 2001). In another study, mice were exposed to 2.45 GHz RF at 200 W/m^2 (SAR 21 W/kg) for 8 h/d for 28 d. Incidences of SCE in bone marrow cells of irradiated mice, sham-irradiated control mice and standard control mice were compared. No statistically significant differences were detected (McRee et al. 1981).

More recent genetic work has focused on formation of single- or double-strand breaks in the DNA of exposed samples. As seems to be a pattern in much of this work, initial reports indicated an increase in DNA breakage in exposed cultures (Kurt and Milham 1988; Lai and Singh 1995, 1997). Subsequent attempts at replication, however, have failed to confirm the changes initially observed (Malyapa et al. 1998; Hook et al. 1999; Lagroye et al. 2004). Additional work at cell phone frequencies also showed no increased DNA single-strand breakage with exposure (Tice et al. 2002). The large majority of studies using mammalian cell lines as well as isolated human cells indicate a lack of strand breaks associated with MW exposure (Maes et al. 1997; Malyapa et al. 1997a,b; Vijayalaxmi et al. 2000, 2001b; Alekseev and Ziskin 2001; Li et al. 2001; McNamee et al. 2002a,b).

Hamrick (1973), studying the response of mammalian lymphocytes exposed to 2.45 GHz MW (CW), reported that power densities from 67 to 160 W/kg had an effect identical to that of heating on the secondary structure of DNA, as determined by comparison of thermal denaturation curves. In other work, circular DNA plasmids, exposed to 2.55 GHz microwaves at SARs ranging up to 85 W/kg for 20 min exhibited a decrease in supercoiled DNA with an increase of relaxed and linear DNA indicating single- and double-strand breaks (Sagripanti and Swicord 1986). The authors indicated that this could be a result of excitation of internal vibrational modes of the DNA but that a nonspecific or thermal mechanism could not be excluded. Furthermore, the presence of copper electrodes in the DNA solution led the investigators to conclude that DNA breakage was not due to RF exposure (Sagripanti et al. 1987).

Hamnerius et al. (1985) exposed *Salmonella typhimurium* in exponential phase growth to a number of different frequencies in pulse wave (PW) and continuous wave modes producing SARs up to 130 W/kg over exposure periods ranging from 2.5 to 6 h. Temperatures were closely controlled between exposed and control samples. Ames' assays on bacteria showed no elevated mutation frequency. The same group, utilizing similar exposure conditions, studied mutations in a gene controlling eye pigmentation in *Drosophila*. No elevated mutation frequency was found.

The mutagenic potential of microwave energy has been evaluated by various techniques including point mutations in bacterial assays (Anderstam et al. 1983), the dominant lethal test in mammalian systems (Varma et al. 1976), or genetic transmission in *Drosophila* (Pay et al. 1972; Mittler 1976) with inconsistent results.

Anderstam et al. (1983) found no change in mutation induction in *Eschericia coli* or *S. typhimurium* exposure to 0.027 and 2.45 GHz continuous wave (CW) and 3.07 GHz pulsed wave (PW) at SARs from 4 to 100 W/kg. Baranski et al. (1976) were not able to attribute mutagenic effects or metabolic changes in *Physarum polycephalum* or *Aspergillus nidulans* to 2.45 GHz, 100 W/m^2 CW or PW microwaves. Correlli et al. (1977) also reported no mutagenesis after exposure to RF at 2.6 to 4.0 GHz, 20 W/kg. They investigated the effects of RF on colony-forming ability (CFA) and molecular structure (determined by infrared (IR) spectroscopy) of *E. coli* B bacterial cells in aqueous suspension. Cells were exposed for 10 h at SARs of 20 W/kg (equivalent to 5 kW/m^2). No RF-induced effects on either CFA or molecular structure were observed.

Other studies of mutagenic effects of RF in bacteria and yeasts have given negative results. A study by Blackman et al. (1976) involved exposure of *E. coli* WWU to 1.70 or 2.45 GHz RF at 20 to 500 W/m^2 for 3 to 4 h. Dutta et al. (1979) exposed *Saccharomyces cerevisiae* D4 to 2.45 GHz CW RF at 400 W/m^2 or to 8.5 to 9.6 GHz pulsed RF at 10 to 450 W/m^2 for 120 min with negative results.

Saunders et al. (1988) exposed mice to 100 W/m^2 at 2.45 GHz CW for 6 h/d for an 8-week period. Exposed male mice were mated to an unexposed group of female mice and followed. There was no significant reduction in pregnancy rate and cytogenetic analysis at the end of the mating period showed no significant variation from the controls.

Meltz et al. (1987) investigated whether there were any RF-induced alterations in DNA repair in normal human fibroblasts maintained *in vitro* after the DNA was damaged by a selected dose of UV light. Power densities of 10 or 100 W/m^2 (0.350 and 1.2 GHz) caused no perturbation of the DNA repair process. Brown et al. (1981) treated mice with streptozocin, a mutagenic and carcinogenic agent known to damage DNA in the rodent liver, and exposed the mice to 400 MHz RF to determine if excision repair of the DNA would be inhibited. Power densities of 16 and 160 W/m^2 (SAR 0.29 and 2.9 W/kg) did not alter the level of excision repair.

Although Mickey (1963) reported increased mutagenesis in *Drosophila* exposed to a 0.02 GHz field, no effects on mutagenesis were observed by Pay et al. (1972) using 2.45 GHz or by Mittler (1976) with 0.003 and 0.150 GHz RF energy.

Varma and Traboulay (1976) reported increased mutagenesis, using the dominant lethal test with male mice exposed to 1.7 GHz continuous wave at 100 and 500 W/m^2 for 90 and 30 min, respectively. Mice exposed to 2.45 GHz CW at 1 kW/m^2 for 10 min and at 500 W/m^2 three times, 10 min each, within 1 d also showed increased mutagenesis. Mice subjected to four 10-min exposures at 500 W/m^2 over a period of 2 weeks, showed no increase in dominant lethality above control levels. Saunders et al. (1983) found no evidence of change in dominant lethality in mice exposed to 2.45 GHz CW, 43 W/kg.

Berman et al. (1980) exposed male rats daily to 2.45 GHz CW (day 12 of gestation to 90 d of age, 100 W/m^2, 4 h/d) or 2.45 GHz CW at 50, 100, or 280 W/m^2 from day 6 of gestation to 90 d of age, 4 or 5 h/d. No significant evidence of germ cell mutagenesis or alteration in reproductive efficiency was detected. In a recent study, Roti Roti et al. (2001) found no evidence of neoblastic transformation in C3H10T1/2 cells after exposure to 835 MHz frequency division multiple access (FDMA) and 847 MHz code division multiple access (CDMA) signals.

It is known that the rate of induction of mutations will increase with increasing temperature. It is possible that artifacts or thermal stress could be factors in some of the reported studies. Baranski and Czerski (1976) noted that there is no satisfactory evidence of microwave-induced genetic effects at low to modest power densities. Additional experimental studies in rodents (Varma et al. 1976) have confirmed earlier findings that microwave exposure at power densities below 100 W/m^2 is not mutagenic in these organisms. Other studies (Liu et al. 1979) have failed to find effects that differ from those resulting from RF heating. Janiak and Szmigielski (1977) likewise reported no significant differences in the sequence and time course of cell membrane injury between cells treated in a water bath and those heated with 2.45 GHz microwaves.

Theories have been proposed that absorption of energy by the DNA from the MW field might result in chemical- or reactive species-mediated damage to the molecule (Swicord and Davis 1982; Edwards et al. 1984, 1985; Davis et al. 1986; Sagripanti et al. 1987). In the actual experimental work in several laboratories, definitive conclusions are less clear. In the interaction of millimeter waves with biological media, a number of investigators reported sharp, distinct resonances in the absorption of the RF energy by various biochemical and biological preparations (Webb and Booth 1969, 1971; Stamm et al. 1974; Dardanoni et al. 1976; Lee and Webb 1977) or that exposure to millimeter waves produces biological effects which exhibit a sharp-frequency dependence (Berteaud et al. 1975; Stamm et al. 1975; Grundler and Keilmann 1978, 1983). These studies suggest sharp millimeter wave, frequency-dependent lethal and mutagenic effects on microorganisms, on metabolic control of growth of cells, on oncogenic viruses, viability properties of cells, protective effects on x-irradiated and cytotoxic drug-treated animals, etc. Because of poor documentation of a number of these studies and failure to replicate the observation of resonances in studies, which were adequately documented (Gandhi et al. 1980) there is little confidence in the validity of such resonances. Experiments in two other laboratories have failed to support either the theoretical propositions or the earlier observations (Foster et al. 1987; Gabriel et al. 1987, 1989).

Absorption resonances were also reported for DNA in solution for the range 8 to 12 GHz (Swicord and Davis 1982). These observations were later retracted by the authors, who identified artifacts of the experiment (Rhee et al. 1988). An attempt to identify such resonances in another laboratory was not successful (Gabriel et al. 1987).

In summary, there is no solid evidence that exposure to RF radiation induces mutations in bacteria, yeasts, or fruit flies. The results of some studies suggest that RF radiation induces mutations in mammals. Critical review has cast doubt on these findings. Other studies have shown no mutagenic effects of RF on mammals. Evidence for cytogenetic effects of RF is mixed. The lowest power density at which cytogenetic effects have been reported is 200 W/m^2 (Stodolnik-Baranska 1974) but these results are contradicted by Chen et al. (1974), who failed to find cytogenetic effects at 2 to 5 kW/m^2.

3.4.2 Kinetic or Functional Changes and Membrane Effects

Many investigations were performed in the 1980s and early 1990s on the potential for MW irradiation to produce functional changes and membrane effects in exposed cell cultures. However, there is very little work reported with such a focus since the last edition of this handbook. The conclusions stated then have not changed; that RF effects were observed; however, consistent changes appeared generally to be associated with accompanying temperature increases and not due to a direct RF field influence. Kinetic or functional changes demonstrated in cells exposed to RF radiation have been reported, as reviewed below.

Dutta et al. (1992) exposed neuroblastoma cells in culture to varying powers of RF at carrier frequencies of 915 and 147 MHz, amplitude modulated at 16 Hz. Enhanced release of calcium ions from the cells was found at two different power densities, between which no enhancement was found. There was enhanced acetylcholinesterase activity under the same exposure conditions. Saffer and Profenno (1989, 1992) found an increase in beta-galactosidase expression exposed to RF between 2 and 4 GHz at 10 W/kg for 4.5 h with temperature changes not exceeding 0.1°C. Control cultures were adjusted to identical temperatures. The authors concluded that the effect was not thermal but possibly caused by thermal gradients due to RF radiation different from conductive heating produced by ambient conditions.

D'Inzeo et al. (1988) studied acetylcholine-activated single channel openings of cultured myotubes from chick embryos by patch clamp techniques. During exposure to RF at 10.75 GHz, 0.5 mW/m^2 at the surface of a bathing solution, the frequency of channel openings decreased but channel-opening time and conductance were not affected. The authors speculated that exposure to microwave energy decreased channel-opening probability via an alteration of intracellular enzymatic processes through a nonthermal mechanism.

Rotkovska et al. (1987) studied stem cell colony-forming units (CFUs) derived from bone marrow cells of female mice utilizing a spleen exocolony method. At 2.45 GHz CW at a range of temperatures there was a significant enhancement of the number of CFUs in RF-exposed systems versus temperature-matched controls.

Rotkovska et al. (1993) also attempted to assess the role of a membrane receptor in the enhancement effect of microwave energy described above. Bone marrow preparations were incubated with a beta-receptor blocker, trimepranol (TMP), during microwave exposure and controlled with preparations without TMP; in addition, comparable samples were exposed only to ambient heat. The microwave stimulatory effect appeared to be blocked in the presence of TMP. The authors stated that activation of the beta-receptors results in enhancement of cAMP synthesis and an increase of Ca^{2+} ions in the cell caused by increased membrane permeability. This is followed by reactions of the adenylate cyclase system, resulting in the enhancement described above.

Cleary et al. (1990a) exposed cultured glioma cells to 0.027 and 2.45 GHz CW RF radiation for 2 h at temperatures maintained at 37°C. At SARs up to 50 W/kg, there was significantly more incorporation of tritiated thymidine and uridine into exposed versus control cultures. These changes persisted for up to 5-d postexposure. Kinetic changes of cell populations exposed to RF energy are also described in Section 3.10.

However, Parker et al. (1988) exposed rodent cell lines to 2.45 GHz microwave radiation at SARs up to 103 W/kg for varying lengths of time at temperatures from 37 to 45°C, using DNA probes for oncogenes, heat protein and long terminal repeat sequences, hybridized to mRNA in exposed and control cells. No significant differences in mRNA expression were observed after microwave exposure.

Brown and Marshall (1986) reported that microwave energy produces no perturbation in cultures of murine erythroleukemic cells undergoing erythroid differentiation in response to hexamethylene bisacetamide (HMBA). Exposures were for 48 h at 1.18 GHz at SARs up to 73 W/kg with temperature maintained at 37.4°C.

Effects of RF on cell membrane dynamics have also been reported. Liburdy and Vanek (1985) exposed rabbit erythrocytes to 2.45 GHz microwaves under a number of conditions and observed a change in membrane permeability to sodium ions within the temperature range 17.7 to 19.5°C. This change was described as linear with the internal electric field strength inside the sample, saturating at 600 V/m. The same investigators studied the effects of microwave energy on liposomes carrying cytosine arabinofuranoside (ARA-C) (Liburdy and Magin 1985). Below the membrane phase transition temperature of 41°C, liposome membranes are normally not leaky. In buffered saline exposure to 2.45 GHz microwaves at a SAR of 60 W/kg ARA-C is released at 33°C. In plasma release occurs at 27°C. Release is enhanced by oxygen and attenuated by antioxidants.

Phelan et al. (1992) studied the effect of microwave energy on membrane fluidity using a B16 melanoma cell line exposed to 2.45 GHz pulsed waves 10 μs wide at 100 pulse per second (pps) for 1 h at 100 W/m^2. An increase in membrane ordering as measured by an electron paramagnetic resonance (EPR) technique, reflecting a shift from a more fluid-like phase to a more solid-like phase, was found. A similar result with artificially prepared liposomes constituted with melanin was discovered. Melanotic cells rendered amelanotic did not show the change. The ordering process was inhibited by superoxide dismutase, implicating oxygen radicals as a cause of the membrane changes. The authors proposed that microwaves stimulate melanin to free radical production and that reactive oxygen species may subsequently produce membrane alterations via lipid peroxidation resulting in the ordering effect.

Sandblom and Thenander (1991) studied the effects of microwaves on the kinetics of gramicidin A channels in lipid bilayer membranes. Artificial membranes into which the decapentapeptide, gramicidin A, was inserted were exposed to 10 GHz pulsed microwaves, pulse width 1 μs at a pulse repetition rate of 1000 per second in waveguide for 1 min at an average output power of 300 W, peak power 300 kW. Temperature excursions were up to 5.1°C and estimated average SAR was 350 W/kg. Although microwave energy did not affect single channel conductance of channel lifetime compared to matched temperature controls, the rate of channel formation was decreased during exposure. The authors stated that this is opposite of the effect expected for a purely thermal effect and offer theories for a nonthermal mechanism.

Philippova et al. (1988, 1994) observed that microwave radiation decreased specific ^{3}H-camphor binding to rat olfactory membrane. Exposures were achieved in a waveguide using 0.9 GHz fields pulsed modulated at frequencies at 1–100 Hz. SAR was varied from 0.5 to 18 W/kg and steady-state temperatures of 4°C maintained. The inhibition of binding appeared to decrease with increasing SAR. Inhibition did not depend on pulse modulation rate. The shedding of protein during microwave exposure of erythrocytes and lymphocytes had been previously reported by Liburdy and Penn (1984).

Allis and Sinha-Robinson (1987) investigated the ATPase activity in human erythrocyte membranes *in vitro* exposed to 2.45 GHz CW microwave radiation for a period of 20 min at a SAR of 6 W/kg and stabilized temperatures from 17 to 31°C. At 25°C, but not at other temperatures, there was a 35% decrease in the activity of the enzyme in microwave-exposed samples. The authors proposed that a transition state of the protein–lipid interaction within the membrane is available at the transition temperature and not at other temperatures.

The activity of a specific enzyme, ornithine decarboxylase, is increased in certain cell lines following exposure to amplitude-modulated microwave fields (Byus et al. 1988). There are many more studies that have documented membrane fluidity and ion transport effects of exposure to thermal levels of RF (Portela et al. 1979; Olcerst et al. 1980; Webber et al. 1980; Friend et al. 1981; Pickard and Barsoum 1981; Barsoum and Pickard 1982a,b; Arber and Lin 1983; Shnyrov et al. 1984; Kim et al. 1985; Liu and Cleary 1988; Sandweiss 1990; Saalman et al. 1991; Bergquist et al. 1994; Neshev and Kirilova 1994; Orlando et al. 1994; Fesenko and Gluvstein 1995; Weaver 1995; Eibert et al. 1999).

3.5 Reproduction, Growth, and Development

Developing fetuses, as well as perinatal organisms, are acknowledged to be particularly sensitive to changes in environmental conditions. On this basis and because of general concern for potential effects on reproduction and development from MW exposures, a large number of research efforts were focused in this area from the early 1970s through the early 1990s. The results from that body of research indicate that little evidence exists for MW exposure effects on either reproduction or physiological processes in the developing organism. Primarily due to the generally observed lack of effects, little new information is available other than a few studies specifically targeting cell phone frequency RF exposure.

3.5.1 Reproduction

MWs have been investigated extensively for possible effects on the testes (Ely et al. 1964; Goud et al. 1982; Kowalczuk et al. 1983). Exposure of the scrotal area at high power densities (>500 W/m^2) results in varying degrees of testicular damage, such as edema, enlargement of the testis, atrophy, fibrosis, and coagulation necrosis of seminiferous tubules in rats and rabbits, exposed to 2.45, 3, and 10 GHz radiation. Saunders and Kowalczuk 1981) exposed the caudal area of anesthesized mature male mice to 2.45 GHz RF in a waveguide system for 30 min. Half-body SARs ranging from 18 to 75 W/kg were estimated from measurements of forward, reflected, and transmitted powers. The corresponding colonic temperatures at the end of exposure ranged from 35.3 to 42.2°C. Other anesthetized mice were sham-exposed. Their mean colonic temperature was 32.6°C, 4 to 5°C lower than unanesthetized mice. For comparison, the caudal areas of other anesthetized mice were inserted for 30 min in a copper well heated by a water bath to 37, 41, 43, or 45°C, which resulted in colonic temperature from 36.4 to 40.7°C. Six days after treatment, sections of testes were scored for cell damage and enumeration of sperm. Extensive degeneration of the spermatogenic epithelium was evident for RF exposure at 75 W/kg and for direct heating to 45°C. At SARs of 57 and 46 W/kg (43 and 41°C), marked depletion of spermatids and spermatocytes, but not spermatogonia, was observed. At the lower SARs (37, 30, and 0 W/kg) or a temperature of 37°C, no effects were seen.

Temperature-sensing probes were also implanted in the testes of other groups of mice, and testicular temperatures were related to SAR values (Berman et al. 1980; Lebovitz and Johnson 1987a). Such measurements of testicular temperature indicated the existence of a threshold of about 39°C for depletion of spermatocytes and of about 41°C for 50% cell death after 6 d of RF exposure or direct heating. The corresponding SARs for these two thresholds were 20 and 30 W/kg, respectively.

Lebovitz and Johnson (1983, 1987a) exposed rats for 8 h to 1.3 GHz CW radiation at a whole body averaged SAR of 9 W/kg. This produced a deep rectal temperature elevation of 4.5°C. Subsequent sequential histologic examination, measurements of daily sperm production, circulating follicle-stimulating hormone (FSH), and leutinizing hormone (LH) were all normal, suggesting that the level of acute hyperthermia was not sufficient to produce disruption for any of the indicated measures of testicular function.

Lebovitz et al. (1987b) compared pulse-modulated 1.3 GHz MW, 1 μs pulse width, 600 pps, to conventional heating in rats. Anesthetized animals were sacrificed over a 13 d cycle of seminiferous epithelium after exposure. At 7.7 W/kg, whole body average SAR for 90 min, there was a modest decline in daily sperm production, primarily from effects on primary spermatocytes. Exposure at 4.2 W/kg produced no change. Temperatures in the first exposure reached 40°C and in the second 38°C. Above 7.7 W/kg, all germ cell types were destroyed. Conventional heating in excess of 39°C for 60 min produced a significant decrease in daily sperm production. The authors stated all damage could be explained by heating.

Dasdag and colleagues exposed rats to cell phone RF and observed a reduction in seminiferous tubule diameter (Akdag et al. 1999; Dasdag et al. 1999). However, further study in the same laboratory was unable to replicate the earlier findings and found no effects on measures of testicular structure and function (Dasdag et al. 2003). With temperature controlled to 37°C, *in vitro* experiments, where RF exposure was to relatively high power (>50 W/kg), resulted in a reduction in sperm fertility, although a mechanism was not presented (Cleary et al. 1989). Additionally, an *in vivo* exposure study suggested a reduction in fertility in rats exposed to 27 MHz without an increase in body temperature (Brown-Woodman et al. 1989). This study has not been independently confirmed and appears to be inconsistent with the larger body of evidence that indicates reproductive effects from RF exposure occur only when temperature in the tissues is elevated.

Exposure to 3 GHz radiation, 80 W/m^2 did not affect mating of mice or rats (Miro et al. 1985). Pituitary gonadotropic function was preserved in female mice exposed to 3 GHz radiation, 100 W/m^2, twice daily for 5 months (Bereznitskaya and Kazbekov 1974).

Reports on reproductive effects of MW exposure in humans include workers using radars, video display units (VDUs), RF heat sealers, and devices used for magnetic resonance imaging (MRI) and medical diathermy. These reports show principally negative results although some positive results were also identified. Whereas a slightly increased risk of infertility (Smith et al. 1997) and a small elevation in risk of miscarriage

(Goldhaber et al. 1988; McDonald et al. 1988) have been reported, no association between exposure to VDUs and pregnancy outcomes was observed in the large majority of studies (Michaelson 1983; Nurminen and Kurpa 1988; Bryant and Love 1989; Taskinen et al. 1990; Larsen 1991; Schnorr et al. 1991). No elevation in risk for adverse reproductive effects was observed in female MRI workers (Evans et al. 1993; Kanal et al. 1993).

Miscarriage in physical therapists exposed to medical diathermy units at 915 and 2450 MHz was reported to be slightly elevated (Ouellet-Hellstrom and Stewart 1993). Unlike many studies showing positive results, a weak dose–response relationship of effects was observed in this study; however, significant questions have arisen concerning the validity of the association between the reported effect and the absorbed RF energy (Hocking and Joyner 1995). Although low birth weight of offspring was reported to be associated with exposure of pregnant physiotherapists to shortwave equipment (Lerman et al. 2001), in another study of a comparable group of female workers, no evidence of malformations or spontaneous abortions was found in their offspring (Taskinen et al. 1990).

Lower sperm concentration, motility and number of normal sperm were observed in RF workers (Lancranjan et al. 1975), although others could not replicate the finding of reduced sperm numbers (Weydant et al. 1996; Schrader et al. 1998). Fertility, measured as semen quality and hormone levels in male heat-sealer workers, was not adversely affected with exposure (Grajewski et al. 2000).

3.5.2 Embryonic Development and Teratology

A recent review of MW exposures and embryonic development has been published by Heynick and Merritt (2003). There are a few reports that suggest particular combinations of frequency, duration, and power density produce effects on embryonic development and postnatal growth. Alterations in development have been reported in insects (Carpenter and Livstone 1971), chick embryos (Van Ummersen 1961), and rodents (Rugh et al. 1975).

Green et al. (1979) and Olsen (1982) have provided a thermal basis for an explanation of the MW-induced teratogenic effects in the mealworm, *Tenebrio molitor*, reported by Carpenter and Livstone (1971), Lindauer et al. (1974), and Liu et al. (1975). Olsen (1977) noted the threshold of teratogenesis in *Tenebrio* to be associated with a rise of pupal temperature in excess of 10°C.

Van Ummersen (1961) reported inhibition of growth and development of chick embryos exposed to 2.45 GHz (CW), 200 to 400 W/m^2, 4.5 min to 5 h. Since all embryos in which effects were produced experienced a significant temperature increase, the observed deleterious effects were ascribed to heating. In other studies in chick eggs, 2450 MHz exposures at 2.9 W/kg showed no effect on hatchability (Braitewaite et al. 1991) and more recently Talau et al. (2003) investigated temperature changes in chicken eggs exposed to 1250 MHz fields of 1.45 10.44 W/kg.

An extensive series of MW exposure studies at 2450 MHz have been completed to examine embryo development in Japanese quail eggs (Hamrick and McRee 1975, 1980; McRee et al. 1975, 1983; Hamrick and Fox 1977; Galvin et al. 1980, 1981; Inouye et al. 1982; Clark et al. 1987; Gildersleeve et al. 1987a,b, 1988a,b).

Quail embryos, exposed during the first 12 d of development to 2.45 GHz MWs at an incident power density of 50 W/m^2 and SAR of 4.03 W/kg, showed no gross deformities in the exposed quail when examined and sacrificed at 24 to 36 h after hatch. No significant changes in the total body weight or weights of the heart, liver, gizzard, adrenals, and pancreas were found in the treated birds. Hematological parameters were also measured in the study. The results showed a statistically significant increase in hemoglobin and decrease in monocytes in birds exposed to MWs. No statistically significant changes in hematocrit, RBCs, total white blood cells (WBCs), lymphocytes, neutrophils, basophils, or eosinophils were detected (McRee and Hamrick 1977).

Although effects were observed in isolated parameters when exposures were of sufficient intensity to produce elevation in temperature, in exposures under 4 W/kg no changes were reported in development of the heart (Hamrick and McRee 1975; Galvin et al. 1980), immune response (Galvin et al. 1981; Gildersleeve et al. 1987a), mortality, egg production, fertility, hatchability of eggs, and reproductive performance (Gildersleeve et al. 1987b).

In additional quail egg studies, Spiers and Baummer (1991) exposed eggs from day 1 through 15 of incubation to sham or 2.45 GHz microwave radiation, 8 h/d at SARs from 3.3 to 15.2 W/kg at a number of ambient temperatures. It was concluded that microwave radiation can be used to increase egg temperature and embryonic growth rate at ambient temperatures below the normal incubation level without altering basic metabolic processes of the embryo.

A number of studies, at various MW frequencies, have been concerned with effects of exposure on mammalian embryonic and fetal development. In many of these studies, a single acute exposure was administered at a power density that causes an increase in body temperature. Some investigators, however, have reported studies in which protracted exposures have been given at power density levels, which apparently do not cause a significant increase in colonic temperature. These and other studies and results are described below.

Much of the work examining development in MW-exposed animals has been performed at 2450 MHz. Rugh et al. (1975) reported abnormalities in mouse fetuses exposed in an environmentally controlled waveguide on gestation days (GDs) 7 to 13 to 2450 MHz (CW) in the range of 84 to 112 W/kg. Hemorrhages, resorptions, exencephaly, stunting, and fetal death were observed. In a further study, Rugh and McManaway (1976) demonstrated that lowering the body temperature of the dam with pentobarbital anesthesia could prevent the teratogenic effects of thermal loading with MWs, thus demonstrating the thermal influence in the earlier studies.

Employing a multimodal cavity at 2.45 GHz, Chernovetz et al. (1975) irradiated pregnant mice with a single intense dose of 38 W/kg for 10 min (22.8 J/g) on GD 11, 12, 13, or 14. This dose resulted in 10% maternal lethality. The exposure during late organogenesis or early fetal stage caused no change in fetal mortality or morbidity when compared to shams. In another study, Chernovetz et al. (1977) noted an increased rate of resorptions in rats after an absorbed dose of 30 W/kg for 20 min once on GDs 10 to 16.

Berman et al. (1978) exposed CD-1 pregnant mice to 2.45 GHz at 34, 136, or 280 W/m^2, GD 6 through 17 for 100 min daily. Estimates of mean dose rate ranged from 2.0 to 22.2 W/kg. No significant increase in colonic temperature was reported. Another group was similarly sham-exposed. The mice in half of each group were examined on GD 18. The incidence of pregnancy, the number of live, dead, and resorbed fetuses, and the total number of fetuses were found to be similar for the exposed and sham-exposed mice. The mean body weight of the live fetuses in the RF-exposed group, however, was significantly smaller (by 10%) than those in the sham-exposed group, a finding consonant with their previous results. In addition, ossification of sternal centers was significantly delayed in the RF-exposed mice. The mice in the other half of each group were permitted to come to term. Significant fetal growth retardation occurred at the highest power density level. A significant increase in CNS anomalies occurred, but only if the data from three irradiated groups were combined. At 7 d of age, the mean body weight of the suckling mice of the RF-exposed group was also significantly smaller (by 10%) than that for the sham-exposed group. The survival rate was not affected. The authors concluded that MW radiation at a frequency of 2.45 GHz was embryopathic at a power density of 280 W/m^2 (22.2 W/kg).

The embryofetal toxicity and teratogenicity of 2.45 GHz at different intensities were investigated in the CD-1 mouse by Nawrot et al. (1981). Mice were exposed on GD 1 to 15 at an incident power density of 50 W/m^2 (SAR of 5 W/kg), and either on GDs 1 to 6 or 6 to 15 to 210 W/m^2 (SAR of 28.14 W/kg) or to 300 W/m^2 (SAR of 40.2 W/kg) for 8 h daily. Exposure either on GD 1 to 6 or 6 to 15 to a power density of 210 or 300 W/m^2 caused an increase in colonic temperature of exposed dams of 1 and 2.3°C, respectively. To distinguish between thermal and nonthermal effects of 210 or 300 W/m^2, groups of mice were also exposed to elevated ambient temperature to raise their body temperature to the level of the animals exposed to MW. Ambient temperatures of 30 and 31°C increased the deep colonic temperature to that obtained with the 210 and 300 W/m^2 MW exposure, respectively. The mice exposed to higher ambient temperature were handled in exactly the same manner as the MW-exposed mice. A significant decrease in implantation sites per litter and reduction in fetal weight was noted in the group exposed at 300 W/m^2 during GDs 1 to 6 compared to that of mice exposed to elevated ambient temperature. Exposure of mice at 300 W/m^2 (GDs 6 to 15) resulted in a slight but significant increase in the percentage of malformed fetuses, predominantly with cleft palate, when compared to all other groups.

In studies of effects of *in utero* exposure of Long-Evans rats to 2.45 GHz CW microwaves, Michaelson et al. (1976) found no adverse effects on the dam or offspring when gestation length or litter size was examined. The rats were exposed at 100 or 400 W/m^2 for 1 h on GD 9 (organogenesis) or day 16 of gestation. Enhanced maturation, as indicated by adrenocortical response, was suggested.

Laskey et al. (1970) observed an increase in resorption rate and a decrease in term fetal weight in offspring of rats exposed to 2.45 GHz for 8 to 13 min on GD 2, 2 and 5, 8, or 15 at 1 kW/m^2. Maternal body temperature increased from 3 to 4°C during exposure. Smialowicz et al. (1979a) exposed rats to 2.45 GHz MWs daily for 4 h at a power density of 40 W/m^2 *in utero* from GD 6 d through 40 d of age. The SAR was determined by twin well calorimetry for several ages of the animals. Pregnant dams weighing 300 to 350 g had a mean SAR of 0.7 W/kg; animals that were 1 to 5 d of age and weighed 6 to 10 g absorbed approximately 4.7 W/kg. There was no significant difference between the mean body weights of the males (female offspring were not used in this experiment) in the 12 sham-irradiated litters when compared to the mean body weights of the males in the 12 MW-irradiated litters.

Jensh et al. (1983a,b) reported on the effects of protracted prenatal exposure of Wistar-derived rats to 2.45 GHz for 8 h daily throughout gestation. The mean exposure time to incident energy at 200 W/m^2 was 115 h. No statistically significant differences were observed between control and exposed animals for maternal mass, embryonic and fetal resorption rate, abnormality rate, and the term fetal and placental masses. These authors (Jensh et al. 1982a,b) also reported similar experiments in which 915 MHz MW were employed. They exposed Wistar rats at 100 W/m^2 for 8 h daily throughout the period of gestation for an average of 110 h of exposure. Postexposure colonic temperatures of dams showed no increase over baseline rectal temperatures. No differences were observed in embryonic or fetal death, abnormalities, fetal mass, litter size, placental mass, fetal sex ratio, maternal mass, or maternal gain of mass.

In additional experiments, rats exposed during gestation to high intensity 2450 MHz fields (16.6 to 22 W/kg) showed lower brain weight (Shore et al. 1977; Berman et al. 1984). However, long-term continuous exposure of rats during gestation at lower SAR (0.4 W/kg) produced no effects in development or fetal and brain weight (Merritt et al. 1984). Furthermore, in a study of brain development at 15, 20, 30 and 40 d of age following prenatal and postnatal 2.45 GHz exposure, no effects on brain development were detected histologically (Inouye et al. 1983). In these studies, the brain SAR was 9.5 W/kg or more. In an additional study, Galvin et al. (1986) exposed rats prenatally between GD 5 and 20 and postnatally between days 2 and 20. Exposed animals exhibited larger body mass and less endurance at 30, but not at 100 d of age. SARs in the fetal rats were 4 W/kg and in the postnatal exposures were 5.5–16.5 W/kg. Jensh and Ludlow (1980) reported that exposure at 2.45 GHz to 3.6–5.2 W/kg did not alter growth or development and no changes were observed in five of six behaviors when the animals were tested as adults.

RF fields at SARs of 0.2, 1.0, and 5.0 W/kg demonstrated no developmental effects on rat embryos grown in culture (Klug et al. 1997). Studies have also been conducted at frequencies other than 2.45 GHz and the results are essentially the same. Lin et al. (1979b) studied the effects of repeated exposure of C3H mice to 48 MHz. The animals were exposed to 5 W/m^2 in a transmission electron microscopy (TEM) exposure chamber for 1 h/d, 5 d/week, beginning on the 4th to 7th day of postpartum, for 10 weeks. The formed elements in the blood were not affected by the exposure. The means of body mass of the irradiated and control animals were comparable. No significant differences in lesion onset, incidence, prevalence, extent, or type were observed when animals repeatedly exposed to RF were compared with sham control groups. During the period of the study, no cataracts were noted. Fertility differences among irradiated and control animals were not detected. Body growth patterns did not differ among sham- and RF-exposed animals.

Bereznitskaya and Kazbekov (1974) exposed mice to 3 GHz, 100 W/m^2, which resulted in increased fetal wastage. No definite abnormalities or inborn genetic defects were found. Neonatal mice exposed to 19.27, 26.6, or 105 MHz pulsed waves in a magnetic field of 55 A/m and an electric field of 8000 V/m, 40 min/d for 5 d did not show any evidence of alteration in growth and development.

Lin et al. (1979b) exposed C3H mice in a rectangular coaxial exposure system at 148 MHz at 5 W/m^2 for 1 h daily from GD 2 through day 19, corresponding to a SAR of 0.013 W/kg. The experiment was conducted as three separate replications. In each of the experiments, some of the dams were allowed to come to term and the fetuses were assessed for weight at birth and again at 60 d of age, whereas in others the uterus was extirpated on GD 19 of gestation. No differences in percentage of resorbed, stillborn, or abnormal fetuses were observed in any of the three experiments.

Dietzel and coworkers exposed pregnant rats using a 27.12 MHz diathermy unit at 55, 70, or 100 W, which was sufficient to raise the animals' colonic temperatures to 39, 40, or 42°C (Dietzel et al. 1972; Dietzel 1975). The rats were exposed once between GDs 1 and 16. The fetuses were examined near term. The peak incidence of anomalies was found to occur on GDs 13 to 14, when 16% of the fetuses were abnormal. This fetal wastage, as in the studies of Rugh et al. (1975), was clearly associated with a general body temperature increase in the dam.

Brown-Woodman and Hadley (1988) exposed pregnant rats at GD 9 to thermalizing levels of RF at 27.12 MHz and found increasing developmental abnormalities with increasing core temperature. Maximum temperature elevation was 5°C. Only a few seconds at this temperature produced abnormalities. Rats heated to comparable levels and for comparable durations in a water bath demonstrated similar levels of abnormalities, suggesting that the teratogenicity of RF radiation is related to hyperthermia.

A series of experiments was conducted on rats to assess the teratologic effects of 27 MHz because of the prevalence of this frequency in industries using RF heat sealers (Conover et al. 1979; Lacy et al. 1980a,b) At various days of gestation, the investigators used energy levels sufficient to cause and sustain high body temperatures (up to 43°C, 20 to 40 min). These experiments showed that GD 9 was the most sensitive day for the rat. A variety of malformations, including decreased body weight, was found. Thermal (maternal) threshold was 41.5°C. Gross microscopic, teratogenic effects were equivalent when increased maternal temperature was caused by MW or hot-water immersion. The threshold was approximately 12 W/kg for 30 min.

Conover et al. (1978) reported on single 20- and 30-min exposures of fetal rats on GD 10, 12, or 14 to 27.12 MHz at SARs of 17 to 35 W/kg. Preliminary results suggested that fetal wastage occurred after exposure on the 10th or 14th day of gestation. Grossly observable malformations occurred in animals exposed during the 10th day for 30 min, while visceral abnormalities occurred after exposure on the 10th and 14th days. Fetal mass as well as crown–rump length was smaller numerically in the fetuses exposed to MWs. The same investigators reported on eight groups of 16 to 28 gravid rats exposed to 27 MHz. The facility for exposure was an RF near field synthesizer operating in the dominant magnetic field mode at a field strength of 55 A/m. The dams were exposed on GDs 2, 4, 8, 10, 14, or 16 for 20 to 40 min at an average dose rate of 125 W/kg until their colonic temperature reached 43°C. Eight groups were sham-irradiated for 30 min and one group of 29 rats served as cage controls. Rats irradiated on GDs 8 to 16 (organogenesis) had a significant increase in gross malformations and a significant decrease in fetal weight and fetal crown–rump length.

Boak et al. (1932) administered shortwave radiation (1 MHz) to rabbits from the 29th day of life through several matings and pregnancies. The total exposure time ranged between 30 and 75 h, during which the temperatures of the animals were raised to 41 to 42°C. There was no interference with mating, fertilization, or development of the young *in utero*. Litter sizes were not significantly different from those of the control animals.

Roux et al. (1986) reported a study of rats exposed to 434 MHz radiation and ionizing radiation, indicating that the two forms of radiation potentiate teratogenic effects in the rat. In a study designed to examine possible effects of chronic RF exposure on mother–offspring behavioral patterns and the electroencephalogram (EEG), Kaplan et al. (1982) exposed 33 female squirrel monkeys near the beginning of the second trimester of pregnancy to 2.45 GHz RF in multimode, mode-stirred MW cavities at whole body SARs of 0.034, 0.34, or 3.4 W/kg, the last value equivalent to about 100 W/m^2 of plane wave RF, for 3 h/d, 5 d/week, until parturition. Eight pregnant monkeys were sham-exposed for the same periods. After parturition, 18 of the RF-exposed dams and their offspring were exposed to RF for an additional 6 months; then the offspring were exposed without the dams for another 6 months. No differences were found between RF- and sham-exposed dams in the numbers of live births or in the growth rates of the offspring. The major difference between RF- and sham-exposed offspring was that four of the five exposed at 3.4 W/kg both prenatally and after birth unexpectedly died before 6 months of age. They apparently had developed a form of pneumonia. Although the numbers of animals used in the behavioral and EEG studies were adequate, the mortality values were too small to place much confidence in statistical inferences.

O'Connor (1980) noted that with respect to basic design, procedure, and variables assessed, the teratogenic studies reported were more diverse than decisive. Wide variation in exposure parameters makes it difficult to compare the results; additional difficulty is generated because many of the reports do not contain information on critical variables such as the manner in which the day of gestation was timed. The day on which the animal is sperm or sperm-plug positive can be timed as day 0, although a more common procedure is to consider this day 1. The manner in which control animals were treated is often not given. Many kinds of controls have been employed including passive cage controls, sham-exposed controls, heat (IR-irradiated) controls, and historic controls. In some reports, probability statements have been substituted for data from concurrently studied control animals. Multiple control procedures in single experiments have not been used extensively. In addition, many of the investigators who have reported defects have employed acute, intense irradiation, which has placed a thermal burden on the exposed subject. Some studies attempt to control heat by including controls heated by means of IR. Of importance in this regard are the comparisons with the literature on heat stress. Many of the teratogenic studies of RF exposure have been performed without a sufficient number of animals, and others lack sophisticated design and thus do not allow for observation of low-probability events. Statistical analyses are usually not given in enough detail to permit evaluation (O'Connor 1980).

However, if attention is focused, not on procedural questions, but only on similarities in the results of the studies, several trends are apparent. The most common result from fetal exposure to MWs suggests that below thermal levels of exposure few changes are observed. An occasional nonspecific, general response of reduced or retarded gain of body mass may occur. However, further study would be required to know if such an effect is maintained after birth.

A more general deleterious response seen in the mammal is increased rate of fetal resorption. The increase may be indicative of malformed fetuses, but the resorbed nature of the fetal material precludes a more fine-grained analysis and thus identification of which, if any, specific structure was damaged. The increased rate of resorption and the range of exposures within which it occurs is remarkably similar to the effects of heat stress. Particularly in the rat, the resorption rate appears to increase within a rather narrow thermal window, the other side of which is fetal death. The majority of defects have been observed following high level, acute exposures with obvious thermal effect (O'Connor 1980).

It does appear that abnormalities in small animal fetuses can be produced with MWs in conjunction with systemic hyperthermia of 2.5 to 5°C above the normal temperature for the species during a time period of specific critical development. From a survey of the literature, it appears that it is the temperature rise in the fetus, irrespective of the manner in which it is produced, that causes damage. It is important to realize that in all species there is a dynamic pattern of maturation during gestation, and in the rat this continues during the first 3 weeks of postnatal life. In interpreting the few teratological effects of MWs that have been reported, as with any agent, it is important to realize that many fetal defects such as hemorrhage, resorption, stillbirth, and exencephaly also occur spontaneously in mice (O'Connor 1980).

Thermal stress, therefore, appears to be the primary mechanism by which RF energy absorption exerts a teratogenic action. Chernovetz et al. (1975) and others have pointed out evidence that indicates increases in mortality and resorption are probably related to peak body temperature and its duration, regardless of the method by which the temperature elevation is elicited. The teratogenic effects of hyperthermia, regardless of the source, have been well-documented (e.g., Edwards 1968, 1969; Pleet et al. 1981). Rugh and McManaway (1976) were able to prevent the increase in incidence of teratogenic activity, which they had previously reported, by lowering the maternal body temperature through controlled use of pentobarbital anesthesia.

Concepts of thermoregulation and thermal stress have been discussed by Michaelson (1982) and Way et al. (1981). The absence of a core temperature change does not necessarily indicate that physiologic adjustments induced by exposure to RF energy are not heat-related. The organism has numerous thermoregulatory mechanisms by which it can maintain homeostasis such as modification of skin blood flow, alterations in peripheral conductance, changes in blood temperature, and alterations in evaporative loss mechanisms. This does not imply that the organism is not stressed. The stress induced by maintaining thermal homeostasis may result in abnormal growth and development in the offspring.

In humans, infections such as rubella, influenza, and smallpox, occurring during early pregnancy, are known to cause abortions and fetal malformations (Wilson 1959). In general, it appears that any infection, giving rise to fever in early stages of pregnancy in humans or animals, is capable of producing fetal malformation or abortion. It is well known that induction of fever can lead to the early termination of pregnancy (Cameron 1943). There are numerous reports of abnormalities from the induction of systemic hyperthermia of 2.5 to 5°C above the normal temperature for the species for 1 h or longer during specific critical developmental stages of the fetus by exposure of the pregnant animals to elevated temperatures and humidity in environmental chambers. Fetal resorption, growth retardation, microphthalmia, and malformations affecting the CNS, musculoskeletal system, and other organs have been observed in mammalian species. In experiments on rats, hyperthermia of 4 to 4.5°C for 40 to 60 min during specific developmental stages produces increased fetal resorptions, retardation of growth, microphthalmia, anencephaly, and defects of tails, limbs, toes, palate, and body, depending upon the gestational stage at which hyperthermia occurred (Edwards 1968). These results indicate that the occurrence of fetal malformations in mammals in early pregnancy is probably related, not as much to the viral or bacterial toxemia, but to the fever, hyperthermia occurring at a particular critical stage of organogenesis. The threshold appears to be an elevation of 2.5 to 5.0°C above the normal temperature of the species, sustained for an hour or more.

As noted by Marston and Voronina (1976) from the standpoint of public health, one must consider the difficulty of extrapolating data from experimental teratology to the human fetus. Such an extrapolation becomes feasible only after detailed analysis of the fine mechanisms of teratogenesis. Also, of great importance is the need for appropriate scaling factors to permit extrapolation of experimental data obtained on small animals, to the human.

The reports of effects of MW exposure on early development have been reviewed by Baranski and Czerski (1976), who concluded that no serious effects are to be expected at power densities below 100 W/m^2 under usual exposure conditions. They further note that defects, when observed, are the result of hyperthermia. The more recent review by Heynick and Merritt (2003) offers no evidence to refute this conclusion. In fact, additional work performed supports the conclusion that no effects on early development are due to MW exposures that do not produce temperature increases in the biological target. There are numerous reports of abnormalities from the induction of systemic hyperthermia of 2.5 to 5°C above the normal temperature for the species, by exposure of the pregnant animal to elevated temperatures at specific critical developmental stages of the fetus (Edwards 1968, 1969). It would thus appear that in the reports of MW-induced developmental abnormalities, it is the temperature rise in the fetus, irrespective of the manner in which it was produced, that caused the damage.

RF energies must be applied at high SAR (>15 W/kg) to rodents, approaching lethal levels, for teratogenesis to result. High maternal body temperatures are known to be associated with birth defects. There appears to be a threshold for the induction of experimental birth defects when a maternal colonic temperature of 41 to 42°C is reached. Any agent capable of producing elevated internal temperatures in this range is a potential teratogen.

Most studies involving prenatal exposures have not shown effects on growth and development. Temperature in the testes of >45°C induced by any modality can cause permanent sterility; from 37 to 42°C, mature sperm may be killed with a temporary loss of spermatogenic epithelium. Changes in reproductive efficiency have not been directly associated with RF exposure.

3.5.3 Postnatal Development and Behavior

Few studies have been reported in which postnatal psychophysiologic parameters have been examined in animals exposed prenatally to MWs. Johnson et al. (1977) observed behavioral alterations in rats subsequent to protracted (20 h daily for 19 d during gestation) prenatal exposure to 0.918 GHz at a power density level of 50 W/m^2. Such exposure resulted in a significantly lower level of achievement in the conditioned avoidance response test. Jensh et al. (1982a) were not able to detect significant behavioral changes associated with protracted 915 MHz MWs at 100 W/m^2 in Wistar rats, approaching the power density level required to increase maternal colonic temperature. Chernovetz et al. (1975) subjected mice to a single 10-min exposure to 2.45 GHz radiation at an SAR of 38 W/kg, without observing any postnatal functional alterations.

In another study, gravid Wistar rats were exposed 8 h daily throughout the pregnancy to 915 MHz MWs at 100 W/m^2 (Jensh and Ludlow 1980). The average exposure time was 109 h. In this study, dams were allowed to deliver their offspring. Within 3 d after birth, the pups were given three tests for reflexes. After the pups were weaned, a series of performance tests was given. The postweaning tests began on postpartum day 60 and were completed by the 90th day. No detrimental effects were observed on performance in a water T-maze, avoidance behavior, open-field behavior, forelimb hanging, or performance in a 24-h activity wheel. Jensh (1984a,b) also reported no observable teratology, growth, eye opening, or postexposure behaviors (open-field tests and water-maze performance) in rats exposed throughout gestation to 350 W/m^2 MWs at 6 GHz. The SARs in pre- and perinatal exposures that are associated with behavioral effects are consistent with the generally accepted threshold for such effects as about 4 W/kg. Prenatal exposure of rats to cell phone RF signals had no effects in behavioral tests on the adult animals (Bornhausen and Scheingraber 2000).

Japanese quail, irradiated continuously in ovo during the first 12 d of embryogenesis at 50 W/m^2, SAR 4.03 W/kg, core temperature 37.5°C for exposed and controls, were followed to reproductive maturity. The reproductive histories of the exposed hatchlings were not different from the controls (Gildersleeve et al. 1987a).

3.6 Effects on the Nervous System

Signal transmission occurs in biological organisms through complex electrical and chemical events. The very nature of the electrical communication within animals and humans suggests a potential interactive target for influences from external EMFs. Thus the intense interest in the early investigations of possible MW exposure effects on the nervous system (see review by Elder and Cahill, 1984). More recent concerns on cellular telephone safety, particularly from exposure of the CNS to such devices have revived wide-ranging research efforts investigating MW and RF exposures to the nervous system. In this section, results will be presented from studies in the 1960s through 1980s at a variety of frequencies, but also discussed are more recent studies using RF exposures principally at cell phone frequencies.

Transient changes in CNS function have been reported following MW exposure. Although some reports describe the thermal nature of MW energy absorption, others implicate nonthermal or specific MW effects at the molecular and cellular level. The first report of the effect of MW energy in the centimeter range on the conditional response activity of experimental animals was made by Gordon et al. (1955). In subsequent years, study of nonthermal effects of MWs gradually occupied the central role in electrophysiological studies in the former Soviet Union (Novitskiy et al. 1971).

Yakovleva et al. (1968) reported that single and repeated exposures of rats to MWs, 50 to 150 W/m^2, weakened the excitation process and decreased the functional mobility of cells in the cerebral cortex. Edematous changes were most often noted throughout the cortex. The greatest number of altered cells was noted with repeated exposures at 150 W/m^2.

Seizure response to noise was transiently suppressed in audiogenic seizure-susceptible mice and rats after exposure to 3 GHz pulsed MWs at an average power density of 100 W/m^2 (Kitsovskaya 1960, 1968). Tolgskaya and Gordon (1960) studied the effects of both CW and PW MWs in rats at 3 and 10 GHz at various intensities. More pronounced morphological changes in the CNS were reported following 3 than 10 GHz at 10 to 100 W/m^2. Pulsed waves were initially considered to be more effective than CW, a phenomenon also reported by Marha (1963). While it is clear that the amount of energy absorbed from MW exposures depends on factors such as wavelength body size and shape, and orientation in the fields (Gandhi 1974; Durney et al. 1986), the overall evidence for differential nervous system or behavioral responses to continuous or pulsed-wave MWs is somewhat tenuous (Frey and Feld 1975). Additional studies comparing pulsed and CW MWs can be found in the monograph by Tolgskaya and Gordon (1973), who noted that exposure to MWs can cause conditioned reflex activity and functional changes in CNS activity were reversible. The functional changes disappeared at the same time that the conditional reflex activity of the animals was restored upon cessation of exposure to MWs. A proliferative reaction of glial cells was also described. Even at high peak powers and ultra-wideband exposures, no evidence was observed of differences between CW and PW (Sherry et al. 1995). A more detailed review of PW investigations is presented by Lu and deLorge (2000) and Pakhomov and Murphy (2000).

Some investigators suggested that MW energy absorption may affect hypothalamic and midbrain function and also affect cerebral, cortical, and reticular system function (Yermakov 1969; Thompson and Bourgeois 1975). According to Gvozdikova et al. (1964), the greatest cortical sensitivity occurs in the meter range, less in the decimeter, and least in the centimeter MW band.

Guy and Chou (1982) and D'Andrea et al. (1994) studied response thresholds in rats for short, high-intensity microwave pulses. The energies were substantial in excess of those producing auditory responses (500 μs wide or less, 5 to 180 mJ/kg in brain; see Section 3.13.1). Rats were subjected to local exposures to the head and especially the brain of a single 915 MHz pulsed field sufficient to produce a local SAR as high as 4×10^5 W/kg and with pulse widths from 1 μs to 360 ms. Animals displayed no reaction other than that due to the hearing effect until the specific absorption exceeded 28 kJ/kg. This correlated with a temperature rise in the brain of 8°C. Seizures lasted for 1 min after exposure followed by a 4- to 5-min unconscious state. Rats then recovered without apparent effect from the exposure. Brain histology revealed some demyelination of neurons 1 d after exposure and focal gliosis 1 month after exposure. Similar studies were performed by Petin et al. (2000).

By contrast Brown et al. (1994), also looking at high power pulses in intervals from 25 ms to 3.2 s, found effects at lower energies. Utilizing a 1.25 GHz source, comparing pulses at a repetition rate of 80 per second to equivalent CW for each episode of exposure, varying peak powers and total duration for different episodes of exposure, a threshold was found for an evoked body movement, an involuntary movement measured by a motion sensor attached to the tail of the rat. No differences were observed between PW and CW (or wide pulses) at any given SAR and specific absorption (SA, total energy absorbed per unit mass in J/kg). When the SAR was held constant at 7.3 kW/kg under either mode and the SA was allowed to increase by varying the total exposure time from 0 to 3.2 s, the percentage of animals displaying the body movement increased proportionally up to 0.9 kJ/kg and then leveled off at 84% of the animals, reflecting a dependency on SA. On the other hand, holding SA constant at 0.9 kJ/kg and varying SAR revealed that above 900 W/kg most animals responded. However, lowering SAR (by lengthening the interpulse interval at constant SA) produced a proportionate decrease in percentage of responding animals out to 3.2 s, reflecting the ability of the thermoregulatory mechanisms to adjust to these lower SARs. At the highest energies, brain temperatures rose by 0.4°C. Skin and subcutaneous temperatures facing the energy source at the higher energies were measured and found to be at levels known to produce a nonpainful thermal sensation. The authors were uncertain as to the mechanism underlying the evoked movement. In further work of Brown et al. (unpublished), skin and subcutaneous regions of the snout and surrounding areas were anesthetized and animals subjected to pulsed energy as before. Thresholds and nature of responses were identical to those of the previous exposures, suggesting that cutaneous mechanisms did not mediate the evoked movement. Wachtel, reflecting on his earlier work with *Aplysia ganglia* (Wachtel et al. 1982), has proposed that the evoked movement reported by Brown et al. (personal communication) may result from the rate of temperature rise rather than the temperature change *per se* in a direct coupling with nervous tissue. The issue is, however, unresolved. Akyel et al. (1991) demonstrated that the evoked body movement produced by PW microwaves is differently affected by a number of neuropharmacologic drugs from that of the classic startle response. Additional work on high peak, low average power acute, and short-term effects is summarized in Section 3.7.

3.6.1 Electroencephalographic Changes

Several investigators have reported that MW exposure produces alterations in the EEG (Baldwin et al. 1960; Livanov et al. 1960; Kolodov 1966; Baranski and Edelwejn 1967; Bawin et al. 1973; Servantie et al. 1975; Suvorov and Kukhtina 1984). Chizhenkova (1988) exposed unanesthetized rabbits to 2.4 GHz for 1 min at 400 W/m^2 and observed an increase in the number of slow waves in the EEG, and a change in the discharge frequency of neurons in the visual cortex. Evoked response of visual cortex neurons to light was enhanced and was regarded as the most sensitive indicator of an MW effect. Baldwin et al. (1960) exposed the heads of rhesus monkeys to 225 to 400 MHz CW in a resonant cavity and noted a progressively generalized slowing and some increase in amplitude of EEG patterns accompanied by signs of agitation, drowsiness, akinesia, and nystagmus, as well as autonomic sensory and motor abnormalities. There were signs of diencephalic and mesencephalic disturbances: alternation of arousal and drowsiness, together with confirming EEG signs. The response depended on orientation of the head in the field and reflections from the surrounding enclosure. In this report from 1960, the animals were described as exposed in a copper mesh resonant cavity into which a quarter wave antenna was inserted at the top. No direct dosimetry is described, but instead temperature measurements were performed in a vessel filled with water the size of an average monkey head. Temperature changes were merely described as trivial. Monkeys were fitted with metallic electrodes driven into the outer table of the skull for EEG measurements. It is a reasonable speculation that greater amounts of energy than suspected produced localized heating around electrodes affecting the results. Additional studies where effects of MW on EEGs were observed include rabbits exposed at 30 MHz, 0.5 to 2 kV/m (Takishima et al. 1979); cats at 147 MHz, unspecified SAR (Bawin et al. 1973); rats at 945 MHz, unspecified SAR (Vorobyov et al. 1997); rats at 900 MHz, 1.3 W/kg (Thuroczy et al. 1994). In the range of studies that have been performed the types of changes in EEG are not consistent across studies.

In a joint U.S. and former USSR study, Mitchell et al. (1989) studied rats exposed for 7 h at 2.45 GHz and 100 W/m^2; in comparing results between U.S. and former USSR investigators, he found statistically significant effects in the power spectral analysis of EEG frequency, but the two groups found such effects at different frequencies. EEG tracings in rabbits exposed to 3 GHz (PW) 50 W/m^2 showed slight desynchronization from the motor region; at 200 W/m^2 variations in the amplitude were observed; 300 MHz had a greater effect than 3 GHz. It was suggested that PW MWs produced a greater effect than CW MWs.

Baranski and Edelwejn (1975) reported that rabbits exposed to 10 GHz (pulsed), 40 W/m^2 (single exposure) showed no changes in EEG tracings, but exposure to 3 GHz, 70 W/m^2, 3 h/d for 60 d, produced functional and morphological changes. Changes in conditional reflexes were also reported in rats exposed to 70 MHz, 150 V/m, 60 min/d for 4 months.

A review of the literature on EEG effects requires awareness of certain deficiencies in methodology and interpretation. The EEG is difficult to quantify due to its time-varying waveform. The use of metallic electrodes either implanted in the brain or attached to the scalp also makes many of the reports on EEG or evoked responses (ER) questionable. Johnson and Guy (1972) have pointed out that such metallic electrodes grossly perturb the field and produce greatly enhanced absorption of the energy in the vicinity of the electrodes. Such enhancement produces artifacts in the biological preparation under investigation. Recording artifacts also result from pickup of fields by the electrodes and leads during the recording of EEGs while the animal is exposed. For example, Tyazhelov et al. (1977) have pointed out that even for the coaxial electrode described by Frey (1968), diffraction of EM waves is still a major source of error because of the metallic nature and large dimensions of the electrode.

Bawin and Adey (1977) reported that electromagnetic energy of 0.147 MHz, amplitude modulated at brain wave frequencies, influenced spontaneous and conditioned EEG patterns in the cat at 10 W/m^2. These amplitude modulated 147 MHz fields induced changes only when the amplitude modulation frequency approached that of physiological bioelectric function rhythms; no effects were seen at modulation frequencies either below 8 Hz or above 16 Hz.

Human studies of RF fields have sometimes reported increases in parts of the EEG spectra (Reiser et al. 1995; von Klitzing 1995) often around the 10 Hz band (Borbély et al. 1999; Huber et al. 2000; Lebedeva et al. 2000, 2001). Additional reports failed to show the increases observed in above studies and reported either decreases (Croft et al. 2002) or no effects of exposure on spontaneous, awake EEG assessments (Kim 1998; de Seze 2000; Heitanen et al. 2000).

In studies examining RF mobile phone signals a shortening of sleep onset time and changes in EEG during sleep were observed (Mann and Roschke 1996; Mann et al. 1998a). However, in follow-up studies, the changes reported earlier were not statistically altered, even at greatly increased (100 to 250 times) exposure levels (Roschke and Mann 1997; Wagner et al. 1998, 2000; Mann and Roschke 2004). Huber et al. (2000) examined the spectral power of the EEG in humans for 900 MHz exposures. They found a short increase in the spectral power of nonrapid eye movement (NREM) sleep during the first 30 min but no effects on sleep latency, sleep stages, or rapid eye movement (REM) sleep spectrum. In a subsequent study, the same group observed that EMF alters waking regional cerebral flow and pulse modulation of the MW exposure is necessary to induce waking and sleep EEG changes (Huber et al. 2002).

A few studies of human event-related brain potentials have reported varied effects following RF exposure (Eulitz et al. 1998; Freude et al. 1998, 2000; Kellenyi et al. 1999).

Other studies looking at similar issues have reported no effects (Kim 1998; Urban et al. 1998; Hladký et al. 1999). Early findings by Krause et al. (2000a,b) of event-related potential changes during working memory tasks were not repeatable in replication efforts.

3.6.2 Calcium Efflux

In 1975, Bawin et al. (1975) reported changes in the calcium released from chick brain tissue following *in vitro* exposure to amplitude-modulated RF. The levels used appeared to preclude RF heating as a mechanism for the effect. Furthermore, effective calcium changes were dependent on amplitude modulation (AM), with significant effects reported for 6, 9, 11, 16, and 20 Hz. The effect also appeared to be power-dependent (Sheppard et al. 1979; Blackman et al. 1980a), suggesting the occurrence of power and frequency windows of effectiveness. Following these initial reports, a variety of studies were performed to investigate the calcium efflux effects of RF exposure (Bawin and Adey 1977; Bawin et al. 1978; Blackman et al. 1979, 1980a,b, 1982, 1985a,b, 1988, 1989, 1991; Joines and Blackman 1980; Albert et al. 1987).

In addition to chick brain, other neurological tissues have been examined following exposure to RF fields. These studies include cats showing irregular calcium efflux (Adey et al. 1982); increased efflux in rats exposed *in vivo* (Paulraj et al. 1999); no effect on rat brain tissue exposed *in vitro* (Shelton and Merritt 1981; Merritt et al. 1982); increased calcium efflux in rat brain synaptosomes exposed *in vitro* (Lin-Liu and Adey 1982); and human and rodent neuroblastoma cell lines, which showed effects similar to chick brain when exposed (Dutta et al. 1984, 1989). An overall evaluation indicates generally inconsistent results in calcium efflux from both *in vitro* and *in vivo* experiments. A detailed review of this literature is available (NRPB 2001).

3.6.3 Histopathology

According to some authors, cellular changes have been found in the nervous system of small animals following MW exposure at 100 W/m^2 (Tolgskaya 1959; Tolgskaya and Gordon 1960, 1964). Albert and Sherif (1988) exposed 1 and 6 d old rats in the far field to 2.45 GHz, CW microwaves at 100 W/m^2 for 5 consecutive days, 7 h/d (SAR 2W/kg). Animals were sacrificed 1 d after exposure cessation, and histology revealed evidence of widespread cell death with associated mononuclear cell infiltration.

Degeneration of neurons in the cerebral cortex and tissue changes in the kidney and myocardium of rabbits have been produced by exposure to 200 MHz. Accinni et al. (1988) found evidence for a condition resembling immune complex glomerulonephritis in rabbits exposed to 27.12 MHz with a diathermy apparatus sufficient to elevate rectal temperatures 2 to 3°C per episode of exposure. The exposures occurred daily for 3 weeks up to 20 min/d.

Exposure of cats for 1 h to 10 GHz, 4 kW/m^2 resulted in injury to cerebral and spinal cord nerve cells; changes occurred in tigroid substance (Nissl bodies) and other components of nerve cells (Bilokrynytsk'ky 1966). On the other hand, rabbits exposed to 10 GHz (pulsed), 40 W/m^2 showed no evidence of morphological damage to the brain, but comparable exposure to 3 GHz did produce such changes (Baranski and Edelwejn 1967).

Tolgskaya and Gordon (1960) have investigated the influence of PW and CW 3 and 10 GHz on the morphology of nervous tissue in rats and rabbits. With exposure to 3 GHz (1.1 kW/m^2 and 400 W/m^2), severe symptoms of overheating were observed, often leading to death. Severe vascular disorders such as edema and hemorrhages in the brain and internal organs were prominent. In repeated but less prolonged exposure, vascular disorders and degenerative changes in internal organs and the nervous system were less severe. With repeated exposures, the animals were better able to withstand successive exposures; they continued to gain weight; body temperature after irradiation quickly recovered; and temperature increase was not evident. Such a response resembles thermal adaptation on a physiological level.

At high field intensities, when death is a result of hyperthermia, the vascular changes are those of hyperemia, hemorrhage, and acute dystrophic manifestations (Dolina 1961; Minecki and Bilski 1961; Baranski et al. 1966). At low field intensities, the changes are of a more general dystrophic character, and proliferation of the glia and vascular changes are not as prominent.

Albert and DeSantis (1975) reported morphological changes in the brains of Chinese hamsters following exposure to 2.45 GHz CW MWs at power densities of 250 and 500 W/m^2. Exposure durations varied from 30 min to 14 h/d for 22 d. Both light and electron microscopic findings revealed alterations in the hypothalamus and subthalamic structures of exposed animals, whereas other regions of the brain appeared unaltered. It should be noted that extremely high local SAR could be present under these conditions. Peak SAR could reach 40 to 200 W/kg in selected brain regions.

In subsequent studies at 1.76 and 2.45 GHz, 100 and 200 W/m^2, Albert (1977, 1978) described cytoplasmic vacuolization of neurons, irregular swelling of axons, and decrease in dendritic spines of cortical neurons. The axonal swelling and spine changes were seen only in chronic exposures, whereas neuronal changes were observed in acute exposure. In all studies, no signs of permanent degenerative changes were recorded and reversibility was noted 2 h after exposure (Albert 1977). The author concluded that while it was possible that higher exposure levels (250 and 500 W/m^2) could have resulted in thermal effects, it was unlikely that 100 W/m^2 would result in significant thermalization of the whole brain. He did not, however, rule out the possibility of hot spots. Exposure of rats to 100 W/m^2 of 2.45 or 2.8 GHz RF resulted in average hypothalamic temperature increases of 0.4°C or less. This increase is lower than hypothalamic temperature increases observed during the normal activity of an animal (Albert et al. 1977, 1978).

Albert et al. (1981b) looked at the effects of RF exposure pre- and postnatally on the Purkinje cells of the rat cerebellum. In one experiment, Sprague–Dawley rats were exposed *in utero* to 100 MHz CW RF at 460 W/m^2 (SAR 2.81 W/kg) for 6 h/d on GDs 16 to 21, and then for 4 h/d for 97 d after birth. Four exposed and four sham-exposed animals were sacrificed 14 months after cessation of irradiation. Quantitative assessment of the cerebella showed that the relative number of Purkinje cells was significantly smaller (12.7%) in experimental animals than in control animals. In another experiment, Sprague–Dawley rats were exposed *in utero* to 2.45 GHz CW RF at 100 W/m^2 (SAR 2 W/kg) for 21 h/d on GD 17 to 21. Power density measured was variable from 40 to 300 W/m^2 because of group exposure conditions. Half of the litters were used shortly after delivery and the other half 40 d after cessation of irradiation to assess effects of the exposure on cerebellar Purkinje cells. Because of the immaturity of the neonates, the Purkinje cell layer was not clearly displayed and quantitative results could not be obtained. Experimental animals sacrificed 40 d postexposure showed significantly fewer (25.8%) cells than did the controls. In a final experiment, rat pups were exposed to 2.45 GHz RF at 100 W/m^2 for 7 h/d on postnatal days 6 to 10. Half the pups were sacrificed at the end of exposure and half at 40 d postexposure. Only those sacrificed immediately showed a significant decrease in the relative number of Purkinje cells as compared with sham-exposed controls. Thus, exposure to two frequencies of RF at similar SAR values (2.8 and 2 W/kg) yielded a reduction in the relative number of Purkinje cells for fetuses and newborn rats. Although the change appeared permanent for rats exposed *in utero*, it appeared to be reversible for those exposed postnatally.

In a related study, Albert et al. (1981a) examined the effects of RF exposure on the Purkinje cells of squirrel monkey cerebella. Pregnant squirrel monkeys were exposed to 2.45 GHz pulsed RF at an equivalent power density of 100 W/m^2 (SAR 3.4 W/kg) 3 h/d starting in the first trimester of pregnancy. The offspring were exposed similarly for the first 9.5 months after birth. At the end of the irradiation period, seven exposed and seven sham-exposed animals were sacrificed and their cerebella examined. There were no statistically significant differences between control and exposed animals in any of the Purkinje cell parameters examined. Several factors were suggested by Albert et al. (1981b) to explain the discrepancy between these results and those with the rat. Factors that might have contributed were differences in geometrical configurations of the head, exposure methods, and daily exposure durations, as well as variations in gestational periods and species differences. It is quite possible that high local SAR values in the rat brain, but not in the squirrel monkey brain, at comparable whole body SARs, may be the important factor in explaining these differences.

3.6.4 Effects on the Blood–Brain Barrier

The existence of a BBB in most regions of the brain normally provides resistance to movements of large molecular weight, fat-insoluble substances (e.g., proteins or polypeptides) from the blood vessels into the surrounding cerebral extracellular fluid, presumably to protect the brain from invasion by various blood-borne pathogens and toxic substances. Several investigators have reported that RF can increase the permeability of the BBB to certain substances of large molecular weight, as described below. However, a majority of studies have been unable to confirm such effects.

The concept of the BBB is replete with definitional problems even preceding the interpretation of experiments. It is known that normal physiologic mechanisms of substance transport can occur by simple diffusion, facilitated diffusion, active transport or pinocytosis, and vesicular transport. Different substances are subjected to different mechanisms, making comparisons of differing substances as markers an uncertain process. The concept of an intrinsic permeability, which can be separated from other factors such as local cerebral blood flow (LCBF), has proved to be especially difficult to characterize. Changes in temperature, which are inevitable concomitants of RF experiments, alter several processes simultaneously making it difficult to identify changes in an intrinsic permeability.

The initial report in Western literature of possible impacts of MW exposure on the BBB was of a study of 2.45 GHz at an SAR of approximately 1 W/kg (Frey et al. 1975). Their report of increased BBB permeability was followed up by reports from Oscar and Hawkins (1977) that supported the original findings. They reported that a single exposure of rats to 1.3 GHz pulsed MWs for 20 min (1 to 20 W/m^2) induces a temporary change in the permeability of small inert polar molecules across the BBB of rats. Increases in permeability were observed for mannitol and inulin, but not for dextran, both immediately and 4 h after exposure, but not 24 h after exposure. They found statistically significant changes in the brain uptake index (BUI), also known as the Oldendorf (1970) technique, at average power densities less than 30 W/m^2. They also found that, depending on the specific pulse characteristics used, pulsed RF could be more or less effective in altering BBB permeability than CW RF of the same average power density. For pulses of long duration and high pulse power density, but only a few pulses per second, mannitol permeation could be induced at average power densities as low as 0.3 W/m^2. The authors considered the possibility of local heating due to hot spots since the greatest BBB alteration occurs in the cerebellum and medulla or close to the neck region of rats.

The Oldendorf technique involves injecting a measured amount of a ^{14}C-labeled test substance and ^{3}H water arterially, allowing one pass of the bolus through the brain, and following by decapitation and measurement of radioactivities in specific parts of the brain. A measurement of the ratio ^{14}C–^{3}H in brain tissue is normalized to the identical ratio in the injectate. The technique assumes that tritiated water freely diffuses between the brain and its vascular system. In fact, the passage of tritiated water is a function of LCBF, which itself varies, and therefore distorts the interpretation of the overall ratio.

Preston et al. (1979) and Preston and Prefontaine (1980), using methods similar to those of Oscar and Hawkins (1977), attempted to determine whether exposure to 2.45 GHz CW RF increased BBB permeability to ^{14}C-labeled D-mannitol. They exposed rats to 1, 5, 10, or 100 W/m^2 and compared with sham-exposed rats for controls, they found no evidence that RFR exposure increased the permeability of the BBB for mannitol. In a second series, rats were exposed to 3, 10, 30, 100, and 300 W/m^2. Again, there were no differences between results from exposed and sham-exposed animals. Like Oscar et al. (1981), Preston et al. (1979) believed changes in LCBF confounded the results of earlier studies.

Following up on earlier studies and in light of Preston's suggestion, Oscar et al. (1981) subsequently used a technique employing ^{14}C iodoantipyrine to measure LCBF in rats. Male rats weighing 250 to 300 g were individually sham-exposed for 5, 30, and 60 min, or individually exposed for 5, 15, 30, 45 or 60 min to pulsed RF at 150 W/m^2 average power density. Carrier frequency was 2.8 GHz, pulse rate was 500 pps, and pulse width was 2 μs. Within 5 min after sham or RF exposure, the previously venous-catheterized, but conscious, animals were infused with isotonic saline containing 5 μCI/mL of ^{14}C-iodoantipyrine. The rats were decapitated 50 s after the start of the infusion. Brain regions were dissected out and assayed for radioactivity by routine liquid scintillation counting. LCBF was then calculated by established procedures. The results show that MW exposure caused a significant increase in LCBF (minimum 39% to >100%) in all 17 brain regions sampled. Because of these findings, the authors indicated that their earlier reported ratio measurements (Oscar and Hawkins 1977) may be an overestimate (Gruenau et al. 1982).

Attempts to duplicate the findings of Oscar and Hawkins (1977) have yielded equivocal results, as Chang et al. (1982) found, or failure as reported in a series of studies by Williams et al. (1984a–d). They (Williams et al. 1984e) concluded that studies reporting increased permeability of the BBB in many cases were subjected to technically derived artifact invalidating the conclusion. The authors found decreased entry of horseradish peroxidase (HRP) and ^{14}C sucrose into the microvessel endothelium of hyperthermic rats at cerebral temperatures up to 43°C. They stated that it is possible that temperatures in excess of 43°C might disrupt the barrier by overt thermal injury as has been reported in clinical studies. They concluded that hyperthermia produces disruption of membrane functions including pinocytosis, but pointed out that other modes of transport, such as active and facilitated transport mechanisms, have not been evaluated. Others have also been unsuccessful in attempts to replicate Oscar and Hawkins (Ward et al. 1982; Ward and Ali 1985). Studies in mice exposed for either 1 h or after a lifetime of exposure at 4 W/kg also demonstrated no permeability changes in the BBB (Finnie et al. 2001, 2002).

Merritt et al. (1978) exposed rats for 30 min to 1.2 GHz pulsed RF at peak power densities in the range of 20 to 750 W/m^2 and 0.5 duty cycle, corresponding to average power densities of 10 to 380 W/m^2, or for 35 min to 1.3 GHz pulsed or CW RF at average power densities in the range of 1 to 200 W/m^2. They examined brain slices under UV light for transfer of fluorescein and under white light for transfer of Evans blue dye (a visual tracer) across the BBB, and chemically analyzed various brain regions for fluorescein content. They also measured the brain uptake of ^{14}C-labeled D-mannitol and determined the BUI values. To validate these detection methods, they used hypertonic urea, known to alter the BBB, as an alternative agent to RF. Finally, sham-exposed rats were heated for 30 min in a 43°C oven to approximate the hyperthermia obtained at 380 W/m^2. In their examination of brain slices, Merritt et al. (1978) found no evidence of enhanced fluorescein or Evans blue dye transfer across the BBB of RF-exposed rats, whereas penetration of the BBB was apparent for rats treated with urea instead of RF. The analyses of fluorescein content corroborated these findings. However, fluorescein uptake was high for the sham-exposed rats that were heated in the oven, an indication that hyperthermia of the brain is necessary to alter BBB permeability. In the ^{14}C-mannitol study of the various brain regions, there were no significant differences in BUI between RF and sham-exposed rats, whereas BUI changes were evident for rats treated with urea. Also, the results showed no evidence of the power density window reported by Oscar and Hawkins (1977). No change in the BBB occurred unless brain temperature was increased by 4°C.

Early on work relevant to MW exposure at high levels was conducted in which the effects of heating were examined on the integrity of the BBB. Sutton et al. (1973) used 2.45 GHz to produce selective hyperthermia of the brain in rats. He then studied the integrity of the BBB with HRP, a protein tracer that can be detected both morphologically and quantitatively. Brains were heated to 40, 42, and 45°C. Barrier integrity was disrupted after heating for more than 45 min at 40°C. Animals with brains heated to 45°C survived for only 8 to 15 min. The most common site of vascular leakage was the white matter adjacent to the granular cell layer of the cerebellum. Sutton concluded that to prevent BBB disruption, brain temperatures must not exceed 40°C in the absence of body core hypothermia.

Albert et al. (1978) also used HRP as a tracer and reported regions of leakage in the microvasculature of the brains of Chinese hamsters exposed to 2.45 GHz MWs at 100 W/m^2 for 2 to 8 h. In control animals, extravascular reaction product was found only in brain regions normally lacking a BBB. In a later paper, Albert (1979) reported that continuation of these earlier studies indicated that a partial restoration of the BBB's impermeability may have occurred within 1 h after exposure ceased, and that restoration was virtually complete within 2 h. Albert believed that the transient changes may be clinically subacute and probably cause no lasting ill effects. It is important to note, however, that such leakage of the microvasculature of the brain occurs irregularly; this was observed in approximately 50% of exposed and 20% of control animals studied by Albert and DeSantis (1975).

In another study, Albert (1979) exposed 52 animals (34 Chinese hamsters and 18 rats) to 2.8 GHz RF for 2 h at 100 W/m^2. Of these, 30 were euthanized immediately, 11 at 1 h after exposure, and 11 at 2 h after exposure. Another 20 animals (12 hamsters and 8 rats) were sham-exposed. Leakage of HRP in some brain regions was reported for 17 of the 30 animals euthanized immediately after RF exposure and for 4 of the 20 sham-exposed animals. Fewer areas of increased BBB permeability were evident for animals euthanized 1 h after RF exposure, and except for one rat, virtually no leakage of HRP was seen for the animals euthanized 2 h after RF exposure. These results indicate that increased BBB permeability due to RF exposure at levels insufficient to denature brain tissue is a reversible effect. Albert suggested that such BBB changes may be clinically subacute and would probably cause no lasting ill effects. However, the increased BBB permeability seen in 4 of 20 sham-exposed animals may indicate that factors other than RFR in the experimental procedure could alter the BBB. One possible confounding point in the use of injected HRP as a tracer is the existence of endogenous peroxidase, the detection of which could yield false positive results. Also, no positive (BBB-altering) control agent was used in these studies for comparative purposes.

Comparable results were reported by Albert and Kerns (1981) on 51 Chinese hamsters exposed to 2.45 GHz CW RF for 2 h at 100 W/m^2. There were 39 original sham-irradiated controls. Of the RF-exposed animals, 12 were allowed to recover for either 1 or 2 h prior to

HRP injection and subsequent fixation. This study appears to be an extension (for Chinese hamsters) of the work reported by Albert (1979) for 2.8 GHz RF exposure, with the same conclusions as that study.

Neubauer et al. (1990) reported that 2.45 GHz PW microwave radiation at 100 W/m^2, SAR approximately 2 W/kg, pulse width 10 μs at 100 pps, resulted in enhanced uptake of rhodamine–ferritin complex by capillary endothelial cells of the cerebral cortex in rats, and was related to exposure duration and power density. Brain temperatures were not reported.

By using a small, dielectrically loaded coaxial applicator, Lin and Lin (1980, 1982) were able to irradiate only the heads of anesthetized adult male Wistar rats with pulsed 2.45 GHz RF (10 μs, 500 pps) for 20 min at average power densities of 5 W/m^2 to 30 kW/m^2. The distribution of absorbed MW energy inside the head was determined by thermographic procedures, and average SARs were found to range from 0.4 to 240 W/kg. Evans blue dye was injected into a catheterized femoral vein following sham or RFR exposure, and 5 min later the animal was perfused via the left ventricle with normal saline. The brain was removed, examined, and scored for degree of tissue staining by the tracer. For average power densities up to and including 26 kW/m^2 (200 W/kg), staining was not significantly different between exposed and control animals. For exposures at 30 kW/m^2 (240 W/kg), extravasation of Evans blue dye could be seen in the cortex, hippocampus, and midbrain. The degree of staining decreased with increasing time to euthanasia postexposure, indicating that the effect was reversible. No alteration in the BBB was evident unless brain temperature was increased by 4°C. Neilly and Lin (1986) found that ethanol inhibits microwave-induced permeation of the BBB through reduced heating of the brain. Other groups have obtained results consistent with thermal effects (Goldman et al. 1984; Moriyama et al. 1991; Ohmoto et al. 1996; Fritze et al. 1997).

Chang et al. (1982) used a technique involving ^{131}I-labeled albumin to investigate alterations of the BBB in dogs. The heads of dogs were exposed at various average power densities between 20 and 2000 W/m^2. In general, no statistically significant differences were found between exposed and sham-exposed animals, but the number of animals used in this study was too small to ascribe a high level of statistical confidence.

One group has reported increases in BBB permeability in rats exposed to CW 915 MHz fields independent of SAR, from 0.02 to 8.3 W/kg. With modulated RF effects were also not SAR-dependent with the confusing result of significant effects at low SARs of 0.0004 to 0.008 W/kg but no effect at the highest SARs (1.7 to 8.3 W/kg) (Persson et al. 1992, 1997; Salford et al. 1993a, 1994, 2003).

It is important to realize that the methods used to investigate BBB permeability are still somewhat controversial. Permeability changes in cerebral blood vessels occur under various conditions, including those that produce heat necrosis (Rozdilsky and Olszewski 1957). Most techniques used to measure BBB permeability in fact measure the net influence of several variables on brain uptake and do not differentiate among the effects of changes in the vascular space, alterations of blood flow, and membrane permeability.

The uncertainty in earlier research on the influence of RF on the BBB is no doubt related to significant artifacts introduced by the kinds of biological techniques used. Several investigators have indicated that exposure to RF may alter the size of vascular and extravascular volumes and cerebral blood flow rate, thereby yielding changes in the BUI that are not necessarily related to BBB permeability alterations. Blasberg (1979) reviewed many of the methods previously used for investigating BBB changes and the problems associated with these methods. Rapoport et al. (1979) developed a method for measuring cerebrovascular permeability to ^{14}C-labeled sucrose that yields results independent of cerebral blood flow rate. As already noted, Oscar et al. (1981) confirmed experimentally that LCBF is increased in the rat brain by exposure to pulsed RFR at 150 W/m^2 average power density.

There is still not a complete concensus among students of the BBB question. There is a widespread view that little confidence can be placed in the results of early experiments on RF-induced BBB alterations. The concensus is that hyperthermic levels of RFR, producing temperatures >43°C, can alter the permeability of the BBB, but in the majority of BBB studies at levels of MW exposure under 4 W/kg, exposure did not affect permeability of the BBB in animals.

3.6.5 Combined Effect of Radio Frequency and Drugs, and Radio Frequency Effects on Neurotransmitters

A pharmacodynamic approach has been taken by some investigators in the study of MW-exposure effects on the CNS and behavior (Baranski and Edelwejn 1968; Edelwejn 1968; Thomas and Maitland 1979). Following exposure to 3 GHz PW MWs, altered sensitivity to neurotropic drugs was noted. Decreased tolerance of rabbits to pentylenetetrazole and increased tolerance to strychnine were observed after a single exposure to 200 W/m^2. Repeated exposures at 70 W/m^2 produced a decreased tolerance to pentylenetetrazole, strychnine, and acetophenetidin (Lin and Lin 1982). Servantie et al. (1976) reported that exposure of rats to 3 GHz at 50 W/m^2 for several days resulted in an altered reaction to pentylenetetrazole. Using curare-like compounds, a neuromuscular site of action for this MW effect was implicated. Edelwejn (1968) observed alterations in the effects of chlorpromazine and D-tubocurarine on EEG recordings in rabbits repeatedly exposed at 70 W/m^2. The author concluded that synaptic structures at the level of the brain stem are affected by MWs. It should be noted that drugs such as pentylenetetrazole and chlorpromazine influence physiologic thermoregulation.

In some cases, drugs and microwave irradiation might have parallel effects, suggesting effects via a common mechanism. Lai, for example, has suggested that the parallel effects of MW and endogenous opioids on choline metabolism may be due to activation of endogenous opioids by MWs. Lai et al. (1987a), studying the effect of RF on central cholinergic activity, measured sodium-dependent, high-affinity choline uptake in different parts of the rat brain *in vivo* exposed to 2.45 GHz microwave radiation, 100 W/m^2, CW and PW (100 W/m^2, 2 μs pulses, 500 pps), whole body average SAR 0.6 W/kg. PW was found to decrease choline uptake in the hippocampus and frontal cortex. There was no effect on the hypothalamus, striatum, and inferior colliculus. Pretreatment with a narcotic antagonist blocked the effect of PW on hippocampal choline uptake but did not alter the effect on the frontal cortex. CW decreased uptake in the frontal cortex but not in the hippocampus, striatum, or hypothalamus. These effects have been further characterized by the same group of investigators (Lai et al. 1987b, 1988, 1989, 1991, 1992a,b; Lai 1992). The authors' conclusion from this series of experiments is a suggestion that the effect of RF on central acetylcholine metabolism is probably a secondary effect of endogenous opioids, which may result from the character of RF as a stressor. The auditory effect of pulsed RF could act as such a stressor. According to Lai, microwave-induced changes in central high-affinity choline uptake may also be mediated by corticotropin-releasing factor (Lai et al. 1990).

Microwave energy can affect the action of psychoactive drugs. Modulation of the action of barbiturates and related effects on the EEG has been reported. It has been reported that warming an animal with microwaves during anesthesia can attenuate the effects of pentobarbital (Bruce-Wolfe and Justesen 1985). Microwave energy can prolong the narcolepsy and hypothermia induced by pentobarbital. The role of local absorption patterns was found to be important.

A number of drugs can modify the thermal responses to microwaves. For example, chlorpromazine can attenuate the thermal effect of microwaves in ketamine-anesthetized rats (Jauchem et al. 1985a). On the other hand, chlorpromazine was reported to cause an increase in susceptibility to microwave heating in the dog (Michaelson et al. 1961).

Anticonvulsant drugs such as chloral hydrate can decrease susceptibility of rats to MW-induced convulsions (Lobanova 1985). The subject of drug modifications of thermal responses to microwaves has been reviewed by Jauchem (1985).

Pulsed microwaves have been reported to enhance the hypothermic effects of acetylcholinesterase inhibitors in the CNS (Ashani et al. 1980). Quock et al. (1987) reported that microwave exposure facilitates the antagonism of domperidone- to apomorphine-induced stereotypic climbing in mice. Quock et al. (1986) reported that microwave exposure facilitates the antagonism of methylatropine to central cholinomimetic drug effects.

Lai et al. (1984) found that microwave radiation enhances ethanol consumption but attenuates ethanol hypothermia at low, nonthermalizing levels in the rat. Hjeresen et al. (1988, 1989) further characterized the ethanol effect. These investigators found, utilizing low-intensity exposures (0.3 W/kg, 45 min exposures, 2.45 GHz), that attenuation of ethanol-induced hypothermia by microwaves is more effective at lower doses of ethanol than at higher doses. They argued for a nonthermal effect of microwaves, possibly an effect on neurotransmitter systems based on similarities to pentobarbital-induced hypothermia.

Neilly and Lin (1986) studied the possible impact of ethanol on the BBB in rats. Varying amounts of ethanol were administered followed by high intensity (30 kW/m^2 for 15 min) microwaves. Rectal temperature remained constant at 37°C and brain temperatures were monitored. After radiation, Evans blue dye in saline was injected and degree of staining in brain parenchyma was examined. The steady-state brain temperatures were found to be the highest (48°C) in animals receiving saline or the lowest doses of alcohol. The degree of staining decreased with increase in the amount of alcohol administered. The authors suggested that ethanol inhibits microwave-induced permeation of the BBB through reduced heating of the brain.

The direct effect of microwave energy on neurotransmitters has also been studied (Lai 1992). In one investigation, a decrease in brain concentration of acetylcholine was seen in mice exposed to pulsed microwaves, which produced a brain temperature increase of 2 to 4°C. This was believed to be due to an increased release of the transmitter (Modak et al. 1981).

Gandhi and Ross (1987) investigated effects of nonionizing radiation on neurotransmitters. Rats were subjected to 0.700 GHz radiation at 150 W/m^2, raising core temperature by 2.5°C. Six brain regions were investigated, but only the hypothalamus showed significant changes in receptor states. The study concluded that both norepinephrine and acetylcholine are released in response to heat. A case was made that norepinephrine suppresses efferent impulses for heat production, whereas the cholinergic system initiates the heat loss or heat dissipation mechanism.

Inaba et al. (1992) investigated the effect on biogenic amines in the rat brain exposed to 2.450 MHz microwaves at 50 and 100 W/m^2 for 1 h at ambient temperatures of 21 to 23°C. At the lower intensity, rectal temperatures stabilized after rising 2.3°C and at the upper intensity 3.4°C. Under these conditions, norepinephrine content was reported to be significantly reduced. 5-Hydroxyindoleacetic acid was increased and serotonin was reported to be unchanged. These and other results suggested that microwaves can affect the function of some neurotransmitters. Browning and Haycock investigated the effect of low level MWs on a neuron-specific phosphoprotein, synapsin I, and found no effect on neurotransmitter levels or brain function at the intensity of microwaves studied (Browning and Haycock 1988). Gandhi and Ross (1989) investigated the effect of low level MWs on rat brain synaptosomes and found that MWs alter the metabolism of inositol phospholipids by enhancing their turnover. It was suggested that this may affect transmembrane signaling in the nerve endings. Further information on combined nonionizing radiation and drugs effects can be found in the review of Lai et al. (1987c).

3.6.6 Neurophysiologic Effects *In Vitro*

Although this section is concerned with effects on the CNS, it seems appropriate to summarize work on the direct effect of RF on transmission of the nerve impulse, even though the bulk of this work has been done on peripheral nerves and ganglia *in vitro*.

Chou and Guy (1978) studied the effects of CW and PW 2.45 GHz MW on frog sciatic, cat saphenous, and rabbit superior cervical ganglia and vagus nerves. These investigations revealed that there is no direct production of an action potential by CW or pulsed microwaves (up to pulsed RF at 220 kW/kg peak or CW RF at 1500 W/kg). Compound action potentials were excited electrically in the presence and absence of RF. No changes in conduction velocity or amplitude were detected except with temperature elevations of at least 1°C in the solution. Identical changes could be produced by ambient changes in temperature.

Wachtel et al. (1975) and Seaman and Wachtel (1978), studying the effect of 1.5 or 2.45 GHz CW RF on abdominal ganglia from *Aplysia*, reported that patterns of pacemaker activity could be altered by MW and that comparable temperature elevations by ambient heating did not produce such changes. Studies by McRee and Wachtel (1982) on frog sciatic nerves suggested that MWs could produce diminished vitality of nerve function independent of temperature changes.

Khramov et al. (1991) investigated the effects of 34 to 78 GHz MWs on the spontaneous firing rate of the stretch-receptor neurons of the crayfish under thermally controlled conditions. The firing rates slowed in the presence of RF and temperature increases from 0.1 to 0.2°C, but not in the absence of temperature elevations. An analysis of the data suggested a thermal rather than direct RF effect.

On the other hand, Bolshakov and Alekseev (1992), studying neurons of *Lymnea stagnalis*, compared the effects of CW and PW at 900 MHz in cells exhibiting a burst-like irregularity in firing. MW above 0.5 W/kg PW (0.5 to 110 pps), but not CW, enhanced the probability of the burst-like irregularity. This was identified as possibly a MW pulse-specific effect, which could have arisen from mechanical oscillations in the exposure system.

Pakhomov et al. (1997) studied the effect of CW RF exposure (40 to 52 GHz, 0.24 to 3.0 mW/cm^2, 10 to 60 min) on isolated frog spinal cord preparations. In the presence of RF-induced heating, effects were identical to conventional heating. Pakhomov et al. (1992) earlier looked at the effects of 915 MHz PWs (peak SAR of 5 to 70 kW/kg) on isolated frog nerves under a variety of conditions. In the exposed groups compound action potential amplitudes decreased significantly over the controls. At the highest SARs microwave heating produced temperature increases of 2.7°C, but comparable conventional heating did not produce the effect. Nerve velocities and refractory characteristics did not change. It was concluded that a particular type of exposure can produce MW-specific changes.

A group has investigated the effects of MW (CW 2.45 GHz, SAR from 12.5 to 125 W/kg on input resistance and action potential in snail neurons (Arber and Lin 1983, 1984, 1985a; Arber et al. 1987; Ginsburg et al. 1992). Although changes occurred, they were not consistent. On the other hand, conventional temperature changes caused consistent changes in input resistance and action potentials. The authors suggested that CW MWs may enhance degenerative effects in neuron metabolism but do not indicate a specific mechanism for MW interaction with neurons (Arber and Lin 1985b).

The hippocampal slice preparation was used by Tattersall et al. (2001) for studying *in vitro* effects of 700 MHz RF exposure. An examination of population spike amplitude showed at low field intensities a 20% potentiation of spike amplitude. At higher intensities, the population spike could change (either increase or decrease) by up to 120 or 80%, respectively. The most significant result was the variability and not the consistency of responses. There were a number of methodological difficulties in this study including use of halothane anesthesia, use of metal-stimulating electrodes, and *post hoc* calculations of dosimetry for the exposures.

3.7 Behavioral Effects

Animal behavioral response to MW exposure has been used by essentially all standard-setting bodies to provide RF-exposure guidelines (NCRP 1986; ICNIRP 1998; D'Andrea 1999; IEEE 1991). Almost all the existing evidence links behavioral responses to MW exposures to the generation of temperature increases in tissues and reinforces the general conclusion that behavioral effects are thermally mediated. These effects have been demonstrated under a variety of conditions and in a range of animals. Parallel studies in humans assessing thermal sensation of MW exposures support the conclusion that changes in behavior of animals exposed to RF are likely to be thermally induced (Vendrik and Vos 1958; Hendler and Hardy 1960; Hendler 1968; Michaelson 1972; Justesen et al. 1982; Justesen 1983, 1988; Brown et al. 1994; Schwan et al. 1996; Blick et al. 1997; Riu et al. 1997; Walters et al. 2000; Adair et al. 2001a; D'Andrea et al. 2003; see also Chapter 4 of this volume). Thermally influenced responses in animals may include perception, aversion, work stoppage or perturbation, reduction in endurance, or even convulsions (Phillips et al. 1975; Justesen 1979; Modak et al. 1981; Guy and Chou 1982; Frei et al. 1995).

Based on thermal responses in animals to MW exposures, behavioral changes show animals to be most sensitive at frequencies close to the resonant frequency ($\sim$2500 MHz for mice, $\sim$6 to 700 MHz for rats, and $\sim$70 MHz for humans). Exposures close to the resonant frequency, particularly at levels sufficient to cause thermal responses, can have significant effects on behavior (D'Andrea et al. 1977; de Lorge and Ezell 1980; Gordon 1983, 1987; Gordon and Ferguson 1984; Durney et al. 1986; Mitchell et al. 1988).

3.7.1 Experimental Behavioral Studies

A large body of data exists and has been extensively reviewed for MW behavioral studies (D'Andrea et al. 2003). A portion of that database is briefly reviewed below (see also Chapter 4 of this volume). Justesen and King (1970) utilized a 2.45 GHz (CW) multimodal resonating cavity to investigate conditioned operant behavior in rats. They used a recurrent cycle of exposure, 5 min on and 5 min off, over a 60 min period at average absorbed energy rates of 3.0, 6.2, and 9.2 W/kg. The performance of the animal usually stopped near the end of the 60 min test period during exposure with an energy absorption rate of 6.2 W/kg; at 9.2 W/kg, this effect occurred much earlier in the test period. Hunt et al. (1975) also, using a multimodal resonating cavity to expose rats to 2.45 GHz PW, found effects on exploratory activity, swimming, and discrimination performance of vigilance task, after a 30 min exposure at about 6 W/kg. Lobanova (1960) exposed rats to 3 GHz pulsed MWs, after which the rats were tested for swimming time. A decrease in endurance was noted after exposure to power–time combinations ranging from 1 kW/m^2 for

5 min to 100 W/m^2 for 90 min. Wang and Lai (2000) also used the swimming test for rats exposed to 2.45 GHz pulsed MW fields. It took exposed rats significantly longer times to locate the submerged platform, suggesting that exposure may have caused a deficit in spatial memory. Variations of the swimming test to assess spatial learning are performed in radial arm mazes (dry land spatial navigation). Testing rats with 900 MHz GSM fields showed no differences between exposed and control animals (Dubreuil et al. 2002). A similar study in mice produced the same results, that is, no evidence of effects on radial arm maze performance indicating the lack of an effect on spatial learning and working memory (Sienkiewicz et al. 2000). In earlier studies, however, Lai et al. (1989, 1994) observed deficits in learning in rats tested in the radial arm memory task following exposure to 2.45 GHz pulsed MWs (average whole body SAR of 0.6 W/kg).

The threshold for observed memory disruption in rats from exposure was demonstrated to be about 10 W/kg in animals even when brain hyperthermia occurred at SARs slightly lower (5, 8.5 W/kg) (Mickley et al. 1994; Mickley and Cobb 1998). In a completely opposite result, Luttges (1980) examined the effects of 3 GHz MW exposure on learning and memory in mice and actually reported an enhancement of performance with exposure at whole body SAR of 13 W/kg. The study, with its unusual result, was repeated by Beel (1983) with the same finding but also with evidence of decreased performance with continued irradiation.

Lin et al. (1977) exposed rats to 0.918 GHz CW at levels of 100, 200, or 400 W/m^2 for 30 min. No effects on response rates were noted at the two lower levels, but at 400 W/m^2 the performance of the animal decreased after 5 min of exposure and ceased after about 15 min of exposure. The average energy absorption rate measured thermographically was 2.1 W/kg incident or 8.4 W/kg absorbed at 400 W/m^2.

Diachenko and Milroy (1975) studied the effects of pulsed and CW MWs on an operant behavior in rats trained to perform a lever-pressing response on a differential reinforcement of low rate (DRL) schedule and tested immediately after 1 h daily exposure to 10, 50, 100, and 150 W/m^2, 2.45 GHz MWs. No behavioral effects were found at these levels. The subjects exposed to 100 W/m^2, while showing no significant decrement in performance, did show signs of heat stress. Thomas et al. (1975, 1982) reported response rate changes in rats exposed between 50 and 200 W/m^2 to 2.86 GHz (CW) and 9.6 GHz PW MWs. Response rates increased in five of ten tests. Decrements in timing of responses were suggested as indicating that low level (estimated at 0.2 to 3.6 W/kg) MWs produce effects when pulsed but not continuous fields were used.

Galloway (1975) studied the performance of four monkeys when using a 2.45 GHz (CW) waveguide applicator (total absorbed power of 10, 15, or 25 W) applied to the head. The duration of the irradiation was 2 min or until convulsions began (20 and 25 W produced convulsions). Because of skin burns, only two subjects completed this series of experiments. There were no performance decrements in a discrimination task that the subjects performed immediately after exposure. Acquisition of a new task during the first ten trials of training was impaired at 25 W, which resulted in convulsions.

Roberti et al. (1975) measured the running time of rats in an electrifiable runway in which each subject was trained to peak performance. Exposure for 185 h at 10.7 GHz (CW), 3 GHz (CW), and 3 GHz pulsed, to 10 W/m^2, caused no performance decrements. No change in baseline performance was noted when rats were irradiated with 3 GHz pulsed for 17 d at a power density of 250 W/m^2.

D'Andrea et al. (1986b) observed significant differences in shuttle box performances and schedule-controlled lever pressing for food pellets in rats exposed to 2.450 GHz CW microwaves at an average power density of 0.05 W/m^2, 7 h/d, 7 d/week, up to 630 h. Similar findings occurred at 2.5 W/m^2 (D'Andrea et al. 1986a). DeWitt et al. (1987) exposed rats 7 h/d for 90 d to CW 2.45 GHz microwaves at an average power density of 0.05 W/m^2 followed by operant testing. In general, no behavioral alterations were found except for a variable effect on a time-related operant task. Galvin et al. (1986) studied long-term exposures of rats prenatally and perinatally (4.45 GHz, 100 W/m^2, 3 h/d, GDs 5 to 20 and days 2 to 20 postnatally) and found altered endurance and changes in gross motor activity in exposed versus control animals.

Mitchell et al. (1988) exposed rats to 2.45 GHz (CW) microwave radiation at 100 W/m^2 for 7 h and then performed a number of behavioral tests. Exposed mice exhibited reduced locomotor activity and diminished responsiveness to acoustic stimuli. Average SAR of the freely moving animals was estimated to be 2.7 W/kg.

Lebovitz (1983) compared CW and PW radiation at levels known to affect behavior. The pulsed source utilized could produce 1 μs wide pulses at 600 pps with a peak pulse power of 750 kW versus the CW source, which could produce a maximum power of 100 W. Rats, trained in an operant task, were exposed during the behavioral sessions for 3 h/d to CW or PW (1.3 GHz) at 3.7, 5.8, and 6.7 W/kg. At the two higher SARs, core temperature was elevated from 0.5 to 1.0°C. The two modes of energy delivery were equally effective in reducing response rates of the operant behavior. From the dosimetry, the auditory effect was probably present.

In some studies, MW exposure was coupled with the administration of various behavior-altering drugs. Thomas and Maitland (1979) investigated the effects of pulsed 2.45 GHz MWs, 1 W/m^2 in combination with dextroamphetamine on behavior in rats. Both acute and repeated exposures modified the normal dose–effect function. The maximum drug effect was obtained at lower MW exposures as the drug dose was increased. Galloway and Waxler (1977) employed a serotonin-depleting drug (fenfluramine) to investigate the effect of 2.45 GHz CW RF applied to the head of 7 to 9 kg rhesus monkeys at dose rates of 2 to 15 W. Combinations of the drug and RF at a dose rate of 15 W resulted in behavioral deficits, whereas the drug or RF alone up to 15 W failed to produce this effect. In respect to drugs such as dextroamphetamine and fenfluramine, their influence on thermal regulation may be significant in these results.

In contrast to the previous studies in which postexposure behavior was measured, de Lorge (1983, 1984) exposed rats and monkeys to 2.45 GHz MWs under far field conditions, while the animals were performing on operant schedules for food reinforcement. Exposure sessions lasted 60 min and were repeated on a daily basis. Stable performance on the operant schedules was disrupted in all three species at power densities positively correlated with the body mass of the animals. The behavior of the squirrel monkeys was disrupted at the middle level of power densities. When the averages of these power densities (280, 450, and 670 W/m^2) are plotted as a function of body mass (0.3, 0.7, and 5 kg), a semilog relationship becomes evident. Extrapolation along the resulting curve could permit prediction of the power densities needed to disrupt ongoing operant behavior in larger animals. The power densities associated with behavioral disruption approximated those power densities that produced an increase in colonic temperature of at least 1°C above control levels in the corresponding animals. These data support the need for scaling factors to extrapolate from small animals to larger animals and it becomes clear that different power densities are necessary to produce equal SARs in different size animals.

A number of studies evaluating the cognitive performance of humans exposed to RF, primarily using mobile phone signals and frequencies, have been recently conducted. Most of the investigations are for short time period exposures of 1 h or less. Several of these acute studies found changes in cognitive performance, in learning and memory tasks, in volunteers exposed to mobile telephone RF (Preece et al. 1999; Koivisto et al. 2000a,b; Lee et al. 2001; Edelstyn and Oldershaw 2002). However, there was a marked lack of consistency across studies, even within the same laboratory, and a number of the studies had methodological problems that precluded drawing conclusions with confidence. A subsequent replication study by Haarala et al. (2003) failed to replicate the positive results of earlier studies in that laboratory (Koivisto et al. 2000a,b), so the determination of cognition effects in humans has not yet been validated for MW exposures.

With new industrial and military applications the issue of high peak, low repetition rate (and low average power) exposure has been studied. The safety standards of the 1980s (ANSI/IEEE 1982; NCRP 1986) did not limit peak power, setting standards only for temporal average powers. Thomas et al. (1982) reported disruption of operant behavior in rats at whole body average SARs of 0.2 to 3.6 W/kg and peak SARs of 0.2 to 3.6 kW/kg, 500 pps, 2 μs pulse widths, 30 min/week exposure just before operant testing. However, Lebovitz (1983) found no effect on operant behaviors of pulsed exposures with whole body average SAR of 6.7 W/kg and peak SAR 11.0 kW/kg, 600 pps, 1 μs pulse width, 3 h exposure daily during behavior. D'Andrea et al. (1989) also found no effects on monkeys, utilizing operant behavioral studies. In these studies in which tested behaviors were concurrent with exposures, whole body average SARS ranged from 0.05 to 0.80 W/kg and peak whole body average SAR was 8.3 W/kg.

In a later study, D'Andrea et al. (1994) exposed rhesus monkeys to high peak power 5.62 GHz microwave pulses during the performance of an operant task. Two different peak powers at each of three different SARs were evaluated. At a whole body average SAR of 2 W/kg and peak powers of 560 W/m^2 or 5.18 kW/m^2, no alterations in behavior were found. At 4 W/kg with 1.28 and 12.7 kW/m^2 peaks, and also at 6 W/kg with 2.77 and 25.2 kW/m^2 peaks, decrements in performance were found. The differences in peak power at the two higher SARs did not produce any differences in behavior. Performance returned to baseline levels following exposure and the animals exhibited no discomfort during the exposure.

Hjeresen et al. (1990), utilizing a source capable of producing peak power densities up to 1.8 kW/cm^2, 10 μs pulse widths, 10 pps, at 1.3 GHz with resultant whole body average SARS from 1.8 to 26.2 W/kg, studied a number of behaviors in rats and concluded that disrupted behaviors correlated only with increased colonic temperature and therefore average SAR. Hjeresen and Umbarger (1989), utilizing a high-power microwave source with frequency range of 2.01 to 2.57 GHz producing peak power densities up to 241 MW/m^2 in pulses 85 ns wide, with 50 such pulses delivered over a 5 min period, produced no interference in behaviors in rats. No aversive properties were found and the authors suggested that peak power is apparently not as important as average power in causing deleterious biological effects.

Akyel et al. (1991), utilizing a high peak power 1.25 GHz source, which could produce a peak SAR of 0.21 MW/kg and SA of 2.1 J/kg per pulse, exposed rats for 10 min and then subjected them to an operant schedule. By varying the pulse repetition rate, a range of whole body average SARs from 0.84 to 213 W/kg could be produced. The highest SAR produced a colonic temperature elevation of 2.5°C. No interference with operant behavior was found at the lower dose levels. At the highest dose, animals failed to respond to the operant schedule. No specific effect of pulse power could be found and the interference with behavior appeared to be of thermal origin.

In further extension of this research, Raslear Akyel et al. (1993) studied high peak power pulsed exposures with short pulse widths and slow repetition rates (irregularly shaped pulse with a 2 MW/m^2 peak and an average of 500 MW/m^2 over the pulse period, 2 W/m^2 average, 80 ns pulse width, repetition rate of 0.125 Hz (200 pulses over 26 min), 3 GHz, average SAR of 0.07 W/kg, peak SAR of 7 MW/kg. These were the highest peak powers and peak SARs found in the literature, and degraded performance of rats was found in a number of test protocols, including Y-maze negotiation and treadmill running.

The authors concluded that at such high peak powers impairments can be found despite the low average SAR. In all of the pulse power studies described, the dosimetry suggests production of the auditory effect. What role it might have played in studies revealing effects is not clear. These studies generally show that pulsed MW are generally, but not always more effective in producing behavioral responses than CW at comparable time averaged, whole body exposure levels. The effects of high peak power pulses have been reviewed by Pakhomov and Murphy (2000) and Pakhomov et al. (2003).

Most of the research on the nervous system and behavior have been carried out in rodents and other lower animals. Behavior among animal species reflects adaptive brain–behavior patterns. Behavioral thermal regulation is seen as an attempt to maintain a nearly constant internal thermal environment. Changes in body temperature bring about not only autonomic drives, but also behavioral drives (Stolwijk and Hardy 1977). The MWs can influence behavioral thermoregulation and has been shown by Stern et al. (1979) and Adair and Adams (1982). This subject is treated in greater detail in Chapter 5 of this volume.

Behavioral responses are not necessarily the manifestations of specific changes in the CNS and may be a function of direct or indirect action of MWs on other body systems. Extrapolation of brain–behavior functions from lower animals to man is thus subjected to many difficulties.

In assessing the significance of the reported behavioral changes, it is important to recognize certain fundamental factors. The resting metabolic rate for rats is approximately 7 W/kg. When this level is exceeded, disruption of behavior could be elicited. In most of the reports, alterations in the behavior of rats were observed with exposures at average energy absorption rates of 5 to 8 W/kg or greater, i.e., at exposure levels similar to those that produce increases in circulating corticosterone (CS) concentrations in rats (Lotz and Michaelson 1978). Behavioral changes may be related to more subtle heat alterations within the body. Heat may produce a general debilitation effect or a decreased motivation for food, since it has been shown that rats maintained in hot environments eat less food (Hamilton 1963) and rats show decreased response and food reinforcement frequency on an operant schedule when environmental temperature is 35°C, but not 25°C. Behavioral responses certainly may be influenced by the interaction of the organism with the environment. Nevertheless, it appears that changes in behavior definitely result from exposure to MWs of varying frequencies and intensities. Yet to be demonstrated with certainty are that observed behavioral responses can occur without the thermal increases associated with the MW exposure.

3.7.2 Behavioral Thermoregulation

The regulation of body temperature can be accomplished by complex thermoregulatory patterns of responses of the skeletal musculature to heat and cold, which modify the rates of heat production and heat loss, for e.g., by exercise, change in body conformation, the thermal insulation of bedding, and by the selection of an environment which reduces thermal stress (Bligh 1975). In humans, optimal temperature in the body is also regulated by thermophysiological responses. Human body temperatures range from 35.5 to 40°C and include exercise, food intake, environmental conditions, and circadian variations. Age can also play a role in thermoregulation (Makrides et al. 1990).

Thermoregulation is part of a complex control system involving circulation, metabolism, and respiration, as well as neural structures. Temperature signals from cutaneous thermoreceptors reach the somatosensory region of the cerebral cortex. The main processing of thermal signals and generation of a controlling signal for the effector part of thermoregulation takes place in the hypothalamus. Adair et al. (1984) found minimal changes in hypothalamic temperature in conjunction with alteration of thermoregulatory behavior in squirrel monkeys exposed to 2.45 GHz, 40 to 200 W/m^2 (SAR 0.2 to 1 W/kg). The whole subject of thermoregulatory behavior in response to energy absorption during MW exposure has been nicely reviewed and summarized by Adair and Black (2003; see also Chapter 5 of this volume).

Successful strategies or mechanisms to maintain body temperatures in a narrow and desirable range in a complex and varying thermal environment are termed thermoregulatory. Although such strategies or mechanisms are found in great variety, they fall into two main categories: voluntary and behavioral adjustments, and involuntary physiological adjustments. The limits of effectiveness of involuntary physiological thermoregulation are rather narrow and we must rely on behavioral methods of thermoregulation over most of the range of environmental temperatures to which we are often exposed. Changes in body temperatures bring about not only autonomic drives, but also behavioral drives (Stolwijk and Hardy 1977).

Thermal motivation arises in situations of thermal stress. On the warm side, it is the uncomfortable feeling of excessive warmth and the desire for temperature reduction; on the cold side, it involves the unpleasant feeling of remaining too cold and the desire for temperature increase. The biological significance of thermal motivation is that, by acting in such a way as to minimize thermal discomfort and maximize thermal comfort, the organism tends to escape from situations of thermal stress and to locate itself in a physiologically neutral thermal environment, thereby solving the problem of physiological temperature regulation (Corbit 1973).

Smialowicz et al. (1981a) made unrestrained, unanesthetized mice hypothermic by injecting them with 5-hydroxytryptamine (5-HT) in a controlled environment of 22°C and 50% relative humidity. Mice injected with saline were used as controls. Colonic temperatures were measured prior to injection. Following injection, groups of mice were exposed to 2.45 GHz CW RF for 15 min at 100, 50, or 10 W/m^2 (equivalent SARs 7.2, 3.6, and 0.7 W/kg) or were sham-exposed, after which their colonic temperatures were measured again. The experiments were performed with BALB/c and CBA/J mice. For saline-injected mice of either type, there were no significant colonic temperature differences between mice exposed at 100 W/m^2 and sham-exposed mice. For BALB/c mice made hypothermic with 5-HT, colonic temperatures were significantly higher for those exposed at all three power densities than those sham-exposed, and the differences increased monotonically with power density. The results for the CBA/J mice were similar, but the increases were statistically significant only at 50 and 100 W/m^2. The investigators conclude that subtle heating by RF can alter the thermoregulatory capacity of hypothermic mice, whereas the colonic temperature of normal (saline-injected) mice was not significantly altered by exposure at 100 W/m^2.

Adair and Adams (1980a, 1982) exposed squirrel monkeys to 2.45 GHz CW RF for 10 or 90 min in ambient temperatures of 15, 20, or 25°C. The power densities ranged from 25 to 100 W/m^2 (SARs 0.4 to 1.5 W/kg). Skin and colonic temperatures were monitored continuously during exposure. The metabolic heat production was calculated from the oxygen deficit in the expired air of each monkey. At all three ambient temperatures, 10 min exposure of two monkeys to a threshold power density of 40 W/m^2 and one monkey to 20 W/m^2 reliably initiated a reduction of their metabolic heat production, and the magnitudes of the reduction were a linear function of the power density above the threshold values. At exposure termination, the metabolic heat production often rebounded sharply and overshot normal levels. For the 90 min exposures at 20°C, the initially large reduction of metabolic heat production gradually diminished towards normal levels, apparently to ensure precise regulation of internal body temperature at the normal value.

Adair (1981) also exposed four squirrel monkeys to 2.45 GHz CW RF in warm ambient temperatures ranging from 32 to 35°C. After an initial equilibration period of 90 min or longer, each monkey was exposed for ten periods to power densities in an increasing sequence from 25 to 200 W/m^2, with sufficient time between exposures for reestablishment of equilibrium. The colonic temperature and the skin temperature at the abdomen, tail, leg, and foot were monitored continuously. As in the previous investigation, the metabolic heat production was determined from the oxygen deficit in the expired air. In addition, thermoregulatory sweating from the foot was determined by sensing the dew point of the air in a special boot over the foot. The results indicate that at ambient temperatures below about 36°C, at which sweating in a sedentary monkey may occur spontaneously, the threshold power density (or SAR) for initiating thermoregulatory sweating from the foot decreased with decreasing ambient temperature.

Adair and Adams (1980b) equilibrated squirrel monkeys for a minimum of 2 h to constant environmental temperatures (22 to 26.5°C), which was cool enough to ensure that the cutaneous blood vessels in the tail and extremities were fully vasoconstricted, an effect produced by the thermoregulatory system to minimize heat loss. The monkeys then underwent 5 min exposures to 2.45 GHz RF at successively higher power densities, starting at 25 to 40 W/m^2, until vasodilation in the tail occurred, as evidenced by an abrupt and rapid temperature increase in the tail skin. For example, a monkey equilibrated to 25°C, exhibited tail vasodilation when exposed to RF at 100 W/m^2 (whole body SAR 1.5 W/kg), whereas it did not when exposed to IR radiation at the equivalent power density, an indication that the effect resulted from stimulation of thermosensitive elements of the thermoregulatory system by the RF rather than from heating of the tail skin. RF exposure at higher power densities was required to cause tail vasodilation in monkeys equilibrated to lower environmental temperatures. Specifically, an increase of 30 to 40 W/m^2 was found necessary for every 1°C reduction in environmental temperature.

3.8 Neuroendocrine Effects

3.8.1 Mechanisms of Interaction

Many of the studies of the endocrine system have involved whole body RF exposures of the rat. The issue has been the level of organization of the system at which thermal interactions exert an effect and the nature of the effect. Some investigators believe that endocrine changes result from stimulation of the hypothalamic–hypophysial system at the hypothalamic or adjacent level of organization; others have advanced evidence that effects are mediated at the hypophysis or the particular endocrine gland or end organ under study. The effects detected are not necessarily pathologic, because the function of the neuroendocrine system is to maintain homeostasis and hormone levels will fluctuate to maintain such organismic stability.

Effects of RF exposure on endocrine function are generally consistent with both immediate and long-term responses to thermal input and nonspecific stress, which can also arise from thermal loading. Changes found in plasma levels of corticosterone (CS) and growth hormone are typical reactions of animals to nonspecific stress; indeed, great care is required in performing experiments to ensure that the changes in hormone level do not result from stress caused by handling of the animals or the novelty of the experimental situation. The topic has been reviewed by Lu et al. (1980b) and also by Black and Heynick (2003).

3.8.2 Hypothalamic–Hypophysial–Adrenal Response

Several investigators have reported biochemical and physiological changes as a result of MW exposure, which suggest an adrenal effect. Indeed, the evidence of increases in plasma steroid levels in rats exposed to MWs is fairly reliable. Furthermore, sufficient research has been performed to determine exposure thresholds for such responses in rats. Demokidova (1974) reported increased adrenal and pituitary gland weight in rats exposed 1.5 months to 69.7 MHz, 12 V/m, 1 h /d and increased adrenal weight in infant rats exposed to 48 V/m, 4 h/d. The same author, however, reported decreased weight of the adrenal glands in infant rats exposed to 0.01488 GHz at 70 V/m. The pituitary gland in female mice retained its gonadotropic function when exposed to 3 GHz (100 W/m^2) twice daily for 5 months, although its activity was reduced in comparison with that of non-exposed animals (Bereznitskaya and Kazbekov 1974). Tolgskaya and Gordon (1960) noted the reversibility of changes in the neurosecretory function of the hypothalamus when exposure was terminated. In another study, however, rats exposed to MWs of varying intensity showed no quantitative changes in CS were found in the adrenals and blood plasma (Mikolajczyk 1972). Prepubescent hypophysectomized rats displayed no differences in adrenal growth rate when treated with pituitary homogenates collected either from rats exposed to MWs or from control rats.

Rats exposed to 2.45 GHz (CW), 100 W/m^2 for 4 h showed no change in adrenal weight, phenylethanolamine-*N*-methyl transferase (PNMT) activity, or epinephrine levels (Parker 1973). After 16 h of exposure, however, decrease in adrenal epinephrine (32%) was significant and PNMT activity was elevated (25%). There were no statistically significant differences in adrenal or plasma CS levels between exposed and sham-exposed animals. It should be noted, however, that similar alterations in epinephrine levels can occur in rats subjected to a stressor such as immobilization or acute exposure to cold.

Rabbits exposed to 3 GHz (500 to 600 W/m^2) 4 h daily for 20 d tended to show a decline in the amount of urinary 17-hydroxycorticosteroids (17-OHCS) at the beginning of exposure, followed by a gradual return to normal (Lenko et al. 1966). No change was evident in the excretion of 17-ketosteroids in the urine. According to Petrov and Syngayevskaya (1970), 3 and 24 h after dogs were exposed to 3 GHz, 100 W/m^2, the serum corticosteroid content had increased by 100 to 150% above the original level. Serum potassium was decreased by 5 to 10% and sodium was increased by the same amount. They also noted that the susceptibility of rats to MW exposure was sharply increased 1 week after bilateral adrenalectomy. Chronic exposure of animals to MWs (CW or PW) was accompanied by reduced cholinesterase activity and an increased amount of 17-ketosteroids in the urine, reduced ascorbic acid, and reduced weight of the adrenal glands.

Increased adrenal function due to MW exposure has been correlated with colonic temperature increase in rats (Lotz and Michaelson 1979). Plasma CS levels in hypophysectomized rats exposed at 600 W/m^2 for 60 min were below control levels. When rats were pretreated with dexamethasone before exposed at 500 W/m^2 for 60 min, the CS response was suppressed. These results suggest that the MW-induced CS response observed in intact rats is dependent on adrenocorticotropic hormone secretion by the pituitary, i.e., the adrenal gland is not the primary endocrine gland stimulated by MW. The evidence obtained in these experiments is consistent with the hypothesis that the stimulation of the adrenal axis in MW-exposed rats is a systemic, integrative process due to a general hyperthermia (Lotz and Michaelson 1978). Consistent with these results were data obtained by Lotz (1985) in which rats exposed to 255 MHz CW for 4 h (SAR 3 to 4 W/kg) showed no change in serum cortisol levels. The boundary between effects and no effects for corticosteroid levels appears to be approximately 4 W/kg in monkeys as well as rats (Lotz 1983). The subject of effects of MW on the adrenal cortex has been reviewed by Lu et al. (1986).

3.8.3 Hypothalamic–Hypophysial–Thyroid Response

The literature offers comparatively few experimental studies of the effect of RF or MWs on the thyroid. No alterations in thyroid structure or function attributable to MW exposure were noted in rats subjected to 2.45 GHz, CW, 10 W/m^2 continuously for 8 weeks or 100 W/m^2, 8 h/d for 8 weeks (Milroy and Michaelson 1972). On the other hand, a stimulatory influence of 50 W/m^2 on the trapping and secretory function of the thyroid gland of rabbits has been reported (Baranski et al. 1972). These functional changes were in agreement with altered histology of the thyroid.

In rats exposed for 16 h to 2.45 GHz (CW) at 100 to 250 W/m^2, tests of thyroid function in general showed no statistically significant deviations from the norm, except that in animals with a 1.0 to 1.7°C increase in colonic temperature there was a reduction in the ability of the thyroid to concentrate iodide (Parker 1973). Decreased thyroid gland weight was noted in infant rats exposed to 0.0697 GHz, 48 V/m, 4 h/d, 1.5 months.

Increased thyroid hormone secretion has been correlated with MW-induced thyroid temperature increase in dogs (Magin et al. 1977). Vetter (1975) found that serum protein levels increased as a function of power density, indicating an alteration of protein synthesis or catabolism; levels of thyroid hormone decreased as power density of 1.45 GHz CW was increased from 50 to 250 W/m^2. In agreement is the finding of Lu et al. (1977) who reported that serum thyroxine levels were transiently elevated after exposure of rats at 10 W/m^2 (2.45 GHz) and were depressed after exposure at 200 W/m^2. None of the reported alterations was irreversible or resulted in morbidity.

Perturbation of the thyroid gland may be the result of an indirect effect, the thermal stress on the body producing a hypothalamic–hypophysial response. This is consistent with MW-induced thermal stimulation of hypothalamic–hypophysial–thyroid (HHT) activity (Michaelson et al. 1961). McLees and Finch (1971) pointed out that temperature elevation and heat stress have been associated with alterations in radioactive iodine (RAI) turnover rate. The HHT axis has been shown to be sensitive to environmental temperature (Michaelson 1961). Differences in rate of temperature change or alteration in thermal gradients could also result in qualitative differences in endocrine response.

In a study described as a delineation of acute neuroendocrine responses to MW exposure, Lu et al. (1980a) exposed gentled rats to the same frequency at power densities ranging from 10 to 700 W/m^2, up to 8 h, at an environmental temperature maintained at 24°C. Sham-exposed rats were used as controls. After treatment, the rats were decapitated, colonic temperatures were taken, and blood was collected for assays of thyroxine (T4), thyrotropin (TSH), growth hormone (GH), and CS. For exposure of 1 h, colonic temperature increased with power density at 200 W/m^2 and higher, but consistent elevation of serum CS did not occur below 500 W/m^2. Lower serum TSH and GH levels also occurred at this and higher power densities. Significant serum T4 elevations were noted at 400 and 700 W/m^2, but they were not consistently related to power density. For sham exposures and exposures at 10 to 200 W/m^2 for longer durations (2 to 8 h), the results were equivocal, presumably, because such exposures encompassed significant portions of the circadian cycle. Specifically, in the sham-exposed rats, the level of T4 did not change significantly under sham conditions with exposure duration; and significant increases of CS and decreases of TSH and GH were seen, so it was difficult to discern consistent differences in these hormones ascribable to RF exposure. The investigators suggest that the divergent responses may be due to two different mechanisms that are dependent on RF intensity and the time of the exposures relative to circadian rhythms.

Abhold et al. (1981) exposed rats to 2.45 GHz CW RF for 8 h (continuously) at 20 or 100 W/m^2 (SAR 0.44 or 2.2 W/kg with the long body axis of the rat parallel to the electric component of the RF). Other rats were sham-exposed for the same period, and still others served as untreated controls. Within 15 min after treatment, the rats were euthanized and their blood was assayed for serum T4, triiodothyronine (T3), and T3 uptake. In addition, serum CS concentrations were measured. The results indicated that concentrations of T4, T3, and T3 uptake were not altered by the treatments. However, the rats that were sham-exposed or exposed to RF at 20 W/m^2 had higher levels of CW than the cage control rats, whereas the rats exposed to 100 W/m^2 had values similar to those of the untreated rats. The changes observed were more likely ascribable to stress and other factors in the experimental protocols than the RF exposure.

In a review article, Lu et al. (1980b) discussed evidence for the existence of threshold intensities for various RF-induced neuroendocrine effects. They indicate that such thresholds are dependent on the intensity and duration of exposure and can be different for each endocrine parameter.

3.8.4 Melatonin Response

Interest has grown in possible effects of MW radiation on circadian function, mediated through the pineal hormone, melatonin. In a number of recent reports, nighttime levels of melatonin have been measured in response to RF exposure in both animals and humans. Burch et al. (2002) reported a marginal increase in melatonin in humans associated with mobile phone use. This report however, is countered by a number of human provocation studies in which no effects in melatonin levels or secretion was observed following RF exposures (de Seze et al. 1998, 1999; Mann et al. 1998b; Radon et al. 2001). Indeed, no changes were observed in exposed humans for a range of hormones as well, including growth hormone, lutinizing hormone, cortisol, melatonin, and others. Research in rats has shown results consistent with these general findings in that no demonstrable effects on hormones, including melatonin, have been reported with low-level RF exposure at 435 MHz (Bonasera et al. 1988; Toler et al. 1988) or at 900 MHz (Vollrath et al. 1997; Heikkanen and Juutilianen 1999).

In summary, although some effects of RF exposure on the endocrine system appear to be relatively straightforward and predictable from physiological considerations, other more subtle effects require further study, notably those related to the interactions among the pituitary, adrenal, thyroid, and hypothalamus and their secretions. Part of the problem in interpreting results appears to arise from uncertainties regarding stressors inadvertently introduced into the experimental design and the response of the animal to such stressors. Animals placed in novel situations are much more prone to exhibit stress responses than animals that have been adapted to the situation. However, there may be large variations in adaptation among animals in a given situation or among experimental situations in different laboratories. Moreover, the use of sham-treated controls may not always reduce the problem. A review of the effects of RF on physiologic regulation has been published by Michaelson (1982).

3.9 Cardiovascular Effects

Low-intensity MW exposures at nonthermalizing levels and clearly thermal levels have been studied. The role of heating and the possibility of unique MW-induced effects even at thermalizing levels of exposure have been addressed. The database is anything but straightforward and not easily distilled into a brief summary. Some aspects of cardiovascular function during exposure to MW are reviewed by Black and Heynick (2003).

Clinicians in the former Soviet Union identified a so-called "microwave neurasthenia," or "microwave radiation sickness," the consequence of chronic occupational exposure to low levels of MW (Sacchikova 1974). This included bradycardia, tachycardia, hypertension, hypotension, T-wave changes, and other neurologic and behavioral changes. Such a syndrome is not considered adequately documented or well defined in Western medicine. The possible hemodynamic changes in the cardiovascular system have been studied in animals ever since. The chronotropic effects (bradycardia or tachycardia) have been of special interest. At nonthermal levels, both bradycardia and tachycardia have been found in various studies of animals but the results have been inconsistent overall. Some investigators have proposed that the heart rate response of rabbits depended on exposure technique, distribution of energy as a function of location, and the possibility of differing regional reflexes producing differing outcomes (Pressman and Levitina 1962). This has not been verified and it is just as likely that biologic variability among animals or the numerous sources of error could account for the findings. Even the distinction between thermal and nonthermal effects is not clear cut; and minor changes in temperature, perhaps less than 1°C, at the right anatomic location could affect hemodynamics by a neural mechanism even though the exposure is reported as nonthermal.

Hyperthermia of non-RF origin produces tachycardia and a decrease in total peripheral resistance caused by vasodilation, a heat-dissipating response to the thermal burden. This is true whether heat is imparted by environmental heating, water bath, or even a general whole body hyperthermia of RF origin (Basset and Kenny 1979).

Evidence that hemodynamic changes produced by MWs might be very sensitive to a precise dosimetric configuration was presented by Lu et al. (1992). Rats were subjected to ventral head–neck exposures in a waveguide at 1.25 GHz, PW and CW at 2.0 and 6.4 W average powers, 10 μs pulses, 400 kW peak power, brain average SAR 9.5 and 30.4 W/kg, neck SAR 34.3 and 110.0 W/kg, for 5 min. Both tachycardia and bradycardia were observed but a decreased stroke volume was also found with consistency. This produced a consistent subnormal cardiac output. Mean arterial pressure was unaltered. From the data, it was possible to conclude that total peripheral resistance had increased. This would be an anomalous or unphysiologic response to the insertion of thermal energy into the body. This kind of a response is not observed in experimental animals subjected to whole body microwave exposure (Jauchem et al. 1984). The investigators concluded that the targeting of the head and neck, even in the presence of generalized hyperthermia, may have elicited specific signals at receptor or effector sites in that area to contravene the normal physiologic response. The baroreceptors were identified as a possible locus of stimulation of efferents, which raised blood pressure by increasing peripheral resistance. A hypothesis that the baroreceptors might respond to pressure waves induced by pulsed MW was not supported, in that both pulse and CW produced the anomalous effect at thermalizing levels.

Rasbury and Seaman (1992) also studied exposures of the head and neck in the rat under different conditions from those of Lu et al. The carotid sinus was surgically exposed and rats exposed to the radiation of a coaxial probe under anesthesia for brief periods, 10 s, and local SARs up to 70 kW/kg. At the highest SARs, where effects were found, approximately half of the animals exhibited decreased blood pressure and heart rate and half the animals exhibited increased blood pressure and heart rate. No changes in colonic or tail temperature occurred for any exposures. The authors could not account for the divergent results, but noted that the inconsistent findings were consistent with other studies on hemodynamic changes associated with MW exposure.

Additional studies in animals at a wide range of MW frequencies have reported definite effects on blood pressure and heart rate, but all at power levels clearly associated with increases in temperature within the body (Phillips et al. 1975; Jauchem et al. 1984, 1985b, 1988; Frei et al. 1988, 1989a,b; Frei and Jauchem 1989). Investigations of ultra-wideband and 94 GHz MW exposures showed no significant changes in the heart rate or mean arterial blood pressure in rats (Jauchem et al. 1999a,b), respectively. However, in another study of rats exposed to ultra-wideband fields, Lu et al. (1999) found a consistent hypotension, which persisted following exposure. This occurred with no changes observed in heart rate.

Cooper et al. (1961, 1962) and Pinakatt et al. (1963, 1965) studied the influence of various drugs in pentobarbital-anesthetized rats exposed to 2.45 GHz, 800 W/m^2 for 10 min, which resulted in a 40.5°C colonic temperature. Pyridoxine and digitoxin did not alter blood pressure or heart rate. Reserpine, vagotomy, and pharmacological ganglioplegia diminished the reaction to MW-induced hyperthermia. Jauchem (1985) also examined the influence of drugs on thermal responses in animals exposed to high-level MW.

A human study reported increases in blood pressure and heart rate in humans exposed to 900 MHz RF (Braune et al. 1998). The same group, however, could not replicate the positive findings in a subsequent study (Braune et al. 2002). Other human studies at much higher intensity RF found no effects of exposure on heart rate (Lu et al. 1992; Adair et al. 2001b). With regard to human exposures, a review by Jauchem (1997) stated that if a tissue is not heated during exposure, current flow appears to be required for significant cardiovascular effects to occur.

3.10 Effects on Hematopoiesis and Hematology

Extensive investigations have been carried out to examine MW exposure effects on the hematopoetic system in animals. The preponderence of this work was performed in the two decades from 1960 to 1980 with the overwhelming evidence that exposed blood cells and blood-forming tissues are not affected by MW exposure except when the target is heated. White cells (leukocytes) are more sensitive than red cells (erythrocytes); however, even the white cell effects are consistent with thermally induced changes. In the absence of significant MW heating or specific electric currents, cardiovascular tissue also does not provide evidence of RF-induced changes, nor is blood pressure changed by MW exposure. Much of this work has been recently reviewed by Black and Heynick (2003).

Several papers from the early literature of Eastern European research, as indicated below, reported effects on blood cells, even at low MW exposures. Kitsovskaya (1964) subjected rats to 3 GHz, 100 W/m^2, 60 min/d for 216 d; 400 W/m^2, 15 min/d for 20 d; 1 kW/m^2, 5 min/d for 6 d. At 400 W/m^2 and 1 kW/m^2, total RBC, WBC, and absolute lymphocytes were decreased; granulocytes and reticulocytes were elevated. At 100 W/m^2, total WBC and absolute lymphocytes decreased, and granulocytes increased. Bone marrow examination revealed erythroid hyperplasia at the higher power densities. Decreased leukocyte count and phagocytic activity had been reported in rats and rabbits exposed to 50 MHz, 0.5 to 6 V/m, 10 to 12 h/d, 180 d (Serdiuk 1969).

Baranski (1971a,b) exposed guinea pigs and rabbits to 3 GHz (PW or CW), 35 W/m^2 for 3 months, 3 h daily. Increases in absolute lymphocyte counts, abnormalities in nuclear structure, and mitosis in the erythroblastic cell series in the bone marrow and in lymphoid cells in lymph nodes and spleen were observed. No alteration in the granulocyte series was noted. Baranski suggests that extrathermal complex interactions seem to be the underlying mechanism for the changes. Unfortunately, minimal information on exposure conditions was provided making determinations on thermal and nonthermal parameters very difficult.

Exposure of mice to 2.45 GHz, 1 kW/m^2 for 5 min resulted in a decrease followed by an increase in ^{59}Fe uptake in the spleen and bone marrow (Vacek 1978). Alteration in ferrokinetics was also found in rabbits and guinea pigs exposed to 3 GHz at 10 or 30 W/m^2, 2 to 4 h daily for 14 to 79 d (Czerski et al. 1974a,b).

Miro et al. (1974) exposed Swiss albino mice to pulsed 3.105 GHz at 200 W/m^2 average power density (4 kW/m^2 peak) for 145 h. Stimulation of splenic lymphopoiesis and increased ^{35}S methionine incorporation in spleen, thymus, and liver were found. The authors interpret these results as a sign of stimulation of cells belonging to the reticuloendothelial system. Possible alteration in the circadian rhythm of bone marrow mitoses in guinea pigs and mice was noted. No effects were seen on precursors of granulocytes and only minimal effects were found on the erythroid series, but pronounced phase shifts were noted in the pool of stem cells. In inbred Swiss albino mice exposed once for 4 h at 5 W/m^2 to pulsed 3 GHz MWs, the diurnal rate of proliferation of the stem cell population was amplified and the phase shifted from that of controls (Czerski et al. 1974b). A comparable phase shift in the circadian rhythm of body temperature was observed in rats exposed to 2.45 GHz, CW, 10 W/m^2, 1 to 8 h, suggesting an interrelated example of physiologic regulation (Lu et al. 1980a).

Subsequent studies, however, have not confirmed such effects. Peterson (1979) investigated both rabbit and human RBCs to 2.45 GHz and demonstrated no hemolysis or changes in exposed samples. Studies of mouse erythroleukemic cells exposed to 1.18 GHz for 48 h at 18.5, 36.3, or 69.2 W/kg also showed no significant differences between exposed and sham control cultures (Marshall 1984). Hyde and Friedman (1968) exposed anesthetized female CF-1 mice to 3 GHz, 200 W/m^2 and 10 GHz, 170, 400, or 600 W/m^2 up to 15 min. No significant effect on total or differential leukocyte count or hemoglobin concentration was noted immediately, 3, 7, or 20 d after exposure. There were no changes in femoral bone marrow other than a variable, but slight, increase in the eosinophil series of the exposed animals, which was not reflected in peripheral blood counts.

Spalding et al. (1971) exposed mice to 800 MHz, 2 h daily for 120 d in a waveguide at an incident power density of 430 W/m^2. RBC and WBC count, hematocrit, hemoglobin, growth, voluntary activity, and life span remained normal. Smialowicz et al. (1981d) conducted long-term continuous exposures of rats for 69 to 70 d to 970 MHz at 2.5 W/kg. No differences were found between exposed and control animals in hemoglobin concentration or hematocrit, mean cell volume of erythrocytes, erythrocyte count, or total and differential leukocyte counts.

In a controlled temperature experiment at 100 MHz, Cleary et al. (1985) exposed rabbit neutrophils for 30 or 60 min and observed no changes due to RF exposure in either viability or phagocytotic ability. Hamrick and Fox (1977) exposed rat lymphocytes to 2.45 GHz MWs and did not demonstrate any significant differences between control and exposed cells.

In dogs exposed whole body to 2.8 GHz pulsed, 1 kW/m^2 for 6 h, there was a marked decrease in lymphocytes and eosinophils (Michaelson et al. 1967). The neutrophils remained slightly increased at 24 h postexposure, while eosinophil and lymphocyte values returned to normal levels. Following 2 h of exposure at 1.65 kW/m^2, there was a slight increase in leukopenia and decrease in neutrophils. When the exposure was of 3 h duration, leukocytosis was evident immediately after exposure and was more marked at 24 h, reflecting the neutrophil response. After exposure to 1.285 GHz pulsed 1 kW/m^2 for 6 h, there was an increase in leukocytes and neutrophils. At 24 h, the neutrophil level was still noticeably increased. Lymphocyte and eosinophil values were moderately depressed initially, but at 24 h slightly exceeded their initial value. An exposure of 6 h to 0.2 GHz (CW), 1.65 kW/m^2 resulted in a marked increase in neutrophils and a mild decrease in lymphocytes. On the following day, the leukocyte count was further increased, and the lymphocytes markedly increased. Such shifts in the WBC picture are consistent with focal thermal lesions to be expected under these exposure conditions.

Work with human cells has demonstrated comparable results. Roberts et al. (1983) exposed human leukocyte cultures for 2 h to 2.45 GHz MW at 0.5, and 4 W/kg. No changes were observed in a number of end points between the groups. In a subsequent study, using cell cultures treated with mitogens or infected with an influenza virus, no significant differences were observed in exposed samples (Roberts et al. 1987). Kiel et al. (1986) also examined human lymphocytes for production of active oxygen species and observed no significant differences for the exposed cells.

In evaluating reports of hematological changes, one must be cognizant of the relative distributions of blood cells in a population of animals or humans and the thermal influence on these alterations. Early and sustained leukocytosis in animals to thermogenic levels of MWs may be related to stimulation of the hematopoietic system, leukocytic mobilization, or recirculation of sequestered cells. Eosinopenia and transient lymphocytopenia with rebound or overcompensation, when accompanied by neutrophilia, may be indicative of increased hypothalamic–hypophysial–adrenal function as a result of thermal stress (Michaelson 1974).

3.11 Effects on the Immune Response

Some interest has continued in the effect of MW exposure on the immune system. A number of articles published in the 1970s and 1980s (Stodolnik-Baranska 1974; Hamrick et al. 1977; Liburdy 1977, 1979, 1980; Wiktor-Jedrzejczak et al. 1977a–c; Smialowicz et al. 1979a,b, 1981b,c 1982a–c; Huang and Mold 1980; Schlagel et al. 1980; Sulek et al. 1980; Roberts et al. 1983; Yang et al. 1983; Smialowicz 1987) were motivated by the possibility of enhancing the immune response with MW exposure with therapeutic goals in mind. By the late 1980s and early 1990s, with a redirection of interest to ELF investigation, the rate of output of studies on immune system effects had declined. Reports of lymphoblastoid transformation by MW were confirmed by some studies and not others (see below). The hope of exploiting such a phenomenon to augment beneficial immune responses in certain disorders was tempered by the concern that there might be a hazard involved if such processes led to autonomous and unregulated lymphoid proliferations. There is no evidence that the latter concern has any validity. On the other hand, no consistent and robust evidence has been elicited to date that MW-induced lymphoblastoid responses can be used to augment immunotherapy of infectious disease or cancer.

3.11.1 Lymphocyte Kinetics

Many reports on effects of MW on lymphocytes have not produced a clear and consistent picture amenable to concise summary. Although lymphocytes can be straightforwardly identified with standard hematologic stains and mature lymphocytes give the appearance of small, uniform, metabolically inert cells, lymphocytes are composed of immunologically distinct subpopulations and, in response to antigenic stimulation or exposure to many chemical substances, can quickly enlarge, exhibiting a diameter two to five times that of the resting cell, with a striking increase in amount of cytoplasm and transformation of the nucleus from a dark, dense mass of chromatin material to fine, lacy, or decondensed structure. This process, known as lymphoblastoid transformation, has been intensively studied immunologically and biochemically. The lymphocyte may or may not proceed to mitosis depending on the nature of the challenge or stimulus. A number of plant lectins have been used to stimulate lymphoblastoid transformation, protein, RNA, and DNA synthesis *in vitro* to study immune status. In immunologically competent animals, unequivocal, vigorous responses can be elicited. Responses to RF have been reported but of a much less robust nature.

Lymphoblastoid transformation, *in vitro*, after free field exposure to 3 GHz pulsed MWs, 70 W/m^2, for 4 h daily and 200 W/m^2 for 15 min daily, 3 to 5 d, has been reported (Smialowicz 1976). At this power density, the temperature of the media increased by 0.5°C after 15 min, and after 20 min the increase was 1°C. Changes in the mitotic index were dependent on the exposure time. Although a 5 min exposure did not influence the proportion of dividing cells, slight differences compared with controls were observed after 10 and 15 min exposures and significant differences were seen following 3- or 4-h exposures at 70 W/m^2. The same research group has examined the proliferative capacity of lymphocytes that are responsible for cellular immune responses (T cells) and humoral immune responses (B cells) following 2.45 GHz exposure *in vitro*. The ability of mouse spleen lymphocytes to undergo blast transformation in response to mitogens that selectively stimulate either T or B cells was measured by the incorporation of ^{3}H-thymidine into DNA. No consistent difference was found between the blastogenic response of exposed (100 W/m^2, 19 W/kg, 1 to 4 h) and control cells. In additional work, Smialowicz et al. (1982a–c) exposed mice to CW and PW MW at 2.45 GHz at SARs up to 8.6 W/kg. No differences in mitogen-stimulated responses of lymphocytes or in primary antibody levels in sensitized animals were observed between exposed and control mice in either the CW or PW exposures.

Czerska et al. (1992) studied human peripheral blood lymphocytes *in vitro*, comparing conventional heating to CW to pulsed (PW) exposures at 2.45 GHz, with each type of exposure evaluated at temperature elevations up to 2°C. Average SARS ranged up to 12.3 W/kg for exposures continuous over 5 d. At nonheating levels, CW had no effect on transformation. At heating levels, both conventional and CW heating produced a small but significant increase in blast transformation; at nonheating levels, PW enhanced transformation significantly and increasingly so at heating levels. The authors speculated that pulse power might act via the larger field amplitudes or by microacoustic shock effects.

Cleary et al. (1990b) presented evidence that lymphocytes are stimulated to take up ^{3}H-thymidine following exposure at 27 MHz or 2.45 GHz *in vitro* under isothermal conditions, with and without phytohemagglutinin (PHA). Maximal uptake occurs at SARs in the range of 25 to 40 W/kg depending on frequency. The authors argue in favor of a nonthermal mechanism.

Kiel et al. (1986) investigated the effects of 2.45 MHz CW MW, SAR of 103.5 W/kg from the point of view of impact on oxidative activity in mononuclear cells, primarily lymphocytes, using a chemiluminescence assay for level of nonphosphorylating oxidative activity. Exposure to MWs at 37°C does not appear to affect oxidative metabolic activity of mononuclear cells.

Huang et al. (1977) reported that lymphocytes from Chinese hamsters exposed to 2.45 GHz CW, from 50 to 450 W/m^2 15 min/d for 5 d showed changes in blast transformation and mitosis. No chromosomal aberrations were evident. These studies noted increased but reversible transformation of lymphocytes without mitogenic stimulation. These authors, noting temperature changes of less than 2°C, were among the first to suggest that hyperthermia could be exploited to manipulate the immune response, assuming the changes were thermally mediated. Further investigations by the same group (Huang and Mold 1980), in which mouse spleen cells were exposed to MW with or without T- or B-cell mitogens, showed responses that varied cyclically with time. The control samples, however, also showed the cyclical variations so it was unclear whether the MW exposure was the causative agent for the observed responses. Zafra et al. (1988) reported that MW at 2.45 GHz and power density of 320 W/m^2 for 15 min enhanced a number of activities of murine macrophages *in vitro* normally elicited in the immune response.

MWs have been reported to induce an increase in the frequency of complement receptor-bearing lymphoid spleen cells in mice (Wiktor-Jedrzejczak et al. 1977c), which may represent a maturation of B lymphocytes. Wiktor-Jedrzejczak et al. (1977a,c) exposed adult male mice to 2.45 GHz CW at an absorbed dose of 12 to 15 W/kg for 30 min in an environmentally controlled waveguide facility and then measured the function of different classes of lymphocytes *in vitro*. Such exposure failed to produce any detectable changes in function of T lymphocytes or increase in DNA, RNA, or protein synthesis, as measured by incorporation of tritiated-thymidine, -uridine, and -leucine by spleen, bone marrow, and peripheral blood lymphocytes *in vitro*. However, the maturation of B lymphocytes from the spleen of exposed mice was stimulated. Consistent with this effect on B lymphocytes are the results reported by Czerski and associates (Czerski et al. 1974a,b; Czerski and Siekierzynski 1975).

Roberts et al. (1987) exposed influenza-virus-infected human mononuclear leukocytes to 2.45 GHz RF, CW or PW at 16 or 60 Hz at an SAR of 4 mW/mL. No alteration by RF of mitogen-stimulated DNA synthesis occurred either in virus-infected leukocytes or uninfected leukocytes.

Mice exposed to 2.45 GHz CW, 50 to 350 W/m^2 (SAR of 4 to 25 W/kg) for 1 to 22 consecutive days (15 to 30 min/d) showed no consistent significant alterations in several parameters such as mitogen-stimulated response of T- and B-splenic lymphocytes, and the primary antibody response of mice to sheep erythrocytes (Smialowicz et al. 1979b). Subsequently, Schlagel et al. (1980) reported that RF-induced complement–receptor-positive (CR+) lymphocyte increase was under genetic control. Smialowicz et al. (1981b) showed, in addition, the age and strain of the mouse, the RF exposure characteristics (waveguide versus far field), and the environmental conditions are all sources of variation that affect the CR+ cell appearance.

Smialowicz et al. (1979a) exposed rats *in utero* and neonatally through 40 d of age (4 h/d, 7 d/week) in a controlled environment to either 2.45 GHz (CW, 50 W/m^2, SAR of 1 to 5 W/kg) or to 0.425 GHz (CW, 100 W/m^2, SAR 3 to 7 W/kg). At 40 d of age significant increases in the response of lymphocytes from exposed rats to *in vitro* stimulation with several mitogens was observed in several experiments. While these results have not been consistently reproduced, the trend in the results suggests that chronic exposure during fetal and neonatal development may change either the frequency or responsiveness of lymphocyte subpopulations. The biological significance of these observed changes is unknown. The mechanism by which these changes are initiated may be related to a thermally induced stress response. Similar responses have been observed in animals following prolonged exposure to nonspecific stressors.

Gildersleeve et al. (1987b) exposed Japanese quail eggs to 2.45 GHz MW at 50 W/m^2 for 12 d. After hatching an immunologic challenge did not elicit any modification of a cell-mediated or humoral immune response.

Hamrick (1973) examined the response of mammalian lymphocytes exposed to 2.45 GHz (CW) in cultures at 200 W/m^2 (7 W/kg) for 48 h. Changes in the stimulation caused by PHA were not affected by MW exposure.

Decrease in natural killer (NK) activity in lymphocytes of hamsters after exposure to 13 W/kg, 2.45 GHz CW for 1 h was reported by Yang et al. (1983). Roberts et al. (1983) examined the effects of exposure to 2.45 GHz continuous wave RF on human mononuclear synthesis after exposure to RF at SARs up to 4 W/kg. They also measured effects on synthesis of specific host defense proteins, namely interferons, and examined for morphological lymphoblastoid transformation as well as changes indicated by incorporation of radiolabeled precursors. Leukocytes were exposed in 37°C chambers without attempts to counteract RF-induced heating, and final culture temperatures were approximately 0.9°C higher than those of sham-exposed cultures, similar to changes induced by Stodolnik-Baranska (1974). Such exposures resulted in no detectable effects on viability or on unstimulated or mitogen-stimulated DNA, RNA, total protein, or interferon synthesis by the human mononuclear leukocytes (Yang et al. 1983).

There can be considerable difficulty in comparing reports, analyzing the literature, and extrapolating from *in vitro* to *in vivo* effects. Immunological responses to MW irradiation have also been recently reviewed by Black and Heynick (2003).

3.11.2 Adaptation

Czerski et al. (1974a,b) have reported that inbred Swiss mice, immunized with sheep RBCs and exposed 2 h/d for 6 or 12 weeks to 5 W/m^2 at 2.95 GHz, showed increased serum hemagglutinin titers and antibody-producing cells in lymph node homogenates. The increase was greater in mice exposed 6 weeks than in those exposed 12 weeks. Similar results were obtained in rabbits exposed 2 h/d to 30 W/m^2 for 6 months with the maximum increase occurring after exposure of 1 or 2 months, and then returning to control values. According to Czerski (1975), this may indicate that after a period of response, the animals become adapted to the MWs. The phenomenon of physiologic adaptation or decreased reaction as a result of repeated exposure to microwaves has also been reported by others (Gordon 1970; Michaelson 1974; Phillips et al. 1975; Varma et al. 1976).

Baranski (1972b) exposed adult guinea pigs 3 h/d for 3 months to 35 W/m^2 at 3 GHz and found an increase in lymphopoiesis over controls, as indicated by increased incorporation of ^3H-thymidine and increased mitotic indices. No differences could be detected between pulsed or CW MWs of the same average power level. Twofold increases in lymphocyte numbers were found in the spleen and lymph nodes.

3.11.3 Influence of Hyperthermia

RF-induced hyperthermia in mice has been associated with transient lymphopenia and neutrophilia, with a relative increase in splenic T and B lymphocytes, and with decreased *in vivo* local delayed hypersensitivity (Liburdy 1977). The latter was not affected by a comparable increase in core temperature produced by warm air. Reduced thymic mass and cell density (Smialowicz et al. 1979b) and suppressed inflammatory response (Phillips et al. 1975; Smialowicz et al. 1979b) have been reported. Such alterations in lymphocyte distribution and function are concomitant with a state of immunosuppression. Qualitatively, similar changes can be induced by administration of synthetic glucocorticoids or CS (Smialowicz et al. 1979b; Roberts et al. 1983). Liburdy (1979) reported elevated plasma CS levels in mice following exposure to RF energy sufficient to cause hyperthermia. A similar response has been reported by Lotz and Michaelson (1979), who showed a correlation between MW-induced body heating and CS levels in the blood of rats. These studies suggest that exposure to MWs of sufficient intensity results in stimulation of the hypophysial–adrenal axis, which could affect the immune system (Liburdy 1979).

It would thus appear that RF exposure initially causes a general stimulation of the immune system, but if the exposure continues, the stimulatory effect disappears, suggesting a phase of adaptation to continued RF exposure. There is a body of literature on the influence of heat *per se* on immunity. Significant influences of microwaves on immune responsiveness would be expected on the basis of the known effects of hyperthermia. Although there has been some uncertainty whether fever enhances host resistance to infection (Bennett and Nicastri 1960; Atkins and Bodel 1972), recent evidence suggests that fever may enhance survival after infection in an animal model (Kluger et al. 1975). Cell-mediated immunity plays a role in defense against facilitative intracellular bacteria, viruses, and certain other infectious agents (Frenkel and Caldwell 1975; Mandell 1975). Roberts and Steigbigel (1977) have shown that increased temperature (38.5°C) enhances human lymphocyte response to mitogen (PHA) and antigen (streptokinase–streptodornase) and enhances, but does not accelerate, certain bactericidal functions of human phagocytic leukocytes. One, therefore, has to be circumspect in assessing the mechanisms of MW exposure related to alterations in immune processes. It may be that if MWs do, in fact, increase the proportion of lymphocytes undergoing transformation, it may not in itself be harmful, but actually beneficial. The immune system has a considerable redundancy and adaptability. Perturbations of the immune system may not have clinical significance.

3.12 Carcinogenesis

There are no laboratory studies in humans with cancer as the direct end point of investigation. Such studies would clearly cross-ethical boundaries and would not be acceptable. Furthermore, RF fields lack sufficient energy to disrupt chemical bonds so there is little theoretical basis for suspecting that such fields would cause mutations or other genotoxic effects. There are a few human studies investigating possible indirect and nongenotoxic effects relevant to cancer that are discussed in the section on endocrine function studies. Additionally, accompanying a rising concern regarding exposure of large numbers of people to mobile telephone signals, a number of recent epidemiologic studies have been performed evaluating RF exposures at cellular phone frequencies.

3.12.1 Long-Term Animal Studies

Long-term animal bioassays, often conducted in two species (usually rats and mice) and in both male and female animals for 2 y, provide a reasonable surrogate for human lifetime exposure. A relatively small number of long-term animal bioassays have been performed exposing rats and mice to RF signals. Unlike many other areas of biological investigation of MW exposures, long-term animal cancer bioassays have almost all been conducted since the last edition of this Handbook was issued. These more recent studies, performed at nonthermal levels, have indicated no pathological or carcinogenic effects. This includes studies with a focus on brain cancer at 836 MHz, 1.1 to 1.6 W/kg (Adey et al.

1999, 2000), 900 MHz at ~1 W/kg (Zook and Simmens 2001), and 1.6 GHz (Anderson et al. 2004). Complete histopathology in life span and hematologic studies were performed at 835/847, 1600, 800, and 2450 MHz at 1.3 W/kg (La Regina and Roti Roti 2003), up to 1.6 W/kg (Anderson et al. 2004); up to 12.9 W/kg (Spalding et al. 1971), and at 0.3 W/kg (Frei et al. 1998), respectively. Some pathological effects have been reported at thermal levels (Roberts and Michaelson 1983). Prausnitz and Susskind (1962) reported the induction of leukemia in mice exposed to 9.27 GHz RF at 1 kW/m^2, 4.5 min/d, 5 d/week for 59 weeks. This report is viewed as a flawed study, principally with regard to interpretation of the histology of the hematopoietic abnormality. Szmigielski et al. (1980) described accelerated development of spontaneous and benzopyrene-induced skin cancer in mice exposed to 2.45 GHz, 150 W/m^2 (6 to 8 W/kg). At 50 W/m^2 (2 to 3 W/kg), the cancer induction was comparable to that which occurs with chronic stress resulting from confinement. Preskorn et al. (1978), however, reported decreased tumor growth and greater longevity in mice exposed to intense levels of 2.45 GHz energy, i.e., sufficient to make the mice hyperthermic.

The only report of an increased tumor incidence with long-term RF exposure at clearly nonthermal levels was reported by Chou et al. (1992). They reported a small increase in overall tumor occurrence in rats exposed for 24 months to 2450 MHz (SAR of 0.15 to 0.4 W/kg). There was no effect in the Chou et al. study on a number of other parameters including metabolism, immune function, hematology, serum chemistry, thyroxine levels, protein parameters, growth or open-field behavior. In another bioassay cancer study at 435 MHz, no overall increase in cancer in animals exposed to 1.0 mW/cm^2 was found (Toler et al. 1997). Some organs (specifically the adrenal glands) did show slight trends toward increased cancer in RF-exposed animals; however, the number of tumors was small and no statistically significant differences could be determined between the exposed and control groups.

Shorter term but repeated exposures of C3H/HeJ mice to ultra-wideband MW pulses showed no effects on the development of mammary tumors (Jauchem et al. 2001). A 2 week exposure experiment at 900 MHz with average whole body SARs of 75 and 270 mW/kg showed no significant effects on tumor development or growth rate or time of tumor onset (Chagnaud et al. 1999).

3.12.2 Radiation or Chemically Initiated and Transgenic Animal Bioassays

Similar to long-term animal bioassays, studies in which tumors have been initiated by means other than RF have been mostly negative. Many different initiation models have been used in these studies in which rodents have been exposed to RFs between 800 and 1500 MHz. The experimental models used include brain tumors initiated in rats with ethyl nitrosourea (836 and 860 MHz, approximately 1 W/kg) (Adey et al. 1999, 2000; Zook and Simmens 2001), benz[*a*]pyrene-initiated sarcomas in rats (900 MHz, 0.075 and 0.27 W/kg) (Chagnaud et al. 1999), dimethyl benzanthracine (DMBA)-initiated rat mammary tumors (900 MHz, SARs from 0.017 to 0.07 W/kg) (Bartsch et al. 2002), diethyl nitrosamine-induced hepatomas in rats (929 and 1500 MHz) (Imaida et al. 1998a,b), and radiation-induced mouse lymphomas (902 MHz, 0.35 W/kg) (Heikkinen et al. 2001). In all of these cited cancer studies, no adverse effects of RF exposure were noted. Mason et al. (2001) studied mice exposed to 94 GHz for 12 weeks. Treatment of both exposed and control mice with DMBA resulted in papilloma development, as expected, but no differences were observed between the groups.

In addition to chemicals and radiation, genetically initiated animal models (transgenic mice) have also been studied in RF carcinogenicity testing. No effects of RF exposure on mutagenicity or tumor development were found using pKZ-1 transgenic mice (900 MHz,

4 W/kg) (Sykes et al. 2001). Another study by Repacholi et al. (1997) in Pim-1 mice reported an association between long-term RF exposure (900 MHz, 0.13 to 1.4 W/kg) and mortality from a certain subtype of lymphoma. A subsequent study performed at multiple dose levels with more uniform and more fully characterized exposure (900 MHz, multiple levels to 4.0 W/kg) did not confirm the positive effects reported in the original study (Utteridge et al. 2002). Chronic exposure of C3H/HeJ mammary tumor-prone mice has also been investigated. The mice were exposed or sham-exposed to 1 mW/cm^2 at 435 MHz for 21 months. No significant effects were observed in any of the tumor-related indices (Toler et al. 1997).

3.12.3 Tumor Cell Injection Bioassays

A few studies of tumor progression, using nonthermal RF exposure levels, have been conducted by injecting tumor cells into mice and determining growth rate, survival, and metastatic progression. Although increased survival of the host, as well as inconsistent evidence of either augmentation or suppression of immune function, has been reported in response to thermal levels of RF exposure, no such effects were observed in studies using lower levels of exposure: at 915 MHz, up to 8.3 W/kg (Salford et al. 1993b) and at 836 and 847 MHz, 0.75 W/kg (Higashikubo et al. 1999).

3.12.4 Summary of Animal Cancer Bioassays

Most animal bioassay studies have not demonstrated increased cancer risk resulting from long- or short-term RF exposure at nonthermal levels. In the very few studies at thermal levels of MW exposure, only inconsistent evidence of effects have been reported, and those have not been confirmed in similar or replicate studies. Recent reviews have been published on cancer studies in MW-exposed animals (Elder 2003; Heynick et al. 2003).

3.13 Other Specialized Organ Response

3.13.1 Auditory Response

A phenomenon called "microwave hearing" or "RF hearing" is a well-established biological effect resulting from exposure of the head to pulsed microwaves. RF hearing is sensed in the head at rather low-power densities (Frey 1961, 1962). In and of itself this phenomenon, characterized by the sound of clicking, buzzing, knocking, or chirping in the head, has no known adverse health consequences (Frey 1985; Roschmann 1991; Elder and Chou 2003). The fundamental MW frequencies that are involved vary from 2.4 MHz to 10 GHz (Frey 1962, 1963; Putthoff et al. 1977; Cain and Rissmann 1978; Khizhnyak et al. 1980). However, the MWs must be modulated to produce the hearing phenomenon, generally in the 7 to 10 kHz range (Chou et al. 1977, 1982b; Watanabe et al. 2000), or 8 to 15 kHz (Lin 1977a,c).

The mechanism responsible for eliciting such an effect appears to be thermoelastic expansion (Foster and Finch 1974; Chou et al. 1975, 1982a; Guy et al. 1975a; Lin 1976, 1977a; see also Chapter 10 of *BBA*). It is probable that RF-generated waves of thermoelastic expansion result in cochlear hair cell distortions, mimicing the sonic waveforms of acoustic stimuli (Foster and Finch 1974; Taylor and Ashleman 1974; Chou et al. 1975,

1985; Guy et al. 1975a; Lebowitz and Seaman 1977a,b; Lin 1977b; Chou and Guy 1979; Lin et al. 1979a, 1982, 1988; Olsen and Lin 1981, 1983). There are studies, however, that indicate the RF-hearing phenomenon can also occur after the middle ear has been ablated (Taylor and Ashleman 1974; Guy et al. 1975b; Chou and Galambos 1979; Wilson et al. 1980). Other studies have described thresholds for this RF response in animals (Guy et al. 1975a; Lebovitz and Seaman 1977a,b; Cain and Rissmann 1978; Seaman and Lebovitz 1989).

Comprehensive reviews are available discussing the MW-hearing effect and research directed to an examination of the mechanisms involved in producing such effects (Chou et al. 1976, 1982a; Lin 1978, 1981, 1989, 1990, 2001, 2002; Elder and Cahill 1984; Postow and Swicord 1996; Stewart 2000; Elder and Chou 2003; see also Chapter 10 of *BBA*).

3.13.2 Ocular Effects

During the past 40 y, numerous investigations in animals and several surveys among human populations have been devoted to assessing the relationship of MW exposure to the subsequent development of cataracts. It is significant that of the many experiments on rabbits by several investigators using various techniques, a power density above 1 kW/m^2 for 1 h or longer appears to be the lowest time–power threshold in the tested frequency range of 0.2 to 10 GHz. In other species of animals such as dogs and nonhuman primates, the threshold for experimental MW-induced cataractogenesis appears to be even higher. This threshold is a time–power threshold, that is, the higher the power density, the shorter is the time threshold and vice versa, down to a certain minimum power density. All of the reported effects of MW radiation on the lens can be explained on the basis of thermal injury.

3.13.2.1 Threshold for Opacity in Rabbits

The most extensive investigations in this area were performed by Carpenter and Van Ummersen (1968) and Carpenter et al. (1960). Later work by Guy et al. (1975b) and Kramar et al. (1975, 1987) has shown that in rabbits exposed to 2.45 GHz, the threshold for cataract production is 1.5 kW/m^2 for 100 min or $\geq$150 W/kg for $\geq$30 min. The data suggest that an intraocular temperature of at least 41°C (Guy et al. 1975b) must be obtained to induce cataracts, although exceptions may be found in the literature. These investigators also found that single potentially cataractogenic exposures will not injure the eye under conditions of controlled general hypothermia, and exposure to 1 kW/m^2, 2 h/d, for 4 to 9 d produced no cataracts, as evidenced by periodic examinations for 6 months after exposure. Guy et al. (1974) exposed rabbits to 0.918 GHz, 4.66 kW/m^2 for 15 min and 1.17 W/m^2 for 100 min with no evidence of cataract and concluded that the threshold for cataractogenesis is higher for this frequency in comparison to 2.45 GHz. Williams et al. (1975) discussed the ultrastructural changes that occur in rabbit eyes when RF exposure fields are intense. Additional studies using high intensity exposures were conducted in the eyes of the dog (Daily et al. 1950; Ballie 1970; Ballie et al. 1970). At the same average power, both pulsed and continuous RF exposures were equally effective in producing cataracts in rabbits (Birenbaum et al. 1969). These results support the indication of a thermal mechanism for the damage.

Appleton et al. (1975) exposed rabbits to 3 GHz, 1 or 2 kW/m^2 for 15 to 30 min. Examining daily for 14 d, weekly for 1 month, and monthly for a year revealed no ocular changes. At power densities of 3, 4, or 5 kW/m^2 for 15 min, acute ocular changes involving the conjunctiva and iris occurred during exposure. In a comparable study by Hirsch et al. (1977), rabbits were repeatedly exposed to 3 GHz once daily for a month.

Clinical examinations were carried out for 1 y afterward. No changes occurred at power densities under 3 kW/m^2.

3.13.2.2 Biochemical Changes

Reports of decreased enzymatic activity (Kinoshita et al. 1966; Carpenter and Van Ummersen 1968) may quite likely be due to thermal inactivation with resultant alterations in metabolism. Decreases in ascorbic acid concentration in the lens have been cited as the first biochemical indication of opacity formation (Merola and Kinoshita 1961; Weiter et al. 1975); and progressive clouding of the lens is associated with decreases in ascorbic acid below 60 μg/g of lens tissue. All of these effects could be fully reparable until the altered metabolism has produced a permanent opacification of the lens. Latent periods and time–power thresholds would be in agreement with a mechanism of this nature.

3.13.2.3 Thermal Aspect of Microwave Cataractogenesis

Guy et al. (1974, 1975a,b), Kramar et al. (1975), and Taflove and Brodwin (1975) have computed the MW energy deposition and induced temperatures in the eyes of rabbits and a model of the human eye. They have indicated that a distinct hot spot exceeding 40.4°C probably occurs deep within the eye at a frequency of 1.5 GHz, when the power density would be cataractogenic (i.e., greater than 2 kW/m^2).

Paulsson (1976) measured the absorption of 0.915, 2.45, and 9.0 GHz MW energy in a model of the human head. The absorbed power showed an essentially exponential decrease with the distance from the cornea at 9.0 GHz. At 2.45 and 0.915 GHz, maximum absorbed power occurred within the eyeball.

3.13.2.4 Pulse Wave Exposures

Kues and Monahan (1992; Kues et al. 1985, 1992) investigated effects of pulsed MW on the eyes of monkeys *in vivo*. Exposures were at 2.45 GHz (10 μs, 100 pps) at an average incident power of 100 W/m^2 (SAR of 2.65 W/kg) on three consecutive days for 4 h/day. Damage to the corneal endothelium and an increase in iris vascular permeability was found persisting for more than 72 h after exposure. This report suggests that PW exposures might have significantly lower thresholds for damage than CW exposures, especially high peak power exposures with low SAR resulting from a sufficiently slow repetition rate. Johnson et al. (1992) also exposed monkeys to 1.25 GHz MW, 0.5 μs wide pulses, 16 pps at an average ocular SAR of 3.5 to 4.0 W/kg, seven 4-h exposures spread over several weeks. At this high peak, low repetition rate exposures retinal damage was found by electroretinography, suggesting cone photoreceptor damage. However, Kamimura et al. (1994) failed to confirm the corneal lesions reported by Kues et al. after CW exposure, and Lu et al. (2000) did not replicate the PW effects of exposure. Furthermore, no functional changes in vision were observed (McAfee et al. 1983). Kues et al. (1999) did not see corneal damage or changes in vascularity permeability of the iris or lens opacity in rabbit or monkey eyes exposed to 60 GHz fields at 100 W/m. Nor were histological changes observed at 2250 W/m^2 (Williams and Finch 1974). Creighton et al. (1987) have also found evidence of an enhanced effect of pulsed over continuous MW at equivalent average power level by observing cataract formation in the rat lens *in vitro*.

3.13.2.5 Cumulative Effect

The possibility of a cumulative effect on the lens from repeated subthreshold exposures of the rabbit eyes to MW was proposed. Such an effect presumably would result from clinically inapparent damage from any given exposure, which would eventually manifest clinically after repeated exposures.

McAfee et al. (1979) trained monkeys to face an RF source and then exposed them to 9.3 GHz pulsed RF at an average power density of 1.5 kW/m^2. Each of 12 monkeys was exposed for up to 20 min/d for 30 to 40 sessions over several months. A total of 75 monkeys, nonexposed and sham-exposed, served as controls. No cataracts or corneal lesions were seen in any of the 12 exposed animals up to 12 months after exposure.

Chou et al. (1982b) exposed two groups of six rabbits to 2.45 GHz CW/PW (10 μs pulses, 100 pps, peak power density 150 kW/m^2) at 150 W/m^2 for 2 h/d over several months; a third group was sham-irradiated. Periodic eye examinations for cataract formation yielded no statistically significant differences among the three groups. Guy et al. (1980) and Chou et al. (1982b, 1983) exposed rabbits to 2.45 GHz RF at 100 W/m^2 (maximum whole body SAR of 17 W/kg) 7 d/week, 23 h/d for 6 months. For controls, other rabbits were sham-exposed for the same durations. Periodic eye examinations with a slit lamp microscope showed normal aging changes in the lenses and no significant differences between the exposed and control groups.

Although it is difficult to identify the time–power threshold precisely, varying the times and powers to produce such a curve, a general reading of the literature suggests that no one has been able to produce cataracts in animal models at power densities below 800 to 999 W/m^2 even by repetitive exposures. These values are well above those of the most widely used safety standards (ANSI/IEEE 1982; NCRP 1986).

In general, the results from animal studies indicate that RF cataractogenesis is essentially a gross thermal effect that has a threshold power density at which the difference between the rates of heat generation by RF and heat removal is large enough to result in damage to the lens of the eye. The case of very high peak powers at low repetition rates may prove to be an exception.

3.13.2.6 Extrapolation to the Human

After 40 y of studies of the effects of MWs on the ocular lens, primarily in the rabbit and monkey, the principal conclusions are:

- The acute thermal insult from high intensity MW fields is cataractogenic in the rabbit if intraocular temperature is greater than 43°C.
- The MW exposure threshold is between 1 and 1.5 kW/m^2 applied for about 60 to 100 min.
- There does not appear to be a cumulative effect from MW exposure unless each single exposure is sufficient to produce some irreparable degree of injury to the lens.

That opacity of the ocular lens can be produced in rabbits by exposure to MW is well-established. Extrapolating results from animal studies to humans is difficult, because the conditions, durations, and intensities of exposure are usually quite different. Reports of adverse reactions to exposures are necessarily uncontrolled and exposure conditions are generally impossible to determine or reconstruct. It is also difficult to relate cause and effect, because lens imperfections do occur in otherwise healthy individuals, especially with increasing age. Numerous drugs, industrial chemicals, and certain metabolic diseases are associated with cataracts. Because of the many uncertainties, workmen's compensation and negligence cases are usually settled out of court with the details of the settlement withheld from the public. Lipman et al. (1988) have reviewed the conditions under which MW might produce injury to the lens.

There have been a few epidemiological studies that specifically address the issue of MW exposure effects in the human eye. The principal end point has been risk of cataract formation or other opacities in the eye. Difficulties arise in almost all of these studies in that exposure frequencies are seldom specified, nor are exposure levels known. These studies are reviewed in Elder (2003) with the strong conclusion that no evidence of changes in the eye are apparent. One might assume that at MW exposure levels high enough to produce heating in the eye (as in animal studies) that one might observe opacities develop. However, only two cancer assessment studies in humans show effects. Holly et al. (1996) found a twofold risk increase for uveal melanoma in RF-exposed workers. Stang et al. (2001) also reported an increased risk of eye cancer from a case–control study of mobile phone users. The limitations of the methodologies used, however, prevented clear interpretation of the data. Johansen et al. (2001, 2002) attempted to replicate these observations in a Danish study and found no association between mobile phone use and eye (or brain or other) cancers.

3.13.3 Hypersensitivity

A number of studies have investigated claims by individuals of an association between exposure to electrical fields and a number of subjective end points, including dizziness, perspiration, irritability, and other bioeffects (COMAR 2002). This condition has been termed "hypersensitivity" and has been reported to occur with exposure to video display terminal (VDT) instruments (Sandstrom et al. 1995; Stenberg et al. 1995), with occupational RF fields (Bini et al. 1986), or other external RF sources (Choy et al. 1986; Flodin et al. 2000; Sandstrom et al. 2001). A study of RF-exposed workers in Poland claimed health effects but study design limitations make interpretation of results difficult (Bielski 1994). Some carefully conducted research has reported no association between mobile telephony exposure and hypersensitivity in both provocation studies (Kovisto et al. 2000a,b) and self-identified sensitive subjects (Heitanen et al. 2002). However, other studies report links between hypersensitivity and cell phone exposure (Hocking 1998; Hocking and Westerman 2000). Kimata (2002) observed increased allergic reactions in sensitive subjects with cell phone exposure for 60 min. A follow-on study, however, indicated that the actual field exposure may not be the causative factor (Kimata 2003).

3.14 Critique

Elucidation of the biological effects of RF exposure requires study and analysis of the available literature. This literature has been accumulating, in a sustained way, since the late 1940s with the development and application of radar in World War II. The literature is not particularly user-friendly in that comparisons of studies must often rest on assumptions of uncertain validity. Over the last half century, there has been a steady improvement in measurement capabilities, dosimetric measurements, and standards of quality in the performance of investigations. This improvement has been so dramatic that it makes contemporary investigators mistrustful of the older literature, resting as it did on a less-sophisticated technology and approach. The older literature also contained a major contribution from the closed societies of Eastern Europe, which was hobbled by barriers to free and open communication and which may have been tainted by political interference. It was commonly the case that reports from countries dominated by the former

Soviet Union were inadequately documented and the results slanted to conform to expectations of authorities. This may have had the effect of endowing RF effects at low power with greater and perhaps more sinister efficacy in producing specific biological effects, especially on the nervous system, than the contemporary consensus would support.

The general improvement in the quality of research has produced areas of agreement on effects and potential hazards, but also left areas of disagreement. The issue of window effects produced by modulated RF on calcium exchange in the nervous system is still permeated with disagreement. The issue of an increase in intrinsic permeability of the BBB at nonthermal levels of RF has been resolved to the satisfaction of a majority of students of the subject but some dissent still remains. General consensus is that there is no such increase at nonthermal levels and that brain temperature must rise to 41 to 43°C, rapidly producing tissue necrosis and morbidity, in order to see transfer of hydrophilic molecules into brain parenchyma. Studies of effects on the neuroendocrine system always pose interpretative problems because of the labile response of the system to stress. Even the most gentled animals may be responding to the stress of handling and this may potentiate an RF effect which otherwise may not be observed.

Even if the effects can be validated and the dose–response worked out quantitatively in a manner to pass the most stringent review, there remains the problem of what constitutes a hazard. One objective definition of injury is an irreversible change in biological function producing a functional deficit as observed at the organ or system level. With this definition, it is possible to define a hazard as a probability of injury on a statistical basis. It is important to distinguish between the intensities of RF energy at which injury may be sustained and those that produce nonharmful effects with no further pathophysiologic consequences. All effects are not necessarily hazards. Some effects may have beneficial application under appropriately controlled condition. For that reason therapeutic application of RF energy by use of diathermy technology to produce frank heating is excepted from safety standards prescribing permissible exposure limits. In this case, risk–benefit analyses are applied.

It is important to determine whether an observed effect is irreparable, transient, or reversible, disappearing when the electromagnetic field is removed either immediately or after some interval of time. Of course, even reversible effects are unacceptable if they transiently impair the ability of the individual to function properly or to perform a required task.

Although there is a much greater appreciation in recent years of the need for rigorous experimental design, the challenges are still formidable. An investigator may think he is observing a low level or nonthermal effect in one animal because the incident power is low, while in actuality the animal may be exposed to as much absorbed power in a specific region of the body due to resonance or focusing effects as another larger animal is with much higher incident power at certain frequencies. A detailed understanding of interspecies scaling factors is essential.

The review of the literature as a whole has its own problem. The process of comparing studies is usually not straightforward, especially if the studies seem to contradict each other. Even an investigation, which has as a specific goal the confirmation or nonconfirmation of a previous study, is inevitably different in some ways. The significance of the differences is debated at scientific meetings and these debates are usually not reflected in the literature. If a follow-up study does not use exactly the same frequency, the question of the ability to extrapolate from one frequency to the other arises.

Pickard and Foster (1987) proposed that it may be time to limit further RF biological effects research by formulating guidelines to restrict further work. They advanced this proposal because of the inherent noisiness of the field and the apparent inability to resolve controversies in a timely manner. In subsequent meeting forums many scientists took issue with the proposal, arguing that more (and more careful) research was needed, not less.

As mentioned in the introduction of this chapter, the rate of RF research output declined significantly during the 1990s. The U.S. Defense Department, the major underwriter of RF research prior to about 1995, dramatically reduced its support of MW research. Public concerns through the late 1980s and early 1990s directed more support to ELF research. However, in the mid-1990s concerns that the RF output of cellular telephones might play a role in the initiation of tumors caused some return of funding to RF effects research. In assessing the literature, it will be evident throughout this chapter that research conducted since the last edition of this Handbook has concentrated in particular areas. A focus on potential RF effects on cancer, certain neurophysiological effects such as investigations of the BBB, and further studies at the cellular level predominate in the recent literature.

Although the public policy rationale undergoes periodic shifts, overall support for continuing research in electromagnetic effects will undoubtedly continue, motivated substantially by safety considerations and exploration of potential medical use of MWs in diagnoses and therapies.

Acknowledgment

L.E.A. acknowledges the important efforts made by Dr. Sol M. Michaelson and Dr. Edward C. Elson in the earlier editions of this work. It is hoped that the updating of their chapter has contributed in a measurable way to the usefulness of the current edition.

References

Abhold R.H., Ortner M.J., Galvin M.J., McRee D.I., 1981. Studies on acute *in vivo* exposure of rats to 2450-MHz microwave radiation: II. Effects on thyroid and adrenal axes hormones, *Radiat. Res.* 88:448–455.

Accinni L., De Martino C., Mariutti G., 1988. Effects of radiofrequency radiation on rabbit kidney: a morphological and immunological study, *Exp. Mol. Pathol.* 49:22–37.

Adair E.R., 1981. Microwaves and thermoregulation, in: *USAR Radiofrequency Radiation Bioeffects Research Program—A Review*, Mitchell, J.C., Ed., Review 4-81, USAF School of Aerospace Medicine, San Antonio, TX.

Adair E.R., 1995. Thermal physiology of radiofrequency radiation (RFR) interactions in animals and humans, in: *Radiofrequency Standards*, Klauenberg B.J., Erwin D.N., Grandolofo M., Eds., Plenum Press, New York, NY, pp. 403–433.

Adair E.R., Adams B.W., 1980a. Microwaves modify thermoregulatory behavior in squirrel monkey, *Bioelectromagnetics* 1:1–20.

Adair E.R., Adams B.W., 1980b. Microwaves induce peripheral vasodilation in squirrel monkey, *Science* 207:1381–1383.

Adair E.R., Adams B.W., 1982. Adjustments in metabolic heat production by squirrel monkeys exposed to microwaves, *J. Appl. Physiol.: Respir. Environ. Exercise Physiol.* 50:1049–1058.

Adair E.R., Black D.R., 2003. Thermoregulatory responses to RF energy absorption, *Bioelectromagnetics* (Suppl. 6):S17–S38.

Adair E.R., Adams B.W., Akel G.M., 1984. Minimal changes in hypothalamic temperature accompany microwave-induced alteration of thermoregulatory behavior, *Bioelectromagnetics* 5:13–30.

Adair E.R., Mylacraine K.S., Cobb B.L., 2001a. Human exposure to 2450-MHz CW energy at levels outside the IEEE C95.1 standard does not increase core temperature, *Bioelectromagnetics* 22:429–439.

Adair E.R., Mylacraine K.S., Cobb B.L., 2001b. Partial-body exposure of human volunteers to 2450-MHz pulsed or CW fields provokes similar thermoregulatory responses, *Bioelectromagnetics* 22:246–259.

Adey W.R., Bawin S.M., Lawrence A.F., 1982. Effects of weak amplitude-modulated microwave fields on calcium efflux from awake cat cerebral cortex, *Bioelectromagnetics* 3:295–307.

Adey W.R., Byus C.V., Cain C.D., Higgins R.J., Jonès R.A., Jean C.J., Kuster N., MacMurray A., Stagg R.B., Zimmerman G., Phillips J.L., Haggren W., 1999. Spontaneous and nitrosourea-induced primary tumors of the central nervous system in Fischer 344 rats chronically exposed to 836 MHz modulated microwaves, *Radiat. Res.* 152:293–302.

Adey W.R., Byus C.V., Cain C.D., Higgins R.J., Jones R.A., Jean C.J., Kuster N., MacMurray A., Stagg R.B., Zimmerman G., 2000. Spontaneous and nitrosourea-induced primary tumors of the central nervous system in Fischer 344 rats exposed to frequency-modulated microwave fields, *Cancer Res.* 60:1857–1863.

Akdag M.Z., Celik M.S., Ketani A., Nergiz Y., Deniz M., Dasdag S., 1999. Effect of chronic low-intensity microwave radiation on sperm count, sperm morphology, and testicular and epididymal tissues of rats, *Electro- Magnetobiol.* 18:133–145.

Akyel Y., Hunt E.L., Gambill C., Vargas C., 1991. Immediate post-exposure effects of high-peak-power microwave pulses on operant behavior of Wistar rats, *Bioelectromagnetics* 12:183–195.

Albert E.N., 1977. Light and electron microscopic observations on the blood brain barrier after microwave irradiation, in: *Symposium on Biological Effects and Measurement of RF/MW: Proceedings of an FDA Conference*, HEW Publication (FDA) 77-8026, pp. 294–304.

Albert E.N., 1978. Ultrastructural pathology associated with microwave induced alterations in blood–brain barrier permeability, in: *Proceedings Biological Effects of Electro Magnetic Waves*, August 1978, 19th Gen. Assembly, Int. Union Radio Sci., Helsinki, Finland.

Albert E.N., 1979. Reversibility of microwave induced blood brain barrier permeability, *Radio Sci.* 14:323–327.

Albert E.N., DeSantis M., 1975. Do microwaves alter nervous system-structure? *Ann. N.Y. Acad. Sci.* 247:87–108.

Albert E.N., Kerns J.M., 1981. Reversible microwave effects on the blood–brain barrier, *Brain Res.* 230:153–164.

Albert E.N., Sherif M., 1988. Morphologic changes in cerebellum of neonatal rats exposed to 2.45 GHz microwaves, in: *Electromagnetic Fields and Neurobehavioral Function*, O'Connor M.E., Lovely, R.H., Eds., Alan R. Liss, Inc., New York, NY, pp. 135–152.

Albert E.N., McCullars G., Short M., 1974. The effect of 2450 MHz microwave radiation on liver adenosine triphosphate (ATP), *J. Microwave Power* 9:205.

Albert E.N., Grau L., Kerns J., 1977. Morphologic alterations in hamster blood–brain barrier after microwave irradiation, *J. Microwave Power* 12:43–44.

Albert E.N., Brainard D.L., Randal J.D., Jannata F.S., 1978. Neuropathological observations on microwave-irradiated hamsters, in: *Proceedings Biological Effects of Electro Magnetic Waves*, August 1978, 19th Gen. Assembly, Int. Union Radio Sci., Helsinki, Finland.

Albert E.N., Sherif M.F., Papadopoulos N.J., Slaby F.J., Monahan J., 1981a. Effects of nonionizing radiation on the purkinje cells of the rat cerebellum, *Bioelectromagnetics* 2:247–257.

Albert E.N., Sherif M.F., Papadopoulos N.J., 1981b. Effect of nonionizing radiation on the purkinje cells of the uvula in squirrel monkey cerebellum, *Bioelectromagnetics* 2:241–246.

Albert E.N., Slaby F.J., Loftus J., 1987. Effect of amplitude-modulated 147 MHz radiofrequency radiation on calcium ion efflux from avian brain tissue, *Radiat. Res.* 109:19–27.

Alekseev S.I., Ziskin M.C., 2001. Distortion of millimeter wave absorption in biological media due to the presence of thermocouples and other objects, *IEEE Trans. Biomed. Eng.* 48:1013–1019.

Allis J.W., 1975. Irradiation of bovine serum albumin with a crossed-beam exposure-detection system, *Ann. N.Y. Acad. Sci.* 247:312–322.

Allis J.W., Sinha-Robinson B.L., 1987. Temperature-specific inhibition of human red cell Na^+/K^+ ATPase by 2,450 MHz microwave radiation, *Bioelectromagnetics* 8:203–212.

Anderson L.E., Sheen D.M., Wilson B.W., Grumbein S.L., Creim J.A., Sasser L.B., 2004. Two year chronic bioassay study of rats exposed to a 1.6 GHz radiofrequency signal, *Radiat. Res.* 162:201–210.

Anderstam B., Hamnerius Y., Hussain S., Ehrenberg L., 1983. Studies of possible genetic effects in bacteria of high frequency electromagnetic fields, *Hereditas* 98:11–32.

Anne A., Saito M., Salati O.M., Schwan H.P., 1962. Relative microwave absorption cross sections of biological significance, in: *Proc. 4th Annu. Tri-Service Conf. Biol. Effects of Microwave Radiating Equipment; Biological Effects of Microwave Radiation*, Peyton M.F., Ed., Plenum Press, New York, pp. 153–176.

ANSI/IEEE, 1982. Safety Levels with Respect to Radio Frequency Electromagnetic Fields, 300 kHz to 100 GHz, Report No. ANSI C95.1-1982, American National Standards Institute, The Institute of Electrical and Electronic Engineers, Inc., New York, NY.

Appleton B., Hirsch S., Kinion R.O., Soles M., McCrossan G.C., Neidlinger R.M., 1975. Microwave lens effects in humans: II. Results of five-year survey, *Arch. Ophthalmol.* 93:257–258.

Arber S.L., Lin J.C., 1983. Microwave enhancement of membrane conductance in snail neurons: role of temperature, *Physiol. Chem. Phys. Med. NMR* 15:259–260.

Arber S.L., Lin J.C., 1984. Microwave enhancement of membrane conductance: effects of EDTA, caffeine and tetracaine, *Physiol. Chem. Phys. Med. NMR* 16:469–475.

Arber S.L., Lin J.C., 1985a. Extra-cellular calcium and microwave enhancement of membrane conductance in snail neurons, *Radiat. Environ. Biophys.* 24:149–156.

Arber S.L., Lin J.C. 1985b. Microwave-induced changes in nerve cells: effects of modulation and temperature, *Bioelectromagnetics* 6:257–270.

Arber S.L., Neilly J.P., Lin J.C., Kriho V., 1987. The effect of 2450 MHz microwave radiation on the ultrastructure of snail neurons, *Physiol. Chem. Phys. Med. NMR* 18:243–249.

Ashani Y., Henry F.H., Catravas G.N., 1980. Combined effects of anticholinesterase drugs and low-level microwave radiation, *Radiat. Res.* 84:469–503.

Atkins E., Bodel P., 1972. Fever, *N. Engl. J. Med.* 286:27–34.

Baillie H.D., 1970. Thermal and nonthermal cataractogenesis by microwaves, in: *Biological Effects and Health Implications of Microwave Radiation*, Cleary S.F. Ed., U.S. Department of Health, Education, and Welfare, HEW Publication BRH/DBE 70-2, Washington, D.C., pp. 59–65.

Baillie H.D., Heaton H.G., Pal D.K., 1970. The dissipation of microwaves as heat in the eye, in: *Biological Effects and Health Implications of Microwave Radiation*, Cleary S.F. Ed., U.S. Department of Health, Education, and Welfare, HEW Publication BRH/DBE 70-2, Washington, D.C., pp. 85–89.

Baldwin M.S., Bach S.A., Lewis S.A., 1960. Effects of radiofrequency energy on primate cerebral activity, *Neurology* 10:178–187.

Baranski S., 1971a. Effect of chronic microwave irradiation on the blood forming system of guinea pigs and rabbits, *Aerosp. Med.* 42:1196–1199.

Baranski S., 1971b. Effect of microwaves on the reaction of the leukocytic system, *Acta Physiol. Pol.* 22:898.

Baranski S., 1972a. Histological and histochemical effects of microwave irradiation on the central nervous system of rabbits and guinea pigs, *Am. J. Physiol. Med.* 51:182–190.

Baranski S., 1972b. Effect of microwaves on the reactions of the WBC system, *Acta Physiol. Pol.* 23:619–629.

Baranski S., Czerski P., 1976. *Biological Effects of Microwaves*, Dowden, Hutchinson & Ross, Stroudsburg, PA.

Baranski S., Edelwejn Z., 1967. Electroencephalographic and morphological investigations on the influence of microwaves on the central nervous system, *Acta Physiol. Pol.* 18:517–532.

Baranski S., Edelwejn Z., 1968. Studies on the combined effect of microwaves and some drugs on bioelectric activity of the rabbit central nervous system, *Acta Physiol. Pol.* 19:31–41.

Baranski S., Edelwejn Z., 1975. Experimental morphologic and electroencephalographic studies of microwave effects on the nervous system, *Ann. N.Y. Acad. Sci.* 247:109–116.

Baranski S., Czekalinski L., Czerski P., Haduch S., 1966. Experimental research on fatal effect of micrometric wave electromagnetic radiation, *Rev. Med. Aeronaut.* (*Paris*) 2:108.

Baranski S., Ostrowski K., Stodolnik-Baranska W., 1972. Functional and morphological studies of the thyroid gland in animals exposed to microwave irradiation, *Acta Physiol. Pol.* 23:1029–1039.

Baranski S., Debiec H., Kwarecki K., Mezykowski T., 1976. Influence of microwaves on genetical processes of *Aspergillus nidulans*, *J. Microwave Power* 11:146.

Barsoum Y.H., Pickard W.F., 1982a. Radio-frequency rectification in electrogenic and nonelectrogenic cells of chara and nitella, *J. Membr. Biol.* 65:81–87.

Barsoum Y.H., Pickard W.F., 1982b. The vacuolar potential of characean cells subjected to electromagnetic radiation in the range 200–8,200 MHz, *Bioelectromagnetics* 3:393–400.

Bartsch H., Bartsch C., Seebald E., Deerberg F., Dietz K., Vollrath L., 2002. Chronic exposure to a GSM-like signal (mobile phone) does not stimulate the development of DMBA-induced mammary tumors in rats: results of three consecutive studies, *Radiat. Res.* 157:183–190.

Basset M.A., Kenny R.A., 1979. Cardiac response to whole-body heating, *Aviat. Space Environ. Med.* 50:387–389.

Bawin S.M., Adey W.R., 1977. Calcium binding in cerebral tissue, in: *Symposium on Biological Effects and Measurement of Radio Frequency/Microwaves*, Hazzard D.G., Ed., U.S. Department of Health, Education, and Welfare, HEW Publication (FDA) 77-8026, Washington, D.C., pp. 305–313.

Bawin S.M., Gavalas-Medici R.J., Adey W.R., 1973. Effects of modulated very high frequency fields on specific brain rhythms in cats, *Brain Res.*, 58:365–384.

Bawin S.M., Kaczmarek L.K., Adey W.R., 1975. Effects of modulated VHF fields on the central nervous system, in: *Biological Effects of Nonionizing Radiation*, Tyler P.W., Ed., *Ann. N.Y. Acad. Sci.* 247:74–81.

Bawin S.M., Adey W.R., Sabbot I.M., 1978. Ionic factors in release of $45Ca^{2+}$ from chicken cerebral tissue by electromagnetic fields, *Proc. Nat. Acad. Sci. USA* 75:6314–6318.

Beel J.A., 1983. Posttrial microwave effects on learning and memory in mice, *Soc. Neurosci. Abstr.* 9:644.

Belkhode M.L., Johnson D.L., Muc A.M., 1974. Thermal and athermal effects of microwave radiation on the activity of glucose-6-phosphate dehydrogenase in human blood, *Health Phys.* 26:45–51.

Bennett I.L., Jr., Nicastri A., 1960. Fever as a mechanism of resistance, *Bacteriol. Rev.* 24:16–34.

Bereznitskaya, A.N. Kazbekov, I.M., 1974. Studies on the reproduction and testicular microstructure of mice exposed to microwaves, in: *Biological Effects of Radiofrequency Electromagnetic Fields*, Gordon, A.V., Ed., No. 4, Moscow, Russia (JPRS63321), pp. 221–229.

Bergqvist B., Arvidsson L., Pettersson E., Galt S., Saalman E., Hamnerius Y., Norden B., 1994. Effect of microwave radiation on permeability of liposomes, evidence against non-thermal leakage, *Biochim. Biophys. Acta* 1201:51–54.

Berman E., Kinn J.B., Carter H.B., 1978. Observations of mouse fetuses after irradiation with 2.45 GHz microwaves, *Health Phys.* 35:791–801.

Berman E., Carter H.B., House D., 1980. Tests of mutagenesis and reproduction in male rats exposed to 2,450-MHz CW microwaves, *Bioelectromagnetics* 1:65–76.

Berman E., Carter H.B., House D., 1984. Growth and development of mice offspring after irradiation *in utero* with 2,450-MHz microwaves, *Teratology* 30:393–402.

Berteaud A.J., Dardalhon M., Rebeyrotte N., Averbeck D., 1975. Action d'un rayonnement electromagnetique a longueur d'onde milimetrique sur la croissance bacterienne, *C.R. Acad. Sci. Ser. D.* 281:843–846.

Bielski J., 1994. Bioelectrical brain activity in workers exposed to electromagnetic fields, *Ann. N.Y. Acad. Sci.* 724:435–437.

Bilokrynytsk'ky V.S, 1966. Changes in the tigroid substance of neurons under the effect of radio waves, *Fiziol. Zh.* 12:70.

Bini M., Checcucci A., Ignesti A., Millanta L., Rubino N., Camici S., Manao G., Tamponi G., 1978. Analysis of the effects of microwave energy on enzymatic activity of lactate dehydrogenase (LDH), *J. Microwave Power* 13:95–100.

Bini M., Checcucci A., Ignesti A., Millanta L., Olmi R., Rubino N., Vanni R., 1986. Exposure of workers to intense rf electric fields that leak from plastic sealers, *J. Microwave Power* 21:33–40.

Birenbaum L., Kaplan I.T., Metlay W., Rosenthal S.W., Schmidt H., Zaret M.M., 1969. Effect of microwaves on the rabbit eye, *J. Microwave Power* 4:232–243.

Bisht K.S., Moros E.G., Straube W.L., Baty J.D., Roti Roti J.L., 2002. The effect of radiofrequency radiation with modulation relevant to cellular phone communication (835.62

FDMA and 847.74 MHz CDMA) on the induction of micronuclei in C3H 10T1/2 cells, *Radiat. Res.* 157:506–515.

Black D.R., Heynick L.N., 2003. Radiofrequency (RF) effects on blood cells, cardiac, endocrine, and immunological functions, *Bioelectromagnetics* (Suppl. 6):S187–S195.

Blackman C.F., Surles M.C., Benane S.G., 1976. The effects of microwave exposure on bacteria mutation reduction, in: *Symp. Biol. Eff. of E.M. Waves*, Vol. 1, Publ. (FDA) 77-8010, U.S. Department of Health, Education and Welfare, Rockville, MD, pp. 406–413.

Blackman C.F., Elder J.A., Weil C.M., Benane S.G., Eichinger D.C., House D.E., 1979. Induction of calcium-ion efflux from brain tissue by radio-frequency radiation: effects of modulation frequency and field strength, *Radio Sci.* 14:93–98.

Blackman C.F., Benane S.G., Elder J.A., House D.E., Lampe J.A., Faulk J.M., 1980a. Induction of calcium-ion efflux from brain tissue by radio-frequency radiation: effect of sample number and modulation frequency on the power-density window, *Bioelectromagnetics* 1:35–43.

Blackman C.F., Benane S.G., Joines W.T., Hollis M.A., House D.E., 1980b. Calcium-ion efflux from brain tissue: power density versus internal field-intensity dependencies at 50-MHz RF radiation, *Bioelectromagnetics* 1:277–283.

Blackman C.F., Benane S.G., Kinney L.S., Joines W.T., House E.E., 1982. Effects of ELF fields on calcium-ion efflux from brain tissue *in vitro*, *Radiat. Res.* 92:510–520.

Blackman C.F., Benane S.G., House D.E., Joines W.T., 1985a. Effects of ELF 1-120 Hz and modulated 50 HZ RF fields on the efflux of calcium ions from brain tissue, *Bioelectromagnetics* 6:1–11.

Blackman C.F., Benane S.G., Rabinowitz J.R., House D.E., Joines W.T., 1985b. A role for the magnetic field in the radiation-induced efflux of calcium ions from brain tissue *in vitro*, *Bioelectromagnetics* 6:327–337.

Blackman C.F., Benane S.G., Elliot D.J., House D.E., Pollock M.M., 1988. Influence of electromagnetic fields on the efflux of calcium ions from brain tissue *in vitro*: a three-model analysis consistent with the frequency response up to 510 Hz, *Bioelectromagnetics* 9:215–227.

Blackman C.F., Kinney L.S., House D.E., Joines W.T., 1989. Multiple power-density windows and their possible origin, *Bioelectromagnetics* 10:115–128.

Blackman C.F., Benane S.G., House D.E., 1991. The influence of temperature during electric- and magnetic-field-induced alteration of calcium-ion release from *in vitro* brain tissue, *Bioelectromagnetics* 12:173–182.

Blasberg R.G., 1979. Problems of quantifying effects of microwave irradiation on the blood–brain barrier, *Radio Sci.* 14 (6S):335–344.

Blick D.W., Adair E.R., Hurt W.D., Sherry C.J., Walters T.J., Merritt J.H., 1997. Thresholds of microwave-evoked warmth sensations in human skin, *Bioelectromagnetics* 18:403–409.

Bligh J., 1975. Physiologic responses to heat, in: *Fundamental and Applied Aspects of Nonionizing Radiation*, Michaelson S., et al., Eds., Plenum Press, New York, NY, pp. 143.

Boak R.A., Carpenter C.M., Warren S.L., 1932. Studies on the physiological effects of fever temperatures. II. The effect of repeated short wave (30 meter) fevers on growth and fertility of rabbits, *J. Exp. Med.* 56:725–739.

Bolshakov M.A., Alekseev S.I., 1992. Bursting responses of Lymnea neurons to microwave radiation, *Bioelectromagnetics* 13:119–130.

Bonasera S., Toler J., Popovic V., 1988. Long-term study of 435 MHz radio-frequency radiation on blood-borne end points in cannulated rats—Part I: engineering considerations, *J. Microwave Power Electromagn. Energy* 23:95–104.

Borbély A.A., Huber R., Graf T., Fuchs B., Gallmann E., Achermann P., 1999. Pulsed high-frequency electromagnetic field affects human sleep and sleep electroencephelogram, *Neurosci. Lett.* 275:207–210.

Bornhausen M., Scheingraber H., 2000. Prenatal exposure to 900 MHz cell-phone electromagnetic fields had no effect on operant-behavior performances of adult rats, *Bioelectromagnetics* 21:566–574.

Braithwaite L., Morrison W., Otten L., Pei D., 1991. Exposure of fertile chicken eggs to microwave radiation 2.45 GHz, CW during incubation: technique and evaluation, *J. Microwave Power Electromagn. Energy* 26:206–214.

Braune S., Wrocklage C., Raczek J., Gailus T., Lucking C.H., 1998. Resting blood pressure increase during exposure to a radio-frequency electromagnetic field, *Lancet* 351:1857–1858.

Braune S., Riedel A., Schulte-Monting J., Raczek J., 2002. Influence of a radiofrequency electromagnetic field on cardiovascular and hormonal parameters of the autonomic nervous system in healthy individuals, *Radiat. Res.* 158:352–356.

Brown R.F., Marshall S.V., 1986. Differentiation of murine erythroleukemic cells during exposure to microwave radiation, *Radiat. Res.* 108:12–22.

Brown R.F., Marshall S.V., Hughes C.W., 1981. Effects of RFR on excision-type DNA repair *in vivo*, in: *USAF Radiofrequency Radiation Bioeffects Research Program—A Review*, Mitchell J.C., Ed., *Aeromed. Rev.* 4–18, Report No. SAM-TR-81-30, pp. 184.

Brown D.O., Lu S.T., Elson E.E., 1994. Characteristics of microwave evoked body movements in mice, *Bioelectromagnetics* 15:143–161.

Browning M.D., Haycock J.W., 1988. Microwave radiation, in the absence of hyperthermia, has no detectable effect on synapsin I levels or phosphorylation, *Neurotoxicol. Teratol.* 10:461–464.

Brown-Woodman P.D., Hadley J.A., 1988. Studies of the teratogenic potential of exposure of rats to 27.12 MHz pulsed shortwave radiation, *J. Bioelectricity* 7:57–67.

Brown-Woodman P.D., Hadley J.A., Richardson L., Bright D., Porter D., 1989. Evaluation of reproductive function of female rats exposed to radiofrequency fields 27.12 MHz near a shortwave diathermy device, *Health Phys.* 56:521–525.

Bruce-Wolfe V., Adair E.R., 1985. Operant control of convective cooling and microwave irradiation by the squirrel monkey, *Bioelectromagnetics* 6:365–380.

Bruce-Wolfe V., Justesen D.R., 1985. Microwaves retard the anesthetic action of pentobarbital, *Bioelectromagnetics Soc. Abstr.* 7:47.

Bryant H.E., Love E.S., 1989. Video display terminal use and spontaneous abortion risk, *Int. J. Epidemiol.* 18:132–138.

Burch J.B., Noonan C.W., Ichinose T., Bachand A.M., Koleber T.L., Yost M.G., 2002. Melatonin metabolite excretion among cellular telephone users, *Int. J. Radiat. Biol.* 11:1029–1036.

Byus C.V., Kartum K., Pieper S., Adey W.R., 1988. Increased ornithine decarboxylase activity in cultured cells exposed to low energy modulated microwave fields and phorbol ester tumor promoters, *Cancer Res.* 48:4222–4226.

Cain C.A., Rissmann W.J., 1978. Mammalian auditory responses to 3.0 GHz microwave pulses, *IEEE Trans. Biomed. Eng.* 25:288–293.

Cameron J.A., 1943. Termination of early pregnancy by artificial fever, *Proc. Soc. Exp. Biol. Med.* 52:76.

Carpenter R.L., Livstone E.M., 1971. Evidence for nonthermal effects of microwave radiation: abnormal development of irradiated insect pupae, *IEEE Trans. Microwave Theory Tech.* 19: 173–178.

Carpenter R.L., Van Ummersen C.A., 1968. The action of microwave radiation on the eye, *J. Microwave Power* 3:3–19.

Carpenter R.L., Biddle D.K., Van Ummersen C.A., 1960. Opacities in the lens of the eye experimentally induced by exposure to microwave radiation, *IRE Trans. Med. Electron.* 7:152–157.

Chagnaud J.L., Moreau J.M., Veyret B., 1999. No effect of short-term exposure to GSM-modulated low-power microwaves on benzo(a)pyrene-induced tumours in the rat, *Int. J. Radiat. Biol.* 75:1251–1256.

Chang B.K., Huang A.T., Joines W.T., Kramer R.S., 1982. The effect of microwave radiation 1.0 GHz on the blood–brain barrier in dogs, *Radio Sci.* 17:165–168.

Chen K.M., Samuel A., Hoopingarner R., 1974. Chromosomal aberrations of living cells induced by microwave radiation, *Environ. Lett.* 6:37–46.

Chernovetz M.E., Justesen D.R., King N.W., Wagner J.E., 1975. Teratology, survival, and reversal learning after fetal irradiation of mice by 2450-MHz microwave energy, *J. Microwave Power* 10:391–409.

Chernovetz M.E., Justesen D.R., Oke A.F., 1977. A teratological study of the rat: microwave and infrared radiations compared, *Radio Sci.* 12:191–197.

Chizhenkova R.A., 1988. Slow potentials and spike unit activity of the cerebral cortex of rabbits exposed to microwaves, *Bioelectromagnetics* 9:337–345.

Chou C.K., Galambos R., 1979. Middle-ear structures contribute little to auditory perception of microwaves, *J. Microwave Power* 14:321–326.
Chou C.K., Guy A.W., 1978. Effects of electromagnetic fields on isolated nerve and muscle preparations, *IEEE Trans. Microwave Theory Tech.* 26:141–147.
Chou C.K., Guy A.W., 1979. Microwave-induced auditory responses in guinea pigs: relationship of threshold and microwave-pulse duration, *Radio Sci.* 14:193–197.
Chou C.K., Galambos R., Guy A.W., Lovely R.H., 1975. Cochlear microphonics generated by microwave pulses, *J. Microwave Power* 10:361–367.
Chou C.K., Guy A.W., Galambos R., 1976. Microwave-induced auditory response: cochlear microphonics, in: *Biological Effects of Electromagnetic Waves*, Vol. 1, Johnson C.C., Shore M.L., Eds., HEW Publication (FDA) 77-8010, Rockville, MD, pp. 89–103.
Chou C.K., Guy A.W., Galambos R., 1977. Characteristics of microwave-induced cochlear microphonics, *Radio Sci.* 12:221–227.
Chou C.K., Guy A.W., Galambos R., 1982a. Auditory perception of radiofrequency electromagnetic fields, *J. Acoust. Soc. Am.* 71:1321–1334. Review.
Chou C.K., Guy A.W., McDougall J.B., Han L.F., 1982b. Effects of continuous and pulsed chronic microwave exposure on rabbits, *Radio Sci.* 17:185–193.
Chou C.K., Guy A.W., Borneman L.E., Kunz L.L., Kramar P., 1983. Chronic exposure of rabbits to 0.5 and 5 mW/sq-cm 2450-MHz CW microwave radiation, *Bioelectromagnetics* 4:63–77.
Chou C.K., Yee K.C., Guy A.W., 1985. Auditory response in rats exposed to 2,450 MHz electromagnetic fields in a circularly polarized waveguide, *Bioelectromagnetics* 6:323–326.
Choy R.V., Monro J.A., Smith C.W., 1986. Electrical sensitivities in allergy patients, *Clin. Ecol.* 4:93–102.
Chou C.K., Guy A.W., Kunz L.L., Johnson R.B., Crowley J.J., Krupp J.H., 1992. Long-term low-level microwave irradiation of rats, *Bioelectromagnetics* 13:469–496.
Ciaravino V., Meltz M., Erwin D.N., 1987. Effects of radiofrequency radiation and simultaneous exposure with mitomycin c on the frequency of sister chromatid exchanges in Chinese hamster ovary cells, *Environ. Mutagen.* 9:393–399.
Ciaravino V., Meltz M., Erwin D.N., 1991. Absence of a synergistic effect between moderate-power radio-frequency electromagnetic radiation and adriamycin on cell-cycle progression and sister-chromatid exchange, *Bioelectromagnetics* 12:289–298.
Clark M.W., Gildersleeve R.P., Thaxton J.P., Parkhurst C.R., McRee D.I., 1987. Leukocyte numbers in hemorrhaged Japanese quail after microwave irradiation in ovo, *Comp. Biochem. Physiol. A* 87:923–932.
Cleary S.F., Liu L.M., Garber F., 1985. Viability and phagocytosis of neutrophils exposed *in vitro* to 100-MHz radiofrequency radiation, *Bioelectromagnetics* 6:53–60.
Cleary S.F., Liu L.M., Graham R., East J., 1989. *In vitro* fertilization of mouse ova by spermatozoa exposed isothermally to radio-frequency radiation, *Bioelectromagnetics* 10:361–369.
Cleary S.F., Liu L.M., Merchant R.E., 1990a. Glioma proliferation modulated *in vitro* by isothermal radiofrequency radiation exposure, *Radiat. Res.* 121:38–45.
Cleary S.F., Liu L.M., Merchant R.E., 1990b. *In vitro* lymphocyte proliferation induced by radio-frequency electromagnetic radiation under isothermal conditions, *Bioelectromagnetics* 11:47–56.
Cleary S.F., Cao G., Liu L.M., Egle P.M., Shelton K.R., 1997. Stress proteins are not induced in mammalian cells exposed to radiofrequency or microwave radiation, *Bioelectromagnetics* 18:499 507.
COMAR, 2002. Technical information statement: electromagnetic hypersensitivity, *IEEE Eng. Med. Biol.* Sept/Oct:173–175.
Conover D.L., Lary J.M., Foley E., 1978. Induction of teratogenic effects in rats by 27.12 MHz RF radiation, Presented at the *Symp. on Electromagnetic Fields in Biological Systems*, Ottawa, June 17–30, 1978.
Conover D.L., Lary J.M., Hanser P.L., 1979. Thermal threshold for teratogenic response in rats irradiated at 17.12 MHz, *Bioelectromagnetics* 1:204.
Cook H.F., 1952. A physical investigation of heat production in human tissues when exposed to microwaves, *Brit. J. Appl. Phys.* 3:1–6.
Cooper T., Pinakatt T., Richardson A.W., 1961. Effect of microwave induced hyperthermia on the cardiac output of the rat, *Physiologist* 4:21.

Cooper T., Jellinek M., Pinakatt T., Richardson A.W., Cooper T., 1962. Effects of adrenalectomy, vagotomy and ganglionic blockade on the circulatory response to microwave hyperthermia, *Aerosp. Med.* 33:794–798.

Corbit J.D., 1973. Thermal motivation in neural control of motivated behavior. *Neurosci. Res. Prog. Bull.* 11:4.

Correlli J.C., Gutmann R.J., Kohazi S., Levy J., 1977. Effects of 2.6–4.0 GHz microwave radiation on *E. coli* B, *J. Microwave Power* 12:141–144.

Creighton M.O., Larsen L.E., Stewart-DeHaan P.J., Jacobi J.H., Sanwal M., Baskerville J.C., Bassen H.E., Brown D.O., Trevithick J.R., 1987. *In vitro* studies of microwave-induced cataract. II. Comparison of damage observed for continuous wave and pulsed microwaves, *Exp. Eye Res.* 45:357–373.

Croft R.J., Chandler J.S., Burgess A.P., Barry R.J., Williams J.D., Clarke A.R., 2002. Acute mobile phone operation affects neural function in humans, *Clin. Neurophysiol.* 113:1623–1632.

Czerska E.M., Elson E.C., Davis C.C., Swicord M.L., Czerski P., 1992. Effects of continuous and pulsed 2450-MHz radiation of spontaneous lymphoblastoid transformation of human lymphocytes *in vitro*, *Bioelectromagnetics* 13:247–259.

Czerski P., 1975. Microwave effects on the blood-forming system with particular reference to the lymphocyte, *Ann. N. Y. Acad. Sci.* 247:232–242.

Czerski P., Siekierzynski M., 1975. Analysis of occupational exposure to microwave radiation, in: *Fundamental and Applied Aspects of Non-Ionizing Radiations*, Michaelson S.M., Miller M.W., Magin R., Carstensen E.L., Eds., Plenum Press, New York, NY, pp. 367–377.

Czerski P., Paprocka-Slonka E., Siekierzynski M., Stolarska A., 1974a. Influence of microwave radiation on the hematopoietic system, in: *Biologic Effects and Health Hazards of Microwave Radiation*, Czerski P., et al., Eds., Polish Medical Publishers, Warsaw, Poland, pp. 67–74.

Czerski P., Paprocka-Slonka E., Stolarska A., 1974b. Microwave irradiation and the circadian rhythm of bone marrow cell mitosis, *J. Microwave Power* 9:31–37.

Daily L., Wakim K.J., Herrick J.F., Parkhill E.M., Benedict W.L., 1950. The effects of microwave diathermy on the eye, *Am. J. Opthalmol.* 33:1241–1254.

D'Ambrosio G., Massa R., Scarfi M.R., Zeni O., 2002. Cytogenetic damage in human lymphocytes following GMSK phase modulated microwave exposure, *Bioelectromagnetics* 23:7–13.

D'Andrea J.A., 1999. Behavioral evaluation of microwave irradiation, *Bioelectromagnetics* 20:64–74.

D'Andrea J.A., Gandhi O.P., Lords J.L., 1977. Behavioral and thermal effects of microwave radiation at resonant and nonresonant wavelengths, *Radio Sci.* 12:251–256.

D'Andrea J.A., DeWitt J.R., Emmerson R.Y., Bailey C., Stensaas S., Gandhi O.P., 1986a. Intermittent exposure of rats to 2450 MHz microwaves at 2.5 mW/cm^2: behavioral and physiological effects, *Bioelectromagnetics* 7:315–328.

D'Andrea J.A., DeWitt J.R., Gandhi O.P., Stensaas S., Lords J.L., Nielson H.C., 1986b. Behavioral and physiological effects of chronic 2,450-MHz microwave irradiation of the rat at 0.5 mW/cm^2, *Bioelectromagnetics* 7:45–56.

D'Andrea J.A., Cobb B.L., de Lorge J.O., 1989. Lack of behavioral effects in the rhesus monkey: high peak microwave pulses at 1.3 GHz, *Bioelectromagnetics* 10:65–76.

D'Andrea J.A., Thomas A., Hatcher D.J., 1994. Rhesus monkey behavior during exposure to high-peak-power 5.62-GHz microwave pulses, *Bioelectromagnetics* 15:163–176.

D'Andrea J.A., Adair E.R., de Lorge L.O. 2003. Behavioral and cognitive effects of microwave exposure. *Bioelectromagnetics Supp*, 6:S39–S62.

Dardanoni L., Torregrossa V., Tamburello C., Zanforlin L., Spalla M., 1976. Biological effects of millimeter waves at spectral singularities, in: *Electromagnetic Compatibility*, Wydawnictwo Politechniki, Wroclawskiej, Breslau, Poland, pp. 308–313.

Dasdag S., Ketani M.A., Akdag Z., Ersay A.R., Sari I., Demirtas O.C., Celik M.S., 1999. Whole-body microwave exposure emitted by cellular phones and testicular function of rats, *Urol. Res.* 27:219–223.

Dasdag S., Zulkuf Akdag M., Aksen F., Yilmaz F., Bashan M., Dasdag M.M., Celik M.S., 2003. Whole body exposure of rats to microwaves emitted from a cell phone does not affect the testes, *Bioelectromagnetics* 24:182–188.

Davis C.C., Edwards G.S., Swicord M.L., Sagripanti J., Saffer J., 1986. Direct excitation of internal modes of DNA by microwaves, *Bioelectrochem. Bioenerg.* 16:63–76.

de Lorge J.O., 1983. The thermal basis for disruption of operant behavior by microwaves in three animal species, in: *Microwaves and Thermoregulation*, Adair E.R. Ed., Academic Press, New York, NY, pp. 379–399.
de Lorge J.O., 1984. Operant behavior and colonic temperature of *Macaca mulatta* exposed to radio frequency fields at and above resonant frequencies, *Bioelectromagnetics* 5:233–246.
de Lorge J.O., Ezell C.S., 1980. Observing-responses of rats exposed to 1.28- and 5.62-GHz microwaves, *Bioelectromagnetics* 1:183–198.
Demokidova N.K., 1974. The effects of radiowaves on the growth of animals, in: *Biological Effects of Radiofrequency Electromagnetic Fields*, Gordon A.V., Ed., U.S. Joint Publications Research Service No. 63321, Arlington, VA, pp. 237–242.
de Seze R., 2000. Effects of radiocellular telephones on human sleep, *J. Sleep Res.* 9 (Suppl. 1):18.
de Seze R., Fabbro-Peray P., Miro L., 1998. GSM radiocellular telephones do not disturb the secretion of antipituitary hormones in humans, *Bioelectromagnetics* 19:271–278.
de Seze R., Ayoub J., Peray P., Miro L., Touitou Y., 1999. Evaluation in humans of the effects of radiocellular telephones on the circadian patterns of melatonin secretion, a chronological rhythm marker, *J. Pineal Res.* 27:237–242.
DeWitt J.R., D'Andrea J.A., Emmerson R.Y., Gandhi O.P., 1987. Behavioral effects of chronic exposure to 0.5 mW/cm^2 of 2,450-MHz microwaves, *Bioelectromagnetics* 8:149–157.
Diachenko J.A., Milroy W.C., 1975. The Effects of High Power Pulsed and Low Level CW Microwave Radiation on an Operant Behavior in Rats, Naval Surface Weapons Center, Dahlgren Laboratory, Dahlgren, VA.
Dietzel F., 1975. Effects of nonionizing electro-magnetic radiation on implantation and intrauterine development of the rat, *Ann. N.Y. Acad. Sci.* 247:367–376.
Dietzel F., Kern W., Steckenmesser R., 1972. Deformity and intrauterine death after short-wave therapy in early pregnancy in experimental animals, *Muench. Med. Wochenschr.* 114:228–230.
D'Inzeo G., Bernardi P., Eusebi F., Grassi F., Tamburello C., Zani B.M., 1988. Microwave effects on acetylcholine induced channels in cultured chick myotubes, *Bioelectromagnetics* 9:363–372.
Dolina L.A., 1961. Morphological changes in the central nervous system due to the action of centimeter waves on the organism, *Arkh. Patol.* 23:51.
Dubreuil D., Jay T., Edeline J.-M., 2002. Does head-only exposure to GSM-900 electromagnetic fields affect the performance of rats in spatial learning tasks? *Behav. Brain Res.* 129:203–210.
Dumansky Y.D., Rudichenko V.F., 1976. Dependence of the functional activity of liver mitochondria with super-high frequency radiation, *Hyg. Sanit.* 4:16–19.
Dumansky Y.D., Serdyuk A.M., Litvinova C.I., Tomashevskaya L.A., Popovich V.M., 1972. Experimental research on the biological effects of 12-centimeter low-intensity waves, in: *Health in Inhabited Localities*, 2nd ed., Kiev, Urkraine, pp. 29.
Durney C.H., Massoudi H., Iskander M.F., 1986. *Radiofrequency Radiation Dosimetry Handbook*, 4th ed., Report USAFSAM-TR-85-73, USAF School of Aerospace Medicine, Brooks AFB, TX.
Dutta S.K., Nelson W.H., Blackman C.F., Brusick D.J., 1979. Lack of microbiol genetic response to 2.45 GHz CW and 8.5 to 9.6 GHz pulsed microwaves, *J. Microwave Power* 14:275–280.
Dutta S.K., Subramoniam A., Ghosh B., Parshad R., 1984. Microwave radiation-induced calcium ion efflux from human neuroblastoma cells in culture, *Bioelectromagnetics* 5:71–78.
Dutta S.K., Ghosh B., Blackman C.F., 1989. Radiofrequency radiation-induced calcium ion efflux enhancement from human and other neuroblastoma cells in culture, *Bioelectromagnetics* 10:197–202.
Dutta S.K., Das K., Ghosh B.G., Blackman C.F., 1992. Dose dependence of acetylcholinesterase activity in neuroblastoma cells exposed to modulated radio-frequency electromagnetic radiation, *Bioelectromagnetics* 13:317–322.
Edelstyn N., Oldershaw A., 2002. The acute effects of exposure to the electromagnetic field emitted by mobile phones on human attention, *NeuroReport* 13:317–322.
Edelwejn Z., 1968. An attempt to assess the functional state of the cerebral synapses in rabbits exposed to chronic irradiation with microwaves, *Acta Physiol. Pol.* 19:897.
Edwards M.J., 1968. Congenital malformations in the rat following induced hyperthermia during gestation, *Teratology* 1:173–178.

Edwards M.J., 1969. Congenital defects in guinea pigs: fetal resorptions, abortions, and malformations following induced hyperthermia during early gestation, *Teratology* 2:313–328.
Edwards G.S., Davis C.C., Saffer J.D., Swicord M.L., 1984. Resonant microwave absorption of selected DNA molecules, *Phys. Rev. Lett.* 53:1284–1287.
Edwards G.S., Davis C.C., Saffer J.D., Swicord M.L., 1985. Microwave-field-driven acoustic modes in DNA, *Biophys. J.* 47:799–807.
Eibert T.F., Alaydrus M., Wilczewski F., Hansen V.W., 1999. Electromagnetic and thermal analysis for lipid bilayer membranes exposed to RF fields, *IEEE Trans. Biomed. Eng.* 46:1013–1021.
Elder J.A., 2003. Survival and cancer in laboratory mammals exposed to radiofrequency energy, *Bioelectromagnetics* (Suppl. 6):101–106.
Elder J.A., Cahill D.F., 1984. Biological Effects of Radiofrequency Radiation, EPA Report (EPA-600/8-83-026F). [Available from National Technical Information Service, 5285 Port Royal Road, Springfield, VA 22161 (Report PB-85-120-848)], pp. 1–269.
Elder J.A., Chou C.K., 2003. Auditory response to pulsed radiofrequency energy, *Bioelectromagnetics* (Suppl. 6):S162–S173.
Ely T.S., Goldman D., Hearon J.Z., Williams R.B., Carpenter H.M., 1964. Heating characteristics of laboratory animals exposed to ten centimeter microwaves, U.S. Naval Medical Research Institute (Res. Rep. Proj. NM 001-056.13.02), Bethesda, MD, *IEEE Trans. Biomed. Eng.* 11:123–135.
Environmental Protection Agency (EPA), 1984. *Biological Effects of Radiofrequency Radiation,* EPA-600/8-83-026F, Health Effects Research Laboratory, United States Environmental Protection Agency, Research Triangle Park, NC, pp. 5–76.
Eulitz C., Ullsperger P., Freude G., Elbert T., 1998. Mobile phones modulate response patterns of human brain activity, *NeuroReport* 9:3229–3232.
Evans J.A., Savitz D.A., Kanal E., Gillen J., 1993. Infertility and pregnancy outcome among magnetic resonance imaging workers, *J. Occup. Med.* 35:1191–1195.
Fesenko E.E., Gluvstein A.Y., 1995. Changes in the state of water induced by radiofrequency electromagnetic fields, *Fed. Eur. Biochem. Soc. Lett.* 367:53–55.
Finnie J.W., Blumbergs P.C., Manavis J., Utteridge T.D., Gebski V., Swift J.G., 2001. Effects of global system for mobile communication (GSM)-like radiofrequency fields on vascular permeability in mouse brain, *Pathology* 33:338–340.
Finnie J.W., Blumbergs P.C., Manavis J., Utteridge T.D., Gebski V., Davies R.H., 2002. Effect of long-term mobile communication microwave exposure on vascular permeability in mouse brain, *Pathology* 34:344–347.
Flodin U., Seneby A., Tegenfeldt C., 2000. Provocation of electric hypersensitivity under everyday conditions, *Scand. J. Work Environ. Health* 26:93–98.
Foster K.R., Finch E.D., 1974. Microwave hearing: evidence for thermoacoustic auditory stimulation by pulsed microwaves, *Science* 185:256–258.
Foster K.R., Epstein B.R., Gealt M.A., 1987. Resonances in the dielectric absorption of DNA? *Biophys. J.* 52:421–425.
Frei M.R., Jauchem J.R., 1989. Effects of 2.8 GHz microwaves on restrained and ketamine anesthetized rats, *Radiat. Environ. Biophys.* 28:155–164.
Frei M., Jauchem J., Heinmets F., 1988. Physiological effects of 2.8 GHz radio-frequency radiation: a comparison of pulsed and continuous-wave radiation, *J. Microwave Power Electromagn. Energy* 23:85–93.
Frei M., Jauchem J.R., Heinmets F., 1989a. Thermoregulatory responses of rats exposed to 9.3-GHz radiofrequency radiation, *Radiat. Environ. Biophys.* 28:67–77.
Frei M., Jauchem J.R., Padilla J.M., 1989b. Thermal and physiological changes in rats exposed to CW and pulsed 2.8 GHz radiofrequency radiation in E and H orientations, *Int. J. Radiat. Biol.* 56:1033–1044.
Frei M.R., Ryan K., Berger R., Jauchem J.R., 1995. Sustained 35 GHz radiofrequency irradiation induces circulatory failure, *Shock* 4:289–293.
Frei M.R., Jauchem J.R., Dusch S.J., Merritt J.H., Berger R.E., Stedham M.A., 1998. Chronic, low-level (1.0 W/kg) exposure of mice prone to mammary cancer to 2450 MHz microwaves, *Radiat. Res.* 150:568–576.

Frenkel J.K., Caldwell S.A., 1975. Specific immunity and nonspecific resistance to infection: listeria, protozoa, and viruses in mice and hamsters, *J. Infect. Dis.* 131:201–209.

Freude G., Ullsperger P., Eggert S., Ruppe I., 1998. Effects of microwaves emitted by cellular phones on human slow brain potentials, *Bioelectromagnetics* 19:384–387.

Freude G., Ullsperger P., Eggert S., Ruppe I., 2000. Microwaves emitted by cellular telephones affect human slow brain potentials, *Eur. J. Appl. Physiol.* 81:18–27.

Frey A.H., 1961. Auditory system response to radio-frequency energy, *Aerosp. Med.* 32:1140–1142.

Frey A.H., 1962. Human auditory system response to modulated electromagnetic energy, *J. Appl. Physiol.* 17:689–692.

Frey A.H., 1963. Some effects on human subjects of ultra-high-frequency radiation, *Am. J. Med. Electron.* 2:28–31.

Frey A.H., 1968. A coaxial pathway for recording from the cat brain during illumination with UHF energy, *Physiol. Behav.* 3:363–364.

Frey A.H., 1985. Psychophysical analysis of microwave sound perception, *J. Bioelectricity* 4:1–14.

Frey A.H., Feld S.R., 1975. Avoidance by rats of illumination with low-power nonionizing electromagnetic energy, *J. Comp. Physio. Psych.* 89:183–188.

Frey A.H., Feld S.R., Frey B. 1975. Neural function and behavior: defining the relationship, in: *Biological Effects of Nonionizing Radiation*, Tyler P.W., Ed., pp. 433–439, *Ann. N.Y. Acad. Sci.* 247:433–439.

Friend A.W., Gartner S.L., Foster K.L., Howe H., 1981. The effects of high power microwave pulses on red blood cells and the relationship to transmembrane thermal gradients, *IEEE Trans. Microwave Theory Tech.* 29:1271–1277.

Fritze K., Sommer C., Schmitz B., Mies G., Hossmann K.A., Kiessling M., Wiessner C., 1997. Effect of global system for mobile communication (GSM) microwave exposure on blood–brain barrier permeability in rat, *Acta Neuropathol.* 94:465–470.

Gabriel C., Grant E.H., Tata R., Brown P.R., Gestblom B., Noreland E., 1987. Microwave absorption in aqueous solutions of DNA, *Nature* 328:145–146.

Gabriel C., Grant E.H., Tata R., Brown P.R., Gestblom B., Noreland E., 1989. Dielectric behaviours of aqueous solutions of a plasmid DNA at microwave frequencies, *Biophys. J.* 55:29–34.

Galloway W.D., 1975. Microwave dose–response relationships on two behavioral tasks, *Ann. N.Y. Acad. Sci.* 247:410–416.

Galloway W.D., Waxler M., 1977. Interaction between microwave and neuroactive compounds, in: *Symp. on Biological Effects and Measurement of Radio Frequency/Microwaves*, Hazzard D., Ed., Publ. (FDA) 77-8026, U.S. Department of Health, Education and Welfare, Rockville, MD, pp. 62–68.

Galvin M.J., McRee D.I., Lieberman M., 1980. Effects of 2.45-GHz microwave radiation on embryonic quail hearts, *Bioelectromagnetics* 1:389–396.

Galvin M.J., McRee D.I., Hall C.A., Thaxton J.P., Parkhurst C.R., 1981. Humoral and cell-mediated immune function in adult Japanese quail following exposure to 2.45-GHz microwave radiation during embryogeny, *Bioelectromagnetics* 2:269–278.

Galvin M.J., Tilson H.A., Mitchell C.L., Peterson J., McRee D.I., 1986. Influence of pre- and postnatal exposure of rats to 2.45-GHz microwave radiation on neurobehavioral function, *Bioelectromagnetics* 7:57–71.

Gandhi O.P., 1974. Polarization and frequency effects on whole animal energy absorption of RF energy, *Proc. IEEE* 62:1171–1175.

Gandhi O.P., 1980. State of knowledge for electromagnetic absorbed dose in man and animals, *Proc. IEEE* 68:24–32.

Gandhi O.P., 1990. Electromagnetic energy absorption in humans and animals, in: *Biological Effects and Medical Applications of Electromagnetic Energy*, Gandhi O.P., Ed., Prentice Hall, Englewood Cliffs, NJ, pp. 174–195.

Gandhi V.C., Ross D.H., 1987. Alterations in alpha-adrenergic and muscarinic cholinergic receptor binding in rat brain following nonionizing radiation, *Radiat. Res.* 109:90–99.

Gandhi C.R., Ross D.H., 1989. Microwave induced stimulation of 32Pi incorporation into phosphoinositides of rat brain synaptosomes, *Radiat. Environ. Biophys.* 28:223–234.

Gandhi O.P., Hunt E.L., D'Andrea J.A., 1977. Deposition of electromagnetic energy in animals and in models of man with and without grounding and reflector effects, *Radio Sci.* 12 (6S):39–45.

Gandhi O.P., Hagmann M.J., Hill D.V., Partlow L.M., Bush L., 1980. Millimeter-wave absorption spectra of biological samples, *Bioelectromagnetics* 1:285–298.

Gildersleeve R.P., Galvin M.J., McRee D.I., Thaxton J.P., Parkhurst C.R., 1987a. Reproduction of Japanese quail after microwave irradiation 2.45 GHz CW during embryogeny, *Bioelectromagnetics* 8:9–21.

Gildersleeve R.P., Thaxton J.P., Parkhurst C.R., Scott T.R., Galvin M.J., McRee D.I., 1987b. Leukocyte numbers during the humoral and cell mediated immune response of Japanese quail after microwave irradiation in ovo, *Comp. Biochem. Physiol. A* 87:375–380.

Gildersleeve R.P., Bryan T.E., Galvin M.J., McRee D.I., Thaxton J.P., 1988a. Serum enzymes in hemorrhaged Japanese quail after microwave irradiation during embryogeny, *Comp. Biochem. Physiol. A* 89:531–534.

Gildersleeve R.P., Satterless D.G., McRee D.I., Bryan T.E., Parkhurst C.R., 1988b. Plasma corticosterone in hemorrhaged Japanese quail after microwave irradiation in ovo, *Comp. Biochem. Physiol. A* 89:415–424.

Ginsburg K.S., Lin J.C., O'Neill W.D., 1992. Microwave effects on input resistance and action potential firing of snail neurons, *IEEE Trans. Biomed. Eng.* 39:1011–1021.

Girirajan K.S., Young L., Prohofsky E.W., 1989. Vibrational free energy, entropy, and temperature factors of DNA calculated by a helix lattice approach, *Biopolymers* 28:1841–1860.

Goldhaber M.K., Polen M.R., Hiatt R.A., 1988. The risk of miscarriage and birth defects among women who use visual display terminals during pregnancy, *Am. J. Ind. Med.*13:695–706.

Goldman H., Lin J.C., Murphy S., Lin M.F., 1984. Cerebrovascular permeability to rb86 in the rat after exposure to pulsed microwaves, *Bioelectromagnetics* 5:323–330.

Gordon, Z.V., 1970. Biological Effect of Microwaves in Occupation Hygiene, Izd. Med., Leningrad, TT 70-50087, NASA TT F-633.

Gordon C.J., 1983. Behavioral and autonomic thermoregulation in mice exposed to microwave radiation, *J. Appl. Physiol.: Respir. Environ. Exercise Physiol.* 55:1242–1248.

Gordon C.J., 1987. Normalizing the thermal effects of radiofrequency radiation: body mass versus total body surface area, *Bioelectromagnetics* 8:111–118.

Gordon C.J., Ferguson J.H., 1984. Scaling the physiological effects of exposure to radiofrequency electromagnetic radiation: consequences of body size, *Int. J. Radiat. Biol.* 46:387–397.

Gordon Z.V., Lobanova Y.A., Tolgskaya M.S., 1955. Some data on the effect of centimeter waves (experimental studies), *Gig. Sanit.* (*USSR*) 12:16.

Goud S.N., Usha Rani M.V., Reddy P.P., Reddy O.S., Rao M.S., Saxena V.K., 1982. Genetic effects of microwave radiation in mice, *Mutat. Res.* 103:39–42.

Grajewski B., Cox C., Schrader S.M., Murray W.E., Edwards R.M., Turner T.W., 2000. Semen quality and hormone levels among radiofrequency heater operators, *J. Occup. Environ. Med.* 42:993–1005.

Green D.R., Rosenbaum F.J., Pickard W.F., 1979. Intensity of microwave irradiation and the teratogenic response of *Tenebrio molitor*, *Radiol. Sci.* 14:165–171.

Gruenau S.P., Oscar K.J., Folker M.T., Rapoport S.I., 1982. Absence of microwave effect on blood–brain barrier permeability to C14-sucrose in the conscious rat, *Exp. Neurobiol.* 75:299–307.

Grundler W., Keilmann F., 1978. Nonthermal effects of millimeter microwaves on yeast growth, *Z. Naturforsch.* 33C(1/2):15–22.

Grundler W., Keilmann F., 1983. Sharp resonances in yeast prove nonthermal sensitivity to microwaves, *Phys. Rev. Lett.* 51:1214–1216.

Guy A.W., Chou C.K., 1982. Effects of high-intensity microwave pulse exposure of rat brain, *Radio Sci.* 17:169–178.

Guy A.W., Lin J.C., Kramar P.O., Emery A.F., 1974. Measurement of absorbed power patterns in the head and eyes of rabbits exposed to typical microwave sources, in: *Proceedings 1974 Conference on Precision Electromagnetic Measurements*, London, pp. 255.

Guy A.W., Chou C.K., Lin J.C., Christensen D., 1975a. Microwave-induced acoustic effects in mammalian auditory systems and physical materials, in: *Biological Effects of Nonionizing Radiation*, Tyler P.W., Ed., pp. 194–218, *Ann. N.Y. Acad. Sci.* 247.

Guy A.W., Lin J.C., Kramar P.O., Emery A.F., 1975b. Effect of 2450-MHz radiation on the rabbit eye, *IEEE Trans. Microwave Theory Tech.* 23:492–498.

Guy A.W., Webb M.D., Sorensen C.C., 1976. Determination of power absorption in man exposed to high frequency electromagnetic fields by thermographic measurements on scale models, *IEEE Trans. Biol. Med. Eng.* 23:361–371.

Guy A.W., Kramar P.O., Harris C.A., Chou C.K., 1980. Long-term 2450-MHz CW microwave irradiation of rabbits: methodology and evaluation of ocular and physiologic effects, *J. Microwave Power* 15:37–44.

Gvozdikova Z.M., Annanyev V.M., Zenina I.N., Zak V.I., 1964. Sensitivity of the rabbit central nervous system to a continuous (nonpulsed) ultrahigh frequency electromagnetic field, *Byull. Eksp. Biol. Med.* (*Moscow*) 58:63.

Haarala C., Bjornberg L., Ek M., Laine M., Revonsuo A., Koivisto M., Hamalainen H., 2003. Effect of a 902 MHz electromagnetic field emitted by mobile phones on human cognitive function: a replication study, *Bioelectromagnetics* 24:283–288.

Hamilton C.L., 1963. Interactions of food intake and temperature regulation in the rat, *J. Comp. Physiol. Psychol.* 56:476–488.

Hamnerius Y., Rasmuson A., Rasmuson B., 1985. Biological effects of high-frequency electromagnetic fields on *Salmonella typhimurium* and *Drosophila melanogaster*, *Bioelectromagnetics* 6:405–414.

Hamrick P.E., 1973. Thermal denaturation of DNA exposed 2450 MHz CW microwave radiation, *Radiat. Res.* 56:400.

Hamrick P.E., Fox S.S., 1977. Rat lymphocytes in cell culture exposed to 2450 MHz (CW) microwave radiation, *J. Microwave Power* 12:125–132.

Hamrick P.E., McRee D.I., 1975. Exposure of the Japanese quail embryo to 2.45 GHz microwave radiation during the second day of development, *J. Microwave Power* 10:211–220.

Hamrick P.E., McRee D.I., 1980. The effect of 2450 MHz microwave irradiation on the heart rate of embryonic quail, *Health Phys.* 38:261–268.

Hamrick P.E., McRee D.I., Thaxton P., Parkhurst C.R., 1977. Humoral immunity of Japanese quail subjected to microwave radiation during embryogeny, *Health Phys.* 33:23–33.

Hardy J.D., 1978. Regulation of body temperature in man—an overview, in: *Energy Conservation Strategies in Buildings*, Stolwijk J.A.J., Ed., University Printing Service, New Haven, CT, pp. 14–37.

Heikkanen P., Juutilianen J., 1999. Chronic exposure to 50-Hz magnetic fields or 900-MHz electromagnetic fields does not alter nocturnal 6-hydroxymelatonin sulfate secretion in CBA/S mice, *Electro- Magnetobiol.* 18:33–42.

Heikkinen P., Kosma V.M., Hongisto T., Huuskonen H., Hyysalo P., Komulainen H., Kumlin T., Lahtinen T., Lang S., Juutilainen J., 2001. Effects of mobile phone radiation on x-ray-induced tumorigenesis in mice, *Radiat. Res.* 156:775–785.

Heller J.H., 1970. Cellular effects of microwave radiation, in: *Biological Effects and Health Implications of Microwave Radiation, Symp. Proc.*, Cleary S.F., Ed., Public Health Service BRH/DBE 70-21, U.S. Department of Health, Education, and Welfare, Washington, D.C., pp. 116–121.

Hendler E., 1968. Cutaneous receptor response to microwave irradiation, in: *Thermal Problems in Aerospace Medicine*, Hardy J.D., Ed., Unwin Bros. Ltd., Surrey, U.K., pp. 149–161.

Hendler E., Hardy J.D., 1960. Infrared and microwave effects on skin heating and temperature sensation, *IRE Trans. Med. Electron.* 7:143–152.

Herrick J.F., Krusen F.H., 1953. Certain physiologic and pathologic effects of microwaves, *Electron. Eng.* 72:239 244.

Heynick L.N., Merritt J.H., 2003. Radiofrequency fields and teratogenesis, *Bioelectromagnetics* (Suppl. 6):S174–S186.

Heynick L.N., Johnston S.A., Mason P.A., 2003. Radio frequency electromagnetic fields: cancer, mutagenesis, and genotoxicity, *Bioelectromagnetics* (Suppl. 6):S74–S100.

Hietanen M., Kovala T., Hamalainen A.M., 2000. Human brain activity during exposure to radiofrequency fields emitted by cellular phones, *Scand. J. Work Environ. Health* 26:87–92.

Hietanen M., Hamalainen A.M., Husman T., 2002. Hypersensitivity symptoms associated with exposure to cellular telephones: no causal link, *Bioelectromagnetics* 23:264–270.

Higashikubo R., Culbreth V.O., Spitz D.R., LaRegina M.C., Pickard W.F., Straube W.L., Moros E.G., Roti Roti J.L., 1999. Radiofrequency electromagnetic fields have no effect on the *in vivo* proliferation of the 9L brain tumor, *Radiat. Res.* 152:665–671.

Hirsch S.E., Appleton B.S., Fine B.S., Brown P.V., 1977. Effects of repeated microwave irradiations to the albino rabbit eye, *Invest. Ophthalmol. Visual Sci.* 16:315–319.

Hjeresen D.L., Umbarger K.O., 1989. Lack of Behavioral Effects of High-Peak-Power Microwave Pulses from an Axially Extracted Virtual Cathode Oscillator, USAFSAM-TR-89-24, November, United States Air Force School of Aerospace Medicine, Brooks Air Force Base, TX.

Hjeresen D.L., Francendese A., O'Donnell J.M., 1988. Microwave attenuation of ethanol-induced hypothermia: ethanol tolerance, time course, exposure duration, and dose response studies, *Bioelectromagnetics* 9:63–78.

Hjeresen D.L., Francendese A., O'Donnell J.M., 1989. Microwave attenuation of ethanol-induced interactions with noradrenergic neurotransmitter systems, *Health Phys.* 56:767–776.

Hjeresen D.L., Hoeberling R.F., Kinross-Wright J., Umbarger K.O., 1990. Behavioral Effects of 1300-MHz High-Peak-power-Microwave Pulsed Irradiation, USAFSAM-TR-90-6, August, United States Air Force School of Aerospace Medicine, Brooks Air Force Base, TX.

Hladk} A., Musil J., Roth Z., Urban P., Blazkova V., 1999. Acute effects of using a mobile phone on CNS functions, *Central Eur. J. Pub. Health* 7:165–167.

Ho H.S., Edwards, W.P., 1977. Oxygen-consumption rate of mice under differing dose rates of microwave radiation, *Radio Sci.* 12 (6S):131–138.

Ho H.S., Edwards W.P., 1979. The effect of environmental temperature and average dose rate of microwave radiation on the oxygen-consumption rate of mice, *Radiat. Environ. Biophys.* 16:325–338.

Hocking B., 1998. Preliminary report: symptoms associated with mobile phone use, *Occup. Med. (Oxford)* 48:357–360.

Hocking B., Joyner K., 1995. Re: Miscarriages among female physical therapists who report using radio- and microwave-frequency electromagnetic radiation [Letter], *Am. J. Epidemiol.* 141:273–274.

Hocking B., Westerman R., 2000. Neurological abnormalities associated with mobile phone use, *Occup. Med.* 50:366–368.

Holly E.A., Aston D.A., Ahn D.K., Smith A.H. 1996. Intreocular melanoma linked to occupations and chemical exposures. *Epidemiol.* 7:55–61.

Hook G.J., Vasquez M., Clancy J.J., Blackwell D.M., Dkonner E.M., Trice R.R., McRee D., 1999. Genotoxicity of Radio Frequency Fields Generated by Analog, TDMA, CDMA, and PCS Cellular Technologies Evaluated Using the Single Cell Gel Electrophoresis (SCGE) and the Cytochalasin B Micronucleus (CB_MN) Assay, Abstract 1–4, Twenty-first Annual Meeting of the Bioelectromagnetics Society, Long Beach, CA June 20–24.

Huang A.T., Mold N.G., 1980. Immunologic and hematopoietic alterations by 2.450-MHz electromagnetic radiation, *Bioelectromagnetics* 1:77–87.

Huang A.T., Engle M.E., Elder J.A., Kinn J.B., Ward T.R., 1977. The effect of microwave radiation (2450 MHz) on the morphology and chromosomes of lymphocytes, *Radio Sci.* 12 (6S):173–177.

Huber R., Graf T., Cote K.A., Wittmann L., Gallmann E., Matter D., Schuderer J., Kuster N., Borbely A.A., Aschermann P., 2000. Exposure to pulsed high-frequency electromagnetic field during waking affects human sleep EEG, *NeuroReport* 11:3321–3325.

Huber R., Treyer V., Borbely A.A., Schuderer J., Gottselig J.M., Landolt H.-P., Werth E., Berthold T., Kuster N., Buck A., Achermann P., 2002. Electromagnetic fields, such as those from mobile phones, alter regional cerebral blood flow and sleep and waking EEG, *J. Sleep Res.* 11:289–295.

Hunt E L., King N.W., Phillips R.D., 1975. Behavioral effects of pulsed microwave radiation, *Ann. N.Y. Acad. Sci.* 247:440–453.

Hyde A.S., Friedman J.J., 1968. Some effects of acute and chronic microwave irradiation of mice, in: *Thermal Problems in Aerospace Medicine*, Hardy J.D., Ed., Unwin, Surrey, England, pp. 163.

ICNIRP (International Commission on Non-Ionizing Radiation Protection), 1998. Guidelines for limiting exposure to time-varying electric, magnetic, and electromagnetic fields (up to 300 GHz), *Health Phys.* 74:494–522.

IEEE C-95-1991 (1999 Ed.). *Standard for Safety Levels with Respect to Human Exposure to Radio Frequency Electromagnetic Fields, 3 kHz to 300 GHz,* IEEE Standards Coordinating Committee 28, Institute of Electrical and Electronics Engineers, Inc., New York, NY.

Imaida K., Taki M., Watanabe S., Kamimura Y., Ito T., Yamaguchi T., Ito N., Shirai T., 1998a. The 1.5 GHz electromagnetic near-field used for cellular phones does not promote rat liver carcinogenesis in a medium-term liver bioassay, *Jpn. J. Cancer Res.* 89:995–1002.

Imaida K., Taki M., Yamaguchi T., Ito T., Watanabi S., Wake K., Aimoto A., Kaminurra Y., Ito N., Shirai T., 1998b. Lack of promoting effects of the electromagnetic near-field used for cellular phones (929.2 MHz) on rat liver carcinogenesis in a medium-term liver bioassay, *Carcinogenesis* 19:311–314.

Inaba R., Shishido K., Okada A., Moroji T., 1992. Effects of whole body microwave exposure on the rat brain contents of biogenic amines, *Eur. J. Appl. Physiol.* 65:124–128.

Inouye M., Galvin M.J., Jr., McRee D.I., 1982. Effects of 2.45 GHz microwave radiation on the development of Japanese quail cerebellum, *Teratology* 25:115–121.

Inouye M., Galvin M.J., McRee D.I., 1983. Effect of 2,450 MHz microwave radiation on the development of the rat brain, *Teratology* 28:413–419.

Janiak M., Szmigielski S., 1977. Injury of cell membranes in normal and SV40-virus transformed fibroblasts exposed *in vitro* to microwave (2450 MHz) or water-bath hyperthermia (43°C), in: *Abstr. 1977 Int. Symp. on the Biological Effects of Electromagnetic Waves*, Airlie, VA.

Jauchem J.R., 1985. Effects of drugs on thermal responses to microwaves, *Gen. Pharmacol.* 16:307–310.

Jauchem J.R., Frei M.R., Heinmets F., 1984. Increased susceptibility to radiofrequency radiation due to pharmacological agents, *Aviat. Space Environ. Med.* 55:1036–1040.

Jauchem J.R., Frei M.R., Heinmets F., 1985a. Effects of doxapram on body temperature of the rat during radiofrequency irradiation, *Clin. Exp. Pharmacol. Physiol.* 12:1–8.

Jauchem J.R., Frei M.R., Heinmets F., 1985b. Effects of psychotropic drugs on thermal responses to radiofrequency radiation, *Aviat. Space Environ. Med.* 56:1183–1188.

Jauchem J.R., Frei M.R., Heinmets F., 1988. Thermal responses to 5.6-GHz radiofrequency radiation in anesthetized rats: effect of chlorpromazine, *Physiol. Chem. Phys. Med. NMR* 2:135–143.

Jauchem J.R., Frei M.R., Heinmets F., 1997. Exposure to extremely-low-frequency electromagnetic fields and radiofrequency radiation: cardiovascular effects in humans, *Int. Arch. Occup. Environ. Health* 70:9–21.

Jauchem J.R., Frei M.R., Ryan K.L., Merritt J.H., Murphy M.R., 1999a. Lack of effects on heart rate and blood pressure in ketamine-anesthetized rats briefly exposed to ultra-wideband electromagnetic pulses, *IEEE Trans. Biomed. Eng.* 46:117–120.

Jauchem J.R., Ryan K.L., Frei M.R., 1999b. Cardiovascular and thermal responses in rats during 94 GHz irradiation, *Bioelectromagnetics* 20:264–267.

Jauchem J.R., Ryan K.L., Frei M.R., Dusch S.J., Lehnert H.M., Kovatch R.M., 2001. Repeated exposure of C3H/HeJ mice to ultrawideband electromagnetic pulses: lack of effects on mammary tumors, *Radiat. Res.* 155:369–377.

Jensh R.P., 1984a. Studies of the teratogenic potential of exposure of rats to 6000-MHz microwave radiation—I. morphologic analysis at term, *Radiat. Res.* 97:272–281.

Jensh R.P., 1984b. Studies of the teratogenic potential of exposure of rats to 6000-MHz microwave radiation—II. Postnatal psychophysiologic evaluations, *Radiat. Res.* 9:282–301.

Jensh R.P., Ludlow J., 1980. Behavioral teratology: application in low dose chronic microwave irradiation studies, Chapter 8, in: *Advances in the Study of Birth Defects*, Vol. 4, Persand T.V.N., Ed., MTP Press, Lancaster, England.

Jensh R.P., Vogel W.H., Brent R.L., 1982a. Postnatal functional analysis of prenatal exposure of rats to 915 MHz microwave radiation, *J. Am. Coll. Toxicol.* 1:73–90.

Jensh R.P., Weinberg I., Brent R.L., 1982b. Teratologic studies of prenatal exposure of rats to 915-MHz microwave radiation, *Radiat. Res.* 92:160–171.

Jensh R.P., Vogel W.H., Brent R.L., 1983a. An evaluation of the teratogenic potential of protracted exposure of pregnant rats to 2450-MHz microwave radiation: II. Postnatal psychophysiologic analysis, *J. Toxicol. Environ. Health* 11:37–59.

Jensh R.P., Weinberg I., Brent R.L., 1983b. An evaluation of the teratogenic potential of protracted exposure of pregnant rats to 2450-MHz microwave radiation: I. Morphologic analysis at term, *J. Toxicol. Environ. Health* 11:23–35.

Johansen C., Boice J.D., McLaughlin J.K., Olsen J.H., 2001. Cellular telephones and cancer—a nationwide cohort study, *J. Natl. Cancer Inst.* 93:203–207.

Johansen C., Boice J.D., McLaughlin J.K., Christensen H.C., Olsen J.H., 2002. Mobile phones and malignant melanoma of the eye, *Br. J. Cancer* 86:348–349.

Johnson C.C., 1975. Recommendations for specifying EM wave irradiation conditions in bioeffects research, *J. Microwave Power* 10:249–250.

Johnson C.C., Guy, A.W., 1972. Non-ionizing electromagnetic wave effects in biological materials and systems, *Proc. IEEE* 60:692–718.

Johnson R.B., Mizumori S., Lovely R.H., 1977. Adult behavioral deficit in rats exposed prenatally to 918-MHz microwaves, in: *Developmental Toxicology of Energy-Related Pollutants,* DOE Symp., Ser. 47, Department of Energy, Washington, D.C., pp. 281–289.

Johnson M.A., McLeod D.S., Lutty G.A., D'Anna S.A., Perry C.R., Kues H.A., 1992. Microwave induced retinal damage in the primate, *Invest. Ophthalmol. Visual Sci.* 33:1310.

Joines W.T., Blackman C.F., 1980. Power density, field intensity, and carrier frequency determinants of RF-energy-induced calcium-ion efflux from brain tissue, *Bioelectromagnetics* 1:271–275.

Justesen D.R., 1979. Behavioral and psychological effects of microwave radiation, *Bull. N.Y. Acad. Med.* 55:1058–1078.

Justesen D.R., 1983. Sensory dynamics of intense microwave irradiation: a comparative study of evasive behaviors by mice and rats, in: *Microwaves and Thermoregulation,* Adair E.A., Ed., Academic Press, New York, NY, pp. 203–230.

Justesen D.R., 1988. Microwave and infrared radiations as sensory, motivational, and reinforcing stimuli, in: *Electromagnetic Fields And Neurobehavioral Function,* O'Connor M.E., Lovely R.H., Eds., Alan R. Liss, Inc., New York, pp. 235–264.

Justesen D.R., King N.W., 1970. Behavioral effects of low level microwave irradiation in the closed space situation, in: *Biological Effects and Health Implications of Microwave Radiation,* Cleary, S.F., Ed., Symp. Proc., Public Health Service BRH/DBE 70-2, U.S. Department of Health, Education and Welfare, Washington, D.C., pp. 154–179.

Justesen D.R., Adair E.R., Stevens J.C., Bruce-Wolfe V., 1982. A comparative study of human sensory thresholds: 2450-MHz microwaves vs far-infrared radiation, *Bioelectromagnetics* 3:117–125.

Kamimura Y., Saito K., Saiga T., Amemiya Y., 1994. Effect of 2.45 GHz microwave irradiation on monkey eyes [Letter], in: *IEICE Trans. Commun., Special Issue on Biological Effects of Electromagnetic Fields,* Vol. E77-B, No. 6, pp. 762–765.

Kanal E., Gillen J., Evans J.A., Savitz D.A., Shellock F.G., 1993. Survey of reproductive health among female MR workers, *Radiology* 187:395–399.

Kaplan J., Polson P., Rebert C., Lunan K., Gage M., 1982. Biological and behavioral effects of prenatal and postnatal exposure to 2450-MHz electromagnetic radiation in the squirrel monkey, *Radio Sci.* 17:135–144.

Kellenyi L., Thuroczy G., Faludy B., Lenard L., 1999. Effects of mobile GSM radiotelephone exposure on the auditory brainstem response (ABR), *Neurobiology* 7:79–81.

Kerbacher J.J., Meltz M.M., Erwin D.N., 1990. Influence of radiofrequency radiation on chromosome aberrations in CHO cells and its interaction with DNA-damaging agents, *Radiat. Res.* 123:311–319.

Khizhnyak E.P., Shorokhov V.V., Tyazhelov V.V., 1980. Two types of microwave auditory sensation and their possible mechanisms, in: *Proc. URSI Int. Symposium on Electromagnetic Waves and Biology,* Paris, France, pp. 101–103.

Kholodov Y.A., 1966. The Effect of Electromagnetic and Magnetic Fields on the Central Nervous System, NASA TT-F-465, Nauka Press, Moscow.

Khramov R.N., Sosunov E.A., Koltun S.V., Ilyasova E.N., Lednev V.V., 1991. Millimeter-wave effects on electric activity of crayfish stretch receptors, *Bioelectromagnetics* 12:203–214.

Kiel J.L., Erwin D.N., 1984. Microwave and thermal interactions with oxidative hemolysis, *Physiol. Chem. Phys. Med. NMR* 16:317–323.

Kiel J.L., Erwin D.N., 1986. Microwave radiation effects on the thermally driven oxidase of erythrocytes, *Int. J. Hyperthermia* 2:201–212.

Kiel J.L., Wong L.S., Erwin D.N., 1986. Metabolic effects of microwave radiation and convection heating on human mononuclear leukocytes, *Physiol. Chem. Phys. Med. NMR* 18:181–187.

Kiel J.L., McQueen C., Erwin D.N., 1988. Green hemoprotein of erythrocytes: methemoglobin superoxide transferase, *Physiol. Chem. Phys. Med. NMR* 20:123–128.

Kim Y.S., 1998. Characteristics of EEG and AEP in human volunteers exposed to RF, *Kor. J. Environ. Health Soc.* 24:58–65.

Kim Y.A., Fomenko B.S., Agafonova T.A., Akoev I.G., 1985. Effects of microwave radiation 340 and 900 MHz on different structural levels of erythrocyte membranes, *Bioelectromagnetics* 6:305–312.

Kimata H., 2002. Enhancement of allergic skin wheal responses by microwave radiation from mobile phones in patients with atopic eczema/dermatitis syndrome, *Int. Arch. Allergy Immunol.* 129:348–350.

Kimata H., 2003. Enhancement of allergic skin wheal responses in patients with atopic eczema/dermatitis syndrome by playing video games or by a frequently ringing mobile phone, *Eur. J. Clin. Invest.* 33:513–517.

Kinoshita J.H., Merola L.O., Dikmak E., Carpenter R.L., 1966. Biochemical changes in microwave cataracts, *Docum. Ophthal.* 20:91–103.

Kitsovskaya I.A., 1960. An investigation of the interrelationships between the main nervous processes in rats on exposure to SHF fields of various intensities, *Tr. Gig. Tr. Prof. AMN SSR* 1:75.

Kitsovskaya I.A., 1964. The effect of centimeter waves of different intensities on the blood and hemopoietic organs of white rats, *Gig. Tr. Prof. Zabol.* 8:14.

Kitsovskaya I.A., 1968. The effect of radiowaves of various ranges on the nervous system (sound stimulation method), in: *On the Biological Effect of Radio-Frequency Electromagnetic Fields*, Moscow, pp. 81.

Klug S., Hetscher M., Giles S., Kohlsmann S., Kramer K., 1997. The lack of effects of nonthermal RF electromagnetic fields on the development of rat embryos grown in culture, *Life Sci.* 61: 1789–1802.

Kluger M.J., Ringler D.H., Anver M.R., 1975. Fever and survival, *Science* 188:166.

Koivisto M., Krause C.M., Revonsuo A., Laine M., Hamalainen H., 2000a. The effects of electromagnetic field emitted by GSM phones on working memory, *NeuroReport* 11:1641–1643.

Koivisto M., Revonsuo A., Krause C., Haarala C., Sillanmäki L., Laine M., Hamalainen H., 2000b. Effects of 902 MHz electromagnetic field emitted by cellular telephones on response times in humans, *NeuroReport* 11:413–415.

Kowalczuk C.I., Saunders R.D., Staple H.R., 1983. Sperm count and sperm abnormality in male mice after exposure to 2.45 GHz microwave radiation, *Mutat. Res.* 122:155–161.

Kramar P.O., Guy A.W., Lin J.C., 1975. The ocular effects of microwaves on hyperthermic rabbits: a study of microwave cataractogenic mechanisms, in: *Biological Effects of Nonionizing Radiation*, Tyler P.W., Ed., pp. 155–163, *Ann. N.Y. Acad. Sci.*, 247.

Kramar P., Harris C., Guy A.W., 1987. Thermal cataract formation in rabbits, *Bioelectromagnetics* 8:397–406.

Krause C.M., Sillanmaki L., Koivisto M., Haggqvist A., Saarela C., Revonsuo A., 2000a. Effects of electromagnetic fields emitted by cellular phones on the electroencephalogram during a visual working memory task, *Int. J. Radiat. Biol.* 76:1659–1667.

Krause C.M., Sillanmäki L., Koivisto M., Häggqvist A., Saarela C., Revonsuo A., Laine M., Hamalainen H., 2000b. Effects of electromagnetic field emitted by cellular phones on the EEG during a memory task, *NeuroReport* 11:761–764.

Kues H.A., Monahan J.C., 1992. Microwave-induced changes to the primate eye, *J. Johns Hopkins APL Tech. Digest* 13:244–255.

Kues H.A., Hirst L.W., Lutty G.A., D'Anna S.A., Dunkelberger G.R., 1985. Effects of 2.45-GHz microwaves on primate corneal endothelium, *Bioelectromagnetics* 6:177–188.

Kues H.A., Monahan J.C., D'Anna S.A., McLeod D.S., Lutty G.A., Koslov S., 1992. Increased sensitivity of the non-human primate eye to microwave radiation following ophthalmic drug pretreatment, *Bioelectromagnetics* 13:379–393.

Kues H.A., D'Anna S.A., Osiander R., Green W.R., Monahan J.C., 1999. Absence of ocular effects after either single or repeated exposure to 10 mW/cm^2 from a 60 GHz source, *Bioelectromagnetics* 20:463–473.

Kurt T.L., Milham S., 1988. Re: Increased mortality in amateur radio operators due to lymphatic and hematopoietic malignancies [Letter and Reply], *Am. J. Epidemiol.* 128:1384–1385.

Lacy K.K., Desesso J.M., Lary J.M., 1980a. A comparison of the teratogenic effects of radiofrequency radiation and hyperthermia: gross evaluation, *Teratology* 21:51A.

Lacy K.K., Desesso J.M., Sadler T.W., Lary J.M., 1980b. A comparison of the teratogenic effects of radiofrequency radiation and hyperthermia: light microscopic evaluation, *Teratology* 21:52A.
Lagroye I., Anane J.R., Wettring B.A., Moros B., Straube W.L., LaRegina M., Niehoff M., Pickard W.F., Roti Roti J.L., 2004. Measurement of DNA damage after acute exposure to pulsed wave 2450 microwaves in rat brain cells by two alkaline comet assay methods, *Int. J. Radiat. Biol.* 80:11–21.
Lai H., 1992. Research on the neurological effects of nonionizing radiation at the University of Washington, *Bioelectromagnetics* 13:513–526.
Lai H., Singh N.P., 1995. Acute low-intensity microwave exposure increases DNA single-strand breaks in rat brain cells, *Bioelectromagnetics* 16:207–210.
Lai H., Singh N.P., 1996. Single- and double-strand DNA breaks in rat brain cells after acute exposure to radiofrequency electromagnetic radiation, *Int. J. Radiat. Biol.* 69:513–521.
Lai H., Singh N.P., 1997. Melatonin and a spin-trap compound block radiofrequency electromagnetic radiation-induced DNA strand breaks in rat brain cells, *Bioelectromagnetics* 18:446–454.
Lai H., Horita A., Chou C.K., Guy A.W., 1984. Ethanol-induced hypothermia and ethanol consumption in the rat are affected by low-level microwave irradiation, *Bioelectromagnetics* 5:213–220.
Lai H., Horita A., Chou C.K., Guy A.W., 1987a. Low-level microwave irradiation affects central cholinergic activity in the rat, *J. Neurochem.* 48:40–45.
Lai H., Horita A., Chou C.K., Guy A.W., 1987b. Effects of low-level microwave irradiation on hippocampal and frontal cortical choline uptake are classically conditionable, *Pharmacol. Biochem. Behav.* 27:635–639.
Lai H., Horita A., Chou C.K., Guy A.W., 1987c. A review of microwave irradiation and actions of psychoactive drugs, *IEEE Eng. Med. Biol.* 6:31–36. Review.
Lai H., Horita A., Guy A.W., 1988. Acute low-level microwave exposure and central cholinergic activity: studies on irradiation parameters, *Bioelectromagnetics* 9:355–362.
Lai H., Carino M.A., Horita A., Guy A.W., 1989. Low-level microwave irradiation and central cholinergic systems, *Pharmacol. Biochem. Behav.* 33:131–138.
Lai H., Carino M.A., Horita A., Guy A.W., 1990. Corticotropin-releasing factor antagonist blocks microwave-induced decreases in high-affinity choline uptake in the rat brain, *Brain Res. Bull.* 25:609–612.
Lai H., Carino M.A., Wen Y.F., Horita A., Guy A.W., 1991. Naltrexone pretreatment blocks microwave-induced changes in central cholinergic receptors, *Bioelectromagnetics* 12:27–33.
Lai H., Carino M.A., Horita A., Guy A.W., 1992a. Single vs. repeated microwave exposure: effects on benzodiazepine receptors in the brain of the rat, *Bioelectromagnetics* 13:57–66.
Lai H., Carino M.A., Horita A., Guy A.W., 1992b. Opioid receptor subtypes that mediate the microwave-induced decreases in central cholinergic activity in the rat, *Bioelectromagnetics* 13:237–246.
Lai H., Horita A., Guy A.W., 1994. Microwave irradiation affects radial-arm maze performance in the rat, *Bioelectromagnetics* 15:95–104.
Lancranjan I., Maicanescu M., Rafaila E., Klepsch I., Popescu H.I., 1975. Gonadic function in workmen with long-term exposure to microwaves, *Health Phys.* 29:381–383.
La Regina M., Roti Roti J.L., 2003. The effect of chronic exposure to 835.62 MHz FMCW or 847.74 MHz CDMA on the incidence of spontaneous tumors in rats, *Radiat. Res.* 160:143–151.
Larsen A.I., 1991. Congenital malformations and exposure to high-frequency electromagnetic radiation among Danish physiotherapists, *Scand. J. Work Environ. Health* 17:318–323.
Laskey J., Dawes D., Howes M., 1970. Progress report on 2450 MHz irradiation of pregnant rats and the effect on the fetus, in: *Radiation Bioeffects*, Summary Report PHS, Publ. BRH/DBE-70, U.S. Department of Health, Education and Welfare, Rockville, MD, pp. 167.
Lebedeva N.N., Sulimov A.V., Sulimova O.P., Kotrovskaya T.I., Gailus T., 2000. Cellular phone electromagnetic field effects on bioelectric activity of human brain, *Crit. Rev. Biomed. Eng.* 28:323–337.
Lebedeva N.N., Sulimov A.V., Sulimova O.P., Korotkovskaya T.I., Gailus T., 2001. Investigation of brain potentials in sleeping humans exposed to the electromagnetic field of mobile phones, *Crit. Rev. Biomed. Eng.* 29:125–133.
Lebovitz R.M., 1983. Pulse modulated and continuous wave microwave radiation yield equivalent changes in operant behavior of rodents, *Physiol. Behav.* 30:891–898.

Lebovitz R.M., Johnson L., 1983. Testicular function of rats following exposure to microwave radiation, *Bioelectromagnetics* 4:107–114.

Lebovitz R.M., Seaman R.L., 1977a. Single auditory unit responses to weak, pulsed microwave radiation, *Brain Res.*126:370–375.

Lebovitz R.M., Seaman R.L., 1977b. Microwave hearing: the response of single auditory neurons in the cat to pulsed microwave radiation, *Radio Sci.* 12:229–236.

Lebovitz R.M., Johnson L., 1987a. Acute, whole-body microwave exposure and testicular function of rats, *Bioelectromagnetics* 8:37–43.

Lebovitz R.M., Johnson L., Samson W.K., 1987b. Effects of pulse-modulated microwave radiation and conventional heating on sperm production, *J. Appl. Physiol.* 62:245–252.

Lee R.A., Webb S.J., 1977. Possible detection of in vivo viruses by fine-structure millimeter microwave spectroscopy between 68 and 76 GHz, *IRCS Med. Sci.* 5:222.

Lee T.M., Ho S.M., Tsang L.Y., Yang S.Y., Li L.S., Chan C.C., 2001. Effect on human attention of exposure to the electromagnetic field emitted by mobile phones. *Neuro Report* 12:729–731.

Lenko J., Dolatowski A., Gruszecki L., Klajman S., Januszkiewicz L., 1966. Effect of 10-cm radar waves on the level of 17-ketosteroids and 17-hydroxycorticosteroids in the urine of rabbits, *Pregl. Lek.* 22:296–299.

Lermanth Y., Jacubovich R., Green M.S., 2001. Pregnancy outcome following exposure to short-aeases among physiotherapists in Israel. *Am J. Indus. Med* 39:499–504.

Li L., Bisht K.S., LaGroye I., Zhang P., Moros E.G., Roti Roti J.L., 2001. Measurement of DNA damage in mammalian cells exposed *in vitro* to radiofrequency fields at SARs of 3–5 W/kg, *Radiat. Res.* 156:328–332.

Liburdy R.P., 1977. Effects of radio-frequency radiation on inflammation, *Radio Sci.* 12:179–183.

Liburdy R.P., 1979. Radiofrequency radiation alters the immune system: modulation of T- and B-lymphyocyte levels and cell-mediated immunocompetence by hyperthermic radiation, *Radiat. Res.* 77:34–46.

Liburdy R.P., 1980. Radiofrequency radiation alters the immune system: II. Modulation of *in vivo* lymphocyte circulation, *Radiat. Res.* 83:66–73.

Liburdy R.P., Magin R.L., 1985. Microwave-stimulated drug release from liposomes, *Radiat. Res.* 103:266–275.

Liburdy R.P., Penn A., 1984. Microwave bioeffects in the erythrocyte are temperature and PO_2 dependent: cation permeability and protein shedding occur at the membrane phase transition, *Bioelectromagnetics* 5:283–291.

Liburdy R.P., Vanek P.F., 1985. Microwaves and the cell membrane II. Temperature, plasma, and oxygen mediate microwave-induced membrane permeability in the erythrocyte, *Radiat. Res.* 102:190–205.

Liburdy R.P., Vanek P.F., 1987. Microwaves and the cell membrane. III. Protein shedding is oxygen and temperature dependent: evidence for cation bridge involvement, *Radiat. Res.* 109:382–395.

Liburdy R.P., Rowe A.W., Vanek P.F., 1988. Microwaves and the cell membrane. IV. Protein shedding in the human erythrocyte: quantitative analysis by high-performance liquid chromatography, *Radiat. Res.* 114:500–514.

Lin J.C., 1976. Microwave auditory effect—a comparison of some possible transduction mechanisms, *J. Microwave Power* 11:77–87.

Lin J.C., 1977a. On microwave-induced hearing sensation, *IEEE Trans. Microwave Theory Tech.* 25:605–613.

Lin J.C., 1977b. Further studies on the microwave auditory effect, *IEEE Trans. Microwave Theory Tech.* 25:938–943.

Lin J.C., 1977c. Theoretical calculation of frequencies and thresholds of microwave-induced auditory signals, *Radio Sci.* 12:237–242.

Lin J.C., 1978. Microwave-evoked brainstem auditory responses, in: *Proc. Diego Biomed. Symp.*, Vol. 17, Academic Press, New York, NY, pp. 451–466.

Lin J.C., 1981. Microwave hearing effect, in: *Biological Effects of Nonionizing Radiation*, Illinger K.H., Ed., Am. Chem. Soc. Symp. Series 157, pp. 317–330.

Lin J.C., 1989. Pulsed radiofrequency field effects in biological systems, in: *Electromagnetic Interaction with Biological Systems*, Lin J.C., Ed., Plenum Press, New York, NY, pp. 165–177.

Lin J.C., 1990. Auditory perception of pulsed microwave radiation, in: *Biological Effects and Medical Applications of Electromagnetic Energy*, Gandhi O.P., Ed., Prentice Hall, Engelwood Cliffs, NJ, pp. 277–318.

Lin J.C., 2001. Hearing microwaves: the microwave auditory phenomenon, *IEEE Antennas Propagation Mag.* 43:166–168.

Lin J.C., 2002. Radio frequency radiation safety and health, *Radio Sci. Bull.* 303:37–39.

Lin J.C., Lin M.F., 1980. Studies on microwave and blood–brain barrier interaction, *Bioelectromagnetics* 1:313–323.

Lin J.C., Lin M.F., 1982. Microwave hyperthermia-induced blood–brain barrier alterations, *Radiat. Res.* 89:77–87.

Lin J.C., Guy A.W., Caldwell L.R., 1977. Thermographic and behavioral studies of rats in the near field of 918-MHz radiations, *IEEE Trans. Microwave Theory Tech.* 25:833–836.

Lin J.C., Meltzer R.J., Redding F.K., 1979a. Microwave-evoked brainstem potentials in cats, *J. Microwave Power* 14:291–295.

Lin J.C., Nelson J.C., Ekstrom M.E., 1979b. Effects of repeated exposure to 148 MHz radiowaves on growth and hematology of mice, *Radio Sci.* 14:173–179.

Lin J.C., Meltzer R.J., Redding F.K., 1982. Comparison of measured and predicted characteristics of microwave-induced sound, *Radio Sci.* 17:159–163.

Lin J.C., Su J.L., Wang Y., 1988. Microwave-induced thermoelastic pressure wave propagation in the cat brain, *Bioelectromagnetics* 9:141–147.

Lin-Liu S., Adey W.R., 1982. Low frequency amplitude modulated microwave fields change calcium efflux rates from synaptosomes, *Bioelectromagnetics* 3:309–322.

Lindauer G.A., Liu L.M., Skewes G.W., Rosenbaum F.J., 1974. Further experiments seeking evidence of nonthermal biological effects of microwave radiation, *IEEE Trans. Microwave Theory Tech.* 22:790–793.

Lipman R.M., Tripathi B.J., Tripathi R.C., 1988. Cataracts induced by microwave and ionizing radiation, *Surv. Ophthalmol.* 33:200–210.

Liu L.M., Cleary S.F, 1988. Effects of 2.45-GHz microwave and 100-MHz radiofrequency radiation on liposome permeability at the phase transition temperature, *Bioelectromagnetics* 9:249–257.

Liu L.M., Rosenbaum F.J., Pickard W.F., 1975. The relation of teratogenesis in *Tenebrio molitor* to the incidence of low-level microwaves, *IEEE Trans. Microwave Theory Tech.* 23:929–931.

Liu L.M., Nickless F.G., Cleary S.F., 1979. Effects of microwave radiation on erythrocyte membranes, *Radio Sci.* 14 (6S):109–115.

Livanov M.N., Tsypin A.B., Grigoriev Y.G., Kruschev U.G., Stepanov S.M., Anenyev A.M., 1960. The effect of electromagnetic fields on the bioelectric activity of cerebral cortex in rabbits, *Byull. Eksp. Biol. Med.* 49:63.

Lloyd D.C., Saunders R.D., Finnon P., Kowalczuk C.I., 1984. No clastogenic effect from *in vitro* microwave irradiation of G0 human lymphocytes, *Int. J. Radiat. Biol.* 46:135–141.

Lloyd D.C., Saunders R.D., Moquet J.E., Kowalczuk C.I., 1986. Absence of chromosomal damage in human lymphocytes exposed to microwave radiation with hyperthermia, *Bioelectromagnetics* 7:235–237.

Lobanova Y.A., 1960. Survival and development of animals at various intensities and duration of the influence of UHF, in: *The Biological Action of Ultrahigh Frequencies*, Letavet A.A., Gordon Z.V., Eds., Academy of Medical Science USSR, Moscow, USSR (English translation, JPRS 12471; OTS 62-19175, US Joint Publications Research Service, Washington, D.C.), pp. 60–63.

Lobanova Y.A., 1985. Investigations on the susceptibility of animals to microwave (MW) irradiation following treatment with pharmacologic agents, in: *Biologic Effects of Radiofrequency Electromagnetic Fields*, Gordon A.V., Ed., Joint Publications Reprints Service, Washington, D.C., pp. 201.

Logani M.K., Ziskin M.C., 1996. Continuous millimeter-wave radiation has no effect on lipid peroxidation in liposomes, *Radiat. Res.* 145:231–235.

Lotz W.G., 1983. Influence of the circadian rhythm on body temperature on the physiological response to microwaves: day vs. night exposure, in: *Microwaves and Thermoregulation*, Academic Press, New York, NY, pp. 445–460.

Lotz W.G., 1985. Hyperthermia in radiofrequency-exposed rhesus monkeys: a comparison of frequency and orientation effects, *Radiat. Res.* 102:59–70.
Lotz W.G., Michaelson S.M., 1978. Temperature and corticosterone relationships in microwave-exposed rats, *J. Appl. Physiol.: Respir. Environ. Exercise Physiol.* 44:438–445.
Lotz W.G., Michaelson S.M., 1979. Effects of hypophysectomy and dexamethasone on rat adrenal response to microwaves, *J. Appl. Physiol.: Respir. Environ. Exercise Physiol.* 47:1284–1288.
Lu S.T., deLorge J.O., 2000. Biological effects of high peak power radio frequency pulses, in: *Advances in Elelctromagnetic Fields in Living Systems*, Vol. 3, Lin J., Ed., Plenum Press, New York, NY, pp. 207–264.
Lu S.T., Lebda N., Michaelson S.M., Pettit S., Rivera D., 1977. Thermal and endocrinological effects of protracted irradiation of rats by 2450-MHz microwaves, *Radio Sci.* 12:147–156.
Lu S.T., Lebda N., Pettit S., Michaelson S.M., 1980a. Delineating acute neuroendocrine responses in microwave-exposed rats, *J. Appl. Physiol.: Respir. Environ. Exercise Physiol.* 48:927–932.
Lu S.T., Lotz W.G., Michaelson S.M., 1980b. Advances in microwave-induced neuroendocrine effects: the concept of stress, *Proc. IEEE* 68:73–77.
Lu S.T., Pettit S., Lu S.J., Michaelson S.M., 1986. Effects of microwaves on the adrenal cortex, *Radiat. Res.* 107:234–249.
Lu S.T., Brown D.O., Johnson C.E., Mathur S.P., Elson E.C., 1992. Abnormal cardiovascular responses induced by localized high power microwave exposure, *IEEE Trans. Biomed. Eng.* 39:484–492.
Lu S.T., Mathur S.P., Akyel Y., Lee J.C., 1999. Ultrawide-band electromagnetic pulses induced hypotension in rats, *Physiol. Behav.* 67:753–761.
Lu S.T., Mathur S.P., Stuck B., Zwick H., D'Andrea J.A., Ziriax J.M., Merritt J.H., Lutty G., McLeod D.S., Johnson M., 2000. Effects of high peak power microwaves on the retina of the rhesus monkey, *Bioelectromagnetics* 21:439–454.
Luttges M.W., 1980. Microwave Effects on Learning and Memory in Mice, NTIS Document No. AD-A094, 788/7.
Maes A., Collier M., Slaets D., Verschaeve L., 1996. 954 MHz microwaves enhance the mutagenic properties of mitomycin C, *Environ. Mol. Mutagen.* 28:26–30.
Maes A., Collier M., Van Gorp U., Vandoninck S., Verschaeve L., 1997. Cytogenetic effects of 935.2-MHz (GSM) microwaves alone and in combination with mitomycin C, *Mutat. Res.* 393:151–156.
Maes A., Collier M., Verschave L., 2000. Cytogenetic investigations on microwaves emitted by a 455.7 MHz car phone, *Folia Biol. (Praha)* 46:175–180.
Maes A.M., Collier M., Verschaeve L., 2001. Cytogenetic effects of 900 MHz (GSM) microwaves on human lymphocytes, *Bioelectromagnetics* 22:91–96.
Magin R.L., Lu S.-T., Michaelson S.M., 1977. Stimulation of dog thyroid by local application of high intensity microwaves, *Am. J. Physiol.* 233:E363–E368.
Makrides L., Heigenhauser G.J., Jones N.L., 1990. High-intensity endurance training in 20- to 30- and 60- to 70-yr-old healthy men, *J. Appl. Physiol.* 69:1792–1798.
Malyapa R.S., Ahern E.W., Straube W.L., Moros E.G., Pickard W.F., Roti Roti J.L., 1997a. Measurement of DNA damage after exposure to 2450 MHz electromagnetic radiation, *Radiat. Res.* 148:608–617.
Malyapa R.S., Ahern E.W., Straube W.L., Moros E.G., Pickard W.F., Roti Roti J.L., 1997b. Measurement of DNA damage after exposure to electromagnetic radiation in the cellular phone communication frequency band (835.62 and 847.74 MHz), *Radiat. Res.* 148:618–627.
Malyapa R.S., Ahern E.W., Straube W.L., LaRegina M., Pickard W.F., Roti Roti J.L., 1998. DNA damage in rat brain cells after *in vivo* exposure to 2.450 MHz electromagnetic radiation and the various methods of euthanasia, *Radiat. Res.* 149:637–645.
Malyshev V.T., Tkachenko M.I., 1972. Activity of ferments on the mucous membrane of the small intestine under the influence of an SHF field, in: *Physiology and Pathology of Digestion*, Kishenev, pp. 186.
Mandell G.L., 1975. Effect of temperature on phagocytosis by human polymorphonuclear neutrophils, *Infect. Immunity* 12:221–223.
Manikowska E., Luciani J.M., Servantie B., Czerski P., Obrenovitch J., Stahl A., 1979. Effects of 9.4 GHz microwave exposure on meiosis in mice, *Experientia* 35:388–390.

Mann K., Roschke J., 1996. Effects of pulsed high-frequency electromagnetic fields on human sleep, *Neuropsychobiology* 33:41–47.

Mann K., Roschke J., 2004. Sleep under exposure to high-frequency electromagnetic fields, *Sleep Med. Rev.* 8:95–107.

Mann K., Röschke J., Connemann B., Beta H., 1998a. No effects of pulsed high-frequency electromagnetic fields on heart rate variability during human sleep, *Neuropsychobiology* 38: 251–256.

Mann K., Wagner P., Brunn G., Hassan F., Hiemke C., Röschke J., 1998b. Effects of pulsed high-frequency electromagnetic fields on the neuroendocrine system, *Neuroendocrinology* 67:139–144.

Marha K., 1963. Biological effects of rf electromagnetic waves, *Prac. Lek.* (*Prague*) 15:238–241.

Marston LV., Voronina V.M., 1976. Experimental study of the effect of a series of phosphoorganic pesticides (Dipterex and Imidan) on embryogenesis, *Environ. Health Perspect.* 13:121.

Massoudi H., Durney C.H., Johnson C.C., 1977. Long-wavelength electromagnetic power absorption in ellipsoidal models of man and animals, *IEEE Trans. Microwave Theory Tech.* 25:47–52.

Mason P.A., Walters T.J., Digiovanni J., Jauchem J.R., Merritt J.H., Murphy M.R., Ryan K.L., 2001. Lack of effect of 94 GHz radio frequency radiation exposure in an animal model of skin carcinogeneses. *Carcinogeneses* 22:1701–1708.

McAfee R.D., Longacre A., Bishop R.R., Elder S.T., May J.G., Holland M.G., Gordon R., 1979. Absence of ocular pathology after repeated exposure of unanesthetized monkeys to 9.3-GHz microwaves, *J. Microwave Power* 14:41–44.

McAfee R.D., Ortiz-Lugo R., Bishop R., Gordon R., 1983. Absence of deleterious effects of chronic microwave radiation on the eyes of rhesus monkeys, *Ophthalmology* 90:1243–1245.

McDonald A.D., McDonald J.C., Armstrong B., Cherry N., Nolan A.D., Robert D., 1988. Work with visual display units in pregnancy, *Br. J. Ind. Med.* 45:509–515.

McLees B.D., Finch E.D., 1971. Analysis of the Physiologic Effects of Microwave Radiation, U.S. Naval Medical Research Institute, Proj. MF12 24.015-0001B, Report No. 3, Bethesda, MD.

McNamee J.P., Bellier P.V., Gajda G.B., Lavallée B.F., Lemay E.P., Marro L., Thansandote A., 2002a. DNA damage in human leukocytes after acute *in vitro* exposure to a 1.9 GHz pulse-modulated radiofrequency field, *Radiat. Res.* 158:534–537.

McNamee J.P., Bellier P.V., Gajda G.B., Miller S.M., Lemay E.P., Lavallee B.F., Marro L., Thansandote A., 2002b. DNA damage and micronucleus induction in human leukocytes after acute *in vitro* exposure to a 1.9 GHz continuous-wave radiofrequency field, *Radiat. Res.* 158:523–533.

McRee D.I., Hamrick P.E., 1977. Exposure of Japanese quail embryos to 2.45-GHz microwave radiation during development, *Radiat. Res.* 71:355–366.

McRee D.I., Wachtel H., 1982. Pulse microwave effects on nerve vitality, *Radiat. Res.* 91:212–218.

McRee D.I., Hamrick P.E., Zinkl J., 1975. Some effects of exposure of the Japanese quail embryo to 2.45-GHz microwave radiation, in: *Biological Effects of Nonionizing Radiation*, Tyler P.W., Ed., pp. 377–390, *Ann. N.Y. Acad. Sci.*, 247.

McRee D.I., MacNichols G., Livingston G.K., 1981. Incidence of sister chromatid exchange in bone marrow cells of the mouse following microwave exposure, *Radiat. Res.* 85:340–348.

McRee D.I., Thaxton J.P., Parkhurst C.R., 1983. Reproduction in male Japanese quail exposed to microwave radiation during embryogeny, *Radiat. Res.* 96:51–58.

Meltz M.L., Walker K.A., Erwin D.N., 1987. Radiofrequency microwave radiation exposure of mammalian cells during UV-induced DNA repair synthesis, *Radiat. Res.* 110:255–266.

Merola L.O., Kinoshita J.H., 1961. Changes in the ascorbic acid content in lenses of rabbit eyes exposed to microwave radiation, in: *Biological Effects of Microwave Radiation*, Vol. 1, Peyton, M.F., Ed., Plenum Press, New York, pp. 285–291.

Merritt J.H., Chamness A.F., Allen S.J., 1978. Studies on blood–brain barrier permeability after microwave-radiation, *Rad. Environ. Biophys.* 15:367–377.

Merritt J.H., Shelton W.W., Chamness A.F., 1982. Attempts to alter $45Ca^{++}$ binding to brain tissue with pulse-modulated microwave energy, *Bioelectromagnetics* 3:475–478.

Merritt J.H., Hardy K.A., Chamness A.F., 1984. *In utero* exposure to microwave radiation and rat brain development, *Bioelectromagnetics* 5:315–322.

Michaelson S.M., 1972. Human exposure to nonionizing radiant energy—potential hazards and safety standards, *Proc. IEEE* 60:389–421.

Michaelson S.M., 1974. Effects of exposure to microwaves: problems and perspectives, *Environ. Health Perspect.* 8:133–155.

Michaelson S.M., 1982. Physiologic regulation in electromagnetic fields, *Bioelectromagnetics* 3:91–104.

Michaelson S.M., 1983. Thermoregulation in intense microwave fields, in: *Microwaves and Thermoregulation*, Adair E.R., Ed., Academic Press, New York, NY, pp. 283–295.

Michaelson S.M., Thomson R.A.E., Howland J.W., 1961. Physiological aspects of microwave irradiation of mammals, *Am. J. Physiol.* 201:351–356.

Michaelson, S.M., Thomson, R.A.E., Howland, J.W., 1967. Biological Effects of Microwave Exposure, Griffiss Air Force Base, Rome Air Development Ctr. (ASTIA Doc. No. AD 824-242), Rome, NY.

Michaelson S.M., Guillet R., Catallo M.A., Small J., Inamine G., Heggeness F.W., 1976. Influence of 2450 MHz microwaves on rats exposed *in utero*, *J. Microwave Power* 11:165–166.

Mickey G.H., 1963. Electromagnetism and its effect on the organism, *N.Y. State J. Med.* 63: 1935–1942.

Mickley G.A., Cobb B.L., 1998. Thermal tolerance reduces hypoerthermia-induced disruption of working memory: a role for endogenous opiates? *Physiol. Behav.* 63:855–865.

Mickley G.A., Cobb B.L., Mason P.A., Farrell S., 1994. Disruption of a putative working memory task and selective expression of brain c-*fos* following microwave-induced hyperthermia, *Physiol. Behav.* 55:1029–1038.

Mikolajczyk H., 1972. Hormone reactions and changes in endocrine glands under influence of microwaves, *Med. Lotnicza* 39:39–51.

Milroy W.C., Michaelson S.M., 1972. Thyroid pathophysiology of microwave radiation, *Aerosp. Med.* 43:1126–1131.

Minecki L., Bilski R., 1961. Histopathological changes in internal organs of mice exposed to the action of microwaves, *Med. Pr.* (*Poland*) 12:337.

Miro L., Loubiere R., Pfister A., 1974. Effects of microwaves on the cell metabolism of the reticuloendothelial system, in: *Biologic Effects and Health Hazards of Microwave Radiation*, Czerski P., et al., Eds., Polish Medical Publishers, Warsaw, Poland, pp. 89–97.

Miro L., Loubiere R., Pfister A., 1985. Studies of visceral lesions observed in mice and rats exposed to UHF waves; a particular study of the effects of these waves on the reproduction of these animals, *Re. Med. Aeronaut.* (*Paris*) 4:37.

Mitchell C.L., McRee D.I., Peterson N.J., Tilson H.A., 1988. Some behavioral effects of short-term exposure of rats to 2.45-GHz microwave radiation, *Bioelectromagnetics* 9:259–268.

Mitchell C.L., McRee D.I., Peterson N.J., Tilson H.A., Shandala M.G., Rudnev M.I., Varetski V.V., Navakatikyan M.I., 1989. Results of a United States and Soviet Union joint project on nervous system effects of microwave radiation, *Environ. Health Perspect.* 81:201–209.

Mittler S., 1976. Failure of 2 and 20 meter radio waves to induce genetic damage in *Drosophila melanogaster*, *Environ. Res.* 11:326–330.

Modak A.T., Stavinoha W.B., Deam A.P., 1981. Effect of short electromagnetic pulses on brain acetylcholine content and spontaneous motor activity of mice, *Bioelectromagnetics* 2:89–92.

Moriyama E., Salcman M., Broadwell R.D., 1991. Blood–brain barrier alteration after microwave-induced hyperthermia is purely a thermal effect: I. Temperature and power measurements, *Surg. Neurol.* 35:177–182.

Nawrot P.S., McRee D.I., Staples R.E., 1981. Effects of 2.45 GHz CW microwave radiation on embryofetal development in mice, *Teratology* 24:303–314.

NCRP, 1981. Radiofrequency Electromagnetic Fields: Properties, Quantities and Units, Biophysical Interactions and Measurements, NCRP Report No. 67, National Council on Radiation Protection and Measurements, Washington, D.C.

NCRP, 1986. Biological Effects and Exposure Criteria for Radiofrequency Electromagnetic Fields, NCRP Report No. 86, National Council on Radiation Protection and Measurements, Bethesda, MD.

Neilly J.P., Lin J.C., 1986. Interaction of ethanol and microwaves on the blood–brain barrier of rats, *Bioelectromagnetics* 7:405–414.

Neshev N.N., Kirilova E.I., 1994. Possible nonthermal influence of millimeter waves on proton transfer in biomembranes, *Electro-Magnetobiol.* 13:191–194.

Neubauer C., Phelan A.M., Kues H., Lange D.G., 1990. Microwave irradiation of rats at 2.45 GHz activates pinocytotic-like uptake of tracer by capillary endothelial cells of cerebral cortex, *Bioelectromagnetics* 11:261–368.

Nikogosyan S.V., 1962. Influence of UHF on the cholinesterase activity in the blood serum and organs in animals, in: *The Biological Action of Ultrahigh Frequencies*, Letavet A.A., Gordon A.V., Eds., JPRS 12471.

Novitskiy Y.I., Gordon Z.V., Presman A.S., Kholodov Y.A., 1971. Radio Frequencies and Microwaves, Magnetic and Electrical Fields, National Aeronautics and Space Administration (NASA TT F-14.021), Washington, D.C.

NRPB, 2001. Possible Health Effects From Terrestrial Trunked Radio (TETRA), Report of an Advisory Group on Non-ionising Radiation, Vol. 12, No. 2, National Radiological Protection Board, Chilton, Didcot, Oxon OX11 0RQ, England.

Nurminen T., Kurpa K., 1988. Office employment, work with video display terminals, and the course of pregnancy, *Scan. J. Work Environ. Health* 14:293–298.

O'Connor M.E., 1980. Mammalian teratogenesis and radiofrequency fields, *Proc. IEEE* 68:56–60.

Ohmoto Y., Fujisawa H., Ishikawa T., Koizumi H., Matsuda T., Ito H., 1996. Sequential changes in cerebral blood flow, early neuropathological consequences and blood–brain barrier disruption following radiofrequency-induced localized hyperthermia, *Int. J. Hyperthermia* 12:321–334.

Olcerst R.B., Rabinowitz J.R., 1978. Studies on the interaction of microwave radiation with cholinesterase, *Radiat. Environ. Biophys.* 15:289–295.

Olcerst R.B., Belman S., Eisenbud M., Mumford W.W., Rabinowitz J.R., 1980. The increased passive efflux of sodium and rubidium from rabbit erythrocytes by microwave radiation, *Radiat. Res.* 82:244–256.

Oldendorf W.H., 1970. Measurement of brain uptake of radiolabeled substances using a tritiated water internal standard, *Brain Res.* 24:372–376.

Olsen R.G., 1977. Insect teratogenesis in a standing-wave irradiation system, *Radio Sci.* 12:199–207.

Olsen R.G., 1982. Constant-dose microwave irradiation of insect pupae, *Radio Sci.* 17:145–148.

Olsen R.G., Lin J.C., 1981. Microwave pulse-induced acoustic resonances in spherical head models, *IEEE Trans. Microwave Theory Tech.* 29:1114–1117.

Olsen R.G., Lin J.C., 1983. Microwave-induced pressure waves in mammalian brains, *IEEE Trans. Biomed. Eng.* 30:289–294.

Orlando A.R., Mossa G., D'Inzeo G., 1994. Effect of microwave radiation on the permeability of carbonic anhydrase loaded with unilamellar liposomes, *Bioelectromagnetics* 15:303–313.

Oscar K.J., Hawkins T.D., 1977. Microwave alteration of the blood–brain barrier system of rats, *Brain Res.* 126:281–293.

Oscar K.J., Gruenau S.P., Folker M.T., Rapoport S.I., 1981. Local cerebral blood flow after microwave exposure, *Brain Res.* 204:220–225.

Ouellet-Hellstrom R., Stewart W.F., 1993. Miscarriages among female physical therapists who report using radio- and microwave-frequency electromagnetic radiation, *Am. J. Epidemiol.* 138:775–786.

Pakhomov A., Murphy M.R., 2000. A comprehensive review of the research on biological effects of pulsed radio frequency radiation in Russia and the former Soviet Union, In: *Advances in Electromagnetic Fields in Living Systems*, Vol. 3, Lin J., Ed., Plenum Press, New York, NY, pp. 265–290.

Pakhomov A.G., Dubovick B.V., Kolupayev V.E., Degtyariov I.G., Pronkevich A.N., 1992. Search for microwave-sensitive structures and functions in the nervous system, in: *Fourteenth Annual IEEE/EMBS*, Paris, France, pp. 297.

Pakhomov A.G., Prol H.K., Mathur S.P., Akyel Y., Campbell C.B.G., 1997. Search for frequency-specific effects of millimeter-wave radiation on isolated nerve function, *Bioelectromagnetics* 18:324–334.

Pakhomov A.G., Doyle J., Stuck B.E., Murphy M.R., 2003. Effects of high-power microwave pulses on synaptic transmission and long-term potentiation in hippocampus, *Bioelectromagnetics* 24: 174–181.

Parker J.E., Kiel J.L., Winters W.D., 1988. Effect of radiofrequency radiation on mRNA expression in cultured rodent cells, *Physiol. Chem. Phys. Med. NMR* 20:129–134.

Parker L.N., 1973. Thyroid suppression and adrenomedullary activation by low-intensity microwave radiation, *Am. J. Physiol.* 224:1388–1390.

Paulraj R., Behari J., Rao A.R., 1999. Effect of amplitude modulated RF radiation on calcium ion efflux and ODC activity in chronically exposed rat brain, *Indian J. Biochem. Biophys.* 36:337–340.

Paulsson L.E., 1976. Measurements of 0.915, 2.45, and 9.0 FHz absorption in the human eye, Presented at the 6th European Microwave Conf., Rome, Italy.

Pay T.L., Beyer E.C., Reichelderfer C.F., 1972. Microwave effects on reproductive capacity and genetic transmission in *Drosophila melanogaster*, *J. Microwave Power* 7:75–82.

Persson B.R., Salford L.G., Brun A., Malmgren L., 1992. Increased permeability of the blood–brain barrier induced by magnetic and electromagnetic fields, *Ann. N.Y. Acad. Sci.* 649:356–358.

Persson B.R., Salford L.G., Brun A., 1997. Blood–brain barrier permeability in rats exposed to electromagnetic fields used in wireless communication, *Wireless Network* 3:455–461.

Peterson D.J., 1979. An investigation of the thermal and athermal effects of microwave irradiation on erythrocytes, *IEEE Trans. Biomed. Eng.* 26:428–436.

Petersen R.C., 1991. Radiofrequency/microwave protection guides, *Health Phys.* 61:59–67.

Petin V.G., Zhurakovskaya G.P., Kalugina A.V., 2000. Microwave dosimetry and lethal effects in laboratory animals, in: *Radio Frequency Radiation Dosimetry*, Klauenberg B.J., Miklavcic D., Eds., Kluwer Academic Publishers, Dordrecht, pp. 375–382.

Petrov I.R., Syngayevskaya V.A., 1970. Endocrine glands, in: Influence of Microwave Radiation on the Organism of Man and Animals, Petrov I.R., Ed., (NASA TT F0-708), Meditsina Press, Leningrad, pp. 31–41.

Phelan A.M., Lange D.G., Kues H.A., Lutty G.A., 1992. Modification of membrane fluidity in melanin-containing cells by low-level microwave radiation, *Bioelectromagnetics* 13:131–146.

Philippova T.M., Novoselov V.I., Bystrova N.F., Alekseev S.I., 1988. Microwave effect on camphor binding to rat olfactory epithelium, *Bioelectromagnetics* 9:347–354.

Philippova T.M., Novoselov V.I., Alekseev S.I., 1994. Influence of microwaves on different types of receptors and the role of peroxidation of lipids on receptor–protein binding, *Bioelectromagnetics* 15:183–192.

Phillips R.D., Hunt E.L., Castro R.D., King N.W., 1975. Thermoregulatory, metabolic, and cardiovascular response of rats to microwaves, *J. Appl. Physiol.* 38:630–635.

Pickard W.F., Barsoum Y.H., 1981. Radio-frequency bioeffects at the membrane level: separation of thermal and athermal contributions in the characeae, *J. Membr. Biol.* 61:39–54.

Pickard W.F., Foster K.R., 1987. The risks of risk research, *Nature* 330:531.

Pinakatt T., Cooper T., Richardson A.W., 1963. Control of ouabain on the circulatory response to microwave hyperthermia in the rat, *Aerosp. Med.* 34:497.

Pinakatt T., Richardson A.W., Cooper T., 1965. The effect of digitoxin on the circulatory response of rats to microwave irradiation, *Arch. Int. Pharmacodyn. Ther.* 156:151–160.

Pleet H., Graham J.M., Smith D.W., 1981. Central nervous system and facial defects associated with maternal hyperthermia at four to 14 weeks gestation, *Pediatrics* 67:785–795.

Portela A., Guardado M.I., de Xammar Oro J.R., Brennan M., Trainotti V., Stewart P.A., Perez R.J., Rodrigues C., Gimeno A., Rozell T., 1979. Quantitation of effects of repeated microwave radiation on muscle-cell osmotic state and membrane permselectivity, *Radio Sci.* 14:127–139.

Postow E., Swicord M.L., 1996. Modulated fields and "window" effects, in: *CRC Handbook of Biological Effects of Electromagnetic Fields*, Polk C., Postow E., Eds., CRC Press, Boca Raton, FL, pp. 535–580.

Prausnitz S., Susskind C., 1962. Effects of chronic microwave irradiation on mice, *IRE Trans. Bio-Med. Electron.* 9:104–108.

Preece A.W., Iwi G., Davies-Smith A., Wesnes K., Butler S., Lim E., Varey A., 1999. Effect of a 915 MHz simulated mobile phone signal on cognitive function in man, *Int. J. Radiat. Biol.* 75:447–456.

Preskorn S.H., Edwards W.D., Justesen D.R., 1978. Retarded tumor growth and greater longevity in mice after fetal irradiation by 2450-MHz microwaves, *J. Surg. Oncol.* 10:483–492.

Pressman A.S., Levitina N.A., 1962. Nonthermal action of microwave on cardiac rhythm. Communication I. A study of the action of continuous microwaves, *Bull. Exp. Biol. Med.* 53:36–39.

Preston E., Prefontaine G., 1980. Cerebrovascular permeability to sucrose in the rat exposed to 2,450-MHz microwaves, *J. Appl. Physiol.: Respir. Environ. Exercise Physiol.* 49:218–223.

Preston E., Vavasour E.J., Assenheim H.M., 1979. Permeability of the blood–brain barrier to mannitol in the rat following 2450 MHz microwave irradiation, *Brain Res.* 174:109–117.

Putthoff D.L., Justesen D.R., Ward L.B., Levinson D.M., 1977. Drug-induced ectothermia in small mammals: the quest for a biological microwave dosimeter, *Radio Sci.* 12:73–80.

Quock R.M., Fujimoto J.M., Ishii T.K., Lange D.G., 1986. Microwave facilitation of methylatropine antagonism of central cholinomimetic drug effects, *Radiat. Res.* 105:328–340.

Quock R.M., Kouchich F.J., Ishii T.K., Lange D.G., 1987. Microwave facilitation of domperidone antagonism of apomorphine-induced stereotypic climbing in mice, *Bioelectromagnetics* 8:45–56.

Radon K., Parera D., Rose D.M., Jung D., Wollrath L., 2001. No effects of pulsed radio frequency electromagnetic fields on melatonin, cortisol, and selected markers of the immune system in man, *Bioelectromagnetics* 22:280–287.

Rapoport S.I., Ohno K., Fredricks W.R., Pettigrew K.D., 1979. A quantitative method for measuring altered cerebrovascular permeability, *Radio Sci.* 14 (6S):345–348.

Rasbury J.A., Seaman R.L., 1992 (October). Transient cardiovascular changes in rats with microwave exposure of carotid sinus region, in: *Proceedings of the 14th Annual IEEE EMBS Conference*, Paris, France, p. 392.

Raslear T.G., Akyel Y., Bates F., Belt M., Lu S.-T., 1993. Temporal bisection in rats: the effects of high-peak power pulsed microwave irradiation, *Bioelectromagnetics* 14:459–478.

Reiser H.P., Dimpfel W., Schober F., 1995. The influence of electromagnetic fields on human brain activity, *Eur. J. Med. Res.* 1:27–32.

Repacholi M.H., Basten A., Gebski V., Noonan D., Finnie J., Harris A.W., 1997. Lymphomas in transgenic mice exposed to pulsed 900 MHz electromagnetic fields, *Radiat. Res.* 147:631–640.

Revutsky E.L., Edelman F.M., 1964. Effects of centimeter and meter electromagnetic waves in the content of biologically active substances in human blood, *Philos. J. Ukr. Acad. Sci.* 10:379–383.

Rhee K.W., Lee C.F., Davis C.C., Sagripanti J.L., Swicord M.L., 1988. Further Studies of the Microwave Absorption Characteristics of Different Forms of DNA in Solution, 10th Annual Meeting of the Bioelectromagnetics Society, Stamford, CT, June 19–23.

Riu P.J., Foster K.R., Blick D.W., Adair E.R., 1997. A thermal model for human thresholds of microwave-evoked warmth sensations, *Bioelectromagnetics* 18:578–583.

Roberti B., Heebels G.H., Hendricx J.C.M., De Greef A.H.A.M., Wolthuis O.L., 1975. Preliminary investigations of the effects of low-level microwave radiation on spontaneous motor activity in rats, *Ann. N.Y. Acad. Sci.* 247:417–424.

Roberts N.J., Michaelson S.M., 1983. Microwaves and neoplasia in mice: analysis of a reported risk, *Health Phys.* 44:430–433.

Roberts N.J., Jr., Steigbigel R.T., 1977. Hyperthermia and human leukocyte functions: effects on response of lymphocytes to mitogen and antigen and bactericidal capacity of monocytes and neutrophils, *Infect. Immunity* 18:673–679.

Roberts N.J., Lu S.-T., Michaelson S.M., 1983. Human leukocyte functions and the U.S. safety standard for exposure to radiofrequency radiation, *Science* 220:318–320.

Roberts N.J., Jr., Michaelson S.M., Lu S.T., 1987. Mitogen responsiveness after exposure of influenza virus-infected human mononuclear leukocytes to continuous or pulse-modulated radiofrequency radiation, *Radiat. Res.* 110:353–361.

Roschke J., Mann K., 1997. No short-term effects of digital mobile radio telephone on the awake human electroencephalogram, *Bioelectromagnetics* 18:172–176.

Roschmann P., 1991. Human auditory system response to pulsed radiofrequency energy in RF coils for magnetic resonance at 2.4 to 170 MHz, *Magn. Reson. Med.* 21:197–215.

Rotkovska D., Vacek A., Bartonichkova A., 1987. Effects of microwaves on the colony-forming ability of haemopoietic stem cells in mice, *Acta Oncol.* 26:233–236.

Rotkovska D., Bartonickova A., Kautska J., 1993. Effects of microwaves on membranes of hematopoietic cells in their structural and functional organization, *Bioelectromagnetics* 14:79–85.

Roti Roti J.L., Malyapa R.S., Bisht K.S., Ahern E.W., Moros E.G., Pickard W.F., Straube W.L., 2001. Neoblastic transformation of C3H10T1/2 cells after exposure to 835.62 MHz FDMA and 847.74 CDMA radiations, *Radiat. Res.* 155:219–231.

Roux C., Elefant E., Gaboriaud G., Jaullery C., Gardette J., Dupuis R., Lambert D., 1986. Association of microwaves and ionizing radiation: potentiation of teratogenic effects in the rat, *Radiat. Res.* 108:317–326.

Rozdilsky B., Olszewski J., 1957. Permeability of cerebral blood vessels studies by radioactive iodinated bovine albumin, *Neurology* 7:270–279.

Rugh R., McManaway M., 1976. Can electromagnetic waves cause congenital anomalies? in: *Int. IEEE/AP-S USN/URSI Symp.*, Amherst, MA, p. 143.

Rugh R., Ginns E.I., Ho H.S., Leach W.M., 1975. Responses of the mouse to microwave radiation during estrous cycle and pregnancy, *Radiat. Res.* 62:225–241.

Saalman E., Norden B., Arvidsson L., Hamnerius Y., Hojevik P., Connell K.E., Kurucsev T., 1991. Effect of 2.45 GHz microwave radiation on permeability of unilamellar liposomes to 56-carboxyfluorescein. Evidence of non-thermal leakage, *Biochim. Biophys. Acta* 1064:124–130.

Sacchikova M.N., 1974. Clinical manifestations of reactions to microwave irradiation in various occupational groups, in: *Biological Effects and Health Hazards of Microwave Irradiation*, Czerski P., Ostrowski K., Silverman C., et al., Eds., Polish Medical Publishers, Warsaw, Poland, pp. 261–267.

Saffer J.D., Profenno L.A., 1989. Sensitive model with which to detect athermal effects of non-ionizing electromagnetic radiation, *Bioelectromagnetics* 10:347–354.

Saffer J.D., Profenno L.A., 1992. Microwave-specific heating affects gene expression, *Bioelectromagnetics* 13:75–78.

Sagripanti J.L., Swicord M.L., 1986. DNA structural changes caused by microwave radiation, *Int. J. Radiat. Biol.* 50:47–50.

Sagripanti J.L., Swicord M.L., Davis C.C., 1987. Microwave effects on plasmid DNA, *Radiat. Res.* 110:219–231.

Salford L.G., Brun A., Eberhardt J.L., Persson B.R., 1993a. Permeability of the blood–brain barrier induced by 915 MHz electromagnetic radiation, continuous wave and modulated at 8, 16, 50, and 200 Hz, *Bioelectrochem. Bioenerg.* 30:293–301.

Salford L.G., Brun A., Persson B.R., Eberhardt J.L., 1993b. Experimental studies of brain tumour development during exposure to continuous and pulsed 915 MHz radiofrequency radiation, *Bioelectrochem. Bioenerg.* 30:313–318.

Salford L.G., Brun A., Sturesson K., Eberhardt J.L., Persson B.R., 1994. Permeability of the blood–brain barrier induced by 915 MHz electromagnetic radiation, continuous wave and modulated at 8, 16, 50, and 200 Hz, *Microsc. Res. Tech.* 27:535–542.

Salford L.G., Brun A.E., Eberhardt J.L., Malgren L., Persson B.R., 2003. Nerve cell damage in mammalian brain after exposure to microwaves from GSM mobile phones, *Environ. Health Perspect.* 111:881–883.

Sandblom J., Thenander S., 1991. The effect of microwave radiation on the stability and formation of gramicidin-a channels in lipid bilayer membranes, *Bioelectromagnetics* 12:9–20.

Sanders A.P., Schaefer D.J., Joines W.T., 1980. Microwave effects on energy metabolism of rat brain, *Bioelectromagnetics* 1:171–181.

Sanders A.P., Joines W.T., Allis J.W., 1985. Effects of continuous-wave, pulsed, and sinusoidal-amplitude-modulated microwaves on brain energy metabolism, *Bioelectromagnetics* 6:89–98.

Sandstrom M., Mild K.H., Stenberg B., Wall S., 1995. Skin symptoms among VDT workers and electromagnetic fields a case referent study, *Indoor Air* 5:29 37.

Sandstrom J., Wilen J., Oftedal G.G., Mild K.H., 2001. Mobile phone use and subjective symptoms. Comparison of symptoms experienced by users of analogue and digital mobile phones, *Occup. Med.* 51:25–35.

Sandweiss J., 1990. On the cyclotron resonance model of ion transport, *Bioelectromagnetics* 11:203–205.

Saunders R.D., Kowalczuk C.I., 1981. Effects of 2.45 GHz microwave radiation and heat on mouse spermatogenic epithelium, *Int. J. Radiat. Biol.* 40 (6):623–632.

Saunders R.D., Darby S.C., Kowalczuk C.I., 1983. Dominant lethal studies in male mice after exposure to 2.45 GHz microwave radiation, *Mutat. Res.* 117:345–356.

Saunders R.D., Kowalczuk C.I., Beechey C.V., Dunford R., 1988. Studies of the induction of dominant lethals and translocations in male mice after chronic exposure to microwave radiation, *Int. J. Radiat. Biol.* 53:983–992.

Savage J.R.K., 1977. Use and abuse of chromosomal aberrations as an indicator of genetic damage, *Int. J. Environ. Stud.* 1:233.
Schlagel C.J., Sulek K., Ho H.S., Leach W.M., Ahmed A., Woody J.N., 1980. Biological effects of microwave exposure. II. Studies on the mechanisms controlling susceptibility to microwave-induced increases in complement receptor positive spleen cells, *Bioelectromagnetics* 1:405–414.
Schnorr T.M., Grajewski B.A., Hornung R.W., Thun M.J., Egeland G.M., Murray W.E., Conover D.L., Halperin W.E., 1991. Video display terminals and the risk of spontaneous abortion, *N. Engl. J. Med.* 324:727–733.
Schrader S.M., Langford R.E., Turner T.W., Breitenstein M.J., Clark J.C., Jenkins B.L., Lundy D.O., Simon S.D., Weyandt T.B., 1998. Reproductive function in relation to duty assignments among military personnel, *Reprod. Toxicol.* 12:465–468.
Schwan H.P., Anne A., Sher I., 1996. Heating of Living Tissue, Report NAEC-ACEL-534, U.S. Naval Eng. Center, Philadelphia, PA.
Seaman R.L., Lebovitz R.M., 1989. Thresholds of cat cochlear nucleus neurons to microwave pulses, *Bioelectromagnetics* 10:147–160.
Seaman R.L., Wachtel H., 1978. Slow and rapid responses to CW and pulsed microwave radiation by individual *Aplysia* pacemakers, *J. Microwave Power* 13:77–86.
Serdiuk A.M., 1969. Biological effect of low-intensity ultrahigh frequency fields, *Vrach. Delo* 11:108–111.
Servantie B., Servantie A.M., Etienne J., 1975. Synchronization of cortical neurons by a pulsed microwave field as evidenced by spectral analysis of electrocorticograms from the white rat, in: *Biological Effects of Nonionizing Radiation*, Tyler P.W., Ed., pp. 82–86, *Ann. N.Y. Acad. Sci.*, 247.
Servantie G., Gillord J., Servantie A.M., Obrenovitch J., Berthanion G., Peeren J.C., Gretin B., 1976. Comparative study of the action of three types of microwave fields upon the behavior of the white rat. *J. Micro Power* 11:145–146.
Shelton W.W., Merritt J.H., 1981. *In vitro* study of microwave effects on calcium efflux in rat brain tissue, *Bioelectromagnetics* 2:161–167.
Sheppard A.R., Bawin S.M., Adey W.R., 1979. Models of long-range order in cerebral macromolecules: effects of sub-ELF and of modulated VHF and UHF fields, *Radio Sci.* 14:141–145.
Sherry C.J., Blick D.W., Walters T.J., Brown G.C., Murphy M.R., 1995. Lack of behavioral effects in non-human primates after exposure to ultrawideband electromagnetic radiation in the microwave frequency range, *Radiat. Res.* 143:93–97.
Shnyrov V.L., Zhadan G.G., Akoev I.G., 1984. Calorimetric measurements of the effect of 330-MHz radiofrequency radiation on human erythrocyte ghosts, *Bioelectromagnetics* 5:411–418.
Shore M.L., Felten R.P., Lamanna A., 1977. The effect of repetitive prenatal low-level microwave exposure on development in the rat, in: *Symposium on Biological Effects and Measurements of Radio Frequency/Microwaves*, Hazzard D.G., Ed., U.S. Department of Health, Education, and Welfare, HEW Publication (FDA) 77-8026, Washington, D.C., pp. 280–289.
Sienkiewicz Z.J., Blackwell R.P., Haylock R.G., Saunders R.D., Cobb B.L., 2000. Low-level exposure to pulsed 900 MHz microwave radiation does not cause deficits in the performance of a spatial learning task in mice, *Bioelectromagnetics* 21:151–158.
Sliney D.H., Wolbarsht M.L., Muc A.M., 1985. Differing radiofrequency standards in the microwave region-implications for future research, *Health Phys.* 49:677–683.
Smialowicz R.J., 1976. The effect of microwaves (2450 MHz) on lymphocyte blast transformation *in vitro*, in: *Biological Effects of Electromagnetic Waves*, Johnson C.C., Shore M.L., Eds., Publ. (FDA) 77-8010, U.S. Department of Health, Education and Welfare, Rockville, MD, pp. 472–483.
Smialowicz R.J., 1987. Immunologic effects of nonionizing electromagnetic radiation, *IEEE Eng. Med. Biol. Mag.* 6:47–51. Review.
Smialowicz R.J., Kinn J.B., Elder J.A., 1979a. Perinatal exposure of rats to 2450-MHz CW microwave radiation: effects on lymphocytes, *Radio Sci.* 14:147–153.
Smialowicz R.J., Riddle M.M., Brugnolotti P.L., Sperazza J.M., Kinn J.B., 1979b. Evaluation of lymphocyte function in mice exposed to 2450 MHz (CW) microwaves, in: *Proc. 1978 Symp. on Electromagnetic Fields in Biological Systems*, Stuchly S.S. Ed., Ottawa, Canada, pp. 122–152.
Smialowicz R.J., Ali J.S., Berman E., Bursian S.J., Kinn J.B., Liddle C.G., Reiter L.W., Weil C.M., 1981a. Chronic exposure of rats to 100-MHz CW radiofrequency radiation: assessment of biological effects, *Radiat. Res.* 86:488–505.

Smialowicz R.J., Brugnolotti B.L., Riddle M.M., 1981b. Complement receptor positive spleen cells in microwave 2450-MHz-irradiated mice, *J. Microwave Power* 16:73–77.
Smialowicz R.J., Riddle M.M., Brugnolotti P.L., Rogers R.R., Compton K.L., 1981c. Detection of microwave heating in 5-hydroxytryptamine-induced hypothermic mice, *Radiat. Res.* 88:108–117.
Smialowicz R.J., Weil C.M., Marsh P., Riddle M.M., Rogers R.R., Rehnberg B.F., 1981d. Biological effects of long-term exposure of rats to 970-MHz radiofrequency radiation, *Bioelectromagnetics* 2:279–284.
Smialowicz R.J., Riddle M.M., Rogers R.R., Stott G.A., 1982a. Assessment of immune function development in mice irradiated *in utero* with 2450-MHz microwaves, *J. Microwave Power* 17:121–126.
Smialowicz R.J., Riddle M.M., Weil C.M., Brugnolotti P.L., Kinn J.B., 1982b. Assessment of the immune responsiveness of mice irradiated with continuous wave or pulse-modulated 425-MHz radio frequency radiation, *Bioelectromagnetics* 3:467–470.
Smialowicz R.J., Weil C.M., Kinn J.B., Elder J.A., 1982c. Exposure of rats to 425-MHz CW radiofrequency radiation: effects on lymphocytes, *J. Microwave Power* 17:211–221.
Smith E.M., Hammonds-Ehlers M., Clark M.K., Kirchner H.L., Fuortes L., 1997. Occupational exposures and risk of female infertility, *J. Occup. Environ. Med.* 39:138–146.
Spalding J.F., Freyman R.W., Holland L.M., 1971. Effects of 800 MHz electromagnetic radiation on body weight, activity, hematopoiesis, and life span in mice, *Health Phys.* 20:421–424.
Spiers D.E., Baummer S.C., 1991. Thermal and metabolic responsiveness of Japanese quail embryos following periodic exposure to 2,450 MHz microwaves, *Bioelectromagnetics* 12:225–240.
Stamm M.E., Winters W.D., Morton D.L., Warren S.L., 1974. Microwave characteristics of human tumor cells, *Oncology* 29:294–301.
Stamm M.E., Warren S.L., Rand R.W., et al., 1975. Microwave therapy experiments with B-16 murine melanoma, *IRCS Med. Sci.* 3:392–393.
Stang A., Anastassiou G., Ahrens W., Bromen K., Bornfeld N., Jockel K.H., 2001. The possible role of radiofrequency radiation in the development of uveal melanoma, *Epidemiology* 12:7–12.
Stenberg B., Eriksson N., Mild K.H., Hoog J., Sandstrom M., Sundell J., Wall S., 1995. Facial skin symptoms in visual display terminal VDT workers. A case-referent study of personal, psychosocial, building- and VDT-related risk indicators, *Int. J. Epidemiol.* 24:796–803.
Stern S., Margolin L., Weiss B., Lu S.T., Michaelson S.M., 1979. Microwaves: effect on thermoregulatory behavior in rats, *Science* 206:1198–1201.
Stewart W., 2000. Mobile Phones and Health, Report by the U.K. Independent Expert Group on Mobile Phones, U.K. National Radiological Protection Board, Chilton, Didcot, Oxon OX11 0RQ, England, pp. 1–160.
Stodolnik-Baranska W., 1974. The effects of microwaves on human lymphocyte cultures, in: *Biologic Effects and Health Hazards of Microwave Radiation*, Czerski P., et al., Eds., Polish Medical Publishers, Warsaw, Poland, pp. 189–195.
Stolwijk J.A., Hardy J.D., 1977. Control of body temperature, in: *Handbook of Physiology. Section 9. Reaction To Environmental Agents*, Lee D.H.K., Ed., American Physiological Society, Bethesda, MD, pp. 45–68.
Stuchly M.A., 1978. *Health Aspects of Radiofrequency and Microwave Radiation Exposure*, Part 2, Department of National Health and Welfare, Ottawa, Canada.
Sulek K., Schlagel C.J., Wiktor Jedrzecjzak W., Ho H.S., Leach W.M., Ahmed A., Woody J.N., 1980. Biologic effects of microwave exposure. I. Threshold conditions for the induction of the increase in complement positive (CR+) mouse spleen cells following exposure to 2450-MHz microwaves, *Radiat. Res.* 83:127–137.
Sutton C.H., Nunnaly R.L., Carroll F.B., 1973. Protection of the microwave-irradiated brain with body-core hypothermia, *Cryobiology* 10:513–514.
Suvorov N.B., Kukhtina G.V., 1984. Spatial dynamics of brain electrical activity during prolonged contact with physical factors, *Human Physiol.* 10:395–401.
Swicord M.L., Davis C.C., 1982. Microwave absorption of DNA between 8 and 12 GHz, *Biopolymers* 21:2453–2460.
Swiecicki W., Edelwejn Z., 1963. The influence of 3 and 10 cm microwave irradiation in blood proteins in rabbits, *Med. Lotnicza* 11:54.

Sykes P.J., McCallum B.D., Bangay M.J., Hooker A.M., Morley A.A., 2001. Effect of radiofrequency exposure on intrachromosomal recombination in mutation and cancer, *Radiat. Res.* 156:495–502.

Szmigielski S., Szudzinski A., Pietraszek A., Bielec M., 1980. Acceleration of cancer development in mice by long-term exposition to 2450 MHz microwave fields, in: *Proc. URSI Int. Symp. on Electromagnetic Waves and Biology*, Paris, France, pp. 165–169.

Taflove A., Brodwin M.E., 1975. Computation of the electromagnetic fields and induced temperatures within a model of the microwave-irradiated human eye, *IEEE Trans. Microwave Theory Tech.* 23:888–896.

Takashima S., Onaral B., Schwan H.P., 1979. Effects of modulated RF energy on the EEG of mammalian brains, *Radiat. Environ. Biophys.* 16:15–27.

Talau H.P., Raczek J., Marx B., Homback V., Cooper J., 2003. Temperature changes in chicken embryos exposed to a continuous wave 1.25 GHz radiofrequency electromagnetic field, *Radiat. Res.* 159:685–692.

Taskinen H., Kyyronen P., Hemminki K., 1990. Effects of ultrasound, shortwaves, and physical exertion on pregnancy outcome in physiotherapists, *J. Epidemiol. Commun. Health* 44:196–201.

Tattersall J.E.H., Scott I.R., Wood S.J., Nettell J.J., Bevir M.K., Wang Z., Somasiri N.P., Chen X., 2001. Effects of low-intensity radiofrequency electromagnetic fields on electrical activity in rat hippocampal slices, *Brain Res.* 904:43–53.

Taylor E.M., Ashleman B.T., 1974. Analysis of central nervous system involvement in the microwave auditory effect, *Brain Res.* 74:201–208.

Thomas J.R., Maitland G., 1979. Microwave radiation and dextroamphetamine: evidence of combined effects on behavior of rats, *Radio Sci.* 14 (6S):253–258.

Thomas J.R., Finch E.D., Fulk D.W., Burch L.S., 1975. Effects of low-level microwave radiation on behavioral baselines, in: *Biological Effects of Nonionizing Radiation*, Tyler P.W., Ed., pp. 425–432, *Ann. N.Y. Acad. Sci.*, 247.

Thomas J.R., Schrot J., Banvard R.A., 1982. Comparative effects of pulsed and continuous-wave 2.8-GHz microwaves on temporally defined behavior, *Bioelectromagnetics* 3:227–235.

Thompson W.D., Bourgeois A.E., 1975. Effects of Microwave Exposure on Behavior and Related Phenomena, Primate Behavior Lab., Aeromedical Research Lab. Report (ARL-TR-65-20; AD 489245), Wright-Patterson AFB, Ohio.

Thuroczy G., Kubinyi G., Bodo M., Bakos J., Bakos L.D., 1994. Simultaneous response of brain electrical activity (EEG) and cerebral circulation (REG) to microwave exposure in rats, *Rev. Environ. Health* 10:135–148.

Tice R.R., Hook G.G., Donner M., McRee D.I., Guy A.W., 2002. Genotoxicity of radiofrequency signals, I. Investigation of DNA damage and micronuclei induction in cultured human blood cells, *Bioelectromagnetics* 23:113–126.

Toler J., Popovich V., Bonasera S., Popovich P., Honeycutt C., Sgoutas D., 1988. Long-term study of 435 MHz radio-frequency radiation on blood-borne end points in cannulated rats—Part II: methods, results and summary, *J. Microwave Power Electromagn. Energy* 23:105–136.

Toler J.C., Shelton W.W., Frei M.R., Merritt J.H., Meltz M.L., Shelton M.A., 1997. Long-term, low-level exposure of mice prone to mammary tumors to 435 MHz radiofrequency radiation, *Radiat. Res.* 148:227–234.

Tolgskaya M.S., 1959. Morphological changes in animals exposed to 10 cm microwaves, *Vopr. Kurortol. Fizioter. Lech. Fiz. Kult.* 1:21.

Tolgskaya M.S., Gordon, Z.V., 1960. Changes in the receptor and interoreceptor apparatuses under the influence of UHF, in: *The Biological Action of Ultrahigh Frequencies*, Letavet A.A., Gordon Z.V., Eds., Academy of Medical Science, Moscow, pp. 104.

Tolgskaya M.S., Gordon Z.V., 1964. Comparative morphological characterization of action of microwaves of various ranges, *Tr. Gig. Tr. Prof. AMN SSR* 2:80.

Tolgskaya M.W., Gordon Z.V., 1973. *Pathological Effects of Radio Waves*, Meditsina Press, Moscow, 1971; transl. Consultants Bureau, New York, NY.

Tyazhelov V.V., Tigranian R.E., Khizhniak E.P., 1977. New artifact-free electrodes for recording of biological potentials in strong electromagnetic fields, *Radio Sci.* 12 (6S):121–123.

Urban P., Lukas E., Roth Z., 1998. Does acute exposure to the electromagnetic fields emitted by a mobile phone influence visual evoked potentials? *Central Eur. J. Pub. Health* 6:288–290.

Utteridge T.D., Gebski V., Finnie J.W., Vernon-Roberts B., Kuchel T.R., 2002. Long-term exposure of Eμ-pim1 transgenic mice to 898.4 MHz microwaves does not increase lymphoma incidence, *Radiat. Res.* 158:357–364.

Vacek D.R.A., 1978. Effect of high-frequency electromagnetic field upon haemopoietic stem cells in mice, *Folia Bio. (Praha)* 18:292–297.

Van Ummersen C.A., 1961. The effect of 2450 mc radiation on the development of the chick embryo, in: *Biological Effects of Microwave Radiation*, Vol. 1, Peyton M.F., Ed., Plenum Press, New York, NY, pp. 201–219.

Varma M.M., Traboulay E.A., 1976. Evaluation of dominant lethal test and DNA studies in measuring mutagenicity caused by nonionizing radiation, in: *Biological Effects of Electromagnetic Waves*, Vol. 1, Publ. (FDA) 77-8010, Johnson C.C., Shore, M.L., Eds., U.S. Department of Health Education, and Welfare, Rockville, MD, pp. 386–396.

Varma M.M., Dage E.L., Joshi S.R., 1976. Mutagenicity induced by nonionizing radiation in Swiss male mice, in: *Biological Effects of Electromagnetic Waves*, Vol. 1, Johnson C.C., Shore, M.L., Eds., Publ. (FDA) 77-8010, U.S. Department of Health, Education, and Welfare, Rockville, MD, pp. 397–405.

Vendrik A.J., Vos J.J., 1958. Comparison of the stimulation of the warmth sense organ by microwave and infrared, *Int. Appl. Physiol.* 13:435–444.

Vetter R.J., 1975. Neuroendocrine response to microwave irradiation, *Proc. Natl. Electron. Conf.* 30:237–238.

Vijayalaxmi, Frei M.R., Dusch S.J., Guel V., Meltz M.L., Jauchem J.R., 1997a. Frequency of micronuclei in the peripheral blood and bone marrow of cancer-prone mice chronically exposed to 2450 MHz radiofrequency radiation, *Radiat. Res.* 147:495–500.

Vijayalaxmi, Mohan N., Meltz M.L., Wittler M.A., 1997b. Proliferation and cytogenetic studies in human blood lymphocytes exposed *in vitro* to 2450-MHz radiofrequency radiation, *Int. J. Radiat. Biol.* 72:751–757.

Vijayalaxmi, Frei M.R., Dusch S.J., Guel V., Meltz M.L., Jauchem J.R., 1998. Correction of an error in calculation in the article 'Frequency of micronuclei in the peripheral blood and bone marrow of cancer-prone mice chronically exposed to 2450 MHz radiofrequency radiation,' *Radiat. Res.* 149:308–308.

Vijayalaxmi, Leal B.Z., Szilagyi M., Prihoda T.J., Meltz M.L., 2000. Primary DNA damage in human blood lymphocytes exposed *in vitro* to 2450 MHz radiofrequency radiation, *Radiat. Res.* 153: 479–486.

Vijayalaxmi, Bisht K.S., Pickard W.F., Meltz M.L., Roti Roti J.L., Moros E.G., 2001a. Chromosome damage and micronucleus formation in human blood lymphocytes exposed *in vitro* to radiofrequency radiation at a cellular telephone frequency (847.74 MHz, CDMA), *Radiat. Res.* 156:430–432.

Vijayalaxmi, Pickard W.F., Bisht K.S., Leal B.Z., Meltz M.L., Roti Roti J.L., Straube W.L., Moros E.G., 2001b. Cytogenetic studies in human blood lymphocytes exposed *in vitro* to radiofrequency radiation at a cellular telephone frequency (835.62 MHz, FDMA), *Radiat. Res.* 155:113–121.

Vijayalaxmi, Sasser L.B., Morris J.E., Wilson B.W., Anderson L.E., 2003. Genotoxic potential of 1.6 GHz wireless communication signal: *in vivo* two-year bioassay, *Radiat. Res.* 159:558–564.

Vollrath L., Spessert R., Kratzsch T., Keiner M., Hollmann H., 1997. No short-term effects of high-frequency electromagnetic fields on the mammalian pineal gland, *Bioelectromagnetics* 18:376 387.

von Klitzing I., 1995. Low-frequency pulsed electromagnetic fields influence EEG of man, *Phys. Med.* 11:77–80.

Vorobyov V.V., Galchenko A.A., Kukushkin N.I., Akoev I.G., 1997. Effects of weak microwave fields amplitude modulated at ELF on EEG of symmetric brain areas in rats, *Bioelectromagnetics* 18:293–298.

Wachtel H., Seaman R., Jolnes W., 1975. Effects of low-intensity microwaves on isolated neurons, in: *Biological Effects of Nonionizing Radiation*, Tyler P.W., Ed., pp. 46–62, *Ann. N.Y. Acad. Sci.*, 247.

Wachtel H., Adey G.R., Chalker R., Barnes F., 1982. Temperature Rise Rate as a Causal Factor of Rapid Neural Responses to Microwave Absorption, Fourth Annual Scientific Session, Bioelectromagnetics Society, Los Angeles, CA, p. 34.

Wagner P., Röschke J., Mann K., Hiller W., Frank C., 1998. Human sleep under the influence of pulsed radiofrequency electromagnetic fields: a polysomnographic study using standardized conditions, *Bioelectromagnetics* 19:199–202.

Wagner P., Roschke J., Mann K., Fell J., Hiller W., Frank C., Grozinger M., 2000. Human sleep EEG under the influence of pulsed radio frequency electromagnetic fields. Results from polysomnographies using submaximal high power flux densities, *Neuropsychobiology* 42:207–212.

Walters T.J., Blick D.W., Johnson L.R., Adair E.R., Foster K.R., 2000. Heating and pain sensation produced in human skin by millimeter waves: comparison to a simple thermal model, *Health Phys.* 78:259–267.

Wang B., Lai H., 2000. Acute exposure to pulsed 2450 MHz microwaves affects water-maze performance of rats, *Bioelectromagnetics* 21:52–56.

Wangemann R.T., Cleary S.F., 1976. The *in vivo* effects of 2.45 GHz microwave radiation on rabbit serum components and sleeping times, *Radiat. Environ. Biophys.* 13:89–103.

Ward T.R., Ali J.S., 1985. Blood–brain barrier permeation in the rat during exposure to low-power 1.7-GHz microwave radiation, *Bioelectromagnetics* 6:131–143.

Ward T.R., Allis J.W., Elder J.A., 1975. Measure of enzymatic activity coincident with 2450 MHz microwave exposure, *J. Microwave Power* 10:315–320.

Ward T.R., Elder J.A., Long M.D., Svendsgaard D., 1982. Measurement of blood–brain barrier permeation in rats during exposure to 2450-MHz microwaves, *Bioelectromagnetics* 3:371–383.

Watanabe Y., Tanaka T., Taki M., Watanabe S., 2000. FDTD analysis of microwave hearing effect, *IEEE Trans. Microwave Theory Tech.* 49:2126–2132.

Way W.I., Kritikos H., Schwan H., 1981. Thermoregulatory physiologic responses in the human body exposed to microwave radiation, *Bioelectromagnetics* 2:341–356.

Weaver J.C., 1995. Electroporation in cells and tissues: a biophysical phenomenon due to electromagnetic fields, *Radio Sci.* 30:205–221.

Webb S.J., Booth A.D., 1969. Absorption of microwaves by microorganisms, *Nature* 222:1199–1200.

Webb S.J., Booth A.D., 1971. Microwave absorption by normal and tumor cells, *Science* 174:72.

Webber M.M., Barnes F.S., Seltzer L.A., Bouldin T.R., Prasad K.N., 1980. Short microwave pulses cause ultrastructual membrane damage in neuroblastoma cells, *J. Ultrastruct. Res.* 71:321–330.

Weiter J.J., Finch E.D., Schultz W., Frattali V., 1975. Ascorbic acid changes in cultured rabbit lenses after microwave irradiation, in: *Biological Effects of Nonionizing Radiation*, Tyler P.W., Ed., pp. 175–181, *Ann. N.Y. Acad. Sci.*, 247.

Weyandt T.B., Schrader S.M., Turner T.W., Simon S.D., 1996. Semen analysis of military personnel associated with military duty assignments, *Reprod. Toxicol.* 10:521–528.

Wiktor-Jedrzejczak W., Ahmed A., Czerski P., Leach W.M., Sell K.W., 1977a. Immune response of mice to 2450-MHz microwave radiation: overview of immunology and empirical studies of lymphoid splenic cells, *Radio Sci.* 12:209–219.

Wiktor-Jedrzejczak W., Ahmed A., Czerski P., Leach W.M., Sell K.W., 1977b. Increase in the frequency of fc receptor fcr bearing cells in the mouse spleen following a single exposure of mice to 2450 MHz microwaves, *Biomedicine* 27:250–252.

Wiktor-Jedrzejczak W., Ahmed A., Sell K.W., Czerski P., Leach W.M., 1977c. Microwaves induce an increase in the frequency of complement receptor-bearing lymphoid spleen cells in mice, *J. Immun.* 118:1499–1502.

Williams R.J., Finch E.D., 1974. Examination of the cornea following exposure to microwave radiation, *Aerosp. Med.* 45:393–396.

Williams R.J., McKee A., Finch E.D., 1975. Ultrastructural changes in the rabbit lens induced by microwave radiation, in: *Biological Effects of Nonionizing Radiation*, Tyler P.W., Ed., pp. 166–174, *Ann. N.Y. Acad. Sci.*, 247.

Williams W.M., del Cerro M., Michaelson S.M., 1984a. Effect of 2450 MHz microwave energy on the blood–brain barrier to hydrophilic molecules. B. Effect on the permeability to HRP, *Brain Res. Rev.* 7:171–181.

Williams W.M., Lu S.T., del Cerro M., Michaelson S.M., 1984b. Effect of 2450 MHz microwave energy on the blood–brain barrier to hydrophilic molecules. D. Brain temperature and blood–brain barrier permeability to hydrophilic tracers, *Brain Res. Rev.* 7:191–212.

Williams W.M., Hoss W., Formaniak M., Michaelson S.M., 1984c. Effect of 2450 MHz microwave energy on the blood–brain barrier to hydrophilic molecules. A. Effect on the permeability to sodium fluorescein, *Brain Res. Rev.* 7:165–170.

Williams W.M., Platner J., Michaelson S.M., 1984d. Effect of 2450 MHz microwave energy on the blood–brain barrier to hydrophilic molecules. C. Effect on the permeability to [C-14] sucrose, *Brain Res. Rev.* 7:183–190.

Williams W.M., Lu S.-T., Del Cerro M., Hoss W., Michaelson S.M., 1984e. Effects of 2450 MHz microwave energy on the blood–brain barrier: an overview and critique of past and present research, *IEEE Trans. Microwave Theory Tech.* 32:808–818.

Wilson G.J., 1959. Experimental studies on congenital malformations, *J. Chron. Dis.* 10:111–130.

Wilson B.S., Zook J.M., Joines W.T., Casseday J.H., 1980. Alterations in activity at auditory nuclei of the rat induced by exposure to microwave radiation: autoradiographic evidence using [C-14] 2-deoxy-D-glucose, *Brain Res.* 187:291–306.

Yakovleva M.I., Shlyafer I.P., Tsvetkova I.P., 1968. On the question of condition cardiac reflexes, the functional and morphological state of cortical neurons under the effect of superhigh-frequency electromagnetic fields, *Zh. Vyssh. Nervn. Deyat.* (*USSR*) 18:973.

Yang H.K., Cain C.A., Lockwood J., Tompkins W.A., 1983. Effects of microwave exposure on the hamster immune system. I. Natural killer cell activity, *Bioelectromagnetics* 4:123–139.

Yao K.T.S., 1976. Cytogenetic consequences of microwave incubation of mammalian cells in culture, *Genetics Abstr.* 83 (Suppl.):584.

Yao K.T.S., 1978. Microwave radiation-induced chromosomal aberrations in corneal epithelium of Chinese hamsters, *J. Hered.* 69:409–412.

Yao K.T.S., 1982. Cytogenetic consequences of microwave irradiation on mammalian cells incubated *in vitro, J. Hered.* 73:133–138.

Yermakov Y.V., 1969. On the mechanism of developing astheno-vegetative disturbance under the chronic effect of a SHF-field, *Voyenno-Medit. Zh.* (*USSR*) 3:42.

Zafra C., Pena J., de la Fuente M., 1988. Effect of microwaves on the activity of murine macrophages *in vitro, Int. Arch. Allergy Appl. Immunol.* 85:478–482.

Zook B.C., Simmens S.J., 2001. The effects of 860 MHz radiofrequency radiation on the induction or promotion of brain tumors and other neoplasms in rats, *Radiat. Res.* 155:572–583.

4

*Behavioral and Cognitive Effects of Electromagnetic Field Exposures**

Sheila A. Johnston and John A. D'Andrea

CONTENTS

*Disclaimer: The views expressed in this chapter are those of the authors and do not reflect the official policy or position of the Department of the Navy, Department of Defense, or the U.S. Government. The trade names of materials and products or nongovernmental organizations are cited as necessary for precision. These citations do not constitute official endorsement or approval of the use of such commercial materials or products.

4.1 Introduction

The human brain is comprised of more than 100 billion individual nerve cells that are functionally interconnected into systems. These systems construct our perceptions and cognitions of the world, and control how we react to the external world (Kandel et al., 2000). This chapter presents an overview of the interaction of nonionizing electromagnetic fields (EMFs) as external stimuli to the nervous systems and the behavior of humans and laboratory animals. This overview of the scientific literature specifically includes the detection of EMFs and effects of EMFs on behavioral performance and cognition. EMF detection by humans and responses of laboratory animals suggests that detection could motivate changes in behaviors. The chapter is divided into two parts, which are distinguished simply by frequency: microwave (MW) radio frequency and extremely low-frequency (ELF). This chapter extends the findings of recent reviews of the animal and human research literature (Hermann and Hossmann, 1997; D'Andrea, 1999; Cook et al., 2002; D'Andrea et al., 2003a,b; Hossmann and Hermann, 2003).

Biological effects of MWs occur due to absorption of energy in the body. Microwave absorption by water molecules has a resonant frequency in the region of 2.45 GHz that causes heating by molecular vibration and is the result of the interaction of MWs and tissue of the body. Water is the molecular component of tissue known to have a resonant frequency that could contribute to the absorption of MW energy. Heating (absolute temperature rise) is the best understood mechanism for the effective transfer of MW energy into tissue. Other tentative mechanisms for the transfer of energy from MWs into human tissue are not established as potential sources for health effects or the modification of human behavior.

The accepted unit of measurement is specific absorption rate (SAR). The IEEE (Std C95.1-2005) defines "specific absorption rate (SAR) as the time derivative of the incremental energy absorbed by (dissipated in) an incremental mass contained in a volume element of given density." The ICNIRP (1998) defines "specific energy absorption rate (SAR) as the rate at which energy is absorbed in body tissues, in watt per kilogram (W/kg). SAR is the dosimetric measure that has been widely adopted at frequencies above about 100 kHz." Absolute rise in temperature in the tissues of the body is also a measure of MW absorption of energy.

Biological effects of ELF are known to be due to electrostimulation which is defined as "the induction of a propagating action potential in excitable tissue by an applied electrical stimulus; electrical polarization of pre-synaptic processes leading to a change in post synaptic cell activity" (Reilly, 1998; IEEE, Std C95.6-2002). The unit of measurement is slightly different for IEEE and ICNIRP. For the IEEE, the unit is the induced *in situ* electric field stimulation of the nervous system (volts per meter: V/m) (IEEE, Std C95.6-2002) and for the ICNIRP the unit is the induced current density (ampere per square meter: A/m^2) that results in stimulation of the nervous system (ICNIRP, 1998).

The body generates thermal noise and also electrical noise and against that background the input of low power MW stimulation may be indistinguishable from the endogenous biological thermal noise while core temperature remains stable (Adair, 2003) and the electrical power stimulation may be indistinguishable from the endogenous electrical noise below the threshold for the generation of action potentials of nerves (Reilly, 1998). Establishing scientific evidence of the time course and dose effects for MW-induced temperature rise and ELF-induced electrical stimulation of nerves are essential. Reviewed in this chapter, these are reflected in the animal and human gradients from biological sensory detection of MWs and ELF exposures up to the level of adverse health effects on behavioral and cognitive responses.

EMF may affect a biological system, causing a biological effect, without necessarily causing an adverse change in health. We feel it is very important and relevant in this chapter to see the full picture that is the gradient of established biological effects up to adverse effects following the international criteria for scientific evidence. Several papers, chapters of books, and Web sites are available that list the research criteria required for evaluation of the scientific literature (Repacholi, 1998; Erdreich and Klauenberg, 2001; Gajšek et al., 2002; ICNIRP, 2003; Johnston, 2003, 2004/2005; Kheifets et al., 2003). An essential part of the full research picture is establishing scientific evidence of the thresholds for MW* and ELF† adverse effects on behavior and cognition for setting international standards and guidelines to protect our health. The scientific evidence exists in its own right, independent of the standards, and reciprocally the standards review bodies review all the scientific evidence, not just adverse effects.

Confirmed and replicated scientific results of MW and ELF exposure are essential to establish scientific evidence. The results of single studies, even when they are of good quality, are not considered as scientific evidence on their own because single results may fail to be reproduced. Only when the results are repeated and confirmed or replicated independently or when the results from well-conducted studies are consistent with the weight of evidence from other good quality research data, are they accepted as scientific evidence.

Within the scientific literature, it is important to distinguish between replication and confirmation studies because they may have different implications for results that fail to be reproduced. Often published studies have incomplete or poor protocols to test the hypothesis and thus only confirmation studies could be considered to investigate the same hypothesis. In this instance, the lack of confirmation could be due to different research designs testing different underlying mechanisms. The failure to confirm single results in this instance would be inconclusive and difficult to resolve. This is especially true when the underlying mechanism for a reported single effect is unknown. But, it is feasible that confirmation studies with slightly different protocols could test the same known underlying mechanism. One such example is different maze protocols that are known to test spatial memory (Morris et al., 1982; Gray and McNaughton, 1983; Kandel et al., 2000; Maguire et al., 2000, 2003a,b; Burgess et al., 2002; see Section 4.2.4). Such studies that test the same hypothesis with slightly different protocols but report the same result are considered as confirmation studies. Single experiments with well-defined protocols, when replicated independently, suggest both experiments are testing the same underlying mechanism. In this instance, if replication fails then the single result is rejected. Well-conducted and published confirmation and replication studies that produce the same result contribute to the weight of scientific evidence. These central concepts will be referred to in this chapter.

This chapter explores the failure of U.S. scientists to replicate some of the MW exposure effects on the nervous system, the blood–brain barrier (BBB), and cognition as reported by

---

*Established adverse MW effects are associated with whole body heating at levels that usually increase temperature by approximately 1°C or more (IEEE Std C95.1-2005). "Available experimental evidence indicates that the exposure of resting humans for approximately 30 minutes to MWs producing a whole-body SAR of between 1 and 4 W/kg results in a body temperature increase of less than 1°C". [ICNIRP, 1998]. Animal data indicate a threshold for behavioral responses in the same SAR range [ICNIRP, 1998; D'Andrea et al., 2003b; IEEE Std C95.1-2005]. These data form the basis for an occupational exposure restriction of 0.4 W/kg, which provides a 10× margin of safety [ICNIRP, 1998; IEEE Std C95.1-1999]. The public standard has an additional 5x margin of safety (0.08 W/kg) for both the IEEE Std C95.1-2005 and ICNIRP guidelines [1998].

†Harmonisation of the ELF basic restrictions between IEEE and ICNIRP is in progress [Reilly, 2005; ICNIRP, 2003]. For public exposure at 60 Hz for the brain the IEEE exposure limit is set at 0.018 V/m and for ICNIRP at 0.01 V/m and for other tissue for IEEE at 0.70 V/m and ICNIRP at 0.01 V/m. The ICNIRP current density was converted to E field, based on $\sigma = 0.2$ S/m [Reilly, 2005].

Russian scientists. Within the area of behavioral and cognitive effects investigated are changes in attention, simple and complex memory tasks, sleep and awake electroencephalogram (EEG), and changes in the BBB (Kandel et al., 2000). This overview explores the past research and extends the findings of recent reviews of the animal and human behavioral and cognitive research literature (Hermann and Hossmann, 1997; D'Andrea, 1999; Cook et al., 2002; D'Andrea et al., 2003a,b; Hossmann and Hermann, 2003) to newly published papers in the MW and ELF fields. One question considered is whether cognitive performance disruption might be more "sensitive," that is have a lower threshold for disruption by MW and ELF exposure, than performance disruption of simple tasks.

The ICNIRP (1998) and IEEE (Std C95.1-1991/1999, Std C95.1-2005) safety standards that limit human exposure to MWs differ tremendously from those of the former Soviet Union (FSU), which are based on Soviet study results on fixed durations of exposure (Hossmann and Hermann, 2003). Since the 1970s, the maximum permissible exposure in the Soviet block countries has been reported to be 1000 times lower than most other countries. Due to these large differences in MW safety standards, a program was established in 1975 for the collaborative study of the biological effects of low-level MW exposures. A leading area of study was the effect of nonionizing radiation on the central nervous system (CNS) and behavior (Shandala et al., 1979). The presumption was that microwave fields might alter ongoing electrical activity in the brain by induction of fields and currents that might alter the resting membrane potential of neurons, synapses, and sensory organs. However, Chou and Guy (1978) failed to replicate the result of Kamenskii (1964) of a MW effect on the isolated nerves exposed in air and parallel to the E-field. When the nerve was kept in a temperature-controlled waveguide filled with physiological solution, no effect on action potentials was observed other than those of thermal origin, even if the nerve was exposed to very high pulsed power levels with peak SAR up to 220 kW/kg (Chou and Guy, 1978). Support for thermal effects on action potentials reported by Chou and Guy (1978) comes from electrophysiological recordings by Andersen and Moser (1995), who reported that "various bioelectrical signals are more sensitive during warming, axonal conduction is speeded up and stimulus elicited transmitter release becomes faster and more synchronized." The effects in the original study (Kamenskii, 1964) appeared to be due to the particular *in vitro* exposure condition. When the proper temperature control was applied, the effects disappeared.

However, at that time most research efforts were focused on electroencephalography, evoked responses, and behavioral changes. D'Andrea et al. (2003b) reviewed the details of many of these studies recently. After initial experiments in the United States and FSU failed to mutually confirm low-level effects, an agreement was made to use similar experimental designs and research protocols (McRee et al., 1979). The results of the joint project, which lasted only a few years, found some alteration in spectral power of the EEG, but these results were not the same between the experiments in both countries (Mitchell et al., 1989). The collaborative project ended without mutual confirmation by both research groups of low-level, microwave-induced biological effects. This outcome illustrates the paramount importance of replication of experimental studies. Replication is important because some estimate of natural variation is needed in order to know if any observed effect can be attributed to the experimental treatment or due to just random change. Any experiment should have sufficient statistical power to detect an effect that is biologically meaningful.

An example of low-level exposure studies that have often yielded positive results, but were not confirmed in replication attempts, is a study by D'Andrea et al. (1986b), which was designed to test previously reported low-level behavioral effects. They exposed rats intermittently to 0.5 mW/cm^2 2450 MHz MWs for 90 d and reported changes in time-related lever-pressing behavior. However, a replication experiment by DeWitt et al. (1987) reported different behavioral changes and failed to replicate the lever-pressing findings reported in the D'Andrea et al. (1986b) study. Interestingly, both of these experiments did not replicate the earlier behavioral changes following low-level MW exposures reported by Rudnev et al. (1983) and Shandala et al. (1979). D'Andrea et al. (2003b) concluded that these experiments were all below the threshold for reliable effects and could not be considered as established scientific evidence. However D'Andrea et al. (1986a) at 2.5 mW/cm^2 did show effects at a higher SAR although this study was never replicated. Other low-level MW exposure studies were reviewed in detail by D'Andrea et al. (2003b). Because these Russian low-level chronic exposure studies remain controversial and are not established as international scientific evidence, they have not served a role in health risk assessment and in safety standard setting for the international standard setting bodies.

The WHO EMF Project (1996; see also Veyret and Lagroye personal communication) in collaboration with ICNIRP and the IEEE have renewed efforts to evaluate the Russian literature and the Russian scientific approach for setting safety standards used in the FSU and allied European countries. As a result of these meetings, international scientists have translated and identified Russian BBB research of the Vinogradov group for replication. Because breaches of the BBB could lead to brain cell death and cognitive impairment, BBB studies are reviewed later in this chapter.

4.2 Microwave Effects

4.2.1 Detection of Microwaves

Humans and laboratory animals have shown the capability to detect fairly low levels of MW exposure. There are two well-established thermal mechanisms that may convey detection of MWs by mammals. One is the auditory hearing effect. MW-induced auditory sensations are thought to be thermal events based on very small areas of heat-related tissue expansion created on the surface of the brain by pulsed MWs (Foster and Finch, 1974). The thermoelastic expansion, created by each MW pulse propagates as a small acoustic wave along the surface of the brain, couples to the skull and is thought to be conducted to the cochlea of the inner ear as sound, transmitted to the auditory cortex and heard as a click (Chou et al., 1982; Elder and Chou, 2003). At first, the reports that persons standing next to a radar antenna could hear the radar pulses drew skepticism (Airborne Instruments Laboratory, 1956). However, the first scientific study by Frey (1961) confirmed the MW-hearing effect when experimental subjects reported hearing "clicks" or "buzzing" during delivery of the microwave pulses. In the case of microwave hearing, biophysical mechanisms (Foster and Finch, 1974) have been confirmed to predict the experimental data (Airborne Instruments Laboratory, 1956; Frey, 1961; Chou et al., 1982; Elder and Chou, 2003) (see also Chapter 10 in *BBA*).

The second well-established thermal mechanism is the heating of tissues by MW exposure that can be detected by thermal receptors in the skin and elsewhere in the body and CNS. The identification of a family of transient receptor potential (TRP) ion channels that are gated by specific temperatures has been an important advance in the elucidation of the molecular mechanisms of thermosensitivity. Research has revealed a family of TRP proteins that sense heat and cold at the cellular level (Patapoutian et al., 2003). These proteins show sensitivity in the range of degrees. A warmth-sensitive

(~34°C) TRP channel is concentrated in skin cells, suggesting that the skin itself senses heat and passes the message to neurons. There is also literature to suggest that mammals have neurons sensitive to temperature changes of fractions of a degree in the brain stem and spinal cord along the walls of the 3–4th ventricles and spinal cord canal (Hensel, 1981). These CNS neurons may be implicated in the autonomic rapid response sweating to MW exposure (100 and 220 MHz) SAR above the guideline limits (Adair et al., 2003, 2005; Allen et al., 2003, 2005; see also Chapter 5 in this volume) when no whole body temperature changes are evident. Further research in this area is ongoing at Brooks Air Force Base. Molecular mechanisms of action associated with MW stimulation of sweating need to be investigated further (Patapoutian et al., 2003).

A variety of studies have shown that humans detect warming by microwave heating at fairly low levels. Adair and Black (2003) described several archival studies that have documented thresholds for detection of microwave warming (Vendrik and Vos, 1958; Hendler and Hardy, 1960; Hendler et al., 1963; Hendler, 1968). The duration of microwave heating in all of these studies was 10 s or less and involved exposure of an area of the forehead or forearm. For example when a small area (37 cm^2) of forehead was exposed for 4 s, the mean absolute threshold of warmth varied across frequency. At 3 GHz this threshold was 33.5 mW/cm^2, at 10 GHz it was 12.6 mW/cm^2, and at frequencies for far infrared (IR) it was 4.2 mW/cm^2.

Justesen et al. (1982) studied the detection on the forearm of IR and 2450 MHz. The authors used a 15-cm diameter aperture in a wall of microwave-absorbent material to limit the exposed area of skin. Using a double-staircase psychophysical method, the subjects were given 10-s exposures. Thresholds of 26.7 and 1.7 mW/cm^2 were documented for 2450 MHz and IR exposures, respectively. The authors noted that a high correlation within subjects between thresholds for detection of just-noticeable warming by IR and MW irradiation ($r = 0.97$) allowed them to conclude that the same set of superficial thermoreceptors was stimulated.

A recent study by Blick et al. (1997) measured detection thresholds on the skin in the middle of the back of human subjects. They used long duration (10 s), large area (327 cm^2) stimuli to minimize temporal or spatial summation. Frequencies of 2.45, 7.5, 10, 35, and 94 GHz as well as IR were tested. From 2.45 to 94 GHz, the detection thresholds decreased more than an order of magnitude. The 2450 MHz threshold (63.1 mW/cm^2) and the IR detection threshold (5.34 mW/cm^2) were roughly 2.5 times the thresholds found by Justesen et al. (1982). However, as Adair and Black (2003) pointed out, the total surface area of the exposures and duration of exposure as well as location of stimulation are important and easily account for the differences noted here.

Animal studies have also determined thresholds for microwave detection. A unique study by King et al. (1971) studied the detection of MWs by rats. They used the technique of conditioned suppression to determine the threshold of detection of 2450 MHz MWs by rats exposed in a multimode cavity. The rats were given dose rates of 0.5 to 6.4 W/kg. The authors described how several sources of artifactual cueing were controlled in this study and that MWs were reliably detected at dose rates as low as 0.6 W/kg. They pointed out that MWs lack the saliency of an auditory stimulus but were, nevertheless, readily detectable.

Ziriax et al. (1996) determined detection thresholds for 94 GHz MWs on the back of a nonhuman primate model (*Macaca mulatta*) and found a threshold of 7.7 mW/cm^2. Likewise, D'Andrea et al. (1999) determined threshold detection for exposure of the face of rhesus monkeys at 94 GHz to be 8 mW/cm^2. These are slightly higher than the power densities detected on the back of humans at 94 GHz (4.5 mW/cm^2) by Blick et al. (1997) but as stated above procedural differences easily account for the discrepancies.

4.2.2 Microwave Performance Disruption

The studies of warmth detection by humans and responses of laboratory animals to similar exposures provide confidence that behavioral changes in animal studies are indeed thermally motivated. Thresholds to disrupt simple operant behaviors during acute whole body exposure were determined to be near ~4 W/kg (de Lorge, 1976, 1979, 1983; D'Andrea et al. 1977; de Lorge and Ezell, 1980; see D'Andrea et al., 2003b). Several studies have shown that the disruption of ongoing behavior during acute MW exposure is generally associated with 1°C or greater increase of body temperature. This is a statistically reliable measure that is associated with whole body SARs in a narrow range between 3.2 and 8.4 W/kg, across differences in carrier frequency (225 MHz–5.8 GHz), species (rodents to rhesus monkeys), and exposure parameters (near and far field, CW and pulse-modulated). These changes in behavior are highly associated with the thermal increase produced by the MW exposures and have withstood the test of replication. Several of the studies cited above form an important part of the core database of scientific evidence from which safe whole body exposure standards have been derived (see IEEE, Std C95.1-2005). To date, however, none of the published MW research on performance disruption has been conducted on humans. To truly understand the nature of a MW hazard equivalent, behavioral experiments should be conducted on human volunteers. Such experiments would confirm or change the threshold hazard level of 4 W/kg currently used in safety standards.

4.2.3 Microwave Cognitive Effects: Animal Studies

4.2.3.1 Microwave Cognitive Performance Disruption in Animals

Nelson (1978) determined the effects of pulsed MWs on the learning of new behaviors by male squirrel monkeys in a repeated acquisition task. The estimated threshold whole body SAR range was 3.2–3.6 W/kg. The results of this experiment were compatible with the hypothesis that behavioral changes were directly related to hyperthermia in the monkey. D'Andrea et al. (2003b) operationally defined performance disruption as a significant rate change in behavior from a well-defined baseline. Work stoppage is simply the point at which the animal ceases performing the trained behavior for a predetermined time period. The performance disruption and work stoppage paradigms, described above, have provided a threshold (4 W/kg) for potential hazards of MW radiation absorption. However, cognitive components have been employed in many of the nonhuman primate studies. For example, the behavioral tasks used by de Lorge (1976) and D'Andrea et al. (1994) are vigilance tasks and involve attention processes by monkeys measured by simple reaction time or choice reaction time to either auditory or visual stimuli. In most cases reaction time was altered by MW exposures, which produced a whole body SAR of ~4 W/kg and above.

D'Andrea (1999) postulated that performance of cognitively mediated tasks might be disrupted at levels of exposure lower than that required to cause performance disruption: "Unlike disruption of performance of a simple task, a disruption of cognitive functions could lead to profound errors in judgment due to alteration of perception, disruption of memory processes, attention, and/or learning ability, resulting in modified but not totally disrupted behavior. Any of these hypothesized deficits can compromise personnel operating in critical industrial or military occupations. Such deficits, which may be subclinical, can be amplified depending on the requirements of the occupation."

However, a "weight of evidence" approach indicates that the animal studies of cognitive performance disruption that have been conducted, to date, have not confirmed the implied concept of more "sensitive" to MW exposure than performance disruption of a simple task (see D'Andrea, 1999). For example, Mickley et al. (1994) and Mickley and Cobb (1998) investigated memory deficits in rats exposed to MWs and determined that the threshold for memory effects was 10 W/kg. Similarly, Luttges (1980) evaluated the effects of MW exposure on learning and memory in mice and found an enhancement of performance (estimated whole body SAR 13 W/kg). The MW memory facilitation was found in both automated active avoidance testing and in single trial, passive avoidance tests. Beel (1983) repeated the study by Luttges (1980) and again found significant enhancement of learning and memory following 15 min exposure, both with five consecutive days of multiple trial, active avoidance training and with single trial, passive avoidance training.

Raslear et al. (1993) exposed rats to high peak power pulsed MWs produced by the TEMPO (Transformer Energized Megavolt Pulsed Output) virtual cathode oscillator at a whole body average SAR of 0.072 W/kg. The authors concluded that significant effects on cognitive function in rats were observed, particularly in the decision-making process. But in a subsequent experiment, Raslear and colleagues reported that the microwave pulse inhibition and enhancement of startle were similar to previously reported effects of sensory stimuli delivered at similar lead times, indicating the possibility that action was mediated by sensory stimulation (Seaman et al., 1994).

From the descriptions of the studies described above, an overall conclusion can be made that measures of cognitive-based performance disruptions in animals are not more sensitive to MW exposure (see D'Andrea, 1999) than simple performance tasks. However, the different cognitive and simple performance tasks, different exposure systems, modulation parameters, and differences in irradiation frequency make a more detailed conclusion difficult. This is true also because of the different test animals and exposure durations that make easy interpretation of this sparse literature difficult. However, there is support for this conclusion from basic neuroscience. The results are in line with the demonstrated conservation of the memory mechanisms across simple tasks and more complex tasks such as spatial memory (Kandel et al., 2000, p. 1265). This means that neurophysiologically simple and complex memory mechanisms are similar, and as such we would expect them to be similarly disrupted by the same rises in temperature induced by similar levels of SAR.

4.2.4 Animals: Spatial Memory Replications and Confirmations

4.2.4.1 Introduction

When considering the effects of MW exposure on spatial memory experiments in rats, it is relevant and essential to understand the established scientific evidence in the spatial memory field over the past three decades (e.g., O'Keefe and Dostrovsky, 1971; O'Keefe and Nadel, 1978; Morris et al., 1982; Andersen and Moser, 1995; Kandel et al., 2000; Bast et al., 2005; O'Keefe and Burgess, 2005). The established scientific evidence in the spatial memory field includes the homology of spatial memory in the hippocampus of rats and humans, a common neural hippocampal component for spatial memory in various maze paradigms and activation of the same molecular mechanism for establishing spatial memory. There is a homology between the use of the hippocampus for spatial memory in rats in a maze and humans performing similar navigation tasks (Maguire et al., 2000, 2003a; Burgess et al., 2002). The importance of this homology is far-reaching and suggests a conservation of spatial memory across species (Kandel et al., 2000; Burgess et al., 2002). Thus there is strong support for the extrapolation of MW spatial memory results from animals to humans.

In animal spatial memory experiments, there are many versions of the maze experimental paradigm. Although the spatial paradigms and apparatus such as the T-maze,

TABLE 4.1
Effects of MW Exposure on Spatial Memory Performance in Animals

EMF Effect	Species	SAR (W/kg)	Frequency (MHz)	Modulation	Intensity (mW/cm^2)	Duration	Ref.
No overall effect for 20-min exposure but more errors each day in a 12-arm radial maze at 45-min exposure	SD Rats	Avg. whole body 0.6 W/kg	2450 MHz	2 μs pulses, 500 pps	1 mW/cm^2 but very high peak power	20 min/d and 45 min/d exposure for 10 d	Lai et al. (1989)
Deficit in spatial working memory function reversed by pretreatment with physostigmine or naltrexone	SD Rats	Avg. whole body 0.6 W/kg	2450 MHz	2 μs pulses, 500 pps	1 mW/cm^2 but very high peak power	45-min exposure	Lai et al. (1994)
These results show that acute exposure to pulsed MWs caused a deficit in the spatial "reference" memory in the rat	SD Rats	Avg. whole body 1.2 W/kg	2450 MHz	Pulsed width 2 μs, 500 pps	Avg. power density 2 mW/cm^2	For 1 h in a circular waveguide system	Wang and Lai (2000)
Low-level exposure to pulsed 900 MHz MW radiation does not cause deficits in the performance of a spatial learning task in mice	Male mice C57BL/6J	Whole body 0.05 W/kg	900 MHz	Pulsed at 217 Hz		45 min/d for 10 d	Sienkiewicz et al. (2000)
No deficits in spatial learning after "head-only" exposure of rats to GSM electromagnetic fields in the two spatial learning tasks	SD Rats	Head only 1 or 3.5 W/kg	900 MHz EMF	GSM modulation at 217 Hz, 1/8 duty factor		45 min over 10–14 d	Dubreuil et al. (2002)
Head-only exposure to GSM 900 MHz electromagnetic fields does not alter rat's memory in spatial and nonspatial tasks	SD Rats	Head only 1 and 3.5 W/kg	900 MHz EMF	GSM modulation at 217 Hz, 1/8 duty factor		45 min	Dubreuil et al. (2003a)
No significant effects on memory performance were found due to RF, avg. SAR within the brain was 7.4 W/kg and the whole body average SAR was 1.4 W/kg	SD Rats	1: Brain (B) 7.4 W/kg and whole body (WB) 1.4 W/kg, 2: Brain 25 W/kg and WB 4.5 W/kg	1439 MHz	PDC (Japan)		I: 1 h/day for 4 d; II: 45 min/day for 4 d	Yamaguchi et al. (2003)

continued

TABLE 4.1 (continued)

Effects of MW Exposure on Spatial Memory Performance in Animals

EMF Effect	Species	SAR (W/kg)	Frequency (MHz)	Modulation	Intensity (mW/cm^2)	Duration	Ref.
The results included no significant exposure effect, no significant drug effect, and no significant interactions between those two factors	SD Rats	Avg. whole body 0.6 W/kg	2450 MHz	Pulsed circular polarized 2 μs pulses, 500 pps	1 mW/cm^2 but very high peak power	45 min/d for 10 d	Cobb et al. (2004) (Lai et al., 1994 replication)
Response of MWs and temporally incoherent 60 mG on spatial learning in a water maze was similar to that of the sham-exposed animals	SD Rats	Avg. whole body 1.2 W/kg	2450 MHz	CW circular polarized	2 mW/cm^2	1 h CW; 1 h CW + 60 mG	Lai (2004)
Whole body exposure to 2.45 GHz electromagnetic fields does not alter radial maze performance in rats.	SD Rats	Avg. whole body 0.6 W/kg	2450 MHz	Pulsed circular polarized 2 μs pulses, 500 pps	1 mW/cm^2 but very high peak power	45 min	Cassel et al. (2004) (Lai et al., 1994 replication)
Whole body exposure to 2.45 GHz does not alter anxiety responses in rats: a plus maze study including test validation	SD Rats	Avg. whole body 0.6 W/kg	2450 MHz	Pulsed circular polarized 2 μs pulses, 500 pps	1 mW/cm^2 but very high peak power	45 min before testing	Cosquer et al. (2005a)
Whole body exposure to 2.45 GHz electromagnetic fields does not alter 12-arm radial maze with reduced access to spatial cues in rats	SD Rats	Avg. whole body 0.6 W/kg	2450 MHz	Pulsed circular polarized 2 μs pulses, 500 pps	1 mW/cm^2 but very high peak power	45 min before testing	Cosquer et al. (2005b)
MW exposure had no effects on learning in a radial arm maze.	Adult male mice C57BL/6J mice	1. Whole Body 0.1, 0.6 or 3 W/kg 2. headmainly exposure 0.1, 0.4 or 2.2 W/kg	1. 900 MHz	1. GSM talk		1.1 h/d over 15 days	Jones et al., 2005

radial arm maze, and water maze differ, spatial learning in these various maze paradigms requires a common hippocampal memory component since hippocampal lesions impair spatial learning in all types of mazes (Morris et al., 1982; Gray and McNaughton, 1983).

In the field of spatial memory, the underlying established molecular mechanism of spatial memory has been demonstrated to be located in the CA1 region of the hippocampus and requires glutamate neurotransmission in the various spatial memory maze paradigms (Morris et al., 1982; Gray and McNaughton, 1983; Kandel et al., 2000; Maguire et al., 2000, 2003a,b; Burgess et al., 2002). Because spatial memory confirmation and replication paradigms address the same molecular mechanism in the hippocampus their results are comparable. It is well established that spatial memory requires glutamate neurotransmission in Schaffer collateral fibers released to glutamate* receptors of CA1 pyramidal cells of the hippocampus. Long-term potentiation in the CA1 cells results in synapse alteration and spatial memory and learning. Even though brain along with body temperatures fluctuates 2–3°C in normal physiological conditions, spatial memory is quite robust and occurs at brain temperatures between 30 and 39°C in rats (Andersen and Moser, 1995).

Research over the years on Alzheimer's disease, which is associated with memory loss and destruction of acetylcholine (ACh) neurons in the basal forebrain, has focused on attempts to find the association between cholinergic systems and spatial memory. ACh neurotransmission has not been established as a mechanism for spatial memory storage. There is no evidence that new synapses have formed on ACh neurons during spatial learning (Hoh et al., 2003). One hypothesis is that place (spatial) learning occurs in the hippocampus through glutamate release and response learning occurs later through ACh release from the striatum and may supersede place learning (Chang and Gold, 2003). Hippocampal ACh is derived entirely from projections of cholinergic neurons of the diagonal band and medial septum (Woolf and Butcher, 1981; Woolf, 1991; Calabresi et al., 2000). ACh neurotransmission may modulate the use of spatial memory possibly through the perforant fiber pathway in the dentate region of the ventral hippocampus (Levin, 1988; Levin et al., 2003, 2005; Dong et al., 2005). It is important to keep in mind that this key established evidence on spatial memory when reviewing the microwave literature on spatial memory.

4.2.5 Microwave Spatial Memory Experiments

Effects of MW exposure on spatial memory have been investigated in animal replication and confirmation experiments in six laboratories ((i) Lai et al., 1989, 1994; Wang and Lai, 2000; Lai, 2004; (ii) Sienkiewicz et al., 2000; Jones et al., 2005; (iii) Dubreuil et al., 2002, 2003a,b; (iv) Yamaguchi et al., 2003; (v) Cobb et al., 2004; (vi) Cassel et al., 2004; Cosquer et al., 2005a–c; see also Table 4.1). These experiments were undertaken to establish scientific evidence on the effects of MW exposure on spatial memory.

It is notable that the Lai et al. (1989, 1994), Cobb et al. (2004), Cassel et al. (2004), and Cosquer et al. (2005b,c) groups used a 12-arm radial maze and Dubreuil et al. (2002) and Sienkiewicz et al. (2000) used an eight arm radial maze. Glassman (1999) pointed seven as the number of different digits or radial arms, etc., we can normally hold in short term memory (Miller, 1956). Eight is within this region. However, the 12-arm radial maze suggests a memory task beyond the "normal" short-term retention

*The glutamate receptors are identified as NMDA (N-methyl-D-Aspartate) and alpha-amino-3-hydroxy-5-methyl-4-isoxazolepropionic acid (AMPA) receptors [see reviews by Kandel et al., 2000, Chapters 62–63; Day et al., 2003; Bast et al., 2005].

range. Maguire et al. (2003b) confirmed this when they investigated the superior and normal mnemonic memory of humans. However, they found the superior memory persons used a spatial learning strategy that also employed the hippocampus. This reaffirms the scientific evidence that performance in the 12-arm radial maze requires the same underlying spatial memory molecular mechanism in the hippocampus as performance in the eight arm radial maze (see Section 4.2.4.1).

The MW dosimetry of the experiments of Lai et al. (1989, 1994), Cobb et al. (2004), Cassel et al. (2004), and Cosquer et al. (2005a,b) are similar. They used the circular waveguide (Guy et al., 1979) to expose the whole body of rats to MWs at 0.6 W/kg. The circular waveguide (Guy et al., 1979) was designed initially for experiments on animals in the MW oven frequency range (2450 MHz) with whole body exposures like those in MW ovens. The exposure is from two sources of "circularly polarized, guided waves that provide a relatively constant and easily quantifiable coupling of microwave energy to each animal, regardless of their position, posture or movement" (Guy et al., 1979). For the same average incident power density, the average SARs in the heads of these rats were about two times higher in the circular waveguide (0.6 W/kg $\times$ 2) than for other exposures, such as in an anechoic chamber (Chou et al., 1985). In the circular waveguide, Guy et al. (1979) and Cassel et al. (2004) have estimated the brain average SAR to be 0.8 W/kg $\pm$4 dB when the whole body SAR is 0.6 W/kg in rats of 250–300g. Wang and Lai (2000) and Lai (2004) also used the circular waveguide (Guy et al., 1979) to expose the whole body of rats to MWs, but used a whole body SAR of 1.2 W/kg. As well, Cosquer et al. (2005c) used the same exposure chamber but used a whole body SAR of 2 W/kg.

This circularly polarized wave, dual source exposure used in the experiments above is in contrast to the plane wave exposure from a single source, typical of the mobile telecommunications industry, used to expose animals in the other spatial memory experiments ((i) Sienkiewicz et al., 2000 (GSM 900 MHz average whole body SAR of 0.05 W/kg); Jones et al., 2005 (900 MHz GSM whole body exposure SARs of 0.1, 0.6, or 3 W/kg, or exposure mainly of the head, local SARs of 0.1, 0.4, or 2.2 W/kg); (ii) Dubreuil et al., 2002, 2003a (GSM 900 MHz, head only exposure, SAR 1 and 3.5 W/kg); (iii) Yamaguchi et al., 2003 (pulsed 1439 MHz time division multiple access (TDMA) brain average SAR 7.5 W/kg or 25 W/kg)).

The MW exposure experiments on spatial memory of Lai and colleagues over the years have used two maze paradigms, the 12-arm radial maze (Lai et al., 1989, 1994), and more recently the water maze (Wang and Lai 2000; Lai, 2004) and unlike the other five laboratories, they have consistently reported positive results. Lai et al. (1989, 1994) investigated the effects of MW exposure (circular waveguide paradigm, *2450* MHz, 2 s pulses, 500 pulse per second (pps), whole body average SAR 0.6 W/kg) on learning in the 12-arm radial arm maze. Rats were trained in the maze to obtain food reinforcements immediately after 20 or 45 min of MW exposure. Lai et al. (1994) did not report on the blindedness of their procedures. Exposure to MWs for 20 min prior to training had no significant main effect on maze learning but affected the shape of the learning curve, whereas 45-min exposure significantly retarded the performance of the rats. Wang and Lai (2000) and Lai (2004) found that acute exposure to pulsed 2450 MHz MWs affected water maze performance of rats. Lai et al. (1989, 1994) and Wang and Lai (2002) maze studies have been reported to have statistical problems and the exposed and sham animals exhibited no apparent interaction between testing days and treatment (IEGMP, 2000; Sienkiewicz, 2002; Dubreuil et al., 2002).

Cobb et al. (2004) attempted to replicate the results of Lai et al. (1994) that showed a working memory deficit in rats exposed to 2450 MHz pulsed MW fields (2 ms, 500 pps, whole body average SAR of 0.6 W/kg) and also attempted that this deficit was reversed by a pretreatment with physostigmine, an acetylcholinesterase inhibitor, or with naltrexone hydrochloride, an opioid antagonist that acts on both CNS and peripheral receptors, but not with the peripheral opioid antagonist, naloxone methiodide. Cobb et al. (2004) study was conducted with a double blind procedure. Cobb's analyses of error rates revealed no significant exposure effect, no significant drug effect, and no significant interaction between the two main factors. There was a significant difference in test days, as expected, with repeated test–trial days, which indicates that learning was accomplished. Cobb et al. (2004) failed to replicate the work of Lai et al. (1994). They concluded that there is no evidence from their current study that exposure to MWs under parameters examined caused decrements in the ability of rats to learn the spatial memory task.

Replication studies by Cassel et al. (2004) and Cosquer et al. (2005b) also failed to confirm the MW exposure effects of Lai et al. (1994) on spatial learning in the 12-arm radial maze. Cosquer et al. (2005a) reported that similar MW exposure did not alter anxiety responses assessed in the elevated plus maze. These elevated plus maze, anxiety responses are hypothesized to be associated with changes in benzodiazepine receptors (Lai et al., 1992: "benzodiazepine receptors in the brain are responsive to anxiety and stress"). Cosquer et al. (2005a) failed to confirm the work of Lai et al. (1992) about 2450 MHz exposure altering anxiety and the hypothetically associated benzodiazepine receptors in the rat brain.

Lai (2005a) has written a letter suggesting that the rats of Cobb et al. (2004) overlearned the spatial memory task, while his rats did not, and thus she failed to replicate his work. He also suggested that there was a "typo" in his 1994 paper and that his dose of physostigmine was 0.1 mg/kg, i.p. instead of 1 mg/kg i.p. as reported in 1994, and as such Cobb et al. (2004) who used 1 mg/kg did not replicate his work. Lai (2005a) has theorized that his reported microwave-induced learning deficit was caused by a decrease in cholinergic functions in the brain. Jauchem (2005) and Cassel (2005) have replied to Lai's (2005a) comments and Lai (2005b) has replied to them. Cassel (2005) presented a graphed comparison of the radial maze performance curves of Cobb et al. (2004) and Lai et al. (1994), showing parallel trends, especially after the third day; and Cassel also pointed out on the last day Cobb et al.'s (2004) rats' mean errors approached 2, whereas Lai et al.'s (1994) rats' mean errors were less than 0.5. Lai (2005b) continues to contend that no one has replicated his results because both Cobb and Cassel gave the rats more chance to learn the maze than he did. Lai (2005b) does concede that Cassel et al. (2004) replicated Cobb et al.'s (2004) results.

There is no confirmatory (see below) or replicated evidence to support Lai and his colleagues' papers (1989, 1994, 2000, 2004), suggesting MW exposure effects on spatial memory. Although Lai (2005a,b) has indicated that other experiments were not exact duplications, his contentions are dismissed by Cassel (2005). Indeed if we refer to the scientific evidence (see Section 4.2.4.1) we know that no matter what the spatial memory paradigm, spatial memory always involves glutamate neurotransmission in the CA1 pyramidal cells of the hippocampus. Therefore slight differences in training procedure in the 12-arm radial maze would not make a difference to the location in the brain of spatial memory or the molecular mechanism forming spatial memory and thus, the possible effects of MW exposure on spatial memory. Furthermore, there is no established neuroanatomical location or molecular mechanism for Lai's cholinergic hypothesis for his MW-induced spatial memory deficit (see Section 4.2.4.1).

As for Lai's (2005b) contention that there may be task or MW-related genetic differences within the substrains of the Sprague-Dawley rat species he used (1994) and those used by Cobb et al. (2004) and Cassel et al. (2004), this seems unlikely since spatial memory occurred in all three experiments in the rats during control conditions. Also the groups of rats of both Cobb et al. (2004) and Lai et al. (1994) show a parallel trend in the graphed curves of their radial maze performance for both control and MW conditions

(Cassel, 2005). The lack of a MW effect on spatial memory is replicated or confirmed in possibly five different sources of Sprague-Dawley rat subspecies ((i) Dubreuil et al., 2002: Charles River Laboratories, IFFA, Credo, France; (ii) Yamaguchi et al., 2003: Charles River, Yokohama, Japan; (iii) Cassel et al., 2004: R. Janvier-St. Berthevin, France; (iv) Cobb et al., 2004: Charles River Laboratories, Portage, MI; (v) Cosquer et al., 2005a–c: R. Janvier Le Genest St. Isle, France). Perhaps the MW positive results of Lai and his colleagues have to do with undisclosed methodological shortfalls or those suggested previously (IEGMP, 2000; Dubreuil et al., 2002; Jauchem, 2005).

There were three groups that investigated spatial memory using plane wave exposures. Sienkiewicz et al. (2000) failed to find any memory deficits in mice repeatedly exposed to 900 MHz fields in a GHz Transverse electromagnetic (TEM) cell. This may be due to the fact that the average whole body SAR was low (0.05 W/kg). Dubreuil et al. (2002) found no significant spatial memory difference between MW- and sham-irradiated rats that were exposed at relatively high local SARs of 1 and 3.5 W/kg. They exposed the heads of the rats in a TEM cell at 900 MHz.

Yamaguchi et al. (2003) reported that pulsed TDMA fields (1439 MHz, brain average 7.5 and 25 W/kg) affected performance of rats in a T-maze task only when body temperature was elevated (at 25 W/kg). Sprague-Dawley rats were exposed for either 1 h daily for 4 d or for 4 weeks to a pulsed 1439 MHz TDMA field in a carousel-type exposure system. When the brain, average SAR was 7.5 W/kg, the whole body average SAR was 1.7 W/kg. Other subjects were exposed at the brain average SAR of 25 W/kg with the whole body average SAR of 5.7 W/kg for 45 min daily for 4 d. Learning and memory were evaluated by reversal learning in a food-rewarded T-maze, in which rats learned the location of food (right or left) by using environmental cues. The animals exposed to MW with the brain average SAR of 25 W/kg for 4 d showed statistically significant decreases in the transition in number of correct choices in the reversal task, compared to sham-exposed or cage control animals. However, rats exposed at the brain average SAR of 7.5 W/kg for either 4 d or for 4 weeks showed no T-maze performance impairments. Intraperitoneal temperatures, as measured by a fiber optic thermometer, increased in the rats exposed to the brain average SAR of 25 W/kg but remained the same for the brain average SAR of 7.5 W/kg. These results suggest that the exposure to a TDMA field at levels about four times stronger than emitted by cellular phones (2 W/kg limit) does not affect the learning and memory processes when there is no whole body temperature rise.

4.2.6 Conclusions: Animals: Spatial Memory Replications and Confirmations

The weight of evidence from the six research groups using two species of mammals (Sprague-Dawley rats and C57BL/6J mice) is that there is no effect on spatial learning as tested by six spatial learning paradigms at whole body or local (head) MW exposures within guideline limits and also well above the whole body and partial body limits, when whole body temperature did not measurably rise. Five groups ((i) Sienkiewicz et al., 2000; Jones et al., 2005; (ii) Dubreuil et al., 2002, 2003a,b; (iii) Yamaguchi et al., 2003; (iv) Cobb et al., 2004; (v) Cassel et al., 2004; Cosquer et al., 2005a–c), using spatial memory animal experiments confirmed or replicated each others' results indicating that MW exposures at frequencies from 900 to 2450 MHz had no effect on spatial memory. The positive results of Lai and colleagues (Lai et al., 1989, 1994; Wang and Lai, 2000; Lai, 2004) are not supported by the weight of scientific evidence. Sienkiewicz's group (Jones et al., 2005) have recently reconfirmed and extended their results on spatial memory in mice exposed to MWs lending further support to this weight of scientific evidence conclusion.

Extensive neurophysiological and neuroanatomical research (see Kandel et al., 2000) has established that the critical neurotransmission for spatial memory to occur is glutamate binding to glutamate receptors of pyramidal cells in the CA1 region of the hippocampus. Newer research in the literature continues to reconfirm this (Maguire et al., 2000, 2003a,b; Burgess et al., 2002; Bast et al., 2005).

If the strong evidence for homology of memory is accepted at this time, there does not seem to be any further requirement for studies in this area in other strains and species of animals and possibly in man (Kandel et al., 2000; Burgess et al., 2002; Maguire et al., 2000, 2003a,b). However, it is generally preferred to have confirmatory MW exposure, spatial memory experiments in humans for health risk assessment. Thus, we recommend testing spatial memory in humans during MW exposure. Human cognitive studies are reviewed in the following section.

In conclusion, MW exposure appears to disrupt spatial memory in rats at a dose at or above the whole body threshold of 4 W/kg in a fashion similar to simple tasks of perception. This is in line with the mechanism of whole body temperature increases of 1°C or higher.

4.2.7 Microwave Cognitive Effects: Human Studies (Table 4.2)

In recent years, the mobile phone has become extremely popular and is widely used by about 1.4 billion members of society. This rapid increase in use of a relatively new technology has resulted in concerns over alleged effects because of exposure of the head to MW radiation from the phone antenna. During the period 1999–2005, a variety of studies have been done, both to measure the dose rate to the head and to evaluate the effects of mobile phone irradiation on cognitive processes in humans (see Table 4.2) (Preece et al., 1999, 2005; Van Leeuwen et al., 1999; Koivisto et al., 2000a,b; Krause et al., 2000a,b, 2004; Lee et al., 2001, 2003; Croft et al., 2002; Edelstyn and Oldershaw, 2002; Hamblin and Wood, 2002; D'Andrea et al., 2003b; Haarala et al., 2003b, 2004, 2005; Curcio et al., 2004; Maier et al., 2004; Hamblin et al., 2004; Besset et al., 2005). The studies that examine human memory processes and mobile phone usage appear to show no established evidence of memory deficits. Rather, some results have been reported of memory and attention facilitation. Preece et al. (1999) reported an improvement in reaction time of humans during exposure to cell phone radiation at 915 MHz while performing a "working" memory task. This finding could be viewed as a cognitive enhancement. However Preece et al. (2005) have failed to replicate this work in 18 children of 10–12 y of age.

A Finnish group has performed a number of experiments testing human cognitive effects. Koivisto et al. (2000a,b) reported that exposure to GSM 902 MHz radio frequency energy emitted by cellular telephones was associated with a facilitation of brain functioning during memory tasks, including faster response times, thus suggesting an improvement in cognitive function. Haarala et al. (2003b, 2004) recently reported a failure to replicate the effects (Koivisto et al., 2000a,b) of GSM 902 MHz EMF emitted by cellular telephones on response times and short-term memory in humans, using double blind conditions in two independent laboratories. Indeed Haarala et al. (2005) have done a similar double blind study on adolescents (32 children: 16 boys, 16 girls) of 10–14 y old (mean 12.1 y, SD 1.1) and found no effect of MW exposure on cognition.

Krause et al. (2004) reported that exposure to similar fields did not alter resting EEG in humans, but caused a decrease in cortical activity in human subjects performing auditory memory tasks. One interpretation of this finding is that more efficient brain function was attained with less effort. Krause et al. (2000b) found that such exposure altered event-related desynchronization and synchronization responses of all EEG frequency bands studied during the performance of a visual working memory task. The biological significance of this finding, if any, was not stated. But recently, the term "desynchronized" for

TABLE 4.2
Effects of MW Exposure on Human Cognition

EMF Effect	Species	SAR (W/kg)	Frequency (MHz)	Modulation	Intensity (mW/cm^2)	Duration of Exposure	Ref.
They calculated a maximum rise in brain temperature of 0.11°C for an antenna with an average emitted power of 0.25 W, the maximum value in common mobile phones, and indefinite exposure	Human bioheat head model	Max SAR 0.91 W/kg for 10 g	916 MHz		0.25 W	Indefinite exposure	Van Leeuwen et al. (1999)
There was evidence of a significant increase with the analog but not with the digital simulation, in choice reaction time only. No effects on memory	Humans		915 MHz and CW	217 Hz pulsed 12.5% duty cycle	1 W mean power	25–30 min	Preece et al. (1999)
No effect of 902 MHz mobile phone transmission on cognitive function in children. Double blind	18 children 9 m., 9 f.; 10.2–12.2 y	Estimate worst case 0.28 W/kg	902 MHz GSM Nokia 3110	217 Hz pulse width 577 μs	*X*: 0 W, *Y*: 0.2 W peak (0.025 W mean), *Z*: 2 W peak (0.25 W mean)	30–35 min over 2 out of 3 d, (*Y*,*Z*) left side	Preece et al. (2005)
No effect on nine tasks and 11 comparisons choice reaction tasks. Facilitatory effect on tasks requiring attention and manipulation of information in working memory. Single blind	48: 24 m., 24 f.; 18–49 y, avg. 26 y, R. handed		GSM 902 MHz, Nokia 6110	217 Hz pulse width 577 μs	0.25 W mean power	About 1 h/d 2 d left ear	Koivisto et al. (2000a)
Memory load was varied from 0 to 3 items in an n-back test. Significantly speeded up response times on three items (on increased memory load). Single blind	Humans	0.683 W/kg avg. over 10 g	GSM 902 MHz, Nokia 6110	217 Hz pulse width 577 μs	0.25 W mean power	30 min left ear	Koivisto et al. (2000b) (SAR measurement for Koivisto et al., 2000b done in replication Haarala et al., 2004)

902 MHz mobile phone exposure had no effect on performance of nine human cognitive tasks: Failed replication of Koivisto et al. (2000a) in Finland and Sweden double blind	Humans: 32 m., 32 f.	0.88 W/kg avg. over 10 g	GSM 902 MHz, Nokia 6110	217 Hz pulse width 577 μs	0.25 W mean power	65 min left ear	Haarala et al. (2003b)
A PET study showed no rCBF changes were in the area of maximum 902 MHz mobile phone SAR. During scanning, the subjects performed a visual working memory task	14 R. handed healthy males		GSM 902 MHz, Nokia 6110	217 Hz pulse width 577 μs	0.25 W mean power		Haarala et al. (2003b)
902 MHz mobile phone does not affect short-term memory in humans. Failed replication in Finland and Sweden: Koivisto (2000b). Double blind	64: 32 m., 32 f.; 20–42 y, Finland avg. 24.19 y. Sweden avg. 29.34 y	0.99 W/kg avg. 10 g	GSM 902 MHz, Nokia 6110	GSM	0.25 W mean power	Mean duration 65 min/d over 2 d left ear	Haarala et al. (2004)
No effect of 902 MHz mobile phone transmission on eight cognitive function tests in children. Double blind. Temperature of face skin was same with and without MW transmission. No subject could detect if the phone was on or off above chance	32 children 16 m., 16 f.; 10–14 y; R. handed; mean 12.1 y	Mean 0.99 W/kg over 10 g. Peak 2.07 W/kg	GSM 902 MHz	217 Hz pulse width 577 μs	0.25 W mean power	60 min over 2 d; tested same time each day. Left side	Haarala et al. (2005)
On three measures of attention mobile phone users performed significantly better than controls on one test (trail making)	72 bright students 16 y		Mobiles Hong Kong	GSM		Median 3712.5 min of use	Lee et al. (2001)
The differential effect of the duration of exposure to the electromagnetic field emitted by mobile phones was on only one of two tests of human attention	78 univ. students R. handed		Mobiles Hong Kong	GSM		30 min	Lee et al. (2003)

continued

TABLE 4.2 (continued)

Effects of MW Exposure on Human Cognition

EMF Effect	Species	SAR (W/kg)	Frequency (MHz)	Modulation	Intensity (mW/cm^2)	Duration of Exposure	Ref.
Significant effects were found for digit span forward, spatial span, and serial subtraction tasks, three of six tests for exposed versus unexposed conditions	38 humans 21 y avg.		900 MHz			30 min left ear	Edelstyn and Oldershaw (2002)
Modulated 450 MHz decreased variance in errors in two visual memory tasks and increased error in two others	100: 37 f., 63 m.		450 MHz	7 Hz	0.158 W/cm^2	10–20 min	Lass et al. (2002)
Mobile phone use facilitates word recall memory in male, but not female, subjects only in the short term. Single blind	62: R. handed 33 m., 29 f.; avg. 26.5 y, 18–53 y	0.79 W/kg	Ericcson A2618s 1800 MHz			Left ear 15 min	Smythe and Costall (2003)
A pilot study on pulsed field interference with auditory discrimination task. Single blind	11 m., 11 f; 23–48 y, 39.1 avg.		GSM 902 MHz Motorola 920	Pulsed at 217 Hz pulse width 577 μs	1 mW/cm^2	50-min exposure then testing	Maier et al. (2004)
Effect of time course (25–30 min) of GSM 902.40 exposure on repeated simple and choice reaction times and increase of local tympanic temperature. No effect on speed and accuracy measures, visual search task, and descending subtraction task. Double blind	20: 10 m., 10 f.; 22–31 y; mean 26.4 y, R. handed	Max 0.5 W/kg	GSM 902.40 MHz Motorola Timeport 260	Pulsed at 217 Hz pulse width 577 μs	Max power 1 W; avg. power 0.25 W/cm^2	45 min left ear	Curcio et al. (2004)
No effect on cognitive function from daily mobile phone use tested after 13-h rest periods. Double blind	55: 27 m., 28 f.; avg. 24.3 y, 18–40 y	0.54 W/kg	GSM 900 MHz	217 Hz pulse width 577 μs		2 h/day over 5 d/week, total 27 d (preferred hand/side)	Besset et al. (2005)
No effects of mobile phone emitted EMFS on human event-related potentials and performance	120 volunteers		895 MHz	217 Hz, pulse width 577 μs	250 mW	N100, P300	Hamblin et al. (2005)

the activated states of waking and REM has been rendered obsolete by the discovery of highly synchronized gamma frequency (30–80 Hz) activity in these states (Hobson and Pace-Schott, 2002). Krause et al. (2004) have recently failed to replicate their 2000 study. All eight of the significant changes in the earlier study were not significant in the present double blind replication.

Among other groups that have studied this problem, Edelstyn and Oldershaw (2002) found that human volunteers exposed to the fields of GSM mobile phones showed improvement in the performance of three cognitive function tests: (i) immediate verbal memory capacity, (ii) immediate visual spatial working memory capacity, and (iii) sustained attention. Lee et al. (2001) performed three measures of attention to compare responses of mobile phone users (Hong Kong, PCS 1900 MHz) with those of nonusers. Mobile phone users performed better on one of the measures, a ''trail making test.'' The authors noted that the results could have been confounded by self-selection. Lee et al. (2003) repeated their previous results of enhanced attention following mobile phone exposure and discovered that duration of the exposure may be important for the development of differential effects on attention processes.

Curcio et al. (2004) reported an improvement of both simple and choice reaction times that was accompanied by an increased temperature of the area irradiated during exposure from a GSM mobile phone. They reported a positive relationship with duration of MW exposure and changes in temperature and reaction time performance. They stated that a minimum of 25 min of EMF exposure is needed to show appreciable changes. These studies (Lee et al., 2001, 2003), however, did not observe increased error rates on the attention and cognitive tests. Also, increases in the temperature of the head during exposure (Curcio et al., 2004) may be largely due to the local insulation of the head by the physical proximity of the phone, rather than the MWs emitted from the phone. This confounding heat needs to be controlled for in further research (Van Leeuwen et al., 1999; Straume et al., 2005).

Besset et al. (2005) investigated the effects of daily exposure to GSM 900 MWs on cognitive function in a double-blinded design. Fifty-five subjects (27 males and 28 females, 18–33 y) were divided into two groups: a group with mobile phones switched on and a group with them switched off. The groups were matched according to age, gender, and IQ. Over 45 d, there were three testing periods: baseline (3 d), exposure (28 d), and recovery (14 d). Subjects were exposed or sham-exposed for 2 h/d, 5 d/week. The neuropsychological test battery composed of 22 tasks screened four cognitive categories: information processing, attention capacity, memory function, and executive function. For the most part, these 22 tasks are well-validated in the scientific literature. This neuropsychological battery was performed four times, once during baseline, twice during exposure, and once during recovery. Their results indicate that daily mobile phone (MP) use has no effect on cognitive function after a 13-h rest period.

Maier et al. (2004) in their pilot study found an increase in discrimination time on an auditory discrimination task after 50 min of pulsed GSM fields (902 MHz) exposure from a programmable mobile phone as compared to sham exposure. Maier used an auditory discrimination task based on the auditory order threshold (OT). They defined OT as the minimum time needed (a) to recognize that two auditory stimuli are presented just separately and (b) to decide without error on what side the first of both stimuli was presented. Out of 11 MW-exposed subjects, two showed shorter discrimination time, and nine showed an increase in discrimination time. There was no indication of the normal variation in discrimination time and no clear verification of the reliability and validity of the measurement. No other laboratory has reported MW exposure results using this paradigm.

Hamblin et al. (2004) studied auditory event-related potentials (ERP) during an auditory task performed during 1 h of exposure to a GSM mobile phone (SAR at the head 0.87 W/kg). In this pilot study with 12 subjects (eight males and four females, 19–44 y), they found reduced amplitudes and latency for the N100 component (stimulus detection) and only delayed latency for the P300 component (thought to represent cognitive processes). Differences during the auditory task were found at cortical sites closest to the active mobile phone. The authors questioned that whether this could be due to stronger MW at those sites or whether these sites were the most MW sensitive to the performance of the task. The authors reported that reaction times were significantly longer for the exposed condition than for sham. These preliminary results suggest that exposure may affect neural activity. The ERP literature was previously reviewed (D'Andrea et al., 2003b), leaving doubt as to the validity of the various ERP experimental paradigms used. That conclusion still holds in relation to the Hamblin pilot paper (Hamblin et al., 2004). Hamblin et al. (2006) at a conference presented a large, double blind, replication study of 120 subjects and concluded that no MW effects on ERP were observed. Any effects they saw were reported as weak and were dismissed as chance. Hamblin et al.'s (2006) study appears to have been conducted with improved methodology; and perhaps firm conclusions about ERP effects, at the low SARs produced by mobile phones, may now be drawn.

The studies that examine human cognitive processes and mobile phone usage appear to show no established evidence of memory deficits. Replication studies with standardized protocols, larger samples (multicentered replications; Haarala et al., 2003b), better experimental controls, double blind conditions, and Bonferroni or other statistical corrections for multiple comparisons appear to have eliminated the "false positives."

4.2.8 Microwave Effects on the Electroencephalogram

Included in cognitive effects are the EEG and BBB studies that were previously reviewed by D'Andrea et al. (2003b). The reviews by D'Andrea et al. (2003a,b) concluded that the evidence for mobile phone effects on human cognitive performance was very weak. And they concluded that until more studies could be conducted with improved methodology and standardized protocols, firm conclusions about cognitive effects at the low SARs produced by mobile phones could not be drawn. Since D'Andrea et al.'s (2003a,b) review of EEG publications (see Table 4.3), there are a few new studies of note including Huber et al. (2003, 2005), showing effects of pulsed signals on EEG, and Hinrikus et al. (2004), reporting the normal EEG is too variable to be able to measure possible small MW effects.

The Borbély, Huber, and Ackerman group papers (Borbély 1999; Huber et al., 2000, 2002, 2003, 2005; Achermann et al., 2005; see Table 4.3) on the effects of MW exposure on nighttime and daytime sleep EEG, heart rate, and positron emission tomography (PET) scans are very difficult to follow because they publish preliminary studies (1999, 2000, 2002), followed years later by further analysis and reanalysis of previous results without addressing discontinuities in methodological details, and in the results. The preliminary sleep EEG studies in 1999 (Borbély et al., 1999) on the effects on EEG (and electrocardiogram) of subjects' nighttime sleep episode due to an intermittent exposure schedule (by three antennas, 900 MHz; SAR 1 W/kg consisting of alternating 15-min on and 15-min off intervals) and in 2000 (Huber et al., 2000) on the effects on EEG of a unilateral 900 MHz GSM exposure for 30 min prior to 3 h of daytime sleep, were reanalyzed in the data given by Huber et al. (2003), finding MW effects on different sleep bands after reanalysis. Borbély et al. (1999) reported effects on nonrapid eye movement (NREM) episodes 1–3, but Huber et al. (2003) reported only effects on the first NREM episode (see Table 4.3).

TABLE 4.3

Effects of MW Exposure on Electroencephalography (EEG)

EMF Effect	Species	SAR (W/kg)	Frequency (MHz)	Modulation	Intensity (mW/cm^2)	Duration	Ref.
Increase number of spindle shape activity firings and slow waves; enhance evoked responses in the visual cortex neurons to single light flashes	Rabbits (unanesthetized)		2400	CW	40	1 min	Chizhenkova et al. (1988)
Significant effects in the power spectral analysis of EEG frequency, but not at the same frequency	Rats	2.7	2450	CW	10	7 h	Mitchell et al. (1989)
Significant elevations of EEG asymmetry in 10–14 Hz range during the first 20 s	8 Rats		945 MHz	(AM-4 Hz) 2 ms pulse duration	0.1–0.2 mW/cm^2 antenna 28–30 cm above rat	1-min on–1-min off alternating for 10 min	Vorobyov et al. (1997)
Increase total power of spectra for (ii) 30 mW/cm^2 condition only	40 Rats	WB	2450 CW	CW	(i) 10 (ii) 30	10 min	Thuroczy et al., 1994
Increase in the power of the delta waves (thermal)	40 Rats	42 mW/g	4000 CW	AM 16 Hz		30 min	Thuroczy et al., 1994
Increase in the power of beta waves	40 Rats	8.4 mW/g	4000	AM 16 Hz		30 min	Thuroczy et al., 1994
Increase in cerebal blood flow	40 Rats	8.4 mW/g	4000 CW	CW		30 min	Thuroczy et al., 1994
Magnetic fields mimic the behavior effects of REM sleep deprivation in humans, enhancement of motivational and drive related behaviors	Human with two cases multiple sclerosis and Parkinsons		2 Hz		7.5 pT	6 min	Sandyk et al. (1992)

Borbély/Huber/Achermann Group (1994–2005)

EMF Effect	Species Human	Antenna	SAR (W/kg)	Frequency (MHz)	Modulation	Intensity (mW/cm^2)	Exposure Duration	Electrodes	Time Duration of EEG Recording	Ref.
Significant sleep-inducing effect	52: 32 f., 20 m. 18–53 y, median 24 y	A LEET Low energy emission therapy device intrabuccally	<10 in oral mucosa: 100 mW/kg in brain	27.12 MHz	Intrabuccal amplitude-modulated 42.7 Hz		3-s on 1-s off for 15 min	16 recoding electrodes eyes closed in a chair in a dark room	Between 15:00 and 20:00 EEG for 15-min post-EMF	Reite et al. (1994)

continued

TABLE 4.3 (continued)

Effects of MW Exposure on Electroencephalography (EEG)

EMF Effect	Species Human	Antenna	SAR (W/kg)	Frequency (MHz)	Modulation	Intensity (mW/cm^2)	Exposure Duration	Electrodes	Time Duration of EEG Recording	Ref.
Significantly reduced sleep onset and spectral power of NREM slow wave sleep increase in the 10–11 and 13.5–14 Hz bands in initial part of sleep, declined but still significant in 2nd and 3rd NREM episodes	24 R. handed men 20–25 y; mean 22.6 y	Three λ/2 dipole antennas 30 cm away from back of head	Spatial peak avg. 1 W/kg over 10 g	GSM 900 MHz	Linear polarized 2, 8, 217, 1736 Hz 87.5% duty cycle base station cocktail		1 Night 11 p.m.–7 a.m.; 15-min on–15-min off	EEG electrodes were placed according to international 10–20 system	23:00–7:00	Borbély et al. (1999)
Spectral power of the EEG in NREM sleep was increased in first 30 min of sleep only $p < .01$. This was significant for the entire 8-h sleep. The maximum rise occurred in the 9.75–11.25 Hz and 12.5–13.25 Hz bands in first 30 min of NREM sleep (alpha-band). No effect on sleep latency or sleep stages or REM sleep spectrum. Same effect whether exposure was from the left or right side. Double blind	16 R. handed men 20–25; avg. 22.3 y	Two planar antennas, unilateral left or right side 11 cm from the ear	Spatial peak avg. 1.0 W/kg over 10 g. Hemi-brain 0.14 W/kg	900 MHz: GSM	Linear polarized 2, 8, 217, 1736 Hz 87.5% duty cycle cocktail. Synthesized base station-like signals (cocktail)		30 min prior to 3-h sleep begin at 9:45 or 10:15 (night before only 4-h sleep)	EEG (F3, C3, P3, O1, F4, C4, P4, O2) 30-min preceding	3-h daytime sleep episode and night before	Huber et al. (2000)

EMF Effect	Species Human	Antenna number	SAR (W/kg)	Freq. (MHz)	Modulation H_2	Intensity (mW/cm^2)	Exposure Duration h	Electrodes location	Time Duration of EEG Recording: h	Reference
1. Reanalysis of 1999 (Borbély et al.). Effects on sleep stages: duration of waking after sleep onset reduced. Spectral power of NREM slow wave sleep increase in the 11.5–12.25 and 13.5–14 Hz bands in initial part of sleep only.	24 R. handed men 20–25 y; mean 22.6 y	Three λ/2 dipole antennas 30 cm away from back of head	Spatial peak avg. 1 W/kg over 10 g	GSM 900 MHz	Linear polarized 2, 8, 217, 1736 Hz 87.5% duty cycle base station cocktail		1 Night 11 p.m.–7 a.m.; 15-min on–15-min off	EEG electrodes were placed according to international 10–20 system	23:00–7:00	Huber et al. (2003). Part 1: Reanalysis of Borbély (1999) data

Notice the change. Waking EEG increase in power of 11–11.5 Hz								
2. Reanalysis of 2000 (Huber et al.). No effects on sleep variables. The sleep EEG in first 30 min of NREM sleep increase in power more pronounced in the left hemisphere for 9–13.5 Hz and 12.5–13.25 Hz bands irrespective of exposure side $p < .05$. Waking EEG reduction in power in 10.5–11 Hz and 18.75–19.5 Hz bands $p < .05$	16 R. handed men 20–25; avg. 22.3 y	Two planar antennas, unilateral left or right side 11 cm from the ear	Spatial peak avg. 1.0 W/kg over 10 g. Hemi-brain 0.14 W/kg	900 MHz: GSM	Linear polarized 2, 8, 217, 1736 Hz 87.5% duty cycle cocktail. Synthesized base station-like signals (cocktail)	30 min prior to 3-h sleep begin at 9:45 or 10:15 (night before only 4-h sleep)	EEG (F3, C3, P3, O1, F4, C4, P4, O2) 30-min preceding	3-h daytime sleep episode and night before	Huber et al. (2003). Part 2: Reanalysis of Huber et al. (2000) data
3A. Analysis of heart rate data from 1999 (Borbély et al.). Heart rate was not affected. But heart rate variability (spectra of RR intervals) was. But none of the all night spectra was significantly different that controls	24 R. handed men 20–25 y; mean 22.6 y	Three λ/2 dipole antennas 30 cm away from back of head	Spatial peak avg. 1 W/kg over 10 g	GSM 900 MHz	Linear polarized 2, 8, 217, 1736 Hz 87.5% duty cycle. Base station cocktail	1 Night 11 p.m.–7 a.m.; 15-min on–15-min off	ECG was recorded with a polygraphic amplifier	23:00–7:00	Huber et al. (2003). Part 3A: Analysis of Borbély et al. ECG (1999) data
3B. Analysis of heart rate data from 2000 (Huber et al.). During the 30 min of MW exposure no difference in heart rate between the three conditions was observed. Heart rate variability was altered during the 3-h sleep spectra but not during first 30-min NREM sleep. Observed effects were weak	16 R. handed men 20–25; avg. 22.3 y	Two planar antennas, unilateral left or right side 11 cm from the ear	Spatial peak avg. 1.0 W/kg over 10 g. Hemi-brain 0.14 W/kg	900 MHz: GSM	Linear polarized 2, 8, 217, 1736 Hz 87.5% duty cycle cocktail. Synthesized base station-like signals (cocktail)	30 min prior to 3-h sleep begin at 9:45 or 10:15 (night before only 4-h sleep)	ECG was recorded with a polygraphic amplifier	3-h daytime sleep episode and night before	Huber et al. (2003). Part 3B: Analysis of Huber et al. ECG (2000) data

continued

TABLE 4.3 (continued)

Effects of MW Exposure on Electroencephalography (EEG)

EMF Effect	Species Human	Antenna	SAR (W/kg)	Frequency (MHz)	Modulation	Intensity (mW/cm^2)	Exposure Duration	Electrodes	Time Duration of EEG Recording	Ref.
4A. EMF exposure dosimetry of Borbély et al. (1999), including the assessment of the exposure variability and uncertainties (SAR variations up to 40% higher)	Head and body phantoms numerical model	Three λ/2 dipole antennas 30 cm away from back of head	DASY 3 Measurement in free space and inside head and I body phantoms	GSM 900 MHz	Linear polarized 2, 8, 217, 1736 Hz 87.5% duty cycle base station cocktail		During testing	Measurements of induced currents of electrodes on surface of phantom shell with diff. orientations (SAR variations up to 40% higher)		Huber et al. (2003). Part 4: Analysis dosimetry. Borbély et al. (1999)
EMF exposure dosimetry of the brain and body areas of Huber et al. (2000) including the assessment of the exposure variability and uncertainties. Measurements of induced currents of electrodes on surface of Phantom shell (8%) and ear and eye electrodes inside the shell (35% higher, 1 g)	Head and body phantoms numerical model	Planar antennas, unilateral left or right side 11 cm from the ear	DASY 3 Measurement in free space and inside dielectric phantoms	900 MHz: GSM	Linear polarized 2, 8, 217, 1736 Hz 87.5% duty cycle cocktail. Synthesized base station-like signals (cocktail)		During testing	Measurements of induced currents of electrodes on surface of phantom shell (8%) and ear and eye electrodes inside the shell (35% higher, 1 g)		Huber et al. (2003). Part 4: Reanalysis dosimetry. Huber et al. (2000)
5. RE electrosensitivity: subjective assignment of experimental condition was not better than chance	16 R. handed men 20–25; avg. 22.3 y	Two planar antennas, unilateral left or right side 11 cm from the ear	Spatial peak avg. 1.0 W/kg over 10 g. Hemi-brain 0.14 W/kg	900 MHz: GSM	Linear polarized 2, 8, 217, 1736 Hz 87.5% duty cycle cocktail. Synthesized base station-like signals (cocktail)		30 min prior to 3-h sleep begin at 9:45 or 10:15 (night before only 4-h sleep)	EEG (F3, C3, P3, O1, F4, C4, P4, O2) 30-min preceding	3-h daytime sleep episode and night before	Huber et al. (2003). Part 5: Huber et al. (2000)

RECOMMENDATIONS: For exposure design of human studies addressing health risk evaluations of mobile phones important exposure parameters are the signal, field distribution, field strength, the setup & dosimetry. State uncertainty on all measurements	Cover entire half cortex, Double blind Temp load, Minimize intersubject variability	Patch antenna for local unilateral exposure R & L Minimize exposure variability modeling	Worst case. 10 W/kg	Mobile phone in use	Worst case cocktail & CW & intermittant exposures	Maximize spectral power	Not specified	Model electrodes increases in current density interference etc	Not specified	Kuster et al. 2004

EMF Effect	Species Human	Antenna Number/ type	SAR (W/kg)	Frequency (MHz)	Modulation Hz	Intensity (mW/cm^2)	Exposure Duration h	Electrodes location	Time Duration of EEG Recording h	Reference
The present results seem to show that (i) PM-EMF alters waking rCBF and (ii) pulse modulation of EMF is necessary to induce waking and sleep EEG changes. Double blind	16 Human male R. handed (20–25 y)	Planar antennas, unilateral left side 11 cm from the ear	Spatial peak 1 W/kg over 10 g (4× higher than base site ex.)	Example 1. 900 MHz handset signal Example 2. 900 MHz CW	Examples 1 and 2 12.5% duty cycle, 2, 8, 217, 1736 Hz and harmonic					Huber et al. (2002)
(1) Pet awake. Compared with the sham condition, PM-EMF exposure increased relative rCBF in the dorsolateral prefrontal cortex of the left hemisphere, the side of prior EMF exposure ($p < 0.01$, two-tailed paired t-test) but not on the contralateral side. EMF-induced changes in the activity of dorsolateral prefrontal cortex may underlie EMF effects on working memory. On the other hand, an interaction between the counting task and EMF exposure cannot be excluded due to the lack of a performance measure of the counting task	16 Human male R. handed (20–25 y)mean 22.5 y reduced to $n = 13$	Planar antennas Unilateral left side 11 cm from the ear. Subject on a chair, head BTW two plates	SAR 1 W/kg 4× higher peak SAR but maintain same time avg. SAR as base station-like signals	900 MHz handset signal	Pulse-modulated 2, 8, 217, 1736 Hz and harmonics in awake subjects. Provide higher spectral power of the 2 and 8 Hz modulation components		Exposure for 30 min to left side of head then 10-min delay before PET scan	PET scan 3 scans 60 s, intervals of 10 min, using 300–350 mBq $H_2{}^{15}O$ slow bolus. End 10-min transmission scan: re-photon attenuation correction	Between 8:00–14:00 h during scanning count from 1 to 60 slowly-cognitive activity	Huber et al. (2002) Part 1

continued

TABLE 4.3 (continued)

Effects of MW Exposure on Electroencephalography (EEG)

EMF Effect	Species Human	Antenna	SAR (W/kg)	Frequency (MHz)	Modulation	Intensity (mW/cm^2)	Exposure Duration	Electrodes	Time Duration of EEG Recording	Ref.
(2) Sleep study. Spectral power of the EEG sleep onset revealed that the alpha-frequency range was increased in the PM-EMF conditions in comparison with the sham condition. This effect was not present for the CW EMF condition. The sleep EEG was also modified after PM EMF exposure. In sleep stage 2, power in the 12.25–13.5 Hz range was increased relative to sham exposure. There tended to be a decrease, which was statistically significant in a single, slightly higher frequency bin. The extended duration of PM EMF-induced changes in the sleep EEG appeared to increase in the course of the night. Spindle amplitude was increased in the PM EMF condition compared with the CW EMF and sham conditions ($p < 0.05$, two-tailed paired t-test). The duration and number of sleep spindles did not differ. The NREM sleep EEG was modified, but the REM sleep EEG was not affected and no differences were seen in sleep architecture, REM sleep latency or the	16 Human male R. handed (20–25 y) mean age 22.3 reduced to $n = 13$	Planar antennas unilateral left side 11 cm from the ear. Subject on a chair, head BTW two plates	SAR 1 W/kg 4 higher peak SAR but maintain same time avg. SAR as base station-like signals	1. 900 MHz hand set signal. 2. 900 MHz CW	1. Pulse modulated 2, 8, 217, 1736 Hz and harmonics in awake subjects. Provide higher spectral power of the 2 and 8 Hz modulation components		Exposure for 30 min to left side of head	EEG recordings C3A4, C4A1 derivations only	Exposure for at 22:20 then 10-min delay before sleep start at 23:00–07:00	Huber et al. (2002) Part 2

duration of NREM–REM sleep cycles										
Reanalysis of the PET studies (2002): Only 12 subjects analyzed. 1. Comp PET after handset versus sham-exposed. Significant increase in relative cerebral blood flow (rCBF) left dorsolateral prefrontal cortex (LDPFC). 2. Comp PET after Bstat versus sham-exposed. Not significant. 3. Comp PET handset versus base station. Significant increase rCBF LDPFC. Physiological effect may be dependent on spectral power in the amplitude modulation of MW carrier of handset signal	16 Human male R. handed (20–25 y) mean 22.5 y reduced to $n = 12$	Planar antennas, unilateral left side 11 cm from the ear. Subject on a chair, head BTW two plates	SAR 1 W/kg 4× higher peak SAR for handset signal but maintain same time avg. SAR as base station-like signals	1.900 MHz handset signal 2.900 MHz base station signal	1. Handset 1. 2, 8, 217, 1736 Hz and harmonics in awake subjects. Duty cycle 12.5% provide higher spectral power of components. 2. Base station 2. 2, 8, 217, 1736 Hz, 87.5% duty cycle cocktail	Ratio between pulse peak power and time avg. power differed by 4×. 1.2. Base station and 4.8 for handset signal	Exposure for 30 min to left side of head then 10 min delay before PET scan	PET scan 3 scans 60 s, intervals of 10 min, using 300–350 mBq $H_2{}^{15}O$ slow bolus. End 10 min transmission scan: re-photon attenuation correction	Between 8:00–1400 h during scanning count from 1 to 60 slowly-cognitive activity	Huber et al. (2005). Reanalysis of Huber et al. (2002) Part 2

EMF Effect	Species	SAR (W/kg)	Frequency (MHz)	Modulation	Intensity (mW/cm^2)	Duration	Ref.
Scalp electrodes (occipital) O1/O2. During and after exposure for some hours the O2 position alpha-wave is altered (increased energy)	Human 17 men and women; 20–27 y		150 MHz (magnetic field coils at the neck, 10^{-8} T)	217 Hz, pulse width 4.6 ms, interrupts 10 μs	At brain 6 cm depth: 1 μW/cm^2	Exposed 2–3× for 15 min	von Klitzing et al. (1995)
In relaxed awake subjects there was increase in EEG power in α2: 9.75–12.5; β1 and 2 with a delay of 15 min after exposure (mega-wave caused increase in EEG power during and after exposure in α2 9.75–12.5; β1 and 2)	Humans 36 males and females	SAR similar to 0.25 W phone	902.4 MHz GSM 8 W; (and 150 MHz MEGA Wave therapy device)	217 Hz pulse frequency (80 μs 9.6 Hz)	(Magnetic flux density in the range of 400 pT)	1 h 6 min	Reiser et al. (1995)

continued

TABLE 4.3 (continued)

Effects of MW Exposure on Electroencephalography (EEG)

EMF Effect	Species	SAR (W/kg)	Frequency (MHz)	Modulation	Intensity (mW/cm^2)	Duration	Ref.
During MW exposure, significant suppressive effect on REM sleep and increased REM EEG spectral power (alpha mainly affected) density and shortening of sleep onset (NREM)	Human 14 males (mean 27 y)	?	900 GSM mobile phone	217 Hz, pulse width 580 μs	0.05 at 40 cm	8 h, 3 night	Mann and Röschke (1996)
No short-term effects of digital mobile phone telephone on the awake closed eye EEG with special attention to the spectral power density of the alpha (8–13 Hz) EEG	Human 34 males (mean 27 y)		900 GSM mobile phone	217 Hz, pulse width 580 μs	0.05 at 40 cm	3.5 min	Röschke and Mann (1997)
No (CNS-mediated) effects heart rate variability during human sleep	Human 12 males 21–34 y	?	900 GSM mobile phone	217 Hz, pulse width 580 μs	0.05 at 40 cm	8 h, 1 night	Mann et al. (1998)
Suppression of REM sleep as well as a sleep inducing effect, previous results could not be replicated, might be due to dose-dependent effects of the EMF on the human sleep profile	Human 24 healthy males 18–37 y mean 26 y	0.3 to Max 0.6 W/kg	circular polarized 900 MHz GSM	217 Hz, pulse width 577 μs	0.2 W/m^2	8 h, 1 night 11 p.m.–7 a.m.	Wagner et al. (1998)
No effects on human EEG activity recorded in an awake, closed eyes situation. Exposure to one of the phones caused a statistically significant change in the absolute power of the delta recording probably due to statistical chance	Human 10 men 28–48 y and 9 women 32–57 y	?	5 cellular phones (analysis and digital 900 MHz or 1800 MHz).	900 NMT, GSM; PCN 1800 MHz	peak power 1–2 W, transmit at max power	20 min;	Hietanen et al. (2000)
Significant difference on the alpha frequency range for the EEG derivation between occupational and nonoccupational group exposure. No significant difference on EEG for long term user, short-term user, and nonuser	98 healthy male volunteers 25–45 y		Cellular phones Korea			30-min field exposure	Kim et al.
Effect with task-relevant target stimuli in the EEG band 18.75–31.25 Hz (Beta/gamma/awake). No effect with irrelevant standard stimuli	Human 13 healthy males 21–27 y		916.2 MHz	217 Hz pulse frequency pulse width 577 μs radiated power of aerial 2.8 W		Around 10 min with exposure to left posterior temporal region	Eulitz et al. (1998)

Significant decrease in preparatory slow brain potentials (SP) in awake visual monitoring task at central and temporo-parietal–occipital regions but not frontal. No effects on finger movement and contingent negative variation task	Human 16 healthy R. handed males 21–26 y	SAR 0.882 mW/g over 10 g	916.2 MHz	217 Hz pulse frequency pulse width 577 μs radiated power of aerial 2.8 W		Around 10 min: telephone at left ear	Freude et al. (1998)
Significant decrease in awake slow brain potentials (SP) in a complex visual task. No effects on finger movement and contingent negative variation task	Human 20 healthy R. handed males 21–30 y	SAR 0.882 mW/g over 10 g	916.2 MHz	217 Hz pulse frequency pulse width 577 μs radiated power of aerial 2.8 W		Around 10 min: telephone at left ear	Freude et al. (2000)
1-h exposure to mobile phone MWs EMG has no effect on auditory brain stem responses (ABRs) and distortion products otoemission (DPOE) recordings in the conditions of their protocol	10 men, 10 women, 20–30 y		900 MHz	217 Hz GSM	Full power peak power 2 W/8 duty cycle 1/8	60-min exposure	Thimonier et al. (1999)
Effects of MWs (900 MHz) on the cochlear receptor: exposure systems and preliminary results. No effect of exposure was found on otoacoustic emissions from the cochlea	S-D rats $n = 8$ per group	(a) Head 0.2 or 1.0 W/kg (b) 1.0 W/kg	(a) 950 MHz (b) 936 MHz	CW	far field exposure	(a) 3 h/day for 3 d or (b) 3 h/d for 5 d	Marino et al. (2000)
Results suggest that the exposure to EMF does not alter the resting EEG *per se* but increases EEG relative power of 8–10 Hz significantly during auditory memory tasks (words) during retrieval but not during resting	8 males and 8 females right handed mean 22 y		Digital 902 MHz GSM	217 Hz, pulse width 577 μs, 0.25 W	Normal use position right side of head	Dimmed room, in chair, 30-min exposure right side	Krause et al. (2000a)
There was no main effect of EMF at any frequency band between the event-related synchronous (ERS) and desynchronous responses (ERP elicited by targets and non targets, reaction time or accuracy (a) visual sequential letter task n-back)	24 right handed adults M/F 20–30 mean 23 y		Digital 902 MHz GSM	217 Hz, pulse width 577 μs, 0.25 W	Normal use position right side of head	Dimmed room, in chair, 30-min exposure right side	Krause et al. (2000b)
MP MWs may suppress the excessive sleepiness and improve performance while solving a monotonous cognitive task requiring sustained attention and vigilance	22 narcolepsy-cataplexy pts mean 48 y	0.06 W/kg over 10 g	900 MHz Motorola d520 MP	217 Hz, pulse width 577 μs, 2.8 Hz 0.25 W		45 min right ear	Jech et al. (2001)

continued

TABLE 4.3 (continued)

Effects of MW Exposure on Electroencephalography (EEG)

EMF Effect	Species	SAR (W/kg)	Frequency (MHz)	Modulation	Intensity (mW/cm^2)	Duration	Ref.
Mobile phone exposure decreased 1–4 Hz and 4–8 Hz activity and increased 8–12 and 30–45 Hz activity as a function of exposure duration recording an auditory discrimination task. Single blind	24 16 m, 8 f., 19–48 y. avg. 27.5 y. 20 R. handed		900 MHz Nokia 5110 and EMF attenuator	217 Hz pulse width 577 μs		20 min 19 scalp electrodes	Croft et al. (2002)
EMF effects on the EEG and on the performance on memory tasks were variable and were not replicated from Krause et al. (2000a) for unknown reasons	24 normal subjects	0.648 mW/kg	Digital 902 MHz GSM Nokia 6110	217 Hz, pulse width 577 μs, 0.25 W	Normal use position right side of head	Dimmed room, in chair, 30-min exposure right side	Krause et al. (2004)
Increase in error rate effects of 450 MHz exposure on human performance on visual memory tasks were reported by large variance between scores. Single blind	100, 63/37 m/f	0.35 W/kg est.	450 MHz	7 Hz 50% duty cycle	0.158 mW/cm^2	10–20 min	Lass et al. (2002)
Cell phones exposure gives transient abnormal slow waves in EEG of awake persons	awake subjects		Mobile phones			16 channel telemetric EEG (ExpertTM)	Kramarenko and Tan (2003)
During sham and MW exposure changes in human EEG varied strongly between subjects. No effect of MWs was evident. Single blind	20, 11/9 m/f; 19–23 y	0.35 W/kg est.	450 MHz	7 Hz 50% duty cycle	0.16 mW/cm^2 at the scalp	60-s on–60-s off for 10 cycles	Hinrikus et al. (2004) (Lass et al., 1999, 2002)
Effects of GSM signals on correlation coefficients of the spectra of auditory evoked responses were observed but it was difficult to deduce relation to human health	9 Healthy and 6 epileptic people		Various GSM MWs			13–14 selected out of 32 electrodes	Maby et al. (2005)
GSM 1800 did not affect verbal memory encoding tested by recoding Event-related magnetic fields	12,10 f., 18–30 avg. 23.6	0.16 W/kg (over 10 g)	GSM 1870	Pulsed at 217 Hz pulse width 577 μs		30 min left ear	Hinrichs and Heinze (2004)

Huber et al.'s (2002) preliminary results on the effect of 900 GSM on waking regional cerebral blood flow (rCBF) and on waking and sleep EEG in humans (see Table 4.3) are reanalyzed by Huber et al. (2005). Although it is not stated in the 2002 paper that they did the base station-like exposures and the subsequent associated PET scans in 2002, which were analyzed in the 2005 paper, we deduce this from the same age range (20–25 y) and average age (22.5 y) of the subjects in the 2005 paper. If the base station-like exposures and PET scans had been done in 2005 with the same subjects they would have been 3 y older. As such they have presented incomplete PET methodology in both the 2002 and 2005 papers, making interpretation difficult. The coordinates of significant results from the first analysis on the PETs after handset-like exposure in 2002 do not match the coordinates in 2005 at all. The authors fail to explain the variance of their PET scores or details of how they computed them and how they controlled for scientific uncertainty in their measurements, and thus this data cannot be interpreted. Presently, PETs are considered unreliable on their own and can present spurious results, so interpretation of PETs must be accompanied by other brain scans such as the computer tomography scans (Wager and Smith, 2003). Computed tomography (CT) scans allow for cross-sectional views of brain tissues. Presently, we have no validation that Huber et al. (2002, 2005) are measuring what they say they are measuring.

Two conclusions can be made. Firstly, the normal EEG appears to be too variable to be able to be a reliable measure of possible small MW effects. For instance, changes in EEG during the night could be attributed to uncontrolled changing brain temperatures that are known to change spike latency of potentials (Moser et al., 1994). And the PET scans used to locate the brain site of significant effects of MWs on EEG are known to be unreliable for this purpose. Unless EEG recording and analysis methodology is improved and standardized and the scientific uncertainty in EEGs is carefully defined and controlled for, it appears that further studies should not be considered.

4.2.9 Blood–Brain Barrier Studies

The BBB prevents high molecular weight substances in the blood from getting into the brain. This barrier protects the brain from foreign toxic substances, but allows the molecules that are necessary for metabolism to enter. Controversy has followed reports that MW exposure caused leakage through the BBB. In many follow-up studies, most researchers could not replicate the low-level BBB permeability changes or could show the effect only at high intensity levels, when the heating of the brain tissue was obvious. Now many researchers believe that the permeability change is associated with an increase in temperature-induced blood flow. D'Andrea et al. (2003b) reviewed 24 BBB studies since 1982 (see Table 4.4) and concluded that the question of MW-induced BBB permeability changes was still subjected to debate and required further experimentation. Presently, remaining BBB concerns related to MW exposures at low levels of SAR that require further study are the reported increase in binding of immunoglobulins to brain extracts (0.6 W/kg; see papers of the Vinogradov group, 1975–1993 in Table 4.4 and cited below), increased BBB permeability, and the occurrence of dark neurons in the brain (Salford et al., 1993, 1994, 2003; 2, 20, and 200 mW/kg). Replication and confirmation BBB experiments are currently in progress to address these uncertainties.

The Vinogradov group (1975–1993) (Vinogradov and Dumanskij, 1975; Dumanskii and Rudichenko, 1976; Vinogradov et al., 1981, 1983, 1985, 1991; Shandala et al., 1983, 1985; Vinogradov, 1993) has shown that exposure of rats to MW radiation led to an increased permeability of the BBB and a modification of the brain structure leading to the

TABLE 4.4

Effects of MW Exposure on the Blood–Brain Barrier

Effect	Species	SAR (W/kg)	Frequency (MHz)	Modulation	Intensity (mW/cm^2)	Duration	Ref.
The pre-1982 (13) publications showed the specific barrier function against nonlipid-soluble macromolecules of the BBB, is permeated above 42–43°C							Oldendorf (1970), Polyaschuk (1971), Sutton et al. (1973), Frey et al. (1975), Oscar and Hawkins (1977), Merrit et al. (1978), Preston et al. (1979), Sutton and Carroll (1979), Albert (1979), Frey (1979), Lin and Lin (1980), Albert et al. (1981), Oscar et al. (1981)
Increase BBB permeability by Evans blue dye, which is related to intense MW hyperthermia	Rats	3.0	2450	PW		20 min	Lin and Lin (1982)
No significant increase in BBB leakage correcting for thermal effects of the MW, using [^{14}C] and [^{3}H] as tracers molecules	CD albino rats	0, 2, 4, or 6	2450	CW	0, 10, 20, or 30	30 min	Ward et al. (1982)
No increase in permeation effects on the BBB were observed, ^{131}I albumin as a tracer molecule	Mongrel dogs		1000	CW	2, 4, 10, 50, or 200	20 min	Chang et al. (1982)
Attempts to alter $^{45}Ca^{2+}$ binding to brain tissue with pulse-modulated MW energy	Rat brain *in vivo* and *in vitro*	1.9–2.9	1000	PW	1	20 min	Merritt et al. (1982)

Absence of MW effects on BBB permeability to [^{14}C] sucrose in the conscious rat		0.3	2450 and 2800	CW and pulsed at 500 pps	10–40 CW 1–15 PW	30 min	Gruenau et al. (1982)
Increase BBB permeability to sodium fluorescein was found only in the brain made considerably hyperthermic by exposure to ambient heat or MW energy	Fisher-344 rats	13 (>41°C)	2450	CW	65	30 or 90 min	Williams et al. (1984a)
No increase in BBB permeability to HRP following exposure to ambient heat or MW, actually, a reduced uptake of the tracer by the brain was observed	Fisher-344 rats	13	2450	CW	0, 20, or 65	30, 90, 180 min	Williams et al. (1984b)
No increase in BBB permeability to tracer [^{14}C]	Rats	13	2450	CW	20 or 65	30 or 90 min	Williams et al. (1984c)
No change in uptake of either [^{14}C] or [^{3}H] tracer was found in any of the eight brain regions as compared with those of sham exposed animals	CD rats	0.1	1700	CW and PW		30 min	Ward and Ali (1985)
Increase of BBB permeability to [^{86}Rb] are associated with intense, MW-induced hyperthermia, and that the observed changes are not due to field-specific interaction	Wistar-derived rats	Avg. 3, a peak SAR: 240	2450	PW		5, 10, or 20 min	Goldman et al. (1984)
Ethanol inhibits MW-induced permeation of the BBB through reduced heating of the brain	Wistar rats	3.0	3150	CW		15 min	Neilly and Lin (1986)
MRI increases the BBB permeability to ^{153}Gd diethylenetriaminepentaacetic acid	Rats		6.25			23 min	Prato et al. (1996)
Increase uptakes of an intravascular molecule (Rh–F complex) through the BBB	Albino rats	~2	2450	PW	10	30–120 min	Neubauer et al. (1990)
Opening the BBB is due to MW-induced hyperthermia, not related to the nonthermal effect of MW	S-D rats	42.5 or 44.3°C	2450	CW		30 or 60 min	Moriyama et al. (1991)
Both CW and PW MW are able to open up the BBB for albumin passage	Fisher-344 rats	3.3	915	CW and PW 8, 16, 50 and 200/s			Salford et al. (1993)

continued

TABLE 4.4 (continued)

Effects of MW Exposure on the Blood–Brain Barrier

Effect	Species	SAR (W/kg)	Frequency (MHz)	Modulation	Intensity (mW/cm^2)	Duration	Ref.
Extravasation was independent of SAR <2.5 W/kg but rose significantly for higher SAR and was not significantly different between pulse and CW exposure conditions	Fisher-344 rats	0.16–5 W/kg	915	CW and PW 8, 16, 50 and 200/s		120 min	Salford et al. (1994)
Nerve cell damage in mammalian brain after exposure to MW from GSM mobile phones	Fischer 344 m. and f. 12–26 weeks	WB 2, 20 and 200 mW/kg	862–960	GSM 217 Hz	Peak power densities 0.24, 2.4, and 24 W/m^2	2 h	Salford et al. (2003)
Extravasation of Evans blue was observed in the regions where the temperature reached 43°C and above, but not in the areas where the temperature was 42°C and below	20 Wistar rats	<43°C	8	CW		30 min	Ohmoto et al. (1996)
Immunohistology against rat's own albumin: extravasation only at 7.5 W/kg	Rat	0.3–7.5 Brain	900	GSM		4 h	Fritze et al. (1997)
Immunohistology against rat's own albumin and fibrinogen. Avidin–biotin: extravasation at all SAR levels	Fischer m. and f. 344 rats 630 exposed various modulation and frequency and 372 controls	0.0012–12 WB	915 CW and GSM	217 Hz with 0.57 ms pulse width, or at 50 Hz with 6.6 ms pulse width		From 2 min to 960 min	Persson et al. (1997)
Immunohistology against rat's own albumin. Avidin–biotin Evan's blue: No extravasation	Rat	2 Brain	1439	GSM TDMA		1 h/d, 2–4 weeks	Tsurita et al. (2000)
In vitro model of BBB immunohistology against the cytosolic peripheral membrane proteins ^{14}C. sucrose flux: increased permeability	Mice	0.3	1800	GSM		4 d	Schirmacher et al. (2000)

In vivo immunohistology against mouse's own albumin. No extravasation	Mice	4.0 WB	898.4	GSM		60 min	Finnie et al. (2001)
Effect of long-term mobile communication MW exposure on vascular permeability in mouse brain. No extravasation	Mice	0.25, 1.0, 2.0 and 4.0 WB	898.4	GSM		60 min, 5 days/week for 104 weeks	Finnie et al. (2002)
New papers since D'Andrea review (2003)							
No leakage effect of 915 MHz exposure on the integrity of the BBB (Salford failed replication) except under conditions of heat. Double blind	S-D Fischer rats	20, 2, 0.2, 0.02, 0.002 W/kg (CW). 0.2, 0.02, 0.002 (pulsed 16). 2.5, 0.25, 0.025, 0.0025 (pulsed 217).	915	CW; Pulsed at 16 Hz and 217 Hz		30 min	McQuade et al. (2005)
Lack of effect of 1439 MHz local exposure on the BBB in immature and young rats	Rats	0, 2, 6 W/kg	1439	TDMA		90 min/d for 1–2 weeks	Shirai et al. (2005)
No BBB leakage due to exposure to 2.45 GHz as evidenced by no effects of scopolamine methylbromide on spatial memory during MW exposure or after MW exposure	S-D rats	Whole body 2.0 W/kg. Brain avg. 3.0 W/kg	2450	500 pps 2 μs		45 min	Cosquer et al. (2005c)
GSM 1800 MHz did not alter blood–brain barrier permeability to sucrose in BBB models *in vitro*	Porcine endothelial cell culture	Avg. 0.3 W/kg	1800	GSM		1–5 d	Franke et al. (2005a)
No evidence of UMTS field-induced disturbance of the BBB cultures quantified by ^{14}C-sucrose, serum albumin permeation, transendothelral electrical resistence, permeation of transporter substrates and integrity of the tight-junction proteins	Porcine endothelial cell culture	Max 1.8 W/kg		UMTS	Avg. 3.4–34 V/m	84 h	Franke et al. (2005b)

increase in binding of immunoglobulins to brain extracts. Transfer of these brain extracts to other animals led to strong immune responses. The Vinogradov group reported this effect on the immune system in rats exposed to a radar-like signal at around 2700 MHz (pulse recurrence frequency of 400 Hz, 40 ms duration of a train of pulses). These Russian papers have been translated and are reviewed by Lagroye and Veyret (personal communication). They reported "the critical effect was found at 500 $\mu W/cm^2$ i.e., 5 W/m^2. In rats, this incident power corresponds to ca. 0.6 W/kg. With a reduction factor of 10 the Russian workers limit is set at 50 $\mu W/cm^2$ and 15 $\mu W/cm^2$ for sensitive population" (Lagroye and Veyret, personal communication). Since the Vinogradov group studies have methodological limitations, replication studies are performed at PIOM, France where they have the exposure systems and the ability to test the immune response.

Because of the recent claims of Salford et al. (1993, 1994, 2003) of leaks in the BBB and the presence of dark neurons, reported to result from low MW exposures (whole body: 2, 20, and 200 mW/kg), replications of the Salford group studies are underway at the U.S. Air Force Research Laboratory (Mason), in Japan (Ohkubo) and within the French research program (Veyret).

Overall, the new BBB references in this chapter (Cosquer et al., 2005c; Franke et al., 2005a,b; McQuade et al., 2005; Shirai et al., 2005; see Table 4.4) indicate no effects of MW exposure in a wide range of exposures above and below the present ICNIRP and IEEE limits for either whole body or local exposure on the BBB where the body temperature does not rise above 1°C. McQuade et al. (2005), in a replication of the work of Salford et al. (1993, 1994, 2003), have reported at a conference no leakage in the BBB after low SARs (2, 0.2, 0.02, 0.002 W/kg). At the same conference, Shirai et al. (2005) reported the lack of an effect of 1439 MHz local exposure on the BBB in immature rats. A published paper by Cosquer et al. (2005c) showed no BBB leakage after 2450 MHz MW exposure (2 μs pulse width, 500 pps, whole body SAR of 2.0 W/kg, ± 2 dB and brain averaged SAR of 3.0 W/kg, ± 3 dB) as evidenced by no spatial memory deficit after scopolamine methylbromide, administered i.p. This muscarinic antagonist substance does not cross the BBB unless there is leakage in the BBB. If it does, it is known to disrupt performance of spatial memory.

Franke et al. (2005a) had previously reported that GSM 1800 exposure increased sucrose permeation across the BBB *in vitro* (Schirmacher et al., 2000). The cell culture BBB model used in their previous study was of rat astrocytes in coculture with porcine brain microvascular endothelial cells. In this study (Franke et al., 2005a), after optimization of cell culture conditions, the GSM 1800-related effects on BBB permeability (Schirmacher et al., 2000) were no longer evident. Cell cultures were exposed for 1–5 d at an average SAR of 0.3 W/kg. They reported that they could not confirm the enhanced permeability of the BBB *in vitro* after the MW exposure of Schirmacher et al. (2000) since the *in vitro* barrier tightness in the Franke et al. (2005a) experiments was now more like that of the *in vivo* situation. Frank et al. (2005b) used this same optimized *in vitro* model of the BBB (Frank et al., 2005a) to investigate the effects of a generic Universal Mobile Telecommunications System (UMTS) signal on BBB tightness, transport processes, and the morphology of the porcine brain microvascular endothelial cells. This *in vitro* model of the BBB was exposed continuously for up to 84 h at an average electric field strength of 3.4–34 V/m (maximum 1.8 W/kg). They reported that they did not find any evidence of the UMTS field-induced disturbances of the function of the BBB.

The BBB area of research remains open while we are awaiting the publication of results of further BBB replication studies.

4.2.10 Electrosensitivity in Humans

The Netherlands Organization for Applied Scientific Research (TNO) study, known colloquially as COFAM, on the effects of GSM and UMTS signals on well-being and cognition, was released to the public on the internet at the Health Council of the Netherlands home page on September 9, 2003 (Zwamborn et al., 2003). COFAM suggested that GSM and UMTS base site-like signals might have a negative influence on "well-being" and cognition. One of the two groups of subjects was self-reported to be "electrosensitive." It was found shortly after release, when international critical scientific comments on the statistical analyses were addressed through reanalyses of their data that the only effects were on "well-being" from UMTS exposure.

Following this, the Dutch Health Council EMF Committee was asked to carry out a critical review of the COFAM research methods (Roubos et al., 2004). The consultant psychometrist advised that there was no empirical evidence to indicate the metrological qualities of the subscales and tests of the "well-being" questionnaire and the cognitive tests used with the particular TNO subject groups under the conditions that prevailed for the TNO study. Since there was no measure of their validity, there was no proof to suggest the "well-being" questionnaire and the cognitive tests measured what they claimed to measure. The Dutch EMF Committee concluded that there was no objective evidence to justify mitigating measures. It is essential that, along with dosimetric and statistical expertise, international expertise in psychometrics be included in replication and follow-up studies (Roubos et al., 2004).

The Swiss Government authorized a TNO "replication" and expansion study in September 2004, to be completed in September 2005 (Achermann et al., 2005). Two similar studies are underway in the U.K. in the MTHR program, and several other European research groups including Denmark will carry out confirmation or replication studies of the TNO-COFAM study. The COFAM study is not published in a peer-reviewed scientific journal and hence is criticized by groups such as the WHO EMF Project as not fulfilling its criteria for inclusion in the research base for health risk assessment. The consequent public concern about the COFAM results has slowed down roll out of UMTS in Europe (BBC NEWS, 2003).

There are several studies of subjects' self-reported symptoms (see Table 4.5; Bergqvist and Vogel, 1997; Hansson Mild et al., 1998; Hocking, 1998; Chia et al., 2000; Oftedal et al., 2000; Koivisto et al., 2001; Sandstrom et al., 2001; COMAR, 2002; Hietanen et al., 2002; Zwamborn et al., 2003; Schmid et al., 2005; Balik et al., 2005; Balikci et al., 2005; Belyaev et al., 2005; Markova et al., 2005; Rubin et al., 2005; Seitz et al., 2005). The scientific consensus is that this is an idiopathetic environmental intolerance (Staudenmayer et al., 2003). There is no demonstrated relation between the self-reported sensitivity to MWs electrohypersensitivity (EHS) and the exposures of these people. However, more research is required to better characterize this self-reported EHS condition. Research is presently ongoing concerning subjects' self-reported symptoms of EHS.

4.2.11 Future Microwave Mobile Signals and Further Research

The SAR values are guaranteed by manufacturers to be within guideline limits (ICNIRP, 1998; IEEE, Std C95.1-2005). With frequency and power modulations different from the GSM signal and other "second generation" cell phone signals in third and fourth generation signals of MW telecommunications (Andersen, 2005), must we once again do biological tests for next generation telecommunications to investigate if the new modulations could have adverse biological effects? This is an important question. If the only established biological effects are from thermal mechanisms, then the existing standards

TABLE 4.5

Effects of MWs on Self-Reported Human Hypersensitivity Ideopathic Environmental Intolerance

This review was unable to establish a relationship between low- or high-frequency fields and electromagnetic hypersensitivity or with symptoms typically occurring among such afflicted individuals	138 centers and 15 EU countries	Limit 2 W/kg or less	Mixed mostly, Analog signals in public use	Mostly CW	Questionnaires self-report	Bergqvist et al. (1997)
GSM users reported symptoms (headaches) warmth sensation on the ear and behind or around the ear less frequently than NMT users	Questionaires 6379 GSM users; 5613 NMT users	Limit 2 W/kg or less	Analog and GSM phones in public use	Analog (CW) GSM (217 pulse width 577 μs)	Self-report	Mild et al. (1998)
MW sickness. The syndrome involves the nervous system and includes fatigue, headaches, dysaesthesia, and various autonomic effects in MW radiation workers	40 self-reporting persons		Australia	Mobile phones	Self-report survey	Hocking (1998)
Subjective sensations of warmth around the ear. No effect on memory	Humans		Sweden 900 MHz	CW, GSM	Normal use	Oftedal et al. (2000)
Comparison of subjective symptoms of users of analog and GSM phones with adjusted odds ratios did not indicate any increased risk for GSM users with his larger sample (changed result from Mild (1998) above)	Sweden 6379 GSM users; 5613 NMT. Norway 2500 GSM and 2500 NMT		900 MHz	GSM NMT	Questionaires, self-report	Sandstrom et al. (2001)
A review of EMF hypersensitivity. The symptoms reported by self-reported EMF hypersensitive (EHS) individuals—headache, fatigue, stress are common and nonspecific. EHS bears close resemblance to idiopathic environmental intolerances (IEI)						COMAR (2002)

Subjective symptoms in humans were not related to their MW exposure. Single blind	1. 48: 24 m., 24 f., 18–49 y; avg. 26 y. 2. Same 18–34 y avg. 23.2 y all R. handed		GSM 902 MHz, Nokia 6110	217 Hz pulse width 577 μs	0.25 W mean power	30–60 min left ear on and off skin temp. about 35°C	Koivisto et al. (2001)
Hypersensitivity symptoms associated with exposure to cellular telephones: no causal link	20: 7 m., 13 f.,		1. NMT 900 MHz. 2. GSM 900 MHz. 3. GSM 1800 MHz	1. CW, 2–3. GSM, 217 Hz pulse width 577 μs	1. 1 W, 2. 2 W, avg. 0.25, 3. 1 W, avg. 0.125	30 min × 4 sessions over 1 d	Hietenan et al. (2002)
Headache was the most prevalent symptom among hand-held cellular telephone users in Singapore as compared to nonusers. Significant increase in headache with increasing duration of usage (minute/day)	808 m. and f., age 12–70 y; mobile users	GSM 900 or 1800 MHz				Self-report questionnaire	Chia et al. (2000)
Hypersensitivity symptoms associated with exposure to cellular telephone: No causal link for GSM. But UMTS-like exposure lead to significant number of feelings of unwellness on non-validated tests with only negative items. Replications underway	Sensitive 36: m. 11, f. 25; 31–74 y, avg. 55.7 y. Normal 36: m. 22, f. 14; 18–72 y; avg. 46.6 y	0.045 (900); 0.082 (1800); 0.64, 0.078 (2100)	GSM 900, 1800; UMTS 2100	GSM and UMTS-like modulations		? possibly 20 min a session during the Taskomat test?	Zwamborn (2003)
No influence on selected parameters of human visual perception of 1970 MHz UMTS-like exposure. Double blind	58: 29 m., 29 f.	High 0.37 W/kg 10 g. avg. range 0.16–0.84 W/kg Low 0.037 W/kg	1.97 GHz	UMTS 5 Hz wide UMTS-FDD signal 1500 Hz, 8–12 Hz components		4 tests × 3 exposure condition approximately 2 h left side	Schmid et al. (2005)
A survey found some subjective ocular symptoms and sensations experienced by LT (1–4 y) users of mobile phones	695: 193 f.; 157: 36 use–non. 502 m.; 392: 110 Use–nonElazigy, Turkey		MPs			LT users (1–4 y)	Balik et al. (2005)

continued

TABLE 4.5 (continued)

Effects of MWs on Self-Reported Human Hypersensitivity Ideopathic Environmental Intolerance

A self-report pilot survey on symptoms of LT users (2–4 y) of mobile phones found no effect on dizziness, shaking in hands, speaking falteringly, and neuropsychological discomfort, but effects for headache, extreme irritation, increase in the carelessness, forgetfulness, decrease of the reflex, and clicking in the ears	695: 193 f.; 157: 36 use–non. 502 m; 392: 110 Use–nonElazigy, Turkey	MPs		LT users (2–4 y)	Balikci et al. (2005)
A review. Idiopathic environmental intolerance IEI is a belief characterized by an overvalued idea of toxic attribution of symptoms and disability, fulfilling criteria for a somatoform disorder, and a functional somatic syndrome					Staudenmayer et al. (2003)
Thirty-one experiments testing 725 electromagnetically hypersensitive (EHS) participants were identified. Metaanalyses found no evidence of an improved ability to detect EMF in hypersensitives	725 EHS persons	161 MP EHS			Rubin et al. (2005)
Thirteen EHS studies re MW of mobile phones were reviewed. They concluded no valid evidence for an association between impaired well-being and exposure to mobile phone radiation. The limited quantity and quality of research do not exclude long-term health effects	13 EHS studies	MPs			Seitz et al. (2005)

are sufficient to insure no adverse cognitive effects occur within the guideline limits. To date, all the literature reviews by expert panels have considered all the reported effects (ICNIRP, 1998; NRPB, 2004; IEEE, Std 95.1-2005) and concluded that none could be replicated and established as an adverse effect, except absolute temperature rise above the guideline limits. Present evidence does not indicate that modulation is biologically significant, apart from a few special cases such as intense pulses (ICNIRP, 1998; NRPB, 2004; IEEE, Std C95.1-2005); if the case were otherwise, the entire rationale for MW exposure guidelines would need revision (Foster and Repacholi, 2004).

There are two ways to proceed with future research and research is proceeding on both fronts. One way is to continue to investigate each modulation for biological effects without a known biophysical mechanism to support this research (Foster and Repacholi, 2004). It is paramount in this uncertain area below guideline limits that the quality of telecommunications research become the highest possible caliber using standardized protocols, double blind procedures, multicentered replications, well-validated psychometrics, and the proper statistics to rule out chance effects and experimental artifacts as we search for other as yet unknown biophysical mechanisms.

The other way to proceed with future research is to build on what we know, that is, to better define thermal effects of MWs. It is paramount that research proceeds to investigate the area where the known mechanism is heat, measured as absolute whole body temperature rise, to better define the effects of MW heat interactions with the body and better define the thresholds more clearly where heat effects becomes adverse to animal behavior and human cognition (Andersen and Moser, 1995; Adair et al., 1998, 2001a,b, 2003, 2005; Adair and Black, 2003; Allen et al., 2003, 2005; Hancock and Vasmatzidis, 2003; Hirata and Shiozawa, 2003; Kheifets et al., 2003; Foster and Adair, 2004; IEEE, Std C95.1-2005; Section C.2.2.2.1.2). Validated thermal models could be used to predict the cognitive effects of MW signals more accurately for new technologies. Due to changing MW exposures of new technologies (Andersen, 2005) and a large number of cognitive and behavioral factors that come into play during temperature changes that are just now beginning to be investigated (Hancock and Vasmatzidis, 2003), the thermal mechanism area remains open to further research.

4.2.12 Microwave Effects: Overall Conclusion

The studies of warmth detection by humans and responses of laboratory animals to similar exposures have provided confidence that behavioral changes in animal studies are indeed thermally motivated. D'Andrea (1999) postulated that performance of cognitively mediated tasks might be disrupted at levels of exposure lower than that required to cause performance disruption. The results from numerous performance and cognitive tasks of varying difficulty do not support this hypothesis.

Hancock and Vasmatzidis (2003), on reviewing the literature on the effects of heat stress on cognitive performance, found two trends: "First, heat stress affects cognitive performance differentially, depending on the type of cognitive task. Secondly, it appears that a relationship can be established between the effects of heat stress and deep body temperature." They concluded, "that much remains to be understood before a (temperature rise) limit becomes universally acceptable."

Several studies have shown that the disruption of ongoing behavior, including spatial memory in rats during acute MW exposure, is generally associated with a 1°C or greater increase of body temperature. The studies that examined human cognitive processes and mobile phone usage appear to show no established evidence of associated memory deficits. Replication studies with standardized protocols, larger samples (e.g., multicentered replications; Haarala et al., 2003b), better experimental controls, double blind conditions, and Bonferroni or other statistical corrections for multiple comparisons have failed to replicate the initial positive results in this area. Further high-quality multicentered research in human cognitive studies including human spatial memory may be needed to strengthen these results.

Because the normal EEG appears to be too variable to be a reliable measure of possibly small MW effects, further EEG studies should not be considered until methodologies improve. The BBB research area remains open, while we are awaiting the publication of results of further BBB replications. The scientific consensus is that MW hypersensitivity is an idiopathetic environmental intolerance with no demonstrated relation between the self-reported sensitivity to MWs and the exposures of these people. Further studies are underway to better characterize this intolerance.

The final conclusion must be that disruption of ongoing behavior during acute MW exposure is generally associated with 1°C or greater increase of body temperature. It seems counterintuitive to expect a chronic effect of MW exposure when no acute effect below 1°C is evident, however, there is limited cognitive information on chronic exposure of humans to low-level emissions.

4.3 Extremely Low-Frequency Effects

Exposure to a time-varying electric field will induce current flow within the biological body relative to the external field strength. When exposed to a time-varying magnetic field, an electric field within the biological body will be induced leading to subsequent current flow. If the induced fields and currents exceed the stimulation thresholds of excitable tissues, then neural activity may be initiated (see Reilly, 1998). These basic effects are related to the interaction between the induced electric fields and voltage-gated ion channels of a variety of excitable tissues (see NRPB, 2004; Kandel et al., 2000).

Detection, perception, behavioral and cognitive effects from weak ELF EMFs have been studied in a variety of animal and human experiments, measuring many different biological end points. Animal and human studies have determined the thresholds at which EMFs are perceptible, disrupt behavioral performance, or produce effects on cognitive processes.

4.3.1 Detection of Extremely Low-Frequency Electric and Magnetic Fields

Studies by Kalmijn (1971) have shown that some species, such as the dogfish, *Scyliorhinus canicula,* have an elegant apparatus, Ampullae of Lorenzini, for detection of ELF fields at extraordinarily low intensities around 5 nV/cm. Many studies have evaluated the use of earth magnetic fields by migratory bird species as a means of navigation (Keeton, 1971). These naturally occurring magnetic fields are of very low intensity and demonstrated to have effects on bacteria as well.

Mammals are not known to possess electrosensory receptors for detection of ELF fields at extraordinarily low intensities. Mammals perceive ELF fields by other mechanisms that stimulate other types of sensory receptors such as light, touch, mechanoreceptors and pain, which result in nervous stimulation at much higher ELF intensities than electrosensory fish with specialized and highly sensitive electroreceptors (Bullock, 1986).

In the ELF range, the effects of magnetic fields that penetrate the body and result in induced electric fields have been the main focus of behavioral research. Smith et al. (1994) evaluated the ability of rats to detect magnetic field levels using a conditioned suppression

TABLE 4.6

Effects of ELF Fields on Cognition and Behavior

Effect	Species	Frequency (Hz)	Electric Field (V/m)	Magnetic Field (A/m)	Modulation	Duration	Ref.
Slowing of event-related potentials (ERP) recorded during behavioral performance	Humans, $n = 21$ and 18 in two experiments	50		100 μT_{rms}	15 s on/off cycles	Three 30-min sessions	Crasson et al. (1999)
No effect of magnetic fields, bright light, or sham exposure on ERP or cognitive performance	Humans, $n = 18$	50		100 μT_{rms}	15 s on/off cycles	Three 30-min sessions	Crasson and Legros (2005)
High-voltage direct current static electric fields did not produce taste aversion learning	Male Long Evans rats, $n = 56$	DC	+75 to −75 kV/m		None		Creim et al. (1984)
Rats could not detect the MFs and were unable to acquire the two-choice discrimination	Male, specific pathogen-free (SPF) Sprague–Dawley rats	60 and DC		50 mT 60 Hz, 26 mT static fields	None		Creim et al. (2002)
Results of this study do not support the hypothesis that ELF MF affects human cognitive performance	Male humans 20–30 y old, $n = 32$	50		20 and 400 mT		65 min	Delhez et al. (2004)
Retarded learning in 12-arm radial maze test of spatial memory	Rats	60		0.75 mT		45 min	Lai (1996)
Magnetic field exposure causes a deficit in spatial "reference" memory in the rat	Male Sprague-Dawley rats, $n = 24$	60		1 mT		60 min	Lai et al. (1998)
Thresholds of phosphene induction for different background illuminations levels and wavelengths	Humans, $n = 11$ (exp 1) $n = 8$ (exp 2) $n = 6$ (exp 3) $n = 9$ (exp 4)	Threshold was lowest at 20–30 Hz		0–40 Hz tested Threshold lowest at 10–12 mT			Lövsund et al. (1980b)

continued

TABLE 4.6 (continued)

Effects of ELF Fields on Cognition and Behavior

Effect	Species	Frequency (Hz)	Electric Field (V/m)	Magnetic Field (A/m)	Modulation	Duration	Ref.
Impairment of spatial memory observed in rats and decreased motivation by patterned magnetic fields	Male Wistar rats, $n = 28$			200 nT or 500 nT	Patterned magnetic fields with inter-stimulus intervals of 0, 2,000, 4,000, or 10,000 ms		McKay and Persinger (2000)
No work stoppage on a delayed match-to-sample task could be achieved	10 baboons (*Papio cynocephalus*)	60	6 to 30 kV/m	50 mT and 100 mT	None		Orr et al. (1995a,b)
The average EF detection threshold was 12 kV/m; the range of means among subjects was 5–15 kV/m	6 baboons (*Papio cynocephalus*)	60	5 to 15 kV/m				Orr et al. (1995a)
No effects on visual duration discrimination or recognition memory	30 male and 50 female human subjects	50		100 μT		1 s on and 1 s off	Podd et al. (2002)
No effects on human reaction time	12 human subjects	Continuous 0.1 Hz, 0.2 Hz, or sham field		1.1 mT			Podd et al. (1995)
Removal of ambient magnetic field induces analgesia in mice. Response may be opiod related as morphine	Swiss CD-1 mice adult male, $n = 4$			Absence of ambient geophysical			Prato et al. (2005)

mimics effects and naloxone (antagonist) blocks effects	to $n = 48$ in different experiments			MF by Mu-metal enclosures		
Baboon response rates on a two component operant schedule showed work stoppage when initially exposed to strong electric fields	12 naive baboons (*Papio cynocephalus*)	60	30 to 60 kV/m			Rogers et al. (1995)
Results magnetic fields affected spatial learning in rats by modifying rate of aquisition	Adult male C57BL/6J mice	60		0.75 mT	45 min	Sienkiewicz et al. (1998a)
Exposure to MF showed dose-dependent effects on spatial learning task in an eight-arm radial maze	Adult male C57BL/6J mice	60		7.5, 75, 0.75, or 7.5 mT	45 min	Sienkiewicz et al. (1998b)
Rats would suppress responding during MF exposure	Rats	7, 16, 30, 60, and 65.1		1900 μT at 7 Hz to 200 μT at 65.1 Hz	3 min	Smith et al. (1994)
Threshold of detection was between 4 and 10 kV/m. Probability of detection increased with increasing field strength	Rats	60	4 to 10 kV/m		3–4 s trials	Stern et al. (1983)
No effects on a performance, a two-alternative, forced-choice, duration-discrimination task	100 male and female human subjects ranging in age from 18 to 48 y of age	50		100 μT	9 min exposures	Whittington et al. (1996)

behavioral technique. They determined that rats were able to detect magnetic fields as low as 200 μT, although this has been questioned by Stern and Justesen (1995) because adequate sham tests were not done.

Recently, Creim et al. (2002) used a two choice discrimination learning task in which the presence of magnetic fields was tested as a discriminative stimulus (SD) to cue food-deprived rats as to which of two directions food could be found. They reported that nine rats trained for 64 trials (eight trials per day) could acquire a two choice discrimination in a modified radial arm maze based on a specified SD of a change in ambient illumination ($p < .001$), whereas nine rats similarly trained could not acquire a two choice discrimination based on the SD of a combination of sinusoidal and static magnetic field exposure (50 μT, 60 Hz, 26 μT static fields) and any cues attendant to energizing the coils that produced the MF exposure. The selection of the field (50 μT, 60 Hz, 26 μT static field) corresponded to the ion resonant conditions for calcium ions (Liboff et al., 1989), since spatial learning and memory tasks have been previously determined to be causally dependent upon the movement of free calcium in the animal cortex (Liboff et al., 1989; Colombo et al., 1997; Antoni et al., 1998; Gnegy, 2000). Recent research on calcium underlines the importance of intracellular calcium in the molecular mechanisms of learning (Kandel et al., 2000). In conclusion, detection thresholds for magnetic fields in animals are less clear and show greater variability than those for electric fields (ICNIRP 2003; NRPB, 2004).

In animal studies, perception refers to the behavioral response to the detection of an ELF electric field as a cue or discriminative stimulus. Orr et al. (1995a) used operant methods to train baboons (*Papio cynocephalus*) to perform a detection task for the presence of an ELF EMF. The animals responded for food reward and showed that the average ELF detection threshold for electric field was 12 kV/m with a range of detection for the different subjects from 5 to 15 kV/m. The ELF detection threshold of nonhuman primates is similar to thresholds reported for electric fields in rats in the range of 3–15 kV/m with probability of detection increasing as a function of electric field intensity (Stern et al., 1983; Creim et al., 1984; Stern and Laties, 1985). Animals may detect the presence of low-frequency electric fields, possibly as a result of surface charge effects, including hair stimulation (Weigel et al, 1987; Stell et al., 1993).

In humans, the CNS has generally been viewed as the chief target for EMF-induced effects because the CNS and peripheral nervous system consist of excitable tissues and are responsive to electric fields and the currents induced within the body by external exposure of ELF fields. This has implied that the CNS may be responsive to very low-intensity fields. One of the biological effects resulting from lower field strengths has been the generation of phosphenes, small sensations of light during exposure to electric and magnetic fields. Visual system phosphenes can be generated at the retina by magnetic fields of near 15 mT oscillating at 50–60 Hz, producing electric fields in tissue on the order of 1 V/m (Reilly, 1998). Although other estimates of electric fields needed to generate phosphenes in the retina vary around 10–60 mV/m at 20 Hz, there is large uncertainty in these values (NRPB, 2004). When a suprathreshold stimulus is recurrent, the force on innervated muscle increases in the case of motor neurons (McNeal and Bowman, 1985) and perceived sensory intensity increases in the case of sensory neurons. Reported adverse reactions to CNS electrostimulation above phosphene excitation thresholds consist of tiredness, headaches, after images (up to 10 min) and spasms of the eye muscles (Lövsund et al., 1980a,b; Silny, 1986), headaches (Aurora et al., 2003), and hyperpolarization (Aurora et al., 2003; Pollen, 2004). One question that has been recently addressed is the strength of the association between headaches and phosphene stimulation. Recently Aurora et al. (2003) have demonstrated that there is a direct neurophysiological correlate for clinical observations, which have inferred hyperexcitability of the occipital cortex in migraineurs. This has been confirmed electrophysiologically by Pollen (2004).

Another case for additional concern is the possible differences in sensitivity in spinal cord and autonomic nervous systems (ANS). However the ANS questions appear to be covered, for instance, by studies of the nervous sensitivity of the heart muscle and CNS responses related to phosphenes. The ANS is predominantly an efferent system transmitting impulses from CNS to peripheral organ systems. Simple autonomic reflexes are completed entirely within the organ concerned, whereas more complex reflexes are controlled by the higher autonomic centers in the CNS, principally the hypothalamus (Bakewell, 1995).

The effects are easily discernable at higher field intensity levels such as those found in magnetic resonance imaging (MRI), transcranial magnetic stimulation (TMS), and electroconvulsive therapy (ECT), which have been well-documented (see Shellock and Crues, 2004; Wasserman, 1998; Sackheim, 1991, respectively). For example, brain stimulation during TMS creates sites of focal neuronal activity and a functional lesion useful during studies of cognitive functions; it has a threshold of near 20 V/m (Reilly, 1998). A recent paper by Nadeem et al. (2003) calculated the electric fields near the head surface for ECT at 1.6–80 kV/m, while deep in the brain they are in the range of 100–500 mV/m for both TMS and ECT. The biological effects of ECT and TMS have been intensively studied and well-documented. At high power, magnetic fields can stimulate nerve tissue directly through the induction of circulating eddy currents (Ueno et al., 1986) but no effects on nerve function are seen within guideline limits. Electrophysiological studies (evoked potentials and ERP) carried out during exposure include the possibility that the electrodes used to record the responses can be the source of field-induced artifact (Angelone et al., 2004).

Recently Shupak et al. (2004b) have published results on perception of pulsed magnetic fields by humans; they confirmed that within their paradigm they did not appear to be perceived. They investigated magnetic field exposure on sensory and pain thresholds following experimentally induced warm and hot sensations, using a double blind design. The magnetic field was established by the presence of three nested orthogonal pairs of Helmholtz-like coils, 2 m square $\times$ 1 m separation, 1.751 m $\times$ 0.875 m separation, and 1.5 m $\times$ 0.75 m separation. A constant current amplifier excited via a digital-to-analog converter was used to drive each of the three coil pairs. The pulsed magnetic field had a frequency content between 0 and 500 Hz as determined by a Fourier analysis and a magnitude of $\pm$200 μT peak in the vertical direction. The ambient static magnetic field (MF) present in both sham and exposed conditions was 11.6 μT in the horizontal direction and 50 μT in the vertical direction. The ambient 60 Hz field for both sham and magnetic field exposed groups was 0.2 μT_{rms}. Their results indicate that within their experimental setup, the magnetic field exposure did not affect basic human perception, but can increase pain thresholds in a manner indicative of an analgesic response. But there is an apparent involvement of the placebo effect that is under investigation.

While strong fields in tissues, as described above, can produce biological effects, laboratory studies at low levels (<500 μT) have produced only subtle and transient effects on neurobehavioral responses and conditions necessary to elicit these are not clearly defined (Cook et al., 2002). There are few laboratory data from human or laboratory animal experiments that would even suggest that ELF-EMF, at or below current safety standards, pose a health risk.

Human detection sensitivity for electric fields is evident in stimulation of the peripheral nervous system in the skin (Bailey and Nyenhuis, 2005). Reilly (1998, p. 354) described research on detection of electric fields by 122 men and 8 women volunteers. He pointed out one study in which men and women were tested at a high voltage facility with overhead transmission lines, and as they walked through the electric fields they rated their own response as no feeling, perception, or annoyance (Reilly, 1978). A few individuals rated perception in remarkably low fields (<2 kV/m). For the women, the median threshold for perception with the hands raised was 17.5 kV/m, whereas median perception for men with hands raised was lower at 6.7 kV/m. Responses to electric fields depend on relative humidity, but these effects were reported to represent a wide range of environmental conditions.

Leitgeb and Schröttner (2003) investigated electrosensibility characterized by perception threshold and its standard deviation in a sample of the general population of 708 adults, including 349 men and 359 women aged between 17 and 60 y. They used an isolated stimulator producing currents of less than 5 μA on the lateral side of the forearms of volunteers. They reported that electrosensitivity and its range of variability change from day to day (see also ICNIRP, 2003).

Spark discharges from people to grounded objects may occur and can be annoying and sometimes painful (Reilly, 1998). The magnitude of the spark will depend on how well the person is insulated from ground, the charge voltage, and impedances involved. Spark discharge perception threshold is complicated by many factors which are involved such as "frequency, induced open circuit voltage and capacitance between the conducting object and exposed person, temperature, speed of making or breaking contact, bodily location where contact is made, and other variables" (IEEE, Std C95.1-1991/1999).

In conclusion, electrophysiological studies remain difficult to interpret, but none suggest that ELF effects within the guideline limits are hazardous (see ICNIRP, 2003; Foster, 2003). Despite the difficulties, research continues in this area but focused for the most part on medical applications, where therapeutic exposure levels are often well above the guideline limits.

4.3.2 Extremely Low-Frequency Cognitive Effects: Animals

Adverse reactions to CNS electrostimulation below excitation thresholds ($B = 0.75$ mT at 50 Hz) may result in transitory reduced acquisition of spatial learning in mice (Sienkiewicz et al., 1998a,b). Other laboratories have also reported effects on spatial memory of rats, mice, and voles at power frequencies at 100 μT and above (Kavaliers et al., 1993, 1996; Lai, 1996). Other behavioral (Creim et al., 2002) effects tested and the various complexities of the signals (Thomas and Persinger, 1997; McKay and Persinger, 2000) make a conclusion difficult. Further replication and confirmation studies are required of the effects of 0.75 mT at 50 Hz field exposure on acquisition and retention of spatial memory in two species (rats and mice) before such an effect could be considered as established evidence.

The mechanism of an ELF effect of a 0.75 mT exposure at 50 Hz on memory has not been established. The possibility that it may be the result of electrostimulation in the CA1 region of the hippocampus needs to be further explored (Tattersall et al., 2001; Pakhomov et al., 2003) (for background neurobiology of the hippocampus, see Section 4.2.4.1). Hippocampal pyramidal cells fire in synchronized assemblies or networks (O'Keefe and Nadel, 1978). The possibility that electrostimulation from an outside ELF source such as 0.75 mT at 50 Hz could thus have a coordinated effect on the firing of an assembly of hippocampal cells above the endogenous biological noise in the system needs further investigation (Saunders, 2003).

Other animal researchers (Kavaliers et al., 1985; Lai, 1996; Thomas and Persinger, 1997; Lai and Carino, 1998), using various paradigms including the spatial memory paradigm of the radial arm maze, have suggested the ELF-associated learning deficits could result from changes in cholinergic system triggered by ELF interactions with the opioid system in the animal. However, as reviewed in Section 4.2.4.1 above, this result is not supported by the established neurophysiological evidence and molecular mechanism for spatial memory (see Kandel et al., 2000).

Recently in support of the research on ELF modulation of the opioid system, Prato's group has investigated magnetic and electric fields' ability to modulate the opioid system in animals. Prato et al. (2005) reinvestigated magnetic and electric fields' ability to modulate the opioid system in CD-1 mice (Choleris et al., 2001; Shupak et al., 2004a). They extended the previous findings from studies performed independently in Pisa, Italy and London, Ontario, Canada (Choleris et al., 2001) to elucidate what effects the absence of the geomagnetic field may have on nociceptive sensitivity and pain sensation. Choleris et al. (2001) did a series of experiments which repeatedly showed that a 2 h stay in a Mu-metal box, shielding the animal from ambient magnetic and electric environment, resulted in a significant decrease in response latencies, which indicated a significant increase in nociceptive sensitivity and reduction in stress-induced analgesia (SIA) in C57 mice. The effect of the Mu-metal box was specific to the shielding of the ambient magnetic field, as a 2 h exposure within a copper box, which shields only the electric field, did not affect SIA. Prato et al. (2005) targeted several objectives. They tested whether 10 d of repeated 1 h acute exposures of mice to a near zero magnetic field environment (less than 1 μT) produced by Mu-metal shielding would affect nociception. They validated the earlier finding that the effect is due to shielding of the magnetic field rather than only the ambient electric field by performing identical experimental procedures to the Mu-metal trial in a thinly clad copper box. They found that the opioid mediation only partly explains the interaction with repeated exposure to the shielded condition. Other variables in the experiment could be the source of variance such as light and moisture in the shielded environment. The authors noted that further research is required (Prato et al., 2005).

4.3.3 Extremely Low-Frequency Animal Cognitive Studies: Conclusion

The ELF exposure conditions and the associated effects on cognition are too variable to draw any conclusions (ICNIRP, 2003; NRPB, 2004; Sienkiewicz et al., 2005). There is no support for changes in the cholinergic system triggered by ELF interactions with the opioid system in the established evidence on the molecular mechanism of spatial memory. Indeed these differences in outcome may depend on other variables including the timing and arousal and motivation, lighting, and duration of exposure relative to learning (McKay and Persinger, 2000; NRPB, 2004; Prato et al., 2005). These results suggest that the neural processes underlying the performance of cognitive tasks may be vulnerable to the effects of magnetic fields. One possible mechanism for the vulnerability of cognitive tasks that remains to be investigated is CNS electrostimulation below excitation thresholds. Overall, what appear to be the transient small biological effects on memory do not suggest harmful disruption of memory (NRPB, 2004; Sienkiewicz et al., 2005).

4.3.4 Extremely Low-Frequency Cognitive Studies: Humans

Previously magnetic and electric field-induced effects on attention, vigilance, memory, and other information-processing functions have been reviewed by NIEHS (1998), Cook et al. (2002), IEEE (Std C95.6-2002), Crasson (2003), NRPB (2004), ICNIRP (2003), and Sienkiewicz et al. (2005). Few field-dependent changes have been observed within guideline limit values. Newer research reviewed for this paper, published 2002–2005, reinforces the conclusion previously made by expert bodies that this type of study seems particularly susceptible to environmental and individual noise factors that could decrease the power to detect small field-dependent effects (Foster, 2003; NRPB, 2004).

Podd et al. (2002) investigated both the direct and delayed effects of a 50 Hz, 100 μT magnetic field on a visual duration discrimination task, with 40 subjects exposed to the field and 40 subjects sham-exposed. The delayed effects of this field were also examined in a recognition memory task that followed immediately upon completion of the discrimination task. Unlike their earlier studies, they were unable to find any effects of the field on reaction time and accuracy in the visual discrimination task. They concluded that after many years of experimentation, finding a set of magnetic field parameters and human performance measures that reliably yield magnetic field effects is proving elusive (Podd et al., 1995, 2002; Whittington and Podd, 1996; Whittington et al., 1996; Preece et al., 1998; Crasson et al., 1999). A recent review by Cook et al. (2002) evaluated cognitive effects and found incomplete and negative findings that ELF MF can alter human cognition or electrophysiology at <500 μT.

Crasson et al. (1999) performed two double blind studies to examine MF exposure effects and to determine the impact of temporal variation (continuous versus intermittent exposure) of 100 μT_{rms} 50 Hz magnetic field diurnal exposure on psychological and psychophysiological parameters in healthy humans. Three cephalic exposure sessions of 30 min, i.e., sham, continuous, and intermittent (15 s on–off cycles) magnetic field conditions were involved. They demonstrated event-related brain potential (ERP) latency and reaction time slowing in the oddball paradigm, a visual discrimination task, after real magnetic field exposure. These results also indicate that a low-level 50 Hz magnetic field may have a slight influence on ERP and reaction time under specific circumstances of sustained attention.

Crasson (2003) reviewed the effect of 50–60 Hz weak electric, magnetic, and combined electric and magnetic field exposure on cognitive functions such as memory, attention, information processing, and time perception. Studies overall are inconsistent and difficult to interpret with regard to possible health risks. Statistically significant differences between field and control exposures, when found, are small, subtle, transitory, without any clear dose–response relationship and difficult to confirm.

The Crasson group (Delhez et al., 2004) designed a study to carefully address several remaining concerns by exploring the possible effects of exposure at two field strengths usually encountered in everyday life through the use of domestic or industrial appliances. They examined whether a dose–response relationship between field strengths and cognitive effects might be established. They maximized the statistical power of their study by using a within-subject design and chose various measures of performance that might be sensitive to ELF magnetic field effects according to previous results. They investigated cognitive effects of a continuous, vertical ELF magnetic field of 20 and 400 μT 50 Hz in healthy young men during performance on cognitive tests (divided attention, flexibility, memory updating, digit span, digit span with articulary suppression, and time perception). Thirty-two volunteers (mean 22.6 $\pm$ 2.2 y) participated in this double blind study. The total duration of the exposure was 65 min. No effect of MF exposure was observed on performance. Having addressed the critical mass of experimental concerns in this area and after a thorough review of the relevant literature (Crasson, 2003), the group concluded that their results do not support the hypothesis that ELF magnetic fields affect human cognitive performance at 20 and 400 μT 50 Hz in healthy young men during performance on the cognitive tests administered.

4.3.5 Extremely Low-Frequency Human Cognitive Studies: Conclusion

New human and animal studies are continuing to explore the thresholds at which electric and magnetic ELF fields are perceptible, disrupt performance, or produce effects on cognitive performance (Saunders, 2003). Presently, there is no other known mechanism to explain ELF effects on perception, disruption of performance, or effects on cognitive performance above threshold limits than electrostimulation based on nerve stimulation, muscle excitation, cardiac excitation (contractions and adverse), synaptic activity alteration including phosphenes, and visually evoked potentials (headaches). Within the ELF guideline limits as set out by expert groups such as the IEEE (Std C95.1-1999/1999, Std C95.6-2002) and ICNIRP (1998), no established adverse effects on perception, disruption of performance, or cognition occur (ICNIRP/WHO, 2003); however, research is continuing to investigate possible effects on cognition below guideline limits. New methods such as genomics, proteomics, and metabolomics and free radical research are used (Rushton and Elliott, 2003), for instance in research funded by the EMF Biological Trust Foundation in the U.K., looking for any mechanism that could explain any of the postulated effects (EMF National Grid, 2005), including altered learning, memory (for a review of established molecular mechanism, see Kandel et al., 2000) and attention, and pain perception (for a review of the opioid hypothesis, see Sienkiewicz et al., 2005). Though funded by industry, the Trust is fully independent in its funding decisions. The *in vitro* results when completed will be followed up by *in vivo* research to verify any positive findings.

Acknowledgment

The work was sponsored by award from Office of Naval Research, Department of the Navy, United States, to D'Andrea (Work Unit Nos. 601153N.MRO4508.518-60285 and 601153N.M4023.60182).

References

Achermann P, Röösli M, Kuster N. 2005. Effects of UMTS radio-frequency fields on well-being and cognitive functions in human subjects with and without subjective complaints. Mobile Communication Research Foundation, Zurich Switzerland. www.mobile-research.ethz.ch;http://www.mobile-research.ethz.ch/var/TNO_abstract_E.pdf

Adair RK. 2003. Biophysical limits on athermal effects of RF and microwave radiation. *Bioelectromagnetics* 24:39–48.

Adair ER, Black DR. 2003. Thermoregulatory responses to RF energy absorption. *Bioelectromagnetics* (Suppl. 6):S17–S38.

Adair ER, Kelleher SA, Mack GW, Morocco TS. 1998. Thermophysiological responses of human volunteers during controlled whole-body radio frequency exposure at 450 MHz. *Bioelectromagnetics* 194:232–245.

Adair ER, Mylacraine KS, Cobb BL. 2001a. Partial-body exposure of human volunteers to 2450 MHz pulsed or CW fields provokes similar thermoregulatory responses. *Bioelectromagnetics* 224:246–259.

Adair ER, Mylacraine KS, Cobb BL. 2001b. Human exposure to 2450 MHz CW energy at levels outside the IEEE C95.1 standard does not increase core temperature. *Bioelectromagnetics* 226:429–439.

Adair ER, Mylacraine KS, Allen SJ. 2003. Thermophysiological consequences of whole body resonant RF exposure 100 MHz in human volunteers. *Bioelectromagnetics* 247:489–501.

Adair ER, Blick DW, Allen SJ, Mylacraine KS, Ziriax JM, Scholl DM. 2005. Thermophysiological responses of human volunteers to whole body RF exposure at 220 MHz. *Bioelectromagnetics* 26:448–461.

Airborne Instruments Laboratory. 1956. An observation on the detection by the ear of microwave signals. *Proc IRE* 44:2A.

Allen SJ, Adair ER, Mylacraine KS, Hurt W, Ziriax J. 2003. Empirical and theoretical dosimetry in support of whole body resonant RF exposure 100 MHz in human volunteers. *Bioelectromagnetics* 247:502–509.

Allen SJ, Adair ER, Mylacraine KS, Hurt W, Ziriax J. 2005. Empirical and theoretical dosimetry in support of whole body radio frequency (RF) exposure in seated human volunteers at 220 MHz. *Bioelectromagnetics* 26:440–447.

Albert EN. 1979. Current status of microwave effects on the blood-brain barrier. *J Microw Power* 14:281–285.

Albert EN, Kerns JM. 1981. Reversible microwave effects on the blood–brain barrier. *Brain Res* 230:153–164.

Andersen JB. 2005. Signal forms in wireless applications. In: *Do sinusoidal versus non-sinusoidal waveforms make a difference?* and 8th MCM, Zurich, 02/03/2005. http://www.cost281.org/documents.php?node = 96&dir_session

Andersen P, Moser EI. 1995. Brain temperature and hippocampal function. *Hippocampus* 5:491–498.

Angelone LM, Potthast A, Segonne F, Iwaki S, Belliveau JW, Bonmassar G. 2004. Metallic electrodes and leads in simultaneous EEG-MRI: specific absorption rate (SAR) simulation studies. *Bioelectromagnetics* 25:285–295.

Antoni FA, Palkovits M, Simpson J, Smith SM, Leitch AL, Rosie R, Fink G, Paterson JM. 1998. Ca^{2+}/calcineurin-inhibited adenylyl cyclase, highly abundant in forebrain regions, is important for learning and memory. *J Neurosci* 18:9650–9661.

Aurora SK, Welch KM, Al-Sayed F. 2003. The threshold for phosphenes is lower in migraine. *Cephalalgia* 23(4):258–263.

Bailey WH, Nyenhuis JA. 2005. Thresholds for 60 Hz magnetic field stimulation of peripheral nerves in human subjects. *Bioelectromagnetics* 26:462–468.

Bakewell S. 1995. The autonomic nervous system, practical procedures. Cambridge: Addenbrooke's Hospital, 5:1. http://www.nda.ox.ac.uk/wfsa/html/u05/u05_010.htm

Balik HH, Turgut-Balik D, Balikci K, Ozcan IC. 2005. Some ocular symptoms and sensations experienced by long term users of mobile phones. *Pathol Biol* (*Paris*) 53:88–91.

Balikci K, Cem Ozcan I, Turgut-Balik D, Balik HH. 2005. A survey study on some neurological symptoms and sensations experienced by long term users of mobile phones. *Pathol Biol* (*Paris*) 53:30–34.

Bast T, da Silva BM, Morris RG. 2005. Distinct contributions of hippocampal NMDA and AMPA receptors to encoding and retrieval of one-trial place memory. *J Neurosci* 25:5845–5856.

BBC NEWS World Edition, U.K. 2003. 3G masts 'cause health problems' BBC internet home page, October 2, 2003. http://news.bbc.co.uk/2/hi/health/3157676.stm

Beel JA. 1983. Posttrial microwave effects on learning and memory in mice. *Soc Neurosci Abstr* 9:1.644.

Belyaev IY, Hillert L, Protopopova M, Tamm C, Malmgren LO, Persson BR, Selivanova G, Harms-Ringdahl M. 2005. 915 MHz microwaves and 50 Hz magnetic field affect chromatin conformation and 53BP1 foci in human lymphocytes from hypersensitive and healthy persons. *Bioelectromagnetics* 26:173–184.

Bergqvist U, Vogel E, Eds. 1997. Possible health implications of subjective symptoms and electromagnetic fields: a report prepared by a European group of experts for the European Commission DG V. Solna, Sweden: European pp. S-171–S-184.

Besset A, Espa F, Dauvilliers Y, Billiard M, de Seze R. 2005. No effect on cognitive function from daily mobile phone use. *Bioelectromagnetics* 26:102–108.

Sandyk R, Tsagas N, Anninos PA, Derpapas K. 1992. Magnetic fields mimic the behavioral effects of rem sleep deprivation in humans. *Int J Neurosci* 65:61–68.

Blick DW, Adair ER, Hurt WD, Sherry CJ, Walters TJ, Merritt JH. 1997. Thresholds of microwave-evoked warmth sensation in human skin. *Bioelectromagnetics* 18:403–409.

Borbély AA, Huber R, Graf T, Fuchs B, Gallmann E, Achermann P. 1999. Pulsed high-frequency electromagnetic field affects human sleep and sleep electroencephalogram. *Neurosci Lett* 275:207–210.

Bullock TH. 1986. *Electroreception*. New York: Wiley, 722 pp.

Burgess N, Maguire EA, O'Keefe J. 2002. The human hippocampus and spatial and episodic memory. *Neuron* 35(4):625–641. Review.

Calabresi P, Centonze D, Gubellini P, Pisani A, Bernardi G. 2000. Acetylcholine-mediated modulation of striatal function. *Trends Neurosci* 23:120–126. Review.

Cassel JC. 2005. Letter to the Editor concerning "Radial arm maze performance of rats following repeated low-level microwave radiation exposure" by Cobb et al. (*BEMS*, 2004, 25:49–57) and Letter to the Editor by Lai (*BEMS*, 2005, 26(2):81). *Bioelectromagnetics* 26:526–527.

Cassel JC, Cosquer B, Galani R, Kuster N. 2004. Whole-body exposure to 2.45 GHz electromagnetic fields does not alter radial-maze performance in rats. *Behav Brain Res* 155(1):37–43.

Chang Q, Gold PE. 2003. Switching memory systems during learning: changes in patterns of brain acetylcholine release in the hippocampus and striatum in rats. *J Neurosci* 23:3001–3005.

Chia SE, Chia HP, Tan JS. 2000. Prevalence of headache among handheld cellular telephone users in Singapore: a community study. *Environ Health Perspect* 108:1059–1062.

Reite M, Higgs L, Lebet JP, Barbault A, Rossel C, Kuster N, Dafni U, Amato D, Pasche B. 1994. Sleep inducing effect of low energy emission therapy. *Bioelectromagnetics* 15:67–75.

Choleris E, Del Seppia C, Thomas AW, Luschi P, Ghione G, Moran GR, Prato FS. 2001. Shielding, but not zeroing of the ambient magnetic field reduces stress-induced analgesia in mice, *Proc R Soc London Ser B* 269:193–201.

Chou C-K, Guy AW. 1978. Effects of electromagnetic fields on isolated nerve and muscle preparation. *IEEE Trans Microwave Theory Tech* 26(3):141–147.

Chou C-K, Guy AW, Galambos R. 1982. Auditory perception of radio-frequency electromagnetic fields. *J Acoust Soc Am* 71:1321–1334.

Chou C-K, Guy AW, McDougall JA, Lai H. 1985. Specific absorption rate in rats exposed to 2,450-MHz microwaves under seven exposure conditions. *Bioelectromagnetics* 6:73–88.

Cobb BL, Jauchem JR, Adair ER. 2004. Radial arm maze performance of rats following repeated low level microwave radiation exposure. *Bioelectromagnetics* 251:49–57.

Colombo PJ, Wetsel WC, Gallagher M. 1997. Spatial memory is related to hippocampal subcellular concentrations of calcium-dependent protein kinase C isoforms in young and aged rats. *Proc Natl Acad Sci USA* 94:14195–14199.

COMAR. 2002. Electromagnetic hypersensitivity—a COMAR technical information statement. *IEEE Eng Med Biol Magazine* 21:173–175. http://www.ewh.ieee.org/soc/embs/comar/Hypersensitivity.htm

Cook CM, Thomas AW, Prato FS. 2002. Human electrophysiological and cognitive effects of exposure to ELF magnetic and ELF modulated RF and microwave fields: a review of recent studies. *Bioelectromagnetics* 23:144–157.

Cosquer B, Galani R, Kuster N, Cassel JC. 2005a. Whole-body exposure to 2.45 GHz electromagnetic fields does not alter anxiety responses in rats: a plus-maze study including test validation. *Behav Brain Res* 156:65–74.

Cosquer B, Kuster N, Cassel JC. 2005b. Whole-body exposure to 2.45 GHz electromagnetic fields does not alter 12-arm radial-maze with reduced access to spatial cues in rats. *Behav Brain Res* 161:331–334.

Cosquer B, Vasconcelos AP, Frohlich J, Cassel JC. 2005c. Blood–brain barrier and electromagnetic fields: effects of scopolamine methylbromide on working memory after whole-body exposure to 2.45 GHz microwaves in rats. *Behav Brain Res* 161:229–237.

Crasson M. 2003. 50–60 Hz electric and magnetic field effects on cognitive function in humans: a review. *Radiat Prot Dosimetry* 106:333–340.

Crasson M, Legros JJ. 2005. Absence of daytime 50 Hz, 100 microT(rms) magnetic field or bright light exposure effect on human performance and psychophysiological parameters. *Bioelectromagnetics* 26:225–233.

Crasson M, Legros JJ, Scarpa P, Legros W. 1999. 50 Hz magnetic field exposure influence on human performance and psychophysiological parameters: two double-blind experimental studies. *Bioelectromagnetics* 20:474–486.

Creim JA, Lovely RH, Kaune WT, Phillips RD. 1984. Attempts to produce taste-aversion learning in rats exposed to 60-Hz electric fields. *Bioelectromagnetics* 5:271–282.

Creim JA, Lovely RH, Miller DL, Anderson LE. 2002. Rats can discriminate illuminance, but not magnetic fields, as a stimulus for learning a two-choice discrimination. *Bioelectromagnetics* 237:545–549.

Croft RJ, Chandler JS, Burgess AP, Barry RJ, Williams JD, Clarke AR. 2002. Acute mobile phone operation affects neural function in humans. *Clin Neurophysiol* 113:1623–1632.

Curcio G, Ferrara M, De Gennaro L, Cristiani R, D'Inzeo G, Bertini M. 2004. Time-course of electromagnetic field effects on human performance and tympanic temperature. *NeuroReport* 151:161–164.

D'Andrea JA. 1999. Behavioral evaluation of microwave irradiation. *Bioelectromagnetics* 20(Suppl. 4): 64–74.

D'Andrea JA, Gandhi OP, Lords JL. 1977. Behavioral and thermal effects of microwave radiation at resonant and nonresonant wavelengths. *Radio Sci* 12:251–256.

D'Andrea JA, DeWitt JR, Emmerson RY, Bailey C, Stensaas S, Gandhi OP. 1986a. Intermittent exposure of rats to 2450 MHz microwaves at 2.5 mW/cm^2, behavioral and physiological effects. *Bioelectromagnetics* 7:315–328.

D'Andrea JA, DeWitt JR, Gandhi OP, Stensaas S, Lords JL, and Nielson HC. 1986b. Behavioral and physiological effects of chronic 2,450-MHz microwave irradiation of the rat at 0.5 mW/cm^2. *Bioelectromagnetics* 7:45–56.

D'Andrea JA, Thomas A, Hatcher DJ. 1994. Rhesus monkey behavior during exposure to high-peak-power 5.62-GHz microwave pulses. *Bioelectromagnetics* 15:163–176.

D'Andrea JA, Hatcher DJ, Walters TJ, Ziriax JM, Hurt WD, Kosub KR, Weathersby F. 1999. Facial detection of 94 GHz millimeter waves by the nonhuman primate. Paper presented at the Bioelectromagnetics Society Annual Meeting, Long Beach, CA, June 20–24.

D'Andrea JA, Adair ER, and de Lorge JO. 2003a. Behavioral and cognitive effects of microwave exposure. *Bioelectromagnetics* (Suppl. 6):S39–S62.

D'Andrea JA, Chou C-K, Johnston SA, Adair ER. 2003b. Microwave effects on the nervous system. *Bioelectromagnetics* (Suppl. 6):S107–S147.

Day M, Langston R, Morris RG. 2003. Glutamate-receptor-mediated encoding and retrieval of paired-associate learning. *Nature* 424:205–209.

Delhez M, Legros J, Crasson M. 2004. No influence of 20 and 400 μT, 50 Hz magnetic field exposure on cognitive function in humans. *Bioelectromagnetics* 25:592–598.

de Lorge JO. 1976. The effects of microwave radiation on behavior and temperature in rhesus monkeys. In: Johnson CC, Shore ML, Eds., *Biological Effects of Electromagnetic Waves*, U.S. Dept. of Health, Education, and Welfare, Washington, D.C.: HEW Publication. FDA 77-8010, Vol. 1. pp. 158–174.

de Lorge JO. 1979. Operant behavior and colonic temperature of squirrel monkeys during microwave irradiation. *Radio Sci* 14:217–225.

de Lorge JD. 1983. The thermal basis for disruption of operant behavior by microwaves in three animal species. In: Adair ER, Ed., *Microwaves and Thermoregulation*. New York: Academic Press, pp. 379–399.

de Lorge JO, Ezell CS. 1980. Observing responses of rats exposed to 1.28 and 5.62 GHz microwaves. *Bioelectromagnetics* 1:183–198.

DeWitt JR, D'Andrea JA, Emerson RY, Gandhi OP. 1987. Behavioral effects of chronic exposure to 0.5 mW/cm^2 of 2,450-MHz microwaves. *Bioelectromagnetics* 8:149–157.

Dong H, Csernansky CA, Martin MV, Bertchume A, Vallera D, Csernansky JG. 2005. Acetylcholinesterase inhibitors ameliorate behavioral deficits in the Tg2576 mouse model of Alzheimer's disease. *Psychopharmacology (Berl)* DOI 10.1007/s00213-005-2230-6.

Dubreuil D, Jay T, Edeline JM. 2002. Does head-only exposure to GSM-900 electromagnetic fields affect the performance of rats in spatial learning tasks? *Behav Brain Res* 129:203–210.

Dubreuil D, Jay T, Edeline JM. 2003a. Head-only exposure to GSM 900-MHz electromagnetic fields does not alter rat's memory in spatial and non-spatial tasks. *Behav Brain Res* 145:51–61.
Dubreuil D, Tixier C, Dutrieux G, Edeline JM. 2003b. Does the radial arm maze necessarily test spatial memory? *Neurobiol Learn Mem* 79:109–117.
Dumanskii YD, Rudichenko VF. 1976. The dependence of the functional activity of liver mitochondria on superhigh frequency radiation. *Gig Sanit* 16–19.
Edelstyn N, Oldershaw A. 2002. The acute effects of exposure to the electromagnetic field emitted by mobile phones on human attention. *NeuroReport* 13:119–121.
Elder JA, Chou C-K. 2003. Review: auditory response to pulsed radiofrequency energy. *Bioelectromagnetics* (Suppl. 6):S162–S173.
EMF National Grid, U.K. 2005. Biological Trust Foundation, Internet Home Page [EMF BRT projects (http://www.emfs.info/sci_trustproj.asp) and publications (http://www.emfs.info/sci_trustpubli.asp)].
Erdreich LS, Klauenberg BJ. 2001. Radio frequency radiation exposure standards: considerations for harmonization, *Health Phys* 80:430–439.
Eulitz C, Ullsperger P, Freude G, Elbert T. 1998. Mobile phones modulate response patterns of human brain activity. Neuro-Report 9:3229–3232.
Foster KR. 2003. Mechanisms of interaction of extremely low-frequency electric fields and biological systems. *Radiat Prot Dosimetry* 106:301–310. Review.
Foster KR, Adair ER. 2004. Modeling thermal responses in human subjects following extended exposure to radiofrequency energy. BioMedical Engineering Online 3:4 http://www.biomedical-engineering-online.com/content/3/1/4
Foster KR, Finch ED. 1974. Microwave hearing: evidence for thermoacoustic auditory stimulation by pulsed microwaves. *Science* 185:256–258.
Foster KR, Repacholi MH. 2004. Biological effects of radiofrequency fields: does modulation matter? *Radiat Res* 162:219–225.
Franke H, Ringelstein EB, Stogbauer F. 2005a. Electromagnetic fields (GSM 1800) do not alter blood–brain barrier permeability to sucrose in models *in vitro* with high barrier tightness. *Bioelectromagnetics* 26:529–535.
Franke H, Streckert J, Bitz A, Goeke J, Hansen V, Ringelstein EB, Nattkamper H, Galla HJ, Stogbauer F. 2005b. Effects of Universal Mobile Telecommunications System (UMTS) electromagnetic fields on the blood–brain barrier. *Radiat Res* 164:258–269.
Frey AH. 1961. Auditory system response to radio frequency energy. *Aerospace Med* 32:1140–1142.
Freude G, Ullsperger P, Eggert S, Ruppe I. 2000. Microwaves emitted by cellular telephones affect human slow brain potentials. Eur J Appl Physiol 81:18–27.
Freude G, Ullsperger P, Eggert S, Ruppe I. 1998. Effects of microwaves emitted by cellular phones on human slow brain potentials. Bioelectromagnetics (Brief Communication) 19:384–387.
Fritze K, Sommer C, Schmitz B, Mies G, Hossmann KA, Kiessling M, Wiessner C. 1997. Effect of global system for mobile communication (GSM) microwave exposure on blood-brain barrier permeability in rat. Acta Neuropathol (Berl) 94:465–470.
Gajšek, P, Pakhomov AG, Klauenberg BJ. 2002. Electromagnetic field standards in Central and Eastern European countries: current state and stipulations for international harmonization. *Health Phys* 821:173 183.
Glassman RB. 1999. A working memory "theory of relativity": elasticity in temporal, spatial, and modality dimensions conserves item capacity in radial maze, verbal tasks, and other cognition. *Brain Res Bull* 485:475–489. Review.
Gnegy ME. 2000. Ca^{2+}/calmodulin signaling in NMDA-induced synaptic plasticity. *Crit Rev Neurobiol* 14:91–129.
Goldman H, Lin JC, Murphy S, Lin MF. 1984. Cerebrovascular permeability to 86Rb in the rat after exposure to pulsed microwaves. Bioelectromagnetics 5:323–330.
Gray JA, McNaughton N. 1983. Comparison between the behavioural effects of septal and hippocampal lesions: a review. *Neurosci Biobehav Rev* 7:119–188.
Guy AW, Wallace J, McDougall JA. 1979. Circularly polarized 2450-MHz waveguide system for chronic exposure of small animals to microwaves. *Radio Sci* 14(6S):63–74.

Haarala C, Aalto S, Hautzel H, Julkunen L, et al. 2003a. Effects of a 902 MHz mobile phone on cerebral blood flow in humans: a PET study. *NeuroReport* 14(16):2019–2023.

Haarala C, Bjornberg L, Ek M, Laine M, Revonsuo A, Koivisto M, Hämäläinen H. 2003b. Effect of a 902 MHz electromagnetic field emitted by mobile phones on human cognitive function: a replication study. *Bioelectromagnetics* 24:283–288.

Haarala C, Ek M, Bjornberg L, Laine M, Revonsuo A, Koivisto M, Hamalainen H. 2004. 902 MHz mobile phone does not affect short term memory in humans. *Bioelectromagnetics* 25(6):452–456.

Haarala C, Bergman M, Laine M, Revonsuo A, Koivisto M, Hämäläinen H. 2005. Electromagnetic field emitted by 902 MHz mobile phones shows no effects on children's cognitive function. *Bioelectromagnetics* DOI 10.1002/bem.20142.

Hamblin DL, Wood AW. 2002. Effects of mobile phone emissions on human brain activity and sleep variables. *Int J Radiat Biol* 78:659–669.

Hamblin DL, Wood AW, Croft RJ, Stough C. 2004. Examining the effects of electromagnetic fields emitted by GSM mobile phones on human event-related potentials and performance during an auditory task. *Clin Neurophysiol* 1151:171–178.

Hamblin DL, Croft RJ, Wood AW, Spong J-L, Stough C. 2005. The sensitivity of human event-related potentials and performance to mobile phone emitted EMFS. BEMS Abstract, Dublin, pp. 116–117. http://bioelectromagnetics.org/bioem2005/bioem2005-abstracts-small. pdf? PHPSESSID = 78d98d6fc2c81667e31b2f9184e163db

Hamblin DL., Croft RJ., Wood AW., Stough C., Spong J. 2006. The sensitivity of human eccent-related potentials and reaction time to mobile entitled potentials and reactions time to mobile entitled electromagnetic fields. *Bioelectromagnetic* 27: 265–273.

Hancock PA, Vasmatzidis I. 2003. Effects of heat stress on cognitive performance: the current state of knowledge. *Int J Hyperthermia* 19:355–372.

Mild K, Oftedal G, Sandstrom M, Wilen J, Tynes T, Haugsdal B, Hauger E. 1998. Comparison of symptoms experience by users of analogue and digital mobile phones. A Swedish-Norwegian epidemiological study (Swed) Arbetslivsrapport Investigation Report No. 23), p. 74.

Hendler E. 1968. Cutaneous receptor response to microwave irradiation. In: Hardy JD, Ed., *Thermal Problems in Aero-Space Medicine*. Surrey: Unwin, Ltd., pp. 149–161.

Hendler E, Hardy JD. 1960. Infrared and microwave effects on skin heating and temperature sensation. *IRE Trans Med Electron Med Electron* 7:143–152.

Hendler E, Hardy JD, Murgatroyd D. 1963. Skin heating and temperature sensation produced by infrared and microwave irradiation. In: Hertzfield CM, Ed., *Temperature: Its Measurement and Control in Science and Industry*. NewYork: Reinhold, pp. 211–230.

Hensel H. 1981. Thermoreception and temperature regulation. *Monogr Physiol Soc* 38:1–321.

Hermann DM, Hossmann KA. 1997. Neurological effects of microwave exposure related to mobile communication. *J Neurol Sci* 152(1):1–14.

Hietanen M, Kovala T, Hämäläinen A-M. 2000. Human brain activity during exposure to radio-frequency fields emitted by cellular phones. *Scand J Work Environ Health* 26:87–92.

Hietanen M, Hämäläinen A-M, Husman T. 2002. Hypersensitivity symptoms associated with exposure to cellular telephones: no causal link. *Bioelectromagnetics* 23:264–270.

Hinrichs H, Heinze HJ. 2004. Effects of GSM electromagnetic field on the MEG during an encoding-retrieval task. *NeuroReport* 15:1191–1194.

Hinrikus H, Parts M, Lass J, Tuulik V. 2004. Changes in human EEG caused by low level modulated microwave stimulation. *Bioelectromagnetics* 25:431–440.

Hirata A, Shiozawa T. 2003. Correlation of maximum temperature increase and peak SAR in the human head due to handset antennas. *IEEE Trans Microwave Theory Tech* 51:1834–1841.

Hobson JA, Pace-Schott EF. 2002. The cognitive neuroscience of sleep: neuronal systems, consciousness and learning. *Nat Rev Neurosci* 3:679–693.

Hocking B. 1998. Preliminary report: symptoms associated with mobile phone use. *Occup Med (London)* 48:357–360.

Hoh TE, Kolb B, Eppel A, Vanderwolf CH, Cain DP. 2003. Role of the neocortex in the water maze task in the rat: a detailed behavioral and Golgi–Cox analysis. *Behav Brain Res* 138:81–94.

Hossmann KA, Hermann DM. 2003. Effects of electromagnetic radiation of mobile phones on the central nervous system. *Bioelectromagnetics* 24:49–62.

Huber R, Graf T, Cote KA, Wittmann L, Gallmann E, Matter D, Schuderer J, Kuster N, Borbély AA, Achermann P. 2000. Exposure to pulsed high-frequency electromagnetic field during waking affects human sleep EEG. *NeuroReport* 11:3321–3325.

Huber R, Treyer V, Borbély AA, Schuderer J, Gottselig JM, Landolt H-P, Werth E, Berthold T, Kuster N, Buck A, Achermann P. 2002. Electromagnetic fields, such as those from mobile phones, alter regional cerebral blood flow and sleep and waking EEG. *J Sleep Res* 11:289–295.

Huber R, Schuderer J, Graf T, Jütz K, Borbély AA, Kuster N, Achermann P. 2003. Radio frequency electromagnetic field exposure in humans: estimation of SAR distribution in the brain, effects on sleep and heart rate. *Bioelectromagnetics* 24:262–276.

Huber R, Treyer V, Schuderer J, Berthold T, Buck A, Kuster N, Landolt HP, Achermann P. 2005. Exposure to pulse-modulated radio frequency electromagnetic fields affects regional cerebral blood flow. *Eur J Neurosci* 21:1000–1006.

ICNIRP. 1998. International Commission on Non-Ionizing Radiation Protection Guidelines for limiting exposure to time-varying electric, magnetic, and electromagnetic fields up to 300 GHz. Health Physics 74:494–522.

ICNIRP. 2003. Matthes R, McKinlay AF, Bernhardt JH, Vecchia P, Veyret B, editors. Exposure to static and low-frequency electromagnetic fields, biological effects and health consequences (0–100 kHz). München, Germany: Märkl-Druck. ICNIRP: 13/2003.

ICNIRP/WHO Proceedings International Workshop 2003. Repacholi MH, McKinlay AF, editors. Weak Electric Field Effects in the Body. *Radiat Prot Dosimetry* 106:4.

IEEE. Std C95.1-1991/1999. IEEE Standard for safety levels with respect to human exposure to radio frequency electromagnetic fields, 3 kHz to 300 GHz. 1999 ed. New York: The Institute of Electrical and Electronic Engineers. http://shop.ieee.org/ieeestore/Results.aspx

IEEE. Std C95.6-2002. Standard for safety levels with respect to human exposure to electromagnetic fields, 0 to 3 kHz. IEEE Standard C95.6-2002. New York: The Institute of Electrical and Electronic Engineers. http://shop.ieee.org/ieeestore/Results.aspx

IEEE Std C95-1™ 2005, IEEE Standard for safety levels with respect to human exposure to Radio frequency electromagnetic fields, $3KH_3$ to 300 GHz. 2005 ed. New York: The institute of electrical and electronic engineers, http://shopieeorg/ieeestore/Resultsaspx.

Independent Expert Group on Mobile Phones (IEGMP). 2000. Mobile Phones and Health, NRPB, Chilton, Didcot, Oxon OX11 0RQ U.K. http://www.iegmp.org.uk/report/index.htm

Jauchem JR. 2005. Letter to the editor concerning Lai's letter on "Radial arm maze performance of rats following repeated low-level microwave radiation exposure" [*BEMS*, 2004, 25:49–57]. *Bioelectromagnetics* 26:525.

Jech R, Sonka K, Ruzicka E, Nebuzelsky A, Bohm J, Juklickova M, Nevsimalova S. 2001. Electromagnetic field of mobile phones affects visual event related potential in patients with narcolepsy. Bioelectromagnetics 22:519–528.

Johnston SA. 2003. Mechanisms modelling, biological effects, therapeutic effects, international standards, exposure criteria. In: Stavroulakis P, Ed., *Biological Effects of Electromagntic Fields*, Heidelberg: Springer-Verlag, pp. 733–777.

Johnston SA. 2004/2005. EMF Dosimetry Handbook: International guidelines for quality EMF (RF or ELF) research. http://www.emfdosimetry.org/

Jones N, Bottomley A, Haylock R, Saunders R, Kuster N, Sienkiewicz Z. 2005. Perform B: Behavioral studies with mice. BEMS Abstract, pp. 184–85. http://bioelectromagnetics.org/bioem2005/bioem2005-abstracts-small.pdf?PHPSESSID = 78d98d6fc2c81667e31b2f9184e163db

Justesen DR, Adair ER, Stevens JC, Bruce-Wolfe V. 1982. A comparative study of human sensory thresholds: 2450 MHz vs. far-infrared radiation. *Bioelectromagnetics* 3:117–125.

Kalmijn AJ. 1971. The electric senses of sharks and rays. *J Exp Biol* 55:371–383.

Kamenskii YI. 1964. Effect of microwaves on the functional state of the nerve. *Biophysics* 9:758–764.

Kandel E, Schwartz J, Jessell M, Eds. 2000. *Principles of Neural Science*, 4th ed. London: McGraw Hill.

Kavaliers M, Eckel LA, Ossenkopp K-P. 1993. Brief exposure to 60 Hz magnetic fields improves sexually dimorphic spatial learning performance in the meadow vole, Microtus pennsylvanicus. Journal of Comparative Physiology A: Sensory, Neural, and Behavioral Physiology (Historical Archive) 173:241–248.

Kavaliers M, Ossenkopp KP, Prato FS, Innes DG, Galea LA, Kinsella DM, Perrot-Sinal TS. 1996 Spatial learning in deer mice: sex differences and the effects of endogenous opioids and 60 Hz magnetic fields J Comp Physiol [A]. 179:715–724.

Kavaliers M, Teskey GC, Hirst M. 1985. The effects of aging on day-night rhythms of κ opiate-mediated feeding in the mouse. Psychopharmacology (Berl). 87:286–291.

Keeton WT. 1971. Magnets Interfere with Pigeon Homing. Proc Natl Acad Sci USA 68:102–106.

Kheifets L, Repacholi M, Saunders R. 2003. Thermal stress and radiation protection principles. *Int J Hyperthermia* 19:215–224.

King NW, Justesen DR, Clarke RL. 1971. Behavioral sensitivity to microwave radiation. *Science* 172:398–401.

Koivisto M, Krause CM, Revonsuo A, Laine M, Hamalainen H. 2000a. The effects of electromagnetic field emitted by GSM phones on working memory. *NeuroReport* 11:1641–1643.

Koivisto M, Revonsuo A, Krause C, Haarala C, Sillanmaki L, Laine M, Hamalainen H. 2000b. Effects of 902 MHz electromagnetic field emitted by cellular telephones on response times in humans. *NeuroReport* 11:413–415.

Koivisto M, Haarala C, Krause CM, Revonsuo A, Laine M, Hämäläinen H. 2001. GSM phone signal does not produce subjective symptoms. *Bioelectromagnetics* 22:212–215.

Kramarenko AV, Tan U. 2003. Effects of high-frequency electromagnetic fields on human EEG: a brain mapping study. *Int J Neurosci* 113:1007–1019.

Krause CM, Sillanmäki L, Koivisto M, Häggqvist A, Saarela C, Revonsuo A, Laine M, Hämäläinen H. 2000a. Effects of electromagnetic field emitted by cellular phones on the EEG during a memory task. *NeuroReport* 11:761–764.

Krause CM, Sillanmäki L, Koivisto M, Häggqvist A, Saarela C, Revonsuo A, Laine M, Hämäläinen H. 2000b. Effects of electromagnetic fields emitted by cellular phones on the electroencephalogram during a visual working memory task. *Int J Radiat Biol* 76:1659–1667.

Krause CM, Haarala C, Sillanmaki L, Koivisto M, Alanko K, Revonsuo A, Laine M, Hämäläinen H. 2004. Effects of electromagnetic field emitted by cellular phones on the EEG during an auditory memory task: a double blind replication study. *Bioelectromagnetics* 251:33–40.

Kuster N, Schuderer J, Christ A, Futter P, Ebert S. 2004. Guidance for exposure design of human studies addressing health risk evaluations of mobile phones. *Bioelectromagnetics* 25:524–529.

Lai H. 1996. Spatial learning deficit in the rat after exposure to a 60 Hz magnetic field. *Bioelectromagnetics* 17:494–496.

Lai H. 2004. Interaction of microwaves and a temporally incoherent magnetic field on spatial learning in the rat. *Physiol Behav* 82(5):785–789.

Lai H. 2005a. Comment on "Radial arm maze performance of rats following repeated low level microwave radiation exposure." *Bioelectromagnetics* 26:81.

Lai H. 2005b. Responses to Jauchem and Cassel. *Bioelectromagnetics* 26:528.

Lai H, Carino M. 1998. Intracerebroventricular injection of mu- and delta-opiate receptor anatagonists block 60 Hz magnetic field-induced decreases in cholinergic activity in the frontal cortex and hippocampus of the rat. *Bioelectromagnetics* 19:432–437.

Lai H, Carino MA, Horita A, Guy AW. 1989. Low-level microwave irradiation and central cholinergic systems. *Pharmacol Biochem Behav* 33:131–138.

Lai H, Carino MA, Horita A, Guy AW. 1992. Single vs. repeated microwave exposure: effects on benzodiazepine receptors in the brain of the rat. *Bioelectromagnetics* 131:57–66.

Lai H, Horita A, Guy AW. 1994. Microwave irradiation affects radial-arm maze performance in the rat. *Bioelecromagnetics* 15:95–104.

Lai H, Carino MA, Iishijima I. 1998. Acute exposure to 60 Hz magnetic field affects water-maze performance in the rat. *Bioelectromagnetics* 19:117–122.

Lass J, Tuulik V, Hinrikus H. 1999. Modulated microwave effects on EEG alpha-rhythms. *Med Biol Eng Comput* 37:s105–s108.

Lass J, Tuulik V, Ferenets R, Riisalo R, Hinrikus H. 2002. Effects of 7 Hz-modulated 450 MHz electromagnetic radiation on human performance in visual memory tasks. *Int J Radiat Biol* 78:937–944.

Lee TMC, Ho SMY, Tsang LYH, Yang SYC, Li LSW, Chan CCH. 2001. Effect on human attention of exposure to the electromagnetic field emitted by mobile phones. *NeuroReport* 12:729–731.
Lee TM, Lam PK, Yee LT, Chan CC. 2003. The effect of the duration of exposure to the electromagnetic field emitted by mobile phones on human attention. *NeuroReport* 18:1361–1364.
Leitgeb N, Schröttner J. 2003. Electrosensibility and electromagnetic hypersensitivity. *Bioelectromagnetics* 24:387–394.
Levin ED. 1988. Psychopharmacological effects in the radial-arm maze. *Neurosci Biobehav Rev* 12:169–175.
Levin ED, Petro A, Caldwell DP. 2003. Nicotine and clozapine actions on pre-pulse inhibition deficits caused by *N*-methyl-D-aspartate (NMDA) glutamatergic receptor blockade. *Prog Neuropsychopharmacol Biol Psychiat* 29:581–586.
Levin ED, Petro A, Caldwell DP. 2005. Nicotine and clozapine actions on pre-pulse inhibition deficits caused by *N*-methyl-D-aspartate (NMDA) glutamatergic receptor blockade. *Prog Neuropsychopharmacol Biol Psychiat* 29:581–586.
Liboff AR, Thomas JR, Schrot J. 1989. Intensity threshold for 60 Hz magnetically induced behavioral changes in rats. *Bioelectromagnetics* 10:111–113.
Lin JC, Lin MF. 1980. Studies on microwave and blood–brain barrier interaction. Bioelectromagnetics 1:313–323.
Lövsund P, Öberg PA, Nilsson SE. 1980a. Magneto- and electrophosphenes: a comparative study. *Med Biol Eng Comput* 18:758–764.
Lövsund P, Öberg PA, Nilsson SE, Reuter T. 1980b. Magnetophosphenes: a quantitative analysis of thresholds. *Med Biol Eng Comput* 18:326–334.
Luttges MW. 1980. Microwave effects on learning and memory in mice. NTIS Document No. AD-A094, 788/7.
Maby E, Jeannes RL, Faucon G, Liegeois-Chauvel C, De Seze R. 2005. Effects of GSM signals on auditory evoked responses. *Bioelectromagnetics* 26:341–350.
Maguire EA, Gadian DG, Johnsrude IS, Good CD, Ashburner J, Frackowiak RS, Frith CD. 2000. Navigation-related structural change in the hippocampi of taxi drivers. *Proc Natl Acad Sci USA* 97:4398–4403. http://www.pubmedcentral.nih.gov/articlerender.fcgi?tool = pubmed&pubmedid = 10716738
Maguire EA, Spiers HJ, Good CD, Hartley T, Frackowiak RS, Burgess N. 2003a. Navigation expertise and the human hippocampus: a structural brain imaging analysis. *Hippocampus* 132:250–259.
Maguire EA, Valentine ER, Wilding JM, Kapur N. 2003b. Routes to remembering: the brains behind superior memory. *Nat Neurosci* 61:90–95.
Maier R, Greter SE, Maier R. 2004. Effects of pulsed electromagnetic fields on cognitive processes—a pilot study on pulsed field interference with cognitive regeneration. *Acta Neurol Scand* 110:46–52.
Mann K, Röschke J. 1996. Effects of pulsed high-frequency electromagnetic fields on human sleep. Neuropsychobiology 33:41–47.
Mann K, Röschke J, Connemann B, Beta H. 1998. No effects of pulsed high-frequency electromagnetic fields on heart rate variability during human sleep. Neuropsychobiology 38:251–256.
Mann K, Röschke J. 1997. No short-term effects of digital mobile radio telephone on the awake human electroencephalogram. Bioelectromagnetics 18:172–176.
Marino C, Cristalli G, Galloni P, Pasqualetti P, Piscitelli M, Lovisolo G. 2000. Effects of microwaves (900 MHz) on the cochlear receptor: exposure systems and preliminary results. Radiat Environ Biophys 39:131–136.
McKay BE, Persinger MA. 2000. Application timing of complex magnetic fields delineates windows of posttraining–pretesting vulnerability for spatial and motivational behaviors in rats. *Int J Neurosci* 103:69–77.
McNeal DR, Bowman BR. 1985. Peripheral neuromuscular stimulation. In: Mykelbust JB, Cusick JF, Sances A, Larsons SJ, Eds., *Neural Stimulation*, Vol. II. Boca Raton, FL: CRC Press, pp. 95–118.
McQuade JS, Merritt JH, Rahimi O, Miller SA, Scholin T, Salazar AL, Cook MC, Mason PA. 2005. Effects of 915 MHz exposure on the integrity of the blood brain barrier. *Bioelectromagnetics* 15(3):171–172.

McRee DI, Elder JA, Gage MI, Reiter LW, Rosenstein LS, Shore ML, Galloway WD, Adey WR, Guy AW. 1979. Effects of nonionizing radiation on the central nervous system, behavior, and blood: a progress report. *Environ Health Perspect* 30:123–131.

Merrit JH, Chamness AF, Allen SJ. 1978. Studies on blood-brain barrier permeability after microwave-radiation. *Radiat Environ Biophys* 15:367–377.

Merritt JH, Shelton WW, Chamness AF. 1982. Attempts to alter $45Ca^{2+}$ binding to brain tissue with pulse-modulated microwave energy. Bioelectromagnetics 3:475–478.

Mickley GA, Cobb BL. 1998. Thermal tolerance reduces hyperthermia-induced disruption of working memory: a role for endogenous opiates? *Physiol Behav* 63:855–865.

Mickley GA, Cobb BL, Mason PA, Farrell S. 1994. Disruption of a putative working memory task and selective expression of brain c-fos following microwave-induced hyperthermia. *Physiol Behav* 55:1029–1038.

Miller GA. 1956. The magical number seven, plus or minus two: some limits on our capacity for processing information. *Psychol Rev* 63:81–97.

Mitchell CL, McRee DI, Peterson NJ, Tilson HA, Shandala MG. 1989. Results of a United States and Soviet Union joint project on nervous system effects of microwave radiation. *Environ Health Perspect* 81:201–209.

Moriyama E, Salcman M, Broadwell RD. 1991. Blood–brain barrier alteration after microwave-induced hyperthermia is purely a thermal effect: I. Temperature and power measurements. *Surg Neurol* 35:277–282.

Ohmoto Y, Fujisawa H, Ishikawa T, Koizumi H, Matsuda T, Ito H. 1996. Sequential changes in cerebral blood flow, early neuropathological consequences and blood-brain barrier disruption following radiofrequency-induced localized hyperthermia in the rat. Int J Hyperthermia 12:321–334.

Morris RG, Garrud P, Rawlins JN, O'Keefe J. 1982. Place navigation impaired in rats with hippocampal lesions. *Nature* 297:681–683.

Moser EI, Moser M-B, Andersen P. 1994. Potentiation of dentate synapses initiated by exploratory learning in rats: dissociation from brain temperature, motor activity, and arousal. *Learn Mem* 1:55–73.

Nadeem M, Thorlin T, Gandhi OP, Persson M. 2003. Computation of electric and magnetic stimulation in human head using the 3-D impedance method. *IEEE Trans Biomed Eng* 50:900–907.

Nelson TD. 1978. Behavioural effects of microwave irradiation on squirrel monkey Saimiri sciureus performance of a repeated acquisition task. NTIS Document No. AD A055 953/4GA.

Neilly JP, Lin JC. 1986. Interaction of ethanol and microwaves on the blood–brain barrier of rats. Bioelectromagnetics 7:405–414.

Neubauer C, Phelan AM, Kues H, Lange DG. 1990. Microwave irradiation of rats at 2.45 GHz activates pinocytotic like uptake of tracer by capillary endothelial cells of cerebral cortex. Bioelectromagnetics 11:261–268.

NIEHS. 1998. National Institute of Enviromental Health Sciences working group report. In: Portier CP, Wolfe MS, Eds., *Assessment of Health Effects from Exposure to Power-line Frequency Electric and Magnetic Fields*. Research Triangle Park NC, National Institute of Health, NIH Publication No 98-3981.

NRPB. 2004. Review of the scientific evidence for limiting exposure to electromagnetic fields (0–300 GHz). Documents of the NRPB 15:3.

Oftedal G, Wilén J, Sandström M, Mild KH. 2000. Symptoms experienced in connection with mobile phone use. *Occup Med* 50:237–245.

Ohmoto Y, Fujisawa H, Ishikawa T, Koizumi H, Matsuda T, Ito H. 1996. Sequential changes in cerebral blood flow, early neuropathological consequences and blood-brain barrier disruption following radiofrequency-induced localized hyperthermia in the rat. Int J Hyperthermia 12:321–334.

O'Keefe J, Burgess N. 2005. Dual phase and rate coding in hippocampal place cells: theoretical significance and relationship to entorhinal grid cells. DOI: 10.1002/hipo.20115.

O'Keefe J, Dostrovsky J. 1971. The hippocampus as a spatial map. Preliminary evidence from unit activity in the freely moving rat. *Brain Res* 34:171–175.

O'Keefe J, Nadel L. 1978. *The Hippocampus as a Cognitive Map*. Oxford: Clarendon Press. http://www.cognitivemap.net/HCMpdf/HCMComplete.pdf

Oldendorf WH. 1970. Measurement of brain uptake of radiolabeled substances using a tritiated water internal standard. *Brain Res* 24:372–376.

Orr JL, Rogers WR, Smith HD. 1995a. Detection thresholds for 60 Hz electric fields by nonhuman primates. *Bioelectromagnetics* (Suppl. 3):23–34.

Orr JL, Rogers WR, Smith HD. 1995b. Exposure of baboons to combined 60 Hz electric and magnetic fields does not produce work stoppage or affect operant performance on a match-to-sample task. *Bioelectromagnetics* (Suppl. 3):61–70.

Oscar KJ, Hawkins TD. 1977. Microwave alteration of the blood-brain barrier system of rats. *Brain Res* 126:381–393.

Pakhomov AG, Doyle J, Stuck BE, Murphy MR. 2003. Effects of high power microwave pulses on synaptic transmission and long term potentiation in hippocampus. *Bioelectromagnetics* 24:174–181.

Patapoutian A, Peier AM, Story GM, Viswanath V. 2003. Thermo TRP channels and beyond: mechanisms of temperature sensation. *Nat Rev Neurosci* 4:529–539.

Persson BRR, Salford LG, Brun A. 1997. Blood–brain barrier permeability in rats exposed to electromagnetic fields used in wireless communication. *Wireless Network* 3:455–461.

Podd JV, Whittington CJ, Barnes GR, Page WH, Rapley BI. 1995. Do ELF magnetic fields affect human reaction time? *Bioelectromagnetics*16:317–323.

Podd J, Abbott J, Kazantzis N, Rowland A. 2002. Brief exposure to a 50 Hz, 100 microT magnetic field: effects on reaction time, accuracy, and recognition memory. *Bioelectromagnetics* 3:189–195.

Pollen DA. 2004. Brain stimulation and conscious experience. *Conscious Cogn* 13:626–645.

Polyashuck L. 1971. Changes in permeability of histo-hematic barriers under the effect of microwaves. *Dokl Akad Naeuk Ukrain* 8:754–758.

Prato FS, Kavaliers M, Carson JJ. 1996. Behavioural evidence that magnetic field effects in the land snail, *Cepaea nemoralis*, might not depend on magnetite or induced electric currents. *Bioelectromagnetics* 17:123–130.

Prato FS, Robertson JA, Desjardins D, Hensel J, Thomas AW. 2005. Daily repeated magnetic field shielding induces analgesia in CD-1 mice. *Bioelectromagnetics* 26:109–117.

Prato FS, Frappier JRH, Shivers RR, Kavaliers M, Zabel P, Drost D, Lee TY. 1990. Magnetic resonance imaging increases the blood–brain barrier permeability to 153-gadolinium diethy·lenetriaminepentaacetic acid in rats. *Brain Res* 523:301–304.

Preece AW, Wesnes KA, Iwi GR. 1998. The effect of a 50 Hz magnetic field on cognitive function in humans. *Int J Radiat Biol* 74:463–470.

Preece AW, Iwi G, Davies-Smith A, Wesnes K, Butler S, Lim E, Varey A. 1999. Effect of a 915-MHz simulated mobile phone signal on cognitive function in man. *Int J Radiat Biol* 75:447–456.

Preece AW, Goodfellow S, Wright MG, Butler SR, Dunn EJ, Johnson Y, Manktelow TC, Wesnes K. 2005. Effect of 902 MHz mobile phone transmission on cognitive function in children. *Bioelectromagnetics* (Suppl. 7) DOI: 10.1002/bem.20128.

Preston E, Vavasour EJ. Assenheim HN. 1979. Permeability of the blood-brain barrier to mannitol in the rat following 2450 MHz microwave irradiation. *Brain Res* 174:109–117.

Raslear TG, Akyel Y, Bates F, Belt M, Lu ST. 1993. Temporal bisection in rats: the effects of high-peak-power pulsed microwave irradiation. *Bioelectromagnetics* 145:459–478.

Reilly JP. 1978. Electric and magnetic coupling from high voltage AC power transmission lines—classification of short-term effects on people. *IEEE Trans Power Appar Syst* 97(6): 2243–2252.

Reilly JP. 1998. *Applied Bioelectricity*. New York: Springer.

Reilly JP. 2005. An analysis of differences in the low-frequency electric and magnetic field exposure standards of ICES and ICNIRP. *Health Phys* 89:71–80.

Thuroczy G, Kubinyi G, Bodo M, Bakos J, Szabo LD. 1994. Simultaneous response of brain electrical activity (EEG) and cerebral circulation (REG) to microwave exposure in rats. *Rev Environ Health* 10:135–148.

Repacholi MH. 1998. Low-level exposure to radiofrequency electromagnetic fields: health effects and research needs. *Bioelectromagnetics* 19:1–19. Review.

Rogers WR, Orr JL, Smith HD. 1995. Initial exposure to 30 kV/m or 60 kV/m 60 Hz electric fields produces temporary cessation of operant behavior of nonhuman primates. *Bioelectromagnetics* (Suppl. 3):35–47.

Roubos EW, van Aernsbergen LM, Brussaard G, Havenaar J, Koops FBJ, van Leeuwen FE, Leonhard HK, van Rhoon GC, Sitskoorn MM, Swaen GMH, van de Weerdt DHJ, Zwamborn APM, van Rongen E (Zwamborn APM, Vorst HCM, invited consultants). 2004. The TNO study on effects of GSM and UMTS signals on well-being and cognition: Review and recommendations for further research. Health Council of the Netherlands, Electromagnetic Fields Committee, Publication No. 2004/13E. http://www.gr.nl/pdf.php?ID = 1042

Rubin GJ, Das Munshi J, Wessely S. 2005. Electromagnetic hypersensitivity: a systematic review of provocation studies. *Psychosom Med* 67:224–232.

Rudnev MI, Tarasyuk NYe, Kulikova AD. 1983. Effect of low-intensity superhigh-frequency energy on respiration and oxidative phosphorylation of organ mitochondria and activity of some blood enzymes. In: *Effects of Nonionizing Electromagnetic Radiation*, USSR Report No. 10, JPRS 83745, pp. 5–8.

Rushton L, Elliott P. 2003. Evaluating evidence on environmental health risks. *Br Med Bull* 68:113–128.

Sackheim HA. 1991. Stimulus intensity, seizure threshold, and seizure duration, *Electroconvulsive Ther* 14: 803–843.

Salford LG, Brun A, Eberhardt JL, Persson RR. 1993. Permeability of the blood–brain barrier induced by 915 MHz electromagnetic radiation, continuous wave and modulated at 8, 16, 50, and 200 Hz. *Bioelectrochem Bioenerg* 30:293–301.

Salford LS, Brun A, Sturesson K, Eberhardt JL, Persson BRR. 1994. Permeability of the blood–brain barrier induced by 915 MHz electromagnetic radiation, continuous wave and modulated at 8, 50, and 200 Hz. *Microscopy Res Tech* 27:535–542.

Salford LG, Brun AE, Eberhardt JL, Malmgren L, Persson BRR. 2003. Nerve cell damage in mammalian brain after exposure to microwaves from GSM mobile phones. *Environ Health Perspect* 111:881–883.

Sandstrom M, Wilen J, Oftedal G, Hansson Mild K. 2001. Mobile phone use and subjective symptoms. Comparison of symptoms experienced by users of analogue and digital mobile phones. *Occup Med* (*London*). 51:25–35.

Vorobyov VV, Galchenko AA, Kukushkin NI, Akoev IG. 1997. Effects of weak microwave fields amplitude modulated at ELF and EEG of symmetric brain areas in rats. *Bioelectromagnetics* 18:293–298.

Saunders RD. 2003. Rapporteur report: weak field interactions in the central nervous system. *Radiat Prot Dosimetry* 106:357–361.

Schirmacher A, Winters S, Fischer S, Goeke J, Galla HJ, Kullnick U, Ringelstein EB, Stogbauer F. 2000. Electromagnetic fields (1.8 GHz) increase the permeability to sucrose of the blood–brain barrier *in vitro*. *Bioelectromagnetics* 21:338–345.

Schmid G, Sauter C, Stepansky R, Lobentanz IS, Zeitlhofer J. 2005. No influence on selected parameters of human visual perception of 1970 MHz UMTS-like exposure. *Bioelectromagnetics* 26:243–250.

Seaman RL, Beblo DA, Raslear T. 1994. Modification of acoustic and tactile startle by single microwave pulses. *Physiol Behav* 55:587–595.

Seitz H, Stinner D, Eikmann T, Herr C, Roosli M. 2005. Electromagnetic hypersensitivity (EHS) and subjective health complaints associated with electromagnetic fields of mobile phone communication—a literature review published between 2000 and 2004. *Sci Total Environ* 349:45–55.

Shandala MG, Dumanskii ID, Rudnev MI, Ershova LK, Los IP. 1979. Study of nonionizing microwave radiation effects upon the central nervous system and behavior reactions. *Environ Health Perspect* 30:115–121.

Shandala MG, Vinogradov G, Rudnev MI, Rudakova SF. 1983. Effect of microwave radiation on cell immunity in conditions of chronic exposure. *Radiobiologiia* 23:544–546.

Shandala MG, Dumanskii ID, Tomashevskaia LA, Soldatchenkov VN. 1985. Hygienic standardization of intermittent-pulse electromagnetic energy of ultra-high 2750 MHz frequency in the environment. *Gig Sanit* 26–29.

Shellock FG, Crues JV. 2004. MR procedures: biological effects, safety, and patient care. *Radiology* 232:635–652.

Shirai T, Kuribayashi M, Wang J, Fujiwara O, Doi Y, Nabae K, Tamano S, Ogiso T, Asamoto M. 2005. Lack of effects of 1439 MHz electromagnetic near field exposure in the blood brain barrier in immature and young rats. BEMS Abstract, Dublin, p. 173. http://bioelectromagnetics.org/bioem2005/bioem2005-abstracts-small.pdf?PHPSESSID = 78d98d6fc2c81667e31b2f9184e163db

Shupak NM, Hensel JM, Cross-Mellor SK, Kavaliers M, Prato FS, Thomas AW. 2004a. Analgesic and behavioral effects of a 100 micro T specific pulsed extremely low-frequency magnetic field on control and morphine treated CF-1 mice. *Neurosci Lett* 354:30–33.

Shupak NM, Prato FS, Thomas AW. 2004b. Human exposure to a specific pulsed magnetic field: effects on thermal sensory and pain thresholds. *Neurosci Lett* 363:157–162.

Sienkiewicz Z. 2002. Biological effects of microwaves: animal studies. In: Stone WR, Wilkinson P, Eds., *The Review of Radio Science 1999–2002*. New York: IEEE Press, Wiley, pp. 943–964.

Sienkiewicz ZJ, Haylock RGE, Bartrum R, Saunders RD. 1998a. Deficits in spatial learning after exposure of mice to a 50 Hz magnetic field. *Bioelectromagnetics* 19:79–84.

Sienkiewicz ZJ, Haylock RGE, Saunders RD. 1998b. 50 Hz magnetic field effects on the performance of a spatial learning task by mice. *Bioelectromagnetics* 19:486–493.

Sienkiewicz ZJ, Blackwell RP, Haylock RG, Saunders RD, Cobb BL. 2000. Low-level exposure to pulsed 900 MHz microwave radiation does not cause deficits in the performance of a spatial learning task in mice. *Bioelectromagnetics* 21:151–158.

Sienkiewicz Z, Jones N, Bottomley A. 2005. Neurobehavioural effects of electromagnetic fields. *Bioelectromagnetics* July 29, DOI 10.1002/bem.20141.

Silny J. 1986. The influence threshold of the time-varying magnetic field in the human organism. In: Bernhardt JH, Ed., *Biological Effects of Static and Extremely Low Frequency Magnetic Fields*. München, Germany: MMV Medzin Verlag, pp. 105–115.

Smith RF, Clarke RL, Justesen DR. 1994. Behavioral sensitivity of rats to extremely low-frequency magnetic fields. *Bioelectromagnetics* 155:411–426.

Smythe JW, Costall B. 2003. Mobile phone facilitates memory in male, but not female subjects. *Neuro Report* 14:243–246.

Staudenmayer H, Binkley KE, Leznoff A, Phillips S. 2003. Idiopathic environmental intolerance: Part 2: A causation analysis applying Bradford Hill's criteria to the psychogenic theory. *Toxicol Rev* 22:247–261.

Stell M, Sheppard AR, Adey WR. 1993. The effect of moving air on detection of a 60 Hz electric field. *Bioelectromagnetics* 14:67–78.

Stern S, Justesen DR. 1995. Comments on "Do rats show a behavioral sensitivity to low-level magnetic fields?" letter and reply. *Bioelectromagnetics* 16:335–338.

Stern S, Laties VG. 1985. 60-Hz electric fields: detection by female rats. *Bioelectromagnetics* 6:99–103.

Stern S, Laties VG, Stancampiano CV, Cox C, de Lorge JO. 1983. Behavioral detection of 60-Hz electric fields by rats. *Bioelectromagnetics* 43:215–247.

Straume A, Oftedal G, Johnsson A. 2005. Skin temperature increase caused by a mobile phone: a methodological infrared camera study. *Bioelectromagnetics* 26:510–519.

Sutton CH, Nunnaly RL, Carroll FB. 1973. Protection of the microwave-irradiated brain with body-core hypothermia. Cryobiology 10:513.

Sutton CH, Carroll FB. 1979. Effects of microwave-induced hyperthermia on the blood-brain barrier of the rat. Radio Science 14: 329–334.

Tattersall JE, Scott IR, Wood SJ, Nettell JJ, Bevir MK, Wang Z, Somasiri NP, Chen X. 2001. Effects of low intensity radio frequency electromagnetic fields on electrical activity in rat hippocampal slices. *Brain Res* 904:43–53.

Thimonier C, Chabert R, Ayoub J, de Seze R, Lallemant J-G. Miro L, Fabbro-Peray P. 1999. No effect in humans of microwaves emitted by GSM mobile phones on the auditory brainstem responses (ABRs) and distortion products of otoemission (DPOE). In: Bersani F, editor. Proceedings of the 2nd world congress for electricity and magnetism in biology and medicine. New York, NY: Kluwer Academic Publishers.

Thomas AW, Persinger MA. 1997. Daily post-training exposure to pulsed magnetic fields that evoke morphine-like analgesia affects consequent motivation but not proficiency in maze learning in rats, *Electro-Magnetobiol* 16, 33–41.

Chizhenkova RA. 1988. Slow potentials and spike unit activity of the cerebral cortex of rabbits exposed to microwaves. *Bioelectromagnetics* 9:337–345.

Tsurita G, Nagawa H, Ueno S, Watanabe S, Taki M. 2000. Biological and morphological effects on the brain after exposure of rats to a 1439 MHz TDMA field. *Bioelectromagnetics* 21:364–371.

Ueno S, Lövsund P, Öberg PA. 1986. Effect of time-varying magnetic fields on the action potential in lobster giant axon. *Med Biol Eng Comput* 24:521–526.

Van Leeuwen GMJ, Lagendijk JJW, Van Leersum BJAM, Zwamborn APM, Hornsleth SN, Kotte ANTJ. 1999. Calculation of change in brain temperatures due to exposure to a mobile phone. *Phys Med Biol* 44:2367–2379.

Vendrik AJH, Vos JJ. 1958. Comparison of the stimulation of the warmth sense organ by microwave and infrared. *J Appl Physiol* 13:435–444.

Veyret B., Zogroye L., (personal communication) partly available in the: WHO Research Agenda for radiofrequency fields 2006. http://www.udo-int/pch-emf/research/ref_research_agenda_2006.pdf

Vinogradov G. 1993. The phenomenon of autoimmunity from the effects of nonionizing microwave radiation. In: Blank M, Ed., *Electricity and Magnetism in Biology and Medicine*. San Francisco: San Francisco Press, Inc., pp. 649–650.

Vinogradov GI, Dumanskij JD. 1975. Sensitization effect of ultra high frequency electromagnetic fields. *Hyg Sanit* 9:31–35.

Vinogradov GI, Naumenko GM. 1986. Experimental simulation of autoimmune reactions by exposure to nonionizing microwave radiation. *Radiobiologiia* 26:705–708.

Vinogradov GI, Gonchar NM, Belonozhko NG, Zhelezniak AA, Vinarskaia EI. 1981. Immunological and hematological effects of low-intensity extremely high frequency electromagnetic fields. *Hyg Residential Areas* 29–33.

Vinogradov GI, Naumenko GM, Vinarskaja EI, Gonchar NM, Zhelezniak AA. 1983. Influence of a low intensity electromagnetic field in the microwave range on the development of the body's immunological reactivity. *Hyg Human Habitats Issue* 22, *Kiev Health* 31–33.

Vinogradov GI, Batanov GV, Naumenko GM, Levin AD, Trifonov SI. 1985. Influence of nonionizing microwave radiation on autoimmune reactions and the antigenic structure of serum proteins. *Radiobiologiia* 25:840–843.

Vinogradov GI, Andrienko LG, Naumenko GM. 1991. The phenomenon of adaptive immunity in exposure to nonionizing microwave radiation. *Radiobiologiia* 31:718–721.

Von Klitzing L. 1995. Low frequency pulsed electromagnetic fields influence EEG of man. *Physica Medica* 11:77–80.

Wagner P, Roschke J, Mann K, Hiller W, Frank C. 1998. Human sleep under the influence of pulsed radiofrequency electromagnetic fields: a polysomnographic study using standardized conditions. *Bioelectromagnetics* 19:199–202.

Wager TD, Smith EE. 2003. Neuroimaging studies of working memory: a meta-analysis. *Cogn Affect Behav Neurosci* 3:255–274.

Wang B, Lai H. 2000. Acute exposure to pulsed 2450-MHz microwaves affects water-maze performance of rats. *Bioelectromagnetics* 21:52–56.

Wasserman E. 1998. Risk and safety of repetitive transcranial magnetic stimulation: report and suggested guidelines from the International Workshop in the Safety of Repetitive Transcranial Magnetic Stimulation. *Electroencephalogr Clin Neurophysiol* 108:1–16.

Weigel RJ, Jaffe RA, Lundstrom DL, Forsythe WC, Anderson LE. 1987. Stimulation of cutaneous mechanoreceptors by 60 Hz electric fields. *Bioelectromagnetics* 8:337–350.

Whittington CJ, Podd JV. 1996. Human performance and physiology: a statistical power analysis of ELF electromagnetic field research. *Bioelectromagnetics* 17:274–278. Review.

Whittington CJ, Podd JV, Rapley BR. 1996. Acute effects of 50 Hz magnetic field exposure on human visual task and cardiovascular performance. *Bioelectromagnetics* 17:131–137.

WHO EMF Project. 1996. http://www.who.int/peh-emf/project/en/

Woolf NJ. 1991. Cholinergic systems in mammalian brain and spinal cord. *Prog Neurobiol* 37:475–524. Review.

Woolf NJ, Butcher LL. 1981. Cholinergic neurons in the caudate–putamen complex proper are intrinsically organized: a combined Evans blue and acetylcholinesterase analysis. *Brain Res Bull* 5:487–507.

Yamaguchi H, Tsurita G, Ueno S, Watanabe S, Wake K, Taki M, Nagawa H. 2003. 1439 MHz pulsed TDMA fields affect performance of rats in a T-maze task only when body temperature is elevated. *Bioelectromagnetics* 24:223–230.

Zwamborn APM, Vossen SHJA, van Leersum BJAM, Ouwens MA, Makel WN. 2003. Effects of Global Communication system radio-frequency fields on well being and cognitive functions of human subjects with and without subjective complaints. COFAM. Netherlands Organisation for Applied Scientific Research (TNO) Physics and Electronics Laboratory, TNO-report FEL-03-C148. http://www.tno.nl/instit/fel/

Ziriax JM, MacCallum M, Hurt W. 1996. Detection of 94.5 GHz radio frequency fields by rhesus monkeys. *Bioelectromagnetics Abstract*, Victoria, B.C.

5

Thermoregulation in the Presence of Radio Frequency Fields

David Black

CONTENTS

5.1 Introduction

Much of the research on thermoregulation in the presence of absorbed radio frequency (RF) energy has used commonly available laboratory animals. These, like all mammals, are endotherms, with a dependence for body temperature on controlled-heat production by metabolism, but there are important limitations. These arise from their size and consequent surface-to-volume ratio and also differences in thermoregulatory mechanisms, such as sweating, which are species dependant. In general, the function of most mammalian tissues and organs depends on the maintenance of temperature within a range of some 5°C, although this will vary to some extent as a result of changes in activity, hormonal and psychological factors, as well as intake of food and physiologically active substances, such as pharmaceuticals and alcohol.

Most tissues are tolerant to low temperatures, at least insofar as potential for recovery, but since there is relatively low tolerance of overheating, organisms have evolved various sets of sophisticated control mechanisms to avoid this. The controls that are used in human physiological systems may be broadly categorized as behavioral and physiological. For behavioral control, the whole organism acts consciously or subconsciously to alter its environment with regard to radiated or convected heat energy or inhaled air. This provides gross control, which also serves to guard against entering or remaining in adverse thermal conditions, if they can be avoided. In humans this is connected to and mediated by much more complex behaviors, including anxiety, which attempts to predict an adverse environment, resulting in sophisticated avoidance behaviors. These mechanisms provide coarse control, and more primitive variants of such behaviors are also found in ectotherms such as reptiles. The added mechanism available to mammals, where body temperature is maintained to a set point, are the physiological automatic thermoregulatory systems, which are largely controlled by the autonomic nervous system, requiring a sophisticated network of sensors and a variety of control mechanisms that have to work in concert with other autonomic functions such as the metabolic demands of circulation.

5.2 Maintenance of Heat Equilibrium

A generalized heat balance equation as used by thermal physiologists was originally proposed by Pennes [1] in 1948. In simpler terms, [2] a generalized heat balance equation using units of power density (W/m^2) provides that

$$M \pm W = R \pm C \pm E \pm S \ [\mathrm{W/m^2}]$$

where M is the rate of metabolic heat production, W is the rate of work expended (+) or received (−), R is the radiant exchange (+ for gain to the body), C is the convective exchange, E is the energy lost to latent heat of vaporization caused by surface evaporation, and S is the rate of heat storage (+ for net gain).

Essentially this is no more or less than an expansion of the first law of thermodynamics, which states that the change in internal energy equals the heat added to the system minus work done by the system, $\Delta U = Q + W$. (In chemistry, it is more common to write the law with a positive sign for work done on systems, whereas physics and engineering usually use a negative sign in the equation if the work done by the system is positive.)

A variation of this to deal with temperature rises associated with electromagnetic fields is described in detail in Chapter 10 of *BBA*, where numerical methods are described which are of practical value in analyzing the ultimate fate of thermal energy lost to or gained from the environment by the living tissue of an endotherm.

5.3 Endogenous Heat Production (*M*)

Basal metabolic rate (BMR) of a human is the term used to describe the heat production of a healthy human at rest in a thermal environment, which does not add or subtract energy from the body. The standard rate for a man is about 84 W or 0.8 MET. MET is the traditional unit for measuring metabolism, equivalent to 58.15 W/m^2 of body surface. A normal adult has a surface area of 1.7 m^2 and so a person in thermal comfort with an activity level of 1 MET would have a heat loss of about 100 W. Human metabolism is at its lowest during sleep (0.8 MET) and at its highest during vigorous physical activity when increases of 10- to 15-fold are frequently reached. The basal metabolic rate can be altered by changes in the body mass, diet, or by variations in endocrine levels but it is probably not reset to accommodate living in a hot climate [2].

In the resting state, most human heat is generated in the central viscera and neurological tissues, particularly the brain and this is conducted away by the circulatory system to more peripheral tissues. Heat is eventually dissipated by the exchange mechanisms of convection, radiation or evaporative loss from the skin, which affords an autonomic mechanism for control of the rate. Under cold conditions, an increase in muscle tone and activation of skin hair (piloerection) can increase metabolic heat products (M) by about 35%. Active shivering increases this to several fold. However, this is temporary; constant heat production under cold conditions will no more than double M [3].

In order to lose excessive heat, after the opportunities for behavioral alterations in environment are exhausted, the autonomic changes in vasodilatation (the increase in caliber of superficial blood vessels) is both effective and relatively rapid. This is described as the vasomotor state. Sweating also provides a further and more efficient method of carrying away significant amounts of heat, lost to the latent heat of vaporization (sudomotor activity). Sweating is activated by both rising heat and by various physiological mechanisms during exercise. Sweating can be activated by ambient temperature rises above about 31°C or by a rise in core temperature above 37°C, providing an emergency mechanism to dump relatively large amount of heat into the environment. A core temperature rise cannot be allowed to go more than a few degrees before damage might occur.

5.4 Thermoregulatory Control

The thermoregulatory system operates with a closed-loop feedback, which maintains a "set" temperature at a reference level determined in the hypothalamus. There are recognized disorders, which are compatible with continued life but have significant effects on the well-being and energy levels of the individual, in which this setting is abnormally low [4]. There are thermal sensors throughout the body in different tissue planes, which monitor the temperature of adjacent tissues and report back to the central control unit, which is in the preoptic/anterior hypothalamic area of the brain stem. The receptors are found elsewhere in the brain and spinal cord, in deep abdominal structures and most importantly in the skin. The autonomic control mechanisms are most sensitive and act pre-emptively to changes in the environment. These generally obviate the need for significant control from more central heating, although the latter can produce relatively powerful responses, particularly in response to hyperthermia (rising heat), which might be reasonably interpreted as a sign of the failure of the systems. Receptors at different levels would be expected to respond in different ways, as alterations in the environment have different implications in the maintenance of a steady state in the type of environment in which the organism has evolved. This must be borne in mind when considering the potential for effect of an agent, such as RF energy, which may not follow the same rules of deposition as sources of radiant and convected heat expected by the organism in the natural environment in which it has evolved.

5.4.1 Tolerance to Added Heat

Before taking into account the ability of various thermal sensors to initiate mechanisms of heat loss, the fundamental question during the gain of heat energy deposited by the dissipation of electrical currents deposited by RF energy is whether there are sufficient mechanisms (vasomotor, sudomotor in addition to convection and radiation) to carry away all of the unwanted heat. Without this, cooling cannot occur and the core body temperature will continue to rise. The extent of heat stress can be predicted by assessing the percentage of the skin surface that is wet with sweat. If this is less than 20% there is generally a state of thermal comfort, but higher values are indicative of erosion of available tolerance [5]. The same ratio is also known as the heat stress index (HSI) [6]. HSI values greater than 30% are uncomfortable whereas levels over 60% become intolerable to severe.

5.4.2 Heat Generated during Exercise

Additional heat is delivered to the circulatory system when muscles are used and under these conditions there are both autonomic changes and mechanical factors, which result in increased blood flow. The autonomic system activates vasodilatation in peripheral skin so that heat is lost from the body surface. Under conditions of equilibrium during constant exercise there may be a small increase in internal body temperature, resulting in a heat gradient into the environment and a stabilized rate of heat loss [7].

5.4.3 Exposure to Radio Frequency

If the body is exposed to RF energy, either by entering electric or magnetic fields or by intercepting an electromagnetic wave, the resultant currents that flow in conductive tissue may heat the various tissues at different rates, which depend on the frequency, amplitude, temporal and other characteristics of the signal and the electrical conductivity and dielectric properties of the absorbing tissue. This results in a markedly nonuniform distribution of heat, even from a collimated wave of constant density. This is not entirely different from the situation with endogenously generated heat, as heat-producing tissues are unevenly distributed throughout the body. However, the difference may be that the circulatory system is adapted to these intrinsic characteristics of the body, whereas the manner in which electromagnetic radiation or fields distribute energy is not anticipated by the adapted physiology. In 1965 Nielsen and Nielsen [8] described equivalence of thermophysical responses during exercise and passive heating using shortwave diathermy to deliver heat directly into deep tissues of the trunk of human subjects. Subsequently it has been argued that passive heating by RF and heat generated by active exercise produces similar thermal loads in the body as a whole, although this might not take into account the tissues that are physiologically protected from heating under normal circumstances. This has been clarified by more recent work by Adair et al. [9] from human studies at two frequencies, 450 and 2450 MHz. In general, these studies show that, despite the differences in depths of energy absorption of the two frequencies, overall absorption of energy varies as expected, and the thermoregulatory system was selectively mobilized to deal efficiently with comparable thermal burdens in the steady state.

5.4.4 Fever

In febrile states, which generally result from some pathology, there is a rise in the "set point," which is maintained by homeostatic mechanisms in a manner which parallels the controls of normal temperature in an afebrile organism. There are important differences in the way the body behaves during fever under different conditions of ambient temperature. Under warm conditions, vasoconstriction will prevent heat loss. If this is inadequate then heat production is activated. Under cold conditions, vigorous shivering, pathognomonic of certain diseases, is induced. Adair et al. [10] in 1997 hypothesized that a febrile animal might use energy from RF exposure to generate a fever in response to an injection of pyrogen into the hypothalamus, thus sparing metabolic energy stores. Four squirrel monkeys were injected through Delrin injection cannulae and reentrant tubes in the medial hypothalamic area; it was found that during the chill and plateau phases of a fever cycle or energy delivered by RF absorption did substitute for metabolic energy in both the generation and the maintenance of fever. However, as the fever fell, as it would during resolution of a febrile illness, RF exposure tended to exacerbate the fever and interfere with mechanisms of heat loss.

5.5 Human Experimental Data

Homo sapiens are very well equipped to deal with a wide range of environmental temperature conditions and to deal with relatively high levels of endogenously generated heat. Humans can tolerate rates of metabolic heat production (*M*) of up to 10 to 15 times the resting rate by a combination of behavioral and physiological adaptations. Recent experiments using human volunteers exposed to moderate levels of RF energy have confirmed expected physiological responses, all of which have been shown to be nonharmful and reversible.

5.5.1 Overexposure

There is limited data on human overexposure. In China, Chiang and Shao [11] reported on experiments using "hundreds of male volunteers" who had their testes exposed to microwaves at 2450 and 450 MHz CW energy once a month for 30 min. It was found that if the testes were heated to 40 to 42°C, sperm counts were reduced significantly and did not recover. In 1991 a similar attempt to use RF energy to achieve sterility employed powers of 20 to 30 W, which raised the scrotal skin temperature by 10°C. Considerable damage was seen on biopsy samples of testicular tissue although this was more resistant than expected.

Mitchell in 1985 [12] indicated that over the preceding 10 y, 300 overexposure incidents were investigated, although only 1 in 5 was confirmed to be over the then-ruling permissible exposure power density limit of 10 mW/cm^2 for any 6 min period. Clinical findings of these subjects were inconsistent although few significant changes were observed.

In 1998 Adair et al. [13] reported the first in a series of studies of human volunteers exposed to plane wave RF fields at controlled power densities in a highly controlled environment. Transducers were mounted on the subjects to measure thermoregulatory responses of heat production and heat loss during 45 min dorsal exposures with 450 MHz continuous wave transmission. The boresight of the antenna was centered on the subjects' back and peak power densities were measured to be 18 and 24 mW/cm^2. The local normalized peak specific absorption rate (SAR) estimated at 0.32 W/kg. Subjects were tested in each of three ambient temperatures, 24, 28, and 31°C. No change in metabolic heat production was seen during any of these exposure conditions. Vigorous increases in local sweating rate on the back and the chest, which were influenced by both ambient temperature and power density, provided skin cooling and maintained regulation of core temperature, which was measured with an oesophageal probe, to within 0.1°C of the normal level. At the highest local power density (24 mW/cm^2) the normalized peak surface SAR was 7.7 W/kg. This level is in excess of the 20 mW/cm^2 allowed for partial body exposure by the IEEE/ANSI Standard [9,14] for a controlled environment, which is six times the level generally permitted for the general public. Under all conditions, the autonomic heat loss response mechanisms of the subjects, principally sweating, maintained thermal homeostasis.

In a later study, Adair et al. [9] compared these results with those collected from a second group of human volunteers at the same frequency. The protocol was identical, as were the ambient temperatures and response measures, but local skin blood flow was measured at three sites on the body. In this study, it was found that the increase in skin temperature was directly related to frequency but local sweating rates on the back and chest were related more to the ambient temperature and SAR. In a third study in the series Adair et al. [15] compared responses of human volunteers at 2540 MHz continuous and pulsed waves of equal average power density. No convincing difference was found between the two. Later studies extended the peak power density to 35 mW/cm^2, with a normalized peak local SAR of 15.4 W/kg, well over current standards limits for partial body exposure, and with the whole body SAR over double the limit for a controlled environment in most standards (0.4 W/kg). Data from sites measured followed similar trends, as expected, with no change in oesophageal temperature or metabolic heat production; and reports of discomfort in the subjects indicated that there are strong warnings available to an over-exposed individual before homeostatic mechanisms for regulation are overcome.

5.6 Magnetic Resonance Imaging

There is a considerable literature relating to the use of RF energy at 64 MHz coming from the alternating fields in the RF coils used in magnetic resonance imaging (MRI) procedures. It is accepted that these frequently exceed the current standards. Shellock in 1992 [16] reported that tissue temperature changes are small and well below the hazardous levels, and in a 1992 review Gordon [17] concluded that whole body SARs up to 2 W/kg may produce significant elevations in skin temperatures. These are still relatively slight and without evidence of adverse effect.

There have been some laboratory studies of humans exposed to RF in the course of MRI scans. In 1992 Adair and Berglund [18] reported tests on two normal male subjects who underwent a series of 20 minute MRI scans with a 1.5 Tesla static magnet of the trunk and were exposed to a whole body SAR of 1.2 W/kg. Reports of thermal sensation and discomfort were recorded throughout the tests. In both the subjects, skin temperatures of chest and thigh are increased by 1.1°C; however, oesophageal temperature was stable throughout nearly 3 h of testing, at the maximum excursion of 0.3°C.

5.7 Thermal Perception and Nociception of Radio Frequency Energy

It is generally accepted that it is the stimulation of nerve endings in the superficial 0.6 mm of human skin that provides the perception of temperature [19]. Whether a RF signal

TABLE 5.1

Dermal Warmth Detection Thresholds as a Function of Frequency

Frequency (GHz)	Threshold (mW/cm^2) (mean ± SEM)	Tissue Penetration[a] (l/e^2 Depth, mm)	Wavelength (m)
2.4	63.1 ± 6.7	32	122
7.5	19.5 ± 2.9	6.3	40
10	19.6 ± 2.9	3.9	30
35	8.8 ± 1.3	0.8	8.6
94	4.5 ± 0.6	0.4	3.2
$1–3 \times 10^5$	5.34 ± 1.1	<0.1	about 0.002

[a]Skin depth (δ) is given by: $\delta = [(67.52/f)\sqrt{(\varepsilon')^2 + (\varepsilon'')^2} - \varepsilon']^{-1/2}$, where f is frequency in GHz, ε' and ε'' are the real and imaginary part of the relative permittivity, respectively. (Durney et al., 1986). Relative permittivity (ε^*) of skin is given (Hurt, 1996) by:

$$\varepsilon^* = \varepsilon' - i\varepsilon'' = 4.3 - 3.19i/f + 105/(1 + if/0.05) + 32/(1 + if/20).$$

Source: Blick, D.W. et al., *Bioelectromagnetics*, 18 (6), 407, 1997. With permission.

produces a sensation of warmth depends on the intensity and duration of exposure, as well as the frequency, which affects the depth of deposition and absorption. Absolute thresholds for detection have been studied in the past by Schwan et al. in 1966 [20] and Eijkman and Vendrik in 1961 [21]. These studies generally involved brief exposures to restricted areas and have reported that shorter wavelengths generally require less energy to produce a threshold of detection, defined as the threshold perception of warmth. The greater the intensity, the shorter the time of exposure for which the effect is determined by intensity alone; this is called the "critical duration" [22]. In 1997 Blick et al. [23] studied the thresholds across a range of five frequencies from 2.54 to 94 GHz as well as for infrared at about 3000 GHz. (see Table 5.1)

The threshold at 2.54 GHz proved to be more than an order of magnitude larger than that at 94 GHz although this was not significantly different from the infrared threshold. In general, these thresholds were proportional to the skin depth at each frequency. The theoretical analysis presented by Riu et al. in 1997 [24] suggested that a constant temperature increase of 0.07°C at or near the surface of the skin would result in perception. Riu also suggests that the depth of the sensors is not in itself important, as long as they are within 0.3 mm of the skin surface. Pain sensation is proportional to temperature and occurs about 46°C [25]. Cook's theoretical explanations were based on theoretical flow of heat energy, which was satisfactory for short exposure. However, that long exposures must also involve physiological responses in the skin and superficial tissues was recognized by Riu in 1997 [24].

5.8 Behavioral Responses to Excessive Heat Load

Both exotherms and endotherms exhibit voluntary activity, which affect the absorption of heat from their environment. There is a relative paucity of human data but substantial animal experimentation. Reptiles exhibit thermotropic reactions, seeking out and basking in sources of warmth including microwaves. Endothermic mammals exhibit more sophisticated responses and experiments show that there is a threshold to initiate the behavior, which then continues until there is equilibrium or change in the environment [26]. Behavioral adaptation can be quite effective; at whole body SARs that are the equivalent of double the resting metabolic rate changes in thermoregulatory behavior can regulate body temperature to normal [27]. Typical values for these effects are given below in Section 5.9 with respect to metabolic (5.9.1) and vasomotor (5.9.2) responses. If a part of the body is exposed to RF energy, behavorial adaptations are determined by the integrated effect of energy accumulation by the whole body [28]. The result of this research has been generalized to occupational exposure of sources of absorbed RF energy and are useful in providing workers and employers with the information about warning signs for early recognition of thermal overload.

5.9 Data from Animal Studies

A thermoregulatory response to an added head load from RF exposure can be reliably predicted under specified environmental and exposure conditions. By definition, there will be no effect below the lower threshold where the energy does not result in any perceptible or behavioral changes sensed by an experimental animal and noticed by an observer, or reported by an experimental human subject. At higher levels, various physiological responses will occur in a hierarchy of metabolic adjustment (5.9.1) and then vasomotor adjustment (5.9.2), followed by evaporative adjustment (5.9.3). Adair [29] provides a detailed discussion of these concepts and their implications for thermoregulation in the presence of RF/Electromagnetic (EM) fields.

5.9.1 Metabolic Responses

A threshold SAR must be exceeded before a reliable reduction in M occurs. In a nonhuman primate this has been shown to be between 0.5 and 1.5 W/kg [30–32], but it has not been systematically explored in other species. The maximum whole body absorption of RF energy occurs at the so-called resonant frequency, when the long axis of the body is parallel to the electric field vector and the length of the body is about 0.4 of the wavelength of the signal. However, exposure of nonhuman primates at their resonant frequency results in somewhat less efficient thermoregulation than exposure to other frequencies [33,34]. While the threshold for reduction of M may be lower at resonance [31], the response change may be less at a given SAR than it is at nonresonance and so the body temperature may rise. This means that, as would be expected, there is no physiologically important difference between exposure at resonance than at other frequencies for equivalent energy absorption. Recent experiments by Adair et al. [35] in which seated human adults undergo 45 min RF whole body exposures at frequencies near resonance (100 MHz) showed that no increase in core temperature occurred at power densities eight times higher than the reference level in the prevailing IEEE/ANSI C95 Standard 1991/2 [14]. It is also clear that if a part of the body is exposed to incident RF energy, the magnitude of the change in M reflects the total absorbed energy integrated over the whole body [28]. It follows that if an endotherm is exposed to RF energy at an SAR greater than that which will reduce M to the resting level, then thermoregulation will be accomplished by mobilization of the next phase in the hierarchy outlined in the preceeding paragraph.

5.9.2 Vasomotor Responses

When RF energy absorption is sufficient to induce vasodilatation, this response progresses as a direct function of the total heat load. Conversely, when the field is removed there is rapid vasoconstriction. In nonhuman primates both the threshold and the degree of vasodilatation depend on the frequency, although there are differences in species. Experimental results suggest that vasomotor control results from a combination of central and peripheral neural inputs and that changes in the caliber of deep blood vessels increase dramatically when the temperature of heated tissue exceeds about 41°C. This phenomenon has been employed to good effect in the treatment of localized malignancies [36] but the possibility of unwanted effects cannot be ignored in predicting effects of unusually high localized exposure.

5.9.3 Evaporative Adjustments during Radio Frequency Exposure

Dry heat loss is at its greatest when vasodilatation is complete and further disposal of energy must be initiated by another mechanism. Evaporation is highly effective and therefore mechanisms for this become mobilized during the phase of the upper levels of peripheral vasodilatation. Depending on species, evaporative loss may be from panting, sweating, or both. Humans have a particularly large capacity to lose body heat by sweating [37].

5.9.4 Intense or Prolonged Exposure

Michaelson [38] showed the characteristic triphasic change in internal body temperature with whole body exposure of dogs at a power density of 1 kW/m^2 of a 2.8 GHz pulsed wave for periods of 2 to 3 h (SAR 6.1 W/kg) or 6 h (SAR 3.7 W/kg). The three phases were described as an initial increase in temperature, a hypothermic plateau, and thermoregulatory collapse. It was presumed that the mobilization of heat loss by panting counteracted the initial effects of RF energy absorption, but that this eventually failed. The ambient temperature was found to be particularly important. Adaptation was observed in the dogs through the series of experiments, resembling the acclimatization to a hot environment seen in humans [2].

Candas [39], experimenting with squirrel monkeys, investigated whether there might be a power density ceiling beyond which further changes in response did not occur. SARs of 5.3 W/kg were found necessary to reduce the elevated *M* to resting level within 10 min. At a lower SAR of 4.3 W/kg resting *M* was reached within 20 min. The resulting change in body temperature was held to about 1°C.

Walters [40] used microwaves to examine whether fatigue during exertional heat stress occurred at a critical body temperature independent of that at the start of the exercise. Rats were exposed to microwaves at 2.1 GHz at 100 mW/cm^2. Using a treadmill, run time exhaustion was significantly reduced after preheating. These results support the idea that a critical temperature exists that limits exercise under hot conditions. The relevance of intrinsic core to skin thermal gradients has also been investigated by Ryan [41,42]. It was concluded that altered core to skin thermal gradient during RF exposure does provide enhanced potential for cardiovascular impairment.

Recently there has been interest in exposure to very high peak power nanosecond, (after nanosecond) pulses composed of an ultra wide band (UWB) of frequencies. It has been hypothesized that these may pose special health risks. Walters et al. [43] reported results from single 2 min exposures of 23 rats to UWB, but found nothing particularly referable to this exposure.

5.9.5 Thermal Hotspots

Thermographic studies on tissue equivalent models have repeatedly shown that wrists, ankles, and the neck are areas where high local absorption and consequent elevation of tissue temperature may be expected. This has been investigated experimentally [33,34] with variable results but somewhat lesser effects than theoretically predicted from a heat deposition model, probably because of the cooling effect of circulating blood.

5.10 Summary

Thermoregulation of the human body in the presence of RF fields follows the laws of thermodynamics in which heat and work are balanced and conserved. More specialized adaptation of the relevant equations have been developed by thermal physiologists, and the non-SI unit MET is widely used for measuring metabolism and does not have an SI equivalent. Thermal comfort at rest equates to about 1 MET, but endogenously generated heat energy can be increased by physiologically generated heat by a factor of more than ten. The autonomic nervous system controls mechanisms for conserving and losing heat to maintain body temperature. These include shivering, vasodilatation, and facilitation of evaporative loss at the air interface. Thermoregulatory control is managed by the body's "central computer" which has a set temperature, maintained at a reference level in specialized tissue in the hypothalamus. The results of monitoring are fed back by thermal sensors in a number of tissues, both central and superficial. Absorption of RF energy, which results in internal fields, resolves into electric currents in conductive tissues, the ultimate fate of energy being heat deposition. Deposition is highly nonuniform, irrespective of the nature of the wave or field, because of variations in penetration, conduction, and current flow. Distribution of heat deposition inevitably affects the pattern of exposed thermoreceptors and thus the feedback to the hypothalamus used to form an impression of overall body temperature. Increases in the set point can result from naturally occurring fever and experimental comparisons have been made between this condition and heat gain caused by RF.

The effect of heat on tissue has been studied in animal experiments and to a limited and lesser extent on humans to the point of irreversible tissue damage. However, in recent experiments using human volunteers exposed to substantial levels of energy, heating to the point of discomfort has shown no elevation of core temperature. Thus, it is generally accepted that behavioral responses to excessive heat provide a generally reliable early warning.

References

1. Pennes, H.H., Analysis of tissue and arterial blood temperatures in the resting human forearm, *J. Appl. Physiol.*, 1, 93–122, 1948.
2. Adair E.A. and Black D.R., Thermoregulatory response to RF energy absorption. *Bioelectromagnetics Supp.* 6:S17-S38, 2003.
3. Iampietro, P.F., Vaughan, J.A., Goldman, R.F., Kreider, M.B., Masucci, F., and Bass, D.E., Heat production from shivering, *J. Appl. Physiol.*, 15, 632–634, 1960.

4. Kloos, R.T., Spontaneous periodic hypothermia, *Medicine* (*Baltimore*), 74 (5), 268–280, 1995.
5. Gagge, A.P., A new physiological variable associated with sensible and insensible perspiration, *Am. J. Physiol.*, 120 (277), 1937.
6. Belding, H.S. and Hatch, T.F., Index for evaluating heat stress in terms of resulting physiological strain, *Heat. Piping Air Cond.*, 27 (129), 1955.
7. Robergs, R.A. and Roberts, S.O., *Exercise Physiology: Exercise, Performance and Clinical Applications*, Mosby, St Louis, MO, 1997, pp. 18–21.
8. Nielsen, B. and Nielsen, M., Influence of passive and active heating on the temperature regulation of man, *Acta Physiologica Scandinavica*, 64, 323–331, 1965.
9. Adair, E.R., Cobb, B.L., Mylacraine, K.S., Kelleher, S.A., Human exposure at two radio frequencies (450 and 2450 MHz): Similarities and differences in physiological response, *Bioelectromagnetics*, 20 (Suppl. 4), 12–20, 1999b.
10. Adair, E.R., Adams, B.W., Kelleher, S.A., and Streett, J.W., Thermoregulatory responses of febrile monkeys during microwave exposure, *Ann. NY Acad. Sci.*, 813, 497–507, 1997.
11. Chiang, H. and Shao, B., Biological responses to microwave radiation: Reproduction, development and immunology, in *Electromagnetic Interaction with Biological Systems*, Lin J.C., Ed., Plenum Press, New York, 1989, pp. 141–163.
12. Mitchell, D. and Laburn, H.P., Pathophysiology of temperature regulation, *Physiologist* 28 (6), 507–517, 1985.
13. Adair, E.R., Kelleher, S.A., Mack, G.W., and Morocco, T.S., Thermophysiological responses of human volunteers during controlled whole-body radio frequency exposure at 450 MHz, *Bioelectromagnetics*, 19 (4), 232–245, 1998.
14. IEEE_Standard, *IEEE Standard for safety levels with respect to human exposure to radio frequency electromagnetic fields, 3 kHz to 300 GHz*, 1999.
15. Adair, E.R., Myalcrainc, K.S., Cobb, B.L., Partial-body exposure of human volunteers to 2450 MHz pulsed or CW fields provokes similar thermoregulatory responses, *Bioelectromagnetics*, 22 (4), 246–259, 2001a.
16. Shellock, F.G., Thermal responses in human subjects exposed to magnetic resonance imaging, in *Biological Effects and Safety Aspects of Nuclear Magnetic Resonance Imaging and Spectroscopy*, Magin, R.L., Liburdy, R.P., and Persson, B., Eds., *Ann. NY Acad. Sci.*, 1992, pp. 260–272.
17. Gordon, C.J., Local and global thermoregulatory responses to MRI fields, *Ann. NY Acad. Sci.*, 649, 273–284, 1992.
18. Adair, E.R. and Berglund, L.G., Predicted thermophysiological responses of humans to MRI fields, in *Biological Effects and Safety Aspects of Nuclear Imaging Resonance Imaging and Spectroscopy*, Magin, R.L., Liburdy, R.P., and Persson, B., Eds., *Ann. NY Acad. Sci.*, 1992, pp. 188–200.
19. Hardy, J.D. and Oppel, T.W., Studies in temperature sensation III. The sensitivity of the body to heat and the spatial summation of the warmth sense organ responses, *J. Clin. Invest.*, 16, 533–540, 1937.
20. Schwan, H.P., Anne, A., and Sher, I., Report No. NAEC-ACEL-534, 1966.
21. Eijkman, E. and Vendrik, A.J., Dynamic behavior of the warmth sense organ, *J. Exp. Psychol.*, 62, 403–408, 1961.
22. Stevens, J.C., Thermal sensation: Infrared and microwaves, in *Microwaves and Thermoregulation*, Adair, E.R., Ed., Academic Press, New York, 1983, pp. 191–201.
23. Blick, D.W., Adair, E.R., Hurt, W.D., Sherry, C.J., Walters, T.J., and Merritt, J.H., Thresholds of microwave-evoked warmth sensations in human skin, *Bioelectromagnetics*, 18 (6), 403–409, 1997.
24. Riu, P.J., Foster, K.R., Blick, D.W., and Adair, E.R., A thermal model for human thresholds of microwave-evoked warmth sensations, *Bioelectromagnetics*, 18 (8), 578–583, 1997.
25. Cook, H.F., The pain threshold for microwave and infra-red radiations, *J. Physiol.*, 118 (1), 1–11, 1952b.
26. Adair, E.R. and Adams, B.W., Behavioural thermoregulation in the squirrel monkey: adaptation processes during prolonged microwave exposure, *Behav. Neurosci.*, 97, 49, 1983.
27. Adair, E.R., Report No. USAFSAM-TR-87–7, 1987.
28. Adair, E.R., Microwave challenges to the thermoregulatory system, in *Electromagnetic Waves and Neurobehavioural Function*, O'Connor, M.E., Lovely, R.H., Eds., Alan R. Liss, New York, 1988, p. 179.

29. Adair, E.R., Thermophysiological effects of electromagnetic radiation, *IEEE Eng. Med. Biol.*, 6, 37, 1987.
30. Lotz, W.G. and Saxton, J.L., Metabolic and vasomotor responses of rhesus monkeys exposed to 225 MHz radiofrequency energy, *Bioelectromagnetics*, 8 (1), 73, 1987.
31. Lotz, W.G. and Saxton, J.L., Thermoregulatory responses in the rhesus monkey during exposure to a frequency (225-MHZ) near whole-body resonance, in *Electromagnetic Fields and Neurobehavioural Function*, Lovely, R.H. and O'Connor, M.C., Eds., Alan R. Liss, New York, 1988, p. 203.
32. Adair, E.R., Adams, B.W., and Hartman, S.K., Physiological interaction processes and radiofrequency energy absorption, *Bioelectromagnetics*, 13 (6), 497, 1992.
33. Krupp, J.H., *In vivo* measurement of radio-frequency radiation absorption, in *Proceedings of a Workshop on the Protection of Personnel Against Radiofrequency Electromagnetic Radiation*, Mitchell, J.C., Ed., School of Aerospace Medicine, Brooks AFB, TX, 1981.
34. Krupp, J.H., *In vivo* temperature measurements during whole body exposure of Macaca mulatta to resonant and non-resonant frequencies, in *Microwaves and Thermoregulation*, Adair, E.R., Ed., Academic Press, New York, 1983, p. 95.
35. Adair, E.R., Myalcraine, K.S., and Allen, S.J., Human thermophysiological responses to whole-body RF exposure (100MHz CW) regulate the body temperature efficiently, in *BEMS 24th Annual Meeting Abstracts*, 2002, p. 34.
36. Guy, A.W. and Chou, C.K., Electromagnetic heating for therapy, in *Microwaves and Thermoregulation*, Adair, E.R., Ed., Academic Press, New York, 1983, pp. 57–93.
37. Wenger, C.B., Circulatory and sweating responses during exercise and heat stress, in *Microwaves and Thermoregulation*, Adair E.R., Ed., Academic Press, New York, 1983, pp. 251–276.
38. Michaelson, S.M., Thermal effects of single and repeated exposures to microwaves—a review, in *Biological Effects and Health Hazards of Microwave Radiation*, Czerski, P., Silverman, C., Ostrowski, K., Suess, M.J., Shore, M.L., and Shalmon, E., Eds., Polish Medical Publishers, Warsaw, 1974, pp. 1–14.
39. Candas, V., Adair, E.R., and Adams, B.W., Thermoregulatory adjustments in squirrel monkeys exposed to microwaves at high power densities, *Bioelectromagnetics*, 6 (3), 221–234, 1987.
40. Walters, T.J., Ryan, K.L., Tate, L.M., and Mason, P.A., Exercise in the heat is limited by a critical internal temperature, *J. Appl. Physiol.*, 89, 799–806, 2000b.
41. Frei, M.R., Ryan, K.L., Berger, R.E., and Jauchem, J.R., Sustained 35-GHz radiofrequency irradiation induces circulatory failure, *Shock*, 4, 289–293, 1995.
42. Ryan, K.L., Frei, M.R., Berger, R.E., and Jauchem, J.R., Does nitric oxide mediate circulatory failure induced by 35-GHz microwave heating? *Shock*, 6, 71–76, 1996.
43. Walters, T.J., Mason, P.A., Sherry, C.J., Steffen, C., and Merritt, J.H., No detectable bioeffects following acute exposure to high peak power ultra-wide band electromagnetic radiation in rats, *Aviat. Space Environ. Med.*, 66, 562–567, 1995.

6

Epidemiologic Studies of Extremely Low-Frequency Electromagnetic Fields

Leeka Kheifets and Riti Shimkhada

CONTENTS

6.1 Introduction

Given the ubiquitous nature of extremely low-frequency (ELF) electromagnetic fields (EMFs), there is concern regarding their potential to adversely affect the health. Numerous health effects have been studied in relation to the EMF exposure: cancer, reproductive disorders, as well as neurodegenerative and cardiovascular diseases. Cancer, especially childhood cancer, has received the most attention.

A number of reviews on the potential of EMF to cause damage to health have been published [1–4]. The general consensus is that cellular effects do not occur with exposures below 100 μT. Also, except for a very few animal studies that suggest adverse effect of EMF, these studies have been largely negative.

The EMFs are imperceptible, ubiquitous, have multiple sources, and can vary greatly over time and short distances [5]. In the absence of a biological mechanism to implicate one or more specific field parameters, the exposure assessment of EMF has varied over the years. The epidemiologic studies in the last decade have employed improved exposure assessment methods. Most of the epidemiologic studies discussed below use the time-weighted average (TWA) measurements to characterize the exposure. Furthermore, with the technological advances and increased sample size studies, the higher exposures, i.e., >0.4 μT, are being explored. Although the epidemiologic evidence is not conclusive, it is generally agreed that the possibility of a causal association between the EMF and the adverse health outcomes cannot be excluded and that the epidemiologic studies of childhood leukemia provide the strongest evidence of an association.

The epidemiologic evidence is a major contributor to the understanding of the potential effects of EMF on health. The International Agency for Research on Cancer (IARC) classified EMF as a Group 2B or possible human carcinogen [1]; this classification was mostly based on the consistent epidemiologic evidence of an association between exposure to these fields and childhood leukemia and of laboratory studies in animals and cells which were not supportive of exposure to EMF causing cancer. Although the body of evidence is always considered as a whole, on the basis of the weight of evidence approach and incorporating different lines of scientific enquiry, epidemiologic evidence, as the most relevant, is given the maximum weight.

The epidemiologic data is routinely critically assessed to shed light on the potential of EMF to cause harm to health. This chapter provides an up-to-date review of the epidemiologic evidence and accompanying methodological concepts.

6.2 Leukemia

6.2.1 Background

The leukemias are cancers of the blood and bone marrow. The classification of leukemias is conventionally based on the origin of the cell types (lymphocytes, myelocytes, monocytes) and rate at which the disease progresses (acute and chronic).

The age distribution of leukemias is bimodal, with a first peak occurring at 4 y, a decline at 15–29 y, and then a slow rise throughout the rest of life. Leukemia is the most common childhood malignancy, constituting more than half of all childhood cancers. Acute lymphocytic leukemias (ALLs) account for 75% of all the cases of childhood leukemia; the most common types in adults are acute myeloid leukemia (AML) and chronic lymphocytic leukemia (CLL).

The age-standardized rate of leukemia for children younger than 15 y has been estimated to be 3.5 per 100,000 per year for females and 4.2 per 100,000 per year for males in the developed world and 2.2 per 100,000 per year for females and 2.9 per 100,000 per year for males in the developing world [6]. For adults, the rates for each type of leukemia have changed very little over the past 20 years. For children, however, incidence rates have risen slightly; similar or more pronounced increases have occurred for other childhood cancers [7]. Some of these increases might be attributable to the improved diagnosis.

The environmental, occupational, and genetic factors have been associated with one or more types of leukemia. Increased risks have been associated with radiation exposure (all types of leukemia except CLL), cigarette smoking (parental smoking for childhood leukemia), and exposure to the human T-cell leukemia virus type one (HTLV-1) (associated with adult T-cell leukemia and lymphoma). Occupational exposures associated with increased risk are exposure to benzene (AML), manufacturing processes of styrene and butadiene production (lymphoid leukemia), and petroleum refining (AML and to a lesser extent CLL). The genetic risk factors include the chromosomal and congenital abnormalities of Fanconi's anemia, along with Down's, Bloom's, Kleinfelter's, and trisomy G syndromes [8].

In general, the relationship between socioeconomic status (SES) and childhood leukemia is weak and inconsistent [9]. The causes of leukemia, especially childhood leukemia, are not well understood. In addition, the presently known or suspected risk factors are likely to account for only a small proportion of all cases [10].

6.2.2 Childhood Leukemia

6.2.2.1 Residential Exposures

6.2.2.1.1 Wire Codes

In the first study by Wertheimer and Leeper [11] in 1979, each subject's exposure was categorized on the basis of (a) the type of electric utility wiring adjacent to the residence, i.e., transmission lines, primary and secondary distribution lines of various types, and (b) the distance from that wiring to the residence. This exposure is referred to as the "wire code" (see Table 6.1 for a list of the major epidemiologic studies using wire codes).

TABLE 6.1

Epidemiologic Studies of the Association between EMF and Childhood Leukemia

Ref.	Study Population	Study Design	Exposure	RR (95%CI) High Category
Wertheimer and Leeper [11]	Cases: deaths between 1950 and 1973, <19 y old. Controls: birth certificates. Denver, Colorado	Case–control: 155 cases, 155 controls	Wire code	HCC 3.0 (1.8–5.0)
Fulton et al. [184]	Cases: <20 y old. Controls: birth certificates. Rhode Island	Case–control: 119 cases, 240 controls	Wire code	HCC 1.0 (—)
Tomenius [21]	Cases: <19 y old. Controls: birth certificates. Stockholm, Sweden	Case–control: 243 cases, 212 controls	Measured field (front door)	≥0.3 μT 0.3(—)
Savitz et al. [13]	Cases: <15 y old. Controls: random digit dialling. Denver, Colorado.	Case–control: Wire code—97 cases, 259 controls	Wire Code	HCC 2.7 (0.9–8.0)
		Measured field—36 cases, 207 controls	Measured field (spot, child's bedroom, low power)	≥0.25 μT 1.9 (0.7–5.6)
Myers et al. [185]	Cases: <15 y old. Controls: birth register. Yorkshire, England	Case–control: 243 cases, 212 controls (all cancers)	Distance (home to overhead line)	≤25 meters 1.1 (0.5–2.6)
			Estimated field strength	≥0.1 μT 0.4 (0.04–4.3)
London et al. [15]	Cases: <10 y old. Controls: friends and random digit dialling. Los Angeles, California	Case–control: Wire code—211 cases, 205 controls	Wire Code	VHCC 2.1 (1.1–4.3)
		Measured field—162 cases, 143 controls	Measured field (24 h bedroom)	≥0.125 μT 1.2 (0.5–2.8)
Feychting and Ahlbom [16]	Residents within 300 m of power line. Cases: <15 y old. Controls: random from cohort. Sweden	Nested Case–control: 38 cases, 554 controls	Historically calculated fields	≥0.3 μT 3.8 (1.4–9.3)
Olsen et al. [61]	Cases: <10 y old. Controls: Central Population Registry. Denmark	Case–control: 833 cases, 1666 controls	Historically calculated fields	≥0.4 μT 6.0 (0.8–44)
Verkasalo et al. [18]	Residents within 500 m of power line. <17 y, 1974–1990. Finland	Cohort: 35 cases	Historically calculated fields	≥0.2 μT 1.6 (0.3–4.5)
Tynes and Haldorsen [17]	Residents in ward with high-voltage power lines. Cases: <15 y old. Controls: random from cohort. Norway	Nested Case–control: 139 cases, 546 controls	Historically calculated fields	≥0.14 μT 0.3 (0.0–2.1)

Linet et al. [26]	Cases: <15 y old. Controls: random digit dialling. U.S. residents in 9 Midwestern and mid-Atlantic States	Case–control: Wire code—402 cases, 402 controls (acute lymphoblastic leukaemia)	Wire Code	VHCC 0.9 (0.5–1.6)
		Measured field—162 cases, 143 controls (acute lymphoblastic leukaemia)	Measured field (time-weighted average)	≥0.3 μT 1.2 (0.9–1.8)
Michaelis et al. [28]	Cases: <15 y old. Controls: govt. office residents' registry. Residents of Northwest Germany and Berlin	Case–control: 176 cases, 414 controls	Measured field (24 h bedroom)	≥0.2 μT 2.3 (0.8–6.7)
Dockerty et al. [33]	Cases: <15 y old. Controls: birth certificate. New Zealand	Case–control: 115 cases, 117 controls	Measured field (24 h bedroom)	≥0.2 μT 15.5 (0.3–7.6)
McBride et al. [27]	Cases: <15 y old. Controls: health insurance records. Canada	Case–control: Wire code—303 cases, 309 controls	Wire code	VHCC 0.8 (0.4–1.6)
		Measured field (48 h personal)—293 cases, 339 controls	Measured field (48 h personal)	≥0.2 μT 1.0 (0.7–1.6)
		Measured field (24 h bedroom)—272 cases, 304 controls	Measured field (24 h bedroom)	≥0.2 μT 1.3 (0.7–2.3)
Green et al. [20]	Cases: <15 y old. Controls: telephone marketing lists. Canada	Case–control: Wire code—79 cases, 125 controls	Wire code	OHCC ± VHCC 1.5 (0.3–8.7)
		Measured field (48 h personal)—88 cases, 133 controls	Measured field (48 h personal)	≥0.4 μT 1.1 (0.3–4.1)
		Measured field (spot)—21 cases, 46 controls	Measured field (spot)	≥0.4 μT 4.5 (1.3–15.9)
UKCSS [30]	Cases: <15 y old. Controls: Family Health Services Authority. England, Wales, Scotland	Case–control: 1094 cases, 1096 controls	Measured field (Phase I: 90 min in family room + spot measurement in bedroom; Phase II: 48 h bedroom and spot measurement in school only for highest 10% in Phase I.	≥0.4 μT 1.7 (0.4–7.1)
Bianchi et al. [19]	Cases: <15 y old and controls in Varese province identified by Lombardy Cancer Registry, Italy	Case–control: 101 cases, 412 controls	Historically calculated fields	≥0.1 μT 4.5 (0.9–23.2)
Schuz et al. [186]	Cases: <15 y old German Childhood Cancer Registry. Controls: population registration files. West Germany	Case–control: 514 cases, 1301 controls	Measured field (24 h residential)	≥0.4 μT 5.9 (0.8–44.1)

Wire Codes: VLCC, very low-current configuration; OLCC, ordinary low-current configuration; OHCC, ordinary high-current configuration; HCC, high-current configuration.

Wertheimer and Leeper [11] initially used a dichotomous exposure classification wherein subjects were classified as living in either high- or low-current configuration (HCC or LCC) homes. Later they introduced a finer degree of resolution into their wire code classification system: very low-current configuration (VLCC), ordinary low-current configuration (OLCC), ordinary high-current configuration (OHCC), and very high-current configuration (VHCC) [12]. Savitz et al. [13] and Kaune and Savitz [14] further refined the original wire code configuration classification scheme.

Wertheimer and Leeper [11] surmised that across an entire living space the outdoor electric lines and the associated grounding system were the dominant source of long-term magnetic field exposures; and unless the lines were modified over time or homes were demolished they provided an estimate of the past exposures. Wire coding did not require resident's participation; thus, the exposure assessment without intrusion was possible for cancer cases and controls (cancer-free comparison subjects) with verified addresses, and the study size was therefore maximized. On the basis of cancer cases between 1950 and 1973, they reported that childhood leukemia mortality was associated with residence in the HCC homes. However, methodological problems, such as nonblind exposure assessment, reliance on death certificates, and lack of evaluation of confounding, rendered this study difficult, if not impossible, to interpret. It, nevertheless, generated a hypothesis that has been followed up by several other studies.

A childhood cancer investigation by Savitz et al. [13], conducted in the same geographical area, was designed to replicate the Wertheimer and Leeper [11] study and to improve upon it through comprehensive ascertainment of incident cases, assessment of potential confounders, and blind exposure assessment. In addition to wire coding, they included point-in-time or spot magnetic field measurements in all available residences. In general, higher magnetic fields were registered in the OHCC and especially VHCC residences, but the wire code or the magnetic field correlation was weak. Savitz et al. [13] reported positive associations of wire code with childhood leukemia. The risk for spot measurements was lower and less precise, partially owing to the reduced number of homes available for measurement. The results for wire code and leukemia were basically consistent with those of Wertheimer and Leeper [11], with the additional evidence that wire code may serve as a weak indicator of contemporary measured magnetic fields.

In 1991, London et al. [15] conducted an incidence study of childhood leukemia similar in its basic design to the study by Savitz et al. [13], but with 24 h bedroom measurements added to the exposure assessment. Similar results emerged: (a) the residential magnetic field strength correlated weakly with wire code and (b) a trend of increased leukemia risk across wire code, with elevated leukemia risk among the VHCC sample relative to the referent.

6.2.2.1.2 *Exposure Based on Distance and Calculated Fields*

There are a handful of studies on childhood leukemia using distance from residence to power lines as the exposure of interest. The results of these studies have not been wholly consistent. The studies using calculated field measurements seek to capture historical exposure levels. The exposure to EMF is most often assessed by using power line load data and specifications for power lines that are specific to the time period of interest. Both Feychting and Ahlbom [16] and Tynes and Haldorsen [17] used a nested case–control study approach to identify leukemia cases within defined cohorts during specific time periods and calculated the field measurements. Feychting and Ahlbom [16] showed that the risk of childhood leukemia was elevated for children exposed to 0.3 μT or greater; however, Tynes and Haldorsen [17] showed that there was no elevated risk of leukemia among those exposed to 0.14 μT and greater. No analysis for higher cut points was presented. A study from Finland using similar design showed a nonsignificant elevated risk of childhood leukemia among those exposed to 0.2 μT and greater [18]. A recent study by Bianchi et al. [19] demonstrated no increased risk at low exposures of 0.1 μT and greater.

6.2.2.1.3 *Household and Personal Exposures*

The majority of epidemiologic studies on EMF and childhood leukemia use residential EMF exposure measurements; they have been either spot measurements [13,20,21] or area (mostly bedroom) magnetic field measurements of 24 h and longer [20,22–30]. Although most of these studies observed elevated odds ratios (ORs), for many, the confidence intervals (CIs) around these estimates of effect are large due to small numbers of highly exposed subjects. Given the rarity of both the disease and the high exposure, securing enough highly exposed cases, needed for statistically stable estimates, is quite difficult.

Only two studies employed the use of personal EMF meters to measure exposure [20,27]. In both of these studies, there was no indication of elevated risk of leukemia due to high personal exposure. Although there is evidence that there is a close correlation between household exposure and personal measurements [30,31], use of personal measurements in case–control studies might be biased as disease status might effect personal exposure measurements. Personal measurements are even more problematic in case–control studies of children, as exposure at young age is quite age dependent, and measurements made few years after the etiologically relevant time period might be different than during the time period of interest.

6.2.2.1.4 *Pooled Analyses*

Two pooled analyses represent the most powerful attempt so far to provide a cohesive assessment of the epidemiologic data of EMF and childhood leukemia [22,32]. These analyses, although focusing on a largely overlapping but distinct set of studies, come to similar conclusions.

In the pooled analysis by Greenland et al. [32], 12 studies using measured or calculated fields were identified; the study included a total of 2656 cases and 7084 controls. For this analysis, the metric of choice was the time-weighted average. The estimated OR for childhood leukemia was 1.68 (95% CI 1.23–2.31) for exposures greater than 0.3 μT as compared with exposures less than 0.1 μT, controlling for age, sex, and study.

Using more stringent inclusion criteria, Ahlbom et al. [22] included nine studies using measured and calculated fields. There were a total of 3203 cases and 10,338 controls in the pooled sample. Using the geometric mean as the metric of choice, the estimated OR for childhood leukemia was 2.00 (95% CI 1.27–3.13) for exposures greater than or equal to 0.4 μT as compared with exposures less than 0.1 μT, controlling for age, sex, SES (in measurement studies only), and East and West (in German study only).

6.2.2.2 Electric Appliance Exposure

Studies that have evaluated the risks of childhood leukemia [15,33–35] associated with the use of electric appliances, used interviews of mothers to assess exposure information. Positive findings were observed in only a couple of these studies. There was an elevated risk of leukemia, but no evidence of dose–response relationship, among children watching black-and-white television [15] in one study; another study observed a rise in leukemia with increasing number of hours children watched television, regardless of the child's distance from the television set [34]. Risks were increased for postnatal exposure to a few other appliances in one study [34], with no dose–response relationship. Postnatal use of electric blankets [33–35] and hair dryers [15,34] was linked with modestly elevated risks in more than one study, but there was no evidence of dose–response relationships. Overall, the small number of studies, evaluation of one appliance at a time leading to a very large misclassification, and the absence of good measurement data within the studies preclude the straightforward interpretation of results.

6.2.2.3 Parental Exposure

Of the studies that have examined parental occupation and childhood cancer [36] none have found any association with parental exposure to EMF. London et al. [15] reported an association between a mother's exposure to nonionizing radiation during pregnancy and the risk of childhood leukemia. However, the exposure question did not focus on EMF and is likely to reflect exposures to higher than power frequencies.

Two studies [34,35] observed small increase in the risk of childhood leukemia with the prenatal use of electric blankets. In general, on the basis of these studies, there is little evidence of elevated risk of leukemia in offspring associated with mothers' prenatal use of other types of electric appliances.

6.2.3 Adult Leukemia

6.2.3.1 Residential Exposures

The first study of residential exposures and adult cancer was also conducted by Wertheimer and Leeper [12]. Although the focus was on total cancers, leukemia did not contribute to the elevated risk found for all cancers combined.

Several other studies have attempted to examine the risk of adult leukemia in populations residing near power lines [37–40]. These studies are based on small numbers, low potential exposures, and very crude exposure estimation methods. The overall results are negative, with a few hints of a small, nonsignificant elevation of risk in some studies.

Since then, other studies on adult leukemia and residential EMF exposure have shed more light on the plausible relationship [41,42], with a follow-up study by Feychting et al. [43] and studies by Verkasalo and coworkers [44,45] and Li et al. [46]. A small increased risk for all leukemia was seen in some of these studies.

6.2.3.2 Electric Appliance Exposure

In a case–control study of adult AML and CML in Los Angeles County that used cases and matched neighborhood controls [47], use of electric blankets was not related to leukemia risk. Lovely et al. [48] examined adult leukemia risk and personal appliance use, on the basis of the study of Severson et al. [42] mentioned above. Most noteworthy in this analysis were a small elevation in risk for ever/never use of electric razors (relative risk (RR) 1.3, 95% CI 0.8–2.2), and a larger increase in risk with duration of daily use (RR 2.4, 95% CI 1.1–5.5 for the highest relative to the lowest category). However, it appears that this finding is due to bias in the proxy-reported information and that there is in actuality no association between leukemia risk and either use versus no use or duration of use of electric razors [49].

6.2.3.3 Occupational Studies

Studies on occupational exposure to EMF and adult leukemia have job titles mostly linked with cancer incidence or mortality. In a meta-analysis of this literature by Kheifets et al. [50], there was a small increased risk of leukemia associated with work in electric occupations. However, jobs thought to have higher exposure (welders, electricians, linemen, and power plant operators) did not have higher risks than electric workers who generally have lower exposures (installers, engineers, and television or radio repairmen). Several occupational cohort and case–control studies have been published since this meta-analysis. Overall, the results are negative with excess risks reported only in some subgroups [51–56]. Of note is a small case–control study on electric workers from New Zealand [51], which constructed a job-exposure matrix based on measurements obtained in a different study in New Zealand. Based on the exposures for 0.50–1.0 μT the OR was 2.9 (95% CI 0.7–11.4) and for >1.0 μT the OR was 3.2 (95% CI 1.2–8.3), both in reference to exposures less than 0.21 μT.

6.2.4 Summary

In summary, there is enough consistency in the results from the large body of high-quality research on childhood leukemia to reach a general agreement that postnatal exposures above 0.4 μT are associated with an elevated risk of childhood leukemia as compared with exposures less than 0.1 μT; this RR has been estimated at around 2, on the basis of two large pooled analyses. This is unlikely to be due to chance but may be partly due to bias. For other exposure measures and for adult leukemia the evidence is considerably weaker.

6.3 Brain Cancer

6.3.1 Background

Of all nervous system cancers, 90% occur in or around the brain. Cancers of the brain and nervous system (brain cancer) are rapidly fatal but fairly rare, with an incidence rate of 6 per 100,000 people per year in the United States [57]. Glioblastoma and astrocytoma are two subtypes of brain cancer that have been examined in relation to EMF exposure. Brain cancers are more common among males than females. Malignant neoplasms of the brain and other parts of the nervous system are the second most common form of cancer in children aged 0–19 y, representing more than one-fifth of all their malignancies.

There are few clearly recognized risk factors for brain cancer. In addition to ionizing radiation, occupational exposures that have been linked to adult brain cancer risk in some studies including, organic solvents, and pesticides. Excess risks have been found among farmers and painters, but the specific agents responsible for these risks have not been identified. Brain cancers have also been linked to exposure to *N*-nitroso compounds and halomethanes. People with a family history of cancer, certain genetic diseases, or head injury may have an elevated risk of brain tumors. Tobacco and alcohol use have not been linked to brain cancer risk. Thus, the causes of brain cancer are not well understood. Presently known or suspected factors appear to account for only a small proportion of all the cases.

6.3.2 Childhood Brain Cancer

6.3.2.1 Residential Exposures

Several of the childhood cancer studies described earlier examined brain cancer in addition to leukemia and other childhood tumors (see Table 6.2 for a list of the major studies). Two studies based in Denver, Colorado [11,13] found an association between wire codes and brain tumors. In the study by Feychting and Ahlbom [16], however, none of the exposure metrics were significantly associated with brain cancer.

TABLE 6.2

Epidemiologic Studies of the Association between EMF and Childhood Brain Tumors

Ref.	Study Population	Study Design	Exposure	RR (95%CI) High Category
Wertheimer and Leeper [11]	Cases: deaths between 1950 and 1973, <19 y old. Controls: birth certificates. Denver, Colorado	Case–control: 66 cases, 66 controls	Wire code	HCC 2.4 (1.0–5.4)
Tomenius [21]	Cases: <19 y old. Controls: birth certificates. Stockholm, Sweden	Case–control: 294 cases, 253 controls	Measured field (front door)	≥0.3 μT 3.7 (—)
Savitz et al. [13]	Cases: <15 y old. Controls: random digit dialling. Denver, Colorado.	Case–control: Wire code—59 cases, 259 controls	Wire Code	HCC 1.9 (0.5–8.0)
		Measured field—25 cases, 207 controls	Measured field (spot, child's bedroom, low power)	≥0.25 μT 1.0 (0.2–4.8)
Feychting and Ahlbom [16]	Residents within 300 m of power line. Cases: <15 y old. Controls: random from cohort. Sweden	Nested Case–control: 38 cases, 554 controls	Historically calculated fields	≥0.3 μT 1.0 (0.2–3.9)
Verkasalo et al. [18]	Residents within 500 m of power line. <17 y, 1974–1990. Finland	Cohort: $N = 39$	Historically calculated fields	≥0.2 μT 2.3 (0.8–5.4)
Preston-Martin et al. [64]	Cases: <20 y old. Controls: random digit dialling. Los Angeles, California	Case–control: 833 cases, 1666 controls	Wire Code	VHCC 1.2 (0.6–2.2)
Gurney et al. [59]	Cases: <20 y old. Controls: random digit dialling. Washington State	Case–control: 120 cases, 240 controls	Wire Code	VHCC 0.5 (0.2–1.6)
Tynes and Haldorsen [17]	Residents in ward with high-voltage power lines. Cases: <15 y old. Controls: random from cohort. Norway	Nested Case–control: 144 cases, 599 controls	Historically calculated fields	≥0.14 μT 90.7 (0.2–2.1)
UKCSS [30]	Cases: <15 y old. Controls: Family Health Services Authority. England, Wales, Scotland	Case–control: 390 cases, 393 controls	Measured field (Phase I: 90 min in family room + spot measurement in bedroom; Phase II: 48 h bedroom and spot measurement in school only for highest 10% in Phase I.	≥0.4 μT 0 cases and 2 controls

Wire Codes: VLCC, very low-current configuration; OLCC, ordinary low-current configuration; OHCC, ordinary high-current configuration; HCC, high-current configuration.

In the most comprehensive study to date of childhood brain tumors, Preston-Martin et al. [58] found that brain cancer risk was not related to measured fields or high (HCC or VHCC) wire codes. A companion study [59] also found no association between high wire code and occurrence of brain tumors. Other studies [21,60,61] have addressed the potential relationship between magnetic fields and childhood brain cancer with apparently inconsistent results.

In general, in the studies of childhood brain tumors and residential EMFs, no consistent relationship has been reported [62]. However, these studies have generally been small and of low quality; a formal pooled analysis for brain cancer has yet to be completed.

6.3.2.2 Electric Appliance Exposure

Four studies examined brain tumor occurrence and use of appliances by children or their mother during pregnancy [35,59,63,64]. No consistent or remarkable associations were reported in any of the studies. Positive associations of certain appliances found in individual studies were generally not replicated in other studies.

6.3.2.3 Parental Exposure

Six case–control studies have examined the occurrence of childhood brain cancer or neuroblastoma among the children of fathers whose work presumably exposed them to EMF [65–70]. As in most occupational studies, exposure assessment was based on job titles, a method that may not be very precise. Elevated RRs (above 2.0) reported in the first two studies were not confirmed in the later studies. Predominance of positive findings in earlier reports could be due to selective reporting.

Some additional information regarding parental occupations and childhood brain tumor is provided by analyses of the childhood brain cancer studies [64,71]. These studies further support the absence of any association.

6.3.3 Adult Brain Cancer

6.3.3.1 Residential Exposures

Few studies have examined the potential association between brain cancer and residential exposure in adults, and none have looked at appliance use. Cancer of the brain or nervous system has been examined as a subtype in several studies of adult cancers with unconvincing results [12,38,39,44,72]. Most notable is a study by Feychting and Ahlbom [41], which showed no evidence of an association between residential EMF exposure and adult brain cancer, regardless of the method of estimating exposure. In the Feychting et al. [43] study, the residential and occupational exposures were combined by incorporating the estimates of occupational magnetic field exposure into Feychting and Ahlbom's earlier residential study [41]; it found no association between occupational or residential exposure and central nervous system tumors in adults exposed to higher levels of magnetic fields both at home and at work. Although the study has limitations, particularly that the occupational magnetic field measurements were from a different study and hence their applicability is questionable, it is important in that it attempts to incorporate both residential and occupational exposures. Results from a large study [73] of adult glioma and residential EMF exposure assessed through spot measurements, wire codes, and distance from electric facilities did not support an association. For residences with measurements above 0.3 μT, there was a suggestion of increased risk; however, the number of cases and controls in this category was small.

6.3.3.2 Occupational Studies

Many epidemiologic studies have examined the possible association between occupational exposures to EMF and brain cancer. In 1995, Kheifets et al. [74] conducted a review of 52 occupational studies; of these, 29 represented original research published in English and provided enough information to be included in a meta-analysis. These studies examined populations in 12 different countries, although most were conducted in the United States and Scandinavia. Although most studies presented a small elevation in risk, there was considerable heterogeneity in the results. An inverse-variance weighted pooling of all the data resulted in a small but significant overall increase in risk among individuals employed in the broadly defined group of electric occupations. The relative risk was higher for some specific jobs and for gliomas. The findings of this meta-analysis were not affected by inclusion of unpublished data, influence of individual studies, weighting schemes, and model specification. Kheifets [75] also reviewed the studies published since the 1995 meta-analysis [43,76–87] and concluded that these studies mainly showed no risk.

Although positive associations have been reported in some studies with employment in electric occupations, these studies have often had methodological limitations, such as reliance on a single job title for exposure assessment in early studies. More rigorous studies on workers in the electricity supply industries, with a specific focus on occupations with high potential for EMF exposure workers [88–90], have had fewer of these limitations. In a comparative analysis by Kheifets et al. [91], conclusions based on these rigorous studies were compatible with a weak association between magnetic fields and brain cancer; on the basis of a combined analysis of data, the RR per 10 μT-y was 1.12 (95% CI 0.98–1.28). However, results were also compatible with chance fluctuation. This analysis concluded that what previously appeared to be important differences in results across studies are small and not a result of analytic methods.

6.3.4 Summary

Brain cancer studies have shown inconsistent results. A pooled analysis of childhood brain cancer studies may be very informative and can inexpensively provide insight into existing data, including possibility of a selection bias, and, if appropriate (i.e., if studies are sufficiently homogeneous), come up with the best estimate of risk.

6.4 Breast Cancer

6.4.1 Background

Breast cancer is the most commonly occurring malignancy in women in the United States. A considerable body of epidemiologic research has identified numerous factors that affect the risk of developing breast cancer in females. The disease occurs most frequently among whites, women of upper social class, women without children or with few children, and those who had their first child at a late age. Other risk factors include early age of menarche (menses), late age of menopause, obesity for postmenopausal women, proliferative fibrocystic disease, and a first degree relative with breast cancer, especially if it was diagnosed at a young age. Considerably less is known about male breast cancer but indications are that genetic and environmental factors including obesity, familial history, and endocrine factors play a causative role. Occupational studies indicate elevated rates of breast cancer among men in such jobs as newspaper printing, soap and perfume manufacturing, and health care.

Breast cancer incidence rates are highest in North America and northern Europe and lowest in Asia and Africa. Until recently, the search for possible explanations of this pattern had focused on the differences in dietary and reproductive patterns of women in societies with different degrees of industrialization. However, the role of diet in the etiology of breast cancer remains uncertain and reproductive risk factors apparently account for only a fraction of the excess disease reported in the modernized societies.

It has been proposed that one factor contributing to the greater occurrence of breast cancer in industrialized compared with nonindustrialized societies is the use of electric power and higher exposures to light at night or to magnetic fields. Stevens [92] hypothesized that EMF and light at night can affect breast cancer through suppression of melatonin. Epidemiologic investigations addressing the potential link between breast cancer and exposure to magnetic fields include occupational exposures and residential studies that examine breast cancer risk in relation to proximity to electric installations and the use of electric blankets.

6.4.2 Female Breast Cancer

6.4.2.1 Residential Exposures

Several studies of residential exposures and other adult cancers examined the risk of breast cancer in populations residing near power lines. One of these studies found an association between pre- but not postmenopausal breast cancer and wiring configuration [12]. The other two studies [38,39] did not detect any associations. Three recently completed large studies on breast cancer found no association with exposure to electric or magnetic fields [93–95].

6.4.2.2 Electric Blankets

Two early case–control studies [96,97] did not support the hypothesis that the use of electric blankets increases the risk of post- or premenopausal breast cancer. Vena et al. [96] combined these two studies and found elevated risk among women who reported some use of electric blankets throughout the night (RR 1.5, 95% CI 1.1–1.9) for the past 10 years; however, the risk was not the highest among the highest exposure group, that is, those who reported daily use of the blankets in season and continuously throughout the night for 10 years (RR 1.20, 95% CI 0.8–1.9).

6.4.2.3 Occupational Studies

Few occupational studies of electric workers included sufficient numbers of females to address the potential association of occupational EMF exposure and development of breast cancer. Albeit based on small numbers, four studies found no elevation in risk of breast cancer among females working in occupations with potential exposure to EMF as compared with low-exposure occupations [98–101].

One large study [102] used computerized mortality files from the National Center for Health Statistics for the years 1985 to 1989. Death certificates included the occupation and industry in which the decedent usually worked, coded according to the 1980 U.S. Census. Excluded were women whose occupation was listed as homemaker and those whose death certificate provided no occupational data; these two groups made up more than half of the database. Seven electric occupations used in previous studies were included, along with seven other occupations, such as computer programmers and telephone operators, presumed to have a large number of female workers and some potential for above-background EMF exposure. All other occupations were considered unexposed. Among 27,882 breast cancer cases and 110,949 controls, 68 cases and 199 controls had been employed in traditional electric occupations. The RR for breast cancer among those employed in electric occupations was 1.4 (95% CI 1.0–1.8). In a more detailed analysis, the association was the strongest for the managerial and professional class and for those of 45–54 years of age. The RRs for the other occupations with potential exposure were around 1.0 or lower. In a separate analysis of the same data, but with a different approach to exposure grouping, Cantor et al. [103] did not find an association of female breast cancer and potential workplace exposure to EMF.

Epidemiologic studies based on death certificates alone have many limitations and are considered to be of a preliminary nature. In such studies, the population at risk is unknown. Thus, an apparent increase in the proportion of deaths from breast cancer among electric workers may be due to an increase in the proportions of deaths from other causes in the control group. The validity of the study results depends upon the accuracy of the occupational information reported on death certificates and the extent to which job titles alone reflect exposure to magnetic fields. Also, the authors did not take into account many risk factors known to be associated with breast cancer, including reproductive and family histories. Working in male-dominated jobs may have favored nulliparity, delayed childbearing, or other characteristics related to risk of breast cancer. The authors attempted to control for some of these characteristics by adjusting for social class. However, it is not clear how effectively social class was determined, and at best it can serve only as a partial control for known risk factors of breast cancer.

Although some earlier registry-based studies provided some support for a possible association between EMF exposure and female breast cancer [104], the most recent very large study, which incorporated exposure measurements in female workers, did not find an association [105].

6.4.3 Male Breast Cancer

6.4.3.1 Occupational Studies

As described earlier, many studies have examined cancer among workers in electric occupations. As part of that examination, many considered breast cancer as one of the outcomes. Male breast cancer is so rare and most of the studies are not based on sufficiently large populations, so estimates of risk for male breast cancer often are not included in tables of results unless an excess risk has been observed [100,101,106–111]. This makes it difficult to evaluate the potential for the excess risk of male breast cancer. Nevertheless, several reports [98,112–115] are suggestive of a positive association, although negative results were reported by Loomis [116] and Cammarano [117]. The large studies of electric workers [88–90] did not identify any excess of male breast cancer.

6.4.4 Summary

For male and female breast cancer, the research began with a biological hypothesis confirmed by some early studies. More rigorous epidemiologic studies that followed showed no effect, thus it appears that EMF fields are not involved in the development of breast cancer.

6.5 Lung Cancer

Of the many studies [89,90,100,106,108–111,117] that have examined the possible association between electric occupations or power frequency magnetic field exposure and lung cancer, only two [101,107] report some elevation in risk.

Armstrong et al. [76] reported an association between lung cancer and cumulative high-frequency electromagnetic transients, or pulsed electromagnetic fields (PEMFs). In this study of EMF exposure and lung cancer it is possible that some residual confounding from smoking remains, because (a) smoking is a strong risk factor for lung cancer and (b) the length of employment is likely to be strongly correlated with both cumulative PEMF exposure and the amount and length of smoking rather than just current smoking status.

In summary, presently available evidence does not suggest an association between power frequency EMF exposure and lung cancer. The study of lung cancer and PEMFs needs replication.

6.6 Other Cancers

Sporadic reports of elevated risks for other cancers have appeared in the literature, most notably for malignant melanoma, but none have been sufficiently suggestive to warrant presentation here. Finally, cancers of the prostate and pituitary gland could be of interest because of their roles in the production of hormones. Patterns of risk for these cancers in relation to jobs with exposure to magnetic fields remain largely unexplored.

6.7 Reproductive Outcomes

6.7.1 Maternal Exposure

6.7.1.1 Video Display Terminals

Video display terminals emit both extremely low frequency fields as well as higher frequencies of up to 100 kHz. The health risks associated with use of such devices, particularly adverse birth outcomes during pregnancy, have been examined in several studies [118–123]. For the most part, studies have not shown increased risks for spontaneous abortion, low birth weight, preterm delivery, intrauterine growth retardation, or congenital abnormalities.

6.7.1.2 Electric Blankets and Heated Beds

Electric blankets and heated beds can be a major source of ELF EMF; it is estimated that exposure from water beds can be up to 0.5 μT and electric blankets produce fields up to 2.2 μT [124–126].

The findings of the studies on electric blankets or heated beds and reproductive health effects have been mostly negative (see Table 6.3). Furthermore, a related outcome, urinary tract anomalies, has also been shown to be unrelated to EMF exposure from such electric devices [127].

6.7.1.3 Residential and Occupational Exposure

Residential EMF exposures have also been studied for possible association with reproductive outcomes (see Table 6.4). Studies using wire codes or proximity to power lines as EMF exposure proxy-measurements have suggested no association. Studies that have

TABLE 6.3

Epidemiologic Studies on Electric Blanket and Heated Bed Exposure and Adverse Reproductive Outcomes

Ref.	Study Population	Design	Exposure	RR (95%CI) High Category
Miscarriage				
Belanger et al. [187]	Women receiving care at 11 private obstetric practices and 2 health maintenance organizations in New Haven, Connecticut area	Prospective: $N = 2967$	Electric blanket use at:	
			Conception	1.74 (1.0–3.2)
			Conception, high setting	1.65 (0.6–4.9)
			Interview	1.61 (0.8–3.2)
			Interview, high setting	2.05 (0.7–4.7)
			Heated bed use at:	
			Conception	0.59 (0.3–1.1)
			Conception, high setting	0.59 (0.3–1.1)
			Interview	0.63 (0.4–1.1)
			Interview, high setting	0.49 (0.2–1.1)
Lee et al. [188]	Women whose first prenatal appointment at one of three California Kaiser Permanente Medical Care Program facilities, 1990–1991	Prospective: $N = 5144$	Electric blanket	0.8 (0.6–1.2)
			Electric blanket, high	1.6 (0.6–3.3)
			setting	1.0 (0.7–1.3)
			Heated bed	1.0 (0.7–1.5)
			Heated bed, high setting	
Birth defects				
Dlugosz et al. [189]	Cases from New York State Congenital Malformations Registry, born in 1983–1986. Controls selected at random from birth registrations	Case–control: 535 cases, 535 controls	Neural tube defect:	
			Electric blanket	0.9 (0.5–1.6)
			Heated bed	1.1 (0.6–1.9)
			Oral cleft defect	
			Electric blanket	0.7 (0.4–1.2)
			Heated bed	0.7 (0.4–1.1)
Milunsky et al. [190]	Women attending obstetric practices	Cohort: $N = 23{,}491$	Neural tube defect: electric blanket	1.2 (0.5–2.6)
Shaw et al. [123]	Cases from liveborn infants and fetuses, 1989–1991. Controls randomly selected from each area hospital	Study 1 case–control: 538 cases, 539 controls	Neural tube defect:	1.8 (1.2–2.6)
			Electric blanket	1.2 (0.8–1.8)
			Heated bed	1.2 (0.6–2.3)
		Study 2 case–control: 265 cases, 481 controls	Neural tube defect:	1.2 (0.8–1.9)
			electric blanket	1.3 (0.5–3.4)
			Heated bed	1.8 (1.0–3.2)
		Study 3 Case–control: 662 cases, 734 controls	Oral cleft defect (multiple):	
			Electric blanket	
			Heated bed	

Low birth weight and growth retardation				
Bracken et al. [124]	Women receiving care at 11 private obstetric practices and 2 health maintenance organizations in New Haven, Connecticut area	Prospective: $N = 2967$	Low birth weight—Electric blanket or heated bed: Low setting, 3rd trimester High setting, 3rd trimester Intrauterine growth retardation— Electric blanket or heated bed: Low setting, 3rd trimester High setting, 3rd trimester	1.2 (0.5–2.8) 1.2 (0.6–2.5) 0.8 (0.4–1.7) 1.6 (1.0–2.6)
Urinary tract anomalies				
Li et al. [127]	Cases identified from the Washington Birth Defects Registry. Controls randomly selected from among infants born in five large hospitals	All women: 118 cases, 369 controls Subfertile women: 37 cases, 85 controls	Electric blanket Heated bed Electric blanket	1.1 (0.5–2.3) 0.8 (0.3–2.7) 4.4 (0.9–23)

TABLE 6.4

Epidemiologic Studies on Residential EMF Exposure and Adverse Reproductive Outcomes

Ref.	Study Population	Design	Exposure[a]	RR (95%CI) High Category
Miscarriage				
Savitz and Ananth [191]	Residents Denver, Colorado Standard Metropolitan Statistical area	Case–control: $N = 257$	Home spot measurement $\geq$ 0.2 μT	0.8 (0.3–2.3)
Belanger et al. [187]	Women receiving care at 11 private obstetric practices and 2 health maintenance organizations in New Haven, Connecticut area	Prospective: $N = 2967$	Very high wire code	0.4 (0.2–1.1)
Li et al. [129]	Pregnant women within northern California Kaiser Permanente medical care system	Prospective: $N = 969$	Personal TWA $\geq$ 0.3 μT	1.2 (0.7–2.2)
			Personal 24 h maximum $\geq$ 1.6 μT	1.8 (1.2–2.7)
Lee et al. [130]	Members of the northern California Kaiser Permanente medical care system	Nested case–control: 155 cases, 509 controls	Very high wire code	1.2 (0.7–2.1)
			Home spot measurement $\geq$ 0.2 μT	1.1 (0.5–2.2)
			Personal TWA $\geq$ 0.128	1.7 (0.9–3.2)
			Personal 24 h maximum $\geq$ 3.51 μT	2.3 (1.2–4.4)
			Personal rate-of-change $\geq$ 0.094 μT	3.1 (1.6–6.0)
			Home spot measurement $\geq$ 0.2 μT	3.1 (1.0–9.7)
			Personal TWA $\geq$ 0.2μT	1.9 (0.6–6.1)
		Prospective: $N = 219$	Personal 24 h maximum $\geq$ 2.69 μT	2.6 (0.9–7.6)
			Personal rate-of-change $\geq$0.069 μT	2.4 (0.9–6.6)
			Early pregnancy loss	1.1 (0.6–2.3)
			Home spot measurement $\geq$ 0.25 μT	5.1 (1.0–26)
Juutilainen et al. [120]	Work and fertility study of women attempting to become pregnant	Case–control: 89 cases, 102 controls	Home spot measurement $\geq$ 0.63 μT	
Low birth weight and growth retardation				
Bracken et al. [124]	Women receiving care at 11 private obstetric practices and 2 health maintenance organizations in New Haven, Connecticut area	Prospective: $N = 2967$	Low birth weight:	
			Very high wire code	0.8 (0.3–2.1)
			Personal TWA $\geq$ 0.2 μT	1.3 (0.3–6.1)
			Intrauterine growth retardation:	
			Very high wire code	0.7 (0.4–1.6)
			Personal TWA $\geq$ 0.2 μT	1.2 (0.4–3.1)
All anomalies				
Robert et al. [192]	Cases from population-based registry in Central-East France, 1988–1991. Controls, matched for birth year and municipality.	11 cases	Distance to power line $\leq$ 100 m	0.9 (0.5–3.2)

[a]Table reports highest exposure category for measured fields.

used measurement-based exposure assessment have shown mixed results. One study using spot measurement at the front door suggested an association between early pregnancy loss and front door fields of above 0.63 μT [120]. Although spot measurements have been shown to be correlated with personal exposure, there is concern that they result in large misclassification [128].

Using personal measurements, there was no association between exposure to time-weighted fields above 0.2 μT and low birth weight or intrauterine growth retardation [124]. Most recently, a similar prospective study found no association between TWA personal measures and spontaneous abortion [129]. There was, however, a significantly increased risk (OR 1.8, 95% CI 1.2–2.7) when exposure above 1.6 μT was used as the exposure metric of interest. In the same year, a nested case–control study using personal exposure measurements [130] showed that TWA magnetic fields were not associated with miscarriage, but there were in fact statistically significant dose–response trends for two different exposure metrics, maximum exposure and magnetic field rate-of-change.

Thus, there is some evidence for increased miscarriage risk associated with high residential EMF exposure; however methodological problems make interpretation of these studies difficult due to possible information bias. There is little evidence, of increased risks of adverse reproductive outcomes other than miscarriage.

6.7.2 Paternal Exposure

Paternal magnetic field exposure has also been studied in relation to adverse reproductive outcomes. The earliest study was by Buiatti et al. [131], in which persons with infertility were more likely to report radioelectric work than controls. However, Lundsberg et al. [132] observed no association between job titles with high EMF exposure and semen abnormalities. Similarly, there was no raised risk of abnormal birth outcome among offspring of power-industry workers [133].

An excess of birth defects was observed among children of fathers who were electronic equipment operators [134]. Similarly, an increased frequency of congenital malformations and fertility difficulties was observed among the offspring of high-voltage switchyard workers [135]. It has also been suggested that there is an association between EMF exposure and decreased male to female ratio in the offspring [136,137]. However, the results of the studies on paternal exposure and adverse reproductive outcomes are unconvincing.

6.8 Cardiovascular Disease

Concerns about cardiovascular changes resulting from exposure to EMFs originated from descriptions in the 1960s and early 1970s of the symptoms among Russian high-voltage switchyard operators and workers [138]. Although these reports remain unconfirmed [139], more recent investigations have focused on direct cardiac effects of EMF exposure, mostly related to heart rate variability and subsequent acute cardiovascular events.

On basis of the idea put forth by Sastre et al. [140], Savitz et al. [141] hypothesized an association between exposure to EMF and acute cardiovascular disease. This hypothesis was based on two independent lines of evidence: The first was experimental data in which intermittent 60 Hz magnetic fields were found to reduce the normal heart rate variability [140]. The second came from several prospective cohort studies, which have indicated that reductions in some components of the variation in heart rate increase the risk for (a) heart disease [142–144], (b) overall mortality rate in survivors of myocardial infarction [145–147], and (c) risk for sudden cardiovascular death [148]. Thus, Savitz et al. [141] postulated that occupational exposure to EMFs will increase the risk for cardiac arrhythmia-related conditions and acute myocardial infarction (AMI), but not for chronic cardiovascular disease. As postulated, they observed an increased risk of AMI and arrhythmia-related death with high exposure, but not for chronic cardiovascular disease. Other studies were not able to replicate their findings [149,150]. Similarly, using morbidity as the outcome measure Johansen et al. [151] and Ahlbom et al. [152] did not observe any association (see Table 6.5 for a summary of the major studies).

Hakansson et al. [153] observed an increased risk of AMI with high exposure, though nonsignificant. Only mortality studies of the association between occupational exposure to EMF and cardiovascular diseases have suggested an association. However, studies that use mortality records are questionable due to possible inaccuracies of the diagnosis of AMI on the death certificates. In general, it can be concluded that the evidence does not support a plausible etiologic relation between EMF exposure and cardiovascular disease.

6.9 Neurodegenerative Disease

6.9.1 Alzheimer's Disease

Alzheimer's disease (AD) is characterized clinically by progressive loss of memory and other cognitive abilities (e.g., language, attention). Its onset is thought to be heralded by a phase of mild cognitive impairment in which cognition is not normal but not severe enough to warrant a diagnosis of dementia. The exact duration of mild cognitive impairment is unclear, but is likely to last at least a few years. Many persons with Alzheimer's disease also develop motor, behavioral, and affective disturbances. In particular, parkinsonian signs, hallucinations, delusions, and depressive symptoms are present in half or more of persons with the disease.

When evaluated across all the studies, there appears to be an association between estimated EMF exposure and disease risk (see Table 6.6). However, this result is mainly confined to the first two studies [154,155] and it is not clearly confirmed by the later studies [156–160]. The exception might be a study by Hakansson et al. [161]. The two studies [154,155] that show excess risk may have been affected by selection bias. Because the study populations are undefined, there is no way to determine the extent to which the controls are representative with respect to exposure of the population from which the cases originated. On the other hand, many of the negative studies are based on death certificates, which is particularly problematic because this diagnosis is often not reported as an underlying cause and is underrepresented as a contributing cause as well. Better studies of Alzheimer's are warranted.

6.9.2 Amyotrophic Lateral Sclerosis

Amyotrophic lateral sclerosis (ALS) is characterized clinically by progressive motor dysfunction, including painless muscle wasting and spasticity. Persons with ALS may develop cognitive and autonomic dysfunction. In particular, a frontal lobe dementia and hypotension may develop in persons with the disease. As the disease progresses, pulmonary function and dysphagia result in the need for artificial respiratory support and the insertion of feeding devices to maintain life.

TABLE 6.5

Epidemiologic Studies of General Cardiovascular Events in Relation to EMF

Ref.	Study Population	Study Design	Exposure	Category
Circulatory disease mortality				
Baris et al. [179]	Workers employed in electric company between 1970 and 1988	Cohort: $N = 21{,}744$	Job-exposure matrix	Highest exposure category; SMR: 0.63 (0.53–0.74); RR: 0.91 (0.73–1.14)
CVD				
Kelsh and Sahl [83]	Utility workers, employed between 1960 and 1991, followed to 1992	Cohort: $N = 40{,}335$	Occupational categories	SMR: 0.62 (0.59–0.65); Linemen, RR: 1.42 (1.18–1.71)
Johansen and Olsen [164]	Male utility workers, employed between 1990 and 1993, followed 1974 and 1993	Cohort: $N = 21{,}236$	Classification of workplaces based on measurements	AMI—High exposure; SMR: 1.0 ($O = 160$)
Savitz et al. [141]	Male utility workers, employed between 1950 and 1986, followed to 1988	Cohort: $N = 138{,}903$	Duration of work in jobs with elevated EMF	AMI—Highest μT-y group; RR: 1.62 (1.45–1.82)
Sahl et al. [149]	Utility workers, employed between 1960 and 1991, followed to 1992 (same as Kelsh and Sahl, 1997)	Cohort: $N = 35{,}391$	Duration of work in jobs with elevated EMF	Chronic CHD—RR: 1.0 (0.86–1.77) AMI—Highest μT-y group; RR: 0.99 (0.65–1.51)
Hakansson et al. [153]	Swedish twins responding to job questionnaire in 1967 or 1973	Cohort: $N = 27{,}790$	Job-exposure matrix	Chronic CHD—RR: 1.19 (0.79–1.77) AMI—Highest exposure group; RR: 1.3 (0.9–1.9)
Sorahan et al. [150]	Utility workers, employed between 1973 and 1982, followed to 1997	Cohort: $N = 79{,}972$	Duration of work in jobs and locations with elevated EMF	AMI—Highest μT-year category; RR: 1.03 (0.88–1.21) Chronic CHD—RR: 0.92 (0.73–1.16)
Pacemaker implantation				
Johansen et al. [151]	Male utility workers, employed between 1990 and 1993, followed 1974 and 1993 (same as Johansen and Olsen, 1998)	Cohort: $N = 24{,}056$	Classification of workplaces based on measurements	Highest exposure category; SMR: 1.00 (0.6–1.5)
AMI morbidity				
Ahlbom et al. [152]	Male population of Stockholm 1992–1993	Population based case–control: 695 cases, 1133 controls	Job titles, job exposure matrix	Highest exposure category; RR: 0.57 (0.36–0.89)

AMI, acute myocardial infarction; CHD, coronary heart disease.

TABLE 6.6

Epidemiologic Studies of the Association between Alzheimer's Disease and Dementia and EMF

Ref.	Study Population	Study Design	Exposure	RR[a](95%CI) High Category
Alzheimer's Disease				
Sobel et al. [155]	Patients at University of Helsinki, Finland, 1982–1985; Koskela Hospital, Helsinki, 1977–1978; and University of Southern California, 1984–1993	Case–control: all sites 387 cases, 465 controls	Work in occupations with exposure	0.2–1 μT or > 1 μT Intermittent to high > 1 μT or > 10 μT Intermittent exposure: 3.0 (1.6–5.4)
Sobel et al. [154]	Patients from an Alzheimer's disease treatment and diagnostic center	Case–control: 326 cases, 152 controls	Occupational exposure: medium and high EMF	3.9 (1.4–10.6)
Savitz et al. [156]	U.S. death certificates of males with occupational coding	Case–control: 256 cases, 768 controls	Electric occupation	1.2 (1.0–1.4)
Feychting et al. [158]	Population-based Swedish twin register	Case–control: 55 cases, 2 control groups: (i) 228 controls; (ii) 238 controls	Primary occupation Last occupation	≥0.2 μT (i) 0.9 (0.3-2.8) (ii) 0.8 (0.3-2.3) ≥0.2 μT: (i) 2.4 (0.8–6.9) (ii) 2.7 (0.9–7.8)
Savitz et al. [157]	Cohort of electric utility workers	Cohort: $N = 139{,}905$	Occupational exposure	Cumulative career—2.1–4.7 μT: 2.0 (0.6–7.0)
Feychting et al. [159]	Economically active individuals in the Swedish 1980 census followed from 1981 through 1995	Cohort: $N = 4{,}812{,}646$	Job-exposure matrix	≥ 0.5 μT: 1.3 (1.0–1.7)
Håkansson et al. [161]	Cohort of Swedish engineering industry workers 1985–1996	Cohort: $N = 718{,}221$	Job-exposure matrix	≥0.53μT: 4.0 (1.4–11.7)
Qiu et al. [160]	Cohort of age 75 y and older in Stockholm, Sweden followed 1987–1989 until 1994–1996	Cohort: $N = 931$	Job-exposure matrix, measurement on historical equipment, and expert estimation high exposure	≥ 0.18μT: 1.1 (0.7–1.5)
Dementia				
Feychting et al. 1998 [158]	Population-based Swedish twin register	Case–control: 77 cases, 228 and 238 controls	Primary occupation Last occupation	≥0.2 μT: 1.5 (0.6–4.0) and 1.2 (0.5–3.2) ≥0.2 μT: 3.3 (1.3–8.6) and 3.8 (1.4–10.2)
Johansen and Olsen [164]	Men employed in utility companies between 1900 and 1993 in Denmark	Standardized mortality study: $N = 21{,}236$	Job-exposure matrix, employment records	≥0.1 μT: 0.6 (nonsignificant) senile dementia

[a]Relative risk for μT-y cumulative exposure.

The earlier studies on ALS examined the possible association of electric shocks and ALS. After the first study in 1964 suggested a plausible relationship [162], two subsequent studies from Japan, where the prevalence of electric work (as recorded in the medical history) and of electric shock was low, failed to provide any support for the hypothesis [163]. Although some of the later studies did report analyses that linked electric shocks to ALS [164–166], most of these studies do not allow examination of confounding of magnetic field effects from electric shock. It is conceivable that exposure to electric shocks increases ALS risk; and, clearly work in the utility industry carries a risk of experiencing electric shocks.

Numerous studies of the relationship between electric work and the experience of electric shocks have been published since the first reports (see Table 6.7 for a review of these studies). In sum, there is modest epidemiologic evidence to suggest that employment in electric occupations may increase the risk of ALS, however, separating the increased risk due to receiving an electric shock from the increased exposure to EMFs is difficult. Studies of ALS that can distinguish between electric shocks and magnetic fields would be of value.

6.9.3 Parkinson's Disease and Other Neurodegenerative Diseases

Parkinson's disease is characterized clinically by progressive motor dysfunction, including bradykinesia, gait disturbance, rigidity, and tremor. Many persons with Parkinson's disease develop cognitive, behavioral, and autonomic signs.

Of the three neurodegenerative diseases that have been considered, Parkinson's disease has received the least attention in epidemiology. Occupation has been considered as a possible cause of Parkinson's disease in several studies. The study by Wechsler et al. [167] included jobs likely to involve relatively high exposures to EMF and reported 3 of 19 affected men were welders against 0 out of 9 controls and that 2 other affected men had worked as electricians or electrical engineers. However, Savitz et al. [157] found very little evidence of an increased risk in electric workers. Overall the OR derived from the occupations of 168 men dying from Parkinson's disease and 1614 controls was 1.1 (95% CI 0.9–1.2).

In the Danish cohort study, the standardized mortality ratio (SMR) for parkinsonism was 0.8 based on 14 deaths and even lower for the more heavily exposed men (0.5). In the study of Savitz et al. [156], positive relationships were observed with both cumulative cancer exposure and exposure more than 20 years before death, neither of which were, however, statistically significant (RR 1.03 per μT-y (95% CI 0.90–1.18), and 1.07 per μT-y, (95% CI 0.91–1.26). Noonan et al. [168] reported a positive association with an OR of 1.50 (95% CI 1.02–2.19) for the highest exposure category for Parkinson's disease and magnetic field exposure in electric workers.

Feychting et al. [159] found no increased risk for vascular dementia, senile dementia, presenile dementia, Parkinson's disease, multiple sclerosis, or epilepsy for either men or women. Hakansson et al. [161] and colleagues also found no increased risk for Parkinson's disease or multiple sclerosis and they observed a decreased RR for epilepsy.

In sum, there is no evidence linking EMF and Parkinson's disease and only very weak evidence to suggest that they affect Alzheimer's disease. The evidence that people employed in electric occupations have an increased risk of developing ALS is substantially strong, but this could be because such employees have an increased risk of experiencing an electric shock rather than any long-term exposure to the fields *per se*.

TABLE 6.7

Epidemiologic Studies of the Association between Amyotrophic Lateral Sclerosis and EMF

Ref.	Study Population	Study Design	Exposure	RR[a] (95%CI) High Category
Gunnarsson et al. [166]	Male population of Sweden 1970–1983	Case–control: 1067 cases (32 exposed), 1005 controls	Job title in census 1960: electric worker	1.5 (0.9–2.6)
Gunnarsson et al. [193]	Male population of central and southern Sweden in 1990	Case–control: 58 cases, 189 controls	Questionnaire: electric work and exposure to MF	0.6 (0.2–2.0)
Davanipour et al. [194]	Patients at outpatient clinic	Case–control: 28 cases, 32 controls	75th percentile of case distribution average exposure	2.3 (0.8–6.6)
Savitz et al. [156]	U.S. death certificates of males with occupational coding	Case–control: 114 cases	Electric occupation	1.3 (1.1–1.6)
Savitz et al. [157]	Cohort of electric utility workers	Cohort: $N = 139{,}905$	Occupational exposure	Cumulative career— 1.1–15.4 μT: 1.2 (0.4–3.3)
Johansen and Olsen [164]	Men employed in utility companies between 1900 and 1993 in Denmark	Standardized mortality study: $N = 21{,}236$	Job-exposure matrix, employment records	≥ 0.3 μT: 2.5 (1.1–4.8)
Feychting et al. [159]	Economically active individuals in the Swedish 1980 census followed from 1981 through 1995	Cohort: $N = 4{,}812{,}646$	Job-exposure matrix	≥ 0.5μT: 0.8 (0.6–1.0)
Håkansson et al. 2003 [161]	Cohort of Swedish engineering industry workers 1985–1996	Cohort: $N = 718{,}221$	Job-exposure matrix	≥ 0.53μT: 2.1 (1.0–4.7)

[a]Relative risk for μT-y cumulative exposure.

6.10 Depression and Suicide

6.10.1 Depression

There have been few of studies on depression and EMF (see Table 6.8). The early studies on the effects of EMF on depression did not use validated scales for identification of depressive symptoms [169,170], hence limiting the inferences based on the results [171]. A number of recent studies, however, have used validated depression scales. Although one of these studies demonstrated an association between proximity to power lines and depression [172], others [173,174] have not provided support for such an association. Similarly, Savitz et al. [175] did not find an association in their study of U.S. veterans. In sum, the findings on depressive symptoms and EMF are largely inconsistent.

6.10.2 Suicide

The earliest studies relating EMF exposure and suicide showed that cases came from homes with significantly higher exposure levels than controls [176,177]. Studies since then have improved exposure assessment techniques; they have used different approaches, such as distance between home and power lines, measurements, and job titles [38,164,178–181]. The most recent occupational study, based on job titles recorded on death certificates, suggested a weak association with suicide [182] (see Table 6.9 for details on these studies). In sum, the recent studies on workers, particularly the study by van Wijngaarden et al. (2000) [181], suggests an excess risk of suicide in electric workers, but the data is sparse.

6.11 Discussion

6.11.1 Challenges

As mentioned above, consistent epidemiologic evidence carries a good deal of weight when considering the potential health effects of EMF. Consistency in results is key because epidemiologic studies of EMF are difficult to design, conduct, and interpret for a number of methodological reasons. In addition to the usual methodological difficulties of designing and carrying out almost any good epidemiologic investigation, studying effects of EMF exposure poses unique and substantial difficulties. The difficulties that arise are in connection with three attributes of epidemiologic studies: outcome (disease), assessment of exposure, and interpretation of findings.

6.11.2 Outcome

Leukemias and malignant brain tumors have received the most attention in both residential and occupational EMF studies. While they are the most common childhood malignancies, in absolute terms they still are quite rare, and even rarer in adults. Thus, a major challenge in EMF epidemiology is the small number of cases available in any given study and the necessity for retrospective study designs. Furthermore, the etiologies of these diseases are poorly understood, making a search for confounding as an explanation for any observed association difficult.

TABLE 6.8

Epidemiologic Studies on Depression and EMF

Ref.	Study Population	Study Design	Exposure	RR (95%CI) High Category
Dowson et al. [169]	Persons living near 132 kV line and persons who lived 3 miles away.	Cross-sectional: $N = 226$	Distance from overhead line	Association between depression and proximity to overhead power line
Perry et al. [170]	Persons discharged with depression from hospital and controls from electoral list	Case–control: $N = 359$	Measurements at front door	Average measurement: Cases = 0.23 μT Controls = 0.21 μT
Poole et al. [172]	Residents along a transmission line right-of-way. Center for Epidemiologic Studies—Depression (CES-D) scale used	Cross-sectional: $N = 382$	Distance from power line	Near versus Far: 2.8 (1.6–5.1)
Savitz et al. [175]	Male veterans who served in the U.S. army. The Diagnostic Interview Schedule and the Minnesota Personality Inventory were used	Cross-sectional: $N = 4044$	Present job and duration	Electric worker: 1.0 (0.5–1.7)
McMahon et al. [173]	Neighborhood near a transmission line; sample of homes near a power line and one block away from the power line. CES-D scale used.	Cross-sectional: $N = 152$	Measurements at front door	Average for homes on easement, 0.486 μT and one block away, 0.068 μT: 0.9 (0.5–1.9)
Verkasalo et al. [174]	Persons from Finnish Twin Cohort Study who answered the Beck Depression Inventory	Cross-sectional: $N = 12{,}063$	Calculated fields	≥0.1 μT: 0.5 (0.1–2.3)—mild depression 1.3 (0.2–9.4)—moderate depression 15.3 (3.5–66.5)—severe depression

TABLE 6.9

Epidemiologic Studies on Suicide and EMF

Ref.	Study Population	Study Design	Exposure	RR (95%CI) High Category
Reichmanis et al. [177] and Perry et al. [176]	Suicide cases and controls in England	Case–control: $N = 589$	Distance from power lines Measurements at the homes of subjects	Higher estimated and measured fields at case homes
McDowall [38]	Persons resident in the vicinity of transmission facilities	Standardized mortality study: $N = 8$	Distance from power lines	Home within 50 m from substation or 30 m from overhead line 0.75 (nonsignificant)
Baris and Armstrong [178]	Deaths in England and Wales during 1970–1972 and 1979–1983	Proportional mortality study: $N = 495$	Job titles on death certificates	No increase for electric workers
Baris et al. [179,180]	Male utility workers. Cases: deaths from suicide in mortality registry. Controls: 1% random sample from the cohort	Case–control: 49 cases, 215 controls	Job-exposure matrix	Very high exposure ($\geq$1.56 μT): 1.38 (0.38–5.0) Blue collar electric workers only
Johansen and Olsen [164]	Male employees in utility companies observed during 1974–1993. Cases: deaths from suicide in mortality registry	Standardized mortality study: $N = 21{,}236$	Employment records and job-exposure matrix	$\geq$0.1 μT: 1.4 (nonsignificant)
van Wijngaarden [181]	Male electric utility workers	Case–control: 36 cases, 5348 controls	Jobs and indices of cumulative exposure based on measurement survey	Electrician: 2.18 (1.25–3.80) Line worker: 1.59 (1.18–2.14)
van Wijngaarden [182]	United States death certificate files for the years 1991 and 1992	Case–control: 11,707 cases, 132,771 controls	Occupation code; usual occupation and industry on the death certificates	1.3 (1.2–1.4)

EMF does not appear to be genotoxic, but biological evidence suggests that it might influence cellular function and proliferation. Therefore, it could act as a promoter or growth enhancer in carcinogenesis. Epidemiologic studies that attempt to consider such secondary events are rare and methodological developments in this area are needed.

6.11.3 Exposure

That the assessment of exposure is a major weakness of epidemiologic studies of EMF is not surprising, because several factors make assessment of EMF exposure more difficult than assessment of many other environmental exposures. Magnetic fields are variable in time and space, and our understanding of the contributions of the multitude of different sources to total exposure is limited. EMF exposure is ubiquitous, but neither detectable nor memorable in most circumstances. The difficulties in exposure assessment are further exacerbated by the retrospective nature of most EMF epidemiologic research, as many of diseases have long latency periods. To quantify past exposure that was unnoticed and unmeasured, epidemiologists rely on surrogate measurements or indicators of exposure. The surrogates used to study EMF have included wire codes, occupational job titles, questions regarding appliance use, and present day measurements. Further, some studies must rely on information provided by proxy respondents, especially for cancer and Alzheimer's disease, if the study subject has died or has been incapacitated.

Although occupational exposures are generally much higher than exposures encountered elsewhere, they are usually fleeting. The highly exposed worker generally encounters high fields not for hours but for seconds or minutes at a time while working. When we consider EMF exposure integrated over time, the brevity of high exposures in most work places and the large amount of time the individual spends in nonoccupational environments combine to wash out the distinctions between the supposedly highly exposed occupational groups and the general population. Thus, often we might not have enough separation between high and low-exposure groups to detect an effect of EMF exposure if we rely on TWAs.

Because EMF exposures are complex, numerous parameters have been used to characterize them, including transients, harmonic content, resonance conditions, peak values, as well as average levels. It is not known which of these parameters or what combinations of parameters, if any, are biologically relevant. If there were a known biological mechanism of interaction for carcinogenesis, it might be possible to identify critical parameters of exposure, including the relevant period or timing of exposure. Furthermore, environmental EMF is not detectable by the exposed person, nor is it memorable. Because it is ubiquitous, exposure assessment has to separate the more exposed from the less exposed, a much more difficult task than simply delineating the exposed from the nonexposed. There is also a considerable degree of variability in exposure in both the short- and long-terms, both of which are influenced by the variability in exposure over space, for example occupational versus household exposure.

All of these difficulties with EMF exposure assessment are likely to have led to substantial exposure misclassification, which is likely, in turn, to interfere with detection of an association between exposure and disease, if indeed such an association exists. In particular, if the true association is small or moderate, it will be difficult to detect with this amount of measurement error.

6.11.4 Interpretation

Three circumstances make it difficult to judge whether EMF poses a cancer risk: (a) difficulties in the appraisal of small risks (b) inconsistencies among and within studies, and (c) lack of knowledge of other risk factors for some diseases that should be considered as potential confounding variables.

Small risks are notoriously hard to evaluate, both because it is difficult to achieve enough precision to distinguish a small risk from no risk and because small risks are more vulnerable to subtle confounding and biases that can go undetected. Study quality remains an important consideration in evaluating epidemiologic evidence of EMF. As noted earlier, the quality of recent studies has improved in a number of important ways. However, earlier studies, many of which were fraught with methodological difficulties, still represent a substantial proportion of available evidence. Any attempt to summarize the available evidence, whether narrative or meta-analytic in its approach, must explicitly consider study date and study quality to arrive at a summary risk estimate that is not unduly influenced by less well-done studies. Given the difficulties with EMF exposure assessment and the other methodological difficulties outlined above, it is not surprising that many results of EMF studies are inconsistent. Nevertheless, an assessment of the internal and external consistency of studies is paramount to our understanding of the underlying risks.

Another impediment to a clear interpretation of these epidemiologic studies is the absence of a dose–response relationship. Distortion of the underlying dose–response relationship by substantial exposure misclassification [183] or by a nonmonotonic relationship between exposure and disease, has been offered as potential explanation. However, both possibilities remain speculative. A well-characterized and reproducible relationship between the most relevant aspect of exposure and disease is needed for the association to be scientifically acceptable.

6.12 Conclusions

Overall, with over two decades of epidemiologic investigation on the relation of EMF to the risk of various diseases, we have learned a great deal; however there remain a number of uncertainties. Among all the outcomes evaluated in epidemiologic studies of EMF, childhood leukemia in relation to postnatal exposures above 0.3–0.4 μT is the one for which there is most evidence of an association. Further study is warranted only if investigations are of high methodological quality, of sufficient size and with sufficient numbers of highly exposed subjects, and studies must include appropriate exposure groups and sophisticated exposure assessment. Particularly for childhood leukemia, little can be gained from further repetition of investigation of risks at moderate and low exposure levels, unless such studies can be designed to test specific hypotheses, such as selection bias or aspects of exposure not previously captured. Investigations of major diseases, such as breast cancer and cardiovascular disease, although initially biologically driven, did not confirm biological hypotheses or early positive studies. There is good evidence that these diseases are not associated with EMF exposure. Further work on neurodegenerative diseases is warranted as current evidence indicates a potential association, but is based on studies with numerous methodological limitations.

References

1. IARC, Non-Ionizing Radiation, Part 1: Static and Extremely Low-Frequency (ELF) Electric and Magnetic Fields, *IARC Monograph*, Vol. 80, 2002, International Agency for Research on Cancer, Lyon, France.

2. National Institute of Environmental Health Sciences (NIEHS), NIEHS Report on Health Effects from Exposure to Power-Line Frequency Electric and Magnetic Fields, NIH Publication No. 99-4493, 1999, National Institute of Environmental Health Sciences, National Institutes of Health, Research Triangle Park, NC.
3. National Radiological Protection Board (NRPB), Health effects from radiofrequency electromagnetic fields, in *Report of an Independent Advisory Group on Non-Ionizing Radiation*, 2003, National Radiological Protection Board.
4. National Research Council (NRC), *Possible Health Effects of Exposure to Residential Electric and Magnetic Field*, 1997, National Research Council, National Academy Press, Washington D.C.
5. Bracken, T., Kheifets, L., and Sussman, S., Exposure assessment for power frequency electric and magnetic fields (EMF) and its application to epidemiologic studies, *J. Expo. Anal. Environ. Epidemiol.*, 1993, 3(1): 1–22.
6. IARC, *Globocan 2000: Cancer Incidence, Mortality and Prevalence Worldwide*, 2000, International Agency for Research on Cancer: Lyon, France.
7. Steliarova-Foucher, E., Stiller, C., Kaatsch, P. et al., Geographical patterns and time trends of cancer incidence and survival among children and adolescents in Europe since the 1970s (the ACCIS project): an epidemiological study, *Lancet*, 2004, 364(9451): 2097–2105.
8. Mezei, G. and Kheifets, L., Clues to the possible viral etiology of childhood leukemia, *Technology*, 2002, 9: 3–14.
9. Poole C., Greenland, S., Luetters, C. et al., Childhood leukemia and socioeconomic status: a systematic review, Int.J. Epidemiol. 2006. 35(2):370–384.
10. Kheifets, L. and Shimkhada, R., Childhood leukemia and EMF: review of the epidemiologic evidence, *Bioelectromagnetics*, 2005, 26(Suppl. 7): S51–S59.
11. Wertheimer, N. and Leeper, E., Electrical wiring configurations and childhood cancer, *Am. J. Epidemiol.*, 1979, 109(3): 273–284.
12. Wertheimer, N. and Leeper, E., Adult cancer related to electrical wires near the home, *Int. J. Epidemiol.*, 1982, 11(4): 345–355.
13. Savitz, D., Wachtel, H., Barnes, F. et al., Case–control study of childhood cancer and exposure to 60-Hz magnetic fields, *Am. J. Epidemiol.*, 1988, 128(1): 21–38.
14. Kaune, W.T. and Savitz, D.A., Simplification of the Wertheimer-Leeper wire code, *Bioelectromagnetics*, 1994, 15(4): 275–282.
15. London, S., Thomas, D., Bowman, J. et al., Exposure to residential electric and magnetic fields and risk of childhood leukemia, *Am. J. Epidemiol.*, 1991, 134(9): 923–937.
16. Feychting, M. and Ahlbom, A., Magnetic fields and cancer in children residing near Swedish high-voltage power lines, *Am. J. Epidemiol.*, 1993, 138(7): 467–481.
17. Tynes, T. and Haldorsen, T., Electromagnetic fields and cancer in children residing near Norwegian high-voltage power lines, *Am. J. Epidemiol.*, 1997, 145(3): 219–226.
18. Verkasalo, P., Pukkala, E., Hongisto, M. et al., Risk of cancer in Finnish children living close to power lines, *Br. Med. J.*, 1993, 307(6909): 895–899.
19. Bianchi, N., Crosignani, P., Rovelli, A. et al., Overhead electricity power lines and childhood leukemia: a registry-based, case–control study, *Tumori*, 2000, 86(3): 195–198.
20. Green, L., Miller, A., Villeneuve, P. et al., A case–control study of childhood leukemia in Southern Ontario, Canada, and exposure to magnetic fields in residences, *Int. J. Cancer*, 1999, 82(2): 161–170.
21. Tomenius, L., 50-Hz electromagnetic environment and the incidence of childhood tumors in Stockholm County, *Bioelectromagnetics*, 1986, 7(2): 191–207.
22. Ahlbom, A., Day, N., Feychting, M. et al., A pooled analysis of magnetic fields and childhood leukaemia, *Br. J. Cancer*, 2000, 83(5): 692–698.
23. Coghill, R., Steward, J., and Philips, A., Extra low frequency electric and magnetic fields in the bedplace of children diagnosed with leukaemia: a case–control study, *Eur. J. Cancer Prev.*, 1996, 5(3): 153–158.
24. Dockerty, J., Elwood, J., Skegg, D. et al., Electromagnetic field exposures and childhood leukemia in New Zealand, *Lancet*, 1999, 354(9194): 1967–1968.
25. Kabuto, M., Hiroshi, N., Yamamoto, S. et al., A Japanese case–control study of childhood leukaemia and residential power-frequency magnetic fields, Unpublished, 2002.

26. Linet, M., Hatch, E., Kleinermann, R. et al., Residential exposure to magnetic fields and acute lymphoblastic leukemia in children, *N. Engl. J. Med.*, 1997, 337(1): 1–7.
27. McBride, M., Gallagher, R., Thériault, H. et al., Power-frequency electric and magnetic fields and risk of childhood leukemia in Canada, *Am. J. Epidemiol*, 1999, 149(9): 831–842.
28. Michaelis, J., Schüz, J., Meinert, R. et al., Childhood leukemia and electromagnetic fields: results of a population-based case–control study in Germany, *Cancer Causes Control*, 1997, 8(2): 167–174.
29. Schuz, J., Grigat, J., Brinkmann, K. et al., Residential magnetic fields as a risk factor for childhood acute leukaemia: results from a German population-based case–control study, *Int. J. Cancer*, 2001, 91(5): 728–735.
30. UK Childhood Cancer Study Investigators, Exposure to power-frequency magnetic fields and the risk of childhood cancer, *Lancet*, 1999, 354(9194): 1925–1931.
31. Kaune, W.T. and Savitz, D.A., Simplification of the Wertheimer–Leeper wire code, *Bioelectromagnetics*, 1994, 15(4): 275–282.
32. Greenland, S., Sheppard, A., Kaune, W. et al., A pooled analysis of magnetic fields, wire codes, and childhood leukemia, *Epidemiology*, 2000, 11(6): 624–634.
33. Dockerty, J., Elwood, J., Skegg, D. et al., Electromagnetic field exposures and childhood cancers in New Zealand, *Cancer Causes Control*, 1998, 9(3): 299–309.
34. Hatch, E., Linet, M., Kleinerman, R. et al., Association between childhood acute lymphoblastic leukemia and use of electrical appliances during pregnancy and childhood, *Epidemiology*, 1998, 9(3): 234–245.
35. Savitz, D.A., John, E.M., and Kleckner, R.C., Magnetic field exposure from electric appliances and childhood cancer, *Am. J. Epidemiol.*, 1990, 131(5): 763–773.
36. Savitz, D.A. and Chen, J.H., Parental occupation and childhood cancer: review of epidemiologic studies, *Environ. Health Perspect.*, 1990, 88: 325–337.
37. Coleman, M.P., Bell, C.M., Taylor, H.L. et al., Leukaemia and residence near electricity transmission equipment: a case–control study, *Br. J. Cancer*, 1989, 60(5): 793–798.
38. McDowall, M.E., Mortality of persons resident in the vicinity of electricity transmission facilities, *Br. J. Cancer*, 1986, 53(2): 271–279.
39. Schreiber, G.H., Swaen, G.M., Meijers, J.M. et al., Cancer mortality and residence near electricity transmission equipment: a retrospective cohort study, *Int. J. Epidemiol.*, 1993, 22(1): 9–15.
40. Youngson, J.H., Clayden, A.D., Myers, A. et al., A case/control study of adult haematological malignancies in relation to overhead powerlines, *Br. J. Cancer*, 1991, 63(6): 977–985.
41. Feychting, M. and Ahlbom, A., Magnetic fields, leukemia, and central nervous system tumors in Swedish adults residing near high-voltage power lines, *Epidemiology*, 1994, 5(5): 501–509.
42. Severson, R.K., Stevens, R.G., Kaune, W.T. et al., Acute nonlymphocytic leukemia and residential exposure to power frequency magnetic fields, *Am. J. Epidemiol.*, 1988, 128(1): 10–20.
43. Feychting, M., Forssen, U., and Floderus, B., Occupational and residential magnetic field exposure and leukemia and central nervous system tumors, *Epidemiology*, 1997, 8(4): 384–389.
44. Verkasalo, P.K., Magnetic fields and leukemia—risk for adults living close to power lines, *Scand. J. Work Environ. Health*, 1996, 22(Suppl. 2): 1–56.
45. Verkasalo, P.K., Pukkala, E., Kaprio, J. et al., Magnetic fields of high voltage power lines and risk of cancer in Finnish adults: nationwide cohort study, *Br. Med. J.*, 1996, 313(7064): 1047–1051.
46. Li, C.Y., Theriault, G., and Lin, R.S., Residential exposure to 60-Hertz magnetic fields and adult cancers in Taiwan, *Epidemiology*, 1997, 8(1): 25–30.
47. Preston-Martin, S., Peters, J.M., Yu, M.C. et al., Myelogenous leukemia and electric blanket use, *Bioelectromagnetics*, 1988, 9(3): 207–213.
48. Lovely, R.H., Buschbom, R.L., Slavich, A.L. et al., Adult leukemia risk and personal appliance use: a preliminary study, *Am. J. Epidemiol.*, 1994, 140(6). 510–517.
49. Sussman, S.S. and Kheifets, L.I., Adult leukemia risk and personal appliance use: a preliminary study, *Am. J. Epidemiol.*, 1996, 143(7): 743–744 [Letters to Editor, Reply].
50. Kheifets, L.I., Afifi, A.A., Buffler, P.A. et al., Occupational electric and magnetic field exposure and leukemia. A meta-analysis, *J. Occup. Environ. Med.*, 1997, 39(11): 1074–1091.
51. Bethwaite, P., Cook, A., Kennedy, J. et al., Acute leukemia in electrical workers: a New Zealand case–control study, *Cancer Causes Control*, 2001, 12(8): 683–689.

52. Björk, J., Albin, M., Welinder, H. et al., Are occupational, hobby, or lifestyle exposures associated with Philadelphia chromosome positive chronic myeloid leukaemia? *Occup. Environ. Med.*, 2001, 58(11): 722–727.
53. Hakansson, N., Floderus, B., Gustavsson, P. et al., Cancer incidence and magnetic field exposure in industries using resistance welding in Sweden, *Occup. Environ. Med.*, 2002, 59(7): 481–486.
54. Oppenheimer, M. and Preston-Martin, S., Adult onset acute myelogenous leukemia and electromagnetic fields in Los Angeles County: bed-heating and occupational exposures, *Bioelectromagnetics*, 2002, 23(6): 411–415.
55. van Wijngaarden, E., Savitz, D., Kleckner, R. et al., Mortality patterns by occupation in a cohort of electric utility workers, *Am. J. Ind. Med.*, 2001, 40(6): 667–673.
56. Willett, E., McKinney, P., Fear, N. et al., Occupational exposure to electromagnetic fields and acute leukaemia: analysis of a case–control study, *Occup. Environ. Med.*, 2003, 60(8): 577–583.
57. Linet, M.S., Ries, L.A., Smith, M.A. et al., Cancer surveillance series: recent trends in childhood cancer incidence and mortality in the United States, *J. Natl. Cancer Inst.*, 1999, 91(12): 1051–1058.
58. Preston-Martin, S., Gurney, J.G., Pogoda, J.M. et al., Brain tumor risk in children in relation to use of electric blankets and water bed heaters. Results from the United States West Coast Childhood Brain Tumor Study, *Am. J. Epidemiol.*, 1996, 143(11): 1116–1122.
59. Gurney, J.G., Mueller, B.A., Davis, S. et al., Childhood brain tumor occurrence in relation to residential power line configurations, electric heating sources, and electric appliance use, *Am. J. Epidemiol.*, 1996, 143(2): 120–128.
60. Myers, A., Cartwright, R.A., Bonnell, J.A. et al., Overhead power lines and childhood cancer, IEE International Conference on Electric and Magnetic Fields in Medicine and Biology, December 4–5, 1985, London.
61. Olsen, J., Nielsen, A., and Schulgen, G., Residence near high voltage facilities and risk of cancer in children, *Br. Med. J.*, 1993, 307(6909): 891–895.
62. Kheifets, L.I., Electric and magnetic field exposure and brain cancer: a review, *Bioelectromagnetics*, 2001, (Suppl. 5): S120–S131.
63. McCredie, M., Maisonneuve, P., and Boyle, P., Perinatal and early postnatal risk factors for malignant brain tumours in New South Wales children, *Int. J. Cancer*, 1994, 56(1): 11–15.
64. Preston-Martin, S., Navidi, W., Thomas, D. et al., Los Angeles study of residential magnetic fields and childhood brain tumors, *Am. J. Epidemiol.*, 1996, 143(2): 105–119.
65. Bunin, G.R., Ward, E., Kramer, S. et al., Neuroblastoma and parental occupation, *Am. J. Epidemiol.*, 1990, 131(5): 776–780.
66. Johnson, C.C. and Spitz, M.R., Childhood nervous system tumours: an assessment of risk associated with paternal occupations involving use, repair or manufacture of electrical and electronic equipment, *Int. J. Epidemiol.*, 1989, 18(4): 756–762.
67. Nasca, P.C., Baptiste, M.S., MacCubbin, P.A. et al., An epidemiologic case–control study of central nervous system tumors in children and parental occupational exposures, *Am. J. Epidemiol.*, 1988, 128(6): 1256–1265.
68. Spitz, M.R. and Johnson, C.C., Neuroblastoma and paternal occupation. A case–control analysis, *Am. J. Epidemiol.*, 1985, 121(6): 924–929.
69. Wilkins, J.R., III and Hundley, V.D., Paternal occupational exposure to electromagnetic fields and neuroblastoma in offspring, *Am. J. Epidemiol.*, 1990, 131(6): 995–1008.
70. Wilkins, J.R., III and Koutras, R.A., Paternal occupation and brain cancer in offspring: a mortality-based case–control study, *Am. J. Ind. Med.*, 1988, 14(3): 299–318.
71. Feingold, L., Savitz, D.A., and John, E.M., Use of a job-exposure matrix to evaluate parental occupation and childhood cancer, *Cancer Causes Control*, 1992, 3(2): 161–169.
72. Wertheimer, N. and Leeper, E., Magnetic field exposure related to cancer subtypes, *Ann. N. Y. Acad. Sci.*, 1987, 502: 43–54.
73. Wrensch, M., Yost, M., Miike, R. et al., Adult glioma in relation to residential power frequency electromagnetic field exposures in the San Francisco Bay area, *Epidemiology*, 1999, 10(5): 523–527.
74. Kheifets, L.I., Afifi, A.A., Buffler, P.A. et al., Occupational electric and magnetic field exposure and brain cancer: a meta-analysis, *J. Occup. Environ. Med.*, 1995, 37(12): 1327–1341.
75. Kheifets, L.I., Electric and magnetic field exposure and brain cancer: a review, *Bioelectromagnetics*, 2001, (Suppl. 5): S120–S131.

76. Armstrong, B., Theriault, G., Guenel, P. et al., Association between exposure to pulsed electromagnetic fields and cancer in electric utility workers in Quebec, Canada, and France, *Am. J. Epidemiol.*, 1994, 140(9): 805–820.
77. Beall, C., Delzell, E., Cole, P. et al., Brain tumors among electronics industry workers. *Epidemiology*, 1996, 7(2): 125–130.
78. Fear, N.T., Roman, E., Carpenter, L.M. et al., Cancer in electrical workers: an analysis of cancer registrations in England, 1981–87, *Br. J. Cancer*, 1996, 73(7): 935–939.
79. Heineman, E.F., Gao, Y.T., Dosemeci, M. et al., Occupational risk factors for brain tumors among women in Shanghai, China, *J. Occup. Environ. Med.*, 1995, 37(3): 288–293.
80. Harrington, J.M., McBride, D.I., Sorahan, T. et al., Occupational exposure to magnetic fields in relation to mortality from brain cancer among electricity generation and transmission workers, *Occup. Environ. Med.*, 1997, 54(1): 7–13.
81. Guenel, P., Nicolau, J., Imbernon, E. et al., Exposure to 50-Hz electric field and incidence of leukemia, brain tumors, and other cancers among French electric utility workers, *Am. J. Epidemiol.*, 1996, 144(12): 1107–1121.
82. Kaplan, S., Etlin, S., Novikov, I. et al., Occupational risks for the development of brain tumors, *Am. J. Ind. Med.*, 1997, 31(1): 15–20.
83. Kelsh, M.A. and Sahl, J.D., Mortality among a cohort of electric utility workers, 1960–1991, *Am. J. Ind. Med.*, 1997, 31(5): 534–544.
84. Miller, R.D., Neuberger, J.S., and Gerald, K.B., Brain cancer and leukemia and exposure to power-frequency (50- to 60-Hz) electric and magnetic fields, *Epidemiol. Rev.*, 1997, 19(2): 273–293.
85. Johansen, C. and Olsen, J.H., Risk of cancer among Danish utility workers—a nationwide cohort study, *Am. J. Epidemiol.*, 1998, 147(6): 548–555.
86. Cocco, P., Dosemeci, M., and Heineman, E.F., Occupational risk factors for cancer of the central nervous system: a case–control study on death certificates from 24 U.S. states, *Am. J. Ind. Med.*, 1998, 33(3): 247–255.
87. Rodvall, Y., Ahlbom, A., Stenlund, C. et al., Occupational exposure to magnetic fields and brain tumours in central Sweden, *Eur. J. Epidemiol.*, 1998, 14(6): 563–569.
88. Sahl, J.D., Kelsh, M.A., and Greenland, S., Cohort and nested case–control studies of hematopoietic cancers and brain cancer among electric utility workers, *Epidemiology*, 1993, 4(2): 104–114.
89. Theriault, G., Goldberg, M., Miller, A.B. et al., Cancer risks associated with occupational exposure to magnetic fields among electric utility workers in Ontario and Quebec, Canada, and France: 1970–1989, *Am. J. Epidemiol.*, 1994, 139(6): 550–572.
90. Savitz, D.A. and Loomis, D.P., Magnetic field exposure in relation to leukemia and brain cancer mortality among electric utility workers, *Am. J. Epidemiol.*, 1995, 141(2): 123–134.
91. Kheifets, L.I., Gilbert, E.S., Sussman, S.S. et al., Comparative analyses of the studies of magnetic fields and cancer in electric utility workers: studies from France, Canada, and the United States, *Occup. Environ. Med.*, 1999, 56(8): 567–574.
92. Stevens, R.G., Electric power use and breast cancer: a hypothesis, *Am. J. Epidemiol.*, 1987, 125(4): 556–561.
93. Davis, S., Mirick, D.K., and Stevens, R.G., Residential magnetic fields and the risk of breast cancer, *Am. J. Epidemiol.*, 2002, 155(5): 446–454.
94. London, S.J., Pogoda, J.M., Hwang, K.L. et al., Residential magnetic field exposure and breast cancer risk: a nested case–control study from a multiethnic cohort in Los Angeles County, California, *Am. J. Epidemiol.*, 2003, 158(10): 969–980.
95. Schoenfeld, E.R., O'Leary, E.S., Henderson, K. et al., Electromagnetic fields and breast cancer on Long Island: a case–control study, *Am. J. Epidemiol.*, 2003, 158(1): 47–58.
96. Vena, J.E., Freudenheim, J.L., Marshall, J.R. et al., Risk of premenopausal breast cancer and use of electric blankets, *Am. J. Epidemiol.*, 1994, 140(11): 974–979.
97. Vena, J.E., Graham, S., Hellmann, R. et al., Use of electric blankets and risk of postmenopausal breast cancer, *Am. J. Epidemiol.*, 1991, 134(2): 180–185.
98. Guenel, P., Raskmark, P., Andersen, J.B. et al., Incidence of cancer in persons with occupational exposure to electromagnetic fields in Denmark, *Br. J. Ind. Med.*, 1993, 50(8): 758–764.
99. Kelsh, M.A. and Sahl, J.D., Sex differences in work-related injury rates among electric utility workers, *Am. J. Epidemiol.*, 1996, 143(10): 1050–1058.

100. Vagero, D., Ahlbom, A., Olin, R. et al., Cancer morbidity among workers in the telecommunications industry, *Br. J. Ind. Med.*, 1985, 42(3): 191–195.
101. Vagero, D. and Olin, R., Incidence of cancer in the electronics industry: using the new Swedish Cancer Environment Registry as a screening instrument, *Br. J. Ind. Med.*, 1983, 40(2): 188–192.
102. Loomis, D.P., Savitz, D.A., and Ananth, C.V., Breast cancer mortality among female electrical workers in the United States, *J. Natl. Cancer Inst.*, 1994, 86(12): 921–925.
103. Cantor, K.P., Stewart, P.A., Brinton, L.A. et al., Occupational exposures and female breast cancer mortality in the United States, *J. Occup. Environ. Med.*, 1995, 37(3): 336–348.
104. Kheifets, L.I. and Matkin, C.C., Industrialization, electromagnetic fields, and breast cancer risk, *Environ. Health Perspect.*, 1999, 107(Suppl. 1): 145–154.
105. Forssen, U.M., Feychting, M., Rutqvist, L.E. et al., Occupational and residential magnetic field exposure and breast cancer in females, *Epidemiology*, 2000, 11(1): 24–29.
106. Guberan, E., Usel, M., Raymond, L. et al., Disability, mortality, and incidence of cancer among Geneva painters and electricians: a historical prospective study, *Br. J. Ind. Med.*, 1989, 46(1): 16–23.
107. Milham, S., Jr., Mortality in workers exposed to electromagnetic fields, *Environ. Health Perspect.*, 1985, 62: 297–300.
108. Olin, R., Vagero, D., and Ahlbom, A., Mortality experience of electrical engineers, *Br. J. Ind. Med.*, 1985, 42(3): 211–212.
109. Pearce, N., Reif, J., and Fraser, J., Case–control studies of cancer in New Zealand electrical workers, *Int. J. Epidemiol.*, 1989, 18(1): 55–59.
110. Spinelli, J.J., Band, P.R., Svirchev, L.M. et al., Mortality and cancer incidence in aluminum reduction plant workers, *J. Occup. Med.*, 1991, 33(11): 1150–1155.
111. Tornqvist, S., Norell, S., Ahlbom, A. et al., Cancer in the electric power industry, *Br. J. Ind. Med.*, 1986, 43(3): 212–213.
112. Demers, P.A., Thomas, D.B., Rosenblatt, K.A. et al., Occupational exposure to electromagnetic fields and breast cancer in men, *Am. J. Epidemiol.*, 1991, 134(4): 340–347.
113. Floderus, B., Tornqvist, S., and Stenlund, C., Incidence of selected cancers in Swedish railway workers, 1961–79, *Cancer Causes Control*, 1994, 5(2): 189–194.
114. Matanoski, G.M., Elliott, E.A., Breysse, P.N. et al., Leukemia in telephone linemen, *Am. J. Epidemiol.*, 1993, 137(6): 609–619.
115. Tynes, T. and Andersen, A., Electromagnetic fields and male breast cancer, *Lancet*, 1990, 336(8730): 1596.
116. Loomis, D.P., Cancer of breast among men in electrical occupations, *Lancet*, 1992, 339(8807): 1482–1483.
117. Cammarano, G., Crosignani, P., Berrino, F. et al., Cancer mortality among workers in a thermoelectric power plant, *Scand. J. Work Environ. Health*, 1984, 10(4): 259–261.
118. Brent, R.L., Gordon, W.E., Bennett, W.R. et al., Reproductive and teratologic effects of electromagnetic fields. *Reprod. Toxicol.*, 1993, 7(6): 535–580.
119. Delpizzo, V., Epidemiological studies of work with video display terminals and adverse pregnancy outcomes 1984–1992, *Am. J. Ind. Med.*, 1994, 26(4): 465–480.
120. Juutilainen, J., Matilainen, P., Saarikoski, S. et al., Early pregnancy loss and exposure to 50-Hz magnetic fields, *Bioelectromagnetics*, 1993, 14(3): 229–236.
121. Parazzini, F., Luchini, L., La Vecchia, C. et al., Video display terminal use during pregnancy and reproductive outcome—a meta-analysis, *J. Epidemiol. Community Health*, 1993, 47(4): 265–268.
122. Shaw, G.M. and Croen, L.A., Human adverse reproductive outcomes and electromagnetic field exposures: review of epidemiologic studies, *Environ. Health Perspect.*, 1993, 101(Suppl. 4): 107–119.
123. Shaw, G.M., Nelson, V., Todoroff, K. et al., Maternal periconceptional use of electric bed-heating devices and risk for neural tube defects and orofacial clefts, *Teratology*, 1999, 60(3): 124–129.
124. Bracken, M.B., Belanger, K., Hellenbrand, K. et al., Exposure to electromagnetic fields during pregnancy with emphasis on electrically heated beds: association with birthweight and intrauterine growth retardation, *Epidemiology*, 1995, 6(3): 263–270.
125. Florig, H.K. and Hoburg, J.F., Power-frequency magnetic fields from electric blankets, *Health Phys.*, 1990, 58(4): 493–502.
126. Kaune, W.T., Stevens, R.G., Callahan, N.J. et al., Residential magnetic and electric fields, *Bioelectromagnetics*, 1987, 8(4): 315–335.

127. Li, D.K., Checkoway, H., and Mueller, B.A., Electric blanket use during pregnancy in relation to the risk of congenital urinary tract anomalies among women with a history of subfertility, *Epidemiology*, 1995, 6(5): 485–489.
128. Eskelinen, T., Keinanen, J., Salonen, H. et al., Use of spot measurements for assessing residential ELF magnetic field exposure: a validity study, *Bioelectromagnetics*, 2002, 23(2): 173–176.
129. Li, D.K., Odouli, R., Wi, S. et al., A population-based prospective cohort study of personal exposure to magnetic fields during pregnancy and the risk of miscarriage, *Epidemiology*, 2002, 13(1): 9–20.
130. Lee, G.M., Neutra, R.R., Hristova, L. et al., A nested case–control study of residential and personal magnetic field measures and miscarriages, *Epidemiology*, 2002, 13(1): 21–31.
131. Buiatti E., Barchielli, A., Geddes, M. et al., Risk factors in male infertility: a case–control study, *Arch. Environ. Health*, 1984, 39(4): 266–270.
132. Lundsberg, L.S., Bracken, M.B., and Belanger, K., Occupationally related magnetic field exposure and male subfertility, *Fertil. Steril.*, 1995, 63(2): 384–391.
133. Tornqvist, S., Paternal work in the power industry: effects on children at delivery, *J. Occup. Environ. Med.*, 1998, 40(2): 111–117.
134. Schnitzer, P.G., Olshan, A.F., and Erickson, J.D., Paternal occupation and risk of birth defects in offspring, *Epidemiology*, 1995, 6(6): 577–583.
135. Nordstrom, S., Birke, E., and Gustavsson, L., Reproductive hazards among workers at high voltage substations, *Bioelectromagnetics*, 1983, 4(1): 91–101.
136. Irgens, A., Kruger, K., Skorve, A.H. et al., Male proportion in offspring of parents exposed to strong static and extremely low frequency electromagnetic fields in Norway, *Am. J. Ind. Med.*, 1997, 32(5): 557–561.
137. Mubarak, A.A. and Mubarak, A.A., Does high voltage electricity have an effect on the sex distribution of offspring? *Hum. Reprod.*, 1996, 11(1): 230–231.
138. Asanova, T.P. and Rakov, A.N., Health conditions of workers exposed to electric fields of open switchboard installations of 400–500 kV, *Gig. Tr. Prof. Zabol.*, 1966, 10(5): 50–52 [Preliminary report].
139. Baroncelli, P., Battisti, S., Checcucci, A. et al., A health examination of railway high-voltage substation workers exposed to ELF electromagnetic fields, *Am. J. Ind. Med.*, 1986, 10(1): 45–55.
140. Sastre, A., Cook, M.R., and Graham, C., Nocturnal exposure to intermittent 60 Hz magnetic fields alters human cardiac rhythm, *Bioelectromagnetics*, 1998, 19(2): 98–106.
141. Savitz, D.A., Liao, D., Sastre, A. et al., Magnetic field exposure and cardiovascular disease mortality among electric utility workers, *Am. J. Epidemiol.*, 1999, 149(2): 135–142.
142. Dekker, J.M., Schouten, E.G., Klootwijk, P. et al., Heart rate variability from short electrocardiographic recordings predicts mortality from all causes in middle-aged and elderly men. The Zutphen Study, *Am. J. Epidemiol.*, 1997, 145(10): 899–908.
143. Liao, D., Cai, J., Rosamond, W.D. et al., Cardiac autonomic function and incident coronary heart disease: a population-based case–cohort study. The ARIC Study. Atherosclerosis Risk in Communities Study, *Am. J. Epidemiol.*, 1997, 145(8): 696–706.
144. Martin, G.J., Magid, N.M., Myers, G. et al., Heart rate variability and sudden death secondary to coronary artery disease during ambulatory electrocardiographic monitoring, *Am. J. Cardiol.*, 1987, 60(1): 86–89.
145. Kleiger, R.E., Miller, J.P., Bigger, J.T., Jr. et al., Decreased heart rate variability and its association with increased mortality after acute myocardial infarction, *Am. J. Cardiol.*, 1987, 59(4): 256–262.
146. Lombardi, F., Sandrone, G., Pernpruner, S. et al., Heart rate variability as an index of sympathovagal interaction after acute myocardial infarction, *Am. J. Cardiol.*, 1987, 60(16): 1239–1245.
147. Vaishnav, S., Stevenson, R., Marchant, B. et al., Relation between heart rate variability early after acute myocardial infarction and long-term mortality, *Am. J. Cardiol.*, 1994, 73(9): 653–657.
148. Malik, M., Farrell, T., and Camm, A.J., Circadian rhythm of heart rate variability after acute myocardial infarction and its influence on the prognostic value of heart rate variability, *Am. J. Cardiol.*, 1990, 66(15): 1049–1054.
149. Sahl, J., Mezei, G., Kavet, R. et al., Occupational magnetic field exposure and cardiovascular mortality in a cohort of electric utility workers, *Am. J. Epidemiol.*, 2002, 156(10): 913–918.

150. Sorahan, T., and Nichols, L., Mortality from cardiovascular disease in relation to magnetic field exposure: findings from a study of UK electricity generation and transmission workers, 1973–1997, *Am. J. Ind. Med.*, 2004, 45(1): 93–102.
151. Johansen, C., Feychting, M., Moller, M. et al., Risk of severe cardiac arrhythmia in male utility workers: a nationwide danish cohort study, *Am. J. Epidemiol.*, 2002, 156(9): 857–861.
152. Ahlbom, A., Feychting, M., Gustavsson, A. et al., Occupational magnetic field exposure and myocardial infarction incidence, *Epidemiology*, 2004, 15(4): 403–408.
153. Hakansson, N., Gustavsson, P., Sastre, A. et al., Occupational exposure to extremely low frequency magnetic fields and mortality from cardiovascular disease, *Am. J. Epidemiol.*, 2003, 158(6): 534–542.
154. Sobel, E., Dunn, M., Davanipour, Z. et al., Elevated risk of Alzheimer's disease among workers with likely electromagnetic field exposure, *Neurology*, 1996, 47(6): 1477–1481.
155. Sobel, E., Davanipour, Z., Sulkava, R. et al., Occupations with exposure to electromagnetic fields: a possible risk factor for Alzheimer's disease, *Am. J. Epidemiol.*, 1995, 142(5): 515–524.
156. Savitz, D.A., Loomis, D.P., and Tse, C.K., Electrical occupations and neurodegenerative disease: analysis of U.S. mortality data, *Arch. Environ. Health*, 1998, 53(1): 71–74.
157. Savitz, D.A., Checkoway, H., and Loomis, D.P., Magnetic field exposure and neurodegenerative disease mortality among electric utility workers, *Epidemiology*, 1998, 9(4): 398–404.
158. Feychting, M., Pedersen, N.L., Svedberg, P. et al., Dementia and occupational exposure to magnetic fields, *Scand. J. Work Environ. Health*, 1998, 24(1): 46–53.
159. Feychting, M., Jonsson, F., Pedersen, N.L. et al., Occupational magnetic field exposure and neurodegenerative disease, *Epidemiology*, 2003, 14(4): 413–419; discussion 427–428.
160. Qiu, C., Fratiglioni, L., Karp, A. et al., Occupational exposure to electromagnetic fields and risk of Alzheimer's disease, *Epidemiology*, 2004, 15(6): 687–694.
161. Hakansson, N., Gustavsson, P., Johansen, C. et al., Neurodegenerative diseases in welders and other workers exposed to high levels of magnetic fields, *Epidemiology*, 2003, 14(4): 420–426; discussion 427–428.
162. Haynal A and Regli F, Zusammenhang der amyotrophischen lateralsclerose mit gehäufter elektrotraumata, *Confin. Neurol.*, 1964, 24: 189–198.
163. Kondo, K. and Tsubaki, T., Case–control studies of motor neuron disease: association with mechanical injuries, *Arch. Neurol.*, 1981, 38(4): 220–226.
164. Johansen, C. and Olsen, J.H., Mortality from amyotrophic lateral sclerosis, other chronic disorders, and electric shocks among utility workers, *Am. J. Epidemiol.*, 1998, 148(4): 362–368.
165. Deapen, D.M. and Henderson, B.E., A case–control study of amyotrophic lateral sclerosis, *Am. J. Epidemiol.*, 1986, 123(5): 790–799.
166. Gunnarsson, L.G., Lindberg, G., Soderfeldt, B. et al., Amyotrophic lateral sclerosis in Sweden in relation to occupation, *Acta Neurol. Scand.*, 1991, 83(6): 394–398.
167. Wechsler, L.S., Checkoway, H., Franklin, G.M. et al., A pilot study of occupational and environmental risk factors for Parkinson's disease, *Neurotoxicology*, 1991, 12(3): 387–392.
168. Noonan, C.W., Reif, J.S., Yost, M. et al., Occupational exposure to magnetic fields in case-referent studies of neurodegenerative diseases, *Scand. J. Work Environ. Health*, 2002, 28(1): 42–48.
169. Dowson, D.I., Lewith, G.T., Campbell, M. et al., Overhead high-voltage cables and recurrent headache and depressions, *Practitioner*, 1988, 232(1447): 435–436.
170. Perry, S., Pearl, L., and Binns, R., Power frequency magnetic field; depressive illness and myocardial infarction, *Public Health*, 1989, 103(3): 177–180.
171. International Commission on Non-Ionizing Radiation Protection (ICNIRP), *Exposure to Static and Low Frequency Electromagnetic Fields, Biological Effects and Health Consequences (0–100 kHz)*, Matthes, R. et al., Eds., 2003, ICNIRP Publications.
172. Poole, C., Kavet, R., Funch, D.P. et al., Depressive symptoms and headaches in relation to proximity of residence to an alternating-current transmission line right-of-way, *Am. J. Epidemiol.*, 1993, 137(3): 318–330.
173. McMahan, S., Ericson, J., and Meyer, J., Depressive symptomatology in women and residential proximity to high-voltage transmission lines, *Am. J. Epidemiol.*, 1994, 139(1): 58–63.
174. Verkasalo, P.K., Kaprio, J., Varjonen, J. et al., Magnetic fields of transmission lines and depression, *Am. J. Epidemiol.*, 1997, 146(12): 1037–1045.

175. Savitz, D.A., Boyle, C.A., and Holmgreen, P., Prevalence of depression among electrical workers, *Am. J. Ind. Med.*, 1994, 25(2): 165–176.
176. Perry, F.S., Reichmanis, M., Marino, A.A. et al., Environmental power-frequency magnetic fields and suicide, *Health Phys.*, 1981, 41(2): 267–277.
177. Reichmanis, M., Perry, F.S., Marino, A.A. et al., Relation between suicide and the electromagnetic field of overhead power lines, *Physiol. Chem. Phys.*, 1979, 11(5): 395–403.
178. Baris, D. and Armstrong, B., Suicide among electric utility workers in England and Wales, *Br. J. Ind. Med.*, 1990, 47(11): 788–789.
179. Baris, D., Armstrong, B.G., Deadman, J. et al., A mortality study of electrical utility workers in Quebec, *Occup. Environ. Med.*, 1996, 53(1): 25–31.
180. Baris, D., Armstrong, B.G., Deadman, J. et al., A case cohort study of suicide in relation to exposure to electric and magnetic fields among electrical utility workers, *Occup. Environ. Med.*, 1996, 53(1): 17–24.
181. van Wijngaarden, E., Savitz, D.A., Kleckner, R.C. et al., Exposure to electromagnetic fields and suicide among electric utility workers: a nested case control study, *Occup. Environ. Med.*, 2000, 57(4): 258–263.
182. van Wijngaarden, E., An exploratory investigation of suicide and occupational exposure, *J. Occup. Environ. Med.*, 2003, 45(1): 96–101.
183. Dosemeci, M., Wacholder, S., and Lubin, J.H., Does nondifferential misclassification of exposure always bias a true effect toward the null value? *Am. J. Epidemiol.*, 1990, 132(4): 746–748.
184. Fulton, J., Cobb, S., Preble, L. et al., Electrical wiring configurations and childhood leukemia in Rhode Island, *Am. J. Epidemiol.*, 1980, 111(3): 292–296.
185. Myers, A., Clayden, A., Cartwright, R. et al., Childhood cancer and overhead powerlines: a case–control study, *Br. J. Cancer*, 1990, 62(6): 1008–1014.
186. Schuz, J., Grigat, J.P., Stormer, B. et al., Extremely low frequency magnetic fields in residences in Germany. Distribution of measurements, comparison of two methods for assessing exposure, and predictors for the occurrence of magnetic fields above background level, *Radiat. Environ. Biophys.*, 2000, 39(4): 233–240.
187. Belanger, K., Leaderer, B., Hellenbrand, K. et al., Spontaneous abortion and exposure to electric blankets and heated water beds, *Epidemiology*, 1998, 9(1): 36–42.
188. Lee, G.M., Neutra, R.R., Hristova, L. et al., The use of electric bed heaters and the risk of clinically recognized spontaneous abortion, *Epidemiology*, 2000, 11(4): 406–415.
189. Dlugosz, L., Vena, J., Byers, T. et al., Congenital defects and electric bed heating in New York State: a register-based case–control study, *Am. J. Epidemiol.*, 1992, 135(9): 1000–1011.
190. Milunsky, A., Ulcickas, M., Rothman, K.J. et al., Maternal heat exposure and neural tube defects, *JAMA*, 1992, 268(7): 882–885.
191. Savitz, D.A. and Ananth, C.V., Residential magnetic fields, wire codes, and pregnancy outcome, *Bioelectromagnetics*, 1994, 15(3): 271–273.
192. Robert, E., Harris, J.A., Robert, O. et al., Case–control study on maternal residential proximity to high voltage power lines and congenital anomalies in France, *Paediatr. Perinat. Epidemiol.*, 1996, 10(1): 32–38.
193. Gunnarsson, L.G., Bodin, L., Soderfeldt, B. et al., A case–control study of motor neurone disease: its relation to heritability, and occupational exposures, particularly to solvents, *Br. J. Ind. Med.*, 1992, 49(11): 791–798.
194. Davanipour, Z., Sobel, E., Bowman, J.D. et al., Amyotrophic lateral sclerosis and occupational exposure to electromagnetic fields, *Bioelectromagnetics*, 1997, 18(1): 28–35.

7

Epidemiological Studies of Radio Frequency Fields

Maria Feychting

CONTENTS

7.1 Introduction

Historically, exposures to more than minimal radio frequency (RF) fields have been rare and mostly limited to occupational settings. In the general population the exposure sources were distant radio and television transmitters, with fields far below exposure guidelines. With the introduction and widespread use of mobile phone technology, the number of people in the general population exposed to low level RF fields has increased tremendously, not only through the use of the phone itself but also through the exposure from base stations, although fields from base stations are considerably lower than from the phones. The rapid increase in mobile phone use has led to a growing concern among the public about potential harmful effects of these exposures.

Current exposure guidelines are based on acute effects from heating of tissue. For exposure levels below these guidelines there is no known mechanism by which extremely low frequency (ELF) or RF fields might cause health effects such as cancer or different types of symptoms. This chapter will review the epidemiological evidence regarding potential health effects of exposure to RF fields below current guidelines that cannot cause substantial heating of tissue.

7.2 Epidemiology

Epidemiological studies are important in the process of health risk assessment; they study the effects of exposures under real circumstances and in the relevant host (humans). However, epidemiological studies also have some limitations; the most frequently discussed are the possibility of confounding, exposure misclassification, and selection bias. These problems can be handled more or less successfully, depending on how the study is designed and conducted. As with all types of research, we must evaluate the scientific evidence carefully to assess the quality of all the available studies to determine if any, and in what way, potential sources of error might have influenced the results.

The two most commonly used designs of epidemiological studies are cohort studies and case–control studies. In a cohort study, assessment of the studied exposure and potential confounding factors must be made for all persons included in the study population. The study population is followed over a period of time, and all cases of the studied disease are identified. It is essential that the same methods to identify the cases be used in both the exposed and unexposed part of the population. In a case–control study, all cases of the disease are identified, just like in a cohort study, and assessment of the exposure and potential confounding factors is made. If population-based disease registers are not available, cases are often identified directly from one or several hospitals. The study population is then defined as the individuals that would seek care at these hospitals had they contacted the studied disease; often the catchment areas of the hospitals that are included. For the part of the population that did not have the disease, information about the exposure is obtained for a representative sample of the population from which the cases came (controls). If the controls truly are a random sample of the study population, they will provide valid information about the exposure distribution in the population. Sometimes when population registers are not available for identification of controls, other methods are used such as selection of controls from patients with other diseases that are not related to the studied exposure (hospital controls) or among friends or persons living in the same neighborhood as the cases.

Selection bias can occur if the controls in a case–control study are not selected randomly from the study population and if the probability of getting selected as a control is related to the studied exposure, either directly or indirectly, e.g., through socioeconomic status. Selection bias might also occur if not all cases are participating in the study and if the probability of participation is related to the exposure.

Exposure misclassification occurs to a varying degree in all studies. A very crude exposure estimate is, for example, the use of job title information from censuses to estimate occupational exposure; obtaining detailed information about a person's job history, including specific job tasks and substances used, reduces but does not remove the possibility of exposure misclassification. If misclassification of the exposure is unrelated to the disease, i.e., degree of exposure misclassification is similar for cases and others (nondifferential exposure misclassification), the effect on the risk estimate will be a dilution toward unity. If the probability of exposure misclassification is related to the disease (differential exposure misclassification) the risk estimate can be affected in any direction. If exposure information can be collected independently of the disease, e.g., before disease onset or from historical records, there is no risk for differential exposure misclassification. In case–control studies, however, exposure information is often collected through a questionnaire or interview after occurrence of disease. If cases and controls differ in their ability to remember their exposures, differential exposure misclassification may occur, i.e., recall bias.

Confounding can occur when a risk factor other than the studied exposure is related to both the exposure and the disease. A classic example is a finding of an association between yellow fingers and lung cancer. Here it is obvious that smoking is a confounder and if controlled in the analysis, the association between yellow fingers and lung cancer would disappear. To completely explain an observed association, in the sense that the confounder rather than the exposure can be considered as the cause of the association, a confounder needs to be strongly related to the exposure and at the same time have an association with the disease that is considerably stronger than the observed association between the studied exposure and the disease. As long as information has been collected about exposure to potential confounding variables, adjustments can be made in the analyses to eliminate the effect of the confounders.

Generally, a cohort study is considered to be the most valid study design, because the potential for bias is greater in the case–control studies. This is, however, a somewhat too simplistic assumption. In cohort studies of rare diseases the study population needs to be very large (several hundred thousand individuals), and it is often not possible to do a very detailed exposure assessment for such a large number of individuals. Therefore, many cohort studies use quite crude exposure assessment methods, e.g., register-based data on occupational titles, which hampers the possibility to detect modestly increased risks because of dilution of the effect estimates caused by nondifferential exposure misclassification. A case–control study can use more sophisticated exposure assessment methods; and sometimes it may be possible to combine information obtained independently of cases and controls, e.g., from registries or historical records, with information collected through questionnaires to limit the potential for recall bias. If population registries are available, controls can always be selected randomly from the studied population to avoid potential selection bias.

7.3 Exposure Sources

People may be exposed to different sources of low-level RF exposure, but the most common source today is exposures related to mobile telephones. Currently this technology uses the frequencies from 450 to 2500 MHz, although this is a technology in constant development and therefore the frequency range may change in the future. Radio and television transmitters, which are the other major sources of exposure to the general population, operate at frequencies between 200 kHz and 900 MHz. The third main exposure source is occupational, e.g., RF PVC welding machines, plasma etching, and military and civil radar systems, all operating at different frequencies.

7.4 Cancer

7.4.1 Brain Tumors and Mobile Telephony

For RF exposure during mobile phone use, although below current guidelines, the exposure levels are of the same order of magnitude as the exposure standards. The exposure is concentrated in the part of the head closest to the handset and the antenna, and it declines rapidly with distance to the antenna. Therefore, the use of hands-free equipment while talking on the phone reduces the exposure of the head considerably while usually increasing the exposure of another part of the body that is near the phone's antenna.

Epidemiological studies of health effects related to RF exposure from mobile telephony is a relatively new research area, and only a limited number of studies are available. These have primarily focused on brain tumors, although a few of the studies included other types of tumors. Handheld mobile phones have only been available since the later part of the 1980s and have become common in the general population only during recent years. Today, a vast majority of the population in many countries use mobile phones; in Sweden, for example, over 80% of the population were mobile phone users in 2003, whereas it was less than 10% in the beginning of the 1990s. Considering the relatively short period of exposure it is currently possible to evaluate only the short-term effects of mobile phone exposure.

Seven studies of mobile phone use and brain tumors (excluding acoustic neuroma) have been published so far [1–7]. The majority of these have found no effects on brain tumor risk, but there are a few exceptions. A Finnish case–control study found an increased risk of glioma related to use of analogue phones [1]. Information about the exposure was collected from registries kept by the mobile phone operators, and therefore there was no risk of recall bias. The exposure misclassification is likely to be considerable, as the investigators had no access to information about corporate mobile phone users, who are likely to be among the heaviest mobile phone users. This type of error cannot, however, explain the increased glioma risk as there is no reason to believe that the exposure misclassification would differ between cases and controls. The risk increase was found after only 1 to 2 years of subscription to an analogue phone. This very short time between exposure and effect seems unlikely to be real; it usually takes a longer time for a tumor to develop and be diagnosed. Furthermore, if mobile phone use truly has an effect on glioma risk after such a short duration, the incidence of glioma should have increased during recent years in countries where there has been a rapid increase in mobile phone use in the general population. There is, however, no evidence of such an increase in a study of the glioma incidence in the Nordic countries, where mobile phone use in the general population started relatively early [8].

A Danish study used exposure assessment methods similar to the Finnish one, but was designed as a cohort study [5]. No indications were found of increased risks for any tumor types. The Danish study has the same limitation as the Finnish study in terms of exposure misclassification, and a modest risk increase would probably remain undetected. A more severe limitation in both the Danish and Finnish studies is the very short duration of exposure in the population when the studies were performed, e.g., only 8% of the cohort in the Danish study had used a mobile phone for at least six years.

Two case–control studies from the United States reported no association between brain tumor risk and mobile phone use [4,6]. Specific analyses of the side on which the mobile phone usually was held did not change these findings. These studies are so-called hospital based studies; cases are identified at certain hospitals and controls are selected from other patients at the same hospitals. This type of design assumes that patients with other diseases can be viewed as a representative sample of the population from which the brain tumor cases came and correctly reflects the exposure distribution, i.e., mobile phone use, in this population. The strength of these studies is the rapid case ascertainment, which is important in a study of malignant brain tumors where the poor prognosis of the disease may prevent many cases from participating in the study. A limitation is the short period of mobile phone use in both studies; very few subjects had used a mobile phone more than five years. Therefore, no conclusions can be drawn about potential effects of long-term exposure.

A Swedish case–control study reported no overall increase in brain tumor risk related to mobile phone use [3], but found an increased risk on the side of the head where the mobile phone was usually held [9]. On the opposite side of the head the risk was reduced. Recall bias is a likely explanation for this finding; it seems biologically implausible that mobile phone use would protect against brain tumors on the opposite side of the head. The same group reported similar results in a subsequent larger study [2,10]. In a reanalysis of the larger study using a different method, the risk reduction on the opposite side of the head was not as evident [11]. The studies by Hardell et al. have been criticized for limitations in methods, analysis, and presentation of the study [12].

7.4.2 Acoustic Neuroma and Cancers Outside the Brain

Acoustic neuroma is a slow-growing benign tumor situated in the vestibular portion of the eighth cranial nerve, a part of the head where the RF exposure from the mobile phone is high relative to other sites, though still below exposure guidelines. Currently, eight studies have investigated the association between mobile phone use and acoustic neuroma [2–5,13–16]. The Danish cohort study described above had too few cases with long-term exposure to provide any meaningful information [5], and the study by Warren et al. had a design that was not adapted for studying acoustic neuroma [16] and thus will not be further mentioned here.

The two case–control studies from the United States mentioned in Section 7.4.1 about brain tumors reported no increased risk of acoustic neuroma, regardless of which side of the head the mobile phone was used [4,15]. The problem with the short exposure duration in these studies is even more pronounced for acoustic neuroma because of the slow-growing nature of the tumor, resulting in a long latency period before the tumor is detected. Both studies reported a slightly increased risk of acoustic neuroma in the category of users with the longest duration of use, but the risks were not statistically significant.

A Swedish case–control study that is part of the Interphone study, a large international collaboration coordinated by the International Agency for Research on Cancer (IARC), found an increased risk of acoustic neuroma among subjects who started to use a mobile phone at least 10 years before diagnosis [14]. The risk increase was confined to tumors on the same side of the head as the phone was usually held, and risk estimates on the opposite side were close to unity. There was no indication of increased risks for mobile phone use with less than 10 years' duration, which is consistent with the findings in the U.S. studies. A Danish study that is part of the same international collaboration and used the same study protocol found no risk increase [13], but the number of subjects with long-term mobile phone use in the Danish study was very small. Thus, the power to detect an effect after long-term mobile phone use was severely limited.

In the latest study by Hardell et al. described above, a more than threefold risk increase of acoustic neuroma was found among users of analogue mobile phones, but with no relation to induction period [2], i.e., an increased risk was found also after a short duration of exposure. It seems highly unlikely that a higher incidence of acoustic neuroma would be detected within a few years of mobile phone use, considering that the tumor can be present more than five years before it is diagnosed [17].

Other types of tumors that have been studied are salivary gland tumors and uveal melanoma. Three studies are available on salivary gland tumors, and none of them reported any effects on disease risk [1,5,18]. They are however limited by poor statistical precision especially for analyses of long-term mobile phone use. Two studies have investigated the risk of uveal melanoma but with conflicting results [19,20].

7.4.3 RF Exposure from Transmitters

The levels of RF exposure from mobile phone base stations and radio and television transmitters are several orders of magnitude lower than current guidelines and are only a fraction of the exposure emitted during use of a mobile phone. An important difference compared to mobile phone use is, however, that transmitters expose the whole body and exposure duration is usually considerably longer.

To date, all studies on exposure from transmitters and cancer risk are based on populations living in the vicinity of radio and TV antennas; no studies around mobile phone base stations have been published as yet. Previous reviews have noted major limitations in the available studies, which weaken the conclusions that can be drawn [12,21,22]. One of the most pronounced problems is related to the exposure assessment. All studies have used distance from a broadcasting tower as a proxy for exposure, and no measurements have been taken. There is evidence that the correlation between distance from a transmitter and RF exposure is poor. The broadcast beam is aimed distantly, and the areas closest to the transmitter may have low-exposure levels, whereas more distant areas are those with the highest exposure. Furthermore, no account is taken of ground reflections and signal reductions from buildings, vegetation, and the surrounding landscape. Furthermore, all studies have had an ecological design; exposure has not been determined for each individual, but for the entire population living within certain distance bands from the transmitter. Therefore, no information has been available on potential confounding variables. Yet another problem is that several of the studies were conducted as a response to public concern about an excess of cancer cases in the vicinity, which has influenced the design of the studies. Several of the studies are also based on a small number of exposed cases.

Some studies have suggested an increased risk of leukemia in people living in the vicinity of radio and television transmitters [23,24], whereas another considerably larger study found no risk increase [25]. The limitations discussed above prevent any conclusions from these studies.

The research on exposures to RF fields from transmitters and cancer risk is at a very early stage of development. A prerequisite for an informative study of high quality is that the exposure assessment is improved considerably, and that study designs using individual exposure assessments is applied, including the ability to adjust for potential confounding factors. Development of a meter that can be used in large-scale epidemiological studies is underway and will be of great value.

7.4.4 Occupational Studies

Occupational exposure to RF fields has been studied over more than 20 y, and most investigations have focused on the risk of hematological malignancies, e.g., leukemia, and brain tumors. A variety of occupations have been investigated, including a range of different exposure frequencies, but exposure assessment methods have generally been poor. None of the studies have made measurements of the RF exposure for the subjects included in the study, and often the exposure classification has been based on the job title alone with no actual knowledge about the exposure levels for the workers in the particular occupations. No or only limited control of confounding has been made, and several of the studies have a poor statistical power.

In a study of Norwegian electrical workers [26], based on a categorization of occupational titles as probably exposed to RF fields, an increased risk of leukemia and a risk estimate below unity for brain tumors was reported. It is not clear which occupations these were and if other potentially harmful exposures could be present in the occupations that might possibly explain the findings.

A study of Polish military personnel reported increased risks for both leukemia and brain tumors, as well as for other cancer types [27,28]. This study has severe methodological limitations and an unconventional study design. The authors put much more effort into finding exposures among the cancer cases than among the personnel that had not been diagnosed with cancer, using multiple sources of information for the cases, but not for healthy persons. The bias introduced by this kind of procedure will inevitably lead to findings of increased cancer risks, even if no such associations exist; and no weight can be given these studies in an evaluation of the scientific evidence regarding this question.

A cohort study of radar technicians in the US navy reported no increased leukemia mortality and a risk reduction for brain tumor mortality when compared to the general population [29]. An reduced mortality overall and for many specific diagnostic categories is noticeable in the comparisons with the general population. Internal comparisons within the cohort showed an increased risk of leukemia in the highest exposed group, primarily for the group of aviation electronics technicians. No potential confounding factors were controlled in the analyses, and there were indications that smoking was less common in the cohort compared to the general population. Another US study found a decreased leukemia risk among aviation electronics technicians [30], but based on small numbers.

A cohort study of workers employed in the design, manufacture and testing of wireless devices such as mobile phones reported reduced risks of both leukemia and brain tumor mortality when compared to the general population [31]. A considerably reduced overall mortality was noted, as well as reduced overall cancer mortality. Exposure was assessed on the basis of expert opinion; no personal measurements were available. Internal comparisons within the cohort did not indicate any associations between the exposure and disease risk.

Although some increased cancer risks have been found in certain studies, there is no consistent evidence of risk increases for any cancer sites. The available studies have, however, not convincingly demonstrated that RF exposure has no effect on cancer risk, mainly because of limitations in the exposure assessment which could lead to a dilution of risk estimates, should there be an increased risk. There is also evidence of the so-called healthy worker effect in several of the studies, demonstrating considerably reduced risk estimates in comparisons with the general population. A working population is overall healthier than the general population, and certain occupations have specific demands on good health, e.g., policemen, making comparisons with the general population invalid. The use of mortality as the outcome of interest is also a limitation in several of the studies. A thorough review of all the occupational studies can be found in the Stewart report [21] and the more recent AGNIR report [12].

7.4.5 RF Exposure and Cancer—Conclusions

For brain tumors other than acoustic neuroma, the current evidence does not indicate any increased risks related to mobile phone use. It is important to note, however, that the available evidence only relates to mobile phone use over a short period of time and that conclusions about long-term exposure cannot be drawn based on the data available today. Several of the studies have limitations, especially regarding exposure assessment, which may hamper the possibility to detect a modestly increased risk, should one exist.

For acoustic neuroma there are some findings that indicate that there might be a risk increase associated with mobile phone use over a longer time period. It is, however, not possible to exclude other explanations to the findings, and additional well-designed studies are needed. Several studies included in the international Interphone study are currently under way and will provide additional information. A way to reduce the potential for recall bias would be to combine information from questionnaires or interviews with register-based data about mobile phone subscriptions from the mobile phone operators.

7.5 Subjective Symptoms

Very few epidemiological studies are currently available of subjective or self-reported symptoms, such as headache, dizziness, loss of appetite, fatigue, local or generalized feelings of heat, related to mobile telephony. In addition, all studies available on mobile phone use [32,33] or on exposures from mobile phone base stations [34,35] are cross-sectional, which makes them of limited value for health risk assessment. The subjects themselves reported the occurrence of various subjective symptoms and estimated their own exposure, e.g., distance to the nearest base station or amount of mobile phone use, and no attempt was made to verify the exposure or disease. The studies of base stations have not adequately described how subjects were selected to be participants in the study, and there is a strong possibility that both selection and reporting bias have affected the results. To date, the available epidemiological studies on symptoms do not provide information of sufficient quality for an evaluation of the effect of mobile phones or base stations on the occurrence of different types of symptoms. It is sometimes discussed that there might be a subgroup of the population that are particularly sensitive to electromagnetic field exposure, "electrically hypersensitive" subjects, and that studies of symptoms in the general population would be unable to detect effects in this small subgroup. In studies trying to identify persons with "electrical hypersensitivity," a severe problem is that there are no clearly defined diagnostic criteria and the exposure is part of the disease definition.

7.6 Pregnancy Outcomes

Various pregnancy outcomes have been studied in relation to RF exposure: spontaneous abortions, birth weight, gender ratio, and congenital malformations. Epidemiological studies have focused on occupational exposures, primarily among physiotherapists. A detailed summary of the available studies can be found in Ref. [12].

Findings for spontaneous abortions are mostly negative; an increased risk was reported in association with use of microwave diathermy in one study [36], but three other studies did not support this finding [37–39].

Several studies of female physiotherapists exposed to high-frequency electromagnetic fields have found no effect on the risk of congenital malformations in their offspring [39–41]. An increased risk was, however, reported in a Swedish study [42], although based on small numbers. Work with deep heat therapy was found to be associated with an increased risk of congenital malformations in a Finnish study [37], but with no dose response pattern.

Two studies are available on the gender distribution in offspring to female physiotherapists. A Danish study found a lower proportion of male offspring [38], but this finding was driven by the higher prevalence of female offspring in the unexposed comparison group. A study based on members of the Swiss Federation of Physiotherapists reported no effects on gender ratio in offspring to exposed mothers [43]. This study also analyzed the prevalence of low birth weight ($\leq$2500 g) in offspring born to physiotherapists and found no effects. In contrast, an increased risk of low birth weight was reported in a study of physiotherapists in Israel [40].

The evidence on potential effects of RF exposure on pregnancy outcomes is virtually limited to occupational exposures among physiotherapists; and although some positive findings have been reported, these findings have not been consistent with regard to specific type of malformation or other adverse outcome. Several of the studies have limited statistical power, especially for rare outcomes such as malformations, and there is a potential for recall bias. The available data does not allow conclusions in any direction.

7.7 Discussion and Conclusions

Compared to ELF epidemiology, studies of RF exposure, particularly from mobile phones, form a relatively new and undeveloped research field, especially regarding exposure assessment. Considerable improvements of the exposure assessment methods are to be expected in future studies, with consideration taken of factors related to the adaptive power control used in mobile telephony that makes the mobile phones down-regulate the output power to lower levels when the user is close to a base station, thereby decreasing the exposure up to 1000 times [44]. Relevant variables are, for example, mobile phone use in urban or rural areas, use of the phone inside or outside, or when being stationary or moving. There is also a need to be able to utilize sources of exposure information that is independent of the subjects included in the study, to avoid potential recall bias in studies performed retrospectively, such as case–control studies. This could for example be information from registries kept by the operators, combined with questionnaire data. An alternative would be to perform a prospective cohort study of mobile phone users, where information about mobile phone use is collected from both the operators and the users, but before the occurrence of the disease. A cohort study has the additional advantage that several different outcomes can be studied at the same time.

Occupational studies have been performed over the last 20 years, but these studies are limited to certain occupational groups, generally with no information about the actual exposure in the included occupations. Currently, efforts are being made to improve the knowledge about RF exposure levels in different occupations through actual measurements, but these efforts are not yet available for use in epidemiological studies. A meter suitable for use in large-scale epidemiological studies has been developed and is likely to be available in the near future [45]. This will greatly improve exposure assessment in occupational studies and thereby also the possibility to detect modestly increased risks, should they exist.

There is a public concern about potential health effects of exposure from mobile phone base stations, even if the exposure levels are orders of magnitude lower than the exposure from the phone itself. The feasibility of epidemiological studies of base station exposure is currently being investigated [46], and the newly developed RF field meter may make it possible to perform studies also on base station exposures.

Currently, there is no or only very limited evidence from epidemiological studies suggesting health effects of exposure to RF fields at levels below exposure standards. The findings for acoustic neuroma may suggest an increased risk, but are still highly uncertain [47], and additional studies are needed before any conclusions can be drawn.

References

1. Auvinen A, Hietanen M, Luukkonen R, and Koskela RS. Brain tumors and salivary gland cancers among cellular telephone users. *Epidemiology*, 2002; 13(3):356–359.

2. Hardell L, Hallquist A, Mild KH, Carlberg M, Pahlson A, and Lilja A. Cellular and cordless telephones and the risk for brain tumours. *Eur. J. Cancer Prev.*, 2002; 11(4):377–386.
3. Hardell L, Nasman A, Pahlson A, Hallquist A, and Hansson Mild K. Use of cellular telephones and the risk for brain tumours: A case–control study. *Int. J. Oncol.*, 1999; 15(1):113–116.
4. Inskip PD, Tarone RE, Hatch EE, Wilcosky TC, Shapiro WR, Selker RG, Fine HA, Black PM, Loeffler JS, and Linet MS. Cellular-telephone use and brain tumors. *N. Engl. J. Med.*, 2001; 344(2):79–86.
5. Johansen C, Boice J, Jr., McLaughlin J, and Olsen J. Cellular telephones and cancer—a nation-wide cohort study in Denmark. *J. Natl. Cancer Inst.*, 2001; 93(3):203–207.
6. Muscat JE, Malkin MG, Thompson S, Shore RE, Stellman SD, McRee D, Neugut AI, and Wynder EL. Handheld cellular telephone use and risk of brain cancer. JAMA, 2000; 284(23):3001–3007.
7. Rothman KJ, Loughlin JE, Funch DP, and Dreyer NA. Overall mortality of cellular telephone customers. *Epidemiology*, 1996; 7(3):303–305.
8. Lönn S, Klaeboe L, Hall P, Mathiesen T, Auvinen A, Christensen HC, Johansen C, Salminen T, Tynes T, and Feychting M. Incidence trends of adult primary intracerebral tumors in four Nordic countries. *Int. J. Cancer*, 2004; 108(3):450–455.
9. Hardell L, Mild KH, Pahlson A, and Hallquist A. Ionizing radiation, cellular telephones and the risk for brain tumours. *Eur. J. Cancer Prev.*, 2001; 10(6):523–529.
10. Hardell L, Mild KH, and Carlberg M. Case–control study on the use of cellular and cordless phones and the risk for malignant brain tumours. *Int. J. Radiat. Biol.*, 2002; 78(10):931–6.
11. Hardell L, Mild KH, and Carlberg M. Further aspects on cellular and cordless telephones and brain tumours. *Int. J. Oncol.*, 2003; 22(2):399–407.
12. NRPB. Health effects from radio frequency electromagnetic fields. Report of an independent advisory group on nonionizing radiation. NRPB. Vol. 14, No 2 Vol 14, No. 2: NRPB, 2003.
13. Christensen HC, Schuz J, Kosteljanetz M, Poulsen HS, Thomsen J, and Johansen C. Cellular telephone use and risk of acoustic neuroma. *Am. J. Epidemiol.*, 2004; 159(3):277–283.
14. Lönn S, Ahlbom A, Hall P, and Feychting M. Mobile phone use and the risk of acoustic neuroma. *Epidemiology*, 2004;15:653–659.
15. Muscat JE, Malkin MG, Shore RE, Thompson S, Neugut AI, Stellman SD, and Bruce J. Handheld cellular telephones and risk of acoustic neuroma. *Neurology*, 2002; 58(8):1304–1306.
16. Warren HG, Prevatt AA, Daly KA, and Antonelli PJ. Cellular telephone use and risk of intratemporal facial nerve tumor. *Laryngoscope*, 2003; 113(4):663–667.
17. Thomsen J and Tos M. Acoustic neuroma: clinical aspects, audiovestibular assessment, diagnostic delay, and growth rate. *Am. J. Otol.*, 1990;11(1):12–19.
18. Hardell L, Hallquist A, Hansson Mild K, Carlberg M, Gertzen H, Schildt EB, and Dahlqvist A. No association between the use of cellular or cordless telephones and salivary gland tumours. *Occup. Environ. Med.*, 2004; 61(8):675–679.
19. Johansen C, Boice JD, Jr., McLaughlin JK, Christensen HC, and Olsen JH. Mobile phones and malignant melanoma of the eye. *Br. J. Cancer*, 2002; 86(3):348–349.
20. Stang A, Anastassiou G, Ahrens W, Bromen K, Bornfeld N, and Jockel KH. The possible role of radio frequency radiation in the development of uveal melanoma. *Epidemiology*, 2001; 12(1):7–12.
21. IEGMP Independent Expert Group on Mobile Phones (Chairman: Sir William Stewart). Mobile phones and health. Chilton, Didcot, available at: http://www.iegmp.org.uk/, 2000.
22. Swedish Radiation Protection Authority. Recent Research on Mobile Telephony and Cancer and Other Selected Biological Effects: First annual report from SSI's Independent Expert Group on Electromagnetic Fields. Dnr 00/1854/02, Available at: http://www.ssi.se/english/EMF_exp_Eng_2003.pdf, 2003.
23. Hocking B, Gordon IR, Grain HL, and Hatfield GE. Cancer incidence and mortality and proximity to TV towers. *Med. J. Aust.*, 1996; 165(11–12):601–605.
24. Michelozzi P, Capon A, Kirchmayer U, Forastiere F, Biggeri A, Barca A, and Perucci CA. Adult and childhood leukemia near a high-power radio station in Rome, Italy. *Am. J. Epidemiol.*, 2002; 155(12):1096–1103.
25. Dolk H, Elliott P, Shaddick G, Walls P, and Thakrar B. Cancer incidence near radio and television transmitters in Great Britain. II. All high power transmitters. *Am. J. Epidemiol.*, 1997; 145(1):10–17.

26. Tynes T, Andersen A, and Langmark F. Incidence of cancer in Norwegian workers potentially exposed to electromagnetic fields. *Am. J. Epidemiol.*, 1992; 136(1):81–88.
27. Szmigielski S. Cancer morbidity in subjects occupationally exposed to high frequency (radiofrequency and microwave) electromagnetic radiation. *Sci. Total Environ.*, 1996; 180(1):9–17.
28. Szmigielski S, Sobiczewska E, and Kubacki R. Carcinogenic potency of microwave radiation: overview of the problem and results of epidemiological studies on Polish military personnel. *Eur. J. Oncol.*, 2001; 6:193–199.
29. Groves FD, Page WF, Gridley G, Lisimaque L, Stewart PA, Tarone RE, Gail MH, Boice JD, Jr., and Beebe GW. Cancer in Korean war navy technicians: mortality survey after 40 years. *Am. J. Epidemiol.*, 2002; 155(9):810–818.
30. Garland FC, Shaw E, Gorham ED, Garland CF, White MR, and Sinsheimer PJ. Incidence of leukemia in occupations with potential electromagnetic field exposure in United States Navy personnel. *Am. J. Epidemiol.*, 1990; 132(2):293–303.
31. Morgan RW, Kelsh MA, Zhao K, Exuzides KA, Heringer S, and Negrete W. Radio frequency exposure and mortality from cancer of the brain and lymphatic/hematopoietic systems. *Epidemiology*, 2000; 11(2):118–127.
32. Oftedal G, Wilen J, Sandstrom M, and Mild KH. Symptoms experienced in connection with mobile phone use. *Occup. Med. (Lond.)*, 2000; 50(4):237–245.
33. Chia SE, Chia HP, and Tan JS. Prevalence of headache among handheld cellular telephone users in Singapore: a community study. *Environ. Health Perspect.*, 2000; 108(11):1059–1062.
34. Navarro EA, Segura J, Portolés M, and Gómez-Perretta de Mateo C. The microwave syndrome: a preliminary study in Spain. Electromagnetic biology and medicine 2003; 22:161–169.
35. Santini R, Santini P, Danze JM, Le Ruz P, and Seigne M. Investigation on the health of people living near mobile telephone relay stations: I/Incidence according to distance and sex. *Pathol. Biol. (Paris)*, 2002; 50(6):369–373.
36. Ouellet-Hellstrom R and Stewart WF. Miscarriages among female physical therapists who report using radio- and microwave-frequency electromagnetic radiation. *Am. J. Epidemiol.*, 1993; 138(10):775–786.
37. Taskinen H, Kyyronen P, and Hemminki K. Effects of ultrasound, shortwaves, and physical exertion on pregnancy outcome in physiotherapists. *J. Epidemiol. Community Health*, 1990; 44(3):196–201.
38. Larsen AI, Olsen J, and Svane O. Gender-specific reproductive outcome and exposure to high-frequency electromagnetic radiation among physiotherapists. *Scand. J. Work Environ. Health*, 1991; 17(5):324–329.
39. Cromie JE, Robertson VJ, and Best MO. Occupational health in physiotherapy: general health and reproductive outcomes. *Aust. J. Physiother.*, 2002; 48(4):287–294.
40. Lerman Y, Jacubovich R, and Green MS. Pregnancy outcome following exposure to shortwaves among female physiotherapists in Israel. *Am. J. Ind. Med.*, 2001; 39(5):499–504.
41. Larsen AI. Congenital malformations and exposure to high-frequency electromagnetic radiation among Danish physiotherapists. *Scand. J. Work Environ. Health*, 1991; 17(5):318–323.
42. Kallen B, Malmquist G, and Moritz U. Delivery outcome among physiotherapists in Sweden: is nonionizing radiation a fetal hazard? *Arch. Environ. Health*, 1982; 37(2):81–85.
43. Guberan E, Campana A, Faval P, Guberan M, Sweetnam PM, Tuyn JW, and Usel M. Gender ratio of offspring and exposure to shortwave radiation among female physiotherapists. *Scand. J. Work Environ. Health*, 1994; 20(5):345–348.
44. Lönn S, Forssén UM, Vecchia P, Ahlbom A, and Feychting M. Output power levels from mobile phones in relation to the geographic position of the user. *Occup. Environ. Med.*, 2004; 61:769–772.
45. www.antennessa.com. 9-band isotropic selective personal dosimeter.
46. Neubauer G, Röösli M, Feychting M, Hamnerius Y, Khelfets L, Kuster N, Ruiz I, Schüz J, Überbacher R, and Wiart J. Study of the feasibility of epidemiological studies on health effects of mobile telephone base stations—final report. Seibersdorf Research, Austria, March 2005.
47. Savitz DA. Mixed Signals on Cell Phones and Cancer. *Epidemiology*, 2004; 15(6):651–652.

8

EMF Standards for Human Health

Emilie van Deventer, Dina Simunic, and Michael Repacholi

CONTENTS

8.1 Introduction

The strength of electromagnetic fields (EMF) to which humans are exposed has been increasing gradually with the growth of electric power generation and transmission, the development of new telecommunication systems, and advances in medical and industrial applications. Although the health effects of EMF on humans have been of research interest for several decades, development of standards that incorporate the limits of human exposure to nonionizing radiation (NIR) has occurred more recently for some parts of the electromagnetic frequency spectrum.

Historically, emerging technologies have triggered the development of EMF standards in different parts of the spectrum. In the radio frequency (RF) fields range, limits to protect human health were introduced in the 1950s, soon after the introduction or proliferation of radar and radio and television broadcasting. The need for exposure standards was highlighted by the observation of acute thermal injury in people exposed to high-level RF fields, especially in the microwave range. The United States and the Soviet Union were the first countries to develop RF standards, but their limits were based on different scientific evidence. With the introduction of industrial equipment, such as induction heaters and heat sealers, later standards recognized the vastly expanded use of the electromagnetic spectrum, especially at lower-frequency RF. Early RF standards were aimed at restricting thermal effects seen at high levels of exposure [1]. However, scientists in both the East and the West continue to postulate that nonthermal effects could also be important, and this has been the driving force for RF research for some decades.

In the extremely low frequency (ELF) region, an epidemiological study by Wertheimer and Leeper [2] was the key publication suggesting that there could be a health concern with exposure to ELF magnetic fields. Their studies suggested that exposure to ELF fields from high current configurations in power lines seemed to be associated with an increased incidence of childhood leukemia. This was supported by a few of the subsequent studies, including those by Savitz and coworkers [3] and more recently by two pooled analyzes by Ahlbom et al. [4] and Greenland et al. [5].

Research results drove legislation for both RF and ELF fields. More recently, the well-publicized putative link between ELF magnetic fields and childhood leukemia and the growing exposure to new technologies at home and in the workplace, e.g., computer monitors and televisions, microwave ovens and mobile phones, have triggered public concern and calls for control and regulation limiting public and occupational EMF exposure.

This chapter describes the available types of EMF standards, and includes an overview of the standard-setting bodies involved in NIR at the international and the national level. It also provides a discussion of the methodology for developing standards, which entails both a scientific approach and a political process. Finally, a snapshot on trends in worldwide NIR exposure standards is given.

8.2 Types of Standards

A standard is a general term incorporating both regulations and guidelines and can be defined as a set of specifications or rules to promote the safety of an individual or a group of people.

The ultimate goal of health-based EMF standards is to protect human health. However, there is often confusion about the various types of standards that exist to limit human exposure to EMF. Standards can specify either limits of emissions from a device or limits of human exposure from all devices that emit EMF into a living or working environment. These limits can be developed for the general public or for specific populations such as workers, medical patients, military personnel, children, or the elderly. Also, EMF standards can be classified as to whether they are voluntary measures or legally binding.

Many forms of EMF find application in medical practice, often at exposure levels that are much greater than those subjected to the general population. EMF exposures that are part of medical treatment or diagnosis are usually considered to lie outside the scope of exposure guidelines as they involve different risk-benefit considerations. This chapter discusses various types of EMF standards for emission and exposure limitation applied for nonmedical purposes.

8.2.1 Emission versus Exposure Standards

Emission standards set various specifications for electrical devices and are generally based on engineering considerations, e.g., to minimize electromagnetic interference with other equipment and to optimize the efficiency of the device. They also provide internationally recognized measurement protocols for devices, e.g., phantom measurement for specific absorption rate (SAR) values for mobile phones. Emission standards are usually developed by the International Electrotechnical Commission (IEC), the Institute of Electrical and Electronic Engineers (IEEE), the International Telecommunications Union (ITU), the Comité Européen de Normalization Electrotechnique/European Committee for Electrotechnical Standardization (CENELEC), as well as other independent organizations and national standardization authorities.

Although emission limits are aimed at ensuring, *inter alia*, compliance with exposure limits, they are not explicitly based on health considerations. In general, emission standards aim at ensuring that the aggregate exposure to the emission from a device will be sufficiently low that its use, even in proximity to other EMF emitting devices, will not cause exposure limits to be exceeded (Table 8.1).

Exposure standards that limit human EMF exposure are based on studies that provide information on biological effects from EMF, as well as the physical characteristics and the sources in use, the resulting levels of exposure, and the people at risk. Exposure standards generally refer to maximum levels to which whole or partial body exposure is permitted from any number of sources. This type of standard normally incorporates safety factors and provides the basic guidelines for limiting personal exposure. Such standards have been developed by the International Commission on Nonionizing Radiation Protection (ICNIRP), the IEEE, and many national authorities.

8.2.2 Voluntary versus Mandatory Standards

At present, there are no internationally mandated standards for EMF, such as exists for ionizing radiation (International Basic Safety Standards, IAEA [6]), in large part because

TABLE 8.1

Microwave Ovens and Mobile Phone Base Stations Operate in the Same Microwave Frequency Range, but the Rationale for Their Emission Standards Is Different

Microwave ovens are designed to heat food in an enclosed cavity, so spurious emissions should be limited as much as possible outside the cavity. The internationally agreed power density limit of $5\,mW/cm^2$ at 5 cm from the oven surface was set as an exposure limit to avoid ocular cataract formation from long-term exposure to a person watching the food being cooked. While this upper limit was subsequently found to be very conservative, the limits have remained the same.

Base station antennas for mobile telecommunications are designed to radiate RF energy and are located to provide coverage over a defined area or cell. The emission (or radiated power) from each base station depends on such factors as the size of the cell, the amount of information being transmitted, and the number of calls made. The power emitted from the base station must be optimized, i.e., enough to provide full coverage over the cell while not causing interference with neighboring cells. Thus, emission limits are not necessary because the signal strength is orders of magnitude below health limits and the signal strength is self-regulated by the network.

of the relatively large scientific uncertainty about adverse health effects and the lack of agreement on exposure limits by national authorities. Rather, there are international guidelines, such as those developed by ICNIRP, IEEE, and others, which provide guidance to national agencies but are not legally binding. Guidelines would only become legally binding if a country incorporates them into their own legislation. At the national level, exposure EMF regulations can be broadly categorized as either voluntary or mandatory instruments.

Voluntary instruments include guidelines, instructions, and recommendations that are not legally mandated. However, it is worth checking the national definitions of these terms, as for example, the term "Guideline Document" in Canada, refers to a mandatory, rather than voluntary, instrument.

Mandatory, compulsory, or legally binding instruments include laws, acts, regulations, ordinances, decisions, and decrees, and require a legislative framework. Because many countries do not possess legislation regarding protection standards for EMF, WHO's International EMF Project has developed model legislation on limiting exposure to NIR (http://www.who.int/emf/standards/emf_model/en/).

Procedures should exist to ensure compliance of mandatory standards. For EMF exposure standards, an agency is normally mandated to check compliance through calculations and measurements made in the workplace and other areas. For emission standards, compliance for devices is usually certified by the manufacturer.

8.3 Relevant Authorities

This section provides an overview of the relevant authorities involved in setting EMF standards. A number of international organizations have the mandate to establish EMF exposure and emission standards and have formulated limits for occupational and residential EMF exposure. National governments may regulate NIR protection through specialized legislative and regulatory agencies or ministries. At the regional and local levels, states and municipalities may have jurisdiction over certain aspects of public EMF protection.

8.3.1 Nongovernmental and International Organizations

The International Commission on Non Ionizing Radiation Protection (ICNIRP, http://www.icnirp.org)—ICNIRP is an independent nongovernmental organization (NGO), established in 1992 to evaluate the state of knowledge about the effects of NIR on human health and well being and to provide scientifically based advice on exposure limits and other methods of protection against harmful effects of NIR. It functions as an international scientific advisory body that does not include political, social, and economic considerations. Members of ICNIRP are experts who are not affiliated with a commercial or industrial enterprise. ICNIRP has developed guidelines on limiting exposure to nonionizing radiation from 0 to 300 GHz [7]. These guidelines have been adopted by the majority of countries that regulate NIR in Europe, Asia, and elsewhere.

The International Committee on Electromagnetic Safety (IEEE ICES, http://grouper.ieee.org/groups/scc28)—ICES is a committee established in 2001 by the Institute of Electrical and Electronic Engineers (IEEE). ICES and its predecessor committees have, since 1988, developed standards for safety levels with respect to human exposure to NIR in the frequency range of 0 Hz to 300 GHz. ICES sponsors and oversees the work of five subcommittees, with representatives of all five continents drawn from industry, government, the military, academia, and medicine. Participation in the meetings and deliberations of ICES is open to all. ELF limits were developed in IEEE Std C95.6 [8] and RF limits in IEEE Std C95.1 [9]. ICES guidelines have been adopted as mandatory standards by US governmental agencies as well as by Canada.

The International Electrotechnical Commission (IEC, http://www.iec.ch)—The IEC is an international organization, which consists of IEC National Committee members. IEC publishes standards for all electrical, electronic, and related technologies. IEC standards are developed to ensure interoperability between devices, product and service quality, and better production and delivery efficiency. IEC/TC106 is a technical body that prepares international standards in the field of characterization of the electromagnetic environments, measurement methods, instrumentation and procedures, calculation methods, assessment methods for the exposure produced by specific sources, and assessment of uncertainties with regard to human exposure.

Comité Européen de Normalisation Electrotechnique/European Committee for Electrotechnical Standardization (CENELEC, http://www.cenelec.org)—CENELEC is a European organization, which consists of the European national committees of IEC. It produces voluntary electrotechnical standards relating to measurement techniques for electrical and electronic goods and services, which are harmonized at the European level.

The International Telecommunications Union (ITU, http://www.itu.int)—The ITU is an international organization within which governments and the private sector work together to coordinate the operation of telecommunication networks and services. The activities in the three sectors of the Union—radiocommunication, telecommunication standardization, and telecommunication development—cover all aspects of telecommunication, from setting standards that facilitate interworking of equipment and systems on a global basis to adopting operational procedures for the vast and growing array of wireless services and designing programs to improve telecommunication infrastructure in the developing world. All ITU recommendations are nonbinding, voluntary agreements.

The World Health Organization (WHO, http://www.who.int/emf)—Through the International EMF Project, WHO (i) provides information on worldwide EMF standards in a web-accessible database; (ii) advocates harmonization of standards and facilitates the development of health-based exposure standards through a framework describing the scientific process required to develop them; and (iii) recognizing that some national authorities may want to provide greater protection than the current international guidelines provided in the presence of scientific uncertainty, has published a practical framework for developing such measures. WHO is in formal relation with ICNIRP and formally endorses and promotes ICNIRP's international exposure guidelines.

8.3.2 Multinational and National Bodies

In the past few decades, a number of multinational or regional agreements have been signed to deal with economic and trade issues, e.g., the North American Free Trade Agreement (NAFTA), MERCOSUL in South America, and the European Union (EU). So far, the latter is the only one to provide regulatory guidance on EMF exposure limits to its 25 member states (for more details, see Section 8.5). At the national level, regulations and legislation relating to EMF standards fall within the purview of specialized bodies. Certain governments have set up independent agencies to provide recommendations on EMF. Details on EMF standards are posted on the WHO web site (http://www.who.int/

docstore/peh-emf/EMFStandards/who-0102/Worldmap5.htm) where country-specific information can be found for over 50 countries.

One of the challenges faced at the national level stems from the fact that systems and devices emitting EMF are used for vastly different applications and may cover overlapping parts of the nonionizing radiation spectrum. This often poses confusion about final responsibility regarding protection from EMF exposure. For example, extremely low frequency (ELF) fields from generation, transmission, and distribution of electricity may come under the Ministry of Industry or of Energy, while radio frequency (RF) fields may be regulated by the ministries of Telecommunications (e.g., mobile phones), Transport (radar equipment for air navigation), or Environment. In situations where multiple authorities, sometimes with conflicting interests, are involved, communication and coordination, perhaps through the establishment of interagency committees, is key to evolve commonly accepted measures to protect the population.

The development of technology policy and its regulation must take into account many considerations in addition to health. In some countries, exposure standards are drafted by committees that represent several stakeholders, and the resulting national standards often reflect the compromises needed to satisfy the diverse points of view. Also, regulators need to facilitate the provision of varied and competitive utility services, from electric power to telecommunications, to most of the population. They have to create an environment that supports innovation and technical development and that best utilizes limited resources, whether it is land in the case of power lines or the frequency spectrum allocation for telecommunications devices.

8.3.3 Local Government

In many federated countries, the state or province has had to deal with the issue of EMF regulation. In California for example, an independent EMF research, evaluation and protection program was established in 1994 under the auspices of the state Department of Health (www.dhs.ca.gov). This resulted in state-level recommendations on how to manage EMF exposure to the public, especially in sensitive areas such as schools and kindergartens.

At the local level, municipalities often have authority over land use and building, including installation permits for power lines and mobile telephone base stations. As such, they are often confronted directly by public anxiety and discontent. It is therefore important that local authorities have at least a minimum knowledge of the EMF issue to answer questions from the public or be ready to direct requests to appropriate sources of information. An example of a successful information campaign was the initiative by the Elettra 2000 consortium in Italy to send a brochure on EMF risk management to every mayor in the country in 2003 (http://www.eletretta2000.it).

Some countries have opted for a structured approach to communication on EMF emitting sources. In France, for example, a charter between mayors and the association of mobile operators was developed in 2004 to provide information and dialogue regarding the placement of mobile phone base stations (http://www.afom.fr). This allows for a more transparent and fairer process over the whole territory.

Municipalities sometimes override national regulations and introduce further conservative measures based on political considerations rather than science. The case of Austria is well known, whereby federal legislature adheres to ICNIRP limits, but the city council of Salzburg, backed by its Public Health Department, agreed with mobile telecommunications operators to a precautionary limit value for radiation around 100 times lower than the ICNIRP's exposure limit [10]. This example has brought up criticisms regarding the technical feasibility of setting such limits in GSM and UMTS networks, along with the precedent that such local decisions establish.

8.4 Basis for Exposure Standards

Exposure to EMF may cause different biological effects, with a variety of physiological responses for a human being. Some effects may have beneficial consequences for a person, whereas others result in pathological conditions. Annoyance or discomforts caused by EMF exposure may not be pathological *per se* but, if substantiated, can affect the physical and mental well being of a person and the resultant effect may be considered as a health hazard. A health hazard is thus defined as a biological effect that has health consequences outside the compensatory mechanisms of the human body and is detrimental to health or well-being.

The development of guidelines or standards that limit EMF exposure should ideally be based on established scientific evidence; otherwise the limit values may be either non-protective or unduly restrictive. Once the evidence is properly evaluated, usually through a health risk assessment of all scientific studies, it can then be used to develop and implement policy and standards. However, policy makers must also take public opinion into account and ultimately policy development should include both a scientific approach and a political process to produce accepted and effective outcomes.

8.4.1 Scientific Methodology

Science-based evaluations of any hazards from EMF exposure are an essential part of an appropriate public policy response, and form the basis of international guidelines on exposure limits. Criteria and procedures for determining the exposure limit values are outlined in WHO's Framework for Developing Health-Based EMF Standards (http://www.who.int/emf). A summary of the procedure to derive numerical exposure limits is given in Table 8.2.

TABLE 8.2

Steps in the Development of EMF Exposure Standards

1. Conducting a detailed review of the scientific literature, using those studies having sufficient quality and information to be useful for health risk assessments and a determination of whether the biological effects found in these studies could lead to any adverse health consequence. This normally involves
   (a) using studies published in the peer-reviewed literature that have good dosimetry and experimental methodology, and
   (b) conducting a weight-of-evidence analysis to determine from the many studies showing effects and those not showing effects, whether an adverse health outcome is really likely to occur, at what level of exposure, and whether the effect is biologically plausible.
2. Identifying the lowest EMF exposure level, called threshold level, at which the first established health effect is seen and assessing all aspects of the uncertainty at this threshold level.
3. Applying a safety factor to the threshold level that accounts for all foreseen uncertainties in the threshold level to derive the exposure limits for the populations being protected. Normally an additional safety factor is applied to the exposure limits for more vulnerable populations, such as children, aged, and sick people that will be found in the general population.
4. Using the predominant mechanism(s) of interaction and results from studies conducted at other frequencies to extrapolate the exposure limits over the full frequency range of the population standard. It is important that the same or similar margin of safety in the exposure limits be applied to the whole frequency range for the standard, thus providing the same level of health protection for all frequencies considered.
5. Including an explanatory text with a full scientific rationale to accompany the standard so that subsequent studies, that may challenge the basis for the limits, can be properly assessed with respect to the original scientific justification for the limit values.
6. Performing a periodic review of the scientific literature to ensure that the basis for the standards, and especially the limit values, continue to be scientifically valid.

Some countries have the human and economic resources to undertake their own scientific evaluation of EMF health related effects through a formal health risk assessment process (e.g., the EMF RAPID program in the United States, NIEHS [11]), through an independent advisory committee (e.g., the independent Advisory Group on Non-Ionizing Radiation report in the United Kingdom, AGNIR 2001 [12]), or through a parliamentary report (e.g., the *Téléphonie mobile et santé* Senate report in France, 2002 [13]). Other countries may go through a less formal process to develop science-based guidelines or a variation thereof. Whereas some countries have adopted the ICNIRP guidelines, others use them as the *de facto* standard without giving them a legal basis. This may provide the local manufacturers and equipment users all the guidance they need to set their exposure limits within an installation.

Exposure guidelines in most countries have been based on acute, well-established effects of relatively high fields. Epidemiological evidence for long-term effects has so far been judged insufficient for the formulation of limits. Actual exposures are typically are considerably lower than guideline limits, and research at present mostly focuses on the possibility of long-term effects at low levels of exposure.

8.4.1.1 Types of Studies

The development of exposure standards requires an in-depth evaluation of the established scientific literature related to health effects. Research in this field rests on the conventional approaches used in toxicology, in particular the scientific studies undertaken in the laboratory on human volunteers, animals or cells and those undertaken on human populations by means of epidemiology. All these studies provide useful and valid information, but no single study or type of study can give a definitive answer. The relevance of different studies varies when considering health risks in people. Epidemiological studies of the distribution of disease in populations and the factors that influence this distribution provide direct information on the health of people exposed to an agent and are given the highest "weighting." However, they may be affected by unintentionally biased samples and confounding factors, and their observational nature makes it difficult to infer causal relationships, except when the evidence is strong. Experimental studies using volunteers can give valuable insight into the transient, physiological effects of acute exposure, although for ethical reasons these studies are normally restricted to healthy people. For obvious reasons, they cannot provide information on possible long-term effects.

Studies on animals, tissues, and cell cultures are also important but are given less weight. Animal studies can often be expected to provide qualitative information regarding potential health outcomes, but the data may not be extrapolated to provide quantitative estimates of risk, largely because of differences between species. However, IARC [14] considers that exposure to any biological, chemical, or physical agent is likely to cause cancer in humans if such a risk has been identified in at least two different animal species. Studies carried out at the cellular level are normally used to investigate mechanisms of interaction, but are not generally taken alone as evidence of effects *in vivo* until the effects are also demonstrated *in vivo*. Nevertheless, all types of studies have a role to play in determining the scientific plausibility of any notional health risk.

Both the published and unpublished reports are considered but preference is always given to published data in peer-reviewed scientific journals. Research that is reported to the press before going through the peer-review process poses a problem on two levels. First, when research results are publicized, they may be presented to maximize public interest. This may introduce distortions that raise unfounded public fears. Second, more dangerously, without first attempting to replicate the results, research results may also be used by policy makers to change regulations or policies. Such decisions could have important economic and free trade repercussions and most importantly, could undermine the scientific basis of human exposure guidelines.

8.4.1.2 Multiple Tiers

An important part of the rationale for any exposure standard is the definition of the population to be protected. Exposure standards can have multiple tiers to protect different populations. The ICNIRP guidelines [7] and, more recently, the IEEE standards [8,9] have adopted a two-tier set of limits; one for the general public and one for workers. Occupational standards are intended to protect healthy adults, exposed as a necessary part of their work, who are aware of the occupational risk and who are likely to be subjected to medical surveillance. By contrast, general population standards are based on broader considerations, including people with a wide range of health sensitivities, age, and illness. The general public will not necessarily have any knowledge of their EMF exposure or be able to minimize it. Thus, an additional safety factor is incorporated into the public exposure limits, which also takes into account the possibility of a 24 h exposure compared with occupational exposure of 8 h or whatever the duration of the workday.

8.4.1.3 Safety Factors

Safety factors in health protection standards represent an attempt to compensate for unknowns and uncertainties in the science. Once the threshold exposure level that produces an adverse health effect at the lowest exposure level has been identified, exposure limits may be derived by reducing this threshold level by a safety factor. Examples of sources of uncertainty about threshold levels include the extrapolation of animal data to effects in people, differences in the susceptibility of different groups or individuals, statistical uncertainties in the dose–response function, and the possibility of combined effects of exposures at different frequencies and other environmental factors.

Identification and quantification of various adverse effects of EMF exposure on health are difficult at best, and such judgments require extensive experience and expertise. Generally, acute effects can be quantified with reasonable precision and so derivation of limits to prevent these effects will not require a substantial safety factor below the observed threshold levels. When the uncertainty of the relationship between exposure and adverse outcome is greater, a larger safety factor may be warranted. There is no rigorous basis for determining precise safety factors. The magnitude of the reduction that results in the limit values is a matter of expert judgment, which explains the small differences in exposure limits derived independently by ICNIRP (http://www.icnirp.org) and IEEE (http://grouper.ieee.org/groups/scc28) (see below).

8.4.1.4 Exposure Limits

Limits on EMF exposure are termed basic restrictions and are based directly on established health effects and biological considerations. The physical quantities used in the guidelines reflect the different concepts of ''dose'' relevant to the lowest threshold for a health effect at different frequencies. In the low-frequency range (between 1 Hz and 10 MHz) the current ICNIRP and ICES basic restriction is the current density (J, in A/m^{-2}) for preventing effects in excitable tissues such as nerve and muscle cells; and in the high-frequency range (between 100 kHz and 10 GHz), the basic restriction is the specific absorption rate (SAR) in W/kg, for prevention of whole-body heat stress and local heating. In the intermediate frequency range (between 100 kHz and 10 MHz), restrictions are on both the current density and SAR, whereas in the very high-frequency range (between 10 and 300 GHz) the basic restriction is the power density (S, in W/m^2)

for excessive tissue heating near or at the body surface. Protection against known acute adverse health effects is assured if these basic restrictions are not exceeded. Numerical values of the basic restrictions over the frequency range up to 300 GHz can be found on the websites of ICNIRP and IEEE.

Because basic restrictions are often specified as quantities that may be impractical to measure, other quantities are introduced for practical exposure assessment purposes to determine whether the basic restrictions are likely to be exceeded or not. These reference levels (ICNIRP) or maximum permissible exposure levels (IEEE ICES) correspond to basic restrictions under worst case exposure conditions for one or more of the following physical quantities: electric field strength (E), magnetic field strength (H), magnetic flux density (B), power density (S), limb current (I_L), contact current (I_c), and, for pulsed fields, specific energy absorption (SA). Exceeding the reference levels does not necessarily imply that the basic restrictions are exceeded. However, in this case it is necessary to test compliance with the relevant basic restrictions and to determine whether additional protective measures are necessary. Reference levels (ICNIRP) and maximum exposure levels (IEEE ICES) are compared in Figure 8.1 and Figure 8.2 for general public and occupational exposures up to 300 GHz. As can be seen from Figure 8.1 and Figure 8.2, ICNIRP generally provides more conservative limits than IEEE, especially below 1 MHz. For the RF range, the peak SAR values are now essentially harmonized. Whereas little data is available between 100 and 300 GHz, the ramp for the general public IEEE limits reflects harmonization with the IEEE infrared standard at 300 GHz.

It is worth noting that several large countries have adopted standards based on different quantities. EMF standards in the Russian Federation are based on the energy loading, a concept that is different from SAR (W/kg). Energy loading is defined either as the square of the electric (or magnetic) field multiplied by the duration of exposure, with units expressed in $(V/m)^2 \cdot h$ or $(A/m)^2 \cdot h$, respectively, or as the product of power density

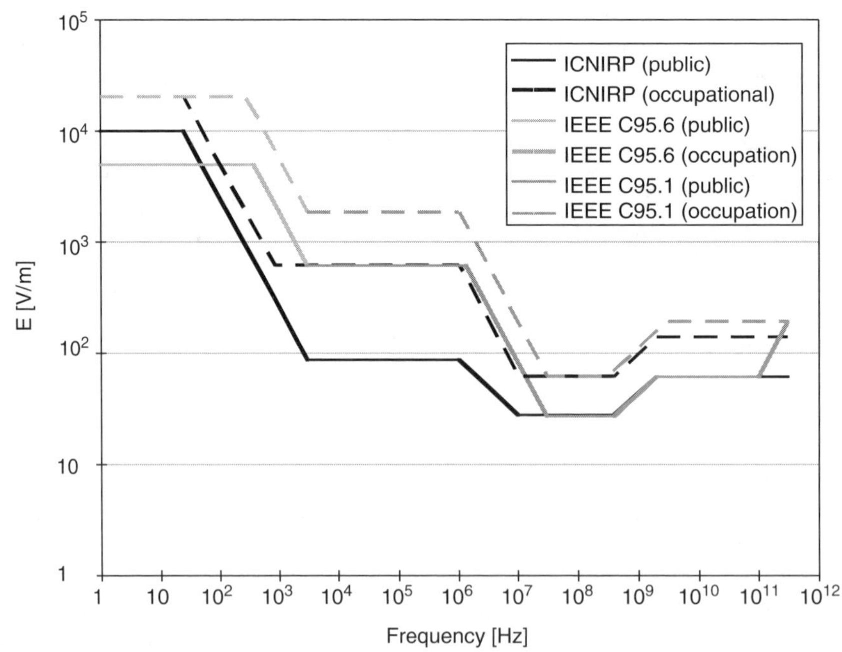

FIGURE 8.1
ICNIRP and IEEE ICES exposure limits for time varying electric fields.

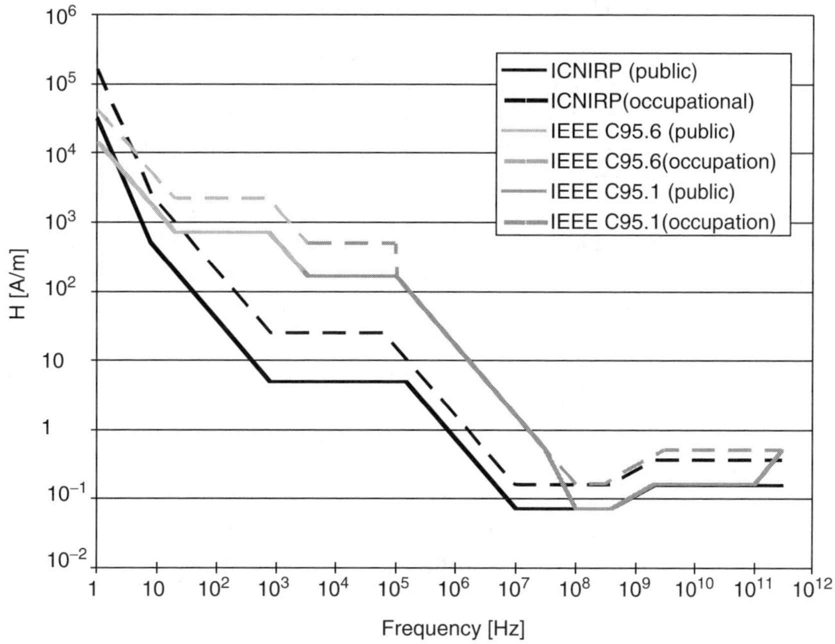

FIGURE 8.2
ICNIRP and IEEE ICES exposure limits for time varying magnetic fields.

and time $(W/cm^2){\cdot}h$. The reference level quantities are the electric and magnetic fields, magnetic flux density, and power density. The People's Republic of China has until now followed the former Soviet Union standard, but is in the process of revising its national RF standard, especially relating to SAR limit for mobile phones.

8.4.1.5 Other Quantitative Measures

EMF standards include mostly field strengths limits, but now many authorities also have guidance on other measures. Examples of such measures for ELF fields from power lines include minimum height of electrical conductors, necessary clearance between a transmission line and buildings (more specifically schools), and the electrical field strength on the ground surface under the line or at the edge of the right-of-way (corridor of clear land along a transmission line, originally used to allow line maintenance). For RF fields, guidance on the location of base stations has been provided by different authorities, including the widely quoted expert committee in the United Kingdom [12].

8.4.1.6 Scientific Uncertainty

There is an increasing awareness of the need to account for uncertainty in the science database. This is traditionally addressed by further research, and the WHO International EMF project addresses these needs through the development of a "research agenda" (http://www.who.int/peh-emf/research/en). However, research programs may take several years to complete, and the long latency associated with diseases such as cancer in people may also preclude a rapid outcome in some studies. Some countries have been addressing the issue of current uncertainty through the development of precautionary measures as discussed below.

8.4.2 Political Process and Precaution

As protecting the health of the population is part of the political process, it is expected that different countries may choose to provide different levels of protection against environmental hazards, responding to their national health policy. Various approaches to protection have been suggested to deal with scientific uncertainty. In recent years, increased reference has been made to cautionary policies and in particular the Precautionary Principle.

The Precautionary Principle is a risk management tool applied in situations of scientific uncertainty where there may be some need to act before there is strong proof of harm. It is intended to justify drafting provisional responses to potentially serious health threats until adequate data are available to develop more scientifically based responses. The Precautionary Principle is mentioned in international law [15,16] and is the basis for European environmental legislation [17]. It has also been referred to in some national legislation, for example in Canada [18], and recommendations of certain EMF exposure situations, for example in Switzerland [19]. The Precautionary Principle and its relationship to science and the development of standards have been discussed by Foster et al. [20].

With regard to the EMF issue, governmental and industry authorities have responded by implementing a wide variety of mandatory and voluntary precautionary approaches, based on cultural, social, and legal considerations. These include the importance given to avoiding a disease that affects mostly children, the acceptability of involuntary, as opposed to voluntary, exposures, and the varying importance given to uncertainties in the decision-making process. Several examples are presented below (Table 8.3).

Prudent avoidance: Prudent avoidance, a version of the Precautionary Principle (although not specifically identified as such), was developed for power-frequency EMF. It is defined as taking steps to lower human exposure to EMF by redirecting facilities and redesigning electrical systems and appliances at low to modest costs [21]. This policy has generally been applied only to new facilities, where minor modifications in design can reduce levels of public exposure at a low cost. Prudent avoidance has been adopted as part of policy in many countries, including Australia, New Zealand, and Sweden. Measures that can be taken at "modest cost" include routing power lines away from schools, phasing and configuring power line conductors to reduce magnetic fields near their rights of way, and optimizing the location of mobile phone base stations within a given area.

Passive regulatory action: This recommendation, introduced in the United States for the ELF issue [11], advocates educating the public and the regulated community on means aimed at reducing exposure, rather than setting up actual measures to reduce exposure.

TABLE 8.3

Examples of Precautionary Approaches

Precautionary Approach	Country	Measures
Prudent avoidance	New Zealand, Australia, and Sweden	Adopt ICNIRP guideline values and add low-cost voluntary measures to reduce exposure
Passive regulatory action	USA	Educate the public on measures to reduce exposure
Precautionary emission control	Switzerland	Adopt ICNIRP guideline values and set lower emission limits if feasible
Precautionary quantitative exposure limits	Italy and Greece	Decrease exposure limits using arbitrary reduction factors

Precautionary emission control: This policy is used in Switzerland to reduce EMF exposure by keeping emission levels "as low as technically and operationally feasible," implying that they have been successfully tested on comparable installations or in trials. Measures to control emissions should be "financially viable" (http://www.umwelt-schweiz.ch/buwal/eng/index.html). Therefore, it is the emission levels from a device or class of devices that is controlled whereas the international exposure limits (ICNIRP) are adopted as the maximum level of exposure to people from all sources of EMF. Such a measure has been taken up by Switzerland regarding base stations emission levels.

Precautionary exposure limits: Some countries have taken precautionary approaches leading to limits on exposure that do not consistently draw on the scientific evidence for adverse health effects*. For example, in 2003, Italy adopted ICNIRP standards but introduced two further limits for EMF exposure: (i) "attention values" of one-tenth of the ICNIRP field reference levels in specific locations, such as children's playgrounds, residential dwellings, and school premises and (ii) further restrictive frequency-dependent "quality goals" that only apply to new sources and new homes.

This diversity of approaches by national authorities led WHO's International EMF Project to develop a "Policy Framework" for rational and cost-effective guidance of policy options in areas of scientific uncertainty (http://www.who.int/emf). Briefly, this aims to develop a set of measures for protecting public health according to the degree of scientific uncertainty and the anticipated severity of the harm that might ensue, taking into account the size of the affected population and the cost. Effective risk communication and consultation between stakeholders are seen as integral parts of this process.

A principal recommendation by WHO is that these types of policies be adopted in such a way that scientific assessments of risk and science-based exposure limits are not ignored. WHO specifically recommends not to reduce the limit values in international standards to some arbitrary level in the name of "precaution" as this undermines the science base on which the limits were based and can introduce an additional cost of compliance for no known health benefit. Also, from a sociological standpoint, there is increasing evidence that making arbitrary reductions in exposure limits lead to increasing public concern, rather than reducing it [24]. Furthermore, arbitrary additional safety factors could lead to limit values so low as to exclude highly beneficial technologies.

8.5 Discussion

Over the past decade, several factors have caused countries to develop or modify their standards for protection against EMF exposure. One of these factors is the technological evolution of telecommunication devices and the overhaul of the electrical grid in many countries. Political changes have also had an impact, notably the break-up of the former

*The European Union (EU) has published Council Recommendation (1999/519/EC) [22] regarding EMF protection for the general public and EC Directive (2004/40/EC) [23] establishing minimum requirements for the protection of workers. Both are based on ICNIRP guidelines. Member States of the EU are required to adopt and incorporate the EC Directives into their legislation. However, in accordance with the Maastricht Treaty, they may provide a higher level of protection than that set out in EC Recommendation or Directive, a prerogative that has been taken up by several European countries on precautionary grounds.

Soviet Union and the enlargement of the EU to include a number of eastern European states, which previously had EMF standards based on Soviet legislation. Furthermore, from a social view point, there has been a rise in public awareness and concern regarding new technologies, from vaccines to genetically engineered plants and mobile phones, and a loss of trust in national responses to public health threats, such as the tainted blood or bovine spongiform encephalopathy (BSE) crises [25].

8.5.1 Technology Leading Health Research

The recent development of new EMF-emitting devices, from novel wireless communication systems to superconducting magnets used in magnetic resonance imaging (MRI), has resulted in a significant increase in human exposure to electromagnetic fields. Given the rapidly expanded use of such devices, any potential health impacts need to be properly assessed, ideally prior to widespread human exposure. Unfortunately, this technological development and its applications usually outpace epidemiological and biological research and the development of proper health risk assessments and consequent standards. For example, a recent review by WHO (in press) indicates that research on exposures to static magnetic fields has been conducted up to about 2 T, but medical and occupational exposures now exceed 3–8 T where no effective studies to determine their safety have been conducted.

8.5.2 Unregulated Portions of the Spectrum

Many countries have now established health protection standards or guidelines. However, in a number of them, only certain parts of the spectrum are regulated. To date, few regulatory exposure standards have been promulgated limiting human exposure to static fields. The greatest interest has been in regulations and guidelines for ELF fields at power frequencies and RF fields at mobile phone frequencies.

8.5.3 Disparities in Standards

Globalization of trade and the rapid expansion in the use of devices emitting EMF have focused attention on differences existing in exposure guidelines or standards limiting EMF exposure. In some cases, these differences are large. Some of the disparities in EMF standards around the world have arisen from the use of only national databases, different criteria for accepting or assessing individual studies, varying interpretations of the scientific data, or different philosophies for public health standards development. Such differences in EMF exposure guidelines might reflect, in part, deficiencies in communications among scientists in different regions as well as certain social differences, at least as far as precautionary measures are concerned. Large disparities between national limits and international guidelines can foster confusion for regulators and policy makers, provide a challenge to manufacturers and operators of communications systems, who need to tailor their products to each market, and increase public anxiety.

8.5.4 Risk Perception and Communication

The lack of worldwide policy harmonization is one of many factors that may exacerbate public anxiety. People's perceptions of a risk depend on personal factors, external factors as well as the nature of the risk [26]. Personal factors include age, sex, and cultural or educational backgrounds while external factors comprise the media and other forms of information dissemination, the current political and economic situation, and opinion movements, as well as the structure of the regulatory process and political decision making in the community. The nature of the risk can also lead to different perceptions, depending on how much control the public has over a situation (e.g., mobile phones versus base stations), fairness and equity aspects in locating EMF sources, and dread of specific diseases (e.g., cancer versus headache). The greater the number of factors adding to the public's perception of risk, the greater is the potential for public concern. Public concern can be reduced through information and communication between the public, scientists, governments, and industry. Effective risk communication is not only a presentation of the scientific calculation of risk, but also a forum for discussion on broader issues of ethical and moral concern [27].

8.5.5 Need for Periodic Evaluation

As new scientific information becomes available, standards should be updated. Certain studies, because of their strength of evidence or because of the severity of the health outcome under study, may be more likely to prompt a reevaluation of the science informing standards than others. Changes to standards or policies should only be made after a proper assessment of the science base as a whole to ensure that the conclusions of the research in a given area are consistent.

Acknowledgment

The authors would like to thank Professor B. Greenebaum and Professor F. Barnes for their valuable suggestions during the elaboration of this chapter.

References

1. Repacholi, M.H., Radiofrequency field exposure standards; current limits and relevant bioeffects data, in *Biological Effects and Medical Applications of Electromagnetic Energy*, Ghandi, O., Ed., Prentice Hall, Englewood Cliffs, NJ, 1990, p. 9.
2. Wertheimer, N. and Leeper, E., Electrical wiring configurations and childhood cancer. *Am. J. Epidemiol.*, 11(4), 345, 1979.
3. Savitz, D.A. et al., Case-control study of childhood cancer and exposure to 60 Hz magnetic fields, *Am. J. Epidemiol.*, 128, 2, 1988.
4. Ahlbom, A. et al., A pooled analysis of magnetic fields and childhood leukemia, *Br. J. Cancer*, 83, 692, 2000.
5. Greenland, S. et al., A pooled analysis of magnetic fields, wire codes, and childhood leukemia, *Epidemiology*, 11, 624, 2000.
6. IAEA (International Atomic Energy Agency), International basic safety standards, 1996 (http://www-pub.iaea.org/MTCD/publications/PDF/SS-115-Web/Start.pdf).
7. ICNIRP (International Commission on Non-Ionizing Radiation Protection), Guidelines for limiting exposure to time varying electric, magnetic and electromagnetic fields (up to 300 GHz). *Health Phys.*, 74(4), 494–522, 1998 (http://www.icnirp.org/).

8. IEEE (Institute of Electrical and Electronics Engineers), C95.6, IEEE standard for safety levels with respect to human exposure to electromagnetic fields in the frequency range 0–3 kHz, International Committee on Electromagnetic Safety (ICES), 2004.
9. IEEE (Institute of Electrical and Electronics Engineers), C95.1, IEEE standard for safety levels with respect to human exposure to radio frequency electromagnetic fields, 3 kHz to 300 GHz, International Committee on Electromagnetic Safety (ICES), 2005.
10. Salzburg Resolution on Mobile Telecommunication Base Stations, International Conference on Cell Tower Siting, Linking Science & Public Health, Salzburg, Austria, June 7–8, 2000.
11. NIEHS (National Institute of Environmental Health Sciences), Assessment of health effects from exposure to power-line frequency electric and magnetic fields, NIEHS Working Group Report, Portier C.J. and Wolfe M.S., Eds., Research Triangle Park, NC, NIH Publication No. 98–3981, 503, 1999 (http://www.niehs.nih.gov/).
12. AGNIR, Independent Advisory Group on Non-Ionizing Radiation, National Radiological Protection Board (http://www.hpa.org.uk/radiation/advisory_groups/agnir/).
13. Lorrain, J-L. and Raoul, D., L'incidence éventuelle de la téléphonie mobile sur la santé, Rapport du Sénat: Rapport 52, Office Parlementaire d'évaluation des choix scientifiques et technologiques, 2002 (http://www.senat.fr/rap/r02-052/r02-052.html).
14. IARC (International Agency for Research on Cancer), IARC Monographs on the evaluation of carcinogenic risks in humans: Preamble. Lyon, International Agency for Research on Cancer, 1995.
15. Treaty of Maastricht, International Legal Materials 31, 1992.
16. United Nations Conference on Environment and Development: Rio Declaration on Environment and Development, *LFNCED document A/CONF*, 151/5/Rev. 1, June 13, 1992.
17. European Commission, Commission of the European Communities, Communication on the Precautionary Principle, Brussels, 2000 (http://europa.eu.int/comm/off/com/health_consumer/precaution.htm).
18. A Canadian Perspective on the Precautionary Approach/Principle, 2003 (http://www.dfo-mpo.gc.ca/ccpa/HTML/pamphlet_e.htm).
19. Swiss Federal Office of Public Health, The Precautionary Principle in Switzerland and Internationally, 2004 (http://www.bag.admin.ch/themen/weitere/vorsorge/e/synthese.pdf).
20. Foster, K., Vecchia, P. and Repacholi, M.H., Science and the precautionary principle, *Science*, 288, 979, 2000.
21. Nair, I., Morgan, M.G., and Florig, H.K. Biological effects of power frequency electric and magnetic fields: Background paper. Washington, DC: Office of Technology Assessment, Congress of the United States. Publ. No. OTA-BP-E-53, 1989.
22. European Commission, 1999/EC/519, Council Recommendation on the limitation of exposure of the general public to electromagnetic fields (0 Hz to 300 GHz), European Council, 1999.
23. European Commission, 2004/40/EC, Directive 2004/40/EC of the European Parliament and of the Council of 29 April 2004 on the minimum health and safety requirements regarding the exposure of workers to the risks arising from physical agents (electromagnetic fields) (18th individual Directive within the meaning of Article 16(1) of Directive 89/391/EC), European Parliament, 2004.
24. Borraz, O., Devigne, M., and Salomon, D., Controverses et mobilisations autour des antennes relais de téléphonie mobile, Centre de Sociologie des Organisations report, 2004.(http://www.cso.edu/upload/pdf_breves/Mobilisations%20Antennes%20Relais%20-%20Rapport%20-final.pdf).
25. Report, evidence and supporting papers of the Inquiry into the emergence and identification of Bovine Spongiform Encephalopathy (BSE) and variant Creutzfeldt-Jakob Disease (vCJD), 2000 (http://www.bseinquiry.gov.uk/report/).
26. Slovic, P., Perception of risk, *Science*, 236, 280, 1987.
27. WHO, *Establishing a Dialogue on Risks from Electromagnetic Fields*, ISBN 92 4 154571 2, 2002 (http://www.who.int/peh-emf/publications/risk_hand/en/index.html).

9

Electroporation

James C. Weaver and Yuri Chizmadzhev

CONTENTS

9.1 Background

Cells and tissues contain multiple, spatially distributed barriers that compartmentalize charged and large molecules. These barriers are largely constructed out of lipids, usually phospholipids. For this reason, only very small molecules with effective high lipid solubility spontaneously penetrate the single or double phospholipid bilayer-based membranes of cells and and their organelles [1]. Of course, these membranes have a large variety of channels and transporters that facilitate transport of particular ions and molecules. Other significant barriers consist of one or more layers of cells connected by tight junctions around bladders and ducts that help retain specialized fluids, and the tough, flexible stratum corneum of mammalian skin that prevents water loss, entry of toxic molecules and infectious agents. Electroporation results in an essentially universal physical reduction of such barriers by creating membrane-spanning aqueous pathways. Aqueous pathways (large dielectric constant $\varepsilon_w \approx 80$) across lipid-containing (small dielectric constant $\varepsilon_l \approx 2$) barriers greatly favor transport of even small, monovalent ions [2,3].

Applied electric field pulses with dominant frequency content below ~ 300 MHz cause concentration of voltage across membranes of isolated cells [4,6], and groups of cells spaced close together in a tissue [7]. If the time-dependent transmembrane voltage, $U_m(t)$, becomes sufficiently large, then stochastic rearrangements of membrane phospholipid molecules are hypothesized to occur at a high rate, such that water-containing defects ("pores") measurably alter the membrane's transport properties. This is electroporation.

The simplest statement is that electroporation "creates new aqueous pathways" through lipid-based barriers. Almost all electroporation studies to date have focused on bilayer membranes (BLMs), both artificial planar BLMs and cell membranes. In BLMs electroporation occurs under biochemically mild conditions, usually with a small temperature rise. For most easily observable phenomena such as cell transfection and molecular uptake, electroporation-related phenomena are believed to depend nonlinearly on the transmembrane voltage, $U_m(t)$. An exposure time-dependent onset within the range $\sim 0.2 < U_m < \sim 1\,V$ is usually found for single pulses with $t_{pulse} > \tau_m$, the membrane charging time. For conventional mammalian cell electroporation this is readily satisfied for widely used pulses that have t_{pulse} in the range 1 μs to 50 ms.

Both dramatic electrical behavior ("reversible electrical breakdown" = REB for cells) and significant molecular transport occur. Most interest to date has focused on the electropermeabilization that coincides with REB conditions, which involves a large increase in membrane permeability for ions and molecules, particularly for longer pulses. Other consequences of electroporation are electrofusion of cells to other cells or tissue and electro-insertion of membrane proteins into cell membranes.

The ability of a lipid-containing sheet to exclude ions and charged molecules is fundamental and is a result of the change in "Born energy" associated with moving a charge from a medium with a large permittivity, e.g., water, to a region with a low permittivity, e.g., the interior of a phospholipid BLM. Here the Born energy is the electrostatic energy of an ionic charge embedded in a medium with permittivity ε, written here in terms of the electric field, E, and with $\varepsilon = K\varepsilon_0$ where K is the dielectric constant and $\varepsilon_0 = 8.85 \times 10^{-12}$ F/m [8]:

$$W_{Born} = \int_{\text{all space but ion}} \frac{1}{2}\varepsilon E^2 dV \tag{9.1}$$

The essential barrier function of cell membranes can be represented by a thin sheet of lipid. This allows the magnitude of the Born energy barrier, ΔW_{Born}, to be computed by the following process: a charged sphere, representing the ion or molecule, is initially located in water far from the lipid sheet ($W_{Born,i}$) and then moved to the center of the sheet ($W_{Born,f}$) which requires the expenditure of energy. The corresponding barrier height, $\Delta W_{Born} = W_{Born,f} - W_{Born,i}$, is large even for small monovalent ions, e.g., Na^+ and Cl^-, is still larger for multivalent charged molecules, and also depends slightly on the membrane thickness, sphere radius, and the amount and distribution of charge within the sphere.

For a single, isolated charge such as a small ion, e.g., Na^+, the largest contribution to ΔW_{Born} arises from the region close to the ion. The small diameter of an ion (solute) of type "s" (typically $2r_s \approx 0.4$ nm) is significantly smaller than a typical membrane thickness of $d_{m,l} \approx 4$ nm) for the lipid hydrocarbon chains. The corresponding full thickness is $d_m \approx 5$ nm, which includes the phospholipid's headgroups. As noted above ΔW_{Born} can be estimated by calculating the change in energy to move the ion from bulk water to bulk lipid. This allows ΔW_{Born} to be estimated by neglecting the membrane thickness and instead considering bulk lipid. This is justified because the greatest contribution to the electric field is in the volume near the ion. This estimate yields

$$\Delta W_{Born} \approx \frac{e^2}{8\pi\varkappa_0 r_s}\left[\frac{1}{K_m} - \frac{1}{K_w}\right] \approx 100\,kT \tag{9.2}$$

where the relevant temperature is $T = 37°C = 310\,K$. Numerical solutions to the electrostatic problem for a thin, low dielectric constant sheet immersed in water yields relatively small corrections. A barrier of this size is surmounted at a negligible rate by thermal fluctuations (spontaneous ion movement). Moreover, a transmembrane voltage, $U_{m,direct}$, which is much larger than physiological values, would be needed to provide this energy. Uncharged molecules can partition into the membrane and then cross the membrane by diffusion; these species are not significantly affected by ΔW_{Born}. Instead, their transport is governed by a passive permeability due to the combined effect of dissolution and diffusion.

Spontaneous barrier crossing is therefore negligible. An early calculation considered not only the "intact sheet" case, but also the case of a fixed cylindrical pore [2,3]. Both aqueous configurations lowered ΔW_{Born}, but the greater reduction was achieved by the pore. The basic pore structure, penetration of the lipid membrane by an aqueous pathway, is of course the essence of channels based on proteins. It is also the basis of the "transient aqueous pore" theory of electroporation [9–19], but with the significant difference that the fluctuating and expandable electroporation aqueous pathways can be created rapidly (time scale of nanoseconds), but are metastable, with inferred pore lifetimes ranging from milliseconds to minutes.

To our knowledge, the first experimental report of electroporation-related phenomena were the irreversible [20] and reversible [21] observations of "breakdown" of the excitable membrane of the node of Ranvier. Almost a decade later nonthermal killing of microorganisms by electric field pulses was reported [22–24], followed a few years later by the observation of a large, field-induced molecular permeability increase in natural vesicles [25]. Increasing numbers of experimental reports involving electrical behavior of field-pulsed cell membranes came in the next few years [26–28]. Artificial planar BLMs [29,30] exhibited dramatic electromechanical behavior, and the first pore-based theory was advanced to explain the fate of pulsed planar membranes [9–15] and then extended in a series of further experimental [31–33] and additional theoretical studies [16,17,34]. Other reports confirmed and extended observation of cell membrane transport due to field-induced permeability increases [35–39]. This included introduction of "inactive" DNA into red blood cells [40], followed by the critically important demonstration of transformation of cells by electrically mediated DNA uptake [41].

An important feature of artificial and cell membranes is that they concentrate electric field pulses with slow rise times (relative to the membrane charging time) because of the large membrane resistance relative to that of the extra- and intracellular media [4–6,42,43]. This form of field amplification can be regarded as voltage concentration due to a spatially distributed voltage divider effect within a single or multicellular system. For the case of current injection (current clamp), field amplification can also arise from current density concentration, arising from multiple cells in close proximity or nearby insulating objects. In the case of cell and organelle membranes providing the predominant barriers and fast pulses with rise times smaller than $\sim 3\,ns$ (significant frequency content above $\sim 300\,MHz$), spatially distributed dielectric voltage division emerges, electric fields tend toward approximate uniformity, and voltage concentration at membranes is much smaller [7].

However, most electroporation studies and applications have utilized pulses with rise times that exceed a typical cell charging time. In the case of artificial planar BLMs the membrane completely blocks the current pathway, so that the entire voltage across the experimental apparatus electrodes appears, after a characteristic charging time, $\tau_m \approx 10^{-6}$ s, across the membrane. In the case of cells, the situation is more complicated, but an approximate guide is obtained for the case of spherical cell for which $U_{m,max} \approx 1.5 E_{app} r_{cell}$, which shows that the change in transmembrane voltage is given approximately by the product of the electric field and the cell size.

Conventional pulse conditions involve large electric field magnitudes and relatively long rise times (relatively low frequencies). Field pulses applied in aqueous extracellular media typically have magnitudes of 0.1 to 1 kV/cm (mammalian cells) and ~ 10 kV/cm (bacteria). As discussed below, most cells have characteristic linear dimensions of $L_{cell} \approx 1$ to 10 μm, although skeletal muscle and nerve cells are typically much larger. Such pulses result in $U_m(t)$ reaching 0.2 to 1.5 V in times of order 1 μs to 10 ms. The widely used "exponential pulse" is often specified only by its decay ($E = E_0 \exp - t/\tau_{pulse}$), with the rise time probably longer than τ_m but usually not reported. Square (rectangular; actually trapezoidal in the approximation that rise and fall times are linear) pulses ($E = E_0$ or 0) are less frequently employed, but usually have pulse widths (durations) of 1 to 100 μs, and sometimes longer. Other waveforms, such as gated RF pulses, bipolar pulses and multiple pulses, have received relatively little attention.

9.2 Pore Models of Planar Lipid Bilayer Electromechanical Behavior

Artificial planar (BLMs are widely used to investigate basic aspects of the bilayer portion of cell membranes [29,30]. Laboratory conditions provide an electric field amplification by the factor $A_{BLM} = L_{elec}/d_m$ where L_{elec} is the electrode separation and d_m is the membrane thickness. Within a charging transient characterized by a time constant $\tau_m = R_e C_m$ (resistance of the electrolyte on both sides and the membrane capacitance) the electric field within a passive (fixed) membrane reaches $E_m = U_m/d_m$. For example, if one-half volt is applied to the electrodes, U_m reaches 0.5 V within about a τ_m of several microseconds. An advantage of such membranes for mechanism studies is that several measurements can be made on the same preparation. These attributes allow short pulse studies to be carried out in which the pulse duration is the same order of magnitude as τ_m or even faster. During the charging transient, the full electrode voltage does not appear across the membrane, and the resistance of the charging pathway should be included. For pulses with rise times shorter than about 3 ns, the dielectric properties of the bathing electrolytes should be included by assigning a capacitance with the dielectric constant of the electrolyte [7] in parallel with R_e.

Creation of artificial planar membranes originally involved use of relatively large amounts of organic solvent [29,30]. Significant improvement was subsequently achieved by using two (different, if desired) previously formed monolayers on aqueous solutions [44]. Unlike their cellular counterparts, these membranes are essentially macroscopic, typically spanning a circular aperture ($D_{ap} \approx 1$ mm; corresponding membrane area is $A_m \approx 7 \times 10^{-7}$ m^2) in an electrically insulating septum. The convenient access to the chambers on both sides of the septum allows electrical measurements to be readily made, and the macroscopic chamber size results in a small electrolyte electrical resistance (sum of resistances on both sides, $R_e \approx 100$ Ω) between the membrane and macroscopic electrodes located about 1 cm away on each side. BLMs have a capacitance per area of about $C_{area} \approx 1$ μF/cm, and therefore a typical BLM has a capacitance of order $C_{BLM} \approx 3 \times 10^{-8}$ F. As a result $\tau_m \approx 3$ μ_s.

The bulk charge relaxation time of an aqueous electrolyte is $\tau_e = \varepsilon_e/\sigma_e = K_e \varepsilon_0 \rho_e$. For physiological saline this basic relaxation time is small compared to the characteristic times of conventional electroporation pulses. Using a resistivity $\rho_{sal} = 0.83$ Ω m and a dielectric constant $K_{sal} = 72$ the relaxation time is $\tau_e \approx 0.5$ ns. This is much less than the charging times of typical BLM (and of cell membranes). For this reason, BLM experiments often have excellent time resolution for electrical measurements.

The measurement of a BLM transmembrane current, $I(t)$, following a step change in the applied electrode voltage, V_0, allows the induced membrane conductance to be followed. In such experiments it is found that during its lifetime a membrane can pass through up to four distinguishable stages. The first stage is a simple charging of the membrane, governed by τ_m given above. The second stage is characterized by a constant voltage and associated (small) current. The third stage has a fluctuating current, but sometimes in this stage the membrane reverts to a quiet, steady current state. If a large voltage is applied to the membrane for a short time, then the intact membrane often makes a transition into a peculiar long-lived (tens of minutes) excited state that is characterized by a large conductance and pronounced current fluctuations, even at low U_m. This peculiar state was termed a "stress state" [45]. A fourth stage is often reached in which $I(t)$ exhibits a drastic irreversible increase to saturation. Visual inspection shows that upon saturation the membrane has ruptured; the membrane material no longer spans the aperture and instead has collected on the aperture rim.

If a new BLM is created and an experiment repeated, then the next $I(t)$ is qualitatively reproduced, but the membrane lifetime, τ_m, the time from step voltage application to onset of a drastic current rise, is usually different. Both the character of the fluctuations and their duration are different. In short, the lifetime exhibits stochastic behavior. This characteristic feature of the membrane response can be explained by the hypothesis of transient aqueous pores.

The dependence of the mean membrane lifetime, $\bar{\tau}_m$, on various factors was studied to clarify the mechanism of membrane rupture [9,31–33]. The effect of varying U_m was most revealing. Over the range 300 to 600 mV the dependence of $\log_{10} \bar{\tau}_m$, on U_m is usually almost linear, e.g., an increase in U_m by 100 mV causes a tenfold decrease in $\bar{\tau}_m$. At larger U_m the decrease in $\bar{\tau}_m$ becomes less pronounced. If, for example, U_m is increased from 200 mV to 1.4 V the mean lifetime changes by more than six orders of magnitude, decreasing to about 10 μs. This highly nonlinear dependence on U_m is not expected for electrostrictive mechanisms and provides key support for the transient aqueous pore hypothesis. Rupture (irreversible breakdown) was also studied by the charge injection technique [31]. In this method, a short (e.g., 400 ns) square pulse of moderate amplitude is applied to the membrane, with the result that $U_m(t)$ rises quickly to about 0.4 V. After 300 to 400 μs, U_m drops to zero over about 50 μs because of rupture. Many features of rupture following charge injection can be described quantitatively by a transient aqueous pore model [9,46,47].

One of the most dramatic aspects of electroporation is "reversible electrical breakdown" (REB), which is actually a high conductance state that acts to protect the membrane through a rapid discharge [46]. Planar membranes made of oxidized cholesterol have been studied using the charge injection method. If such membranes are charged to about 1 V in about 400 ns the membrane resistance reversibly decreases by nearly nine orders of magnitude [31]. In typical experiments U_m reached 0.9 to 1.2 V. Significantly, the value $U_m \approx 1.2$ V cannot be exceeded by further increasing the pulse amplitude. This includes experiments on cell membranes [48–50]. After a rapid discharge, the membrane survives, remains mechanically stable, and can be recharged. This was the first observation of REB in planar BLMs. Later REB was also studied using a voltage-clamp method.

Significantly, REB of cell membranes is similar to the behavior of oxidized cholesterol membranes [31,32] or UO_2^{2+} planar BLMs [51]. An important experimental observation is that such planar membranes can exhibit either rupture or REB, depending only on the charging procedure. A moderate pulse of injected charge leads to destruction, with $U_m(t)$ decreasing in a two-stage, sigmoidal curve: following an initial post pulse decay, $U_m(t)$ begins to level off, but then decays further as the membrane ruptures.

In contrast, for large pulses, U_m (t) reaches significantly higher values and then suddenly drops rapidly. Still bigger pulses led to even more rapid membrane discharge. These very different outcomes can be quantitatively described by a transient aqueous pore model [46,47].

9.3 Cell Membrane Electromechanical Phenomena

A cell membrane is much more complex than an artificial planar BLM. But even if the detailed biochemical composition is neglected, there are still important physical differences. These include the membrane's non-planar shape, the closed membrane topology, the non-zero trans-membrane voltage due to pump activity, and the presence of membrane proteins, and membrane cytoskeleton interactions [52]. Although real cells range from spherical to elongated cylinders to irregular three dimensional shapes, a local membrane area can often be regarded as a small subsystem with a planar geometry. Consider the most commonly treated cell shape (spherical) for the small electric field case. As is well known, for small fields a spherical cell develops a change in transmembrane voltage, ΔU_m, which varies in position over the membrane and also with time:

$$\Delta U_m = \frac{1.5 E_e r_{cell}}{1 + r_{cell} G_m (\rho_{e,\,int} + 0.5\rho_{e,\,ext})} [1 - \exp(-t/\tau_{m,\,cell})] \cos\theta, \qquad (9.3)$$

where r_{cell} is the spherical cell radius, $\rho_{e,int}$ and $\rho_{e,ext}$ are the resistivities of the intra- and extracellular electrolytes, respectively, G_m is the (constant; no electroporation) cell membrane conductance, θ is the angle between the applied electric field, $\vec{E}_{app} = \vec{E}_0 e^{j\omega t}$, and the site on the cell membrane at which U is measured. In this classic case the response field, $\vec{E}_{res}$, differs significantly from $\vec{E}_{app}$, in that $\vec{E}_{res}$ goes around the cell. In contrast $\vec{E}_{app}$ is uniform, as it is the field that would exist if the cell were removed and only aqueous electrolyte was present. Finally, the time constant associated with charging the fixed cell membrane is

$$\tau_{m,\,cell} = \frac{r_{cell} G_m (\rho_{e,\,int} + 0.5\rho_{e,\,ext})}{1 + r_{cell} G_m (\rho_{e,\,int} + 0.5\,\rho_{e,\,ext})} \qquad (9.4)$$

Typical mammalian cells with $r_{cell} \approx 10$ μm have $\tau_m \approx 1$ μs. For $t \gg \tau_m$ and G_m essentially zero, Equation 9.3 reduces to the widely used simple form

$$\Delta U_m = 1.5 E_e r_{cell} \cos\theta \qquad (9.5)$$

However, Equation 9.3 through Equation 9.5 are not valid for electroporation, because these equations do not hold if the membrane conductivity varies with position. Specifically, with electroporation, G_m varies dramatically with time at the sites of large U_m, but hardly at all at other sites. This has been convincingly established experimentally by submicrosecond fluorescence measurements with membrane dyes that respond to the transmembrane voltage [48–50]. Moreover, for suspended cells with significantly different extra- and intracellular medium conductivity, membrane deformation is expected [53–55], and has been observed experimentally [56]. With this in mind theoretical models that use Equation 9.3 and Equation 9.4 under conditions with significant electroporation should be viewed with suspicion.

Two irreversible cell membrane electroporation processes have been identified at the cellular level. First, mechanically bounded small regions of the cell membrane, e.g., regions bounded by cytoskeletal elements and attached to the cytoskeleton by membrane proteins [52], may behave like a small planar membrane and may therefore exhibit prompt rupture even though the plasma membrane typically has a low surface tension [57,58] compared to a artificial planar BLM [9,59]. Surface tension tends to expand a pore once it has formed, so rupture is much less likely in cell plasma membranes unless the cell is osmotically swollen and thus has an elevated tension. In this case the behavior of a BLM is relevant. A topologically closed membrane, e.g., a vesicle, is not expected to rupture globally. Unlike a BLM there is no boundary at which membrane phospholipids can accumulate by expansion of one or a few large pores. The cell membrane surface tension, Γ_{cell}, is also important, and is expected to be small for some cell or vesicle membranes. But for some vesicles and intracellular organelles Γ is large [60].

In the case of BLM the tension is due to the meniscus at the aperture and is a constant during pore expansion. For cells and vesicles, however, the membrane tension is associated with osmotic pressure difference, $\Delta\pi$, through Laplace's law. Following electroporation, $\Delta\pi$ (if it exists) goes to zero, and the tension, Γ, also tends to zero. In this case pore evolution is more complicated and involves the laws of colloid osmotic lysis. In the second case, reversible electroporation may be accompanied by significant molecular transport between the extra- and intracellular volumes, with the resulting chemical imbalance leading to cell stress and eventual lysis. This is an example of the importance of considering biochemical change due to field exposures (see *BBA*, Chapter 5).

Minimum size pores are believed to be small, of order $r_{p,min} \approx 0.8\,\text{nm}$ [61–63] or in erythrocytes about a factor of 2 smaller [36]. However, significantly larger (several nanometer radii) pores are believed to evolve for longer, e.g., 100 μs electroporating pulses that are used to deliver ~1kDa drug molecules such as bleomycin into cells. In this case cell membranes discharge during reversible electrical breakdown (REB) through the pores, and the cellular or vesicular membrane returns to its initial state, except for consequences of net ionic or molecular transport through the transient aqueous pore population. The inability to promptly rupture via critical pores to directly lyse a cell is expected to apply to unswollen cells.

Cell membranes are believed to exhibit REB, i.e., a tremendous increase in electrical conduction which can be inferred from transmembrane voltage measurements [48–50], and which is believed due to ionic conduction through transient aqueous pores. Electric field pulses that cause U_m to rapidly rise into the range 0.5 to 1.5 V cause onset of REB with essentially no dependence on membrane composition. For this reason, the onset of electroporation involving REB is believed to be universal, and it has been observed for a wide variety of cell types. The behavior of the transmembrane voltage, $U_m(t)$, during membrane charging and the subsequent appearance and evolution of a pore population, is intimately connected with the number and size of pores. The success of a transient aqueous pore model in providing a quantitative description of $U_m(t)$ under these conditions gives confidence that electroporation is a valid concept. Significantly, a large increase in molecular transport across cell membranes is generally observed for essentially the same pulse conditions (cell exposures) that cause REB.

9.4 Nonpore Theories of Electromechanical Phenomena

One early approach to explaining membrane destabilization for an elevated U_m involved "punch through" [64] while another was based on electromechanical collapse due to compression of the membrane [65]. Electrocompression models have several important attributes. First, it is deterministic, predicting a critical transmembrane voltage, $U_{m,c}$, above which rupture occurs. However, for realistic values of membrane compressibility for solvent-free membranes this model predicts $U_{m,c} \approx 5\,V$, which is about an order of magnitude too large. Further, the absence of a marked change in membrane capacitance, C_m, before rupture argues strongly against large-scale electrocompression, as an increase in C_m is an inevitable consequence of a decrease in membrane thickness, d_m. Moreover, the observed stochastic nature of rupture is in direct conflict with the concept of a deterministic critical voltage. Neither the stochastic nature nor the strong lifetime dependence on U_m is expected for an electro-mechanical rupture mechanism. Finally, the fate of the membrane in this model depends only on U_m, but experiments also show a dependence on pulse duration.

Electrohydrodynamic models based on viscoelastic behavior have also been advanced [66], but are intimately related to electrocompression theories. The only difference is that the electrocompression model attributes the increase in system energy to elastic compression energy of the membrane, whereas in the electrohydrodynamic model, it corresponds to the work required to form new membrane surface. In both types of deterministic models the development of instability ("irreversible electric breakdown" or "rupture") represents a nonlocal process that occurs simultaneously over a large area of the membrane. In contrast, pore models involve highly localized events that involve only a small fraction of the total membrane area for conventional electroporation used for molecular uptake. At another extreme, however, pulses that cause intracellular effects by organelle electroporation are hypothesized to involve supra-electroporation that involves a very large pore density; see Section 9.11).

9.5 Pore Theories of Electromechanical Phenomena

We first provide a qualitative description of how a membrane is believed to respond to an applied pulse. This is illustrated by the example of $U_m(t)$ given in Figure 9.1. The spontaneous rate of pore formation due to thermal fluctuations is very small. The pore creation rate depends on a Boltzmann factor with an energy contribution that decreases as U_m^2 (Equation 9.6). Even at a resting potential of $U_{m,rest} \approx -60$ mV (nerve membrane) to -200 mV (inner mitochondrial membrane) a minimum size pore will only occasionally appear in a membrane [19,63]. Application of an external electric field pulse begins by charging local membrane areas. Initially displacement currents flow as the membrane charges through the extra- and intracellular electrolyte. In the case of a spherical cell membrane the polar regions (Equation 9.3; θ small) first reach voltages at which pore creation becomes significant. When U_m exceeds $\sim 0.5\,V$, pore creation becomes significant. As more pores rapidly appear whereas U_m rises further, the newly acquired local membrane conductance begins to discharge the membrane through the evolving pore population. When U_m reaches $\sim 1\,V$, there has been a huge increase in the local pore density and U_m begins to decrease even though the applied field may itself still be increasing. This is reversible electrical breakdown (REB), the high conductance state associated with significant electropermeabilization of the local membrane area. If the pulse continues after REB, the transmembrane voltage may continue at about the same value (a plateau in U_m), or may increase again before reaching a plateau. At the end of the pulse, many local areas in the polar region are so conductive that U_m quickly drops to ~ 0. According to what is known, a population of metastable pores remains, with individual pores assumed to vanish stochastically due to local thermal fluctuations.

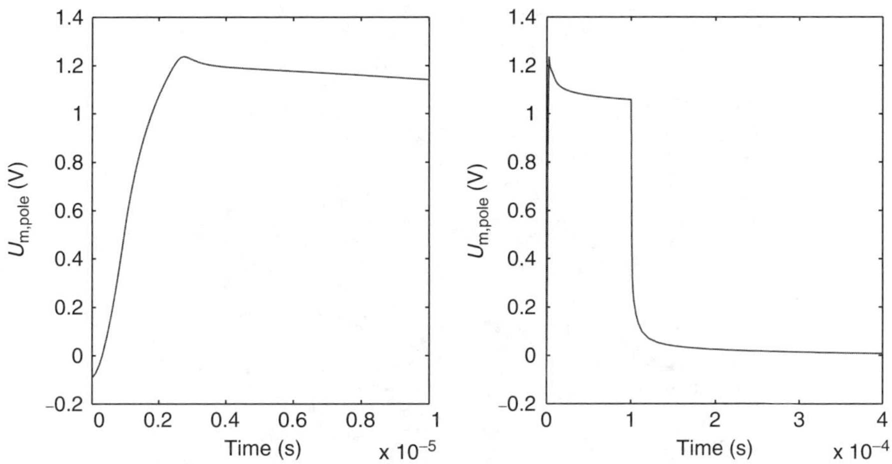

FIGURE 9.1
Illustration of the time-dependent transmembrane voltage, $U_m(t)$, for electroporation. This example shows the behavior at the pole of a two-dimensional cell model for conditions that cause electroporation and reversible electrical breakdown (REB) at $U_m \approx 1.2$ V. A qualitative description of what is believed to occur is given in the text. This example uses a conventional electroporation electric field pulse of magnitude 1100 V/cm and a trapazoidal waveform with 1 μs rise and fall times and a 98 μs flat peak. Left panel: 10 μs timescale showing the initial passive charging of the membrane, followed by the spike at which pore conduction onset is so rapid that it arrests the voltage rise and causes a decrease even though the pulse is still on. Right panel: 400 μs timescale showing REB followed by a transmembrane voltage plateau during the pulse and then a decay when the pulse ends. This model has a distributed resting potential source and membrane resistance that together generate a resting transmembrane voltage of $U_{m,rest} = -90$ mV [69]. This description uses the asymptotic membrane electroporation model assigned to a circular cell with organelle models (Gowrishankar et al., unpublished results).

Molecular transport is believed to occur during the pulse when the pore population changes rapidly, first expanding in pore size and number, and then shrinking mainly in pore size as U_m eventually decreases, with pore number decreasing more slowly. Transient aqueous pore models for artificial planar BLMs [19,46,47] and cell membranes are consistent with this view [67–70].

Initial quantitative theories and models used analytical methods to estimate the creation and destruction rates of pores, based on nucleation theory in which pores are regarded as membrane defects [9,19]. Molecular dynamics (MD) simulations were possible only recently, and these support some of the basic transient aqueous pore concepts [71,76]. For example, a toroidal-like geometry is spontaneously achieved (Figure 9.2).

However, computer limitations presently restrict the model volume to be small, imposing a significant limitation on the concentration of soluble ions and molecules that are candidates for transport through a membrane by passing through a transient pore. Entry of water and aqueous electrolytes (large dielectric constant) into a fluid BLM (small dielectric constant) is increasingly favored as the electric field within the lipid region of the bilayer becomes larger. This is the basis of the transient aqueous pore theoretical model, originally introduced in a series of seven back-to-back papers [9–15]. With similar pore creation mechanisms independently proposed several years later [16,17]. Nucleation theory is based on the absolute rate equation and involves an estimate of the free energy change, $(\Delta W)_p(r_p, U_m)$, associated with formation of a large dielectric constant aqueous pore within the small dielectric constant lipid portion of the membrane. The initial,

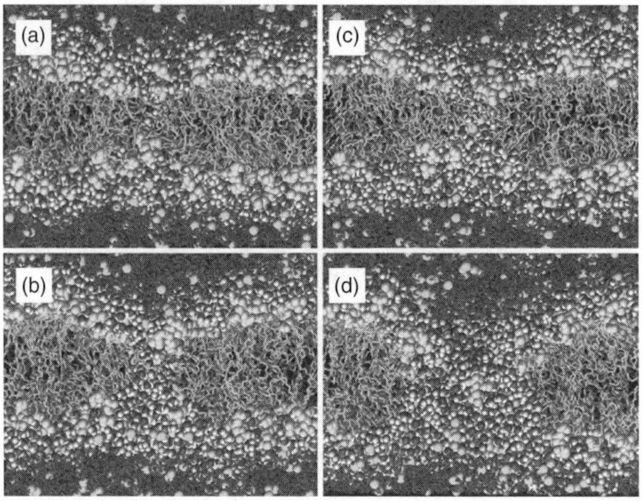

FIGURE 9.2 (See color insert following page 412)
Snapshots of pore creation according to a molecular dynamics simulation. An important, partial test of electroporation was obtained by imposing a constant electric field, $E = 0.5\,\text{V/nm}$ ($5 \times 10^6\,\text{V/cm}$) throughout a small volume that contains aqueous electrolyte and a small area of phospholipid bilayer membrane (Tieleman, D.P. *BMC Biochem.*, 5, 10, 2004). The phospholipid headgroups are white, the hydrocarbon chains are gray, and water is dark gray. The chloride and sodium ions are somewhat difficult to make out in this gray-scale reproduction, so the reader is referred to the original paper for a color depiction. The snapshots are at times (a) 5.33, (b) 5.45, (c) 5.50, and (d) 5.70 ns from starting the simulation. As described in the original paper (Tieleman, D.P. *BMC Biochem.*, 5, 10, 2004) initially water molecules stochastically enter into the bilayer interior, leading to ''formation of single-file like water defects penetrating into the bilayer.'' These and other MD results [71–76] show that pores can form on a nanosecond time scale, and are hydrophilic, with phospholipids present on the interior of the fluctuating pore. These hourglass or toroidal-like pores are reminiscent of the simple drawings used to motivate early models of electroporation. Present MD models are ''noisy'' in the sense that for a typical computational volumes and times only a few transported ions are involved. However this approach has great promise, depending on improved computational power.

simplest form neglected the ''spreading resistance'' (access resistance) and ''Born energy change for ion insertion'' (partitioning) which were included in some later versions, and gave (see Figure 9.3)

$$(\Delta W)_{\text{p}}(r_{\text{p}}, U_{\text{m}}) = 2\pi\gamma r_{\text{p}} - \pi\Gamma r_{\text{p}}^2 - 0.5C_{1\rightarrow\text{w}}U_{\text{m}}^2\pi r_{\text{p}}^2 \tag{9.6}$$

where γ is the line tension associated with a pore edge, Γ is the surface tension of a pore-free planar membrane, and $C_{1\rightarrow\text{w}} = \varepsilon_0[\varepsilon - \varepsilon_1]/d_{\text{m}}$ is the difference in specific capacitance due to replacement of lipid by water. At zero transmembrane voltage $(\Delta W)_{\text{p}}(r_{\text{p}}, U_{\text{m}})$ is a parabola with a maximum of $\sim 100\,kT$, consistent with infrequent spontaneous BLM rupture.

The initial suggestion that pore formation could occur and lead to membrane rupture did not involve electrical behavior. The basic idea was that a ''cookie cutter'' model for a pore formation energy at zero membrane potential, $(\Delta W)_{\text{p}}(r_{\text{p}})$ could be used, based on a gain in ''edge energy,'' $2\pi\gamma$, as a pore is created, and a simultaneous reduction in surface energy, $\pi r_{\text{p}}^2\Gamma$, due to the loss of a circular patch of membrane [77]. The interpretation is simple: a pore-free membrane is envisioned, then a circular region is cut out of the membrane, and the difference in energy between these two states calculated and identified as $(\Delta W)_{\text{p}}(r_{\text{p}})$. The change in pore energy due to an elevated transmembrane voltage is contained in the third term of the right-hand side of Equation 9.6. The membrane is

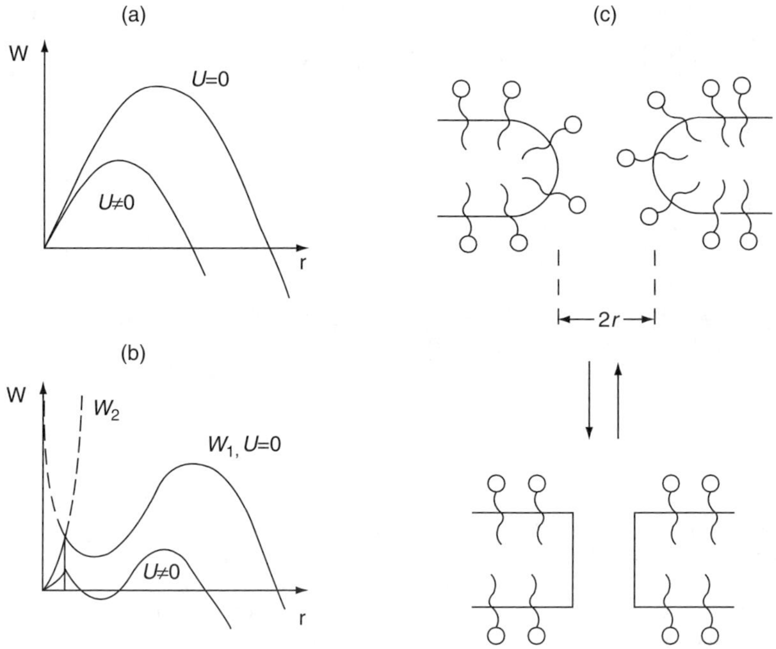

FIGURE 9.3
Drawing of imagined pore structures and associated pore creation energy. Early transient aqueous pore theory was based on two related concepts, hydrophilic pores and hydrophobic pores, with some models considering both and others only hydrophilic pores. (a) Hydrophilic pore creation energy, W, plotted as a function of pore radius, r, for zero transmembrane voltage ($U = 0$) at which W is maximal, and a representative elevated voltages, at which W is decreased (favoring hydrophilic pore formation). (b) More complicated energy landscape for the hypothesis that small hydrophobic pores (curve W_2 at $U = 0$) form first, followed by thermal activation to hydrophilic pores (curve W_1 at $U = 0$). For elevated U the overall energy decreases, with two peaks, which favors increased hydrophobic pore formation and also transitions to hydrophilic pores. Dependent on time and U, hydrophilic pores can surmount the right-most barrier and expand to rupture the membrane (expected for artificial planar bilayer membranes, much less plausible for cell membranes with small surface tension). (c) Envisioned transitions between hydrophobic pores (bottom) and hydrophilic pores (top). In this simple view hydrophobic pores are imagined to be cylindrical, with an interior surface characterized by a hydrocarbon–water interface (large surface energy), and hydrophilic pores are "inverted," with phospholipid head groups lining the pore interior. Recent molecular dynamics simulations show that hydrophilic pores are expected, and the precursor membrane conformations involve water entry, achieving a water chain (analogous to a hydrophobic pore) that penetrates the membrane before a phospholipid-lined pore forms.

regarded as a capacitor, with the U_m-dependent term representing the change in electrical energy resulting from replacing membrane material with dielectric constant ε_l by water with dielectric constant ε_w in a cylindrical pore of radius r_p. In other words, creation of a hydrophilic pore (hereafter simply "pore") is described as a change in specific capacitance, $C_{l \to w}$ in the region occupied by the pore. However, for small hydrophilic pores, even if bulk electrolyte exists within the pores, the permittivity would be $\varepsilon \approx 70\varepsilon_0$, only about ten percent smaller than for pure water. Moreover, the pore interior resistance would still be large, $R_{p,int} \approx \rho_e d_m / \pi r_p^2$ compared to the spreading resistance (see below), so the voltage across the pore would be very close to U_m. Although each of the "ingredients" of the pore model is plausible, if they are combined to represent a complete experimental situation, the overall model is too complex to solve in closed form. Other membrane deformations, conformational changes and hypothesized pore structures are shown in Figure 9.4 [78].

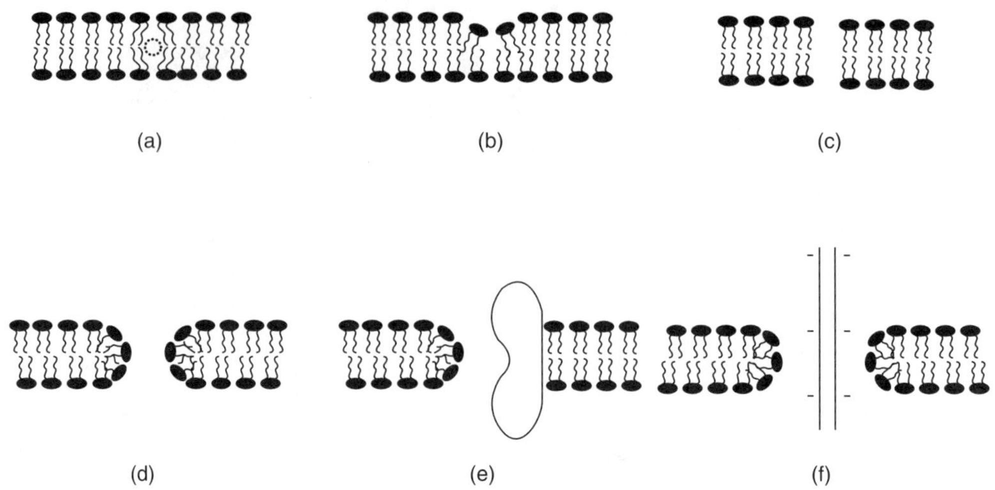

FIGURE 9.4
Hypothesized pore structures. Illustrations of hypothetical structures of both transient and metastable membrane conformations that may be involved in electroporation. Hypothetical bilayer membrane structures related to electroporation: (a) membrane free volume fluctuation; (b) aqueous protrusion into the membrane (''dimple''); (c) hydrophobic pore proposed by Chizmadzhev and coworkers [9]; (d) hydrophilic pore proposed by Litster and by Taupin and coworkers, usually regarded as the ''primary pores'' through which ion and molecules pass during electroporation; (e) composite pore with one or more proteins at the pore's inner edge; (f) composite pore with ''foot-in-the-door'' charged macromolecule inserted into a hydrophilic pore. Although the actual transitions are not known, the transient aqueous pore model assumes that transitions from A → B → C or D occur with increasing frequency as U_m increases. Type E may form by entry of a tethered macromolecule when U_m is large, and then persist after U_m has decayed by pore conduction. These hypothetical structures have not been directly observed, but are consistent with a variety of experimental observations and with theoretical models. (Reproduced from Weaver, *J. Cell. Biochem*. 51, 426–435, 1993. With permission.)

Such models are, however, now routinely solved numerically on computers to generate qualitative and quantitative predictions, and these results can be compared with experimental observations. Solutions for artificial planar BLMs have been developed, and these produce reasonable, but not completely correct, quantitative descriptions of measurable quantities such as the transmembrane voltage, $U_m(t)$ the pore size distribution, and also the membrane conductance, G_m, which increases tremendously during REB. The transport of small ions and small molecules can also be predicted approximately (see Section 9.6). In general agreement with experimental observation [31], a model solved for different pulse magnitudes also predicts the four distinguishable outcomes that are found experimentally for an oxidized cholesterol planar membrane subjected to a 400 ns second square charging pulse: (i) simple, passive charging of the membrane capacitance for small pulses; (ii) charging to about 0.5 V, followed by a sigmoidal decay of U_m as the membrane ruptures for a moderate pulse; (3) incomplete reversible electrical breakdown in which the membrane achieves a high conductance state, but not enough to discharge the membrane fully to zero, for a larger pulse; and (4) reversible electrical breakdown due to a still larger but transient high conductance that discharges the membrane completely [46]. This behavior reflects the interplay between the rapidly changing electrical conductance and the mechanical expansion and contraction of the individual pores in a heterogeneous, dynamic pore population.

An extension of the single artificial planar BLM to two identical membranes in series can be used as a very simple model of a ''cubic cell'' [79]. By using a very low value of Γ the inability of a spherical membrane to rupture can be approximated, and this allows the behavior of cell membranes to longer pulses to be explored. Such an approach is ongoing and generates predictions of $U_m(t)$ and the transport of small charged molecules of type "s", $\bar{n}_s$, that are reasonable. Investigation of a planar membrane has also generated a surprise, the prediction of an approximate intrapulse plateau in U_m as a function of exponential pulse amplitude [47]. Whereas too simple to fully represent cellular electroporation, this prediction may be related to experimentally observed plateaus in charged molecular transport (see Section 9.6).

The various theoretical models support the idea that the reversibility of electroporation in planar membranes is due to (i) the chemical composition of the membrane, (ii) the pulse protocol, and (iii) feedback at the pore level that is associated with the spreading resistance, such that as a pore expands, the local transmembrane voltage across the pore decreases. Overall, considerable progress has been made in devising theoretical models that can quantitatively describe some aspects of electroporation, whereas others are still poorly understood. Mechanical and electrical aspects of electroporation of a relatively simple artificial planar BLM do account for (i) the approximate magnitude of U at which membrane rupture (irreversible breakdown) occurs, (ii) the strong dependence of membrane lifetime on U_m, and (iii) the stochastic nature of rupture. Moreover, a reasonable but not exact quantitative description of U_m is given by a transient aqueous pore theory, and the maximum fractional aqueous area of the membrane during reversible electroporation [47] agrees with limited experimental observations [48,49]. There has been some progress in predicting molecular transport, but much remains to be done, and the mechanisms of membrane recovery, cellular stress, and the determinants of cell survival or death are essentially not understood. Moreover, single pulse electroporation has been emphasized by theories and models to date, and relatively little is known about the importance of pulse shape, repetition and spacing of multiple pulses. The challenge is to understand what changes persist and therefore represent the state of the membrane when second and then additional pulses are applied.

9.6 Molecular and Ionic Transport

Greatly enhanced molecular and ionic transport is the basis of most all applications of electroporation. DNA introduction into cells for transfection was first reported about 25 years ago [41] and "transfection" has almost become synonymous with "electroporation" in some areas of biological research [80–83]. Delivery of drugs, proteins, and fluorescent indicators into cells and tissues is a basic manipulation that is also widely used in research, and is being adopted for clinical treatment of local cancer tumors [84–86]. Accordingly, by now numerous potential applications in biology, biotechnology, and medicine have been suggested, and many other applications are explored. The basic idea is qualitatively simple: application of one or more electrical pulses leads to elevated U_m, which if sufficiently high and long leads to membrane pore formation and subsequent expansion to transport water-soluble molecules under the influence of local driving forces (Figure 9.5). Quantitatively, however, surprisingly little is established theoretically using cell and tissue models. In the case of potential biomedical applications involving *in vivo* electroporation, more research will be needed to understand both efficacy and side effects.

Early studies of transport reported molecular release from biological vesicles due to a field-induced permeability increase [25]. Red blood cells were subsequently used to demonstrate DNA delivery into a cell that was associated with dielectric breakdown of

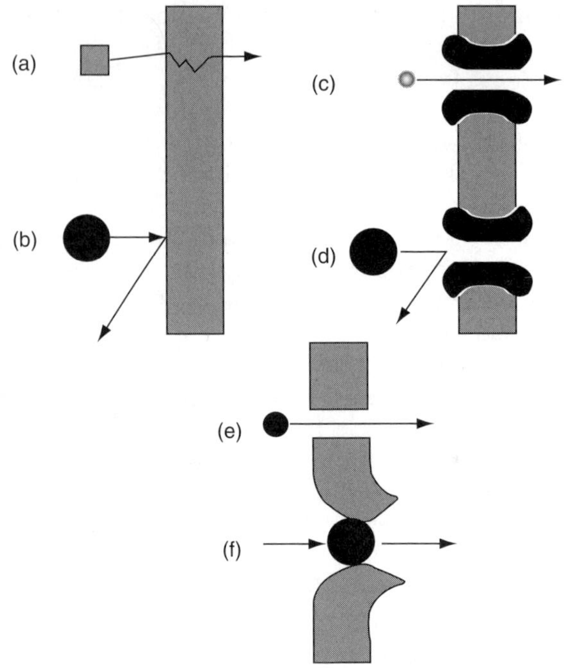

FIGURE 9.5
Schematicized drawing of molecular transport pathways. These pathways are envisioned to exist in the intact or permeabilized lipid bilayer portion of a cell membrane or of an artificial bilayer membrane. The membrane may contain preexisting channels or induced pores. This cartoon can, with some restriction, be considered valid for multilayer structures, e.g., in the skin's stratum corneum. (a) Hydrophobic molecule, which is shown as a gray square can cross the lipid bilayer via partitioning and diffusion. (b) Hydrophilic molecule (large black dot), which is repelled from the lipid phase, can not overcome a hydrophobic barrier. (c) Small ions (gray dot) can cross a membrane through narrow preexisting defects or proteinous channels. (d) Large polar molecules (large black dot) cannot pass through ion channels evolved to regulate small ion transport. (e) Electropores are large enough to provide effective transport of ions and polar particles of moderate size, with the size dependent on evolution of pores during a particular electrical pulse. (f) Bigger charged molecule (large black dot), e.g., plasmid DNA, can be forced by the local electric field to enlarge an electropore and then move across the membrane.

the cell membrane [40]. This experiment did not, however, involve the critical demonstration of cell transfection. This achievement was reported several years later [41,87]. Molecular uptake into cells has now been widely reported, with experiments typically using a large number of cells. These measurements determine the average molecular transport per cell by measuring a total cell population effect. Examples of delivered molecules in such experiments include antibodies [88–90], enzymes [35], and small molecules [91–95]. A few experiments obtained calibrated measurements that yielded that approximate number of molecules delivered into individual cells [91–95], with quantitative results needed for the development of theories and models for transport. Some of these experiments show that there is significant cell–cell variation in molecular transport.

Taken together these studies show that significant molecular transport occurs under approximately the same conditions that cause REB. This is an important finding, but does not address the question of how much molecular transport occurred for various pulsing conditions, nor was the question of cell–cell variation considered. One particularly interesting study showed that uptake of a first fluorescence-labeled molecule (fluorescence-labeled dextran; 19 kDa) was a reliable indicator for the uptake of a second, larger fluorescent molecule (phycoerythrin, 240 kDa) [96]. This report is consistent with the idea that electroporative molecular transport is a physical process (movement through aqueous pathways), which is relatively nonspecific if the pathway is not too close to the molecule's size. As mentioned earlier, studies with artifical planar BLMs show that electroporation depends on the history of the transmembrane voltage, $U_m(t)$. This implies that physical factors which govern field-induced change in U_m over a cell membrane should be important, i.e., the shape, size and orientation of cells with respect to the cells with respect to the local electric field. In addition to isolated cell models, a didactic multicellular model of a hypothetical local tissue region shows that individual cells can experience significantly different changes in U_m over each cell membrane [7].

Biological systems are also well known to be variable, e.g., displaying cell-to-cell variation in many properties ("biological variability"). At a nanometer scale electroporation is believed to arise from fluctuations of water molecules and phospholipid molecules and also nearby ions or molecules that are transported through the membrane in the presence of an average elevated U_m. This view intrinsically arises in molecular dynamics (MD) simulations of artificial membrane electroporation [71–76] and is consistent with stochastic models based on analytic estimates [9,16–19,97]. For both analytic and MD analysis the overall process of forming pores is fundamentally stochastic. Existing continuum models that exhibit REB have significant internal feedback between pore population ionic conduction and $U_m(t)$. This intrinsic regulation of the local transmembrane voltage is missing from the present MD analysis of electroporation.

Pore formation is expected to become more deterministic for pulses significantly shorter than 1 ns or for pores created in spatially restricted regions of a cell membrane, e.g., small phospholipid regions within the mitochondrial inner membrane [98]. However, conditions favoring deterministic pore creation have been uncommon to date. With this in mind, electroporation-related phenomena are expected to exhibit significant cell–cell variability. This in turn supports the idea that quantitative molecular transport determinations at both the population level and the individual cell level are important to understanding the magnitude and mechanisms of molecular transport.

Of the many possible molecular transport measurement methods, optical methods have the best combination of sensitivity, specificity, and speed. Both image analysis (quantitative microscopy) and flow cytometry make quantitative measurements at the individual cell level. This allows measurement of the number of molecules present by light absorbance (least sensitive) and by fluorescence or luminescence (most sensitive). With corrections for background and non-specific binding to external cell surfaces, measurements can determine the number of molecules of type "s" taken up (or released), n_s, from individual cells. Image analysis and flow cytometry are generally complimentary. Flow cytometry makes spatially-unresolved individual cell measurements at rates of 10^2 to 10^4 cell/s, and therefore routinely analyzes large cell populations. Image analysis excels at measurements with spatial resolution (of order 1 μm) but with increasingly improved temporal resolution that is best (fastest) for gated (strobed) image acquisition and generally less (slower) for rastered image acquisition. Both flow cytometric and image-based molecular uptake measurements are at their best when using molecular probes such as propidium iodide (PI) that exhibits major changes in fluorescence upon binding. Such test molecules are attractive because wash steps are avoided and measurement speed is governed by molecular (ionic) transport and binding kinetics combined with optical signal-to-noise ratio considerations. If confocal or two-photon microscopy were used with sufficiently large cells that a portion of the intracellular region can be volumetrically resolved, then molecular transport into an intracellular volume could be quantitatively monitored during the uptake process.

Image analysis also provides information about localization of molecular transport [99–102]. This has allowed observation of the degree of cellular uptake asymmetry. In the case of suspended spherical cells the instantaneous transmembrane voltage, $U_m(t)$, at the two poles (consider $\theta = 0, \pi$ in Equation 9.3 and Equation 9.5 for pre-electroporated cells). These equations suggest that U_m will be increased at one pole and diminished at the other. A mammalian cell's resting potential (transmembrane voltage) is $U_{m,rest} \approx 60$ to 90 mV (cell interior negative with respect to exterior). Qualitatively, it could be expected that $U_{m,rest}$ adds at one pole and subtracts at the other. Quantitatively, this has been addressed by describing U_m during electroporation and finding that the fractional change is small except within the equatorial region [67]. Electroporation asymmetry can also be hypothesized as arising from a metabolically driven ionic current that passes through membrane leaks (the passive cell membrane resistance) to establish $U_{m,rest}$ and to also estimate changes due to electroporation shunting current through electropores [69]. Consideration of U_m alone is not expected to provide a full explanation. Whereas some experiments measure U_m [49,50], other experiments reporting asymmetry involve molecular transport observations [99–102] and in these cases net transport should be understood.

An example of quantitative, individual cell level molecular transport determinations involved yeast cells in a medium containing 80 μM PI (668 Da; charge of $z = +2$). Without electroporation there is negligible uptake. This is expected because of the large Born energy barrier (Equation 9.1) for small charged molecules, and doubly charged PI is a well established "membrane exclusion dye" that is routinely used to test for cell membrane integrity [103]. However a single square electric field pulse within the electrolyte ($E_e = 5 \times 10^3$ V/cm for a pulse with duration $t_{pulse} = 50$ μs) results in a significant average uptake, viz. $\bar{n}_{PI} \approx 10^8$ molecules per cell. Flow cytometric fluorescence measurement is aided by PI binding to doubly stranded nucleic acids within the cell, which significantly increases PI fluorescence [92]. But for typical experimental conditions PI fluorescence is weaker than calcein. Accordingly, in a subsequent example, red blood cell ghosts were suspended in an aqueous electrolyte with the green fluorescent polar tracer molecule, calcein (623 Da; net charge of $z = -4$) at a concentration of 1 mM. Calcein-exposed cells emit only very slight green fluorescence if unpulsed. This control case fluorescence is attributed to weak surface binding and autofluorescence. However, if a single "exponential pulse" $E_e(t) = E_{e,0} \exp -t/\tau_{pulse}$ with $\tau_{pulse} \approx 3$ ms and $E_{e,0} \geq 3 \times 10^3$ V/cm^{-1} then a plateau in molecular uptake $\bar{n}_{Cal} \approx 7 \times 10^4$ molecules per ghost is observed. For the same conditions, but with fluorescence-labeled macromolecules 10 μM lactalbumin (14.5 kDa; $z \approx -15$) and 10 μM bovine serum albumin (BSA; 68 kDa; $z \approx -25$) an uptake plateau is also found, $\approx 7 \times 10^4$ molecules per ghost [104]. Surprisingly, for these single exponential pulse experiments, the plateau uptake per ghost is well below the estimated equilibrium value obtained by multiplying the supplied extracellular concentration and the ghost volume, $\bar{n}_{s,equi} = c_{ext} V_{ghost}$. Extension of these experiments to multiple pulses for BSA uptake shows that the plateau value for $\bar{n}_{BSA}$ increases with pulse number, but remains well below $\bar{n}_{BSA,equi}$. Finally, observation of an uptake plateau well below equilibrium holds for yeast experiments in which the effective partition coefficient between the extracellular medium and cell interior for calcein was estimated experimentally [105] and for other experiments in which calcein uptake and clonal growth were both measured at the individual cell level [95].

The variation in individual cell uptake can also be determined and is significant. Yeast uptake of PI by a 50 μs square pulse and erythrocyte ghost uptake of calcein, fluorescence-labeled lactalbumin, and fluorescence-labeled BSA due to single ~ 1 ms exponential pulses show broad population distributions in the amount of uptake per cell without distinct subpopulations. In contrast, molecular uptake of calcein by a suspension of yeast

(*Saccharomyces cerevisiae*) subjected to a single $\sim$ 1 ms exponential pulse exhibits several subpopulations that change markedly as the pulse amplitude is increased from an approximate threshold $E_{e,0} \approx 1 \times 10^3$ V/cm and above threshold to $\sim 7 \times 10^3$ V/cm [105]. In spite of this significant heterogeneity the average uptake per cell, $\bar{n}_{Cal}$ exhibits plateaus, with separate plateaus found for both a "major uptake subpopulation" and a "minor uptake subpopulation." Surprisingly, for these single exponential pulse experiments with yeast (elongated, budding), the plateau uptake per cell is well below (more than the order of magnitude smaller) than $\bar{n}_{equi}$, even after the intra- or extracellular partition coefficient is included. In summary, several single pulse experiments have found uptake plateaus corresponding to nonequilibrium amounts. These results suggest that the plateau may be a consequence of molecular transport driven by local electric fields, e.g., electrophoresis. An approximate plateau in U_m is predicted by a transient aqueous pore model for an exponential pulse that causes electroporation [47]. If molecular transport is dominated by electrophoresis (or electro-osmosis) through rapidly changing pores during a pulse, then nonequilibrium net transport may therefore result. If instead molecular transport is dominated by diffusion during and after a pulse [106], then net transport can approach equilibrium values or can be less, depending on the number of pores and their lifetime. The total intra- and extracellular volumes of the system, which is usually quite different for *in vitro* and *in vivo* conditions, (Equation 9.7) is also important [107].

Presently quantitative molecular transport measurements at the individual cell level have been made on (i) red blood cell ghosts, for which the uptake per ghost was determined for calcein, lactalbumin, and BSA due to a single exponential pulse with a $\sim$ 1 ms time constant and (ii) intact yeast (*S. cerevisiae) cells*, for which the uptake per cell was determined for PI because of a single 50 μs square pulse, and calcein, lactalbumin, and BSA because of an exponential pulse with a 1 ms time constant. It is therefore not yet established whether nonequilibrium uptake is a general feature, particularly for small charged molecules. Similarly, it is not yet known whether multiple pulses, pulses of other shapes, or certain features of the transported molecules are particularly important. In short, the mechanism of molecular transport is not yet understood in detail. For small molecules, the creation of a pore population probably occurs independently of molecular transport, but for larger molecules, particularly highly charged macromolecules such as DNA, electrically created pores may be significantly enlarged by interaction with approaching molecules driven by the local field (Figure 9.4).

In addition to electrophoresis through pores involving local, inhomogeneous electric fields, and time-dependent diffusion through pores, pulsed cells may be stimulated by presently unknown mechanisms to take up molecules by endocytosis [108]. External additional driving forces such as a pressure difference due to simultaneous or subsequent centrifugation may act to both enlarge primary (electric field-created) pores and to drive flow through the enlarged pores. Quantitative uptake measurements were also reported for mammalian cell experiments with submicrosecond (10 to 100 ns) megavolt-per meter scale (150 kV/cm) pulses in which extracellular media with smaller conductivity were used [56]. For these conditions with low extracellular conductivity media significant cell deformation occurs. Cell deformation increases the challenge of understanding molecular transport through pores. Overall, the strongest evidence to date supports involvement of electrical drift (electrophoresis) for moving small charged molecules through pores, with diffusion contributing to both neutral and charged molecules. The interpretation of experiments that introduce highly charged DNA is consistent with local electrophoresis.

Moreover, a transient aqueous pore theoretical model involving only electrophoretic transport through a dynamic pore population created by a single exponential pulse, predicts some of the plateau behavior found for red blood cell ghosts and yeast cells.

A tentative interpretation is that electroporation involves a rapid interaction between the membrane conductance due to pores, and the transmembrane voltage, $U_m(t)$. A slight increase in U_m results in rapid creation of many pores, which by a voltage divider effect tends to reduce U_m. This tendency to a nearly constant U_m seems to prevail during much, but not all, of the pulse duration, and this is qualitatively consistent with a local transmembrane electrophoretic transport that is independent of the pulse magnitude once a significant number of pores exist. Given the central importance of molecular transport to applications, considerably more development of basic insight into molecular transport mechanism is needed. Both the average and the variation in molecular uptake or release per cell is of basic importance to cellular electroporation, yet has even now received relatively little attention.

9.7 Electrotransfection

The delivery of DNA into cells *in vitro* is the predominant application of electroporation [80–83,109,110] and is extremely important. Molecular transport of DNA into cells by electric field pulsing was first reported in 1976 [40], but only as a demonstration of principle, since it did not alter cell function. Indeed, the first electrically induced gene transfer *in vitro* with subsequent expression was demonstrated in 1982 [41], and this stimulated a large number of papers demonstrating the applicability of electrotransfection to mammalian, plant, bacterial, and yeast protoplasts and cells. Electrotransfection has significant advantages: it is physical and therefore universal, effective, and often is the only method that gives positive results. For this reason, electrotransfection is actively pursued for human gene therapy based on two general approaches: (i) cells treated *in vitro* followed by reimplantation into the body [83,111–113] and (ii) cells transfected either in isolated tissues (e.g., plant tissue) or mammalian tissue *in vivo* [83,110,114–117]. For maximum utility, the efficiency of introduction of active DNA and cell survival must be simultaneously optimized, but understanding mechanisms involved in electroporation and downstream events remains inadequate.

Many investigators have reported a correlation between electroporation and electrotransfection. This could be interpreted as support for the role of electropores as the actual pathways for DNA transport. However, primary pores caused solely by an increased U_m by a short pulse are too small ($r_{p,min} \approx 10\,nm$ [62]) to accommodate DNA. Passive entry and diffusion through a moderately large pore ($r_p \approx 10\,nm$) may allow DNA entry [118]. The frequency-of-occurrence of large pores ($r_p > 10\,nm$) [47,119,120] and the possibility that entering DNA drives enlargement of a pore should also be considered. In any case, DNA uptake is usually not due to passive diffusion. Instead, the following driving forces were considered: (i) flow of water occurring due to the colloid–osmotic cell swelling following electropermeabilization [121], (ii) electro-osmotic flow through a pore with a charged interior surface [122], (iii) electrophoretic transfer of DNA through pores [123,124], and (iv) electrophoretically driven adsorption of DNA onto the membrane, followed by endocytosis which engulfs the DNA [108,125]. Other, relatively recent experiments support the mechanism of electrophoresis by local electric fields focused (by the spreading resistance) through pores, such that DNA is carried into and through a pore that is actively enlarged by the electrically driven DNA [123,124].

The introduction of DNA by electroporation also increases the uptake of dextrans > 20 kDa supplied after pulsing. This occurs even if the dextran is added a few minutes after pulsing, implying a pore lifetime of 10 to 100 s. Moreover, the larger the DNA fragment, the greater the increase in permeability to the dextrans. The interpretation is that the use of a two-pulse technique (a large first pulse to create pores, a second long but much smaller to cause electrophoresis) provides separation of the primary electroporation process and subsequent field-driven transport through long lifetime pores. A DNA molecule may also prevent a DNA-occupied pore from closing, a foot-in-the-door mechanism that may lead to prolonged pore lifetimes. A fluctuating pore temporarily occupied by a macromolecule can allow small molecules to continue to be transported [78]. The introduction of highly charged heparin into the skin's stratum corneum by pulses believed to cause electroporation results in co-transport of small molecules that are consistent with this possibility [126].

An associated theory predicts that DNA in the electroporating electric field exerts high pressure on the membrane, which can then induce structural rearrangements of the membrane [127]. In this view, electrophoresis is the transport mechanism for DNA uptake, as it brings DNA to the cell surface and then drags DNA through the membrane, initially entering preexisting pores, which are then enlarged and deformed in course of translocation. This is an example of a process that leads to secondary pores, which are significantly different than the primary pores that arise solely by the interaction of the electric field with the membrane.

Molecular electroinsertion is distinct from transport of ions or molecules through a membrane. Experiments have shown that certain macromolecules can be inserted stably into cell membranes by pulses associated with electroporation [128–132]. A qualitative picture is that transient openings in the cell membrane provide an opportunity for such macromolecules to partially enter a pore and then become trapped as the pore shrinks. Perhaps one end of a membrane-spanning molecule is loosely bound to part of the membrane, such that the molecule can rotate into a pore, only to become trapped as the pore shrinks and eventually vanishes. Electroinsertion is such a fascinating event that it deserves conceptual emphasis, even though relatively few studies have been made since its discovery.

Electrofusion provides somatic hybridization of cells [111,112,133–139]. This has led to demonstrations of cell-tissue fusion [140,141], hybridomas for antibody production [81,142], individual cell and vesicle manipulation [143,144], tumor vaccine production [145,146], and other applications [80,82,147]. But the mechanism of cell electrofusion is still not fully understood. The understanding we do have comes from different types of studies. Experiments using model systems consisting of two planar lipid bilayers have provided significant insight [59]. After pressure application, the bilayers reach a state of close contact. Close proximity is usually necessary for cell electrofusion. The distance between bilayers is determined by the equilibrium of all acting forces: molecular, electrostatic, hydrostatic pressure, and hydrational. After a lag time, one or more bridges (stalks) form spontaneously between two adjacent monolayers. These bridges then expand, leading to a so-called trilaminar structure, which contains a single bilayer in contact with two bilayers along a perimeter. A trilaminar structure is rather stable, but after a voltage pulse is applied, electroporation occurs, followed by an irreversible rupture of the contact bilayer with its transition to so-called membrane tube. This stage of the process results in fusion because now there are joined lipid phase and water compartments. It is important to note that two different structural rearrangements take place: (i) stalk (or bridge) formation between adjacent monolayers and (ii) pore formation in contact bilayer with measurable time lag. Note that the first stage is essentially spontaneous, and that the electric field is involved only during the second stage of the fusion process.

Additional insight follows from noting that the probability of stalk formation depends on its energy, which in turn is determined by the favorable molecular geometry of phospholipid molecules. From the viewpoint of continuous media mechanics, this implies a dependence on the spontaneous curvature of monolayers. This theory was compared with the results of experiments in which lag times were measured for membranes with different spontaneous curvatures. Good agreement was obtained, supporting the view that the stalk mechanism is involved in lipid bilayers fusion.

Cell fusion is much more complicated. Attempts to find evidence for trilaminar structures in electrically pulsed cell suspensions have not been successful. For this reason, another fusion mechanism and the interaction of coaxial electropores at adjacent membranes have been considered [148]. In this case it was shown that attractive forces due to the Maxwell tension between two parallel bilayers with coaxial pores are large enough to almost bend the membranes and to bring the pores together until a fusion of pore edges is achieved. These two mechanisms differ in their sequence of events: (i) monolayer fusion and followed by electroporation (the first case; see above) and (ii) electroporation followed by fusion of pore edges (the second case). Significantly, it has been experimentally demonstrated [149,150] that fusion can also proceed by using a protocol in which pulsing precedes cell contact. This supports the mechanism of fusion through electropores. Additional understanding of electrofusion should therefore depend on further developments in the understanding of electroporation.

Extension of cell–cell electrofusion *in vitro* to cell–tissue *in vivo* has been briefly described [140,141] and offers the prospect of many applications in research and medicine. In this case critical issues involve making contact between the introduced cells or vesicles to the target tissue and then causing a high probability of fusion while retaining the viability of both the introduced cells and the target tissue.

9.8 Membrane Recovery

Membrane recovery after pulsing is fundamentally important to understanding whether an electroporation protocol is reversible at the membrane level and at the cellular level. Terminology includes "resealing" and "pore lifetime." According to the transient aqueous pore hypothesis, recovery is expected to involve shrinkage and disappearance of pores, with some models further assuming that pores shrink to a minimum size before stochastically vanishing. However, quantitative data that constrain membrane recovery mechanisms remain relatively scarce and are often indirect. For example, molecular uptake observations are often used, which depend on more than pore shrinkage and destruction.

The scientific literature contains reports with widely varied membrane recovery times, from nanoseconds to minutes at room temperature to 37°C and even longer at low temperatures. Here we cite several examples, beginning with several papers with data relevant to membrane barrier recovery. These include molecular and ionic uptake on human erythrocytes ($\tau_p \approx 1$ to 10 h and strongly dependent on temperature, with much longer resealing at 3°C than at 37°C) [36], on isolated rat skeletal muscle cells ($\tau_p \approx 500$ s) [151], PI uptake by electropermeabilized myeloma cells ($\tau_p \approx 100$ s) [152], dye uptake ($\tau_p \approx 120$ s) [153], Ca^{2+} influx into vesicles ($\tau_p \approx 165$ ms) [101], and into adherent cultured cells (τ_p timescale of milliseconds to seconds) [154]. As these examples illustrate, there is a wide range of membrane recovery timescales for molecular uptake, but with artifical membranes often faster than cellular membranes. Electrical measurements can also be used and have the advantage that the transported species (small ions at a high concentration: Na^+, K^+, and Cl^-) are preexisting throughout cells, so they do not require delivery by mixing and diffusion to reach cell membranes. Examples are electrical measurements on isolated frog muscle cell membranes [155] ($\tau_p \approx 500\,s$), on cell suspensions and cell pellets, and on artificial planar BLM. Molecular dynamics simulations of electroporation are relatively recent and suggest pore lifetimes of a only few nanoseconds [75], which is difficult to reconcile with experimental values based on ionic and molecular transport.

Both reversible and irreversible electroporation lead rapidly (typically 1 to 100 μs) to a high conductance state, which discharges the membrane after pulsing or provides a small membrane resistance that participates in a voltage divider effect to reduce the transmembrane voltage, even during the pulse [47]. By definition irreversible behavior has no physical membrane recovery, and recovery by biological repair of significant membrane openings due to large magnitude or longlasting pulses has not yet been investigated. In considering what is known about membrane recovery, it was noted above that both electrical and molecular assessments of membrane barrier function can be made.

Early experiments used electrical measurements and planar membranes to show that after a decrease in the voltage, there is a very rapid (< 5 μs) decrease in the membrane conductance [51,156]. Initially this result was interpreted as reflecting rapid resealing of pores [156]. It was later shown, however, that a rapid decrease in conductance is associated with the nonlinear behavior of a pore's current–voltage curve [45,157,158]. The number and the mean radius of the pores were unchanged on a timescale of microseconds, but in the timescale of 1 to 10 ms the decrease in conductance is governed by a change in the size of the pores. For artificial planar BLMs pores disappear from the membrane on a timescale of milliseconds. For cell membranes the timescale for the disappearance of pores is much longer, ranging from seconds to hours [45].

Electrical assessment of recovery has the advantage of speed and convenience, but testing for restoration of membrane barrier function by probing with molecules of different sizes and charges may fundamentally be more revealing as to the extent and kinetics of recovery. Indeed, several studies have reported the results of delayed addition experiments, with probe molecules supplied to pulsed cells at different times after pulsing. As expected, the qualitative observation from experiments using total population measurements is that significantly less uptake occurs if molecules are added to the extracellular medium after the pulse. Typically the added molecules are small [36,159], but sometimes macromolecules, including DNA, were used [124]. An important finding is that if analytical techniques with individual cell level capability (microscopy, flow cytometry, image analysis) are used, then subpopulations with uptake of delayed addition molecules are observed [105,142,160,161]. Factors affecting membrane recovery kinetics have not yet been thoroughly investigated. However, some experiments show that reduced temperatures lead to much longer membrane recovery times [36,162]. This is not surprising, as the diffusive pore shrinking that is expected as the membrane discharges are expected to have this qualitative temperature dependence. Moreover, most physical and chemical processes of multimolecular systems are characterized by one or more rates that have Boltzmann factors, with their strong, nonlinear reduction in rate at lower temperature.

Pulses that lead to pore formation and evolution are expected to involve a dynamic pore population with time-dependent pore size distributions [46,47]. For this reason the change with time of the membrane's ability to exclude molecules of different sizes and charge should provide additional insight into the membrane recovery process. Indeed, experiments have shown that if molecules are first presented to cells at increasing time delays after pulsing, then a decreasing fraction of cells have a persistent ability to take up molecules [160]. This suggests that a subpopulation of cells has metastable pores with a distribution of pore lifetimes. Enhanced recovery by adding certain surfactants has been demonstrated [163,164]. The discovery of enhancement both challenges our understanding of what membrane recovery is and offers promise as a therapy for electrical injury that involves nonthermal membrane damage by otherwise irreversible electroporation.

9.9 Cell Stress and Survival

Most *in vitro* applications of electroporation benefit from some understanding of cell survival, but for *in vivo* applications it is becoming essential. *In vitro* electroporation often focuses on delivery of molecules into the cytosol, but for some research studies [165–167] related to nonthermal electrical injury [164,168,169] and harvesting of cellular products [170] release of molecules is sought.

A general observation over many experiments is that as the field magnitude is increased for a given pulse shape, phenomena such as molecular uptake become significant at a "fuzzy threshold" (dependent on pulse shape, including pulse duration), and then increase. However, at about the same "fuzzy threshold," cell viability begins to decrease, with almost all cells getting killed as the electric field is increased. Although the mechanism of cell killing by electroporation is not fully established, a plausible hypothesis is that cell survival or death is mainly due to the degree of molecular exchange between cells and their environment and the resulting cell stress due to chemical imbalances [160,171,172]. This is consistent with the observation that cell killing can occur without significant heating [22–24,164,173] and that a tremendous increase in membrane permeability and associated molecular transport occurs.

A transient pore population and also possibly a small number of metastable pores are expected, and all may contribute to stress-inducing transport. Thus, reversible membrane electroporation (complete resealing or pore disappearance) may still lead to net transport (uptake or release). This means that relatively nonspecific molecular exchange may occur between the intra- and extracellular volumes and can lead to stress-inducing chemical imbalances. Depending on the ratio of intra- and extracellular volumes, the composition of the extracellular medium and the cell type, a cell may not recover from the associated biochemical imbalance (stress) and will therefore die. Both reversible and irreversible electroporations result in transient openings (perforations) of the membrane, which are often large enough that the pores are not selective. In this case, molecular transport is expected to be nonspecific. It is plausible that a portion of the cell membrane behaves much like a small planar membrane and undergoes rupture (irreversible electroporation). Just as in the case of reversible electroporation, significant molecular transport between the intra- and extracellular volumes may lead to a significant chemical imbalance. If this imbalance is too large, recovery may not occur, resulting in cell death. It has been hypothesized [172] that the volumetric ratio

$$F_{\mathrm{vol}} = \frac{V_{\mathrm{extracellular}}}{V_{\mathrm{intracellular}}} \tag{9.7}$$

should correlate with cell survival or death. In this equation $V_{\mathrm{extracellular}}$ is the volume of liquid medium outside the cell and $V_{\mathrm{intracellular}}$ is the volume of liquid in the cell interior. The idea is that, for a given cell type and extracellular medium composition, $F_{\mathrm{vol}} \gg 1$ (typical of *in vitro* conditions such as cell suspensions and anchorage-dependent cell culture) should tend to favor cell death. The other extreme is typical of *in vivo* tissue conditions ($F_{\mathrm{vol}} \approx 0.15$, a higher interstitial fluid fraction) and this should favor cell survival. The dilution of released molecules is governed by F_{vol} and partitioning between the extra- and intracellular environments [105]. For example, if the partition coefficient is one, the maximum *in vivo* initial dilution factor in solid tissue should be $1/0.15 \approx 7$, whereas the expected dilution is several orders of magnitude larger for typical *in vitro* conditions. This suggests that for the same degree of electroporation, significantly less damage may occur in tissue than in body fluids or under most *in vitro* conditions. With time, however, released molecules can diffuse within the interstitial space, be cleared by blood perfusion, and then distributed by the systemic circulation. The ultimate fate of released molecules is governed by excretion, metabolism, and storage, which is traditionally described by pharmacokinetics.

Cell viability of culturable cells can be stringently tested by clonal growth assays, whereas diffentiated cells are often assessed by assaying particular functions, such as membrane exclusion dyes. However, caution should be exercised in assessing cell viability after electroporation protocols. After all, there is little *a priori* justification for using membrane-exclusion dyes (e.g., trypan blue [960 D, $z = -4$], PI [668 D, $z = +2$]), because electroporation causes the membrane permeability to significantly increase by unknown amounts for unknown times. Membrane-exclusion dyes are generally larger than some small ions and molecules (e.g., Ca^{2+}). Uptake or release of such ions and molecules may occur even if membrane recovery has progressed to the point that membrane-exclusion dyes are excluded by the membrane. Without validation it cannot be assured that the cell membrane has recovered sufficiently that fatal chemical imbalances are avoided. Instead, relevant functional tests should be considered. If electroporation is to be optimally used, a significantly better understanding of the mechanism of cell death will be needed.

9.10 Tissue Electroporation and *In Vivo* Delivery

A purposeful electroporation of tissue *in vivo* and *in vitro* has been motivated by therapeutic interventions such as tumor treatment by delivery of anticancer drugs [86,174], gene therapy by delivery of DNA, and other genetic material [83,116,117,175], and delivery of various sized molecules into and across the skin [174,176–178]. Undesirable consequences of tissue electroporation have also been identified. A major insight is that tissue electroporation is the source of a nonthermal contribution to electrical injury [164,168,169,173]. Tissue electroporation has also been identified as a side effect to defibrillation interventions [61,179–181]. Finally, tissue electroporation may be relevant to neuromuscular incapacitation (stunning) pulses [182].

Describing the response of many closely spaced cells to an applied pulse is central to understanding electroporation of solid tissue. A fundamental attribute of cells is that applied electric fields result in response fields that are largest in lipid-based membranes, which have an effective high resistivity even in the presence of many ion channels. This is the basis of the amplification associated with Equation 9.3. Although well established at the cellular level [5,6,43], the corresponding concentration of electric fields in tissues has received less attention [7]. Here we emphasize that this voltage-concentrating effect means that electroporation of other lipid-based barriers should occur preferentially. For preferential electroporation, two features should be sought: (i) tissue barriers comprised mainly of lipids and (ii) mechanical deformability (compliance) of membranes comprise of the particular lipids so that the electrostatically favored entry of water into a deformable phospholipid-based membrane results in the creation of aqueous pathways.

In many solid tissues there are significant spaces between most cells, such that for small electric fields the electrical currents at low frequencies flow mainly between the cells. In this case, tissue electroporation consists of electroporation of individual cells whose behavior is expected in many respects to be similar to the electroporation of isolated cells, but with two major differences: (i) the local extracellular electric field depends in a complicated way on many other neighboring cells [7,183] and (ii) the ratio of the extra- to intracellular volumes (Equation 9.7) is usually small, just the opposite of most *in vitro* electroporation conditions. This means that if chemical exchange between the intra- and extracellular volumes is the main cause of cell stress and therefore cell death, tissue electroporation may be intrinsically less damaging than most *in vitro* electroporation conditions.

Tumor tissue is an important example of tissue for which many cells have intercellular aqueous pathways. Even without electroporation there is a significant physiological resistance to entry of anticancer drugs because of limited blood perfusion, elevated interstitial pressure, and relatively large distances to blood vessels [184]. Not only does electroporation greatly reduce the main barrier (that of a cell membrane itself), tissue electroporation should also reduce these physiological resistances. For typical cancer chemotherapy conditions blood perfusion delivers the anticancer drug to sites within a tumor and the drug then diffuses across the avascular regions. Local tissue electroporation should create aqueous pathways that assist drug movement and that may also relieve pressure, but the fourth power dependence of volumetric flow on pathway size implies that significant water flow may be more difficult than diffusion and drift of small drugs. But if the primary pores created electrically subsequently expand, then flow may be successful in reducing the pressure, and perfusion should increase. Diffusional barriers should also be reduced because of the new aqueous pathways. Striking results have, in fact, been reported in experimental tumor systems, mainly using the drug bleomycin, which is a potent drug that ordinarily does not cross the membrane readily [109,185–187]. Importantly, clinical use has been vigorously pursued [84,86,188–190]. Initial results were often obtained by treating surface tumors (nodules) [84]. More generalized use requires electrode access to provide localized electric field pulses to the targeted tumors [191–193]. As a partial guide, the use of electrical impedance measurements or even electrical impedance imaging can be considered [194,195].

Interestingly, there is a significant economic barrier to a general approach to local tumor treatment and ECT-specialized drugs (e.g., molecules with significant charge) that do not passively cross the cell membrane. Almost all approved anticancer drugs (the striking exception is bleomycin) readily cross the cell membrane, actively or passively. Accordingly, such molecules do not enjoy a tremendous increase in delivery by electroporation. Thus, even though it should be fundamentally attractive to use tissue electroporation with cell membrane-impermeant drugs, e.g., an established drug with several charge groups added that are quickly cleaved by intracellular enzymes to greatly reduce side effects, the modified drug will be regarded as new; and the expensive and lengthy approval process for new drugs is a significant disincentive.

Tissues that are comprised of one or more monolayers of cells in which the cells are connected by tight junctions are somewhat favored to experience voltage concentration across their membranes. Such tissues form the linings of organs and other specialized structures, such as the lining of sweat gland ducts [196]. An example of such behavior is found in a didactic, multicellular model [7,183]. Qualitatively, the main barrier of each cell monolayer consists of two cell membranes in series and can therefore be regarded approximately as two planar BLMs in series. In fact, some tissues such as frog skin and toad bladder are plausible approximations to an ideal single cell monolayer tissue, in which a single cell monolayer is the main barrier. For this reason, such tissues provide a convenient experimental system that allows a convenient investigation of electrical behavior of tissue electroporation [197]. In contrast, recent experiments have focused on molecular transport into or through epithelia [198–201]. More complicated and realistic models of regions of a tissue with a layer of cells connected by tight junctions are now becoming available [7,183].

Transdermal drug delivery has long been of interest, with both chemical and physical ''enhancers'' being pursued [178]. Skin electroporation is one of the approaches based on physical intervention [174,176,177]. Mammalian skin is a complicated tissue. It has both the specialized barrier structure, the stratum corneum (SC), and appendages, sweat ducts and hair follicles, which perforate the SC [202–204]. The SC is the skin's main barrier. It is a dead tissue whose barrier properties are often described by using a ''brick wall model'' [205]. The ''bricks'' are corneocytes, which contain cross-linked keratin and increase in volume by up to a factor of 5 if the SC is hydrated, the usual condition for transdermal drug delivery. The ''mortar'' is the intervening lipid, which surrounds the corneocytes, and exists as mostly parallel bilayers made of special (nonphospholipid) lipids, with about five to six bilayers between each of the approximately 20 rows of corneocytes. The SC is mechanically flexible but tough, essentially impenetrable to most infectious microorganisms and very impermeable to water-soluble molecules, particularly charged molecules, as the lipids are present as about 100 bilayers in series. The corneocytes are residual entities, left over from cells that migrated outwards to form the SC, with the multilamellar lipid bilayer providing most of the barrier function of the skin. Clearly this very effective barrier has evolved to provide protection, but it also presents a major barrier to potential medical interventions, viz. transdermal drug delivery [178] and extraction of chemical analytes, e.g., glucose within the subcutaneous interstitial fluid, for minimally invasive sensing [206–208].

The skin's appendages (sweat gland ducts and hair follicles) are usually lined with an extension of the SC, or, deeper within the skin, by a double layer of living cells connected by tight junctions [196]. Both the appendageal barriers and the SC barrier are candidates for electroporation, but at very different transbarrier voltages, U_{barrier}. The double cell lining of sweat ducts should experience electroporation at about $U_{\mathrm{barrier}} \approx 2$ to 4 V, but the approximately 100 multilamellar bilayers of the SC need $U_{\mathrm{barrier}} \approx 50$ to 100 V for pulses with duration of 100 μs to 1 ms, i.e., about 0.5 to 1 V per lipid bilayer [209–212].

Electroporation of skin and other tissues can be studied *in vitro* using a well-established apparatus, a permeation chamber [213], which is a version of the Ussing chamber originally used for molecular flux and electrical studies of epithelial tissues [214]. Two- and four-electrode systems can be used for electrical stimulation and measurement, and in addition, molecular transport measurements can be made by supplying molecules in the donor chamber and removing aliquots at intervals from the acceptor chamber for chemical assay. Experiments of this type with human skin show that if exponential pulses with $U_{\mathrm{SC},0} \approx 50$ to 300 V and time constant $\tau_{\mathrm{pulse}} \approx 1$ ms are applied every 5 s for 1 h, then there is an enhancement by up to a factor of 10^4 in the flux of charged molecules of up to about 1 kDa [176,215]. Companion electrical impedance measurements show a rapid ($\leq$25 ms) decrease in skin resistance [216], and both molecular flux and electrical measurements show that either reversible or irreversible behavior occurs, depending on the transdermal pulse amplitude, $U_{\mathrm{SC},\,0}$. Several *in vivo* experiments show that transdermal delivery can be achieved with minimal damage [176,217–219]. This is consistent with the theoretical estimates of associated heating within the nearby viable epidermis [220].

Extension of this type of *in vitro* method to almost continuous measurement of transdermal molecular flux uses a sampling stream and quantitative, computer-controlled spectrofluorimetry to measure two fluorescent molecule concentrations at alternating intervals, so that with deconvolution, two molecular fluxes and passive electrical properties can be determined for each skin preparation [221]. A faster responding, single flux version of the same method allows the demonstration of both the large magnitude and the rapid onset and cessation of the flux of calcein (623 D, $z = -4$) and supports the view that skin electroporation can deliver small pharmaceuticals with rapid delivery onset [222]. Such methods provide a flexible approach to characterizing new aqueous pathways created by skin electroporation.

Gene therapy based on physical introduction of genetic material into cells *in vivo* is of great and growing interest and includes tissue electroporation [83,116,117,175,223]. Electroporation-based gene therapy is related to electrochemotherapy of tumors by introduction of bleomycin or other anticancer drugs, but with the significant difference that gene therapy requires delivery of large molecules (e.g., DNA) whereas traditional anticancer drugs are small. This is consistent with the use of smaller, longer electroporating pulses in gene therapy [119,120,224]. If the desired therapy does not require that all cells in a tissue be modified, local tissue electroporation can be used to treat some cells. The modified cells can, for example, secrete a therapeutic molecule. An early, partial demonstration was based on subcutaneous injection of DNA followed by surface electrodes for *in vivo* electroporation, leading to transformation of some cells [114]. Major advances have since occurred [83,116,117,175]. As with all new technologies, undesired side effects must be considered and understood along with possible uses.

Thus, the possibility of tissue damage must also be considered. Fortunately, major advances in understanding the role of electroporation in tissue damage has been achieved, through a series of studies motivated by the hypothesis that electroporation can account for a nonthermal permeabilization of cell membranes [164,168,169]. Although electrical injury to tissue has in the past been interpreted as a thermal denaturation phenomenon, more recent studies make a convincing case that electroporation can be a significant factor, particularly for the larger cells such as skeletal muscle cells and nerve cells [173]. Moreover, the demonstration that certain surfactants can significantly accelerate membrane barrier function recovery has offers a basis for therapy after electrical shock injury [163,164,225]. Not only have surfactants such as Polozamer–188 been pursued as a damage-minimizing therapeutic for electrical injury, this polymer is active against damage due to ionizing radiation that also causes cell membrane permeabilization [226] (see chapter 10 in *BMA*). Damaging consequences of tissue electroporation are also relevant to the strong electrical pulses which are widely used to stimulate fibrillating hearts [61,179–181], and this is an active research area. In this case, alteration of electrical behavior has often been emphasized.

Tissue electroporation thus has compelling potential advantages for medical interventions. A general attribute is that it achieves chemical results, transport of molecules into desired sites, without leaving another chemical residue such as the transdermal drug delivery enhancer dimethyl sulfoxide. Perhaps even more important, however, is the fact that electroporation is caused electrically, and the tissue experiencing electroporation can be rapidly assessed electrically, e.g., by impedance measurements. This means that much better control of molecular transport for drug delivery and analyte extraction may be possible. In spite of tremendous progress it is nevertheless true that a better fundamental understanding of electroporation mechanism is needed. The characteristics of such pathways need to be better established quantitatively, in terms of both molecular transport and electrical properties. Also, the potential side effects of possible tissue damage must be confronted, and the trade-off between desired and undesired consequences understood. This is a classic challenge.

Several potential applications of tissue electroporation have now been identified and are vigorously pursued [83,86,116,117,174]. Understanding tissue electroporation thus has several motivations. Unintended exposure to electric fields such as those due to lightning and contact or close approach to electrical power sources or distribution wiring can cause nonthermal tissue damage [164,168,169]. Stunning (neuromuscular incapacitation) devices may also cause tissue electroporation or membrane channel denaturation. Although some acute side effect possibilities have been considered [182,227–230], the spatially distributed fields within the body should be investigated to obtain a thorough understanding of both efficacy and potential side effects.

9.11 Electroporation of Organelles

Most models of cells focus on the outer, plasma membrane. At low frequencies and small applied field amplitudes, the response field is excluded from the cell interior, the site of organelles with their own single or double membranes. However, very-short field pulses have high-frequency content and can lead to displacement currents and associated transient intracellular electric fields [69,70,231]. A recent analysis shows that absolute rate theory can plausibly account for pore creation on a nanosecond timescale at a transmembrane voltage of $U_m \approx 1.2\,V$ [19]. This supports the use of spatially distributed cell models that are based on the asymptotic model of membrane electroporation [69,70]. Such models show that extensive poration occurs for submicrosecond, megavolt-per-meter pulses. Further, the area pore density is exceptionally large. In the plasma membrane this leads to intracellular ionic conduction currents and associated large fields within the cell [50,69,70]. According to initial cell models involving the supraelectroporation hypothesis [69,70] an extremely large pore density (pores per area) is predicted for the cell's outer plasma membrane and also internal organelle membranes. This means that pores are extremely close together, and this in turn suggests that deterministic membrane deformation should be expected [98] in order to create a large membrane conductance that limits the magnitude of U_m to $\sim 1\,V$ [50].

The supraelectroporation hypothesis predicts that for very large intracellular fields the membranes of organelles should also electroporate [70]. Indeed, a striking experimental observation shows that intracellular granules take up calcein from the cytosol [232], with organelle membrane electroporation as the likely cause. Following this report of *in situ* organelle electroporation, there have been a growing number of studies supporting the view that extremely large field pulses cause intracellular effects.

Many of these investigations find evidence for apoptosis, not the prompt necrosis that is consistent with a large increase in plasma membrane permeability by conventional electroporation [233–243]. These experiments involve electric field pulses of durations between 7 and 300 ns and with intensities from 360 to 10 kV/cm. Such pulses cause intracellular effects such as cytochrome *c* release, caspase activation, phosphatidylserine (PS) translocation, disruption of nuclear DNA, vesicle membrane electroporation, and molecular uptake into subcellular granules [232,234–242,244]. It is important to note that apoptosis and necrosis are distinguished as the two major types of cell death [245–249]. Prompt necrosis is known to be an outcome of excessive conventional electroporation when the pulses are sufficiently long, intense, or repeated enough times [98,164,169, 250-252]. Apoptosis (programmed cell death) is usually stimulated by biochemical perturbation of molecular signaling pathways [248,253-257] and not by the application of electrical fields. Indeed, apoptosis has only occasionally been reported to follow conventional electroporation [258,259]. Although significant phosphatidylserine (PS) translocation occurs in the plasma membrane (PM) of red blood cells for conventional electroporation pulses (kV/cm^{-1}) [260–262], some of these recent studies with extremely large ($\sim$ 100 kV/cm^{-1}) field pulses show extensive PS translocation [242,263]. Nevertheless some experiments report insignificant uptake of membrane integrity dyes such as PI, and report that plasma membrane electroporation is minimal or has not occurred [233,235,236,238,240–242,263,264]. This suggests that the PM permeabilization for at least some molecules is only minimally increased by these pulses. Even though the applied fields are extremely strong, their effects are nonthermal due to the limited pulse duration (from 7 to 300 ns). Often, the pulse intensity is adjusted to limit the Joule heating per pulse to $\sigma_e E_{app}{}^2 \Delta t \sim 1.7$ J/ml, where σ_e is the electrolyte conductivity, E_{app} is the applied field intensity, and Δt is the pulse duration. In this case the corresponding adiabatic (assumption of zero heat conduction; a worst case) temperature rise is only $\sim 0.4°C$ per pulse.

The use of these ultrashort (submicrosecond), extreme (10 to 300 k V/cm) pulses is relatively new, but has generated intense interest. The mechanism underlying the observed intracellular effects has not yet been established. One hypothesis is that ultrashort, extreme pulses cause irreversible effects at the membrane level, either membrane rupture or electrically based membrane protein denaturation [74,265–269]. Supraelectroporation is another hypothesis. It predicts that there is a massive creation of transient pores, which expand negligibly because of the short pulse duration, and this leads to full reversibility at the membrane level [69,70]. This is qualitatively consistent with avoiding prompt necrosis, because most ions and molecules should be retained by membranes with minimum size ($\sim$ 0.8 nm) pores. The observed irreversible effects at the cellular level then arise from ionic and molecular transport through pores, which includes phospholipid translocation and other downstream events. As noted above, a contribution to irreversibility may also arise from electro-denaturation (long-lived membrane channel conformational changes induced by the supra-physiologic transmembrane voltage; $\sim$ 0.3 V) [270–272], a magnitude that is achieved by both conventional and supraelectroporation. However, the duration of $U_m \approx 1$ V decreases by several orders of magnitude in going from conventional to supraelectroporation, and the kinetics of the membrane protein conformational changes are not yet known. In any case, the basic idea is that for the largest of these fields, a very large number of pores are created in all of a cell's membranes, such that the field penetrates into the cell interior and also the organelle membranes, with the intracellular electric field nearly equal to the applied electric field, even after displacement currents have decayed [69]. That is, poration is so extensive that ionic current flows through the cell's membrane, in marked contrast to low-field responses in which current flows predominantly around the cell [69,70]. This opens the conceptual possibility of distinct effects due to electroporation of various types of organelles. For example, if pore lifetimes are long (as often inferred from conventional electroporation experiments), then it may be possible to gate open the mitochondrial permeability transition pore (MPTP) [98], which appears to require depolarizing the inner mitochondrial membrane (IMM) for many seconds [273]. However, other mechanism hypotheses are proposed. At the time of this writing, the cause of intracellular effects by extreme field pulses has not yet been established, but is likely to involve electroporation.

Acknowledgment

This study is supported partially by NIH grant RO1-GM63857 and an AFOSR/DOD MURI grant on "Subcellular Responses to Narrowband and Wideband Radio Frequency Radiation," administered through Old Dominion University.

References

1. J. Darnell, H. Locish, and D. Baltimore. *Molecular Cell Biology*. Scientific American Books, New York, 1986.
2. V.A. Parsegian. Energy of an ion crossing a low dielectric membrane: Solutions to four relevant electrostatic problems. *Nature*, 221:844–846, 1969.
3. V.A. Parsegian. Ion-membrane interactions as structural forces. *Ann. NY Acad. Sci.*, 264:161–174, 1975.
4. K.R. Foster and H.P Schwan. Dielectric properties of tissues. In *Handbook of Biological Effects of Electromagnetic Fields*, C. Polk and E. Postow, Eds., 2nd ed., CRC Press, Boca Raton, 1996, pp. 25–102.
5. J. Gimsa and D. Wachner. A polarization model overcoming the geometric restrictions of the Laplace solution for spheroidal cells: Obtaining new equations for field-induced forces and transmembrane potential. *Biophys. J.*, 77:1316–1326, 1999.
6. T. Kotnik and D. Miklavčič. Second-order model of membrane electric field induced by alternating external electric fields. *IEEE Trans. Biomed. Eng.*, 47:1074–1081, 2000.
7. T.R. Gowrishankar and J.C. Weaver. An approach to electrical modeling of single and multiple cells. *Proc. Natl. Acad. Sci. USA*, 100:3203–3208, 2003.
8. M. Zahn. *Electromagnetic Field Theory: A Problems Solving Approach*. Wiley & Sons, New York, 1979.
9. I.G. Abidor, V.B. Arakelyan, L.V. Chernomordik, Yu. A. Chizmadzhev, V. F. Pastushenko, and M. R. Tarasevich. Electric breakdown of bilayer membranes: I. The main experimental facts and their qualitative discussion. *Bioelectrochem. Bioenerg.*, 6:37–52, 1979.
10. V.F. Pastushenko, Yu. A. Chizmadzhev, and V.B. Arakelyan. Electric breakdown of bilayer membranes: II. Calculation of the membrane lifetime in the steady-state diffusion approximation. *Bioelectrochem. Bioenerg.*, 6:53–62, 1979.
11. Y.A. Chizmadzhev, V.B. Arakelyan, and V.F. Pastushenko. Electric breakdown of bilayer membranes: III. Analysis of possible mechanisms of defect origin. *Bioelectrochem. Bioenerg.*, 6:63–70, 1979.
12. V.F. Pastushenko, Y.A. Chizmadzhev, and V.B. Arakelyan. Electric breakdown of bilayer membranes: IV. Consideration of the kinetic stage in the case of the single-defect membrane. *Bioelectrochem. Bioenerg.*, 6:71–79, 1979.
13. V.B. Arakelyan, Y.A. Chizmadzhev, and V.F. Pastushenko. Electric breakdown of bilayer membranes: V. consideration of the kinetic stage in the case of the membrane containing an arbitrary number of defects. *Bioelectrochem. Bioenerg.*, 6:81–87, 1979.
14. V.F. Pastushenko, V.B. Arakelyan, and Y.A. Chizmadzhev. Electric breakdown of bilayer membranes: VI. A stochastic theory taking into account the processes of defect formation and death: Membrane lifetime distribution function. *Bioelectrochem. Bioenerg.*, 6:89–95, 1979.
15. V.F. Pastushenko, V.B. Arakelyan, and Yu. A. Chizmadzhev. Electric breakdown of bilayer membranes: VII. A stochastic theory taking into account the processes of defect formation and death: Statistical properties. *Bioelectrochem. Bioenerg.*, 6:97–104, 1979.
16. J.C. Weaver and R.A. Mintzer. Decreased bilayer stability due to transmembrane potentials. *Phys. Lett.*, 86A:57–59, 1981.
17. I.P. Sugar. The effects of external fields on the structure of lipid bilayers. *J. Physiol. Paris*, 77:1035–1042, 1981.
18. J.C. Weaver and Y.A. Chizmadzhev. Theory of electroporation: A review. *Bioelectrochem. Bioenerg.*, 41:135–160, 1996.
19. Z. Vasilkoski, A.T Esser, T.R. Gowrishankar, D.A. Stewart, and J.C. Weaver. Membrane electroporation: The absolute rate equation and nanosecond timescale pore creation.
20. R. Stämpfli and M. Willi. Membrane potential of a Ranvier node measured after electrical destruction of its membrane. *Experientia*, 8:297–298, 1957.
21. R. Stämpfli. Reversible electrical breakdown of the excitable membrane of a Ranvier node. *Ann. Acad. Brasil. Ciens.*, 30:57–63, 1958.
22. A.J.H. Sale and A. Hamilton. Effects of high electric fields on microoranisms: I. killing of bacteria and yeasts. *Biochem. Biophys. Acta*, 148:781–788, 1967.

23. W.A. Hamilton and A.J.H. Sale. Effects of high electric fields on microorganisms: II. Killing of bacteria and yeasts. *Biochim. Biophys. Acta*, 148:7789–800, 1967.
24. A.J.H. Sale and W.A. Hamilton. Effects of high electric fields on microorganisms: III. Lysis of erythrocytes and protoplasts. *Biochim. Biophys. Acta*, 163:37–43, 1968.
25. E.Neumann and K. Rosenheck. Permeability changes induced by electric impulses in vesicular membranes. *J. Membr. Biol.*, 10:279–290, 1972.
26. U. Zimmermann, J. Schultz, and G. Pilwat. Transcellular ion flow in *Escherichia coli* and electrical sizing of bacterias. *Biophys. J.*, 13:1005–1013, 1973.
27. U. Zimmermann, G. Pilwat, and F. Riemann. Dielectric breakdown of cell membranes. *Biophys. J.*, 14:881–899, 1974.
28. H.G.L. Coster and U. Zimmermann. Dielectric breakdown in the membranes of *Valonia utricularis*: The role of energy dissipation. *Biochim. Biophys. Acta*, 382:410–418, 1975.
29. H.T. Tien. *Bilayer Lipid Membranes (BLM): Theory and Practice*. Marcel Dekker, New York, 1974.
30. H.T. Tien and A. Ottova. The bilayer lipid membrane (BLM) under electric fields. *IEEE Trans. Dielect. Elect. Ins.*, 10:717–727, 2003.
31. R. Benz, F. Beckers, and U. Zimmermann. Reversible electrical breakdown of lipid bilayer membranes: A charge-pulse relaxation study. *J. Membr. Biol.*, 48:181–204, 1979.
32. R. Benz and U. Zimmermann. Relaxation studies on cell membranes and lipid bilayers in the high electric field range. *Bioelectrochem. Bioenerg.*, 7:723–739, 1980.
33. R. Benz and F. Conti. Reversible electrical breakdown of squid giant axon membrane. *Biochim. Biophys. Acta*, 645:115–123, 1981.
34. V.F. Pastushenko and Y.A. Chizmadzhev. Stabilization of conducting pores in BLM by electric current. *Gen. Physiol. Biophys.*, 1:43–52, 1982.
35. U. Zimmermann, F. Riemann, and G. Pilwat. Enzyme loading of electrically homogeneous human red blood cell ghosts prepared by dielectric breakdown. *Biochim. Biophys. Acta*, 436:460–474, 1976.
36. K. Kinosita Jr. and T.Y. Tsong. Formation and resealing of pores of controlled sizes in human erythrocyte membrane. *Nature*, 268:438–441, 1977.
37. K. Kinosita Jr. and T.Y. Tsong. Survival of sucrose-loaded erythrocytes in circulation. *Nature*, 272:258–260, 1978.
38. J. Teissie and T.Y. Tsong. Electric field induced transient pores in phospholipid bilayer vesicles. *Biochemistry*, 20:1548–1554, 1981.
39. U. Zimmermann, P. Scheurich, G. Pilwat, and R. Benz. Cells with manipulated functions: New perspectives for cell biology, medicine and technology. *Angew. Chem. Int. Ed. Engl.*, 20:325–344, 1981.
40. D. Auer, G. Brandner, and W. Bodemer. Dielectric breakdown of the red blood cell membrane and uptake of SV 40 DNA and mammalian cell RNA. *Naturwissenschaften*, 63:391, 1976.
41. E. Neumann, M. Schaefer-Ridder, Y. Wang, and P.H. Hofschneider. Gene transfer into mouse lyoma cells by electroporation in high electric fields. *EMBO J.*, 1:841–845, 1982.
42. H. Fricke. The electric permittivity of a dilute suspension of membrane-covered ellipsoids. *J. Appl. Phys.*, 24:644–646, 1953.
43. H. Pauly and H.P. Schwan. Über die Impedanz einer Suspension von kugelförmigen Teilchen mit einer Schale. *Z. Naturforsch.*, 14B:125–131, 1959.
44. M. Montal and P. Mueller. Formation of bimolecular membranes from lipid monolayers and a study of their electrical properties. *Proc. Natl. Acad. Sci. USA*, 60:3561–3566, 1972.
45. L.V. Chernomordik and Y.A. Chizmadzhev. Electrical breakdown of BLM: phenomenology and mechanism. In E. Neumann, A. Sowers, and C. Jordan, Eds., *Electroporation and Electrofusion in Cell Biology*. Plenum Press, New York, 1989, pp. 83–96.
46. A. Barnett and J.C. Weaver. Electroporation: A unified, quantitative theory of reversible electrical breakdown and rupture. *Bioelectrochem. Bioenerg.*, 25:163–182, 1991.
47. S.A. Freeman, M.A. Wang, and J.C. Weaver. Theory of electroporation for a planar bilayer membrane: Predictions of the fractional aqueous area, change in capacitance and pore–pore separation. *Biophys. J.*, 67:42–56, 1994.
48. Jr. K. Kinosita, I. Ashikawa, N. Saita, H. Yoshimura, H. Itoh, K. Nagayma, and A. Ikegami. Electroporation of cell membrane visualized under a pulsed-laser fluorescence microscope. *Biophys. J.*, 53:1015–1019, 1988.

49. M. Hibino, M. Shigemori, H. Itoh, K. Nagyama, and K. Kinosita. Membrane conductance of an electroporated cell analyzed by submicrosecond imaging of transmembrane potential. *Biophys. J.*, 59:209–220, 1991.
50. W. Frey, J.A. White, R.O. Price, P.F. Blackmore, R.P. Joshi, R. Nuccitelli, S.J. Beebe, K.H. Schoenbach, and J.F. Kolb. Plasma membrane voltage changes during nanosecond pulsed electric field exposures.
51. L.V. Chernomordik, S.I. Sukharev, I.G. Abidor, and Y.A. Chizmadzhev. The study of the BLM reversible electrical breakdown mechanism in the presence of UO_2^{2+}. *Bioelectrochem. Bioenerg.*, 9:149–155, 1982.
52. M. Edidin. Lipids on the frontier: a century of cell-membrane bilayers. *Nat. Rev. Mol. Cell Biol.*, 4:414–418, 2003.
53. M. Winterhalter and W. Helfrich. Deformation of spherical vesicles by electric fields. *J. Colloid. Interface. Sci.*, 122:583–586, 1988.
54. E. Neumann and S. Kakorin. Electrooptics of membrane electroporation and vesicle shape deformation. *Curr. Opin. Colloid Interface Sci.*, 1:790–799, 1996.
55. R.P. Joshi, Q. Hu, K.H. Schoenbach, and H.P. Hjalmarson. Theoretical prediction of electromechanical deformation of cells subjected to high voltages for membrane electroporation. *Phys. Rev. E*, 65:021913-1–021913-10, 2002.
56. K.J. Müller, V.I. Sukhorukov, and U. Zimmermann. Reversible electropermeabilization of mammalian cells by high-intensity, ultra-short pulses of submicrosecond duration. *J. Membr. Biol.*, 184:161–170, 2001.
57. J. Dai and M.P. Sheetz. Regulation of endocytosis, exocytosis, and shape by membrane tension. *Cold Spring Harb. Symp. Quant. Biol.*, 60:567–571, 1995.
58. N. Gov, A.G. Zilman, and S. Safran. Cytoskeleton confinement and tension of red blood cell membranes. *Phys. Rev. Lett.*, 90:228101-1–118101-4, 2003.
59. L.V. Chernomordik, G.B. Milikyan, and Y.A. Chizmadzhev. Biomembrane fusion: A new concept derived from model studies wing two interacting planar lipid bilayers. *Biochim. Biophys. Acta*, 906:309–352, 1987.
60. Y.A. Chizmadzhev, D.A. Kumenko, P.I. Kuzmin, L.V. Chernomordik, J. Zimmerberg, and F.S. Cohen. Lipid flow through fusion pores connecting membranes of different tensions. *Biophys. J.*, 76:2951–2965, 1999.
61. K.A. DeBruin and W. Krassowska. Electroporation and shock-induced transmembrane potential in a cardiac fiber during defibrillation strength shocks. *Ann. Biomed. Eng.*, 26:584–596, 1998.
62. J.C. Neu and W. Krassowska. Asymptotic model of electroporation. *Phys. Rev. E*, 59:3471–3482, 1999.
63. K.C. Melikov, V.A. Frolov, A. Shcherbakov, A.V. Samsonov, Y.A. Chizmadzhev, and L.V. Chernomordik. Voltage-induced nonconductive pre-pores and metastable pores in unmodified planar bilayer. *Biophys. J.*, 80:1829–1836, 2001.
64. H.G.L. Coster. A quantitative analysis of the voltage–current relationships of fixed charge membranes and the associated property of "punch-through." *Biophys. J.*, 5:669–686, 1965.
65. J.M. Crowley. Electrical breakdown of bimolecular lipid membranes as an electromechanical instability. *Biophys. J.*, 13:711–724, 1973.
66. D.S. Dimitrov and R.K. Jain. Membrane stability. *Biochim. Biophys. Acta*, 779:437–468, 1984.
67. K.A. DeBruin and W. Krassowska. Modeling electroporation in a single cell: I. Effects of field strength and rest potential. *Biophys. J.*, 77:1213–1224, 1999.
68. K.A. DeBruin and W. Krassowska. Modeling electroporation in a single cell: II. Effects of ionic concentration. *Biophys. J.*, 77:1225–1233, 1999.
69. D.A. Stewart, T.R. Gowrishankar, and J.C. Weaver. Transport lattice approach to describing cell electroporation: use of a local asymptotic model. *IEEE Trans Plasma Sci.*, 32:1696–1708, 2004.
70. K.C. Smith, T.R. Gowrishankar, A.T. Esser, D.A. Stewart, and J.C. Weaver. Spatially distributed, dynamic transmembrane voltages of organelle and cell membranes due to 10 ns pulses: predictions of meshed and unmeshed transport network models. *IEEE Trans. Plasma Sci.*, 2006. Invited, submitted.
71. D.P. Tieleman, H. Leontiadou, A.E. Mark, and S.-J. Marrink. Simulation of pore formation in lipid bilayers by mechanical stress and electric fields. *J. Am. Chem. Soc.*, 125:6382–6383, 2003.
72. D.P. Tieleman. The molecular basis of electroporation. *BMC Biochem.*, 5:10, 2004.

73. H. Leotiadou, A.E. Mark, and S.J. Marrink. Molecular dynamics simulations of hydrophilic pores in lipid bilayers. *Biophys. J.*, 86:2156–2164, 2004.
74. Q. Hu, S. Viswandham, R.P. Joshi, K.H. Schoenbach, S.J. Beebe, and P.F. Blackmore. Simulations of transient membrane behavior in cells subject to a high-intensity ultrashort electric pulse. *Phys. Rev. E*, 71:03194-1–03194-9, 2005.
75. M. Tarek. Membrane electroporation: A molecular dynamics simulation. *Biophys. J.*, 88:4045–4053, 2005.
76. Q. Hu, R.P. Joshi, and K.H. Schoenbach. Simulations of nanopore formation and phosphatidylserine externalization in lipid membranes subjected to high-intensity, ultrashort electric pulse. *Phys. Rev. E*, 72:031902-1–031902-10, 2005.
77. J.D. Litster. Stability of lipid bilayers and red blood cell membranes. *Phys. Lett.*, 53A:193–194, 1975.
78. J.C. Weaver. Electroporation: A general phenomenon for manipulating cells and tissue. *J. Cell. Biochem.*, 51:426–435, 1993.
79. T.E. Vaughan and J.C. Weaver. A theoretical model for cell electroporation: A quantitative description of electrical behavior. In F. Bersani, Ed., *Electricity and Magnetism in Biology and Medicine*, Plenum, New York, 1999, pp. 433–435.
80. J.A. Nickoloff. Ed. *Methods in Molecular Biology, Electroporation Protocols for Microorganisms*, Vol. 47. Humana Press, Totowa, 1995.
81. J.A. Nickoloff, Ed. *Animal Cell Electroporation & Electrofusion Protocols*, (Methods in Molecular Biology) Vol. 48. Humana Press, Totowa, 1995.
82. J.A. Nickoloff, editor. *Plant Cell Electroporation & Electrofusion Protocols*, (Methods in Molecular Biology) Vol. 55. Humana Press, Totowa, 1995.
83. M.J. Jaroszeski, R. Gilbert, and R. Heller, Eds. *Electrically Mediated Delivery of Molecules to Cells: Electrochemotherapy, Electrogenetherapy and Transdermal Delivery by Electroporation*. Humana Press, Totowa, 2000.
84. M. Belehradek, C. Domenge, B. Luboinski, S. Orlowski, J. Belehradek Jr., and L.M. Mir. Electrochemotherapy, a new antitumor treatment. first clinical phase I-II trial. *Cancer*, 72:3694–3700, 1993.
85. R. Heller, M. Jaroszeski, L.F. Glass, J.L. Messina, D.P. Rapport, R.C. DeConti, N.A. Fenske, R.A. Gilbert, L.M. Mir, and D.S. Reintgen. Phase I/II trial for the treatment of cutaneous and subcutaneous tumors using electrochemotherapy. *Cancer*, 77:964–971, 1996.
86. A. Gothelf, L.M. Mir, and J. Gehl. Electrochemotherapy: results of cancer treatment using enhanced delivery of bleomycin by electroporation. *Cancer Treat. Rev.*, 29:371–387, 2003.
87. T.K. Wong and E. Neumann. Electric field mediated gene transfer. *Biochem. Biophys. Res. Commun.*, 107:584–587, 1982.
88. I. Uno, K. Fukami, H. Kato, T. Takenawa, and T. Ishikawa. Essential role for phosphatidylinositol 4,5-bisphosphate in yeast cell proliferation. *Nature*, 333:188–190, 1988.
89. D.L. Berglund and J.R. Starkey. Isolation of viable tumor cells following introduction of labelled antibody to an intracellular oncogene product using electroporation. *J. Immunol. Methods*, 125:79–87, 1989.
90. M. Rui, Y. Chen, Y. Zhang, and D. Ma. Transfer of anti-TFAR19 monoclonal antibody into HeLa cells by *in situ* electroporation can inhibit apoptosis. *Life Sci.*, 71:1771–1778, 2002.
91. L.M. Mir, H. Banoun, and C. Paoletti. Introduction of definite amounts of nonpermeant molecules into living cells after electropermeabilization: Direct access to the cytosol. *Exp. Cell Res.*, 175:15–25, 1988.
92. D.C. Bartoletti, G.I. Harrison, and J.C. Weaver. The number of molecules taken up by electroporated cells: Quantitative determination. *FEBS Lett.*, 256:4–10, 1989.
93. B. Poddevin, S. Orlowski, J. Belehradek Jr., and L.M. Mir. Very high cytotoxicity of bleomycin introduced into the cytosol of cells in culture. *Biochem. Pharmacol.*, 42, Suppl.:567–575, 1991.
94. M.R. Prausnitz, B.S. Lau, C.D. Milano, S. Conner, R. Langer, and J.C. Weaver. A quantitative study of electroporation showing a plateau in net molecular transport. *Biophys. J.*, 65:414–422, 1993.
95. E.A. Gift and J.C. Weaver. Simultaneous quantitative determination of electroporative molecular uptake and subsequent cell survival using gel microdrops and flow cytometry. *Cytometry*, 39:243–249, 2000.

96. L. Graziadei, P. Burfeind, and D. Bar-Sagi. Introduction of unlabeled proteins into living cells by electroporation and isolation of viable protein-loaded cells using dextran-fluorescein isothiocyanate as a marker for protein uptake. *Anal. Biochem.*, 194:198–203, 1991.
97. K.T. Powell and J.C. Weaver. Transient aqueous pores in bilayer membranes: A statistical theory. *Bioelectrochem. Bioelectroenerg.*, 15:211–227, 1986.
98. J.C. Weaver. Electroporation of biological membranes from multicellular to nano scales. *IEEE Trans. Dielect. Elect. Ins.*, 10:754–768, 2003.
99. W. Mehrle, U. Zimmermann, and R. Hampp. Evidence for asymmetrical uptake of fluorescent dyes through electro-permeabilized membranes of *Avena* meosphyll protoplasts. *FEBS Lett.*, 185:89–94, 1985.
100. E. Tekle, R.D. Astumian, and P.B. Chock. Electroporation using bipolar oscillating electric field: An improved method for DNA transfection of NIH3T3 cells. *PNAS*, 88:4230–4234, 1991.
101. E. Tekle, R.D. Astumian, W.A. Fraiuf, and P.B. Chock. Asymmetric pore distribution and loss of membrane lipid in electroporated DOPC vesicles. *Biophys. J.*, 81:960–968, 2001.
102. M. Golzio, J. Teissie, and M.P. Rols. Direct visualization at the single-cell level of electrically mediated gene delivery. *Proc. Natl. Acad. Sci. USA*, 99:1292–1297, 2002.
103. H.M. Shapiro. *Practical Flow Cytometry*, 3rd ed. Wiley-Liss, New York, 1995.
104. M.R. Prausnitz, C.D. Milano, J.A. Gimm, R. Langer, and J.C. Weaver. Quantitative study of molecular transport due to electroporation: Uptake of bovine serum albumin by human red blood cell ghosts. *Biophys. J.*, 66:1522–1530, 1994.
105. E.A. Gift and J.C. Weaver. Observation of extremely heterogeneous electroporative molecular uptake by *Saccharomyces* cerevisiae which changes with electric field pulse amplitude. *Biochim. Biophys. Acta*, 1234:52–62, 1995.
106. M. Puc, J. Kotnik, L.M. Mir, and D. Miklavčič. Quantitative model of small molecules uptake after *in vitro* cell electropermeabilization. *Bioelectrochemistry*, 60:1–10, 2003.
107. J.C. Weaver. Electroporation of cells and tissues. *IEEE Trans. Plasma Sci.*, 28:24–33, 2000.
108. S. Šatkauskas, M.F. Bureau, A. Mahfoudi, and L. M. Mir. Slow accumulation of plasmid in muscle cells: supporting evidence for a mechanism of DNA uptake by receptor-mediated endocytosis. *Mol. Ther.*, 4:317–323, 2001.
109. L.M. Mir. Therapeutic perspectives of *in vivo* cell electropermeabilization. *Bioelectrochemistry*, 53:1–10, 2001.
110. F. Andre and L.M. Mir. DNA electrotransfer: its principles and an updated review of its therapeutic applications. *Gene Ther.*, 11(Suppl. 1):S33–S42, 2004.
111. E. Neumann, A.E. Sowers, and C.A. Jordan, Ed. *Electroporation and Electrofusion in Cell Biology*. Plenum Press, New York, 1989.
112. D.C. Chang, B.M. Chassy, J.A. Saunders, and A.E. Sowers, Ed. *Guide to Electroporation and Electrofusion*. Academic Press, New York, 1992.
113. L.H. Li, P. Ross, and S.W. Hui. Improving electrotransfection efficiency by post-pulse centrifugation. *Gene Ther.*, 6:364–372, 1999.
114. A.V. Titomirov, S. Sukharev, and E. Kistoanova. *In Vivo* electroporation and stable transformation of skin cells of newborn mice by plasmid DNA. *Biochim. Biophys. Acta*, 1088:131–134, 1991.
115. R.L. Harrison, B.J. Byrne, and L. Tung. Electroporation-mediated gene transfer in cardiac tissue. *FEBS Lett.*, 435:1–5, 1998.
116. T. Goto, T. Nishi, T. Tamura, S.B. Dev, H. Takeshima, M. Kochi, K. Yoshizato, J. Kuratsu, T. Sakata, G.A. Hofmann, and Y. Ushio. Highly efficient electro–gene therpay of solid tumor by using an expression plasmid for the herpes simplex virus thymidine kinase gene. *Proc. Natl. Acad. Sci. USA*, 97:354–359, 2000.
117. D.J. Wells. Gene therapy progress and prospects: electroporation and other physical methods. *Gene Ther.*, 11:1361–1369, 2004.
118. P.-G. de Gennes. Passive entry of a DNA molecule into a small pore. *Proc. Natl. Acad. Sci. USA*, 96:7262–7264, 1999.
119. J.C. Neu, K.C. Smith, and W. Krassowska. Electrical energy required to form large conducting pores. *Bioelectrochemistry*, 60:107–114, 2003.
120. K.C. Smith, J.C. Neu, and W. Krassowska. Model of creation and evolution of stable electropores for DNA delivery. *Biophys. J.*, 86:2813–2826, 2004.

121. H. Stopper, H. Jones, and U. Zimmermann. Large scale transfection of mouse L–cells by electropermeabilization. *Biochim. Biophys. Acta*, 900:38–44, 1987.
122. D.S. Dimitrov and A.E. Sowers. Membrane electroporation-fast molecular exchange by electroosmosis. *Biochim. Biophys. Acta*, 1022:381–392, 1990.
123. V.A. Klenchin, S.I. Sukharev, S.M. Serov, L.V. Chernomordik, and Y.A. Chizmadzhev. Electrically induced DNA uptake by cells is a fast process involving DNA electrophoresis. *Biophys. J.*, 60:804–811, 1991.
124. S.I. Sukharev, V.A. Klenchin, S.M. Serov, L.V. Chernomordik, and Y.A. Chizmadzhev. Electroporation and electrophoretic DNA transfer into cells. the effect of DNA interaction with electropores. *Biophys. J.*, 63:1320–1327, 1992.
125. L.V. Chernomordik, A.V. Sokolov, and V.G. Budker. Electrostimulated uptake of DNA by liposomes. *Biochim. Biophys. Acta*, 1024:179–183, 1990.
126. J.C. Weaver, R. Vanbever, T.E. Vaughan, and M.R. Prausnitz. Heparin alters transdermal transport associated with electroporation. *Biochem. Biophys. Res. Commun.*, 234:637–640, 1997.
127. V.P. Pastushenko and Y.A. Chizmadzhev. Energetic estimations of the deformation of translocated DNA and cell membrane in the course of electrotransformation. *Biol. Mem.*, 6:287–300, 1992.
128. Y. Mouneimne, P-F. Tosi, Y. Gazitt, and C. Nicolau. Electro-insertion of xeno-glycophorin into the red blood cell membrane. *Biochem. Biophys. Res. Commun.*, 159:34–40, 1989.
129. M. Zeira, P-F. Tosi, Y. Mouneimne, J. Lazarte, L. Sneed, D.J. Volsky, and C. Nicolau. Full-length CD4 electro-inserted in the erythrocyte membrane as a long-lived inhibitor of infection by human immunodeficiency virus. *Proc. Natl. Acad. Sci. USA*, 88:4409–4413, 1991.
130. C. Nicolau, Y. Mouneimne, and P.-F. Tosi. Electroinsertion of proteins in the plasma membrane of red blood cells. *Anal. Biochem.*, 1993:1–10, 1993.
131. K.E. Ouagari, J. Teissie, and H. Benoist. Glycophorin a protects K562 cells from natural killer cell attack. *J. Biol. Chem.*, 270:26970–26975, 1995.
132. S. Raffy, C. Lazdunski, and J. Teissie. Electroinsertion and activation of the C-terminal domain of colicin A, a voltage gated bacterial toxin, into mammalian cell membranes. *Mol. Membr. Biol.*, 21:237–246, 2004.
133. P. Scheurich, U. Zimmermann, M. Mischel, and I. Lamprecht. Membrane fusion and deformation of red blood cells by electric fields. *Z. Naturforsch.*, 35:1801–1805, 1980.
134. J. Teissie, V.P. Knutson, T.Y. Tsong, and M.D. Lane. Electric pulse-induced fusion of 3t3 cells in monolayer culture. *Science*, 216:537–538, 1982.
135. U. Zimmermann. Electric field-mediated fusion and related electrical phenomena. *Biochim. Biophys. Acta*, 694:227–277, 1982.
136. U. Zimmermann and G. Küppers. Cell fusion by electromagnetic waves and its possible relevance for evolution. *Naturwissenschaften*, 70:568–569, 1983.
137. A.E. Sowers. Movement of a fluorescent lipid label from a labeled erythrocyte membrane to an unlabeled erythrocyte membrane following electric-field-induced fusion. *Biophys. J.*, 47: 519–525, 1985.
138. D.A. Stenger and S.W. Hui. Kinetics of ultrastructure changes during electrically-induced fusion of human erythrocytes. *J. Membr. Biol.*, 93:43–53, 1986.
139. I.P. Sugar, W. Forster, and E. Neumann. Model of cell electrofusion. membrane electroporation, pore coalescence and percolation. *Biophys. Chem.*, 26:321–335, 1987.
140. J.R. Grasso, R. Heller, J.C. Cooley, and E.M. Haller. Electrofusion of individual animal cells directly to interaction corneal epithelial tissue. *Biochim. Biophys. Acta*, 980:9–14, 1989.
141. R. Heller and R.J. Grasso. Transfer of human membrane surface components by incorporating human cells into interaction animal tissue by cell–tissue electrofusion *in vivo. Biochim. Biophys. Acta*, 1024:185–188, 1990.
142. I. Tsoneva, T.Tomov, I. Panova, and D. Strahilov. Effective production by electrofusion of hybridomas secreting monoclonal antibodies against hc-antigen of *Salmonella. Bioelectrochem. Bioenerg.*, 24:41–49, 1990.
143. D.T. Chiu, C.F. Wilson, F. Ryttsen, A. Stromberg, C. Farre, A. Karlsson, S. Nordholm, A. Gaggar, B.P. Modi, A. Moscho, R.A. Garza-Lopez, O. Orwar, and R.N. Zare. Chemical transformations in individual ultrasmall biomimetic containers. *Science*, 283:1892–1895, 1999.

144. A. Strömberg, F. Ryttsén, D.T. Chiu, M Davidson, P.S. Eriksson, C.F. Wilson, O. Orwar, and R.N. Zare. Manipulating the genetic identity and biochemical surface properties of individual cells with electric-field-induced fusion. *Proc. Natl. Acad. Sci. USA*, 97:7–11, 2000.
145. K.T. Trevor, C. Cover, Y.W. Ruiz, E.T. Akporiaye, E.M. Mersh, D. Landais, R.R. Taylor, A.D. King, and R.E. Walters. Generation of dendritic cell-tumor cell hybrids by electrofusion for clinical vaccine application. *Cancer Immunol. Immunother.*, 53:705–714, 2004.
146. W.T. Lee, K. Shimizu, H. Kuriyama, H. Tanaka, J. Kjaergaard, and S. Shu. Tumor-dendritic cell fusion as a basis for cancer immunotherapy. *Otolaryngol. Head Neck Surg.*, 132:755–764, 2005.
147. S.W. Hui, N. Stoicheva, and Y.-L. Zhao. High-efficiency loading, transfection, and fusion of cells by electroporation in two-phase polymer systems. *Biophys. J.*, 71:1123–1130, 1996.
148. P.L. Kuzmin, V.P. Pastushenko, I.G. Abidor, S.I. Sukharev, A.V. Barbul, and Y.A. Chizmadzhev. Electrofusion of cells: Theoretical analysis. *Biol. Membr.*, 5:600–612, 1988.
149. A.E. Sowers. A long lived fusogenic state is induced in erythrocytes ghosts by electric pulses. *J. Cell. Biol.*, 102:1358–1362, 1986.
150. J.Teissie and M.-P. Rols. Fusion of mammalian cells in culture is obtained by creating the contact between cells after their electropermeabilization. *Biochem. Biophys. Res. Commun.*, 140:258–264, 1986.
151. M. Bier, S.M. Hammer, D.J. Canaday, and R.C. Lee. Kinetics of sealing for transient electropores in isolated mammalian skeletal muscle cells. *Bioelectromagnetics*, 20:194–201, 1999.
152. C.S. Djuzenova, U. Zimmermann, H. Frank, V.L. Sukhorukov, E. Richter, and G. Fuhr. Effect of medium conductivity and composition on the uptake of propidium iodide into electropermeabilized myeloma cells. *Biochim. Biophys. Acta*, 1284:143–152, 1996.
153. E. Neumann, K. Toensing, S. Kakorin, P. Budde, and J. Frey. Mechanism of electroporative dye uptake by mouse B cells. *Biophys. J.*, 74:98–108, 1998.
154. M.N. Teruel and T. Meyer. Electroporation-induced formation of individual calcium entry sites in the cell body and processes of adherent cells. *Biophys. J.*, 73:1785–1796, 1997.
155. M. Bier, W. Chen, T.R. Gowrishankar, R.D. Astumian, and R.C. Lee. Resealing dynamics of a cell membrane after electroporation. *Phys. Rev. E*, 66:062905-1–062905-4, 2002.
156. R. Benz and U. Zimmermann. The resealing process of lipid bilayers after reversible electrical breakdown. *Biochim. Biophys. Acta*, 640:169–178, 1981.
157. L.V. Chernomordik, L.V. Sukharev, I.G. Abidor, and Yu. A. Chizmadzhev. Breakdown of lipid bilayer membranes in an electric field. *Biochim. Biophys. Acta*, 1983.
158. Yu. A. Chizmadzhev and I. Abidor. Bilayer lipid membranes in strong electric fields. *Bioelectrochem. Bioenerg.*, 7:83–100, 1980.
159. G. Pilwat, U. Zimmermann, and F. Riemann. Dielectric breakdown measurements of human and bovine erythrocyte membranes using benzyl alcohol as a probe molecule. *Biochim. Biophys. Acta*, 406:424–432, 1975.
160. J.C. Weaver, G.I. Harrison, J.G. Bliss, J.R. Mourant, and K.T. Powell. Electroporation: High frequency of occurrence of the transient high permeability state in red blood cells and intact yeast. *FEBS Lett.*, 229:30–34, 1988.
161. S. Kwee, H.V. Nielsen, and J.E. Celis. Electropermeabilization of human cultured cells grown in monolayers: incorporation of monoclonal antibodies. *Bioelectrochem. Bioenerg.*, 23:65–80, 1990.
162. K.P. Mishra and A.B. Singh. Temperature effects on resealing of electrically hemolysed rabbit erythrocytes. *Indian J. Exp. Biol.*, 24:737–741, 1986.
163. R.C. Lee, L.P. River, F.-S. Pan, L. Ji, and R.L. Wollmann. Surfactant induced sealing of electropermeabilized skeletal muscle membranes *in vivo*. *Proc. Natl. Acad. Sci. USA*, 89:4524–4528, 1992.
164. R.C. Lee, D. Zhang, and J. Hannig. Biophysical injury mechanisms in electrical shock trauma. *Ann. Rev. Biomed. Eng.*, 2:477–509, 2000.
165. M.R. Prausnitz, J.D. Corbett, J.A. Grimm, D.E. Golan, R. Langer, and J.C. Weaver. Millisecond measurement of transport during and after an electroporation pulse. *Biophys. J.*, 68:1864–1870, 1995.
166. U. Pliquett, M.R. Prausnitz, Y. Chizmadzhev, and J.C. Weaver. Measurement of rapid release kinetics for drug delivery. *Pharm. Res.*, 12:546–553, 1995.
167. P.E. Marszalek, B. Farrell, P. Verdugo, and J. M. Fernandez. Kinetics of release of serotonin from isolated secretory granules. I. Amperometric detection of serotonin from electroporated granules. *Biophys. J.*, 73:1160–1168, 1997.

168. D.C. Gaylor, K. Prakah-Asante, and R.C. Lee. Significance of cell size and tissue structure in electrical trauma. *J. Theor. Biol.*, 133:223–237, 1988.
169. D.L. Bhatt, D.C. Gaylor, and R.C. Lee. Rhabdomyolysis due to pulsed electric fields. *Plast. Reconstr. Surg.*, 86:1–11, 1990.
170. V. Ganeva, B. Galutzov, Eynard N., and J. Teissié. Electroinduced extraction of β-galactosidase from *Kluyveromyces lactis. Appl. Microbiol. Biotechnol.*, 56:411–413, 2001.
171. M.R. Michel, M. Elgizoli, H. Koblet, and Ch. Kempf. Diffusion loading conditions determine recovery of protein synthesis in electroporated p3×63 ag8 cells. *Experientia*, 44:199–203, 1988.
172. J.C. Weaver. Molecular basis for cell membrane electroporation. *Ann. NY Acad. Sci.*, 720:141–152, 1994.
173. R.C. Lee and R.D. Astumian. The physiochemical basis for thermal and nonthermal "burn" injury. *Burns*, 22:509–519, 1996.
174. A.R. Denet, R. Vanbever, and V. Preat. Skin electroporation for transdermal and topical delivery. *Adv. Drug Deliv. Rev.*, 56:659–674, 2004.
175. R. Heller, M. Jaroszeski, A. Atkin, D. Moradpour, R. Gilbert, J. Wands, and C. Nicolau. *In vivo* gene electroinjection and expression in rat liver. *FEBS Lett.*, 389:225–228, 1996.
176. M.R. Prausnitz, V.G. Bose, R. Langer, and J.C. Weaver. Electroporation of mammalian skin: A mechanism to enhance transdermal drug delivery. *Proc. Natl. Acad. Sci.*, 90:10504–10508, 1993.
177. J.C. Weaver and R. Langer. Electrochemical creation of large aqueous pathways: An approach to transdermal drug delivery. *Prog. Dermatol.*, 33:1–10, 1999.
178. M.R. Prausnitz, S. Mitragotri, and R. Langer. Current status and future potential of transdermal drug delivery. *Nat. Rev. Drug Discov.*, 3:115–121, 2004.
179. O. Tovar and L. Tung. Electroporation and recovery of cardiac cell membrane with rectangular voltage pulses. *Am. J. Physiol.*, 263:H1128–H1136, 1992.
180. D.K. Cheng, L. Tung, and E.A. Sobie. Nonuniform responses of transmembrane potential during electric field stimulation of single cardiac cells. *Am. J. Physiol.*, 277:H351–H362, 1999.
181. E.R. Cheek and V.G. Fast. Nonlinear changes of transmembrane potential during electrical shocks: role of membrane electroporation. *Circ. Res.*, 94:208–214, 2004.
182. R.M. Fish and L.A. Geddes. Effects of stun guns and tasers. *Lancet*, 358:687–688, 2001.
183. T.R. Gowrishankar, C. Stewart, and J.C. Weaver. Electroporation of a multicellular system: asymptotoic model analysis. In *Proceedings of the 26th Annual International Conference of the IEEE EMBS*, San Francisco (2004).
184. R.K. Jain. Physiological resistance to the treatment of solid tumors. In B.A. Teicher, Ed., *Drug resistance in oncology* . Marcel Dekker, New York, 1993, pp. 87–105.
185. M. Okino and H. Mohri. Effects of a high-voltage electrical impulse and an anticancer drug on *in vivo* growing tumors. *Jpn. J. Cancer Res.*, 78:1319–1321, 1987.
186. L. . Mir, S. Orlowski Jr., J. Belehradek, and C. Paoletti. *In Vivo* potentiation of the bleomycin cytotoxicity by local electric pulses. *Eur. J. Cancer*, 27:68–72, 1991.
187. L.M. Mir, S. Orlowski, J. Belehradek, J. Teissie, M.P. Rols, G. Sersa, D. Miklavčič, R. Gilbert, and R. Heller. Biomedical applications of electric pulses with special emphasis on antitumor electrochemotherapy. *Bioelectrochem. Bioenerg.*, 38:203–207, 1995.
188. C. Domenge, S. Orlowski, B. Luboinski, T. De Baere, G. Schwaab, J. Belehradek, and L.M. Mir. Antitumor electrochemotherapy: New advances in the clinical protocol. *Cancer*, 77:956–963, 1996.
189. I. Entin, A. Plotnikov, R. Korenstein, and Y. Keisari. Tumor growth retardation, cure, and induction of antitumor immunity in B16 melanoma-bearing mice by low electric field-enhanced chemotherapy. *Clin. Cancer Res.*, 9:3190–3197, 2003.
190. M. Hyacinthe, M.J. Jaroszeski, V.V. Dang, D. Coppola, R.C. Karl, R.A. Gilbert, and R. Heller. Electrically enhanced drug delivery for the treatment of soft tissue sarcoma. *Cancer*, 85:409–417, 1999.
191. R.A. Gilbert, M.J. Jaroszeski, and R. Heller. Novel electrode designs of electrochemotherapy. *Biochim. Biophys. Acta*, 1334:9–14, 1997.
192. D. Sel, S. Mazeres, J. Teissie, and D. Miklavčič. Finite-element modling of needle electrodes in tissue from the perspective of frequent model computations. *IEEE Trans. Biomed. Eng.*, 50:1221–1232, 2003.

193. S.B. Dev, D. Dhar, and W. Krassowska. Electric field of a six-needle array electrode used in drug and DNA delivery *in vivo*: Analytical verus numerical solutions. *IEEE Trans. Biomed. Eng.*, 50:1296–1300, 2003.
194. R.V. Davalos, B. Rubinsky, and D.M. Otten. A feasibility study for electrical impedance tomography as a means to monitor tissue electroporation for molecular medicine. *IEEE Trans. Biomed. Eng.*, 49:400–403, 2002.
195. R.V. Davalos, B. Rubinsky, L.M. Mir, and D.M. Otten. Electrical impedance tomography for imaging tissue electroporation. *IEEE Trans. Biomed Eng.*, 51:761–767, 2004.
196. M.J. Berridge and J.L. Oschman. *Transporting Epithelia*. Academic Press, New York, 1972.
197. K.T. Powell, A.W. Morgenthaler, and J.C. Weaver. Tissue electroporation: Observation of reversible electrical breakdown in viable frog skin. *Biophys. J.*, 56:1163–1171, 1989.
198. E.B. Ghartey-Tagoe, J.S. Morgan, K. Ahmed, A.S. Neish, and M.R. Prausnitz. Electroporation-mediated delivery of molecules to model intestinal epithelia. *Int. J. Pharm.*, 270:127–138, 2004.
199. J.L. Kirby, L. Yang, J.C. Labus, R.J. Lye, N. Hsia, R. Day, G.A. Cornwall, and B.T. Hinton. Characterization of epidymal epithelial cell-specific gene promotors by *in vivo* electroporation. *Biol. Reprod.*, 71:613–619, 2004.
200. H.E. Abud, P. Lock, and J.K. Heath. Efficient gene transfer into the epithelial cell layer of embryonic mouse intestine using low-voltage electroporation. *Gastroenterology*, 126:1779–1787, 2004.
201. E.B. Ghartey-Tagoe, J.S. Morgan, A.S. Neish, and M.R. Prausnitz. Increasing permeability of intestinal epithelial monolayers mediated by electroporation. *J. Controlled Release*, 103:177–190, 2005.
202. L.A. Goldsmith, Ed. *Physiology, Biochemistry, and Molecular Biology of the Skin*, 2nd ed. Oxford University Press, New York, 1991.
203. H. Schaefer and T.E. Redelmeier. *Skin Barrier: Principles of Percutaneous Absorption*. Karger, Basel, 1996.
204. K.C. Madison. Barrier function of the skin "la raison d'etre" of the epidermis. *J. Invest. Dermatol.*, 121:231–241, 2003.
205. A.S. Michaels, S.K. Chandrasekaran, and J.E. Shaw. Drug permeation through human skin: Theory and *in vitro* experimental measurements. *AIChEJ*, 21:985–996, 1975.
206. J.A. Tamada, N.J.V. Bomannon, and R.O. Potts. Measurement of glucose in diabetic subjects using noninvasive transdermal extraction. *Nat. Med.*, 1:1198–1202, 1995.
207. R.T. Kurnik, B. Berner, J. Tamada, and R.O. Potts. Design and simulation of a reverse iontophoretic glucose sensor. *J. Electrochem. Soc.*, 145:4119–4125, 1998.
208. R.O. Potts, J.A. Tamada, and M.J. Tierney. Glucose monitoring by reverse iontophoresis. *Diabetes Metab. Res. Rev.*, 18:S49–S53, 2002.
209. Yu. A. Chizmadzhev, V. Zarnytsin, J.C. Weaver, and R.O. Potts. Mechanism of electroinduced ionic species transport through a multilamellar lipid system. *Biophys. J.*, 68:749–765, 1995.
210. Y. Chizmadzhev, A.V. Indenbom, P.I. Kuzmin, S.V. Galinchenko, J.C. Weaver, and R. Potts. Electrical properties of skin at moderate voltages: Contribution of appendageal macropores. *Biophys. J.*, 74:843–856, 1998.
211. Y. Chizmadzhev, P.I. Kuzmin, J.C. Weaver, and R. Potts. Skin appendageal macropores as possible pathway for electrical current. *J. Invest. Dermatol. Symp. Proc.*, 3:148–152, 1998.
212. J.C. Weaver, T.E. Vaughan, and Y. Chizmadzhev. Theory of electrical creation of aqueous pathways across skin transport barriers. *Adv. Drug Deliv. Rev.*, 35:21–39, 1999.
213. D.R. Friend. *In vitro* permeation techniques. *J. Control. Release*, 18:235–248, 1992.
214. H.H. Ussing. *The alakli metal ions in biology*. Springer-Verlag, Berlin, 1960.
215. D. Bommannan, L. Leung, J. Tamada, J. Sharifi, W. Abraham, and R. Potts. Transdermal delivery of lutenizing hormone releasing hormone: Comparison between electroporation and iontophoresis *in vitro*. *Proceedings of the International Symposium Controlled Release on Bioactive Materials*, Vol. 20, 1993, pp. 97–98.
216. U. Pliquett, R. Langer, and J.C. Weaver. Changes in the passive electrical properties of human stratum corneum due to electroporation. *Biochim. Biophys. Acta*, 1239:111–121, 1995.
217. R. Vanbever, D. Fouchard, A. Jadoul, N. De Morre, V. Preat, and J-P. Marty. *In vivo* noninvasive evaluation of hairless rat skin after high-voltage pulse exposure. *Skin Pharmacol. Appl. Skin Physiol.*, 11:23–34, 1998.

218. R. Vanbever, G. Langers, S. Montmayeur, and V. Preat. Transdermal delivery of fentanyl: Rapid onset of analgesia using skin electroporation. *J. Control. Release*, 50:225–235, 1998.
219. A. Sharma, M. Kara, F.R. Smith, and T.R. Krishnan. Transdermal drug delivery using electroporation: II. Factors influencing skin reversibility in electroporative delivery of terazosin hydrochloride in hairless rats. *J. Pharm. Sci.*, 89:536–544, 2000.
220. G. Martin, U. Pliquett, and J.C. Weaver. Theoretical analysis of localized heating in human skin subjected to high voltage pulses. *Bioelectrochemistry*, 57:55–64, 2002.
221. U. Pliquett and J.C. Weaver. Electroporation of human skin: Simultaneous measurement of changes in the transport of two fluorescent molecules and in the passive electrical properties. *Bioelectrochem. Bioenerg.*, 39:1–12, 1996.
222. M.R. Prausnitz, U. Pliquett, R. Langer, and J.C. Weaver. Rapid temporal control of transdermal drug delivery by electroporation. *Pharm. Res.*, 11:1834–1837, 1994.
223. L.M. Mir, M.F. Bureau, J. Gehl, R. Rangara, D. Rouy, J-M. Caillaud, P. Delaere, D. Brannellec, B. Schwarts, and D. Scherman. High-efficiency gene transfer into skeletal muscle mediated by electric pulses. *Proc. Natl. Acad. Sci. USA*, 96:4262–4267, 1999.
224. J.C. Neu and W. Krassowska. Modeling postshock evolution of large electropores. *Phys. Rev. E*, 67:021915-1–021915-12, 2003.
225. R.C. Lee, E.G. Cravalho, and J.F. Burke, Eds. *Electrical Trauma: The Physiology, Manifestations and Clinical Management*. Cambridge University Press, Cambridge, 1992.
226. G. Greenebaum, K. Blossfield, J. Hannig, C.S. Carrilo, M.A. Beckett, R.R. Weichselbaum, and R.C. Lee. Poloxamer 188 prevents acute necrosis of adult skeletal muscle cells following high-dose irradiation. *Burns*, 30:539–547, 2004.
227. M.N. Robinson, C.G. Brooks, and G.D. Renshaw. Electric shock devices and their effects on the human body. *Med. Sci. Law*, 30:285–300, 1990.
228. S. Anders, M. Junge, F. Schulz, and K. Püschel. Cutaneous current marks due to a stun gun injury. *J. Forensic Sci.*, 48:1–3, 2003.
229. W.C. McDaniel, R.A. Strabucker, M. Nerheim, and J.E. Brewer. Cardiac safety of neuromuscular incapacitating defensive devices. *Pacing Clin. Electrophysiol.*, 28(Suppl. 1):S284–S287, 2005.
230. P.J. Kim and W.H. Franklin. Ventricular fibrillation after stun-gun discharge. *N. Engl. J. Med.*, 353:958–959, 2005.
231. K.R. Foster. Thermal and nonthermal mechanisms of interaction of radio-frequency energy with biological systems. *IEEE Trans. Plasma Sci.*, 28:15–23, 2000.
232. K.H. Schoenbach, S.J. Beebe, and E.S. Buescher. Intracellular effect of ultrashort pulses. *Bioelectromagnetics*, 22:440–448, 2001.
233. S.J. Beebe, P.M. Fox, L.J. Rec, K. Somers, R.H. Stark, and K.H. Schoenbach. Nanosecond pulsed electric field (nsPEF) effects on cells and tissues: apoptosis induction and tumor growth inhibition. *IEEE Trans. Plasma Sci.*, 30:286–292, 2002.
234. S.J. Beebe, P.M. Fox, L.J. Rec, L.K. Willis, and K.H. Schoenbach. Nanosecond high intensity pulsed electric fields induce apoptosis in human cells. *FASEB J.*, 17:1493–1495, 2003.
235. P.T. Vernier, Y. Sun, L. Marcu, S. Salemi, C.M. Craft, and M.A. Gundersen. Calcium bursts induced by nanosecond electric pulses. *Biochem. Biophys. Res. Commun.*, 310:286–295, 2003.
236. J. Deng, K.H. Schoenbach, E.S. Buescher, P.S. Hair, P.M. Fox, and S.J. Bebe. The effects of intense submicrosecond electrical pulses on cells. *Biophys. J.*, 84:2709–2714, 2003.
237. P.S. Hair, K.H. Schoenbach, and E.S. Buescher. Sub-microsecond, intense pulsed electric field applications to cells show specificity of effects. *Bioelectrochemistry*, 61:65–72, 2003.
238. S.J. Beebe, J. White, P.F. Blackmore, Y. Deng, K. Sommers, and K.H. Schoenbach. Diverse effects of nanosecond pulsed electric fields on cells and tissues. *DNA Cell Biol.*, 22:785–796, 2003.
239. M. Stacey, J. Stickley, P. Fox, V. Statler, K. Schoenbach, S.J. Beebe, and S. Buescher. Differential effects in cells exposed to ultra-short, high intensity electric fields: cell survival, DNA damage, and cell cycle analysis. *Mutat. Res.*, 542:65–75, 2003.
240. J.A. White, P.F. Blackmore, K.H. Schoenbach, and S.J. Beebe. Stimulation of capacitive calcium entry in HL-60 cells by nanosecond pulsed electric fields (nsPEF). *J. Biol. Chem.*, 279:22964–22972, 2004.

241. N. Chen, K.H. Schoenbach, J.F. Kolb, R.J. Swanson, A.L. Garner, J. Yang, R.P. Joshi, and S.J. Beebe. Leukemic cell intracellular responses to nanosecond electric fields. *Biochem. Biophys. Res. Commun.*, 317:421–427, 2004.
242. P.T. Vernier, Y. Sun, L. Marcu, C.M. Craft, and M.A. Gundersen. Nanoelectropulse-induced phosphatidylserine translocation. *Biophys. J.*, 86:4040–4048, 2004.
243. K.H. Schoenbach, R.P. Joshi, J.R. Kolb, N. Chen, M. Stacey, P.F. Blackmore, E. S. Buescher, and S. J. Beebe. Ultrashort electrical pulses open a new gateway into biological cells. *Proc. IEEE*, 92:1122–1137, 2004.
244. E. Tekle, H. Oubrahim, S.M. Dzekunov, J.F. Kolb, and K.H. Schoenbach. Selective field effects on intracellular vacuoles and vesicle membranes with nanosecond electric pulses. *Biophys. J.*, 89:274–284, 2005.
245. G. Chen and D.V. Goeddel. TNF-R1 signaling: a beautiful pathway. *Science*, 296:1634–1635, 2002.
246. H. Wajant. The Fas signaling pathway: more than a paradigm. *Science*, 296:1635–1636, 2002.
247. J.-S. Kim, L. He, and J.J. Lemasters. Mitochondrial permeability transition: a common pathway to necrosis and apoptosis. *Biochem. Biophys. Res. Commun.*, 304:463–470, 2003.
248. N.N. Danial and S.J. Korsmeyer. Cell death: critical control points. *Cell*, 116:205–219, 2004.
249. E.E. Varfolomeev and A. Ashkenazi. Tumor necrosis factor: an apoptosis JuNKie? *Cell*, 116:491–497, 2004.
250. R.C. Lee and M.S. Kolodney. Electrical injury mechanisms: Electrical breakdown of cell membranes. *Plast. Reconstr. Surg.*, 80:672–679, 1987.
251. B. Bagriel and J. Teissié. Control by electrical parameters of short- and long-term cell death resulting from electropermeabilization of Chinese hamster ovary cells. *Biochim. Biophys. Acta*, 1266:171–178, 1995.
252. C.R. Keese, J. Wegner, S.R. Walker, and I. Giaver. Electrical wound-healing assay for cells *in vitro*. *Proc. Natl. Acad. Sci. USA*, 101:1554–1559, 2004.
253. D.R. Green and J.C. Read. Mitochondria and apoptosis. *Science*, 281:1309–1312, 1998.
254. M.O. Hengartner. The biochemistry of apoptosis. *Nature*, 407:770–776, 2000.
255. G.I. Evan and K. H. Vousden. Proliferation, cell cycle and apoptosis in cancer. *Nature*, 411:342–348, 2001.
256. S. Orrenius, B. Zhivotosky, and P. Nicotera. Regulation of cell death: the calcium-apoptosis link. *Nat. Rev. Mol. Cell Biol.*, 4:552–565, 2003.
257. L. Scorrano and S.J. Korsmeyer. Mechanisms of cytochrome c release by proapoptotic BCL-2 family members. *Biochem. Biophys. Res. Commun.*, 304:437–444, 2003.
258. F. Hofmann, L.H. Scheller, W. Strupp, U. Zimmermann, and C. Jassoy. Electric field pulses can induce apoptosis. *J. Membr. Biol.*, 169:103–109, 1999.
259. J. Pi nero, M. López-Baena, T. Ortiz, and F. Cortés. Apoptotic and necrotic cell death are both induced by electroporation in HL60 human promyeloid leukaemia cells. *Apoptosis*, 2:330–336, 1997.
260. V. Dressler, K. Schwister, C.W. M. Häst, and B. Deuticke. Dielectric breakdown of the erythrocyte membrane enhances transbilayer mobility of phospholipids. *Biochim. Biophys. Acta*, 732:304–307, 1983.
261. L.Y. Song, J.M. Baldwin, R. O'Reilly, and J.A. Lucy. Relationship between the surface exposure of acidic phospholipids and cell fusion in erythrocytes subjected to electrical breakdown. *Biochim. Biophys. Acta*, 1104:1–8, 1992.
262. C.W.M. Haest, D. Kamp, and B. Deuticke. Transbilayer reorientation of phospholipid probes in the human erythrocyte membrane. lessons from studies on electroporated and resealed cells. *Biochim. Biophys. Acta*, 1325:17–33, 1997.
263. P.T. Vernier, L.I. Aimin, L. Marcu, C.M. Craft, and M.A. Gundersen. Ultrashort pulsed electric fields induce membrane phospholipid translocation and capase activation: differential sensitivities of Jurkat T lymphoblasts and rat glioma C6 cells. *IEEE Trans. Dielect. Elect. Ins.*, 10:795–809, 2003.
264. P.T. Vernier, Y. Sun, L. Marcu, C.M. Craft, and M.A. Gundersen. Nanosecond pulsed electric fields perturb membrane phospholipids in T lyphoblasts. *FEBS Lett.*, 572:103–108, 2004.

265. R.P. Joshi and K.H. Schoenbach. Electroporation dynamics in biological cells subjected to ultrafast electrical pulses: a numerical simulation study. *Phys. Rev. E*, 62:1025–1033, 2000.
266. R.P. Joshi, Q. Hu, R. Aly, K.H. Schoenbach, and H.P. Hjalmarson. Self-consistent simulations of electroporation dynamics in biological cells subjected to ultrashort pulses. *Phys. Rev. E*, 64:011913-1–011913-10, 2001.
267. R.P. Joshi, Q. Hu, K.H. Schoenbach, and H.P. Hjalmarson. Improved energy model for membrane electroporation in biological cells subjected to electrical pulses. *Phys. Rev. E*, 62:041920-1–041920-8, 2002.
268. R.P. Joshi and K.H. Schoenbach. Mechanism for membrane electroporation irreversibility under high-intensity, ultrashort pulse conditions. *Phys. Rev. E*, 66:052901-1–052901-4, 2002.
269. R.P. Joshi, Q. Hu, K.H. Schoenbach, and S.J. Beebe. Energy-landscape-model analysis for irreversibility and its pulse-width dependence in cells subjected to a high-intensity ultrashort electric pulse. *Phys. Rev. E*, 69:051901-1–051901-10, 2004.
270. W. Chen and R.C. Lee. Altered ion channel conductance and ionic selectivity induced by large imposed membrane potential pulses. *Biophys. J.*, 67:603–612, 1994.
271. W. Chen, Y. Han, Y. Chen, and D. Astumiam. Electric field-induced functional reductions in the K^+ channels mainly resulted from supramembrane potential-medicated electroconformational changes. *Biophys. J.*, 75:196–206, 1998.
272. W. Chen. Supra-physiological membrane potential induced conformational changes in K^+ channel conducting system of skeletal muscle fibers. *Bioelectrochemistry*, 62:47–56, 2004.
273. C. Loupatatzis, G. Seitz, P. Schonfeld, F. Lang, and D. Siemen. Single-channel currents of the permeability transition pore from the inner mitochondrial membrane of rat liver and of a human hepatoma cell line. *Cell Physiol. Biochem.*, 12:269–278, 2002.

10

Electrical Shock Trauma

Raphael C. Lee, Elena N. Bodnar, Pravin Betala, and Sigrid Blom-Eberwein

CONTENTS

10.1 Introduction

No engineering achievement has had a greater impact on human culture than electrical power. As electric power gains its significance and vital importance in today's modern society, it poses equal threat to human society in terms of safety. Most citizens have experienced an electric shock at least once in their lifetime. The fear reflex generated by the bad experience of pain usually prevents us from further tampering with electricity. However, no matter how careful we are, accidents do and will occur, especially among electrical workers who have to handle commercial electrical power lines everyday. Recently, use of electrical power has made it into the mainstream of law enforcement. Nonlethal electrical weapons now provide a new intermediate force option for the police. The purpose of this chapter is to provide a basic overview of harmful effects of electrical force from both the engineering and the medical perspectives.

Basically, the range of clinical manifestations of electrical shock is not well documented. Many survivors of accidental electrical shock never seek medical attention. The data that exists is from those who do seek attention or from few case studies of electrical workers.

Furthermore, there is considerable variation in how electrical injury and safety is managed across various countries and cultures. In industrializing countries safety practices are often not the top priority, resulting in high rates of injury. A study of burn injuries by Nursal et al. [1] during a 1 y (2000–2001) period indicated that 21% of burn subjects were victims of electrical injury . In highly industrialized nations, electrical shock rates are on the decline. In the United States, workplace electrocution remains the fifth leading cause of fatal occupational injury with an estimated economic impact of more than $1 billion annually [2]. The rates of injury may be the highest among electrical workers, mostly caused by working on "live" electrical equipment, wiring, light fixtures, and overhead power lines [3]. A study in Virginia suggested that public utilities have the highest rate of fatal electrical injuries among all industrial sectors. More than 90% of these injuries occur in men, mostly between the ages of 20 and 34, with 4 to 8 y of experience on the job [4]. Another source [5] suggests that the average age of victims was 37.5 y and the average years of experience amounted to 11.3 y. In one of the author's own (German) study, the incidence in a 9 y period was 8.3%, 96% of victims were male, and the mean age was 32.7 y [6]. For survivors, the injury pattern is very complex, with a high disability rate due to accompanying neurologic damages and loss of limbs.

Away from the workplace, most electrical injuries are due to either indoor household low-voltage (<1000 V) electrical contact or outdoor lightning strikes [7]. Domestic household 60-Hz electrical shocks are common and usually result in minor peripheral neurological symptoms or occasionally in skin surface burns. However, more complex injuries may result depending on the current path, particularly following oral contact with household appliance cord disclosures or outlets in small children [8]. Compared to a high-voltage shock that usually is mediated by an arc, low-voltage shocks are more likely to produce a prolonged, "no-let-go" contact with the power source. This "no-let-go" phenomenon is caused by an involuntary, current-induced, muscle spasm [9]. For 60 Hz electrical current the "no-let-go" threshold for axial current passage through the forearm is 16 mA for males and 11 mA for females [10,11].

There are roughly 200 human deaths annually in the United States due to lightning strikes and there are three times those many who survive. The range of lightning injury extent is quite broad, depending upon the magnitude of exposure and the condition of the victim. Usually lightning hits result in surface burns, complex neurological damage similar to blunt head trauma, peripheral neurologic injury, and cardiac damage [12]. Radio frequency (RF) and microwave injuries are less common. Nonetheless, they represent an important medical problem to understand. In short, electrical trauma may produce a very complex pattern of injury because of the multiple modes of frequency-dependent tissue–current interactions, the variation in current density along its pathway through the body, as well as variations in body size, body position, and use of protective gear. No two cases are the same.

10.2 Electrical Transport within Tissues

The fundamental bioengineering perspective is that the human body is considered to be a compartmentalized (or lumped element) conducting dielectric. It consists of about 60% of water by weight, in which 33% is intracellular and 27% is extracellular [13]. Body fluid in both the intracellular and the extracellular compartments is highly electrolytic, and these two compartments are separated by a relatively impermeable, highly resistive plasma membrane. Current within the body is carried by mobile ions in the body fluid.

The concentration of mobile ions results in a conductivity of approximately 1.4 S/m in physiological saline. Because electrons are the charge carriers in metallic conductors or electrical arcs, when in contact with the human body the current carrier changes from electrons to ions. This conversion occurs at the skin surface through electrochemical reactions [14].

At low frequencies (i.e., below radio frequencies), the electrical current passing across the body distributes such that the electric field strength is nearly uniform throughout any plane perpendicular to the current path [15–17]. As a consequence, current density distribution depends on the relative electrical conductivity of various tissues and the frequency of the current. Experimental data support this basic concept. Sances et al. [17] measured the current distribution in the hind limb of anesthetized hogs. They found that major arteries and nerves experienced the largest current density because of their higher conductivity. It was also observed that skeletal muscles carried the majority of the current due to their predominant volumetric proportions.

At a more microscopic scale, low-frequency current distribution within tissue is determined by the density, shape, and size of cells. The cell membrane is an insulating ion transport barrier that mostly shields the cytoplasmic fluid from low-frequency electrical current. In addition, the presence of cells diminishes the area available for ionic current and, in effect, makes tissues less conductive. As cell size increases, the membrane has less impact on a cell's electrical properties, because the volume fraction of the cell occupied by the membrane is proportionately decreased [18]. Similarly, the resistivity of skeletal muscle that is measured parallel to the long axis of the muscle cells is less than what is measured perpendicular to the axis. Solid volume fraction of the extracellular matrix can also be important in certain tissues and anatomic locations. For example, the resistivity of cortical bone and epidermis is higher than other tissues because their free water content is lower. This is also evident in the recent work by Kalkan et al. [19].

At higher frequencies, in the RF and microwave ranges, the current distribution is dependent on different parameters. The cell membrane is no longer an effective barrier to current passage, and capacitive coupling of power across the membrane readily permits current passage into the cytoplasm. Frequency-dependent factors like energy absorption and skin-depth effects govern the field distribution in tissues. At the highest frequency ranges, including light and shorter wavelengths, other effects such as scattering and quantum absorption effects become important in governing field distribution in tissues. Table 10.1 provides a categorization of frequency regimes, with corresponding wavelength spectrum, their common applications and their effects on tissues, as a result of electrical injury. Mechanisms of biological effects are different in each regime. A discussion of injury mechanisms must also be separated according to the frequency regime.

TABLE 10.1

Frequency—Wavelength Regimes with General Applications and Harmful Effects

Field Frequency (cycles/sec)	Energy Coupling Mechanism	Tissue Damage
dc to 10^3	Ionic currents	Joule heating
	Forces on cell structures	Membrane poration
10^3 to 10^7	Ionic currents	Joule heating
	Field energy absorption by cells	Cell spinning
10^7 to 10^9	Field energy absorption by proteins	Macromolecular heating
10^9 to 10^{11}	Field energy absorption by water	Microwave heating of water
10^{11} to 10^{15}	Field energy absorption by atomic bonds	Photo-optical protein damage

10.3 Physicochemistry of Tissue Injury

10.3.1 Low-Frequency Electric Shocks

The biophysical mechanism of injuries caused by contacting electrical power sources remain controversial [9]. It has been shown that the pathophysiology of tissue electrical injury is more complex, involving thermal, electroporation (EP), and electrochemical interactions [19–22], and blunt mechanical trauma secondary to thermoacoustic blast from high-energy arc [23]. The various modes of trauma lead to complex patterns of injury which remain incompletely described until today.

In the most general terms, tissue damage exists when proteins and other biomolecules, cellular organelle membranes or water content is altered. Among all the components of the cells and tissues which can be damaged by the electrical shock, it is the thin cell membrane which has the greatest vulnerability. Thus, the cell membrane appears to be the most important determinate of tissue injury accumulation.

The most important function of the cell membrane is to provide a diffusion barrier against free ion diffusion [24]. Because most metabolic energy of mammalian cells is used in maintaining transmembrane ionic concentration differences [25], the importance of the structural integrity of the lipid bilayer is apparent. The conductance of electropermeabilized membranes may increase by several orders of magnitude. ATP production and in turn, ATP-fueled protein ionic pumps, cannot keep pace and lead to metabolic energy exhaustion. If the membrane is not sealed, it results in cell necrosis. Thus, in discussing tissue injury resulting from electrical shock, the principal focus is directed at kinetics of cell membrane injury and the reversibility of that process. A simulation study of the membranes by Tarek [26] explains the EP phenomena in bilayers.

10.3.1.1 Direct Electric Force Damage

A cell within an applied dc or low-frequency electric field will experience electric forces which will act most forcefully across and along the surface of the cell membrane. The forces acting across the membrane can alter membrane protein conformation and disrupt the structural integrity of the lipid bilayer. The magnitude of the forces acting across the membrane is related to the induced transmembrane potential V_m. V_m depends on a variety of factors, such as the intra- and extracellular medium conductivity, cell shape and size, the external electric field strength E as well as how the electric field vector orients with respect to the point of interest on the cell membrane [27–29].

Given that most cells are either spherical or cylindrical in shape, the expressions which describe the relationship between the externally applied electric field and the induced transmembrane potential can be simplified to two simple forms. Considering physiologic conditions, the peak magnitude of induced transmembrane potential V_m (V_m^P) at the electrode-facing poles of spherical cells can be expressed as

$$V_m^P = 1.5R_{cell}\cos(\phi)(1+(f/f_s)^2)^{-1/2}E_{peak}, \tag{10.1}$$

with R_{cell} being the radius of the cell, E_{peak} is the peak field strength in the tissue surrounding the cell, ϕ is the angle of axis from the field direction, f_s is the sub-β-dispersion frequency limit below which the cell charging time is short compared to rate of field change, and f is the field frequency. For cylindrical-shaped cells, such as skeletal muscle and nerve cells, which are aligned in the direction of the field (herein assigned the z coordinate), the induced transmembrane potential takes a different form. Under these

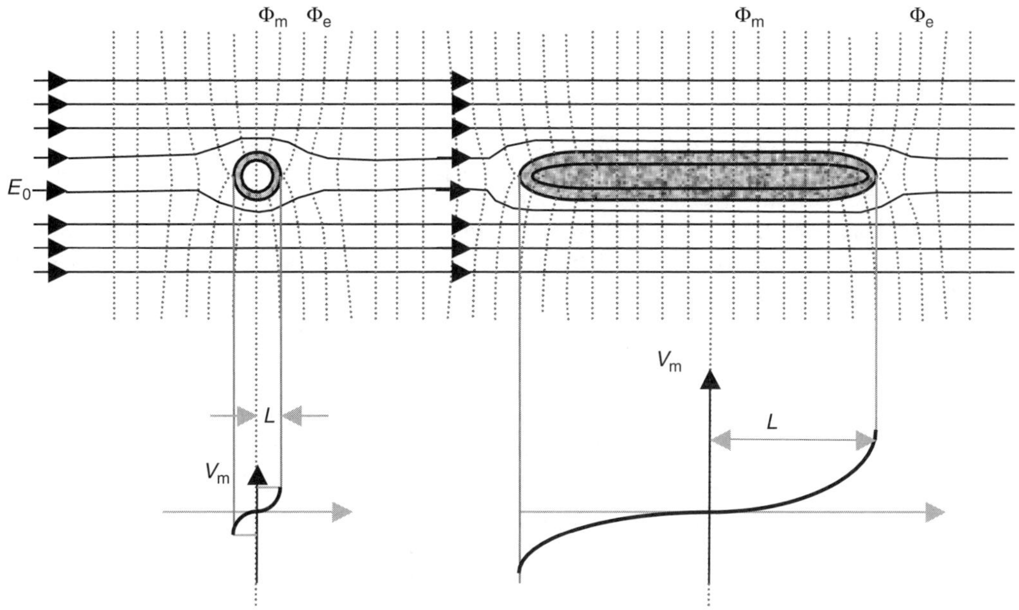

FIGURE 10.1
Dependence of cell size on the spatial variation of the induced transmembrane potential difference when cells are exposed to the same electric field. Electrical current lines are the same as electric field lines (E_0), and constant voltage lines are dashed.

circumstances, an electrical space constant parameter becomes useful in describing the electrical properties of the cell. The induced transmembrane potential can be expressed as a function of z

$$V_m^p(z) \approx A\lambda_m \sinh(z/\lambda_m)(1 + (f/f_s)^2)^{-1/2} E_{peak} \tag{10.2}$$

where λ_m is the electrical space constant of the cell, A is a variable that depends on cell length, the position $z = 0$ corresponds to the mid-point of the cell [28]. Figure 10.1 illustrates schematically the spatial variation of $V_m^p(z)$ on the cell size.

Equation 10.1 and Equation 10.2 are valid as long as the electrical properties of the cell membrane remain constant. However, the major theme of this chapter is that the transport properties of the cell membrane are altered by forces that are much greater than the natural physiologic forces. The natural transmembrane potential of mammalian cells has a magnitude of less than 100 mV [30]. When an artificially imposed potential results in a transmembrane potential magnitude of greater than 200 mV, intra-membrane molecular alterations occur which may lead to membrane damage.

The principal mechanisms of damage are EP of the lipid bilayer and electro-conformational denaturation of the membrane proteins. Electro-conformational damage to membrane proteins has been well documented for voltage-gated membrane protein channels. The processes occur quickly, on the order of milliseconds, after strong fields are applied.

EP is the biophysical process of membrane permeabilization owing to electric-field driven reorganization of lipids in the lipid bilayer by supraphysiologic electric fields [27,31,32]. It has been found useful applications at both single and multicellular levels to (1) introduce foreign DNA into cells; (2) introduce enzymes, antibodies, viruses, and other agents or particles for intracellular assays; (3) precipitate cell fusion; (4) insert or embed macromolecules into the cell membrane; (5) sampling of microenvironments across membranes [33]; (6) gene delivery in human embryonic stem cell [34], gene transfer in whole embryos [35] and gene repair in mammalian cells [36]. Recent reviews and books published have extensively treated this subject [27–29,37–41]. In this chapter, we will briefly examine this phenomenon as it relates to the understanding of electrical injury. (See also Chapter 9 in Vol. 2)

EP can be either transient or stable, depending on the magnitude of the imposed transmembrane potential, duration for which it is imposed, membrane composition, and temperature. The time required for EP ranges from tens of microseconds to milliseconds. The physical state of the lipid bilayer, either liquid crystal or fluidic, is strongly temperature dependent. After application of a brief electroporating field pulse, the transiently electroporated membrane will spontaneously seal. Sealing follows removal of water from the membrane defects. Sealing kinetics are often orders of magnitude slower than the field relaxation because the forces driving the molecular sealing events are not as strong as the electroporating electric field. Sealing of electropores requires reordering of membrane lipids and removal of water molecules from the pore: both time and energy consuming processes [42–44].

The threshold transmembrane potential for induction of membrane EP is remarkably similar across the cell types. The threshold V_m for EP has been found to be in the range from 300 to 350 mV [42–45]. Several authors have developed models to explain the experimentally observed values of V_m required for EP and associated transmembrane aqueous dynamics [46,47]. Using empirical data as parameters in an asymptotic approximation [48], the threshold V_m is predicted to be approximately 250 mV, which is quite consistent with reported experimental data.

Generally, for most media-suspended, isolated cells with a typical diameter of 10–20 μm, the dc field strength threshold for EP is in the range of 1 kV/cm. By comparison, the fields required to alter large cells is much less. Due to their relatively long length, skeletal muscle cells, up to 8 cm long in large animals, and nerve cells, up to 2 m long, have much lower EP thresholds. Therefore, muscle and nerve cell membranes are likely to be damaged with electrical fields as small as 60 V/cm.

The distribution of electropore formation in a cell placed in an applied field was addressed by DeBruin and Krassowska [49,50]. Expanding from previous theoretical models and including the fact that the membrane charging time of about 1 μs is very short compared to a 1 ms field duration, they concluded that supraphysiological V_m at the pole caps is large enough to create pores, and thereby effectively prevent a further increase in V_m in these areas.

This confirms early experimental findings which show a saturation of V_m that is independent of the field strength (for high-voltage shocks, [51–53]). After the effect of ionic concentrations is included in the model, it is even able to confirm asymmetries in V_m observed in respect to the hyperpolarized (anode-facing) and hypopolarized (cathode-facing) pole of a cell [43,44,54]. Although the pore sealing time (time needed for pores to close) in the range of seconds predicted by the model is in agreement with some published experimental results [55], others have found longer sealing times in the range of several minutes [42,43,56]. This might be explained by the fact that [1] this model is based on pure lipid bilayers instead of cell membranes embedded with proteins, and [2] it only considers primary pores formed by V_m (pores formed during shock) and not those formed after the external field pulse ends (secondary pores), which provide transport routes for macromolecules.

Investigation of EP of many cells within a tissue had been initially driven by the need for a better understanding of the pathophysiology of electrical injury [57,58]. In the early 1990s, it was studied in connection with cardiac defibrillation shocks [59,60]. More recently, tissue EP has begun to be envisioned as a potential therapeutic tool in the medical field. It has found use in (1) enhanced cancer tumor chemotherapy (electrochemotherapy [61,62]), (2) localized gene therapy [63,64], (3) transdermal drug delivery and body fluid sampling [65–67]. Computational models of human high-voltage electrical shock suggest that the induced tissue electric field strength in the extremities is high enough to electroporate skeletal muscle and peripheral nerve cell membranes [8,68–70] and to possibly cause electroconformational denaturation of membrane proteins.

Bhatt and coworkers [20] measured EP damage accumulation using isolated, cooled *in vitro* rat *biceps femoris* muscles. After the initial impedance measurement, an electric field pulse was delivered to the muscle using current pulses that setup tissue field pulse amplitudes ranging between 30–120 V/cm, which were thought to be typical forearm field strengths in high-voltage electrical shock. The duration of the dc pulses ranged from 0.5 to 10 ms. These short pulses reduce Joule heating to insignificant levels. Field pulses were separated by 10 s to allow thermal relaxation. The drop in the low-frequency electrical impedance in the muscle tissue following the application of short-duration dc pulses indicated skeletal muscle membrane damage. A decrease in muscle impedance magnitude occurs following dc electric field pulses that exceed 60 V/cm magnitude and 1 ms duration. Thus the field strength, pulse duration, and number of pulses are factors that determine the extent of EP damage.

Based on these results, Block et al. [21] electrically shocked fully anesthetized female Sprague–Dawley rats through cuff-type electrodes wrapped around the base of the tail and one ankle, using a current-regulated dc power supply. The objective was to determine whether EP of skeletal muscle tissue *in situ* could lead to substantial necrosis. The study involved histopathological analysis and diagnostic imaging of an anesthetized animal hind limb. A series of 4 ms dc-current pulses, each separated by 10 s to allow complete thermal relaxation back to baseline temperature before the next field pulse, was applied. The electric field strength produced in the thigh muscle was estimated to range from 37 to 150 V/cm, corresponding to applied currents ranging from 0.5 to 2 A. These tissue fields were suggested to be on the same level as that experienced by many victims of high-voltage electrical shock.

Muscle biopsies were obtained from the injured as well as the collateral control legs 6 h post shock and subjected to histopathological analysis. Sections of electrically shocked muscle revealed extensive vacuolization and hypercontraction-induced degeneration band patterns which were not found in unshocked contra-lateral controls (Figure 10.2). The fraction of hypercontracted muscle cells increased with the number of applied pulses. These results are consistent with the investigators hypothesis that nonthermal electrical effects alone can induce cellular necrosis. The pathologic appearance of the shocked muscle was similar to that seen in the disease malignant hyperthermia indicating that EP may lead to Ca^{2+} influx into the sarcoplasm. A similar muscle injury pattern has been described in a human electrical injury victim published by DeBono in a clinical case report [71]. These results suggested that direct electrical injury of skeletal muscle *in situ* can lead to the commonly seen pattern of injury in electrical shock victims even in the absence of pathologically significant Joule heating.

10.3.1.2 Thermal "Burn" Injury

Passage of electrical current through Ohmic conduction leads to Joule heating, which can lead to severe burn injury in electrical shock victims. Burn injury is used here to specifically refer to tissue injury by damaging supraphysiological temperatures. Burn effects are related to protein lysis of cell membranes and protein denaturation, often followed by recognizable changes in the optical properties of tissue. There are two conceptually

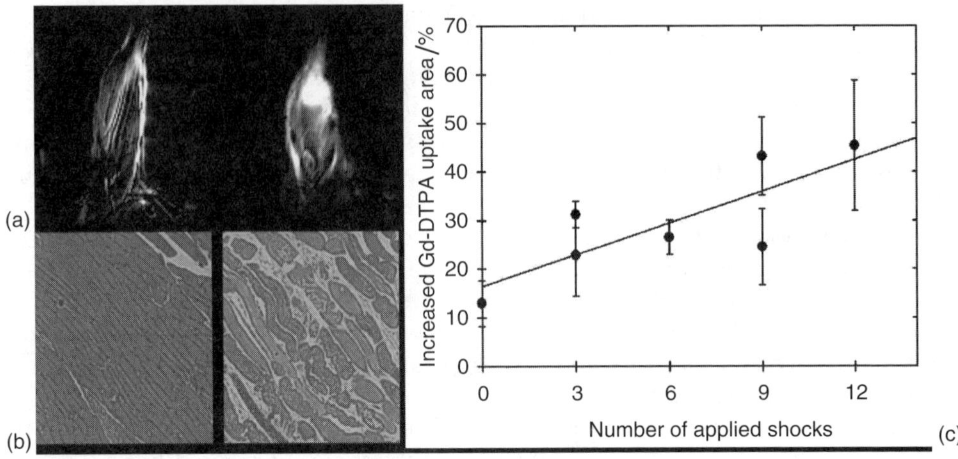

FIGURE 10.2 (See color insert following page 412)
The effects of multiple applied shocks on muscle tissue. (a) T2-weighted images of control (left) and shocked (right) muscles. (b) Histology sections of control (left) and shocked (right) muscle. (c) Relationship between increased Gd uptake with increasing number of applied shocks. The fraction of hypercontracted muscle cells increased with the number of applied pulses.

different potential outcomes for the denatured protein, which depend on the initial molecular structure and configuration. The first possibility occurs when the native folded conformational state of the protein, held by intra-molecular bonds, is different from the most favored conformation when no intramolecular crosslinks are needed to maintain the native folded state (thermodynamically lowest energy level). When this protein is heated, the intramolecular bonds are broken and it denatures to one of several preferred lower energy states from which it will not spontaneously return to the native conformation. Conceivably, if the primary structure of the protein is undamaged, it may be plausible to reconfigure the protein using similar chaperone-assisted mechanisms which establishes its initial folding after biosynthesis. The second possibility occurs when the native folded state of the protein is the same as the most preferred, energetically the lowest protein conformation in the absence of intramolecular crosslinks.

The speed of the transition from natural to denatured states is governed by the Arrhenius rate equation which states that when the kinetic energy of the molecule exceeds a threshold magnitude E_a^i (for activation energy), the transition to the ith state will occur, in this case from natural to denatured state. For a large number of molecules at temperature T the fraction with a kinetic energy above E_a is governed by the Maxwell–Boltzmann relation (Γ) [74],

$$\Gamma^{\iota} = \exp\left(-E_a^i/k_B T\right) \tag{20.4}$$

where k_B is Boltzmann's constant. Because the strength of bonds retaining the folding conformation of macromolecules is very dependent on the nature of the chemical bond, the value of E_a^i is dependent on molecular structure. Despite this complexity the net rate of denaturation of cellular structures containing many different proteins is also often describable in terms of Equation 10.4. For example, the accuracy of this equation in describing thermal damage to cell membranes has been reported [75–77]. Even the thermal injury to intact tissues like human skin is reasonably described by the simple Equation 10.4.

It has been known for more than 50 y that the rate at which damage accumulates in heated skin can be estimated by convolving Equation 10.4 with the temperature history. The resulting expression is called the "heat damage" equation [78],

$$d\Omega/dt = A\Gamma \tag{10.5}$$

where Ω is a parameter reflective of the extent of damage, and A is a frequency factor that describes how often a configuration occurs from which reaction is energetically possible, something which is also very dependent on molecular structure. The shape of the temperature–time curve predicted by Equation 10.5 is indeed the same as the human skin temperature versus time scale burn curve measured by Henriques and Moritz [79]. This temperature–time curve shape has also been obtained for heat damage to isolated cells [77].

Because the lipid bilayer components of the cell membranes are held together only by forces of hydration, the lipid bilayer is the most vulnerable to heat damage [80]. Even at temperatures of only 6°C above normal (i.e., 43°C) the structural integrity of the lipid bilayer is lost [81]. In effect, the warmed lipid bilayer goes into solution, rendering the membrane freely permeable to small ions. At slightly higher temperatures, published reports indicate that the contractile mechanism of muscle cells is destroyed immediately following exposure to 45°C and above [82]. Experiments on fibroblasts demonstrated that heat-induced membrane permeabilization also begins to appear above 45°C [83].

Bischof and coworkers investigated the effect of supraphysiological temperatures on isolated rat muscle cells using a thermally controlled microperfusion stage [84]. Cells were loaded with the membrane permeable fluorescent dye precursor calcein-AM. After entering the cell, the precursor is converted by nonspecific esterases into the membrane impermeable fluorescent Calcein. Using quantitative fluorescent microscopy Bischof et al. measured time-resolved dye leakage from the muscle cells at several supraphysiological temperatures. In addition, using Equation 10.5 and Equation 10.6, the authors determined the activation energy necessary to thermally induce membrane permeabilization in the isolated muscle cells to be 32.9 kcal/mol [84]. Reported activation energy values for thermal damage in other cell types are in the range from 30 to 140 kcal/mol [77].

10.3.1.3 Electro-Conformational Denaturation of Transmembrane Proteins

Imposed supraphysiologic transmembrane potential differences can produce electroconformational changes of membrane proteins, ion channels, and ion pumps. Approximately 30% of cell membrane consists of proteins, some of them embedded into the bilayer, others spanning across the entire membrane. Many of them carry electric charges from amino acids with acidic or basic side groups that can be acted on directly by an intense V_m (by, e.g., charge separation or charge induction through dissociation). In addition, each amino acid has an electrical dipole moment of about 3.5 D (1 Debye $= 3.336\times10^{-30}$ cm), giving the proteins an overall dipole moment that, in the case of an α-helical protein structure, can reach 120 D [85]. In a strong external electric field those molecules will orient themselves and thereby change their conformation to increase the effective dipole moment in the direction of that field.

If the field strength becomes sufficiently intense, those field-induced changes can cause irreversible damage to membrane proteins. In particular ion channels and pumps with their selective, voltage-gated charge transport mechanisms (e.g., a Ca^{2+}-specific channel) are highly sensitive to differences in V_m. Chen and coworkers investigated the effects of large magnitude V_m pulses on voltage-gated Na^+ and K^+ channel behavior in frog skeletal muscle membrane using a modified double vaseline-gap voltage clamp.

They found in both channel types, but more drastically in K^+ channels, reductions of channel conductance, and ionic selectivity by the imposed V_m [86]. Chen et al. were able to demonstrate that these changes are not caused by the huge field-induced channel currents (Joule heating damage) but rather by the magnitude and polarity of the imposed V_m [87]. In the most recent work, Clausen et al. [88] study the effects of shock and EP on the acute loss of force in skeletal muscles and the role of the Na (+) and K (+) pumps in the force recovery after EP. Ionic pumps alone are sufficient to compensate a simple mechanical leakage. They report that EP induces reversible depolarization, partial run-down of Na (+), K (+) gradients, cell membrane leakage, and loss of force. The consequences of this effect may underlie the transient nerve and muscle paralysis in electrical injury victims.

10.3.2 Radio Frequency and Microwave Burns

Every year a few cases of RF or microwave field injuries require medical attention in the United States. The victims are usually industrial workers. Above the low-frequency regime (>10 kHz), tissue response strongly depends upon the field frequency. In the 10–100 MHz RF range, two types of tissue heating occur, Joule and dielectric heating, with Joule heating outweighing dielectric heating. Small molecules like water, when not bound, are able to follow the field up to the gigahertz range [89]. However, at microwave frequencies (100 MHz–100 GHz), dielectric heating is more significant than Joule heating because both bound and free water are excited by microwaves. Molecular dipoles of macromolecules have lower natural frequencies, so that their most efficient induction frequency is in the radio frequency range.

Exposure to ambient microwave fields is known to cause burn trauma. Microwave burns have different clinical manifestations than low-frequency electrical shocks [90–93]. At low frequency the epidermis is a highly resistive barrier, whereas in the microwave regime, electrical power readily passes the epidermis in the form of "capacitive" coupling with very little energy dissipation. Consequently, the epidermis may not be burned unless it is very moist. The microwave field penetration into tissue has a characteristic depth in the range of 1 cm, resulting in direct heating of subepidermal tissue water. The rate of tissue heating is dependent not only on the amplitude of tissue electric field, but also on the density of dipoles. For example, microwave heating is much slower in fatty tissues [94].

10.3.3 Lightning Injury

Lightning arcs result from dielectric breakdown in air caused by build up of free electrical charges on the surface of clouds. The current through an arc can be enormous, but the duration is quite brief (1–10 ms). The primary current is confined to the surface of conducting objects connected by the arc. Peak lightning current ranges between 30,000–50,000 A and is able to generate temperatures near 30,000 K. This abrupt heating generates a high-pressure thermoacoustic blast wave known as thunder.

An individual directly struck by lightning will experience current for a brief period of time. Initially, the surface of the body is charged by the high electric field in the air. This can cause breakdown of the epidermis and several hundred amperes to flow through the body for a 1–10 μs period, which is certainly long enough to induce EP. Following this, a much smaller current persists for several milliseconds, during which time the body is discharging into the ground. The duration of current flow is relatively short, so there is no substantial heating except for a breakdown of the epidermis. However, disruption of cell membrane can wreak havoc on nerve and muscle tissues.

When lightning reaches the ground, it spreads out radially from the contact point. A substantial shock current can be experienced by a person walking nearby, if the feet are widely separated. For example, with an average lightning current of 20,000 A, a step length of 50 cm, and an individual located 10 m away from strike point, the voltage drops between the legs can reach 1500 V. This can induce a 2–3 A current flow through the body between the legs for a 10 μs period.

10.3.4 Diagnostic Imaging of Electrical Injury

Because most of the damage caused by electrical shock occurs beneath the skin, it is important to discuss how to achieve tissue injury detection and localization. There are two basic cellular abnormalities in electrical injury, altered protein structure and disrupted cell membranes. In addition, there can be blood coagulation, tissue edema, elevated tissue pressures and other abnormalities that effect molecular transport. From the clinical perspective, it is important to recognize area of damaged cells and interrupted blood flow.

Magnetic resonance imaging (MRI) methods are particularly well suited for detection of changes in protein folding, disruption of cell membranes and tissue edema. In particular, MRI methods in common hospital settings are typically based upon measurement of water proton behavior. As water behavior will change in the presence of denatured proteins and osmotic swelling will follow disruption of cell membranes, the typical MRI equipment in hospitals is well suited for the task. The MRI images in Figure 10.3 demonstrate the tissue changes seen adjacent to tissue-conductor contact points. The surface contact (220 V) was prolonged, resulting in deep burns and possible brain injury. Patient did not manifest significant neurology abnormality despite anatomic evidence of brain surface injury.

Technetium^{99m}-PYP (pyrophosphate) is widely used as a radiolabel tracer for various forms of soft tissue injury including electrical trauma; it is believed to follow the calcium movement in cellular function [72]. The increased tracer accumulation in the muscle tissue indicates loss of cell membrane integrity, tissue edema, and predictive of tissue. Using the *in vivo* rat hind limb electrical injury model described by Block, Matthews [73] monitored the uptake of Tc^{99m}-PYP in the electrically shocked tissue as a function of the magnitude of the dc current. Either 0.5, 1.0, or 1.85 A of direct current was applied to the rat's hind limb. Intravenous saline infusion was used as sham-treatment. Their results indicated that Tc^{99m}-PYP does accumulate in electroporated tissue and the level of the tracer accumulation is positively correlated to the tissue field pulses applied. This indicates that quantitative imaging of Tc^{99m}-PYP uptake may be developed further as an indicator of the extent of EP or other membrane injury.

10.4 Summary and Conclusions

Given the importance of electrical power to human culture, the problem of electrical injury is one that will continue to exist for the foreseeable future. Electrical injury has been poorly understood and perhaps less than optimally managed in the past. Better understanding of injury mechanisms, anatomical patterns of injury and therapy are required. A prompt, accurate clinical diagnosis of electrical injury is one of the most difficult tasks

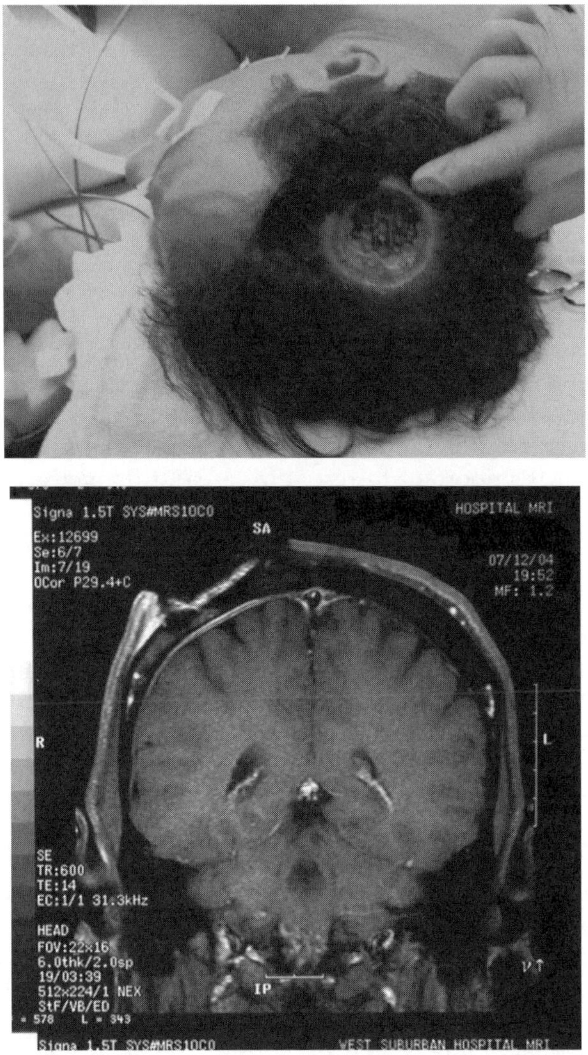

FIGURE 10.3
MRI image of electrical contact wound of scalp with extension through skull into brain surface. Dural and brain cortex damage is clear.

in the medical field [8] because it usually calls upon an understanding of the interaction between electric current and human tissue. Specifically speaking, the difficulty involves the following:

1. The exact tissue damage mechanism and damage level depend on a host of parameters: the characteristics of the power source (dc or ac current, voltage, frequency, etc.), path and duration of closed circuit, area and impedance of contact spot, etc. Correspondingly, there is a whole spectrum of damage characteristics depending on the values of these parameters. The physician needs to do 4-dimensional (spatial plus temporal) detective work in order to arrive at a correct diagnosis [95–114].
2. Electrical damage to the tissues is not easily detectable by visual inspection or physical examination. And often times its sequelae will not manifest themselves for a while: electrically injured tissue may initially appear viable, only to become visibly necrotic at a later point (in a number of days) [100,115,116].

The molecular structure of biological systems can be severely altered by the effects of high-energy commercial frequency electrical power. The mechanisms of damage include cell membrane EP, Joule heating, electroconformational changes of proteins (denaturation), and other macromolecules. Rehabilitation and reconstruction needs in the electrically injured are usually not obvious at the initial evaluation. The reintegration of the individual into their preinjury living situation often becomes a real challenge. Aside from physically obvious impairments (loss of limb, etc.) it is not uncommon for an electrician to develop a phobia toward electricity after being injured. The more in-depth understanding of the underlying mechanism of injury will lead to a more specific treatment regime that may prevent some of the late sequelae of electrical injury.

Acknowledgment

Parts of the research presented were funded by grants from the Electric Power Research Institute (RP WO-2914 and RP WO-9038), the National Institutes of Health (NIGMS 5-R01 GM53113) and Commonwealth Edison.

References

1. Nursal TZ, Yildirim S, Tarim A, Caliskan K, Ezer A, et al. 2003. Burns in southern Turkey: electrical burns remain a major problem. *J. Burn Care Rehabil.*, 24(5):309–314.
2. Occupational Safety & Health Administration Statistics & Data. U.S. Department of Labor, 1999. Source on line: http://www.osha.gov/oshstats
3. McCann M, Hunting KL, Murawski J, Chowdhury R, Welch L. 2003. Causes of electrical deaths and injuries among construction workers. *Am. J. Ind. Med.*, 43(4):398–406.
4. Gourbiere E. 1992. Work-related electrical burns. *Proc. Ann. Int. Conf. IEEE Engineer. Med. Biol. Soc.*, Paris, France.
5. Andrews CJ, Cooper MA, Darveniza M, Mackerras D. 1992. *Lightning Injuries: Electrical, Medical, and Legal Aspects*. Boca Raton, FL: CRC Press, p. 195.
6. Blome-Eberwein S, Jester A, Germann G. 2001. Electrical injuries in an adult patient population—incidence of limb salvage. Presented at the Annual meeting of the American Burn Association 2001 (abstract in Journal of Burn Care and Rehabilitation, March 2001).
7. National Safety Council, 1983.
8. Lee RC. 1997. Injury by electrical forces: Pathophysiology, manifestations, and therapy. *Curr. Probl. Surg.*, 34(9):677–675.
9. Hunt JL. 1992. Soft tissue patterns in acute electrical burns. In *Electrical Trauma: The Pathophysiology, Manifestations, and Clinical Management*, Lee RC, Cravalho EG, Burke JF, Eds., Cambridge, U.K.: Cambridge University Press, pp. 83–104.
10. Dalziel CF. 1943. Effect of frequency on let-go currents. *Trans. Am. Inst. Electr. Eng.*, 62:745–750.
11. Dalziel CF, Lee WR. 1969. Lethal electric currents. *IEEE Spectrum*, 6:44–50
12. Whitcomb D, Martinez JA, Daberkow D. 2002. Lightning injuries. *South Med. J.*, 95(11):1331–1334.
13. Duling BR. 1983. The kidney. In *Physiology*, Berne RM, Levy MN, Eds., St Louis: C.V. Mosby Company, p. 824.
14. Geddes LA, Baker LE. 1967. The specific resistance of biological material—a compendium of data for the biomedical engineer and physiologist. *Med. Biol. Eng.*, 5:271–293.

15. Lee RC, Kolodney MS. 1987. Electrical injury mechanisms: dynamics of the thermal response. *Plast. Reconstr. Surg.*, 80:663–671.
16. Daniel RK, Ballard PA, Heroux P, Zelt RG, Howard CR. 1988. High-voltage electrical injury. *J. Hand Surg.*, 13A(1):44–49.
17. Sances A, Myklebust JB, Larson SJ, Darin JC, Swiontek T. 1981. Experimental electrical injury studies. *J. Trauma*, 21(8):589–597.
18. Gaylor DG, Prakah-Asante A, Lee RC. 1988. Significance of cell size and tissue structure in electrical trauma. *J. Theor. Biol.*, 133:223–237.
19. Kalkan T, Demir M, Ahmed AS, Yazar S, Dervisoglu S, et al. 2004. A dynamic study of the thermal components in electrical injury mechanism for better understanding and management of electric trauma: an animal model. *Burns*, 30(4):334–340.
20. Bhatt DL, Gaylor DC, Lee RC. 1990. Rhabdomyolysis due to pulsed electric fields. *Plast. Reconstr. Surg.*, 86(1):1–11.
21. Block TA, Aarsvold JN, Matthews KL II, Mintzer RA, River LP, et al. 1995. Nonthermally mediated muscle injury and necrosis in electrical trauma. *J. Burn Care Rehabil.*, 16(6):581–588.
22. Lee RC, Astumian RD. 1996. The physicochemical basis for thermal and nonthermal "Burn" injury. *Burns*, 22(7):509–519.
23. Capelli-Schellpfeffer M, Lee RC, Toner M, Diller KR. 1998. Correlation between electrical accident parameters and injury. *IEEE Ind. Appl. Mag.*, 4(2):25–31.
24. Parsegian A. 1969. Energy of an ion crossing a low dielectric membrane: solutions to four relevant problems. *Nature*, 221:844–846.
25. Mandel LJ. 1987. Bioenergetic of membrane processes. In *Membrane Physiology*, 2nd ed. New York: Plenum Medical Book Company, pp. 295–310.
26. Tarek M. 2005. Membrane electroporation: A molecular dynamics Simulation. *Biophys J.*, 88:4045–4053.
27. Weaver JC, Chizmadzhev YA. 1996. Theory of electroporation: a review. *Bioelectroch. Bioener.*, 41:135–160.
28. Ho SY, Mittal GS. 1996. Electroporation of cell membranes: a review. *Crit. Rev. Biotech.*, 16(4):349–362.
29. Neumann E, Kakorin S, Tönsing K. 1999. Fundamentals of electroporative delivery of drugs and genes. *Bioelectroch. Bioener.*, 48:3–16.
30. Cevc G. 1990. Membrane electrostatics. *Biochim. Biophys. Acta*, 1031–1033:311–382.
31. Lee RC, Aarsvold JN, Chen W, Astumian RD, Capelli-Schellpfeffer M, et al. 1995. Biophysical mechanisms of cell membrane damage in electrical shock. *Sem. Neurol.* 15(4):367–374.
32. Chen W, Lee RC. 1994. Altered ion channel conductance and ionic selectivity induced by large imposed membrane potential pulse. *Biophys. J.*, 67(2):603–612.
33. Woods LA, Gandhi PU, Ewing AG. 2005. Electrically assisted sampling across membranes with electrophoresis in nanometer inner diameter capillaries. *Anal. Chem.*, 77(6):1819–1823.
34. Kim JH, Do HJ, Choi SJ, Cho HJ, Park KH, et al. 2005. Efficient gene delivery in differentiated human embryonic stem cells. *Exp. Mol. Med.*, 37(1):36–44.
35. Pierreux CE, Poll AV, Jacquemin P, Lemaigre FP, Rousseau GG. 2005. Gene transfer into mouse prepancreatic endoderm by whole embryo electroporation. *JOP*, 6(2):128–135.
36. Hu Y, Parekh-Olmedo H, Drury M, Skogen M, Kmiec EB. 2005. Reaction parameters of targeted gene repair in Mammalian cells. *Mol. Biotechnol.*, 29(3):197–210.
37. Weaver JC. 1993. Electroporation: a general phenomenon for manipulating cells and tissue. *J. Cell. Biochem.*, 51:426–435.
38. Teissie J, Eynard N, Gabriel B, Rols P. 1999. Electropermeabilization of cell membranes. *Adv. Drug Deliv. Rev.*, 35:3–19.
39. Neumann E, Sowers AE, Jordan CA, Eds. 1989. *Electroporation and Electrofusion in Cell Biology*. New York: Plenum Press.
40. Chang DC, Chassy BM, Saunders JA, Sowers AE, Eds. 1992. *Guide to Electroporation and Electrofusion*. New York: Academic Press.
41. Lynch PT, Davey MR, Eds. 1996. *Electrical Manipulation of Cells*. New York: Chapman & Hall.
42. Bier M, Hammer SM, Canaday DJ, Lee RC. 1999. Kinetics of sealing for transient electropores in isolated mammalian skeletal muscle cells. *Bioelectromagnetics*, 20:194–201.

43. Gabriel B, Teissie J. 1997. Direct observation in the millisecond time range of fluorescent molecule asymmetrical interaction with the electropermeabilized cell membrane. *Biophys. J.*, 73:2630–2637.
44. Gabriel B, Teissie J. 1998. Mammalian cell electropermeabilization as revealed by millisecond imaging of fluorescence changes of ethidium bromide in interaction with the membrane. *Bioelectroch. Bioener.*, 47:113–118.
45. Gowrishankar TR, Chen W, Lee RC. 1998. Non-linear microscale alterations in membrane transport by electropermeabilization. *Ann. NY Acad. Sci.*, 858:205–216.
46. Chizmadzhev AY, Arakelyan VB, Pastushenko VF. 1979. Electric breakdown of bilayer membranes: III. Analysis of possible mechanisms of defect origin. *Bioelectroch. Bioener.*, 6:63–70.
47. Glaser RW, Leikin SL, Chernomordik LV, Pastushenko VF, Sokirko AI. 1988. Reversible electrical breakdown of lipid bilayers: formation and evolution of pores. *Biochim. Biophys. Acta*, 940:275–287.
48. Neu JC, Krassowska W. 1999. Asymptotic model of electroporation. *Phys. Rev. E*, 59(3): 3471–3482.
49. DeBruin KA, Krassowska W. 1999. Modeling electroporation in a single cell. I. Effects of field strength and rest potential. *Biophys. J.*, 77:1213–1224.
50. DeBruin KA, Krassowska W. 1999. Modeling electroporation in a single cell. I. Effects of ionic concentrations. *Biophys. J.*, 77:1225–1233.
51. Hibino M, Shigemori M, Itoh H, Nagayma K, Kinosita K. 1991. Membrane conductance of an electroporated cell analyzed by submicrosecond imaging of transmembrane potential. *Biophys. J.*, 59:209–220.
52. Kinosita K, Hibino M, Itoh H, Shigemori M, Hirano K, et al. 1992. Events of membrane electroporation visualized on a time scale from microseconds to seconds. Guide to Electroporation and Electrofusion, Chang DC, Chassy BM, Saunders JA, Sowers AE, Eds., New York, Academic Press, pp. 29–46.
53. Knisley SB, Grant AO. 1994. Asymmetrically electrically induced injury of rabbit ventricular myocytes. *J. Mol. Cell. Cardiol.*, 27:1111–1122.
54. Hibino M, Itoh H, Kinosita K. 1993. Time courses of cell electroporation as revealed by submicrosecond imaging of transmembrane potential. *Biophys. J.*, 64:1789–1800.
55. Neumann E, Sprafke A, Boldt E, Wolff H. 1992. Biophysical considerations of membrane electroporation. Guide to Electroporation and Electrofusion, Chang DC, Chassy BM, Saunders JA, and Sowers AE, Eds., New York, Academic Press, pp. 77–90.
56. Rols MP, Teissie J. 1990. Electropermeabilization of mammalian cells: Quantitative analysis of the phenomenon. *Biophys. J.*, 58:1089–1098.
57. Lee RC, Gaylor DC, Prakah-Asante K, Bhatt D, Israel DA. 1988. Role of cell membrane rupture in the pathogenesis of electrical trauma. *J. Surg. Res.*, 44(6):709–719.
58. Gaylor DC, Lee RC., In *Electrical Trauma: The Pathophysiology, Manifestations, and Clinical Management*, Lee RC, Cravalho EG, Burke JF, Eds., Cambridge, U.K.: Cambridge University Press, pp. 401–425.
59. Tung L. 1992. Electrical injury to heart muscle cells. In *Electrical Trauma: The Pathophysiology, Manifestations, and Clinical Management*, Lee RC, Cravalho EG, Burke JF, Eds., Cambridge, U.K.: Cambridge University Press, pp. 361–400.
60. Tung L., Tovar O, Neunlist M, Jain SK, O'Neill RJ. 1994. Effects of strong electrical shock on cardiac muscle tissue. *Ann. NY Acad. Sci.*, 720:160–175.
61. Mir LM, Glass LF, Sersa G, Teissie J, Domenge C, et al. 1998. Effective treatment of cutaneous and subcutaneous malignant tumours by electrochemotherapy. *Br. J. Cancer*, 77(12):2336–2342.
62. Heller R, Jaroszeski MJ, Reintgen DS, Puleo CA, DeConti RC, et al. 1998. Treatment of tumors with electrochemotherapy using intralesional bleomycin. *Cancer*, 83:148–157.
63. Aihara H, Miyazaki JI. 1998. Gene Transfer into muscle by electroporation *in vivo*. *Nat. Biotechnol.*, 16:867–870.
64. Mir LM, Bureau MF, Gehl J, Rangara R, Rouy D, et al. 1999. High-efficiency gene transfer into skeletal muscle mediated by electric pulses. *Proc. Natl. Acad. Sci. USA*, 96:4262–4267.
65. Pliquett U, Langer R, Weaver JC. 1995. Changes in the passive electrical properties of human stratum corneum due to electroporation. *Biochim. Biophys. Acta*, 1239:111–121.

66. Prausnitz MR, Lee CS, Liu CH, Pang JC, Singh TP, et al. 1996. Transdermal transport efficiency during skin electroporation and iontophoresis. *J. Control. Release*, 38:205–217.
67. 1999. *Adv. Drug Deliv. Rev.*, 35:1–137.
68. Tropea BI, Lee RC. 1992. Thermal injury kinetics in electrical trauma. *J. Biomech. Eng.*, 114(2):241–250.
69. Diller KR. 1994. The mechanisms and kinetics of heat injury accumulation. *Ann. NY Acad. Sci.*, 720: 38–55.
70. Reilly JP. 1994. Scales of reaction to electric shock: thresholds and biophysical mechanisms. *Ann. NY Acad. Sci.*, 720: 21–37.
71. DeBono R. 1999. A histological analysis of a high voltage electric current injury to an upper limb. *Burns*, 50:541–547.
72. Shen AC, Jennings KB. 1972. Kinetics of calcium accumulation in acute myocardial ischemic injury. *Am. J. Pathol.*, 67:441–452.
73. Matthews KL II, 1997. *Development and application of a small gamma camera.* PhD thesis. University of Chicago, Chicago, IL.
74. Eyring H, Lin SH, Lin SM. 1980. *Basic Chemical Kinetics*. New York, Wiley Interscience.
75. Moussa NA, McGrath JJ, Cravalho EG, Asimacopoulos PJ. 1977. Kinetics of thermal injury in cells. *J. Biomed. Eng.*, 99:155–159.
76. Rocchio CM. 1989. The kinetics of thermal damage to an isolated skeletal muscle cell. SB thesis. Massachusetts Institute of Technology, Cambridge.
77. Cravalho EG, Toner M, Gaylor D, Lee RC. 1992. Response of cells to supraphysiologic temperatures: experimental measurements and kinetic models. In *Electrical Trauma: The Pathophysiology, Manifestations, and Clinical Management*, Lee RC, Cravalho EG, Burke JF, Eds., Cambridge, U.K.: Cambridge University Press, pp. 281–300.
78. Henriques FC. 1947. Studies in thermal injuries V: the predictability and the significance of thermally induced rate processes leading to irreversible epidermal damage. *Arch. Pathol.*, 43:489–502.
79. Henriques FC, Moritz AR. 1944. Studies in thermal injuries I: the conduction of heat to and through skin and the temperature attained therein. *Am. J. Pathol.*, 23:531–549.
80. Gershfeld NL, Murayama M. 1968. Thermal instability of red blood cell membrane bilayers: temperature dependence of hemolysis. *J. Membrane Biol.* 101:62–72.
81. Moussa NA, Tell EN, Cravalho EG. 1979. Time progression of hemolysis of erythrocyte populations exposed to supraphysiologic temperatures. *J. Biomech. Eng.*, 101:213–217.
82. Gaylor D. 1989. Role of electromechanical instabilities in electroporation of cell membranes. PhD thesis. MIT, Cambridge, p. 165.
83. Merchant FA, Holmes WH, Capelli-Schellpfeffer M, Lee RC, Toner M. 1998. Poloxamer 188 enhances functional recovery of lethally heat-shocked fibroblasts. *J. Surg. Res.*, 74:131–140.
84. Bischof JC, Padanilam J, Holmes WH, Ezzell RM, Lee RC, et al. 1995. Dynamics of cell membrane permeability changes at supraphysiological temperatures. *Biophys. J.*, 68(8):2608–2614.
85. Tsong TY, Astumian RD. 1987. Electroconformational coupling and membrane protein function. *Prog. Biophys. Mol. Biol.*, 50:1–45.
86. Chen W, Lee RC. 1994. Altered ion channel conductance and ionic selectivity induced by large imposed membrane potential pulse. *Biophys. J.*, 67:603–612.
87. Chen W, Han Y, Chen Y, Astumian RD. 1998. Electric field-induced functional reductions in the K+ channels mainly resulted from supramembrane potential-mediated electroconformational changes. *Biophys. J.*, 75(1):196–206.
88. Clausen T, Gissel H. 2005. Role of Na,K pumps in restoring contractility following loss of cell membrane integrity in rat skeletal muscle. *Acta Physiol. Scand.*, 183(3):263–271.
89. Chou CK. 1995. Radiofrequency hyperthermia in cancer therapy. In *The Biomedical Engineering Handbook*, Bronzino JD, Ed., Boca Raton, FL: CRC Press, pp. 1424–1430.
90. Sneed PK, Gutin PH, Stauffer P. 1992. Thermoradiotherapy of recurrent malignant brain tumors. *Int. J. Radiat. Oncol. Biol. Phys.*, 23(4):853–861.
91. van Rhoon GC, van der Zee J, Broekmeyer-Reurink MP, Visser AG, Reinhold HS. 1992. Radiofrequency capacitive heating of deep-seated tumours using pre-cooling of the subcutaneous tissues: results on thermometry in Dutch patients. *Int. J. Hyperthermia*, 8(6):843–854.

92. Nicholson CP, Grotting JC, Dimick AR. 1987. Acute microwave injury to the hand. *J. Hand Surg. Am.*, 12(3):446–449.
93. Alexander RC, Surrell JA, Cohle SD. 1987. Microwave oven burns to children: an unusual manifestation of child abuse. *Pediatrics*, 79(2):255–260.
94. Surrell JA, Alexander RC, Cohle SD, Lovell Jr FR, Wehrenberg RA. Effects of microwave radiation on living tissues. *J. Trauma*, 27(8):935–939.
95. Koumbourlis AC. 2002. Electrical injuries. *Crit. Care Med.*, 30(11 Suppl):S424–S430.
96. Peterson RR. 1980. *A Cross-sectional Approach to Anatomy*. Chicago: Yearbook Medical Publishers.
97. Freiberger H. 1933. The electrical resistance of the human body to commercial direct and alternating currents. (Der elektrische Widerstand des menschlichen Körpers gegen technischen Gleich- und Wechselstrom.) *Elektrizitätswirtschaft* (*Berlin*), (2):442–446; 32(17):373–375.
98. Pennes HH. 1948. Analysis of tissue and arterial blood temperatures in the resting human forearm. *J. Appl. Physiol.*, 1:93–122.
99. Jones RA, Liggett DP, Capelli-Schellpfeffer M, Macaladay T, Saunders LF, et al. 1997. Staged tests increase awareness of arc-flash hazards in electrical equipment. *IEEE Ind. Appl. Soc. Proc. PCIC* :313–322 (Abstr.).
100. Baxter CR. 1970. Present concepts in the management of major electrical injury. *Surg. Clin. North Am.*, 50:1401–1418.
101. Hunt JL, Sato RM, Baxter CR. 1980. Acute electric burns: current diagnostic and therapeutic approaches to management. *Arch. Surg.*, 115:434.
102. Zelt RG, Ballard PA, Common AA, Heroux P, Daniel RK. 1986. Experimental high voltage electrical burns: the role of progressive necrosis. *Surg. Forum*, 37:624–626.
103. Hannig J, Kovar DA, Abramov GS, Zhang D, Zamora M, et al. 1999. Contrast enhanced MRI of electroporation injury. *Proc. Jt. BMES/EMBS Conf., 1st*, CD 1078(Abstr.).
104. Isager P, Lind T. 1995. Accidental third-degree burn caused by bipolar electrocoagulation. *Injury*, 26(5):357.
105. Frey FJ. 2004. Microwave-induced heating injury *Ther Umsch.*, 61(12):703–706. [German]
106. DeToledo JC, Lowe MR. 2004. Microwave oven injuries in patients with complex partial seizures., *Epilepsy Behav.*, 5(5):772–774.
107. Sinha AK. 1985. Lightning induced myocardial injury—a case report with management. *Angiology*, 36(5):327–331.
108. Cooper MA. 1980. Lightning injuries: prognostic signs for death. *Ann. Emerg. Med.*, 9:134–138.
109. Ravitch MM, Lane R, Safar P, Steichen FM, Knowles P. 1961. Lightning stroke. *New Engl. J. Med.*, 264:36–38.
110. Cherington M. 2003. Neurologic manifestations of lightning strikes. *Neurology*, 60:182–185.
111. Hooshimand HF, Radfar F, Beckner E. 1989. The neurophysiological aspects of electrical injuries. *Electroencephalogy*, 20:111–120.
112. Moran KT, Thupari JN, Munster AM. 1986. Electric and lightning induced cardiac arrest reversed by prompt cardiopulmonary resuscitation [letter]. *JAMA*, 255:211–257.
113. Amy BW, McManus WF, Goodwin CW, Pruitt BA. 1985. Lightning injury with survival in five patients. *JAMA*, 253:243–245.
114. Taussig HA. 1968. "Death" from lightning—and the possibility of living again. *Ann. Int. Med.*, 68(6):1345–1353.
115. Artz CP. 1967. Electrical injury simulated crush injury. *Surg. Gynecol. Obstet.*, 125:1316–1317.
116. Hammond J, Ward CG. 1994. The use of Tc99m-PYP scanning in management of high-voltage electrical injuries. *Am. Surg.*, 60:886–888.

11

Mechanisms and Therapeutic Applications of Time-Varying and Static Magnetic Fields

Arthur A. Pilla

CONTENTS

11.1 Introduction

It is now commonplace to learn the successful use of weak, nonthermal electromagnetic fields (EMF) in the quest to heal, or relieve the symptoms of a variety of debilitating ailments. This chapter attempts to give the reader an introduction and assessment of EMF modalities that have demonstrated therapeutic benefit for bone and wound repair and chronic and acute pain relief. This chapter will concentrate on the use of exogenous time-varying and static magnetic fields. There is, however, a large body of research, including many clinical studies, describing the successful application of electrical signals via electrodes in electrochemical contact with the skin for pain relief and to enhance wound repair. Consideration of these modalities is beyond the scope of this chapter. The reader is referred to several excellent reviews of such electrical stimulation modalities [1–5]. Electroporation (see Chapter 9 in this volume) [6–8,372], which applies high-amplitude (>100 V/cm), short-duration (≤1 ms) voltage pulses with electrodes in contact with the target, allows controlled transient opening of the cell and other membranes, and has shown promise for gene transvection [9] and treatment of certain cancers [10], and is also beyond the scope of this chapter. Finally radio frequency (RF) (>100 MHz) and microwave signals are also beyond the scope of this chapter because these modalities are rarely utilized to enhance bone or wound repair, but rather for tissue heating, thermal ablation, or as surgical tools. Nonthermal bioeffects at these frequencies have been reported, but there are many controversial findings. Excellent reviews are available for the reader interested in detail [11,373].

As of this writing, there are a considerable number of peer-reviewed publications, which show EMF can result in physiologically beneficial *in vivo* and *in vitro* bioeffects. The number of people who have received substantial clinical benefit from exogenous EMF is certainly in the millions worldwide and is increasing rapidly as new clinical indications emerge. The EMF therapies also present as alternatives to many pharmacologic treatments with virtually no toxicity or side effects. The time-varying EMFs consisting of rectangular or arbitrary waveforms, referred to as pulsing electromagnetic fields (PEMFs), the pulse modulated radio frequency waveforms, particularly in the 15–40 MHz range, referred to as pulsed radio frequency fields (PRFs), and the low-frequency sinusoidal waveforms (<100 Hz) have been shown to enhance healing when used as adjunctive therapies for a variety of musculoskeletal injuries. Indeed, peer-reviewed meta-analyses clearly show that both PEMF and PRF modalities, now approved by regulatory bodies worldwide and widely used on patients to enhance bone and wound repair, are clinically effective [12,13]. Although still not completely elucidated, the mechanism of action of EMF signals at the molecular and cellular level is now much better understood and strongly suggests ion or ligand binding in a regulatory cascade could be the signal transduction pathway [14–28]. Furthermore, *a priori* configuration of physiologically effective waveforms via tuning the electrical properties of the exogenous EMF signal to the endogenous electrical properties of ion binding has recently been reported [29,30].

This chapter will provide a brief overview of the basic and clinical evidence that time-varying magnetic fields (EMF) can modulate molecular, cellular, and tissue function in a physiologically significant manner. The fundamental questions relating to the biophysical conditions under which EMF signals could modulate cell and tissue function will be discussed in detail. Particular attention will be paid to the manner by which signal parameters are related to dosimetry. In other words, the properties that render an EMF signal bioeffective. An attempt is made to correlate dosimetry for weak magnetic field with that for electric field effects. The ratio of signal to (endogenous) thermal noise (SNR) in the target is used in an SNR and Dynamical Systems model which has been successful for the *a priori* configuration of physiologically significant waveforms and which the reader may find useful to decipher the myriad of waveforms that have been utilized. The model may also allow the reader to perform an *a posteriori* analysis of waveforms for dose-related explanations for the presence or absence of a biological effect. Examples of *in vivo* and *in vitro* studies are given, illustrating specific EMF waveforms, including several examples of the use of the model.

11.2 Tissue Repair

11.2.1 Orthopedic Applications

Five million bone fractures occur annually in the United States alone. About 5% of these will become delayed or nonunion fractures with associated loss of productivity and independence [31]. Several techniques are available to treat recalcitrant fractures such as internal and external fixation; bone grafts or graft substitutes, including demineralized bone matrix, platelet extracts, and bone matrix protein; and biophysical stimulation, such as mechanical strain applied through external fixators or ultrasound (US), and EMFs.

The electrical properties of bone tissue have been extensively investigated. Yasuda in Japan hypothesized that endogenous electrical activity observed in bone was the mediator of repair and adaptive remodeling responses to mechanical loading and that an exogenous electrical signal alone could stimulate the response [32,33]. A seminal report soon followed on bone piezoelectric properties from the pioneering work of Fukada and Yasuda [34]. These authors showed a voltage could be obtained upon deformation of dry bone. Several groups, notably led by Becker at the State University of New York, Bassett at Columbia University, and Brighton at the University of Pennsylvania, soon reported the generation of electrical potentials in wet bone on mechanical deformation [35–39]. Similar observations were subsequently made in collagen and cartilaginous tissues [40–43]. The important conclusion from these studies was the revelation that bone and other tissue could respond to electrical signals in a physiologically useful manner. This ultimately led to the use of EMFs to modulate bone repair.

The development of modern EMF therapeutics was stimulated by the clinical problems associated with nonunion and delayed union bone fractures. It started with the pioneering work of Yasuda, Fukada, Becker, Brighton, and Bassett, mentioned above, who responded to the fundamental orthopedic question of how bone adaptively and structurally responds to mechanical input by suggesting that an electrical signal may be involved in the transduction of the mechanical signal to cellular activity. This naturally led to the suggestion that superimposing an exogenous EMF upon the endogenous fields accompanying normal cellular activity could help in the treatment of difficult fractures. The first animal studies employed microampere level direct currents delivered via implanted electrodes. Remarkably, this resulted in new bone formation, particularly around the cathode [44]. As these studies progressed, it became clear that the new bone growth resulted from the chemical changes around the electrodes is caused by electrolysis [45]. However, it has been shown that a mechanical stimulus also plays a role in dc bone stimulation [46]. The first therapeutic devices were based on these early animal studies and used implanted and semiinvasive electrodes delivering dc to the fracture site [47,48]. This was followed by the development of clinically preferable externally applied EMF modalities [49–52]. Subsequent studies concentrated on the direct effects of EMFs leading to modalities that provided a noninvasive, no-touch means of applying an electrical or mechanical signal to a cell or tissue target. Therapeutic uses of these technologies in orthopedics have led to clinical applications, approved by regulatory bodies worldwide, for treatment of recalcitrant fractures and spine fusion [53–59] and recently for osteoarthritis of the knee [60–62]. Additional clinical indications for EMF have been reported in double-blind studies for the treatment of avascular necrosis [63,64] and tendinitis [65]. This spectrum of applications clearly demonstrates the potential of this biophysical modality to enhance musculoskeletal tissue healing.

At present, the clinical modalities in use for bone repair consist of electrodes implanted directly into the repair site or noninvasive capacitive or inductive coupling. Direct current is applied via one electrode (cathode) placed in the tissue target at the site of bone repair and the anode placed in soft tissue. Direct currents of 5–100 μA are sufficient to stimulate osteogenesis [45]. The capacitive coupling (CC) technique utilizes external skin electrodes placed on opposite sides of the fracture site [66]. This requires openings in the cast or brace to allow skin access. Sinusoidal waves of 20–200 kHz are typically employed to induce 1–100 mV/cm electric fields in the repair site [67]. The inductive coupling (PEMF) technique induces a time-varying electric field at the repair site by applying a time-varying magnetic field via one or two electrical coils. The induced electric field parameters are determined by frequency characteristics of the applied magnetic field and the electrical properties of the tissue target [15,30,50,51]. Several waveform configurations have been shown to be physiologically effective. Peak time-varying magnetic fields of 0.1–20 G, (0.01–2 mT in SI units), inducing 1–150 mV/cm peak electric fields in a 3-cm diameter target, have been used [50,68]. The relationship between inductively coupled waveform characteristics and their ability to produce physiologically significant bioeffects will be considered in detail below. One version of the inductive technique utilizes a specific combination of dc and ac magnetic fields (CMFs) that are believed to tune specifically to ion transport processes [17,341].

11.2.1.1 Cellular Studies

The cellular studies have addressed effects of EMFs on both signal transduction pathways and growth factor synthesis. The important overall result from these studies is that EMF can stimulate the secretion of growth factors (e.g., insulin-like growth factor-II, IGF-II) after a short-duration trigger stimulus. The clinical benefit to bone repair is enhanced production of growth factors upregulated as a result of the fracture trauma. The induced electric field thus acts as a triggering mechanism, which modulates the normal process of molecular regulation of bone repair mediated by growth factors.

Studies underlying this working model have shown effects on calcium ion transport [69], a 28% increase in cell proliferation [70], a fivefold increase in IGF-II release [71], and increased IGF-II receptor expression in osteoblasts [72]. Increases of 53 and 93% on IGF-I and IGF-II, respectively, have also been demonstrated in rat fracture callus [73]. Stimulation of TGF-β mRNA by threefold with PEMF in a bone induction model in the rat has been reported [74]. This study also suggests the increase in growth factor production by PEMF may be related to the induction of cartilage differentiation [75]. It also suggests the responsive cell population is most likely mesenchymal cells [76], which are recruited early in the duration of PEMF stimulus to enhance cartilage formation. Upregulation of TGF-β mRNA by 100%, as well as collagen, and osteocalcin synthesis by PEMF has been reported in the human osteoblast-like cell line MG-63 [77,78]. PEMF stimulated a 130% increase in TGF-β1 in bone nonunion cells [79]. That the upregulation of growth factor production may be a common denominator in the tissue level mechanisms underlying electromagnetic stimulation is supported by studies from the Brighton [80,81], Stevens [82], and Aaron [83] groups.

Using specific inhibitors, Brighton et al. suggest that EMF acts through a calmodulin-dependent pathway [81]. This follows reports by the Pilla group [84–90] that specific PEMF and PRF signals, as well as weak static magnetic fields, modulate Ca^{2+} binding to calmodulin (CaM) by a twofold acceleration in Ca^{2+} binding kinetics in a cell-free enzyme preparation. The Stevens group has shown upregulation of mRNA for BMP2 and BMP4 with PEMF in osteoblast cultures. The Aaron group has reported extensively on upregulation of TGF-β in bone and cartilage with PEMF. All of these studies have utilized EMF signals identical to those that have demonstrated clinical success. The ion-binding target pathway has recently been confirmed in other studies using static magnetic fields [91,92]. PEMF has been reported to increase angiogenesis by threefold in an endothelial cell culture [93]. A recent study confirms this and suggests PEMF increases *in vitro* and *in vivo* angiogenesis through a sevenfold increase in endothelial release of FGF-2 [94].

11.2.1.2 Animal Studies

Bassett et al. [95,96] were the first to report a PEMF signal could accelerate bone repair by 150% in a canine tibial osteotomy model. A bilateral cortical hole defect model in the metacarpal bones in horses showed PEMF treated holes produced a statistically significant increase in amount of new bone formation and mineral apposition rate [97,98]. A capacitively coupled signal was shown to prevent osteopenia due to both sciatic-denervation and castration in rat osteopenia models [99,100]. PEMF inhibited bone loss in an ovariectomized canine model [101]. Combined magnetic fields reversed osteopenia in ovariectomized rats [102]. An avian ulna disuse model showed a significant increase in bone formation when treated with PEMF [103]. The frequency dependence of EMF effects was also studied in this model. The results showed maximal response was observed with a 15 Hz sinusoidal waveform producing 10-μV/cm peak electric field in tissue. Experimental models of bone repair show enhanced cell proliferation, calcification, and increased mechanical strength with direct currents [104,105]. Capacitive coupled fields have been reported to improve the mechanical strength of experimental fractures and healing osteotomies [67]. Several studies with PEMF showed increased calcification and enhanced mechanical strength in healing bone [106,107]. Exposure time studies report a linear effect of daily exposure with a 6 hour stimulation being most effective [68]. A series of animal studies reported that dc, CC, and PEMF techniques enhance the formation of bone and improve fusion rates in spinal arthrodeses [108–110]. Direct currents of 10 μA per cm of cathode length showed the best acceleration of spinal fusion [111]. The mechanical strength of late phase osteotomy gap healing in the dog was 35% stronger in PEMF treated limbs [112]. PEMF increased bone ingrowth into hydroxyapatite implants in cancellous bone by 50% [113]. PEMF produced a 10% increase in the diameter of arteriolar microvessels in rat muscle from which the authors suggested increased local blood flow could play a role in the PEMF acceleration of bone repair [114]. The use of *in vivo* micro-computed tomography showed PEMF reduced bone loss in a nonunion fibular model in the rat by threefold [362]. In a related study the effect of PEMF waveform configuration was examined. The results showed callus stiffness in a rat fibular osteotomy was increased twofold by a PEMF signal routinely employed for clinical bone repair, whereas a second PEMF waveform with much higher frequency content was ineffective [368].

11.2.1.3 Clinical Studies

Electromagnetic stimulation modalities have been used clinically to treat fresh fractures, osteotomies, spine fusions, and delayed and nonunion fractures. The efficacy of EMF stimulation on bone repair has been studied in a formal meta-analysis [12]. Twenty randomized control trials were identified. Fifteen trials supported EMF effectiveness and five failed to show effectiveness. Most studies used PEMF. In all cases, the primary outcome measure was bone healing assessed by radiographs and clinical stability test.

Results from pooled trials of 765 cases supported the effectiveness of PEMF stimulation of bone repair. However, because of the inability to pool data from all studies, conclusions regarding PEMF efficacy in bone repair were only suggestive. PEMF significantly accelerated union of femoral and tibial osteotomies in randomized, placebo controlled studies by approximately 50% [115–117].

PEMF, CC, and dc have been used to promote healing of spine fusions for the treatment of chronic back pain from worn or damaged intervertebral discs. This is measured by the increase in successful fusions from 50% to approximately 80% using EMF as adjunctive treatment. This application has also been subjected to meta-analysis [13]. Five randomized, controlled trials and five nonrandomized case controlled studies showed positive results for the enhancement (by 60%) of spine fusion by electrical and electromagnetic stimulation. There are many studies and reviews that show electrical and electromagnetic stimulation is effective in promoting spinal arthrodesis [118–122].

The effectiveness of EMF in promoting healing of recalcitrant fractures has been reviewed [123]. Twenty-eight studies of ununited tibial fractures treated with PMF were compared with 14 studies of similar fractures treated with bone graft with or without internal fixation. The overall success rate for the surgical treatment of 569 ununited tibial fractures was 82%, while that for PMF treatment of 1718 ununited tibial fractures was 81%, suggesting it is significantly more advantageous for the patient to use PEMF rather than submit to invasive surgery for the first bone graft. There are several observational studies suggesting the efficacy of dc, CC, or PEMF techniques in stimulating healing of delayed unions and nonunions [124–132]. Interestingly, and of huge clinical significance, studies comparing PEMF with bone graft show their equivalence in promoting union of delayed union or nonunion fractures [123,133–135]. Finally, there is a promising study on the effects of PEMF on distraction osteogenesis for the correction of bone length discrepancies [136].

Thus, several physical modalities have been shown to effectively manage nonunions and delayed unions of bone. Implantable direct current stimulation is effective as an adjunct in achieving spinal fusion. PEMFs induce weak nonthermal time-varying currents at the fracture site. Inductively and capacitively coupled EMFs appear to be as effective as surgery in managing extremity nonunions and lumbar and cervical fusions. Low-intensity US has also been used to speed normal fracture healing and manage delayed unions. Although these modalities seem vastly different, there appears to be a common mechanism of action. This will be discussed below.

All of the modalities discussed above now constitute the standard armamentarium of orthopedic clinical practice. Since the success rate for these modalities has been reported equivalent to that for the first bone graft, a huge advantage to the patient ensues because PEMF therapy is noninvasive and is performed on an outpatient basis. PEMF therapy also provides significant reductions in the cost of health care since no operative procedures or hospital stays are involved. This also applies for the increased success rate of spinal fusions with EMF. Thus, the clinical effects of EMF on hard tissue repair are physiologically significant and often constitute the method of choice when standard of care has failed to produce adequate clinical results. It is interesting to note that EMF may be the best modulator of the release of the growth factors specific to each stage of bone repair, certainly more so than the exogenous application of the same growth factors [398].

11.2.2 Soft Tissue Applications

Chronic wounds and their treatment are an enormous burden on the healthcare system, both in terms of their cost ($5 billion to $9 billion annually) and the intensity of care required [137]. There is even more cost to society from attendant human suffering and reduced productivity. More than 2 million people suffer from pressure ulcers and as many as 600,000 to 2.5 million more have chronic leg and foot wounds [137]. Diabetic foot ulcers are the most common chronic wounds in western industrialized countries. Of the millions who have diabetes mellitus, 15% will suffer foot ulceration that often leads to amputation (100,000 per annum in the United States alone).

There is an emerging and substantial clinical application of EMF in wound healing. Soft tissue healing has been reported by the use of direct electrode coupled devices delivering waveforms similar to those produced by several transcutaneous electric nerve stimulation (TENS) devices currently approved by the Food and Drug Administration (FDA) [5]. Regulatory and reimbursement issues have prevented more widespread use of PEMF modalities. However, the clear clinical effectiveness of PEMF signals has resulted in significantly increased use [12]. In fact, the Center for Medicare Services (CMS) has now determined PEMF produces sufficient clinical outcome to permit, and reimburse for, use in the off-label application of healing chronic wounds, such as pressure sores and diabetic leg and foot ulcers [138]. In addition, PRF devices have been cleared by the FDA for the relief of acute and chronic pain and the reduction of edema, all symptoms of wounds from postsurgical procedures, musculoskeletal injuries, muscle and joint overuse, as well as chronic wounds.

As for bone repair, application of EMF to soft tissue repair appears to have begun with observations of the electrical events associated with wound repair [139–143]. Injury currents, which develop in the presence of dermal wounds, are postulated to play an important role in the healing process [144,145]. These currents are, however, at least two orders of magnitude larger than the endogenous currents from SGP and are dc, or near dc, currents. Cells involved in wound repair are electrically charged and endogenous direct currents may facilitate cellular migration to the wound area [146,147]. In a manner similar to that for bone repair, the original working hypothesis was that exogenous EMF signals may enhance the endogenous electrical signals to accelerate wound repair. It has also been suggested that externally applied EMF may interact with the current of injury or trigger a relevant growth factor cascade [148,149]. Wound healing can be accelerated 2.5 fold with 200–800 μA direct current [150]. Both dc and PEMF have been reported to reduce edema, increase blood flow, modulate upregulated growth factor receptors, enhance neutrophil and macrophage attraction and epidermal cell migration, and increase fibroblast and granulation tissue proliferation [147,149]. Most wound studies involve arterial or venous skin ulcers, diabetic ulcers, pressure ulcers, and surgical and burn wounds.

11.2.2.1 Cellular and Animal Studies

The PEMF signal currently utilized for bone repair (see Figure 11.2, top) accelerates vascularization by several fold using cells from human umbilical vein and bovine aorta [151]. Studies on human umbilical vein cells showed that endothelial cell migration to a wounded area is accelerated by about 14% if cell cultures are exposed to an induced electrical field similar to the pulse burst currently used for bone repair (2 mT peak, 25 Hz repetition rate) [151]. Chronic stimulation of rat muscles increased blood vessel density by 14–30%, possibly through angiotensin and vascular endothelial growth factor pathways [152]. PEMF produced a significant increase in the rate of growth of the vascular tissue in the rabbit ear chamber [153] showing a dependence on signal configuration (repetitive pulse burst significantly better than repetitive single pulse). Sinusoidal signals (300 Hz) improved microcirculation and stimulated proliferation and differentiation of fibroblasts [154]. Amplitude, frequency, and orientation dependence of EMF modulation of fibroblast protein synthesis has been reported [155]. Inductively coupled sinusoidal fields (0.06–0.7 mT; 50, 60, and 100 Hz) increased chick embryo fibroblast proliferation up to 64% [156]. Human fibroblasts exposed to 20 or 500 mT, 50 Hz sinusoidal signals exhibited no effect on fibroblast proliferation [157]. Fibroblasts exposed to a PRF signal consisting of a 65 μs burst of 27.12 MHz sinusoidal waves repeating at 600/s (1G peak amplitude) showed enhanced cell proliferation by 130–220% [158,159]. Tissue cultures of human foreskin fibroblasts, when exposed to high 2 V/cm induced electric fields at either 1 or 10 Hz, demonstrated a sixfold increase in internal calcium, but excitation at 100 Hz had no significant effect [160]. Recent animal studies have reported that PRF signals produced a statistically significant several fold increase in neovascularization in an arterial loop model, suggesting an important clinical application for angiogenesis [161,162]. PRF signals, configured *a priori* assuming a Ca–CaM transduction pathway, accelerated wound repair in a rat cutaneous wound model by approximately 60% as measured by tensile strength [163]. However, treatment of identical wounds in the rat with PEMF of the type and intensity used for bone healing (see Figure 11.2, top) failed to produce significant increases in soft tissue fibroblast counts or improvement in wound closure [164]. PEMF increased the degree of endothelial cell tubulization and proliferation (threefold) *in vitro* [94]. In the same study PEMF increased fibroblast growth factor β-2 by fivefold from which the authors conclude that PEMF augments angiogenesis primarily by stimulating endothelial release of fibroblast growth factor-2 (FGF-2).

11.2.2.2 Clinical Studies

Nonthermal PRF signals were originally utilized for the treatment of infections in the preantibiotic era [165] and are now widely employed for the reduction of post-traumatic and postoperative pain and edema. Double-blind clinical studies have been reported for chronic wound repair, wherein PRF treated pressure ulcers closed by 84 vs. 40% closure in untreated wounds in one study [166] and 60% closure vs. no closure in the control group in another study [167]; acute ankle sprains, wherein edema decrease was sevenfold vs. the control group [168,169]; and acute whiplash injuries, wherein pain decreased by 50% and range of motion increased by 75% in the treated vs. control patients [170,171]. PRF signals have been reported to enhance skin microvascular blood flow by about 30% in both healthy [172] and diabetic [173] individuals. PRF reduced postmastectomy lymphedema by 56% and increased skin blood flow by fourfold [174]. The PEMF at 600 and 800 Hz, 25 μT mean amplitude, significantly reduced the size of venous ulcers by 63%, and decreased pain by 72%, in a randomized control study [175]. A modulated EMF signal at 10 and 100 Hz relieved the main clinical symptoms of diabetic peripheral neuropathy, improved peripheral nerve conduction by about 40% and the reflex excitability of functionally diverse motoneurons in the spinal cord [176].

A meta-analysis was performed on randomized clinical trials using PEMF on soft tissues and joints [12]. The results showed that both PEMF and PRF were effective in accelerating healing of skin wounds [177–183], soft tissue injury [168–171,184], and hair regrowth [185–187], as well as providing symptomatic relief in patients with osteoarthritis and other joint conditions [61,62]. PEMF has been successfully used in the treatment of chronic pain associated with connective tissue (cartilage, tendon, ligaments, and bone) injury and joint-associated soft tissue injury [188,189].

As for bone repair, EMF clinical effects on soft tissue repair are substantial and often constitute the method of choice when standard of care has failed to produce adequate clinical results. This is particularly true for chronic wounds, which often do not respond to standard of care and can be life-threatening if not resolved. It is interesting to note that

EMF can increase angiogenesis several fold in chronic wounds, significantly more than that achieved to date with growth factors such as vascular endothelial growth factor (VEGF), for which there have been generally disappointing clinical results [397]. This may be the primary reason that EMF is so effective with problem wounds wherein increased blood supply is always one of the primary clinical objectives. It is also interesting to note that EMF can provide an alternative to nonsteroidal anti-inflammatory drugs (NSAIDs) (e.g., ibuprofen, cox-2 inhibitors, etc.) and other pharmacological analgesics for the relief of chronic and acute pain.

11.3 Biophysical Considerations of EMF Therapeutics

11.3.1 Introduction

The above sections have provided an overview of the abundance of *in vitro*, *in vivo*, and clinical evidence that suggest that time-varying magnetic fields of various configurations can produce physiologically beneficial effects for conditions as varied as chronic and acute pain, chronic wounds, and recalcitrant bone fractures. This has all been achieved with low-intensity, nonthermal, noninvasive time-varying EMFs, having many configurations over a very broad frequency range. The reader should be aware that pulsing US and intermittent mechanical loading have also been shown to modulate bone repair and remodeling [190–198]. It has been suggested [199–201] that an ion-binding transduction pathway is common to both mechanical and EMF modalities, which provides useful US signal dosimetry information. This will be further discussed below. There is also compelling evidence that weak static magnetic fields can provide physiologically useful musculoskeletal pain relief [202–209]. Here also, ion binding may be involved in the transduction pathway via the effect of weak magnetic fields on the motion dynamics of the bound ion [19–26,210–216]. This can modulate binding kinetics by accelerating bound ions to preferred active orientations within the binding site or channel. The static magnetic fields as low as 10 μT can be detected in this pathway if the ion remains bound on the order of a second. One model, based upon Larmor precession of the bound ion, predicts static magnetic field effects as well as windows for certain combinations of ac and dc magnetic fields, similar to the ion cyclotron and parametric resonance models, as will be shown below [211–213].

Clearly, many EMF signals appear to have the capacity to achieve a physiologically meaningful bioeffect. Why should such seemingly different doses be effective? Are any signal parameters better than others? Is it the magnetic or electric field, or both? Does the state of the tissue target play a role? Despite the understandable impression that many EMF signals have been chosen in some arbitrary manner, the following sections will attempt to show that EMF dosimetry can have a rigorous quantitative basis based upon relatively simple mechanisms.

The biophysical mechanisms of interaction of weak electric and magnetic fields on biological tissues as well as the biological transductive mechanisms have been vigorously studied. One of the first models was created using a linear physicochemical approach [14,15,29,30,49,50,221,222,223,224,233,256] in which an electrochemical model of the cell membrane was employed to predict a range of EMF waveform parameters for which bioeffects might be expected. This approach was based on the assumption that voltage-dependent processes, such as ion or ligand binding and ion transport at and across the electrified interface of the cell membrane, were the most likely EMF targets. Several elegant studies further quantified this approach using Lorentz force considerations [16–19], including ion resonance and the Zeeman–Stark effect [20]. These suggested combined low-frequency ac and dc magnetic fields could modulate ion or ligand movement in a molecular cleft (binding site) and thereby affect binding kinetics [19–26,210–216,286]. Direct action of the Lorentz force on free electrons in macromolecules such as DNA has also been proposed [217–219].

At present, the most generally accepted biophysical transduction step is ion or ligand binding at cell surfaces and junctions which modulate a cascade of biochemical processes resulting in the observed physiological effect [81,83,220,225]. A unifying biophysical mechanism that could explain the vast range of reported results and allows predictions of which EMF signals and exposures are likely to induce a clinically meaningful physiological effect has been proposed [29,30].

Electromagnetic bioeffects from relatively weak (below heating and excitation thresholds) signals can be produced with a time-varying electric field, $E(t)$, induced from an applied time-varying magnetic field, $B(t)$. A large number of electromagnetic clinical devices in present use (particularly for bone and wound repair) induce 1–100 mV/cm peak E at the treatment site [12,13,68]. A myriad of waveforms have been employed and the fundamental question becomes one of dosimetry. In other words, the relation of waveform configuration to detectability (dose) at the supposed target must be established. The first step is evaluation of the amplitude and spatial dosimetry of the induced EMF within the target site for each exposure system and condition. This has been rigorously carried out for the laboratory dish with coils oriented vertically or horizontally [226–228]. Models have been created for the distribution of induced voltage and current in human limbs [229] and joints [230]. Three-dimensional visualizations of clinical PEMF signals have been reported [361].

11.3.2 Inductively Coupled Clinical EMF Waveforms

The electric field induced via a time-varying magnetic field waveform is directly related to the electrical characteristics of the coil employed and the current waveform applied to the coil. Induced electromotive force (emf) is proportional to the rate of change of current in the coil (dI_{coil}/dt), which produces the shape of the induced electric field. The following is applicable to coils that have been utilized for clinical bone fracture repair.

Coil current $I_{coil}(t)$ for an air-core inductor driven with a voltage step V_o rises exponentially to the limiting current defined by the coil resistance as

$$I_{coil}(t) = (V_o/R_{coil})[1-\exp(-t\, R_{coil}/L)] \tag{11.1}$$

where L is the coil inductance and R_{coil} is the effective coil resistance, including all connecting cable and driving circuit resistances. The waveform of the induced voltage is a direct function of the time derivative of $I_{coil}(t)$:

$$dI_{coil}/dt = V_o/L \exp(-t\, R_{coil}/L) \tag{11.2}$$

The above equation clearly shows that a rectangular-type induced waveform is achieved for a linear rise in coil current if τ_{coil} (L/R_{coil}) is sufficiently greater (e.g., by 10 times) than the rise time of current in the coil. This can be achieved by the proper choice of L and R_{coil}. One modality is to keep L relatively small so that safe driving voltages (<25 V) can be employed. Note that, as given by Equation 11.2, the maximum induced voltage (as $t \to 0$) is inversely proportional to coil inductance for a given V_o. Effective coil resistance can be

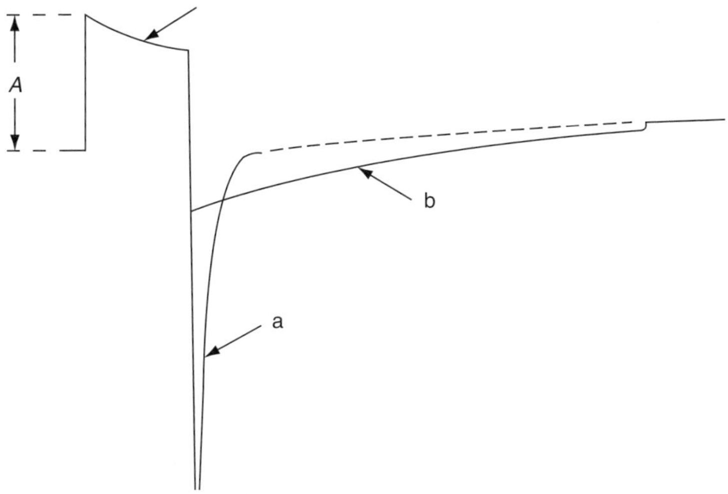

FIGURE 11.1
Schematic illustration of an inductively coupled electric field waveform used for bone and wound repair. The coil time constant (L/R_{coil}) determines the shape of this signal during rise of current in the coil. Collapse of coil current can be diode-limited (b) or determined by the impedance of the electronic driving circuitry (a). (From Pilla, A.A., Sechaud, P., and McLeod, B.R., *J. Biol. Phys.*, 11, 51, 1983. With permission.)

kept small by utilizing heavy magnet wire and connecting cable (e.g., 14–16 B&S gauge). With the above taken into account, it is easy to see that, for a given R_{coil}, the voltage step (V_o) can be applied to the coil for as long as the following relation is approximately valid:

$$(\mathrm{d}I_{coil}/\mathrm{d}t)_{t\to 0} = V_o/L[1-(t\ R_{coil}/L)] \quad (11.3)$$

The above equation shows coil current will rise approximately linearly resulting in an induced voltage waveform in the form of a step having some negative slope, as shown in Figure 11.1.

At the time of coil turn-off, coil current must collapse back to zero. This can be accomplished in either of the two manners. In the first, the coil current is allowed to decay at a rate determined only by the coil and driving circuit impedance (a, Figure 11.1). In the second, the rate of current collapse is controlled by an amplitude-limiting diode (b, Figure 11.1). By using either of these modes, induced waveform patterns having similar or different opposite polarity durations and amplitudes can be achieved.

The waveform shown in Figure 11.1 represents the electric field induced in the cell or tissue target via a coil placed external to the skin surface. Some of the pulse-type–induced electric field waveforms in common clinical use are shown in Figure 11.2. The rationale behind the configuration of these waveforms was based on the assumption that the induced electric field (and associated induced current density) is the primary stimulus. In other words, the magnetic component was considered to be the carrier or coupler, not significantly contributing to the biological effect. That this is correct for many PEMF clinical signals will become evident below. Clinical EMF signals for bone repair are inductively coupled, except for one capacitive coupled 60 kHz sinusoidal signal, for which E dosimetry, estimated from the geometry and dielectric properties of the target [121], is in the mV/cm peak range. There is also the continued and anomalous use of invasive direct currents, despite the accompanying electrolytic effects, which are known to cause bone formation via an inflammatory response [39–41].

The waveforms shown in Figure 11.1 and Figure 11.2 represent the time variation of the electric field signal induced in the extracellular fluid, cell, or tissue complex in a human,

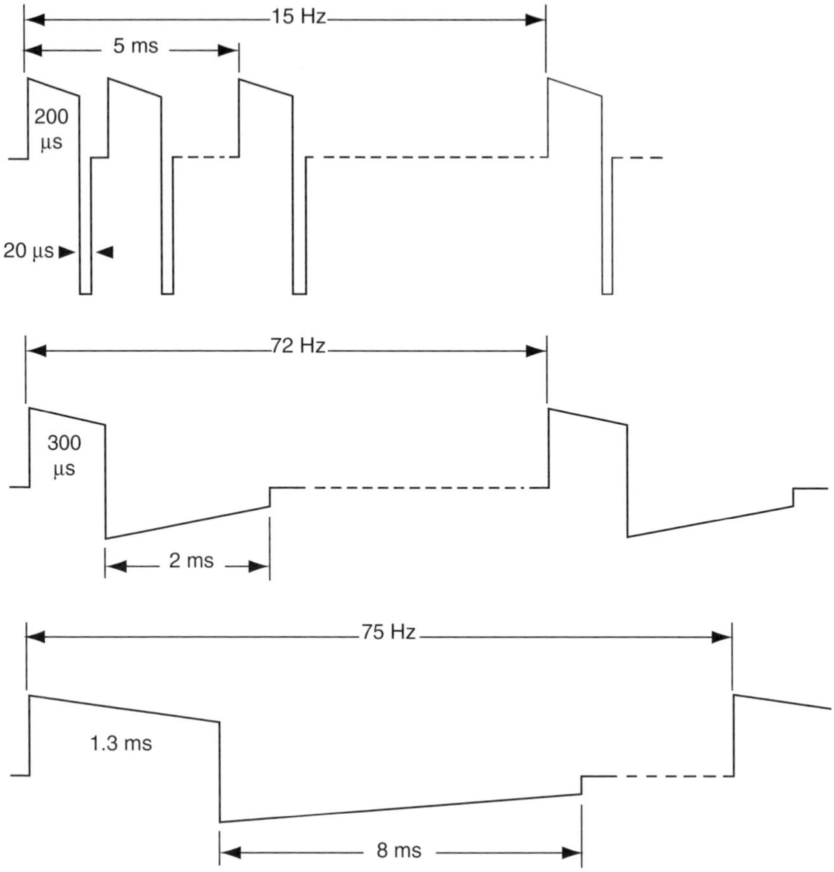

FIGURE 11.2
Induced electric field, $E(t)$, in tissue from the time-varying magnetic fields utilized in EMF devices for clinical applications. The top waveform consists of bursts of asymmetrical pulses; the others are wide asymmetrical single pulses. For all signals peak E is 1–10 mV/cm in a 2 cm target. All are detectable by some tissue targets. Positive clinical effects have been reported for all signals.

animal, or plant target. The distribution of current flow depends upon the geometry of coil and target. The basic rule is that the voltage induced will be defined by the distribution of magnetic flux within the tissue and the electrical properties of the target. To illustrate, consider a pair of Helmholtz aiding circular coils. A plastic culture dish, having cylindrical geometry, can be placed between the coils such that the symmetry (z) axis is aligned with that of the coils (see Figure 11.3). The $z = 0$ plane passes through the midplane of a saline solution, of conductivity σ, placed in the dish. Simple EMF theory predicts, as there is total symmetry in the angular (ϕ) direction, that the angular component of the induced electric field, E_ϕ, varies with radius, r, as

$$E_\phi(t) = -\frac{\mathrm{d}B}{\mathrm{d}t}\frac{r}{2} \tag{11.4}$$

which states the instantaneous amplitude of the induced electric field $E_\phi(t)$, within the area of a cylindrical target, such as a laboratory dish, penetrated by a uniform magnetic field B, is proportional to its rate of change with time $\mathrm{d}B/\mathrm{d}t$, and the radius r of the target. In other words, $E_\phi(t)$ at any point in space and time is dependent upon the rate of change of B and its spatial distribution in the target. The actual waveform of $E_\phi(t)$ depends upon

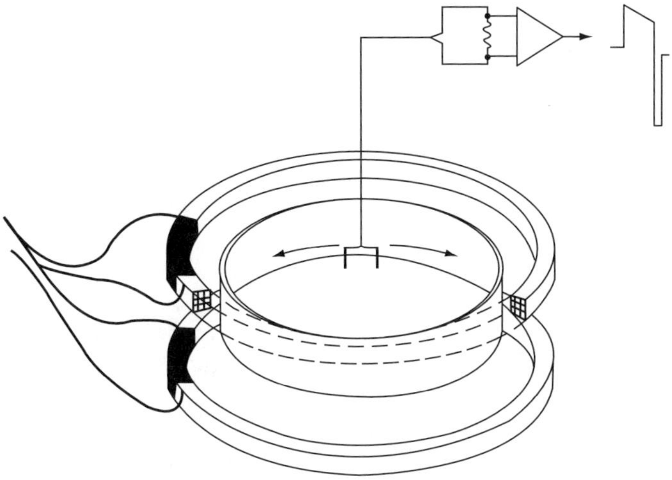

FIGURE 11.3
Schematic illustration of Helmholtz aiding coils with a plastic culture dish placed between the coils in a uniform magnetic field. This is an example of cylindrical geometry for which Equation 11.4 is valid. A special dual electrode probe may be placed at various radii parallel to the induced electric field, allowing E_ϕ, and current density, J_ϕ, vectors to be evaluated. (From Pilla, A.A., Sechaud, P., and McLeod, B.R., *J. Biol. Phys.*, 11, 51, 1983. With permission.)

dB/dt, which defines the applied frequency spectrum. For example, in the case of a clinical PRF signal, which is a sinusoidal wave, $dB/dt = \omega B$, where $\omega = 2\pi f$ (f = frequency, typically 27.12 MHz for PRF devices in the US). As dB/dt can be controlled, the induced electric field waveform may be chosen as a function of the electrical properties of the target (impedance, bandpass).

Equation 11.4 shows that the electric field induced in a homogenous cylindrical target is in circular loops, i.e., is rotatory, orthogonal to the magnetic field lines. In addition, the induced E field will be greater when the magnetic field intercepts a greater cross-sectional area of the sample, i.e., maximum E field in the target depends upon target size. Peak E field and associated current density, J, at a radius of 2 cm is often utilized for dosimetry comparisons. It is also convenient to use dB/dt as a measure of the peak induced electric field, assuming identical target size, for a given EMF signal. For example, a common clinical bone repair signal produces 20 G peak magnetic field in 20 μs. Thus $dB/dt = 10^6$ G/s for which peak $E_\phi(t) = 1$ V/cm $= 10$ mV/cm at a radius of 2 cm in the target, a typical dose metric for EMF bone growth stimulators. Note, however, that dB/dt alone is not sufficient to evaluate as a dose metric for a specific ion-binding target pathway. This will be discussed in Section 11.3.9.

Of course, Equation 11.4 is not valid for complex geometry and nonhomogenous targets; but it is always true, from Faraday's law of induction, that the induced voltage $V(t)$, along the line boundary of the surface S through which the flux ϕ penetrates, is

$$V(t) = -d\phi(t)/dt = -d/dt \int B(t)\, d\bar{S} \tag{11.5}$$

This expression indicates that, for any geometry, the time variation of induced V will also be identical to that of E in air and any other homogeneous nonmagnetic medium contained within the boundaries defined by S.

For most PEMF clinical devices the induced electric field and associated induced currents are small enough such that back emfs (due to the magnetic field from the induced or eddy current itself) are negligible. Thus, measurements of induced fields in air accurately reflect those at the target site for the PEMF devices utilized for bone repair. On the other hand, the induced electric field from PRF devices is orders of magnitude larger at carrier frequencies between 10 and 40 MHz. Therefore, the amplitude of the incident magnetic field (in air) is always perturbed by a tissue or saline load due to the secondary field from the induced currents, which act to cancel the primary magnetic field. Induced field levels and distribution have been evaluated in the presence of a tissue or saline load for PRF signals [249].

11.3.3 Electrochemistry at Cell Surfaces

For a living cell or tissue to respond functionally to an exogenous electric field, it is necessary that the field reaches and be detected at the appropriate molecular, cellular, or tissue site. In contrast to electrical potentials (and associated currents) that are applied directly across the cell's plasma membrane, the situation considered in this section involves the induction of electric fields within the cell or tissue target via electrical pathways originating at the external skin surface. As cells are surrounded by a highly conducting ionic medium, it is clear that an electric field containing relatively low frequencies (<100 MHz) can affect the cell only if current flows. This places certain constraints upon the relationship of the electrical characteristics of both the cell or tissue complex and the applied electric signal.

An important step, therefore, is the characterization of the electrical properties of cells and tissues. There are many such studies and the reader is referred to excellent reviews [231–233] (see also chapter 3 of *BBA*). However, it has also been proposed that a complete description of the electrical properties of cells and tissues should include the electrical equivalents of the electrochemical processes, which could be involved in the signal transduction pathway [50]. The electrical equivalents of electrochemical processes at cell surfaces and junctions and their relevance to EMF therapeutics have been described [14,15,30,49,50,234]. For the purposes of this review it is simply necessary to recall that the bilayer lipid structure of the plasma membrane and the electrical double layer at its inner and outer surfaces make the cell membrane a real capacitor, i.e., a charge storage system [240]. The structure of this capacitor is determined primarily by the interactions of water dipoles and hydrated ions with the charged chemical groups associated with the various lipid, protein, and carbohydrate components of the membrane. In addition, this capacitor is leaky because transmembrane ion transport can occur. These general properties can be described using an electrochemical approach to characterize the passage of current at and across the cell membrane by combining the dielectric and electrochemical properties of the membrane. In this manner not only the passive, but also the functional electrical response of the cell or tissue target may be taken into account.

Accordingly, induced current can affect cell surfaces and junctions via a complex, but readily discernible, set of electrochemical steps that are representative of the cell's real-time response to perturbations in its charged environment for any given functional state. Impedance measurements have reported the relative magnitudes of the time constants (relaxation times) of these processes [235–238]. As expected, dielectric and ion-binding time constants are somewhat smaller (1–100 μs) than those involving membrane transport (1–100 ms). In addition, the steady-state (dc) current pathway across the plasma membrane exhibits a specific resistivity several orders of magnitude above that of the extracellular fluid [239,241]. This means exogenous EMF signals need to contain frequencies well above dc to produce detectable electric field levels in the electrochemical pathways of ion binding and membrane transport. Exogenous direct currents will be mostly extracellular and generally need to be significantly higher than the peak induced current density from typical PEMF therapeutic signals to affect the distribution of charges (receptors) on the cell surface via electrokinetic mechanisms [242].

The electrochemical pathways involved in the transduction of an exogenous EMF signal into a physiologically significant endpoint appear to be operationally similar to the initial gating process involved in the production of the action potential via membrane depolarization [231]. It is therefore appropriate to consider the configuration of EMF waveforms in terms of an informational approach, or trigger, in contrast to one designed to supply energy to drive the biochemical cascade. Examples of the latter would be the use of direct currents large enough to cause cells to move along the electric field in wound repair applications and electroporation, wherein short voltage pulses are applied with sufficient electric field to temporarily cause the cell membrane to become permeable to macromolocules such as DNA or chemotherapeutic agents.

11.3.4 Electrochemical Information Transfer Model

It was proposed by Pilla in 1972 that nonthermal, subthreshold EMFs may directly affect ion binding and transport and possibly alter the cascade of biological processes related to tissue growth and repair [14]. This electrochemical information transfer (EIT) hypothesis postulated the cell membrane as the site of interaction of low level EMFs through modulation of the rate of binding of, e.g., calcium ion to receptor sites as a first step in a biochemical cascade relevant to the desired clinical outcome.

Ionic interactions at electrically charged interfaces of a cell are generally voltage-dependent (electrochemical) processes. Several distinct types of electrochemical interactions can occur at cell surfaces. One includes all of the nonspecific electrostatic interactions involving water dipoles and hydrated (or partially hydrated) ions at the bilayer lipid–aqueous interface of a cell membrane, which make all cell membranes a capacitor [231–233]. The other involves voltage-dependent ion or ligand binding. Here an ion or dipole can effectively compete with water dipoles and hydrated ions for specific membrane sites, which, in turn, can modulate a downstream cascade.

Equivalent electrical circuit models representing these electrochemical processes at cell surfaces and junctions have been derived [14,15,50,234]. Typically, most calculations consider a membrane model that consists of a capacitance, C_d, in parallel with an ionic leak pathway, R_M (see Figure 11.4). Although all membranes exhibit these properties, this simple model does not completely describe the dielectric properties of a functioning

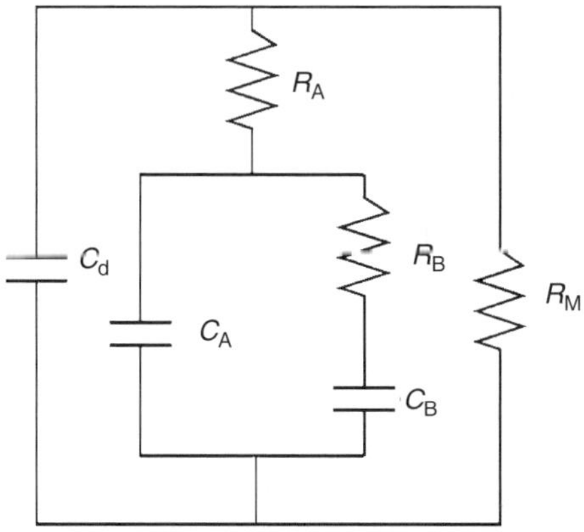

FIGURE 11.4
Electrical equivalent circuit of a cell membrane that exhibits a dielectric membrane capacitance, C_d, a leak resistance, R_M, an ion-binding pathway having an equivalent resistance, R_A, and capacitance, C_A, and a time constant, R_BC_B, representing a coupled surface step, e.g., conformation change. This circuit requires ion binding to occur prior to the follow-up step.

membrane, particularly with respect to the EMF transduction pathway. Impedance measurements on isolated cells have revealed the existence of relaxation processes which appear to reflect the kinetics of ion or ligand binding, as well as follow-up biochemical reactions [235–238]. Thus, a more general description of membrane dielectric properties, which takes into account electrochemical processes relevant to EMF sensitivity, considers an ion-binding step that precedes and possibly triggers a subsequent chemical reaction at the membrane surface. The current, $i_b(\omega)$ that flows into this pathway can be written

$$i_b(s) = q_a s \Gamma_a \Delta\Gamma_a(s) \tag{11.6}$$

where q_a is a coefficient representing the dependence of interfacial charge upon the surface concentration of the bound ion, Γ_a, and s ($=\sigma + \mathrm{j}\,\omega$) is the complex frequency variable of the Laplace transform [243,244]. A similar analysis could, of course, be carried out for $s = \mathrm{j}\,\omega$, i.e., a classical Fourier analysis [240,243].

Equation 11.6 shows the current in this pathway is a function of the change in surface concentration of the binding ion with time, $\Delta\Gamma_a(s)$, which, in turn, is voltage dependent and a function of the change in surface concentration of the product of the follow-up biochemical reaction, $\Delta\,\beta_b(s)$. In order to derive an expression for the impedance, $Z_A(s)$, of this pathway, it is necessary to define relationships between $\Delta\Gamma_a$ and the change in transmembrane voltage (V_M), and $\Delta\beta_b(s)$. This may be written, for first-order linear kinetics as [14,234]

$$\Delta\Gamma_a(s) = \frac{\nu_a}{\Gamma_a s}[-\Delta\Gamma_a(s) + \alpha V_M - \Delta\beta_b] \tag{11.7}$$

where ν_a is the binding rate constant and α is proportional to the potential dependence of binding ($\cong \partial\Gamma/\partial V_M$). The change in surface concentrations of the ion and the biochemical product can also be described by first-order kinetics

$$\Delta\beta_b(s) = \frac{\nu_b}{\beta_b s}[\Delta\Gamma_a(s) - \Delta\beta_b(s)] \tag{11.8}$$

where ν_b is the rate constant for the follow-up chemical reaction (defined as for ν_a) governing the rate of formation (decomposition) of the bound biochemical product after ion binding in a molecular cleft has occurred.

Equation 11.6 through Equation 11.8 allow the electrical impedance of the proposed transduction pathway at the cell membrane, $Z_A(s)$, to be written as

$$Z_A(s) = \frac{1}{\alpha q_a}\left[\frac{1 + \Gamma_a s/\nu_a}{\Gamma_a s} + \frac{1}{\Gamma_a s(1 + \beta_b s/\nu_b)}\right] \tag{11.9}$$

Inspection of Equation 11.9 reveals the existence of two time constants, the parameters of which are identifiable in terms of the rate constant and change in surface concentration of each reaction step. Thus, the equivalent resistance of binding, R_A is

$$R_A = \frac{1}{\alpha q_a \nu_a} \tag{11.10}$$

which shows the equivalent resistance of binding kinetics is, as expected, inversely proportional to the rate constant, ν_a. Correspondingly, the equivalent capacitance of binding, C_A, is directly proportional to the surface concentration of the binding entity, Γ_a as

$$C_A = \alpha q_a \Gamma_a \tag{11.11}$$

Note that the product αq_a ($\alpha \approx \partial\Gamma/\partial V_M$, $q_a = \partial q/\partial\Gamma$) in Equation 11.10 and Equation 11.11 has the dimensions of capacitance, the expected proportionality constant to enable equivalent electric circuit parameters to be related to ion-binding kinetics.

The second time constant, τ_B, in Equation 11.9 relates to the follow-up biochemical reaction

$$\tau_B = R_B C_B = \frac{\beta_b}{\nu_b} \tag{11.12}$$

where R_B and C_B are the equivalent resistance (inversely proportional to reaction rate, as for R_A) and capacitance (proportional to the surface concentration, β_b, of the reaction product) of the follow-up biochemical reaction, respectively. Note the follow-up reaction can be a conformational change as happens when Ca^{2+} is bound to at least three of the four available binding sites in calmodulin [288].

An electrical equivalent circuit, Z_A, which requires ion binding to occur prior to the follow-up reaction, is given in Figure 11.4. The membrane capacitance, C_d, and leak resistance, R_M, are also shown, as it is necessary to include these pathways in the total membrane impedance, Z_M, to fully characterize frequency dependence.

The impedance, $Z_d(s)$, of the C_d/R_M pathway is given by

$$Z_d(s) = R_M + \frac{1}{C_d s} \tag{11.13}$$

The most common representation of subexcitation threshold membrane impedance is via Equation 11.13. However, the electrochemical analysis given above shows this is not complete. Indeed, the pathways depicted in the ion-binding impedance given in Equation 11.9 have proven to be necessary to complete the EIT model. Time constants associated with electrochemical membrane processes have been reported [235–238]. One model system studied was the toad urinary bladder membrane having a single-cell thick epithelial layer with tight junctional electrical contact between cells, thereby affording high-resolution impedance values [235]. Isolated cell impedance studies utilized artificial epithelial layers created by deforming living cells under physiologic hydrostatic pressure into well-defined polycarbonate membrane filters. This technique was applied to melanoma, fibroblast, and osteoblast cells [238]. In all cases the results showed, as expected, a first time constant or relaxation process due to the ubiquitous dielectric capacitance of the lipid–protein bilayer. The time constant for this process is similar for all mammalian cells, in the 1–10 μs range. However, all cells exhibited at least a second time constant, which was characteristic of an ion-binding pathway. The time constant for this pathway was significantly different for each cell type, ranging from 20 μs for human erythrocytes to 200 μs for fibroblasts and osteoblasts. In addition, a longer time constant was often present related to passive ion transport across the cell membrane that could be coupled to the ion-binding step [50,234].

The above summarizes the EIT model that strongly guided the creation of the first clinically effective PEMF signal for recalcitrant fracture repair [51,52]. According to the EIT model, the requirements for an effective waveform could be met if it contained frequency components of sufficient amplitude within the time constant of the proposed target pathway [14,15,50]. Transmembrane ion transport, for which kinetics is in the millisecond range [235], was chosen as the target pathway for bone repair. This, coupled with practical restrictions on the size of the coil for patient use, led to the pulse burst waveform shown in Figure 11.2, top. It was supposed that the cell would ignore the short opposite polarity pulse and respond only to the envelope of the burst that had a duration of 5 ms, enough to induce sufficient amplitude in the kilohertz frequency range. Although the reasoning behind the asymmetric pulse in this waveform was erroneous because the EIT model was not yet complete, requiring a thermal noise analysis as will become apparent below, this signal is nonetheless effective for bone repair. It continues to be part of the standard armamentarium of the orthopedist for the nonsurgical noninvasive treatment of recalcitrant bone fractures.

It is important to note that the role of ions as transducers of information in the regulation of cell structure and function gained widespread acceptance well after the introduction of the EIT model. Ionic regulation mechanisms have now been described in: growth factor activation of the Na K-ATPase enzyme in fibroblasts [246,250]; Ca^{2+} regulation, via CaM, of the cell cycle [251,252]; differential Ca^{2+} requirements of neoplastic vs. nonneoplastic cells [245,253]; Ca^{2+}-dependent adenylate cyclase activation in macrophages [251]; Ca–CaM regulation of growth factor and other cytokine release [247,248,254,255]. EMF could also modulate the distribution of protein and lipid domains in the membrane bilayer, as well as conformational changes in lipid–protein associations by altering the kinetics of binding. Ion or ligand binding represents a coupling or transduction mechanism for exogenous EMFs at biological surfaces and junctions, which can be used to configure quantitatively and predictively the bioeffective EMF waveforms.

11.3.5 Magnetic Field Effects

As stated previously, the first PEMF signals utilized for tissue growth and repair were configured assuming that the induced electric field was the source of the stimulus (information). There is ample evidence that electric field is the dose metric for *in situ* PEMF signals such as those depicted in Figure 11.2. One study showed PEMF enhanced cellular proliferation and extracellular matrix synthesis for bovine articular chondrocytes only when a pair of Helmholtz aiding coils was horizontal to the culture dish. There was no effect when the same coils inducing the same pulsing magnetic field were oriented vertical to the culture dish [262]. Although the magnetic field was exactly the same for both coil orientations, the maximum electric field within the culture dish was more than a factor of 10 lower for vertical vs. horizontal coils, as expected from Faraday's law of induction (Equation 11.5). This is because the height of the medium in a typical culture dish is approximately 2 mm vs. a dish diameter of 35–60 mm. (Refer to Figure 11.3 for a schematic illustration of the horizontal coil configuration.) When coils are oriented vertically current flow is at right angles to that shown in Figure 11.3 and the path is limited to liquid height, not dish diameter.

Liburdy [263] used specially constructed annular culture dishes of differing diameter to show that calcium transport in mitogen stimulated thymic lymphocytes scaled with the induced electric field from a 22 mT, 60 Hz time-varying magnetic field. The magnetic field was identical for each loop diameter suggesting the dose metric in this case followed Equation 11.4, which shows induced electric field amplitude is directly proportional to the target loop radius for the cylindrical geometry of this experiment. The electric field dose metric was also demonstrated *in vivo* by exposure of fibular osteotomies in the rabbit to a clinically effective pulse burst PEMF signal (see Figure 11.2) applied via an external coil or via implanted electrodes [264]. The results showed the biomechanical acceleration of bone repair depended only on the *in situ* electric field and not on the magnetic component produced in the external coil to inductively couple the electric field to the repair site. Another study reported optimization of induced electric field parameters from time-varying magnetic fields to control bone remodeling in an avian model of disuse osteoporosis [265]. Finally, a recent study reported remarkable sensitivity of neutrophil metabolism to induced electric fields as low as 1 μV/cm [266]. The electric vs. magnetic field dose metric was specifically addressed in this study, and it was established that the observed results were indeed due to the induced electric field.

In contrast to the above-mentioned studies, Liboff et al. [267] reported in 1984 that human fibroblasts in culture exhibited enhanced DNA synthesis when exposed to sinusoidally varying magnetic fields over a wide range of frequencies (15–4 kHz) and amplitudes (2–600 μT). The effect appeared to be independent of the time derivative of the magnetic field, suggesting magnetic field was the dose metric for these sinusoidal waveforms. These results were somewhat controversial because the induced electric field at frequencies in the kilohertz range was of the same order as that reported effective for PEMF signals in cellular and clinical studies, and could clearly have contributed to the dose metric. However, this study also reported very low-frequency magnetic fields in the microtesla range could produce a similar effect on DNA synthesis. For these signals the induced electric field was well below that of the clinically effective PEMF signals, suggesting a direct magnetic field effect. Thus began the ion cyclotron and parametric resonance era. These models are covered in chapter 9 BBA [372], however a short summary will be useful.

Ion cyclotron resonance (ICR) [19,268–269] described frequency specific combinations of dc and ac magnetic fields, which can increase ion mobility near receptor sites and through ion channels. The Lorentz force equation was used to relate individual influences of both ac and dc magnetic fields to ligand receptor binding and motions of ions or other charged molecules [268 270]. The main objection to ICR is thermal noise [23,24,214,216,271]. Bianco and Chiabrera [272] have provided an elegant explanation of the inclusion of thermal noise in the Lorentz–Langevin model which clearly shows the force applied by a magnetic field on a charge moving outside the binding site is negligible compared to background Brownian motion and therefore has no significant effect on binding or transport at a cell surface or junction.

Because of thermal noise problems in the ICR model Lednev [20] and others [21,22,26,276–278] formulated an ion parametric resonance (IPR) quantum approach that modeled the ion in the binding site of a macromolecule (e.g., calmodulin, CaM) as a charged harmonic oscillator. It was proposed that the presence of a static magnetic field could split the energy level of the bound ion into two sublevels with amplitudes corresponding to electromagnetic frequencies in the infrared band. The difference between these two energy levels is the Larmor frequency (=cyclotron frequency/2). The IPR model, as for the ICR model, requires parallel ac and dc magnetic fields, and that they both be present.

In addition to difficulties with the experimental verification of the IPR model [279,350,351], the main fundamental objection to the IPR model relates to excited-state lifetime for the low-frequency EMF signals involved [280,281]. Indeed, the acceleration of the bound ion oscillating at frequencies of the order of 10^{12} Hz obviously cannot be affected by the negligible perturbations of the ion orbit generated by weak magnetic fields at 10^{10} lower frequencies. Therefore, the transition rate to the ground state cannot be affected by ELF fields and IPR cannot occur [282]. However, the axis of vibration of bound ligands can be affected by weak ELF, as well as dc, the magnetic fields in a classical manner, e.g., Larmor precession, and some orientations may cause enhanced biological effects [23,24,28,30,213,282].

Not withstanding all of the above, a clinical device was created on the basis of the ICR model that is currently used for recalcitrant bone fractures. Clinical results from this device appear to be equivalent to those from other inductive and capacitive coupled devices [31,59,68]. The signal is applied using an external pair of coils oriented parallel to one another. The alternating 40 μT sinusoidal magnetic field is at 76.6 Hz (a combination of Ca^{2+} and Mg^{2+} resonance frequencies). The static (dc) parallel magnetic field is at 20 μT. Since there are no published clinical studies with either the ac or dc component of the magnetic field alone, there is no solid evidence that this combination of ac and dc fields is unique, or, e.g., that either the ac or dc component alone would not have produced the same clinical results. The fact remains, however, that a clinical device, which produces an electrical field too weak to be detected by the tissue target, has demonstrated clinical success. This can only have been achieved via a magnetic, not electric, field.

There is enough additional significant evidence showing both low-frequency sinusoidal magnetic fields, which induce electric fields well below the thermal noise threshold, and weak static magnetic fields, for which there is no induced electric field, can have biologically and clinically significant effects [84–92,202–209,273,283,290,291,360]. In these cases also, the stimulus must clearly be the magnetic field. This was unexpected, particularly for weak dc magnetic fields. There is, however, a promising, and largely overlooked, model, remarkably unhindered by thermal noise, which considers the Lorentz force on a moving charge in a binding site in terms of Larmor precession and its possible effect on reactivity [24,28,210–216]. The Larmor precession model (LPM) is summarized in the following section.

11.3.6 Larmor Precession Model

Larmor precession, which describes the effects of exogenous magnetic fields on the dynamics of one state of ion binding, i.e., ions already bound, has been suggested as a possible mechanism for observed bioeffects due to weak static and alternating magnetic field exposures [210–216,271,288]. A bound ionic oscillator in a static magnetic field will precess at the Larmor frequency in the plane perpendicular to the applied field. This motion will persist in superposition with thermal forces, until thermal forces eventually eject the oscillator from a binding site. For a magnetic field oriented along the z-axis, the precessional motion will be confined to the x–y plane. The LPM proposes that the biochemical reactivity of a bound ion complex may be affected by changes in the spatial orientation of the bound ionic oscillator.

The effect of weak dc and ac magnetic fields according to the LPM can be summarized as follows. The Lorentz–Langevin equation written to describe the motion of an ion bound in a potential well (molecular cleft) subject to a magnetic field oriented along the z-axis in the presence of thermal noise forces is

$$\frac{d^2r}{dt^2} = -\beta\frac{dr}{dt} + \gamma\frac{dr}{dt}x\,B_o\mathbf{k} - \omega^2\mathbf{r} + \mathbf{n} \tag{11.14}$$

where $\mathbf{r}$ is the position vector of the ion; β is the viscous damping coefficient per unit mass (due to molecular collisions in the thermal bath); γ is the ion charge to mass ratio; B_o is the magnitude of the magnetic field vector; $\mathbf{k}$ is the unit vector along the z-axis; ω is the angular frequency of the oscillator; and $\mathbf{n}$ is the random thermal noise force per unit mass [271,288]. It has also been shown that precession is not limited to the case of a linear isotropic oscillator potential but will occur for any central restorative potential [211].

Equation 11.14 describes the motion of an oscillator (ion) in a molecular cleft due to an exogenous magnetic field in the presence of thermal noise. The solution may be written [212]

$$\mathbf{r}(t) \rightarrow \mathbf{e}^{i\omega t}r(t) = \mathbf{e}^{i\omega t}[C(t) + \Psi(t)] \tag{11.15}$$

where $\mathbf{r}(t)$ is the position vector of the bound ion; $C(t)$ is the coherent oscillation of the bound oscillator, and $\Psi(t)$ is the contribution due to thermal noise forces. The ion trajectory thus consists of a coherent part given by

$$C(t) = -\frac{C_0}{\omega} e^{-\frac{\beta}{2}t} e^{-i\omega_L t} \sin(\omega t) \tag{11.16}$$

where C_0 is determined by initial conditions and ω_L $(=\gamma B/2)$ is the Larmor frequency; and a component due to thermal noise

$$\Psi_n(t) = \frac{k_0}{\lambda_2 - \lambda_1} e^{-\frac{\beta}{2}t} e^{-\frac{i\omega_L}{2}t} [N(t)] \tag{11.17}$$

where k_0 is determined by initial conditions; $N(t)$ is the accumulation of the thermal component with time; and λ_1 and λ_2 are the roots and contain the ac and dc magnetic field terms [212].

The thermal component $\Psi(t)$ of the ion trajectory itself thus consists of an harmonic oscillator driven by thermal noise, subject to viscous damping and undergoing precessional motion at the Larmor frequency about the axis defined by the magnetic field. It oscillates at the fundamental frequency of the oscillator potential with amplitude increasing over time, ultimately resulting in ejection from the binding site after a bound lifetime determined by the magnitude of thermal forces.

Both the coherent and thermal components of an ion at a binding site exhibit Larmor precession in the presence of an applied magnetic field. As the amplitude of the thermal component grows the oscillator orientation still precesses at the Larmor frequency in the plane perpendicular to the applied magnetic field direction. This is illustrated in Figure 11.5, which shows the manner in which the amplitude of the oscillator vibration

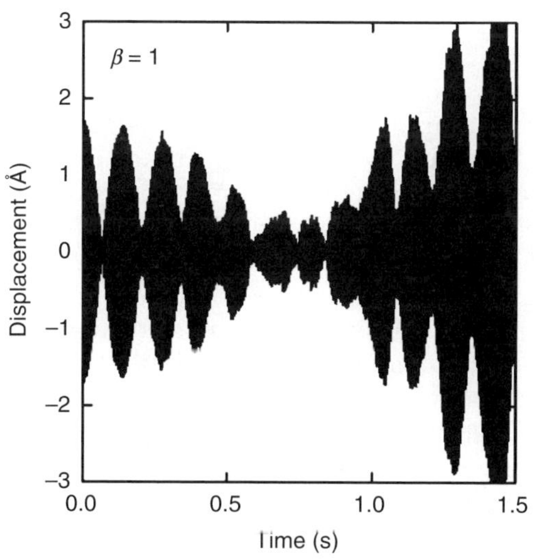

FIGURE 11.5
Overall effect of thermal white noise on precessional motion in a 10 μT static magnetic field. Note that, for this value of the viscous damping coefficient (β), precession is coherent and decays exponentially for approximately 0.5 s. Thermal noise then begins to add to the oscillator vibration amplitude, ultimately leading to ejection of the bound ion from the molecular cleft after a bound time of approximately 1.2 s. Until ejection, the oscillator is still precessing. (From Pilla, A.A., Muehsam, D.J., Markov, M.S., *Bioelectrochem. Bioenerg.*, 43, 241, 1997. With permission.)

(at infrared frequencies) is affected by thermal noise. Note that, even though thermal noise is progressively adding to the amplitude of oscillator vibration, the bound ion continues to precess at the original Larmor frequency until ejected from the binding site. Thus, precession frequency is unaffected by thermal noise while the oscillator is bound. The threshold for LPM is, therefore, determined only by the bound lifetime of the charged oscillator, allowing extremely weak magnetic fields to affect its dynamics.

Although thermal forces will in general be distributed throughout the spherical solid angle available in the binding site, it is important to bear in mind that the bound ion or ligand is not executing random motions in an isotropic region. Rather, it is strongly bound in an oscillator potential, with oscillator frequency in the infrared [211,212]. It is also important to emphasize that an ion bound in a molecular cleft exhibits vibrational and rotational, but not translational, degrees of freedom [393–395]. This means the bound oscillator can precess, but will not retain the ability to move in the random directions permitted in its unbound trajectory.

Larmor precession converts the exogenous magnetic field amplitude into a frequency determined by the gyromagnetic ratio of the target. Thus, for an ion oscillating along the z-axis the Larmor frequency ω_L is

$$\omega_L = -\Gamma B_{x,y} \tag{11.18}$$

where $\Gamma = q/2m$ for a simple unhydrated ion. Equation 11.18 illustrates that precession frequency scales with magnetic field provided the magnetic field has components perpendicular to the axis of oscillation. Note that cyclotron frequency $= 2\omega_L$. Of course, the minimum detectable magnetic field is determined by the contribution of thermal energy to the bound oscillator amplitude as described above.

According to LPM, each Larmor precession frequency determines the minimum time for the bound oscillator to reach reactive orientations at the binding interface. LPM predicts that a bound oscillator will accelerate faster to preferred orientations in the binding site with increasing static magnetic field strength. This can increase binding rate with a resultant acceleration in the downstream biochemical cascade. According to LPM, static magnetic fields in the 0.1–1 μT range can be detected if the oscillator remains bound for the order of a second [271]. To illustrate, the Larmor frequency for calcium in a 50-μT static magnetic field is approximately 18 Hz, suggesting a value for the damping coefficient, β, of about 35 or less is necessary in order for the oscillator to maintain a substantial amplitude over the period of one or more precessional orbits. The geometry of the binding site can create a locally hydrophobic region, from which dipolar molecules such as water are repelled [285], although at least one water dipole remains in a Ca–CaM binding site [287]. Thus, the binding site is a region in which a bound ion would experience very few collisions, resulting in a viscosity significantly below that of bulk water, accounting for the long bound times reported for Ca–CaM.

Weak static magnetic fields have been reported to accelerate Ca–CaM-dependent myosin light chain kinase (MLCK) and protein kinase C (PKC)-dependent processes up to twofold [84–92,273], although one report failed to show effects in the MLCK system [274] and another, which examined Ca^{2+} binding to CaM directly using a fluorescence technique, also failed to show effects [275]. Two studies reported the rate of Ca^{2+} binding to CaM was increased twofold with 2 G static magnetic field [90,92]. Static magnetic fields as low as 6 G increased cell survival by reducing stress-induced apoptosis threefold via a twofold increase of Ca^{2+} influx [352]. Weak dc alone (1 G) caused conformational changes in chromatin of *E. coli* bacteria similar in magnitude to ac–dc combinations chosen according to ICR or IPR models [353,354].

11.3.7 Resonance in Larmor Precession Model

There are credible reports of *in vitro* studies that demonstrate resonance behavior for certain combinations of weak ac and dc magnetic fields [19,278,283,349,350]. However, these do not unequivocally support the predictions of either the ICR or the IPR models [350]. Thus, it is interesting to consider ion resonance in terms of LPM. In order to assess the combined effect of simultaneous weak ac and dc magnetic fields on Larmor precession, it is necessary to recall that the bound charged oscillator will precess when only a dc field is present according to Equation 11.18. Addition of an ac magnetic field to an oscillator already precessing in a binding site will modulate oscillator motion for both perpendicular and parallel orientations with peak effects at multiples of the Larmor frequency.

The addition of an ac frequency in either a perpendicular or parallel direction to the axis of a precessing oscillator will modulate the axis of precession with a resultant effect on the time to reach a reactive orientation. This may be quantified by evaluation of the total excursion of the oscillator, $A(t)$, for a given binding time, t, and for various ac–dc combinations

$$A(t) = C_o \int_0^t \omega_L \, dt = \frac{C_o}{2m} \int_0^t [B_o + B_1 \cos(\omega_{ac} t)] \, dt = \frac{C_o}{2m} \left[B_o t + \frac{B_1}{\omega_{ac}} \sin(\omega_{ac} t) \, dt \right] \tag{11.19}$$

where ω_L is the Larmor frequency; B_o is the amplitude of the dc magnetic field; B_1 and ω_{ac} are the peak amplitude and the frequency of the ac magnetic field, respectively; m is the mass of the bound oscillator; and C_o is a proportionality constant.

Equation 11.19 may be evaluated for any ion or ligand, any combination of ac and dc magnetic fields with any relative orientation. An example is shown in Figure 11.6 in which dc magnetic field amplitude is varied for ac fixed at ω_L for Ca^{2+}. It may be seen that addition of an ac magnetic field to the precessing oscillator either further accelerates or inhibits its time to reach a reactive orientation. This behavior is remarkably similar to reported experimental verifications of IPR [278,350], and supports Larmor precession as a viable alternative mechanism for weak dc and ac magnetic field bioeffects.

It is interesting to speculate whether the parameters of the ac and dc magnetic field components could be chosen to actually prohibit the bound oscillator from reaching a reactive position in the binding site. If this possibility is verified by experiment, a non-invasive treatment for pathologies such as ectopic bone formation (wherein bone grows

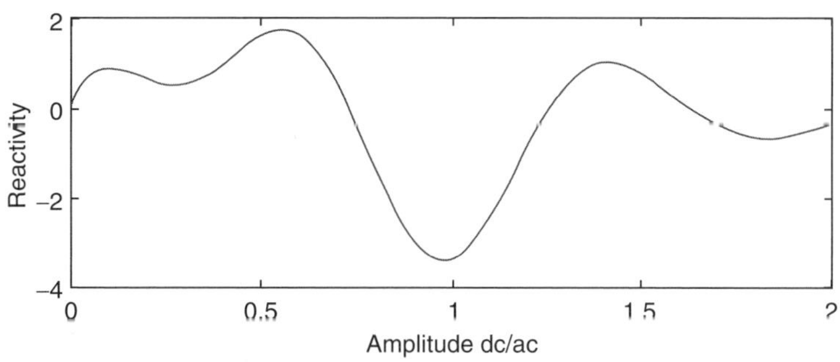

FIGURE 11.6
Effect of scanning dc magnetic field amplitude on a bound oscillator precessing at the Larmor frequency. Reactivity, e.g., Ca–CaM-dependent enzyme activity, can be either increased (vs. dc field alone) or decreased, meaning addition of ac can either further accelerate or inhibit the oscillator excursion to a reactive orientation within the binding site.

where it should not) and malignancies without side effects could emerge. This possibility does not appear to exist for PEMF modalities.

11.3.8 Dynamical Systems Model

Magnetic fields couple with binding kinetics via the Larmor frequency, while the induced electric field couples to the binding process via its potential dependence (see Equation 11.7). This may be modeled by considering the binding process as a dynamical system wherein the particle has two energetically (double potential well), stable points separated by a few kilotesla either bound in the molecular cleft, or unbound in the plane of closest approach to the hydrated surface (Helmholtz plane) at the electrified interface between the molecular cleft and its aqueous environment [288], as shown in Figure 11.7. Ion binding or dissociation is then treated as the process of hopping between these two states driven by thermal noise and EMF effects are measured by modulation of the ratio of time bound (in the molecular cleft) to time unbound (in the Helmholtz plane). This model does not require the ion to move in a trajectory far from the binding site, i.e., the ion is still in a prebound state while in the Helmholtz plane.

According to this model, modulation of the kinetics of binding is considered as either a change in the height or in the bias of the energy barrier between unbound and bound states. This is thus a dynamical system wherein thermal white noise is taken as the driving force for ion binding and dissociation. The reaction coordinate $q(t)$ is subject to inertia, a damping force proportional to dq/dt and a potential energy function $V(q, E_{ind})$

FIGURE 11.7
Cartoon of a molecular cleft illustrating an ion in a bound state that resides in the inner Helmholtz plane, IHP. Kinetics of binding is determined by ratio of time bound to unbound (outer Helmholtz plane, OHP). Bound ion hops between IHP and OHP driven by thermal noise and statistically does not drift into the diffuse double layer where it is subject to mechanical (electrokinetic) perturbation. While bound, the ion possesses rotational degrees of freedom and can precess. Note that water dipoles may remain resident in the binding site while the ion is in the bound state. (From Pilla, A.A., Muehsam, D.J., Markov, M.S., *Bioelectrochem. Bioenerg.*, 43, 241, 1997. With permission.)

dependent upon the induced electric field E_{ind}. The system can be described by the following differential equation

$$\frac{d^2q(t)}{dt^2} + \eta\frac{dq(t)}{dt} + \frac{dV(q, E_{ind})}{dq} = F_{noise} \tag{11.20}$$

where η is the coefficient of damping and F_{noise} is the force due to thermal noise on the bound particle. The force imparted on a charged ion by the induced electric field may be expressed as a perturbation of the potential energy function

$$V(q, E_{ind}) = \alpha_1\frac{q^4}{4} - \alpha_2\frac{q^2}{2} + \varepsilon\, q\, E_{ind} \tag{11.21}$$

where the nonnegative coefficients, α_1 and α_2, are characteristic of the receptor molecule-hydration environment and ε is the effective (hydrated) ion charge.

Equation 11.20 describes a double well potential wherein the potential energy wells correspond to the bound and unbound phases of the binding process. The dynamical system describes binding in a statistical sense. In terms of this model the force due to an induced electric field can modulate the relative depth of the wells thereby affecting the ratio of time bound to time unbound and thus the kinetics of the binding process. A weak magnetic field can indirectly affect the double well potential via Larmor precession by changing the dwell time of the bound vibrating oscillator, which, in turn, modulates the ratio of time bound to time unbound and therefore reaction rate.

The dynamical system model provides a rationale for linking observed static magnetic (LPM) and electric (EIT) field effects in identical molecular and cellular systems. For example, Ca^{2+}–CaM-dependent myosin phosphorylation has been extensively studied with both static magnetic fields [30,90], and with several PRF signals [29]. A comparison of typical results is shown in Figure 11.8, left, which shows a 2-G static field, and a 0.2 G,

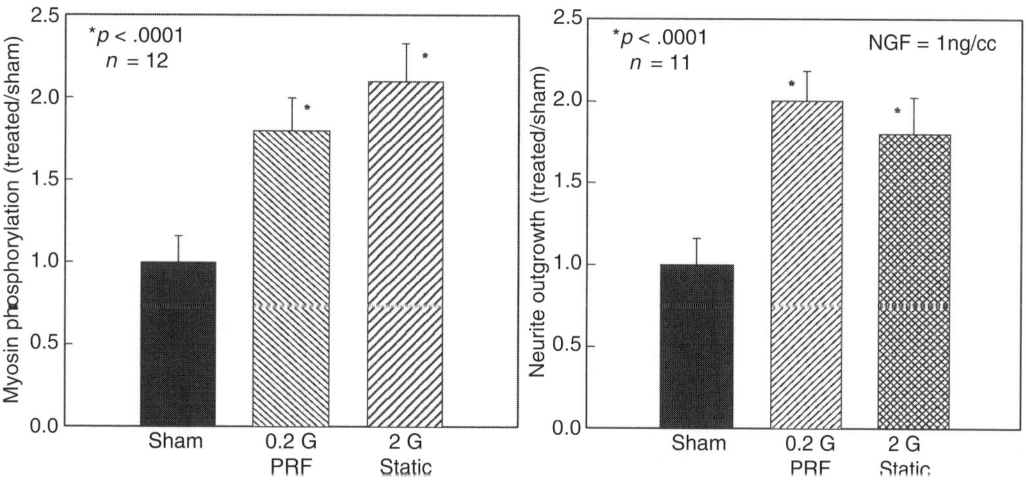

FIGURE 11.8
Comparison of the effect of a 0.2-G PRF signal having a 500-μs burst of 27.12-MHz sinusoidal waves repeating at 1 per s, and a 2-G static magnetic field on Ca^{2+}–CaM-dependent myosin phosphorylation (left), and neurite length from embryonic chick ganglia explants (right). Both signals elicit the same response for identical exposure times, however, the PRF signal couples to the target pathway via induced electric field, whereas the magnet couples via magnetic field.

500 μs burst of 27.12 MHz sinusoidal waves at 1-burst/s PRF signal, both accelerate phosphorylation nearly twofold in the presence of suboptimal free Ca^{2+} concentration. Similar behavior has been obtained for dendrite outgrowth from embryonic chick ganglia in the presence of suboptimal nerve growth factor (NGF) concentration (Figure 11.8, right). Dendrite length was also increased approximately twofold with both static and pulsing magnetic fields.

In both of these systems, the PRF signal induced a time-varying electric field and negligible magnetic field (vs. ambient), and the magnet produced only a static magnetic field, approximately 7X ambient. Dosimetry, therefore, depended upon the characteristics of the induced electric field for the PRF signal and on the magnetic field amplitude for the magnet. The common target pathway for both weak electric and magnetic fields, as proposed in the dynamical systems model, is ion or ligand binding. The common target pathway in both cases has been proposed to be the kinetics of Ca^{2+} binding, for which signal detection (SNR) can be estimated, as shown below, for the PRF signal, or for which the LPM can be employed to make predictions of effective, as well as saturation, static magnetic field strengths.

11.3.8.1 Calcium–Calmodulin-Dependent Myosin Phosphorylation as a Dynamical System

The EMF effects on Ca^{2+}–calmodulin-dependent myosin phosphorylation have been reported to occur only for Ca^{2+}-depleted conditions and during the nonequilibrium phase of the reaction [84–92]. For these depleted conditions, enzyme kinetics favor the bound state according to $k_{on}/k_{off} \approx 10^2$–$10^3$ [288], the instantaneous exchange reaction rate $v(t)$ is dependent upon the instantaneous free Ca^{2+} concentration $[Ca^{2+}(t)]$, and myosin phosphorylation increases for increasing $[Ca^{2+}(t)]$. Enzyme kinetics are interpreted here in a statistical sense wherein $[Ca^{2+}(t)]$ is taken to be proportional to the ratio of the mean time the ion is free to the mean time bound

$$[Ca^{2+}(t)] = \rho \frac{t_{free}}{t_{bound}} \tag{11.22}$$

where ρ is a proportionality constant.

The instantaneous exchange reaction rate $v(t)$ for myosin phosphorylation is proportional to the concentrations of free ions and CaM in the linear phase of the reaction and the reaction rate is thus a function of ion-binding dynamics

$$\nu(t) \propto [Ca^{2+}CaM] \leftrightarrow \rho \frac{t_{free}}{t_{bound}} + [CaM] \tag{11.23}$$

Ion-binding kinetics determines the sensitivity of this system to both induced electric field and weak magnetic fields. Dosimetry for electric fields is determined by detectability in the binding pathway. This requires sufficient voltage in the binding pathway to increase net $[Ca^{2+}]$ above that due to thermal noise. The requirement is to configure the induced electric field waveform to produce sufficient amplitude in the binding pathway within the frequency range defined by binding kinetics. This will be described in detail in the next section. Direct magnetic field effects are linked via Larmor precession to ion-binding kinetics. In fact, the mean residence time of Ca^{2+} in the CaM binding site must be a multiple of the binding time constant. Reported values for this time constant range between 10^{-2} and 10^{-3} s [289]. Assuming a factor of 3, for simple exponential behavior, the static magnetic field effect on binding should be detectable at approximately 10 μT

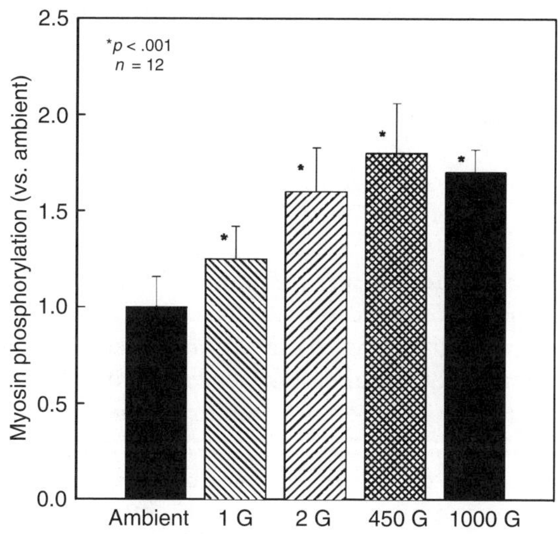

FIGURE 11.9
Effect of static magnetic field amplitude on Ca–CaM-dependent myosin phosphorylation. Acceleration of Ca^{2+} binding appears to saturate at approximately 200 μT, as predicted by the Larmor precession model.

above ambient (≈50 μT) and saturate at 200–300 μT. Results reported to date [90,271] support this prediction. See also Figure 11.9.

It should also be noted that Ca^{2+} has been implicated in the EMF transduction pathway independent of its interactions with CaM. The most commonly reported are those related to Ca^{2+} influx, efflux, or oscillations [27,257–261].

11.3.9 Dosimetry for Induced Electric Fields

The biophysical lore prevailing until the late 1980s and lingering to this day is that, unless the amplitude and frequencies of an applied electric field were sufficient to trigger an excitable membrane (e.g., heart pacemaker), produce tissue heating, or move an ion along a field gradient there could be no effect. This was a formidable obstacle in the quest for therapeutic applications of weak EMF signals. However, this position had to be changed as the evidence for weak (nonthermal) EMF bioeffects became overwhelming. The clinical evidence offered by many double-blind clinical studies coupled with the database of hundreds of thousands of successful treatments of delayed and nonunion bone fractures registered with the FDA simply could no longer be ignored. Noninvasive PEMF treatment is actually as successful as the first bone graft to the huge benefit of the patient. The task now was to provide solid testable models for the biophysical mechanism of weak electric field bioeffects.

The underlying problem for any model, which claims to describe the biophysical mechanism of weak EMF bioeffects, relates to signal detection at the molecular, cellular, or tissue target in the presence of thermal noise, i.e., SNR. Considering the cell membrane as the target, the burden of proof is to show the induced voltage is not buried in thermal and other voltage noise, i.e., that the applied signal is detectable. One of the first simple models assumed the EMF target to be a spherical cell of approximately 10 μm radius. The dielectric properties of the cell membrane in this model were limited to a simple membrane capacitance, with no attempt to take specific ion-binding pathways into account. These calculations often lead to an unfavorable SNR for many EMF signals, which have otherwise demonstrated biological effect [292–294]. Other attempts to account for the thermal noise problem assume unsubstantiated signal processing, such as signal averaging or rectification [295], by the target pathway. Stochastic resonance in which increasing noise strength can increase SNR for a signal having frequency components at a characteristic frequency (e.g., Larmor frequency) has also been proposed [296]. Finally, metabolic amplification via the out of equilibrium state of the ligand–receptor system due to basal cell metabolism has been proposed [297]. None of these models have been proven experimentally.

Without resorting to signal processing or metabolic amplification, it is still necessary to attempt to understand the remarkable sensitivity of biological systems to weak electric fields. In terms of target geometry, certainly the spherical cell model is oversimplified and cannot represent the geometric complexity of cellular and tissue EMF targets. For example, the successful outcome of a healing fracture, wherein bone tissue differentiates both functionally and spatially, is a clinically relevant illustration of cell–cell communication [298]. This suggests that the target for the PEMF signals used to affect nonunions and delayed unions of bone is a highly organized ensemble of cells. In fact, all organized tissue is developed and maintained by an ensemble of complex geometry cells, which have coordinated activity [299]. The most prevalent cell shape in living system tissue is elliptical and flattened, with processes extending in at least two directions. For example, human fibroblasts can typically exceed 100 μm when attached to a substrate (connective tissue) [300]. In addition, nerve axons can be tens of centimeters in length [364]. Gap junctions provide pathways for ions and molecular intercellular communication [301]. They are present in all tissues including bone. The role of cooperative organization in the EMF sensitivity of biological systems has been qualitatively considered [363]. Gap junctions provide ionic coupling and metabolic cooperation, without which disorders in growth control and tissue repair, as well as neoplastic transformations, could occur [301]. Functional modification of gap junctions by modulated microwave fields, as well as EMF signals has been reported [303,304]. There have been several recent reports of modulation of gap junction activity by PEMF signals [305–309].

As shown in Sections 11.3.4 and Section 11.3.5, EMF dosimetry requires knowledge of the electrical properties of the cell or tissue target. Several reports have emerged showing how induced electric field dosimetry can be altered by the presence of gap junctions [310–317,363–367]. A distributed-parameter electrical analog (cable) is often employed to represent the electrical properties of an ensemble of cells in gap junction contact. Most of the cable models represent the cell by a simple membrane capacitance and leak resistance [318,367,368]. However proper dosimetry requires the complete electrochemical state of a functioning cell membrane to be taken into account. A summary of this approach, which incorporates electrochemical processes in the electrical equivalent circuits of the cell membrane, is presented below.

11.3.10 Cell Array Model

Gap junctions allow a group of organized cells, as in a tissue, to present a larger antenna to detect the induced electric field from an exogenous EMF. This increases EMF sensitivity well beyond that due to a molecule or single cell, with a resultant increase in target sensitivity. A useful model is the distributed-parameter linear electrical analog (transmission line) allowing induced transmembrane voltage, V_M, to be evaluated as a function of frequency and position along the array. This is similar to the electrophysiological models which have been proposed for current spread in electrotonically coupled tissues, the dc model proposed to account for tissue sensitivity to the weak electric currents commonly

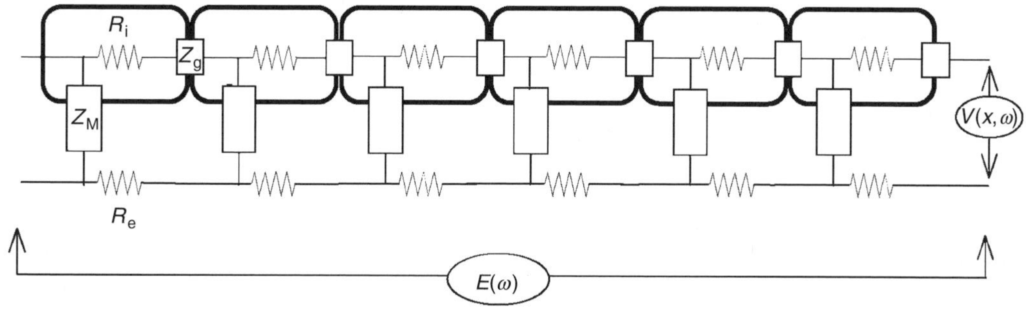

FIGURE 11.10
Distributed-parameter cell array model showing cells in gap junction contact via impedance Z_g. This model behaves electrically identical to a transmission line wherein the applied electric field $E(\omega)$ propagates throughout the array causing progressively higher changes in induced transmembrane voltage $V(x, \omega)$. Z_M, R_i, and R_e are the membrane impedance, and intracellular and extracellular resistances per unit length, respectively. Z_g represents the gap junction impedance. (From Pilla, A.A., Nasser, P.R., Kaufman, J.J., *Bioelectrochem. Bioenerg.*, 35, 63, 1994. With permission.)

found in developing and regenerating tissues [314,315], for direct currents in wound repair [364] and for nonlinear electroporation [365]. Obviously the complete model should be three-dimensional, however the linear model presented here serves as a good first approximation to illustrate the array effect on EMF sensitivity.

The transmission line model is a lossy ladder network, or cable, of finite length, having different internal and external conductance. The steps of this ladder are constituted by the membrane impedance per unit length, which provides a pathway for the internal current to exit the cell. Details are available elsewhere [312], however it is useful to show how the presence of gap junctions affects the induced electric field detected in the cell array when electrochemical processes that represent the EMF transduction pathway are taken into account. The electrical equivalent circuit is shown in Figure 11.10.

Induced transmembrane voltage $V_M(x, s)$ at any position x in the cable shown in Figure 11.10 representing the cell array and for any induced electric field waveform $E(s)$ is

$$V_M(x, s) = -E(s)\,\frac{1}{\gamma}\,\frac{\sin h(\gamma x)}{\cos h(\gamma L)} \tag{11.24}$$

where

$$\gamma = \sqrt{\frac{R_e + R_i + R_g}{Z_M(s)}} \tag{11.25}$$

and R_e and R_i are, respectively, the extracellular and intracellular resistances per unit length along the x-axis, R_g is the gap junction resistance per unit length, and $Z_M(s)$ is the membrane impedance per unit length.

Evaluation of $V_M(x, s)$ has been reported for realistic values of the specific electrical parameters related to the cell array. Typical values for R_e and R_i are 10^{10} Ω/m. These values would be expected to be of the same order, given the cell volume percentage for a typical tissue (50%). The values for R_g range from 10^{10} to 10^{15} Ω/m, representing the limiting electrical conditions of a completely open or completely closed gap junction, respectively [312]. The exact form of Z_M depends upon the impedance assumed. In the most simple form a membrane consists of a capacitance, C_d, representing the real capacitor made up of the lipid bilayer and its electrified interfaces, in parallel with an ionic leak pathway, R_M, a real resistor, as discussed in Section 11.3.5.

When an ion-binding pathway is added (Figure 11.4) the membrane admittance per unit length is

$$Y_M(\omega) = \frac{1}{R_M} + C_d\, j\,\omega + \frac{1}{\left[R_A + \frac{1}{C_A j\omega}\right]} \tag{11.26}$$

where R_A, the equivalent binding resistance, ranges between 10 to 10^3 Ω-m, and C_A, the equivalent binding capacitance, ranges from 10^{-6} to 10^{-5} F/m. Typical values for R_M range from 10^3 to 10^5 Ω-m, and for C_d from 10^{-7} to 10^{-6} F/m. Equation 11.22 describes two time constants, the first due to the membrane capacitance C_d and the membrane leak resistance R_M. The second represents the ion or ligand binding process, the parameters of which are identifiable in terms of rate constants and changes in surface concentration, as shown in Section 11.3.4 and Section 11.3.5.

Equation 11.22 and Equation 11.24 have been employed to evaluate the effect of signal frequency and array length, L, on induced transmembrane voltage. The example shown in Figure 11.11, for a membrane at which ion binding occurs, is typical of models of the effect of gap junctions or long cells on induced transmembrane voltage [310–313]. As shown, there is a substantial increase in $V_M(L, \omega)$ as L increases which can render this target more sensitive to exogenous EMF signals. This is simply due to the increased size of the target. However, overall frequency response is also significantly affected. The frequency response for a single cell (L = 10 μm) indicates that V_M is maximum between 10^5 and 10^6 Hz, as expected. In contrast, for a 1 mm cell array, V_M is about 10^2 higher than for a single cell, but only at frequencies below 100 Hz. Note the presence of the second time constant due to ion binding between 10^3 and 10^4 Hz. Interestingly, the ion-binding time constant does not contribute substantially to the overall frequency response of the single isolated cell. The significance of this will become apparent below.

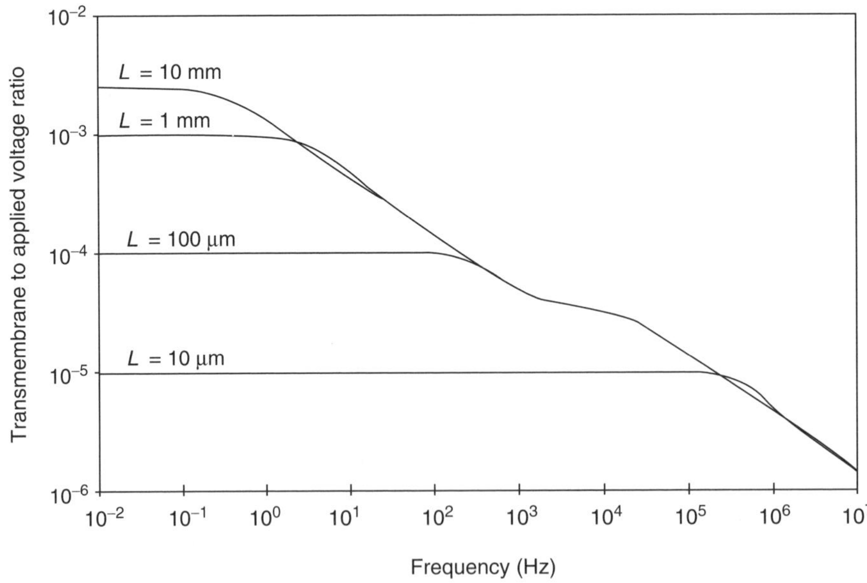

FIGURE 11.11
Frequency dependence of induced transmembrane voltage V_M for various cell array lengths. As predicted by the cell array model, there is a substantial increase in transmembrane voltage as array length L increases, but at significantly lower frequencies vs. that for a single molecule or cell, reflecting the increased propagation time (low pass filter behavior) for longer array lengths.

The presence of gap junctions in the cells of an organized or organizing (repairing) tissue causes the induced transmembrane voltage to be substantially higher than that for the same cell in isolation for the same applied EMF. The frequency range in which increased V_M occurs vs. that for a single isolated cell is shifted toward a substantially lower range. This places different frequency requirements on the induced electric field waveform dependent upon whether the target is a macromolecule, single cell, or tissue. As array length increases beyond 1 mm, the rate of increase in V_M diminishes because of the dissipation of intracellular current via R_M. In the case of myelinated nerve axons, R_M is substantially higher and array lengths above 1 cm can provide further significant increases in V_M [363].

The effect of gap junction impedance on signal amplification resulting from the cell array structure may be assessed by choosing physically meaningful values for R_g corresponding to the state of conductivity of the gap junction. When the gap junction allows macromolecules, as well as ions to pass freely ($R_g \rightarrow 0$), its resistance is negligible compared to R_i and R_M [299]. This represents a realistic lower limit for R_g. When the gap junction is completely closed, its resistance can increase to that of an insulating membrane ($R_g \rightarrow 10^{15}$ Ω/m for an artificial lipid bilayer, nearly a perfect insulator). Under these conditions, all cells in the array are still physically connected, i.e., the long antenna still exists, but has much higher dc resistance. The maximum value of R_g for a cell with a typical membrane resistance of $R_M = 10^5$ Ω-m^2 is of the order of 10^{13} Ω/m.

Quantitatively, the effect of R_g is assessed using Equation 11.24. The results are given in Figure 11.12, wherein Z_M ($=1/Y_M$) is defined by Equation 11.26, for a 1 mm array. As may be seen, signal amplification decreases as R_g increases. However, it is important to note

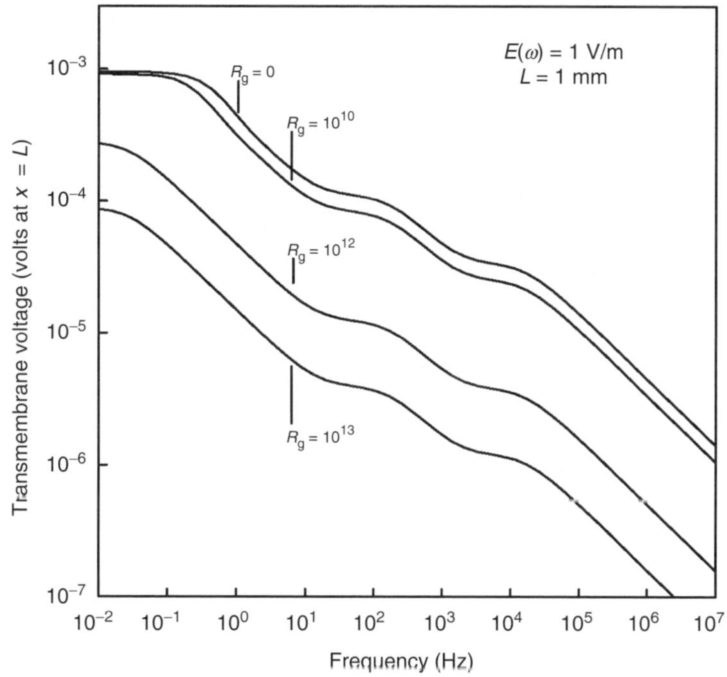

FIGURE 11.12
Effect of gap junction impedance, R_g, on spatial amplification for cells at the end of a 1 mm array. An increase of many orders of magnitude in R_g has relatively little effect on the induced transmembrane voltage. When the value of R_g is equivalent to that of the transmembrane leak, a physiologically realistic limit, spatial amplification is reduced by only 10X. (From Pilla, A.A., Nasser, P.R., Kaufman, J.J., *Bioelectrochem. Bioenerg.*, 35, 63, 1994. With permission.)

that, even when R_g approaches the value of the membrane leak resistance (R_M), i.e., increases by a factor of 10^{13}, V_M decreases only by approximately one order. In fact, the array effect is negated, i.e., V_M is reduced to that of a single cell, only when R_g is significantly larger than the membrane leak resistance, a physiologically unrealistic limiting condition. These results strongly suggest that tissue structures, in which the cellular junctions are physically intact may always be significantly more sensitive than single cells to weak EMF, even though the resistance of the gap junction may change by several orders of magnitude throughout the functional cycles of the living cell.

Examination of Equation 11.24 shows reasoning similar to that given above will hold for the effects of increasing intracellular resistance, R_i. It is also interesting to consider the effect of increasing transmembrane leak resistance, R_M. This is equivalent to increasing the insulating properties of the membrane, an increase of which would allow the signal to propagate a further distance along the array before substantial leakage into the extracellular compartment. An example of such an array is the nerve axon. The myelin sheath that covers the outer cell membrane surface has very high resistance, which allows electrical signals to propagate tens of centimeters before the advantage of array length becomes negligible [363].

The effect of gap junctions upon EMF response for cells in a culture dish has been reported [318]. This study showed alkaline phosphatase activity increased in an osteosarcoma cell line only when cells were electrically connected via gap junctions. The signal employed was 30 Hz sinusoidal at 1.8 mT peak-to-peak amplitude. Interestingly, in the absence of gap junctions, cell division was inhibited for the same signal suggesting the magnetic component of this signal had a different biological effect. Another study [319] found that magnetic fields over a frequency range from 30 to 120 Hz and field intensities up to 12.5 G decreased gap junction intercellular communication in preosteoblastic cells during their proliferative phase of development in a dose-dependent manner. Identical exposure conditions did not affect gap junction communication in well-differentiated osteoblastic cell line and when the preosteoblastic cells were more differentiated. The authors conclude this signal may affect only less differentiated or preosteoblasts and not fully differentiated osteoblasts. Consequently, EMFs may aid in the repair of bone by effects exerted only on osteoprogenitor or preosteoblasts. This is supported by the vast clinical experience that fully differentiated and remodeled bone is not affected in a physiologically significant manner by the same EMF signal that accelerates a repairing fracture, as well as by many cellular studies. This will be reviewed in a later section.

11.3.11 Resonance with Electric Field Signals

As for magnetic field bioeffects, resonance is one manner by which a cellular target can have increased sensitivity to an induced electric field over a restricted frequency range. This will depend upon the structure of the membrane impedance, as can be illustrated by adding a voltage-dependent Hodgkin–Huxley K^+-conduction pathway [320–324] to the simple membrane. Because the conductance is related to the K^+ pathway resistance, R_n, by $g_n = 1/R_n$ and the time constant τ_n has the properties of an inductance, i.e., $\tau_n = L_n/R_n$, the K^+-conduction pathway may be described via the electric circuit analog shown in Figure 11.13. The corresponding voltage-dependent admittance, $YK(V)$, written in terms of the analogous electrical circuit components for the membrane and the K^+ pathway, is thus

$$YK(V) = \frac{1}{R_M} + i\omega C_d + \frac{g_n}{1 + i\omega\tau_n} \tag{11.27}$$

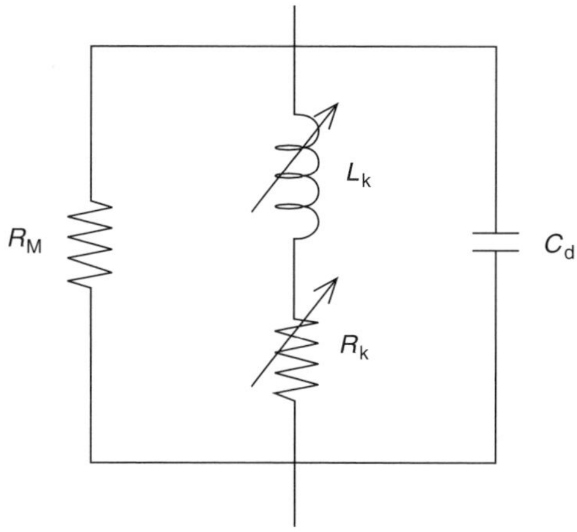

FIGURE 11.13
Potassium conductance is described in Hodgkin–Huxley formulation via addition of resistance–inductance branch to the simple membrane model. R_k and L_k are the equivalent voltage-dependent (arrows) resistance and inductance representing the voltage-dependent kinetics of K^+ transport across the cell membrane. The characteristics of the time constant $\tau_n = L_k/R_k$ of the inductive branch lead to resonance-type frequency response of cell array in a low-frequency range. (From Muehsam, D.S. and Pilla, A.A., *Bioelectrochem. Bioenerg.*, 48, 35, 1999. With permission.)

where g_n and τ_n are the voltage-dependent K^+ parameters given by Hodgkin and Huxley [320,321] and Fishman et. al. [321–324], and the other parameters represent the base membrane impedance.

An electric circuit analog for the K^+-conduction pathway may be formed by associating the conductance g_n with a pathway resistance R_k via $R_k = 1/g_n$ and time constant τ_n to an inductance via $\tau_n = L_k / R_k$ [310]. The corresponding admittance can be written as

$$YK(V) = \frac{1}{R_M} + i\omega C_d + \frac{1}{R_k + i\omega L_k} \tag{11.28}$$

The membrane model, thus, contains a series resistance–inductance pathway in parallel to the membrane resistance and capacitance pathway as shown in Figure 11.13.

The effect of the K^+ conduction membrane pathway on the frequency response of the cell array may be illustrated by holding the membrane-resting potential constant (voltage clamp conditions). Using reported values for R_k and L_k in the Hodgkin–Huxley formulation, the addition of a series resistance–inductance pathway to the membrane model results in a local minimum in the admittance at low frequencies. The frequency response is shown in Figure 11.14 wherein V_M reaches a peak at approximately 16 Hz, for a flat input ($E(\omega) = 10$ mV/cm, a typical therapeutic PEMF) and with the membrane voltage clamped at 40 mV. The introduction of a linear inductive element, according to the Hodgkin–Huxley formulation, to the membrane model produces a broad resonance response to applied fields and thus increases sensitivity to induced electric fields having frequency components in the low-frequency range of this resonance. Note, however, that resonance is only significant for long cells or cells arrays, not for single cells.

Thus, in addition to resonances that may occur via Larmor precession for applied magnetic fields, there is also the possibility of a resonance due simply to the electrical characteristics of the target pathway. In the example given, the presence of a resonance-type response due to K^+ membrane transport depends upon the initial state of membrane polarization. The stimulus to the membrane under these conditions is well below

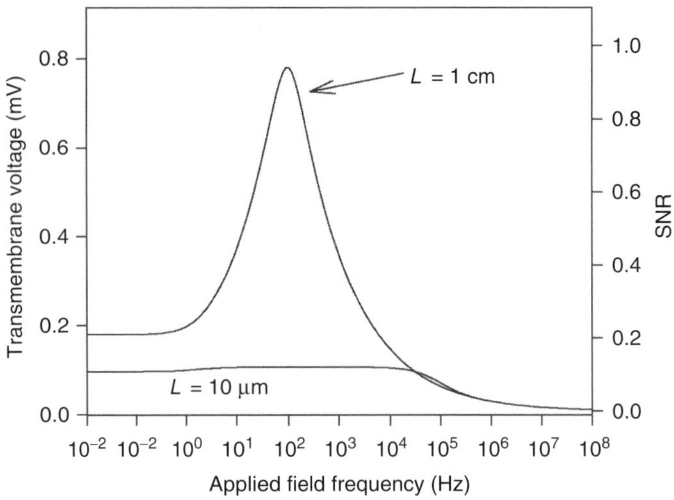

FIGURE 11.14
Response of cell array with K^+ conductance membrane model to 10 mV/cm electric field at a given transmembrane-resting voltage. Frequency response of array exhibits a wide resonance response for applied field frequencies in the 16 Hz range. K^+ conductance at a given voltage is described in the Hodgkin–Huxley formulation as a series resistance–inductance branch in the membrane model. Note the broad resonance at frequencies in the 10–100 Hz range. SNR was calculated according to the method described in Section 11.3.11. Note that resonance is not significant for a single cell, only for a long cell or cell array.

excitatory threshold and the primary response is K^+ membrane transport for which the equivalent electric circuit contains an inductor. This, coupled with the normal capacitive properties of the cell membrane naturally leads to resonance behavior.

11.3.12 Signal to Thermal Noise

In order for an induced electric field to modulate a cellular target, the first requirement is that it be detected by the target. This means the waveform must be configured to produce sufficient voltage at, e.g., a binding site to significantly affect the t_{free}-to-t_{bound} ratio for the binding ligand as defined in Equation 11.18 (Section 11.3.8.1). There are several sources of transmembrane noise in biological membranes. The most common are due to thermal, flicker ($1/f$), shot, and conductance fluctuations [325]. The latter three usually relate to ion transport and their interpretation is model dependent. Thermal noise is present in all voltage-dependent membrane processes and represents the minimum requirement to establish adequate SNR, a front-line measure of detectability. SNR can be evaluated, assuming only the presence of thermal noise (no signal processing or other enhancements which are model dependent) via [326]

$$\text{SNR} = \frac{|V_{\text{target}}(\omega)|}{\text{RMS}_{\text{noise}}} \tag{11.29}$$

where $|V_{\text{target}}(\omega)|$ is the maximum amplitude of the induced voltage in the target pathway and $\text{RMS}_{\text{noise}}$ is the root mean square of the noise voltage

$$\text{RMS}_{\text{noise}} = \left[4kT \int_{\omega_1}^{\omega_2} \text{Re}Z_{\text{target}}(x, \omega)\, d\omega \right]^{1/2} \tag{11.30}$$

where represents the real part of the total impedance of the target Z_{target} and the limits of integration (ω_1, ω_2) are determined by the bandpass of the target, typically 10^{-2}–10^7 rad/s.

SNR, as defined above, can be a powerful indicator of whether a given EMF signal can be expected to have a physiologically significant effect. To illustrate, consider a target in which the pathway is Ca–CaM-dependent. This includes most growth factors in every stage of bone and wound repair [69–83,146,147,151–153,], as well as nitric oxide synthase (NOS) that modulates NO, a signaling molecule in many neurological and cardiac pathologies and in bone and cartilage repair [327,328,388–392]. To evaluate SNR for this target, the quantity of interest is the effective voltage, $E_b(\omega)$ induced across the equivalent binding capacitance, C_A (see Figure 11.4), which is directly proportional to [Ca^{2+}] as defined in Equation 11.18, and given by [329]

$$E_b(\omega) = \frac{X_C\ E(\omega)}{(R_A^2 + X_C^2)^{1/2}} \tag{11.31}$$

where $X_C = 1/\omega C_A$ and R_A is the equivalent binding resistance. Equation 11.18 describes the relation between the frequency response of the target, $E_b(\omega)$, and applied field waveform $E(\omega)$, illustrating clearly EMF response is dependent upon applied waveform parameters.

SNR is evaluated using $E_b(\omega)/\text{RMS}_{noise}$; however, the binding time constant $\tau_A = R_A C_A = 1/k_b$, must be known or be estimable. Consider a Ca^{2+}–CaM-dependent process, for which free [Ca^{2+}] concentration is the EMF-sensitive rate-limiting factor [90,92]. Linearized Michaelis–Menton kinetics describing Ca^{2+} binding to CaM is [90]

$$V_{max}/\nu = [1 + K_D/[Ca^{2+}] \tag{11.32}$$

where V_{max} is the maximal reaction rate, i.e., the slope of the corresponding Lineweaver–Burke plot, the reaction velocity v is given by $v = k_b\,[Ca^{2+}CaM]$ and K_D is the equilibrium constant.

The Michaelis–Menton relation thus determines k_b, i.e., defines the binding time constant for use in Equation 11.27

$$k_b = \frac{V_{max}}{[Ca^{2+}CaM](1 + K_D/[Ca^{2+}])} \tag{11.33}$$

Employing numerical values for which EMF sensitivity has been reported, $V_{max} = 10^{-6}$ -10^{-7} per s, [Ca^{2+}] = 1–3 μM, K_D = 20–40 μM, [Ca^{2+}CaM] = K_D([Ca^{2+}]+[CaM]), yields τ_A between 1 and 5 ms [90].

Evaluation of SNR for Ca^{2+} binding in the manner outlined above may be performed for molecular, cellular, or tissue targets. An interesting example is wound repair. A common model is the full thickness linear incision performed through the skin down to the fascia on the dorsum of adult Sprague Dawley rats [331]. Acceleration of wound repair is assessed by tensile strength measurements at 21 days postoperative. At this time point untreated (control) strength is approximately one-third of that of the fully healed wound. One study used the PEMF signal commonly employed for bone repair (Figure 11.2, top) and reported no effect [331]. A second, more recent study, used a PRF signal specifically configured to enhance Ca^{2+} binding to CaM with the specific goal of enhancing growth factor release and reported a 59% increase in tensile strength vs. controls at 21 days, $p < .001$ [163,332]. SNR analysis for the signals used in these studies is shown in Figure 11.15. It is clear that the induced electric field produced by the PEMF bone repair

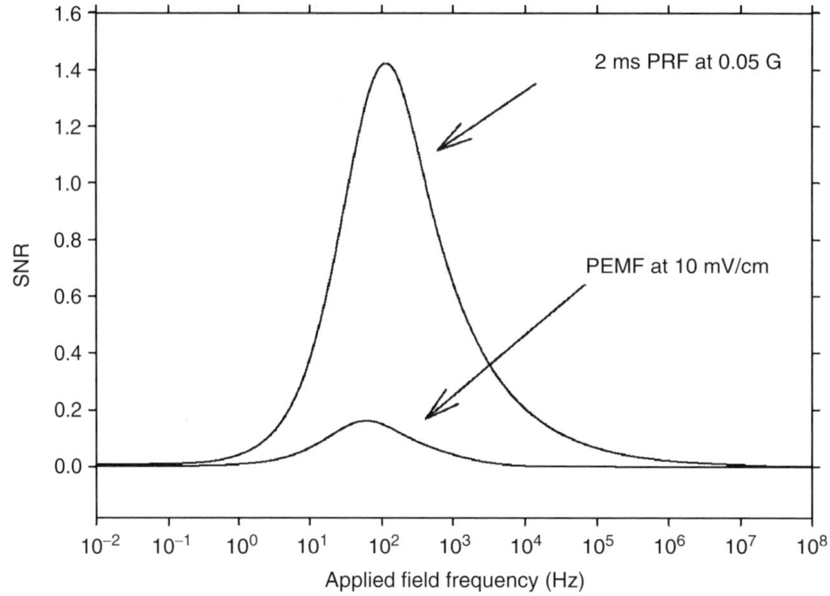

FIGURE 11.15
SNR in a Ca–CaM pathway for PEMF and PRF waveforms used in a rat cutaneous wound model. The asymmetrical repetitive pulse burst PEMF bone repair signal produced low (below detectability threshold) SNR and had no effect on wound repair. The 27.12 MHz repetitive sinusoidal burst produced sufficient SNR for detectability in the Ca–CaM pathway and enhanced tensile strength by 59% at 21 days.

signal, which consists of a 5 ms burst of bipolar pulses (200/20 μs asymmetrical duration) repeating at 15 per s (see Figure 11.2, top), and inducing a gross peak electric field of 1 mV/cm ($dB/dt = 10^6$ G/s), produced very low induced voltage across C_A (Figure 11.4) in the Ca–CaM pathway. The resultant SNR was below the detectability threshold. In contrast, the PRF signal which consisted of a 2 ms burst of 27.12 MHz sinusoidal waves repeating at 1 per s, $dB/dt = 10^7$ G/s, produced a significantly larger induced voltage across C_A with a larger effect on Ca^{2+} binding. It is to be noted that increasing the gross induced electric field produced by the PEMF bone repair signal (Figure 11.2, top) to 50 mV/cm (5X) would have increased SNR by the same factor and allowed this signal to be equally detectable in the Ca^{2+} binding target pathway. Unfortunately, such an amplitude comparison was not performed in this study.

A recent study compared the effects of the PEMF bone repair signal utilized in the example above ($dB/dt = 10^6$ G/s) with a 65 μs burst of rectangular pulses of 4 and 12 μs duration per polarity repeating at 1.5 bursts/s ($dB/dt = 10^4$ G/s) on bone repair in a rat osteotomy model [369]. In this study the standard clinical bone repair PEMF signal produced a twofold increase in new woven bone and callus stiffness, whereas the 4/12 μs signal was ineffective. SNR, calculated as above for each signal, assuming a Ca–CaM target pathway, reveals peak SNR > 1 for the clinical PEMF signal and peak SNR $\ll 1$ for the 4/12 μs signal. Note that modulation of the Ca–CaM pathway requires frequency components of sufficient amplitude in the 10^2–10^4 Hz range and neither of these signals was configured accordingly.

Further support for the SNR and Dynamical Systems model for weak electric field bioeffects has been reported wherein bioeffective waveforms based upon the PRF signal were configured *a priori* for Ca–CaM-dependent myosin phosphorylation, neurite outgrowth from embryonic chick dorsal root ganglia and for bone repair in a rabbit model [29,30]. Growth factor production in the latter two pathways is Ca–CaM-dependent

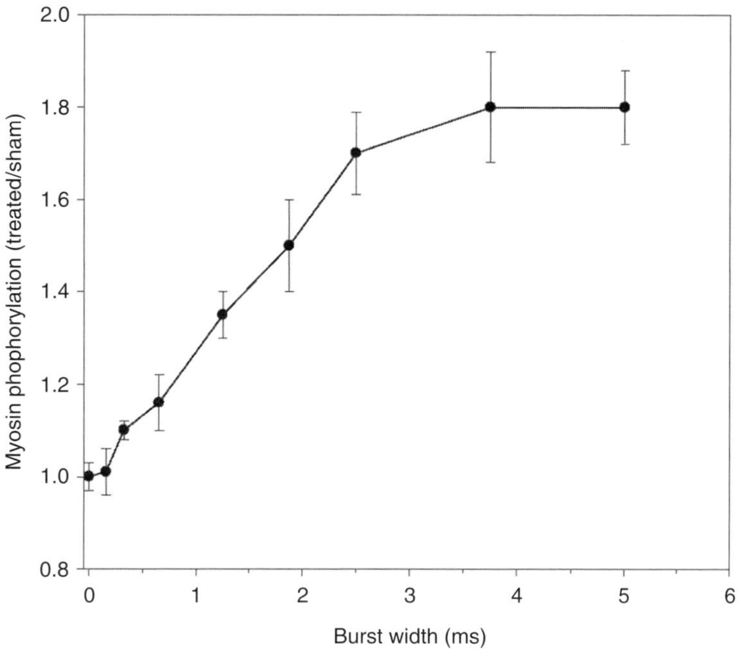

FIGURE 11.16
PRF signal modulates Ca^{2+} binding kinetics (CaM) with increased effect vs. burst duration at constant amplitude (0.05 G) and repetition rate (1 per s). SNR analysis predicted a dose dependence on burst width, with a plateau at approximately 3 ms, in good agreement with experiment. The number of frequency components having detectable amplitude at the Ca^{2+} binding time constant increases with burst duration, thereby increasing SNR.

[81,336]. To illustrate signal prediction, PRF modulation of Ca–CaM-dependent myosin phosphorylation is reviewed here. Specifically, SNR was evaluated for the Ca–CaM pathway for burst durations in the 0.1–5 ms range with constant burst repetition (1 per s) and peak amplitude (0.05 G), and compared to experiment. Recall that the clinical PRF signal consists of repetitive bursts of a 27.12 MHz sinusoidal wave. SNR analysis showed peak SNR > 1 only for burst durations above 0.3 ms, with a plateau at approximately 3 ms. The experimental results shown in Figure 11.16 suggest no significant PRF effect on myosin phosphorylation for burst durations below 0.5 ms and no further significant effect beyond approximately 4 ms, in good agreement with the SNR model.

11.4 Ultrasound for Tissue Repair

As indicated in earlier sections, connective and endothelial tissues, particularly bone, adapt to the mechanical environment by remodeling to accommodate the magnitude and direction of the applied stress [373,374]. Mechanical signals exist in functionally loaded bone [375] and represent strong regulatory signals to skeletal tissue [378], even during fracture healing [377]. Indeed, controlled movement during fracture repair has been shown to enhance healing [378]. There is also considerable evidence that low-intensity US (<100 mW/cm^2 Spatial Average Temporal Average, SATA) can accelerate healing of fresh fractures and established nonunions of bone [379]. Indeed, US stimulation is now part of the standard armamentarium of the orthopedist. In addition, basic studies have demonstrated that US can modulate each of the three key stages of the healing process (inflammation, repair, and remodeling) because it can enhance angiogenic [382], chondrogenic [381], and osteogenic [382] activity as for PEMF signals.

Although it is well known that many cells have mechanoreceptors allowing direct response to a mechanical signal [383], a direct electrical effect, linking US and PEMF signals, has been proposed [199–201]. The absorption of US in a tissue target gives rise to the phenomenon of microstreaming, i.e., the movement of fluid across surfaces. If the fluid contains ions and the surfaces are charged (cell membranes), streaming causes ions in the diffuse electrical double layer to be displaced from their equilibrium or resting positions and the electrokinetic phenomenon of streaming potentials occurs. This could be termed a mechanically induced electrophoretic effect. US and other mechanical inputs such as controlled weight bearing during fracture repair, walking, jumping, hitting a tennis ball, etc., cause rapid flow of fluids past cell surfaces thereby generating a time-varying, mechanically induced, electric field. Should this electric field have sufficient SNR in a proposed transduction pathway, it could act in a similar manner to exogenous PEMF.

The low-intensity US signal most commonly employed clinically is a 500 μ burst of 1.5 MHz sinusoidal acoustic waves repeating at 200 Hz and at an amplitude of 30–50 mW/cm^2 SATA. The sound pressure wave causes fluid flow during each burst, which, in turn, produces a repetitive time-varying voltage. This streaming potential has the form of a distorted trapezoid-like waveform with millisecond rise times and longer relaxation times, similar to the waveforms observed when bone is rapidly deformed. It has been reported that the clinical US signal produces 1–10 mV/cm peak electric field amplitude in tissue fluid [199]. This waveform may be analyzed using the SNR model given above, but it is necessary to define the target pathway.

Low-intensity pulsed US appears to act at the cellular level via biochemical pathways which are remarkably similar to those reported for EMF. Thus, US has been reported to induce a threefold increase in prostaglandin E2 (PGE2) production in murine osteoblasts through the upregulation of cyclooxygenase-2 (COX-2). COX-2 and PGE2 are known mediators in a bone forming response to external stimuli [384]. Pulsed US induced the transient expression of the early response gene c-fos and elevated mRNA levels for IGF-I in bone-marrow–derived stromal cells. Osteocalcin, bone sialoprotein, and bone matrix proteins were also modulated [385,386]. The US stimulation of primary rat chondrocytes elevated the intracellular concentration of Ca^{2+}. Chelating or removing Ca^{2+} from the medium inhibited the stimulatory effects of US on proteoglycan synthesis, suggesting US-stimulated synthesis of cell matrix proteoglycan, associated with accelerated fracture healing, is mediated by intracellular calcium signaling [387]. The US significantly increased VEGF mRNA. Early inhibition of nitric oxide (NO) production, but not calcium or PGE2, significantly reduced US-enhanced VEGF levels. Osteoblasts responded to US treatment by increasing NO production and NOS catalytic activities. Inhibition of NOS activity by *N*-nitro-*L*-arginine methyl ester (L-NAME) reduced VEGF levels [388].

The latter report is strong evidence that pulsed US acts via a pathway that is remarkably similar to that for PEMF. Indeed, PEMF has been reported to enhance proteoglycan synthesis in human cartilage cells via early stimulation of NO production [391]. NO is regulated by NOS, which, in turn, is activated by Ca–CaM or even by Ca^{2+} binding directly to NOS [392]. This provides strong support for the validity of comparing SNR for the US and PEMF signals in the Ca^{2+} binding pathway of CaM. The results are shown in Figure 11.17, wherein it may be seen the PEMF clinical signal (see Figure 11.2, top), as well as the streaming potential induced by the US signal, has sufficient amplitude to be detectable in the frequency range corresponding to Ca^{2+} binding kinetics.

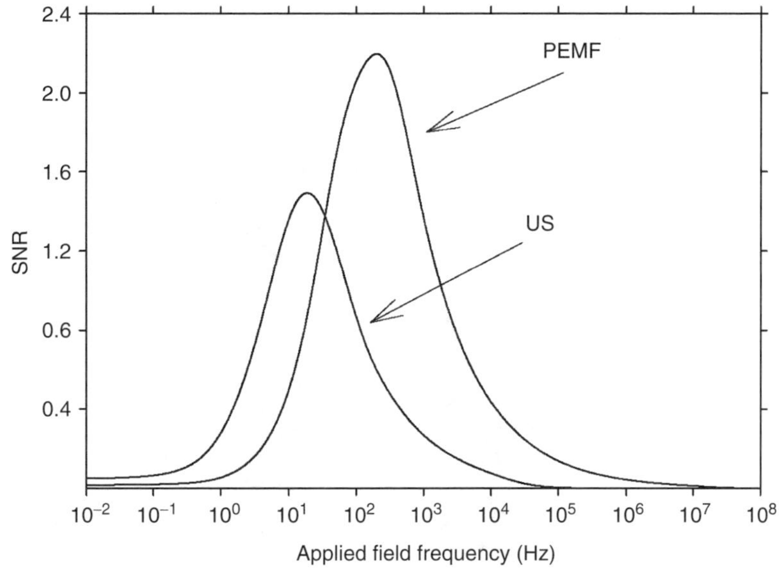

FIGURE 11.17
SNR in a Ca–CaM transduction pathway for a standard bone repair PEMF (see Figure 11.2, top) and the streaming potential from a pulsing US signal, also in use for bone repair (see text for details). These curves show there is sufficient SNR produced by both signals in a transduction pathway which may be common to both signals. SNR peak for the US signal is shifted toward the lower frequency range reflecting the lower frequency content of the mechanically induced time-varying electric field.

This analysis may be tested by examination of results reported in the literature for the effects of PEMF and US signals on identical cellular targets. This is illustrated in Figure 11.18, which shows both signals accelerated proteoglycan synthesis approximately twofold for chondrocytes in culture.

It is clear that both mechanical and EMF stimuli can modulate bone repair and affect a variety of cellular processes. The proposal that both modalities act through the EIT mechanism is reviewed here using the tools of the SNR and Dynamical Systems model. This may further help the reader to assess the myriad of signals and treatment modalities reported in the rapidly expanding literature in this area.

11.5 Effect of Initial Cell or Tissue State on EMF Sensitivity

Clinical experience, particularly in bone and wound repair, as well as numerous *in vitro* studies, suggests that the initial conditions of the EMF-sensitive target pathway determine whether a physiologically meaningful bioeffect can be achieved. Thus, surrounding normal bone tissue, which receives the same PEMF dose as a fracture site, does not respond in a physiologically significant manner, whereas fracture repair may be accelerated [123]. Local peripheral blood flow is not affected by EMF in a healthy subject, but is increased when a musculoskeletal injury or other exogenous stimulus is present [172]. Ca^{2+} binding to CaM is modulated by EMF only under depleted Ca^{2+} conditions [90]. Human fibroblasts in culture exhibit maximum increase in DNA synthesis with sinusoidal EMF during the *S* phase of the cell cycle [267]. Dendritic outgrowth in a nerve

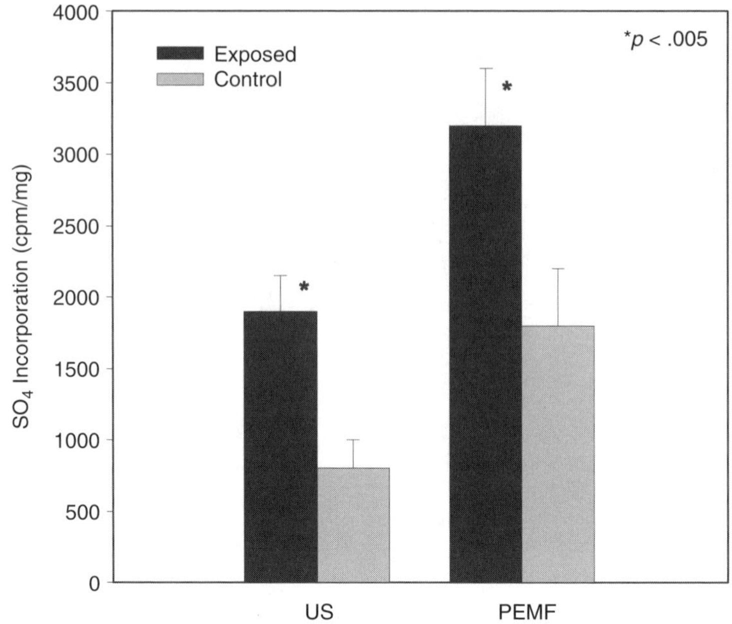

FIGURE 11.18
Comparison of the effects of US and PEMF signals employed clinically for bone repair on chondrocytes in culture. Both signals accelerated proteoglycan synthesis approximately twofold suggesting the transduction pathway may be similar or even identical.

regeneration model is modulated by EMF only in the presence of subsaturation concentrations of NGF [333–335]. PEMF effects on osteoblasts depend upon their maturation stage [337].

This behavior is consistent with the SNR and Dynamical Systems model because it is known that the electrical properties of a tissue target depend upon its initial state. Indeed, SNR calculations depend upon the electrical characteristics of the EMF-sensitive target which can be substantially different when the target is at rest or when reacting to a sudden change in its environment such as a fracture or other musculoskeletal injury [338]. Thus, for an identical EMF stimulus, the impedance of the EMF-sensitive pathway may alter sufficiently in the presence of an injury or pathology to enable a signal, which has otherwise been ignored (not detected), to be detected.

An extension of the SNR and Dynamical System model that examines the variation of the electrical characteristics of the voltage-dependent K^+ conduction model (see Section 11.3.12) with membrane polarization has been proposed [339]. Both conductance g_n and time constant τ_n (see Equation 11.24) are voltage dependent and resting or baseline transmembrane voltage can affect membrane impedance. The electrical characteristics of the target pathway can depend upon the membrane-resting potential, which, in turn, is defined by the baseline activity of the cell via, e.g., up- and down-regulation of receptors involved in the biochemical cascades relevant to baseline activity or response to injury.

The voltage-dependent characteristics of the Hodgkin–Huxley K^+ conduction conductance g_n and time constant τ_n produce the dependence of admittance upon membrane-resting potential and frequency shown in Figure 11.19, left-hand plot. For this example, the Hodgkin–Huxley values for the K^+ conduction pathway are employed and the time constant τ_n reaches a maximum at a membrane-resting potential of about 12 mV, corresponding to the K^+ activation voltage. The admittance drops rapidly with increasing frequency, reaching its lowest value at $\approx$100 Hz.

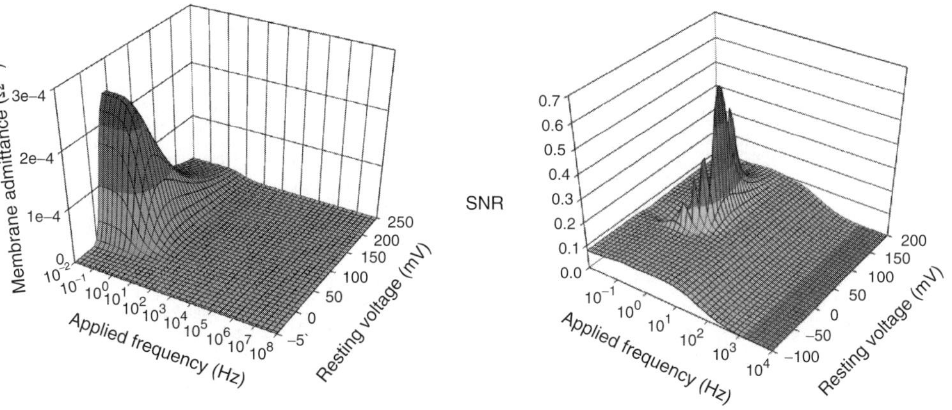

FIGURE 11.19 (See color insert following page 412)
(Left) Membrane admittance depends upon frequency and membrane potential for Hodgkin–Huxley K+-conduction pathway. Region of higher admittance (reduced impedance) corresponds to greater K^+ flux across membrane and admittance drops rapidly with increasing frequency. (Right) Response of Hodgkin–Huxley K^+-conduction membrane pathway to exogenous electric field. Voltage induced in K^+ pathway varies with membrane potential and frequency. Thus, EMF sensitivity depends upon the target initial state.

Figure 11.19, right-hand plot, shows the response to a flat input, $E(\omega) = 10\,\text{mV/cm}$, of a 1 mm (50–100 cells) cell array is determined by the membrane-resting voltage. In a manner similar to that shown in Figure 11.13 for the fixed-voltage K^+ pathway, the voltage-dependent K^+ pathway also exhibits a resonance-like response to applied field frequencies in the 16 Hz range. However, the width of the resonance frequency region changes with changing membrane-resting potential. For membrane-resting voltages less than approximately -10 mV, the array responds equally to all frequencies below approximately 16 Hz. The admittance approaches zero for the depolarized membrane, so that for this condition the response of the K^+ pathway is identical to that of the simple membrane (R_M, C_d). Thus, as the membrane-resting potential increases to 40 mV or more, the array (target) begins to exhibit a preferential response, producing sufficient SNR only for applied field frequencies in the 16-Hz range.

An *in vitro* example of the effect of cell initial conditions upon EMF sensitivity is shown in Figure 11.20 [339]. Jurkat cells were exposed to the PEMF signal shown in Figure 11.2, top. The results show the PEMF signal could only achieve sufficient SNR to produce physiologically significant results in growth stage II. Cells in early and late log growth phase were insensitive to the identical PEMF signal. The authors also report PEMF amplified the antiproliferative effects of anti-CD3 (antibody to the T cell receptor). This was interpreted as a potential clinically significant anti-inflammatory property stemming from PEMF down-regulation of T cells that are activated at the T cell receptor during inflammation. In the absence of inflammation, T cell receptors are most likely not upregulated, i.e., not EMF-sensitive and the identical PEMF signal could not achieve sufficient SNR to be detected in this pathway. In terms of the cell array model this translates to a substantially different electrical impedance at this pathway, dependent upon whether the T cell receptor is up- or down-regulated.

It may thus been seen the complete dosimetry picture for weak EMF modulation of tissue growth and repair clearly involves the changing electrical properties of the target with functional state. Realization of this may provide an explanation for the sometimes-elusive repeatability of EMF studies.

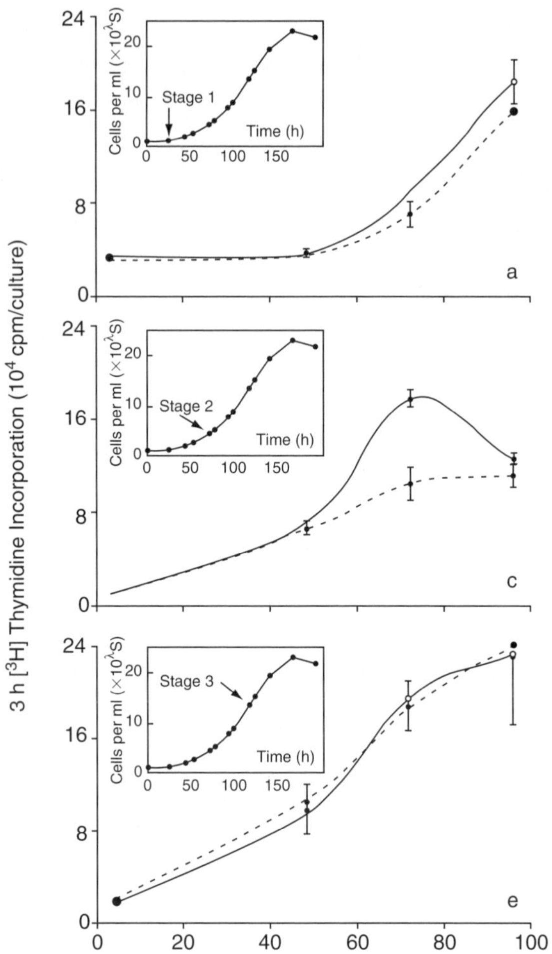

FIGURE 11.20
Illustration of the effect of cell growth stage (initial condition) upon EMF sensitivity. This study was performed on a lymphocyte cell line (Jurkat) and clearly demonstrates how an identical signal can have a large or no effect dependent upon the initial state of the cell. Pilla [50] interpreted these results as down-regulation, by EMF, of the T cell receptor that is apparently not upregulated in growth stages 1 or 3. (From Nindl, G., Swezb, J.A., Millera, J.M., Balcavage, W.X., *FEBS Lett.*, 414, 501, 1997. With permission.)

11.6 The Future

This chapter has attempted to provide the reader with enough information to show there is an abundance of experimental and clinical data demonstrating that exogenous EMFs of surprisingly low levels can have a profound effect on a large variety of biological systems. Both electrical and electromagnetic devices have been demonstrated to positively affect the healing process in fresh fractures, delayed and nonunions, osteotomies, and spine fusion in orthopedics and for chronic and acute wound repair. These clinical results have been validated by well-designed and statistically powered double-blind clinical trials and have survived meta-analyses. The FDA has approved labeling for these biophysical devices, limited at present to these indications. EMF stimulation technologies provide an additional arm to current treatment management strategies for these pathologies. However, the potential clinical applications of EMF therapeutics extend far beyond those considered here and the clinical rewards are certain to be huge. Great strides have been made in the use of PEMF for chronic and acute wound repair. It is often the only effective treatment for chronic wounds and has been chosen as the treatment of choice for postoperative pain and edema reduction in plastic and reconstructive surgery. The advent of new more effective signals may even expand applications in bone repair. There is a significant emerging application for the treatment of osteoarthritis [61,356–358] and rheumatoid arthritis [369,370].

The state of knowledge in EMF therapeutics has significantly advanced in the past decade. The mechanism of action is much better understood, so much so it is often possible to configure, *a priori*, pulsing waveforms for an expected bioeffect. The example is PEMF and PRF signals have been successfully configured for a Ca–CaM ion-binding transduction pathway, which the biologists have established as a primary regulatory pathway. As we continue to learn to properly match dosimetry to pathology, the dreams that many of us had more than 35 years ago may well be realized. Cancer, cardiac muscle regeneration, diabetes, arthritis, and neurological disorders are just some of the pathologies that have already been shown to be responsive to EMF therapy. Successful applications of low-frequency EMFs have been reported for treatment of bronchial asthma, myocardial infarction, and venous and varicose ulcers. There is emerging research on EMF effects on angiogenesis and the manner in which this may increase stem cell survival in the treatment of Altzeimer's and Parkinson's diseases. There are also many studies that point to the possibility of the use of EMF for peripheral nerve regeneration [333–336,360].

There are numerous reports suggesting a role for EMF in the treatment of cancer. In 1979–1981 Larry Norton studied the effect of PEMF (using a 50 ms burst version of the PEMF signal in Figure 11.2, top) on transplanted tumors in mice. Initial results suggested PEMF significantly increased survival time and led to preliminary clinical trials [341,342]. More recent studies on animals provide data that are reliable, and this certainly suggests that clinical applications on humans are possible in the near future [343–348,351,352,355].

EMF therapy modalities are simple, safe, and significantly less costly to the health care system. They offer the ability to treat the underlying pathology rather than simply the symptoms. The time is particularly opportune given the increased incidence of side effects from the use of pharmacological agents. EMF therapeutics will have a profound impact upon health and wellness and their costs worldwide.

Acknowledgments

This is a review to which many people have contributed. I particularly acknowledge Andrew Bassett, without whom I would have never even thought of EMF therapeutics; Jack Ryaby, who made enormous contributions to the bone growth stimulator industry, and without whom there would never have been devices available for Andy's patients; and Alessandro Chiabrera for his enormous contributions and the invaluable and memorable summers he spent in my laboratory. May they all rest in peace. I also deeply acknowledge Robert Siffert who gave me the unique opportunity to explore EMF therapeutics relatively unhindered. There are many people who have collaborated or studied with me, whose immense contributions helped me construct this review. I thank them here: Jonathan Kaufman, Marko Markov, David Muehsam, Philip Nasser, James Ryaby, Betty Sisken, and Berish Strauch.

References

1. Sluka KA and Walsh D. Transcutaneous electrical nerve stimulation: basic science mechanisms and clinical effectiveness. *J. Pain*, 2003; 4: 109–121.
2. Rushton DN. Electrical stimulation in the treatment of pain. *Disabil. Rehab.*, 2002; 24: 407–415.
3. Bjordal JM, Johnson MI, and Ljunggreen AE. Transcutaneous electrical nerve stimulation (TENS) can reduce postoperative analgesic consumption. A meta-analysis with assessment of optimal treatment parameters for postoperative pain. *Eur. J. Pain*, 2003; 7: 181–188.
4. Ojingwa JC and Isseroff RR. Electrical stimulation of wound healing. *J. Invest. Derm.*, 2003; 121: 1–12.
5. Gardner SE, Frantz RA, and Schmidt FL. Effect of electrical stimulation on chronic wound healing: a meta-analysis. *Wound Repair Regen.*, 1999; 7: 495–503.
6. Benz R, Beckers F, and Zimmermann U. Reversible electrical breakdown of lipid bilayer membranes: a charge-pulse relaxation study. *J. Membr. Biol.*, 1979; 48: 181–204.
7. Zimmermann U, Vienken J, Pilwat G. Development of drug carrier systems: electrical field induced effects in cell membranes. *Bioelectrochem. Bioenerg.*, 1980; 7: 553–574.
8. Mir LM. Therapeutic perspectives of *in vivo* cell electropermeabilization. *Bioelectrochemistry*, 2001; 53: 1–10.
9. Ferguson M, Byrnes C, Sun L, Marti G, Bonde P, Duncan M, and Harmon JW. Wound healing enhancement: electroporation to address a classic problem of military medicine. *World J. Surg.*, 2005; 29 (Suppl 1): S55–S59.
10. Gothelf A, Mir LM, and Gehl J. Electrochemotherapy: results of cancer treatment using enhanced delivery of bleomycin by electroporation. *Cancer Treat. Rev.*, 2003; 29: 371–387.
11. Lin JC. Studies on microwaves in medicine and biology: from snails to humans. *Bioelectromagnetics*, 2004; 25: 146–159.
12. Akai M and Hayashi K. Effect of electrical stimulation on musculoskeletal systems: a meta-analysis of controlled clinical trials. *Bioelectromagnetics*, 2002; 23: 132–143.
13. Akai M, Kawashima N, Kimura T, et al. Electrical stimulation as an adjunct to spinal fusion: a meta-analysis of controlled clinical trials. *Bioelectromagnetics*, 2002; 23: 496–504.
14. Pilla AA. Electrochemical information and energy transfer *in vivo*. Proc. 7th IECEC, Washington, DC, American Chemical Society, 1972; 761–764.
15. Pilla A.A. Electrochemical information transfer at living cell membranes. *Ann. NY Acad. Sci.*, 1974; 238: 149–170.
16. Chiabrera A, Grattarola M, and Viviani R. Interaction between electromagnetic fields and cells: microelectrophoretic effect of ligands and surface receptors. *Bioelectromagnetics*, 1984; 5: 173–178.
17. McLeod BR and Liboff AR. Dynamic characteristics of membrane ions in multifield configurations of low-frequency electromagnetic radiation. *Bioelectromagnetics*, 1986; 7: 177–189.
18. Chiabrera A and Bianco B. The role of the magnetic field in the EM interaction with ligand binding. In: *Mechanistic Approaches to Interactions of Electric and Electromagnetic Fields with Living Systems*. Blank M., Findl E, eds., Plenum Press, NY, 1987, pp. 79–90.
19. Liboff AF, Fozek RJ, Sherman ML, McLeod BR, and Smith SD. Ca^{2+}-45 cyclotron resonance in human lymphocytes. *J. Bioelectricity*, 1987; 6: 13–22.
20. Lednev VV. Possible mechanism for the influence of weak magnetic fields on biological systems. *Bioelectromagnetics*, 1991; 12: 71–75.
21. Blanchard JP and Blackman CF. Clarification and application of an ion parametric resonance model for magnetic field interactions with biological systems. *Bioelectromagnetics*, 1994; 15: 217–238.
22. Engstrom S. Dynamic properties of Lednev's parametric resonance mechanism. *Bioelectromagnetics*, 1996; 17: 58–70.
23. Muehsam DS and Pilla AA. Lorentz approach to static magnetic field effects on bound ion dynamics and binding kinetics: thermal noise considerations. *Bioelectromagnetics*, 1996; 17: 89–99.
24. Zhadin MN. Combined action of static and alternating magnetic fields on ion motion in a macromolecule: theoretical aspects. *Bioelectromagnetics*, 1998; 19: 279–292.

25. Chiabrera A, Bianco B, Moggia E, and Kaufman JJ. Zeeman-Stark modeling of the RF EMF interaction with ligand binding. *Bioelectromagnetics*, 2000; 21: 312–324.
26. Binhi VN. Amplitude and frequency dissociation spectra of ion–protein complexes rotating in magnetic fields. *Bioelectromagnetics*, 2000; 21: 34–45.
27. McCreary CR, Thomas AW, and Prato FS. Factors confounding cytosolic calcium measurements in Jurkat E6.1 cells during exposure to ELF magnetic fields. *Bioelectromagnetics*, 2002; 23: 315–328.
28. Zhadin M and Barnes F. Frequency and amplitude windows in the combined action of DC and low-frequency AC magnetic fields on ion thermal motion in a macromolecule: theoretical analysis. *Bioelectromagnetics*, 2005; 26: 323–330.
29. Pilla AA, Muehsam DJ, Markov MS, and Sisken BF. EMF signals and ion/ligand binding kinetics: prediction of bioeffective waveform parameters. *Bioelectrochem. Bioenerg.*, 1999; 48: 27–34.
30. Pilla AA. Weak time-varying and static magnetic fields: from mechanisms to therapeutic applications. In: *Biological Effects of Electromagnetic Fields*, Stavroulakis P, ed., Springer Verlag, 2003, pp. 34–75.
31. Ryaby JT. Clinical effects of electromagnetic and electric fields on fracture healing. *Clin. Orthop.*, 1998; 355(Suppl): 205–215.
32. Yasuda I. Piezoelectric activity of bone. *J. Jpn. Orthop. Surg Soc.*, 1954; 28: 267–271.
33. Yasuda I, Noguchi K, and Sata T. Dynamic callus and electric callus. *J. Bone Joint Surg. Am.*, 1955; 37: 1292–1293.
34. Fukada E and Yasuda I. On the piezoelectric effect of bone. *J. Phys. Soc. Jpn.*, 1957; 12: 121–128.
35. Becker RO. The bioelectric factors in amphibian-limb regeneration. *J. Bone Joint Surg.*, 1961; 43A: 643.
36. Bassett CAL. Biological significance of piezoelectricity. *Calc. Tiss. Res.*, 1968; 1: 252–261.
37. Friedenberg ZB and Brighton CT. Bioelectric potentials in bone. *J. Bone Joint Surg.*, 1966; 48A: 915–923.
38. Shamos MH and Lavine LS. Piezoelectricity as a fundamental property of biological tissues. *Nature*, 1967; 212: 267–268.
39. Williams WS and Perletz L. P–n junctions and the piezoelectric response of bone. *Nat. New Biol.*, 1971; 233: 58–59.
40. Anderson JC and Eriksson C. Electrical properties of wet collagen. *Nature*, 1968; 227: 166–168.
41. Bassett CAL and Pawluk RJ. Electrical behavior of cartilage during loading. *Science*, 1974; 814: 575–577.
42. Grodzinsky AJ, Lipshitz H, and Glimcher MJ. Electromechanical properties of articular cartilage during compression and stress relaxation. *Nature*, 1978; 275: 448–450.
43. Kim Y-J, Bonassar LJ, and Grodzinsky AJ. The role of cartilage streaming potential, fluid flow and pressure in the stimulation of chondrocyte biosynthesis during dynamic compression. *J. Biomechanics*, 1995; 28: 1055–1066.
44. Spadaro JA. Electrically stimulated bone growth in animals and man. *Clin. Orthop.*, 1977; 122: 325–329.
45. Black J. *Electrical Stimulation: Its Role in Growth, Repair, and Remodeling of the Musculoskeletal System*. Praeger: NY, 1987.
46. Spadaro JA. Mechanical and electrical interactions in bone remodeling. *Bioelectromagnetics*, 1997; 18: 193–202.
47. Brighton CT. The treatment of nonunions with electricity. *J. Bone Joint Surg.*, 1981; 63A: 8–12.
48. Friedenberg ZB, Harlow MC, and Brighton CT. Healing of non-union of the medial malleolus by means of direct current. *J. Trauma*, 1971; 11: 8831.
49. Pilla AA. Electrochemical Events in Tissue Growth and Repair. In: *Electrochemical Bioscience and Bioengineering*, Miller I, Salkind A, and Silverman H, (eds.), Electrochem. Society Symposium Series, Princeton: New Jersey, 1973, pp. 1–17.
50. Pilla AA. Mechanisms of electrochemical phenomena in tissue growth and repair. *Bioelectrochem. Bioenerg.*, 1974; 1: 227–243.
51. Bassett CAL, Pawluk RJ, and Pilla AA. Acceleration of fracture repair by electromagnetic fields. *Ann. NY Acad. Sci.*, 1974; 238: 242–262.

52. Basset CAL, Pilla AA, and Pawluk R. A non-surgical salvage of surgically-resistant pseudoarthroses and non-unions by pulsing electromagnetic fields. *Clin. Orthop.*, 1977; 124: 117–131.
53. Bassett C, Mitchell S, and Gaston S. Treatment of ununited tibial diaphyseal fractures with pulsing electromagnetic fields. *J. Bone Joint Surg.*, 1981; 63A: 511–523.
54. Bassett C, Mitchell S, and Schink M. Treatment of therapeutically resistant nonunions with bone grafts and pulsing electromagnetic fields. *J. Bone Joint Surg.*, 1982; 64A: 1214–1224.
55. Bassett C, Valdes M, and Hernandez E. Modification of fracture repair with selected pulsing electromagnetic fields. *J. Bone Joint Surg.*, 1982; 64A: 888–895.
56. Mooney VA. randomized double blind prospective study of the efficacy of pulsed electromagnetic fields for interbody lumbar fusions. *Spine*, 1990; 15: 708–715.
57. Goodwin CB, Brighton CT, Guyer RD, Johnson JR, Light KI, and Yuan HA. A double blind study of capacitively coupled electrical stimulation as an adjunct to lumbar spinal fusions. *Spine*, 1999; 24: 1349–1357.
58. Zdeblick TD. A prospective, randomized study of lumbar fusion: preliminary results. *Spine*, 1993; 18: 983–991.
59. Linovitz RJ, Ryaby JT, Magee FP, Faden JS, Ponder R, and Muenz LR. Combined magnetic fields accelerate primary spine fusion: a double-blind, randomized, placebo controlled study. *Proc. Am. Acad. Orthop. Surg.*, 2000; 67: 376.
60. Nicolakis P, Kollmitzer J, Crevenna R, Bittner C, Erdogmus CB, and Nicolakis J. Pulsed magnetic field therapy for osteoarthritis of the knee—a double-blind sham-controlled trial. *Wien. Klin. Wochenschr.*, 2002; 16: 678–684.
61. Zizic T, Hoffman P, Holt D, Hungerford J, O'Dell J, Jacobs M, et al. The treatment of osteoarthritis of the knee with pulsed electrical stimulation. *J. Rheumatol.*, 1995; 22: 1757–1761.
62. Mont MA, Hungerford DS, Caldwell JR, Hoffman KC, He YD, Jones LC, and Zizic TM. The use of pulsed electrical stimulation to defer total knee arthroplasty in patients with osteoarthritis of the knee. *Am. Acad. Orthop. Surg.*, Mar, 2004.
63. Aaron RK, Lennox D, Bunce GE, and Ebert T. The conservative treatment of osteonecrosis of the femoral head. A comparison of core decompression and pulsing electromagnetic fields. *Clin. Orthop.*, 1989; 249: 209–218.
64. Steinberg ME, Brighton CT, Corces A, Hayken GD, Steinberg DR, Strafford B, Tooze SE, and Fallon M. Osteonecrosis of the femoral head. Results of core decompression and grafting with and without electrical stimulation. *Clin. Orthop.*, 1989; 249: 199–208.
65. Binder A, Parr G, Hazelman B, and Fitton-Jackson S. Pulsed electromagnetic field therapy of persistent rotator cuff tendinitis: a double blind controlled assessment. *Lancet*, 1984; 1: 695–697.
66. Brighton C and Pollack S. Treatment of recalcitrant non-unions with a capacitively coupled electrical field. *J. Bone Joint Surg.*, 1985; 67A: 577–585.
67. Brighton CT, Hozack WJ, Brager MD, Windsor RE, Pollack SR, Vreslovic EJ, and Kotwick JE. Fracture healing in the rabbit fibula when subjected to various capacitively coupled electrical fields. *J. Orthop. Res.*, 1985; 3: 331–340.
68. Aaron RK, Ciombor DMcK, and Simon BJ. Treatment of Nonunions With Electric and Electromagnetic Fields. *Clin. Orthop.*, 2004; 419: 21–29.
69. Fitzsimmons RJ, Ryaby JT, Magee FP, and Baylink DJ. Combined magnetic fields increase net calcium flux in bone cells. *Calcif. Tiss Intl.*, 1994; 55: 376–380.
70. Fitzsimmons RJ, Baylink DJ, Ryaby JT, and Magee FP. EMF-stimulated bone cell proliferation, in *Electricity and Magnetism in Biology and* Medicine. M.J. Blank, ed., San Francisco Press, San Francisco, 1993, pp. 899–902.
71. Fitzsimmons RJ, Ryaby JT, Mohan S, Magee FP, and Baylink DJ. Combined magnetic fields increase IGF-II in TE-85 human bone cell cultures. *Endocrinology*, 1995; 136: 3100–3106.
72. Fitzsimmons RJ, Ryaby JT, Magee FP, and Baylink DJ. IGF II receptor number is increased in TE 85 cells by low amplitude, low-frequency combined magnetic field (CMF) exposure. *J. Bone Min. Res.*, 1995; 10: 812–819.
73. Ryaby JT, Fitzsimmons RJ, Khin NA, Culley PL, Magee FP, Weinstein AM, and Baylink DJ. The role of insulin-like growth factor in magnetic field regulation of bone formation. *Bioelectrochem. Bioenerg.*, 1994; 35: 87–91.

74. Aaron RK, Ciombor DMcK, and Jones AR. Bone induction by decalcified bone matrix and mRNA of TGFb and IGF-1 are increased by ELF field stimulation. *Trans. Orthop. Res. Soc.*, 1997; 22: 548.
75. Ciombor McKD, Lester G, Aaron RK, Neame P, and Caterson B. Low frequency EMF regulates chondrocyte differentiation and expression of matrix proteins. *J. Orthop. Res.*, 2002; 20: 40–50.
76. Aaron RK and Ciombor DMcK. Acceleration of experimental endochondral ossification by biophysical stimulation of the progenitor cell pool. *J. Orthop. Res.*, 1996; 14: 582–589.
77. Aaron RK, Ciombor DMcK, and Jolly G. Stimulation of experimental endochondral ossification by low-energy pulsing electromagnetic fields. *J. Bone Min. Res.*, 1989; 4: 227–233.
78. Lohmann CH, Schwartz Z, Liu Y, Guerkov H, Dean DD, Simon B, and Boyan BD. Pulsed electromagnetic field stimulation of MG63 osteoblast-like cells affects differentiation and local factor production. *J. Orthop. Res.*, 2000; 18: 637–646.
79. Guerkov HH, Lohmann CH, Liu Y, Dean DD, Simon BJ, Heckman JD, Schwartz Z, and Boyan BD. Pulsed electromagnetic fields increase growth factor release by nonunion cells. *Clin. Orthop.*, 2001; 384: 265–279.
80. Zhuang H, Wang W, Seldes RM, Tahernia AD, Fan H, and Brighton CT. Electrical Stimulation induces the level of TGF-beta1 mRNA in osteoblastic cells by a mechanism involving calcium/calmodulin pathway. *Biochem. Biophys. Res. Commun.*, 1997; 237: 225–229.
81. Brighton CT, Wang W, Seldes R, Zhang G, and Pollack SR. Signal transduction in electrically stimulated bone cells. *J. Bone Joint Surg.*, 2001; 83A: 1514–1523.
82. Bodamyali T, Bhatt B, Hughes FJ, Winrow VR, Kanczler JM, Simon B, Abbott J, Blake DR, and Stevens CR. Pulsed electromagnetic fields simultaneously induce osteogenesis and upregulate transcription of bone morphogenetic proteins 2 and 4 in rat osteoblasts *in vitro*. *Biochem. Biophys. Res. Commun.*, 1998; 250: 458–461.
83. Aaron RK, Boyan BD, Ciombor DMcK, Schwartz Z, and Simon BJ. Stimulation of growth factor synthesis by electric and electromagnetic fields. *Clin. Orthop.*, 2004; 419: 30–37.
84. Markov MS, Ryaby JT, Kaufman JJ, and Pilla AA. Extremely weak AC and DC magnetic field significantly affect myosin phosphorylation. In: *Charge and Field Effects in Biosystems-3*. Allen MJ, Cleary SF, Sowers AE, and Shillady DD, eds., Birkhauser, Boston, 1992, pp. 225–230.
85. Markov MS, Wang S, Pilla AA. Effects of weak low-frequency sinusoidal and DC magnetic fields on myosin phosphorylation in a cell-free preparation. *Bioelectrochem. Bioenerg.*, 1993; 30: 119–125.
86. Markov MS and Pilla AA. Ambient range sinusoidal and DC magnetic fields affect myosin phosphorylation in a cell-free preparation. In: *Electricity and Magnetism in Biology and Medicine*. Blank M, ed., San Francisco Press, 1993, pp. 323–327.
87. Markov MS and Pilla AA. Static magnetic field modulation of myosin phosphorylation: calcium dependence in two enzyme preparations. *Bioelectrochem. Bioenerg.*, 1994; 35: 57–61.
88. Markov MS and Pilla AA. Modulation of cell-free myosin light chain phosphorylation with weak low-frequency and static magnetic fields. In: *On the Nature of Electromagnetic Field Interactions with Biological Systems*. Fry AH, ed., RG. Landes Co., Austin, 1994, pp. 127–141.
89. Markov MS, Muehsam DJ, and Pilla AA. Modulation of cell-free myosin phosphorylation with pulsed radio frequency electromagnetic fields. In: *Charge and Field Effects in Biosystems 4*. Allen MJ, Cleary SF, and Sowers AE, eds., World Scientific, New Jersey, 1994, pp. 274–288.
90. Markov MS and Pilla AA. Weak static magnetic field modulation of myosin phosphorylation in a cell-free preparation: calcium dependence, *Bioelectrochem. Bioenerg.*, 1997; 43: 235–240.
91. Engstrom S, Markov MS, McLean MJ, Holcomb RR, and Markov JM. Effects of non-uniform static magnetic fields on the rate of myosin phosphorylation. *Bioelectromagnetics*, 2002; 23: 475–479.
92. Liboff AR, Cherng S, Jenrow KA, and Bull A. Calmodulin-dependent cyclic nucleotide phosphodiesterase activity is altered by 20 mT magnetostatic fields. *Bioelectromagnetics*, 2003; 24: 32–38.
93. Yen-Patton GP, Patton WF, Beer DM, et al., Endothelial cell response to pulsed electromagnetic fields: stimulation of growth rate and angiogenesis *in vitro*. *J. Cell Physiol.*, 1988; 134: 37–39.
94. Tepper OM, Callaghan MJ, Chang EI, Galiano RD, Bhatt KA, Baharestani S, Gan J, Simon B, Hopper RA, Levine JP, and Gurtner GC. Electromagnetic fields increase *in vitro* and *in vivo* angiogenesis through endothelial release of FGF-2. *FASEB J.*, 2004; 18: 1231–1233.

95. Bassett CAL, Pawluk RJ, and Pilla AA. Acceleration of fracture repair by electromagnetic fields. *Ann. NY Acad. Sci.*, 1974; 238: 242–262.
96. Bassett CAL, Pawluk RJ, and Pilla AA. Augmentation of bone repair by inductively coupled electromagnetic fields. *Science*, 1974; 184: 575–578.
97. Cane V, Botti P, Farnetti P, and Soana S. Electromagnetic stimulation of bone repair: a histomorphometric study. *J. Orthop. Res.* 1991; 9: 908–917.
98. Cane V, Botti P, and Soana S. Pulsed magnetic fields improve osteoblast activity during the repair of an experimental osseous defect. *J. Orthop. Res.*, 1993; 11: 664–670.
99. Brighton CT, Katz MJ, Goll SR, Nichols CE, and Pollack SR. Prevention and treatment of sciatic denervation disuse osteoporosis in the rat tibia with capacitively coupled electrical stimulation. *Bone*, 1985; 6: 87–97.
100. Brighton CT, Luessenhop CP, Pollack SR, Steinberg DR, Petrik ME, and Kaplan FS. Treatment of castration induced osteoporosis by a capacitively coupled electrical signal in rat vertebrae. *J. Bone Joint Surg.*, 1989; 71A: 228–236.
101. Skerry TM, Pead MJ, and Lanyon LE. Modulation of bone loss during disuse by pulsed electromagnetic fields. *J. Orthop. Res.*, 1991; 9: 600–608.
102. Ryaby JT, Haupt DL, and Kinney JH. Reversal of osteopenia in ovariectomized rats with combined magnetic fields as assessed by x-ray tomographic microscopy. *J. Bone Min. Res.*, 1996; 11: S231.
103. McLeod KJ and Rubin CT. The effect of low-frequency electrical fields on osteogenesis. *J. Bone Joint Surg.*, 1992; 74A: 920–929.
104. Connolly J, Ortiz J, Price R, et al. The effect of electrical stimulation on the biophysical properties of fracture healing. *Ann. NY Acad. Sci.*, 1974; 238: 519–529.
105. Petersson C, Holmar N, and Johnell O. Electrical stimulation of osteogenesis: studies of the cathode effect on rat femur. *Acta. Orthop. Scand.*, 1982; 53: 727–732.
106. Bassett C, Valdes M, and Hernandez E. Modification of fracture repair with selected pulsing electromagnetic fields. *J. Bone Joint Surg.*, 1982; 64A: 888–895.
107. Inoue N, Ohnishi I, Chen D, et al. Effect of pulsed electromagnetic fields (PEMF) on late-phase osteotomy gap healing in a canine tibial model. *J. Orthop. Res.*, 2002; 20: 1106–1114.
108. France JC, Norman TL, Santrock RD, et al. The efficacy of direct current stimulation for lumbar intertransverse process fusions in an animal model. *Spine*, 2001; 26: 1002–1008.
109. Nerubay J, Marganit B, Bubis JJ, et al. Stimulation of bone formation by electrical current on spinal fusion. *Spine*, 1986; 11: 167–169.
110. Toth JM, Seim HB, Schwardt JD, et al. Direct current electrical stimulation increases the fusion rate of spinal fusion cages. *Spine*, 2000; 25: 2580–2587.
111. Dejardin LM, Kahanovitz N, Arnoczky SP, et al. The effect of varied electrical current densities on lumbar spinal fusions in dogs. *Spine*, 2001; 1: 341–347.
112. Inoue N, Ohnishi I, Chen D, Deitz LW, Schwardt JD, and Chao EY. Effect of pulsed electromagnetic fields (PEMF) on late-phase osteotomy gap healing in a canine tibial model. *J. Orthop. Res.*, 2002; 20: 1106–1114.
113. Fini M, Cadossi R, Cane V, Cavani F, Giavaresi G, Krajewski A, Martini L, Aldini NN, Ravaglioli A, Rimondini L, Torricelli P, and Giardino R. The effect of pulsed electromagnetic fields on the osteointegration of hydroxyapatite implants in cancellous bone: a morphologic and microstructural *in vivo* study. *J. Orthop. Res.*, 2002; 20: 756–763.
114. Smith TL, Wong-Gibbons D, and Maultsby J. Microcirculatory effects of pulsed electromagnetic fields. *J. Orthop. Res.*, 2004; 22: 80–84.
115. Borsalino G, Bagnacani M, Bettati E, et al. Electrical stimulation of human femoral intertrochanteric osteotomies. *Clin. Orthop.*, 1988; 237: 256–263.
116. Mammi GI, Rocchi R, Cadossi R, et al. The electrical stimulation of tibial osteotomies: a double-blind study. *Clin. Orthop.*, 1993; 288: 246–253.
117. Traina G, Sollazzo V, and Massari L. Electrical stimulation of tibial osteotomies: a double blind study. In: *Electricity and Magnetism in Biology and Medicine*. Bersani F, ed., Kluwer Academic/Plenum: New York, 1999, pp. 137–138.
118. Kahanovitz N. Electrical stimulation of spinal fusion: a scientific and clinical update. *Spine*, 2002; 2: 145–150.

119. Oishi M and Onesti S. Electrical bone graft stimulation for spinal fusion: a review. *Neurosurgery*, 2000; 47: 1041–1056.
120. Rogozinski A and Rogozinski C. Efficacy of implanted bone growth stimulation in instrumented lumbosacral spinal fusion. *Spine*, 1996; 21: 2479–2483.
121. Meril AJ. Direct current stimulation of allograft in anterior and posterior lumbar interbody fusions. *Spine*, 1994; 19: 2393–2398.
122. Goodwin CB, Brighton CT, Guyer RD, et al. A double-blind study of capacitively coupled electrical stimulation as an adjunct to lumbar spinal fusions. *Spine*, 1999; 24: 1349–1356.
123. Gossling HR, Bernstein RA, and Abbott J. Treatment of ununited tibial fractures: a comparison of surgery and pulsed electromagnetic fields (PEMF). *Orthopedics*, 1992; 15: 711–719.
124. Paterson D, Lewis G, and Cass C. Treatment of delayed union and nonunion with an implanted direct current stimulator. *Clin. Orthop.*, 1980; 148: 117–128.
125. Brighton C, Black J, and Friedenberg Z. A multicenter study of the treatment of non-union with constant direct current. *J. Bone Joint Surg.*, 1981; 63A: 2–13.
126. Bassett C, Mitchell S, and Gaston S. Treatment of ununited tibial diaphyseal fractures with pulsing electromagnetic fields. *J. Bone Joint Surg.*, 1981; 63A: 511–523.
127. Heckman J, Ingram A, and Loyd R. Nonunion treatment with pulsed electromagnetic fields. *Clin. Orthop.*, 1981; 161: 58–66.
128. Bassett C, Mitchell S, and Schink M. Treatment of therapeutically resistant nonunions with bone grafts and pulsing electromagnetic fields. *J. Bone Joint Surg.*, 1982; 64A: 1214–1224.
129. Brighton C and Pollack S. Treatment of recalcitrant non-unions with a capacitively coupled electrical field. *J. Bone Joint Surg.*, 1985; 67A: 577–585.
130. Sedel L, Christel P, Duriez J, Duriez R, Evrard J, Ficat C, Cauchoix J, and Witvoet J. Acceleration of repair of non-unions by electromagnetic fields. *Rev. Chir. Orthop. Reparatrice Appar. Mot.*, 1981; 67: 11–23.
131. DeHaas W, Watson J, and Morrison D. Noninvasive treatment of ununited fractures of the tibia using electrical stimulation. *J. Bone Joint Surg.*, 1980; 62B: 465–470.
132. Dunn AW and Rush GA. Electrical stimulation in treatment of delayed union and nonunion of fractures and osteotomies. *South Med. J.*, 1984; 77: 1530–1534.
133. Sharrard W. A double blind trial of pulsed electromagnetic fields for delayed union of tibial fractures. *J. Bone Joint Surg.*, 1990; 72B: 347–355.
134. Scott G and King J. A prospective double blind trial of electrical capacitive coupling in the treatment of non-union of long bones. *J. Bone Joint Surg.*, 1994; 76A: 820–826.
135. Brighton C, Shaman P, and Heppenstall R. Tibial nonunion treated with direct current, capacitive coupling, or bone graft. *Clin. Orthop.*, 1995; 321: 223–234.
136. Fredericks DC, Piehl DJ, Baker JT, Abbott J, and Nepola JV. Effects of pulsed electromagnetic field stimulation on distraction osteogenesis in the rabbit tibial leg lengthening model. *J. Pediatr. Orthop.*, 2003; 23: 478–483.
137. Wysocki AB. Wound fluids and the pathogenesis of chronic wounds. *J. Wound Ostomy Continence Nurs.*, 1996; 23: 283–290.
138. CMS. Decision memo for electrostimulation for wounds (CAG-00068R), 2003. http://www.cms.hhs.gov/mcd/.
139. Burr HS, Taffel M, and Harvey SC. Electrometric study of the healing wound in man. *Yale J. Biol. Med.*, 1940; 12: 483–485.
140. Burrows H, Iball J, and Roe EMF. Electrical changes in wounds and inflamed tissues: Part 1. The bioelectric potentials of cutaneous wounds in rats. *Br. J. Exp. Pathol.*, 1942; 23: 253–257.
141. Barnes TC. Healing rate of human skin determined by measurement of the electrical potential of experimental abrasions. *Am. J. Surg.*, 1945; 69: 82–88.
142. Barker AT. Measurement of direct current in biological fluids. *Med. Biol. Eng. Comput.*, 1981; 19: 507–508.
143. Barker AT, Jaffe LF, and Vanable JW Jr. The glabrous epidermis of cavies contains a powerful battery. *Am. J. Physiol.*, 1982; 242: R358–R366.
144. Foulds IS and Barker AT. Human skin battery potentials and their possible role in wound healing. *Br. J. Dermatol.*, 1983; 109: 515–522.

145. Iglesia DD and Vanable JW. Endogenous lateral electric fields around bovine corneal lesions are necessary for and can enhance normal rates of wound healing. *Wound Rep. Reg.*, 1998; 6: 531–542.
146. Fang KS, Farboud B, Nuccitelli R, and Isseroff RR. Migration of human keratinocytes in electric fields requires growth factors and extracellular calcium. *J. Invest. Dermatol.*, 1998; 111: 751–756.
147. Lee RC, Canaday DJ, and Doong H. A review of the biophysical basis for the clinical application of electric fields in soft-tissue repair. *J. Burn Care Rehabil.*, 1993; 14: 319–335.
148. Carley PJ and Wainapel SF. Electrotherapy for acceleration of wound healing: low intensity direct current. *Arch. Phys. Med. Rehabil.*, 1985; 66: 443–446.
149. Gentzkow GD. Electrical stimulation to heal dermal wounds. *J. Dermatol. Surg. Oncol.*, 1993; 19: 753–758.
150. Vodovnik L, Miklavcic D, and Sersa G. Modified cell proliferation due to electrical currents. *Med. Biol. Eng. Comput.*, 1992; 30: CE21–CE28.
151. Goodman E, Greenebaum B, and Frederiksen J. Effect of pulsed magnetic fields on human umbilical endothelial vein cells. *Bioelectrochem. Bioenerg.*, 1993; 32: 125–132.
152. Amaral SL, Linderman JR, Morse MM, and Greene AS. Angiogenesis induced by electrical stimulation is mediated by angiotensin II and VEGF. *Microcirculation*, 2001; 8: 57–67.
153. Greenough CG. The effects of pulsed electromagnetic fields on blood vessel growth in the rabbit ear chamber. *J. Orthop. Res.*, 1992; 10: 256–262.
154. Nikolaev AV, Shekhter AB, Mamedov LA, Novikov AP, and Manucharov NK. Use of a sinusoidal current of optimal frequency to stimulate the healing of skin wounds. *Biull. Eksp. Biol. Med.*, 1984; 97: 731–734.
155. McLeod KJ, Lee RC, and Ehrlich HP. Frequency dependence of electric field modulation of fibroblast protein synthesis. *Science*, 1987; 236: 1465–1469.
156. Katsir G, Baram SC, and Parola AH. Effect of sinusoidally varying magnetic fields on cell proliferation and adenosine deaminase specific activity. *Bioelectromagnetics*, 1998; 19: 46–52.
157. Supino R, Bottone MG, Pellicciari C, Caserini C, Bottiroli G, Belleri M, and Veicsteinas A. Sinusoidal 50 Hz magnetic fields do not affect structural morphology and proliferation of human cells *in vitro*. *Histol. Histopathol.*, 2001; 16: 719–726.
158. George FR, Lukas RJ, Moffett J, and Ritz MC. *In-vitro* mechanisms of cell proliferation induction: a novel bioactive treatment for accelerating wound healing. *Wounds*, 2002; 14: 107–115.
159. Gilbert TL, Griffin N, Moffett J, Ritz MC, and George FR. The provant wound closure system induces activation of p44/42 MAP kinase in normal cultured human fibroblasts. *Ann. NY Acad. Sci.*, 2002; 961: 168–171.
160. Cho MR, Marler JP, Thatte HS, and Golan DE. Control of calcium entry in human fibroblasts by frequency-dependent electrical stimulation. *Front. Biosci.*, 2002; 7: 1–8.
161. Roland, D, Ferder M, Kothuru R, Faierman T, and Strauch B. Effects of pulsed magnetic energy on a microsurgically transferred vessel. *Plast. Reconstr. Surg.*, 2000; 105: 1371–1374.
162. Weber RV, Navarro A, Wu JK, Yu HL, and Strauch B. Pulsed magnetic fields applied to a transferred arterial loop support the rat groin composite flap. *Plast. Reconstr. Surg.*, 2004; 114: 1185–1189.
163. Strauch B, Patel MK, Navarro A, Berdischevsky M, and Pilla AA. Pulsed magnetic fields accelerate wound repair in a cutaneous wound model in the rat. *Plast. Reconstr. Surg.*, 2006, in press.
164. Glassman LS, McGrath MH, and Bassett CA. Effect of external pulsing electromagnetic fields on the healing of soft tissue. *Ann. Plast. Surg.*, 1986; 16: 287–295.
165. Ginsberg AJ. Ultrashort radiowaves as a therapeutic agent. *Med. Record*, 1934; 140: 651–653.
166. Salzberg CA, Cooper SA, Perez P, Viehbeck MG, and Byrne DW. The effects of nonthermal pulsed electromagnetic energy on wound healing of pressure ulcers in spinal cord-injured patients: a randomized, double-blind study. *Ostomy Wound Management*, 1995; 41: 42–51.
167. Kloth LC, Berman JE, Sutton CH, Jeutter DC, Pilla AA, and Epner ME. Effect of pulsed radio frequency stimulation on wound healing: a double-blind pilot clinical study. In: *Electricity and Magnetism in Biology and Medicine*. Bersani F, ed., Plenum press, NY, 1999, pp. 875–878.
168. Pilla AA, Martin DE, Schuett AM, et al. Effect of pulsed radiofrequency therapy on edema from grades I and II ankle sprains: a placebo controlled, randomized, multi-site, double-blind clinical study. *J. Athl. Train.*, 1996; S31: 53.

169. Pennington GM, Danley DL, Sumko MH, et al. Pulsed, non-thermal, high frequency electromagnetic energy (diapulse) in the treatment of grade I and grade II ankle sprains. *Military Med.*, 1993; 158: 101–104.
170. Foley-Nolan D, Barry C, Coughlan RJ, O'Connor P, and Roden D. Pulsed high frequency (27MHz) electromagnetic therapy for persistent neck pain: a double blind placebo-controlled study of 20 patients. *Orthopedics*, 1990; 13: 445–451.
171. Foley-Nolan D, Moore K, Codd M, et al. Low energy high frequency pulsed electromagnetic therapy for acute whiplash injuries: a double blind randomized controlled study. *Scan. J. Rehab. Med.*, 1992; 24: 51–59.
172. Mayrovitz HN and Larsen PB. Effects of pulsed magnetic fields on skin microvascular blood perfusion. *Wounds: Compend. Clin. Res. Pract.*, 1992; 4: 192–202.
173. Mayrovitz HN and Larsen PB. A preliminary study to evaluate the effect of pulsed radio frequency field treatment on lower extremity peri-ulcer skin microcirculation of diabetic patients. *Wounds: Compend. Clin. Res. Pract.*, 1995; 7: 90–93.
174. Mayrovitz HN, Macdonald J, and Sims N. Effects of pulsed radio frequency diathermy on postmastectomy arm lymphedema and skin blood flow: a pilot investigation. *Lymphology*, 2002; 35(Suppl): 353–356.
175. Kenkre JE, Hobbs FD, Carter YH, Holder RL, and Holmes EP. A randomized controlled trial of electromagnetic therapy in the primary care management of venous leg ulceration. *Fam. Pract.*, 1996; 13: 236–241.
176. Musaev AV, Guseinova SG, and Imamverdieva SS. Application of impulse complex modulated electromagnetic fields in management of patients with diabetic polyneuropathy. *Zh. Nevrol. Psikhiatr. Im. S S Korsakova.*, 2002; 102: 17–24.
177. Goldin JH, Broadbent NRG, Nancarrow JD, and Marshall T. The effects of diapulse on the healing of wounds: a double-blind randomized controlled trial in man. *Br. J. Plast. Surg.*, 1981; 34: 267–270.
178. Itoh M, Montemayor JS Jr, Matsumoto E, Eason A, Lee MH, and Folk FS. Accelerated wound healing of pressure ulcers by pulsed high peak power electromagnetic energy (diapulse). *Decubitus*, 1991; 4: 24–25, 29–34.
179. Seaborne D, Quirion-DeGirardi C, and Rousseau M. The treatment of pressure sores using pulsed electromagnetic energy (PEME). *Physiother. Can.*, 1996; 48: 131–137.
180. Comorosan S, Vasilco R, Arghiropol M, Paslaru L, Jieanu V, and Stelea S. The effect of diapulse therapy on the healing of decubitus ulcer. *Rom. J. Physiol.*, 1993; 30: 41–45.
181. Ieran M, Zaffuto S, Bagnacani M, Annovi M, Moratti A, and Cadossi R. Effect of low-frequency electromagnetic fields on skin ulcers of venous origin in humans: a double blind study. *J. Orthop. Res.*, 1990; 8: 276–282.
182. Stiller MJ, Pak GH, Shupack JL, Thaler S, Kenny C, and Jondreau L. A portable pulsed electromagnetic field (PEMF) device to enhance healing of recalcitrant venous ulcers: a double-blind, placebo-controlled clinical trial. *Br. J. Dermatol.*, 1992; 127: 147–154.
183. Canedo-Dorantes L, Garcia-Cantu R, Barrera R, Mendez-Ramirez I, Navarro VH, and Serrano G. Healing of chronic arterial and venous leg ulcers with systemic electromagnetic fields. *Arch. Med. Res.*, 2002; 33: 281–289.
184. Bentall RHC. Low-level pulsed radiofrequency fields and the treatment of soft-tissue injuries. *Bioelectricity and Bioenergetics* 1986; 16: 531–548.
185. Maddin WS, Bell PW, and James JHM. The biologic effects of a pulsed electrostatic field with specific reference to hair. *Int. J. Dermatol.*, 1990; 29: 446–450.
186. Maddin WS, Amara I, and Sollecito WA. Electrotrichogenesis: further evidence of efficacy and safety on extended use. *Int. J. Dematol.*, 1992; 31: 878–880.
187. Benjamin B, Ziginskas D, Harman J, and Meakin T. Pulsed electrostatic fields (ETG) to reduce hair loss in women undergoing chemotherapy for breast carcinoma: a pilot study. *Psycho-oncology*, 2002; 11: 244–248.
188. Trock DH, Bollet AJ, and Markoll R. The effect of pulsed electromagnetic fields in the treatment of osteoarthritis of the knee and cervical spine. Reports of randomized, double blind, placebo controlled trials, *J. Rheumatol.*, 1994; 21: 1903–1911.
189. Trock DH, Bollet AJ, Dyer RH, Fielding LP, Miner WK, and Markoll R. A double-blind trial of the clinical effects of pulsed electromagnetic fields in osteoarthritis, *J. Rheumatol.*, 1993; 20: 456–460.

190. Dyson M and Brookes M. Stimulation of bone repair by ultrasound. *Ultrasound Med. Biol.*, 1983; 8: 61–66.
191. Duarte LR. The stimulation of bone growth by ultrasound. *Arch. Orthop. Trauma. Surg.*, 1983; 101: 153–159.
192. Pilla AA, Mont MA, Nasser PR, Khan SA, Figueiredo M, Kaufman JJ, and Siffert RS. Non-invasive low-intensity pulsed ultrasound accelerates bone healing in the rabbit. *J. Orthop. Trauma*, 1990; 4: 246–253.
193. Pilla AA, Figueiredo M, Nasser P, Alves JM, Ryaby JT, Klein M, Kaufman JJ, Siffert RS, Kristiansen T, and Heckman J. Acceleration of bone repair by pulsed sine wave ultrasound: animal, clinical, and mechanistic studies. In: *Electromagnetics in Biology and Medicine*. Brighton CT and Pollack SR, eds., San Francisco Press, San Francisco, 1991, pp. 331–341.
194. Tis JE, Meffert CR, Inoue N, McCarthy EF, Machen MS, McHale KA, and Chao EY. The effect of low intensity pulsed ultrasound applied to rabbit tibiae during the consolidation phase of distraction osteogenesis. *J. Orthop. Res.*, 2002; 20: 793–800.
195. Chang WH-S, Sun J-S, Chang S-P, and Lin JC. Study of thermal effects of ultrasound stimulation on fracture healing. *Bioelectromagnetics*, 2002; 23: 256–263.
196. Heckman JD, Ryaby JP, McCabe J, Frey JJ, and Kilcoyne RF. Acceleration of tibial fracture-healing by non-invasive, low-intensity pulsed ultrasound. *J. Bone Joint Surg.*, 1994; 76A: 26–34.
197. Kristiansen TK, Ryaby JP, McCabe J, Frey JJ, and Roe LR. Accelerated healing of distal radial fractures with the use of specific, low-intensity ultrasound. A multicenter, prospective, randomized, double-blind, placebo-controlled study. *J. Bone Joint Surg.*, 1997; 79A: 961–973.
198. Rubin CT, Bolander M. Ryaby JP, and Hadjiargyrou M. The use of low-intensity ultrasound to accelerate the healing of fractures. *J. Bone Joint Surg.*, 2001; 83A: 259–270.
199. Pilla AA and Nasser PR. A unified mechanoelectric approach to Wolff's law for streaming potentials and exogenous EMF, *Trans. Orthop. Res. Soc.*, 1993; 18: 151.
200. Pilla AA, Muehsam DJ, and Nasser PR. A unified mechanoelectric basis for bone repair and remodelling via streaming potentials as the primary cellular messenger. *Trans. Orthop. Res. Soc.*, 1995; 20: 589.
201. Pilla AA. Low-intensity electromagnetic and mechanical modulation of bone growth and repair: are they equivalent? *J. Orthop. Sci.*, 2002; 7: 420–428.
202. Valbona C, Hazlewood CF, and Jurida G. Response of pain to static magnetic fields in post-polio patients: a double-blind pilot study. *Arch. Phys. Med. Rehabil.*, 1997; 78: 1200–1207.
203. Man D, Man B, and Plosker H. The influence of permanent magnetic field therapy on wound healing in suction lipectomy patients: a double-blind study. *Plast. Reconstr. Surg.*, 1999; 104: 2261–2266.
204. Colbert AP, Markov MS, Banerij M, and Pilla AA. Magnetic mattress mad use in patients with fibromyalgia: a randomized double-blind pilot study. *J. Back Musculoskelet. Rehab.*, 1999; 13: 19–31.
205. Alfano AP, Taylor AG, Foresman PA, Dunkl PR, McConnell GG, Conway MR, and Gillies GT. Static magnetic fields for treatment of fibromyalgia: a randomized controlled trial. *J. Alt. Comp. Med.*, 2001; 7: 53–64.
206. Weintraub MI. Magnetic bio-stimulation in painful diabetic peripheral neuropathy: a novel intervention—a randomized double-placebo crossover study. *Am. J. Pain Manag.*, 1999; 9: 8–17.
207. Brown CS, Ling FW, Wan JY, and Pilla AA. Efficacy of static magnetic field therapy in chronic pelvic pain: a double-blind pilot study. *Am. J. Obs. Gyn.*, 2002; 187: 1581–1587.
208. Weintraub MI, Wolfe GI, Barohn RA, Cole SP, Parry GJ, Hayat G, Cohen JA, Page JC, Bromberg MB, and Schwartz SL. Static magnetic field therapy for symptomatic diabetic neuropathy: a randomized, double-blind, placebo-controlled trial. *Arch. Phys. Med. Rehabil.*, 2003; 84: 736–746.
209. Wolsko PM, Eisenberg DM, Simon LS, Davis RB, Walleczek J, Mayo-Smith M, Kaptchuk TJ, and Phillips RS. Double-blind placebo-controlled trial of static magnets for the treatment of osteoarthritis of the knee: results of a pilot study. *Alt. Ther. Health Med.*, 2004; 10: 36–43.
210. Zhadin MN, Fesenko EE. Ionic cyclotron resonance in biomolecules. *Biomed. Sci.*, 1990; 1: 245–250.
211. Edmonds DT. Larmor precession as a mechanism for the detection of static and alternating magnetic fields. *Bioelectrochem. Bioenerg.*, 1993; 30: 3–12.

212. Muehsam DJ and Pilla AA. Weak magnetic field modulation of ion dynamics in a potential well: mechanistic and thermal noise considerations. *Bioelectrochem. Bioenerg.*, 1994; 35: 71–79.
213. Muehsam DS and Pilla AA. Lorentz approach to static magnetic field effects on bound ion dynamics and binding kinetics: thermal noise considerations. *Bioelectromagnetics*, 1996; 17: 89–99.
214. Zhadin MN. Effect of magnetic fields on the motion of an ion in a macromolecule: theoretical analysis. *Biophysics*, 1996; 41: 843–860.
215. Pilla AA, Muehsam DJ, and Markov MS. A dynamical systems/Larmor precession model for weak magnetic field bioeffects: ion-binding and orientation of bound water molecules. *Bioelectrochem. Bioenerg.*, 1997; 43: 239–249.
216. Zhadin MH. Combined action of static and alternating magnetic fields on ion motion in a macromolecule: theoretical aspects. *Bioelectromagnetics*, 1998; 19: 279–292.
217. Blank M and Goodman R. Do electromagnetic fields interact directly with DNA? *Bioelectromagnetics*, 1997; 18: 111–115.
218. Blank M and Goodman R. Electromagnetic fields may act directly on DNA. *J. Cell Biochem.*, 1999; 75: 369–374.
219. Adair RK. Extremely low-frequency electromagnetic fields do not interact directly with DNA. *Bioelectromagnetics*, 1998; 19: 136–138.
220. Seegers JC, Engelbrecht CA, and van Papendorp DH. Activation of signal-transduction mechanisms may underlie the therapeutic effects of an applied electric field. *Med. Hypotheses*, 2001; 57: 224–230.
221. Chiabrera A, Grattarola M, Viviani R, and Braccini C. Modelling of the perturbation induced by low-frequency electromagnetic fields on the membrane receptors of stimulated human lymphocytes. *Stud. Biophys.*, 1982; 90: 77–81.
222. Pilla AA, Schmukler RE, Kaufman JJ, and Rein G. Electromagnetic modulation of biological processes: consideration of cell–waveform interaction. In: *Interactions between Electromagnetic Fields and Cells*. Chiabrera A, Nicolini C, and Schwan HP, eds., Plenum Press, NY, 1985, 423–436.
223. Pilla AA, Kaufman JJ, and Ryaby JT. Electrochemical kinetics at the cell membrane: a physicochemical link for electromagnetic bioeffects. In: *Mechanistic Approaches to Interactions of Electric and Electromagnetic Fields with Living Systems*. Blank M and Findl E, eds., Plenum Press, NY, 1987, 39–62.
224. Pilla AA. State of the art in electromagnetic therapeutics. In: *Electricity and Magnetism in Biology and Medicine*. Blank M, ed., San Francisco Press Inc., San Francisco, 1993, 17–22.
225. Nelson FR, Brighton CT, Ryaby J, Simon BJ, Nielson JH, Lorich DG, Bolander M, and Seelig J. Use of physical forces in bone healing. *J. Am. Acad. Orthop. Surg.*, 2003; 11: 344–354.
226. Pilla AA, Sechaud P, and McLeod BR. Electrochemical and electric current aspects of low-frequency electromagnetic current induction in biological systems. *J. Biol. Phys.*, 1983; 11: 51–57.
227. McLeod BR, Pilla AA, and Sampsel MW. Electromagnetic fields induced by Helmholtz aiding coils inside saline-filled boundaries. *Bioelectromagnetics*, 1983; 4: 357–370.
228. Hart FX. Cell culture dosimetry for low-frequency magnetic fields. *Bioelectromagnetics*, 1996; 17: 48–57.
229. van Amelsfort AMJ. An analytical algorithm for solving inhomogeneous electromagnetic boundary-value problems for a set of coaxial circular cylinders, Ph.D. Thesis, Eindhoven University, The Netherlands, 1991.
230. Buechler DN, Christensen DA, Durney CH, and Simon B. Calculation of electric fields induced in the human knee by a coil applicator. *Bioelectromagnetics*, 2001; 22: 224–231.
231. Plonsey R and Fleming DG. *Bioelectric Phenomena*. McGraw-Hill, NY, 1969.
232. Foster KR, Schwan HP. Dielectric properties of tissues and biological materials: a critical review. *Crit. Rev. Biomed. Eng.*, 1989; 17: 25–104.
233. Gabriel. Dielectric and Magnetic Properties of Biological Materials. Chapter 3, *BBA* of this book.
234. Pilla AA. Electrochemical information transfer at cell surfaces and junctions: application to the study and manipulation of cell regulation. In: *Bioelectrochemistry*. Keyser H and Gutman F, eds., Plenum Press, NY, 1980, pp. 353–396.

235. Pilla AA and Margules G. Dynamic interfacial electrochemical phenomena at living cell membranes: application to the toad urinary bladder membrane system. *J. Electrochem. Soc.*, 1977; 124: 1697–1706.
236. Pilla AA. Membrane impedance as a probe for interfacial electrochemical control of living cell function. *Adv. in Chem., ACS*, 1980; 188: 339–359.
237. Margules G, Doty SB, and Pilla AA. Impedance of living cell membranes in the presence of chemical tissue fixative. *Adv. in Chem., ACS*, 1980; 188: 461–484.
238. Schmukler RE, Kaufman JJ, Maccaro PC, Ryaby JT, and Pilla AA. Transient impedance measurements on biological membranes: application to red blood cells and melanoma cells. In: *Electrical Double Layers in Biology*. Blank M, ed., Plenum Press, NY, 1986, pp. 201–210.
239. Blinks LR. The direct current resistance of Nitella. *J. Gen. Physiol.*, 1930; 13: 495–508.
240. Cole KS and Cole RH. Dispersion and absorption in dielectrics I. Alternating current characteristics. *J. Chem. Phys.*, 1941; 9: 341–351.
241. Kao CY. Changing electrical constants of the Fundulus egg plasma membrane. *J. Gen. Physiol.*, 1956; 40: 107–119.
242. Poo MM, Poo WJ, and Lam JW. Lateral electrophoresis and diffusion of Concanavalin A receptors in the membrane of embryonic muscle cell. *J. Cell Biol.*, 1978; 76: 483–501.
243. Cheng DK. *Analysis of Linear Systems*. Addison-Wesley, London, 1959.
244. Pilla AA. A transient impedance technique for the study of electrode kinetics. *J. Electrochem. Soc.*, 1970; 117: 467–477.
245. Quastel MR and Kaplan JG. Inhibition by ouabain of human lymphocyte transformation induced by phytohaemagglutinin *in vitro. Nature*, 1968; 219: 198–200.
246. Kaplan JG. Membrane cation transport and the control of proliferation of mammalian cells. *Ann. Rev. Physiol.*, 1978; 40: 19–41.
247. Means AR and Rasmussen CD. Calcium, calmodulin and cell proliferation. *Cell Calcium*, 1988; 9: 313–318.
248. Berridge MJ. Calcium signal transduction and cellular control mechanisms. *Biochim. Biophys. Acta*, 2004; 1742: 3–7.
249. Markov MS and Pilla AA. Electromagnetic field stimulation of soft tissue: pulsed radio frequency treatment of post-operative pain and edema. *Wounds*, 1995; 7: 143–151.
250. Boonstra J, Van der Sagg PT, Moolenarr WH, and DeLaat SW. Rapid effects of nerve growth factor on Na+,K+-pump in rat pheochromocytoma cells. *Exp. Cell Res.*, 1981; 131: 452–455.
251. Whifield JF, Boyton AL, MacManus JP, Rixon RH, Sikorska M, Tsong B, Walker RP, and Smierenga SH. The roles of calcium and cyclic AMP in cell proliferation. *Ann. NY Acad. Sci.*, 1981; 339: 216–240.
252. Chafoules JC, Bolton WE, Hidaka H, Boyd AE, and Means HR. Calmodulin involvement in regulation of cell-cycle progression. *Cell*, 1982; 28: 41–50.
253. Boynton AL, Whitfield JF, Isaacs RJ, and Trembley RG. Different extracellular calcium requirement for proliferation of nonneoplastic, preneoplastic and neoplastic mouse cells. *Cancer Res.*, 1977; 37: 2657–2661.
254. Hazelton B and Tupper J. Calcium transport and exchange in mouse 3T3 and SV40–3T3 cells. *J. Cell Biol.*, 1979; 81: 538–542.
255. Gemsa D, Seitz M, Kramer W, Grimm W, Till G, and Resch K. Ionophore A 23187 raises cyclic AMP levels in macrophages by stimulation prostaglandin E formation. *Exp. Cell Res.*, 1979; 118: 55–62.
256. Chiabrera A, Bianco B, Caratozzolo F, Gianetti G, Grattarola M, and Viviani R. Electric and magnetic field effects on ligand binding to the cell membrane. In: *Interactions between Electromagnetic Fields and Cells*. Chiabrera A, Nicolini C, and Schwan HP, eds., Plenum Press, NY, 1985, p. 253.
257. Lyle DB, Wang X, Ayotte RD, Sheppard AR, and Adey WR. Calcium uptake by leukemic and normal t lymphocytes exposed to low-frequency magnetic fields. *Bioelectromagnetics*, 1991; 12: 145.
258. Bawin SM, Kaczmarek LK, and Adey WR. Effects of modulated VHF fields on the central nervous system. *Ann. NY Acad. Sci.*, 1975; 247: 74–91.
259. Blackman CF, Benane SG, Kinney LS, Joines WT, and House DE. Effects of ELF fields on calcium efflux from brain tissue *in vitro. Radiat. Res.*, 1982; 92: 510.

260. Blackman CF, Benane SG, Rabinowitz JR, House DE, and Joines WT. A role for the magnetic field in the radiation-induced efflux of Ca-ions from brain tissue *in vitro. Bioelectromagnetics,* 1985; 6: 327.
261. Wei LX, Goodman R, and Henderson AS. Changes in levels of c-myc and histone H2B following exposure of cells to low-frequency sinusoidal electromagnetic fields: evidence for a window effect. *Bioelectromagnetics*, 1990; 11: 269.
262. Elliott JP, Smith RL, and Block CA. Time-varying magnetic fields: effects of orientation on chondrocyte proliferation. *J. Orthop. Res.*, 1988; 6: 259–264.
263. Liburdy RP. Calcium signaling in lymphocytes and ELF fields: evidence for an electric field metric and a site of ion channel interaction involving the calcium ion channel. *FEBS Lett.*, 1992; 301: 53–59.
264. Pilla AA, Figueiredo M, Nasser PR, Kaufman JJ, and Siffert RS. Broadband EMF acceleration of bone repair in a rabbit model is independent of magnetic component. In: *Electricity and Magnetism in Biology and Medicine*. Blank M, ed., San Francisco Press, San Francisco, 1993, pp. 363–367.
265. Rubin CT, Donahue HJ, Rubin JE, and McLeod KJ. Optimization of electric field parameters for the control of bone remodeling: exploitation of an indigenous mechanism for the prevention of osteopenia. *J. Bone Miner. Res.*, 1993; 8: 573–581.
266. Rosenspire AJ, Andrei L, Kindzelskii AL, Simon BJ, and Petty HJ. Real-time control of neutrophil metabolism by very weak ultra-low-frequency pulsed magnetic fields. *Biophys. J.*, 2005; 88: 3334–3347.
267. Liboff AR, Williams T Jr, Strong DM, and Wistar R Jr. Time-varying magnetic fields: effect on DNA synthesis. *Science*, 1984; 223: 818–820.
268. Liboff AR. Cyclotron resonance in membrane transport. In: *Interactions between Electromagnetic Fields and Cells*. Chiabrera A, Nicolini C, and Schwan HP, eds., Plenum Press, NY, 1985, p. 281.
269. Liboff AR, Smith SD, and McLeod BR. Experimental evidence for ion cyclotron resonance mediation of membrane transport. In: *Mechanistic Approaches to Interactions of Electric and Electromagnetic Fields with Living Systems*. Blank M and Findl E, eds., Plenum Press, NY, 1987, p. 109.
270. Chiabrera A, Bianco B, Kaufman JJ, and Pilla AA. Resonant Phenomena. In: *Interaction Mechanisms of Low Level Electromagnetic Fields in Living Systems*. Royal Swedish Academy of Sciences, Stockholm, 1989, p. 256.
271. Muehsam DJ and Pilla AA. Weak magnetic field modulation of ion dynamics in a potential well: mechanistic and thermal noise considerations, *Bioelectrochem. Bioenerg.*, 1994; 35: 71–79.
272. Bianco B and Chiabrera A. From the Langevin-Lorentz to the Zeeman model of electromagnetic effects on ligand-receptor binding. *Bioelectrochem. Bioenerg.*, 1992; 28: 355–365.
273. Shuvalova LA, Ostrovskaja MV, Sosunov EA, and Lednev VV. Weak magnetic field influence of the speed of calmodulin dependent phosphorylation of myosin in solution. *Dokl. Akad. Nauk. SSSR*, 1991; 217: 227.
274. Coulton LA, Barker AT, Van Lierop JE, and Walsh MP. The effect of static magnetic fields on the rate of calcium/calmodulin-dependent phosphorylation of myosin light chain. *Bioelectromagnetics*, 2000; 21: 189–196.
275. Hendee SP, Faour FA, Christensen DA, Patrick B, Durney CH, and Blumenthal DK. The effects of weak extremely low-frequency magnetic fields on calcium/calmodulin interactions. *Biophys. J.*, 1996, 70: 2915–2923.
276. Blackman CF, Blanchard JP, Benane SG, and House DE. Empirical test of an ion parametric resonance model for magnetic field interactions with PC-12 cells. *Bioelectromagnetics*, 1994; 15: 239–260.
277. Blackman CF, Blanchard JP, Benane SG, and House DE. The ion parametric resonance model predicts magnetic field parameters that affect nerve cells. *FASEB J.*, 1995; 9: 547–551.
278. Blackman CF, Blanchard JP, Benane SG, House DE, and Elder JA. Double blind test of magnetic field effects on neurite outgrowth. *Bioelectromagnetics*, 1998; 19: 204–209.
279. Adair RK. Measurements described in a paper by Blackman, Blanchard, Benane, and House are statistically invalid. *Bioelectromagnetics*, 1996; 17: 510–511.
280. Chiabrera AA, Bianco B, Kaufman JJ, and Pilla AA. Quantum analysis of ion binding kinetics in electromagnetic bioeffects. In: *Electromagnetics in Medicine and Biology*. Brighton CT and Pollack SR, eds., San Francisco Press Inc., San Francisco, 1991, p. 27.

281. Chiabrera A, Bianco B, and Moggia E. Effect of lifetimes on ligand binding modeled by the density operator. *Bioelectrochem. Bioenerg.*, 1993; 30: 35–42.
282. Adair RK. A physical analysis of the ion parametric resonance model. *Bioelectromagnetics*, 1998; 19: 181–191.
283. Liburdy RP and Yost MG. Time-varying and static magnetic fields act in combination to alter calcium signal transduction in the lymphocyte. In: *Electricity and Magnetism in Biology and Medicine*. Blank M, ed., San Francisco Press, 1993, pp. 331–334.
284. Sisken BF, Kanje M, Lundborg G, and Kurtz W. Pulsed electromagnetic fields stimulate nerve regeneration *in vivo* and *in vitro*. *Restor. Neurol. Neurosci.*, 1990; 1: 24–27.
285. Cox JA. Interactive properties of calmodulin. *Biochem. J.*, 1988; 249: 621–629.
286. Chiabrera A, Bianco B, Kaufman JJ, and Pilla AA. Bioelectromagnetic resonance interactions: endogenous field and noise. In: *Interaction Mechanisms of Low-Level Electromagnetic Fields*. Oxford University Press, Oxford, 1992, pp. 164–179.
287. Mehler EL, Pascual-Ahuir J, and Weinstein H. Structural dynamics of calmodulin and troponin C. *Protein Eng.*, 1991; 4: 625–637.
288. Pilla AA, Muehsam DJ, and Markov MS. A dynamical systems/Larmor precession model for weak magnetic field bioeffects: ion binding and orientation of bound water molecules, *Bioelectrochem. Bioenerg.*, 1997; 43: 241–252.
289. Blumenthal DK and Stull JT. Effects of pH, ionic strength, and temperature on activation by calmodulin and catalytic activity of myosin light chain kinase. *Biochemistry*, 1982; 21: 2386–2391.
290. Pilla AA. Weak static magnetic fields reduce musculoskeletal pain, reduce edema and may enhance human sleep. Bioelectromagnetics Society 25th Annual Meeting, Maui, Hawaii, June 22–26, 2003.
291. Muehsam DJ, Pilla AA. Larmor precession, thermal noise and mechanisms for static magnetic field bioeffects and therapeutic applications. Bioelectromagnetics Society 25th Annual Meeting, Maui, Hawaii, June 22–26, 2003.
292. Foster KR and Schwan HP. In: *Handbook of Biological Effects of Electromagnetic Fields*. Polk C and Postow E. eds., CRC Press, Boca Raton, FL, 1986, p. 83.
293. Adair RK. Constraints on biological effects of weak extremely-low-frequency electromagnetic fields. *Phys. Rev. A*, 1991; 43: 1038–1049.
294. Weaver JC and Astumian RD. The response of living cells to very weak electric fields: the thermal noise limit. *Science*, 1990; 247: 459–462.
295. Astumian RD, Weaver JC, and Adair RK. Rectification and signal averaging of weak electric fields by biological cells. *Proc. Natl. Acad. Sci. USA*, 1995; 92: 3740–3743.
296. Kruglikov IL and Dertinger H. Stochastic resonance as a possible mechanism of amplification of weak electric signals in living cells. *Bioelectromagnetics*, 1994; 15: 539–547.
297. Chiabrera A, Bianco B, Moggia E, and Kaufman JJ. Zeeman-Stark modeling of the RF EMF interaction with ligand binding. *Bioelectromagnetics*, 2000; 21: 312–324.
298. Doty SB. Morphological evidence of gap junctions between bone cells. *Calcif. Tissue Int.*, 1981; 33: 509.
299. Loewenstein WR. Junctional intracellular communications: the cell-to-cell membrane channel. *Physiol. Rev.*, 1981; 61: 829–841.
300. McLeod KJ, Lee RC, and Ehrlich HP. Frequency dependence of electric field modulation of fibroblast protein synthesis. *Science*, 1987; 236: 1465–1469.
301. Sheridan JD and Atkinson MM. Cell membranes: physiological roles of permeable junctions: some possibilities. *Ann. Rev. Physiol.*, 1985; 47: 337–353.
302. Adey WR. Cell membranes: the electrochemical environment and cancer promotion. *Neurochem. Res.*, 1988; 13: 671.
303. Fletcher WH, Shiu WW, Haviland DA, Ware CF, and Adey WR. Proc. Bioelectromagnetics Soc., 8th Annual Mtg., Madison, WI, 1986, p. 12.
304. Hu GL, Chiang H, Zeng QL, and Fu YD. ELF magnetic field inhibits gap junctional intercellular communication and induces hyperphosphorylation of connexin43 in NIH3T3 cells, *Bioelectromagnetics*, 2001; 22: 568–573.
305. Lohmann CH, Schwartz Z, Liu Y, Li Z, Simon BJ, Sylvia VL, Dean DD, Bonewald LF, Donahue HJ, and Boyan BD. Pulsed electromagnetic fields affect phenotype and connexin 43 protein

expression in MLO-Y4 osteocyte-like cells and ROS 17/2.8 osteoblast-like cells. *J. Orthop. Res.*, 2003; 21: 326–334.

306. Hopper RA, Mehrara BJ, Lerman OZ, et al. Pulsed electromagnetic fields (PEMF) increase VEGF in cultured osteoblasts. Plastic Surgery Research Council 46th Annual Meeting, Milwaukee, WI; 2001: 104.
307. Andrew Lee W-P. What's new in plastic surgery. *J. Am. Coll. Surg.*, 2002; 194: 324–334.
308. Ramundo-Orlandoa A, Serafinoa A, Schiavoa R, Libertib M, and d'Inzeo G. Permeability changes of connexin32 hemi channels reconstituted in liposomes induced by extremely low-frequency, low amplitude magnetic fields. *Biochim. Biophys. Acta*, 2005; 1668: 33–40.
309. Griffin GD, Khalafl W, Hayden KE, Miller EJ, Dowray VR, Creekmore AL, Carruthers CR, Williams MW, and Gailey PC. Power frequency magnetic field exposure and gap junctional communication in Clone 9 cells. *Bioelectrochemistry*, 2000; 51: 117–123.
310. Pilla AA, Nasser PR, and Kaufman JJ. The sensitivity of cells and tissues to weak electromagnetic fields. In: *Charge and Field Effects in Biosystems-3*. Allen MJ, Cleary SF, Sowers AE, and Shillady DD, eds., Birkhauser, Boston, 1992, pp. 231–241.
311. Pilla AA, Nasser PR, and Kaufman JJ. On the sensitivity of cells and tissues to therapeutic and environmental EMF. *Bioelectrochem. Bioenerg.*, 1993; 30: 161–169.
312. Pilla AA, Nasser PR, and Kaufman JJ. Gap junction impedance, tissue dielectrics and thermal noise limits for electromagnetic field bioeffects. *Bioelectrochem. Bioenerg.*, 1994; 35: 63–69.
313. Muehsam DS and Pilla AA. The sensitivity of cells and tissues to exogenous fields: effects of target system initial state. *Bioelectrochem. Bioenerg.*, 1999; 48: 35–42.
314. Cooper MS. Gap junctions increase the sensitivity of tissue cells to exogenous electric fields. *J. Theor. Biol.*, 1984; 111: 123–130.
315. Cooper MS. Membrane potential perturbations induced in tissue cells by pulsed electric fields. *Bioelectromagnetics*, 1995; 16: 255–262.
316. Hart FX. Cell culture dosimetry for low-frequency magnetic fields. *Bioelectromagnetics*, 1996; 17: 48–57.
317. Fear EC and Stuchly MA. A novel equivalent circuit model for gap-connected cells. *Phys. Med. Biol.*, 1998; 43: 1439–1448.
318. Vander Molen MA, Donahue HJ, Rubin CT, and McLeod KJ. Osteoblastic networks with deficient coupling: differential effects of magnetic and electric field exposure. *Bone*, 2000; 27: 227–231.
319. Yamaguchi DT, Huang J, Ma D, and Wang PK. Inhibition of gap junction intercellular communication by extremely low-frequency electromagnetic fields in osteoblast-like models is dependent on cell differentiation. *J. Cell Physiol.*, 2002; 190: 180–188.
320. Hodgkin AL, Huxley AF, and Katz B. Measurement of current-voltage relations in the membrane of the giant axon of loligo. *J. Physiol.*, 1952; 116: 424–448.
321. Hodgkin AL and Huxley AF. A quantitative description of membrane current and its application to conduction and excitation in nerve. *J. Physiol.*, 1952; 117: 500–544.
322. Fishman HM, Poussart DJM, Moore LE, and Siebenga E. K^+ Conduction from the low-frequency impedance and admittance of squid axon. *J. Memb. Biol.*, 1977; 32: 255–290.
323. Fishman HM, Poussart DJM, and Moore LE. Complex admittance of Na^+ conduction in squid axon. *J. Memb. Biol.*, 1979; 50: 43–63
324. Fishman HM, Leuchtag HR, and Moore LE. Fluctuation and linear analysis of Na—current kinetics in squid axon. *Biophys. J.*, 1983; 43: 293–307.
325. Stevens CF. Inferences about membrane properties from electric noise measurements. *Biophys. J.*, 1972; 12: 1028–1047.
326. DeFelice LJ. *Introduction to Membrane Noise.* Plenum, NY, 1981, pp. 243–245.
327. Daff S. Calmodulin-dependent regulation of mammalian nitric oxide synthase. *Biochem. Soc. Trans.*, 2003; 31: 502–505.
328. Damy T, Ratajczak P, Shah AM, Camors E, Marty I, Hasenfuss G, Marotte F, Samuel JL, and Heymes C. Increased neuronal nitric oxide synthase-derived NO production in the failing human heart. *Lancet*, 2004; 363: 1365–1367.
329. Angerbauer GJ. *Principles of DC and AC Circuits*, 3rd Ed. Delmar Publishers, Albany, NY, 1989, p. 490.

330. Paul RG, Tarlton JF, Purslow PP, Sims TJ, Watkins P, Marshall F, Ferguson MJ, and Bailey AJ. Biomechanical and biochemical study of a standardized wound healing model. *Int. J. Biochem. Cell Biol.*, 1997; 29: 211–220.
331. Glassman LS, McGrath MH, and Bassett CA. Effect of external pulsing electromagnetic fields on the healing of soft tissue. *Ann. Plast. Surg.*, 1986; 16: 287–195.
332. Patel M, Yu H, and Strauch B. Effects of Pulsed Magnetic Field Treatment on the Healing of Linear Incision Wounds. Plastic Surgery Research Council, Toronto, Canada, May, 2005.
333. Yamauchi T, Yoshimura Y, Nomura T, Fujii M, and Sugiura H. Neurite outgrowth of neuroblastoma cells overexpressing alpha and beta isoforms of Ca2+/calmodulin-dependent protein kinase II: effects of protein kinase inhibitors. *Brain Res. Brain Res. Protoc.*, 1998; 2: 250–258.
334. Sisken BF, Kanje M, Lundborg G, and Kurtz W. Pulsed electromagnetic fields stimulate nerve regeneration *in vivo* and *in vitro*. *Restor. Neurol. Neurosci.*, 1990; 1: 303–309.
335. Greenbaum B, Sutton C, Vadula MS, Battocletti JH, Swiontek T, DeKeyser J, and Sisken B. Effects of pulsed magnetic fields on neurite outgrowth from chick embryos. *Bioelectromagnetics*, 1996; 17: 293–302.
336. Sisken BF, Kanje M, Lundborg G, and Kurtz W. Pulsed electromagnetic fields stimulate nerve regeneration *in vivo* and *in vitro*. *Restor. Neurol. Neurosci.*, 1990; 1: 303–309.
337. Diniz P, Shomura K, Soejima K, and Ito G. Effects of pulsed electromagnetic field (PEMF) stimulation on bone tissue like formation are dependent on the maturation stages of the osteoblasts. *Bioelectromagnetics*, 2002; 23: 398–405.
338. Muehsam DS and Pilla AA. The sensitivity of cells and tissues to exogenous fields: effects of target system initial state. *Bioelectrochem. Bioenerg.*, 1999; 48: 35–42.
339. Nindl G, Swezb JA, Millera JM, and Balcavage WX. Growth stage dependent effects of electromagnetic fields on DNA synthesis of Jurkat cells. *FEBS Lett.*, 1997; 414: 501–506.
340. Longo JA. The management of recalcitrant nonunions with combined magnetic field stimulation. *Orthop. Trans.*, 1998; 22: 408–409.
341. Norton L, Pilla A, Regelson W, and Tansman L. *J. Electrochem. Soc.*, 1980; 127: 130C.
342. Norton L, Tansman L, and Regelson W. *J. Electrochem. Soc.*, 1982; 129: 132C.
343. Tofani S, Cintorino M, Barone D, Berardelli M, De Santi MM, Ferrara A, Orlassino R, Ossola P, Rolfo K, Ronchetto F, Tripodi SA, and Tosi P. Increased mouse survival, tumor growth inhibition and decreased immunoreactive p53 after exposure to magnetic fields. *Bioelectromagnetics*, 2002; 23: 230–238.
344. Tofani S, Barone D, Berardelli M, Berno E, Cintorino M, Foglia L, Ossola P, Ronchetto F, Toso E, and Eandi M. Static and ELF magnetic fields enhance the *in vivo* anti-tumor efficacy of cisplatin against lewis lung carcinoma, but not of cyclophosphamide against B16 melanotic melanoma. *Pharmacol. Res.*, 2003; 48: 83–90.
345. Salvatore JR, Harrington J, and Kummet T. Phase I clinical study of a static magnetic field combined with anti-neoplastic chemotherapy in the treatment of human malignancy: initial safety and toxicity data. *Bioelectromagnetics*, 2003; 24: 524–527.
346. Plotnikov A, Fishman D, Tichler T, Korenstein R, and Keisari Y. Low electric field enhanced chemotherapy can cure mice with CT-26 colon carcinoma and induce anti-tumour immunity. *Clin. Exp. Immunol.*, 2004; 138: 410–416.
347. Johnson MT, Waite LR, and Nindl G. Noninvasive treatment of inflammation using electromagnetic fields: current and emerging therapeutic potential. *Biomed. Sci. Instrum.*, 2004; 40: 469–474.
348. Ronchetto F, Barone D, Cintorino M, Berardelli M, Lissolo S, Orlassino R, Ossola P, and Tofani S. Extremely low-frequency-modulated static magnetic fields to treat cancer: a pilot study on patients with advanced neoplasm to assess safety and acute toxicity. *Bioelectromagnetics*, 2004; 25: 563–571.
349. Trillo A, Ubeda A, Blanchard JP, House DE, and Blackman CF. Magnetic fields at resonant conditions for the hydrogen ion affect neurite outgrowth in PC-12 cells. *Bioelectromagnetics*, 1996; 17: 10–20.
350. Baureus Koch CL, Sommarin M, Persson BR, Salford LG, and Eberhardt JL. Interaction between weak low-frequency magnetic fields and cell membranes. *Bioelectromagnetics*, 2003; 24: 395–402.

351. Tuffet S, de Seze R, Moreau JM, and Veyret B. Effects of a strong pulsed magnetic field on the proliferation of tumor cells *in vitro*. *Bioelectrochem. Bioenerg.*, 1993; 30: 151–160.
352. Fanelli C, Coppola S, Barone R, Colussi C, Gualardi G, Volpe P, and Ghibelli L. Magnetic fields increase cell survival by inhibiting apoptosis via modulation of Ca^{2+} influx. *FASEB J.*, 1999; 13: 95–102.
353. Belyaev IY, Matronchik AY, and Alipov YD. 1994. The effect of weak static and alternating magnetic fields on the genome conformational state of *E. coli* cells: the evidence for model of phase modulation of high frequency oscillations. In: *Charge and Field Effects in Biosystems-4*. Allen MJ, ed., World Scientific, Singapore. pp, 174–184.
354. Binhi VN, Alipov YD, and Belyaev IY. Effect of static magnetic field on *E. coli* cells and individual rotations of ion–protein complexes. *Bioelectromagnetics*, 2001; 22: 79–86.
355. Cameron IL, Sun LZ, Short N, Hardman WE, and Williams CD. Therapeutic electromagnetic field (TEMF) and gamma irradiation on human breast cancer xenograft growth, angiogenesis and metastasis. *Cancer Cell Int.* 2005; 5: 23.
356. Fini M, Giavaresi G, Torricelli P, Cavani F, Setti S, Cane V, and Giardino R. Pulsed electromagnetic fields reduce knee osteoarthritic lesion progression in the aged Dunkin Hartley guinea pig. *J. Orthop. Res.*, 2005; 23: 899–908.
357. McCaig CD, Rajnicek AM, Song B, and Zhao M. Controlling cell behavior electrically: current views and future potential. *Physiol. Rev.*, 2005; 85: 943–978.
358. Thamsborg G, Florescu A, Oturai P, Fallentin E, Tritsaris K, and Dissing S. Treatment of knee osteoarthritis with pulsed electromagnetic fields: a randomized, double-blind, placebo-controlled study. *Osteoarthritis Cartilage*, 2005; 13: 575–581.
359. Morris C and Skalak T. Static magnetic fields alter arteriolar tone *in vivo*. *Bioelectromagnetics*, 2005; 26: 1–9.
360. De Pedro JA, Perez-Caballer AJ, Dominguez J, Collia F, Blanco J, and Salvado M. Pulsed electromagnetic fields induce peripheral nerve regeneration and endplate enzymatic changes. *Bioelectromagnetics*, 2005; 26: 20–27.
361. Zborowski M, Midura RJ, Wolfman A, Patterson T, Ibiwoye M, Sakai Y, and Grabiner M. Magnetic field visualization in applications to pulsed electromagnetic field stimulation of tissues. *Ann. Biomed. Eng.*, 2003; 31: 195–206.
362. Ibiwoyea MO, Powella KA, Grabinera MD, Patterson TE, Sakaia Y, Zborowskia M, Wolfmanc A, and Midura RJ. Bone mass is preserved in a critical-sized osteotomy by low energy pulsed electromagnetic fields as quantitated by *in vivo* micro-computed tomography. *J. Orthop. Res.*, 2004; 22: 1086–1093.
363. King RW. Nerves in a human body exposed to low-frequency electromagnetic fields. *IEEE Trans. Biomed. Eng.*, 1999; 46: 1426–1431.
364. Cooper MS, Miller JP, and Fraser SE. Electrophoretic repatterning of charged cytoplasmic molecules within tissues coupled by gap junctions by externally applied electric fields. *Dev. Biol.*, 1989; 132: 179–188.
365. Gowrishankar TR and Weaver JC. An approach to electrical modeling of single and multiple cells. *Proc. Natl. Acad. Sci. USA*, 2003; 100: 3203–3208.
366. Fear EC and Stuchly MA. Modeling assemblies of biological cells exposed to electric fields. *IEEE Trans. Biomed. Eng.*, 1998; 45: 1259–1271.
367. Fear EC and Stuchly MA. Biological cells with gap junctions in low-frequency electric fields. *IEEE Trans. Biomed. Eng.*, 1998; 45: 856–866.
368. Midura RJ, Ibiwoye MO, Powell KA, Sakai Y, Doehring T, Grabiner MD, Patterson TE, Zborowski M, and Wolfman A. Pulsed electromagnetic field treatments enhance the healing of fibular osteotomies. *J. Orthop. Res.*, 2005 Jun 2; [Epub ahead of print].
369. Segal NA, Toda Y, Huston J, Saeki Y, Shimizu M, Fuchs H, Shimaoka Y, Holcomb R, and McLean MJ. Two configurations of static magnetic fields for treating rheumatoid arthritis of the knee: a double-blind clinical trial. *Arch. Phys. Med. Rehabil.*, 2001; 82: 1453–1460.
370. Hinman MR, Ford J, and Heyl H. Effects of static magnets on chronic knee pain and physical function: a double-blind study. *Altern. Ther. Health Med.*, 2002; 8: 50–55.
371. Cho CK. Therapeutic Heating Applications of Radio Frequency Energy. Chapter 12, this volume.

372. Liboff A. The Ion Cyclotron Resonance Hypothesis. Chapter 9, *BBA* of this book.
373. Wolff J. *The Law of Bone Remodeling*. Hirshwald, Berlin, 1892, pp. 17–35. German.
374. Huiskes R, Ruimerman R, van Lenthe GH, and Janssen JD. Effects of mechanical forces on maintenance and adaptation of form in trabecular bone. *Nature*, 2000; 405: 704–706.
375. Fritton SP, McLeod KJ, and Rubin CT. Quantifying the strain history of bone: spatial uniformity and self-similarity of low-magnitude strains. *J. Biomech.*, 2000; 33: 317–325.
376. Huang RP, Rubin CT, and McLeod KJ. Changes in postural muscle dynamics as a function of age. *J. Gerontol. A Biol. Sci. Med. Sci.*, 1999; 54: B352–B357.
377. Goodship AE, Lawes T, and Rubin CT. Low magnitude high frequency mechanical stimulation of endochondral bone repair. *Trans. Orthop. Res. Soc.*, 1997; 22: 234.
378. Goodship AE and Kenwright J. The influence of induced micromovement upon the healing of experimental tibial fractures. *J. Bone Joint Surg. Br.*, 1985; 67: 650–655.
379. Busse JW, Bhandari M, Kulkarni AV, and Tunks E. The effect of low-intensity pulsed ultrasound therapy on time to fracture healing: a meta-analysis. *CMAJ*, 2002; 166: 437–441.
380. Azuma Y, Ito M, Harada Y, Takagi H, Ohta T, and Jingushi S. Low-intensity pulsed ultrasound accelerates rat femoral fracture healing by acting on the various cellular reactions in the fracture callus. *J. Bone Miner. Res.*, 2001; 16: 671–680.
381. Parvizi J, Wu CC, Lewallen DG, Greenleaf JF, and Bolander ME. Low-intensity ultrasound stimulates proteoglycan synthesis in rat chondrocytes by increasing aggrecan gene expression. *J. Orthop. Res.*, 1999; 17: 488–494.
382. Korstjens CM, Nolte PA, Burger EH, Albers GH, Semeins CM, Aartman IH, Goei SW, and Klein-Nulend J. Stimulation of bone cell differentiation by low-intensity ultrasound a histomorphometric *in vitro* study. *J. Orthop. Res.*, 2004; 22: 495–500.
383. Wang N, Butler JP, and Ingber DE. Mechanotransduction across the cell surface and through the cytoskeleton. *Science* 1993; 260: 1124–1127.
384. Kokubu T, Matsui N, Fujioka H, Tsunoda M, and Mizuno K. Low intensity pulsed ultrasound exposure increases prostaglandin E2 production via the induction of cyclooxygenase-2 mRNA in mouse osteoblasts. *Biochem. Biophys. Res. Commun.*, 1999; 256: 284–287.
385. Naruse K, Mikuni-Takagaki Y, Azuma Y, Ito M, Ohta T, Kameyama K-Z, and Itoman M. Anabolic response of mouse bone-marrow-derived stromal cell clonal ST2 cells to low intensity pulsed ultrasound. *Biochem. Biophys. Res. Commun.*, 2000; 268: 216–220.
386. Warden SJ, Favaloro JM, Bennell KL, McMeeken JM, Ng KW, Zajac JD, and Wark JD. Low-intensity pulsed ultrasound stimulates a bone-forming response in UMR-106 cells. *Biochem. Biophys. Res. Commun.*, 2001; 286: 443–450.
387. Parvizi J, Parpura V, Greenleaf JF, and Bolander ME. Calcium signaling is required for ultrasound-stimulated aggrecan synthesis by rat chondrocytes. *J. Orthop. Res.*, 2002; 20: 51–57.
388. Wang FS, Kuo YR, Wang CJ, Yang KD, Chang PR, Huang YT, Huang HC, Sun YC, Yang YJ, and Chen YJ. Nitric oxide mediates ultrasound-induced hypoxia-inducible factor-1alpha activation and vascular endothelial growth factor-A expression in human osteoblasts. *Bone*, 2004; 35: 114–123.
389. De Mattei M, Pasello M, Pellati A, Stabellini G, Massari L, Gemmati D, and Caruso A. Effects of electromagnetic fields on proteoglycan metabolism of bovine articular cartilage explants. *Connect. Tissue Res.*, 2003; 44: 154–159.
390. Li H and Poulos TL. Structure-function studies on nitric oxide synthases. *J. Inorg. Biochem.*, 2005; 99: 293–305.
391. Pitsillides AA, Rawlinson SC, Suswillo RF, Bourrin S, Zaman G, and Lanyon LE. Mechanical strain-induced NO production by bone cells: a possible role in adaptive bone (re)modeling? *FASEB J.*, 1995; 9: 1614–1622.
392. Ignarro LJ and Napoli C. Novel features of nitric oxide, endothelial nitric oxide synthase, and atherosclerosis. *Curr. Diab. Rep.*, 2005; 5: 17–23.
393. Bockris JO and Saluja PPS. Ionic solvation numbers from compressibilities and ionic vibration potentials measurements. *J. Phys. Chem.*, 1972; 76; 2140–2151.
394. Buchanan TJ, Haggis GH, Hasted JB, and Robinson BG. The dielectric estimation of protein hydration. *Proc. Roy. Soc. London*, 1952; A213: 379–391.

395. Conway BE. *Ionic Hydration in Chemistry and Biophysics*. Elsevier, New York, 1981.
396. Termaat MF, Den Boer FC, Bakker FC, Patka P, and Haarman HJ. Bone morphogenetic proteins. Development and clinical efficacy in the treatment of fractures and bone defects. *J. Bone Joint Surg. Am.*, 2005; 87: 1367–1378.
397. Bennett SP, Griffiths GD, Schor AM, Leese GP, and Schor SL. Growth factors in the treatment of diabetic foot ulcers. *Br. J. Surg.*, 2003; 90: 133–146.

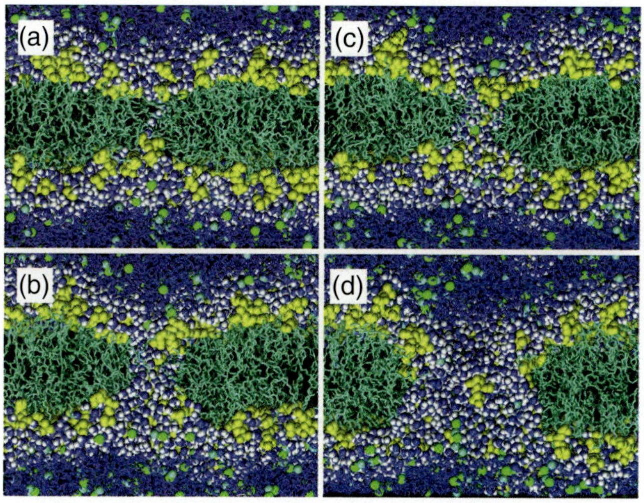

FIGURE 9.2
Snapshots of pore creation according to a molecular dynamics simulation. An important, partial test of electroporation was obtained by imposing a constant electric field, $E = 0.5\,\mathrm{V\,nm^{-1}}$ ($5 \times 10^6\,\mathrm{V\,cm^{-1}}$) throughout a small volume that contains aqueous electrolyte and a small area of phospholipid bilayer membrane (Tieleman, D.P. *BMC Biochem.*, 5, 10, 2004). The phospholipid headgroups are white, the hydrocarbon chains are gray, and water is dark gray. The chloride and sodium ions are somewhat difficult to make out in this gray-scale reproduction, so the reader is referred to the original paper for a color depiction. The snapshots are at times (A) 5.33, (B) 5.45, (C) 5.50, and (D) 5.70 ns from starting the simulation. As described in the original paper (Tieleman, D.P. *BMC Biochem.*, 5, 10, 2004) initially water molecules stochastically enter into the bilayer interior, leading to "formation of single-file like water defects penetrating into the bilayer." These and other MD results [71–76] show that pores can form on a nanosecond time scale, and are hydrophilic, with phospholipids present on the interior of the fluctuating pore. These hourglass or toroidal-like pores are reminiscent of the simple drawings used to motivate early models of electroporation. Present MD models are "noisy" in the sense that for a typical computational volumes and times only a few transported ions are involved. However this approach has great promise, depending on improved computational power.

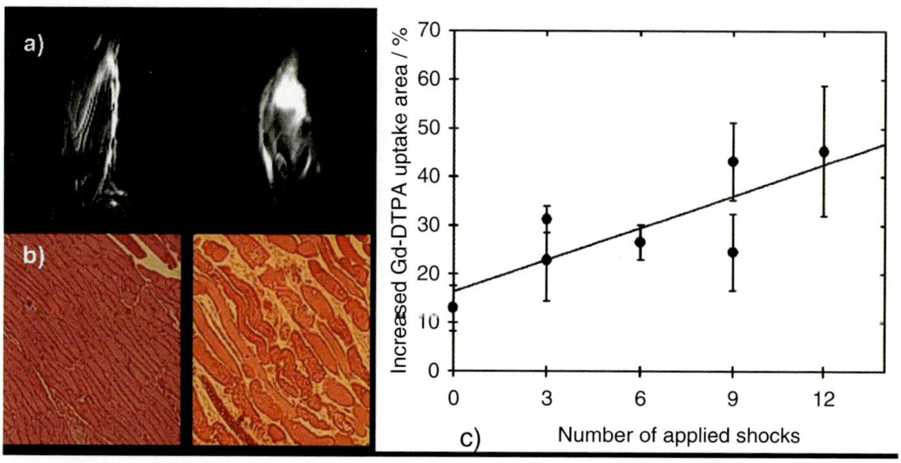

FIGURE 10.2
The effects of multiple applied shocks on muscle tissue. (a) T2-weighted images of control (left) and shocked (right) muscles. (b) Histology sections of control (left) and shocked (right) muscle. (c) Relationship between increased Gd uptake with increasing number of applied shocks. The fraction of hypercontracted muscle cells increased with the number of applied pulses.

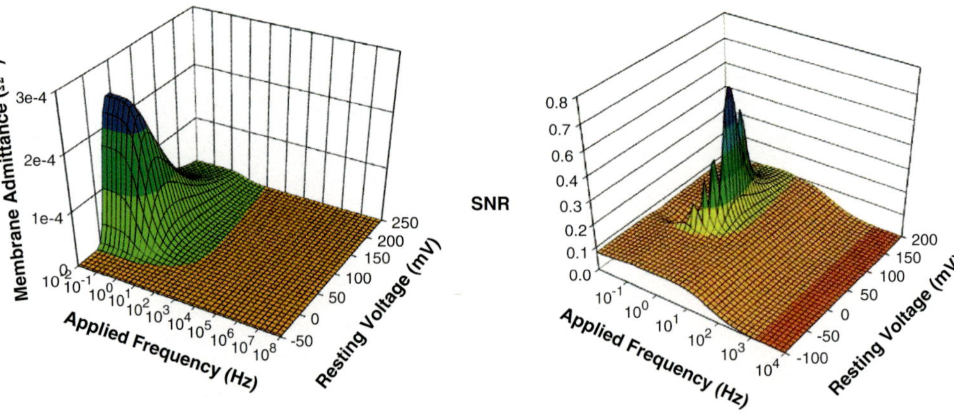

FIGURE 11.19

(Left) Membrane admittance depends upon frequency and membrane potential for Hodgkin–Huxley K+-conduction pathway. Region of higher admittance (reduced impedance) corresponds to greater K^+ flux across membrane and admittance drops rapidly with increasing frequency. (Right) Response of Hodgkin–Huxley K^+-conduction membrane pathway to exogenous electric field. Voltage induced in K^+ pathway varies with membrane potential and frequency. Thus, EMF sensitivity depends upon the target initial state.

12

Therapeutic Heating Applications of Radio Frequency Energy

C-K. Chou

CONTENTS

12.1 Introduction

The Bioelectromagnetics Society recognizes distinguished scientists in the field of bioelectromagnetics with its highest award, the d'Arsonval Award. This award honors a French scientist Jacques-Arsène d'Arsonval who pioneered research on the beneficial effects of radio frequency (RF) energy in the late-19th century and the beginning of the 20th century. Since the 1940s, diathermy ("through heat") has been used in rehabilitation medicine to relieve pain from sprains and strains. RF and ultrasound energy have been used for achieving deep-tissue heating to increase blood flow and collagen tissue extensibility, decrease joint stiffness, and muscle spasm [1]. In the mid-1970s, scientists and engineers started applying hyperthermia (40–45°C heating) in combination with ionizing radiation or chemotherapy to treat localized or metastasized cancer [2,3]. In the 1980s, RF ablation (50–100°C heating) was first used for cardiac arrhythmia and recently in the last ten years, for inducing local tumor necrosis [4,5]. Diathermy, hyperthermia, and ablation all use the same RF energy-heating characteristics to achieve different temperatures to treat various clinical conditions and diseases. Other applications using the propagating characteristics of the RF energy for medical purposes, such as RF telemetry to couple sound signals to implanted hearing devices (cochlear implant or middle-ear hearing device), are not discussed here [6,7].

In this chapter, RF heating mechanisms are first explained, followed by a discussion of the relationship between RF heating and temperature rise. Then, RF hyperthermia and ablation are emphasized and some other heating applications are briefly discussed. Clinical applications published since 1995 are selected as examples, but animal and laboratory studies as well as engineering developments are not included because of page limitation. Also, the use of RF energy frequency current less than 100 kHz that can be used for bone and wound healing, nerve regeneration, and other nonheating applications (wave propagation) are not in the scope of this chapter.

12.2 RF Heating Mechanisms

Most of us have experienced shocks from touching leaky 50–60 Hz household appliances. The human body, although not a good conductor like metals, is not an insulator like plastic or rubber. As the frequency increases to 100 kHz, the electro-stimulatory effect gradually turns into a thermal sensation. Between 100 kHz and 5 MHz, it is still possible to sense the electro-stimulation effect if the RF energy is pulsed. At higher frequencies, many RF engineers have experienced RF burns. To understand how RF energy causes tissue heating, it is necessary to understand how the RF energy interacts with tissues. When RF energy is impinging on a human body, some energy is reflected at the body surface and some penetrates into the body, similar to sunlight shining on a lake. The amount of reflection and penetration depend on a property of the medium called the dielectric property as discussed in one of the previous chapters.

In general, biological tissues can be classified into two major categories. High water-content tissues, such as muscle, skin, kidney, and liver, have higher dielectric constants and conductivity than tissues with low water content, such as fat or bone. Brain, lung, and bone marrow tissues contain intermediate amounts of water and have dielectric constants and conductivities that fall between the values of the other two groups. In the RF range, tissues with high water content conduct electrical current well, because of the ions and the polar water molecules. Tissues with low water content, such as fat and bone, have mostly bound charges that do not easily support conduction currents. Dielectric properties are a function of frequency. With increasing frequency, the dielectric constant decreases whereas the conductivity increases. Camelia Gabriel and colleagues [8] have published extensive dielectric property data for 43 tissues. A calculator for obtaining tissue dielectric constant and conductivity for 10–6000 MHz can be found on the following website from the FCC:http://www.fcc.gov/fcc-bin/dielec.sh.

The action of RF fields on tissues produces two types of effects: (1) the oscillation of free charges on ions and (2) the rotation of polar molecules at the frequency of the applied field [9]. Free charge motion loss is due to the electrical resistance of the medium. At radio frequencies below approximately 10 MHz, the alternating field induces a net movement of ions. Resistive (ohmic) losses associated with the induced current causes "joule heating." This is the mechanism of RF ablation heating using <500 kHz RF current. At frequencies above about 100 MHz, the radiative mode of electromagnetic propagation and dielectric losses in tissue predominate over the conduction current losses. Under these conditions, heating results primarily from friction caused by mechanical interactions between adjacent polar water molecules, which are oscillating in an attempt to maintain alignment with the time-varying electric field. This is the primary heating mechanism of microwave applicators used in hyperthermia treatment of cancer as well as microwave ovens used to cook food. The resistive and friction losses are the basis of RF energy absorption, which produces heating in tissue [10,11].

12.3 Temperature Rise Due to RF Heating

The rate of temperature rise (heating rate, HR) in tissue heated with RF energy is related to the rate of energy absorption (specific absorption rate, *SAR*) by Equation 12.1. The *SAR* is associated with the electrical field in the tissue and tissue conductivity as shown by Equation 12.2. The final tissue temperature is a function of the deposited energy, metabolic heating rate Q_m, power dissipation by thermal conduction $k\nabla^2 T$, and blood flow $c_b w(T - T_a)$, as expressed in the Pennes equation (Equation 12.3).

$$HR = \frac{SAR}{69.77C_H(^\circ\mathrm{C/\,min})} \tag{12.1}$$

$$SAR = \frac{\sigma}{\rho}E^2(\mathrm{W/kg}) \tag{12.2}$$

$$\rho c_H \frac{\mathrm{d}T}{\mathrm{d}t} = k\nabla^2 T + SAR + Q_m - c_b w(T - T_a) \tag{12.3}$$

where *HR* is heating rate in tissue (°C/min); *SAR* is specific absorption rate (W/kg); c_H is the specific heat capacity of tissue (kcal/kg °C); *E* is the root-mean-square value of the induced electric field strength in tissue (V/m); ρ is the density of tissue (kg/m^3); σ is the dielectric conductivity of tissue (siemens/m); *T* is the tissue temperature (°C); *k* is the thermal conductivity of tissue (W/m °C); Q_m is the metabolic heat generation (W/kg); c_b is the blood heat capacity (kcal/kg °C); *w* is the blood perfusion rate (kg/m^3 sec); and T_a is the temperature of incoming arterial blood (°C).

In a typical clinical treatment of soft-tissue tumors using about 50 W output power of an external applicator, the temperature, *T*, will increase as shown in Figure 12.1. Initially, there is a linear increase lasting approximately 3 min. In normal tissue, after this linear increase in temperature, there is a period of nonlinear temperature rise usually lasting another 7–10 min, where *T* becomes high in dissipating the absorbed energy through heat diffusion. In tissues with a low or negligible blood flow, the temperature will monotonically approach a steady state value dictated by the magnitude of *SAR*, as shown on the upper curve in Figure 12.1, with equilibrium reached when $SAR = -k\nabla^2 T$ (heat diffusion). For vascular tissues, a marked increase in blood flow will occur due to vasodilatation when the temperature reaches the range of 42 to 44°C, as shown in Figure 12.1, when $SAR = -k\nabla^2 T + c_b w(T - T_a)$ (heat diffusion and blood flow cooling). This characteristic is used in diathermy treatment to increase blood flow to the treated area.

During hyperthermia treatment, the blood flow in many tumors is vigorous at the periphery and sluggish in the center [12]. Because the blood vessels in tumors are often fully open during normal conditions, there is, in many cases, no vasodilatation during the heat treatment. After steady state conditions are reached, the final temperature of the tumor is higher than that of the surrounding normal tissue. The shaded area in Figure 12.1 indicates the range of temperature rise in tumors. The lower boundary is for the periphery of the tumor and the upper curve is for the center, where no blood flow exists.

The rate of energy absorption, *SAR*, must be sufficiently high so that a therapeutic temperature level can be maintained over the major portion of the treatment period. If too little power is applied, the period or the level of temperature elevation will not be long enough or high enough for any benefit, whereas with too high a level, the temperature of the normal tissue may not be maintained at a safe level by vasodilatation. For diathermy treatment, the pain sensors are a reliable and sensitive means for alerting the patient of an unsafe temperature. If the applied power level is set so that only mild pain or discomfort is experienced by the patient, vasodilatation in normally vascular tissue will be sufficient to limit or even lower the temperature to a level that is both tolerable and therapeutically effective. For RF ablation, *SAR* is very high to cause a large temperature rise (15–60°C) in a short time (<6 min) to destroy tissues. The short-duration heating makes the temperature rise directly proportional to SAR, before significant heat diffusion can take place.

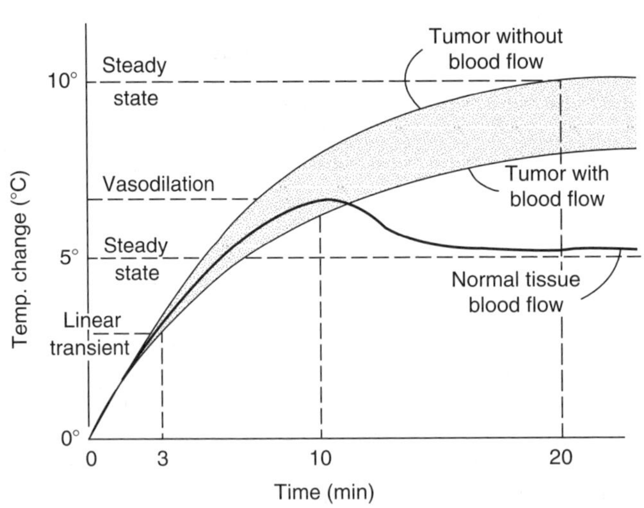

FIGURE 12.1
Temperature change in tumor and normal tissues following application of RF energy. (Reprinted from Guy, A.W. and Chou, C.K., Physical aspects of localized heating by radio frequency waves. In *Hyperthermia in Cancer Therapy*, Storm K., Ed., Boston, G.K. Hall Publisher, 1982, p. 279. With permission.)

12.4 Hyperthermia for Cancer Treatment

Cancer hyperthermia (HT) is a treatment in which the temperature of either local tissue or whole body is raised to a therapeutic level to eradicate tumors. Over the last three decades, much has been learned about the effects of heat on cells and the interactions between heat and ionizing radiation therapy (RT) and chemotherapeutic agents [2,3]. The scientific rationale for its use either alone or combined with other methods is multi-factorial. Heat may be directly cytotoxic to tumor cells or complementary to the effects of ionizing radiation with regard to inhibition of potentially lethal damage and sublethal damage repair, cell-cycle sensitivity, and effects of hypoxia and nutrient deprivation. The synergism of RT and HT is accomplished by the thermal killing of hypoxic and S phase (DNA syntheses) cells, which are resistive to ionizing radiation alone. HT has been used in combination with chemotherapy because heating increases membrane permeability and the potency of some drugs. Increasing drug uptake into cells enhances the amount of DNA damage, inhibits repair of DNA damage, and can reverse drug resistance. Recent developments in gene therapy may also apply HT for targeted, localized induction of gene therapy using the heat shock promoter [14–17].

Although cell and animal studies provide strong biologic rationale for the clinical application of HT, implementation remains difficult. Numerous factors affect the results of HT. The foremost problem in HT, however, is the generation and control of heat in tumors. Current heating methods for whole body heating use hot wax, hot air, hot water suits, or infrared radiation, and for partial body heating, utilize ultrasound, heated blood, fluid perfusion, RF fields, or microwaves. The effective temperature range of HT is very narrow, only 42–45°C. At lower temperatures, the effect is minimal. At temperatures higher than 45°C, normal cells are damaged. Because of this narrow temperature range, the response is highly dependent on to what extent is the tumor tissue heated to a therapeutic level. The clinical use of HT has been hampered by a lack of adequate equipment to effectively deliver heat to deep-seated and even large superficial lesions and by a lack of thermometry that provides reliable information on heat distribution in the target tissues. In RF hyperthermia, the final temperature of tumors is mainly dependent on energy deposition. When RF heating methods are used, the energy deposition is a complex function of the frequency, intensity, and polarization of the applied fields, the applicator's size and geometry, as well as the size, depth, geometry, and dielectric property of the tumor [18]. The material, thickness, and construction of a cooling bolus also influence the amount of energy deposition [19]. Although there are tremendous technical difficulties, there are now eight randomized trials of HT in human cancer patients. The majority demonstrate a survival advantage with the addition of HT to RT. Dewhirst et al. [3] pointed out that as technology improves, the benefits of this form of therapy will only become more visible.

12.5 Methods of RF Hyperthermia

The electromagnetic energy used in HT is usually classified by frequency as either RF or microwave energy. Strictly speaking, the RF spectrum is between 3 kHz and 300 GHz, but for hyperthermia it generally refers to frequencies below 300 MHz. Common RF frequencies are 13.56 and 27.12 MHz, which have been widely used in diathermy. Microwaves occupy the frequency band between 300 MHz and 300 GHz. The most commonly used microwave frequencies in HT are 433, 915, and 2450 MHz, which are designated ISM (industrial, scientific, and medical) frequencies in the United States and Europe (433 MHz in Europe only). Frequencies higher than 2450 MHz have no practical value for HT due to their limited penetrations. At lower frequencies, field penetration is deeper but the applicator must be larger and focusing is difficult.

For superficial (2–5 cm deep) and localized heating, 915 MHz is suitable. To keep the skin below 43°C and minimize pain, a deionized-water cooling bolus is generally placed between the applicator and patient body, although air-cooled method can also be used. For the treatment of tumors deeper than 5 cm, more penetrating lower frequencies must be used to deliver RF energy to the tumor. Applying 5–30 MHz capacitively coupled RF current, heating can be induced in tumors deep inside the body by varying the electrode sizes to control the current density and therefore the heating pattern in patients. Because the highest current-density occurs at the electrodes, cooling boluses must be used to protect the skin. Since fat heats significantly because of its low specific heat and low dielectric constant, RF capacitive heating is applicable only for thin patients. To produce deep-focused heating, a phased array method has been used (Figure 12.2). The system consists of a ring array of antennas to treat tumors inside torsos or legs. Utilizing antennas with RF fields of varying amplitude and phase, due to constructive or destructive interferences, heating can be steered to the desired locations. Noninvasive temperature monitoring of this hyperthermia and magnetic resonance hybrid system has been validated in a heterogeneous phantom [20]. Details of RF heating methods for local, regional, and whole body hyperthermia are available [2,3,13,18–23].

It is impossible for a single piece of equipment to fulfill all of the clinical requirements for patient treatments. Depending on the location and vascularity of the tumor and adjacent tissues and the general physical condition of the patient, the HT practitioners should have the option of choosing the most appropriate equipment. HT is a complicated technique and should be applied only by well-trained individuals. Due to the complexity of RF energy coupling to human tumors, careful heating pattern studies should be performed on all exposure geometries and contingencies prior to treatment to assure the best treatment conditions for the patient. The clinical situation is complicated by the reality that HT in combination with high-energy RT cannot be repeated after the tumor receives a maximal dosage of ionizing radiation. Accurate thermometry is particularly important in all phases of clinical hyperthermia, especially when the patient is anesthetized.

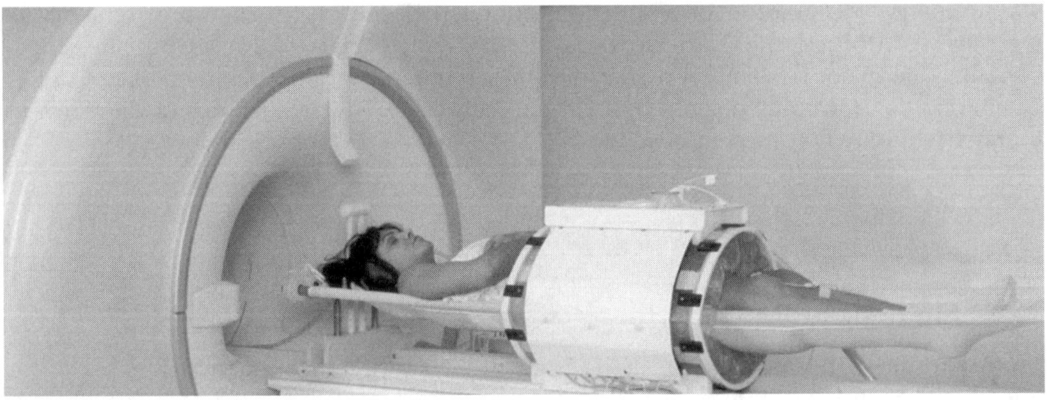

FIGURE 12.2
BSD-2000/3D/MR Hyperthermia System, Combining 3D treatment therapy and MRI treatment monitoring, provides the capability to monitor changes in perfusion, temperature, necrosis, and chemistry. (Courtesy of BDS Medical, Salt Lake City, Utah.)

12.6 Hyperthermia Clinical Phase III Trials

Dewhirst et al. [3] recently reviewed the results of major published Phase III trials on advanced disease and primarily superficial malignancies. The following sections are condensed from that review.

12.6.1 Breast Cancer

Five Phase III trials of breast cancer chest wall recurrence were combined in an international collaborative study [24]. Patients were randomized to either RT alone or RT with HT. In some respects, HT techniques differed between institutions but are well documented, including information concerning temperature distributions and thermal dosimetry. The five trials demonstrated a significant overall improvement in the complete response rate for patients receiving HT + RT compared with RT alone (59% in the former group and 41% in the latter with an odds ratio of 2.3 ([95% Confidence Interval, 1.4 to 3.8]). The greatest effect was observed in patients with recurrent lesions in previously irradiated areas where further irradiation was limited to low doses; however, survival advantages were not apparent.

12.6.2 Other Superficial Malignancies

In a Radiation Therapy Oncology Group (RTOG) study published in 1991 [25], 307 patients with tumors were treated with RT alone or with RT + HT. Approximately, half of the cases comprised head and neck tumors, one third were breast carcinoma (chest wall recurrences), and the remaining patients were a variety of superficial malignancies. Complete response rates were not significantly different between the two groups (RT + HT = 32%; RT alone = 30%); however, subgroup analysis revealed significant improvements in duration of local control in patients with tumors smaller than 3 cm and in those with breast or chest wall recurrences. For lesions larger than 3 cm there was no difference in outcome, which is reasonable since large tumors are more difficult to heat. This trial was also characterized by highly variable heating techniques and thermal dosimetry that led to the development of subsequent RTOG guidelines for performing hyperthermia [26].

12.6.3 Head and Neck Cancer

There are two randomized studies for this category. The first study randomized 65 patients to RT alone versus RT and HT [27]. Patients in stage III who received the combined treatment had a 58% complete response (CR) as compared with 20% in the RT-alone group. Similarly, patients in stage IV receiving combined treatment of RT and HT achieved a CR of 38% as compared with 7% for those receiving RT alone. There was no benefit for more than 90% of patients in stages I and II of the disease; these patients achieving a CR with either treatment. A second trial from Italy restricted evaluation to metastatic cervical lymph nodes [28]. This study randomized 41 patients with advanced local regional squamous cell carcinoma of the head and neck to treatment with either RT alone or RT combined with HT. The CR rate was 41% in the RT-alone arm versus 83% in the combined arm. Five-year survival was 0 in RT-alone patients and 53% in the RT + HT group. These statistically significant differences clearly indicate that HT is beneficial.

12.6.4 Esophagus

Two randomized studies demonstrated an advantage for the addition of HT to chemoradiotherapy or chemotherapy in the treatment of esophageal cancer [29,30]. In the first study, 53 patients with squamous cell carcinoma of the thoracic esophagus were randomized to preoperative HT, RT, and chemotherapy, compared with only chemoradiotherapy. Clinical CR and pathologic responses were significantly improved in the trimodality arm—pathologic CR of 26% combined in the trimodality group versus 8% in the only-chemoradiotherapy group. In a follow-up study, an additional 40 patients were treated with chemotherapy only (bleomycin and cisplatin) or combined with HT. Again, an improvement in histopathologic response was noted favoring the HT group (41% versus 19%).

12.6.5 Malignant Melanoma

Seventy patients with 134 metastatic or recurrent malignant melanoma lesions were randomized to receive RT or RT followed by HT [31]. There was a significant overall benefit for the addition of HT, with a two-year local control of 46% in the combined group compared with 28% for those receiving RT alone. Quality assurance was an issue in this trial, with only 14% of treatments achieving the protocol objective of 43°C for 60 min.

12.6.6 Glioblastoma Multiforme

Seventy-nine patients with glioblastoma multiforme were randomized following external beam RT and chemotherapy to receive brachytherapy alone or combined with HT [32]. Both time-to-tumor progression and survival were significantly improved for the HT patients as compared with those treated with brachytherapy alone. Two-year survivals in the two groups were 31 and 15%, respectively. Toxicity was slightly greater in the HT patients, with seven grade 3 toxicities reported as compared with one reported grade 3 toxicity in the brachytherapy alone group. Good heating was achieved in most patients, but no good correlation was evident between thermal dose and response.

12.6.7 Carcinoma of the Cervix

This study randomized 361 patients with locally advanced, previously untreated, pelvic tumors to receive RT alone or RT + HT [33]. The patients were approximately equally divided between rectal, bladder, and cervix carcinoma. Complete response rates were 39% after RT alone and 55% after RT + HT ($p = <0.001$). The duration of local control was also significantly improved with RT + HT as compared with RT only. Most of the benefit appeared to occur in patients with cervical carcinoma, the CR rate following RT + HT was 83% compared with 57% after RT only. Three-year survival was 27% in the RT-only group of cervix cancer patients and 51% in the RT + HT group. The difference is statistically significant.

12.6.8 Summary of Clinical Trials

The results of the above-published Phase III trials are intriguing; apart from the RTOG trials, they all appear to demonstrate substantial benefits from the addition of HT to RT. All of the trials were associated with significant design and implementation problems including relatively small numbers of patients, inadequate thermal dosimetry information, thermal goals that were often not achieved, and control arms that may not have represented optimal standard therapy. Despite these difficulties, Dewhirst et al. [3] concluded that the generally positive character of these trials suggest the future of HT is quite promising and that technological advances may well lead the way to even more positive results.

12.7 RF Ablation

Radio frequency ablation (RFA) is a modified electrocautery technique that destroys tissue by means of electric current. In surgery, high-frequency electric current is used to cauterize small blood vessels to stop bleeding. Since 1985, there have been numerous reports of the use of RF current as an alternative energy source for cardiac ablation for correcting arrhythmias [34]. RFA has evolved to become a method for creating thermally induced coagulation necrosis in tumors using either a percutaneous approach with image-guided or direct surgical placement of electrodes into tissues. The necrotic tumor becomes smaller and disappears over time to achieve ablation. Goldberg [4] and Rhim et al. [35] reviewed the principles and techniques of tumor ablation with RF energy. A review by Friedman et al. [5] summarized clinical applications for treating a host of malignant diseases throughout the body. Dr. Bradford Wood of the National Institute of Health and Dr. Nahum Goldberg of Harvard University are two leading scientists in this field. Most of the following materials are from their reviews [4,5] with some updates added. Readers who are interested in details can refer to their original articles.

12.7.1 RFA Methods

In contrast to hyperthermia, which elevates tumor temperature to less than 45°C, RFA applies RF electric fields (375–500 kHz) to produce ionic currents in tissue and cause resistive heating to at least 50°C. At 50–52°C, cells undergo coagulation necrosis in 4–6 min; for temperatures greater than 60°C, coagulation necrosis is within seconds. However, temperature greater than 100°C is avoided to prevent tissue boiling, vaporization, and carbonization. These phenomena cause impedance increases and result in reduced ablation. Therefore, the goal of RFA is to achieve and maintain a temperature of 50–100°C throughout the target volume.

Rhim et al. [35] collected essential technical tips for a successful RFA from five international experts. To introduce RF energy into tumor, electrodes are inserted at the center of the tumor. For adequate destruction of tumor tissue, the entire volume of a lesion must be subjected to cytotoxic temperatures (50–100°C for 4–6 min). However, the relatively slow thermal conduction from the electrode through the tissues increases the duration of application to 10–30 min. With single monopole electrodes, coagulated tissue diameters are limited to about 1.6 cm. Many investigators have explored and several commercial companies have manufactured new RFA devices that have increased heating efficacy. Goldberg [4] also reviewed a variety of methods for increasing coagulation volume with RFA. The techniques are briefly summarized below. In addition to an increase in the generator power to 150 W, expandable, multitined electrodes permit the deposition of this energy over a larger volume (e.g., see Figure 12.3). Temperature changes are monitored via embedded thermocouples to avoid overheating. Another method to prevent overheating is to directly measure impedance changes for power control. For single electrode

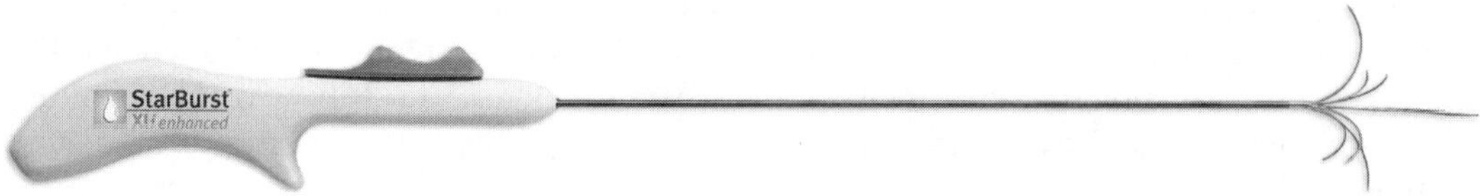

FIGURE 12.3
The StarBurst Xli—Enhances RFA device incorporates controlled infusion of normal saline to promote complete ablation in soft tissue up to 7 cm in diameter. The device features real-time, multipoint temperature feedback system from the seven active arrays. (Courtesy of RTIA Medical Systems, Mountain View, California.)

application, a cooled electrode design can minimize carbonization and gas formation around the needle tip by eliminating excess heat near the electrode, and therefore increases the treatment volume. Pulsing of RF energy has been also shown to be useful. For example, a combined approach that involves use of a multiprobe cluster of internally cooled electrodes with pulsing has also been claimed to create larger lesions than those achieved by any method alone [36]. These technological developments can be used to create an ablation lesion with a maximum diameter of 5 cm.

Although there are improvements from engineering and technical manipulations, it is still very difficult to achieve large lesions, because of the low tumor conductivity and the cooling effect from nearby blood vessels. Therefore other methods to modify the biology or tumor characteristics have also been investigated. These include (1) improved heat conduction by injection of saline and other compounds to increase tissue conductivity; (2) saline injection during RFA; (3) reduced blood flow by occlusion; and (4) drug modulated blood flow. Combination with chemotherapy or chemoembolization to improve tumor response has been shown to be more effective than RFA alone. For details of the RFA techniques, please refer to Goldberg [4] and Rhim et al. [35].

12.7.2 RFA Clinical Applications

12.7.2.1 Liver Tumors

Surgical resection is the only curative treatment for liver cancer patients but only 10–20% of patients with liver tumor have resectable disease [37]. For this reason, RFA has become a popular procedure [38]. Since this method was used 15 years ago [39,40], there has been extensive work on liver tumor RFA [41–43]. Studies of over 3000 RFA-treated patients have shown the efficacy of percutaneous RFA for small (<3 cm) liver tumors. Complete local response averaged 70–75% with tumors between 3 and 5 cm, and drops to 25% in larger tumors. With successful ablation, 5-year survival rates of 40–50% have been reported. This pattern of excellent local control for small liver tumors with significant long-term nonlocal recurrence is also true in RFA of hepatic metastases. Currently, there are no randomized prospective controlled studies comparing liver RFA to standard surgical resection in any population. Livraghi et al. [44] concluded that RFA is a relatively low-risk procedure for focal liver tumor treatment based on results with 3554 lesions.

12.7.2.2 Kidney

Standard radical or partial nephrectomy may be excessively invasive for small kidney-tumors, and RFA can offer an alternative, minimally invasive treatment [45]. Similar to its use in treating liver tumor, RFA is most effective in treating kidney tumors less than 3 cm in diameter [46–50]. With increasing size, there is an increased risk of local recurrence. In general, the number of patients treated is relatively small with short follow-up. However, early short-term results suggest a 70–90% small kidney-tumor success rate.

12.7.2.3 Bone

RFA has been used for more than ten years to treat osteoid osteoma, a benign, slow-growing painful lesion [51]. With RFA, a probe is placed through a bone-penetration cannula into the lesion and activated for 4–6 min at 90°C [52,53]. Success rates for a single ablation range from 91 to 94% with long-term follow-up, and most recurrences can be completely ablated in a second procedure [54]. Cioni et al. [55] did a follow-up of 38 patients for 12–66 months. Their results show primary and secondary clinical success rates were 78.9% (30/38 patients) and 97% (35/36 patients), respectively. Treatment of osteomas by RFA is highly effective, safe, and allows early return to function while minimizing morbidity.

12.7.2.4 Pain

RFA may provide rapid pain relief (in hours to days) for patients with cancer pain resistant to conventional forms of palliation in comparison to the radiation therapy, which palliates over the course of 7–21 days [56]. Treatments of low back-pain [57–59], trigeminal neuralgia [60], and other neurolysis have just started. Currently, the National Institute of Health is conducting trials on painful soft-tissue neoplasms and painful bone tumors resistant to traditional techniques.

12.7.2.5 Lung

For patients with inoperable lung cancer, percutaneous RFA under computer tomography (CT) guidance represents an alternative and minimally invasive treatment [61,62]; however, there are few studies [63–65] and the efficacy of lung cancer ablation is yet to be determined.

12.7.2.6 Breast

Breast cancer RFA is also in the early stage of development [66–68]. The largest study to date has 26 patients with T1 or T2 breast tumors [67]. Coagulation necrosis was confirmed by histology in 96% of the patients following surgical lumpectomy. Several breast RFA clinical trials are underway that will help determine whether RFA may be an alternative breast-conserving treatment in the future.

12.7.2.7 Other Tumors

Other potential applications of RFA include treatment of tumors in the adrenal [69] and prostate [70].

12.7.2.8 Miscellaneous Applications of RFA

Li et al. [71] evaluated the effect of RF treatment of 22 patients' palates on speech, swallowing, taste, sleep, and snoring. After a mean follow-up of 14 months, no adverse effect was reported. The success of RF volumetric reduction of the palate diminishes with time, as with other palate surgical procedures. However, the minimal invasiveness of RF provided high patient-retreatment acceptance and snoring relapse improvement.

Hultcrantz and Ericsson [72] compared two techniques for pediatric tonsil surgery with respect to pain and postoperative morbidity. The two methods were the partial tonsil resection using RF technique versus traditional tonsillectomy. They concluded that RF appears to be a safe and reliable method for tonsil surgery with much less postoperative morbidity than regular tonsillectomy.

Islam et al. [73] described a study on six pediatric patients using RF treatment of the lower esophageal sphincter and showed it is a potentially successful modality to treat recurrent gastroesophageal reflux disease in children.

12.7.2.9 Summary of RFA Clinical Studies

RFA has been studied for about ten years and initial results are promising. Multicenter trials, long-term follow-up studies, further technique refinements, and RFA combined with adjuvant therapies are all still in progress.

12.8 Other Applications

12.8.1 Cosmetics

Narins and Narins [74] used an RF device to induce tightening of skin via uniform volumetric heating into deep dermis, resulting in a "nonsurgical facelift." Twenty treatment areas in 17 patients were treated to evaluate the efficacy and safety of RF treatment to the brow and jowls. The technique was reported to produce gradual skin tightening in most patients with no adverse effects.

Fitzpatrick et al. [75] studied 86 subjects who were evaluated for six months after treatment with an RF tissue-tightening device. A single treatment produced objective and subjective reductions in peri-orbital wrinkles and measurable changes in brow position. These changes were indicative of a thermally induced early tissue-tightening effect followed by additional tightening over a time course consistent with thermal wound-healing response.

Sadick and Shaoul [77] treated 40 adult patients (skin phenotypes II–V) with varied facial and nonfacial hair colors with a combination of laser and RF energy. Maximum hair reduction was observed at 6–8 weeks after each treatment. An average clearance of 75% was observed in all body locations at 18 months. No significant adverse sequelae were reported. They concluded that combined laser and RF energy with contact cooling is a safe and effective method of long-term hair reduction in patients of diversified skin types and varied hair colors.

Chess [76] evaluated laser and RF combination therapy on 25 patients with a total of 35 sites (0.3–5.0 mm vessel diameters). At 1 and 6 months after the final treatment, approximately 77% of treatment sites exhibited 75–100% vessel clearance, and 90% had 50–100% vessel clearance.

12.9 Concluding Remarks

Although there are controversies concerning potential health effects of low-level RF exposure, many beneficial effects using high-level RF energy to heat tissue or destroy tumors are utilized in clinics as shown by the examples in this chapter. Diathermy treatment has a long history as it has been used for more than half a century and millions of patients have benefited from treatment. More recent research on hyperthermia and ablation for cancer treatment has demonstrated the efficacy of RF heating. The literature includes reports of the use of therapeutic RF heating devices as alternative techniques for surgery on the palate, tonsils, and esophagus. Other applications include cosmetic procedures. Complicated RF dosimetry makes the heat delivery an art. Interdisciplinary team work between medical, engineering, and physical science is vital to successful medical applications of RF technology.

Acknowledgments

I would like to thank Dr. Mark Dewhirst of Duke University and Paul Stauffer of the University of California, San Francisco, and Dr. Maxim Itkin of the University of Pennsylvania for providing valuable information on hyperthermia and RF ablation. John McDougall and Dr. Joe Elder's comments on this chapter are highly appreciated.

References

1. Lehmann, J.F., *Therapeutic Heat and Cold*, 4th ed., Baltimore, Williams and Wilkins, 1990.
2. Wust, P., Hildebrandt, B., Sreenivasa, G., Rau, B., Gellermann, J., Riess, H., Felix, R., and Schlag, P.M., Hyperthermia in combined treatment of cancer, *Lancet Oncol.*, 3, 487, 2002.
3. Dewhirst, M.W., Jones, E., Samulski, T., Vujaskovic, Z., Li, C., and Prosnitz, L., Hyperthermia, in *Cancer Medicine* 6th ed., Section 9: Radiation Oncology, Hamilton, London, BC Decker Inc., 2004, p. 623, chap. 41.
4. Goldberg, S.N., Radiofrequency tumor ablation: principles and techniques, *Eur. J. Ultrasound*, 13, 129, 2001.
5. Friedman, M., Mikityansky, I., Kam, A., Libutti, S.K., Walther, M.M., Neeman, Ziv., Locklin, J.K., and Wood, B.J., Radiofrequency blation of cancer, *Cardiovasc. Intervent. Radiol.*, 27, 427, 2004.
6. Garverick, S.L., Kane, M., Ko, W.H., and Maniglia, A.J., External unit for a semi-implantable middle ear hearing device. *Ear Nose Throat J.*, 76, 397, 1997.
7. Zierhofer, C.M., Hochmair, I.J., and Hochmair, E.S., The advanced combi 40+ cochlear implant, *Am. J. Otol.*, 18 (6 Suppl), 37, 1997.
8. Gabriel, C., The dielectric properties of biological materials, in *Radiofrequency Radiation Standards*, Klauenberg B.J., Grandolfo, M., and Erwin, D.N., Eds., New York, Plenum Press, 1995, p. 187.
9. Johnson, C.C. and Guy, A.W., Nonionizing electromagnetic wave effects in biological materials and systems, *Proc. IEEE*, 60, 692, 1972.
10. Guy, A.W. and Chou, C.K., Electromagnetic heating for therapy. *Proceedings of symposium on microwaves and thermoregulation, at the John Pierce Foundation*, New Haven, 1981, Adair, E., Ed., New York, Academic Press, 1983, p. 57.
11. Stauffer, P.R., Thermal therapy techniques for skin and superficial tissue disease, in *A Critical Review, Matching the Energy Source to the Clinical Need*, Ryan TP, Ed., Bellingham, *SPIE Optical Engineering Press*, 327, 2000.
12. Song, C.W., Blood flow in tumors and normal tissues in hyperthermia, in *Hyperthermia in Cancer Therapy*, Storm F.K., Ed., GK Hall Medical Publisher, Boston, 187, 1983.
13. Guy, A.W. and Chou, C.K., Physical aspects of localized heating by radio frequency waves. In *Hyperthermia in Cancer Therapy*, Storm K., Ed., Boston, G.K. Hall Publisher, 1982, p. 279.
14. Huang, Q., Hu, J.K., Lohr, F., et al., Heat-induced gene expression as a novel targeted cancer gene therapy strategy. *Cancer Res.*, 60, 3435, 2000.

15. Borrelli, M., Schoenherr, D., Wong, A., et al., Heat-activated transgene expression from adenovirus vectors infected into human prostate cancer cells. *Cancer Res.*, 61, 1113, 2001.
16. Brade, A.M., Ngo, D., Szmitko, P., et al., Heat-directed gene targeting of adenoviral vectors to tumor cells. *Cancer Gene Ther.*, 7, 1566, 2000.
17. Braiden, V., Ohtsuru, A., Kawashita, Y., et al., Eradication of breast cancer xenografts by hyperthermic suicide gene therapy under the control of the heat shock protein promoter. *Hum. Gene Ther.*, 11, 2453, 2000.
18. Chou, C.K. and Ren, R., Radio frequency hyperthermia in cancer therapy, in *Biomedical Engineering Handbook*, 2nd ed., Bronzino, J., Ed., Boca Raton, FL, CRC Press, Boca Raton, 2000, p. 1, chap. 94.
19. Chou, C.K., Evaluation of microwave hyperthermia applicators. *Bioelectromagnetics*, 13, 581, 1992.
20. Gellermann, J., Wlodarczyk, W., Ganter, H., Nadobny, J., Ahling, H.F., Seebass, M., Felix, R., and Wust, P., A practical Approach to thermography in a hyperthermia/magnetic resonance hybrid system: Validation in a heterogeneous phantom, *Int. J. Radiat. Oncol., Biol., Phys.*, 61, 267, 2005.
21. Seegenschmiedt, M.H., Fessenden, P., and Vernon, C.C., *Thermoradiotherapy and Thermochemotherapy, Clinical Applications,* New York, Springer-Verlag, 1996, p. 2.
22. Stauffer, P.R., Jacobsen, S., and Neuman, D., Microwave array applicator for radiometry controlled superficial hyperthermia, in *Thermal Treatment of Tissue: Energy Delivery and Assessment*, Ryan T.P., Ed., *Proceedings of SPIE*, San Jose, 2001, p. 19.
23. Straube, W.L., Klein, E.E., Moros, E.G., Low, D.A., and Myerson, R.J., Dosimetry and techniques for simultaneous hyperthermia and external beam radiation therapy, *Int. J. Hyp.*, 17, 48, 2001
24. Vernon, C.C., Hand, J.W., Field, S.B., et al., Radiotherapy with or without hyperthermia in the treatment of superficial localized breast cancer: results from five randomized controlled trials, International Collaborative Hyperthermia Group, *Int. J. Radiat. Oncol., Biol., Phys.*, 35, 731, 1996.
25. Perez, C.A., Pajak, T., Emami, B., et al., Randomized phase III study comparing irradiation and hyperthermia with irradiation alone in superficial measurable tumors, final report by the Radiation Therapy Oncology Group, *Am. J. Clin. Oncol.*, 14, 133, 1991.
26. Dewhirst, M.W., Phillips, T.L., Samulski, T.V., et al., RTOG quality assurance guidelines for clinical trials using hyperthermia, *Int. J. Radiat. Oncol., Biol., Phys., 18,* 1249, 1990.
27. Datta, N.R., Bose, A.K., Kapoor, H.K., and Gupta, S., Head and neck cancers: results of thermoradiotherapy versus radiotherapy, *Int. J. Hyp.*, 6, 479, 1990.
28. Valdagni, R. and Amichetti, M., Report of long-term follow-up in a randomized trial comparing radiation therapy and radiation therapy plus hyperthermia to metastatic lymph nodes in stage IV head and neck patients, *Int. J. Radiat. Oncol., Biol., Phys.*, 28, 163, 1994.
29. Sugimachi, K., Kitamura, K., Baba K., et al., Hyperthermia combined with chemotherapy and irradiation for patients with carcinoma of the oesophagus—a prospective randomized trial, *Int. J. Hyp.*, 8, 289, 1992.
30. Sugimachi, K., Kuwano, H., Ide, H., et al., Chemotherapy combined with or without hyperthermia for patients with oesophageal carcinoma: a prospective randomized trial, *Int. J. Hyperthermia*, 10, 485, 1994.
31. Overgaard, J., Gonzalez Gonzalez, D., Hulshof, M.C., et al., Randomised trial of hyperthermia as adjuvant to radiotherapy for recurrent or metastatic malignant melanoma, European Society for Hyperthermic Oncology, *Lancet*, 345, 540, 1995.
32. Sneed, P.K., Stauffer, P.R., McDermott, M.W., et al., Survival benefit of hyperthermia in a prospective randomized trial of brachytherapy boost $\pm$ hyperthermia for glioblastoma multiforme, *Int. J. Radiat. Oncol., Biol., Phys*, 40, 287, 1998.
33. van der Zee, J., Gonzalez Gonzalez, D., van Rhoon, G.C., et al., Comparison of radiotherapy alone with radiotherapy plus hyperthermia in locally advanced pelvic tumours: a prospective, randomised, multicentre trial, Dutch Deep Hyperthermia Group, *Lancet*, 355, 1119, 2000.
34. Huang, S.K., Radio-frequency catheter ablation of cardiac arrhythmias: appraisal of an evolving therapeutic modality, *Am. Heart J.*, 118, 1317, 1989.
35. Rhim, H., Goldberg, S.N., Dodd, G.D. III, Solbiati, L., Lim, H.K., Tonolini, M., Cho, O.K., Essential techniques for successful radio-frequency thermal ablation of malignant hepatic tumors, *Radiographics*, 21, S17, 2001.

36. Goldberg, S.N., Solbiati, L., Hahn, P.F., et al., Large volume tissue ablation with radio frequency by using a clustered, internally cooled electrode technique: laboratory and clinical experience in liver metastases. *Radiology*, 209:371, 1998.
37. Bowles, B.J., Machi, J., and Limm, W.M., Safety and efficacy of radio frequency thermal ablation in advanced liver tumors, *Arch. Surg.*, 136, 864, 2001.
38. Decadt, B. and Siriwardena, A.K., Radiofrequency ablation of liver tumours: systematic review, *Lancet Oncol.*, 5, 550, 2004.
39. McGahan, J.P., Browning, P.D., Brock, J.M., and Tesluk, H., Hepatic ablation using radiofrequency electrocautery, *Invest. Radiol.*, 25, 267, 1990.
40. Rossi, S., Fornari, F., Pathies, C., and Buscarini, L., Thermal lesions induced by 480 kHz localized current field in guinea pig and pig liver, *Tumori*, 76, 54, 1990.
41. Curley, S.A., Izzo, F., Delrio, P., Ellis, L.M., Granchi, J., Vallone. P., Fiore, F., Pignata, S., Daniele, B., and Cremona, F., Radiofrequency ablation of unresectable primary and metastatic hepatic malignancies: results in 123 patients, *Ann. Surg.*, 230, 1, 1999.
42. Solbiati, L., Livraghi, T., Goldberg, S.N., Ierace, T., Meloni, F., Dellanoce, M., Cova L., Halpern, E.F., and Gazelle, G.S., Percutaneous radio-frequency ablation of hepatic metastases from colorectal cancer: long-term results in 117 patients, *Radiology*, 221, 159, 2001.
43. Lencioni, R., Crocetti, L., Cioni, D., Della Pina, C., and Bartolozzi, C., Percutaneous radiofrequency ablation of hepatic colorectal metastases: technique, indications, results, and new promises, *Invest. Radiol.*, 39, 689, 2004.
44. Livraghi, T., Solbiati, L., Meloni, M.F., Gazelle, G.S., Halpern, E.F., and Goldberg, S.N., Treatment of focal liver tumors with percutaneous radio-frequency ablation: complications encountered in a multicenter study, *Radiology*, 226, 441, 2003.
45. Hines-Peralta, A. and Goldberg, S.N., Review of radiofrequency ablation for renal cell carcinoma, *Clin. Cancer Res.*, 10 (18 Pt 2), 6328S, 2004.
46. Matlaga, B.R., Zagoria, R.J., Woodruff, R.D., Torti, F.M., and Hall, M.C., Phase II trial of radio frequency ablation of renal cancer: evaluation of the kill zone, *J. Urol.*, 168, 2401, 2002.
47. Ogan, K., Jacomides, L., Dolmatch, B.L., Rivera, F.J., Dellaria, M.F., Josephs, S.C., and Cadeddu, J.A., Percutaneous radiofrequency ablation of renal tumors: technique, limitations, and morbidity, *Urology*, 60, 954, 2002.
48. Gervais, D.A., McGovern, F.J., Arellano, R.S., McDougal, W.S., and Mueller, P.R., Renal cell carcinoma: clinical experience and technical success with radio-frequency ablation of 42 tumors, *Radiology*, 226, 417, 2003.
49. Kam, A.W., Littrup, P.J., Walther, M.M., Hvizda, J., and Wood, B.J., Thermal protection during percutaneous thermal ablation of renal cell carcinoma, *J. Vasc. Interv. Radiol.*, 15, 753, 2004.
50. Lewin, J.S., Nour, S.G., Connell, C.F., Sulman, A., Duerk, J.L., Resnick, M.I., and Haaga, J.R., Phase II clinical trial of interactive MR imaging-guided interstitial radiofrequency thermal ablation of primary kidney tumors: initial experience. *Radiology*, 232, 835, 2004.
51. Pinto, C.H., Taminiau, A.H, and Vanderschueren, G.M., Technical considerations in CT-guided radiofrequency thermal ablation of osteoid osteoma: tricks of the trade, *AJR Am. J. Roentgenol.*, 179, 1633, 2002.
52. Woertler, K., Vestring, T., and Boettner, F. Osteoid osteoma: CT-guided percutaneous radiofrequency ablation and follow-up in 47 patients, *J. Vasc. Interv. Radiol.*, 12, 717, 2001.
53. Lindner, N.J., Ozaki, T., Roedl, R., Goshcger, G., Winkelmann, W., and Wortler, K. Percutaneous radiofrequency ablation in osteoid osteoma, *J. Bone Joint Surg. Br.*, 83, 391, 2001.
54. Vanderschueren, G.M., Taminiau, A.H., and Obermann, W.R., Osteoid osteoma: clinical results with thermocoagulation, *Radiology*, 224, 82, 2002.
55. Cioni, R., Armillotta, N., Bargellini, I., Zampa, V., Cappelli C., Vagli, P., Boni, G., Marchetti, S., Consoli, V., and Bartolozzi, C., CT-guided radiofrequency ablation of osteoid osteoma: long-term results, *Eur. Radiol.*, 14, 1203, 2004.
56. Wu, J.S., Wong, R., and Johnston, M., Meta-analysis of dose-fractionation radiotherapy trials for the palliation of painful bone metastases, *Int. J. Radiat. Oncol., Biol., Phys.*, 55, 594, 2003.
57. Forouzanfar, T., van Kleef, M., and Weber, W.E., Radiofrequency lesions of the stellate ganglion in chronic pain syndromes: retrospective analysis of clinical efficacy in 86 patients, *Clin. J. Pain*, 16, 164, 2000.

58. Pevsner, Y., Shabat, S., Catz, A., Folman, Y., and Gepstein, R., The role of radiofrequency in the treatment of mechanical pain of spinal origin, *Eur. Spine J.*, 12, 602, 2003.
59. Oh, W.S. and Shim, J.C., A randomized controlled trial of radiofrequency denervation of the ramus communicans nerve for chronic discogenic low back pain, *Clin. J. Pain*, 20, 55, 2004.
60. Kanpolat, Y., Savas, A., Bekar, A., and Berk, C., Percutaneous controlled radiofrequency trigeminal rhizotomy for the treatment of idiopathic trigeminal neuralgia: 25-year experience with 1,600 patients, *Neurosurgery*, 48, 524, 2001.
61. Lim, Y.S., Primary and secondary lung malignancies treated with percutaneous radiofrequency ablation: evaluation with follow-up helical CT, *Am. J. Roentgenol*, 183, 1013, 2004.
62. Kim, C.S., Percutaneous radiofrequency ablation for inoperable non-small cell lung cancer and metastases: preliminary report. *Radiology*, 230, 125, 2004.
63. Highland, A.M., Mack, P., and Breen, D.J., Radiofrequency thermal ablation of a metastatic lung nodule, *Eur. Radiol.*, 12 (Suppl 3), S166, 2002.
64. Thanos, L., Mylona, S., Pomoni, M., Kaliora,s V., Zoganas, L., and Batakis, N., Primary lung cancer: treatment with radio-frequency thermal ablation, *Eur. Radiol.*, 14, 897, 2004.
65. Akeboshi, M., Yamakado, K., Nakatsuka, A., Hataji, O., Taguchi, O., Takao, M., and Takeda, K., Percutaneous radiofrequency ablation of lung neoplasms: initial therapeutic response, *J. Vasc. Interv. Radiol.*, 15, 463, 2004.
66. Jeffrey, S.S., Birdwell, R.L., Ikeda, D.M., Daniel, B.L., Nowels, K.W., Dirbas, F.M., and Griffey, S.M, Radiofrequency ablation of breast cancer: first report of an emerging technology, *Arch. Surg.*, 134, 1064, 1999.
67. Izzo, F., Thomas, R., and Delrio, P., Radiofrequency ablation in patients with primary breast carcinoma: a pilot study in 26 patients, *Cancer*, 92, 2036, 2001.
68. Fornage, B.D., Sneige, N., Ross, M.I., Mirza, A.N., Kuerer, H.M., Edeiken, B.S., Ames, F.C., Newman, L.A., Babiera, G.V., and Singletary, S.E., Small (< or = 2-cm) breast cancer treated with US-guided radiofrequency ablation: feasibility study, *Radiology*, 231, 215, 2004.
69. Wood. B.J., Abraham, J., Hvizda, J.L., Alexander, H.R., and Fojo, T., Radiofrequency ablation of adrenal tumors and adrenocortical carcinoma metastases. *Cancer*, 97, 554, 2003.
70. Selli, C., Scott, C.A., Garbagnati, F., De Antoni, P., Moro, U., Crisci, A., and Rossi, S., Transurethral radiofrequency thermal ablation of prostatic tissue: a feasibility study in humans, *Urology*, 57, 78, 2001.
71. Li, K.K., Powell, N.B., Riley, R.W., Troell, R.J., and Guilleminault, C., Radiofrequency volumetric reduction of the palate: an extended follow-up study, *Otolaryngol. Head Neck Surg.*, 122, 410, 2000.
72. Hultcrantz, E. and Ericsson, E., Pediatric tonsillotomy with the radiofrequency technique: less morbidity and pain, *Laryngoscope*, 114, 871, 2004.
73. Islam, S., Geiger, J.D., Coran, A.G., and Teitelbaum, D.H., Use of radiofrequency ablation of the lower esophageal sphincter to treat recurrent gastroesophageal reflux disease, *J. Pediatr. Surg.*, 39, 282–6, 2004.
74. Narins, D.J. and Narins, R.S., Non-surgical radiofrequency facelift, *J. Drugs Dermatol.*, 2, 495, 2003.
75. Fitzpatrick, R., Geronemus, R., Goldberg, D., Kaminer, M, Kilmer, S., and Ruiz-Esparza, J., Multicenter study of noninvasive radiofrequency for periorbital tissue tightening, *Lasers Surg. Med.*, 33, 232, 2003.
76. Chess, C., Prospective study on combination diode laser and radiofrequency energies (ELOS) for the treatment of leg veins, *J. Cosmet. Laser Ther.*, 6, 86, 2004.
77. Sadick, N.S. and Shaoul, J., Hair removal using a combination of conducted radiofrequency and optical energies—an 18-month follow-up, *J. Cosmet. Laser Ther.*, 6, 21, 2004.

Index